CELLULAR IMMUNOLOGY

Selected Readings
and
Critical Commentary

CELLULAR IMMUNOLOGY

Selected Readings and Critical Commentary

Compiled by
Vicki L. Sato

Department of Cell and
  Developmental Biology
Harvard University

Malcolm L. Gefter

Department of Biology
Massachusetts Institute of Technology

1981

Addison-Wesley Publishing Company
Advanced Book Program/World Science Division
Reading, Massachusetts

London · Amsterdam · Don Mills, Ontario · Sydney · Tokyo

Library of Congress Cataloging in Publication Data

Main entry under title:

Cellular immunology.

   Bibliography: p.
   1. Cellular immunity—Addresses, essays, lectures.
I. Sato, Vicki L.   II. Gefter, Malcolm L.   [DNLM:
1. Immunity, Cellular—Collected works.   QW 568 C3935]
QR185.5.C44   591.2'9   81-12826

ISBN 0-201-10434-2   AACR2

Original text reproduced by Addison-Wesley Publishing Company, Inc.,
Advanced Book Program/World Science Division, Reading, Massachusetts, from camera-ready
copy prepared by the office of the editors.

ABCDEFGHIJ–AL–8987654321

Contents

Preface xi

Chapter I   Early Studies on Cell Cooperation in the Immune Response . . .   1
    Introduction   3

  I.1   J.F.A.P. Miller
     Immunological Function of the Thymus   9

  I.2   H.N. Claman, E.A. Chaperon, and R.F. Triplett
     Thymus-Marrow Cell Combinations. Synergism in Antibody
     Production 11

  I.3   A.J.S. Davies, E. Leuchars, V. Wallis, R. Marchant, and
     E.V. Elliott
     The Failure of Thymus-Derived Cells to Produce Antibody 16

  I.4   G.F. Mitchell and J.F.A.P. Miller
     Cell to Cell Interaction in the Immune Response. II. The
     Source of Hemolysin-Forming Cells in Irradiated Mice Given
     Bone Marrow and Thymus or Thoracic Duct Lymphocytes 25

  I.5   N.A. Mitchison
     The Carrier Effect in the Secondary Response to Hapten-
     Protein Conjugates. II. Cellular Cooperation 42

  I.6   M.C. Raff
     Role of Thymus-Derived Lymphocytes in the Secondary Humoral
     Immune Response in Mice 52

Chapter II B Cells 53
    Introduction 55

  II.1   M.C. Raff, M. Sternberg, and R.B. Taylor
     Immunoglobulin Determinants on the Surface of Mouse
     Lymphoid Cells 61

  II.2   G.L. Ada and P. Byrt
     Specific Inactivation of Antigen-reactive Cells with
     ^{125}I-labelled Antigen 63

  II.3   N.L. Warner, P. Byrt, and G.L. Ada
     Blocking of the Lymphocyte Antigen Receptor Site with
     Anti-Immunoglobulin Sera *in vitro* 65

  II.4   M.H. Julius and L.A. Herzenberg
     Isolation of Antigen-Binding Cells from Unprimed Mice 67

  II.5   I. Scher, A.D. Steinberg, A.K. Berning, and W.E. Paul
     X-Linked B-Lymphocyte Immune Defect in CBA/N Mice.
     II. Studies of the Mechanisms Underlying the Immune Defect . . 84

  II.6   D.G. Sieckmann, I. Scher, R. Asofsky, D.E. Mosier, and
     W.E. Paul
     Activation of Mouse Lymphocytes by Anti-Immunoglobulin.
     II. A Thymus-Independent Response by a Mature Subset of
     B Lymphocytes 98

Chapter III   Functional Diversity of T Lymphocytes 115
        Introduction 117

    III.1   H. Cantor and E.A. Boyse
             Functional Subclasses of T Lymphocytes Bearing Different
             Ly Antigens. I. The Generation of Functionally Distinct
             T-Cell Subclasses is a Differentiative Process Independent
             of Antigen 121

    III.2   J. Jandinski, H. Cantor, T. Tadakuma, D.L. Peavy, and
             C.W. Pierce
             Separation of Helper T Cells from Suppressor T Cells
             Expressing Different Ly Components. I. Polyclonal
             Activation: Suppressor and Helper Activities are
             Inherent Properties of Distinct T-Cell Subclasses 135

    III.3   B. Huber, H. Cantor, F.W. Shen, and E.A. Boyse
             Independent Differentiative Pathways of Ly 1 and
             Ly 23 Subclasses of T Cells 143

    III.4   R. Kiessling, E. Klein, and H. Wigzell
             "Natural" Killer Cells in the Mouse. I. Cytotoxic Cells
             with Specificity for Mouse Moloney Leukemia Cells.
             Specificity and Distribution According to Genotype 149

    III.5   S. Gillis, N.A. Union, P.E. Baker, and K.A. Smith
             The *in vitro* Generation and Sustained Culture of
             Nude Mouse Cytolytic T-Lymphocytes. 155

Chapter IV   Suppression of Antibody Synthesis by T Cells 173
        Introduction 175

    IV.1   R.K. Gershon and K. Kondo
            Infectious Immunological Tolerance 181

    IV.2   T. Takemori and T. Tada
            Properties of Antigen-Specific Suppressive T-Cell Factor
            in the Regulation of Antibody Response of the Mouse.
            I. *in vivo* Activity and Immunochemical Characterizations . . 193

    IV.3   K. Okumura, T. Takemori, T. Tokuhisa, and T. Tada
            Specific Enrichment of the Suppressor T Cell Bearing
            I-J Determinants. Parallel Functional and Serological
            Characterizations 206

    IV.4   T. Tada, M. Taniguchi, and C.S. David
            Suppressive and Enhancing T-Cell Factors as *I*-Region Gene
            Products: Properties and the Subregion Assignment 218

    IV.5   M. Taniguchi, I. Takei, and T. Tada
            Functional and Molecular Organization of an Antigen-
            Specific Suppressor Factor from a T-Cell Hybridoma 227

    IV.6   M. Taniguchi and T. Tokuhisa
            Cellular Consequences in the Suppression of Antibody
            Response by the Antigen-Specific T-Cell Factor 229

IV.7   L.A. Herzenberg, K. Okumura, H. Cantor, V.L. Sato,
        F.-W. Shen, E.A. Boyse, and L.A. Herzenberg
        T-Cell Regulation of Antibody Responses: Demonstration of
        Allotype-Specific Helper T Cells and their Specific
        Removal by Suppressor T Cells   240

IV.8   J.A. Kapp, C.W. Pierce, F. De La Croix, and B. Benacerraf
        Immuno-Suppressive Factor(s) Extracted from Lymphoid Cells
        of Nonresponder Mice Primed with $_L$-Glutamic Acid^{60}-$_L$-
        Alanine^{30}-$_L$-Tyrosine^{10} (GAT). I. Activity and Antigenic
        Specificity   255

Chapter V   Tolerance   261
        Introduction   263

V.1   R.E. Billingham, L. Brent, and P.B. Medawar
        'Actively Acquired Tolerance' of Foreign Cells   267

V.2   J.M. Chiller, G.S. Habicht, and W.O. Weigle
        Kinetic Differences in Unresponsiveness of Thymus and
        Bone Marrow Cells   271

V.3   G.J.V. Nossal and B.L. Pike
        Evidence for the Clonal Abortion Theory of B-Lymphocyte
        Tolerance   274

V.4   M. Venkataraman, M. Aldo-Benson, Y. Borel, and D.W. Scott
        Persistence of Antigen-Binding Cells with Surface Tolerogen:
        Isologous *Versus* Heterologous Immunoglobulin Carriers . . .   287

V.5   C.L. Sidman and E.R. Unanue
        Receptor-Mediated Inactivation of Early B Lymphocytes . . .   291

V.6   E.S. Vitetta, J.C. Cambier, F.S. Ligler, J.R. Kettman,
        and J.W. Uhr
        B Cell Tolerance. IV. Differential Role of Surface IgM
        and IgD in Determining Tolerance Susceptibility of Murine
        B Cells   294

Chapter VI   Macrophages and the Issue of Antigen Presentation   299
        Introduction   301

VI.1   A.S. Rosenthal and E.M. Shevach
        Function of Macrophages in Antigen Recognition by Guinea
        Pig T Lymphocytes. I. Requirement for Histocompatible
        Macrophages and Lymphocytes   309

VI.2   D.W. Thomas and E.M. Shevach
        Nature of the Antigenic Complex Recognized by T Lymphocytes.
        I. Analysis with an *in vitro* Primary Response to Soluble
        Protein Antigens   328

VI.3   C.W. Pierce, J.A. Kapp, and B. Benacerraf
        Regulation by the *H-2* Gene Complex of Macrophage-Lymphoid
        Cell Interactions in Secondary Antibody Responses *in vitro*. .   339

VI.4   J. Sprent
Effects of Blocking Helper T Cell Induction *in vivo* with
Anti-Ia Antibodies. Possible Role of I-A/E Hybrid
Molecules as Restriction Elements 350

VI.5   I. Gery and B.H. Waksman
Potentiation of the T-Lymphocyte Response to Mitogens.
II. The Cellular Source of Potentiating Mediator(s) 364

VI.6   M. Pierres and R.N. Germain
Antigen-Specific T Cell-Mediated Suppression. IV. Role
of Macrophages in Generation of $_L$-Glutamic Acid^{60}-$_L$-
Alanine^{30}-$_L$-Tyrosine^{10} (GAT)-Specific Suppressor T
Cells in Responser Mouse Strains 377

Chapter VII   Antigen Recognition by T Cells 387
Introduction 389

VII.1   H. Binz and H. Wigzell
Shared Idiotypic Determinants on B and T Lymphocytes
Reactive Against the Same Antigenic Determinants.
I. Demonstration of Similar or Identical Idiotypes on
IgG Molecules and T-Cell Receptors with Specificity
for the Same Alloantigens 395

VII.2   K. Eichmann and K. Rajewsky
Induction of T and B Cell Immunity by Anti-Idiotypic
Antibody 410

VII.3   S.J. Black, G.J. Hammerling, C. Berek, K. Rajewsky,
and K. Eichmann
Idiotypic Analysis of Lymphocytes *in vitro*.
I. Specificity and Heterogeneity of B and T Lymphocytes
Reactive with Anti-Idiotypic Antibody 416

VII.4   U. Krawinkel, M. Cramer, T. Imanishi-Kari, R.S. Jack,
K. Rajewsky, and O. Mäkelä
Isolated Hapten-Binding Receptors of Sensitized Lymphocytes.
I. Receptors from Nylon Wool-Enriched Mouse T Lymphocytes
Lack Serological Markers of Immunoglobulin Constant Domains
but Express Heavy Chain Variable Portions 431

VII.5   R.M. Zinkernagel and P.C. Doherty
H-2 Compatibility Requirement for T-Cell-Mediated Lysis
of Target Cells Infected with Lymphocytic Choriomeningitis
Virus. Different Cytotoxic T-Cell Specificities are
Associated with Structures Coded for in *H-2K* or *H-2D* 439

VII.6   J. Sprent
Role of the *H-2* Complex in Induction of T Helper Cells
in vivo. I. Antigen-Specific Selection of Donor T Cells
to Sheep Erythrocytes in Irradiated Mice Dependent upon
Sharing of *H-2* Determinants between Donor and Host 449

VII.7   J. Sprent
Restricted Helper Function of F_1 Hybrid T Cells Positively
Selected to Heterologous Erythrocytes in Irradiated
Parental Strain Mice. I. Failure to Collaborate with B Cells
of the Opposite Parental Strain Not Associated with Active
Suppression 460

VII.8   M.J. Bevan
Killer Cells Reactive to Altered-Self Antigens Can Also Be
Alloreactive 476

Chapter VIII   Thymic Education 481
   Introduction 483

VIII.1   H. von Boehmer, L. Hudson, and J. Sprent
Collaboration of Histo-Incompatible T and B Lymphocytes
Using Cells From Tetraparental Bone Marrow Chimeras 491

VIII.2   M.J. Bevan
In a Radiation Chimaera, Host H-2 Antigens Determine
Immune Responsiveness of Donor Cytotoxic Cells 500

VIII.3   R.M. Zinkernagel, G.N. Callahan, A. Althage, S. Cooper,
P.A. Klein, and J. Klein
On the Thymus in the Differentiation of "H-2 Self-
Recognition" by T Cells: Evidence for Dual Recognition? . . 502

VIII.4   P. Matzinger and G. Mirkwood
In a Fully H-2 Incompatible Chimera, T Cells of Donor
Origin Can Respond to Minor Histocompatibility Antigens
in Association with Either Donor or Host H-2 Type 517

VIII.5   H. von Boehmer, W. Haas, and N.K. Jerne
Major Histocompatibility Complex-Linked Immune
Responsiveness is Acquired by Lymphocytes of Low-
Responder Mice Differentiating in Thymus of High-
Responder Mice 526

Chapter IX   Regulation of Antibody Synthesis: The Network Theory . . . 531
   Introduction 533

IX.1   N.K. Jerne
Towards a Network Theory of the Immune System 539

IX.2   D.A. Hart, A.-L. Wang, L.L. Pawlak, and A. Nisonoff
Suppression of Idiotypic Specificities in Adult Mice by
Administration of Antiidiotypic Antibody 556

IX.3   K. Eichmann, A. Coutinho, and F. Melchers
Absolute Frequencies of Lipopolysaccharide-Reactive
B Cells Producing A5A Idiotype in Unprimed, Streptococcal
a Carbohydrate-Primed, Anti-A5A Idiotype-Sensitized and
Anti-A5A Idiotype-Suppressed A/J Mice 564

IX.4   H. Cosenza
Detection of Anti-Idiotype Reactive Cells in the
Response to Phosphorylcholine 578

IX.5   F.L. Owen, S.-T. Ju, and A. Nisonoff
        Presence on Idiotype-Specific Suppressor T Cells of
        Receptors that Interact with Molecules Bearing the
        Idiotype 581

IX.6   M.H. Dietz, M.-S. Sy, B. Benacerraf, A. Nisonoff,
        M.I. Greene, and R.N. Germain
        Antigen- and Receptor-Driven Regulatory Mechanisms.
        VII. H-2-restricted Anti-Idiotypic Suppressor Factor
        from Efferent Suppressor T Cells 589

Chapter X   The Molecular Biology of the Immune System 603
    Introduction 605

    X.1   N. Hozumi and S. Tonegawa
           Evidence for Somatic Rearrangement of Immunoglobulin
           Genes Coding for Variable and Constant Regions 611

    X.2   E.E. Max, J.G. Seidman, and P. Leder
           Sequences of Five Potential Recombination Sites Encoded
           Close to an Immunoglobulin κ Constant Region Gene 616

    X.3   S. Tonegawa, A.M. Maxam, R. Tizard, O. Bernard,
           and W. Gilbert
           Sequence of a Mouse Germ-Line Gene for a Variable
           Region of an Immunoglobulin Light Chain 621

    X.4   M.M. Davis, K. Calame, P.W. Early, D.L. Livant,
           R. Joho, I.L. Weissman, and L. Hood
           An Immunoglobulin Heavy-Chain Gene is Formed by at
           Least Two Recombinational Events 626

    X.5   J. Schilling, B. Clevinger, J.M. Davie, and L. Hood
           Amino Acid Sequence of Homogeneous Antibodies to Dextran
           and DNA Rearrangements in Heavy Chain V-Region Gene
           Segments 633

    X.6   G. Köhler and C. Milstein
           Continuous Cultures of Fused Cells Secreting Antibody
           of Predefined Specificity 639

    X.7   C.-P. Liu, P.W. Tucker, J.F. Mushinski, and F.R. Blattner
           Mapping of Heavy Chain Genes for Mouse Immunoglobulins
           M and D 642

Preface

    Selected Readings represents the combined attempts of one former molecular biologist (MG) and one former plant photobiologist (VS) to grapple with some of the fundamental issues of their new discipline, cellular immunology. We wished to share some of what we have learned, both about cellular immunology and about the flavor of the field, with students of immunology and fellow scientists in other disciplines. Our approach has been to present a series of papers on individual subjects which collectively contributed to some central idea in cellular immunology. This allows the reader to acquaint himself with the nature of immunological experimentation as well as with the ideas being developed. We have included brief editorial analyses which are designed to offer some guidelines and are mostly a reflection of our own views on these subjects. The selected papers are not meant to be a comprehensive survey of a particular topic but were chosen for their contribution to the development of an idea. Unfortunately, constraints of length, reprint permissions, and availability of original papers have added to our inability to include many noteworthy contributions; obviously the number of relevant publications far exceeds the scope of this book. In some instances, we have tried to compensate for our inability to include papers by listing an additional bibliography.

    The book inevitably represents the bias of recent but enthusiastic converts to cellular immunology. The field lured us from our respective disciplines because it seemed to offer a fascinating breadth of biological research. The problems of the immune response begin with issues of gene structure and expression, proceed through developmental biology and the questions of biological recognition, and culminate in wide-ranging issues about the nature of information handling and storage. For all of our enthusiasm about this new discipline, our initial encounters with immunology were not without their problems. As we began reading the literature and attending seminars several years ago, we all-too-often found ourselves in a mysterious, jargon-befuddled world that made our goal of understanding the state of the art seem almost unattainable. Confusion was reaching a peak about the time of the Cold Spring Harbor Symposium on Lymphocyte Diversity. We both attended the symposium in 1976 whereupon one of us had the following experience.

    "Applying some of the skills acquired during a scientific career, I sat in the front row of the auditorium between two of the most prominent scientists in immunology. Obviously the hope was that a simultaneous translation from immunology into simple prose would be the key to my successful acquisition of useful information. The experts listened to the speakers with an attentiveness designed to confirm the wisdom of my plan of action, but their solemn concentration discouraged me from

stage-whispering questions during the talks. I held them for the inter-
mission and took copious notes. By the time coffee break came, I had
predictably forgotten the questions that I had about the early speakers
of the session but had managed to hold on to a few pertaining to the
last of the speakers (in actual fact, my notes on the early speakers
were sufficiently incomprehensible so that mere recollection of the
topic, let alone incisive questions, eluded me). Before I could even
ask one question, one of my neighbor-experts asked if I could please
decipher the points of interest in the previous session. At that
point, I decided it was futile to keep taking notes and directed all
my attention to a simple assimilation of as much information as possible
in the hope that 'it would all make sense later.' Needless to say,
sudden enlightenment was not forthcoming."

Comparing experiences after our return to Cambridge, we decided to try a
different approach and teach an introductory course in immunology. Not to venture
into the abyss alone each of us sought a buddy. One of us (VS) set out to teach
immunobiology at Harvard in collaboration with another traveller down the immunological road, Wally Gilbert, and the other (MG) managed to find a place in the
immunology course at MIT taught by Lisa Steiner and Herman Eisen. Staying up all
night before class and writing out the entire lecture, after struggling with the
literature ourselves, made us barely able to convey the subject matter to the
uninitiated. The task during the first semester of teaching was obviously a
matter of staying one lecture ahead of the students while keeping just enough
information in reserve to intimidate any student sufficiently confident to ask
pressing questions.

About this time we each passed through a phase of self-affirmation and
decided that surely our difficulties could not lie with us but must lie with the
field. Apparently, the study and language of immunology were sufficiently different in style from molecular genetics and bioenergetics that practicing members
of the disciplines could not communicate. (This continues to be a general problem.
Our colleagues in other biochemical and biological disciplines still bemoan the
seminars given by immunologists, and the two of us, ironically enough, have become
the simultaneous translators.) In an attempt to break down the barriers, we began
to meet once a week to read and discuss a variety of papers, some of which we found
useful in teaching. It seemed to us that a collection of some critical papers in
cellular immunology would be a useful companion to the numerous introductory texts
currently available. Such a collection could serve for undergraduates and graduate
students and as a resource for other scientists entranced but a bit intimidated by
cellular immunology.

Of course, the more we read, the more we found that fascinating threads of
consistency and logic emerged from the morass of often conflicting data, confusing
terminology, and vastly different experimental systems. Exhausted but definitely
addicted, we emerged from the task pleased to confirm our initial instincts that
cellular immunology ranks as one of the most exciting fields in modern biology.

We extend our gratitude to our many colleagues who have helped us in assembling Selected Readings. In particular, we acknowledge Mike Bevan, Herman Eisen,
Ed Golub, and Eli Sercarz for helpful discussions; Nancy Basore, John Douhan III,
Ann Marshak-Rothstein, Leslie Serunian, and Lisa Shinefeld for reading and criti-
cizing the text; Lisa Steiner for the generous loan of her journals to the publishers; Audrey Childs for her work on the early drafts, and especially Neenyah
Ostrom for typing the final copies and offering editorial expertise in the final
hectic stages. Special thanks from VS go to the Herzenbergs and from MG to
Matthew Scharff for taking us into their laboratories and introducing us to
immunology.

Vicki Sato and Malcolm Gefter

ACKNOWLEDGMENTS

The readings in this book have been reprinted by permission from the following sources. Bibliographic citations are given in the list of included papers at the end of the relevant chapter introduction. In the acknowledgments below, Roman numerals refer to chapters and Arabic numbers refer to papers within the chapters.

Papers I.6, II.1, II.2, II.3, IV.5, V.1, V.5, VIII.2, X.4, X.5, X.6 were originally published in Nature. Copyright © 1970, 1970, 1969, 1970, 1980, 1953, 1975, 1977, 1980, 1980 and 1975, respectively, by Macmillan Journals Limited.

Papers I.4, II.4, II.5, II.6, III.1, III.2, III.3, III.5, IV.2, IV.3, IV.6, IV.7, V.3, V.6, VI.1, VI.2, VI.3, VI.4, VI.5, VII.1, VII.3, VII.5, VII.6, VII.7, VIII.1, VIII.3, VIII.4, IX.2, IX.3, IX.5, IX.6 were originally published in The Journal of Experimental Medicine, The Rockefeller University Press.

Papers I.5, III.4, VII.2, VII.4, IX.4 were originally published in European Journal of Immunology, Verlag Chemie GMBH.

Papers IV.8, V.4, VI.6 were originally published in Journal of Immunology. Copyright © 1976, 1977 and 1978, respectively, by the Williams and Wilkins Company. Reproduced by permission.

Papers VII.8, VIII.5, X.1, X.2, X.3 were originally published in Proceedings of the National Academy of Sciences.

Papers V.2, X.7 were originally published in Science. Copyright © 1971 and 1980, respectively, by the American Association for the Advancement of Science.

Paper IV.1 was originally published in Immunology.

Paper IV.4 was originally published in Cold Spring Harbor Symposium on Quantitative Biology. Copyright © 1976 by Cold Spring Harbor Laboratory.

Paper I.2 was originally published in Proceedings of the Society for Experimental Biology and Medicine.

Paper I.3 was originally published in Transplantation. Copyright © 1976 by the Williams and Wilkins Company. Reproduced by permission.

Paper I.1 was originally published in The Lancet.

Paper IX.1 was originally published in Annales d'Immunologie, Masson, S.A.

Chapter I

Early Studies on Cell Cooperation in the Immune Response

Chapter I

Early Studies on Cell Cooperation in the Immune Response

The essence of immunity lies in the ability to distinguish self from nonself. It is upon the ability to make this distinction that the organism has built a biological defense system geared to fend off pathological invaders. The hallmarks of this defense are its DIVERSITY, its SPECIFICITY, and its MEMORY. The diversity of the immune response is reflected in the organism's apparent ability to recognize and respond to a large universe of intrusive foreign antigens. The fine regulation of that potential diversity results in an extraordinarily specific response--highly restricted elements in the immune system are activated by equally restricted antigenic stimuli. In addition, the immune system displays memory--not only the ability to recall past experiences but also the capability to modify behavior upon a subsequent encounter with the initial antigen.

The workings of one area of the immune system, the humoral response, are conducted by a series of white blood cells, primarily macrophages and lymphocytes. The latter cells come in two forms: thymus-dependent (T) lymphocytes and thymus-independent, bone marrow derived (B) lymphocytes. These three cell types work together in ways that are not yet completely understood to elaborate antibodies that are directed against the challenging antigen. These specific antibodies are released into circulation by antibody-producing plasma cells. The other aspect of immunity does not depend on secreted antibodies but rather through the direct action of antigen-specific lymphocytes. Cell-mediated immune responses seem to rely on direct contact between target cells and effector lymphocytes, the latter usually T cells. The two broad classes of lymphocytes for humoral and cellular immunity are distinguished by a series of cell surface antigens, described in subsequent chapters and also clearly reviewed and catalogued in Katz (1977).

This chapter is concerned first with the initial descriptions by Gowans (paper 1) and Miller (paper 2) that small recirculating lymphocytes are fundamentally important in the immune response, and that at least some of these lymphocytes are dependent upon thymic influence for development. Subsequent papers included here present data that support the notion that the cells responsible for the immune response are heterogeneous, both with regard to function and developmental lineage. Finally the experiments of Claman, Davies, Miller and Mitchell, and Mitchison lay the foundation for the concept of T cell-B cell cooperation in the making of a successful humoral response.

Although much of the pioneering work of Gowans in demonstrating the critical role of the small lymphocyte cannot be included here (Gowans and Knight, 1964), his contribution in laying the foundations of modern cellular immunology cannot be underestimated. It was not until his demonstration in 1962 that depletion of the recirculating pool of small lymphocytes resulted in severe immune deficiency that the fundamental importance of the small lymphocyte was realized. Its apparently quiescent nature and rather innocuous appearance made it a far less attractive candidate for mediation of host defenses than that of its dramatic, phagocytic macrophage counterpart. Paper 1, which is but one of several important works, demonstrated that an animal could be rendered immunologically compromised by removal of its recirculating lymphocytes and that immunocompetence could be restored by addition of the lymphocytes removed by thoracic duct cannulation. The paper also demonstrates that addition of lymphocytes obtained from an animal rendered immunologically unresponsive (tolerant) to a particular antigen was unable to restore immune competence to an irradiated recipient.

Extirpation experiments, primarily by Glick (1956) and Cooper (1963-1966) in the chicken and by Miller in the mouse (paper 2) extended the observations by Gowans and revealed that not all of the lymphocytes within an individual were functionally equivalent. The logic behind these experiments was admirably straightforward: to assess the contribution of a particular lymphoid organ (and the cells which it produces) by removing it an an early stage in the host's development and determining how the host fared when faced with an immunological challenge.

The results of these experiments in chickens revealed that removal of the Bursa of Fabricius in newborn chickens resulted in severe depression of antibody production with no apparent effect on cell-mediated immune responses like graft rejection. Neonatal thymectomy affected both humoral and cell-mediated immune responses. Miller confirmed that neonatal thymectomy of a mouse seriously compromised its ability to mediate graft rejection. This data is presented in paper 1. Also presented is the demonstration that any delay in thymectomy results in a failure to remove the immunocompetent cells. The implication, then, is that while the thymus is crucial in contributing to the pool of reactive lymphocytes, its influence in this regard decreases with age. That is, immunocompetent cells are generated which are no longer directly dependent upon the thymus for survival or ability to function. Subsequent work by Miller in a paper not included here (1962) extended these observations to show that the neonatally thymectomized animals subsequently challenged with the antigen sheep red blood cells were far less able to make anti-sheep cell antibodies than their sham-thymectomized or untreated fellows.

Thus, the evidence from extirpation experiments led to the following view of the immune system:

a. Depletion of B lymphocytes by bursectomy results in an inability to mount humoral immune responses (i.e., make antibody) without seriously affecting cell-mediated immunity.

b. Depletion of T lymphocytes by neonatal thymectomy results in an inability to mount both humoral and cell-mediated responses.

A simple description of the immune system in which B cells provide humoral immunity, and T cells provide cell-mediated immunity is not sufficient to explain the available data. The remaining papers in this chapter will deal with resolving why humoral responses were affected by removal of either T or B lymphocytes. These papers are concerned with establishing two basic facts in humoral immunity: 1) T and B cells must interact before antibody can be produced in response to a specific antigenic stimulus, and 2) B cells are directly responsible for synthesis and secretion of antibody while T cells provide essential regulatory signals.

The first direct evidence that T and B lymphocytes act synergistically in the humoral response to antigens came from a simple and elegant experiment performed by Henry Claman and coworkers. Taking advantage of the newly developed transfer method of Playfair, Papermaster, and Cole (1965), Claman transferred thymus and/or bone marrow cells from a naive donor into a recipient which had been rendered immunoincompetent by a large dose of radiation. Using this irradiated host essentially as a test tube, Claman introduced thymocytes and bone marrow cells, either alone or in combination, together with an immunizing dose of sheep red blood cells (SRBC) and assessed the ability of the donor cells to produce anti-SRBC antibodies several days later. Briefly summarized, the results indicated that a mixture of bone marrow and thymus cells were able to mount an immune response comparable to that of splenic lymphocytes, a naturally occurring mixture of T and B cells. Bone marrow cells alone were incapable of significant levels of antibody production, and thymocytes alone were only slightly more effective. (The greater efficiency of antibody production by isolated thymocytes was most likely due to the small but clearly significant numbers of mature B cells found in this organ.) On the basis of these data Claman points out that the simplest explanation is one in which one cell type makes antibody while the other acts as an "auxiliary" partner, although these particular experiments were unable to distinguish which cell was responsible for which function.

Davies and coworkers, working in London, attempted to resolve this issue by a very clever experiment exploiting histocompatibility antigens. Basically, the approach was to use a transfer system similar to that employed by Claman but adding the fillip of eliminating either the T cells or the B cells after transfer and determining the effect on antibody synthesis. Davies began by essentially creating a mouse whose B cells and T cells differed by some minor histocompatibility anti-gens; thus B cells bore minor histocompatibility antigens of strain A, while T cells bore homologous antigens of strain A'. This "chimera" was constructed by grafting bone marrow of strain A and thymus of strain A' onto an irradiated, thym-ectomized recipient. The differences between A and A' were not significant enough to prevent the T cells and B cells from working together after they had matured but were significant enough for a strain A mouse immunized with strain A' cells to make antibody against A'. Now, after these constructed chimeras had developed to an immunocompetent state with their A-type B cells and A'-type T cells, they were immunized with SRBC and transferred to A or A' recipients which had been irra-diated. Here they were boosted with SRBC and assayed for anti-SRBC antibodies. Control groups in which A-type B cells and A'-type T cells were injected and chal-lenged in either irradiated A or A' recipients gave respectable antibody titers. But now the fun begins. Instead of using only plain, run-of-the-mill A or A' recipients, Davies used some A mice which had been previously immunized with A' cells and some A' mice previously immunized with A cells. These mice were produc-ing antibodies against A' and A, respectively. Irradiation of these recipients would deplete immunocompetent cells but not serum antibody. Thus, the A mouse with anti-A' antibodies would provide an environment specifically hostile to A' cells while the A' mouse with anti-A antibodies would be hostile to A cells. Introduction of a mixture of A-derived B cells and A'-derived T cells into such recipients would (if all goes as expected) result in specific depletion of either T or B lymphocytes. The basic results of this tour de force of an experiment are given in a schematic form below:

Group	Donor cells	Recipient	Expected environment of donor	Antibody to SRBC
1	A B cells + A' T cells	A	friendly	+++++
2	A B cells + A' T cells	A'	friendly	+++++
3	A B cells + A' T cells	AαA'	hostile to T cells	++
4	A B cells + A' T cells	A'αA	hostile to B cells	+/-

Thus, elimination of donor B cells had a more serious effect on antibody formation than elimination of T cells; B cells left alone can synthesize respectable levels of antibody while T cells alone cannot. The Claman observation that B cells and T cells together are better than either alone is also confirmed by these experi-ments. Unfortunately, the simple conclusion that B cells make antibody and T cells don't cannot be drawn from these studies. It is conceivable, for example, that both T and B cells can make antibodies but T cells can only do it when B cells are around while B cells function reasonably well alone. In this case, B cells would be responsible for synthesizing X amount of antibody directly and for inducing T cells to make Y amount of antibody. Thus B + T would yield more antibody than B or T alone. The flaw in Davies' experiment lies in the fact that it is designed to

assess precursor function, not effector function. That is, one is basically looking at how effective the surviving pool of cells is at making an immune response,
not at which cell type actually secretes the antibody at the culmination of the
response.

Mitchell and Miller (paper 3) looked at effector cells, also by exploiting
histocompatibility antigens. In this experiment, chimeras expressing A histocompatibility antigens on their B lymphocytes and $(A \times B)F_1$ antigens on their T cells
were selected. These chimeras were immunized with SRBC and their spleens harvested
seven days later. At this time the lymphocytes were treated with antibody and
complement to eliminate either T cells (anti-B-type antibody) or T cells and B
cells (anti-A-type antibody). The results show that elimination of T cells has
almost no effect on the number of antibody secreting cells. Taken with Davies'
experiment, these studies suggest that B cells are responsible for giving rise
to antibody secreting cells, and that Claman's notion that B cells make antibody
while T cells play an auxiliary role was correct.

A more refined picture of how T and B cells interact in the making of a successful humoral response came from important experiments by Mitchison. These
experiments were de nged to determine whether T cells and B cells recognize the
same or different determinants on a complex antigen. Ever since the immunochemical work of Landsteiner allowed the construction of immunogenic carriers bearing
specific haptenic determinants, it had been known that haptens by themselves are
antigenic but not immunogenic. They must be physically coupled to immunogenic
substances before an anti-hapten response can be elicited. In more specific terms,
injection of the hapten dinitrophenol, DNP, will not elicit an anti-DNP response,
while injection of DNP coupled to a protein carrier bovine serum albumin (BSA) <u>will</u>
elicit an anti-DNP response. Once elicited, the anti-DNP antibodies will recognize
and bind to the DNP group alone, or coupled to a number of other "carriers." The
hint that the carrier protein is itself recognized in the course of making anti-DNP
antibodies comes from the observation by Ovary and Benacerraf (1963) that a secondary response to a specific hapten can only be stimulated if the animal is boosted
with the hapten coupled to the same carrier to which it was attached during the
primary injection. In other words, an animal immunized with DNP-BSA will give a
secondary anti-DNP response only if it is boosted with DNP-BSA; boosting with DNP-
OVA (ovalbumin) will not elicit a secondary response. This "carrier effect" was a
key piece of evidence implicating the recognition of two different determinants
(the carrier and the hapten) during the course of a humoral response.

Mitchison, taken by the studies demonstrating that two different cells are
important in making an immune response, tested the idea that these two different
cells each recognized distinct determinants on the same molecule. The approach of
this experiment was to prime T cells and B cells to different determinants and
then to see if these cells would act synergistically when confronted with the
respective determinants joined on the same molecular complex. Using an adoptive
transfer system similar in principle to the one employed by Claman and Davies,
Mitchison primed one group of animals to hapten I coupled to carrier I (H_I-C_I)
and another group to carrier II (C_{II}) alone. He then transferred a mixture of
cells from each group into an irradiated recipient with a challenging dose of
various antigens. The results are summarized below in schematic form adapted
from paper 5:

Cells from donor I Sensitized with	Cells from donor II	Boosting antigen	Anti-hapten response
H_1-C_1	--	H_1-C_1	++++
H_1-C_1	--	C_2	--
H_1-C_1	--	H_1-C_2	--(carrier effect)
--	C_2	H_1-C_1	--
--	C_2	H_1-C_2	--
H_1-C_1	C_2	H_1-C_2	++++
H_1-C_1	C_2	H_1-C_3	--
H_1-C_1	C_2	$H_1-C_3 + C_2$	--

These results indicated that 1) different populations of cells were responsible for the recognition of haptenic and carrier determinants, 2) these cells could act synergistically when confronted with an appropriate hapten-carrier conjugate, and 3) that the hapten and carrier must be physically linked for anti-hapten antibody synthesis to result.

The critical experiment demonstrating that T cells recognize carrier determinants was done by Raff (paper 6). Exploiting the θ antigen (now renamed Thy 1), which is expressed on the membranes of T but not B lymphocytes, he showed that elimination of T cells from the carrier-primed population eliminated the adoptive anti-hapten response, while elimination of T cells from the hapten-primed population had no effect:

Cells from donor I Sensitized to:	Cells from donor II	Boosting antigen	Anti-hapten response
H_1-C_1	C_2	H_1-C_2	++++
H_1-C_1	C_2 ($\alpha-\theta$ treated)	H_1-C_2	--
H_1-C_1 ($\alpha-\theta$ treated)	C_2	H_1-C_2	++++

These papers are but a few of the many that have contributed to the development of our current view that successful antibody synthesis requires cooperation between T and B lymphocytes. They laid the essential groundwork for what has become an increasingly complex picture of how lymphocytes interact. More than ten years after the initial work of Claman we still do not understand fully how T and B cells cooperate in making an immune response. Papers in subsequent chapters will deal with what we have learned about the rules regulating some of the interactions and about possible mechanisms which underlie this complex physiological response.

Papers Included in this Chapter

1. Miller, J.F.A.P. (1961) Immunological function of the thymus. The Lancet (September 30), 748.

2. Claman, H.N., Chaperon, E.A., and Triplett, R.F. (1966) Thymus-marrow cell combinations. Synergism in antibody production. Soc. Exp. Biol. & Med. <u>122</u>, 1167.

3. Davies, A.J.S., Leuchars, E., Wallis, V., Marchant, R., and Elliott, E.V. (1967) The failure of thymus-derived cells to produce antibody. Transplantation <u>5</u>, 222.

4. Mitchell, G.F. and Miller, J.F.A.P. (1968) Cell to cell interaction in the immune response. II. The source of hemolysin-forming cells in irradiated mice given bone marrow and thymus or thoracic duct lymphocytes. J. Exp. Med. <u>128</u>, 821.

5. Mitchison, N.A. (1971) The carrier effect in the secondary response to hapten-protein conjugates. II. Cellular cooperation. Eur. J. Immunol. <u>1</u>, 18.

6. Raff, M.C. (1970) Role of thymus-derived lymphocytes in the secondary humoral immune response in mice. <u>Nature</u> <u>226</u>, 1257.

References Cited in this Chapter

Cooper, M.D., Peterson, R.D.A., Smith, M.A., and Good, R.A. (1966) The functions of the thymus and bursa system in the chicken. J. Exp. Med. <u>123</u>, 75.

Glick, B., Chang, T.S., and Jaap, R.G. (1956) The Bursa of Fabricius and antibody production. Poultry Sci. <u>35</u>, 224.

Gowans, J.L. and Knight, E.J. (1964) The route of recirculation of lymphocytes in the rat. Proc. Roy. Soc. B. <u>159</u>, 257.

Katz, D.H. (1977) <u>Lymphocyte</u> <u>Differentiation</u>, <u>Recognition</u>, <u>and</u> <u>Regulation</u>. Academic Press, New York.

Miller, J.F.A.P. (1962) Effect of neonatal thymectomy on the immunological significance of the thymus. Proc. Roy. Soc. B. <u>156</u>, 145.

Ovary, Z. and Benacerraf, B. (1963) Immunological specificity of the secondary response with dinitrophenylated proteins. Proc. Soc. Exp. Biol. (N.Y.) <u>114</u>, 72.

Playfair, J.H., Papermaster, B.W., and Cole, L.J. (1965) Focal antibody production by transferred spleen cells in irradiated mice. Science <u>149</u>, 998.

IMMUNOLOGICAL FUNCTION OF THE THYMUS

J. F. A. P. Miller

IT has been suggested that the thymus does not participate in immune reactions. This is because antibody formation has not been demonstrated in the normal thymus,[1] and because, even after intense antigenic stimulation plasma cells (the morphological expression of active antibody formation) and germinal centres have not been described in that organ.[2] Furthermore, thymectomy in the adult animal has had little or no significant effect on antibody production.[3]

On the other hand, there are certain clinical and experimental observations in man and other animals which suggest that the thymus may somehow be concerned in the control of immune responses. Thus, in acute infections, when presumably the need for antibody production is great, the thymus undergoes rapid involution; in patients with acquired agammaglobulinæmia the simultaneous occurrence of benign thymomas has been described,[4] and in fœtal or newborn animals, at a time when responsiveness to antigenic stimulation is deficient, the thymus is a very prominent organ.

The apparent contradiction between these two sets of observations may be partly explained by recent work,[5][6] which suggests that the thymus does not respond to circulating antigens because these cannot reach it owing to the existence of a barrier between the normal gland and the blood-stream. If the barrier is broken, for instance by local trauma, the histological reactions of antibody formation take place in the thymus.

In this laboratory, we have been interested in the role of the thymus in leukæmogenesis.[7] During this work it has become increasingly evident that the thymus at an early stage in life plays a very important part in the development of immunological response.

METHODS AND RESULTS

In the preliminary experiments mice of the C3H and Ak strains and of a cross between T_6 and Ak were used. The thymus was removed 1–16 hours after birth. Alternate littermates were used as sham-thymectomised controls—i.e., they underwent the full operative procedure, including excision of part of the sternum, but their thymuses were left intact. Mice in another group had thymectomy at 5 days of age. Wounds were closed with a continuous black silk suture and the baby mice were returned immediately to their mothers. No anti-

1. Fagreus, A. *J. Immunol.* 1948, **58**, 1.
2. Fagreus, A. *Acta med. scand.* 1948, suppl. 204, p. 3.
3. MacLean, L. D., Zak, S. J., Varco, R. L., Good, R. A. *Transplant. Bull.* 1957, **4**, 21.
4. Good, R. A., Varco, R. L. *Lancet*, 1955, i, 245.
5. Stoner, R. D., Hale, W. M. *J. Immunol.* 1955, **75**, 203.
6. Marshall, A. H. E., White, R. G. *Lancet*, 1961, i, 1030.
7. Miller, J. F. A. P. *Nature, Lond.* 1961, **191**, 248.

biotics were administered at any time either to the operated mice or to their mothers.

Mortality during and immediately after the operation ranged between 5 and 15% (excluding deaths due either to neglectful mothers or to cannibalism). Mortality in the thymectomised group was, however, higher between the 1st and 3rd month of life and was attributable mostly to common laboratory infections. This suggested that neonatally thymectomised mice were more susceptible to such infections than even sham-thymectomised littermate controls. When thymectomised and control groups were isolated from other experimental mice and kept under nearly pathogen-free conditions, the mortality in the thymectomised group was significantly reduced.

Absolute and differential white-cell counts were performed on tail blood at various intervals after thymectomy. The significant results of these estimations are summarised in fig. 1. In sham-thymectomised animals the lymphocyte/

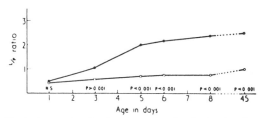

Fig. 1—Average lymphocyte:polymorph ratio of mice thymectomised in the neonatal period compared with sham-thymectomised controls. Statistical differences indicated.

○————○ thymectomised mice.
●————● sham-thymectomised mice.

polymorph ratio rose progressively in the first 8 days of life to reach the normal adult ratio of 2.5 ± 0.08. In the animals whose thymus was removed on the 1st day of life the ratio did not increase significantly and was only 1.0 ± 0.10 at 6 weeks of age.

Histological examination of lymph-nodes and spleens of thymectomised animals at 6 weeks of age revealed a conspicuous deficiency of germinal centres and only few plasma cells (figs. 2 and 3).

At 6 weeks of age, groups of thymectomised, sham-thymectomised, and entirely normal mice were subjected to skin grafting, Ak mice receiving C3H grafts and vice versa, and $(AkXT_6)F_1$ mice receiving C3H grafts. The median survival time of skin grafts in intact mice, sham-thymectomised mice, and mice thymectomised at 5 days of age ranged from 10 to 12 days. In more than 70% of mice whose thymus was removed on the 1st day of life the grafts were established and grew luxuriant crops of hair. Most of these grafts were tolerated for periods ranging from 6 weeks to 2 months and

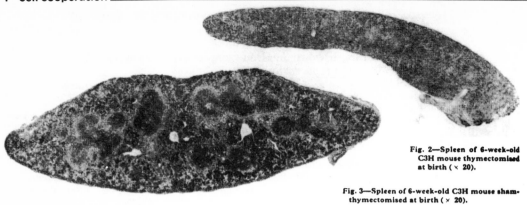

Fig. 2—Spleen of 6-week-old
C3H mouse thymectomised
at birth (× 20).

Fig. 3—Spleen of 6-week-old C3H mouse sham-
thymectomised at birth (× 20).

some for even longer than 2 months. Thereafter they gradually diminished in size, lost their hair, and eventually disappeared. A small group of C3H mice thymectomised immediately after birth, which were 3 weeks later grafted with thymuses taken from C3H fœtuses towards the end of the gestation period, did not tolerate Ak skin grafts for more than 15 days (see table).

DISCUSSION

The above results indicate that thymectomy in the immediate neonatal period is associated with severe depletion in the lymphocyte population and serious immunological defects in the mature animal. Several hypotheses might account for these results. One is that the thymus, particularly in early life, regulates lymphocyte production, not only by being the main producer of such cells, but also by secreting a factor, such as Metcalf's lymphocytosis-stimulating factor,[8] which, after birth, stimulates lymphopoiesis in other lymphoid organs. In mice thymectomised at birth, the deficiency of lymphocytes would simply weaken the host's immunological defence as a whole. Another hypothesis attributes to the thymus a more direct role in the development of immunological response. During embryogenesis the thymus would produce the originators of immunologically competent cells many of which would have migrated to other sites at about the time of birth. This would suggest that lymphocytes leaving the thymus are specially selected cells, and this might possibly be correlated with their epithelial (rather than mesenchymal) origin[9] during embryogenesis. In accordance with elective theories of antibody formation,[10][11] genetically distinct clones of cells might differentiate at various stages during thymic morphogenesis. In parallel with experience gained from experiments on classical immunological tolerance,[12] one might predict that the originators of those cells capable

8. Metcalf, D. *Ann. N.Y. Acad. Sci.* 1958, **73**, 113.
9. Auerbach, R. *Developmental Biol.* 1961, **3**, 336.
10. Burnet, F. M. The Clonal Selection Theory of Acquired Immunity. London, 1959.
11. Lederberg, J. *Science*, 1959, **129**, 1649.
12. Billingham, R. E., Brent, L. *Proc. roy. Soc. B.* 1956, **146B**, 78.

SURVIVAL OF ALLOGENIC SKIN GRAFTS IN MICE THYMECTOMISED IN THE NEONATAL PERIOD

Group	Age at operation	Strain of mice	Skin graft	Number grafted	Number tolerant	Median survival time of graft (days)
Thymectomised	1–16 hours	C3H	Ak	7	5	45 − 101*
		Ak	C3H	6	4	41 − 90 *
		(AkXT6)F₁	C3H	8	8	50 − 118*
Thymectomised	5 days	C3H	Ak	5	0	11 ± 0·7
Thymectomised and thymus-grafted 3 weeks later	5 hours	C3H	Ak	5	0	11 − 15
Sham-thymectomised	1–16 hours	C3H	Ak	6	0	11 ± 0·6
		Ak	C3H	3	0	10 ± 0·8
Intact		C3H	Ak	61	0	11 ± 0·6
		Ak	C3H	45	0	10 ± 0·9
		(AkXT6)F₁	C3H	10	0	11 ± 0·1

* These figures apply to the tolerant mice.

of reacting with the more distantly related antigens would differentiate earlier in embryonic life than the originators of those cells concerned with more closely related antigens.

If this hypothesis is to be tested experimentally, there are obvious technical difficulties to be surmounted. It has been possible, however, to thymectomise and keep alive mice born by cæsarean section 1 day before the end of the normal gestation period. Even within this short period one might be able to obtain sufficient data to correlate the degree of tolerance of skin grafts of increasingly greater " foreignness " with the time at which thymectomy was performed.

I wish to thank Prof. A. Haddow, F.R.S., for his interest in this work. This investigation has been supported by grants to the Chester Beatty Research Institute (Institute of Cancer Research: Royal Cancer Hospital) from the Medical Research Council, the British Empire Cancer Campaign, the Jane Coffin Childs Memorial Fund for Medical Research, the Anna Fuller Fund and the National Cancer Institute of the National Institutes of Health, U.S. Public Health Service.

Thymus-Marrow Cell Combinations. Synergism in Antibody Production.* (31353)

HENRY N. CLAMAN, EDWARD A. CHAPERON, AND R. FASER TRIPLETT
(Introduced by D. W. Talmage)
Departments of Medicine and Pediatrics, University of Colorado Medical Center, Denver

The source of potentially immunocompetent cells and the regulatory mechanisms involved in their maturation are unsolved problems of great interest. Recently considerable evidence has accumulated indicating that the thymus plays an important role in lymphocytopoiesis and in the development and maintenance of the immune system(1). Although small amounts of antibody are synthesized within the intact thymus(2), a more important role for this structure may be to provide potentially competent lymphoid cells which migrate to other lymphoid structures, such as the spleen and lymph nodes, there to differentiate into immunologically active cells(3). Data also exist showing that the thymus may produce a diffusible hormone-like product, influencing the multiplication and maturation of lymphoid cells in the peripheral lymphoid tissues(4). These two mechanisms are by no means mutually exclusive.

The experiments reported here sought to test for the esistence of potentially immunocompetent cells in the thymus, according to the method recently described for spleen cells by Playfair, Papermaster, and Cole(5). This method permits the study of *potentially* immunocompetent cells by demonstrating discrete clones of active, antibody-producing cells arising from a unit number of non-sensitized progenitors which are transferred to irradiated hosts and stimulated with antigen. In the course of these experiments, we found that combinations of normal thymus and bone marrow cells were far more active in producing hemolysins than cells of either type alone. Graft vs. host activity of cells was eliminated since the donor and recipient cells were of the same inbred strain.

Methods. Nine-to twelve-week-old LAF$_1$ or CBA/J recipient male mice were given a

single exposure to 650-750 r (in air) 250 kvp x-rays (30 ma, 0.5 mm CU, 1.0 mm Al, 100 cm) followed by intravenous injection of syngeneic cells (spleen, thymus, marrow or thymus plus marrow). Donor spleen or thymus cell suspensions were obtained by forcing tissue through a stainless steel screen into cold Saline F. Marrow was extruded from femurs by forcing saline through with a syringe after cutting off the epiphyses, and suspensions were made by forcing marrow clumps through a 26 gauge needle. Cells were washed in cold Saline F and counted in a hemocytometer. Concentrations were adjusted so that each animal received the desired number of cells in 0.5 ml of Saline F. When mice received both thymus and narrow cells, the 2 cell populations were combined shortly before injection in a total volume of 0.5 ml. All recipients were injected within 2 hours following irradiation.

In the first group of experiments, the mice were injected iv with 0.2 ml of a 10% suspension of washed sheep erythrocytes on the first day following irradiation. Five days following irradiation, the mice were killed and their spleens removed, cut into fragments, and plated after the manner of Playfair *et al*(5), except that red pulp was not trimmed from the fragments. After 2 hours, incubation at 37°C, 2 ml of 33% guinea pig complement was added to each plate. The plates were incubated for 30 minutes more, fixed by pouring 10% neutral formalin over the surface, then coded and read as unknowns and scored by 2 methods. The experimental design is illustrated in Fig. 1.

The spleen fragments with significant surrounding hemolysis ("active pieces") were not randomly distributed but tended to occur in clusters or "active areas." Playfair *et al*(5) have shown that an active area is probably derived from a unit number of precursor cells, but when there are many active

* Supported by USPHS grants AI 04152, 5 T1 AI 13, HD, 65-04 and AM 07529.

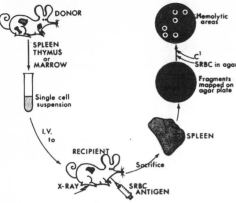

FIG. 1. Experimental design.

pieces, the delineation of discrete active areas is difficult since they tend to overlap. For this reason, the percentage of spleen pieces with hemolysis was also calculated and designated "specific activity." This was found to be linearly related to the number of active areas and permitted quantitative scoring of a larger range of activities. In evaluation of the data ± two standard errors were considered to be the 95% confidence interval.

To determine the percentage of cells reaching recipient spleens, separate suspensions of CBA spleen and thymus cells were incubated for 6½ hours with H^3-thymidine. The cells were washed and the percent labelling determined by autoradiographs of aliquots. Irradiated recipients were injected iv with 10^6 thymus or spleen cells and were sacrificed one hour later. The recipient spleens were made into single cell suspensions and the total number of cells counted. Smears of measured aliquots were made and the number of labelled cells determined by autoradiography. The total number of labelled cells in the spleen was calculated to be 6.5% of those labelled spleen cells injected, and 2.9% of those labelled thymus cells injected. Assuming that the distribution of labelled cells reflected the distribution of all donor cells, roughly twice as many spleen as thymus cells reached the recipient spleens.

Results. Table I shows the results of the first group of experiments using normal and immune spleen and thymus cells. When 5×10^6 spleen cells from non-immunized CBA/J donors were injected into irradiated syngeneic recipients, the mean number of active areas in the spleen 5 days later was found to be 2.5 times that found when 2×10^6 spleen cells were injected. Approximately 2.2×10^6 cells produced one active area in both cases. Five hundred thousand spleen cells from donors injected 3 days previously with sheep red cells (labelled immune) produced as many active areas as two million non-immune cells. Thymus cells from normal and immunized donors did not produce significantly more active areas than controls given no cells.

In the next group of experiments the x-rayed recipients received injections of antigen on days 1 and 4 and were sacrificed on day 8 following irradiation (Table II). Some animals were given 10^7 sygeneic marrow cells either alone or together with thymus cells. Spleens of mice receiving 5×10^7 thymus cells plus 10^7 marrow cells had more and strikingly larger zones of lysis than animals receiving either cell type alone. Marrow alone showed no activity above background, but thymus alone at 8 days showed slight activity.

LAF_1/J mice were used in all subsequent experiments because of their greater resistance to x-rays. Fig. 2 presents pooled data from 6 experiments showing specific activity as related to the dose of thymus cells. When mice received 10^7 marrow cells (upper line), the response curve was linear at least through 5×10^7 thymus cells, with 5.5×10^6 cells producing one active area and 10% specific

TABLE I. Production of Active Areas of Hemolysis in spleens of CBA Mice Which Were Irradiated and Injected with Spleen or Thymus Cells on Day 0, Given Sheep Erythrocytes on Day 1 and Sacrificed on Day 5.

Cells injected	No. of cells ($\times 10^6$)	Active areas per spleen	Mean
Normal spleen	2	00111112	.87
" "	5	111122222222	
		334444	2.3
Immune spleen	.5	00113	.80
" "	5	****	*
Normal thymus	20	00000002	.25
" "	50	0000	.0
" "	90	00011	.40
Immune thymus	100	000	.0
Control	0	0000000000	
		0000001111	.2

* Too many active areas to count accurately.

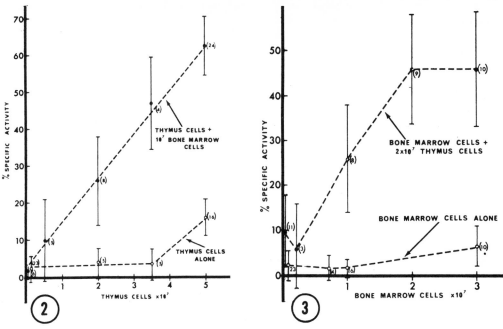

FIG. 2. Percent specific hemolytic activity of recipient spleen fragments are related to number of thymus or thymus plus 10^7 marrow cells injected. Numbers in parentheses represent numbers of mice for each data point. Range indicates ± two standard errors.

FIG. 3. Percent specific hemolytic activity of recipient spleen fragments as related to number of marrow or marrow plus 2×10^7 thymus cells injected. Numbers in parentheses represent numbers of mice for each data point. Range indicates ± two standard errors.

activity. Thymus cells alone (lower line) were much less effective and approximately 4.3×10^7 cells were required to produce one active area (10% specific activity).

In the reciprocal experiment (Fig. 3), the hemolytic response was evaluated as a function of the number of marrow cells injected. In the groups of mice receiving varying num-bers of marrow cells plus 2×10^7 thymus cells (upper line), the response was linear between 2×10^6 and 2×10^7 marrow cells, but leveled out with higher doses of marrow. Recipients of marrow cells alone (lower line) were never significantly more active than controls given no cells.

Discussion. The results confirm the work

TABLE II. Hemolytic Activity in Spleens of CBA Mice Which Were Irradiated and Injected with Spleen, Thymus, and/or Marrow Cells on Day 0, Given Sheep Erythrocytes (iv on Day 1 and ip on Day 4) and Sacrificed on Day 8.

Cells injected	No. of cells ($\times 10^6$)	Active areas per spleen	Mean active areas	Mean specific activity (% ± S.E.)
Thymus	35-50	{ 00011111 { 2333	1.3	12.3 ± 3.9
Marrow	10	{ 0000000000 { 01111122	.5	1.6 ± .5
Thymus + marrow	50 } 10 }	{ 1334** { ******	>3.0	70.7 ± 3.3
Spleen	5	{ 12334* { *****	>3.0	66.1 ± 9.0
None		{ 000000000 { 111112	.5	7.4 ± 2.9

* Too many active areas to count accurately.

of Playfair *et al*(5) by showing the production of clusters of antibody-producing cells in the spleens of irradiated syngeneic mice given normal spleen cells and then challenged with antigen. Over the range tested, the mean number of active areas was directly proportional to the number of cells injected. Five $\times 10^6$ spleen cells produced more active areas in 8-day experiments than in 5-day experiments. Since 2.2×10^6 normal spleen cells produced one active area, and presumably contain one precursor cell, and since 6.5% of tranferred spleen cells (labelled *in vitro*) appear in the recipient spleen, then one precursor is contained in 143,000 normal spleen cells. (Although it is possible that more cells may eventually localize in the spleen following temporary trapping in the lungs, the value presented here agrees quite well with the 4% reported by Playfair *et al*(5) 24 hours following transfer.) Since a normal mouse spleen contains about 10^8 nucleated cells, it would contain 700 precursor cells. This is considerably more than the 50 precursors estimated by Makinodan and Albright(6) or than the 100 precursors estimated by Jerne *et al*(7).

The thymus does not appear to contain similar potentially immunocompetent cells since thymus cells from normal or immunized donors did not produce significant hemolytic activity in recipient spleens at 5 days. At 8 days there was a small amount of activity in recipients of thymus cells, but none in recipients of marrow cells. The combination of thymus and marrow cells, however, produced more active areas and greater specific activity than can be accounted for by simple summation of the activities of the two donor populations.

The simplest interpretation is that one cell population contains cells capable of making antibody ("effector cells"), but only in the presence of cells from the other population ("auxiliary cells"). These data do not establish which cell suspension contains either effector or auxiliary cells nor how these cells interact(8).

Other transfer experiments have shown antibody production by thymus cells from immunized donors, but the cells have generally been transferred many weeks after donor immunization began(9). The known antibody-producing activity of marrow from immunized donors(10) together with evidence that marrow cells migrate through the thymus(11) make it possible that the immunocompetence of transferred thymus from immunized donors is due to the presence of marrow-derived cells within the thymus. A growing body of data shows that the immunocompetence of marrow cells depends upon the presence of the thymus(12). On the basis of all these data, we feel that it is most likely that the effector cell in our experiments is marrow-derived, and that the thymus provides the auxiliary cells.

Summary. Suspensions containing normal thymus, spleen, or marrow cells were injected into irradiated syngeneic mice which were subsequently given antigen. Normal spleen cells produced discrete areas of antibody production in recipient spleens, and the number of areas was proportional to the number of donor cells. Mice receiving both marrow and thymus cells produced more centers of hemolytic activity in their spleens than mice receiving cells of either type alone. Normal and immunized thymus cells produced little or no hemolytic activity, and normal marrow was also inactive.

We are grateful to Dr. David W. Talmage for advice and discussion, to Dr. Carlos Garciga for aid in irradiating mice, and to Miss Jean Baughman for excellent technical assistance.

1. Good, R. A., Gabrielsen, eds., The Thymus in Immunobiology, Hoeber-Harper, New York, 1964.

2. Landy, M., Sanderson, R. P., Bernstein, M. T., Lerner, E. M., Science, 1965, v147, 1591.

3. Nossal, G. J. V., Ann. N. Y. Acad. Sci., 1964, v120, 171.

4. Osoba, D., Miller, J. F. A. P., J. Exp. Med., 1964, v119, 177; Levey, R. H., Trainin, N., Law, L. W., J. Nat. Cancer Inst., 1963, v31, 199; Metcalf, D. M., Brit. J. Cancer, 1956, v10, 442.

5. Playfair, J. H. L., Papermaster, B. W., Cole, L. J., Science, 1965, v149, 998.

6. Makinodan, T., Albright, J. F., J. Cell. Comp. Physiol., 1962, v60, (suppl. 1), 129.

7. Jerne, N. K., Nordin, A. A., Henry, C., in Cell-bound Antibodies, B. Amos, H. Koprowski, eds., Wistar Inst. Press, Philadelphia, 1963, p109.

8. Globerson, A., J. Exp. Med., 1966, v123, 25; Auerbach, R., in Organogenesis, R. L. DeHaan, H.

Ursprung, eds., Holt Press, New York, 1965; Fishman, M., Nature, 1959, **v183, 1200.**

9. Stoner, R. D., Bond, V. P., J. Immunol., 1963, v91, 185.

10. Gengozian, N., Makinodan, T., Shekarchi, I. C., ibid., 1961, v86, 113.

11. Harris, J. E., Ford, E. E., Nature, 1964, v201, 884.

12. Miller, J.F.A.P., Leuchars, E., Cross, A. M., Dukor, P., Ann. N. Y. Acad. Sci., 1964, v120, 205; Feldman, M., Globerson, A., ibid., 1964, v120, 182.

THE FAILURE OF THYMUS-DERIVED CELLS TO PRODUCE ANTIBODY[1]

A. J. S. Davies, E. Leuchars, V. Wallis, R. Marchant and
E. V. Elliott

*Chester Beatty Research Institute, Institute of Cancer Research,
Royal Cancer Hospital, Fulham Road, London, England*

SUMMARY

The capacities of cells of thymus and bone marrow origin taken from radiation chimaeras to produce antibody during a secondary immune response have been tested in an in vivo transfer system. Although it could be shown that thymus-derived cells respond vigorously by mitosis to antigenic stimulation they were not capable of antibody production. By contrast bone marrow-derived cells did not in the first 3 days respond mitotically to antigenic stimulation but they were capable of limited antibody production. Most antibody was found when *both* cell populations were allowed to react to antigenic stimulation.

INTRODUCTION

In both adult irradiated and neonatal mice the development of immunological responsiveness to certain antigenic stimuli is dependent upon the presence of the thymus (*12, 13*). The defects demonstrable after deprivation of the thymus can, however, be repaired by implantation of thymus grafts (*9, 12*). Although it is not at present possible to be sure that these thymus grafts function as does a thymus in the intact animal, it seems a reasonable assumption that they do, since the spectrum of immunological responsiveness in a reconstituted animal has not so far been shown to be different from that of an intact animal. Thus an experimental schedule of thymus deprivation and reconstitution provides a model for investigating both the manner in which a thymus functions and the property of immunological responsiveness.

Two lines of approach to the problem of thymic function are here relevant. The first is based on the findings of Osoba and Miller (*15*) that thymus grafts in Millipore chambers can restore immunological competence to neonatally thymectomized mice. This suggests that the thymus can function as an endocrine organ liberating a humoral factor which, in some way, potentiates the immune response. That this is not the full story is suggested by the fact that

[1] This investigation has been supported by grants to the Chester Beatty Research Institute (Institute of Cancer Research, Royal Cancer Hospital) from the Medical Research Council and the British Empire Cancer Campaign for Research, and by the U.S. Public Health Service Research Grant No. CA-03188-08 from the National Cancer Institute, National Institutes of Health, and the International Atomic Energy Agency, Research Contract No. 313/R1/RB.

the level of immunological responsiveness which is restored by a thymus graft in a Millipore chamber seems to be less than that restored by a thymus graft not encapsulated. This point is however not yet fully documented. Further, full restoration of peripheral blood lymphocyte counts is not achieved by grafting thymuses in Millipore chambers at least in rats and mice (*1, 15*), though the situation may be different in rabbits (*16*). It could, however, be argued that the Millipore chamber experiments do indeed indicate the manner in which the thymus functions but that the chamber itself is not an appropriate milieu for the full expression of thymic function.

The second approach was based on some preliminary evidence of Leuchars and her co-workers (*10*) that cells derived from a thymus graft, used as a reconstituent in a thymectomized irradiated syngeneic chimaera, were mitotically responsive in the spleen to an antigenic stimulus. These experiments were subsequently amplified (*5*) to show that thymus-derived cells had a well-defined mitotic pattern of response to either sheep red cells or skin homografts. These authors reiterated the view (*4*) that cellular traffic from the thymus, when it can occur, may well be a significant component of thymic function, leaving open the question of whether a humoral factor is also operative. Although the experiments of Davies *et al.* (*5*) showed that thymus-derived cell mitosis was a feature of the complex of changes which followed contact with antigen, on the basis of the evidence presented the exact role, if any, played by these cells in the immune response could not be determined. It was critical to attempt to determine whether the thymus-derived cell populations observed dividing, in the first few days after injection with antigen, went on to differentiate into antibody-producing cells. Relevant experiments are reported here.

METHODS

Design of Experiment

The two strains of mice involved in these experiments are: (1) CBA/Cbi, an inbred strain of mouse which has been maintained in the Chester Beatty Research Institute for many years and which has no marker chromosomes; (2) the CBA/H-*T6T6* strain which is an inbred strain derived from the M.R.C. Radiobiological Research Unit, Harwell, and which has two marker chromosomes. In addition to the chromosome difference between these two strains of mice there is also a minor histocompatibility difference which has been documented in detail elsewhere (*3*). Recognition of both the cytological and antigenic differences is essential to an understanding of the present experimental design.

The plan was to make chimaeras as before (*5*) in which CBA/H-*T6T6* cells derive from a thymus graft and CBA/Cbi cells from a bone marrow graft. These (primary host) mice would be immunized and their spleen cells transferred into irradiated secondary host mice which, before irradiation had been rendered isoimmune in such a manner as to kill or impede the growth of cells derived either from the bone marrow or from the thymus graft. The antigenic difference between CBA/Cbi and CBA/H-*T6T6* is a prerequisite of the success of such a scheme. If the secondary hosts are injected with antigen and the immune responses subsequently determined by titration of antibody it should be possible to determine whether either or both of the two relevant cell populations are capable of antibody production. This plan was put into effect as follows.

Primary hosts. Thymectomized syngeneic chimaeras of the constitution CBA/Cbi → CBA/Cbi carrying a CBA/H-*T6T6* thymus graft were prepared as before (*5*). Thirty days after irradiation these mice were given a single i.v. injection of 5×10^8 sheep red blood cells in 0.2 ml Alsever solution. Fourteen days later this injection was repeated and after a further 7 days the animals were bled and sera were titrated for their production of agglutinating antibody to sheep red cells. The titration procedure was as before (*5*). Mice with high titres (negative $\log_2 12$) were selected as donors and their spleens were removed and stripped of cells by teasing in a small volume of medium 199. Cell counts were made on individual spleens and appropriate dilutions were made so that the secondary hosts could each be injected i.v. with 5×10^6 spleen cells in 0.4 ml of medium 199. Approximately 60×10^6 cells were obtained from any one spleen which thus provided material for the injection of up to 12 secondary hosts.

Secondary hosts. Secondary host mice were male or female CBA/Cbi, or CBA/H-*T6T6* mice 10–15 weeks of age. Some mice of each substrain were made isoimmune against the other substrain by two i.p. injections of spleen cells with 14 days between the first and second injection. The mice were irradiated (850R total body dose) 7–10 days after the second injection and used as recipients for spleen cells from the primary hosts. Immediately after injection of the spleen cells some of the secondary hosts received a single i.p. injection of sheep red blood cells (5×10^8 cells as before).

Thus secondary hosts were of eight different types, comprising Groups 1–8 as follows. Group 1. CBA/Cbi, isoimmune anti-CBA/H-*T6T6*, injected with sheep cells—expectation: no thymus-derived cell activity. Group 2. As Group 1, but not injected with sheep cells—expectation: no immune response measurable in 6 days. Group 3. CBA/Cbi, not isoimmune anti-CBA/H-*T6T6* injected with sheep red cells—expectation: both thymus-derived and other cells will react. Group 4. As Group 3, but not injected with sheep red cells—expectation: no immune respone measurable in 6 days. Group 5. CBA/H-*T6T6*, isoimmune anti-CBA/Cbi, injected with sheep red cells—expectation: only thymus-derived cells will react. Group 6. As Group 5, but not injected with sheep red cells—expectation: no immune response measurable in 6 days. Group 7. CBA/H-*T6T6* not isoimmune anti-CBA/Cbi, injected with sheep red cells—expectation: both thymus-derived and other cells will react. Group 8. As Group 7, but not injected with sheep red cells—expectation: no immune response measurable in 6 days.

A number of descriptions of the secondary host mice were made as follows.

Cytological analysis. Three days after irradiation and transfer of cells, spleens were harvested from three mice in each group and half of each spleen was prepared for cytological analysis (5). Discrimination was made between radiation-damaged cells, CBA/H-*T6T6* cells and CBA/Cbi cells. It was assumed that the radiation-damaged cells were native cells of the secondary host, that CBA/H-*T6T6* cells were in the main derived from the thymus graft in the primary host, and that CBA/Cbi cells were derived (mainly) from the bone marrow graft of the primary host. These assumptions are likely to be correct on the basis of what is known about the development of chimaeric states after irradiation and injections of viable haematopoietic cells (*3, 11*). As far as possible, 100 mitoses were scored on two slides prepared from a sample of any one half-spleen. All scoring was done without the observer being aware of the group of mice involved. No two slides from any one mouse were scored by the same person, neither were observers aware of any previous scores. Results were expressed initially as proportions of each cell type which were finally converted to numbers as recounted in the following section.

Histological observations. The other half of each spleen (the remainder having been taken for cytological analysis as indicated above) was fixed in Carnoy and after transverse sectioning was stained with methyl green and pyronin. Three sections from any one spleen were mounted on one slide. The slides were scored for the numbers of mitoses observable along the two longest axes at right angles of the transverse sections. Each spleen was scored independently by three different observers with the same safeguards as employed for the cytological analysis. During the scoring for mitoses a rather more subjective score of +, ++, +++, or ++++ was made of the number of large pyroninophilic cells in the specimen. The numbers of mitoses counted were reconciled with the frequencies of the various cell types derived from the cytological analysis to give an estimate of the numbers of cells of each type likely to have been dividing in any one spleen sample (*vide supra*). The drawbacks of this method have been discussed previously (5) but it is hoped that it provides a picture of the composition of each spleen which is independent of variation in mitotic index. From notes made by the different observers some idea of the localisation of mitoses and of the large pyroninophilic cells was obtained.

Jerne plaque counts. On each of days 4, 5, and 6 after cell transfer three mice from Groups 1, 2, 5, and 6 were bled (0.1 ml) and their spleens prepared for Jerne plaque analysis using sheep red blood cells as target cells. Sera were titrated both for hemolysins and hemagglutinins in saline.

Titrations. About 10 mice in each group were titrated for their serum content of agglutinating antibody to sheep red cells 6 days after irradiation and transfer of cells.

RESULTS

Cytological and Histological Analysis

The results expressed in Table 1 give three facets of the same picture. The numbers of cells seen of any chromosomal phenotype were the numbers of such cells scored on cytological preparations on two slides made from a single spleen. These numbers could form a basis of comparison between mice if larger sam-

TABLE 1

Cytological analysis of secondary host mice 3 days after their irradiation and injection with spleen cells from primed primary hosts

Parameter	Secondary CBA/Cbi hosts				Secondary CBA/H-T6T6 hosts			
	Isoimmune anti-CBA/H-*T6T6*		Not isoimmune		Isoimmune anti-CBA/Cbi		Not isoimmune	
Injected with sheep cells or not	+	−	+	−	+	−	+	−
Exp. Group number	1	2	3	4	5	6	7	8
No. of radiation-damaged cells seen	26, 4, 12	6, 5, 1	4, 4	3, 2, 0	15, 3, 7	7, 1, 14	3, 2, 8	5, 8
No. of CBA/Cbi cells seen	14, 3, 5	2, 2, 4	8, 5	4, 2, 1	2, 0, 0	1, 0, 1	53, 48, 26	94, 32
No. of CBA/H-*T6T6* cells seen	1, 0, 0	2, 1, 1	88, 35	2, 0, 0	75, 44, 23	0, 0, 3	45, 50, 15	1, 0
Mean % radiation-damaged cells	63.7	39.6	6.6	27.8	15.4	88.5	7.1	12.5
Mean % CBA/Cbi cells	35.5	45.8	9.7	64.8	0.7	6.0	51.2	87.0
Mean % CBA/H-*T6T6* cells	0.8	14.6	83.7	7.4	83.9	5.5	41.7	0.5
Estimated No. damaged cells/spleen sample[a]	18, 18, −	−, 7, 2	5, 7	14, 1, 0	14, 5, 7	14, 3, 14	13, 20, 7	12, 17
Estimated No. CBA/Cbi cells/spleen sample	10, 14, −	−, 3, 7	10, 8	18, 1, 24	2, 0, 0	2, 0, 1	232, 164, 66	220, 70
Estimated No. CBA/H-*T6T6* cells/spleen sample	1, 0, −	−, 2, 2	113, 59	9, 0, 0	68, 71, 21	0, 0, 3	196, 170, 38	2, 0
Large pyroninophilic cells	±	−	+++	+	+++	−	++++	++

[a] Approximately 10,000 cells.

ples had been taken but it is not in fact practicable to make anything like a complete survey of a spleen-half by this method. The mean percentages of any one cell type give an idea of the predominant cell type in any one spleen but again are not a satisfactory basis for comparison between mice because of the very wide variation in mitotic rate. The estimated numbers of cells per spleen sample, calculated by multiplying the proportion of cells of any one type by the numbers of mitosing cells observed in about 10,000 cells, dividing and non-dividing, in spleen sections do, however, form a basis of comparison which as previously pointed out makes allowance for variation in mitotic rate and to some extent for variation in spleen size. On this basis it can be seen firstly in Groups 1 and 2 and 5 and 6 that very few cells of the genotype against which the mice in these groups had been isoimmunized were in division at the time of sampling. Secondly, little mitotic activity of any kind was present in Groups 1, 2, 4, and 6. As far as Group 1 is concerned this indicates that bone marrow-derived cells evince little or no mitotic response 3 days after contact with antigen. In Groups 3 and 5 antigen injection is required to give the effect observed (i.e., enhanced mitosis of CBA/H-*T6T6* thymus-graft derived cells). This is also true of thymus-derived cells in Groups 7 and 8 but clearly not of bone marrow-derived cells. An unexpected result was the large number of bone marrow-derived cells brought into division in Groups 7 and 8 irrespective of injection of antigen.

The histological preparations were used in relation to the cytological analysis as indicated above but in addition the numbers of large pyroninophilic cells were assessed as shown in Table 1. The large pyroninophilic cells had large

TABLE 2

The immune responses of secondary host mice at various times after irradiation and transfer of spleen cells from primed primary hosts

Assay[a]	Gp. 1	Gp. 2	Gp. 3	Gp. 4	Gp. 5	Gp. 6	Gp. 7	Gp. 8
4-day Jerne	5, 11, 11	10, 13, 11			23, 23, 11	11, 6, 19		
4-day lysin titre	— — —	— — —			— — —	— — —		
5-day Jerne	135, 255, 75	32, 28, 25			8, 24, 13	6, 12, 5		
5-day lysin titre	3, 2, 1	— — —			— — —	— — —		
6-day Jerne	96, 195, 559	41, 37, 23			32, 11, 11	24, 9, 10		
6-day lysin titre	4, 1, 5	— — —			— — —	— — —		
6-day agglutinin titre (±S.D.)	5.3 ± 1.2 (n = 14)	1.5 ± 0.9 (n = 8)	9.0 ± 0.6 (n = 10)	1.6 ± 1.0 (n = 11)	1.1 ± 0.1 (n = 17)	1.4 ± 1.06 (n = 8)	10.0 ± 1.3 (n = 12)	6.8 ± 1.1 (n = 13)

[a] Jerne plaque assay values = total plaques/spleen; lysin titres are negative $\log_2$ of final dilution showing macroscopically positive.

nucleoli, large nuclei and abundant pyroninophilic cytoplasm similar to those described by Gowans (7). They were located almost entirely in the lymphoid follicles and could be seen to be a mitotically active cell population. It seems that significant numbers of such cells were always observed when, and (with the exception of Groups 7 and 8) *only* when, a thymus-derived cell reaction was in evidence. In Group 7 the reacting cell population was about half thymus-derived and half bone marrow-derived and there were very many large pyroninophilic cells. In Group 8 large pyroninophilic cells were associated with a bone marrow-derived cell population which was not reacting to sheep red cells but to some other stimulus.

The Immune Response

The immune responses of the secondary host mice were measured by the Jerne technique and by lysin titrations on days 4, 5, and 6 and by titration of hemagglutinins on day 6 as shown in Table 2. Taking the findings group by group it is possible to describe the results in Tables 1 and 2 as follows.

Group 1. Bone marrow-derived cells on their own have the capacity to react to previously encountered antigens. The mitotic component of this reaction is not however obvious by 3 days after its initiation.

Group 2. The reactions seen in Group 1 both mitotic and immunological are dependent on the use of a recall injection of antigen.

Group 3. Bone marrow and thymus-graft-derived cells reacting together give a more vigorous anamnestic response than (by reference to Group 1) do bone marrow-derived cells alone ($P < 0.001$). The early mitotic response to antigenic stimulus is of thymus-derived cells only.

Group 4. The reactions seen in Group 3 appear to depend upon the use of a recall injection of antigen and do not simply reflect carry over and amplification of an existing immune condition.

Group 5. Despite the fact that thymus-derived cells are mitotically reactive to antigenic stimulation they are not capable of antibody production, nor on their own of giving rise to antibody-producing cells.

Group 6. Thymus-derived cells, whether or not they respond mitotically to a recall injection of antigen, are incapable of antibody production or on their own of giving rise to antibody-producing cells.

Group 7. The immune response is similar to that of Group 3 and indicates that bone marrow and thymus-derived cells reacting together give rise to a vigorous immune response. However, in contrast to Group 3 the mitotically reactive cell populations early after antigenic stimulation are of both cell types.

Group 8. This was an unexpected result indicating that under certain circumstances a primed cell population can effect a vigorous immune response without further known contact with the priming antigen. The mitotically active component of such a reaction seems in the early stages to be bone marrow-derived, a behaviour pattern in contrast to that observed by bone marrow-derived cells in Group 1. The most obvious possibility is that the immunological reaction observed in Group 8 in some way represents a reaction on the part of primed CBA/Cbi (bone marrow-derived) cells to contact with the antigens which distinguish CBA/H-*T6T6* mice from CBA/Cbi mice. This "cross-reaction" would probably be suppressed in Groups 5 and 6 by the isoimmune status of the secondary host CBA/H-*T6T6* mice.

DISCUSSION

The technique of quantitation of the secondary immunological responsiveness of cell populations by their transfer to irradiated animals is well known. It is usually assumed that the immune responses tested early after irradiation are a property of the transferred cells, anamnestic, and antigen dependent (*6, 14*). There is no reason a priori to assume that isoimmunity is likely to affect the validity of these basic assumptions. Model experiments established that the residual isoimmunity after irradiation was sufficient to suppress the immunological activities of 5×10^6 primed CBA/Cbi spleen cells reacting to antigen in isoimmune CBA/H-*T6T6* secondary host mice and vice versa (*8*). Doria and his colleagues (*6*) used a very similar system for analysing which of the cellular components of an allogeneic chimaera, donor or host, was immunologically responsive.

These characteristics of the test system must be borne in mind in relation to the conclusions firstly, that thymus-derived cell mitosis though not a prerequisite of antibody production during the course of a secondary immune response nevertheless seems to have some adjuvant effect, and secondly, that on their own thymus-derived cells are incapable of antibody production. It must be allowed that antibody may have been produced by cells of thymic origin but not in sufficient quantities to be detected by the present methods; further, thymus-derived cells may in some way have mediated the immune responses determined for Group 1 (secondary host isoimmune anti thymus-type antigens) but in a way that did not require that they should divide. If, however, thymus-derived cells do make a contribution to the secondary immune response they must do so by in some way influencing the antibody-producing population of cells. Superficially there are two ways that it might be envisaged that this influence could be exerted.

Firstly, the thymus-derived cells might elaborate information, specifically related to the antigenic stimulus, which they convey to other cells that then produce the antibody. The reacting thymus cell populations would seem to contain unusually large amounts of RNA on the basis of their staining characteristics and this compound in some form is the obvious candidate for transfer of the hypothetical message. Such a transfer has been postulated as part of the immune response in other experimental systems (4).

Secondly, the thymus-derived cells may, by reacting to the antigenic stimulus, encourage another population of cells to react not by providing information specifically related to the antigen but by providing a non-specific stimulus to mitosis which affects the speed with which the other cell population responds. Any such arguments must necessarily be tentative on the basis of the present evidence. Others have however commented on the existence of a mitotically active non-antibody-producing cell population (2). The present evidence suggests that a proportion of this population may be of thymic origin. It is not at present possible to decide whether thymus-derived cells brought into mitosis by antigenic stimulation are thereby pre-empted for further activity only in relation to the exciting antigen.

The results obtained in Groups 7 and 8 require special consideration. Here an immune response could be observed which though stronger when thymus-derived cells were activated by antigenic stimulation (Group 7) was nevertheless not dependent on these cells nor on the administration of antigen (Group 8). The evidence is clear that in Group 8 only CBA/Cbi cells were reacting mitotically and because a similar reaction was not observed in either Group 3 or 4 it may be supposed that the reaction is a response to the antigens which distinguish CBA/H-*T6T6* mice from CBA/Cbi mice. (Such a reaction would supposedly have been suppressed in Groups 5 and 6.) In effect the reaction to one antigen seems to have led to the production of antibody to a different previously administered antigen. This finding prompted an experiment to ascertain whether the apparent reaction between CBA/Cbi spleen cells and CBA/H-*T6T6* antigens was really non-reciprocal and also whether it depended on the use of chimaeras as primary hosts. The results quoted in Table 3 make it clear that primed CBA/Cbi spleen cells do react to CBA/H-*T6T6* antigens in such a manner that antibodies which agglutinate sheep red cells are pro-

TABLE 3

Agglutinin titres of CBA/Cbi *and* CBA/H-T6T6 *host mice 6 days after 850r and i.v. injection of either* CBA/Cbi *or* CBA/H-T6T6 *spleen cells*[a]

Donor	Host	Host injected with sheep RBC	Host not injected with sheep RBC
CBA/H-*T6T6*	CBA/H-*T6T6*	7.2 ± 1.6, n = 10	1̄.7 ± 0.8, n = 8
CBA/H-*T6T6*	CBA/Cbi	5.3 ± 2.1, n = 6	1̄.66 ± 0.8, n = 3
CBA/Cbi	CBA/H-*T6T6*	6.13 ± 1.5, n = 16	5.28 ± 1.7, n = 18
CBA/Cbi	CBA/Cbi	3.9 ± 2.0, n = 8	— n = 3

[a] 5×10^6 cells per host from donor mice which had been primed with sheep red cells 7 and 21 days before they were killed; some host mice were injected with sheep cells others were not.

duced. This reaction is not reciprocal and is not dependent on the use of chimaeras as primary hosts and although it complicates the present series of experiments it does not seem to obscure the main conclusions. The sera of secondary hosts were not usually titrated before irradiation and injection of spleen cells and there remains the possibility that some of the mice in Groups 1 and 2 had agglutinating antibodies as a result of isoimmunization (CBA/Cbi anti CBA/H-*T6T6*). However a control experiment set up later to check this point revealed that repeated isoimmunization only rarely led to the production of antibody which would agglutinate sheep cells and then only in low titres. Thus, although in future experiments it will be expedient to check secondary host titres before their use it seems unlikely that their slight heterogeneity affects the present series of experiments.

Finally, consideration must be given to the meaning of "thymus-derived". Throughout this paper only cells of CBA/H-*T6T6* chromosomal phenotype have been described as of thymic origin. But, as has been shown elsewhere (*11*), cells of bone marrow origin do enter the thymus graft and it is difficult to see why if they subsequently left the graft they could not reasonably be called thymus-derived. The existence of such a population of thymus-activated cells of bone marrow origin would seriously interfere with the interpretation of the experiments presented and in particular it would suggest that the conclusion that thymus derived cells fail to produce antibody should be qualified. For example, if cells of bone marrow origin can later become thymus-derived then it must be allowed that the Group 1 titres were possibly brought about by such thymus-derived cells. There is however no evidence in the cytological analysis of a population of cells of bone marrow origin with the same early mitotic reactivity in response simply to stimulation by sheep red blood cells as is shown by cells of known thymic origin. Thus, either the "native" thymus cells and immigrant bone marrow cells have subsequent to their emergence from the thymus got difference potentialities for reaction, or few if any cells of bone marrow origin have been processed by and left the thymus graft. The evidence so far is inadequate for any firm conclusions to be made but the latter alternative seems preferable at the moment.

Acknowledgments. The authors are grateful for the advice and encouragement of Professor P. C. Koller during the course of this work.

REFERENCES

1. Aisenberg, A. C.; Wilkes, B. 1965. Nature *205*: 716.
2. Balfour, B. M.; Cooper, E. H.; Meek, E. S. 1965. J. Reticuloendothelial Soc. *2*: 379.
3. Davies, A. J. S.; Doak, S. M. A.; Leuchars, E. 1963. Nature *200*: 1222.
4. Davies, A. J. S.; Doe, B.; Cross, A. M.; Elliott, E. V. 1964. Nature *203*: 1039.
5. Davies, A. J. S.; Leuchars, E.; Wallis, V. J.; Koller, P. C. 1966. Transplantation *4*: 438.
6. Doria, G.; Goodman, J. W.; Gengozian, N.; Congdon, C. 1962. J. Immunol. *88*: 20.
7. Gowans, J. L. 1962. Ann. N.Y. Acad. Sci. *99*: 432.
8. Leuchars, E. 1966. Ph.D. thesis. University of London.
9. Leuchars, E.; Cross, A. M.; Dukor, P. 1965. Transplantation *3*: 28.
10. Leuchars, E.; Cross, A. M.; Davies, A. J. S.; Wallis, V. J. 1964. Nature *203*: 1189.
11. Koller, P. C.; Davies, A. J. S.; Leuchars, E.; Wallis, V. J. *In* Proceedings of Bristol lymphocyte symposium. Edward Arnold, London (in press).
12. Miller, J. F. A. P. 1961. Lancet *2*: 748.
13. Miller, J. F. A. P.; Doak, S. M. A.; Cross, A. M. 1963. Proc. Soc. Exp. Biol. Med. *112*: 205.
14. Makinodan, T.; Perkins, E. M.; Shekarchi, I. C.; Gengozian, N. 1960. p. 182. *In* M. Holub and L. Jarasková (eds.). *Mechanisms of antibody formation.* Czech. Acad. Sci., Prague.
15. Osoba, D.; Miller, J. F. A. P. 1963. Nature *199*: 653.
16. Trench, C. A. H.; Watson, J. W.; Walker, F.; Gardner, P. S.; Green, C. A. 1966. Immunology *10*: 187.

Received 17 June 1966.

CELL TO CELL INTERACTION IN THE IMMUNE RESPONSE

II. The Source of Hemolysin-Forming Cells in Irradiated Mice Given Bone Marrow and Thymus or Thoracic Duct Lymphocytes[*], [‡]

By G. F. MITCHELL and J. F. A. P. MILLER, M.B.

(From The Walter and Eliza Hall Institute of Medical Research, Melbourne, Australia)

(Received for publication 3 June 1968)

An interaction between thymus or thoracic duct cells and antibody-forming cell precursors has been implicated in the response of mice to sheep erythrocytes (1). The possibility was raised that the thymus contributes cells to the circulating pool of lymphocytes which recognize antigen (ARC) and which influence the differentiation of antibody-forming cell precursors (AFCP) to hemolysin-forming cells. The experimental system employing reconstituted neonatally thymectomized mice failed to detect the existence of AFCP in inoculated lymphoid cell suspensions.

Previous work from this laboratory (2) had demonstrated that thoracic duct and spleen cells were capable of affecting the appearance of hemolytic foci in the spleens of heavily irradiated recipients injected with sheep erythrocytes. If such foci represent clusters of antibody-forming cells, then either the inoculum must contain cells potentially capable of producing hemolysins, or the host must provide AFCP which are extremely radio-resistant.

In this paper, we present the results of experiments designed to determine whether both ARC and AFCP are present in populations of cells from thymus, thoracic duct lymph, or bone marrow.

Materials and Methods

Animals.—Mice of the highly inbred strains CBA and C57BL, and F_1 hybrids from crosses between these strains were used. The origin and maintenance of these mice have been described in the previous paper (1).

Cell suspensions were prepared as before (1).

Operative Procedures.—Thymectomy or sham operation was performed in young adult mice, 4–6 wk old, as previously described (3). Checks for the presence of thymus remnants

* This is publication 1248, from The Walter and Eliza Hall Institute of Medical Research.

‡ Supported by the National Health and Medical Research Council of Australia, the Australian Research Grants Committee, the Damon Runyon Memorial Fund for Cancer Research, the Jane Coffin Childs Memorial Fund for Medical Research, and the Anna Fuller Fund.

were made at autopsy and mice with such remnants were discarded from the experiments. Thoracic duct cannulation was performed as described in the previous paper (1).

Irradiation.—Intact or adult thymectomized mice were exposed to total-body irradiation in a Perspex box. In the case of thymectomized mice, the irradiation was performed 1–3 wk after thymectomy. The dose given was 800 rads to mid-point with maximum back scatter conditions and the machine operated under conditions given in the previous paper (1). The focal skin distance was 50 cm and the absorbed dose rate was 170 rads per minute. When thymectomized irradiated mice were to be protected with bone marrow, they received an intravenous injection of 2–5 million cells in 0.1–0.2 ml Dulbecco's phosphate-buffered saline 1–3 hr postirradiation. All irradiated mice were given oxytetracycline[1] (100 mg per liter) or penicillin[2] (600,000 units per liter) in the drinking water.

Assays for Hemolysin Plaque-Forming Cells and Hemolytic Foci.—Hemolysin plaque-forming cells were assayed according to the technique of Jerne et al. (4). The hemolytic focus assay used in the present experiments was a modification of the techniques described by Kennedy et al. (5) and Playfair et al. (6). The spleens were not frozen but serially sectioned into 250-μ segments by means of a tissue chopper[3] and sequentially transferred with forceps to an agar plate. These plates were then heated to 37°C and double the quantity of the top, soft agar layer as used in the Jerne assay, was poured over the segments. This layer contained 3.8 ml of 1.4% agar in double strength Eagle's medium, 2 mg dextran, and 0.2 ml of 20% sheep erythrocytes. After incubation for 1–2 hr, 1:10 guinea pig serum was added and the plates incubated for a further hour. The serum was then removed and the plates cooled to 4°C. The red cells were stained with benzidine[4] for 1–2 min to increase the contrast between the areas of hemolysis which had developed around the segments and the intact red cells. Immediately after staining, the hemolytic areas were counted with the aid of a dissecting microscope.

Preparation of Anti-H2 Sera and Incubation of Plaque-Forming Cells.—The methods for preparing specific anti-H2 sera and for incubating plaque-forming cells were given in the preceding paper (1).

Statistical Analysis.—The standard errors of the means were calculated and P values determined by Student's t test.

RESULTS

Effect of Syngeneic Thymus, Thoracic Duct, or Bone Marrow Cells in Irradiated Hosts.—10 million CBA thoracic duct cells, thymus cells, or bone marrow cells were injected together with 10^8 SRBC[5] into groups of syngeneic recipients that had been exposed to 800 rads total-body irradiation. The number of PFC per spleen was determined 7 days later. As seen in Table I, 10 million thoracic duct cells produced on the average 1270 PFC whereas 10 million thy-

---

[1] Terramycin, Pfizer Pty Ltd., Sydney, Australia.

[2] Crystalline Penicillin G, Evans Medical Australia (Pty) Ltd., Sydney, Australia.

[3] McIlwain's tissue chopper (7), Mickle Laboratory Engineering Co., Mill Works, Gomshall, Surrey, England.

[4] 1 part of 5% hydrogen peroxide to 9 parts of 0.556% benzidine in 12.5% acetic acid kept at 4°C.

[5] The following abbreviations are used: SRBC, sheep erythrocytes; PFC, hemolysin plaque-forming cells; AFCP, antibody-forming cell precursors; ARC, antigen-reactive cells; and SE, standard error.

mus cells did not increase the number of PFC above that obtained in irradiated mice given SRBC alone. Increasing the number of thymus cells to 50 million increased the average number of PFC to only 45. Injection of 25 million thoracic duct cells in the absence of antigen did not cause any elevation above the background number of 15 PFC per spleen. Similarly, bone marrow cells were ineffective.

During investigations of the hemolytic focus–producing capacity of thoracic duct and spleen cell inocula (2) it was noted that the foci produced by thoracic duct cells involved only one or two consecutive segments of the sliced spleens. With spleen cell inocula, hemopoietic regeneration occurred in the irradiated spleens and individual hemolytic foci often extended over three or four seg-

TABLE I

PFC Produced in the Spleens of Heavily Irradiated CBA Mice after Injection of SRBC and Syngeneic Thymus, Bone Marrow, or Thoracic Duct Cells

Cells inoculated	No. of mice	Average PFC per spleen 8 days postirradiation ($\pm$ SE)
SRBC only	16	15 ± 6.1
10×10^6 thymus cells + SRBC	14	15 ± 3.0
50×10^6 thymus cells + SRBC	8	45 ± 5.0
10×10^6 bone marrow cells + SRBC	16	27 ± 5.8
25×10^6 thoracic duct cells only	5	17 ± 4.4
10×10^6 thoracic duct cells + SRBC	23	$1270 \pm 338*$

* $P < 0.05 - < 0.01$ when this value is compared with those in all other groups.

ments. To investigate whatever effect hemopoietic regeneration might have on the size of hemolytic foci produced by thoracic duct cells, bone marrow cells were injected simultaneously. In Table II, the results of two experiments using (CBA $\times$ C57BL)F$_1$ mice as both donors and recipients, are shown. A mixed inoculum of one million thoracic duct cells and 10 million bone marrow cells did not increase the number of hemolytic foci per spleen over that produced by one million thoracic duct cells alone but greatly increased the number of PFC per spleen.

A time course study of the production of hemolytic foci and PFC per spleen was performed in heavily irradiated CBA mice injected with SRBC and one million CBA thoracic duct cells, 10 million CBA bone marrow cells or a mixed inoculum of both cell types. 4–9 days after irradiation and cell inoculation, half the number of spleens were assayed for their content of PFC and the other half were serially sectioned to determine the number of hemolytic foci. The results are shown in Figs. 1 and 2. The majority of heavily irradiated CBA mice of our colony had died by 9 days unless they were given hemopoietic cells. Hence adequate data was not available at 9 days in the case of irradiated mice injected with SRBC and thoracic duct cells alone. At day 8, one million thoracic duct cells produced on the average about 100 PFC and five hemolytic foci per spleen. Addition of bone marrow cells to the inoculum of thoracic duct cells significantly increased the number of PFC per spleen to about 900 ($P <$ 0.01) but did not increase the number of hemolytic foci. As the foci presumably represent discrete clusters of PFC, it can be calculated that, at the peak of the response, the foci resulting from inoculation of thoracic duct cells contain about 20 PFC. The addition of bone marrow cells increased this number to

TABLE II

Hemolytic Foci and PFC Produced in the Spleens of Heavily Irradiated (CBA $\times$ C57BL)F_1 Mice after Injection of SRBC and Syngeneic F_1 Thoracic Duct and F_1 Bone Marrow Cells

Cells inoculated	Hemolytic foci 8 days postirradiation		PFC 8 days post-irradiation	
	No. of mice tested	No. of foci per spleen	No. of mice tested	Average No. of PFC per spleen ($\pm$ SE)
SRBC only	10	0, 0, 0, 0, 0, 0, 0, 1, 1, 1	8	12 $\pm$ 3.5
10×10^6 bone marrow cells + SRBC	4	0, 0, 0, 1	4	31 $\pm$ 2.9
10^6 thoracic duct cells + SRBC	12	3, 3, 3, 4, 5, 5, 5, 6, 6, 7, 8, 8	13	44 $\pm$ 12.3
10^6 thoracic duct cells + 10×10^6 bone marrow cells + SRBC	13	2, 2, 3, 3, 3, 6, 6, 6, 7, 7, 7, 7, 10	16	527 $\pm$ 75.4

approximately 160 PFC. The 7–9 day PFC response of irradiated CBA mice injected with 10 million CBA bone marrow cells varied from 27 $\pm$ 5.8 in one series of experiments (Table I) to 110 $\pm$ 21.1 in another (Fig. 1). The number of hemolytic foci in bone marrow–injected mice however never increased significantly above the background number of one hemolytic focus per spleen obtained in irradiated mice injected with either SRBC only or lymphoid cells only. These results indicate that while irradiated mice given SRBC and thoracic duct cells alone can produce PFC, this response is greatly increased by the simultaneous injection of bone marrow cells.

When thymus cells were used instead of thoracic duct cells, it was evident that even 50 million thymus cells failed to achieve the same effect as one million thoracic duct cells in irradiated mice injected with SRBC and bone marrow (Table III). Therefore thoracic duct cells were far superior to thymus cells in this system in contrast to what was observed in neonatally thymectomized mice (1). This finding prompted the following studies in adult thymectomized irradiated mice injected with bone marrow.

Effect of Syngeneic Thymus or Thoracic Duct Cells in Thymectomized Irradiated Bone Marrow-Protected Hosts.—10 million CBA thymus or thoracic duct cells were injected together with SRBC into adult thymectomized mice 2

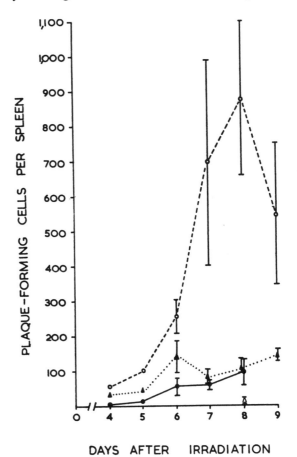

DAYS AFTER IRRADIATION

FIG. 1. PFC produced in the spleens of heavily irradiated CBA mice injected, after irradiation, with SRBC alone (△), SRBC and 1 million syngeneic thoracic duct cells (●——●), SRBC and 10 million syngeneic bone marrow cells (▲....▲), and SRBC and a mixed inoculum of 1 million syngeneic thoracic duct cells and 10 million syngeneic bone marrow cells (○----○). The magnitude of twice the standard errors is shown by the vertical bars. Each point at 4, 5, and 9 days represents the mean of determinations made on 2–5 mice and at 6, 7, and 8 days on 6–13 mice.

wk after irradiation and marrow protection. The number of PFC per spleen was determined 2, 4, 5, 7, and 10 days thereafter. From Fig. 3 it can be seen that while mice given thoracic duct cells produced 72,000 PFC, mice receiving the same number of thymus cells could produce only about 6000 PFC at the

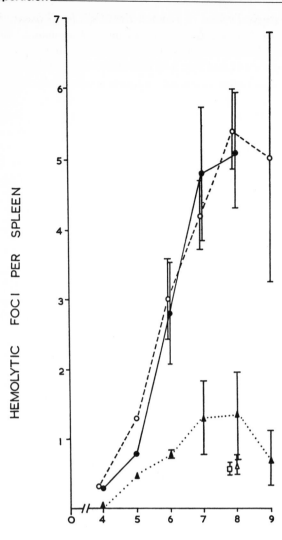

FIG. 2. Hemolytic foci produced in the spleens of heavily irradiated CBA mice injected after irradiation with SRBC alone (△), 1 million syngeneic thoracic duct cells alone (□), SRBC and 1 million syngeneic thoracic duct cells (●——●), SRBC and 10 million syngeneic bone marrow cells (▲....▲), and SRBC and a mixed inoculum of 1 million syngeneic thoracic duct cells and 10 million syngeneic bone marrow cells (○-----○). The magnitude of twice the standard errors is shown by the vertical bars. Each point at 4, 5, and 9 days represents the mean of determinations made on 2–4 mice and at 6, 7, and 8 days on 6–14 mice.

height of the response. Mice injected with SRBC only gave less than 300 PFC. In the case of thoracic duct cell recipients, the number of PFC increased 12-fold between days 4 and 5. When the data for 10 million thoracic duct cells are compared in Table I and Fig. 3, it is evident that far more PFC were produced in the bone marrow–protected thymectomized irradiated hosts than in the irradiated hosts not given bone marrow.

It is apparent, therefore, that thymus cells are not as effective as thoracic duct cells in thymectomized irradiated bone marrow–protected hosts in contrast to the situation (1) in neonatally thymectomized mice. One notable difference between the two hosts was the size of their spleens. Thus, seven mice which had been thymectomized at 6 wk, irradiated and injected with

TABLE III

PFC Produced in the Spleens of Heavily Irradiated CBA Mice after Injection of SRBC and Syngeneic Thymus or Thoracic Duct Cells together with Bone Marrow Cells

Cells inoculated	No. of mice	Average PFC per spleen 8 days post-irradiation ($\pm$ SE)
SRBC	12	13 $\pm$ 1.8
10×10^6 bone marrow cells + SRBC	15	73 $\pm$ 23.5
50×10^6 thymus cells + SRBC	20	52 $\pm$ 19.3
50×10^6 thymus cells + 10×10^6 bone marrow cells + SRBC	13	522 $\pm$ 341
10^6 thoracic duct cells + SRBC	10	97 $\pm$ 34.7
10^6 thoracic duct cells + 10×10^6 bone marrow cells + SRBC	13	877 $\pm$ 218

bone marrow cells had an average of 81 $\pm$ 7.0 million nucleated spleen cells 2 wk after irradiation, which is significantly less than the average number of 148 million cells in the spleens of neonatally thymectomized mice (1). The possibility that this accounts for the different results will be discussed later.

Effect of Allogeneic Thymus, Thoracic Duct, and Bone Marrow Cells in Irradiated Hosts.—In order to identify, by means of anti-H2 sera, the identity of PFC produced in irradiated hosts in response to SRBC and thymus, thoracic duct and bone marrow cells, allogeneic combinations must be used. Accordingly, (CBA $\times$ C57BL)F_1 thoracic duct cells were injected together with SRBC and CBA bone marrow into heavily irradiated CBA mice. The results of two representative experiments are shown in Table IV. When a mixed inoculum of F_1 thoracic duct cells, CBA bone marrow cells, and SRBC was given on the day of irradiation, 800 PFC appeared in the spleens 7 days later. This was only slightly greater than the sum of the mean number of PFC produced in the spleens of mice given either SRBC alone, SRBC and thoracic

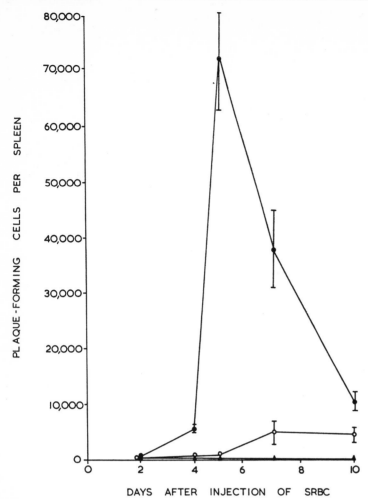

Fig. 3. PFC produced in the spleens of adult thymectomized CBA mice, heavily irradiated and protected with syngeneic bone marrow cells 2 wk previously, and injected with SRBC alone (▲——▲), SRBC and 10 million syngeneic thymus cells (O——O), and SRBC and 10 million syngeneic thoracic duct cells (●——●). The magnitude of twice the standard errors is shown by the vertical bars. Each point represents the mean of determinations made on 3–10 mice.

duct cells, or SRBC and bone marrow cells.[6] These results therefore contrast with those obtained when syngeneic combinations of cells were used (*vide supra*). Attempts were then made to determine whether synergism between

---

[6] Likewise, when irradiated F_1 mice were used, instead of CBA mice, as recipients of a mixed inoculum of SRBC, F_1 thoracic duct cells and CBA bone marrow, no convincing synergistic effect was obtained.

allogeneic cells could take place in irradiated hosts if CBA bone marrow cells were allowed to reside in their hosts for a period of time before introducing F_1 thoracic duct cells and antigen. As shown in Table IV, giving thoracic duct cells and SRBC 2 days after bone marrow did not obviously augment the PFC response. Increasing this interval still further was not attempted in these irradiated hosts since it is well known that immunological recovery takes place in irradiated mice given bone marrow. If adult mice are thymectomized prior to irradiation and bone marrow protection, however, the recovery of immunological capacity is impaired (8). This type of host was therefore chosen to deter-

TABLE IV

PFC Produced in the Spleens of Heavily Irradiated CBA Mice after Injection of SRBC,
(CBA × C57BL)F_1 Thoracic Duct, and CBA Bone Marrow Cells

Exp.	Cells inoculated	No. of mice	Average PFC per spleen 7 days after SRBC
1	6×10^6 F_1 thoracic duct cells + 10×10^6 CBA bone marrow cells + SRBC	6	800
	6×10^6 F_1 thoracic duct cells + SRBC	3	593
	10×10^6 CBA bone marrow cells + SRBC	8	116
	SRBC only	10	19
2	10×10^6 CBA bone marrow cells followed 2 days later by 3×10^6 F_1 thoracic duct cells + SRBC	6	496
	10×10^6 CBA bone marrow cells followed 2 days later by SRBC	8	103
	3×10^6 F_1 thoracic duct cells + SRBC	8	228
	SRBC only	8	43

mine whether synergism between allogeneic cells can take place. 10 million F_1 thymus or thoracic duct cells were injected 2 wk after protecting irradiated adult thymectomized CBA mice with CBA bone marrow. The results of a time course study on the PFC responses of these mice are shown in Fig. 4. Thoracic duct cells increased the response to 19,800 PFC at the peak. Thymus cells failed to produce an increase. A comparison of the results obtained in Figs. 3 and 4 clearly shows that, in thymectomized irradiated bone marrow-protected hosts, syngeneic cells are more effective than semiallogeneic cells.

Identity of the Hemolysin Plaque–Forming Cells.—When spleen cell suspensions containing PFC were incubated with specific isoantisera the number of detectable PFC was reduced by 80–100%. Nonspecific isoantisera or normal mouse sera reduced the number to an insignificant extent (1).

Spleen cells from irradiated CBA mice given F_1 thoracic duct cells and SRBC on the day of irradiation were incubated with normal CBA or C57BL serum,

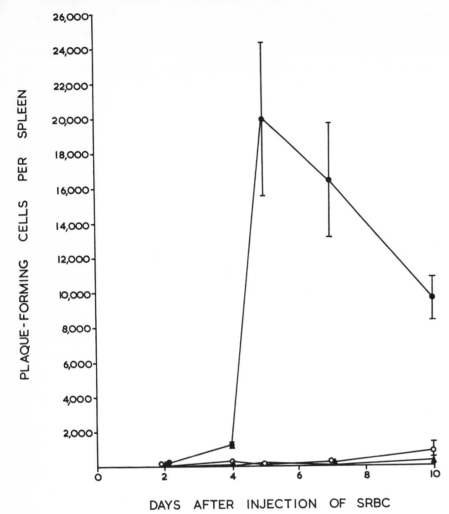

FIG. 4. PFC produced in the spleens of adult thymectomized CBA mice, heavily irradiated and protected with syngeneic bone marrow cells 2 wk previously, and injected with SRBC alone (▲——▲), SRBC and 10 million semiallogeneic (CBA × C57BL)F₁ thymus cells (○——○), and SRBC and 10 million semiallogeneic F₁ thoracic duct cells (●——●). The magnitude of twice the standard errors is shown by the vertical bars. Each point represents the mean of determinations made on 2–10 mice.

anti-CBA serum or anti-C57BL serum. The results of three experiments with 14 mice injected with various numbers of thoracic duct cells are shown in Table V. Anti-CBA serum reduced the number of PFC by 85–93% and anti-C57BL serum reduced the number by 85–95%. It is thus apparent that the PFC produced in these hosts were derived from the inoculated cells.

The number of PFC produced following an inoculation of F_1 thoracic duct cells and SRBC was far greater in adult thymectomized irradiated CBA mice injected with CBA bone marrow than in irradiated mice not receiving bone marrow (cf. Tables V and VI). In Table VI are shown the results of incubating

TABLE V

PFC from Spleens of Reconstituted Irradiated CBA Mice Remaining after Incubation with Isoantisera

Cells used for reconstitution	No. of spleens in pool	Average PFC per spleen at 7 days	No. of PFC per aliquot remaining after incubation with:		
			Normal mouse serum	Anti-CBA serum	Anti-C57BL serum
6×10^6 F_1 thoracic duct cells	3	593	98	14 (85%)*	11 (89%)
10×10^6 F_1 thoracic duct cells	7	822	72	6 (92%)	11 (85%)
30×10^6 F_1 thoracic duct cells	4	5,655	432	32 (93%)	21 (95%)

* Number in brackets refers to per cent reduction.

TABLE VI

PFC from Spleens of Reconstituted Thymectomized Irradiated CBA Mice Remaining after Incubation with Isoantisera

Cells used or reconstitution*	No. of spleens in pool	Average PFC per spleen	No. of PFC per aliquot remaining after incubation with:		
			Normal mouse serum	Anti-CBA serum	Anti-C57BL serum
10×10^6 F_1 thoracic duct cells	5	10,980	80	4 (95%)‡	88 (0%)
10×10^6 F_1 thoracic duct cells	5	18,262§	139	6 (96%)	140 (0%)
35×10^6 F_1 thoracic duct cells	2	69,300	604	20 (97%)	571 (5%)
35×10^6 F_1 thoracic duct cells	2	85,850	1454	70 (95%)	1660 (0%)

* F_1 thoracic duct cells were inoculated 2 wk postirradiation and marrow protection.

‡ Number in brackets refers to per cent reduction.

§ Spleens assayed 7 days after thoracic duct cells were inoculated; all others were assayed 5 days after inoculation.

with isoantisera, aliquots of spleen cells from thymectomized irradiated recipients of F_1 thoracic duct cells and CBA bone marrow. Anti-CBA serum caused a reduction in the number of PFC of 95–97% and anti-C57BL resulted in losses of up to 5%. The majority of PFC produced in these hosts was therefore derived not from inoculated F_1 thoracic duct cells but from CBA cells.

DISCUSSION

Cells capable of producing hemolytic foci in the spleens of heavily irradiated mice, above the background number, were detected in thoracic duct lymphocyte suspensions but not in bone marrow. In irradiated recipients of thoracic duct cells and SRBC, the addition of syngeneic bone marrow did not increase the number of hemolytic foci but increased the number of PFC per spleen to a far greater extent than could be accounted for by summating the activities of either cell population alone. These results can be interpreted in at least two ways. (*a*) By allowing hemopoietic regeneration and possibly creating a favorable environment in the spleen, bone marrow promotes the differentiation and multiplication of PFC precursors present in thoracic duct lymph. Such PFC precursors were described as antigen-sensitive cells by Kennedy et al. (5) who considered that the PFC clustered within an area detectable by the hemolytic focus assay technique, were the direct descendants of an antigen-sensitive cell that had lodged in the spleen at the site where the focus arose. As bone marrow did not produce foci, these investigators concluded that it lacked antigen sensitive cells or PFC precursors according to their definition. (*b*) The second possibility is that bone marrow does provide PFC precursors (the AFCP in our terminology) the differentiation of which can be induced only by the associated presence of some type of cell, presumably derived originally from the thymus and present in thoracic duct lymph (the ARC in our terminology). Clearly, in order to distinguish between these two possibilities, one must identify the source of PFC in irradiated mice receiving bone marrow and thoracic duct or thymus lymphocytes.

In the experiments of Davies et al. (9–11), thymectomized irradiated mice were protected with bone marrow and thymus from donors which were slightly different immunogenetically. When the spleens from these thymus-grafted mice were transferred soon after challenge with SRBC, into irradiated recipients presensitized against either the thymus-donor or the marrow-donor, those capable of rejecting cells of thymus donor–type were able to produce antibody in response to SRBC. By contrast, those immunized against marrow donor–type cells produced much less antibody. These transfer experiments were, however, performed 30 days after irradiation and thymus grafting. It is known that, at that time, the lymphoid cell population of the graft has been replaced entirely by cells derived from the bone marrow (12). Thus hemolysins detected in irradiated mice, presensitized against thymus donor–type cells, may have been produced by bone marrow–derived cells which had repopulated, and migrated from, the thymus graft. Hence the cells producing antibody could have the immunogenetic characteristic of the marrow donor and yet be thymus derived. Thus, "it may be that thymus-derived cells can produce antibody, but only in the presence of cells of bone marrow origin. Equally, cells of bone marrow origin . . . may be the cells whose immunological potential is enhanced by association with cells of thymic origin. These are not problems which the present analysis can resolve" (11).

In the present study, no attempt was made to determine, with anti-H2 sera, the identity of the PFC produced in the spleens of irradiated recipients of a *mixed* inoculum of F_1 thoracic duct lymphocytes, CBA bone marrow cells, and SRBC, because no synergistic effect was observed in this experimental system. This contrasts with the results obtained in the syngeneic situation but is in general agreement with a previous report indicating that allogeneic thymus and bone marrow cells do not interact in irradiated hosts (13). Since it appears that a synergistic effect can be obtained only in syngeneic systems, methods other than those using anti-H2 sera must be devised to determine the source of the PFC. The only technique available so far is the detection of a particular chromosome marker in individual antibody-forming cells (14). This has been applied to PFC obtained from the spleens of heavily irradiated mice injected simultaneously after irradiation with CBA/T6T6 bone marrow cells, CBA thoracic duct cells, and SRBC and the results will be given in the following paper (15).

The reason for the failure of an interaction to take place between allogeneic lymphocytes and bone marrow cells immediately following transfer to irradiated hosts, is unknown but it may be that the phenomenon of allogeneic inhibition (16) operates. If this is so, however, it no longer operates once differentiation of some bone marrow stem cell has taken place, since an excellent synergistic effect was obtained in thymectomized irradiated mice receiving F_1 thoracic duct cells and SRBC 2 wk after an injection of CBA bone marrow. In this system, therefore, techniques using anti-H2 sera were used to determine the identity of the PFC produced. The results indicate that the bone marrow must be a source of AFCP since the PFC were CBA-type and not F_1-type. This conclusion is corroborated by the results obtained in the following study in which chromosome marker techniques were used (15). It seems therefore that bone marrow provides AFCP, but not cells capable of *initiating* the production of hemolytic foci, so that it must lack ARC. The hemolytic focus assay is thus a measure, not of the number of PFC precursors in a given population, but of the number of ARC which settle in the irradiated spleens.

Irradiated recipients of thymus cells and SRBC will produce neither hemolytic foci nor PFC in their spleens (5, 17–19). When bone marrow is added, however, both hemolytic foci and an increase in the number of PFC per spleens can be detected (17–19). One of the differences between the population of cells from thymus and thoracic duct lymph, which could not have been demonstrated in the experimental system used in the previous study (1), is that PFC can be derived directly from some thoracic duct lymphocytes, but not from thymus cells. The evidence for this is that PFC were produced in the spleens of heavily irradiated recipients of syngeneic thoracic duct cells but not of thymus cells, and that the PFC in the spleens of irradiated CBA mice injected with F_1 thoracic duct cells were F_1 type. All the evidence obtained in this and the previous study therefore suggests that the thymus contains only ARC, the bone marrow only AFCP, but the thoracic duct lymph contains both ARC and AFCP. Preliminary studies using density gradient centrifugation techniques indicate that the two cell types in thoracic duct lymph may indeed be separated.

It is evident from a lot of experimental data that the immunological activity of thymus cells is inferior to that of spleen or lymph node cells (20). For instance, mouse thymus cells injected with rat erythrocytes into preirradiated mice enable their hosts to produce much less antibody than a similar number of spleen cells (21). Furthermore, in order to prevent the protective action of allogeneic bone marrow in irradiated mice two to four times as many syngeneic thymus lymphocytes as syngeneic lymph node cells were required (22, 23). In the present experiments, a comparison was made between the peak PFC response of irradiated recipients of thymus and bone marrow on the one hand, and of thoracic duct cells and bone marrow on the other. Far more thymus than thoracic duct lymphocytes were required to cooperate with a certain number of bone marrow cells in order to produce a given number of PFC. The mean PFC response in irradiated recipients of thymus and bone marrow was greater than that of mice given either cell population alone, but there was a considerable variation between individual mice (Table III). The synergistic effect in these experiments was much less impressive than that reported by Claman et al. (17), perhaps because they gave two injections of SRBC to their irradiated recipients whereas only one such injection was given here at the time of cell transfer. Thymus cells were also less effective than thoracic duct cells in thymectomized irradiated, bone marrow–protected mice. This difference between the two cell types was not evident in neonatally thymectomized recipients (1). The spleens of these mice were much larger than those of thymectomized irradiated mice. It is possible that the larger spleens trap more thymus cells since these cells preferentially home there rather than to lymph nodes (24, 25). Thoracic duct cells, however, have a capacity to recirculate and to home to all the lymphoid tissues (26). These features might account for the different results obtained in neonatally thymectomized and in adult thymectomized, bone marrow–protected mice. An alternative possibility is that a particular cell type, which is radiosensitive and required for antigen handling in this system, is present in spleen and thoracic duct lymph, but absent from the thymus lymphocyte population. Neonatal thymectomy would presumably not affect this cell type.

As a general conclusion, it may be said that there are ARC in the thymus but in a proportion that is much less than that found in the cell population of thoracic duct lymph, spleen, and lymph nodes. Perhaps ARC are exported out of the thymus as soon as they are formed so that the thymus would contain a greater proportion of "immature" cells. Perhaps emigrating thymus cells must undergo a period of maturation before they can become ARC. Perhaps this maturation entails interaction with antigen (or macrophage-processed antigen) to give rise to a clone of cells that are specifically reactive to the particular antigen (? memory cells). Perhaps only these cells are capable of recirculating so that the population of cells in thoracic duct lymph would be richer in ARC than that in the thymus. Whatever the case may be, it seems evident that in the 19S immune response to SRBC, some interaction takes place between thymus-derived ARC and bone marrow-derived AFCP and that this allows the differentiation of the AFCP to PFC. Whether a similar interaction takes place in other antigenic systems is not known but it would seem more likely to occur only in those immune responses which are initiated by cells the development of which is under thymus control.

SUMMARY

The number of discrete hemolytic foci and of hemolysin-forming cells arising in the spleens of heavily irradiated mice given sheep erythrocytes and either syngeneic thymus or bone marrow was not significantly greater than that detected in controls given antigen alone. Thoracic duct cells injected with sheep erythrocytes significantly increased the number of hemolytic foci and 10 million cells gave rise to over 1000 hemolysin-forming cells per spleen. A synergistic effect was observed when syngeneic thoracic duct cells were mixed with syngeneic marrow cells: the number of hemolysin-forming cells produced in this case was far greater than could be accounted for by summating the activities of either cell population given alone. The number of hemolytic foci produced by the mixed population was not however greater than that produced by an equivalent number of thoracic duct cells given without bone marrow. Thymus cells given together with syngeneic bone marrow enabled irradiated mice to produce hemolysin-forming cells but were much less effective than the same number of thoracic duct cells. Likewise syngeneic thymus cells were not as effective as thoracic duct cells in enabling thymectomized irradiated bone marrow–protected hosts to produce hemolysin-forming cells in response to sheep erythrocytes.

Irradiated recipients of semiallogeneic thoracic duct cells produced hemolysin-forming cells of donor-type as shown by the use of anti-H2 sera. The identity of the hemolysin-forming cells in the spleens of irradiated mice receiving a mixed inoculum of semiallogeneic thoracic duct cells and syngeneic marrow was not determined because no synergistic effect was obtained in these recipients in contrast to the results in the syngeneic situation. Thymectomized irradiated mice protected with bone marrow for a period of 2 wk and injected with semiallogeneic thoracic duct cells together with sheep erythrocytes did however produce a far greater number of hemolysin-forming cells than irradiated mice receiving the same number of thoracic duct cells without bone marrow. Anti-H2 sera revealed that the antibody-forming cells arising in the spleens of these thymectomized irradiated hosts were derived, not from the injected thoracic duct cells, but from bone marrow.

It is concluded that thoracic duct lymph contains a mixture of cell types: some are hemolysin-forming cell precursors and others are antigen-reactive cells which can interact with antigen and initiate the differentiation of hemolysin-forming cell precursors to antibody-forming cells. Bone marrow contains only precursors of hemolysin-forming cells and thymus contains only antigen-reactive cells but in a proportion that is far less than in thoracic duct lymph.

We wish to thank Miss Winifred House, Miss Susie Bath, Miss Sue Hughes, and Miss Catriona Jelbart for technical assistance.

BIBLIOGRAPHY

1. Miller, J. F. A. P., and G. F. Mitchell. 1968. Cell to cell interaction in the immune response. I. Hemolysin-forming cells in neonatally thymectomized mice reconstituted with thymus or thoracic duct lymphocytes. *J. Exptl. Med.* **128**:801.

2. Miller, J. F. A. P., G. F. Mitchell, and N. S. Weiss. 1967. Cellular basis of the immunological defects in thymectomized mice. *Nature.* **214**:992.

3. Miller, J. F. A. P. 1960. Studies on mouse leukaemia. The role of the thymus in leukaemogenesis by cell-free leukaemic filtrates. *Brit. J. Cancer.* **14**:93.

4. Jerne, N. K., A. A. Nordin, and C. Henry. 1963. The agar plaque technique for recognizing antibody-producing cells. *In* Cell Bound Antibodies. B. Amos and H. Koprowski, editors. Wistar Institute Press, Philadelphia. 109.

5. Kennedy, J. C., L. Siminovitch, J. E. Till, and E. A. McCulloch. 1965. A transplantation assay for mouse cells responsive to antigenic stimulation by sheep erythrocytes. *Proc. Soc. Exptl. Biol. Med.* **120**:863.

6. Playfair, J. H. L., B. W. Papermaster, and L. J. Cole. 1965. Focal antibody production by transferred spleen cells in irradiated mice. *Science.* **149**:998.

7. McIlwain, H., and H. L. Buddle. 1953. Techniques in tissue metabolism. I. A mechanical chopper. *Biochem. J.* **53**:412.

8. Cross, A. M., E. Leuchars, and J. F. A. P. Miller. 1964. Studies on the recovery of the immune response in irradiated mice thymectomized in adult life. *J. Exptl. Med.* **119**:837.

9. Davies, A. J. S., E. Leuchars, V. Wallis, and P. C. Koller. 1966. The mitotic response of thymus-derived cells to antigenic stimulus. *Transplantation.* **4**:438.

10. Davies, A. J. S., E. Leuchars, V. Wallis, R. Marchant, and E. V. Elliott. 1967. The failure of thymus-derived cells to produce antibody. *Transplantation.* **5**:222.

11. Davies, A. J. S., E. Leuchars, V. Wallis, N. R. S. C. Sinclair, and E. V. Elliott. 1968. The selective transfer test. An analysis of the primary response to sheep red cells. *In* Advance in Transplantation. J. Dausset, J. Hamburger, and G. Mathe, editors. Munksgaard, Copenhagen. 97.

12. Dukor, P., J. F. A. P. Miller, W. House, and V. Allman. 1965. Regeneration of thymus grafts. I. Histological and cytological aspects. *Transplantation.* **3**:639.

13. Chaperon, E. A., and H. N. Claman. 1967. Effect of histocompatibility differences on the plaque-forming potential of transferred lymphoid cells. *Federation Proc.* **26**:640.

14. Nossal, G. J. V., K. D. Shortman, J. F. A. P. Miller, G. F. Mitchell, and J. S. Haskill. 1967. The target cell in the induction of immunity and tolerance. *Cold Spring Harbor Symp. Quant. Biol.* **32**:369.

15. Nossal, G. J. V., A. Cunningham, G. F. Mitchell, and J. F. A. P. Miller. 1968. Cell to cell interaction in the immune response. III. Chromosomal marker analysis of single antibody-forming cells in reconstituted, irradiated, or thymectomized mice. *J. Exptl. Med.* **128**:839.

16. Möller, G., and E. Möller. 1966. Interaction between allogeneic cells in tissue transplantation. *Ann. N.Y. Acad. Sci.* **129**:735.

17. Claman, H. N., E. A. Chaperon, and R. F. Triplett. 1966. Thymus-marrow cell combinations—synergism in antibody production. *Proc. Soc. Exptl. Biol. Med.* **122**:1167.

18. Miller, J. F. A. P., and G. F. Mitchell. 1967. The thymus and the precursors of antigen reactive cells. *Nature.* **216**:659.

19. Mitchell, G. F., and J. F. A. P. Miller. 1968. Immunological activity of thymus and thoracic duct lymphocytes. *Proc. Natl. Acad. Sci. U. S.* **59**:296.

20. Miller, J. F. A. P., and D. Osoba. 1967. Current concepts of the immunological functions of the thymus. *Physiol. Rev.* **47**:437.

21. Thorbecke, G. J., and M. W. Cohen. 1964. Immunological competence and responsiveness of the thymus. *In* The Thymus. V. Defendi and D. Metcalf, editors. Wistar Institute Press, Philadelphia. 33.

22. Vos, O., M. J. de Vries, J. S. Collenteur, and D. W. van Bekkum. 1959. Transplantation of homologous and heterologous lymphoid cells in x-irradiated and non-irradiated mice. *J. Natl. Cancer Inst.* **23**:53.

23. Congdon, C. C., and D. B. Duda. 1961. Prevention of bone marrow heterografting. Use of isologous thymus in lethally irradiated mice. *Arch. Pathol.* **71**:311.

24. Fichtelius, K. E., and B. J. Bryant. 1964. On the fate of thymocytes. *In* The Thymus in Immunobiology. R. A. Good and A. E. Gabrielsen, editors. Harper and Row, New York. 274.

25. Parrott, D. M. V., M. A. B. de Sousa, and J. East. 1966. Thymus-dependent areas in the lymphoid organs of neonatally thymectomized mice. *J. Exptl. Med.* **123**:191.

26. Gowans, J. L., and E. J. Knight. 1964. The route of re-circulation of lymphocytes in the rat. *Proc. Roy. Soc. London, Ser. B.* **159**:257.

N. A. Mitchison

National Institute for Medical Research,
London

The carrier effect in the secondary response to hapten-protein conjugates.
II. Cellular cooperation

The adoptive secondary response of mice to conjugates of NIP (4-hydroxy-5-iodo-3-nitro-phenacetyl-) and DNP (2,4-dinitrophenyl-) is here used to elucidate the mechanism of cellular cooperation. The framework into which the experiments fit can be formulated as follows. Priming immunization raises a crop not only of specific antibody-forming-cell-precursors (AFCP) but also of specific helper cells. Upon secondary stimulation the helper cells serve a role as handlers or concentrators of antigen, thus enabling AFCP which would otherwise be incapable of reacting to initiate antibody synthesis. In this act of cooperation both cells recognise antigen; in the system examined here the helpers recognise carrier determinants and the AFCP recognise either the hapten or other carrier determinants.

The first aim of the experiments was to raise populations of helpers and AFCP of distinguishable specificity. Mice were primed with NIP-Ovalbumin (OA) mixed with chicken γ-globulin (CGG) and bovine serum albumin (BSA); in comparison with controls primed with unmixed NIP-OA, their cells after transfer were relatively more sensitive to secondary stimulation with NIP-CGG or NIP-BSA and similar findings were obtained in cross-checks of these carriers. For reasons which are not entirely clear, non-transferred cells did not show the same effect. In further experiments cells primed with one conjugate (e. g. NIP-OA) were mixed with cells primed with another protein (e. g. BSA), transferred and challenged with the hapten conjugated to the second protein (i. e. NIP-BSA). In comparison with controls lacking the protein-primed cells, the mixture regularly showed greater sensitivity to stimulation. NIP and DNP were tested in many of the possible combinations with BSA, OA and CGG with the same result. The mixture system was used in the further analysis.

Tests with allotype-marked protein-primed cells showed that these cells did not participate in the production of the anti-hapten antibody and could therefore properly be regarded as helpers. Tests of specificity showed that physical union of the hapten and carrier were required: cells primed with BSA, for example, would not help NIP-OA-primed cells to make a response to NIP-HSA even when stimulated at the same time with BSA. Transfer of less than one-tenth of the spleen gives a maximum helper effect, whereas AFCP activity continues to rise as larger numbers of cells are transferred. Helper cells are therefore normally present in excess.

Helper activity is more resistant than AFCP activity to irradiation, drugs and semi-allogeneic cell transfer across an H-2 barrier. This suggests that helper cells play a relatively passive role in the immune response.

Several observations indicate that helper cells are thymus-derived mediators of cellular immunity. Passively transferred antibody did not substitute for helper cells. After immunization helper activity developed faster than AFCP activity. Spleen cells obtained from lethally-irradiated, thymocyte-repopulated, immunized donors provided help. Cells from the thymus-derived fraction of thymus/marrow chimeras also appear to provide help.

Thus, the hapten-carrier cooperative response maps onto the well-established synergy of thymus and marrow in the response to foreign erythrocytes.

1. Introduction

The cooperation hypothesis [1, 2, 3] assumes that an antigen becomes more potent if recognised via carrier determinants. Since at the same time the inducing determinant (which in the case of conjugates as studied here is the hapten) must also be recognised, two receptors are assumed to operate. If both receptors are immunoglobulins, and if one cell makes one immunoglobulin, two populations of cells must participate. In the humoral response the receptor which binds the inducing determinant is presumably present on the antibody-forming-cell-precursor (AFCP, a term introduced by Mitchell and Miller, 1969 [4]) while the other cell, which bears the carrier-specific receptor, will here be termed the helper cell.

According to this hypothesis then two cell populations are involved, both of which should be open to priming. One way

Correspondence: Nicholas Avrion Mitchison, National Institute for Medical Research, Mill Hill, London, N. W. 7, G. B.

Abbreviations: AFCP: Antibody-forming-cell-precursors
CGG: Chicken γ-globulin   BSA: Bovine serum albumin
OA: Ovalbumin   NIP: 4-Hydroxy-5-iodo-3-nitro-phenacetyl-

of obtaining separate priming is to immunize an individual against a hapten on one carrier via one route and at one time, by another protein via another route at another time, and subsequently challenge with the hapten conjugated to the second protein [5]. Another way is to immunize separate but syngeneic individuals against the conjugate and the protein and then to combine cells from the two in an irradiated third party prior to challenge [6]. The latter method and its applications are the subject of the present paper.

2. Materials and Methods

2.1. Animals

Mice of the CBA strain were used as before [3], except where specified.

2.2. Cell transfers

Spleen cell transfers were performed as before [3]. In cooperation experiments in which combinations of cells were tested, the two cell suspensions were mixed and injected intraperitoneally through the same needle. Thymus cells were transferred by essentially the same procedure, using normal 3–4 weeks old donors (yield approximately 150×10^6 cells/ donor). Marrow cells were harvested from the tibia and fibia by squirting balanced salt solution vigorously through a 26 g needle rammed into the marrow cavity (yield $15–30 \times 10^6$ cells/donor). The cells were injected intravenously, and to avoid deaths it was found advisable (a) to pass the cell suspension through a fine (16 wires/cm) stainless steel screen immediately before injection and (b) to inject 10 units heparin intraperitoneally 10 min beforehand.

Cell suspensions made from the spleens of lethally irradiated, thymocyte-injected donors contained only 10–20 % viable cells as judged by trypan blue exclusion, an unusually low proportion.

2.3. Miscellaneous

Immunization, antigens, serology, calculations and statistics followed the procedures specified in the preceding paper [3]. Log units (l. u.) are based on the $\log_{10}$ scale; 1 l. u. (anti-hapten-antibody) = 10^{-7} M hapten binding capacity of serum at a free hapten concentration of 10^{-8} M. 1 l. u. (anti-protein-antibody) = $10 \mu g/ml$ antigen binding capacity of serum at a free antigen concentration of $1 \mu g/ml$.

3. Results

3.1. Cooperation when the same population of cells is immunized with both hapten and secondary carrier

3.1.1. With transfer

In experiments of this design, groups of mice were immunized with either $200 \mu g$ $NIP_{12}BSA$ + $200 \mu g$ OA + $200 \mu g$ CGG, or $200 \mu g$ BSA + $200 \mu g$ $NIP_{14}OA$ + $200 \mu g$ CGG, or $200 \mu g$ BSA + $200 \mu g$ OA + $200 \mu g$ NIP_6CGG (as alum precipitate mixed with the standard dose of *Bordetella pertussis*). Spleen cells from each group were then transferred into irradiated hosts and there stimulated with either $NIP_{12}BSA$, $NIP_{14}OA$, or NIP_6CGG in the standard dose range (0.001-1000 μg). Thus, for example, the $NIP_{14}OA$ boost encoun-

tered cells which had been primed with NIP conjugated to BSA or CGG, admixed with cells primed with its own carrier protein, OA (or conceivably individual cells primed with both NIP and OA). The potency of this form of stimulation could then be compared with that obtained with the entirely heterologous carriers, as measured in the preceding paper[3]. The NIP-binding antibody was measured at the standard time after stimulation and the method of calculating relative potency was used as described in the previous paper[3].

Table 1. Increase in relative potency ($\log_{10}$) obtained by carrier-immunization prior to inducing anti-NIP or anti-DNP secondary responses

		(1) BSA	OA	CGG
(2)	BSA	0	+2.3^b (−4.4)	+3.1^b (−4.5)
			+1.6^e (−3.4)^a	+0.8^c (−4.1)
			+1.0^c (−3.5)	+0.6^d (−5.3)
			+2.1^c (−4.4)	+1.3^d (−5.2)
	OA	+2.0^b (−2.6)	0	+3.7^b (−6.0)
				+0.0^c (−3.4)
				−0.1^c (−5.8)
	CGG	+4.9^b (−1.3)	+0.4^b (−2.9)	0
			+1.0^c (−0.2)	

a) Figures in brackets refer to control relative potencies obtained without carrier-immunization. Thus the entry +1.6 (−3.4) here indicates that for cells obtained from mice immunized with NIP-OA in this experiment, NIP-OA is x $10^{3.4}$ as potent as NIP-BSA in stimulating the secondary response; addition of BSA-primed cells increases the potency of NIP-BSA by 1.6 log units, so that now NIP-OA is only x $10^{2.1}$ as potent as NIP-BSA.

b) Experiments in which donors of NIP-sensitive cells were themselves immunized, additionally, with the protein used as a heterologous carrier for the secondary response. The control values here are the means taken from Table 2 in the preceding paper [3].

c) Experiments in which donors separate from the donors of NIP-sensitive cells were immunized with the protein used as heterologous carrier. The control values here were obtained with aliquots of the same cell populations.

d) As c) above, but using DNP-sensitive rather than NIP-sensitive cells.

The results of these experiments are shown in Table 1. In every case priming with the carrier used for secondary stimulation enhanced potency. This design of experiment was not pursued further because (a) it leaves open the possibility, however remote, that carrier- and hapten-primed cells do not comprise entirely separate populations and (b) it precludes separate manipulation of the two hypothetical cell populations.

3.1.2. Without transfer

In two experiments of this design, the secondary response to $NIP_{12}BSA$ was compared in mice immunized with either $NIP_{14}OA$ or $NIP_{14}OA$ + BSA. This is the most straightforward design and it has yielded dramatic results with other haptens in rabbits [5]. Little or no effect of the carrier-priming could be detected in the present experiments of which one is illustrated in Fig. 1. The crucial difference from the preceding design, in which the cells were transferred, may lie in the presence of antibody during stimulation in the non-transfer design; this antibody would be likely to alter the distribution of the injected antigen.

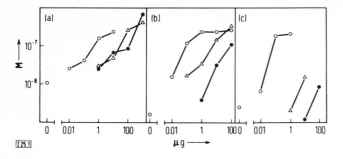

Figure 1. Anti-hapten response to homologous conjugates, compared with heterologous conjugates, with or without help. (a) No cell transfer. Anti-NIP response (serum molar NIP-binding sites) as a function of dose of conjugate (μg/mouse). Open circles: response of $NIP_{13}OA$-primed cells to $NIP_{13}OA$. Closed circles: response of same cells to $NIP_{12}BSA$. Triangles: response of ($NIP_{13}OA$ + BSA)-primed cells to $NIP_{12}BSA$. (b) Cell transfer. Otherwise same as (a), except that triangles mark response of mixture of $NIP_{14}OA$-primed and BSA-primed cells. (c) Cell transfer. Anti-DNP response (serum molar DNP-binding sites) as a function of dose of conjugate (μg/mouse). Open circles: response of $DNP_{10}CGG$-primed cells to $DNP_{10}CGG$. Closed circles: response of same cells to $DNP_{16}BSA$. Triangles: response of same cells to $DNP_{16}BSA$, with addition of BSA-primed cells.

3.2. Cooperation when the population of cells immunized with the secondary carrier is distinct from that immunized with hapten

This design of experiment is illustrated in Fig. 2. One group of mice is immunized with a NIP-conjugate, for example NIP-OA. A second group of mice is immunized with another protein, for example BSA. NIP-OA-primed cells are then transferred and their response measured to stimulation with NIP-BSA, in terms of peak level of anti-NIP antibody. The same response is also measured in other groups of hosts which receive in addition BSA-primed spleen cells from the second group of donors. In the second case the response of the BSA-primed cells can be measured independently, in terms of the peak level of anti-BSA antibody. The control experiments described in the preceding paper [3] would suggest that the NIP-OA-primed cells would not contribute to the anti-BSA response, nor would the BSA-primed cells contribute to the anti-NIP response, but this is a point which requires and will receive confirmation in the mixed-cell design.

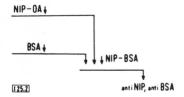

Figure 2. Design used in cooperation experiments in which the population of cells immunized with the secondary carrier is distinct from that immunized with the hapten. Two spleen cell populations are transferred simultaneously into irradiated recipients, which are boosted one day later and bled after another 8–10 days.

The results of two such experiments are illustrated in Fig. 1, one performed with NIP and the other with DNP, and the relative potencies obtained in these and other experiments are included in Table 1. The relative potencies were obtained by the method described in the preceding paper [3], using the same standard slopes and plateau-intercepts as before. The results obtained with DNP-conjugates could be adequately fitted by the slopes and intercepts derived from the corresponding NIP-conjugates.

The effect of adding carrier-primed cells is to shift the area of response without altering the slope, as shown in Fig. 1; a higher peak value was also often obtained, but it should be recalled that the response to the heterologous conjugate did not reach a plateau in the experiments described in the preceding paper [3], probably because of tolerance intervening. The effect of the addition is therefore to increase the relative potency of the heterologous conjugate.

An increase in relative potency was obtained in all experiments, as shown in Table 1, except those in which OA-primed cells were tested. An effect of OA-priming had apparently been obtained in the preceding experiment with mixed OA and NIP_6CGG priming, but the controls then were variable (cf. Table 2 in the preceding paper[3]). Since NIP-OA conjugates previously yielded evidence of a pronounced carrier effect it follows that their effective carrier determinants must have been new determinants induced by conjugation, a contention that seems reasonable for such a readily-denatured protein.

The magnitude of the increase in the other cases is approximately large enough to account for the entire carrier-effect, *i. e.* a "heterologous" carrier, when employed in the presence of its own carrier-primed cells, immunizes approximately as potently as it does when acting as a homologous carrier. The intrinsic potency, as defined in the preceding paper [3], of course remains unaffected by the addition of carrier-primed cells. The fact that approximately complete compensation occurs leaves little room for local environment effects in this form of secondary stimulation.

In these and other experiments a reliable effect was obtained with hapten-sensitive cells taken from mice immunized with NIP-OA, combined with carrier-sensitive cells taken from mice immunized with BSA. This combination was accordingly selected for use in further analysis of the effect. However, CGG is intrinsically a more potent immunogen than OA, as was shown in the preceding paper [3], and accordingly NIP-CGG was used to immunize mice when large amounts of antibody were required.

Under the standard conditions of transfer (40×10^6 cells from each group of donors) and when the number of cells transferred varies over a restricted range, the peak levels of antihapten and anti-carrier antibody is positively correlated; this is illustrated in Fig. 2 in a following paper [7]. The interpretation which should be placed on this correlation is not entirely clear, since it will be argued that those cells active in producing the anti-carrier antibody are not those responsible for increasing the anti-hapten response. Presumably the two cell populations are correlated in their fate after transfer.

3.3. Cell numbers

The effect of varying the number of anti-carrier cells was tested in the two experiments illustrated in Fig. 3. In both experiments small numbers of cells were relatively better at helping the anti-hapten response than in producing their own anti-carrier antibody. As the number of cells was increased the anti-hapten response reached a plateau, at $3-10 \times 10^6$ cells. The experiment shown in Fig. 3 (b) was repeated and yielded curves almost superimposable on those shown, whereas the anti-carrier response continued to rise. The continued rise in antibody production over this range of cell number is in line with previous findings [8]. Since the maximum effect on the anti-NIP response can be obtained with a relatively small number of cells, it follows that the cells responsible for this effect must have been present in excess in the donor spleen. (Note that the average yield of cells from the spleen was 120×10^6).

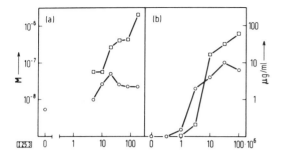

Figure 3. The effect of varying the number of helper cells. (a) Anti-NIP response (serum molar NIP-binding capacity – circles left scale) and anti-BSA response (serum BSA-binding capacity, $\mu g/ml$ – squares, right scale) after transfer of constant number (40×10^6) $NIP_{13}OA$-primed cells mixed with varying number of BSA-primed cells ($0-160 \times 10^6$); all hosts boosted with $100 \mu g$ $NIP_{12}BSA$. (b) Anti-DNP response (serum molar DNP-binding capacity) and anti-BSA response after transfer of constant number (40×10^6) DNP_6CGG-primed cells mixed with varying number of BSA-primed cells ($0-100 \times 10^6$); all hosts boosted with $100 \mu g$ $DNP_{11}BSA$.

3.4. Specificity

The cooperative effect might be ascribed to a non-specific interaction in which a population of cells reacting to one antigenic determinant enhances the reaction of another population to another determinant. According to this hypothesis the addition of a population of protein-primed cells, and their stimulation with their own antigen, should enhance the response of a hapten-primed population to an unrelated heterologous conjugate.

Table 2. Specificity of cooperation: the requirement for physical union of the hapten and carrier

Primed cells	Stimulus ($100 \mu g + 100 \mu g$)	
	$NIP_{12}BSA + HSA$ (l. u.)	$NIP_{12}HSA + BSA$ (l. u.)
$NIP_{13}OA$ – primed cells + BSA – primed cells	1.74	0.40
$NIP_{13}OA$ – primed cells + HSA – primed cells	0.16	1.78
$NIP_{13}OA$ – primed cells alone	0.79	0.31

The prediction was tested in the experiment shown in Table 2. Hapten-sensitive cells were obtained from mice immunized with $NIP_{14}OA$ and carrier-sensitive cells from other mice immunized with either BSA or HSA. When BSA-sensitive cells were employed, stimulation with $NIP_{12}BSA + HSA$ produced an enhanced anti-NIP response, but not stimulation with $NIP_{12}HSA + BSA$. The reverse applied when HSA-sensitive cells were employed. Thus no evidence could be obtained for the hypothetical non-specific interaction. A similar experiment was performed with CGG-sensitive and BSA-sensitive anti-carrier cells and the same result obtained. The experiments also demonstrate the need for a physical union between the two determinants involved in cooperation.

3.5. Do carrier-primed cells produce anti-hapten antibody?

In order to answer this question an experiment was performed with an allotype marker on the carrier-primed cells. The hapten-sensitive cells were taken from the usual CBA donors, immunized with NIP_6CGG, and BSA-sensitive cells from backcross mice, kindly provided by Dr. H. Wortis, histocompatible with the CBA strain at the H-2 locus, bearing the $Ig-1^b$ allotype of C 57 origin in heterozygous form ($H-2^k/H-2^k$, $Ig-1^a/Ig-1^b$). Adequate cooperation was obtained between the two cell populations: addition of the BSA-sensitive cells increased the response obtained with $100 \mu g$ $NIP_{12}BSA$ from $-0.50 \times 10^{-8}M$ hapten binding capacity to $1.05 \times 10^{-8}M$, approximately a 35-fold increase.

Five sera of this response were analysed qualitatively for allotype. Each serum was allowed to form a line of precipitation with anti-$Ig-1^a$ and anti-$Ig-1^b$ antisera. The lines of precipitation were then reacted with radioactively-labeled BSA or hapten and autoradiographs prepared, with the result shown in Fig. 4. In this figure a line of precipitation can be seen with both sera, indicating that the cells from these backcross mice were producing immunoglobulin (a heavier line was obtained with the anti-$Ig-1^b$ antiserum, which happened to be a particularly strong one). Both the $Ig-1^a$ and $Ig-1^b$ lines bound BSA, indicating as expected that the transferred heterozygous cells were producing both allotypes in their antibody. Only the $Ig-1^a$ line bound the hapten, although comparison with the BSA autoradiograph indicates that if any anti-hapten antibody of allotype $Ig-1^b$ had been present it would have shown up strongly. No detectable anti-hapten antibody was therefore produced by the heterozygous cells. Thus the carrier-primed cells appeared to serve a role as helpers but not as precursors of the anti-hapten antibody-forming cells. This experiment formally justifies the distinction already drawn between 'helper cell' and 'AFCP'. Although it severely restricts the possibility of information transmission from helper to precursor a hypothetical transmission is not entirely excluded since the allotype marker used here is located in the Fc region [9] distant from the antigen-combining site.

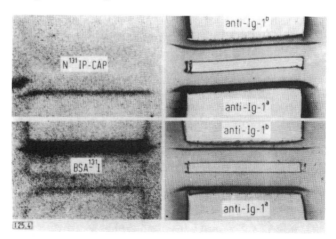

Figure 4. Antiserum from irradiated CBA mouse which received spleen cells from a NIP_6CGG-primed donor ($H-2^k/H-2^k$, $Ig-1^a/Ig-1^a$) and from a BSA-primed donor ($H-2^k/H-2^k$, $Ig-1^a/Ig-1^b$), and were boosted with $NIP_{12}BSA$. On the right, lines of precipitation obtained when the undiluted antiserum was diffused against anti-allotype sera. On the left, autoradiographs obtained after exposing the plates to N^ϵ-($N^{131}IP$-) amino-caproic-acid ($N^{131}IP$-CAP), ($10^{-8}M$) or BSA-^{131}I ($1 \mu g/ml$).

3.6. Failure of secreted antibody to mediate cooperation

In one published experiment antiserum containing 100 µg BSA binding capacity (estimated at 1 µg/ml free BSA) was transferred together with cells sensitized to $NIP_{14}OA$ and was found not to enhance the response of the cells to $NIP_{12}BSA$ [10]. Other experiments performed with fractionated anti-carrier antibody have so far yielded equally negative results. A pool of anti-BSA antiserum was passed over G-150 Sephadex and the ascending portion of the front peak (19 S) collected; of this material 2.7 µg BSA binding capacity (at 1 µg/ml free BSA) was injected intravenously per recipient and an average of 0.6 µg/ml BSA binding capacity was found in the serum one hour later. The 7 S peak was pooled separately and passed through DEAE-cellulose (0.02 M, pH 7.2 phosphate buffer) to yield IgG; of this material 15 and 1.5 µg BSA binding capacity was injected intravenously per recipient and an average of 6.0 µg/ml BSA binding capacity was found in the serum one hour after the higher dose. None of these three treatments detectably enhanced the response of $NIP_{13}OA$-primed cells to $NIP_{12}BSA$.

3.7. The tempo of immunization

Cellular immunity typically precedes in time the appearance of demonstrable antibody production [11] and thymus-derived lymphocytes show a relatively early mitotic response [12]. It is of particular interest, therefore, to ascertain whether helper activity appears relatively early after immunization compared with AFPC activity. Spleen cells obtained soon after immunization with BSA were accordingly compared with others obtained later in respect of their ability (a) to produce anti-BSA antibody after secondary stimulation (AFPC activity) and (b) to help NIP-OA-primed cells make an anti-NIP secondary response (helper activity). The outcome of two such experiments are shown in Table 3.

Table 3. Tempo of immunization: helper and AFCP activity at various intervals after priming

	$NIP_{12}BSA$ boost:	100 µg (l. u.)	1000 µg (l. u.)
Exp. 1			
8 days post BSA-priming[a]	Anti-NIP	−0.04	0.63
	Anti-BSA	−0.33	−0.30
65 days post BSA-priming	Anti-NIP	−0.46	0.73
	Anti-BSA	1.91	2.10
Exp. 2			
7 days post BSA-priming	Anti-NIP	0.39	1.28
	Anti-BSA	−0.57	−0.07
145 days post BSA-priming	Anti-NIP	0.86	1.59
	Anti-BSA	1.74	2.91

a) BSA-primed cells are transferred together with $NIP_{14}OA$-primed cells.

The late cells (65-149 days post-immunization) made of the order of one hundredfold more anti-BSA antibody than the early cells (7-8 days post-immunization) but were only slightly more active as helpers. In interpreting these observations it should be recalled that helper activity in a well-immunized spleen was found to be present in excess; however enough of a dose response effect was observed in the anti-hapten response to make it clear that the two populations of cells were not much different from one another in respect of helper activity. It follows therefore that helper activity develops earlier in response to immunization than does AFPC activity.

3.8. Differential inhibition of helper and AFPC activity:

3.8.1. Irradiation

Cells taken from irradiated donors retain activity as helpers for longer than would be expected from the effectiveness of radiation in suppressing antibody production [10]. The results of three further experiments on the effect of irradiation are shown in Table 4. Doses of radiation of 300 r and 600 r can be seen to be highly effective in preventing antibody production; slightly erratic results in almost completely suppressed groups can be attributed to errors in estimating low levels of antibody. Helper activity is diminished but is still quite appreciable after both doses, although it eventually vanishes after 600 r.

Table 4. Effect of irradiation upon helper and AFCP activity of BSA-primed cells

Treatment of helper cells	Expt. 1 Anti-NIP (l. u.)	Expt. 1 Anti-BSA (l. u.)	Expt. 2 Anti-NIP (l. u.)	Expt. 2 Anti-BSA (l. u.)	Expt. 3 Anti-NIP (l. u.)	Expt. 3 Anti-BSA (l. u.)
No helper cells	−0.52	<1.00	−0.03	<1.00	−0.70	<1.00
Non-irradiated	0.54	0.58	1.40	1.41	0.10	0.04
300 r −2 h[a]	0.22	−0.50	0.01	−0.86		
600 r −2 h	0.37	−0.26	0.04	−0.90	0.02	−0.92
600 r −2 days					−0.09	−0.99
600 r −4 days					−0.72	−0.65

BSA-primed cells transferred together with $NIP_{14}OA$-primed cells: response to stimulation with 100 µg $NIP_{12}BSA$.

a) i. e. Donors irradiated with 300 r two hours prior to cell transfer.

Taken together with the earlier evidence of excess helper activity, these findings suggest but do not prove that radiation differentially spares helper activity. They establish that helper activity is retained after cells have been exposed to substantial doses of radiation.

3.8.2. Drugs

Preliminary experiments involving treatment of the donors of helper cells with cortisol acetate (400 mg/kg, 4, 3, 2, 1 days before transfer), myleran (10 mg/kg, 2, 1 days before transfer) cyclophosphamide (50 mg/kg, 24, 4 h before transfer), mitomycin (20 µg/ml, 20 min in vitro treatment) and colchicine, indicated that the clearest differential inhibition could be obtained with the latter drug. The effect of colchicine treatment proved repeatable and is essentially similar to that of irradiation with perhaps more complete sparing of helper activity (Table 5).

Table 5. Effect of colchicine (10 mg/kg 24 h, 4 h before transfer) upon helper and AFCP activity of BSA-primed cells

	Expt. 1 Anti-NIP (l. u.)	Expt. 1 Anti-BSA (l. u.)	Expt. 2 Anti-NIP (l. u.)	Expt. 2 Anti-BSA (l. u.)	Expt. 3 Anti-NIP (l. u.)	Expt. 3 Anti-BSA (l. u.)
No helper cells	−0.52	<1.00	−0.60	<1.00	−0.11	<1.00
Helper cells from non-colchicine-treated donors	0.54	0.58	1.02	0.84	0.21	0.19
Helper cells from colchicine- treated donors	0.42	−0.68	1.30	−0.91	0.24	−0.54

Table 6. Allogeneic and semi-allogeneic cell transfers

Hosts	Donors NIP-primed cells	BSA-primed cells	Exp. 1 Anti-NIP (l.u.)	Anti-BSA	Exp. 2 Anti-NIP (l.u.)	Anti-BSA	Exp. 3 Anti-NIP (l.u.)	Anti-BSA
CBA	CBA	–	−0.03		−0.78		−0.11	
CBA	CBA	CBA	0.41	0.39	0.43	1.04	0.24	0.19
CBA	CBA	F$_1$ a)	0.34	−0.78	0.67	−0.51	0.11	−0.78
CBA	CBA	C3H	n. d.	n. d.	0.45	1.55	n. d.	n. d.
CBA b)	CBA	C3H	n. d.	n. d.	−0.78	−0.58	n. d.	n. d.
F$_1$	F$_1$	–	0.60		0.55		0.30	
F$_1$	F$_1$	F$_1$	1.79	0.58	1.63	0.80	1.53	1.21
F$_1$	F$_1$	CBA	1.26	0.80	0.31	1.62	0.65	0.99
C57	C57	–	−0.21		n. d.		n. d.	
C57	C57	C57	0.24	−0.60	n. d.	n. d.	n. d.	n. d.
C57	C57	CBA	−0.62	−0.98	n. d.	n. d.	n. d.	n. d.

a) All F$_1$ mice are CBA x C57.
b) Hosts alloimmunized by two i. p. injections of 20 x 10^6 C3H spleen cells 2−6 weeks prior to cell transfer.

3.9. Allogeneic and semi-allogeneic transfers

The activity of cells was studied after allogeneic and semi-allogeneic transfers in the hope of obtaining further evidence of dissociation of helper and AFCP activity, and also in the hope of using allogeneic markers on the two types of cell in the same way as has been done with the response to sheep red blood cells [13, 14]. Donors and hosts were used in the combinations shown in Table 6. In these experiments the hosts as usual received 600 r before cell transfer, so that their capacity to react against allogenic cells was impaired.

In the weak allogeneic combination C3H(H-2^k) → CBA(H-2^k) the results shown in Table 6 indicate that both helper and AFCP activity was normal in unimmunized hosts. Prior immunization with donor cells, however, enabled the hosts to inhibit both activities. This then provides a combination in which the weak alloantigenic difference can be exploited as a marker, as was done with the thymus/marrow chimeras described below and in the *in vitro* combinations studied [15]. Prior immunization was not required for inhibition of both activities in the strong combination CAB → C57(H-2^b), although this experiment was incomplete in so far as the syngeneic transfer C57 → C57 displayed only weak AFCP activity.

In the combination F$_1$ (CBA x C57) → CBA a clear dissociation can be seen in which full helper activity was obtained but AFCP activity was inhibited. This effect parallels those already obtained by treatment of the donor with radiation immunosuppressive drugs and may reflect either a quantitative reduction of a response in which helper cells were present in excess, or inhibition of the multiplicative and productive phases of the immune response without inhibition of antigen presentation, or differential resistance of thymus-derived lymphocytes to radiation and drugs.

In the reverse combination CBA → F$_1$ the transferred cells react to alloantigens without themselves provoking a reaction. Under these circumstances both activities can be detected, although a comparison with the appropriate controls, F$_1$ → F$_1$ for helper and CBA → CBA for AFCP activity, indicates that helper activity is quantitatively impaired and AFCP activity enhanced. Here then a dissociation occurs in which for the first time the balance is shifted towards AFCP activity. This may be because helper cells localise abnormally in an allogeneic environment [16] or because of antigenic competition from alloantigens [17].

3.10. Cooperation after two successive cell transfers

3.10.1. Syngeneic cells

In order to investigate more directly the role of thymus-derived lymphocytes in cooperation, use was made of an experimental design introduced by Mitchell and Miller [4]. Mice were deprived of lymphocytes by lethal irradiation and immediately afterwards received a large number of thymocytes intravenously. Such mice, which lack hemopoietic cells, will not survive for long. Nevertheless, they can be used as a vessel in which to incubate the thymus cells in the presence of antigen. They were therefore immunized with BSA immediately after the intravenous injection. A week later cells were recovered from their spleens and these are referred to in Table 7 as thymus-derived spleen cells.

Their activity was then tested in the standard design of cooperation experiment, in parallel with comparable preparations of spleen-derived spleen cells and spleen cells from conventional mice which had been immunized with BSA for the same time. From the spleen of a mouse which received 950 r plus BSA alum pertussis 7 days earlier approximately 1 x 10^6 cells can be recovered, indicating that the great majority of cells recovered from the mice transferred with thymus or spleen cells originate in the first donor (cf. cell numbers in Table 7). Doubtless the cells which are recovered from the spleen are derived from the minor population of mature lymphocytes present in the thymus [18, 19]. The recovery of viable cells from the spleen of irradiated, transferred mice is poor as judged by trypan blue exclusion (10−20 %); the figures in Table 7, which refer to total nucleated cells, are therefore overestimates.

The results shown in Table 7 indicate that thymus-derived cells of this type are active in cooperation and when due allowance is made for poor viability they appear to be at least as active as comparable populations obtained from the spleen of conventional mice. They could hardly be very much more active, for the mouse spleen normally contains 35 % thymus-derived lymphocytes [20]; the figures in Table 7, which show considerable scatter, are quite compatible with an activity on the part of thymus-derived cells 2.8-fold greater than that of normal spleen cells comparably immunized.

Table 7. Cooperation after two successive cell transfers: syngeneic cells.

Helper cell source	Exp. 1		Exp. 2		Exp. 3	
	Number of helper cells used	Anti-NIP (l.u.)	Number of helper cells used	Anti-NIP (l.u.)	Number of helper cells used	Anti-NIP (l.u.)
Nil		−0.15		−0.47		−0.77
Thymus-derived spleen cells[a]	27×10^6	0.32	10×10^6 30×10^6 90×10^6	−0.10 0.06 0.39	40×10^6	0.04
Spleen-derived spleen cells[b]			10×10^6 30×10^6 90×10^6	−0.66 −0.16 −0.06		
Spleen cells[c]	135×10^6	0.38	10×10^6 30×10^6 90×10^6	−0.06 −0.06 0.25	40×10^6	0.27
AFCP	$NIP_{14}OA$-immunized		$NIP_{14}OA$-immunized		NIP_6CGG-immunized	

Note: Responses elicited throughout by 100 μg $NIP_{12}BSA$. The anti-BSA responses were negligible.

a) Donors received 7 days earlier: 950 r + $90-100 \times 10^6$ thymus cells + BSA alum pertussis [3]
   Yield: $9-17 \times 10^6$ spleen cells per donor.
b) Donors received 7 days earlier: 950 r + 90×10^6 spleen cells + BSA alum pertussis
   Yield: 70×10^6 spleen cells per donor
c) Donors received 7 days earlier: BSA alum pertussis
   Yield: $170-270 \times 10^6$ spleen cells per donor.

3.10.2. Helper activity of chimeric cell populations

Another method of evaluating the activity of thymus-derived lymphocytes has been introduced by Davies, Leuchars, Wallis, Marchant and Elliott [13]. This is to establish chimeras in which marrow- and thymus-derived cells are slightly histincompatible and then to sort out the two cell populations by adoptive transfer into appropriately allo-immunized hosts.

Table 8. Cooperation after two successive transfers: help from allogeneic chimeras

Helper cell source	Secondary hosts	Anti-NIP response to 100 μg $NIP_{12}BSA$	
		Exp. 1 (short-term) (l. u.)	Exp. 2 (long-term) (l. u.)
Nil	CBA	−0.67	−0.59
Nil	C3H		−0.25
T-CBA + M-C3H	CBA	−0.07[b]	−0.15[a]
T-CBA + M-C3H	C3H		−0.05
T-C3H + M-CBA	CBA	−0.48	−0.49
T-C3H + M-CBA	C3H		0.21[b]

a) $0.05 < P < 0.1$
b) $P < 0.05$

Note:
1) For make-up of chimeras, see text.
2) All secondary hosts allo-immunized by two i. p. injections of 20×10^6 CBA or C3H cells 2–6 weeks prior to transfer.
3) All secondary hosts received AFCP from $NIP_{11}OA$-primed donors.

Two experiments of this sort are shown in Table 8. The first of these is a short-term experiment. CBA mice were irradiated with 950 r and then transferred with 50×10^6 thymus cells (T-CBA or T-C3H in Table 8) plus 5×10^6 bone marrow cells (M-CBA or M-C3H in Table 8). In order to reduce contamination of the marrow with thymus-derived lymphocytes the marrow donors received 0.2 ml x 2 injections of anti-lymphocyte antiserum (ALS) on days 1 and 3 before harvest. The ALS used belonged to the 29.5 day mean survival time batch referred to later [21], where justification for the procedure can be found. Its use for preparing marrow-derived cells was suggested by the finding that lymphoid cells from ALS-treated donors can induce tolerance but do not cause runt disease [22]. Four days later the mice were immunized with BSA alum pertussis [3] and eleven days after that their spleen cells were taken and tested for helper activity. The standard design of cooperative experiment was used, except that the hosts ("secondary hosts" in Table 8) were allo-immunized before transfer in order to eliminate cells derived from either the thymus or marrow inoculum.

Because the activity of cells obtained from these short-term chimeras was disappointingly low (Table 8, exp. 1), longer-term chimeras were set up according to the following schedule. On day 0 six weeks-old F_1 (CBA x C3H) mice were thymectomized; in week 4 they received 950 r, followed immediately by 10×10^6 bone marrow cells from CBA x C3H donors treated with ALS as above (M-CBA or M-C3H in Table 8) plus 100×10^6 thymus cells (T-CBA or T-C3H) from the opposite strain intravenously; in week 5 they received BSA on alum without pertussis; in week 6 they received another 100×10^6 thymus cells intravenously from the same source as used previously; in week 7 they received BSA alum pertussis; and in week 15 their spleen cells were tested in the standard cooperation system (Table 8, exp. 2). Prior to transfer these chimeras developed an appreciable level of anti-BSA antibody, approximately 20 % of the level obtained in normal mice after comparable immunization and about 10-fold more than that developed in mice treated similarly but with the thymus cell transfers omitted.

Several batches of both types of mice were tested and Table 8 shows the outcome obtained with the most active representatives. The activity is only just high enough to give meaningful results. The tests show consistently that cooperation takes place when the thymus-derived component of the first transfer can survive in the secondary host. On the other hand, the negative conclusion that marrow-derived cells are inactive cannot be drawn with much confidence.

3.11. Cooperation in the adoptive primary response with hapten-specific helper cells

After transfer of cells from mice primed with one hapten-protein conjugate, stimulation with large doses of the same hapten on a second protein irregularly elicits low levels of antibody to the second protein (-1.0 to 0.0 log units). Since this might provide a system in which cooperation in the primary response could be examined, the experiments shown in Table 9 were performed. Since the titres elicited by a single injection of DNP_9BSA were low and irregular, two injections were given in series as stated in the table. The anti-BSA responses obtained from cells primed originally with DNP on other proteins were consistently better than those of the controls, suggesting that the original priming had generated effective hapten-specific helper cells. However, the responses were feeble and subsequent attempts to exploit the system further yielded inconsistent results: this applied particularly to experiments in which hapten-primed cells from BSA-tolerant mice were tested in combination with non-primed, non-tolerant cells which might have been expected to serve as AFCP.

Table 9. Cooperation in the adoptive primary response with hapten-specific helper cells

Cell source	Exp. 1 Anti-BSA	Exp. 1 Anti-DNP (l. u.)	Exp. 2 Anti-BSA	Exp. 2 Anti-DNP (L u.)
Normal	-0.55	0.10	-0.69	-0.40
$DNP_{10}CGG$-immunized	1.06[a]	1.34[a]	-0.10[a]	0.99[a]
DNP KLH-immunized[c]			-0.26[b]	-0.16[b]

a) $P < 0.05$
b) $0.05 < P < 0.1$ } in comparison with normal cell group.
c) KLH = keyhole limpet hemocyanin.

Inject 1000 μg DNP_9BSA day 1 after cell transfer and 100 μg DNP_9BSA day 20; bleed day 30.

4. Discussion

That cooperation between antigen determinants can occur in the immune response hardly requires further demonstration. A recent review [23] includes 27 examples divided into 17 categories; cooperation has been demonstrated in each one in the sense that immunity to one or more determinant enhances the response to other determinants on a multivalent antigen. Occasionally, however, the reverse occurs and the response to the second determinant is inhibited [24, 25, 26]. In the present experiments cooperation failed to occur in the secondary response when cells were not transferred. Provisionally this can be attributed to the presence of inhibitory humoral antibody. The point may be trivial from a theoretical standpoint, but is important in practice in the design of cooperative experiments. Explanations of these failures or counter-effects have been sought in antigenic competition, steric hindrance and gross (*i. e.* non-determinant-specific) inhibition by humoral antibody. The present experiments suggest two other explanations: use of antigens (*e. g.* polysaccharides) which do not elicit cellular immunity and use of doses of antigen large enough to override the helper effect.

As regards the mechanism of cooperation neither population of cells can play its part in mediating the secondary response unless taken from a primed animal. Both populations of cells respond to immunization, both carry immunological memory and both can therefore presumably recognise antigen. For helper cells the point is made most forcibly by the specificity experiment in which cells not primed against the appropriate carrier would not serve as helpers, although they could evidently do so when tested with another carrier to which they had been primed. For AFCP the same point is made by the allotype experiment: when not taken from a hapten-primed population, cells failed to join in the anti-hapten response, even though at the same time they were evidently capable of producing the anti-BSA antibody for which they had been primed. These two experiments also warrant more far-reaching conclusions about the mechanism of cooperation, as has already been mentioned. The specificity experiment excludes the possibility that cooperation operates through a non-specific or pharmacological step of the kind proposed by Weigle [27]. The allotype experiment places severe restrictions on the possibility of transfer of information from helper cell to AFCP, of the kind discussed by Mitchell and Miller [4].

The present experiments consistently support the hypothesis that helper cells in the hapten-carrier system are derived from the thymus [10]. The evidence is as follows: (1) passively transferred antibody fails to act cooperatively, (2) the tempo of priming for helper activity is characteristic of cellular immunity and thymus-derived cells, (3) thymus-derived cells, immunized in irradiated hosts in the absence of marrow cells, act as helpers, (4) thymus-derived cells, sorted out of chimeric populations, act as helpers. To this list can be added further evidence in support obtained from experiments performed with the marker theta [19] and from the anatomical distribution of helper cells [7]. It is natural to conclude that the helper effect maps onto the thymus-marrow synergy which can be obtained in the response to erythrocyte antigens [12, 28, 29] and to proteins [30].

The question can then be asked whether a mechanism of cell-mediated immunity operates in all instances in which immunity to one determinant of a multideterminant antigen enhances the response to other determinants. This is clearly not so, for we have examples of cooperation which passively transferred antibody has been shown to mediate. Two examples have been analysed in detail [31, 32] and others are listed in a recent survey [33]. In yet another example, which can provisionally be assigned to the same category, cooperation appears to be mediated by an antibody which can be eluted from peritoneal exudate cells [34]. All these examples involve the response to foreign erythrocytes and so it may well be that this kind of mechanism applies particularly to cellular antigens. The response to foreign erythrocytes can be accompanied by gross differences in antigen localisation, *e. g.* as between liver and spleen [35]. A reasonable speculation therefore is that cooperation mediated by secreted antibody depends on gross localisation effects [36].

As regards the mechanism of cooperation between thymus-derived helper cells and AFCP, the present findings argue in favour of a relatively passive role in the response for the helper cells in so far as helper cells were not entirely inactivated by treatment of the animal with irradiation and drugs. In previous work thymus-derived cells were inactivated by this form of treatment [4, 37], but that may be because the response then studied was a primary one in which cell multiplication would be at a greater premium than in the secondary responses studied here.

Most recent discussions of the helper-AFCP interaction emphasise the possibility of an antigen bridge linking the receptor (presumably a normal immunoglobulin) on the AFCP with another receptor (IgX) on the thymus-derived cell [16, 38, 39]. Nothing in the present experiments argues against the possibility. One particular version, in which a matrix of antigenic determinants specified by the IgX matrix on the helper cell fits the matrix of AFCP receptors, has grown more attractive lately as evidence has accrued from other systems in which the stimulation of AFCP can be studied. Other contexts in which stimulation by a matrix may occur include the following: (1) antigen bound to the macrophage surface [10, 40]; (2) polymerised antigens, such as flagellin [41]; (3) the enhancement by antibody of the induction of tolerance [42]; (4) induction by allogeneic cells of cell-mediated immune responses [6, 43]. On the other hand the objection has been raised that rare cells will seldom encounter one another and so the possibility that helper cells transfer their "antibody" to other cells has been invoked [25, 44]. An alternative possibility which avoids the need for an encounter between rare cells would be for helper cells to pick up antigen and transfer it to the dendritic cells of germinal centres (as recirculating lymphocytes can apparently do with soluble complexes of antigens [45], where stimulation of AFCP would then occur: however, one cannot easily imagine how this mechanism could apply to the examples of *in vitro* cooperation cited later [14].

In a sense the suggestion is that cooperation is no more than one of the methods used in the body to construct a matrix of antigen. From the molecular point of view this may be trivial, but not necessarily from a biological one. Cooperation apparently concentrates antigen so that doses which would otherwise be too small to provoke an immune recan do so. It enables immunological memory to be stored in two cell lines: for conventional protective vaccination it may be equally important to establish a specific antigen-concentrating mechanism, operating through thymus-derived helper cells, as to expand the appropriate doses of AFCP. It suggests reasons why antigens may work weakly or not at all in some animal strains and at the same time indicates ways of improving their performance. The reaction against tumour-specific transplantation antigens, for example, may well be open to help [23, 46]. It provides a detailed explanation of the termination of tolerance by cross-reactive antigens [16, 47] and of the difference between states of tolerance induced by high and low doses of protein antigens [48, 49]. It suggests ways whereby an immune response can be provoked against normal body constituents [50, 51], and in doing so, provides a clue to the genesis of auto-immune disease.

I thank Miss Jeanette Dilley and Miss Jenny Bruce for skilled technical assistance.

Received September 6, 1970.

5. References

1 Rajewsky, K. and Rottländer, E., *Cold Spring Harbor Symp. Quant. Biol.* 1967. *32:* 547

2 Mitchison, N. A., *Cold Spring Harbor Symp. Quant. Biol.* 1967. *32:* 431.

3 Mitchison, N. A., *Eur. J. Immunol.* 1971. *1:* 10.

4 Mitchell, G. F. and Miller, J. F. A. P., *Proc. Nat. Acad. Sci. U. S.* 1968. *59:* 296.

5 Rajewsky, K., Schirrmacher, V., Nase, S. and Jerne, N. K., *J. Exp. Med.* 1969. *129:* 1131.

6 Mitchison, N. A. in   (Ed.) *Symp. Int. Soc. Cell Biol. Differentation and Immunology*, 1968. 7: 29.

7 Boak, J. L., Mitchison, N. A. and Pattison, P. H., *Eur. J. Immunol.* (in press).

8 Mäkelä, O. and Mitchison, N. A., *Immunology* 1965. *8:* 539.

9 Mishell, R. and Fahey, J. L., *Science* 1964. *143:* 1440.

10 Mitchison, N. A. in *Organ Transplantation Today.* Excerpta Medica Foundation, Amsterdam 1969, p. 13.

11 Gell, P. G. H. and Benacerraf, B., *Advan. Immunol.* 1961. *1:* 319.

12 Davies, A. J. S., *Transplantation Reviews.* 1969. *1:* 43.

13 Davies, A. J. S., Leuchars, E., Wallis, V., Marchant, R. and Elliott, E. V., *Transplantation* 1967. *5:* 222.

14 Mitchell, G. F. and Miller, J. F. A. P., *J. Exp. Med.* 1968. *128:* 821.

15 Britton, S., Mitchison, N. A. and Rajewsky, K., *Eur. J. Immunol.,* submitted for publication.

16 Mitchison, N. A., Rajewsky, K. and Taylor, R. B. in Sterzl, J. (Ed.) *Prague Symposium on Developmental aspects of antibody formation and structure.* Publishing House of the Czechoslovak Academy of Sciences, Prague 1970. 2: 547.

17 Lawrence, W. L. and Simonsen, M., *Transplantation* 1967. *5:* 1304.

18 Lance, E. M. and Taub, R. N., *Nature* 1968. *221:* 841.

19 Raff, M. C., *Nature* 1970. *226:* 1257.

20 Raff, M. C. and Wortis, H. H., *Immunology* 1970. *18:* 931.

21 Mitchison, N. A., *Eur. J. Immunol.,* (in press)

22 Lance, E. M. and Medawar, P. B., *Proc. Roy. Soc. Lond.,* B., 1969. *173:* 447.

23 Mitchison, N. A., *Transplant. Proc.* 1970. *2:* 92.

24 Brody, N. R. and Siskind, G. W., *J. Exp. Med.* 1969. *130:* 821.

25 Taylor, R. B. and Iverson, G. M., *Proc. Roy. Soc. Lond.* B., 1970. (in press).

26 Mollison, P. L., Hughes-Jones, N. C., Lindsey, M. and Wessely, J., *Vox. Sang.* 1969. *16:* 421.

27 Weigle, W. O. in Sterzl, J. (Ed.) *Prague Symposium on Developmental aspects of antibody formation and structure.* Publishing House of the Czechoslovak Academy of Sciences, Prague 1970. 2: 1020.

28 Claman, H. N., Chaperon, E. A. and Triplett, E. L., *Proc. Soc. Exp. Biol. Med.* 1966. *122:* 1167.

29 Miller, J. F. A. P. and Mitchell, G. F., *Transplantation Reviews.* 1969. *1:* 3.

30 Taylor, R. B., *Nature* 1968. *220:* 611.

31 Henry, C. and Jerne, N. K., *J. Exp. Med.* 1968. *128:* 133.

32 McBride, R. A. and Schierman, L. W., *Science* 1966. *154:* 655.

33 McBride, R. A. and Schierman, L. W., *J. Exp. Med.* 1970. *131:* 377.

34 Kennedy, J. C., Treadwell, P. E. and Lennox, E. S., *J. Exp. Med.* 1970. *132:* 353.

35 Mollison, P. L., *Blood Transfusion in Clinical Medicine.* 3rd Edit., Blackwell Scientific Publications, Oxford 1961.

36 Dennert, G., Pohlit, H. and Rajewsky, K. in Mäkelä, O., Cross, A. M. and Kosunen, T. U. (Eds.) *Cell interactions in immune responses,* Academic Press, London/New York, 1971.

37 Claman, H. N., Chaperon, E. A. and Selner, J. C., *Proc. Soc. Exp. Biol. Med.* 1968. *127:* 462.

38 Mitchison, N. A. and Rajewsky, K. in *Immunological Tolerance.* Academic Press, N. Y. 1969. p. 113.

39 Taylor, R. B., *Transplantation Reviews* 1969. *1:* 114.

40 Unanue, E. R. and Cerottini, J. C., *J. Exp. Med.* 1970. *131:* 247.

41 Shortman, K., Diener, E., Russell, P. and Armstrong, W. D., *J. Exp. Med.* 1970. *131:* 461.

42 Feldman, M. and Diener, E., *J. Exp. Med.* 1970. *131:* 247.

43 Möller, G., Symposium on *"Mediators of Cellular Immunity"* – Brooke Lodge, 1969. (in press)

44 Lachmann, P., *Proc. Roy. Soc. Lond.,* B., 1970. (in press)

45 Brown, J. C., De Jesus, D. G., Holborow, E. J. and Harris, G., *Nature* 1970. *228:* 367.

46 Lindenmann, J. and Klein, P. A., *J. Exp. Med.* 1967. *126:* 93.

47 Boak, J. L., Kölsch, E. and Mitchison, N. A., *Antibiot. Chemother.* 1969. *15:* 98.

48 Rajewsky, K., *Proc. Roy. Soc. Lond.* B., 1970. (in press)

49 Mitchison, N. A., *Third Sigrid Juselius Symposium,* Helsinki. 1970. (in press)

50 Iverson, G. M., *Nature* 1970. *227:* 273.

51 Iverson, G. M. and Dresser, D. W., *Nature* 1970. *227:* 274.

Role of Thymus-derived Lymphocytes in the Secondary Humoral Immune Response in Mice

THERE is now good evidence that the theta (θ) isoantigen of mice[1] is carried only by those peripheral lymphocytes which are thymus-derived[2-4], so that anti-θ antiserum should be useful in elucidating the role of these cells in immunity. In experiments reported here, anti-θ has been used *in vitro* in an adoptive transfer system[5] to study the function of thymus-derived lymphocytes in the secondary humoral immune response to hapten–protein conjugates, particularly in terms of cellular cooperation and carrier effects.

Adult CBA mice were immunized with 4-hydroxy-3-iodo-5-nitrophenyl acetic acid[6]–bovine serum albumin (NIP-BSA), NIP-chicken γ-globulin (NIP-CGG), or BSA. In each case, 400 μg of alum-precipitated antigen was mixed with pertussis and injected intraperitoneally[7]. Eight to twelve weeks later the spleens from these mice were removed into cold Gey's solution, pressed through fine nylon sieves and the equivalent of one-third of a spleen was injected intraperitoneally into each of six syngeneic recipients irradiated with 600 rad 12 h earlier. In some cases, the cells to be injected into a group of six mice were first treated *in vitro* with anti-θ antiserum (anti-θC3H, prepared and tested as previously described[3]) or normal AKR serum (NMS). The cells were sedimented, by centrifugation, resuspended in 1 ml. of the appropriate neat serum and then incubated for 30 min at 37° C. The cells were then washed, resuspended in guinea-pig complement (GPC') diluted 1:3 in veronal buffered saline, and incubated for 30 min at 37° C. After being washed, the cells were injected as outlined. One day later, the hosts received 100 μg NIP-BSA intraperitoneally and were bled after 10 days. The sera were tested for antibodies to NIP and BSA by a modified Farr technique[7,8].

In the experiment of Table 1, NIP-BSA-sensitized spleen cells were treated with anti-θ, NMS or left untreated, and were boosted with NIP-BSA in their adoptive hosts. Another group of mice received untreated cells, but were

Table 1. EFFECT OF ANTI-θ* ON SECONDARY RESPONSE OF NIP-BSA-PRIMED† SPLEEN CELLS TO NIP-BSA

Treatment of cells (*in vitro*)	of host (*in vivo*)	Boost with NIP-BSA (100 μg)	Antibody on day 10‡ Anti-NIP	Anti-BSA
0	0	0	-0.9 ± 0.24	0.51 ± 0.48
0	0	+	1.05 ± 0.15	1.24 ± 0.66
0	Anti-θ§	+	1.09 ± 0.08	1.21 ± 0.61
Anti-θ alone	0	+	1.2 ± 0.28	1.64 ± 0.12
Anti-θ + mouse C'||	0	+	1.08 ± 0.21	1.57 ± 0.24
Anti-θ + GPC'¶	0	+	0.23 ± 0.26	0.55 ± 0.29
NMS + GPC'	0	+	0.97 ± 0.26	1.47 ± 0.16

* Anti-θ C3H used neat (but dilutions of 1 : 2 and 1 : 4 were equally effective).
† NIP-BSA used for priming and boosting had 5–10 moles of NIP bound per mole of BSA.
‡ Expressed as $\log_{10}$ molar binding capacity ($\times 10^{-8}$ M) (see refs. 7 and 8).
§ Intraperitoneally.
|| Neat, fresh CBA serum.
¶ Guinea-pig complement, absorbed with mouse erythrocytes and diluted 1 : 3 in veronal buffered saline.

given 1 ml. of anti-θ intraperitoneally 1 h after cell transfer. Anti-θ produced a marked reduction in the NIP and BSA responses *in vitro*, but was ineffective when given *in vivo*. The effect was complement-dependent and mouse (CBA) complement was ineffective. Similar results have been obtained using 2,4-dinitrophenylated CGG (DNP-CGG), DNP-BSA, and NIP-haemocyanin (NIP-KLH) (unpublished observations of M. C. R.).

Table 2. EFFECT OF ANTI-θ ON CARRIER-PRIMED (BSA) AND HAPTEN-PRIMED (NIP-CGG*) SPLEEN CELLS IN COOPERATIVE SECONDARY ANTI-HAPTEN RESPONSE TO NIP-BSA

NIP-CGG* primed cells	(*In vitro* treatment)	BSA-primed cells	(*In vitro* treatment)	Boost with NIP-BSA† (100 μg)	Anti-NIP response on day 10‡ Exp. 1	Exp. 2
+	0	+	0	0	-0.91 ± 0.20	-0.78 ± 0.32
+	0	0	0	+	0.27 ± 0.48	0.45 ± 0.25
0	0	+	0	+	-0.77 ± 0.33	Not done
+	0	+	0	+	1.0 ± 0.15	Not done
+	0	+	Anti-θ + GPC'	+	-0.27 ± 0.54	0.7 ± 0.29
+	0	+	NMS + GPC'	+	1.05 ± 0.42	1.79 ± 0.25
+	Anti-θ + GPC'	0		+	0.99 ± 0.43	1.78 ± 0.18
+	NMS + GPC'	0		+	1.08 ± 0.16	1.92 ± 0.12

* NIP-CGG used had 6 moles of NIP bound per mole of CGG.
† NIP-BSA used had 5–10 moles of NIP bound per mole of BSA.
‡ Expressed as $\log_{10}$ molar binding capacity ($\times 10^{-8}$ M) (see ref. 7).

In another set of experiments (Table 2) the effect of anti-θ on cellular cooperation was studied using a similar adoptive transfer system[9]. In this system hapten–carrier-primed cells fail to produce a significant anti-hapten response when boosted with the hapten conjugated to a different carrier, but do so if cells primed with the second carrier (helper cells) are given at the same time[9]. In Table 2, the results of experiments using NIP-CGG-primed cells, boosted with NIP-BSA and helped by BSA-primed cells are outlined. In some groups the NIP-CGG-primed cells were treated with anti-θ or NMS; in others the BSA-primed cells were so treated. It can be seen that treating the helper cells (BSA-primed) with anti-θ caused a marked reduction in the anti-NIP response, but treating the NIP-CGG-primed cells had no significant effect.

Because approximately 30–50 per cent of spleen lymphocytes carry the θ antigen and these cells are thymus-dependent[3,4], it is reasonable to assume that anti-θ is exerting its effect in these experiments by killing the thymus-derived lymphocytes with the help of the GPC'. Anti-θ does not kill thymocytes or lymphocytes in cytotcxic testing in the presence of various strains of mouse complement (unpublished observations of M. C. R.) and this may explain its failure to suppress humoral immunity (Table 1) or to prolong skin allograft survival (unpublished observations of R. H. Levey and M. C. R.) when given *in vivo*.

The finding that anti-θ depressed the response to NIP and BSA in the first experiment (Table 1) indicates that thymus-derived cells are important in the secondary humoral response to at least some antigens at certain dose levels. The fact that in the second set of experiments anti-θ had no effect on the NIP-CGG-primed cells (Table 2), the cells which give rise to the anti-NIP antibody in this cooperating system[10], provides further evidence that the role of the thymus-derived cells in the humoral response is one of antigen-specific handling rather than secreting antibody.

It has long been known that immunity against both hapten and carrier is required to elicit a maximum secondary anti-hapten response (carrier effect)[11]. Whereas this phenomenon has previously been attributed to the local environment provided for the hapten by the carrier, there is increasing evidence that carrier effects can best be explained in terms of cellular cooperation[10,12,13]. According to the simplest model, carrier-primed helper cells would interact with antigen (hapten–carrier) through its carrier determinants and present the hapten determinants to hapten-primed cells which are thus induced to produce anti-hapten antibodies[10]. The results in Table 2 indicate that the helper cells are thymus-derived lymphocytes, while the anti-hapten producing cells are thymus-independent (marrow-derived). This is further evidence that carrier effects can be attributed to thymus–marrow synergism[10].

I thank Dr N. A. Mitchison for advice and encouragement and Misses J. Cressey and P. Chivers for technical assistance. I am supported by a post-doctoral fellowship from the National Multiple Sclerosis Society of the USA.

MARTIN C. RAFF

National Institute for Medical Research,
Mill Hill, London NW7.

Received February 16, 1970.

[1] Reif, A. E., and Allen, J. M., *J. Exp. Med.*, **120**, 413 (1964).
[2] Schlesinger, M., and Yron, I., *Science*, **164**, 1412 (1969).
[3] Raff, M. C., *Nature*, **224**, 378 (1969).
[4] Raff, M. C., and Wortis, H. H., *Immunology* (in the press).
[5] Mäkelä, O., and Mitchison, N. A., *Immunology*, **8**, 549 (1965).
[6] Brownstone, A., Mitchison, N. A., and Pitt-Rivers, R., *Immunology*, **10**, 465 (1966).
[7] Brownstone, A., Mitchison, N. A., and Pitt-Rivers, R., *Immunology*, **10**, 481 (1966).
[8] Mitchison, N. A., *Proc. Roy. Soc.*, B, **161**, 275 (1964).
[9] Mitchison, N. A., in *Differentiation and Immunology* (edit. by Warren, K. B.), 29 (Academic Press, New York, 1968).
[10] Mitchison, N. A., Rajewsky, K., and Taylor, R. B., in *Symp. Developmental Aspects of Antibody Formation and Structure, Prague* (in the press).
[11] Ovary, Z., and Benacerraf, B., *Proc. Soc. Exp. Biol. NY*, **114**, 72 (1963).
[12] Rajewsky, K., Schirrmacher, V., Nase, S., and Jerne, N. K., *J. Exp. Med.*, **129**, 1131 (1969).
[13] Taylor, R. B., *Transplant. Rev.*, **1**, 43 (1969).

Chapter II

B Cells

Chapter II

B Cells

As described in Chapter I, two functionally distinct types of lymphocytes are required to manufacture antibodies against most antigens. The B lymphocytes, responsible for the ultimate production of antibody molecules, can be distinguished from the thymus-dependent T lymphocytes by a set of characteristic surface antigens and a distinct path of differentiation. Among the questions we shall be asking and attempting to answer about B cells are:

1. Do B cells "look" different from T cells? What are characteristic surface markers of these lymphocytes?
2. How do B cells recognize antigen?
3. How many different antigens can any single B cell recognize? How many different antibodies will one B cell make?
4. Are all B cells alike or is there functional heterogeneity* in the B cell population?
5. What do B cells do after they recognize antigen?

This chapter will consider several papers which were instrumental in identifying the nature of B cell specific surface antigens, the heterogeneity of the B cell population and some of the consequences of B cell encounter with antigen.

The membrane hallmark of the B lymphocyte is immunoglobulin. In surprising consonance with Paul Ehrlich's predictions (1900) at the turn of the century, B lymphocytes are marked by approximately 100,000 copies of the same immunoglobulin molecule which serve as specific receptors for antigen and harbingers of the antibodies which their plasma cell progeny will ultimately secrete. The initial observation that suggested the presence of immunoglobulin on the surfaces as lymphocytes was (as is often the case in science) a serendipitous consequence of a set of experiments designed to examine something quite different. G. Moller (1961) was using the immunofluorescent technique to detect cell surface histocompatibility antigens on mouse lymphocytes. He first labelled cells with mouse antisera against the appropriate H-2 antigens then exposed these cells to a second antibody labelled with fluorescein and specific for mouse Ig molecules. An important control for these experiments, of course, is to ensure that the fluorescence one detects is a function of the anti-H-2 antibody and not of some other property of the anti-mouse Ig which carries the label. The unexpected finding was that a small number of cells exposed to just the second antibody (rabbit anti-mouse Ig) were fluorescent. Moller correctly thought that this staining must be due to serum

*By "functional heterogeneity" we mean differences in the way different B cells work as opposed to differences in the way B cells might look.

proteins on these cells but failed to deduce the real significance of the finding.

Some ten years later, Raff and coworkers in paper 1 of this chapter reported the critical experiments demonstrating a) the presence of immunoglobulin on some but not all lymphoid cells, b) the ability of these cells to "cap" or redistribute this surface immunoglobulin, and c) the restriction of any single cell to the expression of only one class of Ig. The paper concludes by acknowledging that the question of whether this membrane immunoglobulin functions as an antigen receptor remains unanswered.

It is important to note in considering this paper that the conventional distinction between B and T cells as being surface immunoglobulin (sIg) positive and negative, respectively, began with studies of this sort. The authors however, being aware of the limitations of immunofluorescence as a technique, do suggest that one population of cells may just bear very much less sIg than the other. As will become apparent, the precise role (if any) of membrane-bound immunoglobulin in T cell recognition of antigen is yet to be determined.

Papers 2 and 3 address the functional properties of the sIg on B cells. Does this immunoglobulin act as an antigen receptor? How many different receptors can a single B cell express? Do these sIg$^+$ B cells also secrete antibody? These papers contributed to the establishment of two essential facts in cellular immunology: 1) antigen reacts with specific lymphocytes capable of contributing to the immune response to that antigen only, and 2) the reaction of antigen with specific B lymphocytes is through the receptor activity of their surface immunoglobulin.

Ada and Byrt employed the technique of hot antigen suicide to test the idea that lymphocytes which bind a particular antigen are important in the subsequent synthesis of antibody against that antigen. Their experiments were based on the previous observations of Naor and Sulitzeanu (1967) that approximately one out of every five thousand (0.02%) lymphocytes in a naive, or immunologically virgin, mouse could bind a radioisotopically labelled antigen (BSA). These initial studies, however, did not examine the function of these antigen-specific cells. Ada and Byrt exposed cell suspensions of mouse lymphocytes to heavily radio-iodinated polymerized flagellin under conditions which would destroy any cell which bound sufficient quantities of the antigen. Control groups were exposed to polymerized flagellin which was not labelled with radioactive iodine. Using the adoptive transfer method, these antigen-exposed cell suspensions were transferred to irradiated mice together with a boosting dose of antigen. It was found that the animals receiving cells that had been exposed to the "hot" antigen were specifically unable to respond to the test antigen although they were able to respond to another closely related but distinct antigen. Cells treated with the "cold" antigen preparation responded equally well to both test and control antigens. These important results demonstrated that antigen could react directly with certain lymphocytes which were in some way ultimately responsible for producing antibody to that antigen. In addition, these experiments support the notion that the antigen specificity of a particular set of B cells is fairly limited: elimination of cells binding the test antigen did not eliminate the lymphocytes specific for a closely related antigen. On the other hand, these studies do not prove that B cells are monospecific nor do they address the issue of how the antigen is recognized.

Expanding on these studies, Warner and coworkers (paper 3) determined that surface immunoglobulin acted as the relevant vehicle for the antigen-lymphocyte interaction. In their experiments, lymphocytes were treated with anti-immunoglobulin sera before exposure to radiolabelled antigen. This pretreatment seriously inhibited the subsequent binding of antigen by these cells. In another experiment, these investigators demonstrated that lymphocytes incubated with anti-Ig were less able to mount an immune response to a variety of antigens. Thus, pretreatment of B cells with antibody to immunoglobulin can interfere with antigen binding and subsequent antibody synthesis. The latter experiment also indicates that all B cells, regardless of their specificity, use sIg as the antigen receptor; anti-sIg inhibited the animal's reactivity to several different antigens.

While these experiments do not offer definitive proof that the B cell receptor for antigen is sIg (consider, for example, the possibility that the anti-Ig antibodies, in binding to the sIg, are sterically hindering the real receptor from binding antigen), they were certainly instrumental in laying the groundwork for the way we have come to think about B lymphocyte-antigen interactions.

The foregoing papers have contributed to a picture of the B lymphocyte in which these cells use a display of surface immunoglobulin to recognize and bind antigen. In addition, there is data indicating that any given B cell expresses a single kind of immunoglobulin--restricted with regard to both its antigen specificity and its heavy chain. As such, the immunoglobulin expressed at the surface would function as an accurate forecast of the ultimate antibody secreted during the immune response. Julius and Herzenberg extended this view by demonstrating (paper 4) that the B cell receptor not only reflects the antigen specificity and Ig class of the eventual product but also the relative avidity with which the secreted Ig (receptor) binds the antigen.

In the course of an immune response to antigen X, it had been observed that the antibodies produced late in the response bind antigen more tightly than the antibodies which appeared early in the response. This refinement, if you will, could be due to a preferential selection of B cells bearing high avidity receptors under conditions where antigen concentration is decreasing. This hypothesis was tested by Julius and Herzenberg using a sophisticated direct cell selection method. By exposing spleen lymphocytes to lower and lower concentrations of a fluorescein labelled antigen, and then selecting the cells which persisted in binding antigen under these limiting conditions, they were able to determine that B lymphocytes able to bind antigen at very low concentrations were preferentially able to secrete high avidity antibodies. Thus, the surface immunoglobulin of B lymphocytes is indeed a remarkably accurate forecast of things to come. In the limit, of course, this would mean that the B cell expressing an sIg receptor for antigen X could use exactly the same V region genes to do so as the plasma cell progeny which secrete antibody to antigen X. This is discussed again in Chapter X.

B lymphocytes also express other non-immunoglobulin cell surface markers, including a complement receptor (CR), a receptor for the F_C portion of immunoglobulin molecules (FCR), I-region associated antigens (Ia) and a set of differentiation antigens (Lyb1-7) which are identified by alloantisera and correlated with various aspects of B lymphocyte maturation and activation. These markers have been described in several recent papers and reviewed and discussed by Katz (1977).

Although surface immunoglobulin is the characteristic marker of B lymphocytes, the relative display of these other antigens reflects an important and somewhat unexpected fact: not all B cells are alike. As one probes this heterogeneity further, one discovers that different B cells not only express different surface characteristics, but also behave somewhat differently in making immune responses. This was surprising because it was relatively easy to believe that B cells reflected an elegant uniformity of function: they bind antigen through immunoglobulin receptors and then use T cells to stimulate their differentiation into plasma cells. This notion was especially attractive in view of the seemingly limitless array of interactions indulged in by T cells. Such was not to be, however, and B cells also express a significant level of heterogeneity at the level of both function and surface phenotype.

One of the most important, and revealing, functional differences between B cells is their relative dependence upon T cells in the successful production of antibody. As discussed in Chapter I, the humoral response to most antigens requires both T and B cell participation. Studies with T cell-deprived mice have demonstrated, however, that these immunodeficient animals are capable of producing antibodies to certain kinds of antigens, grouped conveniently if not necessarily accurately, as T-independent antigens.

Paper 5, by Scher and coworkers, presents strong genetic evidence in support of the idea that B cells that are specific for at least some T-independent antigens are a distinct subset within the B cell population. They describe a mutant strain of mouse, CBA/N, which carries an immune defect associated with the X chromosome. The result of this mutation is a decrease in the number of sIg$^+$ cells as well as a depressed immune response to the T-independent antigen DNP-lys-Ficoll. These mice are, however, able to respond to the T-dependent antigen TNP-SRBC, suggesting that the mutation has only affected some B cells. These studies were expanded by Mosier, et al. (1976) who reported by CBA/N mice were still capable of responding to certain other T-independent antigens. This observation caused scientists to group B cells into three functional sets: responsive to T-dependent antigens, responsive to T-independent antigens of type I (DNP-Brucella abortus), and responsive to T-independent antigens of type II (DNP-Ficoll). It is not yet clear whether these groups represent different cell lineages or different stages along one cell lineage.

These genetic studies demonstrating the functional heterogeneity of B cells are supported by serological studies characterizing the Ly b 5.1 antigen (Subbarao, et al., 1979) which turns out to be expressed on B cells responsive to type II T-independent antigens like TNP-Ficoll but not on B cells responsive to type I T-independent antigens like TNP-Brucella abortus. Similar results have been reported by Zitron and coworkers (1977) who report that B cells responsive to type II antigens express surface immunoglobulin bearing the δ heavy chain (IgD) while B cells responsive to type I antigens do not. Thus, two kinds of B cells which differ with regard to antigen reactivity reflect that difference in their surface phenotypes.

In addition to offering some insights into B cell heterogeneity, the study of T-independent responses has also been instrumental in the analysis of B cell activation. The effects that follow antigen binding which are important in the subsequent proliferation and differentiation of the B cell remain one of the major mysteries of immunology. As is true for many hormonal and developmental interactions, lymphocyte triggering requires the translation of a membrane-associated stimulus into a nuclear event. In the immune response, antigen binding apparently triggers two distinct phenomena: proliferation of the appropriate B cell clones and differentiation of those antigen-binding cells into antibody-secreting ones. That these two events must occur is indicated by experiments which show a) that animals which have been immunized to antigen X contain 100 to 1000 fold more antigen-binding cells specific for X than naive animals, and b) that such antigen-binding B cells are themselves incapable of antibody production (for a comprehensive review of early studies directed at answering these questions, see Chapter 9 of Antigens, Lymphoid Cells and the Immune Response, by Nossal and Ada). Thus, antigen encounter and a family of appropriate cell-cell interactions are responsible for first expanding the population of lymphocytes reactive to the intruding antigen and then inducing their maturation into plasma cells capable of delivering secreted antibodies into the serum. This is difficult to dissect in a system requiring cell interaction, as do the immune responses to conventional antigens. T-independent antigens apparently activate B lymphocytes directly and as such offer a simpler system in which to study certain phenomena directly. How T-independent antigens achieve this--whether by substituting for a T-cell derived signal or solely by acting on a quantitatively distinct B cell subset, is not resolved.

The last paper in this chapter is one of a number of studies that attempts to analyze the consequences of the interaction between antigen and surface immunoglobulin by mimicking the interaction with anti-Ig. What is the advantage of such an approach? Unfortunately, there is no advantage--rather, a necessity borne of the diversity of the lymphocyte population. Consider the fact, mentioned before, that approximately 0.02% of the lymphocyte pool binds any given antigen. Clearly, one cannot analyze a specific antigen-receptor interaction in such a heterogeneous pot; it is necessary to either stimulate the entire B cell pool, regardless of specificity, or employ a homogeneous cell line restricted in specificity. As the latter is not yet available, investigators are forced into the former tactic.

Polyclonal stimulation of B cells can be achieved through the use of either anti-Ig reagents or mitogens, specific compounds that induce cells to divide. Since it is likely that mitogens do not act through the immunoglobulin receptor for antigen, the use of anti-Ig reagents offers a more accurate representation of antigen-sIg interaction. Paper 6 by Sieckmann and colleagues demonstrates that anti-Ig can be used to induce proliferation in a population of mature B cells without aid from T cells or macrophages. It also suggests that Ig synthesis cannot be induced by anti-Ig, supporting the idea that two kinds of signals may be necessary to trigger both proliferation and differentiation. This kind of study lays important groundwork for future investigations on the consequences of B cell encounter with antigen.

Papers Included in this Chapter

1. Raff, M.C., Sternberg, M., and Taylor, R.B. (1970) Immunoglobulin determinants on the surface of mouse lymphoid cells. Nature 225, 553.

2. Ada, G.L. and Byrt, P. (1969) Specific inactivation of antigen-reactive cells with ^{125}I-labelled antigen. Nature 222, 1291.

3. Warner, N.L., Byrt, P., and Ada, G.L. (1970) Blocking of the lymphocyte antigen receptor site with anti-immunoglobulin sera in vitro. Nature 226, 942.

4. Julius, M.H. and Herzenberg, L.A. (1974) Isolation of antigen-binding cells from unprimed mice. J. Exp. Med. 140, 904.

5. Scher, I., Steinberg, A.D., Berning, A.K., and Paul, W.E. (1975) X-Linked B-lymphocyte immune defect in CBA/N mice. II. Studies of the mechanisms underlying the immune defect. J. Exp. Med. 142, 637.

6. Sieckmann, D.G., Scher, I., Asofsky, R., Mosier, D.E., and Paul, W.E. (1978) Activation of mouse lymphocytes by anti-immunoglobulin. II. A thymus-independent response by a mature subset of B lymphocytes. J. Exp. Med. 148, 1628.

References Cited in this Chapter

Ehrlich, P. (1900) The Croonian Lecture. On immunity with special reference to cell life. Proc. Royal Soc. London B66, 424.

Katz, D.H. (1977) Lymphocyte Differentiation, Recognition, and Regulation. Academic Press, New York.

Moller, G. (1961) Demonstration of mouse isoantigens at the cellular level by the fluorescent antibody technique. J. Exp. Med. 114, 415.

Mosier, D.E., Scher, I., and Paul, W.E. (1976) In vitro responses of CBA/N mice: Spleen cells of mice with an X-linked defect that precludes immune responses to several thymus-independent antigens can respond to TNP-lipopolysaccharide. J. Immunol. 117, 1363.

Naor, D. and Sulitzeanu, D. (1967) Binding of radioiodinated bovine serum albumin to mouse spleen cells. Nature 214, 687.

Nossal, G.J.V. and Ada, G.L. (1971) Antigens, Lymphoid Cells, and the Immune Response. Academic Press, New York.

Scher, I., Sharrow, S.O., and Paul, W.E. (1976) X-linked B-lymphocyte defect in CBA/N mice. III. Abnormal development of B-lymphocyte populations defined by their density of surface immunoglobulin. J. Exp. Med. 144, 507.

Subbarao, B., Mosier, D.E., Ahmed, A., Mond, J.J., Scher, I., and Paul, W.E. (1979) Role of a nonimmunoglobulin cell surface determinant in the activation of B lymphocytes by thymus-independent antigens. J. Exp. Med. 149, 495.

Taylor, R.B., Duffus, W.P.H., Raff, M.C., and dePetris, S. (1971) Redistribution and pinocytosis of lymphocyte surface immunoglobulin molecules induced by anti-immunoglobulin antibody. Nature New Biology 233, 225.

Zitron, I.M., Mosier, D.E., and Paul, W.E. (1977) The role of surface IgD in the response to thymic-independent antigens. J. Exp. Med. 146, 1707.

Immunoglobulin Determinants on the Surface of Mouse Lymphoid Cells

THERE is increasing evidence that lymphocytes have antigen-specific receptors on their surface[1-3] and it seems reasonable to suppose that the receptors are immunoglobulins (Ig)[1]. The existence of Ig on the surface of lymphocytes has also been inferred from the transforming effect of anti-Ig sera[4] and the opsonic adherence to macrophages of lymphocytes treated with anti-Ig[5]. We report here the demonstration of Ig determinants on the surface of living mouse lymphocytes by the use of immunofluorescence and immunoautoradiography.

Fluorescein-labelled anti-Ig has commonly been used to demonstrate Ig within the cytoplasm of lymphoid cells[6]. In all of these studies the tissues have been frozen and/or fixed, and the fluorescence has been largely confined to the cytoplasm; the majority of cells stained have been plasma cells. Surface staining was first encountered by Möller[7] as an unavoidable background in his study of histocompatibility antigens by indirect immunofluorescence on living lymphocytes. More recently, surface fluorescence has been observed in cultured Burkitt lymphoma cells[8]. We have treated normal living mouse lymphoid cells with either fluorescein-labelled or ^{125}I-labelled rabbit anti-mouse immunoglobulin (fl-anti-M Ig or ^{125}I-anti-M Ig) and report here experiments designed to prove that the staining observed does indicate the presence of Ig on the cell surface and that only a certain proportion of the lymphoid cells become stained—a proportion which is characteristic of the tissue examined.

The anti-M Ig was prepared by immunizing rabbits with an Ig fraction from normal mouse serum for fl-anti-M Ig, and with myeloma protein 5563 (ref. 9) for ^{125}I-anti-M Ig, both in Freund's complete adjuvant. The rabbit antisera were fractionated on DEAE–cellulose and that fraction which was not adsorbed to the resin in 0·02 M phosphate buffer at pH 7·5 was used. It was conjugated with fluorescein isothiocyanate (FITC) by a modification of the dialysis method of Clark and Shepard[10]. The Ig fraction was adjusted to 10 mg/ml. and pH 9·5 in buffered saline (0·05 M phosphate–0·075 M NaCl) and was dialysed for 24 h at 4° C against 10 volumes of FITC solution at 0·125 mg/ml. in the same buffer. The conjugate was dialysed against phosphate buffered saline (PBS) at pH 7·3 until fluorescein was no longer detectable in the dialysate. The conjugate was absorbed with acetone dried mouse liver powder just before use. Iodination was carried out by the method of Hunter and Greenwood[11] as modified by Klinman and Taylor[12]. Details of the specificities of these labelled antisera will be published elsewhere (submitted to *Immunology*).

Cell suspensions were prepared as previously described by Raff[13] for immunofluorescence and by Klinman and Taylor[12] for autoradiography. The only significant difference between the two methods concerns the use of ammonium chloride to lyse the red cells for fluorescence studies. Immunofluorescent labelling of the cells was carried out by the method of Möller[7]. Equal volumes (0·025 ml.) of cells (20 × 10⁶/ml.) in veronal buffered saline (VBS) with 5–10 per cent foetal calf serum (FCS) and fl-anti-M Ig (10 mg/ml.) were mixed and incubated for 15 min at room temperature. The cells were washed twice and resuspended in VBS with 5–10 per cent FCS and examined under a coverslip with ultraviolet light. 100–300 cells were counted and the percentage of cells showing surface fluorescence was determined.

For autoradiographic studies the cells were stained similarly, except that the ^{125}I-anti-M Ig was used at a much lower concentration (approximately 1 μg/ml.). The stained cells were diluted to 1 × 10⁶/ml. and 0·1 ml. was mixed with 0·3 ml. of 0·6 per cent agarose in VBS and spread on an agarose-subbed slide. The slides were fixed for 10 min in 0·25 per cent glutaraldehyde in PBS, washed for 10 min in 0·5 per cent glycine (adjusted to pH 9) to inactivate the glutaraldehyde, and finally washed for 10 min in distilled water and dried in a current of warm air. The slides were then dipped in 1 : 3 dilution of Ilford G-5 emulsion and exposed for 1–7 days. The autoradiographs were developed in Ilford D 19b and stained with methyl green. 100–500 cells were scored and any cell judged to have more than background label was counted as bearing Ig. By 7 days' exposure, the difference between labelled and unlabelled cells became very obvious.

Various lymphoid tissues of normal adult CBA and Balb/c mice were examined and the results are shown in Table 1. The two methods gave similar results and the number of lymphocytes bearing Ig on their surface was approximately 40 per cent in spleen, 20 per cent in lymph node, and 15 per cent in thoracic duct and bone marrow. Titrations were performed with both methods and showed that the percentage of stained cells increased with increasing concentration of anti-M Ig up to a plateau beyond which no more cells could be labelled. The plateau was reached at 0·05 μg/ml. of ^{125}I-anti-M Ig and at 500 μg/ml. of fl-anti-M Ig. The fact that the two methods stained approximately the same percentage of cells in any one tissue, despite the very great difference in sensitivity, suggests that there are at least two distinct populations of lymphocytes, one bearing Ig on their surface and one having very much less or no surface Ig. With immunofluorescence the staining had a striking distribution, most of the labelled cells showing a clearly defined fluorescent "cap" occupying approximately one-fifth to one-third of the cell surface and usually overlying a pseudopod-like structure.

Table 1. **PERCENTAGE OF IMMUNOGLOBULIN-BEARING CELLS IN VARIOUS TISSUES OF NORMAL MICE**

Experiment No.	Spleen	Lymph node	Thoracic duct	Thymus	Bone marrow
A. Autoradiography					
1	41	17	14	0·31	15
2	36	—	—	—	—
3	43	16	—	0·14	—
4	44	18	—	0·21	—
Mean	41	17	14	0·22	15
B. Fluorescence					
1	38	13		0	
2	40	—		—	
3	39	13		—	
*4	36	23		0	
*5	—	16		—	
*6	33	22		0	
*7	—	15		—	
Mean	35	19		0	

* Experiments using Balb/c mice. In all other experiments CBA mice were used.

Control experiments were done to establish that the staining of cells with the anti-M Ig was the result of the presence of Ig on the cell surface. Ig fractions of normal rabbit serum were prepared and labelled with fluorescein and ^{125}I as before and did not stain normal mouse lymphoid cells. With both fluorescence and radiolabelling techniques, the staining could be strikingly inhibited by pretreating the cells with unlabelled rabbit anti-M Ig.

The morphology of the labelled cells has not been studied, but smears of cell suspensions of lymph nodes, spleen, thoracic duct and thymus stained with Wright's stain and methyl green pyronine all show that the great majority of the nucleated cells are lymphocytes[13]. Thus most labelled and unlabelled cells in these tissues must clearly have been lymphocytes. Although it might be expected that macrophages with cytophilic antibody on their surface would become labelled by these methods, they could only account for a small percentage of the labelled cells in most of the tissues examined.

The origin of the immunoglobulin on the surface of lymphocytes is uncertain. The three obvious possibilities are that it is adsorbed from the serum, transmitted from outside the cell to the cell surface by an active process, or it could be a product of the cell which bears it. The first hypothesis seems unlikely, because thymus lymphocytes failed to stain for surface Ig after incubation in normal mouse serum for 30 min, followed by washing, although this does not exclude the possibility of *in vivo* adsorption. Preliminary experiments with class-specific anti-M Ig suggest that the Ig on any one cell belongs to a single class. This result is most easily reconciled with the third hypothesis. The question of whether the surface Ig molecules function as antigen-specific receptors remains to be answered.

We thank Mr R. Lucken, Mr A. Savill and Miss J. Cressey for their technical assistance, and Dr B. A. Askonas for the anti-M Ig. This work was supported in part by the US National Multiple Sclerosis Society.

M. C. RAFF
M. STERNBERG
R. B. TAYLOR

National Institute for Medical Research,
Mill Hill,
London NW7.

Received September 18, 1969.

[1] Mitchison, N. A., *Cold Spring Harbor Symp. Quant. Biol.*, **32**, 431 (1967).
[2] Wigzell, H., and Andersson, B., *J. Exp. Med.*, **129**, 23 (1969).
[3] Sulitzeanu, D., and Noar, D., *Int. Arch. Allergy*, **35**, 564 (1969).
[4] Sell, S., and Gell, P. G. H., *J. Exp. Med.*, **122**, 813 (1965).
[5] Greaves, M. F., Torrigiani, G., and Roitt, I. M., *Nature*, **222**, 885 (1969).
[6] Coons, A. H., Leduc, E. H., and Connolly, J. M., *J. Exp. Med.*, **102**, 49 (1955).
[7] Möller, G., *J. Exp. Med.*, **114**, 415 (1961).
[8] Osunkoya, B. O., Mottram, F. C., and Isoun, M. J., *Intern. J. Cancer*, **4**, 159 (1969).
[9] Askonas, B. A., *Biochem. J.*, **79**, 33 (1961).
[10] Clark, H. F., and Shepard, C. C., *Virology*, **20**, 642 (1963).
[11] Hunter, W. M., and Greenwood, F. C., *Nature*, **194**, 495 (1962).
[12] Klinman, N. R., and Taylor, R. B., *Clin. Exp. Immunol.*, **4**, 473 (1969).
[13] Raff, M. C., *Nature*, **224**, 378 (1969).

Specific Inactivation of Antigen-reactive Cells with ^{125}I-Labelled Antigen

A QUESTION of current interest is whether the initial stages of an immune response—either the induction of antibody formation or of specific tolerance—may involve the reaction of an antigen with a specific lymphocyte. Naor and Sulitzeanu[1] have demonstrated, both *in vivo* and *in vitro*, a reaction of ^{125}I-labelled bovine serum albumin with a small proportion (about 1/5,000) of mouse spleen lymphocytes. We have since shown[2] that both flagellin and polymerized flagellin (*Salmonella adelaide*) and haemocyanin (*Jasus lalandii*) labelled with ^{125}I or ^{131}I react *in vitro* with certain cells from spleens of rats and mice. Of the strongly reactive cells, almost all are mononuclear with a high nuclear/cytoplasmic ratio and 7–12 microns in diameter and, for any one antigen, comprise about 1/5,000 of the total cell population. These reactive cells adsorb between 4,000–40,000 molecules of labelled protein when allowed to react at $0°$ C with labelled protein (about 3×10^{12} molecules/ml.) in 10 per cent foetal calf serum. A similar proportion of such reactive cells was observed in cell suspensions from lymph nodes and thoracic duct lymph, while peritoneal exudate contained a higher proportion and thymus a lower proportion of these cells[2]. The reaction was not inhibited by concentrations of sodium azide which restricted the uptake of labelled antigen by macrophages but was inhibited, using mouse cells, by rabbit anti-mouse globulin serum. Spleen cells from germ-free and conventional mice reacted equally well with ^{131}I-labelled flagellin[2]. We wished to know whether the ability of these cells to react with antigen was immunologically significant. Experiments were devised to distinguish between the following possibilities: that the reactive cells were (1) cells present because of a prior experience of the animal with a related antigen, (2) antigen-reactive cells[3,4] which had not had prior antigenic experience but which were capable of contributing to a specific immune response such as formation and secretion of antibody, and (3) cells coated non-specifically with cytophilic antibody.

We treated normal mouse spleen cells with unlabelled or ^{125}I-labelled polymerized flagellin and stored the cells in appropriate conditions *in vitro* to allow radiation damage to cells coated with the labelled antigen. The cells were then injected into X-irradiated syngeneic hosts which 1 day later were challenged with an immunogenic dose both of the same and a serologically unrelated flagellar antigen. In detail: spleen cells of either male or female CBA/C$_{57}$ mice (6–8 weeks old) were reacted with ^{125}I-labelled polymerized flagellin (moles iodide/40,000 g protein, about 0·5; specific activity, 25 µCi/µg) at $0°$ C for 30 min in the presence of sodium azide (0·015 M) and in the proportion of 5 or 50 µg protein/$1·3 \times 10^8$ cells, present in 6 ml. of 10 per cent foetal calf serum in Dulbecco's[5] medium. The two preparations of polymerized flagellin were Salmonella, SW1338, H antigen, f, g; and Salmonella, SL871, H antigen, 1, 2. The cells were then deposited by centrifugation at low speed, resuspended in 6 ml. medium and centrifuged in 2 ml. lots through 9 ml. of a 50–100 per cent (v/v) foetal calf serum gradient. The supernatant, containing unbound antigen, was removed. About 0·5 per cent of the added radioactivity was associated with the pellet of cells; these were gently resuspended in the medium and kept for 16–20 h at $0°$–$4°$ C, after which they were recovered by centrifugation and suspended in 2–3 ml. of medium. Cells ($1·3 \times 10^7$) in 0·2 ml. were then injected intravenously into X-irradiated (750 r.) syngeneic mice (7–8 weeks old, ten per group). Twenty-four hours later, the mice were injected intraperitoneally with an aqueous solution containing 1 µg of each of the two antigens. Eight days after the injection of antigen, the mice were killed and serum antibody titres measured by the immobilization test[6]. Control experiments yielded the following results: (1) injection of both antigens into normal mice caused high antibody titres; (2) injection of both antigens into X-irradiated mice did not cause detectable antibody formation; (3) injection of cells, pretreated with labelled or unlabelled antigen into X-irradiated mice without a later injection of antigen resulted in either no or only a trace of antibody response; (4) injection of cells, pretreated with either unlabelled antigen (Table 1) or antigen labelled with non-radioactive (carrier) iodide, into irradiated mice, gave the results below.

Smears and autoradiographs were made of cells exposed to labelled antigen before and after the 16–20 h exposure period at $4°$ C. In one experiment, the counts of labelled mononuclear cells were 148 and 49 per 10^6 cells, respectively. The number of grains per cell varied from > 200 to < 30 (lowest counted) which indicated that the average number of disintegrations occurring per labelled cell was very approximately 200 (γ-ray emission) and 230 (β-ray emission)[7].

Table 1. ANTIBODY TITRES OF X-IRRADIATED MICE INJECTED WITH TWO SEROLOGICALLY DISTINCT FLAGELLAR ANTIGENS 1 DAY AFTER RECEIVING SYNGENEIC CELLS PRETREATED WITH ^{125}I-LABELLED OR UNLABELLED ANTIGEN

Experi- ment	Antigenic pretreatment (per $1\cdot3 \times 10^7$ cells)	Mean antibody titre ($\log_2$) from groups of ten mice after challenge with	
		SW1338	SL871
1	0·5 µg SW1338, labelled	0·5	4·2
	0·5 µg SW1338, unlabelled	2·7	5·1
	5 µg SW1338, labelled	< 0·5	5·4
	5 µg SW1338, unlabelled	2·9	4·3
	No antigen	2·3	4·7
2	5 µg SL871, labelled	0·82	< 0·5
	5 µg SL871, unlabelled	1·6	4·8
	No antigen	1·1	4·2
3	5 µg SL871, labelled	2·0	1·1
	5 µg SL871, unlabelled	3·3	4·2
	No antigen	2·4	4·9

The results of these experiments are presented in Table 1. Statistical analysis was by the Rank test[8]. In every case, pretreatment of cells with unlabelled antigen before injection into X-irradiated animals did not significantly diminish the subsequent response either to the same or to the heterologous antigen. In contrast, pretreatment of the cells with labelled antigen either abolished or significantly reduced the subsequent response to the same antigen ($P < 0\cdot01$), without affecting the response to the heterologous antigen. Pretreatment of cells with antigen labelled with carrier iodide did not significantly reduce the subsequent antibody response to the same antigen.

It is clear that many aspects of this phenomenon require further investigation, but at present the following conclusions can be tentatively drawn: (1) at least some of the cells which reacted with antigen take part in an antibody response and, because of their radiosensitivity, may be lymphocytes; (2) at any one time, most if not all cells in the mouse spleen which are capable, on appropriate stimulation, of producing antibody by 8 days, have the ability to react with antigen *in vitro*; (3) the specificity of the reaction shown in these experiments is difficult to reconcile with any theory other than one requiring cell populations with a very restricted potential for antigenic stimulation, such as the clonal selection theory[9]; (4) the ability to inactivate selectively particular cells in mixed populations should prove useful in studies to assess the contribution of such cells in the immune response.

G. L. ADA*
PAULINE BYRT

The Walter and Eliza Hall
Institute of Medical Research,
Royal Melbourne Hospital,
Victoria, 3050, Australia.

Received February 10, 1969.

* Present address: Department of Microbiology, John Curtin School of Medical Research, Australian National University, Canberra.

[1] Naor, D., and Sulitzeanu, D., *Nature*, **214**, 687 (1967).
[2] Byrt, P., and Ada, G. L., *Immunology* (in the press).
[3] Playfair, J. H. L., Papermaster, B. W., and Cole, L. J., *Science*, **149**, 989 (1965).
[4] Kennedy, J. C., Till, J. E., Siminovitch, L., and McCulloch, E. A., *J. Immunol.*, **96**, 973 (1966).
[5] Dulbecco, R., and Vogt, M., *J. Exp. Med.*, **99**, 167 (1954).
[6] Nossal, G. J. V., *Immunology*, **2**, 137 (1959).
[7] Ada, G. L., Humphrey, J. H., Askonas, B. A., McDevitt, H. O., and Nossal, G. J. V., *Exp. Cell Res.*, **41**, 557 (1966).
[8] Moroney, M. J., *Facts from Figures*, 355 (Penguin Books, London, 1965).
[9] Burnet, F. M., *The Impact on Ideas of Immunology, Cold Spring Harbor Symposia*, **32**, 1 (1967).

BIOLOGICAL SCIENCES

Blocking of the Lymphocyte Antigen Receptor Site with Anti-immunoglobulin Sera *in vitro*

ALTHOUGH the events occurring during the induction of an antibody response are not clearly understood, it is evident that antigen must react with some type of receptor which is either free in serum or present at a cell surface. Various experiments[1-3] in which labelled antigen is allowed to react *in vitro* with lymphocytes suggest that a very small proportion of all lymphocytes does in fact have the ability to recognize and bind a given antigen. The specificity of this binding reaction between antigen and lymphocyte was demonstrated in both adoptive transfer experiments[4] and *in vivo*[3]. Electron microscopic studies indicate that the reactive lymphocytes possess localized receptor sites for antigen on their surface[5]. Preliminary studies[3] showed that pretreatment of these cells with a polyvalent anti-immunoglobulin serum inhibited antigen binding and suggested that the antigen receptor site involves some type of immunoglobulin molecule. We now report in more detail on the nature of the receptor sites on lymphocytes and suggest that interaction of antigen with these sites is crucial in the induction of the antibody response.

Normal CBA mouse spleen or peritoneal cell suspensions were preincubated with a 20–50 per cent final concentration of different, specific anti-mouse immunoglobulin sera (or control normal sera) for 2 h in an ice bath, and the cells washed by centrifugation through a foetal calf serum gradient. The cells were then allowed to react with ^{125}I or ^{131}I-labelled antigens at 0° C for a further 30 min, centrifuged through serum gradients and finally smeared on glass slides for the preparation of autoradiographs. The procedures for labelling of antigens, preparation of cell suspensions, autoradiography and counting of labelled cells have been described[3]. Antigens used[3] were flagellin and polymerized flagellin from different *Salmonella* strains and haemocyanin (*Jasus lalandii*).

In a second type of experiment, cell suspensions pretreated as above with anti-globulin sera, were injected into lethally (850 rad) irradiated syngeneic mice (groups of eight per cell suspension) together with an immunogenic dose of unlabelled antigen. The recipient mice were then bled either 8 or 14 days later and serum samples titrated for antibody activity.

Anti-mouse immunoglobulin sera were prepared by repeated immunizations of rabbits with purified mouse myeloma proteins emulsified in Freund's complete adjuvant. The sera were rendered heavy chain specific by appropriate absorptions with either mouse immunoglobulin light chains or myeloma proteins of another immunoglobulin class. The specificity of the absorbed antisera was verified by use of the very sensitive inhibition of precipitation assays of ^{125}I-labelled mouse myeloma proteins as previously described[6].

Table 1. EFFECT OF ANTI-IMMUNOGLOBULIN SERA ON THE ABILITY OF NORMAL MOUSE SPLEEN CELLS TO BIND ^{131}I-LABELLED ANTIGENS

Rabbit antiserum pretreatment (0° C, 2 h)	Percentage of control antigen uptake*	Mean per cent control†
Normal serum	69, 74, 110, 138	98
Anti-human serum albumin	101, 125	113
Polyvalent anti-LH chain	<10, 11, 12, 23	<14
Anti-L-chain	6, 10, 11, 14, 33, 38, 90	29
Anti-L-chain preabsorbed with mouse L chains	100, 170	135
Anti-α-heavy chain	71, 90, 105	89
Anti-γ_1-heavy chain	90, 94, 100	95
Anti-γ_2-heavy chain	72, 99, 150	107
Anti-μ-heavy chain	7, 33, 51	30

* Each value represents an individual experiment performed with either *Salmonella* flagellin or haemocyanin as antigen. Each value is the percentage of antigen uptake for the serum treated cells as compared with the control cells with no serum. Antigen uptake is scored as the number of labelled lymphocytes per 10^4 cells.
† Mean of the individual values.

The results with all antisera are presented in Table 1, in which the percentage of control antigen uptake is shown. All controls (no serum pretreatment) showed the usual level of labelled cells. Each value represents an individual experiment, and the mean values clearly indicate that significant inhibition of antigen uptake by cells was achieved only with antisera containing antibodies either to immunoglobulin light chains or to μ (macroglobulin) heavy chains.

To determine whether these same cells with IgM surface receptors were indeed involved in the primary antibody response, antiserum treated cell suspensions were injected together with various antigens into irradiated mice. The results (Table 2) indicate that marked suppression of antibody formation has been achieved by pretreatment of cells with either anti-light chain serum or an antiserum containing predominantly anti-μ-chain antibodies. This particular anti-μ-chain serum was unabsorbed and contained traces of anti-light-chain activity. In view of the concentration of the anti-light-chain serum required to achieve inhibition, it is unlikely that the trace of anti-light chain antibody present in the anti-μ-chain serum was responsible for its inhibitory activity. Specific anti-γ-chain sera did not inhibit this response.

A preliminary examination has also been made of the nature of antigen binding receptors on lymphocyte-like cells from other sources. Peritoneal cell suspensions (from the normal peritoneum and not oil induced exudates of adult mice) contain a very high proportion of cells which bind antigen. With both antigens used inhibition of binding occurred after treatment of cells with a polyvalent anti-immunoglobulin serum (24 per cent of control uptake) although this reduction was not as marked as with adult spleen cells. A similar degree of inhibition (29 per cent) was observed with newborn mouse spleen cells and a greater inhibition (7 per cent) with spleen

Table 2. EFFECT OF ANTI-IMMUNOGLOBULIN PRETREATMENT OF NORMAL MOUSE SPLEEN CELLS ON THEIR ABILITY ADOPTIVELY TO TRANSFER PRIMARY ANTIBODY PRODUCTION TO IRRADIATED RECIPIENTS

Antigen	Serum preatment (0° C, 3 h)	Antibody titre in irradiated recipients ($\log_2$)* Day 7–9	Day 14	Significance (P value) compared with normal serum or no serum
Polymerized flagellin SW 1338	No serum	2·5	—	
	Preimmune normal serum	2·5	—	
	Anti-μ (trace L)	1·2	—	< 5, > 1%
Polymerized flagellin SL 871	Normal serum	7·3	—	
	Preimmune normal serum	8·2	—	
	Anti-L-chain	3·7	—	1%
Brucella abortus	Preimmune (anti-L) normal	—	8·5	
	Anti-L-chain	—	4·1	1%
	Anti-μ-chain (trace L)	—	5·0	1%
Sheep erythrocytes	Preimmune (anti-L) normal	—	3·5	
	Anti-L-chain	—	1·0	1%
	Anti-μ-chain (trace L)	—	< 1·0	1%

In each experiment the preimmune normal serum is the preimmunization bleed of the anti-μ or anti-L rabbit.

* Bacterial immobilization was used for flagellin, agglutination assay for *Brucella* and sheep erythrocytes.

cells from 3 day old mice. Further studies are in progress to determine whether the incomplete inhibition observed with newborn mouse spleen and adult peritoneal cells is attributable to a phagocytic type of cell not involved in specific antigen recognition.

Our results, combined with related studies[2,3,5], strongly suggest that lymphocytes from unimmunized mice have at their surface IgM macroglobulins which specifically bind antigens. Other studies have indicated that the normal lymphoid cell population contains two discrete populations of cells: thymic derived cells involved primarily in cellular immunity and bursal (or bursal equivalent) derived cells involved in humoral antibody formation[7]. From the result of the *in vivo* cell transfers in these and other studies[3], it is probable that these lymphocyte-like cells are involved in humoral immune responses. Thus differentiation of haemopoietic stem cells towards antibody producing precursor cells would involve the expression of the light and μ-chain immunoglobulin genes, with a resulting localization of the IgM molecules on the cell surface. This is supported by the observation[8] that IgM is the first immunoglobulin synthesized within the bursa of Fabricius itself, which is the first location of immunoglobulin synthesis in chickens.

This suggests that further differentiation toward cells secreting large amounts of antibody is induced by contact of antigen with the surface IgM receptor. Further, if all normal antibody producing precursor cells carry these IgM receptors, indicative of μ-chain gene activity, then the resulting IgG antibody formation which occurs in many responses may reflect an intracellular switch from the μ-chain gene to the γ-chain gene, as has been sug-

gested[9]. Alternatively, different cells may possess IgM, IgG and other immunoglobulin class receptors, but there is a preponderance of cells with IgM receptors.

These results may also be compared with those from a similar approach, involving the use of the same antisera to identify the nature of the antigen recognition site on cells mediating cellular immunity[10]. Cell suspensions pretreated with antisera were assayed for immunological competence in either graft versus host reactions or in the cellular transfer of delayed hypersensitivity. In the latter systems, almost complete suppression of activity was achieved with the anti-light-chain serum, but by contrast with the results now reported, no anti-heavy-chain antibody (including anti-μ-chain antibody) had any inhibitory activity.

Thus the initial contact between antigen and normal unprimed cells involved in either cellular or humoral immunity involves a cell surface bound immunoglobulin molecule as the unit of recognition, and it is intriguing to consider that a different class of immunoglobulin is involved in the two systems. Further studies with ^{125}I-labelled purified anti-immunoglobulin antibodies, now in progress, may demonstrate directly the location of the immunoglobulin receptors on the cell surface.

This work was supported by a US Public Health Service research grant from the National Institute of Arthritis and Metabolic Diseases and by grants from the National Health and Medical Research Council, Canberra, Australia.

NOEL L. WARNER
PAULINE BYRT

The Walter and Eliza Hall Institute
of Medical Research,
Royal Melbourne Hospital,
Victoria 3050.

G. L. ADA

Department of Microbiology,
John Curtin School of Medical Research,
Australian National University,
Canberra.

Received January 30; revised March 9, 1970.

[1] Naor, D., and Sulitzeanu, D., *Nature*, **208**, 500 (1965).
[2] Byrt, P., and Ada, G. L., *Immunology*, **17**, 503 (1969).
[3] Ada, G. L., and Byrt, P., *Nature*, **222**, 1291 (1969).
[4] Humphrey, J., and Keller, H. U., in *Developmental Aspects of Antibody Formation and Structure* (edit. by Sterzl, J., and Riha, M.) (Academic Press, New York, 1970, in the press).
[5] Mandel, T., Byrt, P., and Ada, G. L., *Exp. Cell Res.*, **58**, 179 (1970).
[6] Herzenberg, L. A., and Warner, N. L., in *Regulation of the Antibody Response* (edit. by Cinader, B.), 322 (Thomas, Springfield, 1968).
[7] Warner, N. L., *Folia Biol.*, **13**, 1 (1967).
[8] Thorbecke, G. J., Warner, N. L., Hochwald, G. K., and Ohanian, S. H., *Immunology*, **15**, 123 (1968).
[9] Nossal, G. J. V., Szenberg, A., Ada, G. L., and Austin, C. M., *J. Exp. Med.*, **119**, 485 (1964).
[10] Mason, S., and Warner, N. L., *J. Immunol.* (in the press).

ISOLATION OF ANTIGEN-BINDING CELLS FROM UNPRIMED MICE

Demonstration of Antibody-Forming Cell Precursor Activity and Correlation between Precursor and Secreted Antibody Avidities*

By MICHAEL H. JULIUS AND LEONARD A. HERZENBERG

(*From the Department of Genetics, Stanford University School of Medicine, Stanford, California 94305*)

The time-dependent increase in the average affinity of serum antibody after immunization (1), is said to be a result of antigen-driven selection of those antibody-forming cell precursors having the highest affinity antigen receptors (2). This interpretation infers that the specificity and affinity of antigen-binding receptors on antibody-forming cell precursors are directly correlated to the serum antibody produced by progeny antibody-secreting cells.

It has been found that changes in the affinity of serum antibody with time after immunization reflect changes in the population of antibody-secreting cells. By determining the concentration of antigen required to inhibit individual plaque-forming cells (PFC)[1] assayed at various times after immunization, a time-dependent avidity increase at the antibody-secreting cell level is seen and is comparable to the increase in affinity found in serum antibody (3). Thus, a good correlation between the (intrinsic) affinity of serum antibody and the avidity of the antibody-secreting cell products has been established.

In contradistinction, evidence correlating the avidity of antigen receptors on antibody-secreting cell precursors, with the avidity of antibody secreted by their progeny cells, is meager. Indirect evidence has been reported demonstrating that memory cells with high avidity antigen-binding receptors are required to give rise to high avidity antibody-secreting cells (4). However, the putatively high avidity precursors were not isolated and tested directly in these experiments.

To directly determine the relationship between the avidity of secreted antibody and the avidity of antigen receptors on precursors of antibody-secreting cells requires the isolation of precursor cells with different antigen avidities and measurement of the avidities of antibody-secreting cells derived from these

* This work supported by NIH grants GM 17367, CA 04681, and AI 08917.

[1] *Abbreviations used in this paper:* DNP-MGG, 2,4-dinitrophenyl mouse gamma globulin; FDNP-MGG, fluorescein-conjugated DNP-MGG; FACS, fluorescence-activated cell sorter; FCS, fetal calf serum; F/P, fluorescein per protein; KLH, keyhole limpet hemocyanin; RKLH, rhodamine-conjugated KLH; PBS, phosphate-buffered saline; PFC, plaque-forming cell(s); TNP, 2,4,6-trinitrophenyl.

precursors on antigen stimulation. We have already directly demonstrated that populations of antigen-binding cells from immune (5) and nonimmune (6) animals contain the precursors of antibody-secreting cells. Others have found direct evidence that antigen-binding cells isolated from both nonimmune mice on affinity columns (7) and from immune mice on antigen-derivatized nylon fibers (8) contain antigen-specific precursors required to give adoptive primary and secondary immune responses, respectively. To date, neither of the latter methods have provided adequate purity of precursor activity or enabled purification of precursors with a known range of antigen-binding avidities, both of which are required for a precursor-product correlation study.

The development of a fluorescence-activated cell sorter (FACS) (9) has enabled the isolation of virtually pure populations of viable antigen-binding cells with full functional activity (5, 6). We have isolated cells with low, medium, and high binding avidity for DNP by staining normal spleen cells with fluorescein-ated DNP-mouse gamma globulin (^{F}DNP-MGG) and separation using a FACS.

It has been postulated that precursors of low affinity antibody-secreting cells require higher concentrations of antigen to bind and be stimulated than do precursors of high affinity antibody-secreting cells (1). Therefore, high avidity cells were stained using low concentrations of ^{F}DNP-MGG while low and medium avidity cells were stained using moderately high concentrations in the presence of various concentrations of the univalent competing ligand ϵ-DNP-lysine (10, 11).

The precursor activity of the purified DNP-binding cells was tested by transferring these cells together with DNP-keyhole limpet hemocyanin (KLH) and a source of KLH carrier-primed cooperator cells into irradiated recipients and subsequently assaying the recipient spleens for anti-DNP-plaque-forming cells (PFC). In addition to quantitating the precursor activity of the purified DNP-binding cells, the hemolytic plaque assay was used to study the avidity of single antibody-secreting cells (3). By determining the avidity of anti-DNP-PFC resulting from the transfer of purified DNP-binding cells with varying avidities for ^{F}DNP-MGG, we have found a direct correlation between the avidity of the DNP antigen-binding cells and the anti-DNP antibody secreted by the PFC derived from them.

Materials and Methods

Animals. Male and female mice of the congenic strains BALB/cN and BAB/14 Hz were used at the age of 2–6 mo (BAB/13 were kindly supplied by Dr. M. Potter (NCI, NIH) and BAB/14 were derived in this laboratory by one further back-cross to BALB/cN and subsequent repeated inbreeding of the resulting heterozygotes).

Preparation of Cell Suspensions and Media. Single cell suspensions of spleen were prepared as previously described (5). Cell suspensions in preparation for loading onto nylon wool columns, cell separation, and immunofluorescent staining were prepared in Dulbecco's phosphate-buffered saline (PBS), pH 7.5, (12) supplemented with 5% heat-inactivated fetal calf serum (FCS). Before processing spleen cell suspensions through the FACS, erythrocytes were lysed by incubating the cells for 5 min at 0°C in Gey's balanced salt solution (13) in which the NaCl was replaced with an equimolar concentration of NH$_4$Cl. Spleen cells in preparation for plaquing were prepared in MEM with Hanks' balanced salt solution (14). FCS wherever used was first inactivated at 56°C for 30 min.

Preparation of Carrier-Primed Splenic T Cells, Indirect Immunofluorescence Staining for T and Ig-Bearing Cells and Fluorescence Microscopy. As previously described (5, 15).

Preparation of ᶠDNP-MGG and Staining of DNP-Binding Cells. MGG was purified from BALB/cN normal serum by ion-exchange chromatography and conjugated with 1-fluoro-2-dinitrobenzene (Sigma Chemical Co., St. Louis, Mo.) at pH 8.0 in 0.5 M NaHCO₃ for 2–3 h at room temperature. The resulting DNP-MGG contained an average of 23 DNP groups/molecule of MGG.[2] Fluorescein conjugates of DNP-MGG were prepared using fluorescein isothiocyanate (16). The ᶠDNP-MGG was fractionated by a gradient elution from DEAE-cellulose (17, 18). Fractions with fluorescein per protein (F/P) ratios between 2 and 5 were pooled and concentrated by ultrafiltration. ᶠDNP-MGG with an F/P ratio of 3.5 was used throughout these experiments.

Staining concentrations of ᶠDNP-MGG were varied from 6 to 1,600 μg/ml, 0.1 ml of the appropriate ᶠDNP-MGG concentration was added/2×10^7 pelleted cells. Subsequent to a 20 min incubation at 22–25°C the cells were pelleted through neat FCS and then washed once in an excess of medium.

A modification of the above staining procedure was used to determine the relative avidity of ᶠDNP-MGG-binding cells. A 10- to 100-fold molar excess of ϵ-DNP-lysine (relative to the molarity of DNP present on the ᶠDNP-MGG) was included in the staining mixture. The proportion of binding cells observed by fluorescent microscopy in the presence and absence of ϵ-DNP-lysine was determined and the percent inhibition of staining calculated.

To demonstrate specificity of inhibition, cells were simultaneously stained with ᶠDNP-MGG in the presence of various hapten concentrations and with an unrelated antigen, contrastingly fluorescent rhodamine-conjugated KLH (ᴿKLH) (using 0.1 ml of ᴿKLH at 0.5 mg/ml/2×10^7 cells). The proportion of ᴿKLH-binding cells observed was unaffected by any of the ϵ-DNP-lysine concentrations used.

Purification of ᶠDNP-MGG-Binding Cells by the FACS. The FACS used here allows separation of cells according to fluorescence, light-scattering characteristics, or selected combinations of these two parameters. Details of separation protocol and efficiency have been previously described (5, 6). A typical separation yields two fractions of cells, an enriched fraction containing greater than 85% ᶠDNP-MGG-binding cells and a depleted fraction containing 500-fold fewer ᶠDNP-MGG-binding cells compared to the unseparated spleen.

Antigens. DNP-KLH was prepared by reacting 1-fluoro-2,4-dinitrobenzene (Sigma Chemical Co.) at pH 8.0, in 0.5 M NaHCO₃ with KLH (Pacific Bio-Marine Supply Co., Venice, Calif.) for 2–3 h at room temperature. Molar ratios of DNP/10^5 daltons of KLH were calculated[3] and DNP₈₋₁₀-KLH was used in these experiments. Aliquots of DNP-KLH (prepared as described above) were alum precipitated at pH 6.5 with a 9% solution of ALK(SO₄)₂ for 3 h at room temperature.

Irradiation and Adoptive Transfers. Recipients in adoptive transfer experiments received 600 rads whole-body X irradiation, 16–18 h before receiving cells. A modification of the Mitchison hapten-carrier transfer protocol (19, 20) was used. Various combinations of normal spleen cells, purified ᶠDNP-MGG-binding cells, spleen cells depleted of ᶠDNP-MGG-binding cells and KLH-primed splenic T cells were injected intravenously. 100 μg of alum-precipitated DNP-KLH was given intraperitoneally immediately after injection of cells and 10 μg of aqueous DNP-KLH was given intravenously on day 5. Recipients were bled and sacrificed on day 12, serum and cell suspensions of recipient spleens were assayed for anti-KLH antibody and DNP-PFC, respectively.

DNP-PFC Assay and Inhibition with ϵ-DNP-Lysine for Determination of Relative Hapten Avidities. Anti-DNP antibody-secreting cells (PFC) were measured using a modified version of the hemolytic plaque technique (21). SRBC were conjugated with 2,4,6-trinitrobenzene-sulphonic acid (Eastman Organic Chemicals Div., Eastman Kodak Co., N. Y.) (22). Virtually no PFC were observed using unconjugated SRBC. Fresh frozen guinea pig serum at a final concentration of 1/24 was used as a complement source. In early experiments a polyvalent rabbit antimouse antiserum was used to facilitate hemolysis. Since substantial numbers of indirect PFC were not found, in subsequent experiments only direct PFC were measured. We have detected identical numbers of DNP-PFC in experiments where the same spleens were assayed using either TNP-SRBC (22) or DNP-SRBC (23) as indicators. Due to the ease of preparation, TNP-SRBC were used in subsequent experiments.

The plaque inhibition method of Andersson (3) was used to determine the relative avidity for DNP of the PFC. The term "avidity" rather than "affinity" is used here since it is not clear how accurately

---

[2] Based on $E_{1\,cm}^{1\,\%} = 14$ at 280 nm and $\epsilon_M = 1.74 \times 10^4$ at 360 nm for MGG and DNP, respectively.

[3] Using $E_{1\,cm}^{1\,\%} = 15.5$ at 280 nm and $\epsilon_M = 1.74 \times 10^4$ at 360 nm for KLH and DNP, respectively.

an affinity constant (Ka) (24) can be determined in this way. Concentrations of ϵ-DNP-lysine ranging from 10^{-4} to 10^{-7} M were included in some plaquing chambers. High avidity antibody is saturated at lower concentrations of ϵ-DNP-lysine than low avidity antibody. Therefore, low concentrations of ϵ-DNP-lysine prevent high avidity antibody from lysing TNP-SRBC, whereas higher concentrations of ϵ-DNP-lysine would be required to block hemolysis by low avidity antibody. Strictly, a weighted product of antibody avidity and amount secreted per cell is measured by ϵ-DNP-lysine inhibition. However, since the sizes of uninhibited PFC varied less than about 50% in diameter, while the inhibitions vary over a 10^3 concentration range of inhibitor, the major parameter measured is relative avidity rather than amount per cell. In some experiments for assaying very low avidity DNP-PFC, multivalent DNP (DNP-MGG) was used as an inhibitor and was varied from 10^{-5} to 10^{-8} M DNP in the plaquing chambers. Inhibition of PFC with either ϵ-DNP-lysine or DNP-MGG was specific for DNP-PFC and did not reduce the number of SRBC-PFC detected on day 4 of a primary response to 4×10^8 SRBC.

Titrations. Anti-KLH serum titers were individually determined using the passive microhemagglutination technique (25). KLH was coupled to SRBC using a modification of the glutaraldehyde method (26). Sera were heat inactivated at 56°C for 30 min before titration.

Results

Adoptive Primary Response to DNP-KLH. The DNP precursor activity contained in unprimed spleen was determined using an adoptive syngeneic or congenic hapten-carrier transfer system (19). Graded numbers of unprimed spleen cells and KLH-primed, nylon column-purified (15) T cells were mixed and transferred into irradiated recipients with antigen (see Materials and Methods).

The DNP-PFC response of irradiated recipients reconstituted with normal histocompatible spleen cells is substantially improved by supplementation with carrier-primed cooperator cells (27). Fig. 1 *a* shows the DNP-PFC response obtained on transfer of graded numbers of normal spleen cells with or without addition of a constant number of KLH-primed splenic T cells. Addition of 7.5×10^6 KLH-primed and nylon-purified (15) cooperators results in a greater than 10-fold increased response at each dose of normal spleen tested. Thus in normal spleen, cooperators for the DNP-KLH response are limiting on transfer.

The increase in response of a fixed number of normal spleen cells transferred as a function of the number of carrier-primed T cells added is shown in Fig. 1 *b*. There is about a 10-fold greater than additive increase in response with addition of 5×10^6 KLH-primed splenic T cells. This increase must be due to the expression of DNP-PFC precursors derived from the normal spleen, since the response of 5×10^6 T cells transferred alone could have accounted for only 10% of the response of the mixture of normal spleen and T cells.

The addition of 5×10^6 cooperators allows maximal expression of the DNP-PFC precursors in 5×10^6 normal spleen cells. When 1×10^7 carrier-primed T cells are added (in another experiment), there is no further increase in the number of PFC observed (26×10^3 PFC/spleen with 5×10^6 T cells and 29×10^3 PFC/spleen with 1×10^7 T cells.) In experiments which follow, 7.5×10^6 carrier-primed splenic T cells were added to populations to be tested for precursor activity in order to assure an excess of cooperator activity.

FDNP-MGG-Binding Cells in Normal Mouse Spleen. The proportion of FDNP-MGG-binding cells in normal spleen detectable by fluorescence microscopy increases with the staining concentration of FDNP-MGG (Table I). It reaches a plateau of about 2.5% at a staining concentration of FDNP-MGG between 500 and 1,600 μg/ml.

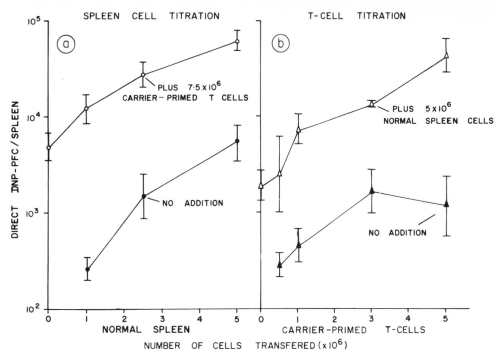

Fig. 1. (a) Adoptive primary dose response to DNP-KLH. Irradiated BAB/14 mice were injected with either normal BALB/cN spleen cells alone or in conjunction with 7.5×10^6 7-day KLH-primed nylon column-purified splenic T cells (obtained from BALB/cN spleen and contained 92% T cells and 4% Ig-bearing cells as assessed by immunofluorescence). Animals received 100 μg alum-precipitated DNP-KLH intraperitoneally on the day of transfer, boosted with 10 μg aqueous DNP-KLH intravenously on day 5, and sacrificed on day 12. Each point represents the geometric mean of the responses of four animals and one SE about the mean is indicated. (b) Dose response of carrier-primed T cells. Irradiated BAB/14 mice were injected with either 2.5-mo KLH-primed nylon column-purified splenic T cells (obtained from BALB/cN spleen and contained 89% T cells and 4% Ig-bearing cells as assessed by immunofluorescence) alone or in combination with 5×10^6 normal BALB/cN spleen cells. See legend Fig. 1 a for immunization regime. Each point represents the geometric mean of the response of four animals and one SE about the mean is indicated.

TABLE I

Effect of ^{F}DNP-MGG Concentration on the Number of Labeled Lymphocytes in Normal Mouse Spleen

Staining concentration* of ^{F}DNP-MGG	No. of lymphocytes‡ counted ($\times 10^2$)	Labeled lymphocytes
μg/ml		%
6	101	0.03
18	51	0.1
55	19	0.3
167	15	1.6
500	15	2.7
1,600	15	2.3

* Normal BALB/cN spleen cells were prepared and stained with ^{F}DNP-MGG as described in the Materials and Methods.

‡ Cell smears were first examined under white-light darkfield illumination and only those cells with intact plasma membranes exhibiting speckled, ringed, or capped membrane-associated fluorescence were considered positive.

The increased proportion of binding cells with the ^{F}DNP-MGG concentration reflects a wide range of avidity in the binding population as well as a certain degree of nonspecificity. High avidity binding cells (i.e., those which stain at low ^{F}DNP-MGG concentrations) are specific for DNP. A 100-fold molar excess of ϵ-DNP-lysine reduced the number of detectable binding cells to 15% of the control (see Table II). However, binding by low avidity cells showed less

TABLE II

Specific Inhibition of ^{F}DNP-MGG-Binding Cells

Staining concentration* ^{F}DNP-MGG	Molarity of DNP‡ on ^{F}DNP-MGG $\times 10^{-6}$ M	Molarity of ϵ-DNP-lysine§ in staining mixture $\times 10^{-5}$ M	^{F}DNP-MGG‖-labeled lymphocytes	Inhibition of binding of ^{F}DNP-MGG
$\mu g/ml$			%	%
167	27	—	1.0	—
		27	0.9	10
		270	0.8	20
55	9	—	0.38	—
		9	0.33	25
		90	0.20	50
18	3	—	0.20	—
		3	0.12	40
		30	0.08	60
6	1	—	0.07	—
		1	0.02	71
		10	0.01	86

* Normal BALB/cN spleen cells were prepared and double stained with ^{F}DNP-MGG and ^{R}KLH as described in the Materials and Methods. 500 $\mu g/ml$ of ^{R}KLH was included in all staining mixtures. The percent of ^{R}KLH-labeled lymphocytes ranged from 0.20 to 0.28% in tubes not containing ϵ-DNP-lysine. None of the concentrations of ϵ-DNP-lysine used in inhibitions significantly decreased the number of ^{R}KLH-binding lymphocytes. The percent of ^{R}KLH-labeled lymphocytes observed in the presence of from 1 to 27 $\times 10^{-5}$ M ϵ-DNP-lysine ranged from 0.20 to 0.26%.

‡ The effective molarity of DNP present in varying concentrations of ^{F}DNP-MGG was calculated on the basis of 23 mol of DNP/mol (150,000 daltons) of MGG.

§ Either a 10- or 100-fold molar excess of ϵ-DNP-lysine (based on the molarity of DNP present in the form of ^{F}DNP-MGG) was added to the staining mixtures for inhibition studies.

‖ Rhodamine-labeled glutaraldehyde-fixed chicken erythrocytes (which fluoresce under both fluorescein and rhodamine illumination conditions) were added to the stained spleen cells in a ratio of 1:200 before smears were prepared. Slides were scanned for ^{F}DNP-MGG-binding cells and the proportion of labeled lymphocytes calculated based on the number of chicken erythrocytes counted. The same procedure was used to determine the proportion of ^{R}KLH-binding lymphocytes. Between 5,000 and 30,000 lymphocytes were scanned on each slide.

specificity in that a 100-fold molar excess of ϵ-DNP-lysine at the highest staining concentration reduced the number of detectable binding cells only to 80% of the control. Recent experiments[4] indicate that a major component of this nonspecificity is due to binding of the MGG moiety of the ^{F}DNP-MGG.

[4] Julius, M. H. Unpublished observations.

The Adoptive Primary Response of Isolated FDNP-MGG-Binding Cells. 2% of normal BALB/cN splenic lymphocytes bound detectable FDNP-MGG at a high-staining concentration (167 µg/ml). Isolation of these cells using the FACS yielded a population containing 91% FDNP-MGG-binding cells and a ninefold depleted population containing 0.3% binding cells (see Table III). These populations were assayed for DNP-PFC precursor activity in adoptive transfer experiments with carrier-primed T cells. Transfer without carrier-primed T cells resulted in low DNP-PFC responses (see Table III).

To test for enrichment of precursor activity, the number of isolated binding

TABLE III

Adoptive Primary Response of Purified FDNP-MGG-Binding Cells

% Fluorescent cells* in fractions			No. of cells transferred ($\times 10^6$)‡				Direct anti-DNP§ PFC/spleen ($\times 10^3$)	Total anti-KLH‖ titer ($\log_2$)
Unfractionated	FDNP-MGG (+)	FDNP-MGG (−)	KLH-primed T¶	Unfractionated	FDNP-MGG (+)	FDNP-MGG (−)		
2.0	91	0.3	—	5	—	—	6 (4–9)	2.0 ± 0
			7.5	—	—	—	6 (6–7)	3.4 ± 0.6
			—	—	0.1	—	1 (0.8–1)	3.5 ± 0.5
			—	—	—	5	1 (0.5–2)	4.0 ± 0
			7.5	5	—	—	59 (52–67)	8.5 ± 2.9
			7.5	—	0.1	—	27 (23–30)	4.0 ± 0
			7.5	—	—	5	28 (24–33)	8.5 ± 2.9

* BALB/cN normal spleen cells were stained with FDNP-MGG at 167 µg/ml. DNP-MGG-positive and -negative cells were isolated using the FACS as described in the Materials and Methods.

‡ Unfractionated, FDNP-MGG (+), or FDNP-MGG (−) cell populations were transferred into 600 R irradiated BAB/14 mice intravenously. Animals received 100 µg alum-precipitated DNP-KLH on day 0, 10 µg aqueous DNP-KLH on day 5, and were bled and sacrificed on day 12.

§ Geometric mean of the responses of four animals. Numbers in parentheses represent one SE about the mean.

‖ Log₂ anti-KLH hemagglutination titer ± SE. Each number represents the arithmetic mean of the titers of four animals.

¶ KLH-primed T cells were isolated from 7-day primed BAB/14 spleens by nylon wool passage and contained 93% T cells and 1.4% Ig-bearing cells as assessed by immunofluorescent staining.

cells transferred from the enriched fraction was adjusted to be roughly equivalent to the number of binding cells found in 5×10^6 unfractionated spleen. Thus, in this experiment 1×10^5 isolated binding cells (at 91% purity) were transferred since the unfractionated population contained 2% binding cells. As seen by comparing lines 5 and 6 of Table III, 10^5 cells of the DNP (+) fraction gave rise to slightly fewer DNP-PFC (actually half as many in this experiment) as 50 times more unfractionated cells. Thus, the precursor activity in the antigen-binding cells is greatly enriched.

To assess depletion of precursor activity, equal numbers of the unfractionated and the DNP-negative cells were transferred along with T cells. Comparison of the last two lines of Table III shows no apparent depletion. However (see below), there is a striking qualitative difference in the PFC obtained from the two populations.

The enrichment for DNP-PFC precursors was specific in that it did not give a concomitant enrichment for anti-KLH precursors. The DNP (+) fraction gave a KLH response not above the background from KLH-primed B cells contaminating the T cells. Furthermore, the depleted fraction contained the same anti-KLH precursor activity as unfractionated spleen (see Table III).

Relationship Between the Avidity of ^{F}DNP-MGG Binding DNP-PFC Precursor Cells and the DNP-PFC Resulting on Transfer of Purified Binding Cells. There is a striking qualitative difference in the DNP-PFC obtained from the depleted fraction and the DNP-PFC obtained from either the isolated binding cells or unfractionated spleen. The avidities of the DNP-PFC in the recipients of the depleted fraction are about 300-fold lower.

The avidities of DNP-PFC were determined by sensitivity to inhibition by either ϵ-DNP-lysine or DNP-MGG. The data in Fig. 2 *a* show that while only 45% of the DNP-PFC in recipients of cells from the depleted fraction are inhibited at the highest concentration of ϵ-DNP-lysine (10^{-4} M), the same concentration inhibited 90% of the DNP-PFC in recipients of unfractionated spleen. By interpolation, 45% of the DNP-PFC in recipients of unfractionated spleen were inhibited at about 3×10^{-7} M ϵ-DNP-lysine, a factor of 300 higher. All of the DNP-PFC in the recipients of cells from the depleted fraction are inhibited by DNP_{23}-MGG indicating that even the lowest avidity PFC are specific for DNP. (The multivalency of DNP on DNP-MGG makes it a more efficient inhibitor

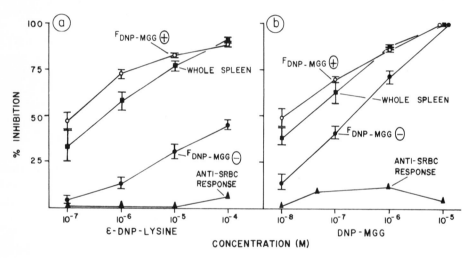

Fig. 2. (*a*) Inhibition profiles of DNP-PFC by ϵ-DNP lysine. Cells are from spleens of mice receiving carrier-primed T cells in addition to either FACS-purified splenic ^{F}DNP-MGG (+) binding cells (stained with ^{F}DNP-MGG at 167 μg/ml), the corresponding depleted fraction (−), or whole spleen cells. Details of adoptive transfer in the Materials and Methods. ϵ-DNP-lysine was included in the plaquing chambers. Each point represents the geometric mean of the percent inhibition of the PFC responses of four animals compared to the geometric mean of four animals for each control group. One SE about the mean is indicated. The mean number of uninhibited PFC/spleen in these groups are given in lines 5–7 of Table III. In addition, the effect of the various concentrations of ϵ-DNP-lysine on a day 4 primary response to SRBC is shown. (*b*) Inhibition profiles of anti-DNP PFC by DNP-MGG. See legend of Fig. 2 *a*.

than ϵ-DNP-lysine.) Neither ϵ-DNP-lysine nor DNP_{23}-MGG inhibit anti-SRBC-PFC (see Fig. 2).

Reduction in the avidity of the DNP-PFC in recipients of cells from the fraction depleted of nearly all cells which bind ^{F}DNP-MGG at a high-staining concentration suggests that low avidity DNP-PFC precursors give rise to low avidity DNP-PFC. This is consistent with the observation that the avidity of DNP-PFC in recipients of cells from the enriched fraction appear to be slightly higher than the DNP-PFC in recipients of unfractionated spleen (see Fig. 2 *a*). The high ^{F}DNP-MGG concentration used here stained nearly all the ^{F}DNP-MGG-binding cells in the unfractionated spleen (see Tables I and II), therefore only the very low avidity precursors are expected to be absent from the enriched fraction. Thus, the higher avidity of the DNP-PFC in the recipients of cells from the enriched fraction reflects this absence.

Isolation of High Avidity ^{F}DNP-MGG-Binding Cells and Assessment of Their Precursor Activity. To determine whether high avidity ^{F}DNP-MGG-binding cells contained the precursors of high avidity PFC, normal BALB/cN splenic lymphocytes were stained at a low concentration of ^{F}DNP-MGG (10 μg/ml) to allow isolation of the highest avidity lymphocytes.

At this staining concentration, 0.2% of the lymphocytes bound detectable (by fluorescence microscopy) ^{F}DNP-MGG and the FACS-enriched fraction contained 85% ^{F}DNP-MGG-binding cells (see Table IV). The depleted fraction contained less than 0.03% fluorescent binding cells. It also should contain those low avidity binding lymphocytes which are detected at higher ^{F}DNP-MGG-staining concentrations (compare Table I).

The avidity of the DNP-PFC in recipients of the stained (high binding avidity) cells are 100-fold higher than the avidity of the DNP-PFC derived from the depleted fraction (Fig. 3). The lines in this figure show that while 75% of the DNP-PFC obtained from the enriched fraction are inhibited by 10^{-6} M ϵ-DNP-lysine, 10^{-4} M ϵ-DNP-lysine is required to inhibit the same proportion of DNP-PFC obtained from the depleted fraction. Thus, enrichment of high avidity ^{F}DNP-MGG-binding cells gives concomitant enrichment for high avidity DNP-PFC precursors.

In the recipients of cells from the depleted fraction, the avidity of the DNP-PFC were 10-fold lower than the DNP-PFC in recipients of unfractionated spleen. This contrasts with the last experiment (see Fig. 2) where the depleted fraction obtained from a population stained with a high concentration of ^{F}DNP-MGG gave rise to DNP-PFC of 300-fold lower avidity compared to unfractionated spleen. Thus, when the majority of the high avidity binding cells are removed, the depleted fraction contains the medium and low avidity DNP-PFC precursors (as well as any contaminating high avidity precursors missed in the separation) which can be removed by increasing the staining concentration. This suggests that over a wide range of avidities there is a direct correlation between the avidity of the ^{F}DNP-MGG-binding precursor cells and the avidity of the DNP-PFC they produce.

Several additional controls relevant to the entire study were included in this experiment. Isolated ^{F}DNP-MGG-binding cells did not contain anti-KLH

TABLE IV
Adoptive Primary Response of Purified High Avidity ᶠDNP-MGG-Binding Cells

% Fluorescent cells* in fractions			No. of cells transferred ($\times 10^6$)‡				Direct anti-DNP§ PFC/spleen ($\times 10^3$)	Total anti-KLH‖ titer ($\log_2$)
Unfrac-tionated	ᶠDNP-MGG (+)	ᶠDNP-MGG (−)	KLH-primed T¶	Unfrac-tionated	ᶠDNP-MGG (+)	ᶠDNP-MGG (−)		
0.2	84	<0.03	—	5	—	—	9 (5–16)	3.0 ± 0.6
			7.5	—	—	—	2 (1–2)	4.7 ± 0.8
			—	—	0.05	—	0.7 (0.5–1)	2.5 ± 0.5
			—	—	—	5	8 (5–16)	2.0 ± 0
			7.5	5	—	—	30 (27–34)	12 ± 4.1
			7.5	0.05	—	—	5 (4–6)	2.7 ± 1.3
			7.5	—	**0.05**	—	20 (17–23)	3.3 ± 0.7
			7.5	—	0.05	—	3 (2–3)**	2 ± 0.5
			7.5	—	—	5	20 (17–22)	10 ± 1.8

* BALB/cN normal spleen cells were stained with ᶠDNP-MGG at 10 µg/ml. ᶠDNP-MGG-positive and -negative cells were isolated using the FACS as described in the Materials and Methods.

‡ Unfractionated, ᶠDNP-MGG (+), or ᶠDNP-MGG (−) cell populations were transferred into 600 R irradiated BAB/14 mice intravenously. All but one group of animals received 100 µg alum-precipitated DNP-KLH on day 0, 10 µg aqueous DNP-KLH on day 5, and were bled and sacrificed on day 12.

§ Geometric mean of the responses of four animals. Numbers in parentheses represent one SE about the mean.

‖ Log₂ anti-KLH hemagglutination titer ± SE. Each number represents the arithmetic mean of the (log₂) titers of four animals.

¶ KLH-primed T cells were isolated from BAB/14 spleens by nylon wool passage and contained 95% T cells and 0.7% Ig-bearing cells as assessed by immunofluorescent staining.

** For this group of animals KLH was substituted for DNP-KLH.

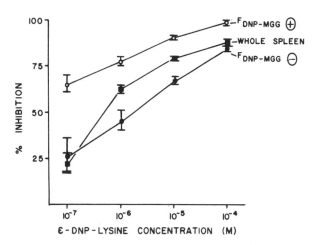

FIG. 3. Same as Fig. 2 *a* except cells were separated after staining at 10 µg/ml ᶠDNP/MGG. The mean number of uninhibited PFC per spleen in these groups are given in lines 5, 7, and 9 of Table IV.

precursor activity, while the depleted fraction contained the same anti-KLH precursor activity as the unfractionated spleen (see Table IV). The hapten specificity of the isolated binding cells was further demonstrated since the DNP-PFC response obtained when recipients of isolated cells are challenged with KLH was 15% of the response obtained when DNP-KLH was used as the immunogen. In addition, transfer of 5×10^4 unfractionated spleen which contained 0.2% binding cells gave rise to only 5×10^3 DNP-PFC emphasizing that enrichment for ^{F}DNP-MGG-binding cells is specific for DNP-PFC precursor activity (see Table IV). To complete the demonstration that the avidity of the antigen-binding precursor cell is directly related to the avidity of the progeny PFC, low avidity ^{F}DNP-MGG-binding cells were isolated directly and shown to be the precursors of low avidity DNP-PFC.

Isolation of Medium and Low Avidity DNP-PFC Precursors. In this experiment, low avidity binding cells were differentially labeled by staining with ^{F}DNP-MGG in the presence of ϵ-DNP-lysine which blocked staining of the high avidity binding cells.

When normal BALB/cN spleen cells were stained with 55 μg/ml ^{F}DNP-MGG in the presence of a 100-fold molar excess (calculated from DNP residues) of ϵ-DNP-lysine, 0.8% of the lymphocytes bound detectable ^{F}DNP-MGG. Passage of this population through the FACS yielded an enriched population containing 90% ^{F}DNP-MGG-binding cells and a depleted fraction containing 0.03% binding cells (see Table V). The isolated binding cells were enriched 100-fold for DNP-PFC precursor activity, while the depleted fraction was depleted twofold for DNP-PFC precursor activity (see Table V).

TABLE V

Adoptive Primary Response of Purified Medium and Low Avidity
^{F}DNP-MGG-Binding Cells

% Fluorescent cells*			No. of cells transferred ($\times 10^6$)‡				Direct anti-DNP§ PFC/spleen ($\times 10^3$)
Unfrac-tionated	^{F}DNP-MGG (+)	^{F}DNP-MGG (−)	KLH-Primed T‖	Unfrac-tionated	^{F}DNP-MGG (+)	^{F}DNP-MGG (−)	
0.8	90	0.03	—	5	—	—	6 (5–8)
			7.5	—	—	—	4 (2–7)
			—	—	0.05	—	0.3 (0.2–0.5)
			—	—	—	5	3 (3–4)
			7.5	5	—	—	47 (41–54)
			7.5	—	0.05	—	40 (38–43)
			7.5	—	—	5	29 (26–31)

* BALB/cN normal spleen cells were stained with ^{F}DNP-MGG at 55 μg/ml in the presence of 8.7×10^{-4} M ϵ-DNP-lysine. ^{F}DNP-MGG-positive and -negative cells were isolated using the FACS as described in the Materials and Methods.

‡ Unfractionated, ^{F}DNP-MGG (+) or ^{F}DNP-MGG (−) cell populations were transferred into 600 R irradiated BAB/14 mice intravenously. All animals received 100 μg alum-precipitated DNP-KLH on day 0, 10 μg aqueous DNP-KLH on day 5, and were sacrificed on day 12.

§ Geometric mean of the responses of four animals. Numbers in parentheses represent one SE about the mean.

‖ KLH-primed T cells were isolated from BAB/14 spleens by nylon wool passage and contained 93% T cells and 1% Ig-bearing cells as assessed by immunofluorescent staining.

The data in Fig. 4 show that the avidity of the DNP-PFC obtained from the isolated medium and low avidity binding cells are 100-fold lower than the avidity of the DNP-PFC obtained from the depleted fraction. While 10^{-5} M ϵ-DNP-lysine inhibited 60% of the DNP-PFC obtained from the enriched fraction, only 10^{-7} M was required to inhibit the same proportion of DNP-PFC obtained from the depleted fraction.

The avidity of the DNP-PFC obtained from enriched fraction in this experiment were slightly lower than the avidity of the DNP-PFC obtained from the unfractionated spleen, which suggests that under these staining conditions,

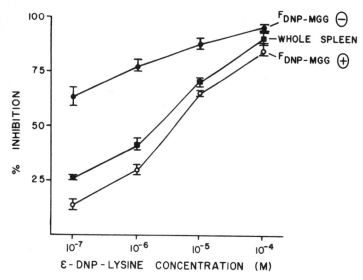

FIG. 4. Same as Figure 2 a except cells were separated after staining at 55 μg/ml ^{F}DNP-MGG in the presence of 8.7×10^{-4} M ϵ-DNP-lysine. The mean number of uninhibited PFC per spleen in these groups are given in lines 5–7 of Table V.

the medium and low avidity DNP-PFC precursors are stained while staining of the highest avidity precursors was blocked by ϵ-DNP-lysine.

Discussion

The concept of clonal selection (28) rests on two basic assumptions: (a) precursors of antibody-forming cells are antigen-binding cells, and (b) antigen receptors on the precursors have the same specificity and affinity as the antibody secreted by the progeny plasma cells. Although much of the data accumulated in cellular immunology is consistent with the predictions of clonal theory, there have been few studies which provide direct evidence to validate these basic assumptions.

To demonstrate that the precursors of antibody-forming cells are antigen-binding cells, Wigzell and Mäkelä (29) passed suspensions of primed or unprimed spleen cells through antigen-coated columns. With this method, they specifically depleted the precursors from the column effluent population but were unable to recover cells enriched for precursor activity from the column. Ada and Byrt (30) also showed that specific precursors are depleted from spleen cell populations when allowed to bind highly radioactive antigen ("suicide" experiments). These experiments gave strong support for the role of antigen-binding cells as precursors but the question still remained open because of the inability to isolate these cells and directly test their function.

The development of the FACS has now allowed the isolation of functional antigen-binding cells. In a previous publication, we showed that the KLH-binding cells isolated from KLH-primed spleen contained all the anti-KLH precursors present in intact spleen (5). We therefore firmly established that the precursors of antibody-forming cells in primed spleen are antigen-binding cells.

In this work we have extended this conclusion to antigen-binding cells in unprimed animals. We have shown that isolated DNP-binding cells are the precursors of anti-DNP antibody-secreting cells. Moreover, we have shown that the avidity of the antibody produced by the antibody-secreting cell reflects the avidity of the receptor on the antigen-binding precursor cell.

In the experiments reported here, high avidity DNP-binding cells gave rise to predominantly high avidity DNP-PFC. Isolation and transfer of medium and low avidity binding cells gave rise to medium and low avidity DNP-PFC. In both cases, transfer of the complementary unstained populations (i.e., the "depleted" fractions) gave rise to DNP-PFC with avidities in the range expected for those precursors intentionally left unstained.

The determination of DNP-PFC avidities is based on their inhibition by ϵ-DNP-lysine as described by Andersson (3) in 1970. Hapten inhibition of PFC has been shown to be a valid measure of antibody avidity in other systems. Yamada et al. (31) have shown that mouse myeloma (MOPC-315) cells, which secrete homogeneous IgA molecules with a uniform binding constant for DNP, form DNP-specific plaques. All of the DNP-PFC were sharply inhibited over a very narrow concentration range of ϵ-DNP-lysine included in the plaquing assay. 50% of the DNP-PFC were inhibited by a concentration of ϵ-DNP-lysine which closely approximated the reciprocal of the binding constant of the myeloma protein.

Although these results indicate the possibility of avidity, or in fact affinity, measurements at the cellular level, certain restrictions to the use of hapten inhibition for avidity determinations merit consideration here. First, it may be used (uncorrected) to determine avidity only when there is a relatively constant rate of antibody secretion among PFC in a given experiment, i.e., when plaque size is relatively constant. Since the avidity measured by this method is a weighted average between the local antibody concentration and the true avidity of the antibody secreted, large differences in rates of antibody secretion will cause apparent avidity differences between two PFC secreting identical antibody molecules. Such differences have been shown experimentally with PFC from the B-cell clone E9 which secretes a homogeneous immunoglobulin specific for DNP when large variation in uninhibited plaque size was observed ($\sim$10-fold) (32). In our experiments, the variation in uninhibited plaque size was small ($\sim$twofold). Therefore, it is likely that the major parameter measured in these experiments is relative avidity of anti-DNP antibody rather than the amount secreted per DNP-PFC.

Secondly, if IgG and IgM PFC are present in the same experiment, the hapten concentrations giving equivalent inhibitions might well be quite different due to the differing valencies of the different antibodies. In these experiments, as will be further discussed below, only IgM PFC are produced.

Thirdly, comparison of avidities for PFC between experiments may be made only when the extent of hapten conjugation to indicator erythrocytes is kept constant. Pasanen and Mäkelä (33) have shown that the ability to inhibit PFC with free hapten decreases as the extent of hapten conjugation to indicator cells increases. Therefore, while comparison of PFC avidity is valid within a given experiment, it must be considered a relative value when comparing experiments unless the hapten conjugation is strictly controlled between experiments. This restriction applies to the data we have reported since each experiment was performed with newly conjugated erythrocytes which may have varied somewhat with respect to the extent of hapten conjugation. Therefore, the data is discussed in terms of the avidities of DNP-PFC from the unfractionated populations in each experiment.

Despite this reservation, there is at least a rough correlation between average avidities measured in different experiments. The average avidities of DNP-PFC obtained from adoptive secondary transfers (i.e., transfer of DNP-KLH-primed spleen) measured in other experiments are 100-fold higher than the average avidities of the DNP-PFC which we find here in an adoptive primary transfer[5] (high avidity PFC obtained in adoptive secondary transfers are predominantly due to IgG antibody, whereas the PFC in these adoptive primary transfers are virtually all due to IgM antibody). This difference in avidities is consistent with higher affinity antibody found in a secondary response (1) and suggests that the hapten inhibition method may be used for comparisons between experiments when there are large avidity differences.

Very few, if any, IgG DNP-PFC are found in the adoptive primary recipients in the experiments reported here, although supplementation of an unprimed precursor population with carrier-primed cooperator cells has been shown to favor the production of IgG PFC in other systems (34, 35). It may be that the numbers of IgG DNP-PFC would have been larger had there been a larger interval between antigen boost and day of assay for PFC. Using a similar adoptive primary transfer system, the peak IgG anti-DNP serum titer occurred at day 30 after transfer, while at day 12 the anti-DNP titer was due completely to IgM antibody.[6]

The absolute primed T-cell dependence of this adoptive primary DNP response both with unfractionated spleen cells and the isolated antigen-binding cells is remarkable in view of the wholly IgM response (at 12 days). This absolute dependence is manifest even when only the highest avidity DNP precursors are used with all the remaining normal spleen cells present, except for low and medium avidity DNP-binding cells hence precursors. This may provide an exquisitely sensitive system for assay of active T cells or T-cell substitutes. The recipients are boosted with rather large amounts of DNP-KLH for the presumed relatively few high avidity precursors. Any explanation of T-cell dependence invoking aid to better antigen binding by precursors would be least applicable, one would expect, to these high avidity cells. It will be interesting to see whether continued stimulation of these precursors will lead to production of IgG DNP-PFC and with higher avidities or whether some suppression of these indirect PFC is occurring.

_____

[5] Metzler, C. M. Unpublished observation.
[6] Segal, S. Personal communication.

Summary

Cells binding DNP groups conjugated to fluoresceinated mouse gamma globulin (ᶠDNP-MGG) were isolated from spleens of unprimed mice using a fluorescence-activated cell sorter (FACS). The isolated cells were specifically enriched at least 100-fold for anti-DNP precursor activity in an adoptive transfer assay as compared to unfractionated spleen. The fraction depleted of binding cells, although depleted of anti-DNP precursor activity, responded as well as unfractionated spleen when assayed for anticarrier (keyhole limpet hemocyanin [KLH]) precursor activity.

High avidity binding cells were stained using low concentrations of ᶠDNP-MGG. Medium and low avidity binding cells were stained using high concentrations of ᶠDNP-MGG in the presence of free hapten which selectively blocked staining of the high avidity binding cells. Cells were supplemented with an excess of carrier-primed (KLH), nylon-purified splenic T cells and transferred to irradiated recipients. DNP-KLH was given at transfer and 5 days later. The anti-DNP plaque-forming cell (DNP-PFC) response and the avidities of the DNP-PFC in the irradiated recipients were measured by hapten inhibition of direct PFC plaque formation 12 days after transfer. At this time, very few indirect PFC were found.

There was a positive correlation between the avidity of the DNP-binding cells and the avidity of the anti-DNP antibody secreted by their progeny. High avidity DNP-binding cells gave rise to predominantly high avidity anti-DNP-PFC. Medium and low avidity binding cells gave rise to medium and low avidity DNP-PFC.

It is a pleasure to acknowledge the excellent technical assistance of R. V. Waters, V. M. Bryan, D. H. Hewgill, T. Knaak, and R. T. Stovel. We also appreciate the devoted editorial assistance of Lee Herzenberg and the manuscript preparation by K. Dalman.

Received for publication 7 May 1974.

Bibliography

1. Eisen, H. N., and G. W. Siskind. 1964. Variations in affinities of antibodies during the immune response. *Biochemistry.* **3:**996.
2. Siskind, G. W., and B. Benacerraf. 1969. Cell selection by antigen in the immune response. *Adv. Immunol.* **10:**1.
3. Andersson, B. 1970. Studies on the regulation of avidity at the level of the single antibody-forming cell. The effect of antigen dose and time after immunization. *J. Exp. Med.* **132:**77.
4. Andersson, B. 1972. Studies on antibody affinity at the cellular level. Correlation between binding properties of secreted antibody and cellular receptor for antigen on immunological memory cells. *J. Exp. Med.* **135:**312.
5. Julius, M. H., T. Masuda, and L. A. Herzenberg. 1972. Demonstration that antigen binding cells are precursors of antibody-producing cells after purification using a fluorescence activated cell sorter. *Proc. Natl. Acad. Sci. U. S. A.* **69:**1934.
6. Julius, M. H., R. G. Sweet. C. G. Fathman, and L. A. Herzenberg. 1974. Fluorescence activated cell sorting and its applications. Los Alamos, N. M., October 17–19, 1973.

Atomic Energy Commission Symposium Series (C.O.N. 73–1007). P. F. Mullaney, D. F. Petersen, and C. R. Richmond, editors. Technical Information Center, U. S. Atomic Energy Commission, Oak Ridge, Tenn. In press.

7. Henry, C., J. Kimura, and L. Wofsy. 1972. Cell separation on affinity columns: the isolation of immunospecific precursor cells from unimmunized mice. *Proc. Natl. Acad. Sci. U. S. A.* **69:**34.

8. Rutishauser, U., P. D'eustachio, and G. M. Edelman. 1974. Immunological function of lymphocytes fractionated with antigen-derivatized fibers. *Proc. Natl. Acad. Sci. U. S. A.* **70:**3894.

9. Bonner, W. A., H. R. Hulett, R. G. Sweet, and L. A. Herzenberg. 1972. Fluorescence activated cell sorting. *Rev. Sci. Instrum.* **43:**404.

10. Davie, J. M., A. S. Rosenthal, and W. E. Paul. 1971. Receptors on immunocompetent cells. III. Specificity and nature of receptors on dinitrophenylated guinea pig albumin-^{125}I-binding cells of immunized guinea pigs. *J. Exp. Med.* **134:**517.

11. Davie, J. M., and W. E. Paul. 1972. Receptors on immunocompetent cells. IV. Direct measurement of avidity of cell receptors and cooperative binding of multivalent ligands. *J. Exp. Med.* **135:**643.

12. Dulbecco, R., and M. Vogt. 1954. Plaque formation and isolation of pure lines with poliomyelitis viruses. *J. Exp. Med.* **99:**167.

13. Gey, G. O., and M. K. Gey. 1936. The maintenance of human normal cells and tumor cells in continuous culture. I. Preliminary report: cultivation of mesoblastic tumors and normal tissue and notes on methods of cultivation. *Amer. J. Cancer.* **27:** 45.

14. Eagle, H. 1959. Amino acid metabolism in mammalian cell cultures. *Science (Wash. D. C.)* **130:**432.

15. Julius, M. H., E. Simpson, and L. A. Herzenberg. 1973. A rapid method for the isolation of functional thymus derived murine lymphocytes. *Eur. J. Immunol.* **3:**645.

16. McKinney, R. M., J. T. Spillane, and G. W. Pearce. 1964. Factors affecting the rate of reaction of fluorescein isothiocyanate with serum proteins. *J. Immunol.* **93:**232.

17. Wood, B. T., S. H. Thompson, and G. Goldstein. 1965. Fluorescent antibody staining. III. Preparation of fluorescein-isothiocyanate-labelled antibodies. *J. Immunol.* **95:**225.

18. Cebra, J. J., and G. Goldstein. 1965. Chromatographic purification of tetramethyl-rhodamine-immune globulin conjugates and their use in the cellular localization of rabbit γ-globulin polypeptide chains. *J. Immunol.* **95:**230.

19. Mitchison, N. A. 1967. Antigen recognition responsible for the induction *in vitro* of the secondary response. *Cold Spring Harbor Symp. Quant. Biol.* **32:**431.

20. Mitchison, N. A. 1969. Cell populations involved in the immune response. *In* Immunological Tolerance. M. Landy and W. Braun, editors. Academic Press, Inc., New York. 149.

21. Cunningham, A. J., and A. Szenberg. 1968. Further improvements in the plaque technique for detecting single antibody-forming cells. *Immunology.* **14:**599.

22. Rittenberg, M. B., and K. L. Pratt. 1968. Antitrinitrophenyl (TNP) plaque assay. Primary response of BALB/c mice to soluble and particulate immunogen. *Proc. Soc. Exp. Biol. Med.* **132:**575.

23. Inman, J. K., B. Merchant, L. Claflin, and S. E. Tacey. 1973. Coupling of large haptens to proteins and cell surfaces: preparation of stable, optimally sensitized erythrocytes for hapten-specific hemolytic plaque assays. *Immunochemistry.* **10:**165.

24. Eisen, H. N., and F. Karush. 1949. Interaction of purified antibody with homologous hapten: antibody valence and binding constant. *J. Am. Chem. Soc.* **71:**363.

25. Boyden, S. V. 1951. The adsorption of proteins on erythrocytes titrated with tannic acid and subsequent hemagglutination by antiprotein sera. *J. Exp. Med.* **93:**197.

26. Avrameas, S., B. Taudou, and S. Chuilon. 1969. Glutaraldehyde, cyanuric chloride

and tetraazotized O-dianisidine as coupling reagents in the passive hemagglutination test. *Immunochemistry*. **6**:67.

27. Katz, D. H., W. E. Paul, E. A. Goidl, and B. Benacerraf. 1970. Carrier function in anti-hapten immune responses. I. Enhancement of primary and secondary anti-hapten antibody responses by carrier preimmunization. *J. Exp. Med.* **132**:261.

28. Burnet, F. M. 1959. The Clonal Selection Theory of Acquired Immunity. Cambridge University Press, London, England.

29. Wigzell, H., and O. Mäkelä. 1970. Separation of normal and immune lymphoid cells by antigen-coated columns. Antigen-binding characteristics of membrane antibodies as analyzed by hapten-protein antigens. *J. Exp. Med.* **132**:110.

30. Ada, G. L., and P. Byrt. 1969. Specific inactivation of antigen reactive cells with ^{125}I-labelled antigen. *Nature* (Lond.). **222**:1291.

31. Yamada, H., A. Yamada, and V. P. Hollander. 1970. 2,4-dinitrophenyl-hapten specific hemolytic plaque-in-gel formation by mouse myeloma (MOPC-315) cells. *J. Immunol.* **104**:251.

32. North, J. R. and B. A. Askonas. 1974. Analysis of affinity of monoclonal antibody responses by inhibition of plaque forming cells. *Eur. J. Immunol.* **4**:361.

33. Pasanen, V. J., and O. Mäkelä. 1969. Effect of the number of haptens coupled to each erythrocyte on hemolytic plaque formation. *Immunology*. **16**:399.

34. Cunningham, A. J., and E. E. Sercarz. 1971. The asynchronous development of immunological memory in helper (T) and precursor (B) cell lines. *Eur. J. Immunol.* **1**:413.

35. Cheers, C., and J. F. A. P. Miller. 1972. Cell-to-cell interaction in the immune response. IX. Regulation of hapten-specific antibody class by carrier priming. *J. Exp. Med.* **136**:1661.

X-LINKED B-LYMPHOCYTE IMMUNE DEFECT IN CBA/N MICE

II. Studies of the Mechanisms Underlying the Immune Defect*

By IRWIN SCHER, ALFRED D. STEINBERG, ALICE K. BERNING, AND WILLIAM E. PAUL

(From the Department of Clinical and Experimental Immunology, Naval Medical Research Institute; the Division of Clinical Immunology, Department of Medicine, National Naval Medical Center; the Arthritis, and Rheumatism Branch, National Institute of Arthritis, Metabolism and Digestive Diseases, and the Laboratory of Immunology, National Institute of Allergy and Infectious Diseases, National Institutes of Health, Bethesda, Maryland 20014)

CBA/N (CN)[1] mice are a subline of CBA mice with a marked defect in the function and a diminution in the number of thymus-independent (B) lymphocytes. These animals are unable to form specific antibody to Type III pneumococcal polysaccharide (SIII), bacterial lipopolysaccharide (LPS) (1), or polyriboinosinic-polyribocytidylic acid (poly I·C) (2). Each of these antigens is known to evoke an antibody response in all conventional mouse strains studied, even if these mice are deprived of thymus-dependent (T) lymphocytes (3-5). The inability of the CN mice to respond to these "T-independent" antigens is inherited as an X-linked recessive trait (1, 2). We have shown that CN mice and F_1 male mice of the CN × DBA/2N (DN) cross (CN × DN F_1 males) have a diminished number of immunoglobulin (Ig)-bearing spleen cells, an impaired response to agents mitogenic for B lymphocytes (B mitogens) and a diminished ability to participate in antibody-dependent cell-mediated cytotoxicity (6).

Although the X-linked immune defect in the CN mice is expressed as a functional abnormality of B lymphocytes, the mechanism underlying the defect has not been studied. In particular, it is not known whether the abnormality of CN B-lymphocyte function represents a defect intrinsic to the B-lymphocyte line or results either from an abnormality in the microenvironment in which CN B lymphocytes differentiate or from abnormal function of CN T lymphocytes. Because of the X-linked nature of the immune defect of CN mice, lymphoid cell

* This work was supported in part by the Bureau of Medicine and Surgery, Navy Department, Work Unit nos. MR041.02.01.0020B2GI and CIC 3-06-132. The opinions or assertions contained herein are the private ones of the authors and are not to be construed as official or reflecting the views of the Navy Department or the naval service at large. The animals used in this study were handled in accordance with the provisions of Public Law 89-54 as amended by Public Law 91-579, "Animal Welfare Act of 1970," and the principles outlined in the "Guide for the Care and Use of Laboratory Animals," U. S. Department of Health, Education and Welfare publication no. (NIH) 73-23.

[1] *Abbreviations used in this paper:* anti-θ, anti-Thy 1.2; B, thymus independent; BRBC, burro erythrocytes; Con A, concanavalin A; CN, CBA/N mice; DN, DBA/2N mice; DNP-lys-Ficoll, 2,4-dinitrophenyl-lysyl-derivative of Ficoll; LPS, bacterial lipopolysaccharide; PFC, plaque-forming cell; poly I·C, polyriboinosinic-polyribocytidylic acid; SIII, type III pneumococcal polysaccharide; T, thymus dependent; TNP, trinitrophenyl.

transfers between phenotypically normal F_1 female mice and their abnormal male littermates can be performed. By immunization of recipients of transferred cells with either of two T-independent antigens, poly I·C and the 2,4-dinitrophenyl-lysyl-derivative of Ficoll (DNP-lys-Ficoll) (7, 8), we have studied the cellular basis of the immune defect of CN mice. Our data indicate that the failure of these mice to respond to T-independent antigens is a result of a deficiency or an intrinsic abnormality of B lymphocytes and/or their progenitors rather than a microenvironmental or T-lymphocyte abnormality.

Materials and Methods

Animals. CN,[2] DN mice, and F_1 animals derived from these strains, (CN × DN) F_1, were obtained from the Rodent and Rabbit Production Section of the National Institutes of Health, Bethesda, Md. All mice were 6–12wk of age at the time of study. The CN mice are a distinct subline of CBA mice and the history of their establishment has been described elsewhere (1, 2). F_1 mice of both sexes were produced by breeding CN females with DN males (CN × DN F_1 males or females).

Lethally irradiated mice (1,000 rads, ^{60}cobalt source, dose rate 40 rads/min) were reconstituted on the day after irradiation by the administration of spleen or bone marrow cells intravenously. CN × DN F_1 males or females were thymectomized at 6 wk of age by aspiration through a sternum-splitting incision. These mice were lethally irradiated 14 days after surgery and reconstituted with $10 × 10^6$ F_1 male or female bone marrow cells which had been treated with AKR anti-Thy 1.2 (anti-θ) plus rabbit complement (C) as has been previously described (9).

Cell Suspensions. Mice were killed by cervical dislocation and their thymuses and/or spleens removed. Spleen and thymus cells were obtained by gentle teasing with a rubber policeman and forceps into RPMI 1640 (Grand Island Biological Co, Grand Island, N. Y.). Bone marrow cells were flushed from the femur and tibial bones using a 25 gauge needle. Cell aggregates were disrupted by passing the cell suspensions through a 26 gauge needle. The single cell suspensions were then washed twice with RPMI 1640 and the number of cells counted.

Antigens. Poly I·C, prepared by annealing the homopolymers polyriboinosinic acid and polyribocytidylic acid, (P-L. Biochemicals, Inc., Milwaukee, Wis.) was dissolved in borate buffer, pH 8, and emulsified with an equal volume of complete Freund's adjuvant (Difco Laboratories, Detroit, Mich.) so that 0.3 ml of the final emulsion contained 100 μg of antigen. Mice were immunized with 0.3 ml of this material by intraperitoneal injection.

DNP-lys-Ficoll was prepared according to the method of Sharon et al. (8). Briefly, Ficoll (Pharmacia Fine Chemicals, Inc., Piscataway, N. J.) was reacted with cyanuric chloride at 4°C, followed by the addition of ϵ-DNP-L-lysine and subsequent reaction at room temperature. The DNP-lys-Ficoll used in these studies had a molar ratio of DNP to Ficoll of 32:1.

Sheep erythrocytes (SRBC) and burro erythrocytes (BRBC) were obtained from a single sheep or burro and were collected sterilely in citric acid-dextrose solution (ACD solution; Abbott Laboratories, Chemical Marketing Div., North Chicago, Ill.). These were washed three times in Hanks' balanced salt solution before use. SRBC were heavily conjugated with the trinitrophenyl (TNP) hapten by reacting sodium 2,4,6-trinitrobenzenesulfonate with SRBC according to the method of Kettman and Dutton (10). Mice were immunized by the intravenous injection of $2 × 10^8$ of these heavily conjugated TNP-SRBC immediately after their preparation.

Antibody and Plaque-Forming Cell Assays. Antibody directed against poly I·C was assayed in sera obtained by orbital sinus puncture of immunized mice by an ammonium sulfate precipitation assay using [^{14}C]poly I·C (Miles Laboratories Inc., Miles Research Div., Elkhart, Ind.) as ligand (11). Data are expressed as the percent binding of 80 ng of [^{14}C]poly I·C (4,000 dpm/μg) by 25 μl of mouse serum. Binding of greater than 20% was considered a positive response; in previous studies we have shown that serum from unimmunized mice do not exceed this degree of binding (2) as is the case for mice immunized with noncross-reacting antigens or adjuvant alone.

Spleen cells from mice immunized with SRBC, heavily substituted TNP-SRBC or DNP-lys-

---

[2] CBA/N mice were formerly designated CBA/HN.

Ficoll were assayed for cells releasing specific antibody by a modification of the Jerne hemolytic plaque technique (12), using SRBC or lightly conjugated TNP-BRBC or TNP-SRBC as indicator cells, respectively. TNP-SRBC and TNP-BRBC were prepared by the method of Rittenberg and Pratt (13). There was less than 1% cross-reactivity between SRBC and BRBC in our system. Plaque-forming cells (PFC) releasing IgG antibodies were detected according to the method of Pierce et al. (14) by inhibition of IgM PFC with goat antimouse μ-chain antibody incorporated in the agar and development of IgG PFC with rabbit polyvalent antimouse γ-chain antibody (both antisera were the gift of Dr. R. Asofsky, NIAID, Bethesda, Md.). Data are expressed as the mean PFC per spleen or the PFC per 10^6 cells $\pm$ SE unless otherwise indicated.

Results

Response of CN $\times$ DN F_1 Male and Female Mice to SRBC and TNP-SRBC. As a basis for comparison with subsequent studies of T-independent antibody responses of male and female CN $\times$ DN F_1 mice, we initially evaluated the response of these animals to two highly thymus-dependent antigens, SRBC and TNP-SRBC. In each of these experiments, the IgM response of F_1 male mice at 4 days after immunization with SRBC or TNP-SRBC was less than those of female mice when considered on the basis of PFC per spleen (Table I). However, since male mice have substantially fewer nucleated cells per spleen (6), it is more instructive to consider the antibody response on the basis of the number of PFC per 10^6 spleen cells. In the first two experiments shown in Table I, the male response, considered on this basis, was approximately half that of the female response and in the third experiment, only slightly less than that of the female. These results are particularly significant in view of our recent finding that the percent of B lymphocytes in the spleen of F_1 male mice is substantially less than that of F_1 female mice (6), and they indicate that F_1 male mice make considerable responses to thymus-dependent antigens. Or on other hand, the IgG anti-SRBC response of F_1 male mice was substanially less than that of F_1 females

TABLE I

IgM and IgG Responses of CN $\times$ DN F_1 Male and Female Mice to TNP-SRBC

Exp.	Sex (no. of mice)	Immuno-globulin	TNP-BRBC*		SRBC*	
			PFC/spleen	PFC/10^6 cells	PFC/spleen	PFC/10^6 cells
1	Male (6)	IgM	—	—	2,000 ± 488	44 ± 6.1
	Female (6)		—	—	7,583 ± 1,450	77 ± 10.6
2	Male (3)	IgM	4,037 ± 678	83.4 ± 10.7	4,175 ± 724	86 ± 11.4
	Female (3)		17,443 ± 2,362	208.2 ± 23.4	15,400 ± 1,584	166 ± 30.0
3	Male (3)	IgM	28,900 ± 8,636	586.0 ± 158.0	9,925 ± 1,615	204 ± 22.4
	Female (3)		59,417 ± 17,774	700.0 ± 76.0	29,933 ± 11,855	287 ± 95.4
3	Male (4)	IgG	—	—	2,858 ± 414	36 ± 8.2
	Female (4)		—	—	18,427 ± 3,080	216 ± 19.2

* The mean number of IgM or IgG PFC measured at 4 or 11 days, respectively, in response to 2×10^8 heavily conjugated TNP-SRBC. Data are expressed as the net mean number of PFC ± SE.

even when considered on a per B-cell basis (Table I, exp. 3).

Response of CN and CN × DN F₁ Mice to DNP-lys-Ficoll. DNP-lys-Ficoll has been recently demonstrated to be a potent T-independent antigen which elicits both IgM and IgG anti-DNP antibody in many strains of mice (8). CN × DN F₁ male and female mice were immunized with varying doses of DNP-lys-Ficoll and their direct (IgM) anti-DNP PFC response was measured at 4 days (Table II). All of the F₁ female mice formed large numbers of PFC after immunization with from 0.1–500 μg of DNP-lys-Ficoll. In contrast to the vigorous responses of the F₁ females, 10 of 12 F₁ males immunized with DNP-lys-Ficoll had fewer PFC than the number seen in unimmunized F₁ males. Two of the F₁ male mice had responses of 800 and 1,600 PFC/spleen after immunization with 0.1 and 1.0 μg of DNP-lys-Ficoll, respectively. These values were the highest responses seen with CN × DN F₁ male mice in our experience and are in

TABLE II

Response of CN × DN F₁ Male and Female Mice to Varying Doses of DNP-Lys-Ficoll

| DNP-lys-Ficoll | TNP-SRBC at day 4* | | | |
| | Male | | Female | |
	PFC/spleen	PFC/10^6 cells	PFC/spleen	PFC/10^6 cells
μg				
0	250 ± 72	6.9 ± 1.4	955 ± 325	12 ± 3.6
0.1	330 ± 235	9.4 ± 7.6	13,600 ± 3,372	156 ± 24.0
1.0	641 ± 480	11.8 ± 9.0	22,108 ± 9,125	357 ± 113.1
20.0	150 ± 68	3.3 ± 0.5	37,000 ± 4,485	340 ± 60.4
500.0	100 ± 14	2.1 ± 0.6	13,616 ± 3,156	169 ± 28.4

* The mean number (three animals per group) of IgM PFC (±SE) measured on day 4 in response to varying doses of DNP-lys-Ficoll.

the same range as the number of background TNP-SRBC plaques seen in CN × DN F₁ females.

Studies of the 4- and 8-day IgM and IgG anti-DNP PFC responses of CN and CN × DN F₁ mice to DNP-lys-Ficoll are shown in Table III. The CN males and females and CN × DN F₁ males had no response (less than background) while the CN × DN F₁ females made large numbers of PFC at both 4 and 8 days. These experiments demonstrate the CBA/N mice have an X-linked defect in responsiveness to another T-independent antigen, DNP-lys-Ficoll. Moreover, since CN × DN F₁ male mice are able to form antibody to the highly cross-reactive hapten (TNP) coupled to erythrocytes, their X-linked immune defect is not a simple absence of the genetic information required to synthesize anti-DNP antibody.

Reconstitution of poly I·C Response in Lethally Irradiated CN × DN F₁ Mice by CN × DN Spleen Cells. In order to study the influence of the CN × DN F₁ male and female environment on the ability of splenic B lymphocytes to respond to T-independent antigens, lethally irradiated F₁ male and female mice were reconstituted with either F₁ male or F₁ female spleen cells and immunized immediately with poly I·C. At 14 days, antibody to poly I·C was found in 9 of 11

TABLE III

IgM and IgG Responses of Male and Female CN and CN × DN F₁ Mice to DNP-Lys-Ficoll (100 μg)

Strain	Sex	TNP-SRBC*			
		IgM (day 4)		IgG (day 7)	
		PFC/spleen	PFC/10⁶ cells	PFC/spleen	PFC/10⁶ cells
CN	Male	325 ± 160	5.8 ± 3.1	125 ± 11.1	2.6 ± 0.2
	Female	125 ± 32	2.9 ± 1.6	13 ± 2.0	0.3 ± 0.1
CN × DN	Male	150 ± 25	3.0 ± 0.2	25 ± 5.0	0.8 ± 0.3
	Female	11,775 ± 1,041	201.0 ± 24.1	3,695 ± 150.0	53.0 ± 10.2

* The mean number (four animals per group) of IgM or IgG PFC ± SE measured at 4 or 7 days, respectively, in response to 100 μg or DNP-lys-Ficoll.

TABLE IV

Response of Lethally Irradiated CN × DN F₁ Mice to Poly I·C After Reconstitution with CN × DN F₁ Spleen Cells

Sex of recipient	[¹⁴C]poly I·C binding*	
	Female donor	Male donor
	%	%
Female	56.1	4.7
	80.6	3.2
	76.9	0.2
	28.5	4.7
Male	21.0	1.7
	0.5	2.4
	14.4	0.2
	71.7	0.0
	21.1	0.0
	71.7	2.4
	50.9	0.0

* Recipient mice were lethally irradiated (1,000 R) and reconstituted with 50 × 10⁶ F₁ male or female spleen cells given intravenously on the day of irradiation. They were immediately challenged with 100 μg of poly I·C in CFA and the percent [¹⁴C]poly I·C binding was assayed on 25 μl of serum obtained 14 days after immunization.

recipients of F₁ female spleen cells. This included four of four F₁ female recipients and five of seven F₁ male recipients (Table IV). In contrast, none of the 11 F₁ mice (either male or female) which had received male F₁ spleen cells formed antibody to poly I·C.

Reconstitution of DNP-Lys-Ficoll Response in Lethally Irradiated CN × DN Mice with CN × DN Bone Marrow.   The previous experiment demonstrated that a population of female cells containing splenic B lymphocytes was able to transfer to irradiated F_1 males the ability to form antibody to poly I·C upon immediate immunization. In order to study the influence of environment on the maturation of these responsive cells, we reconstituted lethally irradiated male and female CN × DN F_1 mice with $10 × 10^6$ bone marrow cells derived from either male or female CN × DN donors and waited 8 wk before immunizing with DNP-lys-Ficoll (Table V). The mice were sacrificed 4 days later and the numbers of nucleated spleen cells and direct TNP-SRBC PFC determined. Both male and female recipients of female bone marrow cells had substantially greater numbers of nucleated spleen cells than male and female recipients of male bone marrow cells. Indeed, the number of nucleated spleen cells in lethally irradiated recipients which received F_1 male cells resembled that of nonirradiated F_1 male animals, while the numbers of nucleated spleen cells in recipients of F_1 female cells resembled that of normal F_1 females (6). Furthermore, male and female

TABLE V

Response of Lethally Irradiated CN × DN F_1 Male and Female Mice to DNP-Lys-Ficoll after Reconstitution with 10 × 10^6 F_1 Male or Female Bone Marrow Cells

Sex of recipient	Sex of donor	No. of nucleated cells per spleen‡	TNP-SRBC*	
			PFC/spleen	PFC/10^6 cells
Male	Male	37.8 ± 4.2	50 ± 25	1.5 ± 1.2
	Female	90.3 ± 4.8	21,875 ± 709	243.7 ± 24.0
Female	Male	58.0 ± 7.5	1,575 ± 368	29.2 ± 9.8
	Female	112.0 ± 8.5	26,217 ± 3,397	210.3 ± 7.9

* Recipient mice were lethally irradiated (1,000 R) at 8 wk of age and reconstituted with $10 × 10^6$ bone marrow cells. 8 wk after reconstitution, they were immunized with 100 μg of DNP-lys-Ficoll. IgM PFC were measured at 4 days.

‡ The number of nucleated cells per spleen ± SE (three animals per group).

recipients of F_1 female bone marrow cells made a vigorous response to DNP-lys-Ficoll while male and female recipients of male cells had very few TNP-SRBC PFC 4 days after immunization. Male recipients of male cells made essentially no response; female recipients of male cells mounted a small response, which can probably be ascribed to the contribution of F_1 female bone marrow cells that escaped or recovered from the effects of irradiation.

These experiments demonstrate that F_1 female spleen and bone marrow cells can transfer to lethally irradiated F_1 male recipients the ability to respond to T-independent antigens and imply that the maturation and function of F_1 female B cells can proceed normally in the environment of the irradiated abnormal host. Similarly, they show that cells from the abnormal donor do not develop normally even when placed in a normal (although irradiated) environment.

Reconstitution of Poly I·C and DNP-Lys-Ficoll Response in Nonirradiated CN × DN F_1 Male Mice. Transfers of spleen cells into irradiated recipients as described above do not rule out the possibility that an abnormal radiation-sensitive regulatory mechanism operates in the F_1 male animals nor do they provide information about the cell type critical for restoring responsiveness upon transfer.

To approach these problems, we transferred cells from F_1 male and female donors to nonirradiated F_1 male and female recipients. In the experiment presented in Table VI, 50×10^6 male or female spleen cells were transferred to nonirradiated F_1 male recipients which were immunized on the same day with 100 μg of poly I·C. Recipients of F_1 female cells made substantial responses to poly I·C, indicating that female cells could function in the environment of an intact F_1 male mouse. When F_1 male cells were transferred to intact F_1 female recipients, no suppression of the anti-poly I·C response was obtained (data not shown) further indicating that abnormal T-cell regulation of B-cell activation was not responsible for the defect of the F_1 male.

TABLE VI

Response of CN × DN F_1 Male Mice to Poly I·C after the Administration of CN × DN F_1 Spleen Cells

Sex of recipient	[^{14}C]poly I·C binding*	
	Female donor	Male donor
	%	%
Male	38.5	0.0
	41.3	0.0
	58.7	13.0
	44.0	0.0
	72.3	0.0
	63.5	5.0
	51.6	2.4

* Recipient CN × DN F_1 male mice received 50×10^6 F_1 male or female spleen cells intravenously. They were immediately challenged with 100 μg of poly I·C in CFA and 14 days later the percent [^{14}C]poly I·C binding by 25 μl of serum was determined.

These results were confirmed and extended by transferring varying numbers of F_1 female spleen cells to F_1 male recipients which were immunized with 100 μg of DNP-lys-Ficoll immediately. As few as 1×10^6 female spleen cells allowed males to make a detectable response; 50×10^6 cells transferred a response comparable, in terms of PFC per 10^6 spleen cells, to that of intact F_1 female mice (Fig. 1). Moreover, treatment of F_1 female spleen cells with anti-θ and C before transfer had no effect on the ability of these cells to reconstitute the response of F_1 male recipients (Fig. 2). The effectiveness of treatment with anti-θ and C in removing T lymphocytes was demonstrated by the marked impairment of the responsiveness of such cells to concanavalin A (Con A). Thus, anti-θ- and C-treated cells gave a net incorporation of [^{3}H] thymidine of 3,081 cpm while cells treated with normal mouse serum and C incorporated 86,869 cpm in response to Con A. This result indicates that F_1 female T lymphocytes are probably not required to reconstitute the responsiveness of F_1 male mice. On the other hand, in vitro irradiation (1,000 R) of F_1 female spleen cells completely abolished the

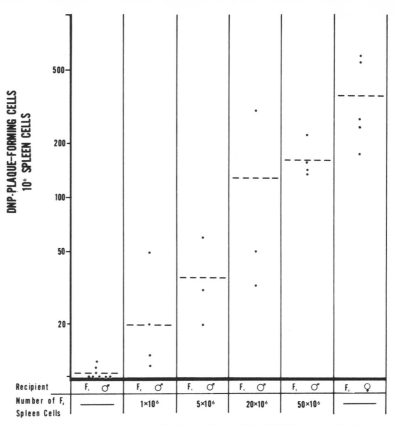

FIG. 1. The number of DNP-PFC/10^6 spleen cells in CN × DN F_1 males which were given different numbers of CN × DN F_1 female spleen cells intravenously and immunized with 100 μg of DNP-lys-Ficoll. The response to DNP-lys-Ficoll was measured on the 6th day after the administration of the female cells and immunization.

ability of these cells to transfer responsiveness to DNP-lys-Ficoll to intact F_1 male recipients. Thus, no PFC (less than background) were seen in recipients of 50×10^6 irradiated F_1 female cells while three F_1 male recipients of nonirradiated F_1 female cells had 275 ± 18 PFC/10^6 spleen cells after immunization with 100 μg of DNP-lys-Ficoll. Further investigation of this transfer model demonstrated that 3×10^6 F_1 female lymph node cells could also transfer responsiveness to DNP-lys-Ficoll to nonirradiated F_1 male recipients, whereas 50×10^6 F_1 female thymocytes were ineffective in initiating a response (Table VII).

Reconstitution of F_1 male mice with spleen cells from F_1 female donors not only allowed responses of the recipients to immediate challenge with DNP-lys-Ficoll but, as shown in separate experiments (Table VIII), allowed responses by the males to primary challenges administered as late as 42 days after cell transfer. This indicates that transferred F_1 female B lymphocytes survive in the recipient or that B lymphocytes develop from F_1 female stem cells present in the spleen cells used for transfer.

Lack of Thymic Influence on the Reconstitution of Responsiveness in Lethally Irradiated CN × DN F_1 Male Mice. The previous studies provide convincing

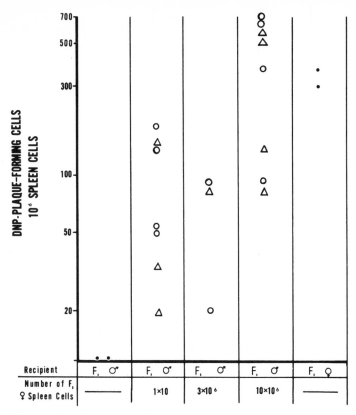

FIG. 2.   The number of DNP-PFC/10^6 spleen cells in CN × DN F_1 males which were given different numbers of CN × DN F_1 female spleen cells which had been previously treated with anti-θ and C ($\triangle$), or NMS and C ($\bigcirc$). The response to DNP-lys-Ficoll was measured on the 6th day after the administration of the female cells and 100 μg of the antigen.

evidence that F_1 female cells can transfer to both intact and irradiated F_1 male recipients the capability to respond to T-independent antigens. Moreover, since anti-θ-treated spleen cells are effective, they provide strong evidence that the defect of the F_1 male is at the B-lymphocyte level. However, it could be argued that F_1 male B lymphocytes develop abnormally in the F_1 male because of an abnormality in the F_1 male T lymphocytes which, in some way, regulate B-cell development or, alternatively, because of an abnormal sensitivity of the F_1 male B cells to normal regulatory effects of F_1 male thymus or T lymphocytes. In order to investigate this point, F_1 male mice were thymectomized at 6 wk of age, irradiated (1,000 R) at 8 wk, and reconstituted with 10 × 10^6 anti-θ-treated male or female bone marrow cells 1 day later. They were held for 10 wk to allow repopulation of their lymphoid system and were then immunized with 100 μg of DNP-lys-Ficoll. Animals reconstituted with anti-θ-treated F_1 female bone marrow made excellent responses to DNP-lys-Ficoll, but those repopulated with anti-θ-treated F_1 male bone marrow were unresponsive to DNP-lys-Ficoll (Table IX). Spleen cells isolated from these F_1 male mice were cultured in the presence of Con A in order to determine if functional T lymphocytes were present. The proliferative response of these spleen cells were all less than 4% of normal F_1

TABLE VII

Response of CN × DN F$_1$ Male Mice to DNP-Lys-Ficoll after the Administration of CN × DN F$_1$ Female Spleen, Lymph Node, or Thymus Cells

| F$_1$ male recipient | F$_1$ female donors | | TNP-SRBC* | |
	Cell source	No. of Cells	PFC/ spleen	PFC/10^6 cells
1	—	—	75	1.5
2	—	—	100	3.2
3	Spleen	3 × 10^6	10,950	267.0
4	Spleen	3 × 10^6	6,450	140.0
5	Spleen	10 × 10^6	11,050	254.0
6	Spleen	10 × 10^6	8,450	275.0
7	Lymph node	3 × 10^6	5,500	149.0
8	Lymph node	3 × 10^6	5,075	108.0
9	Lymph node	10 × 10^6	6,650	155.0
10	Lymph node	10 × 10^6	8,275	194.0
11	Thymus	50 × 10^6	75	1.2
12	Thymus	50 × 10^6	200	3.1
13	Thymus	50 × 10^6	150	2.2

* Recipient F$_1$ male mice were given varying numbers of F$_1$ female spleen, lymph node, or thymus cells intravenously. They were then challenged with 100 μg of DNP-lys-ficoll and IgM TNP-SRBC PFC were assayed on day 6.

TABLE VIII

Response of CN × DN F$_1$ Male Mice to DNP-Lys-Ficoll at Different Times after the Administration of F$_1$ Female Spleen Cells

| Days after F$_1$ female cells | TNP-SRBC* | |
	PFC/spleen	PFC/10^6 spleen
18	1,316 ± 719	35.2 ± 13.5
25	1,550 ± 123	38.5 ± 1.9
39	3,000 ± 177	57.4 ± 1.7
46	1,900 ± 50	45.8 ± 0.5

* Recipient F$_1$ male mice were given 50 × 10^6 F$_1$ female cells intravenously. At 14, 21, 35, and 42 days after the administration of these cells, three F$_1$ male mice were immunized with 100 μg of DNP-lys-Ficoll and their IgM TNP-SRBC PFC response was measured 4 days later. Data are expressed as the mean number of PFC of three mice ±SE.

male mice. Thus, it seems clear that the B-lymphocyte defect exhibited by F$_1$ male mice cannot be ascribed to development under the influence of abnormal T-dependent stimuli or to an abnormal response to normal T-dependent stimuli.

TABLE IX

Response of Thymectomized, Lethally Irradiated, Bone Marrow Reconstituted CN × DN F$_1$ Male Mice to DNP-Lys-Ficoll

Sex of recipient	Sex of donor	TNP-SRBC*	
		PFC/spleen	PFC/10^6 spleen
Female‡	—	60,450	530
Male	Female	51,000	362
Male	Female	27,927	338
Male	Male	25	0.9
Male	Male	75	1.1

* Recipient F$_1$ mice were thymectomized at 6 wk of age, lethally irradiated (1,000 R) at 8 wk and reconstituted with 10×10^6 anti-θ-treated bone marrow cells on the day after irradiation. At 18 wk of age, these mice were immunized with 100 μg of DNP-lys-Ficoll and their IgM TNP-SRBC PFC response was assayed 4 days later.

‡ Nonirradiated normal female immunized with 100 μg of DNP-lys-Ficoll.

Discussion

CN mice and male CN × DN F$_1$ mice have been previously shown to have a profound defect in immune responses to SIII, LPS, and poly I·C (1, 2). We show here that they are essentially unresponsive to DNP-lys-Ficoll. Moreover, in vitro studies confirm the unresponsiveness of F$_1$ male mice to DNP-lys-Ficoll while F$_1$ female mice mount vigorous responses.[3] Thus, CN mice fail to respond to four different T-independent antigens. Our results with DNP-lys-Ficoll are particularly impressive as this T-independent antigen yields very large numbers of IgM and IgG PFC in a variety of normal strains (8).

On the other hand, F$_1$ male CN × DN mice mount substantial responses to SRBC and TNP-SRBC. When considered in terms of PFC per unit of B lymphocytes (rather than as PFC per spleen), the IgM response of F$_1$ males is similar to that of F$_1$ females. This is so because of F$_1$ males generally have half as many nucleated cells per spleen as F$_1$ females (6). In addition, F$_1$ males have only 26% Ig-bearing cells among spleen lymphocytes, whereas 40% of spleen lymphocytes from F$_1$ females bear surface Ig (6). It should be noted that considering responses of F$_1$ males to DNP-lys-Ficoll on this basis does not result in such a normalization, as F$_1$ males make no net TNP-specific PFC as a result of immunization with this T-independent antigen. These results confirm our previous conclusion that the X-linked functional immune defect expressed by CN and male CN × DN F$_1$ mice is principally an inability to mount T-independent responses rather than a global defect in all humoral immune responses (6).

The finding that CN and CN × DN F$_1$ male mice fail to respond to T-independent antigens does not, by itself, establish that the defect of these

---

[3] Cohen, P., I. Scher, and D. Mosier. Manuscript submitted for publication.

animals resides in the B-lymphocyte pool. In order to study this problem, we evaluated the capacity of lymphoid cells obtained from F_1 female donors to transfer responsiveness to F_1 male recipients. Spleen cells from F_1 female mice were able to reconstitute an easily detectable antibody response to poly I·C or DNP-lys-Ficoll in F_1 males after immediate challenge. Moreover, both anti-θ-treated spleen cells and normal lymph node cells from F_1 female donors transferred responsiveness to F_1 males, while irradiated F_1 female spleen cells did not transfer responsiveness. The former indicates that mature T lymphocytes from the F_1 female are not required for the DNP-lys-Ficoll response; the latter experiments suggest that macrophages are not the limiting cell type, since macrophages are less frequent in lymph node cell populations than in spleen cell suspensions and are quite radiation resistant. Indeed, Mosier et al. (7) have shown that the in vitro response to DNP-lys-Ficoll is much less dependent on adherent cells than is the response to SRBC, further suggesting that macrophages are not likely to be the defective cell type in CN and male CN $\times$ DN F_1 mice.

Since cell populations rich in mature B lymphocytes from F_1 female mice will reconstitute the response of F_1 male mice, it is likely that the X-linked functional defect in these mice resides in the B-cell pool. Nonetheless, it could be proposed that the B-cell defect results from a defect in control of B-cell development or function based upon abnormal regulatory cells or abnormal sensitivity of B cells to external regulatory influences. Several of our experiments bear on these points. Firstly, mature F_1 female spleen cells function normally in intact or irradiated males, indicating that an excess of an acutely acting regulatory factor is not present in the F_1 male. Furthermore, nonirradiated F_1 males which have received F_1 female spleen cells mount anti-DNP-responses to DNP-lys-Ficoll administered up to 42 days after transfer, making it unlikely that a suppressive factor active only over a prolonged period limits responsiveness of mature B cells in the F_1 male. Also supporting this contention is the failure of F_1 male spleen cells to inhibit the response of intact F_1 females.

These data strongly indicate that mature B lymphocytes capable of responding to T-independent antigens are absent in CN and CN $\times$ DN F_1 mice. Such an absence could represent a defect in the differentiative potential of B-lymphocyte progenitors or, alternatively, a defect in the microenvironment in which B lymphocytes develop. Our data strongly points to the former rather than the latter. Thus, F_1 female bone marrow cells, when transferred to lethally irradiated F_1 male mice, allow the latter to respond to challenge with DNP-lys-Ficoll administered 8 wk later. This implies that female B-lymphocyte progenitors can develop normally in the environment of the F_1 male. On the other hand, F_1 male bone marrow cells transferred to lethally irradiated F_1 female mice do not reconstitute the response to DNP-lys-Ficoll of the recipients. The small response which these animals mount is equivalent to the number of background TNP-specific PFC in normal females and may represent a response on the part of surviving female B lymphocytes. Similarly, failure in development of male B cells responsive to DNP-lys-Ficoll cannot be ascribed to a regulatory influence of the male thymus or T lymphocytes since anti-θ-treated F_1 male bone marrow cells could not reconstitute responsiveness in thymectomized, lethally irradiated F_1 male recipients.

A final objection to the concept that a functional class of B lymphocytes is absent in CN and F_1 male mice could be raised, if it could be shown that the cells responding to poly I·C or DNP-lys-Ficoll in F_1 males which had received F_1 female cells were the defective F_1 male cells. According to this hypothesis, the presence of the F_1 female cells would allow the F_1 male cells to form antibody against these T-independent antigens. In order to study this problem, we have reconstituted F_1 males with T-lymphocyte-depleted DN spleen cells. In preliminary experiments, we have shown that the cell responding to DNP-lys-Ficoll was derived from the DN strain rather than the F_1 male.

Our data present convincing evidence that the X-linked defect in responsiveness to T-independent antigens by CN mice represents an intrinsic defect in the B-lymphocyte line rather than an abnormal extrinsic regulation of B-lymphocyte development and/or function. We cannot, as yet, determine whether these animals have a deletion in a single line of B lymphocytes (i.e., B cells responsive to T-independent antigens) or, alternatively, a defect in all lymphocytes which affects T-independent responses more profoundly than T-dependent responses. These mice provide a model system in which an analysis of the molecular mechanism of B-lymphocyte activation and a determination of the nature of the functional heterogeneity of such cells should be possible. In addition, they provide an excellent model for understanding the role of X-linked genes in the control of immune responses.

Summary

The mechanisms underlying the X-linked thymus-independent (B) lymphocyte functional defect in the CBA/N (CN) mice and their F_1 progeny were studied. Immune defective mice were unable to respond to the T-independent antigen 2,4-dinitrophenyl-lysyl-derivative of Ficoll (DNP-lys-Ficoll) but were able to form antibody against the highly cross-reactive hapten (trinitrophenyl) when it was coupled to an erythrocyte carrier. Immune defective CN × DBA/2N (DN) F_1 male mice, which do not normally respond to T-independent antigens, were able to respond to both polyribosinic-polyribocytidylic acid and DNP-lys-Ficoll after the adminstration of CN × DN F_1 female spleen cells even if these cells had been depleted of T lymphocytes.

In addition, it was shown that the inability of the CN mice and their F_1 progeny to respond to T-independent antigens was not due to an intrinsic abnormality of their microenvironment or the suppressive actions of a T lymphocyte. Our data present evidence that the X-linked defect in the CN mice is due to an intrinsic defect in B-lymphocyte development.

We would like to acknowledge the excellent editorial assistance of Mrs. Betty J. Sylvester.

Received for publication 27 May 1975.

References

1. Amsbaugh, D. F., C. T. Hansen, B. Prescott, P. W. Stashak, D. R. Barthold, and P. J. Baker. 1972. Genetic control of the antibody response to Type III pneumococcal polysaccharide in mice. I. Evidence that an X-linked gene plays a decisive role in determining responsiveness. *J. Exp. Med.* **136**:931.

2. Scher, I., M. Frantz, and A. D. Steinberg. 1973. The genetics of the immune response to a synthetic double-stranded RNA in a mutant CBA mouse strain. *J. Immunol.* **110**:1396.

3. Andersson, B., and H. Blomgren. 1971. Evidence for thymus-dependent humoral antibody production in mice against polyvinylpyrrolidone and *E. coli* lipopolysaccharide. *Cell. Immunol.* **2**:411.

4. Davies, A. J. S., R. L. Carter, E. Leuchars, V. Wallis, and F. M. Dietrich. 1970. The morphology of immune reactions in normal thymectomized and reconstituted mice. III. Response to bacterial antigens: salmonella flagellar antigen and pneumococcal polysaccharide. *Immunology.* **19**:945.

5. Chused, T. M., A. D. Steinberg, and L. M. Parker. 1973. Enhanced antibody response of mice to polyinosinic-polycytidylic acid by antithymocyte serum and its age-dependent loss in NZB/W mice. *J. Immunol.* **111**:52.

6. Scher, I., A. Ahmed, D. M. Strong, A. D. Steinberg, and W. E. Paul. 1975. X-linked B-lymphocyte immune defect in CBA/HN mice. I. Studies of the function and composition of spleen cells. *J. Exp. Med.* **141**:788.

7. Mosier, D. E., B. M. Johnson, W. E. Paul, and P. R. B. McMaster. 1974. Cellular requirements of the primary in vitro antibody response to DNP-Ficoll. *J. Exp. Med.* **139**:1354.

8. Sharon, R., P. R. B. McMaster, A. M. Kask, J. D. Owens, and W. E. Paul. 1975. DNP-lys-Ficoll: a T-independent antigen which elicts both IgM and IgG anti-DNP antibody secreting cells. *J. Immunol.* **114**:1585.

9. Scher, I., D. M. Strong, A. Ahmed, R. Knudsen, and K. W. Sell. 1973. Specific murine B-cell activation by synthetic single- and double-standed polynucleotides. *J. Exp. Med.* **138**:1545.

10. Kettman, J., and R. W. Dutton. 1970. An in vitro primary immune response to 2,4,6-trinitrophenyl substituted erythrocytes: response against carrier and hapten. *J. Immunol.* **104**:1558.

11. Steinberg, A. D., T. Pincus, and N. Talal. 1971. The pathogenesis of autoimmunity in New Zealand Mice. III. Factors influencing the formation of anti-nucleic acid antibodies. *Immunology.* **20**:523.

12. Mosier, D. E. 1969. Cell interactions in the primary immune response in vitro: a requirement for specific cell clusters. *J. Exp. Med.* **129**:351.

13. Rittenberg, M. B., and C. Pratt. 1969. Anti-trinitrophenyl (TNP) plaque assay. Primary response of BALB/c mice to soluble and particulate immunogen. *Proc. Soc. Exp. Biol. Med.* **132**:575.

14. Pierce, C. W., B. M. Johnson, H. E. Gershon, and R. Asofsky. 1971. Immune responses in vitro. III. Development of primary γM, γG, and γA plaque-forming cell responses in mouse spleen cell cultures stimulated with heterologous erythrocytes. *J. Exp. Med.* **134**:395.

ACTIVATION OF MOUSE LYMPHOCYTES BY ANTI-IMMUNOGLOBULIN

II. A Thymus-Independent Response by a Mature Subset of B Lymphocytes*

By DONNA G. SIECKMANN,‡ IRWIN SCHER, RICHARD ASOFSKY, DONALD E. MOSIER, AND WILLIAM E. PAUL

From the Laboratory of Immunology and Laboratory of Microbial Immunity, National Institute of Allergy and Infectious Diseases, National Institutes of Health; the Department of Experimental Pathology, Naval Medical Research Institute; and the Department of Medicine, Uniformed Services University of the Health Sciences, Bethesda, Maryland 20014

It is now clear that the binding of antigen by specific receptors of thymus-independent (B) lymphocytes is an important step in the physiologic process by which such cells are activated (1, 2). However, the precise role of antigen-specific membrane receptors is still unknown. It has been proposed that the binding of antigens by membrane immunoglobulin directly generates some type of transmembrane signal which is critical to lymphocyte activation (3–7). Others (8–11) have suggested that the interaction of antigen with immunoglobulin receptors, in and of itself, has no effect on the cell but that the receptor-ligand interaction serves to concentrate an intrinsically stimulatory molecule on the cell surface. The latter theory, which has been referred to as the "one non-specific signal" theory (10, 11), predicts that anti-immunoglobulin antibodies, in the absence of "help" from thymus-dependent (T) lymphocytes or from polyclonal activators, should have no effect on B lymphocytes. Consequently, a detailed evaluation of the behavior of B lymphocytes after their exposure to anti-immunoglobulin (anti-Ig)[1] antibodies should provide important information to aid in the choice between these hypotheses.

It has previously been shown that anti-Ig antibodies will induce proliferation by lymphocytes from several species, including rabbits (12), humans (13–16), pigs (17),

---

* Supported in part by the Naval Medical Research and Development Command, Work Unit M0095.PN.001.1030 and Uniformed Services University of the Health Sciences Research grant no. C08310. The opinions or assertions contained herein are those of the authors and are not to be construed as official or reflecting the views of the Navy Department or the naval service at large. The animals used in this study were handled in accordance with the provisions of Public Law 89-54 as amended by Public Law 91-579, the "Animal Welfare Act of 1970," and the principles outlined in the "Guide for the Care and Use of Laboratory Animals," U.S. Department of Health, Education, and Welfare, publication no. 73-23.

‡ Supported by a Postdoctoral Fellowship from the Arthritis Foundation.

[1] *Abbreviations used in this paper:* FATS, anti-thymocyte serum; Con A, Concanavalin A; ACS, fluorescence activated cell sorter; FCS, fetal calf serum; HBSS, Hanks' balanced salt solution; Hepes, N-2-hydroxyl-ethylpiperazine-N-2-ethane sulfonic acid; [^{3}H]TdR, methyl[^{3}H]thymidine; Ig, immunoglobulin; LPS, lipopolysaccharide; 2-ME, 2-mercaptoethanol; *nu/nu*, athymic nude mice; *nu/+*, heterozygous littermates; NWSM, *Nocardia* water-soluble mitogen; NWT, nylon wool T lymphocytes; PFC, plaque-forming cells; PHA, phytohemagglutinin-P; RE, reference endotoxin; SRBC, sheep erythrocytes.

and chickens (18, 19). Recently, conditions have been described under which mouse lymphocytes will proliferate in response to anti-Ig (20–22). Studies from our laboratory (22) indicate that lymphocytes from young adult mice will synthesize DNA when exposed to specifically purified anti-μ or anti-κ antibodies. Both deaggregated antibodies and F(ab')$_2$ fragments were equally effective, and serum is not required for these responses.

In this communication, we show that the response of mouse lymphocytes to anti-μ and anti-γ,κ is a function of B lymphocytes, that it does not require the presence of T lymphocytes, and that adherent cells are not required for the response. These results thus suggest that the activation of B cells by anti-Ig is a direct consequence of its interaction with membrane Ig and is independent of auxilliary signals from either T lymphocytes or macrophages.

Finally, our data also indicate that anti-Ig stimulates a mature subset of B lymphocytes. This is based on the finding that responsiveness to anti-μ or to anti-γ,κ does not appear until 4 wk of age and that lymphocytes from mice with the CBA/N X-linked immune defect in B-lymphocyte function (23) fail to respond to anti-Ig.

Materials and Methods

Animals. (C57BL/6 × DBA/2J)F$_1$ (BDF$_1$/J) and CBA/J mice were purchased from The Jackson Laboratory, Bar Harbor, Maine. (C57BL/6 × DBA/2N)F$_1$ (BDF$_1$/N) and (CBA/N × DBA/2N)F$_1$ mice were obtained from the Division of Research Services, National Institutes of Health. Athymic nude (*nu/nu*) mice and heterozygous littermates (*nu/+*), members of a 10th generation backcross to BALB/c, were obtained at 10 wk of age from Charles River Breeding Laboratories, Wilmington, Mass. All mice were used at 2–3 mo of age unless noted otherwise.

Anti-Immunoglobulin (Anti-Ig) Antibodies. Affinity column-purified goat anti-mouse Ig specific for μ-heavy chains (anti-μ) or for γ-heavy and κ-light chains (anti-γ,κ) were prepared and assayed for specificity as previously described (22).

Cell Culture and Assay for Methyl-[^{3}H]Thymidine ([^{3}H]TdR) Incorporation. Spleen cells were cultured at 5 × 10^5 per culture (in 0.2 ml) for proliferative responses, except where otherwise noted. Cells were cultured in a modified Mishell-Dutton medium (24) containing 10% fetal calf serum (FCS) (Rehatuin, Armour Pharmaceutical Co., Phoenix, Ariz.), 16-mM Hepes buffer, and 5 × 10^{-5} M 2-mercaptoethanol (2-ME), in flat-bottom microtiter plates (Microtest II, Falcon Plastics, Div. of BioQuest, Oxnard, Calif.) and were assayed for [^{3}H]TdR uptake at 48 h as previously described (22). For primary antibody responses to sheep erythrocytes (SRBC), culture conditions similar to those for [^{3}H]TdR incorporation were used, except that 1 × 10^6 cells were added to each well. The number of plaque-forming cells (PFC) for SRBC were measured on day 4.

Depletion of T Cells by Anti-Thy 1.2 Antiserum or Rabbit Anti-Mouse Thymocyte Serum (ATS) Treatment. Spleen cells were suspended in a 1:5 or 1:10 dilution of an AKR anti-C3H (anti-Thy 1.2) antiserum (25) or in a 1:50 dilution of a rabbit anti-mouse thymocyte serum (ATS, batch 14580, Microbiological Associates, Walkersville, Md., generously supplied by Drs. John Kappler and Philippa Marrack, the University of Rochester, Rochester, N. Y.) in Hanks' balanced salt solution (HBSS) at 30 × 10^6 cells/ml and were held on ice for 30 min. The cells were subsequently pelleted and resuspended in a 1:4 dilution of absorbed guinea pig serum (Flow Laboratories, Inc., Rockville, Md.). After a 30-min incubation at 37°C, the cells were washed with HBSS and adjusted to 5 × 10^6 viable cells/ml in medium before plating 0.1 ml/culture. Control spleen cells were held on ice in HBSS, followed by incubation with complement.

Preparation of Splenic T Cells by Nylon Wool Column Passage. Spleen cells (300 × 10^6) in 5 ml HBSS containing 5% FCS were passed into a 3-g column of nylon wool (30-ml vol in a 50-ml syringe; Leuko-pac Leukocyte Filter, Fenwal Laboratories, Deerfield, Ill., 26). After a 45-min incubation at 37°C, the effluent cells were slowly collected from the column while maintaining a 15-ml head volume above the column. The eluted cells contained 0–7% Ig$^+$ cells, as detected by staining for surface Ig using a fluorescein-labeled goat anti-mouse Ig. This T-cell enriched population is referred to as "nylon wool T cells" (NWT).

Preparation of Ig$^+$ and Ig$^-$ Spleen Cell Populations.   Separation of surface Ig$^+$ cells from Ig$^-$ cells was accomplished with a fluorescence activated cell sorter (FACS) (Becton, Dickinson & Co., Mt. View, Calif.) as previously described (27). Briefly, spleen cells from CBA/J mice were treated with ammonium chloride to lyse erythrocytes, or centrifuged on Ficoll-Paque (Pharmacia Fine Chemicals, Piscataway, N. J.) to remove dead cells and erythrocytes, washed in RPMI 1640 (Grand Island Biological Co., Grand Island, N. Y.) containing 10% FCS, and incubated with a fluorescein-conjugated F(ab')$_2$ fragment of a polyspecific goat anti-mouse Ig or a fluorescein-labeled F(ab')$_2$ rabbit anti-mouse Ig under aseptic conditions. After washing at 0°C, the cells were analyzed with the FACS, and a bimodal fluorescence profile was generated. Cells with relative fluorescence intensities of 0–25 were considered Ig$^-$, whereas cells within channels 125–1,000 were considered Ig$^+$. Cells with these fluorescence intensities were separated from each other with the FACS over a period of approximately 4 h. During the sorting process all cells were kept at 0°C to prevent capping. We routinely obtain Ig$^+$ populations that are 94–96% pure and Ig$^-$ populations that are 94–98% pure using this procedure.

Depletion of Adherent and Phagocytic Cell Populations.   Spleen cells (400 × 10^6) were depleted of adherent cells by passage sequentially over two 30-ml columns of Sephadex G-10, (lot 9067, Pharmacia Fine Chemicals; 28) at 37°C. Adherent cells were also removed by incubation of spleen cells at 20 × 10^6 cells/ml in HBSS containing 5% FCS with 20 mg washed carbonyl iron (General Aniline & Film Corp., Easton, Pa.) at 37°C for 30 min (29). Cells that had taken up iron or to which the iron had adhered were removed by three settlings in 60 × 20-mm Petri dishes (no. 3002, Falcon Plastics) placed on a ceramic magnet (35 lb pull, no. 42,098, Edmund Scientific Co., Barrington, N. J.). The depletion of macrophages was tested by phagocytosis of latex beads. Cells (15 × 10^6 in 1.5 ml) were incubated with 10 µl of latex beads (1.091 µm diameter, Uniform Latex Particles, Dow Chemical Co., Midland, Mich.) overnight, rotating at 37°C. The percentage of cells that had phagocytosed latex was measured by light microscopy. Peritoneal cells were prepared by injecting 5 ml HBSS into the peritoneal cavity of a normal mouse and withdrawing the fluid after brief agitation. The cells were immediately exposed to 1,500 rads of γ-irradiation from a Cs-137 source (Gammator M, Isomedix Inc., Parsippany, N. J.), washed, and added to cultures of spleen cells that had been depleted of adherent cells.

Mitogens.   Lipopolysaccharide (LPS) (*Escherichia coli* 0111:B4, Westphal or Boivin, Difco Laboratories, Detroit, Mich.) was used at 50 µg/ml. Reference endotoxin (RE) *E. coli.* 0113:H10 was kindly provided by Dr. Ronald Elin, Clinical Pathology Department, National Institutes of Health. Concanavalin A (Con A) (lot 7001, Pharmacia Fine Chemicals, Uppsala, Sweden) was used at 2 µg/ml. Phytohemagglutinin-P (PHA; lot K1954, Wellcome Research Laboratories, Beckenham, England) was used at 1 µg/ml. *Nocardia* water-soluble mitogen (NWSM; a generous gift of Dr. Constantin Bona, Pasteur Institute, Paris, France) was used at 10 µg/ml.

Results

Tissue Distribution of Cells Responding to Anti-Ig.   It has been previously shown (22) that spleen cells of BDF$_1$ mice are stimulated to a substantial proliferative response when cultured with affinity column-purified goat anti-µ antibody or goat anti-γ,κ antibody. As an initial step in the characterization of the cell type responsive to anti-Ig, lymphoid cell populations from various tissues were tested for responsiveness. Cells (5 × 10^5) from spleen, thymus, bone marrow, mesenteric and peripheral lymph node, and Peyer's patch were cultured for 48 h in the presence of an optimal stimulatory concentration of anti-µ or anti-γ,κ. The results of an experiment of this type, shown in Fig. 1, indicate that spleen, mesenteric lymph node, peripheral lymph node, and Peyer's patch cells given excellent responses to anti-µ; bone marrow cells display an intermediate response, and thymocytes a minimal response. This pattern of responsiveness is similar to that observed for LPS, suggesting the importance of B lymphocytes in the stimulation caused by anti-µ. Furthermore, responsiveness of bone marrow

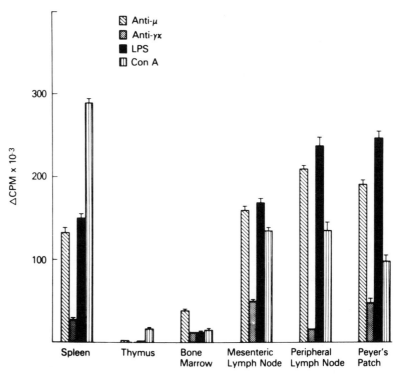

Fig. 1. Anti-Ig stimulation of cells from various lymphoid tissues. Cells were cultured at 5×10^5 per microwell in 0.2 ml medium containing the following stimulants: anti-μ (50 μg/ml), anti-γ,κ, (250 μg/ml), LPS (50 μg/ml), or Con A (2 μg/ml). Responses were measured after 48 h of culture and a 16-h pulse with [^{3}H]TdR.

cells is consistent with a lack of a requirement for T lymphocytes in the anti-Ig response. As noted previously, the response to optimal concentrations of anti-μ is substantially greater than the response to anti-γ,κ. This difference is particularly striking in cultures of peripheral lymph node cells. Although subsequent data suggests that cells responding to anti-μ and anti-γ,κ are not identical, the reason for differences in relative responsiveness has not yet been established. The most obvious explanation is that the relative frequency of the responding cell type varies in different lymphoid tissues.

Depletion of T Cells by Anti-Thy 1.2 and Complement Treatment. To support the hypothesis that B cells are able to respond to anti-Ig in the absence of T cells, spleen cell populations, which were depleted of T cells by treatment with an anti-Thy 1.2 antiserum and complement, were tested for their ability to respond to anti-Ig. Table I shows that, although spleen cells depleted of Thy 1.2-bearing lymphocytes were unresponsive to Con A and PHA, such cells responded normally to anti-γ,κ and LPS. Furthermore, reconstitution of T-depleted spleen cell cultures with NWT cells restored Con A and PHA responsiveness but had no effect on anti-γ,κ or LPS responses. The response anti-μ in these and other experiments showed a slight diminution after anti-Thy 1.2 and complement treatment. The loss in response was somewhat restored by the addition of NWT cells. However, the majority of the anti-μ response was maintained after such treatment, suggesting that the majority of anti-μ-responsive B cells can be stimulated in the absence of T cells. NWT cells alone were unable to

*Effect of Pretreatment with Anti-Thy 1.2 Antiserum and Complement on Anti-Ig-Induced Proliferation**

Mitogen	Concentration	No treatment	Complement alone	Anti-Thy 1.2 + Complement	Anti-Thy 1.2 + Complement with NWT‡	NWT‡ alone
	µg/ml			cpm per culture		
None	—	13,333 ± 461	19,196 ± 1,453	6,410 ± 237	14,141 ± 802	183 ± 41
G102 anti-µ	100	125,947 ± 12,591	187,997 ± 10,151	171,754 ± 4,926	178,582 ± 3,411	894 ± 99
	10	92,700 ± 8,869	108,008 ± 4,542	81,259 ± 1,224	119,353 ± 13,324	408 ± 46
G125 anti-γ,κ	200	23,711 ± 1,511	33,505 ± 1,808	55,306 ± 2,009	54,260 ± 2,676	155 ± 68
	50	19,873 ± 1,823	32,411 ± 2,840	38,745 ± 2,125	50,669 ± 1,920	42 ± 55
LPS	50	161,818 ± 7,293	182,484 ± 5,893	147,721 ± 3,385	143,089 ± 1,878	1,120 ± 153
Con A	2	281,598 ± 2,521	279,039 ± 12,745	8,836 ± 2,475	265,661 ± 4,219	85,912 ± 2,231
PHA	1	156,193 ± 11,268	157,340 ± 3,852	3,792 ± 110	185,547 ± 8,396	83,224 ± 1,027

* 3×10^5 cells per culture.
‡ 1×10^5 NWT.

respond significantly to anti-Ig or LPS, but did respond to the T-cell mitogens. Thus, it appears that the response to anti-µ or anti-γ,κ of spleen cells is largely a response of B lymphocytes which does not require T-lymphocyte help and that T lymphocytes, when cultured alone, do not respond to anti-Ig.

Anti-Ig Stimulation of FACS-Sorted Spleen Cells.   To correlate the response to anti-Ig with the presence of Ig on the membrane of responding cells, spleen cells were stained with fluoresceinated $F(ab')_2$ fragments of a goat anti-Ig reagent and sorted into Ig⁻ and Ig⁺ populations by the FACS. It should be noted that staining of the Ig⁺ B cells with the fluorescein-conjugated $F(ab')_2$ anti-Ig reagent did not cause any stimulation on its own, because there were no differences in background levels of [³H]TdR uptake between nontreated spleen cells and those treated with fluorescein-conjugated anti-Ig. Similarly, staining did not influence a subsequent response to either anti-µ or anti-γ,κ. This is consistent with other data[2] showing an inability to pulse-stimulate spleen cells with either anti-µ- or anti-γ,κ-antibodies.

A profile of the distribution of the amount of Ig per cell in the whole spleen cell population is shown in Fig. 2. Cells with fluorescence intensities detected in channels 0 through 25 (39.9% of total) were classed as Ig⁻, whereas cells with fluorescence intensities detected in channels 125 to 1,000 (56.5% of total) were classed as Ig⁺. After separation, the two populations were cultured at 2×10^5 cells per culture in the presence of anti-µ, anti-γ,κ, and conventional T- or B-cell mitogens. The results in Fig. 3 show that the Ig⁻ pool failed to respond to anti-µ or to anti-γ,κ. These cells responded minimally to LPS, while giving substantial responses to Con A and PHA. In contrast, anti-µ, anti-γ,κ, and LPS stimulated responses in cultures of nonsorted spleen cells and in cultures of Ig⁺ spleen cells. Indeed, the responses of Ig⁺ cells to anti-γ,κ in this experiment were significantly greater than the response of nonsorted cells, suggesting an enrichment for the anti-γ,κ responding cell in the Ig⁺ pool. This enrichment has also been noted for the anti-µ response in the majority of experiments of this type, although it was not seen in the experiment illustrated in Fig. 3.

It was noted that cells in the Ig⁺ pool still retained some responsiveness to Con A and PHA. This suggested the presence of at least some contaminating T lymphocytes in the Ig⁺ population. To overcome this difficulty we attempted to sort spleen cells that had been previously depleted of T lymphocytes by treatment with a rabbit ATS

[2] D. G. Sieckmann and W. E. Paul. Unpublished observations.

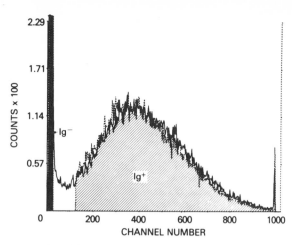

FIG. 2. Fluorescence profile of CBA/J spleen cells stained with fluorescein-labeled F(ab')₂ goat anti-mouse Ig (——). A second profile (– – –) shows Ig⁻ (channels 0–25) and Ig⁺ (channels 125–1,000) pools as they were detected during the sorting process.

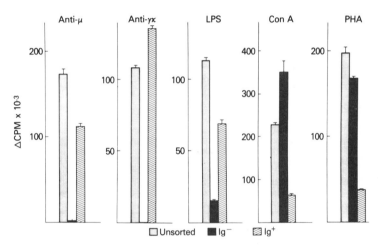

FIG. 3. Anti-Ig stimulation of FACS-sorted spleen cells. Nonsorted, Ig⁺, or Ig⁻ cells were cultured at 2×10^5 cells per microwell with anti-μ (100 μg/ml), anti-γ,κ (250 μg/ml), LPS (50 μg/ml), Con A (2 μg/ml), or PHA (1 μg/ml).

and complement. Preliminary experiments showed that pretreatment of spleen cells with ATS did not alter the anti-Ig staining profile of Ig⁺ cells, when such cells were stained with a fluorescein-labeled rabbit F(ab')₂ anti-mouse Ig. Cells treated in this manner were sorted on the FACS. Ig⁺ cells were collected from channels 451 to 1,000. Such cells, when reanalyzed after sorting, were 98.3% Ig⁺. For comparison, normal cells were also sorted into Ig⁻ (channels 0–45) and Ig⁺ (channels 90–1,000) cells. The results of this experiment (Table II) demonstrate that ATS and complement-treated-Ig⁺-sorted spleen cells, although completely unresponsive to Con A or PHA, were still able to produce a vigorous response to both anti-μ and anti-γ,κ. These responses to anti-Ig were quite enhanced as compared with normal unsorted spleen cells. The response to LPS was also maintained. These results indicate that anti-Ig stimulation

TABLE II

Anti-Ig Proliferative Response of FACS-Sorted Spleen Cells Previously Depleted of T Cells by ATS + Complement Treatment*

Stimulant	Concentration	Normal			ATS + Complement treated	
		Unsorted	Ig⁺	Ig⁻	Unsorted	Ig⁺
	$\mu g/ml$			*cpm per culture*		
None	—	2,536 ± 153	1,329 ± 69	195 ± 7	3,441 ± 233	1,537 ± 120
G102 anti-μ	100	95,201 ± 1,342	100,429 ± 3,236	980 ± 49	132,329 ± 2,608	140,684 ± 4,012
G125 anti-γ,κ	200	12,219 ± 1,497	12,885 ± 1,181	244 ± 34	22,118 ± 2,156	52,213 ± 3,010
LPS	50	114,226 ± 6,515	184,688 ± 16,573	3,388 ± 290	109,244 ± 11,130	169,078 ± 8,691
Con A	2	300,087 ± 7,806	23,272 ± 2,693	155,915 ± 11,869	4,854 ± 977	1,147 ± 211
PHA	1	196,297 ± 5,755	24,371 ± 15	117,266 ± 7,741	1,820 ± 115	387 ± 50

* BDF₁ spleen cells were treated with a rabbit ATS and complement before staining with a fluorescein-labeled rabbit F(ab')₂ anti-mouse Ig. Ig⁺ cells were sorted on the FACS into channels 451–1,000. Normal spleen cells were stained in a similar manner and sorted into Ig⁻ (channels 0–45) and Ig⁺ (channels 90–1,000) cells. Cells were cultured at a density of 1.5 × 10⁵ per microwell.

TABLE III

Stimulation of Athymic Nude (nu/nu) Mice by Anti-Ig*

Mitogen	Concentration	nu/+		nu/nu		NWT‡
		—	1 × 10⁵ NWT	—	1 × 10⁵ NWT	
	$\mu g/ml$			*cpm per culture*		
Experiment I						
None	—	3,382 ± 240	6,160 ± 674	2,236 ± 247	4,966 ± 1,023	300 ± 32
G615 anti-μ	50	94,770 ± 9,600	112,537 ± 9,262	49,225 ± 3,813	68,575 ± 5,103	20,401 ± 561
G125 anti-γ,κ	250	10,516 ± 2,523	18,770 ± 2,947	1,198 ± 142	4,153 ± 116	2,018 ± 114
LPS	50	159,232 ± 7,706	161,331 ± 7,447	128,849 ± 12,828	121,190 ± 10,335	9,981 ± 194
Con A	2	232,894 ± 12,277	213,715 ± 17,280	1,936 ± 197	169,407 ± 1,601	277,692 ± 7,001
Experiment II						
None	—	7,339 ± 2,155	10,343 ± 2,721	2,209 ± 195	5,332 ± 933	190 ± 40
G615 anti-μ	50	95,440 ± 3,585	89,295 ± 1,551	69,660 ± 9,569	77,803 ± 9,771	5,967 ± 139
G125 anti-γ,κ	250	18,675 ± 3,917	30,754 ± 8,625	2,816 ± 564	8,704 ± 2,030	378 ± 16
LPS	50	114,782 ± 6,169	85,981 ± 4,301	82,564 ± 6,680	79,186 ± 6,185	7,384 ± 85
Con A	2	167,554 ± 7,269	146,248 ± 4,899	3,470 ± 623	20,803 ± 6,710	63,035 ± 2,246
Experiment III						
None	—	4,232 ± 340	8,509 ± 2,836	1,840 ± 167	2,464 ± 367	45 ± 7
G125 anti-γ,κ	250	24,923 ± 4,390	32,251 ± 8,758	7,171 ± 1,273	8,027 ± 1,173	81 ± 7
LPS	50	103,203 ± 2,182	97,084 ± 2,931	74,144 ± 14,043	73,684 ± 13,104	296 ± 58
Con A	2	201,111 ± 7,725	166,252 ± 5,929	1,143 ± 139	205,307 ± 16,160	66,590 ± 7,006

* Spleen cells from 10 to 12-week-old athymic (*nu/nu*) mice or heterozygous littermates (*nu/+*) were cultured individually. Each of the values expressed in this table represents the arithmetic mean ± standard error of determinations made on four individual mice.
‡ Nylon wool T cells were cultured alone at 1 × 10⁵ cells/culture (exp. II and III) or at 5 × 10⁵ cells/culture (exp. I).

is dependent upon an Ig⁺ cell and that Ig⁻ cells are not required for responses by Ig⁺ cells.

Stimulation of Athymic (nu/nu) Mice by Anti-Ig. To obtain additional evidence for the T-cell independence of B-cell stimulation by anti-Ig reagents, the ability of spleen cells from *nu/nu* BALB/c mice to respond to anti-Ig was examined (Table III). Spleen cells from *nu/nu* and from *nu/+* BALB/c mice (10–12 wk of age) were cultured individually with various concentrations of anti-μ, anti-γ,κ, and standard concentrations of LPS or Con A. Results are displayed as the arithmetic mean of the maximum response to anti-Ig of the four individual mice in each group. Cultures of *nu/nu* spleen cells responded quite well to anti-μ; however, their responses were only 50–70% of that of normal (*nu/+*) heterozygous littermates. Although not shown in Table III, the lower responsiveness of *nu/nu* spleen cells was apparent at all concentrations of anti-μ tested. *Nu/nu* spleen cells also responded somewhat less well to LPS than did *nu/+* spleen cells. Responses of *nu/nu* spleen cells to anti-μ were only marginally

TABLE IV

Effect of Removal of Adherent Cells from Spleen Cell Populations on Anti-Ig-Induced Proliferation*

Mitogen	Concentration	No treatment	Sephadex G-10 passed	Carbonyl iron treated
	µg/ml		*cpm per culture*	
None	—	7,251 ± 438	7,885 ± 181	9,001 ± 250
G615 anti-µ	50	84,220 ± 4,007	94,260 ± 3,198	78,164 ± 3,405
	10	58,732 ± 3,391	77,756 ± 1,112	68,732 ± 6,329
G125 anti-γ,κ	250	29,912 ± 2,992	67,107 ± 3,841	83,241 ± 4,547
	50	23,971 ± 3,766	43,176 ± 9,285	52,976 ± 2,648
LPS	50	72,665 ± 2,203	72,840 ± 881	66,634 ± 4,163
Con A	2	230,238 ± 8,759	55,271 ± 3,739	119,240 ± 7,859
PHA	1	103,545 ± 2,766	98,970 ± 7,735	111,169 ± 4,419

* BDF_1 spleen cells (15 wks of age) were passed through two sequential Sephadex G-10 columns or treated with carbonyl iron. Nontreated spleen or nonadherent cell populations were cultured at 5×10^5 cells per culture.

enhanced by addition of nylon wool-purified splenic T cells obtained from *nu/+* littermates. In contrast, responsiveness of *nu/nu* spleen cells to anti-γ,κ was quite variable, and most individuals were either nonresponsive or responded only minimally. Again, addition of NWT cells did not restore responsiveness to anti-γ,κ, suggesting that the poor responses by the *nu/nu* spleen cells are not the result of the absence of a functional T-cell population, but reflect an abnormality in the B cells of these mice.

Effect of Removal of Adherent Cells from Spleen Cell Populations on Anti-Ig-Induced Proliferation. We have previously reported that F(ab')$_2$ fragments prepared from anti-µ are as stimulatory as whole molecules in inducing a proliferative response (22). This would suggest that binding to Fc receptors on either lymphocytes or macrophages was not required for stimulation. It was nevertheless of interest to determine whether or not the presence of macrophages was important in B-cell responses to anti-Ig.

Two methods were used for depletion of adherent cell populations: (*a*) sequential passage of spleen cells over two Sephadex G-10 columns (28) or (*b*) carbonyl iron treatment (29). The effectiveness of macrophage removal was monitored by testing for responsiveness to SRBC, which is a macrophage-dependent response, and by testing for phagocytosis of latex particles. The results of one such experiment are shown in Table IV. In this particular experiment, the day 4 primary SRBC response (not shown) was abrogated to the extent of 99 and 90% by G-10 column and carbonyl iron treatment, respectively, and these responses could be partially reconstituted by addition of 2% irradiated peritoneal cells. The starting spleen cell population had 7.1% phagocytic cells by the latex test, G-10 column and carbonyl iron treatment reduced this to 1.8 and 0.9%, respectively. The results in Table IV show that neither anti-µ nor anti-γ,κ responses were affected by either treatment. The LPS and PHA responses similarly were unaffected, whereas the Con A response was considerably diminished, as has been previously reported (30).

The above results suggested that macrophages are not required for anti-Ig-induced proliferation. However, because we had not completely removed all phagocytic cells, as detected by the latex phagocytosis assay, we tested responsiveness of depleted cell populations at limiting cell densities. We reasoned that if small numbers of residual macrophages supported responses in our standard culture condition (5×10^5 cells/culture), they should be diluted out at lower cell concentrations. In this experi-

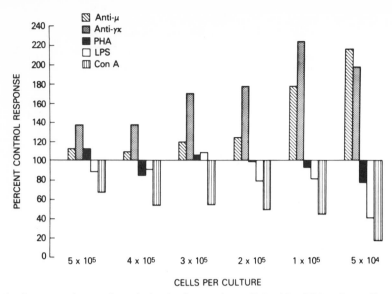

FIG. 4. Response of macrophage-depleted cells at various cell densities. BDF$_1$ spleen cells (15 wk of age) were treated with carbonyl iron and cultured at densities of 5×10^4 to 5×10^5 cells per culture with anti-μ (50 μg/ml), anti-γ,κ (250 μg/ml), LPS (RE, 50 μg/ml), Con A (2 μg/ml), and PHA (1 μg/ml). Results are reported as the response of nonadherent spleen cells ÷ response of control normal spleen cells × 100%.

ment, phagocytic cells were removed by treatment with carbonyl iron and exposure to a magnetic field. This treatment diminished the frequency of cells capable of phagocytosing latex particles from 4.0 to 0.9% and diminished the in vitro primary anti-SRBC antibody response by 80%. The cells were then cultured at concentrations of 5×10^4 to 5×10^5 cells per microwell. The results are presented in Fig. 4 as the ratio of the response of macrophage-depleted cells to that of untreated cells, multiplied by 100. The results show that the relative responsiveness of macrophage-depleted cells to Con A falls as the number of cells cultured is diminished. However, the relative responsiveness of macrophage-depleted cells to anti-μ and to anti-γ,κ actually increases upon cell dilution. Thus, at 5×10^4 cells per microwell, the response to anti-μ and to anti-γ,κ is approximately twofold greater in macrophage-depleted populations than in unseparated cell populations. In every case, addition of irradiated peritoneal cells (2%) to macrophage-depleted populations returned the response to anti-Ig to that of unseparated cells (data not shown). These results provide strong support for the concept that macrophages are not required for anti-Ig stimulation and that they may be suppressive under certain circumstances.

Ontogeny of the Anti-Ig Response. We have previously reported that anti-μ is able to stimulate responses of spleen cells from a wide variety of mouse strains at 8–12 wk of age. A more complete study of the ontogeny of responsiveness was made to determine if there was any correlation between responsiveness to anti-μ and anti-γ,κ, and the development of various B-cell types in the spleen. Cultures of spleen cells from BDF$_1$ mice ranging from 1 to 32 wk of age were stimulated with either anti-μ, anti-γ,κ, LPS, or NWSM (31). The responses are shown in Fig. 5 as the absolute response and as the percent of the average adult response for each mitogen. From these two graphs, it can be seen that spleen cells from mice younger than 4 wk of age do not respond to anti-

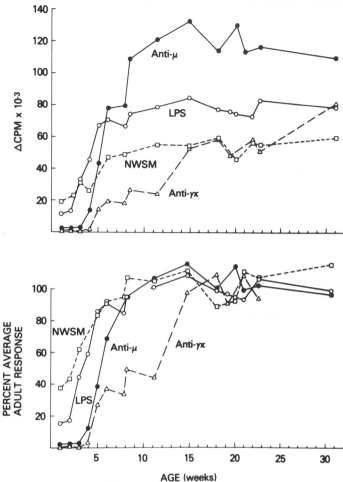

FIG. 5. Ontogeny of the anti-Ig response in BDF₁ mice. BDF₁ spleen cells from two or three mice of various ages were cultured with anti-μ (100 μg/ml), anti-γ,κ (100 μg/ml), LPS (50 μg/ml), and NWSM (50 μg/ml). The results are reported as Δcpm (upper graph) or as a percentage of the average maximum response calculated from data points in the plateau of the age response curve (lower graph). Calculated average maximum adult responses used for percentage calculations were: anti-μ, 114,263 ± 867; anti-γ,κ, 54,229 ± 4,588; LPS, 78,694 ± 4,451; and NWSM, 51,885 ± 6,881.

μ or to anti-γ,κ. Adult levels of responsiveness to anti-μ are not reached until after 8 wk. Responsiveness to anti-γ,κ can usually be observed at 4 wk of age although in the experiment illustrated, responses were first noted at 5 wk. This response, however, does not reach adult levels until after 15 wk of age. The developmental patterns of responsiveness to anti-μ and anti-γ,κ are quite different from those of responsiveness to LPS and NWSM. These mitogens stimulate responses by cells from 1-wk-old mice and give responses that are 60% of adult levels at 4 wk of age, at a time when responses to anti-μ are only 10% of adult levels. These results indicate that the B cells responsive to anti-Ig are members of a late-developing subset.

Anti-Ig Stimulation of Lymphocytes from (CBA/N × DBA/2N)F₁ Mice. The CBA/N mouse strain has been shown to carry an X-linked genetic defect (23), which appears

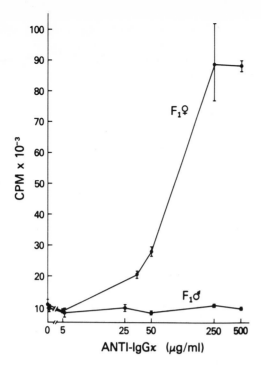

Fɪɢ. 6.　Failure of anti-γ,κ to stimulate (CBA/N × DBA/2N)F$_1$ male spleen cells. Spleen cells from F$_1$ male or female mice (19 wk old) were cultured at 5×10^5 cells/culture with various concentrations of anti-γ,κ.

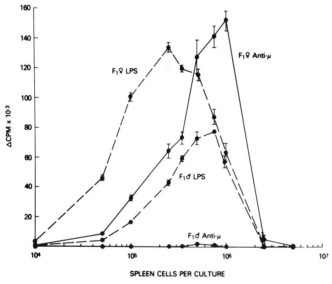

Fɪɢ. 7.　Failure of anti-μ to stimulate (CBA/N × DBA/2N)F$_1$ male spleen cells. Spleen cells of F$_1$ male (17 wk old) or female (19 wk old) mice were cultured with G615 anti-μ antiserum (1:200) or LPS (50 μg/ml).

to be due to the absence or defective function of a mature or late-developing subset of the B-lymphocyte population (32, 33). Fig. 6 shows that spleen cells from (CBA/N × DBA/2N)F_1 male mice, which carry this defect, cannot be stimulated to a proliferative response by anti-γ,κ antibodies over a wide range of antibody concentrations, whereas spleen cells from F_1 female donors, which are phenotypically normal, exhibit a vigorous response. Fig. 7 shows that the unresponsiveness is also seen with anti-μ antibody in cultures of various numbers of spleen cells. In contrast, the F_1-defective male mice respond well to LPS, although higher cell densities are required for maximum responses than are needed for responses of F_1 female cells. The unresponsiveness of lymphocytes from mice that carry the CBA/N B cell defect further suggests that anti-μ and anti-γ,κ selectively activate a mature subset(s) of Ig^+ B lymphocytes, which are functionally absent in these mice and in normal mice of <4 wk of age.

Discussion

In our previous paper (22), we demonstrated that specifically purified goat anti-μ, goat anti-γ,κ, and rabbit anti-κ antibodies could initiate substantial DNA synthetic responses by spleen cells from a wide variety of mouse strains, and we presented evidence indicating that this stimulatory activity depended upon the specificity of the antibody for Ig.

In this paper, we have studied the cellular basis of this response. Our results indicate that Ig^+ cells prepared by sorting for the Ig^+ phenotype or by pretreatment with anti-Thy 1.2 antiserum and complement can respond to anti-Ig, that T lymphocytes are not required for the response, and that thymocytes and nylon wool-passed spleen cells do not respond. The response of cells from nu/nu donors to anti-μ was somewhat lower than that of cells from phenotypically normal littermates, and nu/nu cells often did not respond to anti-γ,κ. However, the addition of nylon wool-purified splenic T cells from a $nu/+$ littermate donor to cultures of nu/nu spleen cells had little effect on the anti-μ response and did not reconstitute the anti-γ,κ response. This suggests that B lymphocytes from nu/nu mice may be abnormal, particularly in the development of the subset of mature B lymphocytes that respond to anti-Ig. It also suggests that maturation of these cells may be under T-cell control. Nonetheless, these results indicate that both anti-μ and anti-γ,κ principally stimulate B lymphocytes and that their response is T independent.

The data presented here strongly suggest that the stimulatory effect of anti-μ and of anti-γ,κ is not exerted on all Ig-bearing lymphocytes. This conclusion is supported by the failure of anti-μ to stimulate proliferative responses of spleen cells from BDF_1 mice until the donors are 4 wk of age. Indeed, it has previously been shown (27, 34) that significant numbers of Ig^+ cells are present in the spleen 3 days after birth and adult frequencies of Ig^+ cells are present by 2 wk of age. In contrast, maximal responsiveness to anti-μ is not reached until 8 wk of age. Responses to anti-γ,κ develop even more slowly. The rate of development of responsiveness to anti-Ig contrasts strikingly with that to LPS and NWSM. The latter stimulate substantial responses by cells from 1-wk-old donors. Finally, mice with the CBA/N immune defect are unresponsive to both anti-μ and to anti-γ,κ. It has previously been shown that these mice lack a subset of mature or late-developing B lymphocytes which bear a distinct phenotypic marker, Lyb 5 (35). It appears that the B cells that proliferate in response to anti-μ and to anti-γ,κ are members of this $Lyb\ 5^+$ subset. Indeed, preliminary experiments[3] indicate that pretreatment with anti-Lyb 5.1 and complement destroys responses to anti-μ and to anti-γ,κ, strongly supporting the idea that Lyb 5$^+$ cells are responsible for the anti-Ig response. The distinctive response patterns to anti-μ and to anti-γ,κ may indicate a degree of heterogeneity within this population.

Our results, indicating that spleen cells from 4-wk-old mice can respond to anti-Ig, are quite different from those of Wiener et al. (21) who have reported that responsiveness to anti-Ig is not achieved until mice are approximately 7 mo of age. Although we cannot explain this difference, it seems unlikely to us that the natural process of lymphocyte differentiation is not completed until mice are 7 mo of age. It is possible that the failure of Wiener et al. to observe responses in young adult mice may reflect differences in the culture conditions used in our laboratories.

Wiener et al. also reported (36) that adult thymectomy speeded the rate at which anti-Ig responsiveness was obtained, whereas we noted that congenitally athymic mice were less responsive to anti-γ,κ than were their normal littermates. This difference may be explicable by the considerable difference in the T-lymphocyte status of these two types of mice. In the congenitally athymic animal, the peripheral environment is exceptionally depleted of T-lymphocyte influence, whereas in the adult thymecto-mized animal, there is preservation of the long-lived T-lymphocyte population. Both sets of observations point to the possibility that T lymphocytes (or thymic influence), although not required for response of mature B lymphocytes to anti-Ig, may play a role in the development of responsive cells.

The response of B lymphocytes to anti-Ig reagents is largely independent of the action of macrophages. Thus, depletion of adherent cells by Sephadex G-10 or by treatment with carbonyl iron and exposure to a magnetic field did not diminish responsiveness to anti-Ig reagents. Indeed, when cell populations depleted of phago-cytic cells were cultured at limiting cell numbers (5×10^4/well), their responses to anti-μ and to anti-γ,κ were greater than those of unseparated spleen cells. These enhanced responses were abolished when irradiated normal peritoneal cells, as a source of macrophages, were added to macrophage-depleted populations. Further-more, in preliminary experiments, we have removed macrophages by carbonyl iron treatment and exposure to a magnetic field and then isolated Ig$^+$ cells from this macrophage-depleted population. Such cells, lacking both macrophages and T cells, responded well to anti-μ and anti-γ,κ. These results strongly suggest that the stimu-latory effect of anti-Ig reagents is independent of both macrophages and T cells. However, one cannot absolutely exclude the need for very small numbers of either cell type. Other types of cell-cell interaction in the activation caused by anti-Ig (e.g. B-B interaction or interaction of nonadherent, nonphagocytic accessory cells with B cells) have not been formally excluded. Nevertheless, our results are most consistent with the concept that anti-Ig leads to the direct activation of mouse B lymphocytes. Inasmuch as F(ab')$_2$ fragments of anti-μ and deaggregated anti-μ are equivalent to control anti-μ preparations in B-cell activation, it does not appear likely that Fc receptors participate in the response. Furthermore, responses to anti-μ can be obtained in serum-free media (22), although such responses do depend on the addition of 2-ME. We would propose that the activation of B-lymphocyte proliferation by anti-μ and anti-γ,κ antibodies is directly due to an interaction with Ig determinants on the B-cell membrane and that the generation of transmembrane signals is a function of

---

[3] D. G. Sieckmann and B. Subbarao. Unpublished observations.

the Ig determinants and not of some postulated non-Ig receptor for a mitogenic principle (10, 11).

Although not addressed in this paper, anti-Ig fails to stimulate Ig synthesis despite the marked uptake of [³H]TdR which occurs in stimulated cells. This is consistent with the need for intrinsically different types of stimuli for activation of the proliferative and differentiative aspects of normal B-lymphocyte responses. As anti-Ig antibody is a globular protein similar in general structure to most thymus-dependent antigens, its interaction with B-lymphocyte Ig may be regarded as comparable to the binding of a thymus-dependent antigen to receptors of specific B lymphocytes. In the absence of T-cell influence, such interactions may lead to priming and proliferation of specific cells but not to antibody synthesis (37–39). Furthermore, Kishimoto and Ishizaka (40) have reported that a T-lymphocyte supernate factor, together with anti-Ig, may cause Ig synthesis by rabbit lymphocytes. Current efforts are underway to find conditions that allow anti-Ig antibodies to stimulate Ig synthesis in mouse cells. This should provide a system in which the distinctive components of B-lymphocyte activation can be precisely analyzed.

Summary

Mouse spleen cells can be stimulated to proliferate in vitro by purified anti-μ or anti-γ,κ antibodies. These responses can be obtained in cell populations bearing membrane immunoglobulin (Ig), purified by the fluorescence activated cell sorter (FACS), but they are not observed in FACS-purified Ig⁻ cell populations. Furthermore, treatment of spleen cell populations with anti-Thy 1.2 and complement does not impair the response, nor does addition of nylon wool-purified T lymphocytes enhance it. These results indicate that B lymphocytes respond to anti-Ig and that their response does not require T cells. On the other hand, cells from athymic nude (*nu/nu*) mice respond slightly less well to anti-μ than do cells from heterozygous littermate (*nu/+*) controls; *nu/nu* cells are almost unresponsive to anti-γ,κ and addition of nylon wool-purified T cells from *nu/+* controls does not restore the response. This suggests that T lymphocytes or the thymus may control the appearance of cells responsive to anti-γ,κ.

Responsiveness of normal mice to anti-μ does not appear until 4 wk of age and does not reach maximum levels until 8 wk of age. Acquisition of full responsiveness to anti-γ,κ is even more delayed. This, together with the failure of mice with the CBA/N B-cell defect to respond to anti-Ig, suggests that cells stimulated to proliferate by anti-Ig are a mature subset of B cells.

Depletion of adherent cells by Sephadex G-10 treatment or by treatment with carbonyl iron and exposure to a magnetic field does not diminish anti-μ or anti-γ,κ responses, suggesting that the responsiveness does not require the presence of macrophages. Thus, activation of B-cell proliferation by anti-Ig appears to be a T-cell independent, macrophage-independent process in which membrane Ig plays a direct role in signal generation.

The authors wish to thank Dr. James Mond for many helpful discussions and Dr. Carol Cowing for assistance in the Sephadex macrophage depletion technique. We also wish to express our gratitude to Robert Habbersett for preparing the FACS-sorted cells, and Ms. Weltha Logan and Iris McCalla for technical assistance.

Received for publication 25 July 1978

References

1. Warner, N. L. 1974. Membrane immunoglobulins and antigen receptors on B and T lymphocytes. *Adv. Immunol.* **17**:67.

2. Schreiner, G. F., and E. R. Unanue. 1976. Membrane and cytoplasmic changes in B lymphocytes induced by ligand-surface immunoglobulin interaction. *Adv. Immunol.* **24**:37.

3. Bretscher, P., and M. Cohn. 1970. A theory of self-nonself discrimination. *Science (Wash. D. C.).* **169**:1042.

4. Diener, E., and M. Feldmann. 1972. Relationship between antigen and antibody-induced suppression of immunity. *Transplant. Rev.* **8**:76.

5. Schrader, J. W. 1973. Mechanism of activation of the bone marrow-derived lymphocyte. III. A distinction between a macrophage-produced triggering signal and the amplifying effect on triggered B lymphocytes of allogeneic interactions. *J. Exp. Med.* **138**:1466.

6. Barton, M. A., and E. Diener. 1975. A new perspective on B cell triggering: Control of immune response by organizational changes in the lipid bilayer. *Transplant. Rev.* **23**:5.

7. Bretscher, P. A. 1975. The two signal model for B cell induction. *Transplant. Rev.* **23**:37.

8. Coutinho, A., and G. Möller. 1974. Immune activation of B cells: Evidence for one nonspecific triggering signal not delivered by the Ig receptors. *Scand. J. Immunol.* **3**:133.

9. Coutinho, A., and G. Möller. 1975. Thymus-independent B-cell induction and paralysis. *Adv. Immunol.* **21**:113.

10. Coutinho, A. 1975. The theory of the "one non-specific signal" model for B cell activation. *Transplant. Rev.* **23**:49.

11. Möller, G. 1975. One non-specific signal triggers B lymphocytes. *Transplant. Rev.* **23**:126.

12. Sell, S., and P. G. H. Gell. 1965. Studies on rabbit lymphocytes *in vitro*. I. Stimulation of blast transformation with an anti-allotype serum. *J. Exp. Med.* **122**:423.

13. Adenolfi, M., B. Gardner, F. Gianelli, and M. McGuire. 1967. Studies on human lymphocytes stimulated *in vitro* with anti-γ and anti-μ antibodies. *Experientia (Basel).* **23**:271.

14. Daguillard, F., D. C. Heines, M. Richter, and B. Rose. 1969. The response of leukocytes of agammaglobulinemic subjects to phytohemagglutinin and anti-immunoglobulin serum. *Clin. Exp. Immunol.* **4**:203.

15. Oppenheim, J. J., G. W. Rogentine, and W. D. Terry. 1969. The transformation of human lymphocytes by monkey antisera to human immunoglobulin. *Immunology.* **16**:123.

16. Gausset, P., G. Delespesse, C. Hubert, B. Kennes, and A. Govaerts. 1976. *In vitro* response of subpopulations of human lymphocytes. II. DNA synthesis induced by anti-immunoglobulin antibodies. *J. Immunol.* **116**:446.

17. Maino, V. C., M. J. Hayman, and M. J. Crumpton. 1975. Relationship between enhanced turnover of phosphatidylinositol and lymphocyte activation by mitogens. *Biochem. J.* **146**:247.

18. Skamene, E., and J. Ivanyi. 1969. Lymphocyte transformation by H-chain specific anti-immunoglobulin sera. *Nature (Lond.).* **221**:681.

19. Kirchener, H., and J. J. Oppenheim. 1972. Stimulation of chicken lymphocytes in a serum-free medium. *Cell. Immunol.* **3**:695.

20. Parker, D. C. 1975. Stimulation of mouse lymphocytes by insoluble anti-mouse immunoglobulins. *Nature (Lond.).* **258**:361.

21. Weiner, H. L., J. W. Moorehead, and H. Claman. 1976. Anti-immunoglobulin stimulation of murine lymphocytes. I. Age dependency of the proliferative response. *J. Immunol.* **116**:1656.

22. Sieckmann, D. G., R. Asofsky, D. E. Mosier, I. M. Zitron, and W. E. Paul. 1978. Activation of mouse lymphocytes by anti-immunoglobulin. I. Parameters of the proliferative response. *J. Exp. Med.* **147**:814.

23. Amsbaugh, D. F., C. T. Hansen, B. Prescott, P. W. Stashak, D. R. Barthold, and P. J. Baker. 1972. Genetic control of the antibody response to type III pneumococcal polysac-

charide in mice. I. Evidence that an X-linked gene plays a decisive role in determining responsiveness. *J. Exp. Med.* **136**:931.

24. Mishell, R. I., and R. W. Dutton. 1967. Immunization of dissociated spleen cell cultures from normal mice. *J. Exp. Med.* **126**:423.

25. Reif, A. E., and J. M. Allen. 1966. Mouse thymic isoantigens. *Nature (Lond.).* **209**:521.

26. Julius, M. H., E. Simpson, and L. A. Herzenberg. 1973. A rapid method for the isolation of functional thymic-derived murine lymphocytes. *Eur. J. Immunol.* **3**:645.

27. Scher, I., S. O. Sharrow, R. Wistar, R. Asofsky, and W. E. Paul. 1976. B-lymphocyte heterogeneity: Ontogenetic development and organ distribution of B-lymphocyte populations defined by their density of surface immunoglobulin. *J. Exp. Med.* **144**:494.

28. Cowing, C., B. D. Schwartz, and H. B. Dickler. 1978. Macrophage Ia antigens. I. Macrophage populations differ in their expression of Ia antigen. *J. Immunol.* **120**:378.

29. Sjoberg, O., J. Andersson, and G. Möller. 1972. Requirement for adherent cells in the primary and secondary immune response *in vitro*. *Eur. J. Immunol.* **20**:123.

30. Habu, S., and M. C. Raff. 1977. Accessory cell dependence of lectin-induced proliferation of mouse T lymphocytes. *Eur. J. Immunol.* **7**:451.

31. Bona, C., C. Damais, and L. Chedid. 1974. Blastic transformation of mouse spleen lymphocytes by a water-soluble mitogen extracted from *Nocardia*. *Proc. Natl. Acad. Sci. U. S. A.* **71**:1602.

32. Scher, I., A. D. Steinberg, A. K. Berning, and W. E. Paul. 1975. X-linked B-lymphocyte immune defect in CBA/N mice. II. Studies of the mechanisms underlying the immune defect. *J. Exp. Med.* **142**:637.

33. Scher, I., S. O. Sharrow, and W. E. Paul. 1976. X-linked B-lymphoyte defect in CBA/N mice. III. Abnormal development of B-lymphocyte populations defined by their density of surface immunoglobulin. *J. Exp. Med.* **144**:507.

34. Gelfand, M. C., G. J. Elfenbein, M. M. Frank, and W. E. Paul. 1974. Ontogeny of B lymphocytes. II. Relative rates of appearance of lymphocytes bearing surface immunoglobulin and complement receptors. *J. Exp. Med.* **139**:1125.

35. Ahmed, A., I. Scher, S. O. Sharrow, A. H. Smith, W. E. Paul, D. H. Sachs, and K. E. Sell. 1977. B-lymphocyte heterogeneity: Development and characterization of an alloantiserum which distinguishes B-lymphocyte differentiation alloantigens. *J. Exp. Med.* **145**:101.

36. Weiner, H. L., D. J. Scribner, and J. W. Moorhead. 1977. Anti-immunoglobulin stimulation of murine lymphocytes. III. Enhancement of the development of responsive B cells by thymic deprivation. *Cell. Immunol.* **31**:77.

37. Roelants, G. E., and B. A. Askonas. 1973. Immunological B memory in thymus deprived mice. *Nature (Lond.).* **239**:63.

38. Davie, J. M., and W. E. Paul. 1974. Role of T lymphocytes in the humoral immune response. I. Proliferation of B lymphocytes in thymus-deprived mice. *J. Immunol.* **113**:1438.

39. Dutton, R. W. 1975. Separate signals for the initiation of proliferation and differentiation in the B cell response to antigen. *Transplant. Rev.* **23**:66.

40. Kishimoto, T., and K. Ishizaka. 1975. Regulation of antibody response *in vitro*. IX. Induction of secondary anti-hapten IgG antibody response by anti-immunoglobulin and enhancing soluble factor. *J. Immunol.* **114**:585.

Chapter III

Functional Diversity of T Lymphocytes

Chapter III

Functional Diversity of T Lymphocytes

T lymphocytes participate in many aspects of the immune response. Helper
and suppressor T cells contribute to the regulation of humoral immunity while
cytotoxic T cells and T cell mediators of delayed-type hypersensitivity (DTH)
are the effectors of cell-mediated immunity. Given this impressive array of
functions, it was reasonable to raise the question of heterogeneity within the
T cell population: do different kinds of T cells mediate different functions,
or are all T cells multi-functional?

The first inkling that functionally distinct populations of T cells did
indeed exist came from studies conducted by Cantor and Asofsky (1970, 1972)
about ten years ago. While those early experiments are not included in this
chapter, subsequent experiments conducted by Cantor and his associates are.
These later studies have formed the basis of many of our current views about
T cell function.

We also include in this chapter one paper on natural killer cells, although
these cells are not currently considered to be bona fide members of the T cell
family. Nonetheless, they are responsible for intriguing and still mysterious
immunological functions, and some evidence does exist to suggest that these cells
may be immature T cells.

The story behind the characterization of functionally distinct T cells
presents a persuasive argument for the utility of antisera capable of identifying differentiation antigens expressed on the surfaces of cells. In the course
of studies on surface antigens expressed by normal and leukemic cells,
Old and Boyse (1969) put forth the idea that the maturational changes of a cell
might be accompanied by changes in the display of molecules at the cell surface.
These molecules are presumably responsible for various new functions assumed by
the developing cell, but the precise function of these surface molecules is less
important for this discussion than the fact that functionally and/or develop-
mentally distinct cells will present unique arrays of these differentiation
antigens. Thus, antisera specific for a particular antigen expressed at a
discrete developmental stage will detect those cells which are at that stage
of differentiation.

Cantor and his colleagues (chief among them, Boyse) exploited a set of
differentiation antigens to dissect the T cell population. This set of antigens
had been described by Boyse and designated the Ly (for lymphocyte) A, B, and C
antigens. They were believed to be expressed on T cells, rather like the Thy 1
antigen. Cantor, using antisera specific for Ly A, B or C antigens, demonstrated
that a) not all T cells expressed all of these antigens and b) that the selective
expression of Ly A, B, or C was correlated to a unique T cell function.

Before describing these experiments in more detail, we should pause to
acknowledge a change in nomenclature (a favorite pastime of immunologists). The

Ly A, B, and C antigens are now designated Lyt (for T lymphocyte) 1, 2, and 3. Two alleles for each of these genes have been identified and are noted as Lyt 1.1 and Lyt 1.2, Lyt 2.1 and Lyt 2.2, and Lyt 3.1 and Lyt 3.2. The Lyt 2 and 3 genes are closely linked on chromosome 6, while Lyt 1 has been mapped to chromosome 19.

Using highly specific alloantisera reacting with the Lyt 1, 2, or 3 antigens, Cantor and Boyse (paper 1) demonstrated that selective displays of the Lyt antigens distinguished precursors of helper T cells from precursors of cytotoxic T cells. The helper T cell precursors express the Lyt $1^+ 2,3^-$ phenotype, while the cytotoxic T cell precursors are Lyt $1^- 2,3^+$. A third population, bearing all three Lyt antigens, may be an immature population of T cells which is not yet committed to either helper or killer activities.

The work by Jandinski et al. (paper 2) expanded these initial findings by demonstrating a) that mitogen-stimulated effector (as opposed to precursor) T cells could also be distinguished and separated on the basis of their Lyt phenotypes, and b) that suppressor T cells bore the Lyt $1^-2,3^+$ phenotype.

As pointed out in this report, the helper activity of the unseparated population of mitogen stimulated cells could not be described since it is masked or dominated by the suppressor activity also present in the population. This helper activity was revealed by treatment of the cells with anti-Lyt 2 and complement, removing the suppressor population. The ability to reveal such activity points to the limitations of conclusions drawn from experiments measuring immune function in unseparated lymphocyte populations.

It is important to point out two things about the utility of Lyt antigens in describing lymphocyte subsets: 1) the Lyt antigens so far described certainly do not include all lymphocyte-specific antigens, and 2) there is no conclusive evidence that proves that a particular Lyt phenotype is absolutely restricted to expression by a particular set of effector T cells.

These cautions notwithstanding, Cantor and his colleagues have continued to derive much useful information about T cells by employing anti-Lyt sera. The report by Huber et al. (paper 3) demonstrates that the Lyt 1^+ and Lyt $2,3^+$ populations are distinct cell sets which are not interconvertible. At least some cells of the Lyt $1,2,3^+$ phenotype may serve as common precursors to the Lyt 1^+ and Lyt $2,3^+$ cells. Current experiments suggest that the Lyt $1,2,3^+$ cells appear first during ontogeny and constitute about 50% of the peripheral T cell population.

Further studies probing the Lyt phenotypes of T cells have revealed that the effector T cells in a delayed-type hypersensitivity response, as well as the cells which respond by proliferation to foreign I region determinants (mixed lymphocyte response) are Lyt $1^+2,3^-$.

Although apparently devoid of easily detectable levels of the Lyt 1,2,3 and Thy 1 antigens, the natural killer (NK) cell can be recognized by its expression of another differentiation antigen, Ly 5. These cells are clearly distinct from cytotoxic T lymphocytes, but they are included in this chapter because of their ability to kill virally-infected cells in the absence of antibody, and the possibility that they may represent cells in the T lineage which have not passed through the thymus. The report by Kiessling (paper 4) addresses primarily the remarkable efficiency of NK cells in killing virally-infected cells upon initial encounter. Unlike resting cytotoxic T cells, these cells seem to be in a perpetual state of activation. Future studies on NK cells will be concerned with the critical issues of what activates these cells and which cell surface structures of the virally-infected targets are recognized by their receptors.

The judicious use of antisera specific for differentiation antigens has contributed a great deal to our understanding of functional diversity within the T cell population. Final answers about the functions of these differentiation antigens, and about the absolute limitations of a single T cell's immunological abilities, will come from the development of the hybridoma and tissue culture technologies. First, high affinity monoclonal antibodies may facilitate the purification of those antigens and also enable the scientist to interfere with whatever cell function (if any) is tied to the expression of the antigens. Second, although still in their infancy, the techniques of cell cloning will allow scientists to work with homogeneous populations of particular kinds of T cells. Cloned lines of antigen-specific helper or cytotoxic T cells will provide new answers about the functional capacities of certain cells and ulti- mately, about the mechanism of function. Although numerous techniques of T cell cloning are being reported in the literature, we have included in this chapter only the paper by Gillis et al. (paper 5). This is not to slight other methods but to emphasize the dependence of these other techniques on the T cell growth factor characterized by Gillis and Smith (Tees and Schreier, 1980).

The combined efforts of scientists developing conditions for the long-term maintenance of normal T cells in culture and those exploiting cell fusion tech- nologies will very shortly allow biochemists and molecular biologists finally to penetrate the secretes of the T cell's receptors and products.

Papers Included in this Chapter

1. Cantor, H. and Boyse, E.A. (1975) Functional subclasses of T lymphocytes bearing different Ly antigens. I. The generation of functionally distinct T-cell subclasses is a differentiative process independent of antigen. J. Exp. Med. 141, 1376.

2. Jandinski, J., Cantor, H., Tadakuma, T, Peavy, D.L., and Pierce, C.W. (1976) Separation of helper T cells from suppressor T cells expressing different Ly components. I. Polyclonal activation: Suppressor and helper activities are inherent properties of distinct T-cell subclasses. J. Exp. Med. 143, 1382.

3. Huber, B., Cantor, H., Shen, F.W., and Boyse, E.A. (1976) Independent differentiative pathways of Ly 1 and Ly 2,3 subclasses of T cells. J. Exp. Med. 144, 1128.

4. Kiessling, R., Klein, E., and Wigzell, H. (1975) "Natural" killer cells in the mouse. I. Cytotoxic cells with specificity for mouse Moloney leukemia cells. Specificity and distribution according to genotype. Eur. J. Immunol. 5, 112.

5. Gillis, S., Union, N.A., Baker, P.E., and Smith, K.A. (1979) The in vitro generation and sustained culture of nude mouse cytolytic T-lymphocytes. J. Exp. Med. 149, 1460.

References Cited in this Chapter

Cantor, H. and Asofsky, R. (1970) Synergy among lymphoid cells mediating the graft-versus-host response. II. Synergy in graft-versus-host reactions produced by Balb/c lymphoid cells of differing anatomic origin. J. Exp. Med. 131, 235.

Cantor, H. and Asofsky, R. (1972) Synergy among lymphoid cells mediating the graft-versus-host response. III. Evidence for interaction between two types of thymus-derived cells. J. Exp. Med. 135, 764.

Raff, M.C. and Cantor, H. (1971) Subpopulations of Thymus Cells and Thymus-Derived Lymphocytes. In: Progress in Immunology (Academic Press, NY), p. 83.

Tees, R. and Schreier, M.H. (1980) Selective reconstitution of nude mice with long-term cultured and cloned specific helper T cells. Nature 283, 780.

Vadas, M.A., Miller, J.F.A.P., McKenzie, I.F.C., Chism, S.E., Shen, F.-W., Boyse, E.A., Gamble, J.R., and Whitelaw, A.M. (1976) Ly and Ia antigen phenotypes of T cells involved in delayed-type hypersensitivity and in suppression. J. Exp. Med. 144, 10.

FUNCTIONAL SUBCLASSES OF T LYMPHOCYTES BEARING
DIFFERENT Ly ANTIGENS

I. The Generation of Functionally Distinct T-Cell Subclasses is a
Differentiative Process Independent of Antigen*

By H. CANTOR and E. A. BOYSE

(From the Department of Medicine, Harvard Medical School, and the Division of Tumor Immunology, Sidney Farber Cancer Center, Boston, Massachusetts 02115 and the Memorial Sloan-Kettering Cancer Center, New York 10021)

T lymphocytes mediate many immunologic functions. For example, they generate cytotoxic responses to alloantigens (1, 2), exert helper (3) and suppressor (4) effects on the production of antibody, and initiate graft-vs.-host responses (5). We do not know whether this diversity of function reflects a functional heterogeneity of T lymphocytes existing before antigen stimulation. If this were the case, then the response of a T-cell clone to stimulation by antigen would be limited to the particular immune function for which it had already been programmed during the differentiation of that T cell. Alternatively, antigen stimulation of a single T cell may induce the formation of progeny that can mediate the complete range of T-dependent responses. These alternatives are illustrated in Fig. 1.

One can pose the question in this way: *is it possible to separate subclasses of T cells from nonimmune animals that are already determined to express, respectively, helper activity or cytotoxic activity before they encounter antigen?* A direct approach to this question could be based upon the use of alloantisera that would define cell surface components expressed selectively on one or another T-cell subclass. Genes coding for such components would most likely be expressed exclusively in T cells.

Because the Ly antigens are said to be reduced on lymphoid cells from neonatally thymectomized mice (6), and have not been detected on the surface of non lymphoid cells (7, 8), they may represent components expressed exclusively on the surface of cells undergoing thymus-dependent differentiation. Each Ly system comprises a genetic locus (*Ly-1* on chromosome 19, *Ly-2* and *Ly-3* closely linked on chromosome 6) each with two alternative alleles, each allele specifying an alternative alloantigen denoted 1 and 2. Thus, the *Ly-1* alleles specify alloantigens Ly-1.1 and Ly-1.2, and all inbred mice express one or the other on their thymocytes; and similarly for *Ly-2 and Ly-3* (9).

---

* This work was supported in part by National Cancer Institute CA-08748, CA-16889, and National Institute of Allergy and Infectious Diseases grant AI-12184.

Shiku and his colleagues observed that effector killer cells, ie. cells that have already responded to antigen, express a distinctive profile of Ly antigens (10). In the study reported here, we have extended the use of the Ly systems to attack the fundamental question posed above, namely whether or not the generation of functional T-cell diversity precedes the encounter with antigen. We find that subclasses of peripheral T cells with different immunological functions and biological characteristics, distinguishable by expression of different Ly alloantigens, pre-exist in mice that have not been immunized. This indicates that diversification of T-cell function, resulting in cells committed to express either helper or killer activity, is a differentiative process that has taken place *before* T cells meet antigen and that these diverse effector cells are derived from different maturational lines of T cells.

Materials and Methods

Notations. The notation for congenic mice and antisera follows Shiku et al. (10).

Mice. C57BL/6 (B6)[1] (Ly phenotype 1.2,2.2,3.2) and B10.D2 were obtained from the Jackson Laboratories, Bar Harbor, Maine, congenic lines B6/Ly-1.1 (Ly-1.1,2.2,3.2), B6/Ly-2.1 (Ly-1.2,2.1,3.2), and B6/Ly-2.1,Ly-3.1 (Ly-1.2,2.1,3.1) from EAB (ref. 12, Table II); B10.T(6R) and AQR were bred by H. Cantor from breeders supplied by Dr. K. Melief of Tufts Medical School; (B6 × BALB)F$_1$ (Ly-1.2,2.2,3.2) was bred by H. Cantor.

Antisera. Anti-Ly-1.2, anti-Ly-2.2 and anti-Ly-3.2, and anti-Thy-1.2 are described in Shiku et al. (ref. 10, Table III). The Ly antisera we used, diluted 1:10, were absorbed once with 120×10^6 syngeneic thymus + LNC/ml to remove autoantibody.

Complement (C)-Dependent Cytotoxicity Assay. 10–40×10^6 cells/ml (^{51}Cr-labeled; 30 min, 100 μCi/ml) were incubated with Ly antiserum diluted in phosphate-buffered saline with 5% fetal calf serum (PBS-FCS) for 1/2 h at 37°C, washed once, brought up in 1 ml of freshly thawed rabbit serum (diluted 1:8 in PBS), and incubated for a further 1/2 h at 37°C. Rabbit sera selected for C were preabsorbed with mouse cells in the presence of EDTA (see 10), which reduced background cytotoxicity for spleen and LNC, under the conditions described above, to <12%.

Sequential Lysis with Different Ly Antisera and C. The proportions of cells displaying one or more Ly antigens were estimated from the lytic effects resulting from sequential exposure to two different Ly antisera; the protocol using B6 LNC as target cells is illustrated (with controls) as follows:

Step 1 (+ C)	Cell lysis*	Step 2§ (+ C)	Lysis of remaining population
	%		%
(a) Anti-Ly-2.2	33	Anti-Ly-1.2	21
(b) NMS	Standard‡	Anti-Ly-1.2	51
(c) Anti-Ly-1.2	52	Anti-Ly-1.2	0
(d) Anti-Ly-2.2	33	Anti-Ly-2.2	0
(e) NMS	Standard‡	NMS	Standard‡

* $\dfrac{\text{cpm (antiserum)} - \text{cpm (NMS)}}{\text{cpm (freeze-thaw)} - \text{cpm (NMS)}} \times 100.$

‡ $\dfrac{\text{cpm (NMS)}}{\text{cpm (freeze-thaw)}} \times 100$ (= 9% in both Steps 1 and 2 in this particular test).

§ Cells from Step 1 were spun down and resuspended in fresh PBS-FCS before Step 2.

---

[1] *Abbreviations used in this paper:* BALB, BALB/c; B6, C57BL/6; C, complement; LNC, lymph node cells; MHC, major histocompatibility complex; MIg, mouse immunoglobulin; MLC, mixed lymphocyte culture; NMS, normal mouse serum; PBS-FCS, phosphate-buffered saline + 5% fetal calf serum; PFC, plaque-forming cells.

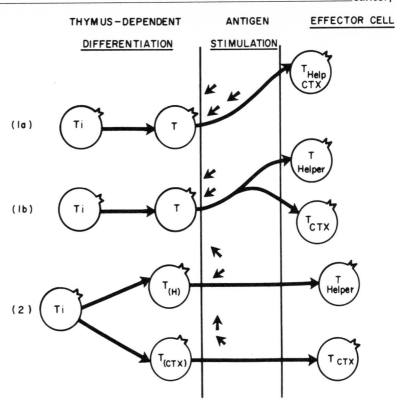

THYMUS—DEPENDENT DIFFERENTIATION   ANTIGEN STIMULATION   EFFECTOR CELL

FIG. 1. Alternative maturation schemes to account for diversity of T-cell function. (1a) After stimulation by antigen, a single T cell mediates all T-cell function. (1b) After stimulation by antigen, a single T cell generates functionally different progeny. (2) Subclasses of T cells are programmed for different T-cell functions during differentiation *before* contact with antigen. (T_i = immature T cell).

This analysis is valid because maximum killing with each Ly antiserum is obtained in one incubation, see (*c*) and (*d*) above, and because pretreatment with NMS + C does not alter susceptibility to subsequent lysis with anti-Ly + C, compare (*b*) Step 2 with (*c*) Step 1. Therefore, all such tests included controls (*b*) to (*e*).

Use of Ly Congenic Mice for Confirming Specificity of Ly Antisera. The specificity of any effects of a given Ly antiserum + C upon the immunologic function of B6 cells was confirmed by examining the effects of the same antiserum on cells from B6 congenic mice which are genetically identical to B6 except for the Ly locus in question (negative controls) as follows: (*a*) anti-Ly-1.2 tested on B6/Ly-1.1 cells, (*b*) anti-Ly-2.2 on B6/Ly-2.1, and (*c*) anti-Ly-3.2 on B6/Ly-2.1,3.1. If there is no effect upon the immune function of these control cells, then the effect of the Ly antiserum on B6 cells must be specific.

Purification of T Lymphocytes. Purification of T cells by elution from Sephadex G200 columns coated with rabbit antimouse immunoglobulin (MIg) according to Schlossman and Hudson (11) permitted virtually 100% T-cell recovery from spleen or LNC; contamination by Ig⁺ cells was approximately

2–5%. In some experiments, T cells were enriched using nylon wool columns according to Julius et al. (12), resulting in a recovery of 60–80% of Thy-1$^+$ cells, and a 5–15% contamination by Ig$^+$ cells; despite this incomplete T-cell recovery, cells purified by nylon wool and rabbit anti-MIg columns had similar functional properties.

Assays of Immunologic Function. (*a*) Sensitization of T lymphocytes and measurement of cytotoxic activity in a 4-h ^{51}Cr release assay is described elsewhere (2). Briefly, 5×10^5 responder cells and 3×10^5 irradiated (2,500 R) BALB or (B6 × BALB)F$_1$ stimulator cells were incubated at 37°C in Falcon 3040 plates for 5 days in 7% CO_2. Cytotoxic activity for each sensitized cell population was calculated thus:

$$\% \text{ Lysis} = \frac{\text{cpm (sensitized cells)} - \text{cpm (unsensitized cells)}}{\text{cpm (freeze-thaw)}} \times 100.$$

(*b*) The ability of cell populations to generate cytotoxicity in vivo was determined by injecting parental (B6) cells (*H-2^b/H-2^b*) into irradiated (900 R) (B6 × BALB)F$_1$ recipients (*H-2^b/H-2^d*) aged 4–6 wk. 4 days later anti-*H-2^d* cytotoxic activity in spleen was measured against ^{51}Cr-labeled LSTRA target cells in a 4-h assay (2). (*c*) In other experiments, proliferative responses of allogeneic mixed cell cultures were measured by incorporation of [^{3}H]T 72–96 h after initiation. These cultures contained 10% human sera and were pulsed with [^{3}H]T according to Nabholz et al. (13). (*d*) Helper activity after treatment with Ly antisera + C was assessed as follows: cell populations passed through rabbit anti-MIg or nylon columns were treated with the different Ly antisera + C and inoculated intravenously into syngeneic irradiated hosts (15×10^6 viable cells/mouse); this inoculum was combined with 10^7 viable spleen cells (pretreated with anti-Thy-1 + C, as a source of B cells). Recipient spleens were tested for anti-SRBC PFC 6 days later.

In both the in vitro and in vivo test systems described above, the concentration of viable cells remaining after pretreatment with antiserum or NMS + C was equalized for all groups.

Results

I. Expression of Ly Antigens on Thymocytes and Peripheral T Cells as Indicated by the C-Dependent Cytotoxicity Assay. Each Ly antiserum (Ly-1.2,Ly-2.2, or Ly-3.2) lysed about 90% of B6 thymocytes, but only 35-60% of cortisone-resistant thymocytes (Fig. 2). One explanation is that although the majority of thymocytes [corresponding to the TL$^+$ phase population (9)] display all three Ly antigens, many cortisone-resistant thymocytes do not. Anti-Thy-1.2 lysed approximately 40% of spleen cells, Ly-1 antiserum 30%, Ly-2 or Ly-3 antiserum 15-20%, and all three Ly antisera together around 35% (Fig. 3 A). The respective figures for spleen suspensions enriched for T cells by passage through a column coated with rabbit anti-MIg were 85% (Thy-1), 65% (Ly-1), 35-45% (Ly-2 or Ly-3), and 80% (Ly-1,2 and 3) (Fig. 3 B).

These data: (*a*) show that the relative proportions of cells expressing different Ly antigens in spleen are similar to those of the cortisone-resistant thymocyte

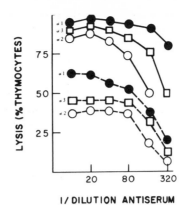

Fig. 2. Complement-dependent lysis of thymocytes by Ly antisera. Lysis of thymocytes from untreated B6 mice (—) and from B6 mice treated with cortisone (2.5 mg cortisone acetate i.p. 48 h previously) (---), by anti-Ly-1.2 (●), anti-Ly-2.2 (○) or anti-Ly-3.2 (□). (For calculations of cells lysed %, see Materials and Methods)

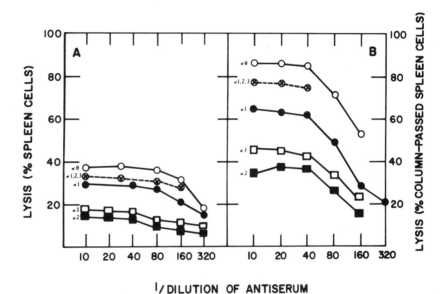

Fig. 3. Complement-dependent lysis of spleen cells by Ly antisera. Lytic activity of anti Thy-1.2 (○), anti Ly-1.2 (●), anti Ly-2.2 (■) and anti Ly-3.2 (□) against pooled B6 spleen cells (A) and B6 spleen cells passed through a rabbit anti-MIg coated column (B). Lysis by all 3 Ly antisera combined is also shown (⊗) (starting dilution of 1/10 for each of the three Ly antisera in the mixture).

population (Fig. 2), and (b) imply that the Ly antisera define subpopulations of Thy-1⁺ lymphocytes in peripheral lymphoid tissues.

II. Estimation of the Proportions of Spleen and LNC Marked by Different Ly Antigens. To estimate the proportions of peripheral cells that displayed different combinations of Ly antigens, the lytic activity of each Ly antiserum was assessed against populations of spleen and LNC that had been pretreated with either NMS or another Ly antiserum in the presence of C (see Materials and

Methods). The results (Table I) indicate that approximately 30% of Thy-1$^+$ cells in peripheral lymphoid tissues display predominantly Ly-1, approximately 7% Ly-23, and approximately 50% all three Ly antigens. These experiments also show that cells resistant to anti-Thy-1.2 serum are not susceptible to subsequent exposure to anti Ly sera; *i.e.*, Ly expression is confined to Thy-1$^+$ cells.

III. Ontogeny of Subclasses of Ly-Bearing Cells in Peripheral Lymphoid Tissues, and their Dependence on the Thymus in Adult Life.   To study the ontogeny of the three subclasses of Ly$^+$ cells defined above (Ly-123$^+$, Ly-1$^+$, Ly-23$^+$), spleen cells from mice of different ages were examined by sequential lysis with different Ly antisera. Virtually all *Ly$^+$* spleen cells 1 wk after birth were Ly-123$^+$ (Fig. 4). Ly-1$^+$ and Ly-23$^+$ cells, although undetectable in neonatal life,

TABLE I

Proportions of Spleen and LN Lymphocytes Bearing One or More
Ly Antigens, Calculated from Results of Sequential Lysis with
Different Ly Antisera*

Ly phenotypes inferred	% of total spleen + LN population		
	Exp. 1	Exp. 2	Exp. 3
For Thy-1$^-$ Cells:			
Ly$^+$	0	1	0
	% of Thy-1$^+$ population		
For Thy-1$^+$ cells:			
Ly-1$^+$2$^+$3$^+$	51	56	49
Ly-1$^+$	32	30	33
Ly-2$^+$3$^+$	6	8	7
Ly-2$^+$	0	0	0
Ly-3$^+$	0	2	1

* See Materials and Methods for details of procedure.

gradually increased in numbers so that by the 10th week of life they together accounted for roughly 20% of the total spleen cell population.

The effect of removing the thymus in adult life upon the concentration of these Ly subclasses in the spleen was then examined. Spleen cells obtained 3 wk after thymectomy or sham thymectomy (performed on 7-wk old mice) were sequentially treated with Ly antisera. These experiments indicate that adult thymectomy resulted in approximately a 50% decrease in the proportion of Ly-123$^+$ cells and a slight increase in the relative proportions of Ly-1$^+$ and Ly-23$^+$ cells as compared with sham-operated controls (Fig. 5).

IV. The Effect of Pretreatment with Ly Antisera and C upon the Capacity of Lymphoid Cells to Develop Helper and Cytotoxic Activities in Irradiated Hosts.   Nylon-column-passed spleen and LNC (pooled) from B6 mice were treated with Ly antisera + C and resuspended to give a standard concentration of viable cells. Half the cells from each group were combined with anti-Thy-

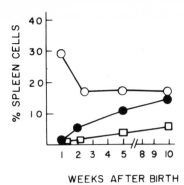

FIG. 4. Ontogeny of subclasses of Ly⁺ cells in spleen. Proportions of Ly-123⁺ (O), Ly-1⁺ (●) and Ly-23⁺ (□) cells at various times after birth, calculated by sequential lysis with the different Ly antisera + C (see Materials and Methods, and Table I).

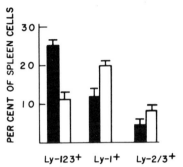

FIG. 5. Effect of adult thymectomy on different subclasses of Ly⁺ lymphocytes. Proportions of Ly-123⁺, Ly-1⁺, and Ly-23⁺ cells in spleen of thymectomized (□) and sham-operated (■) mice. (Spleen cells taken from 10-wk old B6 mice thymectomized, or sham-operated, 3 wk previously.)

1.2-treated B6 spleen cells (source of B cells) and inoculated into B6 (800 R) recipients together with 7×10^6 SRBC. The other half was inoculated i.v. into F_1 hybrid (B6 × BALB) irradiated recipients. After 5 days, spleen cells from the former were assessed for anti-SRBC PFC, and spleen cells from the F_1 recipients for anti H-2^d cytotoxic activity. Removal of Ly-1⁺ cells abolished the subsequent development of PFC activity in irradiated hosts (Table II) but had little effect upon the generation of cytotoxic cells (Fig. 6). By contrast, removal of Ly-2⁺ or Ly-3⁺ cells almost eliminated the generation of cytotoxic cells (Fig. 6) but did not reduce the PFC response (Table II). In control tests, treatment of T cells from congenic B6/Ly-1.1 mice (see Materials and Methods) with anti-Ly-1.2 serum, or cells from B6/Ly-2.1 mice with anti-Ly-2.2 had no effect upon subsequent immune function, indicating the specificity of the results observed for the *Ly* systems named.

V. The Effect of Pretreatment with Ly Antiserum Upon the Generation of Cytotoxic Activity by Lymph Node Cells In Vitro. To analyze further the cellular basis of the production of cytotoxic activity, T cells were sensitized to alloantigens in vitro. B6 cells treated with anti-Ly-2.2 or anti-Ly-3.2 (and thus enriched for Ly-1⁺ T cells) were unable to generate appreciable cytotoxic activity as a result of sensitization in vitro (Fig. 7), thus confirming the effects of these antisera in the circumstances of sensitization in vivo (Section IV above). Pretreatment with anti-Ly-1.2, thus enriching for Ly-23⁺ cells, produced an

TABLE II

The Effect of Pretreatment with Ly Antiserum and C Upon the Capacity of B6 Lymphoid Cells to Develop Helper Function in a Primary Antibody Response In Vivo

B6 cells treated with:	B cells (× 10⁷)*	PFC/spleen (direct/developed)	
		Mean	Range
NMS	+	1,450/2,160	866–2,450/950–2,840
[No cells]	+	11/0	0–30/0
Anti-Thy-1.2	+	42/10	0–95/0–20
Anti-Ly-1.2	+	188/35	20–305/0–65
Anti-Ly-2.2	+	1,890/2,940	1,050–3,920/2,400–3,960
Anti-Ly-3.2	+	1,850/2,360	1,260–2,940/1,960–2,760
NMS	−	25/0	—
Anti-Ly-1.2 (vs B6/Ly-1.1 cells = Ly specificity control)	+	1,050/1,450‡	—

* Obtained by treatment of B6 spleen cells with anti-Thy-1.2 + C.

‡ This control was performed in 2 of the 5 experiments presented, and did not differ significantly from NMS-treated B6 cells in the same experiments.

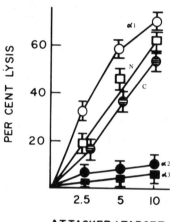

FIG. 6.   Effect of treatment with Ly antiserum and C upon the capacity of parental LN cells to generate killer activity in F₁ hybrid recipients. After treatment of B6 LN cells with either NMS or each of the Ly antisera, equal numbers of viable cells were injected intravenously into (B6 × BALB)F₁ irradiated (900 R) hosts were harvested and assayed for killer activity against LSTRA (*H-2ᵈ*) target cells. The effects of pretreating the B6 donor cells, before in vivo sensitization, with: anti-Ly-1.2 (O), anti-Ly-2.2 (●), anti-Ly-3.2 (■) or NMS (□) are shown. The effect of anti-Ly-2.2 on the ability of cells from B6/Ly-2.1 donors, to generate a cytotoxic response is also shown as a specificity control (⊜) (See Materials and Methods for details). Vertical bars denote the limits of one standard error, based upon triplicate cytotoxic measurements.

increase in the generation of cytotoxic activity by comparison with: (*a*) the activity generated by B6 cells pretreated with NMS, or (*b*) the activity of B6/Ly-1.1 congenic cells pretreated with anti-Ly-1.2 (Ly specificity control). To determine whether the Ly-23+ phenotype of the precursor of the killer cell (as demonstrated by these experiments) is retained by the killer cell which is generated from it, we next examined the effects of treating with Ly antiserum

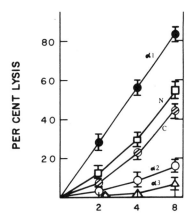

FIG. 7. Effect of pretreatment with Ly antisera upon the capacity of T cells to generate cytotoxic activity to alloantigens in vitro. Nylon passed lymph node cells from B6 mice were treated with Ly antisera, or NMS; equal numbers of remaining viable cells were sensitized to irradiated BALB cells for 5 days in vitro. Cytotoxic responses produced at the end of this sensitization period by cells that had been pretreated with anti-Ly-1.2 (●), anti-Ly-2.2 (○), anti-Ly-3.2 (△) and NMS (□) are shown. The cytotoxic response generated by B6/Ly-2.1 cells treated with anti-Ly-2.2 (Ly specificity control) is also indicated (⊖). Although not shown, pretreatment of lymph node cells with anti-Thy-1.2 + C abolished the ability of these cells to generate a cytotoxic response. Vertical bars denote the limits of one standard error, based upon triplicate cytotoxic measurements.

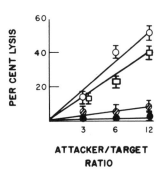

FIG. 8. The effect of Ly antisera + C on killer cell activity. B6 cells that had been sensitized in vitro to irradiated BALB cells were treated with Ly antisera + C, adjusted to equal numbers of viable cells and tested for lytic activity against ^{51}Cr-labeled LSTRA cells. The lytic activity of cells after treatment with anti-Ly-1.2 (○), anti-Ly-2.2 (⊖), anti-Ly-3.2 (●), and NMS + C (□) is shown. Vertical bars denote limits of one standard error.

after sensitization of B6 cells in vitro (i.e., treatment of preformed killer cells) (Fig. 8). Between 70 and 90% of cytotoxic activity was eliminated by anti-Ly-2.2 or anti-Ly-3.2; none by anti-Ly-1.2. Thus, the Ly phenotype of the prekiller cell is stable, and also distinguishes its killer progeny.

VI. The Effects of Pretreatment with Ly Antiserum on the Proliferative Response of Lymphoid Cells to Allogeneic Lymphocytes (Table III). The findings above signify that cytotoxic effector cells originate from Ly-23$^+$ T cells, but do not indicate the relative contributions of the different Ly$^+$ subclasses to the proliferative or recognition phase of the response to alloantigens. For this

TABLE III

Recognition of Different Alloantigens by Subpopulations of T Cells Distinguished by Different Ly Antigens

Antiserum pretreatment*	Incompatibility‡					
	H-2 + non-*H-2*		*H-2*		*I* + *S*	
	E-B	E/B	E-B	E/B	E-B	E/B
NMS	8249	13.31	5310	8.10	6936	10.97
Anti Thy-1.2	331	1.10	321	1.34	129	1.07
Anti Ly-1.2	4931	6.79	3960	4.41	2424	3.08
Anti Ly-2.2	5344	6.88	3485	3.60	11591	12.06
Anti Ly-3.2	5150	6.91	4360	4.95	12116	13.81

	Percent reduction in MLC activity§		
Anti Thy-1.2	96	94	98
Anti Ly-1.2	40	25	65
Anti Ly-2.2	35	34	0 (+67)
Anti Ly-3.2	38	18	0 (+77)

* Following exposure to antiserum + C, cells in each group were adjusted to the same final concentration of viable cells.

‡ *H-2* + non-*H-2* = B10 + BALB (2,500 R); *H-2* = B10+ B10.D2 (2,500 R); *I* + *S* = B10.T(6R) + AQR (2,500 R); E (experimental) = cpm (responder + stimulator: 2,500 R); B (background) = cpm (responder + responder: 2,500 R).

§ Percent reduction in MLC activity = $\dfrac{\text{E-B (anti-Ly + C)}}{\text{E-B (NMS + C)}}$

purpose, we assessed the effects of pretreatment with Ly antisera upon MLC responses. Pretreatment of responder cells with each Ly antiserum reduced subsequent proliferation in MLC by about 50%, suggesting that although Ly-1$^+$ cells did not directly contribute to the production of the cytotoxic response (Section V above), they nonetheless constitute a substantial proportion of the cells that are stimulated to proliferate by alloantigens.

To investigate the possibility that Ly-1$^+$ cells were responding to alloantigens controlled by genes outside the major histocompatibility complex (MHC), which do not normally elicit a cytotoxic response, we tested the effect of pretreatment with Ly antisera upon the MLC response of B10 cells stimulated with irradiated B10.D2 congenic cells. Again pretreatment with anti-Ly-1.2 reduced the MLC response by about 50-60% and so did anti-Ly-2.2 and anti-Ly-3.2. Thus, both Ly-1$^+$ and Ly-23$^+$ cells contribute to the MLC response to antigens of the MHC. Finally, the participation of Ly subclasses in the proliferative response to antigens coded by the $I + S$ region of the MHC is indicated in Table III. Here we tested the effects of pretreatment with Ly antisera upon the MLC response of B10.T(6R) cells stimulated with irradiated AQR cells. These data show that Ly-1$^+$ T cells accounted for almost all the proliferative response to I region ($+ S$) determinants; in fact, pretreatment with anti-Ly-2 or anti-Ly-3 resulted in substantially increased proliferative responses to these antigens.

Discussion

Antigen-stimulated T cells have various immune functions, including secretion of pharmacologic mediators influencing cellular responses (14), amplification or suppression of antibody responses (3, 4), and generation of cells capable of specifically destroying target cells of contact. There is increasing evidence that, after antigenic stimulation has occurred, these individual functions may be performed by T lymphocytes that have different physical and biologic properties (15, 16), and are distinguished by different *Ly* markers (10). But there has so far been no direct evidence as to whether the cells of a single clone develop diverse functions after contact with antigen (Fig. 1; Model 1B) or whether T-cell differentiation involves the evolution of subclasses of functionally different T cells, independently of triggering by antigen (Fig. 1; Model 2).

This is a question of fundamental relevance to the physiology of the immune system, because it involves a choice between, on the one hand, a true differentiative process for which a corresponding regulatory mechanism must be proposed and sought, and on the other hand, a quite different mechanism that would depend on fortuitous antigenic stimulation and so would be excluded from the category of physiological internally regulated differentiative events.

According to our data, for at least two T-cell functions, generation of killer activity and helper activity, respectively, functional commitment is manifest before contact with antigen: (*a*) According to the ^{51}Cr release cytotoxicity assay with Ly antisera, only 50% of Thy-1$^+$ spleen cells are Ly-123$^+$, the rest being Ly-1$^+$ or Ly-23$^+$; (*b*) Lymphoid populations depleted of Ly-1$^+$ T cells (and thus enriched for Ly-23$^+$ cells) are unable to generate appreciable helper responses but their capacity to generate cytotoxic activity in vivo and in vitro is unimpaired; (*c*) Lymphoid cell populations depleted of Ly-23$^+$ cells (and consequently enriched for Ly-1$^+$ cells) produced substantial helper responses but their ability to generate cytotoxic activity in vitro and in vivo was greatly reduced.

Thus, T cells have already been instructed to express exclusively helper function or cytotoxic function *before* they encounter antigen. We have not investigated whether the different Ly components on the T-cell surface contribute directly to the expression of these functions, though Shiku et al. give evidence against this as far as cytotoxic effector function is concerned (10), or whether they perform some other function peculiar to cells that are destined to undertake these activities upon contact with antigen.

In MLC, the Ly-23$^+$ T-cell subclass that gives rise to killer cells responds to MHC antigens coded by the *H-2K* and *H-2D* regions, whereas the MLC response to *I*-region antigens is confined to the Ly-1$^+$ (helper) subclass. This is consistent with reports indicating that cytotoxicity can be generated in response to *H-2K* and *H-2D* region antigens (13), and although total lytic activity may be augmented by additional *I*-region differences (17) the latter are not thought to suffice for triggering cytotoxicity in the systems currently used, and probably do not contribute to the specificity of cytotoxic effector cells (17).

Recognition of *I*-region products (including Ia) by the Ly-1$^+$ helper subclass is also relevant to evidence that these determinants: (*a*) are found mainly on B cells (18), (*b*) are closely associated with or identical to B-cell Fc receptors (19), and (*c*) may play a role in T-B interactions (20). Possibly the proliferative response of Ly-1$^+$ helper cells to Ia antigens on allogeneic B cells may pre-empt their capacity to exert helper activity (20). Recognition of antigen associated with syngeneic Ia molecules on the surface of B cells or macrophages may by contrast favor expression of helper function.

Our data imply that TL$^+$Ly-123$^+$ cells generate three (TL$^-$) subclasses of T cells, denoted Ly-1$^+$, Ly-23$^+$ and Ly-123$^+$, as a normal differentiative process that is independent of exposure to antigen. Whether the Ly-1$^+$ and Ly-23$^+$ subclasses are generated from intermediary TL$^-$Ly-123$^+$ precursors has yet to be decided. Possibly relevant to this question are the findings that: (*a*) all Thy-1$^+$ cells in peripheral tissues in the 1st week of life are Ly-123$^+$, but the proportion of this cell type declines with time in favor of Ly-1$^+$ and Ly-23$^+$ cells, and (*b*) the selective reduction of Ly-123$^+$ cells seen shortly after adult thymectomy may suggest that these cells are transitional peripheral derivatives of TL$^+$Ly-123$^+$ thymocytes. If the Ly-123$^+$ subclass is transitional, the further differentiation of any antigen-specific clone to yield Ly-1$^+$ or Ly-23$^+$ progeny may be regulated by genes within the MHC, i.e., *Ir* genes. Alternatively, the Ly-123$^+$ subclass may be a separately differentiated regulatory population, perhaps capable of exerting immunosuppressive effects after stimulation with antigen. In this latter case, the relative proportions of Ly-123$^+$ and Ly-1$^+$ cells with specificity for a given antigen would determine the net helper effect to that antigen. Isolation of Ly-123$^+$ cells from peripheral tissues may yield definitive evidence for either of these developmental pathways and should permit functional studies of this T-cell subclass.

We have shown that both the prekiller cell and killer cell express the Ly-23 phenotype. Whether Ly-1$^+$ cells, which also recognize certain alloantigens according to the MLC criterion, influence the generation of Ly-23$^+$ killer cells, is dealt with in the following report (21).

Summary

Ly alloantigens coded by two unlinked genetic loci (*Ly-1* and *Ly-2/Ly-3*) are expressed on lymphoid cells undergoing thymus-dependent differentiation. Peripheral Thy-1$^+$ cells from C57BL/6 mice can be divided into three subclasses on the basis of differential expression of Ly-1, Ly-2, and Ly-3; about 50% express all three Ly antigens (Ly-123$^+$), about 33% only Ly-1 (Ly-1$^+$), and about 6–8% Ly-2 and Ly-3 (Ly-23$^+$). Cells of the Ly-123$^+$ subclass are the first peripheral Thy-1$^+$ cells to appear in ontogeny, and are reduced in the periphery shortly after adult thymectomy. In contrast, Ly-1$^+$ and Ly-23$^+$ subclasses appear later in the peripheral tissues than do Ly-123$^+$ cells, and are resistant to the early effects of adult thymectomy.

Peripheral lymphoid populations depleted of Ly-1$^+$ cells and Ly-123$^+$ cells (and thereby enriched for Ly-23$^+$ cells) were incapable of developing significant helper activity to SRBC but generated substantial levels of cytotoxic activity to allogeneic target cells. The same lymphoid populations, depleted of Ly-23$^+$ cells and Ly-123$^+$ cells (and thereby enriched for Ly-1$^+$ cells), produced substantial helper responses but were unable to generate appreciable levels of killer activity.

These experiments imply that commitment of T cells to participate exclusively in either helper or cytotoxic function is a differentiative process that takes place before they encounter antigen, and is accompanied by exclusion of different Ly groups, Ly-23 or Ly-1 respectively, from TL$^+$Ly-123$^+$ T-cell precursors. It is yet to be decided whether the TL$^-$ phase Ly-123$^+$ subclass is a transitional form or a separately differentiated subclass with a discrete immunologic function.

We are grateful for the excellent technical assistance of Ms. Joan Hugenberger and Ms. Linda Gassett, and for the assistance of Ms. Martha Mann and Mr. Austin Segel.

Received for publication 18 December 1974.

References

1. Goldstein, P., H. Wigzell, H. Blomgren, and E. A. J. Svedmeyer. 1974. Cells mediating specific in vitro cytotoxicity. II. Probable autonomy of T lymphocytes for the killing of allogeneic target cells. *J. Exp. Med.* **135**:890.

2. Cantor, H., E. Simpson, V. Sato, G. Fathman, and L. A. Herzenberg. 1975. Characterization of subpopulations of T lymphocytes. Separation and functional studies of peripheral T cells binding different amounts of fluorescent anti Thy-1.2 antibody. *Cell. Immunol.* **15**:180.

3. Transplantation Reviews, Vol. 1. 1969. G. Möller, editor. Williams and Wilkins Co., Baltimore, Md.

4. Gershon, R. K. 1974. T cell suppression. *Contemp. Top. Immunobiol.* **3**:1.

5. Cantor, H. 1972. The effects of anti-theta serum upon graft vs. host activity of spleen and lymph node cells. *Cell. Immunol.* **3**:461.

6. Schlesinger, M. 1972. Antigens of the thymus. *Prog. Allergy.* **16**:214.

7. Boyse, E. A., M. Miyazawa, T. Aoki, and L. J. Old. 1968. Ly-A and Ly-B: two systems of lymphocyte isoantigens in the mouse. *Proc. Roy. Soc. Lond. B. Biol. Sci.* **170**:175.

8. Boyse, E. A., K. Itakura, E. Stockert, C. Iritani, and M. Miura. 1971. Ly-C: a third locus specifying alloantigens expressed only on thymocytes and lymphocytes. *Transplantation.* **11**:351.

9. Itakura, K., J. J. Hutton, E. A. Boyse, and L. J. Old. 1972. Genetic linkage relationships of loci specifying differentiation alloantigens in the mouse. *Transplantation.* **13**:239.

10. Shiku, H., P. Kisielow, M. A. Bean, T. Takahashi, E. A. Boyse, H. F. Oettgen, and L. J. Old. 1975. Expression of T-cell differentiation antigens on effector cells in cell-mediated cytotoxicity in vitro: evidence for functional heterogeneity related to surface phenotype of T cells. *J. Exp. Med.* **141**:227.

11. Schlossman, S. F., and L. Hudson. 1973. Specific purification of lymphocyte populations on a digestible immunoabsorbant. *J. Immunol.* **110**:313.

12. Julius, M., E. Simpson, and L. A. Herzenberg. 1973. A rapid method for the isolation of functional thymus-derived lymphocytes. *Eur. J. Immunol.* **112**:420.

13. Nabholz, I., J. Vives, H. M. Young, T. Meo, V. Miggiano, A. Rijnbeek, and D. C. Shreffler. 1974. Cell mediated cell lysis in vitro: Genetic control of killer cell production and target specificities in the mouse. *Eur. J. Immunol.* **4**:378.

14. David, J. R., and R. A. David. 1972. Cellular hypersensitivity and immunity. Inhibition of macrophage migration and the lymphocyte mediators. *Prog. Allergy* **16**:300.

15. Dennert, G. 1974. Evidence for non-identity of T killer and T helper cells sensitized to allogeneic cells. *Nature (Lond.).* **249**:358.

16. Tigelaar, R. E., and R. M. Gorczynski. 1974. Separable populations of activated thymus-derived lymphocytes identified in two assays for cell-mediated immunity to murine tumor allografts. *J. Exp. Med.* **140**:267.

17. Schendel, D. J., B. J. Alter, and F. H. Bach. 1973. Involvement of LD and SD region differences in MLC and CML in a 3 cell experiment. *Transplant. Proc.* **5**:1651.

18. Shreffler, D. C., and D. S. Chella. 1975. The H-2 major histocompatability complex and the *I* immune response region: genetic variation, function and organization. *Adv. Immunol.* In press.

19. Dickler, H. B., and D. H. Sachs. 1974. Evidence for identity or close association of the Fc receptors of B lymphocytes and alloantigens determined by the Ir region of the *H-2* complex. *J. Exp. Med.* **140**:779.

20. Katz, D. H., M. Graves, M. E. Dorf, H. Dimuzio, and B. Benacerraf. 1975. Cell interactions between histoincompatible T and B lymphocytes. VII. Cooperative responses between lymphocytes are controlled by genes in the *I* region of the *H-2* complex. *J. Exp. Med.* **141**:263.

21. Cantor, H., and Boyse, E. A. 1975. Functional subclasses of T lymphocytes bearing Different Ly antigens. II. Cooperation between subclasses of Ly+ cells in the generation of killer activity. *J. Exp. Med.* **141**:1390.

SEPARATION OF HELPER T CELLS FROM SUPPRESSOR T CELLS EXPRESSING DIFFERENT Ly COMPONENTS

I. Polyclonal Activation: Suppressor and Helper Activities are Inherent Properties of Distinct T-Cell Subclasses*

By J. JANDINSKI, H. CANTOR,‡ T. TADAKUMA,§ D. L. PEAVY,‖ AND C. W. PIERCE¶

(From the Departments of Medicine and Pathology, Harvard Medical School and the Sidney Farber Cancer Center, Boston, Massachusetts 02115)

Peripheral T lymphocytes can be subclassified on the basis of differential expression of Ly components on their surfaces (1–3). Approximately 50% of peripheral T cells manifest all three Ly components analyzed so far (phenotype Ly123), about 33% only Ly1 (phenotype Ly1), and about 5–10% Ly2 and Ly3 (phenotype Ly23). Functional studies indicate that Ly1 T cells can generate helper function during an adoptive primary antibody response, but do not generate appreciable killer activity to alloantigens, whereas the reverse is true of Ly23 T cells. In this report we examine the participation of T-cell subclasses of different Ly phenotypes from the standpoint of their regulatory or suppressive functions (4).

Our approach to this question involves the use of a polyclonal activator, concanavalin A (Con A).[1] Con A stimulates T cells to undergo proliferative and differentiative events similar to those which T cells undergo when they encounter specific antigen (5, 6), and Con A-activated T cells can under appropriate circumstances perform helper, suppressor, and killer functions (5, 6). We analyze here whether each of these functions is confined to a distinct subclass of T cells.

Materials and Methods

Animals. C57BL/6 (B6) mice 10–14 wk of age were obtained from The Jackson Laboratory, Bar Harbor, Maine. The congenic lines B6-*Ly-1*[a] and B6-*Ly-2*[a] (see Klein, reference 7), phenotypes Ly-1.1, 2.2,3.2 and Ly-1.2,2.1,3.2, respectively, were produced and supplied by E. A. Boyse, Memorial Sloan-Kettering Cancer Center, New York.

_____

* This investigation was supported by U. S. Public Health Service Research Grants CA-14723, AI-09897, and AI-12184 from the National Institutes of Health, Bethesda, Md.

‡ Scholar of the Leukemia Society of America.

§ Supported by Keio University School of Medicine, Tokyo, Japan.

‖ Recipient of U. S. Public Health Service Research Fellowship CA-00566 from the National Institutes of Health.

¶ Recipient of U. S. Public Health Service Research Career Development Award AI-70173 from the National Institutes of Health.

[1] *Abbreviations used in this paper:* α, anti; Con A, concanavalin A; B6, C57BL/6; NMS, normal mouse serum; PBS-FBS, phosphate-buffered saline plus 10% fetal bovine serum; PFC, plaque-forming cells; SIRS, soluble immune response suppressor; [³H]TdR, tritiated thymidine.

Antisera. For details of the preparation and use of antisera to Ly-1.2, Ly-2.2, and Thy-1.2 see Shen et al., (8) which also provides a bibliography of the Ly systems.

Isolation of Ly Subclasses. 20–50 × 10⁶ cells/ml were incubated with Ly antiserum at a final dilution of 1:40 in phosphate-buffered saline plus 10% fetal bovine serum (PBS-FBS) for 1/2 h at 20°C. After washing once, the cells were brought up in 1 ml of freshly thawed selected rabbit serum (diluted 1:12 in PBS) as the source of complement (C), and were incubated for another 1/2 h at 37°C. Ly specificity was confirmed for each group of experiments by substituting cells from congenic B6-_Ly-1_ᵃ and _2_ᵃ donors (negative controls) as described previously (1, 3). In each case, anti(α)-Ly activity was absorbed from Ly antiserum only by B6 cells and not by cells of the respective B6-Ly congenic lines.

Polyclonal Activation of T Cells with Con A. B6 spleen cells that had been treated with the various Ly sera were incubated at 10⁷ cells/ml in RPMI 1640 or Eagle's minimal essential medium (MEM) containing 10% FBS (lot no. M26302; Reheis Chemical Co., Kankakee, Ill.) alone (control) or with 1 μg/ml Con A (ICN Nutritional Biochemical Div., International Chemical & Nuclear Corp., Cleveland, Ohio) for 48 h. The cells were harvested, washed four times, and added in graded numbers of viable cells to test cultures to assess their influence on the generation of cytotoxic lymphocytes or plaque-forming cells (PFC). In other experiments, the cells were activated with Con A first, and then treated with the various Ly sera plus C, before addition to test cultures.

Con A-activated T cells produce a factor(s), soluble immune response suppressor (SIRS), which nonspecifically suppresses antibody responses in vitro (9). To determine which T-cell subclass produces SIRS, the different Ly subclasses were prepared from spleen and incubated (in numbers corresponding to 10⁷ of the starting spleen cell population) in MEM containing 2% FBS with or without 1 μg/ml Con A for 48 h. Supernatant fluids were harvested, absorbed with Sephadex G50 to remove residual Con A, sterilized by membrane filtration, and added (at a final dilution of 1:40) to test cultures to assay activity on PFC responses (9).

To assess the proliferative responses of T-cell subclasses to Con A, spleen cells or nylon wool-enriched T cells (1, 2), after treatment with the various Ly sera or NMS, plus C, were incubated at 5 × 10⁶ or 10⁷ viable cells/ml in RPMI 1640 containing 10% FBS with 0, 1, and 2 μg/ml Con A. At 60 h, 1 μCi tritiated thymidine ([³H]TdR) was added, at 72 h the cultures were terminated and [³H]TdR incorporation was assessed.

Test Cultures for Assay of Activity of Con A-Activated T Cells and Their Products. In most experiments, the activities of Con A-activated cells were assayed simultaneously in two laboratories. A modified Mishell-Dutton culture system was used to generate αSRBC PFC (10, 11): 4–8 × 10⁶ lymphocytes were incubated in 16-mm wells (dispose-trays; Linbro Chemical Co., New Haven, Conn.) with 3 × 10⁶ SRBC in 1 ml of RPMI-1640 containing 10% FBS and 2 × 10⁻⁵ 2-mercaptoethanol. Cultures were rocked (8–12 times/min) at 37°C in an atmosphere of 8% CO_2, and fed daily with a nutrient mixture (11) for 5 days, harvested, and assayed for αSRBC PFC activity. PFC responses were measured using either the Cunningham assay or the slide modification of the Jerne localized-hemolysis-in-gel assay (10) and are expressed as PFC per culture. Cytotoxic responses generated during mixed lymphocyte reactions were measured in ⁵¹Cr-release assays as described previously (1). Cytotoxic activity was calculated as follows:

$$\% \text{ Cytotoxicity} = \frac{\text{cpm (sensitized cells)} - \text{cpm (unsensitized cells)}}{\text{cpm (freeze-thaw control)}} \times 100.$$

Ly Subclass Notation. Because there is so far no evidence that Ly2 and Ly3 can be expressed independently of one another, i.e. no cells of Ly-2⁻3⁺ or Ly-2⁺3⁻ phenotypes have yet been identified (1), there are as yet only three well-defined Ly subclasses: Ly1, Ly23 and Ly123. Therefore in the following account we have used the following notations: (_a_) Ly1 signifies the selected T-cell population obtained by treating spleen cells or T cells with αLy-2 or αLy-3, plus C; (_b_) Ly23 signifies the equivalent T-cell population remaining after treatment with αLy-1 plus C.

Results

I. Proliferative Responses of Separate Ly Subclass Populations to Stimulation by Con A (Table I). Ly1 and Ly23 populations were prepared from (_a_) nylon-enriched, and (_b_) unfractionated, spleen cells of B6 mice, and then stimulated in

TABLE I
Thymidine Incorporation by T-Cell Ly Subclass Populations Exposed to Con A

Starting populations from B6 mice	Treated with C and:	T-cell population derived	Thymidine incorporation after 48 h		
			No Con A	Con A, 1 μg/ml	Con A, 2 μg/ml
			cpm		
Nylon-enriched splenic T cells	NMS	All	4,059	25,372	28,646
	αLy-1.2	Ly23	3,426	21,414	26,631
	αLy-2.2	Ly1	3,081	23,702	29,594
	αLy-3.2	Ly1	3,300	23,573	25,959
Unfractionated spleen cells	NMS	All	2,473	25,573	28,658
	αLy-1.2	Ly23	2,376	18,975	24,438
	αLy-2.2	Ly1	2,384	21,167	21,504

vitro with 1 or with 2 μg Con A/ml. According to [³H]TdR incorporation by equal numbers of cells all these populations responded to Con A to roughly the same extent, and to about the same extent as control spleen cells treated only with normal mouse serum (NMS). Thus the responses of the Ly1 and Ly23 populations to the indicated concentrations of Con A are not markedly different from one another or from the T-cell population as a whole.

II. Suppressor Activities of Separate Ly Subclass Populations Stimulated by Con A (Fig. 1). Having established that the Ly1 and Ly23 populations are equally stimulated by Con A (I above), we asked whether these discrete nonspecifically activated populations would exhibit differences in regard to helper-suppressor capabilities.

For this purpose the Ly1 and Ly23 populations from B6 spleen were exposed to Con A (10^7 viable cells/ml culture containing 1 μg/ml Con A), harvested at 48 h, washed four times, and introduced in graded numbers (viable cell count) into cultures of 5×10^6 sheep erythrocyte (SRBC)-stimulated B6 spleen cells. As few as 10^5 Con A-activated spleen cells that had not been fractionated (treated only with NMS plus C before Con A activation) markedly suppressed the PFC response; 10^5 Con A-activated Ly23 cells gave somewhat more suppression; but Con A-activated Ly1 cells gave no demonstrable suppression, in fact their addition in large numbers to normal spleen cell cultures slightly increased the PFC response. Similar results (not shown) were obtained by substituting αLy-3.2 for αLy-2.2 (i.e. once again the Ly2 and Ly3 phenotypes are concordant). Thus polyclonally activated Ly23 cells, but not Ly1 cells, exhibit suppressor activity in the αSRBC system.

III. Ly Phenotypes of Effector-Suppressor Cells Generated by Con A (Fig. 2). The data above indicate that the resting T-cell population, before stimulation by Con A, already contains Ly23 cells that are programmed for suppression (nature unspecified) of the αSRBC response, as well as Ly1 cells that are not so programmed. To decide whether the Ly23 phenotype remains unchanged during the process of activation by Con A, we conducted tests to ascertain the Ly phenotype of the effector-suppressor cell generated by Con A. This entailed activating the whole spleen cell population with Con A, and then testing the

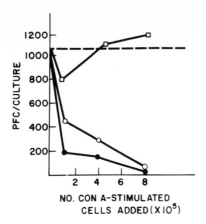

FIG. 1.   Suppressor activity of Ly23 cells activated by Con A. Ly1 and Ly23 populations were isolated from B6 spleen and activated by Con A (see II in Results) and added in graded numbers (abscissa) to fresh spleen cells (5×10^6 per culture; plus 3×10^6 SRBC). The PFC responses were assayed on day 5 (ordinate). (●), Ly23 cells; (□), Ly1 cells; (○), intact spleen population (control; NMS plus C); and (---), no cells added.

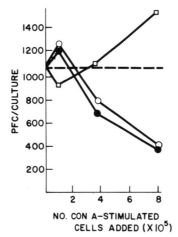

FIG. 2.   Ly23 phenotype of effector-suppressor cells. The test system is the same as that shown in Fig. 1, except that the Ly subclass populations were separated after the intact spleen cell population had been activated by Con A. (●), Ly23 cells; (□), Ly1 cells; (○), intact spleen population (control; NMS plus C); and (---), no cells added.

suppressor capabilities of Ly subclass populations separated from the already activated population.

Again suppression was effected by Ly23 cells but not by Ly1 cells. Thus in the process whereby effector-suppressor cells are generated from antecedent cells already committed to suppressor function, there is no demonstrable change in Ly phenotype.

IV. Elaboration of SIRS by Ly⁺ T-Cell Subclasses (Table II).   Supernatant fluids from Con A-activated spleen cells contain a factor (or factors), SIRS, made by T cells, that inhibits primary αSRBC PFC responses in vitro (9). We there-

TABLE II

Suppressive Effects of Supernates Generated from Treated Spleen Cells

Supernate of cultured spleen population:	PFC/culture	
	IgM	IgG
A. Intact	3,700	770
B. Intact (NMS)	3,780	810
Intact (NMS) + Con A	390	170
C. Ly23	1,490	540
Ly23 + Con A	500	140
D. Ly1	4,090	1,040
Ly1 + Con A	3,340	1,230
E. C + D	3,180	880
C + D + Con A	630	170

Supernates were obtained after 24 h from cultures containing 10^7 viable spleen cells, either intact or after the usual elimination of Ly1 cells (group C) or Ly23 cells (group D). The data show the effects of these supernates at a final dilution of 1:40, upon a primary in vitro αSRBC response.

fore asked whether SIRS was produced by the same subclass of T cells (Ly23) that suppressed the SRBC response in the previous experiments (II and III above). For this purpose Ly1 and Ly23 populations were prepared from B6 spleen cells, at a starting concentration of 10^7/ml, and incubated, without readjustment of viable cell concentration, with or without 1 μg/ml of Con A for 48 h. Supernatant fluids were harvested, absorbed with Sephadex, and tested for SIRS activity in SRBC-stimulated B6 spleen cell cultures. These data indicate that SIRS is produced by cells of the Ly23 T-cell subclass, the same subclass of T cells responsible for suppressor T-cell activity (II and III above).

V. Immunologic Activity of Ly1 Cells Activated by Con A (Fig. 3). The experiments described above show that Ly23 cells, but not Ly1 cells, mediate suppressor activity after polyclonal activation with Con A. We next tested whether Ly1 cells might develop helper activity after polyclonal activation. Ly subclasses were prepared from Con A-activated B6 spleen cells in the usual way. These were added to cultures of B lymphocytes (spleen population treated with αThy-1 plus C) and stimulated with SRBC. Although 1.5×10^6 of the intact spleen cell population incubated for 48 h without Con A expressed significant helper activity (group B), the same number of Con A-activated spleen cells did not (group C). Thus either helper activity was lacking in the intact Con A-activated spleen population, or it was masked by Ly23 suppressor cells. Groups D and E give the answer: Removal of Ly23 cells (group E) reveals helper activity of the intact spleen population. This helper activity was substantially greater than the activity of unstimulated spleen cells (group B). Thus the helper activity of Con A-activated Ly1 cells is masked in the intact population by suppressor Ly23 cells.

VI. The Effect of Con A-Activated Ly Subclasses on the Generation of Killer

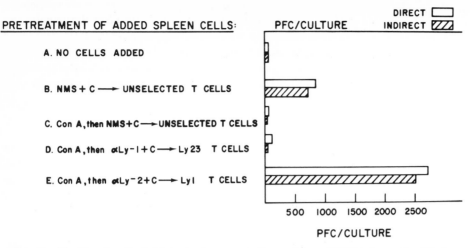

FIG. 3. The Con A-activated intact spleen population contains αSRBC helper cells of Ly1 phenotype. 1.5×10^6 remaining viable cells, after 2 days of incubation with or without 1 μg Con A, were added to cultures of 5×10^6 B cells (spleen cells treated with αThy-1 plus C) and 3×10^6 SRBC; assay for αSRBC on day 5.

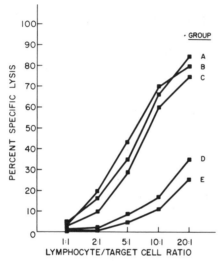

FIG. 4. Suppression of the generation of cytotoxic lymphocytes by Con A-activated Ly23 cells. Spleen cells were incubated for 48 h with 1 μg/ml Con A, and then treated with αLy-2 or αLy-1 as usual to yield the Ly1 and Ly23 subclass populations. 1.5×10^6 viable cells of each population were then added to fresh cultures containing 25×10^6 B6 (H-2^b) spleen cells plus 5×10^6 mitomycin C-treated BALB/c (H-2^d) cells. The figure shows cytotoxicity against ^{51}Cr-labeled P815 (H-2^d) target cells on day 5. Key: (A), no cells added; (B), 0.1 ml medium added; (C), Ly1 cells added; (D) Ly23 added; and (E), NMS-treated cells added.

Cells in vitro (Fig. 4).   Con A-activated T cells have been shown to suppress the generation of cytotoxic lymphocytes in vitro (6). To determine the Ly phenotype of T cells that might suppress this cell-mediated immune response, Ly1 and Ly23 populations were prepared as usual from Con A-activated B6 spleen cells;

1.5×10^6 viable cells of each type were added to cultures of syngeneic responder cells and allogeneic (BALB/c; *H-2^d*) mitomycin C-treated stimulator cells, as detailed in Fig. 4. Cytotoxic responses assayed on ^{51}Cr-labeled P815 (*H-2^d*) target cells 5 days later show that the intact Con A-activated population (NMS plus C; group E) and the Ly23 subclass (group D) suppressed the generation of cytotoxic effector cells. Thus polyclonal activation of T cells gives rise to Ly23 cells that suppress not only humoral but also cellular immune responses.

Discussion

We have asked whether suppressor and helper functions are mediated by the same or different Ly subclasses separated from T cells that have been activated by Con A. The advantage of this approach is that any functional differences between such subclasses cannot be ascribed to preferential activation of particular Ly-distinctive T-cell subclasses by specific antigen.

The possibility that the suppression we observed reflects carry-over of Con A is extremely unlikely because: (*a*) elimination of Ly23 cells, but not of Ly1 cells, before Con A activation abolishes the development of suppressor T-cell activity; (*b*) inclusion of alpha-methyl-mannoside, a competitive inhibitor of Con A activation, does not influence suppressor activity (5), and (*c*) the amount of Con A actually carried over is two orders of magnitude below that required to suppress the generation of PFC to SRBC in vitro (9).

The findings are that: (*a*) Ly1 and Ly23 cells are equally activated by Con A to incorporate [^{3}H]TdR, and (*b*) Ly1 cells help, and Ly23 cells suppress. This rules out the possibility that suppression by Con A-activated T cells is due to some hypothetical mechanism involving an excess of helper activity, because activated Ly1 cells only help, they do not suppress, even in relatively high concentrations. Thus Ly1 and Ly23 cells are programmed for their respective helper and suppressor functions independently of their ability to discriminate and react to specific antigen. Moreover the same results can be obtained with Ly1 and Ly23 cells separated from the resting T-cell population before exposure to Con A. This confirms previous conclusions that programming for discrete T-cell functions, which is linked with the expression of particular Ly profiles, has already taken place before specific or nonspecific cellular activation (1, 2). Expression of those functions may of course depend upon activation.

Our interpretation of these data is that after polyclonal activation, immunosuppression is confined to a subclass of T cells (T$_S$) distinct from helper T cells (T$_H$). The Ly23 phenotype of T$_S$ cells is the same as that expressed on cytotoxic T cells (T$_C$), although whether these two functions are mediated by the same population of Ly23 cells or by different populations of Ly23 cells that may be distinguishable by future immunogenetic analysis, and whether suppressor T cells act by killing the responding T or B cells, are questions yet to be decided.

The central implication is that even before activation, suppression is an obligatory response of a specialized T-cell subclass. We are now faced with the questions (*a*) whether suppression is also generated from this subclass as a normal consequence of immunization with antigen, and (*b*) whether the suppression generated is specific for the inducing antigens. These points are dealt with in the accompanying report (12).

Summary

Concanavalin A, a nonspecific polyclonal activator of T lymphocytes, activates Ly1 and Ly23 subclasses to the same degree. After activation, the Ly23 subclass, but not the Ly1 subclass, has the following properties: (a) Suppression of the antibody response to sheep erythrocytes (SRBC) in vitro. (b) Production of a soluble factor that suppresses the anti-SRBC response in vitro. (c) Suppression of the generation of cell-mediated cytotoxicity to H-2 target cells in vitro. Con A-activated cells of the Ly1 subclass, but not the Ly23 subclass, express helper function in the anti-SRBC response in vitro.

Because the intact Con A-stimulated T-cell population contains both cell types, these cells do not exert detectable helper effects in an anti-SRBC system in vitro, because the helper effect of Ly1 cells is masked by the suppressor effect of the Ly23 cells. Each function is revealed by eliminating one or the other population with the relevant Ly antiserum.

The resting T-cell population, before activation by Con A, also contains already programmed Ly1 and Ly23 cells with similar helper and suppressor potentials, respectively. This is revealed by experiments with Ly subclasses which have been separated from the resting T-cell population and then stimulated by Con A.

Thus helper and suppressor functions, as expressed in these systems, are manifestations of separate T-cell-differentiative pathways and do not depend upon stimulation of the cells by antigen.

Received for publication 13 February 1976.

References

1. Cantor, H., and E. A. Boyse. 1975. Functional subclasses of T lymphocytes bearing different Ly antigens. I. The generation of functionally distinct T-cell subclasses is a differentiative process independent of antigen. *J. Exp. Med.* **141**:1376.

2. Cantor, H., and E. A. Boyse. 1975. Functional subclasses of T lymphocytes bearing different Ly antigens. II. Cooperation between subclasses of Ly$^+$ cells in the generation of killer activity. *J. Exp. Med.* **141**:1390.

3. Shiku, H., P. Kisielow, M. A. Bean, T. Takahashi, E. A. Boyse, H. F. Oettgen, and L. J. Old. 1975. Expression of T-cell differentiation antigens on effector cells in cell-mediated cytotoxicity in vitro. Evidence for functional heterogeneity related to the surface phenotype of T cells. *J. Exp. Med.* **141**:227.

4. Gershon, R. K. 1974. T cell suppression. *Contemp. Top. Immunobiol.* **3**:1.

5. Dutton, R. W. 1975. Suppressor T cells. *Transplant. Rev.* **26**:in press.

6. Pierce, C. W., and J. A. Kapp. 1975. Regulation of immune responses by suppressor T cells. *Contemp. Top. Immunobiol.* **5**:in press.

7. Klein, J. 1973. List of congenic lines of mice. I. Lines with differences at alloantigen loci. *Transplantation (Baltimore)*. **15**:137.

8. Shen, F. W., E. A. Boyse, and H. Cantor. 1975. Preparation and use of Ly antisera. *Immunogenetics.* **2**:591.

9. Rich, R. R., and C. W. Pierce. 1974. Biological expressions of lymphocyte activation. III. Suppression of plaque-forming cell responses *in vitro* by supernatant fluids from concanavalin A-activated spleen cell cultures. *J. Immunol.* **112**:1360.

10. Pierce. C. W., B. M. Johnson, R. Gershon, and R. Asofsky. 1970. Immune responses in vitro. III. Development of primary γM, γG, and γA plaque-forming cell responses in mouse spleen cell cultures stimulated with heterologous erythrocytes. *J. Exp. Med.* **134**:395.

11. Mishell, R. I., and R. W. Dutton. 1967. Immunization of dissociated spleen cell cultures from normal mice. *J. Exp. Med.* **126**:423.

12. Cantor, H., F. W. Shen, and E. A. Boyse. 1975. Separation of helper T cells from suppressor T cells expressing different Ly components. II. Activation by antigen: after immunization, antigen-specific suppressor and helper activities are mediated by distinct T-cell subclasses. *J. Exp. Med.* **143**:1391.

INDEPENDENT DIFFERENTIATIVE PATHWAYS OF Ly1 AND
Ly23 SUBCLASSES OF T CELLS
Experimental Production of Mice Deprived of Selected
T-Cell Subclasses*

By B. HUBER,‡ H. CANTOR,§ F. W. SHEN, AND E. A. BOYSE

(From the Department of Medicine, Harvard Medical School, Farber Cancer Center, Boston, Massachusetts 02115, and the Memorial Sloan-Kettering Cancer Center, New York 10021)

The T-lymphocyte population comprises several subclasses, each programmed to express a particular set of Ly cell surface components and a particular set of immunologic functions. For example, the Ly1 subclass is programmed for the $Ly1^+Ly23^-$ surface phenotype and for helper function. The Ly23 subclass is programmed for the Lyl^-Ly23^+ phenotype and for killer/suppressor function (1, 2).

How are these two subclasses related to one another? Are they two stages of a single line of differentiation, in which case one of them must at some point give rise to the other? Or are they products of separate pathways of thymus-directed differentiation?

We have approached this question by isolating Ly1 and Ly23 cells and allowing them to populate syngeneic 'B mice', i.e. lethally irradiated thymectomized mice restored with T-cell-deficient bone marrow cells, for up to 6 mo.[1] We refer to these recipients as 'B-Ly1 mice' and 'B-Ly23 mice'. Similarly, we call B mice given unselected T cells 'B-T mice'.

Ly Subclass Notation. Because there is so far no evidence that Ly2 and Ly3 can be expressed independently of one another, i.e. no cells of $Ly2^-Ly3^+$ or $Ly2^+Ly3^-$ phenotypes have yet been identified, there are as yet only three well-defined Ly subclasses: Ly1, Ly23, and Ly123. Therefore in this report we use the following notation: Elimination of $Ly2^+$ cells with αLyt-2 (or αLyt-3) plus complement (C) removes the subclasses Ly123 and Ly23 (both being $Ly2^+$), leaving the subclass called Ly1. Elimination of $Ly1^+$ cells with αLyt-1 plus C removes the subclasses Ly123 and Ly1 (both being $Ly1^+$), leaving the subclass called Ly23. Therefore: (*a*) Ly1 signifies the T cells remaining after a given population has been treated with αLyt-2 (or αLyt-3) plus C. Because Ly1 cells selected with αLy2 have so far been indistinguishable from αLyt-3-selected Ly1 cells in regard to helper and killer criteria, only the results obtained by αLyt-2

* Supported by U. S. Public Health Service Grants AI-12184, AI-13600, CA-08748, and CA-16889.

‡ Recipient of an American Cancer Society-Eleanor Roosevelt Fellowship.

§ Scholar of the Leukemia Society of America.

[1] This was suggested by the work of Howard and Wilson, who segregated populations of antigen-stimulated T cells in syngeneic thymectomized recipients (3).

selection are given in this report. (*b*) Ly23 signifies the T cells remaining after a given population has been treated with αLyt-1 plus C.

This terminology, and that for B mice repopulated with Ly1 or Ly23 cells (see below), will require revision if and when new Ly systems, such as Lyt-4 (formerly Ly5) (see 4), necessitate further subdivision of the Ly subclasses of T cells.

Materials and Methods

All Ly serology conformed to the account by Shen et al. (4). Ly subclasses were enumerated according to Cantor and Boyse (1). This paper and reference (2) also give details of all relevant functional assays together with the respective bibliography. Briefly, these procedures are:

Primary Helper Function. Primary helper function was measured by the ability of spleen cells from the different experimental mice to produce anti-sheep red blood cells (SRBC) plaque-forming cells (PFC) after stimulation in vitro by SRBC.

αSRBC Memory Function. αSRBC memory function was measured by the ability of spleen cells from SRBC-primed donors to (*a*) generate anti-SRBC PFC after in vitro stimulation with trinitrophenyl (TNP)-SRBC or (*b*) generate anti-TNP PFC after in vitro stimulation with TNP-SRBC.

Primary Alloreactive Killer Function. Primary alloreactive killer function was measured by the generation of specific cytotoxic activity after stimulation of B6 cells with irradiated BALB/c cells.

Alloreactive Killer Memory. Alloreactive killer memory was measured by the generation of anti-H-2^d-specific cytotoxicity after in vitro "boosting" of cells from B6 mice which have been skin grafted with BALB/c tail skin 3 wk before sacrifice.

Results

I. *Production and Testing of B-Ly1 and B-Ly23 Mice*

All recipients and donors were C57BL/6 (B6) mice.

STEP 1.   Thymectomy at 6 wk of age.

STEP 2.   Lethal irradiation (750 R; ^{137}Cs source) at 8 wk of age, followed immediately by intravenous restoration with 5×10^6 bone marrow (BM) cells pretreated with αThy-1 plus C.

STEP 3.   4–8 wk later: intravenous administration of 2–10×10^6 nylon-enriched T cells from pooled spleen and lymph node cells. Different groups of recipients were given: (*a*) Ly1 cells, making the B-Ly1 mouse, or (*b*) Ly23 cells, making the B-Ly23 mouse, or (*c*) the starting T-cell population untreated with any αLy serum, making the B-T mouse.

STEP 4.   4–28 wk later the mice were killed and their spleen and LNC tested for their ability to generate cytotoxic effector and antibody-forming cells. To determine whether antigenic stimulation might cause subclass conversion, some B recipients were given 5×10^5 SRBC together with the T-subclass cells, or were grafted with BALB/c tail skin 3 wk before reconstitution with T cells. A summary of our analysis of these mice is given in Table I.

II. *Retention of Ly Subclass Constitution by B-Ly1 and B-Ly23 Mice.*

The question is: do B-Ly1 and B-Ly23 mice remain deprived of T cells of the Ly phenotype that was eliminated from the population used for T-cell reconstitution in step 3? This was ascertained by determining the proportions of $Ly1^+$ and $Ly2^+$ cells serologically as already described (1). To exclude radioresistant host T cells from analysis, B6 Ly1 cells were injected into thymectomized-irradiated

TABLE I

Ly Phenotypes and Immune Functions of T Cells from B Mice Repopulated with Ly1 or Ly23 T-Cell Subclasses*

Parameter:		Experimental mice:			
		B	B-T	B-Ly1	B-Ly23
Ly phenotype					
Ly subclasses	Ly1⁺	<10	85	93	<10
(% Thy1⁺ cells)	Ly2⁺	<10	48	<10	72
Helper function:					
Primary helper	SRBC	<5	100‡	137	15
(% B-T PFC)	HRBC	<5	100§	288	16
SRBC memory	Direct	<5	100‖	136	22
(% B-T PFC)	Indirect	<5	100¶	104	19
TNP memory	Direct	<5	100**	95	20
(% B-T PFC)	Indirect	<5	100‡‡	106	17
Alloreactive killer function:					
Primary killer – spleen	2:1§§	ND	53	4	12
(% lysis)	5:1	ND	75	5	30
	10:1	0	88	8	48
Primary killer – LNC	2:1	ND	50	2	22
(% lysis)	5:1	ND	73	2	38
	10:1	0	81	3	47
Killer memory – spleen	2:1	ND	13	4	3
(% lysis)	5:1	ND	30	10	7
	10:1	0	50	13	11

ND, not done.

* Results expressed as percentages based on combined data. The raw data for the B-T mice, and numbers of mice per group, are given in the following footnotes. See text for time intervals between T-cell reconstitution and testing.

‡ Mean direct PFC/culture = 2,115, (range 1,560–2,250); 5 mice/group.

§ Mean direct PFC/culture = 275, (range 117–450); 4 mice/group.

‖ Mean direct PFC/culture = 1,760, (range 1,550–2,000); 4 mice/group.

¶ Mean indirect PFC/culture = 1,905, (range 1,040–2,800); 4 mice/group.

** Mean direct PFC/culture = 700, (range 480–845); 4 mice/group.

‡‡ Mean indirect PFC/culture = 595, (range 411–676); 4 mice/group.

§§ Ratio of attacker cells to target cells; 3–5 mice/group.

B6/Ly2.1,3.1 recipients, while B6 Ly23 cells were injected into thymectomized-irradiated congenic B6/Ly1.1 recipients.

As late as the 20th day after T-cell reconstitution B-Ly1 and B-Ly23 mice yielded T cells only of the Ly subclass originally administered. During this period, Ly1 cells evidently cannot give rise to Ly123 or Ly23 cells, nor can Ly23 cells give rise to Ly123 or Ly1 cells. Thus, according to the criterion of Ly surface phenotypes the genetic programs of Ly1 and Ly23 sets of T cells are independent and not sequential.

III. *Sustained Functional Restrictions of T Cells From B-Ly1 and B-Ly23 Mice.* We have seen that the T-cell subclass constitution of B-Ly1 and B-Ly23 mice is sustained for at least 20 days (II). The question now is: do the T-cell functions of B-Ly1 and B-Ly23 mice continue to be restricted to those that accord with the functions that have been assigned to the Ly1 and Ly23 subclasses in preceding studies?

This was assessed by testing B-Ly1 and B-Ly23 mice at increasing intervals of time for the presence of cells capable of helper and killer functions in vitro. T help is a characteristic function of Ly1 cells, and killing of Ly23 cells.

In the following account, the abbreviation 'interval' is used in reference to the periods elapsing between the administration of T cells to each B mouse and the subsequent testing of the recipient's T cells in the various assays. The data for each group have been combined in Table I because there was no demonstrable change with time.

The B-Ly1 Mouse

PRIMARY HELPER FUNCTION.   In comparison with the standard B-T mouse, primary helper capacity of the B-Ly1 mouse was augmented in both αSRBC and αHRBC systems. Augmentation can be accounted for by the higher proportion of Ly1 cells entering the assay, which relates to the count of all T cells, or to lack of suppression by the missing $Ly2^+$ subclasses (5) (intervals 15, 18, 28, and 64 days).

HELPER MEMORY FUNCTION.   This is measured by inclusion of 5×10^6 SRBC with the T-cell-reconstituting inoculum and subsequent measurement of αSRBC and αTNP-SRBC activities. By both criteria the helper memory function of the B-Ly1 mouse was at least equal to the B-T mouse standard (intervals 48, 51, and 64 days).

PRIMARY ALLOREACTIVE KILLER FUNCTION.   Both the killer-effector cell and the prokiller cell from which the killer is generated belong to the Ly23 subclass (1, 6). No significant killer response could be elicited from T cells taken from B-Ly1 mice and challenged in vitro (intervals 28, 64, and 160 days).

ALLOREACTIVE KILLER MEMORY FUNCTION.   This is tested by priming in vivo with an H-2-incompatible skin homograft and then testing by challenge in vitro (cell-mediated cytotoxicity). No significant memory response was elicited from the T cells of B-Ly1 mice. But neither was there a significant response from B-Ly23 mice for reasons discussed below (interval 160 days).

The B-Ly23 Mouse

PRIMARY HELPER AND HELPER MEMORY FUNCTIONS.   There was little response from T cells of B-Ly23 mice in either the former assay (intervals 15, 18, 28, and 64 days) or the latter (intervals 48, 51, and 64 days).

PRIMARY ALLOREACTIVE KILLER FUNCTION.   Ly1 cells amplify the killer function that is generated from Ly23 pro-killer cells but do not themselves acquire killer function (4). T cells of B-Ly23 mice generated substantial killer function (intervals 28, 64, and 160 days) but not so great as those of B-T mice. The lesser response is in keeping with the known amplification of killer function by Ly1 cells during the initiation phase of immunization, but the present experiments do not exclude a role for Ly123 cells (6).

ALLOREACTIVE KILLER MEMORY FUNCTION.   The weak reaction of T cells from B-Ly23 mice in this assay (intervals 160 days) was no greater than that of B-Ly1 mice and greatly inferior to the B-T mouse.

These findings suggest that, in the absence of other T-cell subclasses, after antigen activation Ly23 cells may (*a*) actively suppress regeneration of a cytotoxic response or (*b*) undergo 'exhaustive differentiation' after in vivo stimulation by the $H-2^d$ graft. We favor the first possibility because of reported evidence that previously activated Ly23 cells can suppress the generation of Ly23 cytotoxic effector cells (2).

Discussion

Evidently B-Ly1 mice are equipped for helper function, primary and remembered, but not for killer function, and B-Ly23 mice are equipped for primary alloreactive killer function but not for helper function. The defective killer memorization of B-Ly23 mice accords with the previously recognized amplification of Ly23 killer activity by Ly1 cells (6) which has been attributed to the recognition of *I*-region antigen by Ly1 cells during the initial phase of immunization. However, the experiments described in the present report do not exclude the possibility that previously activated Ly23 cells can themselves suppress new generation of cytotoxic effector cells, a finding noted previously (2).

All this is compatible with the hypothesis that the Ly1 and Ly23 T-cell phenotypes always signify actual and potential immune functions, because the immune functions of B-Ly1 and B-Ly23 mice concur with the Ly subclasses which they possess.

A further important conclusion follows: Even after prolonged residence of Ly1 cells in hosts that have been deprived of Ly23 cells, and in which therefore all physiological controlling mechanisms must be set to favor expansion of the latter population, there is no appreciable generation of Ly23 from Ly1 cells in B-Ly1 mice. Reversely, there is no appreciable stock of Ly1 cells in B-Ly23 mice. Clearly, Ly1 and Ly23 cells must each have exercised a differentiative option that bars them from giving rise to one another. In other words, these two subclasses belong to different lines of differentiation and are not sequential stages of a single progression.

What bearing does this have on models of T-cell differentiation? Assuming that the TL^+Ly123 thymocyte is the precursor of all three subclasses (which is not finally proven, but is not essential to the point in question) the following models can be discarded:

$$(a) \rightarrow Ly1 \rightarrow Ly23$$
$$\text{or}$$
$$TL^+Ly123 \rightarrow Ly123 \quad (b) \rightarrow Ly23 \rightarrow Ly1$$
$$\text{or}$$
$$(c) \rightarrow Ly1 \leftrightarrow Ly23$$

This leaves two models:

$$[\text{Previous nomenclature} \quad T_0 \quad \rightarrow \quad T_1 \rightarrow \quad T_2 \ (7, 8)]$$

$$(d) \qquad TL^+Ly123 \rightarrow Ly123 \rightarrow \begin{cases} Ly1 \\ Ly23 \end{cases}$$

$$\text{or}$$

$$
\begin{array}{lll}
& \rightarrow Ly123 & (T_1) \\
(e) \qquad TL^+Ly123 \rightarrow & Ly1 & (T_2) \\
& \rightarrow Ly23 & (T_2)
\end{array}
$$

At the moment we favor model (d) in view of recent evidence that at least after antigen stimulation some Ly123 cells $\rightarrow$ Ly23 cells (9). If this model is correct the factors regulating the formation of mature Ly1 and Ly23 T cells from Ly123 precursors must play a central role in generating and regimenting the total population of T cells.

Summary

When B mice are supplied with Ly1 or Ly23 cells they acquire, over the next 6 mo, only the immune functions associated with each of these T-cell subclasses, respectively. The T-cell population of these 'B-Ly1' and 'B-Ly23' mice also remains restricted to the Ly1 and Ly23 subclass phenotypes. Thus the Ly1 and Ly23 populations are derived from two separate lines of differentiation and are not sequential stages of a single differentiative pathway.

Received for publication 19 July 1976.

References

1. Cantor, H., and E. A. Boyse. 1975. Functional subclasses of T lymphocytes bearing different Ly antigens. I. The generation of functionally distinct T-cell subclasses is a differentiative process independent of antigen. *J. Exp. Med.* **141:**1376.
2. Jandinski, J., H. Cantor, T. Tadakuma, D. L. Peavy, and C. W. Pierce. 1976. Separation of helper T cells from suppressor T cells expressing different Ly components. I. Polyclonal activation: suppressor and helper activities are inherent properties of distinct T-cell subclasses. *J. Exp. Med.* **143:**1382.
3. Howard, J. C., and D. B. Wilson. 1974. Specific positive selection of lymphocytes reactive to strong histocompatibility antigens. *J. Exp. Med.* **140:**660.
4. Shen, F. W., E. A. Boyse, and H. Cantor. 1975. Preparation and use of Ly antisera. *Immunogenetics.* **2:**591.
5. Cantor, H., F. W. Shen, and E. A. Boyse. 1975. Separation of helper T cells from suppressor T cells expressing different Ly components. II. Activation by antigen: after immunization antigen specific suppressor and helper activities are mediated by distinct T-cell subclasses. *J. Exp. Med.* **143:**1391.
6. Cantor, H., and E. A. Boyse. 1975. Functional subclasses of T lymphocytes bearing different Ly antigens. II. Cooperation between subclasses of Ly⁺ cells in the generation of killer activity. *J. Exp. Med.* **141:**1390.
7. Cantor, H. 1972. Two stages in the development of T' lymphocytes. Cell Interaction. Third Lepetit Colloquim. L. G. Silvestri, editor. North Holland Publishing Co., Amsterdam. 172.
8. Raff, M. C., and H. Cantor. 1972. Heterogeneity of T cells. *In* Progress in Immunology. Academic Press, Inc., New York. 83.
9. Cantor, H., and E. A. Boyse. 1976. Regulation of cellular and humoral immune responses by T cell subclasses. *Cold Spring Harbor Symp. Quant. Biol.* In press.

R. Kiessling[+], Eva Klein[+] and H. Wigzell[°]

Department of Tumor Biology, Karolinska Institute, Stockholm[+] and Department of Immunology, Uppsala University, Uppsala[°]

"Natural" killer cells in the mouse

I. Cytotoxic cells with specificity for mouse Moloney leukemia cells. Specificity and distribution according to genotype*

In the spleens of young, adult mice there exist naturally occurring killer lymphocytes with specificity for mouse Moloney leukemia cells. The lytic activity was directed against syngeneic or allogeneic Moloney leukemia cells to a similar extent, but was primarily expressed when tested against *in vitro* grown leukemia cells. Two leukemias of non-Moloney origin were resistant and so was the mastocytoma line P815. Although killer activity varied between different strains of mice, the specificity of lysis was the same as indicated by competition experiments using unlabeled Moloney or other tumor cells as inhibitors in the cytotoxic assays. Capacity to compete and sensitivy to lysis by the killer cells were found to be highly positively correlated. Analysis of the kinetics of the cytotoxic assay revealed a rapid induction of lysis within one to four hours, arguing against any conventional *in vitro* induction of immune response. No evidence was found of soluble factors playing any role in the cytolytic assay.

1. Introduction

Cytolytic activity of lymphoid cells from normal, nonimmunized donors has been reported in several *in vitro* systems both in animal [1–2]* and human [3, 4] systems. Some systems would seem to involve specific recognition in the sense that only certain target cells will be lysed by the effector cells [1–4], whereas in other assays no discrimination in killer activity can be demonstrated**. In a few systems there exist indications that the capacity of normal lymphoid cells to function as specific killer cells is induced by antibodies in these s.c. normal individuals, thereby recruiting effector cells in antibody-induced, cell-mediated lysis. This would occur either directly by coating the target cells when the test is carried out in the presence of donor serum, or indirectly via arming the effector cells through *in vivo* coating with antigen-antibody complexes [5]. In the majority of tests, however, the exact cause of "normal" cytolytic activity would seem obscure and different effector cells would seem to be involved in the various tests [1, 2]. The biological relevance of these normally occurring killer cells is hypothetical and has been assumed to be of importance in autoimmune reactions as well as in the reaction against persistant virus infections [2]. Irrespective of hypothetical biological importance the occurrance of such spontaneous killer cells has serious impacts in experimental assays, especially in tumor immune research, when very weak "immune" reactions are to be measured that can be easily overshadowed by the existance of varying spontaneous killer activity in the control material.

In the present article we report on the existence in normal mice of killer cells with rapid cytolytic, specific activity against *in vitro* grown mouse Moloney leukemia cells. We have analyzed the killer activity in relation to genotype of killer or target cells, and report on selective killer deficiencies linked to certain strains of mice. Competition experiments using unlabeled cells as competing targets for killer activity together with labeled targets demonstrate a highly positive correlation between sensitivity to cytolysis and competing capacity. As will be seen from an accompanying article, the killer cells involved would seem to be neither T nor B lymphocytes as classified by conventional markers but rather belong to yet a third group of lymphocytes [6], maybe of the kind described by other workers on the basis of morphology [7].

2. Materials and methods

2.1. Tissue culture medium

Medium F 13 (Grand Island Biological Co., Grand Island N.Y.) containing 10 % heat-inactivated fetal calf serum (Bio-Cult, Glasgow, Scotland) with penicillin (75 units/ml) and streptomycin (50 μg/ml) were used throughout the experiments as diluents and tissue culture media.

2.2. Animals

Animals of our own inbred strains of mice were used throughout most of the experiments. Sex of effector cell donors was not found to be important in this assay and is not documented. Age of cell donors is indicated in each experiment if not between 1 to 2 months of age. In experiments involving genetically athymic mice, such mice were bought from Dr. C.W. Friis (Gl. Bomholtgaard Ltd., Ry, Denmark), and were back-crossed to BALB/c mice on the 6th backcross generation when used. The breeding scheme is described in detail elsewhere [8]. Mice of the same age and sex, homozygous or heterozygous for the nude gene, were used as indicated.

2.3. Tumors

Eight different tumor cell lines were used. Five of these were lymphomas induced by inoculation of the Moloney leukemia virus (MLV) into newborn recipients. The various cell lines are symbolized as follows: the first letter 'Y' stands for Moloney lymphomas. The second letter in 3 letter symbols and the second and third letter in 4 letter symbols indicate

---

* Gomard, E., Leclerc, J.C. and Levy, J.P., submitted for publication.

** Grant, C.K. and Alexander, P., submitted for publication.

[I 875]

* Supported by the Swedish Cancer Society and by NIH contracts NOI-CB-33870 and NOI-CB-33859.

Correspondence: Rolf Kiessling, Department of Tumor Biology, Karolinska Institute, S-104 01 Stockholm 60, Sweden

Abbreviations: MLV: Moloney leukemia virus

the genotype of origin. The symbols correspond to the following strains: 'A' = strain A, 'CA' = (A x ACA)F_1, 'B' = = CBA, 'L' = Leaden. The final letter distinguishes between different MLV tumors of the same genotype. The final numbers denote different sublines of the same tumor.

Three non-MLV tumors were also used. P815-X2, a methylcholanthrene-induced mastocytoma of DBA origin, EL-4, a chemically induced lymphoma of C57BL origin and LAA, a spontaneous lymphoma of A/Sn origin. All tumors were propagated both *in vivo* in syngeneic mice and *in vitro* as established cell lines.

2.4. Target cell labeling

Tumor cells were incubated for 30 min at 37 °C at a concentration of $5 \times 10^6 - 10 \times 10^6$ cells suspended in 0.5 ml of complete medium to which 200 μCi of ^{51}Cr as sodium chromate was added. The cells were washed twice, counted and adjusted to the desired cell density.

2.5. Cytotoxic test

0.1 ml aliquots of various effector cell concentrations were added to glass tubes (11 x 55 mm) in duplicates. A constant number of 4×10^4 ^{51}Cr-labeled target cells in 0.05 ml were added. Target cells were also incubated without lymphocytes in medium only to estimate the level of spontaneous release of ^{51}Cr. This varied between 20–30 % of total label under the experimental conditions described. The tubes were incubated in a 37 °C, 5 % CO_2 incubator on a rocking platform for 12–14 h. Then 1 ml of medium was added to each tube and the tubes were centrifugated. For each tube total radioactivity and radioactivity of half of the supernatant was measured in a gamma counter. Maximum release was determined by treating the target cells with distilled water overnight, which released 70–80 % of the total ^{51}Cr.

The % ^{51}Cr release was calculated according to the following formula:

$$\%^{51}Cr \text{ release} = \frac{2 \times (\text{radioactivity of } 1/2 \text{ of the supernatant})}{\text{total radioactivity}} \times 100$$

Results are expressed as percent lysis according to the following formula:

$$\% \text{ lysis} = \frac{\% \text{ release in test} - \% \text{ spontaneous release}}{\% \text{ maximum release} - \% \text{ spontaneous release}} \times 100$$

The standard deviation between the two duplicates never exceeded ± 5 % lysis.

2.6. Competition studies

1×10^6 effector cells were incubated with various concentrations of "competing" tumor cells in duplicates for 1 h at 37 °C at a total volume of 0.2 ml complete medium. 2×10^4 ^{51}Cr-labeled target cells were added up to a final volume of 1 ml. Thus a constant effector to target cell ratio of 50:1 was used in all experiments, while the ratio of competing tumor cells to target cells was titrated out. The rest of the experiment was carried out as described above. All results are expressed as % inhibition at a certain competitor to target cell ratio according to the following formula:

$$\% \text{ inhibition} = \frac{\% \text{ lysis without competing cells} - \% \text{ lysis with competing cells}}{\% \text{ lysis without competing cells}} \times 100.$$

2.7. Preparation of effector cell suspension

Spleen cell suspensions were prepared in complete F-13 medium. Erythrocytes were lysed by treatment for 10 min at 4 °C with Tris buffered 0.75 % ammonium chloride, pH 7.2. If not otherwise stated, this treatment was done with all spleen cell suspensions.

3. Results

3.1. Strain distribution of normally occurring anti-Moloney leukemia killer cell activity

Spleen cells from normal, adult, nonimmunized mice have a certain cytolytic activity against the *in vitro* line of the A/Sn Moloney leukemia YAC. When analyzing the strain distribution of this "background" cytolytic activity it was found that cells from A/Sn mice were least efficient. Cells of certain other mouse strains were consistently highly cytotoxic as shown in Table 1.

Table 1. Spontaneous cytolytic activity in spleen cells from normal adult mice of different strains tested against YAC *in vitro* line[a]

Strain	No. of Exp.	% lysis at 30/1 ratio Mean	Range
CBA	12	63	34–95
A/Sn	9	16	5–33
C57BL	2	30	23–36
C3H	1	31	
ACA	1	38	
BALB/c	2	48	46–50
A x C57BL	18	37	9–70
A x Leaden	4	23	9–44
A x ACA	3	24	5–48
A x C3H	1	29	
A x CBA	2	30	16–44

a) In each experiment 2–5 spleens are pooled and tested for cytotoxicity. For experimental details see Section 2.

As YAC is of A/Sn origin, we assumed that the low effect of normal A/sn cells might be due to the lack of isoantigenic difference between effector and target cells. However, low spontaneous activity was also observed when the target cell YBA, another MLV leukemia of CBA origin, was tested against the non-H-2 compatible effector cells from A/Sn animals. Furthermore, effector cells of the syngeneic strain (CBA) were more cytolytic than A/Sn cells as could be seen from the experiment in Table 2.

Table 2. Cytolytic activity of A/Sn and CBA spleen cells tested against the YBA *in vitro* line[a]

	Effector cell		
	A/Sn % lysis 50/1		CBA % lysis 50/1
Mouse 1	2	Mouse 1	17
Mouse 2	2	Mouse 2	7
Mouse 3	3	Mouse 3	10
Mouse 4	–2	Mouse 4	29
Mouse 5	0	Mouse 5	28
Mean % lysis	1	Mean % lysis	18

a) Spleens from five A/Sn and five CBA mice of the same age and sex were tested individually in the same experiment against the YBA *in vitro* line.

It would thus seem clear that for yet unknown reasons A/Sn mice have a comparatively low "background" activity of killer cells against these two MLV lymphomas.

3.2. Time kinetic studies on the cytolytic effect on YAC-1 cells by normal spleen cells

In spite of the fact that the effector cells in our system were taken from normal nonimmunized mice, the onset of the *in vitro* lytic activity was remarkably fast. As could be seen in the time course experiment in Fig. 1, a clear effect was obtained by the CBA spleen cells on the YAC target cells already after 4 h of incubation, with an effector-to-target cell ratio of 50/1. Thus, it seemed that a rapid cell lysis independent of known preimmunization or added antibodies was active in this system.

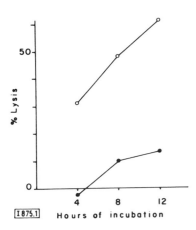

Figure 1. Effect of CBA spleen cells on YAC *in vivo* (●——●) or *in vitro* (○——○) cell lines as a function of incubation time. Spleens from 4 CBA mice were pooled. Spontaneous release after 12 h of incubation is for the *in vitro* line 13 % and for the *in vivo* line 22 % of the total label.

3.3. Origin of mice

In order to rule out the possibility that the activity of the normal spleen cells might depend on "spread" of leukemia virus in our mouse colony, we also investigated mice from other sources where no work on mouse leukemia viruses is carried out. We thus tested CBA mice from the Department of Genetics, Stockholm University and also BALB/c mice from a commercial breeder in Denmark (Friis, see Section 2.2.). We found, however, that spleen cells taken from mice on the day of arrival to the department were as efficient as our own mice of the same strain and age.

3.4. Comparative studies between different tumor cell lines for their sensitivity for lysis by spleen cells from normal animals

In order to establish the nature of this reaction, more tumor cell lines were tested for sensitivity in the present system.

Among the 8 cell lines tested significant effects (> 10 % lysis) have only been seen with three MLV cell lines. Two non-MLV lymphomas as well as two induced by MLV were not sensitive (Table 3). It is of interest that the P815 line, known to be highly sensitive to lymphocyte lysis, was not lysed.

Table 3. Specificity of spontaneous cytolytic activity for Moloney leukemia cells[a]

Exp.	Effector cells	Target cells	% Lysis (ratio 50/1)
I	CBA	YAC-1	63
	CBA	YCAB-1	−4
	CBA	P815	−2
II	CBA	YAC-1	64
	CBA	YCAB-1	8
	CBA	P815	−2
	A x BL	YAC-1	38
	A x BL	YCAB-1	2
	A x BL	P815	0
III	A x BL	YAC-1	47
	A x BL	YAC-301	25
	A x BL	YCAB-1	5
	A x BL	YCAB-6	11
	A x BL	P815	0
	A/Sn	YAC-1	3
	A/Sn	YAC-301	4
	A/Sn	YCAB-1	−2
	A/Sn	YCAB-6	2
	A/Sn	P815	−11
IV	CBA	YAC-1	69
	CBA	YCAB-1	8
	CBA	YBA	17
	CBA	YAA	0
	CBA	EL-4	10
V	CBA	YAC-1	91
	CBA	YBA	26
	CBA	YCAB-1	42
	CBA	YLI	7
	CBA	YAA	6
	CBA	LAA	10
	CBA	P815	−2
	CBA	EL-4	10

a) Comparison between different *in vitro* tumor cell lines for cytolytic sensitivity. Five independent experiments are shown. In each experiment the effector cells were derived from 2–5 spleens from CBA, A/Sn or A x C57BL F_1-hybrids. The spontaneous release is within the same range for all cell lines compared (20–30 % of total label).

3.5. Competition studies

The fact that a target cell is found to be insensitive to lysis does not exclude that the relevant target structures are present on the surface. Some target cells might be intrinsically more resistant to lysis than others, and therefore, competition studies were carried out as a more direct way of defining presence or absence of relevant target structures in different cells [9]. These studies were designed so that different numbers of non-labeled tumor cells were added to a constant number of ^{51}Cr-labeled target cells and effector cells. Thus, a tumor cell with the relevant target structure on its surface may compete for the effector cells, resulting in reduction of lysis of the ^{51}Cr-labeled target cells. Tumor cells with irrelevant surface components would not compete. The results of a typical competition experiment are shown in Fig. 2. Here, as in all experiments, five different ratios of competitor-to-target cell were used. In this experiment, among the 6 lines tested only the YAC and the YAA *in vitro* lines showed competition. We express the competition results as % inhibition (formula see Section 2.6.) at a competitor-to-target cell ratio of 7/1.

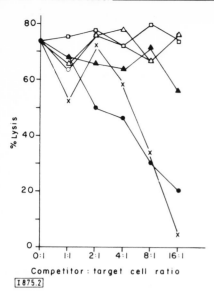

Figure 2. Competition experiment. As effector cells CBA spleen cells at an effector-to-target cell ratio of 50/1 were used. As target cells the YAC-1 *in vitro* line were used. Competing cells are as follows: YAC-1 *in vitro* line (●——●), YAC *in vivo* line (○——○), YAA *in vitro* line (×——×), YCAB-1 *in vitro* line (△——△), EL-4 *in vitro* line (▲——▲) and P815 *in vitro* line (□——□).

3.6. Comparative studies between sensitivity for lysis and competing capacity

Table 4 shows an experiment similar to that described above in which it can be seen that a positive correlation between sensitivity for lysis and capacity to compete exists. The competition experiments seemed to be more sensitive in demonstrating the presence of relevant target structures than direct cytolytic tests on the same cells (see YLI which was significant in competition but insensitive to lysis). As seen in Table 4, both sensitivity to lysis and capacity to compete seems to be confined to tumors induced by MLV (exception YCAB).

Table 4. Positive correlation between cytolytic sensitivity of target cells and capacity to compete in the cytotoxic assay[a]

In vitro line	% lysis (CBA)	% inhibition
YAC-1	46	108
YAA	10	47
YBA	9	69
YLI	0	39
YCAB-1	1	0
LAA	3	8
EL-4	5	0
P815	2	4

a) Experiment comparing the cytolytic sensitivity with competing capacity in different *in vitro* grown tumor cell lines. As effector cells in both systems CBA spleen cells at an effector to target cell ratio of 50/1 were used. As target cells in the competing system the YAC-1 *in vitro* cell line was used. Competing capacity for each tumor cell line is expressed as % inhibition at a competitor-to-target cell ratio of 7/1.

3.7. Differences in competing capacity between MLV- and non-MLV-induced tumor cells

These results prompted us to carry out further competition experiments in order to see if there was a consistent difference between MLV- and non-MLV-induced tumors. In Table 5 the results of four experiments are shown. It can be con-

Table 5. Detailed analysis of the capacity of various *in vitro* tumor lines to compete in cytolysis against YAC[a]

Competing tumor cell	% inhibition[b]			
	Exp. I	II	III	IV
YAC	93	59	113	108
YAA	70	54	125	47
YBA	70	n.d.	138	69
YLI	87	n.d.	106	39
YCAB-1	33	−5	11	0
EL-4	37	3	19	0
P815	22	−8	n.d.[c]	4
LAA	n.d.	n.d.	n.d.	8

a) As effector cells CBA spleen cells at an effector-to-target cell ratio of 50/1 were used. As target cells the YAC-1 *in vitro* line were used.
b) Competing capacity for each tumor cell line is expressed as % inhibition at a competitor-to-target cell ratio of 7/1.
c) n.d. = Not done.

cluded that whereas none of the non-MLV-induced cell lines compete strongly, 3 out of 4 MLV-induced cell lines are efficient competitors.

3.8. Differences between in *vitro* and *in vivo* grown cell lines

During these studies we observed that the *in vivo* grown tumors were considerably less sensitive to the lytic effect of normal spleen cells than their corresponding *in vitro* lines. This difference can be clearly seen in Fig. 1, where both the *in vitro* and the *in vivo* grown YAC cells were used as target cells. The same difference was also seen when comparing *in vivo* and *in vitro* lines of the YBA target cell.

As already mentioned, competition studies may be the most sensitive assay to detect the relevant target structures in the present system. We therefore performed such studies to compare *in vivo* with *in vitro* grown cells. In agreement with the results on cytolytic sensitivity, *in vivo* grown tumor cells did not compete, or competed to a lesser extent than did their corresponding *in vitro* lines, as can be seen from the 3 experiments summarized in Table 6.

It thus seems clear that the structure or antigen which the killer cells recognize is less well expressed on the *in vivo* grown MLV tumor cells.

4. Discussion

In the present article we describe the existance of "naturally" occurring killer cells in mice with selective lytic activity against Moloney leukemia cells. Of the tumor lines tested, only Moloney leukemia cell lines were found to be sensitive to the cytotoxic activity of the killer cells. Between the Moloney leukemia lines there existed significant variations in susceptibility to lysis. Also, *in vitro* grown cells were found to be more sensitive to lysis than *in vivo* grown cells from the same Moloney leukemia cell line. *In vitro* Moloney leukemia cells express more Moloney leukemia-associated antigens than the corresponding cells grown *in vivo* [10] and this might be relevant to our findings. Susceptibility to lysis could be shown to be positively correlated to the capacity of cells of the same tumor line to function as competitive inhibitors when added to the cytolytic test (see Table 4). Any susceptible Moloney leukemia cell line could be used as a selective competitor for lysis of isotope labeled

Table 6. Comparison in competing capacity in the cytotoxic assays against YAC using *in vivo* or *in vitro* grown Moloney leukemia cell lines[a]

Competing tumor cell	Exp. I		II		III		IV	
	in vitro	*in vivo*	*in vitro*	*in vivo*	*in vitro*	*in vivo*	*in vitro*	*in vivo*
YAC	93	27	113	43	59	−3	87	13
YAA	70	37	125	75	n.d.[c]	n.d.	n.d.	n.d.
YBA	80	33	n.d.	n.d.	n.d.	n.d.	n.d.	n.d.
YLI	87	43	106	38	n.d.	n.d.	n.d.	n.d.

% inhibition[b]

a) As effector cells CBA spleen cells at an effector-to-target cell ratio of 50/1 were used. As target cells the YAC-1 *in vitro* line were used.
b) Competing capacity for each tumor cell line is expressed as % inhibition at a competitor to target cell ratio of 7/1.
c) n.d. = Not done.

target cells from another Moloney leukemia line demonstrating that the killer cells, irrespective of strain of origin, had the same specificity for the different Moloney leukemia lines. However, that each tumor line had its own level of susceptibility and each mouse strain had its own level of killer activity as well will be discussed further below.

No evidence for any influence linked to allogeneic inhibition phenomena [11] could be demonstrated in the present system, as the relative killer activity was the same whether tested against syngeneic or allogeneic Moloney leukemia cells. We could thus conclude that the killer cells from these normal mice behave as though specifically immunized against Moloney lymphoma-associated antigens.

The *in vitro* cytolytic activity of the killer cells is very rapid and can be detected within less than 4 h (see Fig. 1) as detected by ^{51}Cr release from the damaged target cells. In fact, when using purified killer cells as obtained by fractionation procedures [6], specific lysis can be found within 1 h after start of incubation. We would thus consider the time periods so short that conventional *in vitro* induction of immune killer cells is excluded and the present test actually demonstrates the existence of *in vivo* generated killer cells with cytotoxic potential for Moloney leukemia cells. As the nature of the killer cell is the topic of an accompanying article [6], let us only state that the effector cell has the morphological appearance of a small lymphocyte. However, it can be clearly separated from both B and T lymphocytes by organ distribution data, as well as by surface markers used in fractionation procedures. The relationship to the lymphocytes described by Stobo and coworkers [7] remains to be clarified. The killing capacity is seemingly mediated by contact, as all attempts to demonstrate lytic activity of supernatants from *in vitro* cultivated killer cells have failed (unpublished observations).

Spleen cells contain the highest killer activity of the lymphoid organs [6], and killer activity peaks around the first 3 months of the life of a mouse. A variation in the efficiency of killer cells present in the spleen of normal mice of different, inbred strains was observed with A/Sn mice containing the lowest killer activity (see Table 1). Of the other strains tested, CBA mice seemed to have the highest killer activity in their spleens.

What is then the cause and origin of the present "spontaneous" killer cells? As already stated, the present killer cells would seem quite distinct from conventional, immune effector cells against Moloney leukemia cells that dominate

the *in vitro* cytotoxic reactions when immune anti-Moloney effector cells are used [12]. This in itself makes a discussion as to the cause of the occurrence of the present killer cells difficult, as the present effector cells would seem to be lymphocytes of a new type with yet unknown biological properties. If we assume that the activity of the killer cells is of immunological nature, a simple explanation would be that they are directed against murine leukemia virus antigens, and as such viruses are known to be quite ubiquitous, this would seem a reasonable assumption. Alternatively, the killer activity might not be Moloney leukemia-specific, but rather of an autoimmune type directed against some organ-associated antigen being expressed on the Moloney leukemia cells but not being directly coded for by the Moloney leukemia virus. Such an argument is partly weakened by the fact that both susceptible Moloney lymphomas and resistant lymphomas such as EL-4 can be shown to carry surface markers such as the theta antigen denoting their common, thymic origin, but only the Moloney leukemia cells are susceptible to lysis. Other workers have reported on the existence of "spontaneous" killer cell activity against tumor cells in the mouse or rat systems, with two systems detecting anti-lymphoma activity [1,*]. In one system, preleukemic AKR mice were found to express an anti-lymphoma specific activity with the effector cells being of probable macrophage type*, whereas in the other test spontaneous cytolytic activity was found in mice against a Rauscher virus-induced leukemia with the effector cells being of lymphocyte nature [1]. Interestingly, in the latter assay the effector cells, although of lymphocyte nature, were found to be resistant to anti-theta serum plus complement treatment in analogy with results in the present system [6].

Received August 5, 1974.

_____

* Gomard, E., Leclerc, J.C. and Levy, J.P., submitted for publication.

5. References

1 Herberman, R., Nunn, M.E., Lavrin, D.H. and Asofsky, R., *J. Nat. Cancer Inst*. 1973. *51*: 1509.

2 Greenberg, A.H. and Playfair, J.H.L., *Clin. Exp. Immunol*. 1974. *16*: 99.

3 Tagasuki, M., Mickey, M.R. and Terasaki, P., *Cancer Res*. 1973. *33*: 2898.

4 Petranyi, G.G., Banczur, M., Onody, C. and Holland, S.R., *Lancet* 1974. *i*: 736.

5 Perlmann, P., Perlmann, H. and Biberfeld, P., *J. Immunol.* 1972. *108*: 558.

6 Kiessling, R., Klein, E., Pross, H. and Wigzell, H., *Eur. J. Immunol.* 1975. *5*: 117.

7 Stobo, J.D., Rosenthal, A.S. and Paul, W.E., *J. Exp. Med.* 1973. *138*: 71.

8 Rygaard, J., *Acta Pathol. Microbiol. Scand.* 1969. 77: 761.

9 Rosenberg, E., McCoy, J., Green, F., Donelly, F., Siwarski, D., Lewine, P. and Herberman, R., *J. Nat. Cancer Inst.* 1974. *52*: 345.

10 Cikes, M., Friberg, S., Jr. and Klein, G., *J. Nat. Cancer Inst.* 1973. *50*: 347.

11 Hellström, K.E. and Hellström, I., *Progr. Exp. Tumor Res.* 1967. *9*: 40.

12 Lamon, E.W., Wigzell, H., Klein, E., Andersson, B. and Skurzak, H., *J. Exp. Med.* 1973. *137*: 1472.

THE IN VITRO GENERATION AND SUSTAINED CULTURE OF NUDE MOUSE CYTOLYTIC T-LYMPHOCYTES*

By STEVEN GILLIS,‡ NANCY A. UNION, PAUL E. BAKER,‡ AND KENDALL A. SMITH

From the Hematology Research Laboratory, Department of Medicine, Dartmouth-Hitchcock Medical Center and The Norris Cotton Cancer Center, Hanover, New Hampshire 03755

We recently reported the development of culture methodologies that allow for the sustained in vitro exponential growth of both murine and human, antigen-specific cytolytic T-lymphocyte lines (CTLL)[1] (1, 2). Long-term culture of CTLL was thoroughly dependent on the continual presence of a T-cell growth factor (TCGF) produced by T-cell mitogen or antigen-stimulated, normal murine, rat or human mononuclear cells. The ability of TCGF to allow for the indefinite culture of differentiated effector T cells prompted an investigation regarding several of the biological characteristics of TCGF.

We found that TCGF production was T-cell specific in that only T-cell mitogenic or antigenic stimulation resulted in TCGF release by mononuclear cells (3). In addition, TCGF production required the presence of both mature T cells and adherent cells. Removal of Thy-1 antigen-positive splenic T cells (3) or adherent splenic cells markedly decreased TCGF production.[2] The proliferative response to TCGF was also found to be T-cell specific. Only cells previously activated by T-cell mitogens or antigens were found to absorb TCGF activity and to proliferate in an indefinite exponential fashion (4). These findings suggested that TCGF served as the second signal in the T-cell immune response and functioned to mediate the proliferative expansion of antigen- or mitogen-activated T-cell clones.

Of particular interest were the observations that murine thymocytes produced relatively little TCGF and exhibited weak proliferative responses after stimulation with concanavalin A (Con A), whereas cortisol-resistant thymocytes produced TCGF and mediated proliferative responses identical to those generated by normal spleen cells.[2] We also observed that, although thymocytes produced little TCGF, they were capable of mounting a Con A-induced proliferative response equal in magnitude to that produced by Con A-stimulated spleen cells, provided TCGF was supplied exogenously.[2] These observations suggested that the limiting factor behind poor Con

* Supported in part by National Cancer Institute grant RO1-17643-04, National Cancer Institute contracts NO1-CB-74141 and NO1-CN-55199, and a grant from the National Leukemia Association, Inc.

‡ Fellows of the Leukemia Society of America.

[1] *Abbreviations used in this paper:* CTLL, cytotoxic T-lymphocyte lines; Con A, concanavalin A; DEAE, diethylaminoethyl; FCS, fetal calf serum; HBSS, Hanks' buffered salt solution; HP-1, helper peak-1; ^{3}H-Tdr, tritiated thymidine; LC, lytic capacity; LE, lytic efficiency; LMC, lymphocyte-mediated cytolysis; LU, lytic unit; MLC, mixed lymphocyte culture; pp, partially purified; SRBC, sheep erythrocytes; TCGF, T-cell growth factor.

[2] Smith, K. A., S. Gillis, F. W. Ruscetti, P. E. Baker, and D. McKenzie. The production and action of T-cell growth factor. Manuscript submitted for publication.

A-induced thymocyte proliferation was a relative lack of those cells capable of TCGF production.

If this supposition were correct, then one might expect that immature thymic precursors, such as those present in the athymic nude mouse, might also respond to immunologic stimuli provided TCGF were present. In this communication, we present the results of experimentation which show that nude mouse spleen, lymph node, and bone marrow cells are capable of T-cell mitogen-induced proliferative responses, provided mitogen sensitization is performed in the presence of TCGF. Furthermore, nude mouse spleen cells, when stimulated in TCGF-supplemented mixed lymphocyte culture (MLC), give rise to Thy-1 antigen-positive effector cell populations capable of mediating significant in vitro cytolysis of alloantigen-specific target cells. Nude mouse cytolytic lymphocytes have been maintained in TCGF-dependent culture for over 3 mo during which time they have continued to demonstrate antigen-specific cytolytic reactivity. The observation that nude mouse pre-T-cell populations can respond to in vitro alloantigen sensitization lends supportive evidence that: (a) antigen-reactive pre-T-cells are present in the spleens of athymic mice; and, (b) a fundamental reason behind the T-cell immunodeficiency of the nude mouse is its inability to produce TCGF.

Materials and Methods

Animals. BALB/c female, athymic, nu/nu (nude) mice, 5–7 wk of age were purchased from ARS Sprague-Dawley, Solon, Ohio. NIH female, nude mice, 5-7 wk of age were purchased from Harlan Industries, Indianapolis, Ind. Normal BALB/c, C57Bl/6, DBA/2, DBA/1, Ajax, and SJL female mice, 4–8 wk of age were purchased from the Jackson Laboratory, Bar Harbor, Maine. Charles River CD rats, 6–10 wk of age were purchased from Charles River Breeding Laboratories, Inc., Wilmington, Mass.

TCGF Production. TCGF for use in the routine maintenance of both normal mouse and nude mouse CTLL was produced by the 48-h stimulation of CD rat spleen cells (1×10^6 cells/ml) with Con A (5 μg/ml, Miles-Yeda Laboratories, Rehoveth, Israel) as previously described (3). In experimentation designed to monitor TCGF produced by normal murine spleen, as well as by nude mouse spleen, bone marrow, and lymph node cells, responding cell populations (1×10^7 cells/ml) were stimulated with Con A (2.5 μg/ml) for various time periods (Results). Cells were cultured at 37°C in RPMI 1640 medium, supplemented with 10% heat-inactivated (56°C for 30 min) fetal calf serum, (FCS, Grand Island Biological Co., Grand Island, N. Y.) 2.5 $\times 10^{-5}$ μM/ml 2-mercaptoethanol, 300 μg/ml fresh L-glutamine (Grand Island Biological Co.), 50 U/ml penicillin-G, and 50 μg/ml gentamicin, in a humidified atmosphere of 5% CO_2 in air. At the conclusion of the culture periods, the cells were removed by centrifugation (1,000 g for 10 min) and the supernates assayed for TCGF activity.

Preparation of Partially Purified (pp) TCGF. Conditioned medium containing TCGF activity was partially purified to remove Con A. Rat TCGF was precipitated by the addition of solid ammonium sulfate to a 70% concentration. The resulting precipitate was applied to a diethylaminoethyl (DEAE) cellulose column at pH 8.2. TCGF activity was found to elute from the column between 15 and 35 mM phosphate salt concentration. These fractions contained less than 1% of the starting protein and radioiodinated Con A was not found to elute in these fractions.[3]

TCGF Microassay. TCGF activity was assayed as previously described (3) using either CTLL 1 or CTLL 2 cells as the indicator cell population (1). The results were quantified by probit analysis (3) and expressed as units of activity based on a standard rat TCGF preparation.

Tritiated Thymidine (^3H-Tdr) Incorporation Assays. Murine cells (100 μl, 1×10^6 cells/ml) were seeded in triplicate into 96-well microtiter plates (No. 3596, Costar, Data Packaging, Cam-

---

[3] McKenzie, D., S. Gillis, W. Culp, and K. A. Smith. Manuscript in preparation.

bridge, Mass.) in Click's medium (Altick Associates, Hudson, Wis.) supplemented with normal mouse serum, 25 μM/ml Hepes buffer (Calbiochem-Behring Corp., American Hoechst Corp., San Diego, Calif.), 16 μM/ml $NaHCO_3$, 300 μg/ml fresh L-glutamine, 50 U/ml penicillin-G, and 50 μg/ml gentamicin. Cell populations included NIH and BALB/c nude mouse spleen, bone marrow, and lymph node cells. The following stimulants were added in 100-μl vol: (*a*) supplemented Click's medium, (*b*) Con A (5 μg/ml), (*c*) pp-TCGF, and (*d*) pp-TCGF plus Con A (5 μg/ml). Microplates were incubated at 37°C in a humidified atmosphere of 5% CO_2 in air. On day 4 of culture, 0.5 μCi of ^{3}H-Tdr (Schwartz/Mann Div., Becton, Dickinson, & Co., Orangeburg, N. Y., sp act, 1.9 Ci/mM) was added to each well and the incubation continued for 4 h. Cultures were harvested onto glass fiber filter strips and ^{3}H-Tdr incorporation determined as previously described (3). Results are expressed as the mean counts per minute $\pm$ 1 SD of triplicate cultures.

CTLL Culture.   Both normal and nude mouse CTLL were seeded at 1 $\times$ 10^4 cells/ml in 50% TCGF/50% supplemented Click's medium (vol/vol) in either 25 cm^2 or 75 cm^2 plastic tissue culture flasks (Nos. 3013, 3024, Falcon Labware Div. of Becton-Dickinson, Inc., Oxnard, Calif.). After 3 or 4 d of culture, when the cells had reached a density of 1–2 $\times$ 10^5 cells/ml, the cultures were subcultured in fresh Click's medium/TCGF (50%/50%, vol/vol) to 1 $\times$ 10^4 cells/ml.

Nude Mouse MLC Stimulation.   MLC were conducted as previously described (5). Briefly, 10 ml of either NIH or BALB/c nude mouse spleen cells (2.5 $\times$ 10^6 cells/ml) were mixed with an equal volume and concentration of x-irradiated (1,500-rad-cobalt source) C57Bl/6 spleen cells. MLC were conducted in 2% FCS-supplemented Click's medium and cultured upright in 30-cm^2 tissue culture flasks (No. 3012 Falcon Labware Div. of Becton-Dickinson, Inc.) in a humidified atmosphere of 5% CO_2 in air at 37°C. On successive days of culture (days 0–4), 10 ml of tissue culture medium was removed from one of five replicate flasks and replaced with 10 ml of TCGF. After 5 d of MLC stimulation, viable effector cells from both TCGF-supplemented and control MLC were harvested and either tested for cytolytic reactivity in a standard 4-h ^{51}Cr-release assay or placed in TCGF-dependent culture. In some instances, effector cells were treated with anti-Thy-1 serum (1/20 dilution, mouse AKR-anti-C3H acites, No. 8301, Bionetics Laboratory Products, Litton Bionetics Inc., Kensington, Md.) and absorbed-rabbit complement (Cedar Lane Laboratories, London, Ontario, Canada) before use in ^{51}Cr-release assays.

Lymphocyte-Mediated Cytolysis (LMC) Assays.   4-h ^{51}Cr-release assays were conducted in 96-well, v-bottom, microplates (1S-MVC-96-TC, Linbro Chemical Co., New Haven, Conn.) using methodology previously described (5). The percentage of specific lysis was determined by using the following equation: % specific lysis = 100 $\times$ (experimental cpm − medium control cpm/ maximum release cpm − medium control cpm). Data are displayed both graphically and in tabular form in terms of both lytic capacity (LC) and lytic efficiency (LE). LC is defined as the number of lytic units (LU)/25 $\times$ 10^6 effector cells. 1 LU is defined as the number of effector cells necessary to mediate 30% specific lysis. LE is defined as the percentage of specific lysis observed at an effector/target cell ratio of 100/1. Target cells used in LMC assays included the FBL-3(Hn) (H-2^b) murine leukemia cell line (5), the AKT-8 (H-2^k) thymoma cell line (obtained from Dr. Janet Hartley, Laboratory of Viral Diseases, National Institute of Allergy and Infectious Diseases, Bethesda, Md.) and the P815 (H-2^d) mastocytoma cell (obtained from Dr. H. Robson MacDonald, Ludwig Institute for Cancer Research, Lausanne, Switzerland). In addition to the above-detailed tumor targets, several normal mouse thymus cell populations were used as targets for nude mouse CTLL. The strains tested included: DBA-2 (H-2^d), DBA-1 (H-2^q), Ajax (H-2^a), and SJL (H-2^s).

Adsorption Of Anti-Thy-1 Serum by CTLL Cells.   To test for cell surface expression of Thy-1 antigen, 100-μl aliquots of anti-Thy-1 serum (1/640) were incubated for 1 h at 37°C in the presence of increasing log$_2$ concentrations of nude CTLL cells, normal murine CTLL cells, and normal C57Bl/6 thymocytes (concentrations ranged from 3 $\times$ 10^5 to 4 $\times$ 10^7 cells/ml). Adsorbed and control antisera were then tested for complement-mediated cytolysis directed against ^{51}Cr-labeled normal C57Bl/6 thymocytes using methodology previously described (5). The antiserum dilution (1/640) chosen for adsorption routinely generated between 70 and 100% complement-mediated cytolysis as assayed on normal mouse thymocytes. Results of adsorption experiments are expressed in terms of the percentage of inhibition of cytotoxicity observed after adsorption at a particular cell concentration.

Direct Membrane Immunofluorescence. Target cells to be assayed (nude CTLL or normal C57Bl/6 thymocytes) were adjusted to a concentration of between 5×10^7 and 2×10^8 cell/ml in cold Hanks' buffered salt solution (HBSS). 50 μl of cells was then incubated for 30 min at 4°C with an equal volume of either of two antisera: (*a*) fluorescein-conjugated, absorbed rabbit-anti-mouse (C3H) brain (anti-Thy-1, No. 8301-63, Bionetics Laboratory products or (*b*) fluorescein-conjugated goat-anti-mouse immunoglobulin (Behring Diagnostics, American Hoechst Corp., Somerville, N. J.). Cells were washed twice with 5-ml vol of cold HBSS, resuspended in two–four drops of cold HBSS and observed with the aid of a Zeiss fluorescence microscope (Carl Zeiss, Inc., New York).

Results

Inability of Immature Prothymocytes to Produce TCGF. Because lymphoid organs from athymic nude mice represent populations of cells which contain appreciable numbers of immature prothymocytes, we questioned whether nude mouse lymphoid cells would produce TCGF in response to T-cell mitogenic stimulation. To assess TCGF production, both NIH and BALB/c nude mouse spleen, lymph node, and bone marrow cells (1×10^7 cells/ml) were cultured with a mitogenic dose of Con A (2.5 μg/ml). At 24-h culture intervals for each of 6 d, culture supernates were harvested and tested for TCGF activity. Spleen cells from normal BALB/c mice were cultured in an identical fashion. As detailed in Table I, nude mouse lymphoid cell populations were not capable of producing TCGF in response to Con A, whereas normal BALB/c spleen cells produced significant quantities of TCGF upon stimulation with Con A.

Response of Nude Mouse Lymphoid Populations To Con A and TCGF. To test whether nude mouse prothymocyte-containing cell populations were capable of responding to stimulation with Con A in the presence of TCGF, BALB/c and NIH nude mouse spleen, lymph node, and bone marrow cells were cultured in the presence of four different culture additives: (*a*) tissue culture medium, (*b*) Con A (2.5 μg/ml), (*c*) pp-TCGF and (*d*) pp-TCGF plus Con A (2.5 μg/ml). After 4 d of culture, cellular proliferation was assayed in terms of resultant ^{3}H-Tdr incorporation. As detailed in Fig. 1, all six responding cell types were capable of mounting a marked proliferative response to stimulation with Con A in the presence of TCGF. However, ^{3}H-Tdr incorporation by cultures treated with either Con A or pp-TCGF alone was no greater than the cellular proliferation witnessed in medium control cultures. The results detailed in Fig. 1 provide further evidence that after activation with Con A, lymphocyte proliferation is mediated solely via *in situ* production of, or exogenous supplementation with TCGF. Most importantly, the data detailed above show that, as was the case with normal mouse thymocytes,[2] nude mouse spleen, lymph node, and bone marrow cells are capable of mounting normal T-cell mitogen responses, as long as exogenous TCGF (which they are incapable of producing *in situ*) is supplied.

TCGF-Dependent In Vitro Generation of Nude Mouse Cytolytic Effector Cells. The observation that nude mouse spleen cells could respond to Con A in the presence of TCGF led us to question whether addition of TCGF to an identical cell population might be able to provoke a response to alloantigen as well. We have previously shown that the in vitro generation of alloantigen-specific, cytolytic effector T lymphocytes is thoroughly dependent upon alloantigen-provoked TCGF production (6). Furthermore, supplementation of cultures aimed at the generation of cytolytic effector cells with additional TCGF greatly increased both the quantity and efficiency of the cytolytic T-cells generated therein (6). Based on these observations, we hypothesized that perhaps addition of TCGF to nude mouse MLC might allow for the proliferative

TABLE I

TCGF Production by Normal and Nude Mouse Lymphoid Cell Populations

Responding culture*	TCGF production culture duration‡			
	24 h	48 h	72 h	96 h
BALB/c normal spleen cells	1.29	0.63	0.41	0.16
NIH nude mouse spleen cells	0.00	0.00	0.00	0.00
NIH nude mouse lymph node cells	0.00	0.00	0.00	0.00
NIH nude mouse bone marrow cells	0.00	0.00	0.00	0.00
BALB/c nude mouse spleen cells	0.00	0.00	0.00	0.00
BALB/c nude mouse lymph node cells	0.00	0.00	0.00	0.00
BALB/c nude mouse bone marrow cells	0.00	0.00	0.00	0.00

* 100 μl of cells (10^7 cells/ml) cultured with an equal volume of Con A (2.5 μg/ml) containing tissue culture medium.

‡ Units of activity contained in culture supernates.

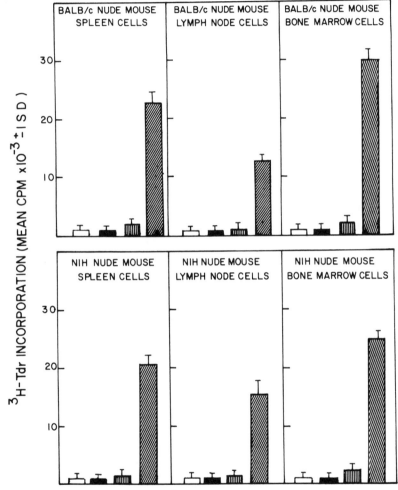

FIG. 1. ^{3}H-Tdr incorporation of 1×10^5 BALB/c and NIH nude mouse lymphocyte populations assayed after 96 h of culture in the presence of tissue culture medium (□); Con A (2.5 μg/ml, ■); pp-TCGF (▥); pp-TCGF plus Con A (2.5 μg/ml, ▨).

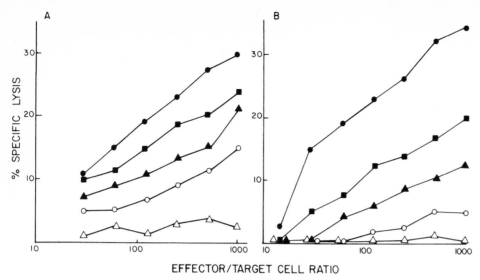

FIG. 2. Cytolytic reactivity of BALB/c (A) and NIH (B) nude mouse × C57Bl/6 MLC effector cells harvested from control cultures (△) and cultures supplemented with TCGF on day 0 (●), day 1 (■), day 2 (▲), and day 3 (○). Cytotoxicity directed against the FBL-3(Hn) (H-2^b) tumor target cell.

expansion of alloantigen-reactive responding cells to the extent where meaningful cytotoxic reactivity might be observed.

To test whether TCGF might initiate nude mouse lymphocyte-alloantigen reactivity, spleen cells from both NIH and BALB/c nude backgrounds were co-cultured in MLC with x-irradiated allogeneic C57Bl/6 spleen cells. On day 0–4, 10 ml of tissue culture medium was removed from one of five replicate flasks and replaced with 10 ml of TCGF. Viable effector cells harvested from 5-d control and TCGF-supplemented cultures, were then tested for cytolytic reactivity directed against a C57Bl/6 tumor target cell FBL-3(Hn) (H-2^b) in a standard 4-h ^{51}Cr-release assay. The ability of alloantigen (H-2^b)-sensitized nude mouse, splenic effector cells to lyse ^{51}Cr-labeled FBL-3(Hn) (H-2^b) target cells is shown in Fig. 2. Effector cells harvested from MLC supplemented with TCGF on day 0 mediated significant levels of cytotoxicity (30% specific lysis at an effector cell/target ratio of 400/1). TCGF supplementation on successively later days of culture resulted in the generation of effector cells which demonstrated a progressively weaker cytotoxic response. It is important to note that MLC stimulation in the absence of TCGF did not result in the generation of cytolytic effector cells. The results depicted in Fig. 2 confirmed that the addition of TCGF to nude mouse MLC resulted in the proliferation of alloantigen-reactive effector cells to the point where demonstrable cytotoxicity could be detected. Furthermore, the inability of nude mouse lymphoid cell populations to produce their own TCGF ensured that non-TCGF-supplemented MLC would be incapable of generating antigen-directed cytolytic effector cells.

Abrogation of Nude Mouse Effector Cell Cytotoxicity by Treatment with Anti-Thy-1 Serum and Complement. As previously detailed, only activated T cells have been shown to be capable of proliferating in response to TCGF (1–3, 6). The TCGF-dependent generation of nude mouse cytolytic effector cells, therefore, intimated that the cytotoxicity observed was mediated by cells derived from the T-cell lineage. A widely accepted

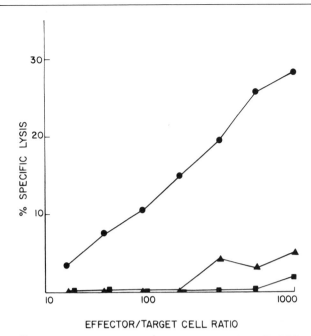

Fig. 3. Lysis of ^{51}Cr-labeled FBL-3(Hn) (H-2^b) tumor target cells by BALB/c nude mouse ✕ C57BL/6 MLC effector cells: control effector cells (■); effector cells harvested from day-zero, TCGF-supplemented MLC (●); day-zero, TCGF-supplemented MLC effector cells treated with anti-Thy-1 serum and complement before use in LMC assay (▲).

demonstration of T-cell mediated cytotoxicity is the elimination of cytolytic reactivity after effector cell treatment with anti-Thy-1 serum and complement (7). The results of anti-Thy-1 serum and complement treatment of nude mouse cytolytic effector cells generated in TCGF-supplemented MLC are displayed in Fig. 3. Such treatment completely abrogated alloantigen-directed cytolysis. In fact, anti-Thy-1 serum-treated effector cells mediated little, if any, more cytolytic reactivity than did effector cells harvested from non-TCGF-supplemented MLC (2–4% lysis at an effector/target cell ratio of 1,000/1). Therefore, it appeared that immediate TCGF supplementation of nude mouse MLC resulted in the proliferative expansion of an allo-reactive prothymocyte (Thy-1-antigen-positive) population which was in turn capable of mediating significant alloantigen-directed cytolytic reactivity.

Creation of Alloantigen-Specific Nude Mouse CTLL. We have demonstrated that both mouse, and human, antigen-specific, cytolytic T cells may be maintained indefinitely in a TCGF-dependent state of exponential proliferation (1, 2). Indeed, two murine, Thy-1-antigen-positive CTLL have remained in culture in the presence of TCGF for over 2.5 yr. The demonstration that cytolytic effector cells generated in TCGF-supplemented nude mouse MLC were also Thy-1-antigen-positive, provided some basis for anticipating that perhaps the continuous culture of nude mouse cytolytic effector cells would also be possible.

In hopes of creating a nude mouse CTLL, effector cells harvested from nude mouse MLC (BALB/c nu/nu vs. C57Bl/6) were seeded in replicate flasks in a solution containing 50% supplemented Click's medium and 50% TCGF. Initial cultures were seeded at concentrations well below 3 ✕ 10^5 cells/ml to ensure that effector cells were

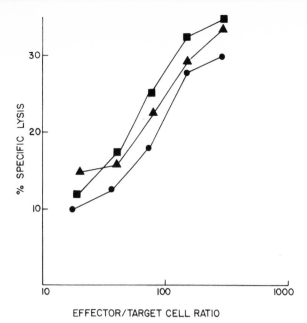

FIG. 4. Cytolysis of FBL-3(Hn) (H-2^b) tumor target cells by BALB/c × C57Bl/6 nude CTLL cells after 3 (●), 7 (■), and 11 (▲) wk of TCGF-dependent culture.

always in a situation of TCGF excess. Throughout the 1st 3 wk of culture, cells growing in suspension were routinely layered over a solution of Ficoll-Hypaque (Ficoll, Pharmacia Fine Chemicals, Div. of Pharmacia, Inc., Piscataway, N. J., Hypaque, Radiopaque Media, Winthrop Laboratories, New York) (2) before subculturing in fresh Click's medium/TCGF, to ensure that subcultured cells contained a minimal amount of cell debris. It is our belief that debris present in early stages of CTLL cell culture are the result of the eventual degradation of non-TCGF-responsive cells. In the case of TCGF-dependent culture of nude mouse MLC effector cells, cell debris present in initial cultures were particularly acute, presumably due to a large number of nonlytic cells (B cells, macrophages).

On the protocol of subculturing to 2×10^4 cells/ml, once cells reached a density of $2–4 \times 10^5$ cells/ml, nude mouse MLC effector cells have remained in a TCGF-dependent state of exponential proliferation for over 14 wk. The cells have the same morphological and growth characteristics (saturation density of $2–4 \times 10^5$ cells/ml, doubling time: every 18–30 h) as previously described CTLL (1, 2, 8–10). Deprivation of TCGF leads to irreversible cell damage and complete death of nude CTLL cultures within 24 h.

Throughout their culture, nude mouse CTLL have continued to mediate the in vitro cytolysis of allogeneic, FBL-3(Hn) (H-2^b), leukemia cells as displayed in Fig. 4. Nude mouse CTLL harvested from 3-, 7-, and 11-wk-old TCGF-dependent cultures demonstrated almost identical in vitro lytic reactivity directed against the FBL-3(Hn) (H-2^b) target cell. It is important to note that nude mouse CTLL mediated 30% specific lysis of the FBL-3(Hn) (H-2^b) target cell at an effector/target cell ratio of ≦150/l; a considerable increase in lytic efficiency as compared to effector cells harvested directly after TCGF-supplemented MLC stimulation (Figs. 2 and 3). As

TABLE II

Specificity of Cytotoxicity Mediated by Nude CTLL Cells

Age of CTLL culture	Cytotoxicity directed against							
	Tumor target cells			Thymocyte target cells				
	FBL-3(Hn) (H-2^b)	AKT-8 (H-2^k)	P815 (H-2^d)	C57Bl/6 (H-2^b)	DBA/2 (H-2^d)	DBA/1 (H-2^q)	SJL (H-2^s)	AJAX (H-2^a)
wk								
8	390/37*	0.8/3	4.1/6	227/26	<0.1/3	0.7/4	<0.1/3	0.4/2
12	250/31	1.1/2	5.0/8	116/21	0.3/6	<0.1/2	<0.1/1	<0.1/0

* Expressed at LC/LE. LE = LU/25 × 10^6 nude CTLL cells. LC is the percentage of specific lysis observed at an effector/target cell ratio of 100/1.

mentioned above, we have hypothesized that the observed increase in lytic effect after prolonged culture may be due to a gradual elimination of non-TCGF-responsive cells and a reciprocal enrichment of the culture population for TCGF-dependent CTLL.

The cytolysis mediated by nude CTLL cells was observed to be alloantigen-(H-2^b)-specific. As detailed in Table II, nude CTLL harvested after 8 and 12 wk of long-term culture were capable of effecting only the lysis of H-2^b target cells (FBL-3(Hn) and C57Bl/6 thymocytes). DBA-2 (H-2^d) thymocyte target populations as well as the P815 (H-2^d) tumor target cell were unaffected. In addition, nude CTLL mediated no cytolytic reactivity whatsoever against the third party tumor target cell, AKT8 (H-2^k) nor were (H-2^b)-reactive nude CTLL capable of killing several third party thymocyte target populations: DBA-1 (H-2^q), SJL (H-2^s), and Ajax (H-2^a).

Throughout their long-term culture, nude CTLL have remained Thy-1-antigen-positive, surface-immunoglobulin-negative, and negative by histochemical stains for specific and nonspecific esterases. Cell surface expression of Thy-1 antigen was determined by two distinct assays: (*a*) adsorption of anti-Thy-1 serum activity and (*b*) direct membrane immunofluorescence. In adsorption tests, increasing numbers of nude CTLL cells were used to adsorb anti-Thy-1 serum cytotoxic reactivity as tested on ^{51}Cr-labeled normal C57Bl/6 thymocytes in standard antibody-dependent, complement-mediated cytolysis assays. As detailed in Fig. 5, 12-wk-old nude CTLL cells were quite efficient in effecting the concentration-dependent adsorption of anti-Thy-1 serum reactivity (50% inhibition of cytotoxicity after adsorption with 2.5 × 10^5 cells). Furthermore, the ability of nude CTLL to adsorb anti-Thy-1 serum activity was essentially identical to that demonstrated by normal murine CTLL cells or normal C57Bl/6 thymocytes.

The results of direct membrane immunofluorescence tests conducted on both normal C57Bl/6 thymocytes and 14-wk-old nude CTLL cells are shown in Fig. 6. Both nude CTLL (Fig. 6 A) and normal thymocyte populations (Fig. 6 B) were found to be 95–100% Thy-1-antigen-positive in tests using fluorescein-conjugated, absorbed rabbit-anti-mouse brain antiserum. With identical direct immunofluorescence techniques and fluorescein-conjugated goat-anti-mouse immunoglobulin, nude CTLL cells were found to be 100% negative for surface-bound immunoglobulin (Fig. 6 C). Therefore, as determined by three separate assays: (*a*) inhibition of cell-mediated cytotoxicity by effector cell treatment with anti-Thy-1 serum and complement; (*b*) adsorption of cytolytic reactivity from anti-Thy-1 serum; and (*c*) direct membrane

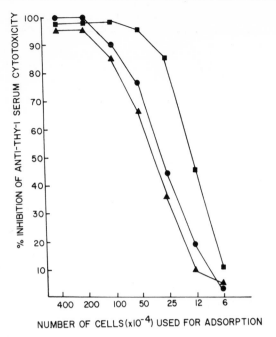

Fig. 5. Concentration-dependent adsorption of anti-Thy-1 serum cytotoxic reactivity by nude mouse CTLL cells (▲), normal murine CTLL cells (●), and C57Bl/6 thymocytes (■). 100 μl of cells incubated for 1 h at 37°C with an equal volume of anti-Thy-1 serum before testing for complement-mediated cytolysis of normal thymocytes. Resultant serum dilution (1/640) routinely promoted 80–100% specific lysis of thymocyte targets.

immunofluorescence, it appears that alloantigen-specific, cytotoxic effector cells generated in TCGF-supplemented nude mouse MLC and maintained in a TCGF-dependent state of exponential proliferation are Thy-1-antigen-positive cytotoxic T lymphocytes.

Discussion

The findings detailed in this report have shown that dual stimulation of nude mouse spleen cells with both alloantigen and TCGF allows for the proliferative expansion of allo-reactive, Thy-1 antigen-positive, cytotoxic lymphocytes. Furthermore, lytic specificity conferred by alloantigen stimulation is retained by nude CTLL throughout 3 mo of subsequent TCGF-dependent culture. These results, compounded with the observation that TCGF also allows for normal T-cell mitogen stimulation of nude mouse spleen, lymph node, and bone marrow cells, may have important ramifications as to our present thinking regarding several aspects of T-cell differentiation; most notably, the functional potential of pre-thymocytes, the role of the thymus in directing T-cell maturation, and, finally, possible mechanisms for overcoming pathological disease states stemming from thymic deficiencies.

Several other investigators have previously reported that nude mouse spleen cells, when used as responding lymphocyte populations in MLC, were not capable of effecting the in vitro cytolysis of ^{51}Cr-labeled allogeneic target cells, even when tested at killer/target cell ratios as high as 400/1 (11, 12). These investigators concluded

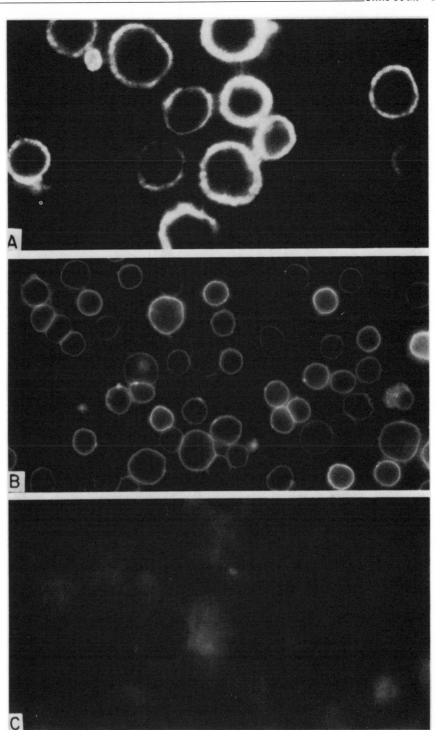

Fig. 6.   Cell surface labeling of nude CTLL cells (A) and normal C57Bl/6 thymocytes (B) via direct immunofluorescence with fluorescein-conjugated anti-Thy-1 serum. Cell surface labeling of nude CTLL cells with fluorescein-conjugated goat-anti-mouse immunoglobulin (C). Approximately ×3,200.

that the lack of cytotoxicity mediated by nude mouse MLC-activated spleen cells was due to a lack of cytotoxic precursors. The experimentation detailed above (Figs. 2 and 3) suggests that the inability of the nude mouse to manage an effective MLC reaction is not due to a dearth of appropriate reactive precursors, but is instead due to the lack of cells which are capable of producing TCGF (Table I). As we have previously described, the in vitro generation of cytolytic T-cells is thoroughly dependent upon the ability of responding lymphocyte populations to produce TCGF as triggered by alloantigen sensitization (6). Because nude mouse cell populations were unable to produce TCGF, it·followed that cells from nude mice would be unable to generate cytolytic cells in classical MLC. However, when TCGF was provided exogenously, nude mouse splenic responder cells were capable of proliferating in response to alloantigen stimulation to the point where demonstrable in vitro cytolysis was observed. It should be noted that the level of cytolysis observed (30% specific lysis at an effector cell/target ratio of 400:1), was less than the efficiency of lysis we routinely observe when normal spleen cells were used as responder cells in an MLC (100% specific lysis at an effector cell/target ratio of 100:1). It remains to be investigated whether this difference is due to relatively fewer alloreactive nude precursor T cells or whether additional factors (perhaps thymic humoral factors) are necessary to effect the maturation of all precursor T cells to a fully TCGF-responsive state. Nonetheless, these results suggested that nude mice contained alloreactive precursor T cells which, when provided with antigen and exogenous TCGF, were capable of mediating a specific immunologic function previously ascribed only to normal, mature, T lymphocytes.

Considerable evidence has been presented detailing the fact that athymic nude mice contain thymic precursor cells which can function to reconstitute lethally irradiated mice (13). Chromosome analyses have shown without question that reconstitution in such cases is due to the evolution of a new T-lymphoid system stemming from the nude donor rather than from some radioresistant subpopulation of the recipient (14). Conversely, nude mice themselves can be reconstituted to normal T-cell function after implantation of a thymus graft (15) or thymic epithelial tissue (16). In both types of studies, the concensus has been that restoration of T-dependent function was the result of some type of interaction between thymus tissue or tissue products and nude mouse precursor T cells. Such an interaction might involve the actual training and differentiation of prothymocytes into antigen-reactive cells which, when later exposed to antigen, are capable of responding in the manner imparted to them during exposure to thymic influence.

Of particular interest with regard to the influence of the thymus on the immunologic capacity of immature T cells, was the recent report by Irle et al. (17). These investigators found that immature thymocytes were capable of mediating normal immunologic, in vitro, responses provided cultures were supplemented with medium conditioned by mitogen-stimulated peripheral T lymphocytes. On the basis of these observations, it was postulated that the thymus was responsible for two steps in the differentiation of immature precursors into mature, immunologically reactive T lymphocytes: (a) the acquisition cf specific mitogen and antigen receptors thought to occur during cell division in the thymic cortex, and (b) the development of the capacity to respond to antigen-receptor triggering by proliferation in the absence of mitogenic factors released by activated lymphoid cells. Irle et al. (17) further concluded that mature peripheral T cells did not require extracellular factors to mediate proliferation in response to mitogenic or antigenic stimulation.

The experimentation described herein, as well as our previous observations (1–3, 6),[2] provides evidence that immature, nude mouse prothymocytes (in the absence of thymic influence) possess mitogen and alloantigen responsiveness, and that prothymocytes, as well as peripheral mature T-lymphocytes, require a second proliferative signal to respond to mitogen/antigen sensitization. Other investigators have described similar studies involving the provocation of T-cell function in the nude mouse which supports the *de novo* presence of antigen-specific pre-T-cells capable of functioning without prior exposure to thymus tissue or tissue products. In fact, these studies, when interpreted with the knowledge of the pivotal role that TCGF plays in the proliferation of activated T lymphocytes, lend further evidence to our contention that the specific influence mediated by a thymus graft in the reconstitution of nude mouse T-lymphoid function, is to provide for the maturation of TCGF-producing cells.

For example, Schimpl and Wecker (18) as well as Kindred and Corley (19) have shown that treatment of nude mice with tissue-culture supernates derived from normal MLC restored the ability to produce antigen-specific antibody directed against sheep erythrocytes (SRBC) or alloantigen. Both sets of investigators suggested that the mode of action of "helper factors" present in the MLC supernate was directly on B cells reactive to the particular antigen involved. However, in light of our findings that MLC as well as mixed tumor lymphocyte cultures produced copius quantities of TCGF (6), it is plausible to hypothesize that the TCGF present in MLC supernates together with antigen presentation allowed for the proliferation of antigen-specific helper pre-T-cells already present in the nude mouse. The TCGF-dependent proliferation of these cells then allowed for specific B-cell activation and antibody production.

Such a hypothesis is similar to that offered by Farrar et al. (20) to explain results they obtained detailing the ability of helper peak 1 (HP-1 isolated from MLC supernates) to reconstitute anti-SRBC reactivity by T-cell deficient spleen cell populations. As opposed to implicating a direct effect of HP-1 on antibody-producing B cells, Farrar et al. (20) acknowledged the possibility that helper molecules might be involved in the activation of pre-T-cells which in turn provided the putative antigen-specific helper signal. The observation that rat pp-TCGF has a mol wt of $\cong$14,000 daltons (similar to the 10,000–15,000 dalton mol wt described for HP-1)[3] lends further support to the hypothesis that MLC supernate-dependent reconstitution of T-dependent, antigen-directed antibody production might have been due to the effects of a TCGF-dependent proliferation of pre-helper T cells already present in T-cell-deficient populations.

The TCGF-driven proliferation of antigen-reactive nude mouse pre-T-cells also provides a means for explaining the results of Piguet and Vassalli (21) who recently reported that intravenous injection of nude mice with allogeneic or xenogeneic purified T-cell populations resulted in the ability of the T-cell treated mouse to reject a skin graft syngeneic to the T-cell donor (via production of graft-specific cytotoxic antibody). These investigators suggested that the ability of nude mice to respond to a foreign graft after treatment with graft strain T cells, represented a unique form of T-B-cell cooperation across both histocompatibility and species barriers in which the grafted T cells recognized nude mouse histocompatibility antigens and in some manner provided a means of helping nude mouse B cells to make antibody against the graft. Once more, the knowledge that TCGF is produced by alloantigen stimulation provides us with a mechanism to explain the above-mentioned data. Donor cells, upon recognizing nude mouse histocompatibility antigens as foreign, produced TCGF which, in turn, allowed for the proliferation of host, graft-antigen-specific, helper pre-T-cells whose action resulted in syngeneic T-B-cell cooperation eventually leading to allo- or xeno-graft rejection. The nude mouse, incapable of producing its own TCGF, was the beneficiary of graft vs. host-generated TCGF. Therefore, it is conceivable that once more, the TCGF-dependent proliferation of already existent antigen-reactive nude mouse pre-T-cells triggered the demonstration of competent T-cell immune function.

The hypothesis that antigen-specific pre-T-cells exist in the nude mouse and are capable of contributing in cell-mediated reactions without additional thymic influence should not be construed as incongruous with the repeated demonstration that exposure of T-cell deficient lymphoid populations to thymic influences either in vitro or in vivo can result in total reconstitution of T-cell function. In fact, co-culture of T-cell-deficient lymphocyte populations or nude mouse lymphoid cell populations with thymic epithelial cells (22–24), supernates from thymic epithelial cells (24, 25), or thymic extracts (26–28) (in particular, thymosin [29, 30]), has been shown to result in several manifestations of normal T-cell function, including: acquisition of T-cell surface markers (31–33); reconstitution of normal proliferative responses to T-cell mitogens (22); and reconstitution of in vitro responsiveness to T-dependent antigens such as SRBC (34), dinitrophenol-protein conjugates, and alloantigen (35). The data presented in this report, in particular, the inability of nude mouse lymphoid cell populations to produce TCGF and still respond to both T-cell mitogen and alloantigen provided exogenous TCGF was present, lead us to hypothesize that the influence of the thymus in the above-cited T-cell function reconstitution experiments was to program the differentiation of prothymocytes into mature T cells capable of producing TCGF. Indeed, it is our belief that a major function of the thymus as a site of T-cell differentiation is to influence the maturation of T lymphocytes capable of cooperating in the production of TCGF.

It should be noted that the pp-TCGF used in the experimentation described in this report might contain more than one molecular entity. In experimentation to be reported elsewhere, we have found that the TCGF activity obtained from DEAE-cellulose chromatography elutes as a single protein moiety after Bio-Gel-p30 (Bio-Rad Laboratories, Richmond, Calif.) gel-exclusion chromatography. Furthermore, active fractions pooled after DEAE chromatography migrate as a single peak of activity on analytical isoelectric focusing columns (isoelectric point = 5.65).[3] Although these results imply that the TCGF activity present in pp-TCGF (obtained via DEAE chromatography) resides in a single molecular entity, it remains possible that additional factors with similar molecular characteristics (present in pp-TCGF preparations) might also be involved in effecting the results we have described.

The biologic activity that distinguishes TCGF from other previously described mitogenic factors (see 36 for review) is its ability to initiate and sustain the continuous proliferation of activated T cells. TCGF production requires mature lymphocytes (either cortisol-resistant thymocytes or peripheral T cells)[2] and is elicited as a consequence of T-cell mitogen or antigen stimulation (3). Furthermore, only activated T cells absorb TCGF activity and respond by continuous proliferation. T-cells which have not been exposed to mitogen or antigen, as well as activated B lymphocytes, fail to absorb TCGF activity and do not proliferate in response to TCGF (4). Therefore, the weight of the evidence indicates that TCGF acts as a second signal effecting the proliferation of mature activated T cells and that prothymocyte populations respond in a similar manner after mitogen or antigen sensitization in the presence of TCGF.

Whether or not the function of the thymus in T-cell differentiation is mediated solely by effecting the maturation of TCGF-producing cells or is intimately associated with conferring antigen-specificity upon nondirected prothymocytes, the fact that TCGF restores normal T-cell responses to nude mouse spleen cells, suggests that TCGF may be capable of alleviating a great deal of the T-cell immunodeficiency presented by the nude mouse. The observations that: (a) nude mouse spleen cells generate cytolytic effector cells after TCGF-supplemented MLC stimulation and (b), antigen-specific nude mouse killer T cells can be maintained in culture indefinitely in the presence of TCGF, provide evidence that TCGF-treatment itself may lead to the development of a new modality for short-circuiting and treating T-cell immune deficiencies. It is our hope that continued studies of the effect of TCGF on nude mouse immune reactivity compounded with further investigation regarding the role of TCGF in thymus-directed T-cell maturation and differentiation, will provide additional insight into the regulation of both normal and deficient T-cell immune reactivity.

Summary

In addition to allowing for the long-term culture of both murine and human cytolytic T lymphocytes, T-cell growth factor (TCGF) functions as the key proliferation-inducing second signal in both T-cell antigen sensitization and mitogenesis. The observation that thymocytes responded normally to T-cell mitogens in the presence of TCGF, prompted the investigation of the effect of TCGF on nude mouse lymphocyte responses in vitro. We found that spleen, lymph node, and bone marrow cells, isolated from nude mice, were incapable of producing TCGF yet responded normally to T-cell mitogen sensitization provided stimulation was conducted in the presence of TCGF. Nude mouse spleen cells were also capable of responding to alloantigen sensitization in mixed lymphocyte cultures (MLC) conducted in the presence of TCGF. Thy-1 antigen-positive cells harvested from TCGF-supplemented nude mouse MLC effectively mediated the cytolysis of alloantigen-specific target cells as tested in standard ^{51}Cr-release assays. Cytolytic nude mouse effector cells have remained in TCGF-dependent culture for over 3 mo during which they have continued to mediate significant levels of alloantigen-specific cytolytic reactivity. These results suggest that prothymocytes present in nude mice are capable of responding to immunologic stimuli by differentiating, in vitro, into cytolytic T lymphocytes and that furthermore, a major function of the thymus may be to effect the maturation of TCGF-producing cells.

The authors gratefully acknowledge the generosity of Dr. William Culp and Dr. Stan Froehner of the Department of Biochemistry, Dartmouth Medical School, for the use of their fluorescence microscope and camera. We would also like to thank Dr. Francis Ruscetti for many helpful conversations and Ms. Linda Waidlich for her help in the preparation of this manuscript.

Received for publication 15 January 1979.

References

1. Gillis, S., and K. A. Smith. 1977. Long-term culture of tumor-specific cytotoxic T-cells. *Nature (Lond.)*. **268:**154.

2. Gillis, S., P. E. Baker, F. W. Ruscetti, and K. A. Smith. 1978. Long-term culture of human antigen-specific cytotoxic T-cell lines. *J. Exp. Med.* **148:**1093.

3. Gillis, S., M. M. Ferm, W. Ou, and K. A. Smith. 1978. T-cell growth factor: parameters of production and a quantitative microassay for activity. *J. Immunol.* **120:**2027.

4. Smith, K. A., S. Gillis, P. F. Baker, D. McKenzie, and F. W. Ruscetti. 1979. T-cell growth factor, mediated T-cell proliferation. *Ann. N. Y. Acad. Sci.* In press

5. Gillis, S., and K. A. Smith. 1977. In vitro generation of tumor-specific cytotoxic lymphocytes. *J. Exp. Med.* **146:**468.

6. Baker, P. E., S. Gillis, M. M. Ferm, and K. A. Smith. 1978. The effect of T-cell growth factor on the generation of cytolytic T-cells. *J. Immunol.* **121:**2168.

7. Cerottini, J. C., A. A. Nordin, and K. T. Brunner. 1970. Specific in vitro cytotoxicity of thymus-derived lymphocytes sensitized to alloantigens. *Nature (Lond.)*. **228:**1308.

8. Nabholz, M., H. D. Engers, D. Collavo, and M. North. 1978. Cloned T-cell lines with specific cytolytic activity. *Curr. Top. Microbiol. Immunol.* **81:**176.

9. Strausser, J. L., and S. A. Rosenberg. 1978. In vitro growth of cytotoxic human lymphocytes. I. Growth of cells sensitized in vitro to alloantigens. *J. Immunol.* **121:**1491.

10. Rosenberg, S. A., S. Schwartz, and P. J. Spiess. 1978. In vitro growth of murine T-cells. II. Growth of in vitro sensitized cells cytotoxic for alloantigens. *J. Immunol.* **121:**1951.

11. Wagner, H. 1972. The correlation between the proliferative and the cytotoxic responses of mouse lymphocytes to allogeneic cells in vitro. *J. Immunol.* **109:**630.

12. Feldman, M., H. Wagner, A. Basten, and M. Holmes. 1972. Humoral and cell mediated responses in vitro of spleen cells from mice with thymic aplasia (nude mice). *Austr. J. Exp. Biol. Med. Sci.* **50:**651.

13. Wortis, H. H., S. Nehlsen, and J. J. Owen. 1971. Abnormal development of the thymus in nude mice. *J. Exp. Med.* **134:**681.

14. Pritchard, H., and H. S. Micklem. 1973. Haemopoietic stem cells and progenitors of functional T-lymphocytes in the bone marrow of nude mice. *Clin. Exp. Immunol.* **14:**597.

15. Pantelouris, E. M. 1971. Observation on the immunobiology of nude mice. *Immunology.* **20:**247.

16. Wortis, H. H. 1975. Pleiotropic effects of the nude mutation. *Birth Defects, Orig. Art. Ser.* **11:**582.

17. Irle, C., P. F. Piguet, and P. Vassalli. 1978. In vitro maturation of immature thymocytes into immunocompetent T-cells in the absence of direct thymic influence. *J. Exp. Med.* **148:**32.

18. Schimpl, A., and E. Wecker. 1972. Replacement of a T-cell function by a T-cell product. *Nature (Lond.)*. **237:**15.

19. Kindred, B., and R. B. Corley. 1977. A T-cell replacing factor specific for histocompatibility antigens in mice. *Nature (Lond.)*. **268:**531.

20. Farrar, J. J., W. J. Koopman, and J. Fuller-Bonar. 1977. Identification and partial purification of two synergistically acting helper mediators in human mixed leukocyte culture supernatants. *J. Immunol.* **119:**47.

21. Piguet, P. F., and P. Vassalli. 1978. Rejection of allo- or xenografts of lymphoid cells by nude mice: T-cell suicide as a result of cooperation between histoincompatible T and B cells. *J. Immunol.* **120:**79.

22. Gershwin, M. E., R. M. Ikeda, W. L. Kruse, F. Wilson, M. Shifrine, and W. Spangler. 1978. Age-dependent loss in New Zealand mice of morphological and functional characteristics of thymic epithelial cells. *J. Immunol.* **120:**971.

23. Waksal, S. D., I. R. Cohen, H. W. Waksal, H. Wekerle, R. L. St. Pierre, and M. Feldman. 1975. Induction of T-cell differentiation in vitro by thymus epithelial cells. *Ann. N. Y. Acad. Sci.* **249:**492.

24. Willis-Carr, J. I., H. D. Ochs, and R. J. Wedgewood. 1978. Induction of T-lymphocyte differentiation by thymic epithelial cell monolayers. *Clin. Immunol. Immunopathol.* **10:**315.

25. Hensen, E. J., E. C. M. Hoefsmit, and J. G. Van den Tweel. 1978. Augmentation of mitogen responsiveness in human lymphocytes by a humoral factor obtained from thymic epithelial cultures. *Clin. Exp. Immunol.* **32:**309.

26. Trainin, N., and M. Small. 1970. Studies on some physicochemical properties of a thymus humoral factor conferring immunocompetence on lymphoid cells. *J. Exp. Med.* **132:**885.

27. Miller, H. C., S. K. Schmiege, and A. Rule. 1973. Production of functional T-cells after treatment of bone marrow with thymic factor. *J. Immunol.* **111:**1005.

28. Lonai, P., B. Mogolner, V. Rotter, and N. Trainin. 1973. Studies on the effect of a thymic humoral factor on differentiation of thymus-derived lymphocytes. *Eur. J. Immunol.* **3:**21.

29. Bach, J. F., J. Dardenne, A. L. Goldstein, A. Guha, and A. White. 1971. Appearance of T-cell markers in bone marrow rosette-forming cells after incubation with thymosin, a thymic hormone. *Proc. Natl. Acad. Sci. U. S. A.* **68:**2734.

30. Goldstein, A. L., T. L. K. Low, M. McAdoo, J. McClure, G. B. Thurman, J. Rossio, C. Lai, D. Chang, S. Wang, C. Harvey, A. H. Ramel, and J. Meienhofer. 1977. Thymosin α_1: Isolation and sequence analysis of an immunologically active thymic polypeptide. *Proc. Natl. Acad. Sci. U. S. A.* **74:**725.

31. Komuro, K., and E. A. Boyse. 1973. Induction of T-lymphocytes from precursor cells in vitro by a product of the thymus. *J. Exp. Med.* **138:**479.

32. Scheid, M. P., M. K. Hoffmann, K. Komuro, U. Hammerling, J. Abbott, E. A. Boyse, G. H. Cohen, J. A. Hooper, R. S. Schulof, and A. L. Goldstein. 1973. Differentiation of T-cells induced by preparations from thymus and by nonthymic agents. The determined state of the precursor cell. *J. Exp. Med.* **138:**1027.

33. Komuro, K., and E. A. Boyse. 1973. In vitro demonstration of thymic hormone in the mouse by conversion of precursor cells into lymphocytes. *Lancet.* **I:**740.

34. Small, M., and N. Trainin. 1967. Increase in antibody-forming cells of neonatally thymectomized mice receiving calf-thymus extract. *Nature (Lond.).* **216:**377.

35. Amerding, D., and D. H. Katz. 1974. Activation of T and B lymphocytes in vitro. IV. Regulatory influence on specific T-cell functions by a thymus extract factor. *J. Immunol.* **114:**1248.

36. Oppenheim, J. J., S. B. Mizel, and M. S. Melt. 1979. Comparison of lymphocyte and mononuclear phagocyte derived mitogenic "amplification" factors. *In* Immunobiology of the Lymphokine. S. Cohen, E. Pick, and J. J. Oppenheim, editors. Academic Press Inc., New York. In press.

Chapter IV

Suppression of Antibody Synthesis by T Cells

Chapter IV

Suppression of Antibody Synthesis by T Cells

The capacity of the immune system to generate an enormously diverse population of antibody molecules with specificities for a universe of different antigens is controlled in a variety of ways so that only a highly specific and appropriate immune response is evoked by a given antigen. In addition to the positive regulation that establishes the specificity of a response there must also be regulation that results in the modulation or abolition of a response. Experiments conducted over the past decade have demonstrated that much of the negative regulation (suppression) of the immune response is mediated by a special set of suppressor T lymphocytes.

The issues that will be addressed in this chapter are the characterization of suppressor T cells, their mode of action and their cellular targets. Before considering the specific papers, let us try to construct a simple model to use as a guide. We know that helper T cells are required to stimulate B cells to divide and to differentiate into antibody secreting cells. For conventional antigens, both the helper T cell and the B cell are specific for the same antigen, although not necessarily for the same determinants. In experimental systems using hapten-carrier conjugates as antigens, the helper T cell is specific for the carrier and the B cell for the hapten. Successful activation of an antibody response also requires the participation of macrophages. Where, then, in this intricate set of interacting cells, could the suppressor T cell exert its influence? Clearly, the suppressor T (T_S) cell could affect regulation by altering the ability of helper T (T_H) cells, B cells, or macrophages to function. For example, suppression could result by a) preventing macrophages from "presenting" antigen to T_H cells, b) inhibiting B or T_H cell proliferation or differentiation, c) actively eliminating reactive B or T_H cell clones. It is conceivable that suppressor T cells might demonstrate sufficient functional heterogeneity to act in a number of different ways.

In addition to the issue of the target cells of suppressor T cell action, one must also consider the detailed mechanism of action of T_S cells: is cell contact essential? Are there freely diffusible substances capable of acting at a distance? How long does suppression last? Is it reversible? Does it require the continued presence of suppressor T cells? How do T_S differ from other clones of T cells? Finally, how do these T cells recognize their respective antigens and targets?

We will explore all of these issues as well as a brief historical account of the suppression phenomenon in this chapter.

The discovery of suppressor T cells is described in the paper by Gershon and Kondo (paper 1). They were exploring the basis of experimentally induced unresponsiveness, or tolerance, in the immune response to sheep red blood cells (SRBC). Conventional wisdom at the time of this discovery held that most, if not all, forms of tolerance were due to the deletion of specific antigen-reactive clones of lymphocytes. If this were true, a mixture of lymphocytes from an animal tolerant to SRBC with lymphocytes from an animal immune to SRBC would be expected to be reactive to SRBC; responsiveness would be dominant over unresponsiveness. Surprisingly, Gershon and Kondo found the reverse to be true: transfer of spleen cells from a tolerant animal into an immune animal resulted in the transfer of tolerance to the recipient. Hence, tolerance was "infectious." Results from further experiments described in paper 1 indicated to Gershon that T lymphocytes were responsible for this "active" form of immune unresponsiveness. This phenomenon is antigen specific in that the recipient test animal can still respond to a closely related antigen (e.g., horse red blood cells).

In an attempt to explore further the mechanisms of suppression, hapten-carrier model systems were developed. In this chapter we have chosen to consider in some depth the regulations of the immune response to the T-dependent hapten-carrier conjugate DNP-KLH, considering the many aspects of "suppressorology" which it represents. We do this partly because a detailed consideration of the many different T-suppressor mediated phenomena is impossible, and learning the details of numerous experimental protocols can be overwhelming. This is not to slight important contributions made by other workers studying other systems, and relevant papers from other labs are included here. A bibliography containing recent review articles on the nature of suppressor T cells is included at the end of the chapter. Hapten-carrier systems provided the means for examining whether or not suppression is effected by direct interference with the antibody forming cell or with the helper T cell needed to trigger the B cell to produce antibody. The following papers by Tada and collaborators demonstrate that suppression of antibody formation is carrier specific, suggesting that suppression can be directed at the helper T cell. In their first paper in the series, not included here (Tada and Takemori, 1974), they demonstrated that T cells derived from animals injected with high doses of a carrier protein can prevent a naive animal from responding to a hapten coupled to the same carrier. These same T cells are ineffective in preventing a response to the hapten when the hapten is coupled to a different carrier. The properties of this carrier specific suppressor T cell are the subject of many papers by Tada and collaborators. In the first of the selected papers by these authors, Takemori and Tada (paper 2) demonstrate that an active soluble factor can be obtained by sonication of spleen cells or thymocytes of animals injected in such a way as to produce suppressor T cells. This "factor" mimicks exactly the action of the suppressor T cells: if it is injected into animals prior to immunization, it prevents anti-hapten antibody formation. As was described before, the suppression is only seen if the hapten is coupled to the same carrier that is used to generate the "factor."

The advantages of obtaining a cell-free product that has biological activity are numerous. As Takemori and Tada show, the factor is at least in part a protein that can bind to antigen. However, it does not have the characteristics of immunoglobulin. The most surprising finding is that the factor contains determinants encoded within the H-2 complex and will act preferentially on H-2 compatible carrier-specific cells. That is, the "factor" will act only to inhibit an immune response if it is derived from an animal that shares the same H-2 associated antigens as the test animal.

The specific T cell responsible for the generation of suppressor factor was identified next by taking advantage of the differential display of surface antigens by functionally distinct T cells. In paper 3, Okumura et al. demonstrate that suppressor T cells (and not helper T cells) bind to the antigen which they recognize, carry the Lyt 2 and Lyt 3 antigens and have no F_C receptor. These cells, as well as the factor derived from them, bear an antigen controlled by a gene within the H-2 complex.

In fact, as described by Tada, Tominuchi and David (paper 4), the antigen responsible for marking suppressor T cells was coded for by a gene shown to map to a previously undescribed region of the H-2 complex, the I-J region. This paper serves as an excellent example of the power of genetic analysis for studying phenomena of the immune system. The study makes use of recombinant inbred strains that show defects in the production of, or in the sensitivity to, suppressor factor.

Having identified the cell type responsible for suppression as being a T cell, the next question to address is the characterization of these specialized cells and their products. This task, however, is not trivial. The number of suppressor T cells specific for a particular antigen in an unresponsive animal is less than 1% of the total splenic lymphocyte pool. The factor derived from the sonication of suppressor T cells is undetectable by ordinary chemical means. An additional complication comes from the likelihood that the "suppressor molecules" of T cells are as heterogeneous and diverse as antibody molecules. That is, recognition of a complex carrier molecule could involve numerous helper T cells, each specific for a distinct determinant. These T cells could each be producing factors which reflect the antigen specificity of the particular receptor and are collectively responsible for the carrier-specific suppression detected. Thus in order to characterize the "factor" in detail, a homogeneous population of T cells (i.e., clones) would have to be obtained. In paper 5, Taniguchi, Takei and Tada describe the production of a continuously growing cloned T cell line which synthesizes a biologically active suppressor factor. The line was produced by fusion of an enriched population of suppressor T cells with a continuous T lymphoma cell line to give rise to a "T cell hybridoma."

Analysis of the active product of this hybridoma revealed that it was composed of two dissociable polypeptides, one having antigen binding properties, the other containing I-J encoded determinants. These results are in accord with what one could expect from the characterization of the previously described crude substance. One intriguing question is raised by the analysis of the activity of the factor (presumably homogeneous): why does administration of a monospecific factor result in the abolition of all of the carrier activity? That is, the KLH carrier is a large, complex protein, bearing many distinct antigenic determinants. It therefore has many monospecific helper T cells recognizing it whose combined activity contributes to an anti-hapten response.

How could the factor operate to prevent help for all carrier determinants? The apparent answer comes from a study on the mechanism of action of the factor. Taniguchi and Takuhisa (paper 6) report that the factor is elaborated in response to a specific antigen interacting with a specific T cell. The target of the factor is not a specific helper T cell however. It appears that the target of the factor is yet another T cell that, in turn, elaborates an antigen-non-specific "factor." It is the second of these factors which results in the final suppression of antibody formation. The antigen-non-specific end point of this chain of reactions explains how a mono-specific suppressor cell could suppress all helper activity.

At this stage there are two additional issues which remain unresolved. First, what is the target cell at the end of the suppressor chain? Second, what is the global effect caused by the antigen non-specific final step?

Experiments addressing the first question give experimental support for the idea that a helper T cell is a direct target of the suppressor T cell.

The paper by Herzenberg et al. employs an entirely different experimental system to analyze suppression (paper 7) and shows that elimination of the suppressor T cell population of a suppressed animal does not result immediately in the resumption of an antibody response. These authors demonstrate that the combined failure of the immune response is due to the complete absence or total inactivation of the specific helper T cell population. In contrast, the resident B cell population in suppressed animals is perfectly functional and can be elicited by presentation of appropriate, active helper T cells.

As is often the case in immunological research, we are left to determine for ourselves whether the results obtained in the Herzenberg system can be generalized and applied to the system studied by Tada and associates.

Regardless of the resolution of this issue which must await additional experimentation, it would appear that suppression can also be mediated (as studied in yet another system) by interference with the B cell. The latter result stems from the analysis of mice that are unable, for reasons of their genetic constitution, to respond to certain antigens. Such mice, injected with the antigen to which they cannot respond, produce suppressor T cells which have an interesting property. Injection of these cells, or factors derived from them, into naive responder animals, prevents the normally responsive animals from responding to the antigen. The same result is obtained even if the test antigen is coupled to a carrier. In the latter case, the antigen is being studied as a hapten by coupling it to a carrier, yet animals that have received suppressor cells cannot respond. Thus if we consider this as a hapten-carrier system, suppressor cells generated to a hapten can prevent an animal from responding to the hapten regardless of the carrier to which that hapten can be coupled. It would appear that suppression in this case is hapten specific, not carrier specific, and is therefore likely to be directed at the B cell. These results are described in the paper by Kapp et al. (paper 7).

There are several differences in the immune responses studied by Kapp and Tada. Tada and coworkers examine carrier specific suppression as assayed in a secondary response while Kapp and associates examine the hapten-directed suppression of a primary response. It is not unreasonable to conclude that both modes of suppression are possible and do occur under normal physiological circumstances.

The student of immunology may well be confused by the complexities and apparent paradoxes presented by the results of the different experimental systems. It is essential to bear in mind that the immune system may not be a collection of cells involved only in binary actions. As discussed in detail in a subsequent chapter on immune networks, a more reasonable model of the immune system may be one that envisions a complex network of interactive cells communicating with each other both by contact and through diffusable molecules. Even the picture given in this chapter may be an over-simplification of suppression. In some laboratories evidence is accumulating (Eardley, et al., 1978; Cantor, et al., 1978) which suggests that many different T cells participate in enormously complex regulatory interactions.

Where do we stand now?

From the collected works of Tada and associates, we know that suppression is brought about by a cascade of events which involves several cell types and soluble substances that ultimately lead to the prevention of antibody formation. Suppressor cells and antigen are at the head of the cascade. We do not yet know what the ultimate target is, although the most likely target seems to be the helper T cell. One puzzling feature of this system is that free antigen would appear to initiate the cascade by activating suppressor T cells. Why would an immune response be turned off in response to free antigen? We obviously need to know where within the entire system this chain of events belongs. At this stage we are able only to catch a glimpse into possible regulatory events within the immune system; the functioning of the entire system still eludes us.

Papers Included in this Chapter

1. Gershon, R.K. and Kondo, K. (1971) Infectious immunological tolerance. Immunology 21, 903.

2. Takemori, T. and Tada, T. (1975) Properties of antigen-specific suppressive T-cell factor in the regulation of antibody response of the mouse. I. In vivo activity and immunochemical characterizations. J. Exp. Med. 142, 1241.

3. Okumura, K., Takemori, T., Tokuhisa, T., and Tada, T. (1977) Specific enrichment of the suppressor T cell bearing I-J determinants. Parallel function and serological characterizations. J. Exp. Med. 146, 1234.

4. Tada, T., Taniguchi, M., and David, C.S. (1976) Suppressive and enhancing T-cell factors as I-region gene products: properties and the subregion assignment. Cold Spring Harbor Symp. Quant. Biol. 50, 119.

5. Taniguchi, M., Takei, I., and Tada, T. (1980) Functional and molecular organization of an antigen-specific suppressor factor from a T-cell hybridoma. Nature 283, 227.

6. Taniguchi, M. and Tokuhisa, T. (1980) Cellular consequences in the suppression of antibody response by the antigen-specific T-cell factor. J. Exp. Med. 151, 517.

7. Herzenberg, L.A., Okumura, K., Cantor, H., Sato, V.L., Shen, F.-W., Boyse, E.A., and Herzenberg, L.A. (1976) T-cell regulation of antibody responses: demonstration of allotype-specific helper T cells and their specific removal by suppressor T cells. J. Exp. Med. 144, 330.

8. Kapp, J.A., Pierce, C.W., de la Croix, F., and Benacerraf, B. (1976) Immunosuppressive factor(s) extracted from lymphoid cells of nonresponder mice primed with L-glutamic acid^{60}-L-alanine^{30}-L-tyrosine^{10} (GAT). I. Activity and antigenic specificity. J. Immunol. 116, 305.

References Cited in this Chapter

Cantor, H., Hugenberger, J., McVay-Boudreau, L., Eardley, D.D., Kemp, J., Shen, F.W., and Gershon, R.K. (1978) Immunoregulatory circuits among T cell sets. Identification of a subpopulation of T-helper cells that induces feedback inhibition. J. Exp. Med. 148, 871.

Eardley, D.D., Hugenberger, J., McVay-Boudreau, L., Shen, F.W., Gershon, R.K., and Cantor, H. (1978) Immunoregulatory circuits among T-cell sets. I. T-helper cells induce other T-cell sets to exert feedback inhibition. J. Exp. Med. 147, 1106.

Tada, T. and Takemori, T. (1974) Selective roles of thymic derived lymphocytes in the anitbody response. I. Differential suppressive effect of carrier-primed T cells on hapten-specific IgM and IgG antibody responses. J. Exp. Med. 140, 239.

Infectious Immunological Tolerance

R. K. Gershon and K. Kondo

*Department of Pathology, Yale University School of Medicine,
New Haven, Connecticut 06510, U.S.A.*

(*Received 11th March* 1971)

Summary. Previous studies have shown that thymectomized lethally irradiated bone marrow grafted mice, reconstituted with thymocytes and pretreated with a large dose of sheep red blood cells (SRBC), are unable to respond to a subsequent immunizing injection of SRBC even after an inoculation of normal thymocytes. If, however, the mice are not thymocyte reconstituted prior to the pretreatment with SRBC, they can respond almost normally to an immunizing injection of SRBC if inoculated with normal thymocytes after the termination of antigen pretreatment.

In the present study the immunosuppressive effect of the presence of thymocytes during the antigen pretreatment was studied by adoptively transferring the spleen cells of the antigen pretreated mice to thymus-deprived chimeras. These spleen cells not only did not co-operate with normal thymocytes in the secondary hosts, but they also prevented the co-operation of normal thymocytes with normal bone marrow derived cells. Untreated spleen cells or treated spleen cells from mice not reconstituted with thymocytes did not affect cell co-operation in the secondary hosts. The abrogation of the co-operation in the secondary host was specific in that the addition of spleen cells did not affect the anti-horse red blood cell response. If the primary host made antibody as a result of the pretreatment, the transfer of their spleen cells did not prevent antibody production in the secondary host.

INTRODUCTION

We have recently reported that co-operation between lymphoid cells may play an important role in the induction of immunological tolerance (Gershon and Kondo, 1970). In those studies tolerance induction was studied in two groups of mice. Both groups were thymectomized, lethally irradiated and bone marrow grafted; one group was also given 1.5×10^7 thymocytes along with the bone marrow. Both groups of mice were then given a large number (2.5×10^{10}) of sheep red blood cells (SRBC) over a 30-day period. Four days after termination of this tolerance induction schedule, neither group of mice could respond to an immunizing dose of antigen. However, if normal thymocytes were innoculated at the time of immunization, the mice that had been pretreated in the *absence* of thymocytes (the thymus-deprived mice) could respond almost as well as non-pretreated controls. On the other hand, mice pretreated in the *presence* of thymocytes (the thymus-reconstituted mice) were totally unable to respond to an immunizing dose of antigen even after the addition of further thymocytes. Thus the presence of thymus-derived lymphocytes during the course of the tolerance induction had created a milieu in the experimental animal where normal thymocytes could not co-operate with pretreated bone-marrow-derived cells (BMDC).

To gain further insight into how the co-operation of these two cell types had been abrogated, we have transferred the spleen cells of pretreated mice into thymectomized lethally irradiated recipients to see if they can co-operate with normal thymocytes in a new environment. This manoeuvre removes the cells from a possible source of residual antigen or other potential immunosuppressive factors that might be circulating in the original host.

We have found that not only will the spleen cells from animals pretreated in the presence of thymocytes not co-operate with normal thymocytes in the new environment, but that they will also prevent the normal thymocytes from co-operating with normal BMDC.

MATERIALS AND METHODS

Mice

Male or female CBA mice were used in these experiments. They were of either strain CBA/H T_6T_6 from our own colony or strain CBA/J from Jackson Laboratories. All experiments were controlled for sex or strain of mouse.

Thymectomy

Thymectomies were performed on adult mice, 7–8 weeks of age, under light ether anaesthesia following the technique of Miller (1960). At the termination of experiments all mice were autopsied and thymic remnants were searched for. None were found in any animals used in these experiments.

Irradiation

A mid-axis dose of 850 R was delivered from a Siemans 250 kV machine at a rate of 85 rad/min.

Cell suspensions

Bone marrow cell suspensions were prepared by washing out the femurs of adult syngeneic mice with cold sterile medium M199. Thymus cell suspensions were prepared by gently teasing thymus glands of syngeneic weanling (4–5 weeks of age) mice between sterile glass slides in cold M199, filtered through gauze, and washed before injection. Spleen cell suspensions were made by the same technique. Counts of viable cells were made in a haemocytometer using the trypan blue dye exclusion method. The cells were inoculated intravenously via the tail vein.

Red blood cells

These were obtained in Alservers solution and washed three times in saline before use.

Bleeding

Blood was collected from the retro-orbital sinus by capillary pipette. Serum was separated and used for titration within 24 hours. Individual mice were ear-marked so that each could be followed serially.

Titrations

All sera were individually titrated by the microhaemagglutination technique of Sever

(1962). Titres are expressed as the $\log_2$ of the last well showing macroscopic agglutination. After the results were recorded, the red cells were resuspended by gentle tapping of the plates and 0·025 ml of 0·1 M 2-mercaptoethanol (ME) was added to each well. The cells were allowed to resettle at room temperature and end-points were read as before. These titres were taken to represent ME resistant (MER) antibody. This method of ME inactivation has been studied at some length and has been shown to produce the same results as more standard techniques (Scott and Gershon, 1970). It was used in these studies in order to minimize the blood loss of experimental animals. MER antibody, although not exactly analogous, is roughly equivalent to 7S antibody under ordinary circumstances (Adler, 1965).

Statistical analysis

Student's *t*-test was used in all statistical analyses.

EXPERIMENTAL PLAN

An outline of the protocol followed is presented in Fig. 1. The method used to induce tolerance was the same as the one previously reported (Gershon and Kondo, 1970). Four days after the termination of tolerance induction, the spleen cells of the treated and control mice were transferred to thymectomized mice that had been lethally irradiated and bone marrow grafted 30 days previously. Some of these secondary recipients were given additional normal thymocytes. The resultant chimeras were then immunized with either a 20 per cent suspension of SRBC, a 20 per cent suspension of horse red blood cells (HRBC) or were unimmunized.

The numbers of thymocytes and spleen cells that were given varied from experiment to experiment, and the numbers for each individual experiment are given in the text.

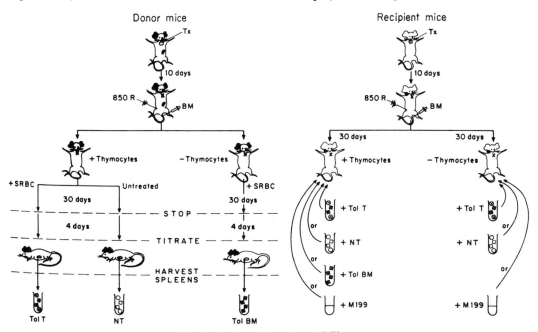

FIG. 1. Outline of Experimental Plan.

RESULTS

The first experiment we report was performed on CBA/HT_6T_6 mice. The donor mice had been reconstituted with 3×10^7 thymocytes and after the termination of tolerance induction 1×10^8 of their spleen cells were transferred to thymus-deprived recipients. Some of these recipients also received an inoculation of $3 \cdot 0 \times 10^7$ normal thymocytes.

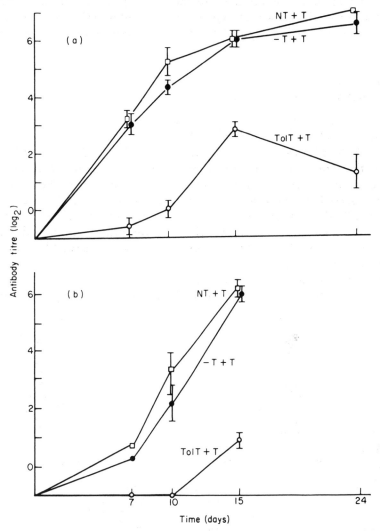

Fig. 2. Anti-SRBC response ($\pm$ S.E.) of thymus-deprived mice given spleen cells from thymus-reconstituted chimeras (see Table 1) and normal thymocytes. (a), Total antibody. (b), MER antibody. (NT+T), Donor mice untreated; normal thymocytes added; ($-$T+T), no donor cells given; normal thymocytes added; (TolT+T), donor mice given SRBC; normal thymocytes added.

In Fig. 2 the anti-SRBC response of three groups of recipient mice, all of which got normal thymocytes, is given. In addition to the normal thymocytes, one group received spleen cells from untreated mice (NT), one received no spleen cells ($-$T), and one received tolerant spleen cells (ToIT). Two points are clearly made: (1) The addition of untreated spleen cells did not significantly affect the immune response (NT *vs* −T).* (2) The addition of tolerant spleen cells significantly decreased the response (ToIT *vs* NT or −T). The inhibition produced by the tolerant spleen cells was much more marked in the MER fraction of antibody (ME inactivation was not performed on the anti-serum on day 24).

Fig. 3 gives the anti-SRBC titres of recipient mice that got the same spleen cells as the mice presented in Fig. 2, without the addition of normal thymocytes. It is clear that the

TABLE 1

PROBABILITY ($P<$) ANALYSES OF THE DATA PRESENTED. SEE VARIOUS FIGURES FOR EXPLANA-
TIONS OF GROUPS

			Day				
			5	7	10	14 or 15	21 or 24
FIG. 2.	NT+T	Total		0·001	0·001	0·001	0·001
	vs						
	ToIT+T	MER		0·02	0·001	0·001	
	−T+T	Total		0·001	0·001	0·001	0·001
	vs						
	ToIT+T	MER		0·05	0·001	0·001	
FIG. 2.	ToIT+T						
vs	*vs*	Total		N.S.	N.S.	N.S.	N.S.
FIG. 3.	ToIT−T	MER*		N.S.	N.S.	N.S.	
	NT+T	Total		0·001	0·001	0·001	0·001
	vs						
	NT−T	MER*		0·05	0·001	0·001	
	−T+T	Total		0·001	0·001	0·001	0·001
	vs						
	−T−T	MER*		0·05	0·01	0·001	
FIG. 4.	ToIT+T	Total		0·05	0·02	0·02	0·001
(HRBC)	*vs*						
	ToIT−T	MER*		N.S.	0·05	0·05	
FIG. 5.	ToIT+T	Total		N.S.	0·05	N.S.	0·02
	vs						
	ToIBM+T	MER		N.S.	N.S.	0·001	0·01
FIG. 6.	ToIT+T	Total	N.S.	0·5	0·001	0·001	
	vs						
	−T+T	MER	N.S.	N.S.	0·01	0·001	
	TMMT+T	Total	N.S.	0·001	0·001	0·01	
	vs						
	−T+T	MER	N.S.	0·001	0·001	0·01	
FIG. 7.	IMMT (high)+T	Total	N.S.	N.S.	N.S.	N.S.	
	vs						
	IMMT (low)+T	MER	N.S.	N.S.	N.S.	N.S.	
	IMMT (low)+T	Total	N.S.	N.S.	N.S.	N.S.	
	vs						
	ToIT+T	MER	N.S.	N.S.	N.S.	N.S.	
	IMMT (high)+T	Total	N.S.	N.S.	0·01	0·01	
	vs						
	ToIT+T	MER	N.S.	N.S.	0·02	0·001	

* Titres not given in figures.

* Table 1 summarizes the *P* values for all the experiments reported herein.

spleen cells of untreated donors did not contain enough thymus-derived lymphocytes to augment the anti-SRBC response (NT *vs* −T). A comparison of Fig. 4 with Fig. 3 reveals that the addition of thymocytes to the recipients significantly augmented their response if they had been given normal donor spleen cells (NT) or no donor spleen cells (−T), but did not if they had been given tolerant spleen cells (TolT).

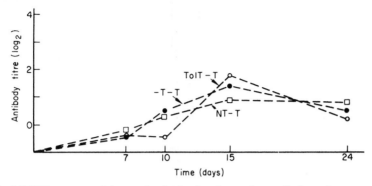

FIG. 3. Anti-SRBC response of thymus-deprived mice given spleen cells from thymus-reconstituted chimeras (see Table 1); no additional normal thymocytes were given. (a) Total Antibody. (NT−T), Donor mice not given SRBC; normal thymocytes not added; (−T−T), no donor cells given; normal thymocytes not added; (TolT−T), donor mice given SRBC; normal thymocytes not added.

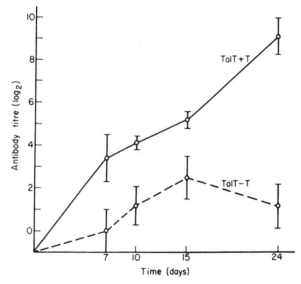

FIG. 4. Anti-HRBC response (±S.E.) of thymus-deprived mice given spleen cells from thymus-reconstituted chimeras (see Table 1); some mice also got normal thymocytes. (a) Total antibody. (TolT+T), Donor mice given SRBC; normal thymocytes added; (TolT−T), donor mice given SRBC; normal thymocytes not added.

A specificity control from this experiment is presented in Fig. 4. It can be seen that the addition of tolerant spleen cells did not prevent thymocytes from augmenting the immune response to HRBC.

Another experiment was done to test whether thymocytes had to be present in the donor spleen for the immunosuppressive effect to occur. In this experiment half the donors were given 3×10^7 thymocytes on the day of irradiation and bone marrow reconstitution. All animals were then given the standard SRBC pretreatment and 4 days after the last injection of SRBC 1×10^8 spleen cells plus 3×10^7 normal thymocytes were transferred to thymus-deprived recipients. This experiment was also done on T_6 mice.

Fig. 5 demonstrates that only the spleen cells of thymus-reconstituted mice produced an immunodepression after adoptive transfer. As in previous experiments, the suppression was most marked in the MER fraction of antibody.

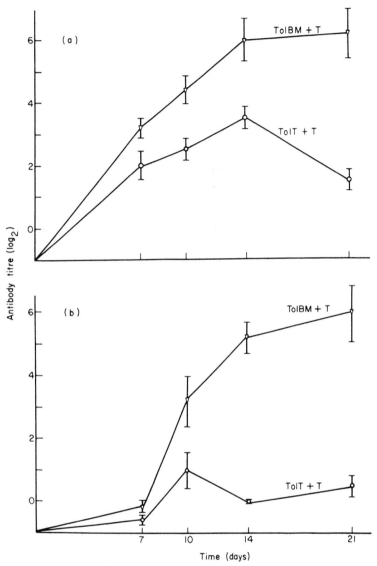

FIG. 5. Anti-SRBC response ($\pm$ S.E.) of thymus-deprived mice given spleen cells from thymus-deprived chimeras (see Table 1) some of which had been reconstituted with thymocytes. All donor mice were pretreated with SRBC. (a) Total antibody. (b) MER antibody. (Tol BM+T), Donor mice not reconstituted with thymocytes; normal thymocytes added; (TolT+T), donor mice reconstituted with thymocytes; normal thymocytes added.

The experiments presented above show that spleen cells from thymus-reconstituted mice that have been pretreated with large amounts of SRBC can prevent normal thymocytes and BMDC from co-operating in a secondary recipient. We present two more experiments below which confirm these findings and which also show the difference in the effect the adoptively transferred spleen cells may have dependent upon whether or not the donor mice make antibody during the pretreatment.

In the first experiment twenty-four CBA/J mice were pretreated after reconstitution with 4×10^7 thymocytes. At the termination of pretreatment eleven mice had antibody titres of $\log_2 1$ or less. These antibodies were all ME sensitive. Ten mice made antibody with a titre of $\log_2 6$ or more, which was mostly MER (mean antibody titre $\log_2$; total 6·7; MER 6·2). Separate spleen cell suspensions were made from these two groups of donor mice and 8×10^7 cells were given to thymus-deprived chimeras along with 3×10^7 normal thymocytes. The recipients were then immunized with SRBC.

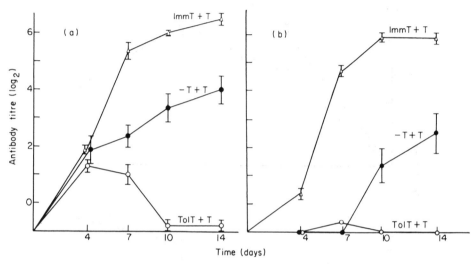

FIG. 6. Anti-SRBC response ($\pm$ S.E.) of thymus-deprived mice given spleen cells from thymus-reconstituted chimeras (see Table 1) and normal thymocytes. All donor mice were pretreated with SRBC. (a) Total antibody. (b) MER antibody. (IMMT+T), Donor mice made antibody; normal thymocytes added; (TolT+T), donor mice did not make antibody; normal thymocytes added; (−T+T), no donor cells given; normal thymocytes added.

The results (Fig. 6) show that the spleen cells from the antibody making mice were immune; recipients of these cells made significantly more antibody than controls (IMMT vs −T). On the other hand recipients that got spleen cells from the mice that made no antibody made significantly less antibody than controls (TolT vs −T). In fact they made no MER antibody at all.

The last experiment we report is one in which more donor mice made antibody than in the previous experiment. These were also CBA/J mice and they had been reconstituted with $1·5 \times 10^7$ thymocytes prior to the antigen pretreatment. Five donor mice made no antibody at all, five made antibody with a $\log_2$ titre between 4 and 5 (total 4·8; MER 4·6) and twenty-seven made antibody with a $\log_2$ titre between 6 and 8 (total 7·7; MER 7·1). Twelve $\times 10^6$ spleen cells from three separate pools were given to thymus-deprived chimeras along with 3×10^7 normal thymocytes.

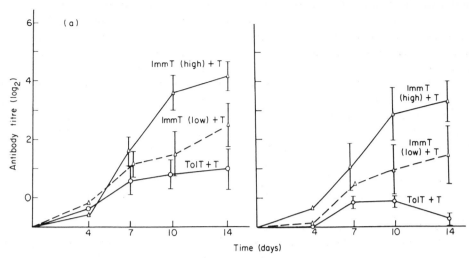

F$_{IG}$. 7. Anti-SRBC response ($\pm$S.E.) of thymus-deprived mice given spleen cells from thymus-reconstituted chimeras (see Table 1) and normal thymocytes. All donor mice were pretreated with SRBC. (a) Total antibody. (b) MER antibody. (IMMT (high)+T), Donor mice made titres > $\log_2$ 6; normal thymocytes added; (IMMT (low)+T), donor mice made titres between $\log_2$ 4–5; normal thymocytes added; (TolT+T), donor mice did not make antibody; normal thymocytes added.

The results (Fig. 7) show that the more antibody the donor mice had made, the more antibody the recipients of their spleen cells made.

DISCUSSION

The results presented above demonstrate that the adoptive transfer of spleen cells from mice made tolerant to SRBC clearly and specifically prevents the co-operation of normal thymocytes and normal BMDC. The results also demonstrate that it is necessary for thymus-derived lymphocytes to be present during the course of tolerance induction for this phenomenon to occur.

We have considered three general mechanisms by which the adoptive transfer of tolerance may be produced. They are: (1) The transfer of antigen, (2) The production of an immunosuppressive substance by the transferred BMDC and (3) The production of an immunosuppressive substance by the transferred thymus-derived cells.

1. *The transfer of free antigen.* This mechanism seems least likely to us since the adoptive transfer of spleen cells from mice treated in the absence of thymus-derived lymphocytes did not produce adoptive tolerance. Thus in order to postulate that antigen was a causative factor one would have to postulate that it was thymus 'processed' antigen, which was particularly tolerogenic.

2. *The effect is caused by a product of the BMDC.* The most well known product of BMDC is antibody (Davies, 1969; Miller and Mitchell, 1969) and it is well established that antibody can interfere with the immune response (see Uhr and Moller, 1968). For the following reasons we think that it is unlikely that conventional antibody is responsible for the effect we have reported. (a) Suppression occurred when neither donor mice nor their transferred spleen cells made any significant amount of detectable antibody. (b) Transferred spleen cells from donor mice that did make antibody did not produce a shut-off effect. The possibility that some exhaustively differentiated antibody making cells (Sterzl, 1966),

released a small amount of antibody and then went no further seems unlikely to us as the donor mice had large pools of antibody-making precursor cells and only small numbers of thymus-derived lymphocytes to activate them. (c) The ability of antibody to interfere with antibody production is related to the affinity of the antibody, with high affinity antibodies being most efficient (Walker and Siskind, 1968). Partially tolerant animals make antibody of low affinity (Theis and Siskind, 1968). Thus it seems unlikely that unmeasurable amounts of low affinity antibody could cause this effect. (d) In the experiments reported above there was always a preferential effect on the MER fraction of antibody. Passive antibody has not been reported to have this effect. Indeed it appears that the shut-off ability of passive antibody may preferentially affect ME sensitive antibodies (Sahiar and Schwartz, 1964; Wigzell, 1966; Morris and Moller, 1968; Uhr and Moller, 1968). There appears to be no more basis for ascribing the effects we have reported to the production of conventional antibodies than for considering this to be the mechanism by which tolerance is generally produced.

3. The shut-off effect is produced by a product of the thymus-derived lymphocytes. By exclusion this mechanism appears most likely to us. The product might be either directly produced by the thymus-derived lymphocytes in the transferred spleens or indirectly produced by the BMDC that have been influenced by them. We hope to be able to distinguish between these two possibilities with the use of an anti-theta anti-serum (Raff, 1969).

The immuno-enhancing effect of thymus-derived lymphocytes has been ascribed to a putative immunoglobulin called IgX (Mitchison, 1968). Following simple algebra, we suggest that the putative immunosuppressive substance be called IgY (Gershon and Kondo, 1970).

There are a number of disparate observations in the literature, some of which are difficult to explain otherwise, which could all be explained or united by an IgY hypothesis. Most striking of these is the observation that *in vitro* incubation, a procedure that kills large lymphocytes, abrogated the tolerance of a population of thoracic duct cells (McGregor, McCullagh and Gowans, 1967). We would theorize that the cells killed were the IgY producing thymus-derived lymphocytes which were responsible for the tolerance.

A number of workers have observed that the adoptive transfer of normal syngeneic immunocompetent cells into tolerant animals does not abrogate the tolerant state (Chase, 1963; Crowle and Hu, 1969; Tong and Boose, 1970; McCullagh, 1970). McCullagh also showed that the transferred cells became unresponsive 3 days after transfer into the tolerant hosts. Although the possibility that residual antigen rendered the adoptively transferred cells tolerant was never completely excluded, the observation by Tong and Boose that even immunized cells were unable to break tolerance, renders this explanation unlikely in our opinion.

Horiuchi and Waksman (1968) injected antigen into the thymuses of normal adult rats and showed that significant amounts of antigen did not escape from the thymus into the circulation. This procedure rapidly rendered the rats partially tolerant. The conclusion that the antigen injected rendered the cells in the thymus tolerant cannot explain the results entirely. Rendering thymus cells tolerant should have no immediate effect on the immunocompetence of an adult animal, unless the tolerance were infectious, as thymectomy itself at that age does not affect the immune response.

Baker, Stashak, Amsbaugh, Prescott and Barth (1970) have noted that thymectomy and ALS treatment increases the immune response of mice to pneumococcal polysaccharide, again suggesting the thymus-derived lymphocyte might make a product which shuts off other cells. Our observation that thymus-derived cells shut-off other cells in antigenic competition is also in line with this idea (Gershon and Kondo, 1971).

An IgY would be of great intellectual comfort in explaining ultra low zone tolerance (Shellam and Nossal, 1968). The amount of antigen used to produce tolerance in this situation can hardly be explained without resorting to some mechanism of amplification.

The fact that immunosuppressive agents can prevent tolerance induction (Claman and Bronsky, 1968) suggests that tolerance is an active process such as might be required for the production of IgY.

Last and perhaps most direct is the observation that the adoptive transfer of antigen pretreated lymphoid cells can abrogate the delayed hypersensitivity response of recipient mice immunized to that antigen (Crowle and Hu, 1969). These authors have suggested that the lymphoid cells make a substance they call 'contrasensitizer†.'

For the above stated reasons we favour the hypothesis that thymus-derived lymphocytes make an immunosuppressive substance, to explain the results we have presented. We would like to emphasize that at the present time we consider this a working hypothesis and that more information is needed before alternate explanations can be ruled out.

If our results do nothing else they vitiate one of the interpretations we have previously made (Gershon and Kondo, 1970). We had favoured the interpretation that the inability of thymocytes to break tolerance was due to tolerance of the BMDC. Since it now appears that the tolerance in our system can be spread from cell to cell this conclusion cannot be validated by our data. Nonetheless, the recent results of Playfair (1969) and of Chiller, Habicht and Weigle (1970) showing specific unresponsiveness of bone marrow cells from tolerant animals, are less likely to have been caused by IgY producing thymus-derived lymphocytes. Nonetheless, a remote possibility that the bone marrow cells were contaminated with small numbers of thymus-derived lymphocytes must be considered. We estimate that no more than several hundred thousand thymus-derived lymphocytes were present in the spleen cell suspensions we transferred. It is possible that very few contaminating cells could produce the effect.

Lastly, we would like to comment on why we have been able to adoptively transfer tolerance while some other workers have not. We believe this is because we transferred cells to minimally reconstituted animals wherein the effect of a small amount of IgY production could be seen. When we transferred spleen cells that could produce an effect in reconstituted chimeras to normal animals, we failed to observe an immunosuppressive effect (Gershon and Kondo, unpublished observations). That the effect is small and difficult to see does not necessarily mean, however, that the same is true for its significance.

ACKNOWLEDGMENTS

We thank Ellen Searle for expert technical assistance.

Work supported by USPHS grant CA-08593 from the NCI. R. K. Gershon is a recipient of a USPHS Career Development Award (CA-10,316).

† More recently several new reports of adoptive tolerance production in immunocompetent mice have appeared. (Asherson, G. L., Zembala, M. and Barnes, R. M. (1971) *Clin. exp. Immunology* 9, 109.) (Terman, D. S., Minden, P. and Crowle, A. J. [1971]. 'Adoptive transfer of neonatal tolerance into normal mice.' *Fed. Proc.*, **30**, 650.)

REFERENCES

ALDER F. L. (1965). 'Studies on mouse antibodies. I. The response to sheep red cells.' *J. Immunol.* **95**, 26.

BAKER, P. J., STASHAK, P. W., AMSBAUGH, D. F., PRESCOTT, B. and BARTH, R. (1970). 'Evidence for the existence of two functionally distinct types of cells which regulate the antibody response to type III pneumococcal polysaccharide.' *J. Immunol.*, **105**, 1581.

CHASE, M. W. (1963). 'Tolerance towards chemical allergens.' La Tolérance Acquise et la Tolérance Naturelle à l'égard de substances antigéniques définies. (Ed. by A. Bussard), p. 139. Centre National de la Recherche Scientifique Paris VIIᵉ.

CHILLER, J. M., HABICHT, G. S. and WEIGLE, W. O. (1970). 'Cellular sites of immunologic unresponsiveness.' *Proc. nat. Acad. Sci. (Wash.)*, **65**, 551.

CLAMAN, H. N. and BRONSKY, E. A. (1965). 'Inhibition of antibody production and acquired immunologic tolerance by actinomycin.' *J. Immunol.*, **95**, 718.

CROWLE, A. J. and HU, C. C. (1969). 'Adoptive transfer of immunologic tolerance into normal mice.' *J. Immunol.*, **103**, 1242.

DAVIES, A. J. S. (1969). 'The thymus and the cellular basis of immunity.' *Transplant. Rev.*, **1**, 43.

GERSHON, R. K. and KONDO, K. (1970). 'Cell interactions in the induction of tolerance: the role of thymic lymphocytes.' *Immunology*, **18**, 723.

GERSHON, R. K. and KONDO, K. (1971). 'Antigenic competition between heterologous erythrocytes. I. Thymic dependency.' *J. Immunol.*, **106**, 1524.

HORIUCHI, A. and WAKSMAN, B. H. (1968). 'Role of the thymus in tolerance. VII. Relative effectiveness of nonaggregated and heat-aggregated bovine γ globulin, injected directly into lymphoid organs of normal rats, in suppressing immune responsiveness.' *J. Immunol.*, **101**, 1322.

MCCULLAGH, P. J. (1970). 'The immunological capacity of lymphocytes from normal donors after their transfer to rats tolerant of sheep erythrocytes.' *Aust. J. exp. Biol. med. Sci.*, **48**, 369–399.

MCGREGOR, D. D., MCCULLAGH, P. J. and GOWANS, J. L. (1967). 'The role of lymphocytes in antibody formation. 1. Restoration of the haemolysin response in x-irradiated rats with lymphocytes from normal and immunologically tolerant donors.' *Proc. roy. Soc. B*, **168**, 229.

MILLER, J. F. A. P. (1960). 'Studies on mouse leukaemia. The role of the thymus in leukaemogenesis by cell-free leukaemic filtrates.' *Brit. J. Cancer*, **14**, 93.

MILLER, J. F. A. P. and MITCHELL, G. F. (1969). 'Thymus and antigen reactive cells.' *Transpl. Rev.*, **1**, 3.

MITCHISON, N. A. (1968). '*Transplantation immunology*.' Excerpta Medica Foundation Symposium on Organ Transplantation. Amsterdam.

MORRIS, A. and MOLLER, G. (1968). 'Regulation of cellular antibody synthesis. Effect of adoptively transferred antibody-producing spleen cells on cellular antibody synthesis.' *J. Immunol.*, **101**, 439.

PLAYFAIR, J. H. L. (1969). 'Specific tolerance to sheep erythrocytes in mouse bone marrow cells.' *Nature (Lond.)*, **222**, 882.

RAFF, M. C. (1969). 'Theta isoantigen as a marker of thymus-derived lymphocytes in mice.' *Nature (Lond.)*, **224**, 378.

SAHIAR, K. and SCHWARTZ, R. S. (1964). 'Inhibition of 19S antibody synthesis by 7S antibody.' *Science*, **145**, 395.

SCOTT, D. W. and GERSHON, R. K. (1970). 'Determination of total and mercaptoethanol resistant antibody in the same serum sample.' *Clin. exp. Immunol.*, **6**, 313.

SEVER, J. L. (1962). 'Application of a micro technique to viral serological investigation.' *J. Immunol.*, **88**, 320.

SHELLAM, G.R. and NOSSAL, G.J.V. (1968). 'Mechanism of induction of immunological tolerance. IV. The effects of ultra-low doses of flagellin. *Immunology*, **14**, 273.

STERZL, J. (1966). 'Immunological tolerance as the result of terminal differentiation of immunologically competent cells. *Nature (Lond.)*, **209**, 416.

TERMAN, D. S., MINDEN, P. and CROWLE, A. J. (1971). 'Adoptive transfer of neonatal tolerance into normal mice.' *Fed. Proc.*, **30**, 650.

THEIS, G. A. and SISKIND, G. W. (1968). 'Selection of cell populations in induction of tolerance: affinity of antibody formed in partially tolerant rabbits.' *J. Immunol.*, **100**, 138.

TONG, J. L. and BOOSE, D. (1970). 'Immunosuppressive effect of serum from CBA mice made tolerant by the supernatant from ultracentrifuged bovine γ-globulin.' *J. Immunol.*, **105**, 426.

UHR, J. W. and MOLLER, G. (1968). 'Regulatory effect of antibody on the immune response.' *Adv. Immunol.*, **8**, 81.

WALKER, J. G. and SISKIND, G. W. (1968). 'Studies on the control of antibody synthesis. Effect of antibody affinity upon its ability to suppress antibody formation.' *Immunology*, **14**, 21.

WIGZELL, H. (1966). 'Antibody synthesis at the cellular level. Antibody induced suppression of 7S antibody synthesis.' *J. exp. Med.*, **124**, 953.

PROPERTIES OF ANTIGEN-SPECIFIC SUPPRESSIVE
T-CELL FACTOR IN THE
REGULATION OF ANTIBODY RESPONSE OF THE MOUSE
I. In Vivo Activity and Immunochemical Characterizations*

By TOSHITADA TAKEMORI AND TOMIO TADA

(From the Laboratories for Immunology, School of Medicine, Chiba University, Chiba, Japan)

Our previous publications (1, 2) indicated that thymocytes and spleen cells of mice that had been immunized with a high dose of carrier antigen, when transferred into syngeneic host, could suppress the antibody response against a hapten coupled to the homologous carrier. This suppressive effect was found to be mediated by a population of thymus-derived lymphocytes, whose activity was completely abrogated by the treatment with anti-θ and complement. Moreover, the specificity of the observed suppression has been established by the fact that neither normal T cells nor those obtained from mice immunized with an unrelated antigen could suppress the antibody response. These results indicate that immunization with a relatively high dose of carrier antigen generated a subpopulation of T cells that specifically inhibits the antibody response against hapten on the same carrier. Similar antigen-specific T-cell-mediated suppression has recently been demonstrated in a number of other experimental systems (3–10), and is now considered to be an important regulatory mechanism in various forms of immune responses (reviewed in 11).

A possible molecular mechanism of the antigen-specific suppression is that certain primed T cells liberate a suppressive factor(s) which then gives rise to an "off" signal to other cell types via combination with antigen. There are several examples of soluble T-cell factors that influence the magnitude and quality of the immune response (12–24). Thus, the above findings prompted us to explore the subcellular component of T cells which can mediate the antigen-specific suppression of the antibody response. We have attempted to separate such a soluble component by simple sonication followed by ultracentrifugation of the thymocytes and spleen cells possessing the suppressor activity, since we were successful in obtaining an antigen-specific suppressive factor from sonicated T-cells of the rat, which had shown a strong inhibitory effect on an ongoing IgE antibody response of the same species (19, 25). This paper will describe some of the properties of the antigen-specific suppressive T-cell factor obtained in mice, which was found to suppress mainly IgG antibody response of syngeneic mice in an in vivo experimental system.

---

* This work was supported by a grant from the Ministry of Education of Japan.

Materials and Methods

Antigens. Keyhole limpet hemocyanin (KLH)[1] was purchased from Calbiochem, San Diego, Calif. Bovine serum albumin (BSA) and bovine gamma globulin (BGG) were obtained from the Nutritional Biochemical Corporation, Cleveland, Ohio. The following 2,4-dinitrophenyl (DNP) conjugates of these proteins were prepared by the method of Eisen et al. (26): DNP_{730}-KLH (assuming the average mol wt of KLH as 7,000,000); DNP_{43}-BGG, and DNP_{34}-BSA. Subscripts refer to the numbers of DNP groups per molecule of carrier protein. *Bordetella pertussis* vaccine was purchased from the Chiba Serum Institute, Chiba, Japan.

Animals. Randomly bred BALB/c AnN and DBA/2 mice were raised in our animal facility. C57BL/6J mice were supplied from the National Institute of Radiological Science, Chiba, Japan. All mice were used at 8- and 12-wk old.

Immunization. To test the primary antibody response, mice were immunized with an intraperitoneal injection of 100 μg of DNP-KLH or DNP-BGG mixed with 1×10^9 *B. pertussis* vaccine. The antibody response was estimated by enumerating the DNP-specific plaque-forming cells (PFC) in their spleen 6 days after the immunization by the method of Cunningham and Szenberg (27) using sheep erythrocytes coated with DNP_{34}-BSA by chromium chloride. Direct PFC were considered to be IgM antibody-producing cells, and indirect PFC only developed with a 1:250 dilution of rabbit antimouse IgG antiserum to be IgG antibody producers.

To obtain carrier-primed thymocytes and spleen cells, mice were immunized with two intraperitoneal injections of 100 μg of soluble KLH or BGG without adjuvant at a 2 wk interval. Animals were killed 2 wk after the second injection, and their thymuses and spleens were removed and processed as below.

Preparation of Cell-Free Extracts from Primed Thymocytes and Spleen Cells. The thymuses and spleens of KLH- or BGG-primed mice were placed in a small quantity of chilled Eagle's minimal essential medium (MEM) in Petri dishes. They were minced and teased with forceps, and then gently pressed between two glass slides to release the cells from fibrous tissues. The cells were washed three times with cold MEM, and were resuspended in a small amount of borate-buffered saline at a concentration of 5×10^8 cells/ml. The cell suspensions were then subjected to sonication for 2 min in ice water. Cell-free extracts were obtained by ultracentrifugation at 40,000 *g* for 1 h. These extracts will be designated as KLH- or BGG-primed thymocyte (T) extract and spleen cell (S) extract.

Antisera. A polyvalent antiserum against mouse immunoglobulins (Ig's) was raised by repeated injections of washed precipitates of diphtheria toxoid and mouse antitoxoid antibodies in rabbits. Antisera specific for mouse IgG, IgM, and Fab fragment were prepared by immunization of rabbits with purified proteins followed by appropriate absorption. The specificity of these antisera was confirmed by immunodiffusion and immunoelectrophoresis.

Antimouse thymocyte serum (ATS) was produced in rabbits by repeated immunizations with washed mouse thymocytes in complete Freund's adjuvant (CFA). The pooled serum was absorbed with mouse erythrocytes and bone marrow cells. The resulting antiserum killed 100% of the thymocytes, 35% of the spleen cells and less than 5% of the bone marrow cells at a dilution of 1:64. The maximal dilution of antiserum for 50% cytotoxicity for thymocytes was 1:256.

Three alloantisera against the *H-2* complex were utilized: antisera against the products of whole $H-2^d$ (B10 anti-B10.D2) and left-hand side (*K* end) of $H-2^d$ (B10.A anti-B10.D2) were kindly provided by Dr. B. Benacerraf of the Department of Pathology, Harvard Medical School, Boston, Mass. An antiserum reactive with only the products of right-hand side (*D* end) of $H-2^d$ [(B10 × LP.RIII)F$_1$ anti-B10.A(5R)] (serum designation, D-13) was provided by the Jackson Laboratories through the courtesy of the National Institute of Health, Bethesda, Md.

Preparation of Immunoadsorbents. Insoluble immunoadsorbents composed of antigens and antibodies were prepared by the method of Axén et al. (28). 2 ml of Sepharose 4B (Pharmacia Fine Chemicals, Inc., Uppsala, Sweden) were activated by cyanogen bromide and allowed to react with 20–40 mg of antigens or gamma globulin fraction of antisera for 6 h. Residual active sites of

---

[1] *Abbreviations used in this paper:* AEF, allogeneic effect factor; ATS, antimouse thymocyte serum; BGG, bovine gamma globulin; BSA, bovine serum albumin; CFA, complete Freund's adjuvant; EA, hen's egg albumin; KLH, keyhole limpet hemocyanin; PFC, plaque-forming cells; S extract, spleen cell extract; T extract, thymocyte extract.

Sepharose were blocked by adding excess amounts of BSA. They were washed thoroughly with borate-buffered saline, and were packed in 0.7 × 15.0 cm columns. The absorption of the T-cell factor was performed by gradually passing the extract through the column at 4°C.

Gel Filtration. The T or S extract was dialysed against borate-buffered saline, pH 8.0, and applied to a column (2.5 × 90 cm × 2) of Sephadex G-200 (Pharmacia Fine Chemicals, Inc.). The elution was performed with a constant upward flow of 10 ml/h at 4°C. Fractions of 2.6 ml were collected and monitored with a spectrophotometer at 280 nm. For an estimation of the molecular weight of the active component, the column was calibrated by passing marker proteins of known molecular weights (mouse IgM, IgG, albumin, and hen's egg albumin [EA]) immediately after the extract had been excluded. Fractions corresponding to the marker proteins were pooled as shown in the results, concentrated by pressure dialysis, and then tested for their activity. The molecular weight ranges of fractions were calculated by the method of Andrews (29).

Digestion with Enzymes. The T extract was digested with DNase (from beef pancreas; 2,000 Kunitz U/mg, Sigma Chemical Co. St Louis, Mo.), RNase (50 Kunitz U/mg, C. F. Boehringer and Sons, Mannheim, Germany), and Pronase (Pronase P from *Streptomyces griseus*, 1,200 tyrosine U/mg; Kaken Chemical Co., Tokyo, Japan) as follows: 1 ml of the T extract obtained from 1×10^9 cells was incubated with 50 μg of DNase or RNase at room temperature for 1 h. The digestion with Pronase was performed at 37°C at a concentration of 100 μg of enzyme/1 ml of the extract. All digested materials were immediately tested for their activity.

Experimental Procedure. The T or S cell extract from carrier-primed mice was injected intravenously into syngeneic (in some cases allogeneic) mice that were concomitantly immunized with 100 μg of DNP-KLH or DNP-BGG plus pertussis vaccine. Direct and indirect PFC responses were estimated 6 days after the immunization at the time when control animals were producing the greatest number of DNP-specific indirect PFC. All the absorbed materials, fractions, and digests were tested for their activity by a similar procedure. The geometric means and standard deviations were calculated from the logarithmically transformed PFC numbers of at least six similarly treated animals. P values were determined by the student's t test.

Results

Suppression of Hapten-Specific Antibody Response by Carrier-Specific T and S Extract. It has previously been shown that the passive transfer of thymocytes and spleen cells from KLH-primed mice into syngeneic normal mice suppresses antibody response of the host against DNP-KLH (1). Thus in the present experiment, the cell-free extracts of thymocytes and spleen cells obtained from KLH-primed BALB/c mice were tested for their suppressive activity in the syngeneic recipient. The extracts corresponding to 1×10^8 original thymus or spleen cells were inoculated intravenously into normal BALB/c mice that were subsequently immunized with DNP-KLH and pertussis vaccine. As a control, groups of mice were given T or S extracts from donors immunized with an unrelated antigen (BGG), and were similarly immunized with DNP-KLH. DNP-specific PFC response was examined 6 days after the immunization, since the previous study indicated that KLH-primed suppressor T cells caused maximal suppression of IgG (indirect PFC) response on day 6 (1).

The upper part of Table I shows the result of one experiment in which one can see a significant suppression of indirect (IgG) PFC response in groups given KLH-primed T and S extracts. BGG-primed T or S extracts produced no such suppression. Direct (IgM) PFC response was not significantly suppressed in repeated experiments.

The lower part of Table I shows the result of a reverse experiment in which DNP-BGG was used as immunizing antigen. In this case, the inoculation of BGG-primed T and S extracts produced a statistically significant ($P < 0.01$)

TABLE I

Suppression of Hapten-Specific Antibody Response by T and S Extract from BALB/c Mice Primed with Carrier

Immunizing antigen	Extract injected	Anti-DNP PFC/spleen*	
		Direct	Indirect
DNP-KLH	—	$7{,}280 \overset{\times}{\div} 1.24$	$11{,}100 \overset{\times}{\div} 1.15$
DNP-KLH	KLH-primed T extract	$7{,}800 \overset{\times}{\div} 1.19$	$929 \overset{\times}{\div} 2.91\ddagger$
DNP-KLH	KLH-primed S extract	$7{,}080 \overset{\times}{\div} 1.65$	$2{,}930 \overset{\times}{\div} 1.45\ddagger$
DNP-KLH	BGG-primed T extract	$8{,}470 \overset{\times}{\div} 1.72$	$11{,}100 \overset{\times}{\div} 1.93$
DNP-KLH	BGG-primed S extract	$9{,}650 \overset{\times}{\div} 1.79$	$13{,}600 \overset{\times}{\div} 1.96$
DNP-BGG	—	$4{,}560 \overset{\times}{\div} 1.77$	$10{,}100 \overset{\times}{\div} 1.79$
DNP-BGG	BGG-primed T extract	$2{,}260 \overset{\times}{\div} 2.14$	$458 \overset{\times}{\div} 3.13\ddagger$
DNP-BGG	BGG-primed S extract	$1{,}310 \overset{\times}{\div} 3.03$	$465 \overset{\times}{\div} 4.23\ddagger$
DNP-BGG	KLH-primed T extract	$3{,}650 \overset{\times}{\div} 1.61$	$10{,}300 \overset{\times}{\div} 1.67$
DNP-BGG	KLH-primed S extract	$3{,}870 \overset{\times}{\div} 1.75$	$10{,}100 \overset{\times}{\div} 1.46$

* Geometric means and standard deviations calculated from six similarly treated mice; $\overset{\times}{\div}$ means "multiply and divide."

‡ $P < 0.01$ as compared with the control response.

suppression of PFC response of the IgG class, whereas KLH-primed T or S extracts displayed no such activity. Thus, the results indicate that the suppression is specific for the carrier of the immunizing antigen.

Affinity of the Suppressive T-Cell Factor for Antigen. Since the suppression of antibody response caused by T and S extracts was found to be carrier specific, the possibility that the suppressive factor possesses an affinity for antigen was examined by absorption with antigens. The T extract from KLH-primed donors was absorbed by passing it through a column of immunoadsorbent composed of KLH or BGG, and the effluent corresponding to the 10^8 original cell number was injected into mice that were subsequently immunized with DNP-KLH. As shown in Table II, the KLH-primed T extract after absorption with KLH immunoadsorbent lost most of its suppressive activity, while that absorbed with BGG immunoadsorbent well retained the suppressive activity.

Absence of Ig Determinants on Suppressive T-Cell Factor. Since the suppressive T-cell factor had specificity and affinity for carrier antigens, it could be an Ig. This possibility was tested by absorbing the T-cell extract with various

TABLE II
Absorption of the Suppressive T-Cell Factor of KLH-Primed Mice with Antigens and Antibodies

Extract injected	Absorbed with[*]:	Anti-DNP PFC/spleen[‡]	
		Direct	Indirect
—	—	$7{,}550 \overset{\times}{\div} 1.75$	$7{,}640 \overset{\times}{\div} 1.60$
KLH-primed T extract	—	$5{,}860 \overset{\times}{\div} 2.02$	$820 \overset{\times}{\div} 3.65$§
KLH-primed T extract	KLH	$7{,}050 \overset{\times}{\div} 1.42$	$5{,}830 \overset{\times}{\div} 1.78$
KLH-primed T extract	BGG	$8{,}360 \overset{\times}{\div} 1.54$	$640 \overset{\times}{\div} 3.77$
KLH-primed T extract	Anti-Igs	$4{,}530 \overset{\times}{\div} 1.99$	$330 \overset{\times}{\div} 5.12$
KLH-primed T extract	Anti-Fab	$4{,}190 \overset{\times}{\div} 2.31$	$910 \overset{\times}{\div} 3.25$§
KLH-primed T extract	Anti-μ	$4{,}560 \overset{\times}{\div} 1.98$	$540 \overset{\times}{\div} 5.03$
KLH-primed T extract	Anti-γ	$3{,}930 \overset{\times}{\div} 2.24$	$200 \overset{\times}{\div} 2.84$

[*] Absorption was carried out by passing the T extract through a column of immunoadsorbent composed of each antigen or antibody.
[‡] Geometric means and standard deviations.
§ $P < 0.01$ as compared with the control response.

immunoadsorbents composed of anti-Ig antibodies. The suppressive T-cell extract was passed through anti-Ig immunoadsorbent columns, and then likewise injected into test animals which then were immunized with DNP-KLH.

As shown in the lower part of Table II, none of the anti-Ig columns, i.e. polyvalent antimouse Ig's, anti-Fab, anti-μ, and anti-γ could remove the suppressive activity. These results exclude the possibility that the suppressive T-cell factor belongs to any known classes or fragments of Ig.

Absorption of Suppressive T-Cell Factor by Antithymocyte Serum and Anti-H-2 Alloantisera. To determine the antigenic characteristics of the suppressive T-cell component, the KLH-primed T extract was absorbed with immunoadsorbents composed of gamma globulin fractions of rabbit ATS and anti-H-2 alloantisera. As shown in Table III, absorption with ATS largely removed the suppressive activity, indicating that the suppressive factor contains antigen which is shared with normal thymocytes. The alloantiserum reactive with the products of the whole *H-2* complex (anti-*H-2^d*) was also capable of absorbing the suppressive activity. Thus, in order to determine whether the suppressive factor contains antigen(s) coded for by genes in the *K* or *D* half of the *H-2* complex, two alloantisera directed to either the *K* or *D* end of *H-2^d* were tested for their absorbing capacity. B10.A anti-B10.D2 contains antibodies against the expression of the *K^d*, *I-A^d*, and *I-B^d* subregions, while (B10 × LP.RIII)F$_1$ anti-

TABLE III

Absorption of the Suppressive T-Cell Factor with ATS and Anti-H-2 Antibodies

Extract injected	Absorbed with*:	Anti-DNP PFC/spleen‡	
		Direct	Indirect
—	—	$7{,}420 \overset{\times}{\div} 2.37$	$15{,}000 \overset{\times}{\div} 2.32$
KLH-primed T extract	—	$8{,}640 \overset{\times}{\div} 1.43$	$2{,}300 \overset{\times}{\div} 2.49$§
KLH-primed T extract	ATS	$7{,}260 \overset{\times}{\div} 2.02$	$13{,}400 \overset{\times}{\div} 1.43$
KLH-primed T extract	Anti-*H-2^d* [B10 anti-B10.D2]	$7{,}080 \overset{\times}{\div} 2.29$	$16{,}000 \overset{\times}{\div} 1.21$
KLH-primed T extract	Anti-*H-2^d* (K^d, I-A^d, I-B^d) [B10.A anti-B10.D2]	$9{,}280 \overset{\times}{\div} 3.32$	$12{,}700 \overset{\times}{\div} 2.18$
KLH-primed T extract	Anti-*H-2^d* (I-C^d, S^d, D^d) [(B10 × LP.RIII)F$_1$ anti-B10.A(5R)]	$6{,}400 \overset{\times}{\div} 2.30$	$936 \overset{\times}{\div} 5.21$‖

* Absorption was carried out by passing the extract through a column of immunoadsorbent composed of gamma globulin fraction of each antiserum.

‡ Geometric means and standard deviations.

§ $P < 0.01$.

‖ $P < 0.02$.

B10.A(5R) contains antibodies reactive with antigens coded for by genes in the *I-C^d* and D^d subregions. The results in Table III clearly indicate that the former antiserum specific for the *K* end of *H-2^d* is capable of removing the suppressive activity while the latter (anti-*D* end) is ineffective in absorbing the activity. Hence, it was concluded that the suppressive factor contains antigen(s) coded for by genes in the left-hand side of the *H-2* complex in addition to an antigen present on mouse T cells in general.

Failure of the Suppressive T-Cell Factor to Suppress the Antibody Response of Allogeneic Mice. Since the suppressive T-cell component contains antigen coded for by genes in the *H-2* complex, we examined whether the factor obtained from one strain of mice could suppress the antibody response of other strains having different *H-2* histocompatibility. KLH-primed suppressive T extract was obtained in BALB/c (*H-2^d*) and C57BL/6 (*H-2^b*) mice. These extracts were capable of suppressing the antibody response against DNP-KLH of syngeneic recipient (Table IV). However, the T extract from C57BL/6 mice failed to suppress the response of BALB/c mice. This cannot be ascribed to the induction of nonspecific stimulation due to allogeneic effect, since the T extract from unprimed C57BL/6 mice produced no such stimulation. Similarly, the extract from BALB/c mice having strong suppressive activity in syngeneic hosts produced no suppression in C57BL/6 mice. In contrast, when the BALB/c T extract was given to DBA/2 mice, which have the same *H-2^d* histocompatibility as BALB/c mice, a statistically significant suppression was observed (Table IV). Thus, it appeared that the suppressive T-cell factor could not act across the barrier of *H-2* histocompatibility.

TABLE IV

Failure of Suppression of Antibody Response against DNP-KLH by the T-Cell Factor Obtained from Histoincompatible Donors

Recipient strain	Donor strain	Primed with*:	Anti-DNP PFC/spleen‡	
			Direct	Indirect
BALB/c	—		$6{,}340 \overset{\times}{\div} 1.59$	$7{,}710 \overset{\times}{\div} 1.38$
BALB/c	BALB/c	KLH	$2{,}030 \overset{\times}{\div} 3.19$	$580 \overset{\times}{\div} 3.34$§
BALB/c	C57BL/6	—	$6{,}530 \overset{\times}{\div} 2.05$	$6{,}970 \overset{\times}{\div} 1.31$
BALB/c	C57BL/6	KLH	$6{,}810 \overset{\times}{\div} 1.27$	$8{,}370 \overset{\times}{\div} 1.26$
C57BL/6	—		$10{,}400 \overset{\times}{\div} 1.81$	$10{,}700 \overset{\times}{\div} 1.67$
C57BL/6	C57BL/6	KLH	$3{,}030 \overset{\times}{\div} 1.58$	$1{,}560 \overset{\times}{\div} 3.63$‖
C57BL/6	BALB/c	—	$14{,}900 \overset{\times}{\div} 1.99$	$11{,}100 \overset{\times}{\div} 2.03$
C57BL/6	BALB/c	KLH	$8{,}830 \overset{\times}{\div} 2.04$	$10{,}300 \overset{\times}{\div} 1.75$
DBA/2	BALB/c	—	$7{,}710 \overset{\times}{\div} 1.69$	$11{,}100 \overset{\times}{\div} 1.25$
DBA/2	BALB/c	KLH	$7{,}740 \overset{\times}{\div} 1.32$	$3{,}520 \overset{\times}{\div} 2.19$¶

* Donor animals were either unprimed or primed with two injections of soluble KLH.
‡ Geometric means and standard deviations.
§ $P < 0.001$.
‖ $P < 0.05$.
¶ $P < 0.01$.

The Molecular Weight of the Suppressive T-Cell Factor. The S extract from KLH-primed mice was fractionated by gel filtration with Sephadex G-200. Fractions corresponding to the known marker proteins were separated as shown in Fig. 1. Fraction I is the protein peak eluted with the void volume, fraction II corresponds to mouse IgG, fraction III to serum albumin, and fraction IV contains the second protein peak which is eluted slightly faster than EA. The pooled fractions were concentrated and likewise tested for their suppressive activity.

As presented in Table V, the suppressive activity was exclusively contained in fraction IV, which was eluted slower than serum albumin and slightly faster than EA. The approximate mol wt range of fraction IV was calculated to be between 35,000 and 60,000, being definitely smaller than the values for usual Ig's.

Chemical Nature of the Suppressive T-Cell Factor. The suppressive S extract was digested with nucleases and proteinase, and then immediately injected into mice that were then immunized with DNP-KLH. As shown in Table

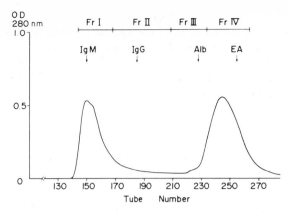

FIG. 1. Elution pattern of S extract obtained from KLH-primed mice from a Sephadex G-200 column. *Arrows* indicate the positions of the marker proteins eluted at the identical condition. Four fractions (Fr) were separated as indicated, and were tested for the presence of suppressive activity after concentration.

TABLE V

Suppressive Effect of Sephadex G-200 Fractions of KLH-Primed S Extract

Fraction* injected	Anti-DNP PFC/spleen‡	
	Direct	Indirect
—	$9,320 \overset{\times}{\div} 1.55$	$7,490 \overset{\times}{\div} 1.38$
Fr. I	$7,680 \overset{\times}{\div} 1.67$	$7,440 \overset{\times}{\div} 1.65$
Fr. II	$14,160 \overset{\times}{\div} 1.33$	$7,130 \overset{\times}{\div} 1.82$
Fr. III	$8,000 \overset{\times}{\div} 1.68$	$6,380 \overset{\times}{\div} 1.56$
Fr. IV	$2,430 \overset{\times}{\div} 4.65$	$252 \overset{\times}{\div} 3.24$§

* Fractions of KLH-primed S extract separated by gel filtration through a Sephadex G-200 column (see Fig. 1).
‡ Geometric means and standard deviations.
§ $P < 0.001$.

VI, the treatment with DNase and RNase did not affect the suppressive activity, while the digestion with Pronase completely destroyed the suppressive activity.

The activity was found to be labile by heating the extract at 56°C for 1 h. When the extract was kept in a refrigerator at 4°C for a week, the suppressive activity disappeared completely (Table VI). These results indicated that the active component is an extremely labile protein.

Discussion

Our previous publications (1, 2) reported that passive transfer of thymocytes and spleen cells from mice primed with a carrier antigen suppresses the anti-

TABLE VI

Stability of the Suppressive T-Cell Factor for Treatments with Enzymes, Heating, and Storage

Extract injected	Treatment	Anti-DNP PFC/spleen*	
		Direct	Indirect
—	—	$5{,}510 \overset{\times}{\div} 1.96$	$9{,}140 \overset{\times}{\div} 1.60$
KLH-primed S extract	—	$6{,}240 \overset{\times}{\div} 2.97$	$720 \overset{\times}{\div} 4.90$‡
KLH-primed S extract	Pronase	$7{,}850 \overset{\times}{\div} 1.91$	$12{,}200 \overset{\times}{\div} 1.89$
KLH-primed S extract	DNase	$5{,}190 \overset{\times}{\div} 1.88$	$850 \overset{\times}{\div} 4.97$§
KLH-primed S extract	RNase	$8{,}970 \overset{\times}{\div} 1.92$	$900 \overset{\times}{\div} 3.56$‡
KLH-primed S extract	56°C 1 h	$14{,}200 \overset{\times}{\div} 1.30$	$13{,}500 \overset{\times}{\div} 1.61$
KLH-primed S extract	4°C 1 wk	$6{,}550 \overset{\times}{\div} 3.32$	$12{,}200 \overset{\times}{\div} 1.70$

* Geometric means and standard deviations.
‡ $P < 0.01$.
§ $P < 0.02$.

body response of normal syngeneic hosts against a hapten coupled to the homologous carrier. This suppression was found to be clearly antigen specific, and mediated by T cells primed by the carrier antigen. We had attempted to separate a subcellular component of such antigen-specific suppressor T cells, since we were successful in obtaining an antigen-specific soluble component from primed T cells of rats which specifically inhibited the IgE antibody response of rats immunized with the same antigen (19, 25).

In the present communication, we have reported the presence of a similar carrier-specific soluble factor that suppresses antibody response of mice against a hapten coupled to the homologous carrier. This factor was extractable from physically disrupted thymocytes and spleen cells of mice immunized with carrier antigen, and upon inoculation into syngeneic hosts, suppressed the antibody response against a hapten on the homologous carrier. Although the suppressive effect was not as strong as that of live suppressor T cells, the injection of the factor invariably caused a statistically significant suppression of indirect (IgG) PFC response of the host.

The specificity of suppression was warranted by two independent experiments. In the present studies, the factor obtained from KLH-primed mice could suppress the response to DNP-KLH, while failing to suppress the response to DNP-BGG. The reverse was true for the factor obtained from BGG-primed mice. Furthermore, the suppressive activity of KLH-primed T-cell extract was completely removed by the immunoadsorbent composed of KLH but not that of BGG. The results clearly indicate that the specificity of suppression is based on the specific binding of the suppressive T-cell factor to the determinant present on the carrier molecule.

Despite its definite specificity and affinity for the carrier molecule, attempts to remove the suppressive T-cell factor with anti-Ig immunoadsorbents were always unsuccessful. None of the antibodies directed to Ig's and their fragments were able to remove the suppressive activity from the extract. The results indicate that the suppressive T-cell factor is not Ig in nature. This conclusion was further supported by its physicochemical properties. The factor has an approximate mol wt between 35,000 and 60,000, which is considerably less than those of conventional Ig's. These immunochemical and physicochemical properties of the mouse T-cell factor are much the same as those of the antigen-specific suppressive T-cell factor of the rat, which had produced a strong suppressive effect on an ongoing IgE antibody response of the species (19, 25).

The suppressive factor described in the present study undoubtedly originated from carrier-primed T cells. It was previously shown that suppressor activity of parental live cells was associated with thymocytes and θ-bearing spleen cells, their activity being abrogated by treatment of the cells with anti-θ or ATS and complement. Furthermore, in the present experiment it was shown that the suppressive activity of thymocyte extract was completely removed by passing it through a column of immunoadsorbent composed of ATS. The latter finding indicates that the suppressive T-cell factor contains a common antigenic determinant shared by the thymocytes and splenic T cells of various mouse strains, since the ATS used in this study is reactive with these cell types of different mouse strains.

The availability of alloantisera directed to the products of genes in the _H-2_ complex in the mouse enabled us to study further the antigenic characteristics of the suppressive T-cell factor. It was found that the alloantiserum against _H-2^d_ could absorb the suppressive activity with great efficiency. Furthermore, alloantisera containing antibodies directed to the _K_ end (_H-2K, I-A,_ and _I-B_) of _H-2^d_ were capable of absorbing the suppressive activity, while the serum reactive only with the _D_ end (_I-C, SsSlp,_ and _H-2D_) had no absorbing capacity. The results suggest that suppressive T-cell factor contains a product of genes in either the _K_ or _I_ region of the _H-2_ complex. It will be reported in another paper[2] that the T-cell factor is, in fact, an _I_ region gene product. It is most interesting that the antibody response to DNP-KLH is definitely regulated by the product of _I_ region genes, since the response to KLH per se is not under _Ir-1_ gene control (30).

In accordance with the above genetic characteristics of the T-cell factor, it appeared that the suppressive factor could only suppress the response of _H-2_ histocompatible strains. The factor derived from BALB/c mice could suppress the response of BALB/c and DBA/2 strains possessing the same _H-2^d_ haplotype, while failing to suppress the response of C57BL/6 (_H-2^b_) mice, in which the syngeneic C57BL/6 factor exerted a significant suppressive influence. This apparent failure of suppression by the _H-2_ histoincompatible T-cell factor is not due to the induction of nonspecific stimulation by the given T-cell extract, since the normal allogeneic T-cell extract did not enhance or suppress the response of the host.

_____

[2] Taniguchi, M., K. Hayakawa, and T. Tada. 1975. Properties of antigen-specific suppressive T-cell factor in the regulation of antibody response of the mouse. II. In vitro activity and evidence for the _I_ region gene product. Manuscript submitted for publication.

The immunochemical and physicochemical properties of the suppressive T-cell factor are much like those of the antigen-specific cooperative T-cell factor reported by Taussig and his associates (22, 31–33), although the effect is quite opposite. Their factor is released from educated T cells by a short-term culture with antigen, and cooperates with bone marrow cells to induce antibody response to the same antigen in irradiated hosts. Although their factor could cooperate with *H-2* histoincompatible bone marrow cells (32), the factor contains antigen coded for by genes in the left-hand side of the *H-2* complex (33). It has specificity and affinity for the immunizing antigen, but is not an Ig in nature (22, 31). The molecular size of their cooperative factor is apparently very close to that of the presently described suppressive T-cell factor (33). Thus, the most important question is to what differences in the molecular structure can the opposite activity be ascribed. Although some important differences in the mode of release and the target of the T-cell factor will be presented in a subsequent paper,[2] it is apparent that both factors may share an identical structure while the other molecular characteristics may be different. Furthermore, the suppressive T-cell factor seems to have considerable structural similarity to the allogeneic effect factor (AEF) described by Armerding and his associates (23, 34) with respect to the physicochemical and antigenic properties, although the AEF has no specificity for immunizing antigen. Elucidation of the interrelationship between these multiple T-cell factors seems to be of crucial importance.

The chemical nature of the suppressive T-cell factor is still unclear. It has only been demonstrated so far that the active factor is a protein with extreme lability. Heating at 56°C for 1 h and storage at 4°C for 1 wk completely destroyed the suppressive activity. This lability as well as the large quantity required for in vivo suppression restricted further analysis in the present experimental system.

Summary

An antigen-specific suppressive T-cell factor was extracted from physically disrupted thymocytes and spleen cells of mice that had been immunized with soluble protein antigens. The factor, when inoculated into syngeneic normal mice, could induce a significant suppression of IgG antibody response against a hapten coupled to the carrier protein by which the donor of the suppressor factor was immunized. The suppressor factor was found only effective in suppressing the antibody response of syngeneic or *H-2* histocompatible recipients. The suppressive T-cell factor was removed by absorption with immunoadsorbent composed of the relevant antigen, but not with any of those of anti-immunoglobulin antibodies. The factor was successfully removed by alloantibodies with specificity for the *K* end (*H-2K*, *I-A* and *I-B*) of the *H-2* complex of the donor strain, but not by those for the *D* end (*I-C*, *SsSlp*, and *H-2D*). The activity was removed by absorption with a heterologous antithymocyte serum. The mol wt of the suppressive T-cell factor was between 35,000 and 60,000 as determined by Sephadex G-200 gel filtration. The suppressive T-cell factor was found to be a heat-labile protein.

The authors wish to express their sincerest thanks to Doctors B. Benacerraf and C. S. David for their generous supply of well-defined alloantisera against mouse *H-2* complex. They are also grateful for the excellent technical and secretarial assistance by Mr. H. Takahashi and Miss Yoko Yamaguchi.

Received for publication 2 July 1975.

References

1. Tada, T., and T. Takemori. 1974. Selective roles of thymus-derived lymphocytes in the antibody response. I. Differential suppressive effect of carrier-primed T cells on hapten-specific IgM and IgG antibody responses. *J. Exp. Med.* **140**:239.

2. Takemori, T., and T. Tada. 1974. Selective roles of thymus-derived lymphocytes in the antibody response. II. Preferential suppression of high-affinity antibody-forming cells by carrier-primed suppressor T cells. *J. Exp. Med.* **140**:253.

3. Gershon, R. K., and K. Kondo. 1971. Infectious immunological tolerance. *Immunology.* **21**:903.

4. Okumura, K., and T. Tada. 1971. Regulation of homocytotropic antibody formation in the rat. VI. Inhibitory effect of thymocytes on the homocytotropic antibody response. *J. Immunol.* **107**:1682.

5. Kapp, J. A., C. W. Pierce, S. Schlossman, and B. Benacerraf. 1974. Genetic control of immune responses in vitro. V. Stimulation of suppressor T cells in nonresponder mice by the terpolymer L-glutamic acid^{60}-L-alanine^{30}-L-tyrosine^{10} (GAT). *J. Exp. Med.* **140**:648.

6. Ha, T.-Y., and B. H. Waksman. 1973. Role of the thymus in tolerance. X. "Suppressor" activity of antigen-stimulated rat thymocytes transferred to normal recipients. *J. Immunol.* **110**:1290.

7. Basten, A., J. F. A. P. Miller, J. Sprent, and C. Cheers. 1974. Cell-to-cell interaction in the immune response. X. T-cell-dependent suppression in tolerant mice. *J. Exp. Med.* **140**:199.

8. Baker, P. J., P. W. Stashak, D. F. Amsbaugh, and B. Prescott. 1974. Regulation of the antibody response to type III pneumococcal polysaccharide. IV. Role of suppressor T cells in the development of low-dose paralysis. *J. Immunol.* **112**:2020.

9. Benjamin, D. C. 1975. Evidence for specific suppression in the maintenance of immunological tolerance. *J. Exp. Med.* **141**:635.

10. Zembala, M., and G. L. Asherson. 1973. Depression of the T cell phenomenon of contact sensitivity by T cells from unresponsive mice. *Nature (Lond.).* **244**:227.

11. Gershon, R. K. 1974. T cell control of antibody production. *Contemp. Top. Immunobiol.* **3**:1.

12. Dutton, R. W., R. Falkoff, J. A. Hirst, M. Hoffmann, J. W. Kappler, J. R. Kettman, J. F. Lesley, and D. Vann. 1971. Is there evidence for a nonantigen specific diffusable chemical mediator from the thymus-derived cell in the initiation of the immune response? *In* Progress in Immunology. D. B. Amos, editor. Academic Press, Inc., New York. 355.

13. Shimple, A., and E. Wecker. 1972. Replacement of T-cell function by a T-cell product. *Nat. New. Biol.* **237**:15.

14. Doria, G.. G. Agarossi, and S. Di Pietro. 1972. Enhancing activity of thymocyte culture cell-free medium on the in vitro immune response of spleen cells from neonatally thymectomized mice to sheep RBC. *J. Immunol.* **108**:268.

15. Gorczynski, R. M., R. G. Miller, and R. A. Phillips. 1972. Initiation of antibody production to sheep erythrocytes in vitro: replacement of the requirement for T-cells with a cell-free factor isolated from cultures of lymphoid cells. *J. Immunol.* **108**:547.

16. Rubin, A. S., and A. H. Coons. 1972. Specific heterologous enhancement of immune responses. III. Partial characterization of supernatant material with enhancing activity. *J. Immunol.* **108**:1597.

17. Feldmann, M., and A. Basten. 1972. Cell interactions in the immune response in vitro. IV. Comparison of the effects of antigen-specific and allogeneic thymus-derived cell factors. *J. Exp. Med.* **136**:722.

18. Sjöberg, O., J. Andersson, and G. Möller. 1972. Reconstitution of the antibody response in vitro of T cell-deprived spleen cells by supernatants from spleen cell cultures. *J. Immunol.* **109**:1379.

19. Tada, T., K. Okumura, and M. Taniguchi. 1973. Regulation of homocytotropic antibody formation in the rat. VIII. An antigen-specific T cell factor that regulates anti-hapten homocytotropic antibody response. *J. Immunol.* **111**:952.

20. Watson, J. 1973. The role of humoral factors in the initiation of in vitro primary immune responses. III. Characterization of factors that replace thymus-derived cells. *J. Immunol.* **111**:1301.

21. Kishimoto, T., and K. Ishizaka. 1973. Regulation of antibody response in vitro. VII. Enhancing soluble factors for IgG and IgE antibody response. *J. Immunol.* **111**:1194.

22. Taussig, M. J. 1974. T cell factor which can replace T cells in vivo. *Nature (Lond.).* **248**:234.

23. Armerding, D., and D. H. Katz. 1974. Activation of T and B lymphocytes in vitro. II. Biological and biochemical properties of an allogeneic effect factor (AEF) active in triggering specific B lymphocytes. *J. Exp. Med.* **140**:19.

24. Zembala, M., and G. L. Asherson. 1974. T cell suppression of contact sensitivity in the mouse. II. The role of soluble suppressor factor and its interaction with macrophages. *Eur. J. Immunol.* **4**:799.

25. Okumura, K., and T. Tada. 1974. Regulation of homocytotropic antibody formation in the rat. IX. Further characterization of the antigen-specific inhibitory T cell factor in hapten-specific homocytotropic antibody response. *J. Immunol.* **112**:783.

26. Eisen, H. N., S. Belman, and M. E. Carsten. 1953. The reaction of 2,4-dinitrobenzensulfonic acid with free amino group of proteins. *J. Am. Chem. Soc.* **75**:4583.

27. Cunningham, A. J., and A. Szenberg. 1968. Further improvements in the plaque technique for detecting single antibody-forming cells. *Immunology*.**14**:599.

28. Axén, R., J. Porath, and S. Ernback. 1967. Chemical coupling of peptides and proteins to polysaccharides by means of cyanogen halides. *Nature (Lond.).* **214**:1302.

29. Andrews, P. 1965. The gel-filtration behaviour of proteins related to their molecular weights over a wide range. *Biochem. J.* **96**:595.

30. Cerottini, J. C., and E. R. Unanue. 1971. Genetic control of the immune response of mice to hemocyanin. I. The role of macrophages. *J. Immunol.* **106**:732.

31. Taussig, M. J., and A. J. Munro. 1974. Specific cooperative T cell factor; removal by anti-H-2 but not by anti-Ig sera. *Nature (Lond.).* **251**:63.

32. Taussig, M. J., E. Mozes, and R. Isač. 1974. Antigen-specific thymus cell factors in the genetic control of the immune response to poly-(tyrosyl, glutamyl)-poly-D,L-alanyl-poly- -lysyl. *J. Exp. Med.* **140**:301.

33. Munro, A. J., M. J. Taussig, R. Campbell, H. Williams, and Y. Lawson. 1974. Antigen-specific T-cell factor in cell cooperation: physical properties and mapping in the left-hand (*K*) half of *H-2*. *J. Exp. Med.* **140**:1579.

34. Armerding, D., D. H. Sachs, and D. H. Katz. 1974. Activation of T and B lymphocytes in vitro. III. Presence of Ia determinants on allogeneic effect factor. *J. Exp. Med.* **140**:1717.

SPECIFIC ENRICHMENT OF THE SUPPRESSOR T CELL
BEARING I-J DETERMINANTS

Parallel Functional and Serological Characterizations*

By KO OKUMURA, TOSHITADA TAKEMORI, TAKESHI TOKUHISA, AND TOMIO TADA

(From the Laboratories for Immunology, School of Medicine, Chiba University, Chiba, Japan)

Recent evidence indicates that determinants controlled by a locus (*Ia-4* locus) mapped in *I-J* subregion of mouse *H-2* major histocompatibility complex are selectively expressed on a functional subpopulation of peripheral T lymphocyte, which is endowed with a role to suppress the antibody and immunoglobulin production (1–3).[1] Unlike other *I*-region associated (Ia) antigens, which are primarily detectable on B cells, the products of *Ia-4* locus are not found on B cells, but are uniquely expressed on T cells (1, 4). In addition, the same locus in *I-J* subregion appears to control the determinants found on the antigen-specific suppressive T-cell factor (3), and thus *I-J* subregion products provide an important clue for studying the nature of both the Ia antigen and the antigen-receptor which are unique to T cells.

However, the presence of such *I*-region determinants on T cells has been mainly determined by functional analyses in which the activity of suppressor T-cell and suppressive T-cell factor is removed by anti-Ia antisera (1, 3, 5, 6), and no direct serological affirmation for the T-cell Ia antigen is yet available using standard cytotoxic assays. This is probably due to the fact that only a very small portion of T cells among total lymphoid cells express such Ia antigens, and this imposes a great limitation in determining the specificity, function, and biochemical structure of T-cell Ia antigens.

We have reported in a preliminary form a simple technique to enrich the antigen-specific suppressor T cell which carries *I-J* subregion determinants (7). We confirmed further that this method is highly effective in obtaining antigen-specific T cells and in analyzing their phenotypic expressions by the usual serological procedures. This report will describe the detailed method to enrich the *I-J*-bearing suppressor T cell, and the serological and functional analyses of the purified suppressor T cell and its products. We also present our recent results concerning the relationship between the I-J determinants, Lyt phenotype, and Fc receptor (FcR)[2] expressed on the specifically purified suppressor T cell.

---

* Supported by grants from the Ministry of Education, Culture, and Science, and the Ministry of Health, Japan.

[1] Hämmerling, G. J. *In* Proceedings of the Third Ir Gene Workshop. H. O. McDevitt, editor. Academic Press, Inc., New York. In press.

[2] *Abbreviations used in this paper:* alum, aluminum hydroxide gel; BAT, brain associated T-cell antigen; C, complement; DNP, 2,4-dinitrophenyl; D'PBS, Dulbecco's phosphate-buffered saline; FACS, fluorescence-activated cell sorter; FcR, Fc receptor; FCS, fetal calf serum; Fr, fraction; KLH, keyhole limpet hemocyanin; MIg, mouse immunoglobulin; PFC, plaque-forming cells; SRBC, sheep erythrocyte; TsF, suppressive T-cell factor.

Materials and Methods

Animals. C3H/He and C57BL/6J mice were purchased from the Shizuoka Agricultural Cooperative Association for Laboratory Animals (Hamamatsu City, Shizuoka, Japan). Strains of B10.A(3R) and B10.A(5R) were kindly provided by Dr. C. S. David of the Department of Genetics, Washington University School of Medicine, St. Louis, Mo., and have been maintained in our animal facility.

Antigens. Keyhole limpet hemocyanin (KLH) was obtained from Calbiochem, San Diego, Calif. Dinitrophenylated KLH (DNP_{770}-KLH) was prepared by the method described previously (8). Egg albumin recrystallized five times was purchased from Nutritional Biochemicals Corporation, Cleveland, Ohio.

Immunization of Mice. As the source of antigen-specific suppressor T cells, mice were immunized with two intraperitoneal injections of 100 μg soluble KLH at a 2-wk interval as described previously (8). Other mice were immunized with 100 μg of DNP-KLH or KLH alone in aluminum hydroxide gel (alum) together with 1×10^9 killed *Bordetella pertussis* vaccine. They were used as the source of DNP-primed B and KLH-primed T cells after appropriate treatments (see below).

Antisera. A polyvalent anti-mouse immunoglobulin antiserum (anti-MIg) was obtained by repeated immunizations of rabbits with normal mouse gamma globulin fraction in complete Freund's adjuvant. The rabbit anti-mouse brain-associated T-cell antigen (anti-BAT) was prepared by the method of Sato et al. (9). The anti-sheep erythrocyte (anti-SRBC) antibody for the Fc rosette assay is the 7S fraction of mouse antiserum against SRBC, which was obtained by gel filtration with Sephadex G-200.

Alloantisera directed at *I-J* subregion of H-2^k and H-2^b haplotypes (anti-I-J^k and anti-I-J^b) were prepared by reciprocal immunization of B10.A(3R) and B10.A(5R) with their lymphoid cells. The antisera were absorbed with syngeneic spleen cells to remove auto-reactive antibodies before use. Anti-Lyt alloantisera were kindly provided by Dr. D. B. Murphy of Stanford University, Stanford, Calif., and anti-Thy-1.2 antiserum produced in Thy-1 congenic mice was the gift of Dr. H. Sato of the Asahikawa Medical School, Hokkaido, Japan.

Preparation of Antibody- and Antigen-Coated Columns. Rabbit anti-MIg antibody was specifically purified by adsorption to and elution from the immunoadsorbent composed of mouse gamma globulin fraction. The purified anti-MIg was then coupled to cyanogen bromide-activated Sephadex G-200 beads (Pharmacia Fine Chemicals, Inc., Uppsala, Sweden) according to the method described by Schlossman and Hudson (10). The conjugation of KLH to Sephadex G-200 was likewise performed by coupling 30 mg of KLH to 20 ml of activated Sephadex G-200. 20 ml of these materials was packed in 20-ml disposable syringes in which about 1 ml of Sephadex G-25 was layered at the bottom as a sieve. The columns were equilibrated with Dulbecco's phosphate-buffered saline (D'PBS) fortified with 5% heat-inactivated fetal calf serum (FCS; Grand Island Biological Co., Grand Island, N.Y.). Detailed methods to use these columns are described in results.

Adoptive Secondary Antibody Response. DNP-primed B cells were obtained by treating DNP-KLH-primed spleen cells with anti-BAT antiserum and guinea pig complement. KLH-specific helper T cells were separated with the nylon wool column according to the method of Julius et al. (11). The mixture of DNP-primed B and KLH-primed T cells was transferred intravenously into lethally (650R) irradiated syngeneic recipients that were subsequently immunized with 10 μg of DNP-KLH. The number of DNP-specific plaque-forming cells (PFC) in the spleen was measured 7 days after the immunization by the method of Cunningham and Szenberg (12).

Cell Culture Technique. A modified Mishell-Dutton culture system was utilized to induce hapten-specific in vitro secondary antibody response. 4×10^6 of DNP-KLH-primed spleen cells were cultured with 0.1 μg/ml of DNP-KLH in RPMI-1640 enriched with 10% FCS in Falcon No. 3008 tissue culture plates (Falcon Plastics, Div. of BioQuest, Oxnard, Calif.). The culture was maintained at 37°C for 5 days, and the anti-DNP antibody response was measured by the PFC assay.

Preparation of the Cell-Free Extract from Fractionated Cells. Cells fractionated by the antigen-coated column (see Results) were suspended in 0.15 M saline at a concentration of 1×10^6/ml. The suspension was subjected to sonication on ice with a Tomy UR-150 Sonicator (Tomy Seiko Co., Ltd., Tokyo, Japan) as described previously (8). The cell-free supernate was obtained by centrifugation at 40,000 g for 1 h, and then dialyzed against saline at 4°C.

Separation of FcR⁺ and FcR⁻ Cells. Rosetting of surface Fc receptor-positive (FcR⁺) cells was performed by the method of Möller (13) except that a mouse 7S anti-SRBC fraction was used instead of rabbit antibody to sensitize SRBC. The rosetting (FcR⁺) cells were separated from nonrosetting (FcR⁻) cells by centrifugation on Isopaque/Ficoll gradient (Isopaque; Nyegaard and Co., Oslo, Norway, Ficoll; Pharmacia Fine Chemicals, Inc., Uppsala, Sweden) according to the method of Parish et al. (14).

Cytotoxic Assay. The ^{51}Cr release assay was used for cytotoxic test (15). ^{51}Cr-labeled 1×10^6 cells/ml were incubated with diluted antiserum at room temperature for 30 min and further at 4°C for 5 min. They were washed and subjected to the further 30-min incubation at 37°C with diluted rabbit complement (C). After centrifugation, radioactivity in the supernate was assayed. Sequential cytotoxic treatments with the combination of alloantisera were performed by the method described by Cantor and Boyse (16).

Analysis of I-J-Bearing Cells by Fluorescence-Activated Cell Sorter (FACS). The cell fractions separated with antigen-coated columns were analyzed by FACS II (Becton, Dickinson Electronics Laboratory, Mountain View, Calif.). The cells were treated with anti-I-J antisera followed by staining with fluoresceinated anti-mouse IgG. The fluorescence profile was analyzed by FACS II after gating out dead cells by the size scatter analysis (17).

Results

Fractionation of KLH-Primed Spleen Cells with Antibody- and Antigen-Coated Columns. Spleen cells from KLH-immunized mice were first passed through a column of Sephadex G-200 coupled with anti-MIg at 4°C to deplete B cells. The medium used in these procedures was D'PBS containing 5% FCS. About 40% of the original spleen cells were harvested in the effluent, which consisted of more than 70% of Thy-1 antigen-positive cells. 3 to 5×10^8 of these enriched T cells were resuspended in 5 ml of warm (37°C) medium. The suspension was then applied to the column of KLH-coated Sephadex G-200 at 37°C. Cells were allowed to penetrate into the column, and were further incubated at 37°C for 30 min in the column. The column was then washed with a warm (37°C) medium by adjusting the flow rate to about 1 ml/min. The elution of nonadherent cells was completed by washing the column with 150–250 ml of warm medium. The column was then placed in the 4°C cold chamber for 30 min. The cells bound to the column were eluted by washing with cold (0–4°C) medium at a flow rate of 1 ml/min.

The elution pattern of cells from the column is depicted in Fig. 1. The total number of cells eluted with the cold medium fraction II (Fr. II) was usually about 0.5% of the original spleen cells. If the spleen cell suspension from the egg-albumin immunized or normal mice were applied to the same KLH-coated column, the recovery of Fr. II did not exceed 0.1% of the original cells under the identical condition.

Functional Analyses of Cell Fractions Separated by Antigen-Coated Column

SUPPRESSOR ACTIVITY. The suppressor activity of column-separated fractions from C57BL/6 mice was assayed by the ability of the mice to suppress specifically an adoptive secondary antibody response of primed syngeneic spleen cells against DNP-KLH. The cells were cotransferred with DNP-primed B cells and KLH-primed nylon-purified T cells into irradiated syngeneic recipients that were subsequently immunized with 10 μg of DNP-KLH. The data in Table I shows that even 0.2×10^6 of Fr. II cells could completely suppress the anti-DNP antibody response, whereas 1×10^7 of Fr. I cells

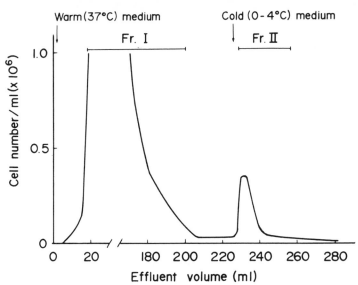

Fig. 1. Elution profile of KLH-primed T cells from the antigen-coated column. 5×10^8 cells were applied to a column of KLH-coated Sephadex G-200 at 37°C. After incubation at 37°C for 30 min, the cells were eluted with warm (37°C) medium (Fr. I), followed by elution at 4°C with cold (0–4°C) medium (Fr. II). Cells were washed and used for experiments. Normal as well as egg-albumin primed spleen cells did not yield the Fr. II peak under the same condition.

TABLE I

Enrichment of Suppressor T Cell with Antigen-Coated Column

KLH-primed suppressor T cell*		Anti-DNP IgG PFC/10^6‡
Fraction	Dose	
——§	——	1,516 (1.54)
Unfractionated	1×10^7	621 (1.68)
Fr. I	1×10^7	1,248 (1.13)
Fr. II	0.2×10^6	<10
B cell only‖		<10

* Suppressor T cells were cotransferred with 5×10^6 DNP-primed B cells and nylon column-purified KLH-primed helper T cells (1×10^6) into irradiated recipients.

‡ Direct PFC subtracted, geometric means and standard deviations calculated from four recipients.

§ Without suppressor T cells.

‖ 5×10^6 DNP-primed B cells without helper T cells.

produced no significant suppression. The T-cell fraction which was obtained by passage only through the anti-MIg column, produced moderate suppression at a dose of 1×10^7, and thus the suppressor activity was found almost exclusively concentrated in Fr. II which bound to the antigen-coated column at 37°C.

HELPER ACTIVITY.   To assess whether the helper T cell could be separated by the same adsorption-elution procedure with the antigen-coated column, the spleen cells of C57BL/6 mice, which had been immunized by a single injection of KLH in alum 6 wk earlier to generate strong helper activity, were fraction-

TABLE II
Helper T Cell Does not Bind to the Antigen-Coated Column

KLH-primed helper T cell*		Anti-DNP IgG PFC/10^6‡
Fraction	Dose	
——§	——	351 (1.29)
Unfractionated	5×10^6	5,631 (1.67)
Fr. I	5×10^6	5,315 (1.67)
Fr. II	0.5×10^6	410 (1.47)
Fr. I + Fr. II	$5 \times 10^6 + 0.5 \times 10^6$	1,823 (1.89)

* KLH-primed helper T cells were cotransferred with 5×10^6 DNP-primed B cells into irradiated recipients.
‡ Direct PFC subtracted, geometric means and standard deviations calculated from four recipients.
§ Without helper T cells.

ated by exactly the same procedure as that used to enrich the suppressor T cell. The yield of Fr. II cells from the helper source was about the same as that from suppressor spleen cells. Each fraction was cotransferred with DNP-primed B cells into irradiated recipients that were then immunized with DNP-KLH. Table II shows that Fr. I cells, which did not bind to the KLH-column exerted a comparable helper effect to that of unseparated T-cell fraction at the same dose. On the contrary, even 0.5×10^6 Fr. II cells, which were obtained from more than 5×10^8 original spleen cells, exhibited very little helper activity. If the same number of Fr. II cells were admixed with 5×10^6 Fr. I cells and cotransferred with DNP-primed B cells to the recipient, the anti-DNP antibody response was much lower than that mounted by Fr. I cells and B cells. This indicates that the Fr. II cells which bind to the antigen-coated column are, in fact, suppressor T cells, even though the cells were derived from animals immunized to generate optimal helper activity. The results collectively indicate that the helper T cell does not bind to the antigen-coated column under the condition in which suppressor T cells are successfully removed. It is also shown that even the spleen cells, which exhibit strong helper activity as a whole, contain a significant number of suppressor T cells which can be separated by binding to the antigen-coated column.

The Suppressive T-Cell Factor Derives from Antigen-Binding T Cells. Since our previous studies indicated that the antigen-specific suppressive T-cell factor (TsF) is obtainable by sonication of suppressor T cells, we have attempted to extract TsF from fractionated cells. The cells of each fraction were disrupted by sonication, and the suppressive activity of the soluble supernate was assessed in the in vitro secondary antibody response. Table III shows the results of an experiment using DNP-KLH-primed C57BL/6 spleen cells as the responding cells to which varying doses of cell-free extracts obtained from syngeneic fractionated cells were added. It is clearly demonstrated that the extract from Fr. II cells produces a significant suppression of anti-DNP PFC response even at a dose corresponding to 10^3 live cells. With the extract corresponding to 10^4 Fr. II cells gave a comparable degree of suppression to that produced by the extract obtained from 100 times (1×10^6) as many unfractionated spleen cells. This

TABLE III

The Suppressive T-Cell Factor Derives from Antigen-Binding T Cells Lacking Fc Receptor

Extract from*	Corresponding cell number	Anti-DNP IgG PFC/culture‡	Suppression
			%
Without extract	——	3,217 ± 423	——
Unfractionated	1×10^6	813 ± 44	75
spleen	1×10^4	3,061 ± 114	5
Fr. I	1×10^6	2,873 ± 673	11
	1×10^4	3,228 ± 523	0
Fr. II	1×10^4	638 ± 365	80
	1×10^3	1,703 ± 356	47
Fr. II (FcR⁺)§	1×10^5	3,413 ± 299	0
Fr. II (FcR⁻)	1×10^5	420 ± 216	87

* Cell-free extract of each fraction corresponding to the given cell number was added to the cultured 4×10^6 DNP-KLH-primed spleen cells in modified Mishell-Dutton culture.

‡ Direct PFC subtracted, arithmetic means and standard deviations calculated from five cultures.

§ FcR⁺ and FcR⁻ cells in Fr. II were separated by the rosetting with antibody-coated SRBC.

indicates that the fractionation procedure resulted in the 100-fold enrichment of suppressor T cell in Fr. II. On the other hand, almost no suppressive activity was observed with the extract from Fr. I cells.

Suppressive T-Cell Factor Derives from FcR⁻ T Cells. Fr. II cells were separated into two fractions by Fc-rosetting technique. About 25% of Fr. II cells were recovered as Fc rosette-forming (FcR⁺) cells. After removal of SRBC by treating the cells with Gey's solution (18), both FcR⁺ and FcR⁻ T-cell fractions were disrupted by sonication, and the supernates were likewise tested for their suppressive activity. As shown in the lower part of Table III, the suppression was induced only by the extract from FcR⁻ T cells, and hence, the producer of TsF was shown to lack surface Fc receptors.

Demonstration of I-J Determinants on Enriched Suppressor T Cells. Column-separated Fr. I and Fr. II cells were analyzed by cytotoxic ^{51}Cr release with various alloantisera and rabbit C. The results obtained with C57BL/6 spleen cells are depicted in Table IV. Both fractions contained 70–75% of Thy-1 antigen-positive cells. About 30% of Fr. II cells were constantly killed by anti-*I-J*^{b} in repeated experiments, whereas no significant killing of Fr. I cells was observed with anti-*I-J*^{k} antiserum. This killing was haplotype-specific, since the reciprocal antiserum (anti-*I-J*^{k}) did not lyse the cells from C57BL/6 mice, whereas it was able to kill Fr. II cells from C3H mice (see below). Another interesting fact is that Fr. II cells contained a larger proportion of cells that were killed by anti-Lyt-2.2, 3.2 antiserum than did Fr. I cells.

Cytotoxic curves with two anti-*I-J* antisera using Fr. II cells from C57BL/6 and C3H as targets are depicted in Fig. 2. The anti-*I-J*^{b} antiserum specifically killed C57BL/6 Fr. II cells, and the anti-*I-J*^{k} only killed C3H Fr. II cells. The titers of both antisera were relatively low showing the steep decline of cytotoxicity at dilutions beyond 1:20.

Relationship between the I-J Expression, Lyt-Phenotype, and Surface Fc Receptor on Enriched Suppressor T Cells. The above studies indicate that

TABLE IV

Cytotoxic Analysis of Cell Fraction of C57BL/6J Mice Separated by
Antigen-Coated Column

Antiserum	Serum dilution	Cytotoxic ^{51}Cr release	
		Fr. I	Fr. II
		%	%
Anti-Thy 1.2*	1:40	70	75
Anti-*I-J^b*‡	1:10	<5	30
Anti-*I-J^k*§	1:10	<5	<5
Anti-Lyt-1.2	1:40	35	20
Anti-Lyt-2.2,3.2	1:40	20	40

* Congenic anti-Thy 1.2.
‡ B10.A(5R) anti-B10.A(3R).
§ B10.A(3R) anti-B10.A(5R).

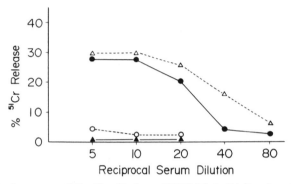

FIG. 2.   Cytotoxic curves of Fr. II cells from C57BL/6J (solid lines) and C3H (broken
lines) with anti-*I-J^b* (●——● ○- - -○) and anti-*I-J^k* (▲——▲ △- - -△) antisera. Cytotoxicity was undetectable with unfractionated cells and Fr. I cells (see text).

the enrichment of antigen-specific suppressor T cells by the antigen-coated
column accompanied with the proportional increase of *I-J^+*, Lyt-2^+,3^+ cells. To
examine whether the *I-J^+* cells in fact belong to the Lyt-2^+,3^+ subclass, the
phenotypic expressions of Fr. II cells were analyzed by the two-step sequential
killing protocol according to the method used by Cantor and Boyse (16). The
^{51}Cr-labeled Fr. II cells from C57BL/6 mice were first treated with various
alloantisera and rabbit C, and the residual live cells were then subjected to the
second cytotoxic killing with different antisera. As shown in Table V, all the *I-
J* determinant-bearing cells belong to the Thy-1 antigen-positive cell population. Only a small proportion of cells were killed by anti-Lyt-1.2 serum, while
nearly 40% of cells were killed by anti-Lyt-2.2,3.2 serum. If the cells were first
treated with anti-Lyt-2.2,3.2 antiserum, the residual cells were not killed by
anti-Lyt-1.2 serum. This suggests that all the Lyt-1^+ cells in Fr. II are carrying
also Lyt-2,3 determinants (Lyt-1^+,2^+,3^+ cells). The results of sequential killing
with the combination of anti-Lyt and anti-*I-J* antisera (lower part of Table V)
indicate that virtually all the I-J determinant-bearing cells possess Lyt-2,3
alloantigens, although the presence of a small number of Lyt-1^+,2^+,3^+ cells in
this *I-J^+* population is not excluded.

TABLE V

Phenotypic Expressions on Specifically Enriched Suppressor T Cells of C57BL/6J Mice

1st killing*		2nd killing‡	
Antiserum	^{51}Cr release	Antiserum	^{51}Cr release
	%		%
Anti-I-J^b	33	—	—
Anti-I-J^k	0	—	—
Anti-Thy 1.2	70	Anti-I-J^b	≤5
Anti-Lyt-1.2	12	NMS§	0
Anti-Lyt-2.2,3.2	38	Anti-Lyt-1.2	≤5
NMS§	0	Anti-Lyt-1.2	12
Anti-Lyt-1.2	12	Anti-I-J^b	25
Anti-Lyt-2.2,3.2	40	Anti-I-J^b	0

* ^{51}Cr-labeled Fr. II cells of C57BL/6J mice were treated with various allo-antisera and rabbit C, and the radioactivity released in the medium was measured.
‡ Live cells after the 1st killing were then treated with antisera and C.
§ Normal C57BL/6J mouse serum absorbed with syngeneic spleen cells.

To learn the relationship between FcR and I-J determinants, the Fr. II cells were separated into FcR$^+$ and FcR$^-$ populations. As stated above, about 25% of Fr. II cells were FcR$^+$. No significant killing with anti-*I-J* antiserum was detectable in FcR$^+$ population, whereas about 30% of FcR$^-$ cells were killed by anti-*I-J* and C. Taken together, the I-J determinant-positive cells bear Lyt-2,3 alloantigens and lack the surface Fc receptor.

Analysis of I-J Determinant-Bearing Cells by Fluorescence-Activated Cell Sorter (FACS). Fr. II cells from C57BL/6 mice were reacted with anti-*I-J^b* or anti-*I-J^k* antiserum followed by staining with fluoresceinated anti-mouse IgG. The fluorescence distribution of cells was analyzed with FACS II after gating out dead cells by size scatter analysis. As can be seen in Fig. 3, the fluorescence profile of the Fr. II cells stained with anti-*I-J^b* showed a bimodal distribution, whereas the same cells treated with irrelevant (anti-*I-J^k*) antiserum showed a single fluorescence-negative peak. The computer analysis of the fluorescence-positive peak indicated that about 30% of the live Fr. II cells carry I-J determinants, the value being consistent with that obtained with cytotoxic ^{51}Cr release.

Discussion

The presence of unique *I* region gene products which are selectively expressed only on certain subsets of T cells has created new problems concerning the role of T-cell Ia antigen in the regulatory cell interactions in the immune response. The cell type which carries I-J determinants is involved in the allotype-, idiotype-, and antigen-specific suppressions (1–3, footnote 1), and in the initiation of T-cell response to concanavalin A (19). In addition, the determinants are found on the molecule which suppresses the antibody response in an antigen-specific fashion. Hence, the functional and biochemical analyses of *I-J* subregion gene products are of obvious importance for the further understand-

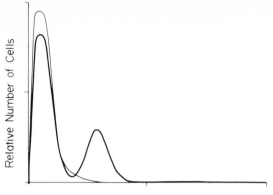

FIG. 3.   Fluorescence distribution of Fr. II cells of C57BL/6J mice labeled with anti-*I-J*ᵇ (bold line) or anti-*I-J*ᵏ (light line) antiserum in FACS analysis. The cells were treated with alloantisera followed by staining with fluoresceinated rabbit anti-mouse IgG. Note the bimodal distribution of cells stained with anti-*I-J*ᵇ but not with anti-*I-J*ᵏ. The cells stained with anti-*I-J*ᵇ comprise about 30% of total Fr. II cells.

ing of the molecular events in the regulatory cell interactions as well as of the nature of antigen receptor of T cells.

There was, however, a great limitation in proceeding with these studies, since the presence of *I-J*-bearing cells was only demonstrable by functional studies in which the activity of suppressor T cells and suppressor factor was removed by anti-*I-J* antisera, and since there was no way to directly detect the *I-J* subregion gene products by serological and immunochemical procedures. This is mainly due to the fact that the number of *I-J*-bearing cells in total spleen cells is too small to allow such an analysis. The present report described a simple method to enrich the antigen-specific suppressor T cell about 100-fold allowing us to study the phenotypic expressions on the cell type with respect to the I-J determinants, Lyt alloantigens, and Fc receptors. Results collectively indicate that I-J determinants are expressed mainly within the $Lyt-2^+,3^+$, FcR^- T-cell population. More recently, Taniguchi and Miller[3] were able to separate the antigen-specific suppressor T cell by adsorption to and elution from antigen-coated Petri dishes. Properties of the purified suppressor T cell are very similar to those of ours despite the fact that their suppressor T cell was derived from tolerant mice.

The present studies together with those reported previously (3, 20) indicate that the antigen-specific suppressor T cell indeed possesses antigen-binding sites together with I-J determinants. This is in sharp contrast to the helper T cell which does not bind to the antigen-coated column under the identical condition. These observations raise an important question as to whether or not the antigen recognition of helper and suppressor T cells are the same. It has been reported that macrophages are required to induce the helper T cell but not the suppressor T cell (21, 22). The optimal conditions to induce helper and

[3] Taniguchi, M., and J. F. A. P. Miller. Enrichment of specific suppressor T cells and characterization of their surface markers. Manuscript submitted for publication.

suppressor T cells are different with respect to the dose of antigen (23, 24), adjuvant vehicle (23), and the form of antigen (25, 26). One possible explanation is that the suppressor T cell recognizes antigen itself and can be stimulated by free antigen, while the helper T cell is only generated by modified antigen presented on macrophages. The other, and perhaps related explanation is that the affinity of antigen-binding sites of helper and suppressor T cells is different. Further biochemical and functional studies will give the final answer.

The method described in this paper appears to have several important applications to future studies on T-cell Ia antigens. In fact, the method allowed us to titrate the anti-*I-J* antibody by the direct cytotoxic assay, which was not possible with the total splenic T cells. We were also able to analyze and separate *I-J*-bearing cells with FACS. These will facilitate the biochemical analysis of the *I-J* subregion gene products with respect to both the possible constant part and antigen-binding site of the molecule, and the serological analysis of specificities of I-J determinants possessed by various haplotype strains. One of the most attractive applications may be to make hybrid cell lines expressing I-J determinants, which will allow the more precise biochemical and genetic studies of I-J molecules having antigen specificity and possibly an idiotype.

Summary

A simple procedure to enrich the antigen (keyhole limpet hemocyanin, KLH)-specific suppressor T cell was described. The suppressor T cell from KLH-immunized mice specifically bound to the KLH-coated Sephadex G-200 column at 37°C, and was eluted from the column by cold (0–4°C) medium. The helper T cell did not bind to the column under the identical condition. The suppressor T cell thus obtained had 100 times as potent suppressor activity as the original spleen cells in in vivo and in vitro secondary antibody responses against a hapten coupled to KLH. This procedure also enriched the cells bearing I-J determinants and Lyt-2,3 alloantigens, allowing us to study the phenotypic expressions on the suppressor T cell by direct serological procedures as well as by the use of the fluorescence activated cell sorter. Parallel functional and serological analyses indicated that the antigen-specific suppressor T cell belongs to a population of $I\text{-}J^+$, Lyt-2^+,3^+ and Fc R$^-$ T cells.

The authors wish to thank Doctors C. S. David, D. C. Shreffler, D. B. Murphy, L. A. Herzenberg, and H. Sato for their generous supply of alloantisera and congenic mice together with a lot of invaluable advice. They are grateful to Doctors I. Morimoto, M. Nonaka, and K. Hayakawa for their devoted collaborations. The excellent technical and secretarial assistance by Mr. H. Takahashi and Ms. Yoko Yamaguchi, respectively, are also gratefully acknowledged. They are indebted to the people at the Fijisawa Medical Supply Co. (Doshomachi, Osaka, Japan) who generously allowed us to use FACS II in this study.

Received for publication 26 July 1977.

References

1. Murphy, D. B., L. A. Herzenberg, K. Okumura, L. A. Herzenberg, and H. O. McDevitt. 1976. A new *I* subregion (*I-J*) marked by a locus (*Ia-4*) controlling surface determinants on suppressor T lymphocytes. *J. Exp. Med.* 144:699.

2. Okumura, K., L. A. Herzenberg, D. B. Murphy, H. O. McDevitt, and L. A. Herzenberg. 1976. Selective expression of *H-2* (*I*-region) loci controlling determinants on helper and suppressor T lymphocytes. *J. Exp. Med.* **144**:685.

3. Tada, T., M. Taniguchi, and C. S. David. 1976. Properties of the antigen-specific suppressive T-cell factor in the regulation of antibody response of the mouse. IV. Special subregion assignment of the gene(s) that codes for the suppressive T-cell factor in the *H-2* histocompatibility complex. *J. Exp. Med.* **144**:713.

4. Tada, T. 1977. Regulation of the antibody response by T cell products determined by different *I* subregion. *In* Regulation of the Immune System. E. Sercarz and L. A. Herzenberg, editors. Academic Press, Inc., New York. In press.

5. Hämmerling, G. J., K. Eichmann, and C. Sorg. 1976. Differential expression of Ia antigens on suppressor T cells, helper T cells and B precursor cells. *In* The Role of Products of the Histocompatibility Gene Complex in Immune Responses. D. H. Katz and B. Benacerraf, editors. Academic Press, Inc., New York. 417.

6. Vadas, M. A., J. F. A. P. Miller, I. F. C. McKenzie, S. E. Chism, F-W. Shen, E. A. Boyse, J. R. Gamble, and A. M. Whitelaw. 1976. Ly and Ia antigen phenotypes of T cells involved in delayed-type hypersensitivity and in suppression. *J. Exp. Med.* **144**:10.

7. Okumura, K., T. Takemori, and T. Tada. 1977. Specific enrichment of suppressor T cells bearing the products of *I-J* subregion. *In* Regulation of the Immune System. E. Sercarz and L. A. Herzenberg, editors. Academic Press, Inc., New York. In press.

8. Takemori, T., and T. Tada. 1975. Properties of antigen-specific suppressive T-cell factor in the regulation of antibody response of the mouse. I. In vivo activity and immunochemical characteristics. *J. Exp. Med.* **142**:1241.

9. Sato, V. L., S. D. Waksal, and L. A. Herzenberg. 1976. Identification and separation of pre T-cell from nu/nu mice: differentiation by preculture with thymic reticuloepithelial cells. *Cell. Immunol.* **24**:173.

10. Schlossman, S. F., and L. Hudson. 1973. Specific purification of lymphocyte populations on a digestible immunosorbent. *J. Immunol.* **110**:313.

11. Julius, M. H., E. Shimpson, and L. A. Herzenberg. 1973. A rapid method for the isolation of functional thymus-derived lymphocytes. *Eur. J. Immunol.* **3**:645.

12. Cunningham, A. J., and A. Szenberg. 1968. Further improvements in the plaque technique for detecting single antibody-forming cells. *Immunology*. **14**:599.

13. Möller, G. 1974. Effect of B-cell mitogens on lymphocyte subpopulations possessing C′3 and Fc receptors. *J. Exp. Med.* **139**:969.

14. Parish, C. R., S. M. Kirov, N. Bowern, and R. V. Blanden. 1974. A one step procedure for separating mouse T and B lymphocytes. *Eur. J. Immunol.* **4**:808.

15. Wigzell, H. 1965. Quantitative titrations of mouse *H-2* antibodies using ^{51}Cr labelled target cells. *Transplantation (Baltimore)*. **3**:423.

16. Cantor, H., and E. A. Boyse. 1975. Functional subclasses of T lymphocytes bearing different Ly antigens. I. The generation of functionally distinct T-cell subclasses is a differentiative process independent of antigen. *J. Exp. Med.* **141**:1376.

17. Loken, M. R., and L. A. Herzenberg. 1975. Analysis of cell populations with a fluorescence-activated cell sorter. *Ann. N.Y. Acad. Sci.* **254**:163.

18. Gey, G. O., and M. K. Gey. 1936. The maintenance of human normal cells and tumor cells in continuous culture. I. Preliminary report: cultivation of mesoblastic tumors and normal tissue and notes on methods of cultivation. *Am. J. Cancer* **27**:45.

19. Frelinger, J. A., J. E. Niederhuber, and D. C. Shreffler. 1976. Effects of anti-Ia sera on mitogenic responses. III. Mapping the genes controlling the expression of Ia determinants on concanavalin A-reactive cells to the *I-J* subregion of the *H-2* gene complex. *J. Exp. Med.* **144**:1141.

20. Taniguchi, M., K. Hayakawa, and T. Tada. 1976. Properties of antigen-specific suppressive T cell factor in the regulation of antibody response of the mouse. II. *In vitro* activity and evidence for the *I* region gene product. *J. Immunol.* **116**:542.

21. Feldmann, M., and S. Kontiainen. 1976. Suppressor cell induction in vitro. II. Cellular requirements of suppressor cell induction. *Eur. J. Immunol.* **6**:302.

22. Ishizaka, K., and T. Adachi. 1976. Generation of specific helper cells and suppressor cells in vitro for the IgE and IgG antibody responses. *J. Immunol.* **117**:40.

23. Kapp, J. A., C. W. Pierce, and B. Benacerraf. 1975. Genetic control of immune responses in vitro. III. Tolerogenic properties of the terpolymer L-glutamic acid[60]-L-alanine[30]-L-tyrosine[10] (GAT) for spleen cells from nonresponder (*H-2*[s] and *H-2*[q]) mice. *J. Exp. Med.* **140**:172.

24. Kontiainen, S., and M. Feldmann. 1976. Suppressor cell induction in vitro. I. Kinetics of induction of antigen-specific suppressor cells. *Eur. J. Immunol.* **6**:295.

25. Basten, A., J. F. A. P. Miller, and P. Johnson. 1975. T cell-dependent suppression of an anti-hapten antibody response. *Transplant. Rev.* **26**:130.

26. Takatsu, K., and K. Ishizaka. 1975. Reaginic antibody formation in the mouse. VI. Suppression of IgE and IgG antibody response to ovalbumin following the administration of high dose urea-denatured antigen. *Cell. Immunol.* **20**:276.

Suppressive and Enhancing T-cell Factors as *I*-region Gene Products: Properties and the Subregion Assignment

T. TADA, M. TANIGUCHI AND C. S. DAVID*

Laboratories for Immunology, School of Medicine, Chiba University, Chiba, Japan 280; *Department of Genetics, Washington University School of Medicine, St. Louis, Missouri 63110

A considerable body of evidence has been accumulated suggesting that cell interactions in the initiation and regulation of immune responses are achieved via complementary interactions of cell-surface molecules expressed on different subsets of immunocompetent cells. Among these molecules, much attention has been focused on the role of products of the major histocompatibility gene complex (MHC) as devices for such cell-to-cell communication (reviewed by Katz and Benacerraf 1975). Thus the diversity of the phenotypic expression of different MHC genes on functionally different subsets of lymphoid cells is indeed a crucial problem in understanding the network of immunocompetent cell interactions. This has been strengthened by recent discoveries of various T-cell factors carrying *I*-region-associated determinants (Taussig and Munro 1974; Armerding and Katz 1974; Takemori and Tada 1975; Erb and Feldmann 1975; Kapp et al. 1976) which enhance or suppress humoral antibody responses in mice. These factors now provide important clues for studying interactions of functionally distinct subpopulations of lymphoid cells at the molecule and gene levels. This paper will describe two distinct antigen-specific T-cell factors having opposite functions in the regulation of antibody response, one suppressive and the other enhancing for a T-cell-dependent antibody response. Both factors are found to have *I*-region gene expressions. Available evidence indicates that they are coded for by genes present in distinct *I* subregions, and that the targets for these factors are probably T cells also carrying *I*-region determinants. Incorporating the new target cell type in the network of immunocompetent cell interactions, possible mechanisms of T-cell-mediated regulation of antibody response will be discussed.

The Suppressive T-cell Factor and Its Properties

Our previous studies indicated the presence of an antigen-specific suppressive T-cell factor that can suppress primary and secondary IgG antibody responses of mice under both in vivo and in vitro conditions (Takemori and Tada 1975; Taniguchi et al. 1976a). The factor was extractable from thymocytes and spleen cells of mice that had been immunized with a relatively high dose of keyhole limpet hemocyanin (KLH) and exerted a suppressive effect on the antibody response against a hapten 2,4-dinitrophenyl (DNP) coupled to the homologous carrier.

The methods used to obtain the suppressive T-cell factors and to test their activity in in vitro-cultured spleen cells have been described in detail in previous publications (Taniguchi et al. 1976a,b). In brief, the antigen-specific T-cell factor was obtained by physical extraction from thymocytes of mice that had been immunized twice with 100 μg of KLH at a 2-week interval. The cell-free extract, which will be designated as T (thymocyte) extract, was added to a culture of primed spleen cells at a dose corresponding to 10^7 original thymocytes at the beginning of the culture. In general, 10^7 spleen cells from mice primed with DNP-KLH plus 10^9 *Bordetella pertussis* vaccine 4 weeks previously were cultured in the Marbrook culture system for 5 days in enriched Eagle's minimal essential medium (MEM) with 0.1 μg/ml of DNP-KLH inside the inner culture chamber.

To test the absorbing capacity of antigens and antibodies for the suppressive T-cell factor, the T extract was passed through a column of Sepharose 4B immunoadsorbent coupled with the antigen or gamma globulin fraction of antiserum at 4°C, and the residual suppressive activity of the effluent was assessed by adding it to the cultured spleen cells. The antibody response was measured by enumeration of indirect (IgG) plaque-forming cells (PFC) detected by DNP-coated sheep erythrocytes, according to the method of Cunningham and Szenberg (1968), after a 5-day culture.

By means of the above experimental procedure, the specificity of this factor has been firmly established on the basis of its binding affinity for KLH but not for unrelated antigens, despite the fact that no immunoglobulin determinants are detectable by absorption with various anti-immunoglobulin antibodies. The molecular weight of the T-cell factor was found to be between 35,000 and 55,000 daltons. Immunochemical and physicochemical properties are shown in Table 1.

Two lines of experimental evidence indicate that the T-cell factor is a product of *I*-region genes:

Table 1. Properties of the Antigen-specific Suppressive T-cell Factor

1. Extractable from carrier-primed thymocytes and spleen cells
2. Has specificity and affinity for carrier
3. Activity dependent on the presence of antigen
4. Suppresses IgG antibody response
5. No Ig determinants
6. Protein in nature
7. Molecular weight: 35,000–55,000
8. Target: probably T cell
9. Possesses *I*-region determinant
10. *H-2* barrier present

It has been clearly shown that the factor is adsorbable with immunoadsorbents composed of alloantibodies directed to the *I* region of the *H-2* complex (see below). Second, the factor cannot act across the major histocompatibility barrier, and strict identity among genes in the *I* region between the donor of the factor and the responding spleen cells is required (Table 2). The latter finding indicates that the acceptor site for the T-cell factor is also an *I*-region gene product, providing evidence that two genes are involved in the induction of T-cell-mediated suppression (Taniguchi et al. 1976b). In genetic analysis, it was generally found that the suppressive T-cell factor of F$_1$ of two different haplotype strains could suppress the responses of both parental stains equally, and the parental factors could suppress the response of F$_1$. Hence, both the suppressor and the acceptor molecules are found to be codominantly expressed on F$_1$ T cells (Table 2).

During the course of this study we have found two peculiar types of genetic defect of the expression of either suppressor or acceptor genes. One is A/J mice, which could not produce the suppressive T-cell factor despite the fact that they could accept the factor produced by other *H-2*-compatible mouse strains (nonproducer). In contrast, all the B10 congenic lines tested so far could produce the T-cell factor, but they could not accept the factor produced by syngenic and *H-2* histocompatible non-B10 congenic lines. The F$_1$ hybrid of A/J × B10.A could both produce and accept the T-cell factor, and thus the expression of both the suppressor and the acceptor molecules was found to be a dominant trait. Because of the occurrence of these two types of defect in gene expression, it was concluded that the suppressor and acceptor molecules are, in fact, coded for by two distinct genes, both of which are present in the *I* region.

In fact, we are aware of the interesting findings reported by Cerottini and Unanue (1971), who showed that the A/J strain is an extremely high responder to KLH, whereas CBA is a low responder, and that this responsiveness to KLH is not linked to the *H-2* complex. In view of the present observation, such a high responsiveness observed in A/J mice might be due to the lack of expression of the KLH-specific suppressor gene, although the response itself is not under H-linked *Ir* gene control. B10 congenic lines were also found to be high responders in our experimental system. It seems possible that the strain differences in responsiveness to various complex protein antigens may be explained at least in part by suppressor and/or ac-

Table 2. Complementary Interaction of the Factor and Acceptor, Both of Which Are Determined by the *I* Region

Donor of T extract	Responding spleen cells	Identities of *H-2* subregion	Suppression (%)
CBA	CBA	K,I,S,D	93
	A/J	K,I	74
	BALB/c	none	0
	(BALB/c × CBA)F$_1$	K,I,S,D	97
(BALB/c × CBA)F$_1$	(BALB/c × CBA)F$_1$	K,I,S,D	91
	BALB/c	K,I,S,D	88
	CBA	K,I,S,D	73
	SJL	none	0
A/J[a]	A/J[a]	K,I,S,D	0
	CBA	K,I	0
	BALB/c	S,D	0
B10.A[b]	B10.A[b]	K,I,S,D	0
	A/J	K,I,S,D	53
	B10.BR[b]	K,I	0
	C3H	K,I	69
(A/J[a] × B10.A[b])F$_1$	(A/J[a] × B10.A[b])F$_1$	K,I,S,D	72
SJL	SJL	K,I,S,D	63
	B10.S[b]	K,I,S,D	0
B10.S[b]	B10.S[b]	K,I,S,D	0[c]
	SJL	K,I,S,D	72

[a] Nonproducer; [b] nonacceptor; [c] enhancement (twofold increase).

ceptor gene expression rather than by *Ir* gene control.

Furthermore, it was frequently found that if the nonacceptor B10 congenic lines were given the syngenic T-cell extract, their IgG antibody responses were greatly enhanced. Such results are also included in Table 2. The enhancement also showed a strict strain specificity (see below).

These results led us to suspect that the T-cell extract obtained from KLH-primed mice might contain two distinct T-cell factors which suppress or enhance the antibody response depending on the presence or absence of the corresponding acceptor sites on target responding cells. In fact, the suppressive activity of T extract was completely absorbed with thymocytes and spleen cells of acceptor strains but not with those of B10 congenic nonacceptor strains. Furthermore, it has recently been demonstrated that spleen cells which absorb the suppressive T-cell factor are nylon-adherent T cells. Neither B cells nor macrophages could absorb the suppressive activity.

If we assume that there are two distinct suppressive and enhancing T-cell factors, several questions arise as to the nature of these T-cell factors and their corresponding acceptor sites: (1) Are the suppressive and enhancing T-cell factors the same? (2) If different, what genes code for these different T-cell factors? (3) Are the target cells for these T-cell factors different or the same? The following sections will summarize our recent findings concerning these points.

I-J Subregion Assignment of the Suppressor Gene

We have already shown that the suppressive activity of the T-cell extract can be successfully removed by certain anti-Ia alloantisera (Takemori and Tada 1975; Taniguchi et al. 1976a; Tada et al. 1976). Thus an attempt was made to define the *I* subregion that determines the suppressive T-cell factor. This was achieved with two different ap-

proaches: (1) by absorbing the suppressive T-cell factor with various well-defined alloantisera of known Ia specificities mapping at different *I* subregions, and (2) by analyzing the identity requirements of genes among different *I* subregions for the induction of suppression in combinations of donor and recipient strains differing at restricted *I* subregions.

Some alloantisera used in these experiments were kindly provided by Drs. B. Benacerraf and D. Sachs, to whom we are greatly indebted. The combinations of the donors and recipients to produce these antisera and the resultant specificities are shown in Tables 3 and 4. In general, T extract corresponding to 5×10^7 original thymocytes was absorbed with one of the immunoadsorbents composed of alloantisera and then tested in five identical cultures. The number of indirect PFC in cultures with absorbed T extract was compared with results obtained with unabsorbed T extract.

Representative results of absorption studies of the $H\text{-}2^k$ suppressive T-cell factor are summarized in Table 3. It is shown that several (but not all) alloantisera against the *I* region of $H\text{-}2^k$ are capable of absorbing the suppressive activity of C3H T extract, but apparently there is no relationship between the absorbing capacity of alloantisera and their anti-Ia specificities. Several important points can be deduced from these experimental results: Since active antiserum (A.TH × B10.HTT)F$_1$ anti-A.TL is only reactive with the products of the *I-A* and *I-B* subregions of $H\text{-}2^k$, the participation of the *I-C* subregion is excluded. Similarly, the antiserum B10.S(7R) anti-B10.HTT directed against the homologous $I\text{-}C^k$ subregion is ineffective in absorbing the $H\text{-}2^k$ factor. On the other hand, antiserum (B10.A(4R) × 129)F$_1$ anti-B10.A(2R), directed against the products of the *I-B* subregion of $H\text{-}2^k$ can absorb the suppressive T-cell factor. These results indicate that the gene coding for the suppressive T-cell factor maps between the *I-A* and *I-C* subregions.

Table 3. Lack of Association between the Absorbing Capacity of Anti-Ia Alloantisera and Their Ia Specificities

| C3H ($H\text{-}2^k$) T extract absorbed with | Specificity | | Anti-DNP IgG PFC/culture | Absorption |
	H-2 subregions	Ia[a]		
Control	—	—	1798 ± 43	—
Unabsorbed	—	—	288 ± 90	—
AQR anti-B10.A	K^k	—	278 ± 111	—
(C3H.H-2° × 129)F$_1$ anti-C3H	$K^k, I\text{-}A^k, I\text{-}B^k, I\text{-}C^k$	1,2	1484 ± 201	+
A.TH anti-A.TL	$I\text{-}A^k, I\text{-}B^k, I\text{-}C^k$	1,2,3,7,15	1777 ± 339	+
(B10.D2 × A.TH)F$_1$ anti-A.TL	$I\text{-}A^k, I\text{-}B^k, I\text{-}C^k$	1,2,3	1621 ± 259	+
(A.TH × B10.HTT)F$_1$ anti-A.TL	$I\text{-}A^k, I\text{-}B^k$	1,2,3,15	1516 ± 329	+
(B10.A(4R) × 129)F$_1$ anti-B10.A(2R)	$I\text{-}B^k, I\text{-}C^d$	(6),7	1479 ± 244	+
B10.S(7R) anti-B10.HTT	$I\text{-}C^k$	7	310 ± 199	—
B10.A(18R) anti-B10.A(5R)	$I\text{-}C^d$	(6),7	1588 ± 75	+
(B10 × LP.RIII)F$_1$ anti-B10.A(5R)	$I\text{-}C^d, D^d$	(6)	2026 ± 115	+
(B10.K × A.TL)F$_1$ anti-B10.A	$I\text{-}C^d$	(6)	241 ± 57	—

[a] Numbers in parentheses indicate Ia specificities not possessed by $H\text{-}2^k$ mice.

Table 4. Possible Presence of a New Subregion *(I-J)* Which Codes for the Suppressive T-cell Factor between the *I-B* and *I-C* subregions

T extract absorbed with[a]	Specificities of *H-2* subregions	Anti-DNP IgG PFC/culture	Absorption
Control	–	1974 ± 87	–
Unabsorbed	–	250 ± 82	–
(A.TH × B10. HTT)F$_1$ anti-A.TL	$I\text{-}A^k, I\text{-}B^k, (I\text{-}J^k)$	1999 ± 134	+
(A.BY × B10.HTT)F$_1$ anti-A.TL	$I\text{-}A^k, I\text{-}B^k, (I\text{-}J^k)$	1975 ± 224	+
(B10.A(4R) × C3H.OH)F$_1$ anti-B10.K	$I\text{-}B^k, (I\text{-}J^k), I\text{-}C^k$	2070 ± 156	+
B10.S(7R) anti-B10.HTT	$I\text{-}C^k$	238 ± 52	–
B10.S(7R) anti-B10.S(9R)	$(I\text{-}J^k), I\text{-}C^d$	1932 ± 268	+
(A.TH × B10.S(9R))F$_1$ anti-A.TL	$I\text{-}A^k, I\text{-}B^k$ $I\text{-}C^k$	329 ± 59	–
B10.A(3R) anti-B10.A(5R)	$(I\text{-}J^k)$	2017 ± 192	+

[a] The $H\text{-}2^k$ (C3H) T extract was absorbed with alloantisera and the residual suppressive activity assayed in the response of C3H spleen cells.

The most interesting fact is that two out of three alloantisera putatively reactive with the $I\text{-}C^d$ subregion product can absorb the suppressive T-cell factor of $H\text{-}2^k$ mice. It has been shown previously that these anti-$I\text{-}C^d$ antisera lack the absorbing capacity of the homologous $H\text{-}2^d$ T-cell factor (Tada et al. 1976). The important point is that these two anti-$I\text{-}C^d$ sera were raised against B10.A(5R), a strain which is a recombinant of $H\text{-}2^a$ and $H\text{-}2^b$. This puzzling phenomenon can be explained only if we assume that the B10.A(5R) strain possesses a chromosomal segment of $H\text{-}2^k$ origin, found between the $I\text{-}B$ and $I\text{-}C$ subregions, originally possessed by the B10.A ($H\text{-}2^a$) strain, and that this segment *(I-J)* accommodates the suppressor gene.

In order to test this, several alloantisera were raised specifically to detect this possible new subregion gene product. These alloantisera were produced at the Department of Genetics, Washington University, St. Louis, Mo. As will be discussed later, this new subregion is designated as "*I-J*," as suggested by Drs. D. C. Shreffler and H. O. McDevitt. The $H\text{-}2^k$ (C3H) T extract was absorbed with these alloantisera, and the residual suppressive activity was likewise tested in the response of DNP-KLH-primed C3H spleen cells.

The results presented in Table 4 are, in fact, most reasonably explained by the presence of such a new locus between the $I\text{-}B$ and $I\text{-}C$ subregions specialized in coding for the suppressive T-cell factor. For example, B10.S(7R) anti-B10.S(9R) could absorb the $H\text{-}2^k$ suppressive T-cell factor, indicating that B10.S(9R) has the $I\text{-}J^k$ subregion between $I\text{-}B^s$ and $I\text{-}C^d$. Conversely, (A.TH × B10.S(9R))F$_1$ anti-A.TL, which theoretically lacks $I\text{-}J^k$ specificity, could not absorb the $H\text{-}2^k$ suppressive T-cell factor, despite the fact that this antiserum is reactive with all other I subregions, i.e., $I\text{-}A^k$, $I\text{-}B^k$, and $I\text{-}C^k$. However, B10.S(7R) anti-B10.HTT was unable to absorb the $H\text{-}2^k$ reactive factor, even though B10.S(9R) and B10.HTT have the same $I\text{-}A^s$ and $I\text{-}B^s$ subregions. This indicates that, unlike B10.S(9R), B10.HTT does not possess the $I\text{-}J^k$ subregion between $I\text{-}B^s$ and $I\text{-}C^k$.

More striking is the fact that antiserum B10.A(3R) anti-B10.A(5R) could effectively absorb the $H\text{-}2^k$ suppressive T-cell factor. Since both B10.A(3R) and B10.A(5R) mice are supposed to have the same haplotype, it is conceivable that the only difference in the $H\text{-}2$ complex between these strains is the $I\text{-}J$ subregion. Furthermore, this antiserum did not show any cytotoxic activity for $H\text{-}2^k$ spleen cells, once again indicating that the suppressor molecule is in fact distinct from known Ia molecules expressed on B cells.

While our work was in progress, the possible differences in the position of crossing-over in certain recombinant strains, (B10.A(3R) and B10.A-(5R), B10.HTT and B10.S(9R)), were found by Murphy et al. (1976), which also indicates that the region coding for an Ia-like molecule on the allotype suppressor T cell maps between the $I\text{-}B$ and $I\text{-}C$ subregions, and that this molecule is different from Ia antigens detected by cytotoxic assays for splenic B cells. The absorption of alloantisera with B cells completely removed the cytotoxic activity for splenic B cells leaving the inhibitory effect on the allotype suppressor T-cell intact (L. A. Herzenberg, pers. comm.). Furthermore, Colombani et al. (1976) have presented evidence by means of cytotoxic assays for B10.A(5R) cells with antiserum (C3H.Q × B10.D2)F$_1$ anti-AQR, that a new subregion *(I-E)* is present between $I\text{-}B$ and $I\text{-}C$. Since antibodies reactive with $H\text{-}2^b$ and $H\text{-}2^d$ haplotypes are negative with this antiserum, the reaction of B10.A(5R) should be due to a region derived from $H\text{-}2^k$. Absorption of this antiserum with B10.A(4R) leaves residual activity for B10.A(2R), B10.A(5R), and B10.S(9R) (C. S. David, unpubl.). These observations strongly suggested that there are at least two loci between $I\text{-}B$ and $I\text{-}C$, one coding for the molecule on the suppressor T cell *(I-J)* and the other for B-cell Ia antigen *(I-E)*.

In order to confirm that the *I-J* subregion is responsible for determining the suppressor molecule, we prepared T extracts from various recombinant mice and tested their activity in the responses of other strains sharing only restricted I subregions. As has been reported elsewhere (Taniguchi et al. 1976b), there is a strict requirement for identity of I-region genes between the factor and responding spleen cells. Combinations of the donor and responding spleen cells are listed in Table 5. It is evident that identities of haplotype in the proposed I-J subregion between the donor and recipient always resulted in an effective suppression. Furthermore, in terms of position of crossing-over in B10.A-(3R), B10.A(5R), B10.HTT, and B10.S(9R), the results are in complete agreement with the findings by Murphy et al. (1976). These results indicate clearly that the gene which determines the suppressor molecule maps in the newly described I-J subregion located between I-B and I-C. Figure 1 illustrates the logical bases for the presence of the I-J subregion at this position.

One further point to be noted in Table 5 is the fact that the combinations of donor and recipient differing in I-J subregion but identical in I-A subregion generally resulted in the enhancement of the antibody response. The identity of the I-C subregion caused neither suppression nor enhancement. The results suggested that an enhancing T-cell factor is present in the same preparation of T extract, and that this may well be another I-region product, since only the strain-specific enhancing effect was observed when the suppressive effect was negated by differing I-J subregions. This postulate prompted us to determine an enhancing T-cell factor in the KLH-primed T-cell extract.

Table 5. Requirement for Identity of I-J Subregions for Induction of Suppression

T extract (A, B, J, E, C)	Responding spleen cells	Identities of I subregion	Suppression (%)
B10 (b b *b* b b)	C57BL/6J	A B *J* E C	68
	A/J	none	0
B10.A (k k *k* k d)	A/J	A B *J* E C	53
	C3H	A B *J* E	69
	BALB/c	C	0
B10.A(4R) (k b *b* b b)	CBA	A	0[a]
	C57BL/6J	B *J* E C	46
B10.A(5R) (b b *k* k d)	C57BL/6J	A B	0[a]
	CBA	*J* E	82
	BALB/c	C	0
B10.A(3R) (b b *b* k d)	C57BL/6J	A B *J*	71
	CBA	E	0
	BALB/c	C	0
B10.S(9R) (s s *k* k d)	SJL	A B	0
	C3H	*J* E	94
	BALB/c	C	0
B10.HTT (s s *s* k k)	SJL	A B *J*	57
	C3H	E C	0
	BALB/c	none	0

[a] Enhancement.

Mapping of the suppressor gene in I-J subregion

Strain	H-2 haplotype	H-2 subregions K	I-A	I-B	(I-J)	I-C	S	D	Suppressible strain
BIO.A	a	k	k	k	(k)	d	d	d	H-2^b ...

Strain	H-2 haplotype	K	I-A	I-B	(I-J)	I-C	S	D	Suppressible strain
BIO.A	a	k	k	k	(k)	d	d	d	H-2^k
BIO	b	b	b	b	(b)	b	b	b	H-2^b
BIO.A(4R)	h4	k	k	b	(b)	b	b	b	H-2^b
BIO.A(5R)	i5	b	b	b	(k)	d	d	d	H-2^k
BIO.A(3R)	i3	b	b	b	(b)	d	d	d	H-2^b

Figure 1. Schematic presentation of the I-J subregion in the H-2 complex of some recombinant strains. Only the response of I-J-compatible strains is suppressed by the factor obtained from these recombinants.

An Enhancing T-cell Factor Determined by the I-A Subregion

Based on the above findings, we have attempted to detect the enhancing T-cell factor. B10 congenic lines are producers of the suppressive T-cell factor but cannot accept the H-2-compatible T-cell factor (see above). Thus the responses of B10 congenic lines were often enhanced by the addition of T extract. Therefore, these mice were found to be convenient strains for studying the effect of the enhancing T-cell factor.

In order to establish the experimental conditions needed to test the enhancing T-cell factor, DNP-KLH-primed spleen cells were cultured with 0.1 μg/ml of DNP-KLH in Marbrook culture bottles, and the syngenic or histocompatible KLH-primed T-cell extract was added on days 0 (at the start of cultivation), 1, 2, and 3. Representative results in the response of B10.BR spleen cells are shown in Figure 2. Both B10.BR and C3H T extracts induced a significant enhancement of the DNP-specific IgG antibody response of B10.BR spleen cells, and the effect was in general most pronounced when the T extract was added on day 2 or day 3. This enhancement was also observed in C3H, the acceptor strain for the suppressive T-cell factor, provided the same T extract was added 1 to 3 days after the start of cultivation of spleen cells with antigen (Fig. 3). The results indicate that the acceptor site for the suppressive T-cell factor is already expressed on responding (C3H) spleen cells before cultivation, whereas the acceptor for the enhancing T-cell factor predominates after cultivation with antigen. As will be discussed later, this sequential expression of acceptor sites for different T-cell factors raises some complicated questions. However, this phenomenon is itself in keeping with the general notion that antigen-nonspecific factors are most effective in enhancing the in vitro antibody response when they are given to the cultured spleen cells at 24 to 48 hours after antigenic stimulation

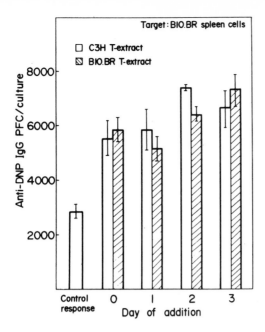

Figure 2. Enhancement of anti-DNP IgG antibody response of B10.BR (nonacceptor strain for the suppressive T-cell factor) spleen cells by addition of T extracts from syngenic and histocompatible (C3H) strains given on different days.

(Dutton 1972; Schimpl and Wecker 1973; Gorczynski et al. 1973; Kishimoto and Ishizaka 1975a,b; Kishimoto et al. 1975).

Utilizing the in vitro culture system described above, we have partially characterized the enhancing T-cell factor as an *I*-region gene product.

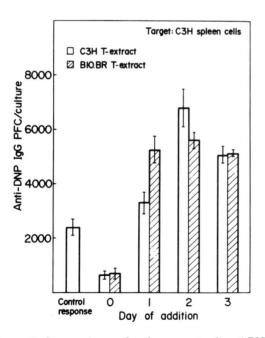

Figure 3. Suppression and enhancement of anti-DNP IgG antibody response of C3H (acceptor strain for the suppressive T-cell factor) by addition of syngenic and histocompatible (B10.BR) T extracts given on different days.

By absorption studies using various immunoadsorbents composed of antigens and anti-immunoglobulin antibodies, it was firmly established that the enhancing T-cell factor also has a definite antigen (KLH) specificity but no immunoglobulin determinants. The activity was also dependent on the presence of relevant antigen (DNP-KLH, but not DNP-egg albumin). Probably the most important point is that the factor is *not* a T-cell-replacing factor, since the addition of the factor to the T-cell-depleted (anti-Thy-1- and C-treated) spleen cells did not induce the production of anti-hapten antibodies of either the IgM or IgG class. Nevertheless, the factor could augment the IgG response of intact, primed spleen cells about two to three times in PFC numbers. Furthermore, we have found that the enhancing T-cell factor can be absorbed with thymocytes and splenic T cells but not with B cells (M. Taniguchi and T. Tada, unpubl.).

Evidence that the enhancing T-cell factor is also an *I*-region gene product was obtained by absorption studies using various alloantisera against restricted *I* subregions. Representative results are shown in Table 6. The extract of KLH-primed CBA thymocytes could enhance the anti-DNP PFC response of DNP-KLH-primed B10.BR spleen cells about 90% over the control. Whereas the absorption of the extract with anti-H-$2K^k$ (AQR anti-B10.A) did not remove the enhancing activity, several, but not all, alloantisera against the *I* region of H-2^k could absorb the enhancing activity. Close analysis of the specificities of alloantisera that can remove the enhancing activity leads to the conclusion that antisera having specificity for the *I-A* subregion are invariably effective regardless of their Ia specificities. In contrast, alloantisera directed to *I-B*, *I-J*, *I-E*, and/or *I-C* subregions were all ineffective in absorbing the enhancing activity (Table 6). The results suggest that the enhancing T-cell factor is an *I-A*-subregion gene product distinguishable from the suppressive T-cell factor, which is an *I-J*-subregion gene product. It was also noted that the absorbing capacities of alloantisera did not correlate with their anti-Ia specificities, suggesting that the factor is not the known Ia molecule, detectable on splenic B cell.

Further evidence for the *I-A*-subregion gene product as the enhancing T-cell factor was obtained by cross-strain experiments. Fortunately, we have two recombinant strains (B10.A(4R) and D2GD) in which crossover has occurred between the *I-A* and *I-B* subregions, and therefore we tested the factors obtained in these strains in the responses of various partner strains. The results are summarized in Table 7. It is clearly shown that the identity of the *I-A* subregion is the prerequisite for effective enhancement, which again indicates that the antigen- and strain-specific enhancing T-cell factor is a product of an *I-A*-subregion gene.

The second important point to be deduced from

Table 6. Absorption of the Enhancing T-cell Factor in H-2^k (CBA) T Extract with Alloantisera Directed to Different I Subregions (Target: B10.BR)

| CBA (H-2^k) T extract absorbed with | Specificity | | Anti-DNP IgG PFC/culture | Absorption |
	H-2 subregion	Ia		
Control	−	−	1492 ± 261	−
Unabsorbed	−	−	2865 ± 225	−
AQR anti-B10.A	K^k	−	2848 ± 113	−
(B10 × LPR.III)F$_1$ anti-B10.A(2R)	K^k, I-A^k, I-B^k, I-J^k, I-E^k, I-C^d	2, 18	1610 ± 219	+
A.TH anti-A.TL	I-A^k, I-B^k, I-J^k, I-E^k, I-C^k	1, 2, 3, 7, 19	1448 ± 256	+
(A.BY × B10.HTT)F$_1$ anti-A.TL	I-A^k, I-B^k, I-J^k	1, 2, 19	1507 ± 230	+
(B10.A(4R) × C3H.OH)F$_1$ anti-B10.K	I-B^k, I-J^k, I-E^k, I-C^k	−	2559 ± 118	−
(B10.A(4R) × HTI)F$_1$ anti-B10.A	I-B^k, I-J^k, I-E^k, I-C^d	7	2994 ± 246	−
B10.S(7R) anti-B10.HTT	I-E^k, I-C^k	7	2504 ± 162	−
B10.A(3R) anti-B10.A(5R)	I-J^k	−	3192 ± 199	−

Table 7. Identity Requirement of I-A Subregions for Effective Enhancement

T extract (A B J E C)	Responding spleen cells	Identities of I subregion	Enhancement (%)
B10.A	B10.A	A B J E C	136
(k k k k d)	B10.BR	A B J E	135
	B10.D2	C	−[a]
B10.A(5R)	B10	A B	144
(b b k k d)	B10.BR	J E	−
	B10.D2	C	−
B10.A(4R)	B10.BR	A	141
(k b b b)	B10	B J E C	−
D2GD	B10.D2	A	77
(d b b b)	B10	B J E C	−
B10.BR	B10.BR	A B J E C	136
(k k k k k)	B10.A	A B J E	141
	B10.A(4R)	A	166
	B10.A(5R)	J E	−

[a] Less than 15% enhancement (not significant).

this experiment is that B10.BR (H-2^k) factor can only enhance the responses of strains sharing the same I-A subregion (e.g., B10.A and B10.A(4R)). This suggests that the acceptor site for the enhancing T-cell factor is also encoded by genes in the I-A subregion. In fact, the enhancing T-cell factor of H-2^k was successfully removed by absorption with spleen cells of B10.BR, B10.A, and B10.A(4R) but not with those of B10, B10.A(5R), and B10.D2. The cells capable of absorbing the enhancing T-cell factor were Thy-1-bearing but adhered to the nylon wool column. In summary, the results presented here indicate that there are two different T-cell factors, determined by different loci in the I region of the H-2 complex, which positively and negatively regulate T-cell-dependent antibody responses.

DISCUSSION

The above results provide evidence that different I-region gene products mediate two opposite regulatory events in T-cell-dependent antibody responses. Of particular interest is that both these T-cell factors are apparently different from previously known Ia antigens detected on B cells. The results are in keeping with observations by Murphy et al. (1976), who showed that the I-J-subregion gene product is expressed on allotype suppressor T cells, and that this determinant is distinct from Ia antigens detectable on splenic B cells by cytotoxic assay. Furthermore, Okumura et al. (1976) have presented evidence that an I-region determinant is detectable on helper T cells, and this determinant is clearly distinguishable from the I-J determinant expressed on suppressor T cells. Since, in their experiments, allotype suppressor T cells remove the helper T-cell activity for Ig-1b B cells (Herzenberg et al. 1975), it is suggested that these I-region gene products are in fact part of devices by which different lymphoid subpopulations communicate. Both the antigen (and in some cases allotype) specificity and I-region specificity observed in these studies suggest that these restrictions are useful in making such cell communications meaningful and unmistakable. How then can we fit all these findings to the concept of collaboration between T and B cells?

Although in the present studies the loci coding for the suppressive and enhancing T-cell factors were determined fairly unambiguously, the mechanisms by which these T-cell factors exert their specific effects are largely unknown. A crucial question remains as to what regulatory mechanisms operate the phenotypic expression of genes in either subregion under different conditions on functionally different subsets of T cells. The question is undoubtedly related to the more fundamental problems concerning the function of specific Ir and Is genes. The proposed locations of the loci coding for the suppressive and enhancing T-cell factors are in close proximity to the Ir and Is loci, but exact relationships between these functionally different genes have not been clearly established. However, the antigen-specific, cooperative T-cell factors reported by Taussig and Munro (1974), Munro et al. (1974), and Mozes et al. (1975) are apparently related to the expression of specific Ir genes. More recently, Kapp et al. (1976) have shown that the unresponsiveness of mice to the synthetic copolymer

L-glutamic acid-L-alanine-L-tyrosine (GAT), which is under Is gene control, is mediated by a T-cell product having properties similar to the KLH-specific factor. Therefore, these results should be integrated so as to make a conceptual framework of T-cell-mediated regulation of antibody response which is accomplished by *I*-region gene products.

However, the problem has been complicated by the demonstration of a third cell type which absorbs suppressive and enhancing T-cell factors. These cells are present in the thymus and spleen but not in the lymph nodes and bone marrow, are Thy-1-bearing, but adhere to the nylon wool column. Since the acceptor sites for the suppressive T-cell factor are not expressed in B10 congenic lines, which are capable of accepting the enhancing T-cell factor, the cells expressing the acceptor sites for suppressive and enhancing T-cell factors may be different, or those acceptor sites are expressed sequentially after antigenic stimulation. Although there are no definite experimental data, the acceptor sites for the factors appear to be determined by distinct *I* subregions (*I-J* and *I-A*). Since the enhancing T-cell factor is not a T-cell-replacing factor, the presence of such a cell type accepting the T-cell factor may be of considerable importance in the transmission of both the suppressor and enhancing effects of T cells. Investigation of the nature of this third cell type with respect to Ly phenotype and antigen specificity is now underway.

Finally, the picture we have described here is far more complicated than the previously proposed collaboration of T and B cells by antigen links and thus provides a new framework for considering the complex processes of immunoregulation. As discussed here, suppression and enhancement both require very strict identity in *I*-region genes, which has not been observed in some other experimental systems (Bechtol et al. 1974; Taussig et al. 1975; von Boehmer et al. 1975). We are also aware of the presence of several antigen-nonspecific factors which can replace the helper T cell and have no *I*-region determinants (Schimpl and Wecker 1973; Gorczynski et al. 1973; Rubin et al. 1973). In an experimental system very similar to ours (Basten et al. 1975), suppressor T cells were shown to act directly on B cells, and Herzenberg et al. (1975) have demonstrated that the allotype suppressor T cell removes helper T-cell activity. In view of the complexity of the immunoregulatory process, including the third transmitter cell type, it is conceivable that certain processes of regulatory cell interactions require *I*-region identity, whereas others require antigen specificity. In addition, it would seem important to know whether or not experiments involve the third cell type for the full expression of T-cell function. In any event, by considering the intermediate transmission process for the suppressor and helper effects, the contro-versy presently existing among different experimental systems can hopefully be resolved.

Acknowledgments

We are grateful to Drs. B. Benacerraf, D. Sachs, R. Schwartz and K. Moriwaki for their generous supply of alloantisera and mouse strains. We thank Drs. T. Takemori, K. Hayakawa and T. Tokuhisa for their excellent technical assistance. Excellent secretarial assistance by Ms. Yoko Yamaguchi is also gratefully acknowledged. This work was supported by a grant from the Ministry of Education, Science and Culture, Japan, and in part by American Cancer Society Grants I-M 80 and I-M 74.

REFERENCES

ARMERDING, D. and D. H. KATZ. 1974. Activation of T and B lymphocytes *in vitro*. II. Biological and biochemical properties of an allogeneic effect factor (AEF) active in triggering specific B lymphocytes. *J. Exp. Med.* **140:** 19.

BASTEN, A., J. F. A. P. MILLER and P. JOHNSON. 1975. T cell-dependent suppression of an anti-hapten antibody response. *Transplant. Rev.* **26:** 130.

BECHTOL, K. B., J. H. FREED, L. A. HERZENBERG and H. O. McDEVITT. 1974. Genetic control of the antibody response to poly-L(Tyr, Glu)-poly-D,L-Ala--poly-L-Lys in C3H ↔ CWB tetraparental mice. *J. Exp. Med.* **140:** 1660.

CEROTTINI, J. C. and E. R. UNANUE. 1971. Genetic control of the immune response of mice to hemocyanin. I. The role of macrophages. *J. Immunol.* **106:** 732.

COLOMBANI, J., M. COLOMBANI, D. C. SHREFFLER and C. S. DAVID. 1976. Separation of anti-Ia from anti-*H-2* antibodies in complex sera, by absorption on blood platelets. Description of three new specificities. *Tissue Antigens* **7:** 74.

CUNNINGHAM, A. J. and A. SZENBERG. 1968. Further improvements in the plaque technique for detecting single antibody-forming cells. *Immunology* **14:** 599.

DUTTON, R. W. 1972. Inhibitory and stimulatory effects of concanavalin A on the response of mouse spleen cell suspensions to antigen. I. Characterization of the inhibitory cell activity. *J. Exp. Med.* **136:** 1445.

ERB, P. and M. FELDMANN. 1975. The role of macrophages in the generation of T helper cells. III. Influence of macrophage-derived factors in helper cell induction. *Eur. J. Immunol.* **5:** 759.

GORCZYNSKI, R. M., R. G. MILLER and R. A. PHILLIPS. 1973. Reconstitution of T cell-depleted spleen cell populations by factors derived from T cells. III. Mechanism of action of T cell-derived factors. *J. Immunol.* **111:** 900.

HERZENBERG, L. A., K. OKUMURA and C. M. METZLER. 1975. Regulation of immunoglobulin and antibody production by allotype suppressor T cells in mice. *Transplant. Rev.* **27:** 57.

KAPP, J. A., C. W. PIERCE, F. DeLa CROIX and B. BENACERRAF. 1976. Immunosuppressive factor(s) extracted from lymphoid cells of non-responder mice primed with L-glutamic acid^{60}-L-alanine^{30}-L-tyrosine^{10} (GAT). *J. Immunol.* **116:** 305.

KATZ, D. H. and B. BENACERRAF. 1975. The function and interrelationships of T-cell receptor, Ir genes and other histocompatibility gene products. *Transplant. Rev.* **22:** 175.

KISHIMOTO, T. and K. ISHIZAKA. 1975a. Regulation of antibody response *in vitro*. IX. Induction of secondary anti-hapten IgG antibody response by anti-immunoglobulin enhancing soluble factor. *J. Immunol.* **114:** 585.

————. 1975b. Immunologic and physicochemical properties of enhancing soluble factors for IgG and IgE antibody responses. *J. Immunol.* **114:** 1177.

KISHIMOTO, T., T. MIYAKE, Y. NISHIZAWA, T. WATANABE and Y. YAMAMURA. 1975. Triggering mechanisms of B lymphocytes. I. Effect of anti-immunoglobulin and enhancing soluble factor on differentiation and proliferation of B cells. *J. Immunol.* **115:** 1179.

MOZES, E., R. ISAC and M. J. TAUSSIG. 1975. Antigen-specific T-cell factor in the genetic control of the immune response to poly(Tyr, Glu)-poly-D,L-Ala--poly-Lys. Evidence for T- and B-cell defects in SJL mice. *J. Exp. Med.* **141:** 703.

MUNRO, A. J., M. J. TAUSSIG, R. CAMPBELL, H. WILLIAMS and Y. LAWSON. 1974. Antigen-specific T-cell factor in cell cooperation: Physical properties and mapping in the left hand (K) half of *H-2. J. Exp. Med.* **140:** 1579.

MURPHY, D. B., L. A. HERZENBERG, K. OKUMURA, L. A. HERZENBERG and H. O. McDEVITT. 1976. A new *I* subregion (*I-J*) marked by a locus (Ia-4) controlling surface determinants on suppressor T lymphocytes. *J. Exp. Med.* (in press).

OKUMURA, K., L. A. HERZENBERG, D. B. MURPHY, H. O. McDEVITT and L. A. HERZENBERG. 1976. Selective expression of *H-2* (*I* region) loci controlling determinants on helper and suppressor T lymphocytes. *J. Exp. Med.* (in press).

RUBIN, A. S., A. B. MacDONALD and A. H. COONS. 1973. Specific heterologous enhancement of immune responses. V. Isolation of a soluble enhancing factor from supernatants of specifically stimulated and allogeneically induced lymphoid cells. *J. Immunol.* **111:** 314.

SCHIMPL, A. and E. WECKER. 1973. Stimulation of IgG antibody formation *in vitro* by T cell replacing factor. *J. Exp. Med.* **137:** 547.

TADA, T., M. TANIGUCHI and C. S. DAVID. 1976. Properties of antigen-specific suppressive T cell factor in the regulation of antibody response of the mouse. IV. Special subregion assignment of the gene(s) which codes for the suppressive T cell factor in the *H-2* histocompatibility complex. *J. Exp. Med.* (in press).

TAKEMORI, T. and T. TADA. 1975. Properties of antigen-specific suppressive T cell factor in the regulation of antibody response of the mouse. I. In vivo activity and immunochemical characterization. *J. Exp. Med.* **142:** 1241.

TANIGUCHI, M., K. HAYAKAWA and T. TADA. 1976a. Properties of antigen-specific suppressive T cell factor in the regulation of antibody response of the mouse. II. In vitro activity and evidence for the *I* region gene product. *J. Immunol.* **116:** 542.

TANIGUCHI, M., T. TADA and T. TOKUHISA. 1976b. Properties of antigen-specific suppressive T cell factor in the regulation of antibody response of the mouse. III. Dual gene control of the T cell-mediated immunosuppression. *J. Exp. Med.* **144:** 20.

TAUSSIG, M. J. and A. J. MUNRO. 1974. Removal of specific cooperative T cell factor by anti-*H-2* but not by anti-Ig sera. *Nature* **251:** 63.

TAUSSIG, M. J., A. J. MUNRO, R. CAMPBELL, C. S. DAVID and N. A. STAINES. 1975. Antigen-specific T-cell factor in cell cooperation. Mapping within the *I* region of the *H-2* complex and ability to cooperate across allogeneic barriers. *J. Exp. Med.* **142:** 694.

VON BOEHMER, H., L. HUDSON and J. SPRENT. 1975. Collaboration of histoincompatible T and B lymphocytes using cells from tetraparental bone marrow chimeras. *J. Exp. Med.* **142:** 989.

Functional and molecular organisation of an antigen-specific suppressor factor from a T-cell hybridoma

Masaru Taniguchi & Izumi Takei

Laboratories for Immunology, School of Medicine, Chiba University, Chiba, Japan

Tomio Tada

Department of Immunology, Faculty of Medicine, University of Tokyo, Japan

Table 1 Presence of the associated and non-associated forms of antigen-binding and I–J coded molecules in the extract and lack of the non-associated form in ascites

Materials added to the culture	Anti-DNP IgG PFC per culture	
	Ascites	Extract
None	$1,774 \pm 182$	$1,211 \pm 269$
Unabsorbed	313 ± 75	146 ± 113
Absorbed with anti-Igs	462 ± 164	131 ± 193
Eluted from anti-Igs	$1,802 \pm 440$	$1,547 \pm 172$
Absorbed with KLH	$2,236 \pm 174$	$1,255 \pm 465$
Eluted from KLH	668 ± 237	219 ± 170
Absorbed with anti-I–J^b	$2,043 \pm 258$	$1,342 \pm 21$
Eluted from anti-I–J^b	580 ± 128	131 ± 193
Combination of materials absorbed with KLH and anti-I–J^b	$2,118 \pm 335$	100 ± 94

The secreted or extracted materials were obtained from suppressor T-cell hybridoma as described previously[16]. Briefly, the hybrids were made by fusion of AKR thymoma cell line BW5147 with the enriched suppressor T cell of C57BL/6 mice specific for KLH. For hybridisation, 5×10^6 enriched suppressor T cells of C57BL/6 origin and $5-50 \times 10^6$ BW5147 cells were washed twice in serum-free Dulbecco's modified Eagle's medium (DMEM) and pelleted together at 400 g. Polyethylene glycol (2 ml: PEG, molecular weight 2,000)–dimethyl sulphoxide (DMSO) solution (1 weight of PEG plus 1 volume of 15% DMSO–DMEM) was added to the cell pellet and mixed gently. A further 2 ml of 50% (W/V) PEG solution without DMSO was poured into the test tube and the mixture was gently stirred with a Pasteur pipette. The cell suspension was then gradually diluted with 16 ml of serum free DMEM and further diluted with 180 ml of DMEM containing 13% fetal calf serum (FCS). It was then incubated at 37 °C for 3–4 h in 5% CO_2. After incubation cells were washed four times with DMEM and cultured for 1–2 weeks in HAT medium (RPMI 1640 supplemented with 10% FCS and hypoxanthine, aminopterin, thymidine (HAT)) on Falcon plates (Falcon 3008).Cells grown in HAT medium were collected and stained with anti-I–J^b anti-serum (B10.A(5R) anti-B10.A(3R)) and fluorescein-conjugated rabbit anti-mouse Igs. The stained cells were then sorted by fluorescence-activated cell sorter, FACS II (Becton-Dickinson), and I–J positive cells were cloned in a multiwell microplate (Falcon 3040) by limiting dilution or single cell manipulation. So far we have established several I–J positive suppressor T-cell hybridomas with specific functions, that is, cell lines 34S-704, 34S-18, 34S-11, 1L-1, 8C-23, and 9F-18, each of which has the same characteristics as the cell line 9F181a described previously[16]. The cell-free extract used was obtained by ultracentrifugation of the frozen and thawed materials of the hybrid cells (34S-704) at 20,000g for 1 h, and the secreted materials were obtained in ascites of hybridoma (34S-704)-bearing F_1 mice. 20 μl of ascites or extracted materials from 3×10^5 hybridoma cells were passed through the immunoadsorbents composed of KLH (5 mg per ml of beads), anti-I–J^b (B10.A(5R)anti-B10.A(3R); 1 ml γ-globulin fraction of antiserum per ml of beads), rabbit anti-mouse Igs (5 mg purified anti-mouse Ig antibodies per ml of beads). The beads used were Sepharose 4B prepared as described previously[16]. The materials absorbed by columns were eluted with 0.175M glycine–HCl buffer pH 3.2 followed by an instant neutralisation with 1 M sodium bicarbonate, dialysed for 4–6 h with 0.01 M sodium phosphate-buffered saline pH 7.2 and concentrated by negative pressure. The effluents and the eluates from the columns, and the mixture of the effluents from each column were added to the culture of 4×10^6 spleen cells of C57BL/6 mice primed with 100 μg DNP–KLH plus 1×10^9 pertussis vaccine in the Mishell-Dutton system (Falcon 3008 plate), in 1 ml RPMI 1640 enriched with 10% FCS in the presence of 0.1 μg DNP–KLH and 2.0×10^{-5}M 2-mercaptoethanol. Anti-DNP IgM and IgG PFC were assayed on day 5 of culture by the method of Cunningham and Szenberg as described previously[3]. Results are expressed as mean numbers of anti-DNP IgG PFC of four cultures ±s.d.

Thymus-dependent (T) lymphocytes have been shown to have antigen specificity. The antigen receptor on T lymphocytes, in contrast to that on B lymphocytes, does not appear to be of the conventional immunoglobulin (Ig) type. Studies on the antigen-specific factors derived from helper and suppressor T cells (Ts) demonstrated that they possess determinants with antigen binding affinity and products of genes in the H-2 complex (MHC)[1-6]. Furthermore, antibodies against the variable region of Ig heavy chains or idiotypes have been shown to react with T-cell antigen receptors as well as antigen-specific helper and suppressor T-cell factors (TsF)[7-15]. It is, therefore, conceivable that at least two gene products are involved in the structural entity of these receptors: one each coded for by genes in either. To establish the molecular nature of the recognition component of T cells we have used homogeneous TsF from a T-cell hybridoma with a specific function. We report here that the antigen binding and I–J coded molecules on TsF are independently synthesised in the cytoplasm, and are secreted as an associated form of the two molecules; this association is required for antigen-specific suppression of antibody response.

The T-cell hybridomas with specific suppressive activity were established 20 months previously by the fusion of BW5147 (H-2^k) and enriched Ts of C57BL/6(H-2^b) mice primed with keyhole limpet haemocyanin (KLH). The hybrids continuously secrete specific suppressor molecules with an antigen binding site and products of genes in the I–J subregion of MHC. The properties and functions of the hybridoma-derived TsF have been reported previously[16]. We attempted to separate the active moiety so as to characterise further the structure of the hybridoma-derived TsF. The suppressive extract or ascitic material from the hybridoma (34S–704) was passed through immunoadsorbent columns of KLH, anti-I–J^b and anti-mouse Igs under the same conditions as described previously[16]. The absorbed materials were eluted from columns with glycine–HCl buffer, pH 3.2 at 0 °C. Both effluents and eluates were tested for their suppressive activity. As shown in Table 1, the activity of TsF from the hybridoma was absorbed by KLH and anti-I–J^b columns but not by the anti-Igs. The activity was successfully recovered in the acid eluate from KLH and anti-I–J^b columns. These results indicate that TsF in the extracted and secreted materials has both antigen-binding sites and I–J determinants, and can be recovered from either the KLH or anti-I–J^b column. These two determinants are also demonstrated on the surface of semipurified Ts[15,17] and hybridomas[18]. In contrast, recent reports[9,10] showed that isolated T-cell receptor molecules do not contain detectable MHC gene products. It is, therefore, of interest to determine whether or not the I–J determinant is in fact associated with antigen-receptor of Ts, and if such association is indeed required for suppression. If these two distinct determinants are present on a single molecule, the mixture of the materials after absorption with the KLH or anti-I–J^b column should not exert any suppressive activity. On the other hand, if they are on separate molecules and combine when they act on target cells, the mixture should reconstitute the suppressor activity. The mixture of the effluents of ascites from the KLH and anti-I–J^b columns gave no suppressive effect (Table 1). Unlike ascites, the mixture of the effluents of the extracted material definitely produced a strong suppression, despite the fact that either effluent alone had no detectable activity. These results indicate that the extract, but not ascites, contains two discrete molecules which would combine to reconstitute suppressor activity. Associated forms of suppressor molecule, however, should be present in the extract, since the activity could be eluted from either KLH or anti-I-J^b column. The secreted material, however, contains only molecules of the associated form.

These experiments do not, however, exclude the possibility that the suppressor activity in the mixture of the two effluents is caused by the additive effect of residual TsF remaining after absorption, even though the effluents alone did not exert detectable suppressor effects. To eliminate this possibility, the extract was passed through the KLH or anti-I-J^b column (first column), and the effluent from each of the columns was successively applied to both KLH and anti-I-J^b columns (second column) (Fig. 1). The effluent from either the KLH or anti-I-J^b column was tested directly for its suppressive activity. In other experiments, the effluent from the KLH column was reabsorbed with another KLH column to ascertain complete absorption. Similar double absorption tests were performed with anti-I-J^b column. The effluents, each after absorption twice with the KLH or anti-I-J^b column, were admixed and tested for the suppressor activity. As shown in Table 2, the suppressive activity in the extract was abrogated by absorption with the KLH or anti-I-J^b column (group III or IV). However, the mixture of the KLH → KLH effluent and anti-I-J^b → anti-I-J^b effluent (group V) showed strong suppressor activity. Effluent from the KLH double columns should not contain KLH-binding molecules while I-J molecules are still present. Similarly, after absorption twice with the anti-I-J^b column, I-J molecules are completely removed leaving KLH-binding molecules in the effluent. Hence, the combination of these two effluents resulted in the mixture of the two molecules which reconstituted the suppressive activity. On the contrary, no suppressive activity was recovered when the effluent from the KLH (or anti-I-J^b) column was successively absorbed with the anti-I-J^b (or KLH) column (the mixture of KLH → anti-I-J^b effluent and anti-I-J^b → KLH effluent; group VI). These results suggest that antigen-binding and I-J coded

Table 2 Requirement of the co-existence of I-J bearing and antigen-binding molecules for the manifestation of suppressor function

Experimental group	Materials after absorption with: first column → second column		Anti-DNP IgG PFC per culture
I	None		2,824±98
II	Unfractionated		525±196
III	KLH	None	1,722±224
IV	Anti-I-J^b	None	2,923±598
V	KLH Anti-I-J^b	→ KLH → Anti-I-J^b	combined 383±150
VI	KLH Anti-I-J^b	→ Anti-I-J^b → KLH	combined 2,507±287

The extracted materials (34S-704) unabsorbed (group II), absorbed with the KLH (group III) or anti-I-J^b (group IV) column, or the mixture of the effluents from the double columns (groups V or VI) as indicated in Fig. 1 were added to the culture of DNP-KLH primed spleen cells from C57BL/6 mice and tested for their suppressive activity. The assay of anti-DNP IgG PFC is the same as in Table 1. Results are expressed as mean numbers of anti-DNP IgG PFC of four cultures ±s.d.

molecules are present independently in the extract, and the association of these two molecules, either of which has no detectable activity by itself, reconstitutes the suppressive activity. Although we have not excluded the possibility that the non-associated form in the extract is due to an imbalance in the synthesis of the two molecules in hybridoma cells, it is concluded that there are at least three molecules in the extracted material: antigen-binding molecule, I-J coded molecule and the associated form of the two molecules. The secreted material consists of the associated form alone. It is, therefore, likely that the two independent molecules in the extract are synthesised in the cytoplasm as a precursor of the suppressor factor which is then secreted after the association.

Our previous studies showed that KLH-TsF specifically suppresses the response of histocompatible spleen cells but not that of allogeneic cells[19]. This genetic restriction is governed by genes in the I-J subregion. The hybridoma-derived TsF also showed rigid genetic restriction[16]. In agreement with this antigen specificity and H-2 restriction of TsF, our present results show firm evidence that the functional structure of TsF is composed of two distinct molecules; one involved in the specific antigen-binding molecule, probably having V$_H$ structure and the other the I-J coded molecule which may serve as a restricting element in the suppressor cell interaction. A definitive answer must await chemical and physiochemical analysis of the hybridoma-derived suppressor molecule.

This work was supported by a grant from the Ministry of Education, Culture and Science, Japan.

Received 28 August; accepted 6 November 1979.

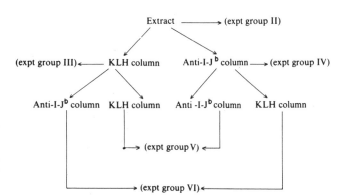

Fig. 1 Procedures of the absorption of the extract with the KLH and anti-I-J^b columns. Extract equivalent of 6×10^5 hybridoma cells (34S-704) was applied to the KLH or anti-I-J^b column (first column). The effluent from either the KLH or anti-I-J^b column was divided into four equal parts, each of which contained the factor equivalent of 1.5×10^5 hybridoma cells. Two of these were used as the effluent of group III or group IV as indicated. The other two were subsequently applied to another KLH and anti-I-J^b column, respectively (second column). The effluent from the KLH → KLH columns and that from the anti-I-J^b → anti-I-J^b columns were mixed (group V), as were the effluent from the KLH → anti-I-J^b columns and that from the anti-I-J^b → KLH columns (group VI). The unabsorbed (group II), the absorbed (groups III and IV) or the mixture of the effluents absorbed with double columns (groups V and VI) were added to the culture and their suppressive activities were tested in comparison to the response of group I without the factor. The materials of each group should contain the factor equivalent of 3×10^5 hybridoma cells.

1. Isac, R., Dorf, M. E. & Mozes, E. *Immunogenetics* **5**, 467-475 (1977).
2. Taussig, M. J., Munro, A. J., Compbell, R., David, C. S. & Staines, N. A. *J. exp. Med.* **142**, 694-700 (1975).
3. Taniguchi, M., Hayakawa, K. & Tada, T. *J. Immun.* **116**, 542-548 (1976).
4. Tada, T., Taniguchi, M. & David, C. S. *J. exp. Med.* **144**, 713-725 (1976).
5. Theze, J., Kapp, J. A. & Benacerraf, B. *J. exp. Med.* **145**, 839-856 (1977).
6. Theze, J., Waltenbaugh, C., Dorf, M. E. & Benacerraf, B. *J. exp. Med.* **146**, 287-292 (1977).
7. Eichmann, K. & Rajewsky, K. *Eur. J. Immun.* **5**, 661-666 (1975).
8. Binz, H. & Wigzell, H. *J. exp. Med.* **142**, 197-211 (1975).
9. Binz, H. & Wigzell, H. *Scand. J. Immun.* **5**, 559-571 (1976).
10. Krawinkel, U. *et al. Cold Spring Harb. Symp. quant. Biol.* **41**, 285-294 (1977).
11. Mozes, E. & Haimovich, J. *Nature* **278**, 56-57 (1979).
12. Kontiainen, S. *et al. Nature* **274**, 477-480 (1978).
13. Germain, R., Shyr-Te, J., Kipps, T. J., Benacerraf, B. & Dorf, M. E. *J. exp. Med.* **149**, 613-622 (1979).
14. Bach, B. A., Greene, M. I., Benacerraf, B. & Nisonoff, A. *J. exp. Med.* **149**, 1084-1098 (1979).
15. Tada, T., Hayakawa, K., Okumura, K. & Taniguchi, M. *Molec. Immun.* (in the press).
16. Taniguchi, M., Saito, T. & Tada, T. *Nature* **278**, 555-558 (1979).
17. Taniguchi, M. & Miller, J. F. A. P. *J. exp. Med.* **146**, 1450-1454 (1977).
18. Taniguchi, M. *Transplantn Proc.* (in the press).
19. Taniguchi, M., Tada, T. & Tokuhisa, T. *J. exp. Med.* **144**, 20-31 (1976).

CELLULAR CONSEQUENCES IN THE SUPPRESSION OF ANTIBODY RESPONSE BY THE ANTIGEN-SPECIFIC T-CELL FACTOR*

By MASARU TANIGUCHI AND TAKESHI TOKUHISA

From the Laboratories for Immunology, School of Medicine, Chiba University, Chiba, Japan

Previous studies from our laboratory indicated that a soluble factor extracted from carrier-primed suppressor T cells (TsF)[1] inhibits the in vitro secondary antibody response against a hapten coupled to the same carrier. The factor was found to possess determinants controlled by a locus (Ia-4) mapped in the I-J subregion of the mouse H-2 histocompatibility complex (1). Unlike other antigen-specific TsF, there has been shown a strict genetic restriction in that TsF derived from one strain of animals can suppress the response of only H-2 histocompatible strains (2, 3).

Furthermore, TsF was shown to be absorbable by splenic T cells, but not by B cells or macrophages of the same H-2 haplotype origin. Such T cells, which were assumed to be the direct targets of TsF, were adherent to a tightly packed nylon-wool column, but were definitely killed by anti-Thy-1 antiserum (2). Thus, the suppression of the antibody response by TsF is mediated by an interaction between the TsF and the acceptor site on the target cells. The most reasonable explanation is that such an acceptor site is controlled by a gene closely linked to that for the TsF within the same H-2 complex, as there have been no exceptional cases in which H-2 histoincompatible TsF can initiate the specific suppression.

Because little is known about the consequences of this initial interaction between TsF and acceptor T cells, we have performed a series of experiments in which subsequent cellular events after the TsF-acceptor interaction were studied. In this communication, we wish to report that the final suppression of antibody response was, in fact, achieved via the intermediary type of the acceptor T cells. Some properties and the mode of action of this cell type are described.

Materials and Methods

Antigens. Keyhole limpet hemocyanin (KLH) was purchased from Calbiochem-Behring Corp., American Hoechst Corp., San Diego, Calif. Hen's egg albumin (EA), recrystallized five times, was obtained by the method of Kekwick and Cannan (4). Dinitrophenylated KLH (DNP_{770}-KLH) and dinitrophenylated EA (DNP_{12}-EA) were prepared by coupling with 2,4-dinitrobenzensulfonic acid under alkaline condition by the method of Eisen et al. (5). *Bordetella pertussis* vaccine (BPV) was purchased from the Chiba Serum Institute, Chiba, Japan.

---

* Supported by a Grant-in-Aid from the Ministry of Education, Science and Culture, Japan.

[1] *Abbreviations used in this paper:* BPV, *Bordetella pertussis* vaccine; C, complement; DNP, 2,4-dinitrophenyl; DNP_{12}-EA, dinitrophenylated hen's egg albumin; DNP_{770}-KLH, dinitrophenylated keyhole limpet hemocyanin; EA, hen's egg albumin; GAT, L-glutamic acid^{60}-L-alanine^{30}-L-tyrosine^{10}; GT, L-glutamic acid^{50}-L-tyrosine^{50}; KLH, keyhole limpet hemocyanin; MEM, Eagle's minimum essential medium; MHC, major histocompatibility complex; TsF, suppressive T-cell factor.

Animals.   Randomly bred C3H/HeJ and BALB/cAnN mice 8–12 wk old were raised in our animal facility.

Immunization of Mice.   Mice were immunized intraperitoneally with 100 μg of DNP$_{770}$-KLH or DNP$_{12}$-EA with 1×10^9 BPV as described previously (6). Spleen cells were obtained 4–8 wk after immunization.

Preparation of Nylon-Wool-adherent and -nonadherent Spleen Cells.   Spleen cells from carrier-primed or unprimed mice were fractionated by the method of Julius et al. (7). Carrier-primed spleen cells were obtained from mice that had been immunized with 100 μg of KLH or EA plus 1×10^9 BPV 4 wk previously. In brief, well-washed, sterile nylon wool of LP-1 Leukopak Leukocyte filters (Fenwal Inc., Walter Kidde & Co. Inc., Ashland, Mass.) was packed into a 100-ml glass syringe to the 40-ml mark. After sterilization, the column was rinsed with 200 ml of Eagle's minimum essential medium (MEM) containing 5% fetal calf serum at 37°C. The column was drained of excess medium and placed in an incubator at 37°C for 1 h before loading the cells. A total of 1×10^9 spleen cells in a vol of 20 ml was applied to the column. The column was left to stand for 45 min at 37°C, and then it was eluted slowly with warm, enriched MEM. After elution, the nylon was placed in a 1-liter beaker and was shaken in 300 ml of cold (4°C), enriched MEM to obtain nylon-wool-adherent cells. Under the conditions stated above, 15–20% of the original cells were recovered in the effluent, which was comprised of >95% Thy-1-positive cells and <5% Ig-bearing cells, of the nylon-wool column. The nylon-wool-adherent population had 60–70% Ig-bearing cells and 10–20% Thy-1-positive cells.

Preparation of Antigen-specific TsF.   The method to obtain antigen-specific TsF was described in detail elsewhere (6). Briefly, mice were immunized intraperitoneally twice with 100 μg of KLH or EA without adjuvant at 2-wk intervals. Their thymuses and spleen cells were obtained and subsequently disrupted by sonication using a Tomy UR 150 sonicator (Tomy Seiko Co., Ltd., Tokyo, Japan). The cell-free supernate was obtained by ultracentrifugation at 40,000 g for 1 h.

Alloantisera.   Anti-Ia antisera were supplied by Dr. C. S. David, Mayo Medical School, Rochester, Minn. The combination of the mouse strains to produce these antisera and resultant specificities were shown in the Results. Anti-Thy-1.2 antisera (AKR anti-C3H) were raised in our laboratory.

Treatment of Cells with Alloantisera.   0.5 ml of 1:5 (anti-Ia) or a 1:10 (anti-Thy-1.2) dilution of alloantisera was added to 0.5 ml of cell suspension containing 1.5×10^7 viable spleen cells. They were incubated at 0°C for 45 min, washed twice with cold MEM, and were then treated with a well-selected rabbit or guinea pig complement (C) diluted to 1:10 at 37°C for 30 min.

Cell Cultures.   The modified Marbrook culture system was utilized. 5×10^6 DNP-primed B cells obtained from DNP$_{770}$-KLH-primed spleen cells treated with anti-Thy-1.2 plus C were cocultured with an equal number of carrier-primed nylon-wool-purified helper T cells with or without the addition of 2.5×10^6 nylon-wool-adherent cells. They were incubated for 5 d with 0.1 μg/ml of appropriate antigen at 37°C in 10% CO$_2$. The number of DNP-specific IgM and IgG plaque-forming cells were assayed using DNP-coupled sheep erythrocytes as described previously (6).

Results

Requirement of Antigen-primed Nylon-Wool-adherent T Cells for the Suppression by TsF.   The DNP-specific in vitro secondary IgG antibody response was effectively induced by cocultivation of DNP-primed B cells with nylon-wool purified KLH-primed helper T cells, whereas low IgG response was elicited with KLH-primed nylon-wool-adherent T cells (Table I). However, the addition of KLH-specific TsF did not significantly suppress the response mounted by B cells and nylon-wool-purified helper T cells.

This suggests that a third cell type presumably present in the adherent population is required for the effect of TsF. To prove this postulate, 2.5×10^6 adherent cells were added to the culture of B and helper T cells. This dose of adherent cells did not affect the net IgG antibody response. However, the addition of TsF into this mixture greatly suppressed the antibody response (Table I).

TABLE I

Requirement of Nylon-Wool-adherent Cells for the Suppression Induced by TsF

Cell mixture‡			KLH-TsF added§	Anti-DNP IgG PFC/culture
DNP-primed B cells*	Nylon-wool-purified T cells‖	Nylon-wool-adherent cells¶		
× 10^6	× 10^6	× 10^6		
5	5	—	—	2,231 ± 245**
5	5	—	+	2,345 ± 350
5	—	5	—	231 ± 23
5	—	5	+	261 ± 70
5	5	2.5	—	2,176 ± 354
5	5	2.5	+	681 ± 151
5	—	—	—	11 ± 4

* DNP-KLH-primed spleen cells treated with anti-Thy-1.2 and guinea pig C.
‡ 5 × 10^6 DNP-primed B cells from C3H spleen cells were cocultured with 5 × 10^6 nylon-wool-purified helper T cells and/or 2.5–5.0 × 10^6 nylon-wool-adherent cells from KLH-primed C3H spleen cells in the presence of 0.1 μg DNP-KLH.
§ KLH-TsF from C3H mice corresponding to 1 × 10^7 viable spleen cells was added at the start of cultivation.
‖ KLH-primed splenic cells passed through the tightly packed nylon-wool column.
¶ KLH-primed splenic cells adherent to the tightly packed nylon-wool column.
** Arithmetic means and standard deviations from five cultures.

TABLE II

Necessity for the Carrier Specificity of Nylon-Wool-adherent Cells in the Induction of the Effective Suppression

Nylon-wool-purified T cells*	Nylon-wool-adherent cells‡	Stimulating antigen	TsF added§	Anti-DNP IgG PFC/culture
KLH	—	DNP-KLH	—	2,448 ± 68‖
KLH	KLH	DNP-KLH	—	2,535 ± 898
KLH	KLH	DNP-KLH	KLH	620 ± 240
KLH	KLH	DNP-KLH	EA	1,798 ± 557
KLH	Unprimed	DNP-KLH	—	1,077 ± 468
KLH	Unprimed	DNP-KLH	KLH	908 ± 312
KLH	Unprimed	DNP-KLH	EA	1,101 ± 165

* 5 × 10^6 DNP-primed B cells from C3H spleen cells treated with anti-Thy-1.2 + C were cocultured with 5 × 10^6 nylon-wool-purified helper T cells and 2.5 × 10^6 nylon-wool-adherent cells from primed or unprimed C3H spleen cells with or without KLH- or EA-TsF.
‡ 2.5 × 10^6 nylon-wool-adherent splenic cells from KLH-primed mice.
§ KLH- or EA-TsF from C3H mice corresponding to 1 × 10^7 viable cells was added at the start of cultivation.
‖ Arithmetic means and standard deviations from five cultures.

The question arises as to whether or not the adherent cells have to be primed with relevant antigen. Thus adherent cells were taken either from KLH-primed or unprimed animals and then added to the mixture of B cells and KLH-primed helper T cells. As shown in Table II, KLH-specific TsF was able to induce suppression in the presence of KLH-primed nylon-wool-adherent cells, whereas normal cells were incapable of inducing the suppressive effect. The observed suppression was found to be antigen-specific, because EA-specific TsF did not suppress the response in the presence of KLH-primed adherent cells.

TABLE III

Evidence that a Nylon-Wool-adherent Cell is the Target of TsF

Culture*			Anti-DNP IgG PFC/culture
Cells‡	Cell number	Treatment§	
Nylon purified	5×10^6	—	
+			$2,405 \pm 117\parallel$
Nylon adherent	2.5×10^6	—	
Nylon purified	5×10^6	KLH-TsF	
+			$2,408 \pm 203$
Nylon adherent	2.5×10^6	—	
Nylon purified	5×10^6	—	
+			401 ± 58
Nylon adherent	2.5×10^6	KLH-TsF	
—	—	—	10 ± 5

* 5×10^6 DNP-primed B cells from C3H spleen cells treated with anti-Thy-1.2 and C were cocultured with 5×10^6 KLH-primed nylon-wool-purified helper T cells and 2.5×10^6 nylon-wool-adherent cells, either of which had been pretreated with KLH-TsF at 0°C for 1 h before the culture.

‡ KLH-primed spleen cells passed through or adherent to nylon-wool columns were treated with KLH-TsF at 0°C for 1 h. They were then washed and added to the culture.

§ KLH-TsF from C3H mice corresponding to 1×10^7 viable cells.

$\parallel$ Arithmetic means and standard deviations from five cultures.

Evidence that Nylon-Wool-adherent T cells are the Direct Target of the Antigen-specific TsF. To confirm that the nylon-wool-adherent cell is the target of TsF, the adherent and nonadherent populations from KLH-primed mice were incubated with KLH-specific TsF at 0°C for 1 h without antigen. Cells were then washed twice with cold MEM and were cocultured with DNP-primed B cells in the presence of DNP-KLH. As shown in Table III, the addition of nylon-wool-adherent cells pretreated with KLH-specific TsF significantly suppressed the response mounted by DNP-primed B cells and nylon-wool-purified helper T cells. The same treatment of nylon-wool-purified helper T cells did not produce any suppressive effect. Because the nylon-wool-adherent population contains mostly B cells and macrophages, but only a small number of T cells, we attempted to define the type of cells which mediated this suppression. KLH-primed nylon-wool-adherent cells were treated with anti-Thy-1.2 and guinea pig C. 2.5×10^6 treated or nontreated nylon-wool-adherent cells were added to the coculture of DNP-primed B cells and KLH-primed nylon-wool-purified helper T cells with or without TsF. As shown in Table IV, the addition of nylon-wool-adherent cells together with TsF gave strong suppression. This suppression was, however, completely abrogated by the treatment of nylon-wool-adherent cells with a higher dilution of anti-Thy-1.2 and C. The same effect was obtained by the treatment with lower dilution of anti-Thy-1 and C. The results suggest that a high density of Thy-1 antigen is present on nylon-wool-adherent T cells involved in this suppression. The treatment with higher concentration of anti-Thy-1.2 was, however, found to eliminate both target cells for TsF and helper T cells (Th$_2$) present in the nylon-wool-adherent cell population which can polyclonally help the B-cell response as described by Tada et al. (8).

Presence of the I-J Subregion Gene Product on the Acceptor Site for TsF. KLH-primed

TABLE IV

Abolishment of the TsF-mediated Suppressive Activity by the Pretreatment of Nylon-Wool-adherent Cells with Anti-Thy-1.2 and C

Adherent cells treated with*	Dilution of antiserum	TsF added‡	Anti-DNP IgG/culture
C§	—	—	1,277 ± 259‖
C	—	+	57 ± 17
Anti-Thy 1.2 + C	1:20	—	526 ± 110
Anti-Thy 1.2 + C	1:20	+	437 ± 201
Anti-Thy 1.2 + C	1:60	—	1,279 ± 126
Anti-Thy 1.2 + C	1:60	+	1,291 ± 302

* C3H nylon-wool-adherent cells were treated with anti-Thy-1.2 and C before the culture. They were then washed thoroughly with medium. 2.5×10^6 of the cells were cocultured with DNP-primed B cells and KLH-primed nylon-wool-purified helper T cells from C3H mice with or without KLH-TsF in the presence of 0.1 µg DNP-KLH.
‡ KLH-TsH from C3H mice corresponding to 1×10^7 viable cells was added to the culture.
§ Guinea pig C.
‖ Arithmetic means and standard deviations from five cultures.

TABLE V

The Presence of I-J Subregion Gene Product on the Acceptor Cells for TsF

Adherent cells treated with*	Subregion specificity	Complement‡	TsF added§	Anti-DNP-IgG PFC/culture
—	—	—	—	1,290 ± 198‖
A.TL anti-A.TH	—	+	—	1,170 ± 236
A.TL anti-A.TH	—	—	+	250 ± 88
		+	+	61 ± 26
A.TH anti-A.TL	A,B,J,E,C	—	+	1,631 ± 578
		+	+	1,344 ± 178
(C3H.Q × B10.D2)F₁ anti-AQR	A,B,J,E	—	+	1,058 ± 206
		+	+	942 ± 286
(A.TH × B10.A(5R))F₁ anti-A.TL	A,B,C	—	+	145 ± 106
		+	+	50 ± 27
B10.A (3R) anti-B10.A(5R)	J	—	+	1,062 ± 58
		+	+	1,612 ± 361
B10.S (7R) anti-B10.HTT	E,C	—	+	142 ± 114
		+	+	93 ± 41

* KLH-primed nylon-wool-adherent cells from C3H mice were treated with various alloantisera with or without C. They were then washed twice with cold MEM, and 2.5×10^6 of the cells were cocultured with 5×10^6 DNP-primed B cells and 5×10^6 KLH-primed nylon-wool-purified helper T cells from C3H mice in the presence or absence of KLH-TsF.
‡ Rabbit C.
§ KLH-TsF from C3H mice corresponding to 1×10^7 viable cells was added at a start of cultivation.
‖ Arithmetic means and standard deviations from five cultures.

nylon-wool-adherent cells were treated with various alloantisera with different I-subregion specificities. The treatment was performed with or without rabbit C. As shown in Table V, the suppressive effect by KLH-specific TsF was completely abrogated by the treatment with anti-Ia antisera having specificity for products of the *I-J* subregion, regardless of other subregion specificities. The treatment with those lacking I-J subregion specificity did not affect the function of the acceptor T cells. Table V, also includes the results of the treatment of nylon-wool-adherent cells with

TABLE VI

Induction of Suppressor T Cells by the Precultivation of Nylon-Wool-adherent Cells with TsF and Antigen

Preculture*			Precultured cells treated with‡	Anti-DNP IgG PFC/culture
Adherent cells	Antigen	KLH-TsF		
+	−	+	−	942 ± 69§
+	KLH	−	−	1,495 ± 400
+	KLH	+	−	231 ± 85
+	KLH	+	Anti-Thy-1.2 + C	1,207 ± 249

* 1×10^7/ml KLH-primed nylon-wool-adherent cells from C3H mice were precultured with or without 0.1 μg KLH and/or KLH-TsF for 48 h in the Marbrook bottle. They were then harvested, washed thoroughly with MEM, and 2.5×10^6 of the cells were further cocultured for 5 d with DNP-primed B cells and KLH-primed helper T cells without KLH-TsF in the presence of 0.1 μg DNP-KLH.
‡ KLH-primed nylon-wool-adherent cells precultured with 0.1 μg KLH and KLH-TsF for 48 h were treated with anti-Thy-1.2 and C before further cultivation with DNP-primed B cells and KLH-primed helper T cells.
§ Arithmetic means and standard deviations from five cultures.

anti-Ia for 1 h at 0°C without C. The intermediary function of the acceptor T cells was completely blocked by the relevant antisera having anti-I-J activity.

Induction of Antigen-nonspecific Suppressor T Cells From Nylon-Wool-adherent Cell Population in the Presence of TsF. To understand the properties of nylon-wool-adherent cells after accepting TsF, KLH-primed adherent cells were cultured with TsF in the presence of 0.1 μg KLH for 48 h. Cells were then washed thoroughly with medium and added to the culture of DNP-primed B cells and KLH-primed helper T cells. In some experiments, the cultured adherent cells were treated with anti-Thy-1.2 and C. The results shown in Table VI, clearly demonstrate that the addition of the adherent cells precultured with TsF and antigen caused strong suppression, whereas the suppressive activity was abrogated by the treatment of the cells with anti-Thy-1.2 and C. These results indicate that TsF generates new suppressor T cells from the KLH-primed nylon-wool-adherent cells in the presence of antigen during the 48-h culture.

The antigen specificity of newly induced suppressor T cells by TsF was investigated. The mixture of DNP-primed B cells and EA-primed nylon-wool-purified helper T cells was cultured with KLH-primed nylon adherent cells together with KLH-specific or EA-specific TsF in the presence of 0.1 μg DNP-EA and 0.1 μg free KLH. Under these conditions, DNP-primed B cells and EA-primed helper T cells effectively cooperated to induce a secondary anti-DNP IgG response. The response against DNP-EA was, however, suppressed by KLH-specific TsF, but not by EA-specific TsF when the KLH-primed adherent cells coexisted (Table VII, upper part). The same conclusion was reached from the reverse experiments as illustrated in the lower part of Table VII. The suppression by EA-specific TsF was induced only in the presence of EA-primed but not KLH-primed nylon-wool-adherent cells. The results imply that KLH-primed nylon-wool-adherent T cells after accepting KLH-specific TsF in the presence of the relevant free carrier (KLH) can suppress the response against unrelated antigen (DNP-EA) even though TsF is not specific for the relevant (EA) carrier to the helper T cells; furthermore, the identity of the antigen specificity between TsF and adherent T cells is required.

Non-Requirement of H-2 Histocompatibility between Acceptor T Cells and Helper T Cells for

TABLE VII

Requirement for the Identity of Antigen Specificity between Adherent Cells and TsF

Nylon-wool-purified T cells primed with	Nylon-wool-adherent cells primed with	Stimulating antigens	TsF added*	Anti-DNP IgG PFC/culture
EA	KLH	DNP-EA + KLH	—	1,259 ± 136‡
EA	KLH	DNP-EA + KLH	KLH	310 ± 120
EA	KLH	DNP-EA + KLH	EA	1,203 ± 88
KLH	EA	DNP-KLH + EA	—	1,848 ± 320
KLH	EA	DNP-KLH + EA	EA	260 ± 31
KLH	EA	DNP-KLH + EA	KLH	1,623 ± 187

5×10^6 DNP-primed BALB/c B cells were admixed with 5×10^6 nylon-wool-purified helper T cells and 2.5×10^6 nylon-wool-adherent cells from BALB/c mice primed with indicated antigen (KLH or EA). The mixture was cultured with either KLH-TsF or EA-TsF in the presence of 0.1 µg DNP-EA and KLH, or 0.1 µg DNP-KLH and EA.

* KLH- or EA-TsF from BALB/c mice corresponding to 1×10^7 viable cells was added at the start of cultivation.

‡ Arithmetic means and standard deviations from five cultures.

TABLE VIII

Non-Requirement of H-2 Compatibility between Induced Suppressor T Cells and Helper T Cells for the Induction of Suppression

Nylon-wool-purified helper T cells from	Adherent cells from	TsF from*	Anti-DNP IgG PFC/culture
C3H	—	—	1,155 ± 142‡
C3H	C3H	—	3,343 ± 103
C3H	C3H	C3H	462 ± 138
C3H	C3H	BALB/c	3,119 ± 77
C3H	BALB/c	—	1,126 ± 143
C3H	BALB/c	C3H	992 ± 24
C3H	BALB/c	BALB/c	145 ± 77

2.5×10^6 KLH-primed nylon-wool-adherent cells either from C3H or BALB/c mice were added to the mixture of 5×10^6 DNP-primed B cells and KLH-primed nylon-wool-purified helper T cells from C3H mice. They were cultured for 5 d with KLH-TsF either from C3H or BALB/c mice in the presence of 0.1 µg DNP-KLH.

* KLH-TsF either from C3H or BALB/c corresponding to 1×10^7 viable cells was added at a start of cultivation.

‡ Arithmetic means and standard deviations from five cultures.

the TsF-mediated Suppression. The requirement of histocompatibility was investigated in the suppressive interaction between acceptor T cells, helper T cells, and TsF. DNP-primed B cells and nylon-wool-passed KLH-primed helper T cells from C3H mice were cocultured with KLH-primed syngeneic (C3H) or allogeneic (BALB/c) adherent cells in the presence of KLH-specific C3H or BALB/c TsF. As shown in Table VIII, the response of the mixture of C3H B cells, nylon-wool-passed helper T cells, and adherent cells was suppressed by C3H TsF but not by BALB/c TsF. BALB/c TsF, however, suppressed the response of C3H B cells and nylon-wool-purified helper T cells only if BALB/c nylon-wool-adherent cells coexisted. This suppressive effect mediated by BALB/c nylon-wool-adherent cells and BALB/c TsF was obviously not a result of the allogeneic effect, because no effect was obtained when BALB/c nylon-wool-adherent cells were added to the culture of C3H B cells and nylon-wool-purified helper T cells together with C3H TsF.

Discussion

A certain mechanism of action of the antigen-specific TsF was studied in the secondary antibody response. Antigen-specific TsF acts on the nylon-wool-adherent T cells which have acceptor sites for TsF, generating an actual effector type of suppressor T cells.

Our previous studies have shown that TsF can be absorbed with syngeneic adherent T cells to the nylon-wool column, but not with B cells, macrophages, nylon-wool-purified T cells or allogeneic T cells, whereas the final target cells for TsF were found to be helper T cells (2). We have indeed been able to demonstrate that the suppressive effect of TsF was mediated only if the KLH-specific nylon-wool-adherent cells were present in the culture of DNP-primed B cells and KLH-primed helper T cells (Tables I and II). This suggested a intermediary function of nylon-wool-adherent cells in the suppressor cell interaction. The adherent cells involved in this suppression were shown to be T cells expressing the high density of Thy-1 antigen, because the effect was abolished by the treatment of the cells with a higher dilution of anti-Thy-1.2 and C (Table IV). The primed state of the adherent cells was demonstrated to be necessary for TsF-acceptor cell interaction (Table III and VII). Thus, the direct target cell possessing the acceptor site for TsF is a primed T cell adherent to the nylon-wool column.

The intermediary function of the acceptor T cell was completely abrogated by the cytotoxic treatment with alloantibodies that had specificity to the product of genes in the I-J subregion of the MHC, but not by those of antibodies lacking this I-J specificity (Table V). More specifically, the function was blocked by the treatment with anti-I-J antiserum in the absence of C. This strongly suggests that the acceptor site for TsF is coded for by genes in the I-J subregion, although the possibility that anti-I-J antibodies sterically interfere with TsF binding to the acceptor site, which is non-I-J-products in nature, is still not excluded. These results, therefore, well explain that the requirement of strict genetic restriction, in which identities exist among genes in the I-J subregion of the MHC between the donor of the factor and the acceptor cells, is necessary for effective suppression (2, 3).

KLH-primed adherent cells precultured with KLH-specific TsF plus antigen (KLH) for 48 h became strongly suppressive when they were added to the mixture of B cells and nylon-wool-purified helper T cells. The suppressive activity was abolished by the treatment of precultured cells with anti-Thy-1.2 and C before the addition to the culture (Table VI). Thus it is apparent that TsF activates acceptor cells to generate new suppressor T cells, acting as an actual effector cell type.

The question now to be asked is whether the induced suppressor T cells in our system have the same properties as the producer of TsF. Table VII shows the antigen specificity of induced suppressor T cells. KLH-primed nylon-wool-adherent T cells, after acceptance of KLH-specific but not EA-specific TsF, suppressed the response to DNP-EA mounted by DNP-primed B cell and EA-primed helper T cells. In other words, antigen-nonspecific suppression induced by KLH-TsF was mediated only when KLH-primed acceptor T cells and specific antigen (KLH) were both present in the response to DNP-EA. Therefore, identity of antigen specificity between TsF and acceptor T cells is required in the induction of this nonspecific suppression. Furthermore, adherent cells, after acceptance of syngeneic but not allogeneic TsF, could suppress both syngeneic and allogeneic antibody responses (Table VIII). The identities among MHC genes between TsF and acceptor T cells are definitely required for the induction of the effector type of suppressor T cells, whereas the final step of suppression can be induced by either syngeneic or allogeneic induced suppressor T cells. These findings imply that there are amplification loops in the regulatory pathways by which initial antigen-specific, genetically restricted regulatory forces are transformed into antigen-nonspecific effects that are exerted across the H-2 barrier.

Taken collectively, it seems that at least two distinct T cells are involved in the suppression of antibody response, one of which produces antigen-specific TsF, and the second of which is activated by this TsF plus antigen to become the effector type of suppressor T cells. The experiments of Feldmann and Kontiainen (9) on T_1-T_2 cell interaction in the antibody suppression support this concept. This is also supported by the experiments of Germain et al. (10), Waltenbaugh et al. (11) and Kapp (12) on the induction of antigen-specific suppressor T cells by L-glutamic acid^{60}-L-alanine^{30}-L-tyrosine^{10} (GAT)-TsF or L-glutamic acid^{50}-L-tyrosine^{50} (GT)-TsF. Germain et al. (10) have shown that the addition of GAT-TsF to the culture of syngeneic spleen cells stimulates the development of GAT-specific suppressor T cells. Waltenbauth et al. (11) and Kapp (12) have also demonstrated that suppressor cells induced by GAT-TsF or GT-TsF in naive mice can inhibit the primary anti-GAT or anti-GT response in an antigen-specific fashion.

Genetic restrictions of cell interaction have been reported in several experimental systems (13–17). In the suppressor cell interaction, the genetic restrictions were mapped in the I-J subregion of MHC (2). GAT-TsF and GT-TsF, however, act across H-2 barriers in the suppressor T cell induction (10–12). Kontiainen and Feldmann (18) have also demonstrated that antigen-specific suppressor factor produced by metabolically active, in vitro induced suppressor T cells inhibits both syngeneic and allogeneic antibody responses. These findings are partially in contrast to our results. Their factors were mostly assayed on the in vitro and in vivo primary responses, whereas KLH-TsF was assayed on secondary response. It is, therefore, possible that the discrepancy between KLH-TsF and other factors is a result of the primed state of the target cells.

The genetic restriction in the suppressor cell interaction must be considered in relation to the biological role of I-J determinants on suppressor molecules, because I-J products have been shown to be present on suppressor T cells and their factors in various systems (2, 3, 10–12, 18). Recent studies from our laboratory on T-cell hybridomas with suppressor function have shown that the antigen-specific suppressor molecule is composed of at least two distinct molecules; antigen-binding and I-J-bearing molecules, and the association of which is required for the antigen-specific I-J-restricted suppressive cell interaction (19, 20). Furthermore, I-J subregion gene products were expressed on T-cell hybridomas and their factors with different functional activities, e.g., antigen-specific, antigen-nonspecific, and no suppressive function (19, 21). I-J products on the antigen-specific T-cell hybridoma with restriction specificity was found to be serologically different from that on the T-cell hybridoma with nonspecific, genetically nonrestricted suppressor function (22). Therefore, heterogeneity of I-J subregion gene products may reflect functional difference between subsets of I-J-bearing suppressor T cells. The potential roles of I-J subregion gene products on the suppressor T cells and their factors can be explored by using I-J positive T-cell hybridomas with different functional specificities.

Although the actual cellular events are still largely unknown, the suppressive mechanism is indeed maintained and amplified by the presence of the novel type of T cells. The close characterization of the intermediary type of T cells and the use of I-J-bearing T-cell hybridomas may lead us to understand the interrelationship between various T-cell factors.

Summary

Cellular events mediated by antigen-specific soluble factor extracted from carrier-primed suppressor T cells (TsF) in the suppressive interaction was studied. Keyhole limpet hemocyanin (KLH)-specific TsF directly acts on KLH-primed, I-J positive, nylon-wool-adherent T cells that have an acceptor site for TsF. The nylon-wool-adherent T cells, after accepting TsF in the presence of specific antigen, generate new suppressor T cells acting as an actual effector cell type. Antigen-specificity and syngeneity at I-J between TsF and acceptor T cells are both required for the induction of new suppressor T cells. Newly induced suppressor T cells, however, suppress both syngeneic and allogeneic responses in an antigen-nonspecific fashion.

We thank Dr. D. C. Shreffler and Dr. C. S. David for a generous supply of alloantisera. We also wish to express our gratitude to Prof. Tomio Tada for his excellent advice and constructive criticism. Furthermore, we extend our sincerest appreciation to Ms. Fujiko Takemoto for her kind secretarial assistance.

Received for publication 18 October 1979.

References

1. Tada, T., M. Taniguchi, and C. S. David. 1976. Properties of antigen-specific suppressive T-cell factor in the regulation of antibody response of the mouse. IV. Special subregion assignment of the gene(s) that codes for the suppressive T-cell factor in the *H-2* histocompatibility complex. *J. Exp. Med.* **144:**713.
2. Taniguchi, M., T. Tada, and T. Tokuhisa. 1976. Properties of antigen-specific suppressive T cell factor in the regulation of antibody response of the mouse. III. Dual gene control of the T-cell-mediated suppression of antibody response. *J. Exp. Med.* **144:**20.
3. Tada, T., M. Taniguchi, and C. S. David. 1977. Suppressive and enhancing T cell factors as I-region gene products: properties and the subregion assignment. *Cold Spring Harbor Symp. Quant. Biol.* **41:**119.
4. Kekwick, R. A., and R. K. Cannan. 1936. The hydrogen ion dissociation curve of the crystalline albumin of the hen's egg. *Biochem. J.* **30:**227.
5. Eisen, H. N., S. Belman, and M. E. Carsten. 1953. The reaction of 2,4-dinitrobenzenesulfonic acid with free amino group of proteins. *J. Am. Chem. Soc.* **75:**4583.
6. Taniguchi, M., K. Hayakawa, and T. Tada. 1976, Properties of antigen-specific suppressive T cell factor in the regulation of antibody response of the mouse. II. In vivo activity and evidence of the I region gene product. *J. Immunol.* **116:**542.
7. Julius, M. H., E. Simpson, and L. A. Herzenberg. 1973. A rapid method for the isolation of functional thymus-derived lymphocytes. *Eur. J. Immunol.* **3:**645.
8. Tada, T., T. Takemori, K. Okumura, M. Nonaka, and T. Tokuhisa. 1978. Two distinct types of helper T cells involved in the secondary antibody response: independent and synergistic effects of Ia⁻ and Ia⁺ helper T cells. *J. Exp. Med.* **147:**446.

9. Feldmann, M., and S. Kontiainen. 1976. Suppressor cell induction in vitro. II. Cellular requirement of suppressor cell induction. *Eur. J. Immunol.* **6**:302.

10. Germain, R. N., J. Théze, J. A. Kapp, and B. Benacerraf. 1978. Antigen-specific T-cell-mediated suppression. I. Induction of L-glutamic acid^{60}-L-alanine^{30}-L-tyrosine^{10} specific suppressor T cells in vitro requires both antigen-specific T-cell suppressor factor and antigen. *J. Exp. Med.* **147**:123.

11. Waltenbaugh, D., J. Théze, J. A. Kapp, and B. Benacerraf. 1977. Immunosuppressive factor(s) specific for L-glutamic acid^{50}-L-tyrosine^{50} (GT). III. Generation of suppressor T cells by a suppressive extract derived from GT-primed lymphoid cells. *J. Exp. Med.* **146**:970.

12. Kapp, J. A. 1978. Immunosuppressive factors from lymphoid cells of nonresponder mice primed with L-glutamic acid^{60}-L-alanine^{30}-L-tyrosine^{10}. IV. Lack of strain restrictions among allogeneic, nonresponder donors and recipients. *J. Exp. Med.* **147**:997.

13. Katz, D. H., T. Hamaoka, and B. Benacerraf. 1973. Cell interactions between histoincompatible T and B lymphocytes. II. Failure of physiologic cooperative interactions between T and B lymphocytes from allogeneic donor strains in humoral response to hapten-protein conjugates. *J. Exp. Med.* **137**:1405.

14. Yano, A., R. H. Schwartz, and W. E. Paul. 1977. Antigen presentation in the murine T lymphocyte proliferative response. I. Requirement of genetic identity at the major histocompatibility complex. *J. Exp. Med.* **146**:828.

15. Erb, P., and M. Feldmann. 1975. The role of macrophages in the generation of T-helper cells. II. The genetic control of the macrophage-T cell interaction for helper cell induction with soluble antigen. *J. Exp. Med.* **142**:460.

16. Rich, S. S., and R. R. Rich. 1976. Regulatory mechanisms in cell-mediated immune responses. III. I-region control of suppressor cell interaction with responder cells in mixed lymphocyte reactions. *J. Exp. Med.* **143**:672.

17. Singer, A., K. S. Hathcock, and R. J. Hodes. 1979. Cellular and genetic control of antibody responses. V. Helper T cell recognition of H-2 determinants on accessory cells but not B cells. *J. Exp. Med.* **149**:1208.

18. Kontiainen, S., and M. Feldmann. 1978. Suppressor-cell induction in vitro. IV. Target of antigen-specific suppressor factor and its genetic relationships. *J. Exp. Med.* **147**:110.

19. Taniguchi, M., T. Saito, and T. Tada. 1979. Antigen-specific suppressive factor produced by a transplantable I-J bearing T cell hybridoma. *Nature (Lond.).* **278**:555.

20. Taniguchi, M., I. Takei, and T. Tada. Functional and molecular organization of an antigen-specific suppressor factor derived from a T cell hybridoma. *Nature (Lond.).* In press.

21. Taniguchi, M., T. Saito, I. Takei, and T. Tada. The establishment of T cell hybridomas with specific suppressive function. *In* T and B Lymphocytes: Recognition and Function. F. Bach, B. Bonavida, and E. Vitetta, editors. Academic Press, Inc., N. Y. In press.

22. Taniguchi, M., I. Takei, T. Saito, K. Hiramatsu, and T. Tada. Functional organization of I-J subregion gene products on T cell hybridomas. *In* Proceedings of the 2nd International Lymphokine Workshop. Academic Press, Inc., N. Y. In press.

T-CELL REGULATION OF ANTIBODY RESPONSES: DEMONSTRATION OF ALLOTYPE-SPECIFIC HELPER T CELLS AND THEIR SPECIFIC REMOVAL BY SUPPRESSOR T CELLS*

By LEONORE A. HERZENBERG, KO OKUMURA, HARVEY CANTOR,‡ VICKI L. SATO,§ FUNG-WIN SHEN,‖ E. A. BOYSE,‖ AND LEONARD A. HERZENBERG

(From the Department of Genetics, Stanford University School of Medicine, Stanford, California 94305)

The existence of T lymphocytes which suppress IgG antibody responses has been well established in a variety of systems (1–6). At least three questions about these cells, however, remain to be answered: (*a*) Are suppressors and cooperators (helpers) different types of T cells, or is suppression only a different manifestation of helper T-cell activity? (*b*) Do suppressor T cells directly attack the B cells responsible for antibody production, or do they suppress indirectly, for example, by interfering with the interaction between B cells and helper T cells? (*c*) How do suppressor T cells recognize their target?

The allotype suppression system provides a useful model for attacking these problems. Highly active suppressor T-cell populations are induced in SJL × BALB/c hybrids by perinatal exposure to antibody to allotype (Ig-1b) determinants present on IgG antibody molecules. These suppressor cells are capable of completely and specifically preventing production of antibody that carries the allotypic determinants both *in situ* and in adoptive transfer assays.

Since the suppressor T cells also suppress allotype production by co-transferred cells from syngeneic nonsuppressed donors in the adoptive transfer assay, suppressor and target populations may be taken from different donors. Thus with a hapten-carrier adoptive secondary assay, independently derived populations of carrier-primed helper cells (Th),[1] hapten-primed B cells (B), and suppressor cells (Ts) may be isolated, manipulated, and combined in various experimental protocols to effectively study the interactions between these three basic components of the humoral antibody response.

* This investigation was supported, in part, by grants from the National Cancer Institute (no. CA-04681), the National Institute of Allergy and Infectious Diseases (nos. AI-08917 and AI-12184), the National Institute of Child Health & Human Development (no. HD-01287), and the National Institute of General Medical Sciences (no. GM-17367).

‡ Sidney Farber Cancer Center, Harvard Medical School, Boston, Mass. 02115.

§ Fellow of the Giannini Foundation. Present address: The Biological Laboratories, Harvard University, Cambridge, Mass. 02138.

‖ Sloan-Kettering Institute for Cancer Research, New York 10021.

[1] *Abbreviations used in this paper:* AFC, antibody-forming cell; FACS, fluorescence-activated cell sorter; GAT, linear copolymer of glutamic acid, alanine, and tyrosine; KLH, keyhole limpet hemocyanin; PFC, plaque-forming cells; Th, helper T cell(s); Ts, suppressor T cell(s); TsF, soluble suppressive factors (from suppressed spleen culture supernate(s)).

Previous studies with this system suggested that the mechanism of allotype suppression might involve the removal or inactivation of Th by Ts rather than a direct attack on B cells committed to production of the suppressed allotype (7). Titration of Th, Ts, and B cells in these experiments showed that Th activity is lost in proportion to the Ts dose, regardless of Th- or B-cell dose (7). This quantitative relationship indicates stoichiometric removal of Th by Ts.[2] It suggests that Th and Ts are functionally different and that Th are the target of Ts.

The studies presented here directly confirm this hypothesis. We show first that Ts and Th are different types of cells distinguished by their Ly surface antigens. Cantor and Boyse (8, 9) have recently shown that Th carry Ly1 but not Ly2 surface antigens, which places Th in a distinct T-cell subclass (Ly1) comprising roughly 30% of peripheral T cells. The Th studied here, consistent with this observation, are killed by treatment with Ly1 antisera (plus complement) but not with Ly2 antisera.

In contrast, Ts used in these studies belong to the Ly2-positive subclass. They are killed by exposure to anti-Ly2 plus complement (C) but not by exposure to anti-Ly1 plus C. Thus Ts fall within the same (Ly2) subclass Cantor and Boyse have shown to contain cytotoxic cells and in a different subclass from (Ly1) Th (8, 9).

The ability to selectively kill Ts with anti-Ly2 while leaving the Th population unharmed allows a direct experimental approach to determining whether suppression of Ig-1b allotype production is due to removal of Th activity. If Ts remove Th, then a completely suppressed mouse should have no detectable Th activity for the suppressed allotype, since the Ts present in the animal would be expected to remove Th on appearance. Therefore killing the Ts (by anti-Ly2 treatment) in spleen cells from carrier-primed suppressed animals should not unmask any memory Th activity capable of helping with Ig-1b hapten-primed B cells. This prediction is confirmed by the data presented here.

This evidence, coupled with the well-documented specificity of allotype Ts for Ig-1b responses (6), suggested an unexpected division among Th which help IgG responses: Since Ts suppress Ig-1b production but do not impair production of other IgG antibodies in the same animal, the target Th must be dedicated to help only those B cells destined to produce Ig-1b antibody; Th which help other IgG B cells must be unable to help the Ig-1b response. Such Th specificity is unprecedented. Kishimoto and Ishizaka have presented evidence (10) indicating that IgE Th show restricted class specificity, but no evidence exists for allotype-specific help. In the studies presented here, however, we show directly that in the strain combination we use, Th which help Ig-1b B cells do not help other B cells and vice versa.

Materials and Methods

Most of the methods and materials used for studies presented here are described in detail in an accompanying publication (11). The following briefly summarizes these methods and adds others unique to studies in this publication.

Mice. A new mouse strain, SJA/Hz congenic with SJL/JHz but carrying the BALB/c (Ig^a) chromosome region was mated with BALB/c to obtain Ig^a homozygous hybrids (SJA × BALB/c) congenic with the (Ig^b/Ig^a) heterozygous (SJL × BALB/c)F$_1$ hybrid to SJL, selecting for progeny carrying the Ig^a chromosome region at each successive backcross. Mice used here for mating (SJA/9) were from the third and fourth generations of an inbred line started with ninth backcross generation progeny.

---

[2] The suppressed response is accurately predicted by the equation: response/B = k(Th − $\alpha \cdot$ Ts); where B, Th, and Ts are given as the number of spleen cells transferred from the appropriate donor and k and α are empirically determined scaling constants (7). The equation is valid so long as residual Th activity, i.e., Th − Ts, does not exceed saturating Th levels.

Allotype Suppressed Donors.   All suppressed donors were (SJL × BALB/c)F₁ mice exposed perinatally to maternal (BALB/c) anti-Ig-1b. Donors were generally over 6 mo of age and always tested for Ig-1b just before transfer. Only donors showing no serum Ig-1b detectable by immunodiffusion (<0.01 mg/ml) were used.

Priming.   Mice were primed with 100 μg of 2,4-dinitrophenyl (DNP) keyhole limpet hemocyanin (KLH) on alum (hapten priming) plus 2×10^9 of heat-killed *Bordetella pertussis* (kindly supplied by American Cyanamid Co., Lederle Laboratories Div., Pearl River, N. Y.) at least 6 wk before use as donors in adoptive transfer. KLH (carrier)-primed mice received 100 μg KLH on alum plus 2×10^9 *B. pertussis* 7 days before use as donors in adoptive transfer.

Adoptive Transfer and Plaquing.   Spleen cells from various donors were suspended in minimum essential medium (MEM) (Grand Island Biological Co., Grand Island, N. Y.) and mixed at appropriate doses just before intravenous injection into BALB/c recipients irradiated (600 R) 18 h previously. Recipients were challenged at time of transfer with 10 μg aqueous DNP-KLH and sacrificed 7 days later for determination of DNP plaque-forming cells (PFC) in spleens. DNP-PFC were measured in Cunningham chambers (12). Indirect DNP-PFC were measured by determining the increase in DNP-PFC in chambers containing the appropriate facilitating antiserum over the response in chambers with no facilitating antisera (direct DNP-PFC). Results are expressed as DNP-PFC/10^6 recipient spleen cells. Spleen size in adoptive recipients did not vary substantially.

PFC Developing Antisera.   Sera used to develop Ig-1a, Ig-1b, and total IgG were prepared as previously described (11). Alloantisera (BALB/c anti C57BL/6 allotype) used to develop Ig-4b DNP-PFC also contained anti-Ig-1b activity; therefore, Ig-4b response was determined by taking the difference between the number of DNP-PFC developed with this antiserum and the number of DNP-PFC developed with a specific anti-Ig-1b. The Ig-4b anti-DNP response determined in this fashion was completely blocked by addition of purified Ig-4b myeloma protein (MOPC-245).

Sera used to develop Ig-4a DNP-PFC were prepared by absorption of an SJL anti-BALB/c allotype serum onto S-8 (Ig-4a) myeloma protein bound to Sepharose, elution of the absorbed antibody, and passage through a Sepharose RPC-5 (Ig-1a) column to remove contaminating antibody. Specificity of all sera was tested in radioimmune assay.

T-Enriched Spleen Cells by Nylon Wool Passage.   Spleen cells were passed through nylon wool columns as previously described (13). Between 20 and 30% of spleen cells were recovered after passage. The recovered cells had greater than 95% T cells but less than 5% B cells (Ig bearing).

Soluble Suppressive Factor (TsF) from Culture Supernates.   Spleen cells from suppressed donors suspended in "Click" medium (14) were incubated for 48 h at 37°C in 5% CO_2 in air (2×10^7 spleen cells in 1 ml of medium/6 cm Falcon tissue culture dish). After incubation cells were spun for 10 min at 290 g to pellet cells. Supernates were then removed and passed through a Swinnex filter (Millipore Corp., Bedford, Mass.) (0.22 μm) to remove residual cells. Spleen cells from nonsuppressed animals were similarly treated to provide control supernates. As an additional control, medium without cells was carried through the entire procedure.

Incubation with Suppressive Supernatant Factor.   1 ml of supernate prepared as above was added to 1 ml of fresh medium containing 10^7 KLH-primed spleen cells from nonsuppressed donors. These suspensions were then incubated as above for 24 h after which the treated cells were harvested by centrifugation, washed three times with MEM (no fetal calf serum), resuspended, and tested for cooperator T-cell activity in the adoptive transfer assay. Harvested cells were between 50 and 70% viable in individual experiments.

Antisera.   Antisera to Ly antigens were prepared (and absorbed) as previously described (15). Anti-Ly1.2 was made by immunizing C3H/An with CE/J thymocytes, anti-Ly2.2 by immunizing (C3H/An × Bb-Ly2.1)F₁ with Bb leukemic cells ERLO, and anti-Thy1.2 by immunizing (A.Thy-1ᵃ × AKR-*H-2ᵇ*)F₁ with A-strain leukemia ASL1.

Cytotoxic Treatment.   Spleen cell suspensions were incubated with Ly1.2 and Ly2.2 or Thy1.2 cytotoxic antisera as previously described (reference 8 and footnote 3). Anti-Ly1.2, anti-Ly2.2, and anti-Thy1.2 were all used at a 1:20 final dilution. Rabbit C was used at a 1:10 or 1:15 final dilution. The Ly phenotype of (SJL × BALB/c)F₁ hybrid mice used in these studies is Ly-1.2,2.2,3.2; their Thy-1 phenotype is Thy-1.2.

---

³ Huber, B., O. Devinski, R. K. Gershon, and H. Cantor. 1976. Delayed type hypersensitivity is mediated by a subclass of T cells. Manuscript in preparation.

TABLE I
Ly Phenotype of Ts

(SJL × BALB/c)F₁ spleen cells transferred (× 10⁶)				Indirect DNP-PFC*		
Exp. no.	DNP-KLH-Primed (Th + B)	Suppressed Ts		Ig-1b	Ig-1a	Total IgG
		No. treated	Cytotoxic treatment‡			
1	10	—	—	220	380	2,760
	10	5	NMS	0	210	2,020
	10	5	Anti-Ly1.2	0	320	2,810
	10	5	Anti-Ly2.2	140	180	2,100
2	10	—	—	1,140	1,240	12,220
	10	5	NMS	40	1,130	10,220
	10	5	Anti-Ly2.2	1,080	1,220	10,620
	10	5	Anti-Ly2.2 (absorbed by Ly2.2 cells)	0	990	9,120
	10	5	Anti-Thy1.2	1,170	1,210	10,720

* Indirect DNP-PFC/10⁶ recipient spleen cells. Direct DNP-PFC (exp. 1, <50; exp. 2, <200) subtracted.

‡ Cells treated with indicated serum plus C. For details see Materials and Methods section. Number of cells transferred = remainder after treatment of indicated cell number.

Results

Allotype Ts Belong to the Ly2 T-Lymphocyte Subclass.   The data in Table I show that Ts belong to the Ly2⁺Ly1⁻ T-cell subclass [as defined by Cantor and Boyse (8, 16)]. All Ts activity, measured as suppression of the Ig-1b allotype-adoptive secondary DNP response mounted by co-transferred syngeneic, non-suppressed DNP-KLH-primed spleen cells, is completely removed from spleen cell suspensions of suppressed mice by cytotoxic pretreatment with antibody to Ly2.2. Cytotoxic treatment of the same cells with normal mouse serum, or with antibody to Ly1.2 determinants expressed on Th cells in the same strain (see below), does not impair suppressive activity, nor does treatment with anti-Ly2.2 previously absorbed with Ly2.2-bearing cells congenic to the antiserum donor. The selective killing of Ts by anti-Ly2.2 but not Ly1.2 shows that Ts belong to the Ly2-bearing subclass of T lymphocytes.

Data in Table II show that Th in SJL × BALB/c, like Th in other strains (8, 16) belong to the Ly1 subclass. Th activity, measured as the ability to help syngeneic hapten-primed B cells (T-depleted spleen cells) to mount an adoptive secondary DNP response, is completely removed by cytotoxic pretreatment with anti-Ly1.2 but not with anti-Ly2.2. Data in the second experiment shown in Table II demonstrate that Ly2.2-treated Th tested at a dose where Th limit the response still show no effect of anti-Ly2.2 treatment. Thus these cells carry Ly1.2 but not Ly2.2 and therefore belong to the Ly1 subclass.

Genetically Determined Specificity of T-Cell Cooperation for Production of Allotype-Marked Antibody.   As indicated earlier (see introduction), previous

TABLE II
Ly Phenotype of Th

Exp. no.	(SJL × BALB/c)F₁ spleen cells transferred (× 10⁶)			Indirect DNP-PFC*		
	DNP-KLH-primed B cells‡	KLH primed		Ig-1b	Ig-1a	Total IgG
		No. treated	Cytotoxic treatment§			
1	5	—	—	10	0	120
	5	10	NMS	120	160	1,420
	5	10	Anti-Ly1.2	10	20	210
	5	10	Anti-Ly2.2	100	120	1,250
2	5	—	—	<10	<10	130
	5	4	Anti-Ly2.2	210	240	2,440
	5	4	NMS	200	220	2,290
	5	8	NMS	320	360	3,960

* Indirect DNP-PFC/10^6 recipient spleen cells, direct DNP-PFC (<40) subtracted.

‡ T Cells were depleted by treatment with anti-Thy-1 plus C.

§ Cells treated with indicated serum plus C. For details see Materials and Methods section. Number of cells transferred = remainder after treatment of indicated cell number.

work suggested that Ts suppress immunoglobulin allotype production by removing Th required for production of antibody carrying the suppressed allotype. The work presented here, which confirms this hypothesis, tests several of its key predictions. The most startling of these derives from the well-documented specificity of allotype suppression. It predicts a heretofore unrecognized specificity of Th for the immunoglobulin allotype commitment of the B cells with which they cooperate. While no precedent for such specificity exists, the studies presented in the following section provide clear evidence that Th populations capable of helping B cells committed to produce antibody carrying one parental allotype in an allotype heterozygote do not help B cells from the same donor which are committed to production of antibody carrying the allelic allotype.

For these studies, we use memory B cells from hapten (DNP)-primed Ig^b/Ig^a (SJL × BALB/c) hybrids. These donors are heterozygous for the C_H-chain allotypes specified by alleles at closely linked loci in the Ig chromosome region (17). They receive the Ig^b chromosome region from SJL and the Ig^a from BALB/c. Since the alleles in this chromosome region are co-dominantly expressed, the heterozygote produces both parental allotypes at each locus.

Priming these Ig^b/Ig^a heterozygotes with DNP-KLH generates a number of distinct populations of hapten-primed memory B cells, each committed with respect to class and allotype. The relevant subsets of these memory B cells relevant for these studies are: Ig-1b and Ig-1a (class γG_{2a}), and Ig-4a and Ig-4b (class γG_1). Priming with DNP-KLH also generates Th capable of helping with each of these DNP memory B-cell subsets. The data in Table III show that in the adoptive secondary response to DNP-KLH, spleen cells from DNP-KLH-primed SJL × BALB/c hybrids give rise to both γG_{2a} and γG_1 DNP-PFC, the γG_{2a} response being roughly one-quarter the γG_1 response. Within each of these

TABLE III
Failure of Ig^a/Ig^a Th to Help Ig-1b Memory B Cells

Activity tested	Ig congenic spleen cells transferred ($\times 10^6$)*			Indirect DNP-PFC‡				
	DNP-KLH-primed	KLH-primed nylon-passed T‖						
	Ig^b/Ig^a B cells§	Ig^b/Ig^a Th	Ig^a/Ig^a Th	Ig-1b	Ig-1a	Ig-4b	Ig-4a	Total IgG
	5			<10	<10	40	40	150
Th	5	2.5		160	170	600	650	1,930
	5	5		310	300	1,000	1,100	3,470
	5	10		380	410	1,240	1,320	4,140
	5		2.5	20	180	560	510	1,620
	5		10	<10	350	860	980	3,110
	5		30	20	320	740	870	3,230
Ts in Ig^a/Ig^a donor	5	2.5	10	120	260	920	1,270	3,030
	5	2.5	30	120	230	1,220	1,190	3,320

* Ig^b/Ig^a donors were (SJL $\times$ BALB/c)F$_1$; Ig^a/Ig^a donors were (SJA $\times$ BALB/c)F$_1$.
‡ Indirect DNP-PFC/10^6 recipient spleen cells. Direct DNP-PFC (<50) subtracted.
‖ Nylon wool column purified T cells were used as source of Th.
§ T cells were depleted by treatment with anti-Thy-1 plus C.

classes, the number of DNP-PFC for each of the two parental allotypes (e.g., Ig-1a and Ig-1b) is essentially equal.

To compare Th from different sources for ability to help B-memory cells, the T cells from the DNP-KLH-primed heterozygous B-cell donor spleen must be removed by treatment with anti-Thy-1 before transfer so that all anti-DNP production will be dependent upon the carrier-primed Th being tested. As the data in Table III show, anti-Thy-1 treatment of these spleen cells before transfer abolishes the response. The response, however, is completely restored by syngeneic (SJL $\times$ BALB/c)F$_1$ KLH-primed splenic T cells, confirming that heterozygous mice have Th capable of cooperating with all four types of memory B cells.

In sharp contrast, the response is not completely restored when the Th from the carrier-primed Ig^b/Ig^a heterozygotes are replaced with similarly primed Th from congenic (SJA $\times$ BALB/c)F$_1$ Ig^a/Ig^a homozygotes. Th from the homozygotes restore only the Ig-1a and IgG$_1$ response. They do not restore the Ig-1b response. Even at 12 times the optimal dose (30 $\times$ 10^6) of nylon-passed T cells, (which is equivalent to approximately 60 $\times$ 10^6 spleen cells), no Ig-1b DNP-PFC are produced (see Table III). Thus although Th from Ig^a/Ig^a homozygotes do help hapten-primed B cells from Ig^b/Ig^a heterozygotes, as is shown by restoration of the Ig-1a and IgG$_1$ response, the Ig^a/Ig^a homozygous Th are unable to interact with Ig-1b memory B cells to produce an Ig-1b response.

This failure of Th from Ig^a homozygous donors to help Ig-1b memory B cells is not due to the presence of Ig-1b Ts which suppress Ig-1b memory cell expression, since 10 or 30 million SJA $\times$ BALB/c T cells co-transferred with a limiting dose of heterozygous Th gives essentially the same Ig-1b response as that number of

Table IV

Absence of Ig-1b Th Activity in Ig-1b Suppressed Mice: Ts Removed by Anti-Ly2.2

(SJL × BALB/c)F$_1$ spleen cells transferred (× 10^6)					Indirect DNP-PFC*		
Activity tested	DNP-KLH-primed B cells‡	KLH-primed Th	KLH-primed suppressed (Th + Ts enriched)§		Ig-1b	Ig-1a	Total IgG
			No. treated	Treatment‖			
	6				<10	<10	20
	6	6			290	310	3,050
Th	6	4			200	250	1,880
	6		4	NMS	<10	370	3,790
	6		4	Anti-Ly2.2	<10	300	3,220
Ts	6	2			100	130	900
	6	2	2.5	Anti-Ly2.2	120	300	3,000
	6	2	2.5	NMS	<10	370	3,690

* Indirect DNP-PFC/10^6 recipient spleen cells. Direct DNP-PFC (<40) subtracted.

‡ T cells were depleted by treatment with anti-Thy-1 plus C.

§ B cells were depleted from spleen cell population by nylon wool passage before treatment. T-enriched population had more than 80% T cells and less than 5% B cells (13).

‖ Cells treated with indicated serum plus C. For details see Materials and Methods section. Number of cells transferred = remainder after treatment of indicated cell number.

heterozygous Th transferred alone (see Table IV). The dose of heterozygous Th used here is set considerably below the saturating Th dose, making this assay highly sensitive for detecting Ts in the homozygous donors. Therefore, the absence of the Ig-1b response when carrier-primed Ig^a/Ig^a homozygotes are used as donors must be due to a genetically determined absence of Th (or Th activity) capable of helping Ig-1b B cells.

Th capable of helping with IgG memory cells also show specificity with respect to the B cells which they help. Although the Ig^a/Ig^a homozygous populations help both Ig-4a and Ig-4b (IgG$_1$ allotypes) the absence of an Ig-1b response in the presence of these Th shows that they do not help Ig-1b memory cells. Thus, at least in the (SJL × BALB/c)F$_1$ and (SJA × BALB/c)F$_1$ congenic pair, Th show specificity for the immunoglobulin commitment of the B cells which they help. This demonstration, then, clears the way for consideration of a mechanism of suppression based on the selective removal of Th capable of helping Ig-1b B cells.

Absence of Ig-1b Th Activity in Carrier-Primed Suppressed Mice.   Before testing for Ig-1b Th activity in spleen cell suspensions from suppressed mice, Ts must be removed. Otherwise, the Ig-1b Th may be masked by the Ts. Demonstrating the complete depletion of Ts, however, is complicated by the fact that the Ig-1b response in adoptive transfer recipients of Th and Ts is determined by the difference between the amounts of Th and Ts present, i.e., DNP-PFC/B = k(Th − $\alpha \cdot$ Ts) (7). There is no problem if Ts exceeds Th, even by a small amount, since the suppressive activity is then detectable by transferring with a low dose of nonsuppressed Th (and hapten-primed B cells) which makes the assay highly sensitive for Ts. If Th exceeds Ts, again there is no problem because the Th will be detectable by transferring with hapten-primed B cells despite Ts presence. But if the amounts of Ts and Th in the test cell suspension are sufficiently close so as to simply neutralize one another, neither Ts nor Th will be detectable and the result will be the apparent absence of Th. To overcome this problem, we used two quite different methodologies for selectively depleting the Ts from the KLH-primed suppressed donors: killing, in the presence of C, with antiserum to Ly2 surface determinants and size separation with the fluorescence-activated cell sorter (FACS) (18).

Data presented earlier with the anti-Ly2 antiserum showed that Ts are killed by treatment with the antiserum and C but that Th in KLH-primed nonsuppressed spleen are not (see Tables I and II). Therefore, spleen cells from KLH-primed Ig-1b-suppressed mice were treated with anti-Ly2 and C and the surviving cells assayed for Th with hapten-primed B cells in the DNP-adoptive secondary assay.

Results from these studies are presented in Table IV. The inability of treated cells to suppress Ig-1b DNP-PFC formation when co-transferred with a low dose of Th from a nonsuppressed primed donor shows the complete removal of Ts by anti-Ly2 treatment. Less than 5% of the original Ts activity is detectable. However, no Ig-1b are unmasked. KLH-primed suppressed spleen cells still are unable to help Ig-1b memory cells when Ts are gone, although, as the data show, Ig-1a and γG_1 (total IgG) Th are unharmed by the treatment with anti-Ly2.

Treatment and transfer of 10 million KLH-primed suppressed spleen cells showed similar results, i.e., complete removal of Ts activity but no detectable Ig-1b help. Since, in the assay used here, Ts activity in 0.3 million spleen cells from unprimed suppressed animals is adequate to significantly suppress the Ig-1b response mounted by the co-transferred nonsuppressed primed cells (7), treatment with anti-Ly2 could leave no more than 3% of the original Ts activity in the treated cell population. This residual Ts activity would be far too small to suppress the response due to masked Ig-1b Th in the primed suppressed donor if these Th were present in the same numbers as in primed nonsuppressed donors.

Similar results are obtained when Ts are removed by size separation (measured by a light-scattering parameter) with the FACS. Preliminary experiments showed that Ts in unprimed suppressed mice are confined to the FACS-separated fraction containing the largest 20% of splenic T cells. Th in the splenic T-cell suspension, however, are found in both large and small cell fractions so that, as the data in Table V show, when splenic T cells from carrier-primed suppressed mice are separated by size with the FACS, a substantial portion of the Th for Ig-1a and γG_1 are found in the small cell fraction. No Ig-1b Th are found in this fraction, however, despite the demonstration that the fraction has no apparent suppressor T-cell activity (see Table V).

Thus, since spleen cell suspensions from KLH-primed suppressed mice have no Ig-1b Th activity when tested after Ts have been removed by two independent methods, we feel reasonably safe in concluding that Ts specifically remove Ig-1b Th activity in intact primed suppressed animals. This conclusion is supported by evidence presented in the following section which shows that pretreatment of spleen cells from KLH-primed, nonsuppressed mice with supernates from cul-

TABLE V

Absence of Ig-1b TH Activity in Ig-1b Suppressed Mice: Removal of Ts by Size Separation with FACS

Activity tested	(SJL × BALB/c)F₁ spleen cells transferred (× 10⁶)				Indirect DNP-PFC*		
	DNP-KLH primed		KLH-primed suppressed (Th + Ts enriched)‡		Ig-1b	Ig-1a	Total IgG
	(Th + B)	B Cells§	Fraction	No.			
		4			<10	<10	20
Th		4	Unseparated	1.5	<10	750	7,020
		4	Small‖	1.5	<10	310	3,240
Ts	6				780	880	9,380
	6		Unseparated	1.5	10	850	8,910
	6		Small‖	1.5	800	900	9,420

* Indirect DNP-PFC/10⁶ recipient spleen cells. Direct DNP-PFC (<100) subtracted.

‡ See footnote §, Table IV.

§ T cells were depleted by treatment with anti-Thy-1 plus C.

‖ Small cell fraction (smallest 70%) was separated by low-angle light scatter with the FACS. For details of separation and transfer, see Materials and Methods section.

tures of unprimed suppressed spleen cells specifically removed Th activity capable of helping Ig-1b B cells.

Ig-1b Helper T-Cell Activity Removal by In Vitro Treatment with Cell-Free Culture Supernate (Factor) from Suppressed Spleen Cells.   The Th capable of helping Ig-1b memory cells are selectively removed from KLH-primed SJL × BALB/c spleen cells by culturing these cells for 24 h with culture medium in which suppressed unprimed spleen cells were first cultured for 48 h. The ability of treated Th to help Ig-1a and γG_1 memory B cells in the adoptive transfer assay is unaffected by the treatment (see Table VI).

Only the TsF is able to deplete the Ig-1b Th. Supernates from nonsuppressed spleen cultures or from culture dishes with no cells added have no effect on Ig-1b help (also Table VI).

Induction of suppressor cells in the TsF-treated carrier-primed spleen, which could subsequently suppress Ig-1b production in the adoptive transfer assay, was ruled out by transferring the TsF-treated cells together with a low dose of untreated carrier-primed spleen (as in the preceding experiment). The data in Table VI show that the TsF-treated cells have no Ts activity. Ig-1b responses were the same with or without addition of the treated cells.

The absence of Ts in the treated Th cultures was further substantiated by exposing the TsF-treated Th to anti-Ly2.2 plus C before testing in adoptive transfer. If Ts had been induced by exposure to TsF these Ts should have been killed by the anti-Ly2.2 and the Th activity of the treated culture restored. As the data in Table VII show, however, Th activity in the TsF-treated cultures was still absent after Ly2.2 treatment. Similar results are obtained if the KLH-primed cells are treated with Ly2.2 before exposure to TsF. These data suggest

TABLE VI

Removal of Ig-1b Th Activity by Treatment with TsF

	(SJL × BALB/c)F$_1$ spleen cells (× 10^6)				Indirect DNP-PFC*		
		KLH-primed					
Activity tested	DNP-KLH-primed T-depleted B‡	Un-treated Th	Treated§		Ig-1b	Ig-1a	Total IgG
			Num-ber	Factor Source			
	5				<10	<10	<10
Th	5		10	Suppressed	20	280	3,420
	5		10	Nonsuppressed	250	260	2,720
	5		10	Medium alone	220	250	2,440
Ts	5	2.5			90	120	1,120
	5	2.5	10	Suppressed	250	260	2,830

* Indirect DNP-PFC/10^6 recipient spleen cells. Direct DNP-PFC (<20) subtracted.

‡ T cells were depleted by treatment with anti-Thy-1 plus C.

§ Culture supernates were obtained by culturing spleen cells of indicated type for 48 h. KLH-primed cells were incubated for 24 h with culture supernates, then washed and tested for Th and Ts activity in adoptive transfer. Number of cells transferred = remainder after treatment of indicated cell number (see Materials and Methods).

TABLE VII

Failure of (TsF) to Generate Ts In Vitro

(SJL × BALB/c)F$_1$ spleen cells (× 10^6)			Indirect DNP-PFC*		
DNP-KLH-primed T-depleted B‡	KLH-primed TsF incubated§		Ig-1b	Ig-1a	Total IgG
	No. treated	Cytotoxic treatment‖			
5	—	—	<10	<10	<10
5	10	NMS	10	230	1,220
5	10	Anti-Ly2.2	0	260	1,230

* Indirect DNP-PFC/10^6 recipient spleen cells. Direct DNP-PFC (<20) subtracted.

‡ T cells were depleted by treatment with anti-Thy-1 plus C.

§ KLH-primed spleen cells were incubated with TsF.

‖ For details of treatment and adoptive transfer, see Materials and Methods section. Cells treated with indicated serum plus C. Number of cells transferred = remainder after treatment of indicated cell number.

that Th activity for Ig-1b is specifically removed by TsF treatment rather than masked by the induction of Ts during the culture.

Discussion

In recent years, a number of cases have been studied where nonresponsiveness occurs because T-cell populations are present which actively suppress antibody production (1–6). We have shown here that in one such system (allotype suppression), the Ts remove Th activity and thus regulate antibody formation by limiting the amount of available Th activity.

We first demonstrated that Ts and Th are different types of T cells which belong to different T-cell subclasses. This was accomplished by showing that Ts carry Ly2 and not Ly1 surface antigens, thus placing Ts within the same $Ly2^+Ly1^-$ T-cell subclass as cytotoxic precursor and effector cells (8, 9). Th, in contrast, express Ly1 and not Ly2 surface antigens and thus belong to the $Ly1^+Ly2^-$ T-cell subclass which helps both humoral (8, 16) and cytotoxic responses (8, 9) and can initiate delayed hypersensitivity reactions.[3]

These findings have considerable bearing on possible mechanisms of suppression. Since cells of the Ly2,3 subclass do not show helper activity, the identification of Ts as belonging to this subclass makes highly unlikely the suggestion that Ts populations suppress by providing an excess of helper activity.[3] Furthermore, since Huber and Cantor have shown that T cells in the Ly1 subclass do not convert to Ly2-positive cells,[4] the data presented here make it unlikely that Ts are modified Th. Instead, these data suggest that Ts and Th are distinct differentiated populations, each with its own role in regulation of the antibody production.

The difference in these roles is shown directly by our studies on the mechanism of suppression. Th are required to help B cells to increase in number and differentiate to antibody-forming cells (AFC). Therefore, Th exert direct control over IgG antibody production. Ts, on the other hand, regulate antibody production indirectly. They remove Th activity and thus reduce the amount of Th activity available to help B cells.

The conclusion that Ts remove Th is based on evidence presented here which shows (a) that carrier-primed allotype suppressed mice have no demonstrable Th activity capable of helping Ig-1b B cells; and (b) that cultured spleen cells from suppressed mice produce soluble factors which interact with carrier-primed nonsuppressed spleen cells to specifically deplete Ig-1b Th activity. Taken together, these studies strongly suggest that Ts exert a direct effect on Th rather than on Ig-1b B cells. We support this conclusion with direct evidence showing that, at least in the strain combination used here, Ig-1b B-memory cells require Ig-1b-specific Th which cannot be replaced by Th which help Ig-1a or IgG_1 B cells. This demonstration is required to explain how allotype Ts can specifically suppress Ig-1b antibody production without interfering with Ig-1a or other IgG production in allotype suppressed mice.

In the accompanying publication (11), we have presented evidence showing that priming and persistence of Ig-1b memory B cells is unimpaired in hapten-primed allotype-suppressed mice. T-depleted spleen cell suspensions from these mice (supplemented with carrier-primed spleen cells from nonsuppressed mice) show the same response in an adoptive secondary assay as supplemented T-depleted spleen cells from hapten-primed nonsuppressed mice. Thus the inability of intact, allotype-suppressed mice to produce Ig-1b antibody appears to be due solely to the removal of Ig-1b Th activity by Ts.

---

[4] Huber, B., and H. Cantor. 1976. The developmental relationship between helper (T_H) and killer (T_{CS}) T cell subclasses. Manuscript in preparation.

The restrictions in inducing allotype suppression in other mouse strains (6) dictate caution in extending our findings to other suppressor systems; nevertheless, there are some suggestions that our findings that Ts suppress by removing Th activity may reflect a general immunoregulatory mechanism. Tada has proposed that KLH Ts interfere with Th function because KLH-specific Ts-soluble factors suppress anti-hapten responses when the hapten is coupled to KLH as a carrier (3). Okumura and Tada's studies on suppression of IgE responses (19) may be similarly interpreted, especially since Kishimoto and Ishizaka have shown that IgE Th appear to be specific for IgE B cells (10). Kapp et al. (20) have shown that removal of Ts which suppress the response to GAT, (a linear copolymer of glutamic acid, alanine, and tyrosine) in GAT-suppressed nonresponder mice does not unmask GAT Th activity. These authors suggest that Th are missing because GAT does not prime T cells in GAT-suppressed mice (20), but it is also possible that the GAT Ts have removed the GAT Th. Thus, Th removal by Ts could prove to be a general mechanism of suppression of T-dependent responses.

In addition to the functional similarities listed above, several surface antigenic similarities exist between allotype Ts and other Ts. Ts in several other systems have now been shown to belong to the Ly2,3 subclass in contrast to Ly1 helper cells (16, 21). We have also shown recently that allotype Ts, like Ts generated in the A5A idiotype suppression system (22) and like soluble suppressive factors in Tada's KLH Ts system (23), carry Ia determinants (references 24 and 25, and footnotes 5 and 6). Data from Tada's and our studies indicate that these determinants map to a previously uncharted segment of the *I* region between *I-B* and *I-C*. Our studies indicate that this new region controls Ia antigens on T rather than B cells.[5,6]

Parallels between allotype and idiotype suppression, while not fully established, offer intriguing ground for speculation on how Ts recognize Th and how Th recognize B cells. Eichmann (26) has shown that exposure to antibody determinants (idiotypes) on immunoglobulin molecules results in the generation of a Ts population which specifically suppresses production of immunoglobulin molecules carrying that idiotype. This closely parallels the allotype-suppression system, where exposure to antibody to allotypic determinants on immunoglobulin generates Ts specific for allotype production.

There are, however, significant differences between allotype and idiotype suppression. Idiotypic determinants are in the variable region located in the Fab portion of the immunoglobulin molecule, while the Ig-1b allotypic determinants used in our studies are found on the Fc portion of the immunoglobulin H chain. Furthermore, idiotype Ts suppress production of idiotype-bearing antibody molecules in all immunoglobulin classes, while allotype Ts suppress production of antibody molecules carrying the Ig-1b allotype regardless of the specificity of the

---

[5] Okumura, K., L. A. Herzenberg, D. B. Murphy, H. O. McDevitt, and L. A. Herzenberg. 1976. Selective expression of *H-2* (*I*-region) loci controlling determinants on helper and suppressor T lymphocytes. *J. Exp. Med.* 144:in press.

[6] Murphy, D. B., L. A. Herzenberg, K. Okumura, L. A. Herzenberg, and H. O. McDevitt. 1976. A new *I* subregion (*I-J*) marked by a locus (*Ia-4*) controlling surface determinants on suppressor T lymphocytes. *J. Exp. Med.* 144:in press.

antibody-combining site. These differences indicate that the two types of Ts affect the expression of different subsets of B cells; but whether the mechanism of suppression, i.e. helper depletion, is the same in both cases remains to be determined.

The demonstration that allotype Ts suppress B-memory expression by removing helper T cells establishes these three types of cells in an expanded network similar to that postulated by Jerne (27, 28). Idiotype and carrier-specific suppression also fit into similar networks. It is possible that these networks contain other cells as well, e.g. helper cells for the development of functional Ts and Th, precursors of Th, Ts, B, etc. (The allotype network could also contain two types of Th, both required for B-cell expression — one for carrier recognition and one for allotype recognition; this is entirely speculative but would avoid having to endow a single Th with the ability to recognize both types of determinants.) Such networks must then interlock with one another since allotype suppression stays within Ig class lines but cuts across antigen-specific responses, whereas idiotype and carrier-specific suppression do the opposite.

It is still too early to obtain a clear insight into the molecular basis of communication between cells in a given network. Tada has shown that carrier-specific suppressive factors (KLH-TsF) have both I-region and carrier recognition determinants (23). We have shown here that spleens from allotype-suppressed mice produce a soluble factor (allo-TsF) which interacts with carrier-primed spleen to remove allotype-specific Th. Since allotype Ts have surface I-region determinants which map quite close to the I-region determinants on KLH-TsF, it is quite possible that allo-TsF also carries I-region determinants. If so, then the communication between Ts and Th may generally utilize I-region determinants as part of a recognition mechanism. This would be consistent with the view that I-region determinants are involved in Th-B collaboration (29), although these determinants need not be the same as those involved in Ts-Th communication. Such considerations, however, do not address the heart of the recognition question posed by the mechanism of suppression presented here, i.e., how allotype Ts recognize Ig-1b Th and how Ig-1b recognize Ig-1b memory B cells. No data are currently available which bear on this point.

Thus far in this discussion we have considered the demonstration of allotype-specific help mainly within the context of the mechanism of allotype suppression. The implications of this unprecedented finding with respect to regulation of antibody production, however, deserve consideration in their own right. The data presented here show directly that Th capable of cooperating with Ig-1a B cells do not cooperate with Ig-1b B cells. The converse is also true, since removal of Ig-1b Th activity does not affect the Ig-1a response at limiting Th doses.

Summary

Allotype suppressor T cells (Ts) generated in SJL × BALB/c mice specifically suppress production of antibodies marked with the Ig-1b allotype. The studies presented here show that allotype Ts suppress by specifically removing helper T cell (Th) activity required to facilitate differentiation and expansion of B cells to Ig-1b antibody-forming cells.

We show first that Ts and Th belong to different T-cell subclasses as defined by Ly surface antigens. Ts are Ly2$^+$Ly1$^-$ and thus belong to the same subclass as cytotoxic precursor and effector cells; Th are Ly1$^+$Ly2$^-$ cells and thus belong to the subclass containing cells which can exert helper functions and initiate delayed hypersensitivity reactions. Placing these cells in these two subclasses shows that Th are different from Ts and suggests that they play different roles in regulating antibody responses. The difference in these roles is defined by the evidence presented here showing that Ts attack Th and regulate the antibody response by specifically regulating the availability of Th activity. We show that in allotype suppressed mice, Ts which suppress Ig-1b antibody production have completely removed the Th activity capable of helping Ig-1b B cells without impairing Th activity which helps other IgG B cells.

These findings imply the existence of allotype-specific Th for Ig-1b cells (Ig-1b Th). We directly establish that Ig-1b cells require such help by showing that carrier-primed spleen cells from Ig^a/Ig^a congenic hybrids help Ig-1a B cells from hapten-primed Ig^b/Ig^a donors but do not help Ig-1b B cells from the same donor in the same adoptive recipient.

The authors wish to express their appreciation to Mr. F. T. Gadus and to Mr. Theta Tsu for their excellent technical assistance. We also wish to thank Ms. Jean Anderson for her help in the preparation of this manuscript.

Received for publication 29 March 1976.

References

1. Gershon, R. K. 1974. T cell control of antibody production. *Contemp. Top. Immunobiol.* 3:1.
2. Benacerraf, B., J. A. Kapp, P. Debre, C. W. Pierce and F. de la Croix. 1975. The stimulation of specific suppressor T cells in genetic non-responder mice by linear random copolymers of L-amino acids. *Transplant. Rev.* 26:21.
3. Tada, T. 1974. The mode and sites of action of suppressor T cells in antigen induced differentiation of B cells. *In* Immunological Tolerance. D. H. Katz and B. Benacerraf, editors. Academic Press, Inc., New York. 471.
4. Basten, A., J. F. A. P. Miller and P. Johnson. 1975. T cell-dependent suppression of an anti-hapten antibody response. *Transplant. Rev.* 26:130.
5. Baker, P. J. 1975. Homeostatic control of antibody response: a model based on the recognition of cell-associated antibody by regulatory T cells. *Transplant. Rev.* 26:3.
6. Herzenberg, L. A., and L. A. Herzenberg. 1974. Short-term and chronic allotype suppression in mice. *Contemp. Top. Immunobiol.* 3:41.
7. Herzenberg, L. A., K. Okumura, and C. M. Metzler. 1975. Regulation of immunoglobulin and antibody production by allotype suppressor T cells in mice. *Transplant. Rev.* 27:57.
8. Cantor, H., and E. A. Boyse. 1975. Functional subclasses of T lymphocytes bearing different Ly antigens. I. The generation of functionally distinct T-cell subclasses is a differentiative process independent of antigen. *J. Exp. Med.* 141:1376.
9. Cantor, H., and E. A. Boyse. 1975. Functional subclasses of T lymphocytes bearing different Ly antigens. II. Cooperation between subclasses of Ly$^+$ cells in the generation of killer activity *J. Exp. Med.* 141:1390.
10. Kishimoto, T., and K. Ishizaka. 1973. Regulation of antibody response *in vitro*. VI. Carrier-specific helper cells for IgG and IgE antibody response. *J. Immunol.* 111:720.
11. Okumura, K., C. M. Metzler, T. T. Tsu, L. A. Herzenberg, and L. A. Herzenberg. 1976. Two stages of B-cell memory development with different T-cell requirements. *J. Exp. Med.* 144:345.

12. Cunningham, A. J., and A. Szenberg. 1968. Further improvements in the plaque technique for detecting single antibody-forming cells. *Immunology.* 14:599.

13. Julius, M. H., E. Simpson, and L. A. Herzenberg. 1973. A rapid method for the isolation of functional thymus derived murine lymphocytes. *Eur. J. Immunol.* 3:645.

14. Click, R. E., L. Benk, and B. J. Alter. 1972. Immune responses *in vitro*. I. Culture conditions for antibody synthesis. *Cell. Immunol.* 3:264.

15. Shen, F. W., E. A. Boyse, and H. Cantor. 1975. Preparation and use of Ly antisera. *Immunogenetics.* 2:591.

16. Cantor, H., F. W. Shen, and E. A. Boyse. 1976. Separation of helper T cells from suppressor T cells expressing different Ly components. II. Activation by antigen: after immunization, antigen-specific suppressor and helper activities are mediated by distinct T-cell subclasses. J. Exp. Med. 143:1391.

17. Herzenberg, L. A. 1964. A chromosome region for gamma$_{2a}$ and beta$_{2a}$ globulin H chain isoantigens in the mouse. *Cold Spring Harbor Symp. Quant. Biol.* 24:455.

18. Loken, M., and L. A. Herzenberg. 1975. Analysis of cell populations using FACS. *Ann. N. Y. Acad. Sci.* 254:163.

19. Okumura, K., and T. Tada. 1974. Regulation of homocytotropic antibody formation in the rat. VI. Inhibitory effect of thymocytes on the homocytotropic antibody response. *J. Immunol.* 107:1682.

20. Kapp, J. A., C. W. Pierce, and B. Benacerraf. 1975. Genetic control of immune response in vitro. VI. Experimental conditions for the development of helper T-cell activity specific for the terpolymer L-glutamic acid^{60}-L-alanine^{30}-L-tyrosine^{10} (GAT) in nonresponder mice. *J. Exp. Med.* 142:50.

21. Jandinski, J., H. Cantor, T. Tadakuma, D. L. Peavy, and C. W. Pierce. 1976. Separation of helper T cells from suppressor T cells expressing different Ly components. I. Polyclonal activation: suppressor and helper activities are inherent properties of distinct T-cell subclasses. J. Exp. Med. 143:1382.

22. Hämmerling, G. J., S. J. Black, S. Segal, and K. Eichmann. 1975. Cellular expression of Ia antigens and their possible role in immune reactions. *Proc. Leucocyte Cult. Conf.* In press.

23. Tada, T., and M. Taniguchi. 1976. Characterization of the antigen-specific suppressive T cell factor with special reference to the expression of I region genes. *In* The Role of the Products of the Histocompatibility Gene Complex in Immune Response. D. H. Katz and B. Benacerraf, editors. Academic Press, Inc., New York. In press.

24. McDevitt, H. O. 1976. Functional analysis of Ia antigens in relation to genetic control of the immune response. *In* The Role of the Products of the Histocompatibility Gene Complex in Immune Response. D. H. Katz and B. Benacerraf, editors. Academic Press, Inc., New York.

25. McDevitt, H. O., T. L. Delovitch, J. L. Press, and D. B. Murphy. 1976. Genetic and functional analysis of the Ia antigens: their possible role in regulating the immune response. *Transplant. Rev.* 30:197.

26. Eichmann, K. 1975. Idiotypic suppression. II. Amplification of a suppressor T cell with anti-idiotype activity. *Eur. J. Immunol.* 5:511.

27. Jerne, N. K. 1971. The somatic generation of immune recognition. *Eur. J. Immunol.* 1:1.

28. Jerne, N. K. 1972. What precedes clonal selection? *In* Ontogeny of Acquired Immunity. R. Porter and J. Knight, editors. Elsevier-Excerpta Medica-North Holland, Amsterdam, The Netherlands. 1.

29. Katz, D. H., M. Graves, M. E. Dorf, H. Dimuzio, and B. Benacerraf. 1975. Cell interactions between histoincompatible T and B lymphocytes. VIII. Cooperative responses between lymphocytes are controlled by genes in the *I* region of the *H-2* complex. *J. Exp. Med.* 141:263.

IMMUNOSUPPRESSIVE FACTOR(S) EXTRACTED FROM LYMPHOID CELLS OF NONRESPONDER MICE PRIMED WITH L-GLUTAMIC ACID⁶⁰-L-ALANINE³⁰-L-TYROSINE¹⁰ (GAT)

I. Activity and Antigenic Specificity[1]

JUDITH A. KAPP, CARL W. PIERCE,[2] FERN De La CROIX, AND BARUJ BENACERRAF

From the Department of Pathology, Harvard Medical School, Boston, Massachusetts 02115

The synthetic terpolymer of L-glutamic acid⁶⁰-L-alanine³⁰-L-tyrosine¹⁰ (GAT) not only fails to elicit a GAT-specific antibody response in nonresponder mice, but also prior injection of GAT specifically decreases the ability of nonresponder mice to develop a GAT-specific antibody response to a subsequent challenge with GAT-MBSA. This inhibition is mediated by GAT-specific suppressor T cells. Further, a suppressive factor can be extracted from lymphoid cells of GAT-primed nonresponder mice that inhibits the development of primary GAT-specific antibody responses to GAT-MBSA and to GAT-PRBC by normal syngeneic mice. The suppressive activity is dose-dependent and absorbed by GAT-Sepharose, but not by BSA-Sepharose. The suppressive activity elutes from a G-100 Sephadex column in the same fraction as ovalbumin, suggesting its m.w. is approximately 45,000 daltons.

The demonstration of suppressive as well as cooperative activities of thymus-derived cells (T cells)[3] suggests that T cells are the critical regulators of the development and expression of antibody responses by B cells (1–3). Suppressor T cells nonspecifically inhibit immunoglobulin synthesis (4, 5) and concanavalin A(Con A) activated suppressor T cells nonspecifically suppress humoral and cell-mediated immune responses (3, 6, 7). In antigenic competition, antigen-induced suppressor T cells can suppress immune responses to a second unrelated antigen administered at the appropriate time (8, 9). In other systems, antigen-induced suppressor T cells inhibit only re-

sponses to the antigen that activated the T cells (3). Although the precise mechanism(s) by which suppressor T cells regulate immune responses are still obscure, suppression via soluble mediators is one possibility that has been demonstrated to be operative in some systems. For example, Con A-activated suppressor T cells elaborate a mediator(s), called soluble immune response suppressor (SIRS), that nonspecifically terminates antibody responses by mouse spleen cells *in vitro* (3, 10). Thomas *et al.* (3, 11) have described another protein released by specifically stimulated T cells that inhibits antibody responses to unrelated antigens by mouse spleen cells *in vitro*. Antigen-specific suppressive factors have also been obtained from T cells. For example, Zembala and Asherson (12) have described a factor elaborated by lymph node cells from picryl sulfonic acid-tolerized mice that specifically inhibits the adoptive transfer of delayed hypersensitivity to picryl chloride. In other systems, suppressor factors that specifically inhibit an ongoing IgE antibody response (13) have been extracted from carrier-primed rat thymocytes, and factors that inhibit the IgG antibody responses to hapten-conjugates of the homologous carrier have been extracted from carrier-primed mouse T cells (13).

Our interest in antigen-induced suppressor T cells grew from studies of the mechanisms(s) by which an H-2-linked, immune response (*Ir*) gene(s) controls the antibody response to the synthetic terpolymer L-glutamic acid⁶⁰-L-alanine³⁰-L-tyrosine¹⁰ (GAT) in inbred strains of mice. GAT fails to elicit a GAT-specific antibody response in nonresponder mice (14). Prior injection of GAT specifically decreases the ability of nonresponder mice to develop a GAT-specific antibody response to a subsequent challenge with GAT bound to methylated bovine serum albumin (GAT-MBSA) (15), a form of GAT that is normally s immunogenic for both responder and nonresponder mice (14). This unresponsiveness is the result of active suppression; spleen cells from GAT-primed nonresponder mice of the H-2$^{p, q, s}$ halotypes specifically suppress the anti-GAT plaque-forming cell (PFC) response to GAT-MBSA by spleen cells from normal syngeneic mice. The suppressor cells are identified as T cells by their sensitivity to anti-θ serum and complement, and by the demonstration that they are not retained by an anti-mouse immunoglobulin column, but are eluted in the fraction containing θ-positive, but no immunoglobulin-bearing cells (16). Furthermore, B cells eluted from the column do not have suppressor activity and develop an anti-GAT PFC response in the presence of helper T cells and GAT-MBSA (16).

Although we have not detected suppressor activity in the supernatant of cultured suppressor T cells (Kapp, unpublished

Submitted for publication September 2, 1975.

[1] This investigation was supported by United States Public Health Service Grants AI-09920 and AI-09897 from the National Institute of Allergy and Infectious Diseases.

[2] Recipient of United States Public Health Service Research Career Development Award 5K4-AI-70173 from the National Institute of Allergy and Infectious Disease.

[3] Abbreviations used in this paper: B cell, precursor of antibody-producing cell; BSA, bovine serum albumin; Con A, concanavalin A; GAT, random terpolymer of L-glutamic acid⁶⁰-L-alanine³⁰-L-tyrosine¹⁰; GAT-MBSA, GAT coupled to methylated bovine serum albumin; GAT-PRBC, GAT coupled to pigeon red blood cells; GAT-SRBC, GAT coupled to sheep red blood cells; GT, copolymer of L-glutamic acid⁵⁰-L-tyrosine⁵⁰; H-2, major histocompatability complex in mice; Ig, immunoglobulin; IgG refers to IgG1 and IgG(2a + 2b); Ir gene; immune response gene; MBSA, methylated bovine serum albumin, MEM, completely supplemented Eagle's minimal essential medium; PBS, phosphate-buffered saline; PFC, plaque-forming cell(s); SIRS, soluble immune response suppressor; SRBC, sheep red blood cells; T cell, thymus-derived cell; θ, surface alloantigen on T cells.

observations), using the technique of Tada *et al.* (13), a suppressive factor has been extracted from lymphoid cells of GAT-primed nonresponder mice. In this communication, we will describe some of the characteristics of this extract including: conditions for stimulation; tissue distribution of cells from which it can be extracted; activity *in vivo* and *in vitro;* the antigenic specificity of its activity; and a preliminary determination of its molecular size.

MATERIALS AND METHODS

Mice. DBA/1 (H-2^q) mice were purchased from Jackson Laboratory, Bar Harbor, Maine. A.SW (H-2^s) mice were bred in our animal facilities. Mice used in these experiments were 2 to 8 months old and were maintained on laboratory chow and acidified-chlorinated water *ad libitum.*

Antigens. GAT, m.w. 32,000, was purchased from Miles Laboratories, Miles Research Division, Kankakee, Ill. MBSA, sheep red blood cells (SRBC), GAT, and GAT-MBSA were prepared as described previously (15). GAT was coupled to pigeon red blood cells (PRBC) as described for GAT-SRBC (15).

Preparation of cell-free extracts. DBA/1 mice were injected with 10 μg GAT in Maalox (Wm. H. Rorer, Inc., Fort Washington, Pa.), or with 10 μg GAT as GAT-MBSA in Maalox, or with Maalox alone, and sacrificed 3 to 5 days later. The spleens, thymuses, and lymph nodes were teased separately into single cell suspensions, washed twice in Hanks' balanced salt solution, and resuspended to 6 $\times$ 10^8 cells/ml in Eagle's minimum essential medium (MEM) supplemented with 4 mM HEPES, 2 mM glutamine, and 50 units each penicillin and streptomycin (Microbiological Associates, Bethesda, Md.)

Cells were sonicated in a Sonifier Cell Disruptor, Model W-140-E (Heat Systems-Ultrasonics Inc., Plainview, N. Y.) using a standard micro-tip. From 50 to 60 watts were delivered for 5 min to 3 to 8 ml samples maintained at 7°C in a continous flow-cooling cell.

The lysate was then centrifuged in a fixed angle rotor type 40 for 1 hr at 40,000 $\times$ G at 4°C in a Beckman Model L3-50 centrifuge (Beckman Instruments, Palo Alto, Calif.). The supernatant was collected and assayed *in vivo* or *in vitro* at concentrations designated in the experimental protocol. Some supernatants were stored at −80°C until use. To date, 18 extracts have been prepared from GAT-primed nonresponder lymphoid cells and all have had suppressive activity.

Preparation of immunoadsorbents. BSA, fraction V (Armour Pharmaceutical Corp., Chicago, Ill.) and GAT were coupled individually to Sepharose 4B (Pharmacia Fine Chemicals, Piscataway, N. J.) by the method of Cuatrecasas (17). Briefly, aminoethyl-Sepharose was prepared by reacting ethylene diamine (Fisher Scientific Corp., Chemical Manufacturing Div., Fair Lawn, N. J.), pH 10, with cyanogen bromide-activated Sepharose; the aminoethyl-Sepharose was washed with H$_2$O until no color developed in the eluate by the sodium 2,4,6-trinitrobenzene sulfonate test (18). GAT or BSA, trace-labeled with ^{125}I, was mixed with the aminoethyl-Sepharose and 1-ethyl-3-(3-dimethylaminopropyl) carbodiimide was added. After 15 to 20 hr at 4°C, the BSA-Sepharose was washed with H$_2$O and the GAT-Sepharose with 0.05N Na (CO$_3$)$_2$ containing 50% v/v dimethyl formamide. The immunoadsorbents were stored in the cold in phosphate-buffered saline (PBS) containing 0.05% sodium azide. They were washed with PBS until the effluent contained no detectable azide by absorption at 230 nm

or radioactivity, indicating that azide and free antigen were not leaking from the column. BSA-Sepharose contained 6.74 mg BSA/ml of packed Sepharose; GAT-Sepharose contained 1.21 mg GAT/ml of packed Sepharose.

Gel chromatography. The cell-free lysate was fractionated by gel chromatography with a 2.5 x 100 cm column of Sephadex G-100 (Pharmacia Fine Chemicals, Piscataway, N. J.), equilibrated with PBS, pH 7.5. 2.5 ml of the extract were applied to the column and eluted with PBS at 4°C under controlled pressure at a constant flow rate of 30.8 ml/hr; the eluate was collected in 7.7 ml fractions. Appropriate fractions were pooled and concentrated to 5.0 ml by a Diaflo PM10 ultrafiltration membrane (Amicon Corp., Lexington, Mass.). Mouse albumin and ovalbumin were used as marker proteins to calibrate the column.

Immunization of mice. Mice were immunized i.p. with 10 μg GAT as GAT-MBSA in Maalox and pertussis vaccine (Eli Lilly Co., Indianapolis, Ind.) or 0.5 ml 10% v/v GAT coupled to PRBC (GAT-PBRC). Extracts, 0.5 ml, were injected i.v. within 2 hr of antigen.

Spleen cell cultures and PFC assay. Suspension of single spleen cells containing 8 $\times$ 10^6 nucleated cells/ml in MEM were incubated according to the method of Mishell and Dutton (19) with modifications previously described (20). Extracts were sterilized by membrane filtration, diluted in MEM, and added at culture initiation. PFC responses were determined 7 days after *in vivo* immunization or 5 days after culture initiation using SRBC, GAT-SRBC or PRBC as indicator cells. Statistical analysis was carried out using the Student *t*-test.

RESULTS

Effect of GAT-primed lymphoid cells and their extracts on PFC responses to GAT-MBSA by nonresponder mice in vivo. We have previously shown that 2 to 8 $\times$ 10^6 spleen cells from GAT-primed DBA/1 mice inhibit the development of GAT-specific PFC responses to GAT-MBSA by 8 $\times$ 10^6 spleen cells from normal DBA/1 mice *in vitro* (16, 21). The experiments summarized in Table I demonstrate that 2 to 5 $\times$ 10^7 spleen cells from DBA/1 mice primed 3 days earlier with 10 μg GAT in Maalox, but not spleen cells from DBA/1 mice primed with

TABLE I

Effect of DBA/1 lymphoid cells and their extracts on the response to GAT-MBSA by DBA/1 mice in vivo

2 to 5 $\times$ 10^7 DBA/1 Cells Transferred	GAT-Specific IgG PFC/Spleen[a]	Inhibition	p Value
	Arith. mean ± S.E.	%	
None	10,200 ± 800	—	—
Control spleen	8,800 ± 2,000	14	0.606
GAT-spleen	3,400 ± 700	67	<0.001
Extract from 30 $\times$ 10^7 DBA/1 cells			
Control spleen	11,800 ± 1,000	−15	0.209
GAT-spleen	3,600 ± 500	64	<0.001
GAT-thymus	1,800 ± 400	82	<0.001
GAT-lymph node	4,100 ± 900	60	0.003
GAT-spleen + thymus[b]	1,900 ± 800	82	<0.001
GAT-MBSA-spleen + thymus[b]	13,400 ± 1,800	−30	0.088

[a] All mice immunized with GAT-MBSA immediately before i.v. transfer of cells or extracts and assayed 7 days later (5 to 29 mice per group).

[b] Extract equivalent to 15 $\times$ 10^7 DBA/1 cells.

Maalox alone, suppressed the development of primary GAT-specific IgG PFC responses by normal DBA/1 mice immunized simultaneously with 10 μg GAT as GAT-MBSA in Maalox-pertussis. In addition, cell-free extracts, equivalent to 30×10^7 spleen cells, thymocytes, or lymph node cells from mice primed 3 days earlier with 10 μg GAT in Maalox, suppressed the antibody response to GAT-MBSA from 60 to 82%. However, the suppressive activity of extracts from lymph node cells was not as great as that from spleen or thymus as determined by limiting dilution analysis (data not shown). Splenic extracts from Maalox-primed control mice did not suppress the response to GAT-MBSA, and, more importantly, extracts from DBA/1 mice primed with GAT-MBSA in Maalox 3 days earlier did not suppress the immune response to GAT-MBSA. Since the suppressive activity of extracts from spleen cells and thymocytes were approximately equal, these cells were pooled before sonication in subsequent experiments.

GAT-specific suppressor cells have also been demonstrated in nonresponder mice with H-2[p and s] haplotypes (Kapp, unpublished observations). We have prepared extracts from pooled spleen cells and thymocytes from mice of the congenic resistant strain, A.SW, which has the H-2[s] haplotype on an A strain background. As shown in Table II, extracts from lymphoid cells of GAT-primed, but not Maalox-primed, A.SW mice suppressed the response to GAT-MBSA by A.SW mice *in vivo*.

Effect of lymphoid cell extracts in vitro. Extracts prepared from Maalox (control) or GAT-primed DBA/1 spleens and thymuses were added to cultures of normal DBA/1 spleen cells stimulated with either SRBC or GAT-MBSA. After 5 days, the PFC response in cultures containing the extracts was plotted as a percentage of the control responses to SRBC or GAT-MBSA. The results of a typical experiment are shown in Figure 1. Dilutions of the extracts of 1:100 or greater were not cytotoxic as determined by viable cell recovery after 5 days (data not shown). At a 1:100 dilution of the extracts, nonspecific

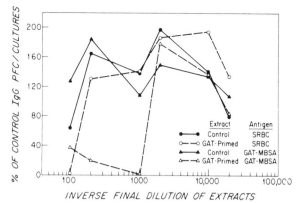

Figure 1. Effect of lymphoid cell extracts on the IgG PFC response to SRBC and GAT-MBSA by normal DBA/1 spleen cells *in vitro.* Extracts were prepared from 6×10^8 spleen and thymus cells from Maalox-primed (control) or GAT-primed DBA/1 mice. Extracts were sterilized by membrane filtration and added at the indicated concentrations to 8×10^6 normal DBA/1 spleen cells at culture initiation. PFC responses were assayed on day 5 and the results are expressed as a percentage of the response in spleen cell cultures that received no extracts.

suppression of responses to SRBC was observed. However, extracts from both control and GAT-primed mice routinely enhanced the response to SRBC at dilutions from 1:200 to 1:10,000. The control extract not only failed to suppress the response to GAT-MBSA at any dilution, but actually enhanced this response. On the other hand, extracts from GAT-primed DBA/1 mice specifically suppressed the responses to GAT-MBSA at dilutions of 1:200 and 1:1000. These dilutions of extracts are equivalent to 1.5×10^6 and 3×10^5 cells, respectively. Dilutions of this extract greater than 1:1000 also enhanced the response to GAT-MBSA. The reasons for the nonspecific suppression or enhancement of immune responses *in vitro* observed with these extracts are not known at present.

Specificity of suppression. The specificity of the suppression mediated by cell-free extracts from pooled spleen and thymuses of GAT-primed DBA/1 mice was exmined further by determining their effect on the IgG anti-PRBC PFC response by DBA/1 mice immunized with GAT-PRBC. The data in Table III demonstrate that the GAT-specific PFC responses of normal DBA/1 mice immunized with either GAT-MBSA or GAT-PRBC were suppressed by the extract. However, the response to PRBC was not significantly decreased.

Antigenic specificity of the extract. Since the suppression mediated by these cell-free extracts was specific for GAT regardless of whether the immunogen was GAT-MBSA or GAT-PRBC, we examined the possibility that the active

TABLE II

Effect of extracts of lymphoid cells from A.SW mice on response to GAT-MBSA by A.SW mice in vivo

Extract from 15×10^7 A.SW Cells	GAT-Specific IgG PFC/Spleen[a]	Inhibition	p Value
	Arith. mean $\pm$ S.E.	%	
None	$13,200 \pm 2,000$	—	—
Control spleen + thymus	$12,500 \pm 4,000$	5	0.860
GAT-spleen + thymus	$2,200 \pm 1,100$	83	<0.001

[a] All mice received GAT-MBSA immediately before i.v. injection of extract and response measured on day 7 (three to seven mice per group).

TABLE III

Specificity of suppression by extracts from GAT-primed DBA/1 mice

Extract from 30×10^7 Cells[a]	Antigen[b]	IgG PFC/Spleen[c] PRBC	IgG PFC/Spleen[c] GAT	GAT-Specific Inhibition	p Value
		Arith. mean $\pm$ S.E.		%	
—	GAT-MBSA		$12,200 \pm 1,000$	—	—
+	GAT-MBSA		$2,800 \pm 700$	77	<0.001
—	GAT-PRBC	$24,300 \pm 10,000$	$6,600 \pm 200$	—	—
+	GAT-PRBC	$20,100 \pm 6,000$	$1,600 \pm 700$	76	<0.001
Inhibition PRBC response:		17%			0.723

[a] Extract from pooled spleen cells and thymocytes at GAT-primed DBA/1 mice.

[b] Mice were immunized immediately before i.v. injection of extract.

[c] Day 7 response (seven mice per group).

component in these extracts had affinity for GAT. An extract from GAT-primed DBA/1 mice was passed over GAT-Sepharose or BSA-Sepharose. These immunoadsorbents were shown independently to bind specifically anti-GAT and anti-BSA antibodies, respectively (data not shown). Titration of the untreated extract from GAT-primed lymphoid cells (Table IV, Groups B, C, D) demonstrated that 0.5 ml of a 1:8 dilution caused significant suppression. Passage of 2 ml of undiluted extract over 5 ml of BSA-Sepharose did not diminish its suppressive activity when tested at a 1:2 dilution (Group E), indicating that BSA-Sepharose did not nonspecifically remove the activity. However, after passage of 2 ml of the extract over 5 ml of GAT-Sepharose, suppressive activity at a 1:2 dilution was significantly decreased (Group F).

Estimate of m.w. We performed two experiments to approximate the m.w. of the suppressive moiety in these extracts. In one experiment, 2.5 ml of extract from GAT-primed DBA/1 mice was fractionated on a G-100 column. The eluate was pooled as shown in Figure 2, concentrated to 5 ml, and tested *in vivo* and *in vitro*. The greatest suppressive activity was in Pool III when tested at a dilution of 1:2 *in vivo* and 1:400 *in vitro*. Ovalbumin (m.w. 45,000 daltons) also eluted in Pool III.

These results were confirmed by passage of another extract through successively smaller Amicon ultrafiltration membranes. When tested undiluted, the immune response to GAT-MBSA *in vivo* was inhibited 77% by material with m.w. of 10,000 to 50,000 daltons. Materials which passed the PM10 membrane or were excluded by the PM50 membrane did not significantly affect the response to GAT-MBSA. More precise determinations of the m.w. of the suppressor factor will depend upon analysis of a more purified product.

DISCUSSION

These data demonstrate that GAT-specific suppression can be adoptively transferred by lymphoid cells *in vivo* and that cell-free extracts of lymphoid cells from two different strains of nonresponder mice primed with GAT can suppress the development of an immune response to GAT-MBSA by normal syngeneic nonresponder mice *in vivo*. Furthermore, the suppressive activity of these extracts can be evaluated *in vitro*. Suppressive activity is not detected in extracts prepared from lymphoid cells of Maalox-primed or GAT-MBSA-primed nonresponder mice. The activity of this extract is GAT-specific and binds to GAT-Sepharose, but not BSA-Sepharose.

TABLE IV

Adsorption of GAT-specific suppressive extract with GAT- or BSA-sepharose

Group	Dilution GAT-primed Extract[a]	GAT-Specific PFC/Spleen[b]	Inhibition	p Value
		Arith. mean ± S.E.	%	
A	None	11,000 ± 1,400	—	—
B	1:2	1,600 ± 1,000	86	<0.001
C	1:4	2,300 ± 500	79	<0.001
D	1:8	4,600 ± 1,000	58	<0.004
E	BSA-Sepharose 1:2	1,300 ± 400	88	<0.001
F	GAT-Sepharose 1:2	7,500 ± 500	32	<0.030

[a] Extract prepared from spleen and thymocytes of GAT-primed DBA/1 mice (30×10^7 cells).

[b] Immune response by DBA/1 mice, 7 days after injection of extract and GAT-MBSA (seven mice per group).

p value of: B:E = 0.8253; B:F = 0.008; D:F = 0.0276.

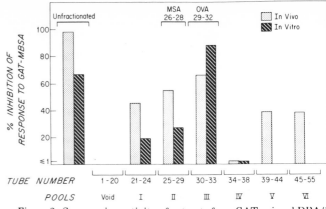

Figure 2. Suppressive activity of extracts from GAT-primed DBA/1 lymphoid cells after passage over Sephaedex G-100 assayed *in vivo* and *in vitro*. An extract was prepared from 6×10^8 GAT-primed DBA/1 spleen and thymus cells and a portion of the extract was fractionated on Sephadex G-100. The eluate was pooled as indicated above, concentrated to 5 ml, and tested for suppressive activity at a dilution of 1:2 *in vivo* and 1:400 *in vitro* on the response to GAT-MBSA by DBA/1 mice. The results are expressed as the percentage of inhibition of the responses to GAT-MBSA in the absence of extracts.

However, it is unlikely that the activity mediator in the extract is a classical antibody since the apparent m.w., 10,000 to 50,000 daltons, is much less than IgG.

This GAT-induced extract is very similar to the suppressive factor extracted from carrier-primed mouse T cells that has been described by Tada (13). The carrier-specific factor has a m.w. of approximately 50,000 daltons, has no determinants that cross-react with antibody specific for immunoglobulins, but immunoadsorbents prepared with alloantisera specific for products encoded by the *I-r* genes of the H-2 complex do remove the suppressive activity. Preliminary studies on the antigenic characteristics of the GAT-induced extract are in agreement with those of the carrier-specific factor reported by Tada *et al.* (13). Furthermore, these antigen-specific suppressive factors are similar to the antigen-specific cooperative T cell factors described by Taussig and Munro (22, 23). Although there is good indirect evidence that antigen-specific helper and suppressor cells are different subpopulations of T cells (3), the great similarities between the factors elaborated by or extracted from these T cells makes it mandatory that in future experiments suppressive factors be assayed for the ability to substitute for helper T cells in the development of an immune response by B cells stimulated with the specific antigen. Likewise, helper factors must be examined for suppressor activities by determining their effect on immunologically competent lymphoid cell populations.

The target cell for the various suppressive factors has not been directly determined. However, based on the ability of the factor extracted from carrier-primed T cells to inhibit antibody responses to haptens coupled to the homologous carrier and the binding of the factor to carrier coupled to Sepharose, Tada *et al.* have postulated that helper T cell function may be inhibited by the factor. The inhibition of anti-GAT antibody responses stimulated by GAT-PRBC by the suppressive extracts could be interpreted as evidence that this extract inhibits GAT-specific B cell function. However, we have recently demonstrated GAT-specific helper T cells in nonresponder mice immunized with GAT-MBSA or macrophages-bearing GAT (24). This raises the possibility that GAT-specific helper T cells are also stimulated by GAT-PRBC and that the

GAT-specific suppressive extract may affect these helper T cells.

The observation that GAT-specific suppressor cells are demonstrable after injection of GAT in three nonresponder strains of mice (H-2$^{p,\ q,\ s}$) and that suppressive factors can be extracted from H-2^q and H-2^s lymphoid cells (H-2^p mice have not been tested) raises the issue of whether genetic nonresponsiveness is always associated with the stimulation of suppressor T cells. In fact, we have recently found that the related synthetic copolymer of L-glutamic acid^{50}-L-tyrosine^{50} (GT), which fails to stimulate an antibody response in 19 inbred strains of mice, suppresses the antibody response elicited by GT complexed to MBSA in some but not all strains of mice (25). Furthermore, these studies indicate that the strain distribution of suppression induced by the two related polymers (GT and GAT) is distinct. The specificity of suppression elicited by GT and GAT is also distinct in SJL (H-2^s) mice, since GAT inhibits only the response to GAT-MBSA, not GT-MBSA, whereas GT inhibits both GAT-MBSA and GT-MBSA responses. If GT-induced suppressive extracts can be obtained, the close relationship between GT and GAT should permit characterization of the antigen-binding sites of suppressive factors.

Acknowledgments. We thank Dr. Zelig Eshhar for preparing the immunoadsorbents, GAT-Sepharose and BSA-Sepharose, and Ms. Deborah Siner for preparation of this manuscript.

REFERENCES

1. Katz, D. H., and B. Benacerraf. 1972. The regulatory influence of activated T cells on B cell responses to antigen . Adv. Immunol. 15:2.

2. Gershon, R. K. 1974. T cell control of antibody production. Contemp. Top. Immunobiol. 3:1.

3. Pierce, C. W., and J. A. Kapp. Regulation of immune responses by suppressor T cells. Contemp. Top. Immunobiol. in press.

4. Waldmann, T. A., S. Broder, H. Diorm, B. Meade, R. Krakauer, M. Blackman, and G. Goldman. 1975. T cell suppression of pokeweed mitogen induced immunoglobulin production. *In* The Roles of Mitogens In Immunobiology. Edited by J. J. Oppenheim and D. L. Rosenstreich. Academic Press, Inc., New York. in press.

5. Herzenberg, L. A., and L. A. Herzenberg. 1974. Short term and chronic allotype suppression in mice. Contemp. Top. Immunobiol. 3:41.

6. Rich, R. R., and C. W. Pierce. 1973. Biological expressions of lymphocyte activation. II. Generation of a population of thymus-derived suppressor lymphocytes. J. Exp. Med. 137:649.

7. Peavy, D. L., and C. W. Pierce. 1974. Cell-mediated immune responses *in vitro*. I. Suppression of the generation of cytotoxic lymphocytes by concanavalin A and concanavalin A-activated spleen cells. J. Exp. Med. 140:356.

8. Pross, H. F., and D. Eidinger. 1974. Antigenic competition: A review of nonspecific antigen-induced suppression. Adv. Immunol 18:133.

9. Liacopoulos, P., and S. Ben-Efraim. 1975. Antigenic competition. Prog. Allergy 18:97.

10. Rich, R. R., and C. W. Pierce. 1974. Biological expressions of lymphocyte activation. III. Suppression of plaque-forming cell

11. Thomas D. W., W. K. Roberts, and D. W. Talmadge. 1975. Regulation of the immune response: Production of a soluble suppressor by immune spleen cells *in vitro*. J. Immunol. 114:1616.

12. Zembala, M., and G. L. Asherson. 1974. T cell suppression of contact sensitivity in the mouse. II. The role of soluble suppressor factor and its interaction with macrophages. Eur. J. Immunol. 4:779.

13. Tada, T., M. Taniguchi, and T. Takemori. 1975. Properties of primed suppressor T cells and their products. Transplant. Rev. 26:106.

14. Kapp, J. A., C. W. Pierce, and B. Benacerraf. 1973. Genetic control of immune response *in vitro*. I. Development of primary and secondary plaque-forming cell responses to the random terpolymer L-glutamic acid^{60}-L-alanine^{30}-L-tyrosine^{10} (GAT) by mouse spleen cells *in vitro*. J. Exp. Med. 138:1107.

15. Kapp, J. A., C. W. Pierce, and B. Benacerraf. 1974, Genetic control of immune response *in vitro*. III. Tolerogenic properties of the terpolymer L-glutamic acid^{60}-L-alanine^{30}-L-tyrosine^{10} (GAT) for spleen cells from nonresponder (H-2^s and H-2^q) mice. J. Exp. Med. 140:172.

16. Kapp, J. A., C. W. Pierce, S. Schlossman, and B. Benacerraf. 1974. Genetic control of immune response *in vitro*. V. Stimulation of suppressor T cells in nonresponder mice by the terpolymer L-glutamic acid^{60}-L-alanine^{30}-L-tyrosine^{10} (GAT). J. Exp. Med. 140:648.

17. Cuatrecasas, P. 1970. Protein purification of affinity chromatograph. Derivatization of agarose and polyacrylamide beads. J. Biol. Chem. 245:3059.

18. Inman, J. K., and H. M. Dintzis. 1969. The derivatization of cross-linked polyacrylamide beads. Controlled introduction of functional groups for the preparation of special-purpose, biochemical adsorbents. Biochemistry 8:4074.

19. Mishell, R. I., and R. W. Dutton. 1967. Immunization of dissociated spleen cell cultures from normal mice. J. Exp. Med. 126:423.

20. Pierce, C. W., B. M. Johnson, H. E. Gershon, and R. Asofsky. 1971. Immune responses *in vitro*. III. Development of primary γM, γG and γA plaque-forming cell responses in mouse spleen cell cultures stimulated with heterologous erythrocytes. J. Exp. Med. 134:395.

21. Kapp, J. A., C. W. Pierce, and B. Benacerraf. 1975. Role of suppressor T cells in an Ir gene controlled immune response. *In* Suppressor Cells in Immunity. Edited by S. K. Singhal and N. R. Sinclair. University of Western Ontario Press, London, Ontario, Canada. p. 84.

22. Taussig, M. J. 1974. T cell factor which can replace T cells *in vivo*. Nature 248:234.

23. Munro, A. J., and M. J. Taussig. 1975. Two agents in the major histocompatibility complex control immune response. Nature 256:103.

24. Kapp, J. A., C. W. Pierce, and B. Benacerraf. 1975. Genetic control of immune response *in vitro*. VI. Experimental conditions for the development of helper T cell activity specific for the terpolymer L-glutamic acid^{60}-L-alanine^{30}-L-tyrosine^{10} (GAT) in nonresponder mice. J. Exp. Med. 142:50.

25. Debre, P., J. A. Kapp, M. E. Dorf, and B. Benacerraf. Genetic control of immune suppression. II. H-2-linked dominant genetic control of immune suppression by the random copolymer L-glutamic acid^{50}-L-tyrosine^{50} (GT). J. Exp. Med. In press.

Chapter V

Tolerance

Chapter V

Tolerance

An individual animal can respond in one of two ways when confronted with an antigen. It can make an immune response, recruiting T and B lymphocytes which proliferate and differentiate into effector cells, or it can make no response and suffer the consequences. Clearly, the failure to make an immune response against a pathogenic invader can have dire results on the host, but the indiscriminate mounting of immune defenses against non-foreign, or self, antigens could have equally dire consequences. It is the ability of the immune response to discriminate between self and non-self which allows an individual to be immunologically tolerant of its own antigens. More generally, tolerance can be defined as the failure of an individual to make an immune response against any specific antigen, be it self or non-self in origin. Loss of tolerance to self-antigens results in autoimmune disease.

Because tolerance to self reflects a fundamental property of the immune system, immunologists are striving to understand how it is achieved. To do this, they have developed experimental systems which render animals tolerant to certain non-self antigens to which they would normally be immune. The assumption is that an understanding of the "experimental tolerance" will provide an insight into the mechanism(s) of self-tolerance. Such an insight would obviously be helpful also in developing rational approaches to organ and tissue transplantation in the treatment of various diseases. The papers in this chapter explore certain aspects of such "experimental" tolerance and suggest that a state of immunological unresponsiveness in individuals can be induced and maintained in a number of different ways.

One of the first clues about the nature of immune tolerance came from an observation made in 1945 by Ray Owen. He noted that fraternal bovine twins had identical blood types. This was an unexpected finding because full siblings, which bear the same genetic relationship to each other as fraternal, or non-identical twins, rarely demonstrated blood type identity. Owen attributed this unusual sharing of blood type antigens to an actual mixing of red blood cells between the twins, who shared a common vascular system during fetal development. Apparently, this natural coexistence of histoincompatible cell types was able to persist through the twins' adult lives.

Burnet and Fenner (1949) used Owen's observations to develop the theory of neonatal tolerance. Basically, this theory proposed that an individual exposed to histoincompatible cells before birth would consider those foreign cells as self for the rest of its life. Hence, the otherwise histoincompatible blood cells were able to coexist in the non-identical bovine twins but not in the full siblings which had not shared a vascular system during development.

In a test of this theory, Billingham, Brent and Medawar (paper 1) performed a now-classic experiment. These immunologists injected mouse fetuses with cells from a histoincompatible donor and waited for the fetuses to develop into adult mice. These adults were then challenged with a skin graft from the same histo-incompatible donor. Billingham and colleagues found that the mice which had received the foreign cells during fetal development were not able to reject the skin grafts. Control mice which had not previously received histoincompatible cells demonstrated a rapid rejection of the grafts. Further experiments revealed that the tolerance displayed by the experimental group of mice was due neither to an alteration in the antigens of the graft nor to a general debility of the host's immune system. The results of these experiments provided support for the model previously proposed by Burnet and Fenner and laid the groundwork for the idea that encounter with any antigen renders a developing animal unresponsive to that anti-gen for life. Thus, immune tolerance toward a foreign antigen would occur in much the same way as tolerance for self molecules. One way in which this tolerance might be achieved would be through the simple elimination of all lymphocytes bear-ing receptors specific for antigens encountered during ontogeny. These would, of course, normally be self antigens.

This model is, however, too simple to explain the phenomenon of immunological tolerance completely. Such a model maintains that an animal has essentially only one chance to develop tolerance toward antigens, and that chance would come at some critical time in development. The fact that adult animals can be rendered toler-ant to foreign antigens suggests either that the model is inadequate or that tolerance to self antigens is quite different from tolerance to foreign antigens. An alternative hypothesis could argue that lymphocyte development is an ongoing process and that immature lymphocytes, in both neonatal and adult animals, respond to antigens by becoming tolerant.

In fact, more recent studies of tolerance have focused on the properties of specific unresponsiveness in adults rather than in newborns. They lend support to the idea that it is the developmental stage of the lymphocyte, not of the animal, which contributes to tolerance. These studies have expanded our knowledge about the many factors which can contribute to making an animal tolerant to specific antigens. They have demonstarted that a) injection of aggregate-free preparations of antigens renders the recipient tolerant (Dresser, 1962); b) both very high and very low doses of antigen result in tolerance, whereas intermediate doses lead to immunity (Mitchison, 1964); and c) a state of tolerance can be maintained through either the B or T cell population (paper 2).

This array of factors which can contribute to developing a state of tolerance probably reflects the fact that an individual can be made immunologically tolerant in a number of different ways. The mechanisms involved could include deletion or inactivation of specific clones (clonal deletion vs clonal abortion), suppression of reactive B or T lymphocytes by suppressor T cells (see the preceding chapter), and inactivation of the receptors by antigen.

Paper 3 is one of a series of papers by Nossal and collaborators presenting evidence for the clonal abortion of immature lymphocytes which encounter antigen. Using a combination of in vivo and in vitro techniques, Nossal and Pike demon-strate that B lymphocytes allowed to mature in the presence of a tolerogenic form (aggregate-free) of the human gamma globulin (HGG) were rendered specifically unresponsive to that antigen. In addition, very low doses of antigen were effec-tive in tolerizing the population of immature (bone marrow) B cells, whereas higher doses were required to render mature (spleen) B cells tolerant. Two further experiments showed that suppressor T cells were not involved in this tolerance to HGG, and that the tolerance was not transferrable by lymphocytes. The data are consistent with, although not proof of, a model of clonal abortion.

In contrast to the clonal abortion hypothesis, the receptor blockade theory (Aldo-Benson and Borel, 1974) postulates that the specific B cells of a tolerant animal are not irreversibly inactivated or deleted by antigen but rather are rever-sibly inhibited by blockade of cell-surface receptors which prevents a stimulating signal from reaching the lymphocyte. The central prediction of this theory is that B cells capable of binding the antigen should be present in tolerant animals. Some evidence for this model is presented in paper 4.

Interaction of antigen with surface receptors is a critical requirement in both the clonal abortion and receptor blockade models of tolerance induction. It has therefore been important to determine if lymphocytes which are easily rendered unresponsive either express different receptors or have receptors which behave differently from normal, reactive lymphocytes. Paper 5 by Sidman and Unanue suggests that neonatal lymphocytes cannot regenerate surface immunoglobulin after these molecules have been capped by anti-immunoglobulin. Adult cells, in contrast, demonstrate rapid regeneration. Paper 6 observes that neonatal cells which are susceptible to tolerance induction express only surface IgM, whereas the majority of adult cells express both IgM and IgD as potential surface receptors for antigen. The authors propose that the presence of IgD is instrumental in making a B cell responsive, as opposed to unresponsive, to antigen.

Although these and other studies have contributed to our understanding of experimental tolerance to exogenous antigens, it is difficult to determine how much of what we have learned is attributable to the issue of self-tolerance. For example, the ability to demonstrate antibodies against self-constituents in certain diseases make the clonal deletion model of self-tolerance fairly untenable, at least as a generally operative mechanism. It is more likely that self-reactive B cell clones exist but are prevented from functioning by antigen blockade or suppressor T cells. In addition, as we learn how critical self-recognition is to the successful functioning of T lymphocytes (Chapters V and VIII), it becomes clear that the simple notion that an individual cannot recognize self-antigens needs some modification. At this point it seems likely that unresponsiveness to self-constituents is induced and maintained by a number of factors. The form and dose of antigen, the developmental state of B cells, and the participation of regulatory T cells must all contribute to a set of interactions that results in immune tolerance.

Papers Included in this Chapter

1. Billingham, R.E., Brent, L., and Medawar, P.B. (1953) "Actively acquired tolerance" of foreign cells. Nature 172, 603.

2. Chiller, J.M., Habicht, G.S., and Weigle, W.O. (1971) Kinetic differences in unresponsiveness of thymus and bone marrow cells. Science 171, 813.

3. Nossal, G.J.V. and Pike, B.L. (1975) Evidence for the clonal abortion theory of B-lymphocyte tolerance. J. Exp. Med. 141, 904.

4. Venkataramen, M., Aldo-Benson, M., Borel, Y., and Scott, D.W. (1977) Persistence of antigen binding cells with surface tolerogen: Isologous vs. heterologous immunoglobulin carriers. J. Immunol. 119, 1006.

5. Sidman, C.L. and Unanue, E.R. (1975) Receptor-mediated inactivation of early B lymphocytes. Nature 257, 149.

6. Vitetta, E.S., Cambier, J.C., Legler, F.S., Kettman, J.R., and Uhr, J.W. (1977) B cell tolerance. IV. Differential role of surface IgM and IgD in determining tolerance susceptibility of murine B cells. J. Exp. Med. 146, 1804.

References Cited in this Chapter

Aldo-Benson, M. and Borel, Y. (1974) Tolerant cells: direct evidence for receptor blockade by tolerogen. J. Immunol. 112, 1793.

Burnet, F.M. and Fenner, T. (1949) The Production of Antibodies. MacMillan, N.Y.

Dresser, D.W. (1962) Specific inhibition of antibody production. II. Paralysis induced in adult mice by small quantities of protein antigen. Immunol. 5, 378.

Mitchison, N.A. (1964) Induction of immunological paralysis in two zones of dosage. Proc. Roy. Soc. B. 161, 275.

'ACTIVELY ACQUIRED TOLERANCE' OF FOREIGN CELLS

By Dr. R. E. BILLINGHAM*, L. BRENT and Prof. P. B. MEDAWAR, F.R.S.

Department of Zoology, University College, University of London

THE experiments to be described in this article provide a solution—at present only a 'laboratory' solution—of the problem of how to make tissue homografts immunologically acceptable to hosts which would normally react against them. The principle underlying the experiments may be expressed in the following terms : that mammals and birds never develop, or develop to only a limited degree, the power to react immunologically against foreign homologous tissue cells to which they have been exposed sufficiently early in fœtal life. If, for example, a fœtal mouse of one inbred strain (say, *CBA*) is inoculated *in utero* with a suspension of living cells from an adult mouse of another strain (say, *A*), then, when it grows up, the *CBA* mouse will be found to be partly or completely tolerant of skin grafts transplanted from any mouse belonging to the strain of the original donor.

This phenomenon is the exact inverse of 'actively acquired immunity', and we therefore propose to describe it as 'actively acquired tolerance'. The distinction between the two phenomena may be made evident in the following way. If a normal adult *CBA* mouse is inoculated with living cells or grafted with skin from an *A*-line donor, the grafted tissue is destroyed within twelve days (see below). The effect of this first presentation of foreign tissue in adult life is to confer 'immunity', that is, to increase the host's resistance to grafts which may be transplanted on some later occasion from the same donor or from some other member of the donor's strain. But if the first presentation of foreign cells takes place in fœtal life, it has just the opposite effect : resistance to a graft transplanted on some later occasion, so far from being heightened, is abolished or at least reduced. Over some period of its early life, therefore, the pattern of the host's response to foreign tissue cells is turned completely upside down. In mice, it will be seen, this inversion takes place in the neighbourhood of birth, for there is a certain 'null' period thereabouts when the inoculation of foreign tissue confers neither tolerance nor heightened resistance—when, in fact, a 'test graft' transplanted in adult life to ascertain the host's degree of immunity is found to survive for the same length of time as if the host had received no treatment at all.

Earlier Work

The literature of experimental embryology is rich in evidence that embryos are fully tolerant of grafts of foreign tissues. It is less well known (though no less firmly established) that embryonic cells transplanted into embryos of different genetic constitutions may survive into adult life, although their hosts would almost certainly have rejected them if transplantation had been delayed until after birth. The transplantation of embryonic melanoblasts[1] provides the most conspicuous evidence of this phenomenon—not because melanoblasts are peculiar in their immunological properties, but simply because their genetic origins are at once betrayed by the

* British Empire Cancer Campaign Research Fellow.

pigmentation of the cells into which they ultimately develop. Unfortunately, experiments with embryonic melanoblasts, having been done with quite different purposes in mind, do not make it possible to decide whether survival into adult life is due to an antigenic adaptation of embryonic cells which have been obliged to complete their development in genetically foreign soil, or whether it is due to a suppression or 'paralysis'[2] of the host's immunological response.

An exactly comparable phenomenon has been described by Owen[3], who found that the majority of dizygotic cattle twins are born with, and long retain, red blood cells of dizygotic origin : each calf contains a proportion of red cells belonging genetically to itself, mixed with red cells belonging to the zygote lineage of its twin. There is no reason to doubt that this is because the cattle twins, being synchorial, exchange blood in fœtal life through the anastomoses of their placental vessels. (This is not a peculiarity of cattle, for a human twin with red cells of dizygotic origin has lately been described[4].) Inasmuch as the provenance of the red cells was revealed by their reactions with specific agglutinins, it is most unlikely that the survival of foreign erythrocyte-forming cells into adult life was made possible by any kind of antigenic adaptation. Moreover, we have found that the majority of cattle twins at birth and for long after are fully tolerant of grafts of each other's skin[5]. Being freshly transplanted, these grafts can have had no opportunity to 'adapt' themselves antigenically to foreign hosts, but they survived nevertheless.

The experiments of Cannon and Longmire[6] have a direct bearing on the phenomenon of actively acquired tolerance. About 5–10 per cent of skin grafts exchanged between pairs of newly hatched chicks of different breeds are tolerated and survive into adult life ; but the percentage of successes falls rapidly as the age at which the chicks are operated increases, and reaches zero by the end of the second week. These results will be referred to later.

Experiments with Mice

A single experiment will be described in moderate detail : the recipients were mice of *CBA* strain, the donors of *A* strain. The data for transplantations between normal mice of these strains are as follows. The median survival time of *A*-line skin grafts transplanted to normal *CBA* adults (regardless of differences of sex, or of age within the interval 6 weeks–6 months) is $11 \cdot 0 \pm 0 \cdot 3$ days[7]. In reacting against such a graft, the host enters a state of heightened resistance ; a second graft transplanted up to sixty days after the transplantation of the first survives for less than six days, and immunity is still strong, though it has weakened perceptibly, after four months. Heightened resistance may be passively transferred to a normal *CBA* adult by the intraperitoneal implantation of pieces of lymph node excised from a *CBA* adult which has been actively immunized against *A*-line skin[8].

In the experiment to be described (Exp. 73), a *CBA* female in the 15–16th day of pregnancy by a

CBA male was anæsthetized with 'Nembutal', and its body wall exposed by a median ventral incision of the skin. The skin was mobilized but not reflected, and particular care was taken not to damage the mammary vessels. By manipulation of the abdomen with damped gauzes, six fœtuses were brought into view through the body wall. Each was injected intra-embryonically with 0·01 ml. of a suspension of adult tissue cells through a very fine hypodermic needle passing successively through the body wall, uterine wall, and fœtal membranes. (The inoculum itself, consisting of a suspension in Ringer's solution of small organized tissue clumps, isolated cells, and cell debris, had been prepared by the prolonged chopping with scissors of testis, kidney and splenic tissue from an adult male *A*-line mouse.) After injection of the fœtuses, the skin was closed with interrupted sutures.

Five healthy and normal-looking young were born four days later; of the sixth fœtus there was no trace. Eight weeks after their birth, when the lightest weighed 21 gm., each member of the litter was 'challenged' with a skin graft from an adult *A*-line donor. The first inspection of the grafts was carried out eleven days later, that is, at the median survival age of *A*-line skin grafts transplanted to normal *CBA* hosts. The grafts on two of the five mice were in an advanced stage of breakdown; the grafts on the other three (one male and two females) resembled autografts in every respect except their donor-specific albinism. Each of these three grafts became perfectly incorporated into its host's skin and grew a white hair pelt of normal density and stoutness. Fifty days later, one of the three mice received a second *A*-line graft from a new donor, and this graft also settled down without the least symptom of an immunological reaction.

The graft on one of the three animals underwent a long-drawn-out 'spontaneous' involution, beginning somewhat before the 75th day after transplantation and ending with complete breakdown shortly after the 91st day. The other two mice were made the subjects of an experiment[9] designed to show that acquired tolerance is due to a failure of the host's immunological response and not to an antigenic adaptation of the grafted cells. When the two grafts were of 77 and 101 days standing, respectively, and still in immaculate condition, their hosts were inoculated intraperitoneally with chopped fragments of lymph nodes from normal *CBA* mice which had been actively immunized against *A*-line skin. The grafts began to deteriorate 2–3 days later, with signs of vascular congestion and stasis; contracture, hardening and discoloration took place progressively, and the grafts were reduced to dry scabs by the fourteenth day after the nodes had been implanted. It follows that tolerant hosts are fully capable of giving effect to a state of immunity which has been elicited by proxy, and that the tolerated grafts have not lost their ability to respond to it.

The fertility of the mice was entirely unimpaired. Both females repeatedly bore litters of normal size by their male litter mate. When they had grown up, two litters (of six and eight respectively) were challenged with grafts of *A*-line skin. Breakdown of all the grafts was far advanced or complete by the eleventh day after their transplantation.

The results of Exp. 73 demonstrate the great (but not necessarily indefinite) prolongation of the life of homografts made possible by a pre-emptive exposure of their hosts to foreign homologous cells. Beyond this, they demonstrate (*a*) that this prolongation of life is not due to an antigenic transformation of the grafts, nor to a competitive absorption of antibodies by, for example, cells of the fœtal inoculum which had survived into adult life; and (*b*) that acquired tolerance is either not transferred to, or is too weak or too ephemeral to make itself evident in, the offspring.

Two of the five mice of Exp. 73 gave no evidence of increased tolerance of *A*-line cells. We suppose that this was because they were imperfectly injected. Only in one experiment so far has every injected fœtus given rise to a tolerant adult.

The more important results obtained from the investigation of other litters may be summarized as follows. (1) The conferment of tolerance is not of an all-or-nothing character; every degree is represented, down to that which gives the test-grafts only a few days of grace beyond the median survival time of their controls. (2) The conferment of tolerance is immunologically specific. Thus a *CBA* mouse made tolerant of *A*-line tissue, or vice versa, retains the ability to react with unmodified vigour against skin from a donor belonging to a third strain, *AU*. (Equally clear evidence of the specificity of acquired tolerance is given in the following section.) But it has so far been our experience that the transplantation of (say) *AU* skin to a *CBA* mouse that is tolerant of *A*-line cells will elicit a reaction which, in addition to destroying the *AU* graft, causes an *A*-line graft already in residence to go through a severe immunological crisis. Although we have evidence of a sharing of tissue antigens between strains *A* and *AU*, this phenomenon is difficult to interpret, for the antigens common to strains *A* and *AU* are merely a sub-group of those to which the *CBA* host is manifestly unresponsive. It may therefore turn out that the continued well-being of a tissue homograft upon a tolerant host depends upon the quiescence of the antibody-forming system, and that if this is awakened by tissue antigens other than those of which the host is tolerant, antibodies directed against its, until then, tolerated graft may be formed as well. (3) A wide histological variety of tissue cells is capable of conferring tolerance to homografts of skin. It is by no means obligatory that skin cells, or even epithelial cells, should be among them. (4) We have inoculated ninety-six new-born mice with adult or fœtal tissue cell suspensions to decide whether tolerance can be conferred by exposure to foreign cells at this relatively advanced stage of development. The majority received a single inoculation of cells as soon as possible after birth; a small subgroup of these was injected with 0·05 mgm. cortisone acetate on the same occasion and several more times during the first week of life in an attempt to delay the maturation of the antibody-forming system[e]. The remainder of the new-born mice received repeated injections of foreign tissue cells in increasing quantities over the period of a month from birth. In all, only nine mice (about 10 per cent: cf. the experiments of Cannon and Longmire referred to above) showed an increase of tolerance when tested with a skin graft from the donor strain, and six of these were members of a single litter of eight which had received a single inoculation of cells, without cortisone, immediately after birth. It is of particular interest that when challenged with skin grafts in adult life the great majority of inoculated new-born mice showed neither tolerance nor enhanced resistance. New-born mice are, in general, too old for a tissue inoculum to confer tolerance, and too young for it to confer immunity; the epoch of birth represents a null period during which the net outcome of exposure to foreign cells is to leave its subjects in a state of unaltered reactivity. (5) Grafts removed from hosts which have tolerated them, and then transplanted to normal mice of the host's strain, survive 2–3 days longer than freshly transplanted homografts of normal skin. Such homografts cannot, however, be compared directly with homografts of normal skin, because a high proportion of the corium of each will probably have been replaced by cells and cellular derivatives of host origin.

Preliminary Experiments with Chickens

Donors and recipients in these experiments were of Rhode Island Red and White Leghorn breeds, respectively. Skin transplanted from two weeks old Rhode Island Red chicks to White Leghorn recipients of the same age, using Cannon and Longmire's methods[6], is completely destroyed within ten days of grafting, to the accompaniment of an inflammatory reaction of conspicuous violence.

The embryonic chick is particularly well suited to experiments which make use of cellular inoculation, because the intravenous route is so easily accessible. Using methods demonstrated to us by Dr. C. Kaplan, whose help has been of the greatest value, we have obtained successful results by transfusing 0·2 ml. unmodified whole blood from an 11–12 day old embryonic Rhode Island Red donor into a chorio-allantoic vein of a White Leghorn embryo of the same age. Fourteen days after hatching, a test-graft of skin was transplanted to the recipient from its original donor. In seven such trials, five grafts showed prolongation of survival; of these, three succumbed within fifty days to the accompaniment of very much subdued inflammatory changes, and two still survive, with normal growth of red feathers, to the present time (125 days).

The success of these experiments has been found to depend on the strict pairing of donors with recipients. The transplantation of skin between our Rhode Island Red chicks when two weeks old showed them to be a highly heterogeneous assembly, in spite of their uniformity of breed characters. It is therefore understandable that chicks made tolerant to grafts from one Rhode Island Red donor will not, in general, accept grafts from another. This strict specificity of acquired tolerance has made it difficult to test the efficacy of grafts transplanted to the chorio-allantois, for an embryonic donor must be killed to provide tissues suitable for grafting. This difficulty has been circumvented by killing a young hatched donor, storing its skin[10] for later use in the test operation, and grafting fragments of a wide variety of its tissues upon the chorio-allantoic membrane of 10–11 day old White Leghorn recipients. Tolerance is conferred less regularly and less completely by this method than by blood transfusion: only three out of seven such experiments have yielded test grafts which survived for longer than twice their normal expectation of life, and those which have done so have given evidence of chronic low-grade inflammatory changes.

Washed blood corpuscles are as effective as whole blood in conferring tolerance; plasma is therefore a dispensable ingredient. If adult blood proves to be capable of conferring tolerance, it should be easy to decide whether red cells or white cells or both are efficacious.

Discussion

It is one of the predictions of Burnet and Fenner's[11] theory of immunity that the exposure of animals to antigens before the development of the faculty of immunological response should lead to tolerance rather than to heightened resistance. The homograft immunity system is particularly well suited to the appraisal of such a hypothesis, because the antigens are at once powerful, innocuous and persistent. For this reason, no great weight should be attached to failure of the experiments of Burnet, Stone and Edney[12] to verify it. It must be emphasized, however, that our experiments do not yet bear upon the fundamental problem of whether the production of antibodies represents an *inherited* derangement of protein synthesis, that is, a transformation which can persist through repeated cell divisions after the disappearance of the antigen originally responsible for it. It would be a highly significant fact if a transient exposure to antigen in foetal life could confer a permanent or very long-lasting tolerance; but in our present experiments there is no reason to doubt that at least some of the cells of the foetal inoculum survived as long as the tolerant state which they were responsible for creating. At all events, any complete theory of antibody formation must be competent to explain two sets of facts: that although embryos do not make antibodies, they respond to antigens in a manner that prejudices their ability to do so in later life; and that acquired tolerance is highly specific, for antibody-forming cells can be prevented from responding to one antigen without impairing their capacity to respond to any other.

A state of tolerance similar in principle to that which we have described may not necessarily depend upon an exposure to antigens in foetal life. Phenomena which occur in adult life and which may be cognate with tolerance induced by foetal inoculation include: (*a*) the highly specific 'immunological paralysis' of mice by the administration of relatively high doses of pneumococcal polysaccharide, made clear by the important experiments of Felton[2]; (*b*) that restraint of certain drug allergies in guinea pigs which, so Chase[13] has shown, may be brought about by the oral administration of the sensitizing chemical—an observation of particular interest, in view of the affinity between transplantation immunity and sensitization reactions of just this type[14]; and (*c*) the enhancement of the growth of tumour homografts by treatment of their prospective hosts with a variety of lyophilized tissue preparations, revealed by the work of Casey, Snell, Kaliss and their colleagues[15]. The relationship between these various phenomena has yet to be determined. We propose to investigate the hypothesis that exposure of the adult antibody-forming system to what may be called a 'tissue hapten' preparation—that is, to tissue ingredients which, although not themselves antigenic, contain the determinant groupings that confer specific activity upon complete antigens—may have the same kind of effect as that produced by an exposure of the immature antibody-forming system of the foetus to complete antigens.

Actively acquired tolerance may not be a wholly artificial phenomenon. We are inquiring into the possibility that it may occur naturally by the accidental incorporation of maternal cells into a foetus during normal development.

In dizygotic twin cattle of unlike sex, the confluence of foetal vessels that leads to red cell chimærism[3] and tolerance of homografts[6] has long been known to be associated with infertility of the female. In our present experiments, tolerance has been conferred upon female mice by inoculation with male cells without affecting their fertility. The two phenomena are therefore separable, although in cattle they go together and share an anatomical pre-requisite in common; but this is an inference that leaves entirely open the question of whether or not the freemartin state is due to a purely humoral influence of the male fœtus upon its synchorial female twin.

The experiments described in this article will in due course be reported on in full.

Summary

(1) Mice and chickens never develop, or develop to only a limited degree, the power to react immunologically against foreign homologous tissue cells with which they have been inoculated in fœtal life. Animals so treated are tolerant not only of the foreign cells of the original inoculum, but also of skin grafts freshly transplanted in adult life from the original donor or from a donor of the same antigenic constitution.

(2) Acquired tolerance is immunologically specific: mice and chickens made tolerant of homografts from one donor retain the power to react against grafts transplanted from donors of different antigenic constitutions.

(3) Acquired tolerance is due to a specific failure of the host's immunological response. The antigenic properties of a homograft are not altered by residence in a tolerant host, and the host itself retains the power to give effect to a passively acquired immunity directed against a homograft which has until then been tolerated by it.

(4) The fertility of tolerant mice is unimpaired.

[1] Cf. Rawles, M. E., *Physiol. Rev.*, **28**, 383 (1948).

[2] Felton, L. D., *J. Immunol.*, **61**, 107 (1949).

[3] Owen, R. D., *Science*, **102**, 400 (1945). Owen, R. D., Davis, H. P., and Morgan, R. F., *J. Hered.*, **37**, 291 (1946). Stone, W., Stormont, C., and Irwin, M. R., *J. An. Sci.*, **11**, 744 (1952).

[4] Dunsford, I., Bowley, C. C., Hutchison, A. M., Thompson, J. S., Sanger, R., and Race, R. R., *Brit. Med. J.*, ii, 81 (1953).

[5] Anderson, D., Billingham, R. E., Lampkin, G. H., and Medawar, P. B., *Heredity*, **5**, 379 (1951). Billingham, R. E., Lampkin, G. H., Medawar, P. B., and Williams, H. Ll., *Heredity*, **6**, 201 (1952).

[6] Cannon, J. A., and Longmire, W. P., *Ann. Surg.*, **135**, 60 (1952).

[7] Billingham, R. E., Brent, L., Medawar, P. B., and Sparrow, E. M. (unpublished work).

[8] Billingham, R. E., Brent, L., and Medawar, P. B. (unpublished work). The passive transfer of transplantation immunity was first demonstrated by Mitchison, N. A., *Nature*, **171**, 267 (1953); *Proc. Roy. Soc.*, B (in the press).

[9] The principle underlying this test was formulated by Chase, M. W., Abstr. 49th Gen. Meeting Soc. Amer. Bacteriol., 75 (1949).

[10] Billingham, R. E., and Medawar, P. B., *J. Exp. Biol.*, **29**, 454 (1952).

[11] Burnet, F. M., and Fenner, F., "The Production of Antibodies" (Melbourne, 1949).

[12] Burnet, F. M., Stone, J. D., and Edney, M., *Aust. J. Exp. Biol. Med. Sci.*, **28**, 291 (1950).

[13] Chase, M. W., *Proc. Soc. Exp. Biol. Med.*, N.Y., **61**, 257 (1946).

[14] Mitchison (ref. 8). Medawar, P. B., Colloquia of the Ciba Foundation (in the press).

[15] Summarized and reviewed by Snell, G. D., *Cancer Res.*, **12**, 543 (1952).

Kinetic Differences in Unresponsiveness of Thymus and Bone Marrow Cells

Abstract. *Both thymus and bone marrow cells of adult mice can be made specifically unresponsive to human gamma globulin, but each cell population displays a distinct kinetic pattern for both the induction and spontaneous loss of the unresponsive state. These kinetics appear to be much slower in the bone marrow cells than in the thymus cells. In addition, the dose of deaggregated human gamma globulin needed to induce unresponsiveness in bone marrow cells is much greater than that needed to induce unresponsiveness in thymus cells. Apparently, unresponsiveness in only one cell type is sufficient for the tolerant state to be exhibited by the intact animal.*

Investigations of the cellular events manifested in the induction process of the state of immunological unresponsiveness have been limited to studies of whole animals or heterogeneous lymphoid cell populations, such as those obtained from the spleen or from the peripheral blood (*1*). These in vivo studies have indicated that this induction period is short, from 24 hours to 4 or 5 days depending on the antigen and the test system used. In view of the collaboration between thymus cells and bone marrow cells in the immune response to a number of different antigens (*2*) and the subsequent demonstration that unresponsiveness can exist at each of these cellular sites (*3*), cellular events in the induction of unresponsiveness now demand new study. We now report that in adult mice there exists a distinct kinetic pattern for both the thymus and the bone marrow cell populations with regard to both the induction and spontaneous loss of the unresponsive state to human gamma globulin (HGG). In addition, our data demonstrate differential susceptibility of thymus cells and bone marrow cells to varying doses of tolerogen (a physical form of antigen that induces tolerance).

Synergism has been observed in the response to HGG when both thymus and bone marrow cells were obtained from normal mice, but not if either cell type came from tolerant mice (*3*). From these data, the prediction can be made that unresponsiveness in one cell type would be sufficient to make an intact animal appear tolerant. To test whether such a cellular dichotomy exists, the kinetics of the induction and spontaneous loss of immunological unresponsiveness to HGG were determined separately for thymus cells and for bone marrow cells.

Groups of adult A/J male mice (*4*) were injected with 2.5 mg of HGG deaggregated by ultracentrifugation (*5*) and each group was killed at a specified time. Suspensions of their thymus cells (90×10^6) or bone marrow cells (30×10^6) were injected intravenously along with normal bone marrow cells (30×10^6) or thymus cells (90×10^6), respectively, into lethally irradiated recipients (*3*). At the same time and again 10 days later, these irradiated mice were challenged with 0.4 mg of aggregated HGG (*6*) and, 5 days after the second injection, their spleens were assayed for plaque-forming cells (PFC) to HGG (*7*). Since very few, if any, direct PFC were found by this system, the data are expressed only as indirect PFC detected with the use of goat antiserum to mouse gamma globulin at a concentration shown to be optimal.

The percentage of suppression of the response obtained in the recipients receiving one cell type from donors injected with tolerogen (deaggregated HGG) and one cell type from normal donors compared to the response ob-

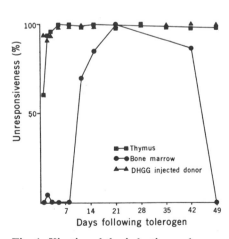

Fig. 1. Kinetics of the induction and spontaneous loss of unresponsiveness in thymus and bone marrow cells. Calculations of the values obtained for unresponsiveness in both cell types are described in (*8*); *DHGG*, deaggregated human gamma globulin.

tained in animals receiving both normal cell types was then calculated (*8*) and plotted as shown in Fig. 1. Thymus cells from tolerogen treated animals were essentially unresponsive by day 2 and remained so for the duration of the experiment (day 49). In contrast, bone marrow cells from the same tolerogen treated donors did not begin to show unresponsiveness until day 11 and were totally unresponsive only after 21 days. In addition, by day 49 bone marrow cells had returned to the responsive state. The induction of unresponsiveness in the intact animal paralleled the induction of unresponsiveness in thymus cells. Thus, the kinetics of the induction of unresponsiveness with this experimental protocol are much faster in thymus cells than those seen in bone marrow cells, and the state of unresponsiveness persists much longer in thymus cells. It is also of interest that an animal with an unresponsive thymus cell population may appear tolerant, although it can possess bone marrow not yet rendered unresponsive (for example, day 5) or having recovered from unresponsiveness (for example, day 49). These data could account for the fact that Taylor (*9*) was unable to demonstrate specific unresponsiveness in bone marrow cells 24 hours after he administered a tolerogenic regimen of bovine serum albumin (BSA).

To determine the effect of varying the dose of tolerogen on the induction of unresponsiveness in thymus and bone marrow cells, mice were injected with 0.1, 0.5, or 2.5 mg of deaggregated HGG. Eleven days later, cell suspensions of their thymus (90×10^6) or bone marrow (30×10^6) were injected along with normal bone marrow (30×10^6) or thymus (90×10^6) cells, respectively, into irradiated recipients. As before, mice were challenged with aggregated HGG on the day of transfer and again 10 days later, and their spleens were assayed for PFC 5 days after the second injection. The results of this experiment show that thymus cells obtained from each tolerogen treated group of mice were equally unresponsive (Fig. 2) although bone marrow cells obtained from the same tolerogen treated donors—that is, injected with 0.1, 0.5, or 2.5 mg of deaggregated HGG—were 9, 56, and 70 percent unresponsive, respectively. In summary, bone marrow cells require higher doses of tolerogen for the induction of unresponsiveness, they require a longer induction period, and in these cells the unresponsive state is of shorter duration than that in thymus cells.

The distinct kinetic pattern for the induction of unresponsiveness in each cell population probably reflects a difference in the mechanism by which the unresponsive state is achieved. It may be that qualitative differences exist to account for the temporal differences seen in the kinetic data. While thymus cells may be rendered unresponsive simply by the passive interaction of tolerogen with cell, bone marrow cells may require an active process, such as cell division, differentiation, or antibody formation, to reach a tolerogen-sensitive stage. On the other hand, the kinetic differences might be quantitative so that more tolerogen is needed to confer unresponsiveness in the bone marrow population compared to the thymus population because either more antigen reactive cells are involved or more tolerogen per antigen reactive cell is required. By the same logic, the maintenance of unresponsiveness in bone marrow would necessitate continued high doses of tolerogen. When the dose falls below the threshold of this effective concentration, the bone marrow would be expected to return to a normal state, whether this recovery involves the de-repression of unresponsive cells or the generation of new antigen reactive cells. The data on dose variation presented above (Fig. 2) and some recent findings by Mitchison (*10*) are compatible with such a hypothesis, as is the observation that there is a paucity of receptor sites on thymus as compared to bone marrow cells (*11*).

These data offer an insight into the cellular mechanisms operating both in the termination of induced immunological unresponsiveness (*12*) and in experimental autoimmunity (the termination of natural unresponsiveness) in rabbits (*13*). Each of these unresponsive states could be terminated by injection with either cross-reacting antigens or altered tolerated antigen, indicating that these tolerant animals still had the capacity to produce antibody specific for the tolerated antigen but lacked the ability to recognize the antigen as an immunogenic molecule. The observation would be explained by the involvement of two cell types: (i) the thymus cell which recognizes the immunogenic cross-reacting molecule and (ii) the bone marrow cell which is responsive to the tolerated antigen and which produces antibody. In view of

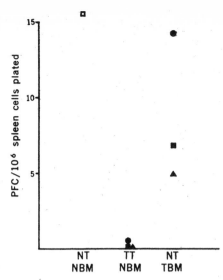

Fig. 2. The effect of tolerogen dose on the unresponsiveness of thymus and bone marrow cells. Each point represents the arithmetic mean of the individual response (PFC) obtained in six mice: NT, NBM; thymus, bone marrow obtained from normal donors. TT, TBM; thymus, bone marrow obtained from tolerogen injected donors which received 0.1 mg ●, 0.5 mg ■, or 2.5 mg ▲ of deaggregated human gamma globulin. ☐ Both thymus and bone marrow cells were from normal donors.

the results from the studies presented above, rabbits in which unresponsiveness to a heterologous serum protein had been induced at birth would be expected to contain at the time of testing (90 days after treatment with tolerogen) thymus cells that were unresponsive but bone marrow cells that had recovered from the unresponsive state. However, since thymus cells containing receptors for cross-reacting antigens would be present in the unresponsive animals, interaction could occur between these thymus cells (via determinants specific to the terminating antigen) and the bone marrow cells (via determinants cross-reacting with the tolerated antigen) resulting in a normal response to such cross-reacting determinants. The stimulation of bone marrow cells in this manner would result in their proliferation, differentiation, and subsequent synthesis of antibody with a specificity analogous to the receptor sites on the cells. It is known that in the case of natural unresponsiveness to thyroglobulin there are low levels of circulating antigen (*14*) which would, in a situation analogous to the dose response described above, be sufficient to render the thymus cells, and thus the entire animal, unresponsive to thyroglobulin but not sufficient to render bone marrow cells unresponsive.

In accord with this hypothesis, cross-reacting or chemically altered thyroglobulins could be recognized as immunogenic by thymus cells and could stimulate the bone marrow cells to produce antibody specific for native thyroglobulin. These observations are compatible with the suggestion that thymus and bone marrow cells each may react with different determinants on an antigen (*15*).

JACQUES M. CHILLER
GAIL S. HABICHT
WILLIAM O. WEIGLE
Scripps Clinic and Research Foundation, 476 Prospect Street, La Jolla, California 92037

References and Notes

1. N. A. Mitchison, *Immunology* **15**, 531 (1968); E. S. Golub and W. O. Weigle, *J. Immunol.* **99**, 624 (1967); P. Matangkasombut and C. V. Seastone, *ibid.* **100**, 845 (1968); J. M. Chiller and W. O. Weigle, *ibid.*, in press.
2. H. N. Claman, E. A. Chaperon, R. F. Triplett, *Proc. Soc. Exp. Biol. Med.* **122**, 1167 (1967); J. F. A. P. Miller and G. F. Mitchell, *Transplant. Rev.* **1**, 3 (1969); R. B. Taylor, *ibid.*, p. 114; G. S. Habicht, J. M. Chiller, W. O. Weigle, in *Developmental Aspects of Antibody Formation and Structure*, J. Sterzl and I. Riha, Eds. (Academia, Prague, 1970), vol. 2, p. 893.
3. J. M. Chiller, G. S. Habicht, W. O. Weigle, *Proc. Nat. Acad. Sci. U.S.* **65**, 551 (1970). In this publication the possibility that tolerogen carry-over could induce unresponsiveness in transferred normal cells was judged unlikely by the inability to detect antigen on thymus or bone marrow cells from tolerogen treated mice. Subsequent experimental data substantiates this conclusion. Irradiated recipients of normal thymus, normal bone marrow, and either bone marrow, thymus, or spleen cells obtained from animals previously treated with tolerogen were capable of responding to aggregated HGG.
4. Purchased from the Jackson Laboratories, Bar Harbor, Maine.
5. Human γ-globulin (Cohn fraction II) was obtained by courtesy of the American Red Cross and further purified by chromatography on DEAE-cellulose at 0.01M phosphate buffer, pH 8.0. The desired fractions were pooled, concentrated, and diluted to a concentration of 30 mg/ml with 0.15M NaCl. The protein was then deaggregated by ultracentrifugation, 100,000g for 150 minutes in a swinging bucket rotor. The HGG contained in the top one-third of each tube was removed and diluted with 0.15M NaCl to 2.5 mg/ml. Protein concentration was estimated spectrophotometrically by absorbance at 280 nm ($E_{1 cm}^{1\%} = 15$). Mice to be made unresponsive were injected intraperitoneally with 2.5 mg of the deaggregated material in a volume of 1 ml.
6. A solution of DEAE-purified HGG, 20 mg/ml in 0.01M sodium phosphate buffer, pH 8.0, was precipitated by heating at 63°C for 25 minutes, with occasional stirring, the mixture was then placed at 0°C for 12 hours. Enough 2.18M Na$_2$SO$_4$ was then added to reach a final concentration of 0.62M, and the mixture was incubated at 0°C for 30 minutes. The precipitate was washed three times with 0.62M Na$_2$SO$_4$, then dissolved in phosphate buffered saline, pH 8.0 (PBS), and dialyzed free of Na$_2$SO$_4$ against excess PBS. Protein nitrogen concentration of the aggregated solution was determined by micro-Kjeldahl analysis. The suspension was then diluted with 0.15M NaCl, pH 7.0, to contain 2 mg of protein per milliliter and stored frozen until used.
7. E. S. Golub, R. I. Mishell, W. O. Weigle, R. W. Dutton, *J. Immunol.* **100**, 133 (1968).
8. At each time interval, individual spleens of the following groups of mice were tested for PFC. Recipients of (i) tolerant thymus (TT

and normal bone marrow (NBM), (ii) normal thymus (NT) and tolerant bone marrow (TBM), and (iii) normal thymus (NT) and normal bone marrow (NBM). The size of each group at individual points varied from four to eight mice. The percentage of suppression in thymus cells or bone marrow cells from tolerogen treated donors was calculated at each point of the curve in the following manner:

$$\% \text{ suppression} = 100 - \left[\frac{\dfrac{\Sigma \text{ PFC in TT, NBM (or NT, TBM)}}{\text{No. mice treated}}}{\dfrac{\Sigma \text{ PFC in NT, NBM}}{\text{No. mice treated}}} \times 100 \right]$$

9. R. B. Taylor, *Nature* **220**, 661 (1968).
10. N. A. Mitchison, in 3rd Sigrid Juselius Symposium, Helsinki, Finland, in press.
11. P. Byrt and G. L. Ada, *Immunology* **17**, 503 (1969); J. H. Humphrey and H. U. Keller, in *Developmental Aspects of Antibody Formation and Structure*, J. Sterzl and I. Riha, Eds. (Academia, Prague, 1970); D. Naor and D. Sulitzeanu, *Israel J. Med. Sci.* **6**, 519 (1970).
12. W. O. Weigle, *Natural and Acquired Immunologic Unresponsiveness* (World, Cleveland, 1967); D. C. Benjamin and W. O. Weigle, *J. Exp. Med.* **132**, 66 (1970).
13. W. O. Weigle, *J. Exp. Med.* **122**, 1049 (1965); ——— and R. M. Nakamura, *J. Immunol.* **99**, 223 (1967).
14. P. M. Daniels, O. E. Pratt, I. M. Roitt, G. Torrigiani, *Immunology* **12**, 489 (1967).
15. N. A. Mitchison, *Cold Spring Harbor Symp. Quant. Biol.* **32**, 431 (1967); K. Rajewsky, V. Schirrmacher, S. Nase, N. F. Jerne, *J. Exp. Med.* **129**, 1131 (1969).
16. We thank S. Cossentine for technical help. This is publication No. 453 from the Department of Experimental Pathology, Scripps Clinic and Research Foundation. Supported by PHS grant AI 07007, American Cancer Society grant T-519, AEC contract AT(04-3)-410, Dernham Fellowship (J-166) of the California Division of the American Cancer Society to J.M.C., and PHS research career award 5-KG-GM-693 to W.O.W.

19 October 1970

EVIDENCE FOR THE CLONAL ABORTION THEORY OF
B-LYMPHOCYTE TOLERANCE*

BY G. J. V. NOSSAL AND BEVERLEY L. PIKE

(*From The Walter and Eliza Hall Institute of Medical Research, Melbourne, Victoria 3050, Australia*)

There is now general agreement (1–3) that different models of experimentally induced immunological tolerance illustrate a variety of cellular mechanisms. Many models relating to antibody synthesis show that the tolerance-inducing antigen exposure has only affected thymus-derived (T) lymphocytes, leaving the compartment of bone marrow-derived progenitors of antibody-forming cells (B lymphocytes) normally reactive (reviewed in 4). This may sometimes reflect the activity of certain T lymphocytes (suppressor T cells), which possess the capacity to suppress the reactivity to antigen of a normal population of B cells (5, 6). In other examples, tolerance has been achieved at the B-cell level. Such models usually involve either the administration of relatively high doses of soluble, aggregate-free and poorly immunogenic antigens (e.g. 7–9), or the use of highly polymeric antigens with multiple antigenic determinants in supra-immunogenic concentrations (10, 11) or in molecular arrangements (12–14) peculiarly favorable for tolerogenesis. In some of these models, there is evidence that B cells have been rendered nonreactive because their receptors have been blockaded by antigen (reviewed in 1), permitting reversibility of the tolerance on removal of the antigen. In other cases, the tolerance appears to be due to some action of antigen on the B cell leading to its permanent inactivation or even elimination (15).

The above cellular mechanisms pertain to the injection of a tolerogen into a mature animal already equipped with competent T and B lymphocytes in its secondary lymphoid organs. The acquisition of self-tolerance to many autologous proteins may be different, as the protein towards which tolerance must be achieved may be present before lymphocytes develop. This consideration opens the possibility that self-tolerance is achieved through the existence of a phase of lymphocyte differentiation during which the cell is particularly sensitive to tolerogenesis on contact with antigen (16–18). To distinguish this concept from that of clonal deletion of competent lymphocytes, we have termed it clonal abortion (19). The clonal abortion theory predicts that lymphocytes, at a particular stage of their differentiation at which some receptors for antigen have already appeared, can be permanently switched off or eliminated if they encounter antigen in appropriate concentration. As for every other phenomenon in cellular immunology, there must be an avidity parameter to the postulate, with cells possessing receptors of low affinity for a particular "self" antigen requiring higher concentrations for clonal abortion.

* This is publication no. 2057 from The Walter and Eliza Hall Institute of Medical Research. This work was supported by the National Health and Medical Research Council, Canberra (Australia), National Institutes of Health grant AI-O-3958, and was in part pursuant to contract no. NIH-NCI-G-7-3889 with the National Cancer Institutes of Health; and by the Volkswagen Foundation grant no. 112147.

Recent studies from our laboratory (20–22) have pin-pointed a stage in B-lymphocyte differentiation that seemed particularly favorable for a test of the postulate. Adult mouse bone marrow was found to be a site in which large numbers of membrane immunoglobulin (Ig)-positive B lymphocytes were generated, with rapid renewal kinetics (20). Furthermore, radioautographic double-labeling studies in which animals received [³H]thymidine in vivo and cells were exposed to [¹³¹I]antiglobulin in vitro showed that mouse bone marrow small lymphocytes exited from the proliferative process (presumably involving large and medium lymphocytes) which generated them, lacking detectable Ig receptors. They acquired these receptors in progressively increasing amounts through a nonmitotic maturation period lasting approximately 2 days (21). When bone marrow was placed in tissue culture, i.e. when escape of mature, Ig-positive lymphocytes by emigration was rendered impossible, the proportion of Ig-positive cells rose (21), as did the performance of the cell population in an adoptive immune system where B-cell reactivity to a T-cell-independent antigen was measured (22). These studies revealed mouse bone marrow as being a major factory for B lymphocytes, and suggested that cells "caught" during the nonmitotic phase of progressive receptor display might be suitable targets for clonal abortion tolerogenesis.

In order to mimic in vivo realities as closely as possible, it was desirable to use monomeric, aggregate-free antigens. For most of the work, we have used deaggregated dinitrophenyl human gamma globulin (DNP-HGG)[1] at the low conjugation ratio of one hapten molecule per protein molecule. This paper presents evidence to show that hapten-specific tolerance in bone marrow cell cultures can be achieved with remarkably low antigen concentrations, the kinetics paralleling that of B-cell neogenesis, and the phenomenon differing crucially from observations made with spleen cell cultures.

Materials and Methods

Animals. CBA/H/Wehi mice aged 8–11 wk, bred and maintained under strict pathogen-free (SPF) conditions behind a sterile barrier, were used for most experiments. Congenitally athymic (nu/nu) mice in their fifth back-cross generation to BALB/c mice were also used.

Antigens. The DNP hapten was conjugated onto various proteins, including polymerized Salmonella flagellin (POL), human gamma globulin (HGG; Commonwealth Serum Laboratories, Melbourne), and, for preliminary experiments, bovine serum albumin (BSA), and fowl gamma globulin (FGG) using 2,4-dinitro-benzene-sulphonic acid (DNBS) (Eastman Kodak Co., Rochester, N.Y.) by Eisen's method (23). The DNP-POL possessed 4.2 mole of DNP per mole of monomeric flagellin (mol weight 40,000) in the POL. Three batches of DNP-HGG were used, bearing 0.9–1.1 mole of DNP per mole of HGG (DNP-HGG). HGG used as a hapten-unsubstituted protein carrier control in tissue culture studies was treated in a manner indentical to that used during dinitrophenylation of HGG, except that DNBS was not added. DNP$_1$-HGG and HGG were deaggregated by centrifugation at 145,000 g for 90 min. After some preliminary experimentation with freshly-deaggregated tolerogen, it was found that the tolerogenic properties were not altered if deaggregated antigens were stored in

---

[1] *Abbreviations used in this paper:* BSA, bovine serum albumin; DNBS, 2,4-dinitro-benzene-sulphonic acid; FCS, fetal calf serum; FGG, fowl gamma globulin; HEM, Eagle's minimal essential medium buffered with Hepes; HGG, human gamma globulin; LPS, *Escherichia coli* lipopolysaccharide endotoxin; NIP, 4-hydroxy-3-iodo-5-nitrophenylacetic acid; PFC, plaque-forming cell; POL, polymerized flagellin from Salmonella strains SW1338 or SW871; SPF, specific pathogen-free.

diluted form at 4°C. The hapten 4-hydroxy-3-iodo-5-nitrophenylacetic acid (NIP) was prepared and conjugated to HGG or POL by standard methods (24, 25). The single batch of NIP-HGG used had a substitution rate of 1.5 mole NIP per mole HGG. The NIP-POL used was at 3.4 mole NIP per mole of monomeric flagellin in the POL. *Escherichia coli* lipopolysaccharide endotoxin (LPS) was obtained from Difco Laboratories, Detroit, Mich.

Preparation of Cell suspensions. The medium used for the preparation of cell suspensions was Eagle's minimal essential medium (MEM) (Catalogue F-15 Grand Island Biological Company, Grand Island, N.Y.) adjusted to mouse osmolarity and buffered with 20 mM Hepes (HEM) (Calbiochem, San Diego, Calif.). Bone marrow cell suspensions were prepared from femoral plugs, ejected with HEM using a syringe and a 21 gauge hypodermic needle, and the plugs were dissociated by repeated gentle aspiration through the needle. Spleen cell suspensions were prepared by gentle dissociation of the spleens with forceps into cold HEM, with 5% vol/vol fetal calf serum (FCS) using a stainless steel mesh sieve. Large cell clumps were removed from both cell suspensions by settling over 1 ml of FCS for 5 min, and the light debris was removed by washing the cells twice through FCS. Nucleated cell counts were performed in a hemocytometer and the viability determined by the eosin dye exclusion method.

Tissue Culture System. A Marbrook (26) system was used for cell culture. The culture medium used was MEM supplemented with 5% FCS and buffered with bicarbonate. 6×10^7 viable nucleated cells from either bone marrow or spleen were placed in an inner well (insert) in 4 ml of culture medium (containing the antigen, if any), and were separated from an outer reservoir of 50 ml of medium by a dialysis membrane. After tissue culture periods of 3 h to 5 days, the flask inserts were vortexed before cell harvest and the dialysis membranes were vigorously rinsed. The cells were washed once through FCS, resuspended in HEM, assessed for viability, and injected for measurement of adoptive immune performance.

Adoptive Immune Assay System. Cells were injected intravenously into lethally X-irradiated (850 rads) syngeneic SPF mice. A Phillips RT 250 X-ray machine was used (Philips-Electronic Instruments, Mount Vernon, N.Y.), operating at 250 KV, 15 ma, and a 0.2 mm Cu filter, (half value layer, HVL, = 0.8 mm Cu). Irradiation rate was 127 rads/min under conditions of maximum backscatter. Cells were injected within 3 h of irradiation, and, almost immediately after cell injection, the host mice received challenge antigen, namely an intraperitoneal injection of 5 μg DNP-POL, frequently mixed with 25 μg of NIP-POL. These challenge doses of the T-cell-independent antigens had previously been found to be the minimal doses giving a maximal plaque-forming cell (PFC) response by adoptively transferred cells. 6–10 (usually 9) days after cell transfer, host mice were killed and the spleen content of hapten-specific PFC determined. Anti-DNP and anti-NIP direct and enhanced PFC were determined by the Cunningham liquid monolayer plaque technique as previously described (22). As SPF mice were used, irradiation mortality in reconstituted mice was negligible. Nonreconstituted irradiated control mice died between 10 and 14 days, and were shown incapable of mounting any antihapten PFC response on antigenic challenge. Furthermore, background PFC numbers in irradiated, nonreconstituted mice were negligible. In a typical experiment, four–six cultures belonging to a treatment group were pooled and transferred to three–five irradiated mice. Most experiments were repeated two–three times.

Anti-θ Treatment of Mouse Bone Marrow Cells. This was as previously described (22), and conditions used were shown to be capable of completely abrogating "helper" or "suppressor" activity of authentic peripheral T-lymphocyte populations.

Results

Preliminary Experiments. In view of the complexities surrounding the adoptive immune response as a quantitative read-out system for the immune capacity of B lymphocytes (22, 27), preliminary experiments were performed to study the nature of the cell dose: PFC response relationship. Using transfers of normal, noncultured marrow, and with killing of host mice 9 days after transfer, the number of PFC rose linearly with cell dose from 2 to 6×10^7 viable cells transferred, after which an appreciably supra-linear increment occurred (see also 22). Similarly, the adoptive PFC response of marrow that had been cultured for 3 days showed an approximately linear relationship to cell dose from $5-40 \times 10^6$ viable cells transferred. Accordingly, 2×10^7 was chosen as the standard dose of viable cultured cells transferred. Preliminary work also showed the absence of a significant cross-reaction between the NIP and DNP haptens under the PFC-revealing conditions used. In a typical 3 day culture, the 6×10^7 input bone marrow cells yielded 3×10^7 recoverable nucleated cells of which 1.6 to 2×10^7 were viable.

Adoptively transferred bone marrow cells did not give rise to significant numbers of "enhanced" PFC at any day of killing between 6 and 10 days post-transfer. With spleen cell transfers, small numbers of enhanced plaques were occasionally found, but only when hapten-protein conjugates were present during the culture period. Accordingly, only the results for direct PFC are presented below.

The Effect of Tissue Culture on Adoptive Immune Performance. The effect of a period of tissue culture on the capacity of bone marrow or spleen cells to give an adoptive immune response, in the absence of any antigen (other than those present in FCS) added to the cultures, was studied. The results (Fig. 1) show that the performance of bone marrow cells rose considerably over the first 3 days of culture, whereas that of spleen cells fell off. Longer periods of culture signifi-

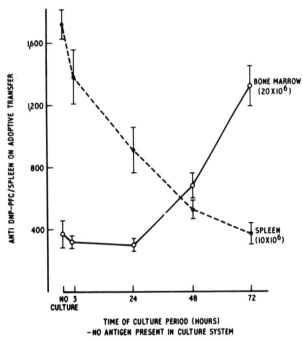

Fig. 1. The effect of tissue culture on adoptive immune performance. Adoptive hosts received 2×10^7 viable nucleated bone marrow cells (O—O) or 10^7 spleen cells (●---●) that had been cultured for periods of 3 h to 3 days, or that had not been cultured. They were immediately challenged with 5 μg of DNP-POL. 9 days later, the anti-DNP direct PFC content of host spleens was determined. Vertical bars here and elsewhere denote standard errors of the mean.

cantly decreased the potential of bone marrow cells. This confirmed previous experience (22).

The Effect of DNP₁-HGG on the Increased Adoptive Immune Performance of Cultured Bone Marrow Cells. Next, the effect of adding 4 µg/ml of DNP_1-HGG (2.5×10^{-8} M hapten) to bone marrow cultures was studied. The results (Fig. 2) show that this low concentration of hapten present in predominantly monomeric form completely abrogated the increment in adoptive immune capacity occurring on culture in the absence of antigen.

Influence of Day of Killing of Adoptive Host on DNP₁-HGG-induced Tolerance of Marrow Cells. To obviate the possibility that some subpopulation of B cells with abnormal maturation kinetics in the adoptive host might have been

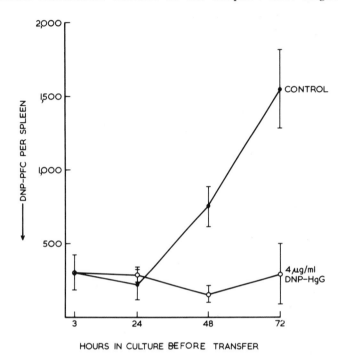

HOURS IN CULTURE BEFORE TRANSFER

FIG. 2. The effect of DNP-HGG on the adoptive immune response of bone marrow cultures. Cells were incubated with or without 4 µg/ml of DNP_1-HGG for 3 h to 3 days before adoptive transfer.

escaping the postulated clonal abortion process, the time of killing of host animals was varied. Bone marrow cultures were held for 3 days with 4 µg/ml DNP_1-HGG, or HGG, or no added antigen. The results (Fig. 3) establish three points. First, control mice killed 9 days after cell transfer yielded the highest PFC numbers, and this day was chosen as the standard for subsequent experiments. Secondly, cultures containing HGG gave PFC numbers equivalent to those containing no added antigen. Thirdly, no matter which day of killing was considered, DNP_1-HGG present in culture substantially reduced the adoptive anti-DNP response.

Antigen Concentration Required for Tolerance Induction in Bone Marrow Cultures. Tissue cultures of mouse bone marrow or spleen cells were held with

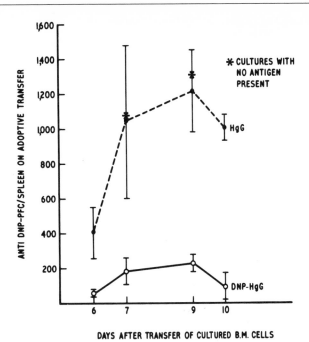

FIG. 3. The effect of varying the day of killing of adoptive hosts on the observed tolerogenesis of bone marrow cell cultures by DNP-HGG. All cultures were maintained for 3 days at 37°C with either no antigen (*) 4 μg/ml HGG (●---●), or 4 μg/ml DNP₁-HGG (O——O). Cells were then adoptively transferred, challenged with 5 μg DNP-POL, and hosts killed at intervals between 6 and 10 days after transfer.

various concentrations of DNP₁-HGG for 3 days of culture. The normalized results of a pool of experiments are presented in Fig. 4. It is shown that this antigen fails to cause tolerance in spleen cell cultures at any concentration up to 400 μg/ml (2.5×10^{-6} M). In other experiments, (not shown) spleen cells were held in culture for 1 day (a period during which better adoptive immune capacity is retained) with concentrations of DNP₁-HGG up to 400 μg/ml and again no tolerance was induced. Higher concentrations were somewhat toxic, as shown by reduction in irrelevant immune responses. In contrast, with bone marrow significant reductions in anti-DNP responsiveness was shown with concentrations as low as 0.4 μg DNP₁-HGG/ml (2.5×10^{-9} M), and marginal reduction may even have been present at 0.04 μg antigen/ml (2.5×10^{-10} M) though this did not reach statistical significance. These concentrations of DNP₁-HGG had no effect on the anti-NIP responsiveness of the cells.

Specificity of Tolerance Induction. To demonstrate the specificity of reduction in adoptive immune capacity, bone marrow cells were cultured for 3 days with either DNP₁-HGG or NIP₁.₅-HGG. The adoptive hosts were challenged with both DNP-POL and NIP-POL. The results (Table I) show that tolerance is specific, the response being reduced only towards the hapten present in culture. This result has been obtained repeatedly. The normal anti-NIP response of DNP-HGG-treated cultures has been included as a specificity control in many experiments, the anti-NIP results not always being shown.

Antigen is Required During most of the Culture Period for Optimal

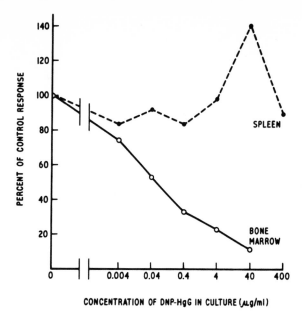

Fɪɢ. 4. Dose-response relationships in tolerogenesis of bone marrow cultures by DNP-HGG. Normalized, pooled results of several experiments. The control values for spleen (●---●) were lower in absolute PFC numbers than for bone marrow (O——O) (see Fig. 1).

Tᴀʙʟᴇ I

Specificity of Tolerance Induced in Cultures of Bone Marrow Cells

Antigen present in culture system	DNP-PFC/spleen	NIP-PFC/spleen
	Mean ± SE	*Mean ± SE*
4 μg/ml DNP$_1$-HGG	315 ± 72	1,133 ± 152
4 μg/ml NIP$_{1.5}$-HGG	1,163 ± 282	248 ± 120

Adoptive transfer recipients were challenged with both DNP-POL and NIP-POL and spleens were assayed for both anti-DNP and anti-DNP PFC.

Tolerogenesis. On the working hypothesis, cultured cells would be maturing through the critical tolerance-sensitive phase asynchronously throughout the 3-day period. Accordingly, if antigen were added only late in the culture, some maturing small lymphocytes would have passed the critical stage in an antigen-free environment, and would have gained anti-DNP reactivity. Thus we reversed the experimental design shown in Fig. 2, maintaining a series of cultures for the full 3 days, but pulsing in antigen at variable times. The results (Fig. 5) show that 4 μg/ml of DNP$_1$-HGG had no significant effect on spleen cell cultures. With bone marrow cell cultures, it caused a partial reduction in response if present for the last 24 h; and it caused the usual 80% reduction if present for the full period of culture. With neither cultured nor noncultured bone marrow cells exposed to 4 μg/ml of DNP$_1$-HGG immediately before transfer, or for very brief culture periods, was a reduction in anti-DNP responses achieved.

The Tolerogenesis is not Dependent on Suppressor T cells. CBA mouse bone

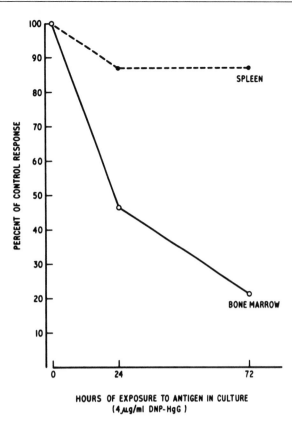

FIG. 5. The effect of addition of antigen at various times after initiation of culture on bone marrow cell tolerogenesis. Cultures of bone marrow to (O——O) or spleen (●----●) cells were maintained for 3 days. 4 µg/ml of DNP$_1$-HGG was added 24 h or 3 days before harvesting the cultures.

marrow only contains small numbers of θ-positive cells (20), but we wished to ensure that suppressor T cells were in no way involved in our system. Therefore, bone marrow cells were treated with anti-θ serum and complement before tissue culture in the presence of 4 µg/ml of DNP$_1$-HGG or HGG. The results (Table II) show that tolerogenesis was in no way affected by the anti-θ treatment.

To further document the irrelevance of T cells in this form of tolerance, cultures of bone marrow cells from partially inbred congenitally athymic mice were performed. The results yielded much lower PFC numbers than with CBA mice, but demonstrated the same order of tolerogenesis by 4 µg/ml of DNP$_1$-HGG.

The B-Cell Tolerance Induced in Bone Marrow Cultures is not "Infectious". To demonstrate whether tolerance might have been due to some active component, e.g. a suppressor B cell, an experiment was performed in which normal or tolerized bone marrow cell cultures were injected into lethally irradiated hosts either alone, or as a mixture. The results (Table III) show that the numbers of PFC appearing in adoptive recipients of cell mixtures did not differ significantly from the expected sums of PFC numbers in recipients of singly-treated cell cultures.

TABLE II

Tolerance Induction in Cultures of Anti-θ Treated Bone Marrow Cells

Antigen present in culture system	Anti-DNP-PFC/spleen	Anti-NIP-PFC/spleen
	Mean ± SE	*Mean ± SE*
4 μg/ml HGG	1,230 ± 95	1,103 ± 145
4 μg/ml DNP₁-HGG	245 ± 124	1,155 ± 110

TABLE III

Adoptive Immune Response of Mixtures of Tolerized and Nontolerized Bone Marrow Cultures

Reconstituting cell population	Anti-DNP-PFC/ spleen	Anti-NIP-PFC/ spleen
	Mean ± SE	*Mean ± SE*
20 × 10⁶ HGG treated	589 ± 71	1,185 ± 120
20 × 10⁶ DNP₁-HGG treated	250 ± 59	1,052 ± 108
20 × 10⁶ HGG treated plus 20 × 10⁶ DNP₁-HGG treated	613 ± 93	2,250 ± 160

Polyclonal B-cell Activators (28) do not Abolish Clonal Abortion. Part of the clonal abortion hypothesis is the concept that the lymphocyte matures from a stage during which it can only be tolerized on contact with antigen to a stage where it can be either immunized or tolerized depending on concentration, molecular form of antigen, and other factors such as co-mitogens. It has been shown (15, 29) that the tolerogenesis induced in mature B lymphocytes by monomeric antigens can be prevented by the addition of adjuvant substances with the capacity for polyclonal B-cell activation (28). Failure of such abrogation of tolerance in the present system would strengthen the probability that immature B cells were reacting in a unique way. Therefore, we investigated the effects of LPS added to bone marrow cultures undergoing tolerogenesis. Preliminary experiments were performed to show that the LPS batch used performed as would be expected in inducing appearance of PFC in antigen-free spleen cell cultures and in causing B-cell blastogenesis. When added to bone marrow cell cultures containing antigen (Table IV), 10 μg/ml of LPS did not diminish tolerogenesis. With 50 μg/ml LPS, the adoptive immune performance of tolerized cells increased but not to a statistically significant extent. Basically similar results were obtained when POL was used as a B-cell activator.

Discussion

The key findings of this study have been as follows: (*a*) Remarkably low molar concentrations, (c. 10^{-9} M) of an essentially monomeric antigen have induced specific immunological nonreactivity in B cells of bone marrow cell but not of spleen cell cultures. (*b*) The tolerogenesis follows a kinetics similar to that of

TABLE IV

Effect of LPS on Tolerance Induction of Bone Marrow Cells in Culture

Treatment in culture	Anti-DNP-PFC/spleen
	Mean ± SE
4 µg/ml HGG	780 ± 120
4 µg/ml DNP$_1$-HGG	167 ± 37
4 µg/ml DNP$_1$-HGG + 10 µg/ml LPS	132 ± 54
4 µg/ml DNP$_1$-HGG + 50 µg/ml LPS	270 ± 84

renewal of Ig-positive small lymphocytes in marrow. Full effects are achieved only if antigen is continuously present throughout culture. (*c*) The in vitro treatment always resulted in a residual immune competence in the cultured cells. In fact, the observed effect of antigen was really to prevent the emergence of new immune competence rather than to abrogate preexisting competence (Fig. 2). (*d*) The nonreactivity, tested through a T-cell-independent antigen challenge in a carefully calibrated adoptive immune model, could be induced as well in the absence of T lymphocytes as in their presence, and could not be transmitted to a population of untreated, cultured bone marrow cells. Thus, active suppression of one population of lymphocytes by another does not appear to be at work.

This series of results is entirely consistent with the predictions of the clonal abortion theory of B-cell tolerogenesis. The bone marrow lymphocyte population is seen as consisting of cells at diverse maturational stages. The large and medium lymphocytes in rapid mitotic cycle, as well as the nondividing small lymphocytes which are their immediate progeny, lack detectable Ig receptors (20, 21) and therefore cannot react with antigen. Over the subsequent 2 days or so, progressively increasing amounts of Ig receptors appear on the maturing small lymphocyte's surface (21), and we postulate that, for any given lymphocyte, all receptors are of identical antibody specificity. We suggest that any contact with an antigen capable of reacting with such receptors above a certain threshold of avidity of binding results in death or permanent functional suppression of that cell. This phase of the cell's maturation, which may last around 2 days, we refer to as the tolerance-sensitive, or obligatorily paralyzable phase. At its completion, the cell emerges as a fully mature, immunologically competent B lymphocyte with the density of surface Ig receptors characteristic of B cells in secondary lymphoid organs. At any given time, bone marrow samples will contain a small proportion of such mature B cells. Ex hypothesi, these show no special sensitivity to tolerogenesis and could account for the residuum of irreducible immune competence present in DNP$_1$-HGG-treated cultures. In culture, marrow cells appear to mature from Ig-negative to Ig-positive at approximately the same rate as in vivo (21), and their immune competence gradually builds up (22). However, in the presence of antigen, specifically reactive lymphocytes may be aborted as their receptors appear. Thus, while irrelevant immune competence increases normally, reactivity to the particular antigen present in culture does not increase.

If clonal abortion is a major physiological mechanism of tolerogenesis, this in no way negates the possibility that B-cell tolerance could also be achieved by other means. For example, recent studies from our own laboratory (29) have shown that in cultures of spleen cells from congenitally athymic mice, oligomeric antigens such as FGG or DNP$_{4.5}$-HGG can cause tolerance, but in 1,000- to 10,000-fold higher threshold concentrations. Highly polymeric antigens can also functionally silence mature B cells (11–14). It may be that for many antigens present in extracellular fluids in concentrations above 10^{-9} M, clonal abortion at both T and B cell levels is the chief physiological guarantee against self-reactivity, and antigen-antibody complex-mediated tolerogenesis (12) may act as a useful "fail-safe" device. Tolerogenesis of mature B cells by antigens such as POL may represent an unphysiological counterpart of the latter phenomenon.

If clonal abortion of B cells indeed occurs in vivo, it is pertinent to ask why so many of the models of tolerance induced by the repeated injection of small amounts of antigen into intact animals have recently been shown to be exclusively at the T-cell level. We believe that, in some cases, such experimental protocols allow a degree of immunization to take place (especially amongst B cells of low avidity for the tolerogen) because of unavoidable presence of some aggregates or of contaminating endotoxinlike materials in proteins. The suppressor mechanisms documented in such experiments, while interesting and valid, may reveal more about feedback mechanisms in immune responses than about self-recognition.

The low concentrations of antigen shown capable of tolerogenesis of bone marrow cultures are striking when it is considered that our assay system measures the behavior of B cells whose progeny secrete relatively low-affinity IgM antibody. Our read-out system as presently constituted does not specifically highlight tolerogenesis of the subset of anti-DNP bone marrow B cells that might, with appropriate T-cell help and challenge, have matured into high affinity IgG-producing anti-DNP PFC. It is quite possible, in line with the speculations of Mitchell (30), that such cells would have been tolerized with even lower concentrations of antigen. As it is, we have documented B-cell tolerance with antigen concentrations usually regarded as more typical for T-cell tolerogenesis (7, 8). Experiments with comparable doses of tolerogens injected into living animals would have failed to induce B-cell tolerance because of the long-lived nature of many peripheral B cells. Any tolerogenesis of newly emerging B cells would have been obscured by the residual normal reactivity of the older B cells, not tolerized by that particular antigen dosage level. The present postulate explains why tolerance may sometimes be more readily induced in newborn animals, where the existing pool of mature lymphocytes is small; and why tolerance may sometimes be maintained (once established) by lower antigen doses than were needed to induce it.

We have not investigated the molecular mechanisms which may be involved in clonal abortion. The theory needs ad hoc explanations for the special sensitivity of immature cells to contact with oligomeric antigen; for example, if tolerogenesis were due to cyclic AMP synthesis, the immature B cell might display a much lower inhibitory threshold. It is worthwhile recording that the nature of the protein carrier in the induction of hapten-specific B-cell tolerance is significant. In vivo, DNP coupled to albumins is not as effective as coupled to globulins (9), and in the tolerization of spleen cell cultures with high concentrations of DNP-proteins (29), the same was true, although the B-cell Fc receptor was probably not involved as $F(ab)_2$ fragments were efficient carriers for hapten tolerogens. In the present bone marrow culture system, the role of the carrier has not been extensively investigated, but preliminary experiments indicate that FGG is as effective, but BSA much less effective, than HGG. A simple methodology[2] for the preparation of functionally enriched populations of DNP- or NIP-reactive spleen cells has recently been developed in our laboratory. If this can be successfully applied to bone marrow lymphocytes, new approaches to the molecular biology of tolerance induction might become possible.

Summary

This paper deals with the behavior of adult mouse bone marrow cells placed in tissue culture with or without antigen, and subsequently assessed for immune competence after adoptive transfer into lethally X-irradiated, syngeneic hosts. Attention was focussed on B lymphocytes through using hapten human gamma globulin (HGG) preparations as putative tolerogens in tissue culture, the T-cell-independent antigens DNP-POL and NIP-POL as challenge injections in adoptive hosts, and numbers of hapten-specific PFC in host spleens for the quantitation of immune competence.

It was found that the capacity of bone marrow cells to mount an adoptive immune response rose by a factor of about fivefold over 3 days in tissue culture. This rise was completely abolished by the presence in the culture of hapten-HGG conjugates with about one mole of hapten per carrier molecule. The prevention of the emergence of immune competence amongst maturing B cells was termed clonal abortion tolerogenesis. Dose-response studies showed the lowest effective antigen concentration to be between 2.5×10^{-10} and 2.5×10^{-9} M, and a standard concentration of 2.5×10^{-8} M was chosen as producing near maximal effects.

The tolerance was antigen-specific and time-dependent, being maximal only when antigen was present continuously as the cultured cells were maturing. It did not depend on the presence of T lymphocytes in marrow, and was not of an "infectious" type. In contrast to tolerogenesis of mature B lymphocytes by high antigen concentrations, it could not be abolished by lipopolysaccharide. We speculate that clonal abortion may be a tolerance mechanism of great physiological significance for self-recognition, and discuss the results in the framework of other recent tolerance models, including those involving receptor blockade and suppressor T cells.

We are grateful for the excellent technical assistance of Ms. Kathy Davern, Lucy Strzelecki, and Susan Wickes.

Received for publication 2 December 1974.

References

1. Nossal, G. J. V. 1974. Principles of immunological tolerance and immunocyte receptor blockade. *Adv. Cancer Res.* **20**:93.
2. Katz, D. H. and B. Benacerraf, Editors. 1974. Immunological tolerance: mechanisms and potential therapeutic applications. Academic Press, Inc., New York.
3. Howard, J. G. 1972. Cellular events in the induction and loss of tolerance to pneumococcal polysaccharides. *Transplant. Rev.* **8**:50.

---

[2] Haas, W., and J. Layton. 1975. Separation of antigen-specific lymphocytes. I. Enrichment of antigen-binding cells. *J. Exp. Med.* **141**: in press.

4. Weigle, W. O. 1973. Immunological unresponsiveness. *Adv. Immunol.* **16**:61.

5. Gershon, R. K., and K. Kondo. 1971. Infectious immunological tolerance. *Immunology.* **18**:723.

6. Basten, A., J. F. A. P. Miller, J. Sprent, and C. Cheers. 1974. Cell-to-cell interaction in the immune response. X. T-cell-dependent suppression in tolerant mice. *J. Exp. Med.* **140**:199.

7. Weigle, W. O., J. M. Chiller, and G. S. Habicht. 1972. Effect of immunological unresponsiveness on different cell populations. *Transplant. Rev.* **8**:3.

8. Mitchison, N. A. 1971. Cell Interactions and Receptor Antibodies in Immune Responses. O. Mäkelä, A. Cross, and T. U. Kosunen, editors. Academic Press, Inc., New York. 249.

9. Golan, D. T., and Y. Borel. 1971. Nonantigenicity and immunologic tolerance: the role of the carrier in the induction of tolerance to the hapten. *J. Exp. Med.* **134**:1046.

10. Howard, J. G., H. Zola, G. H. Christie, and B. M. Courtenay. 1971. Studies on immunological paralysis. *Immunology* **21**:535.

11. Diener, E., and W. D. Armstrong. 1969. Immunological tolerance in vitro: kinetic studies at the cellular level. *J. Exp. Med.* **126**:591.

12. Feldmann, M., and E. Diener. 1971. Antibody-mediated suppression of the immune response in vitro. II. Low zone tolerance in vitro. *Immunology.* **21**:387.

13. Feldmann, M. 1972. Induction of immunity and tolerance in vitro by hapten-protein conjugates. *J. Exp. Med.* **135**:735.

14. Katz, D. H., T. Hamaoka, and B. Benacerraf. 1972. Immunological tolerance in bone marrow-derived lymphocytes. I. Evidence for an intracellular mechanism of inactivation of hapten-specific precursors of antibody-forming cells. *J. Exp. Med.* **136**:1404.

15. Louis, J., J. M. Chiller, and W. O. Weigle. 1973. Fate of antigen-binding cells in unresponsive and immune mice. *J. Exp. Med.* **137**:461.

16. Burnet, F. M. 1959. The clonal selection theory of acquired immunity. Cambridge University Press, New York.

17. Nossal, G. J. V. 1958. The induction of immunological tolerance in rats to foreign erythrocytes. *Aust. J. Exp. Biol. Med. Sci.* **36**:235.

18. Lederberg, J. 1959. Genes and antibodies: (Do antigens bear instructions for antibody specificity or do they select cell lines that arise by mutation?) *Science (Wash. D.C.).* **129**:1649.

19. Nossal, G. J. V., and B. L. Pike. 1975. New concepts in immunological tolerance. Immunological aspects of neoplasia. Proc. 26th Ann. Symp. Fund. Cancer Res. M.D. Anderson Hospital. In press.

20. Osmond, D. G. and G. J. V. Nossal. 1974. Differentiation of lymphocytes in mouse bone marrow: I. Quantitative radioautographic studies of antiglobulin binding by lymphocytes in bone marrow and lymphoid tissues. *Cell. Immunol.* **13**:117.

21. Osmond, D. G. and G. J. V. Nossal. 1974. Differentiation of lymphocytes in mouse bone marrow. II. Kinetics of maturation and renewal of antiglobulin-binding cells studied by double labeling. *Cell. Immunol.* **13**:132.

22. Stocker, J. W., D. G. Osmond, and G. J. V. Nossal. 1974. Differentiation of lymphocytes in the mouse bone marrow. III. The adoptive response of bons marrow cells to a thymus cell-independent antigen. *Immunology.* **27**:795.

23. Eisen, H. N. 1964. Methods in Medical Research. Year Book Medical Publishers, Inc., Chicago, Ill. **10**:94.

24. Brownstone, A., N. A. Mitchison and R. Pitt-Rivers. 1966. Chemical and serological studies with an iodine-containing synthetic immunological determinant 4-hydroxy-3-iodo-5-nitrophenylacetic acid (NIP) and related compounds. *Immunology.* **10**:465.

25. Schlegel, R. A. 1974. Antigen-initiated B lymphocyte differentiation. *Aust. J. Exp. Biol. Med. Sci.* **52**:455.

26. Marbrook, J. 1967. Primary immune response in cultures of spleen cells. *Lancet.* **2**:1279.

27. Stocker, J. W., and G. J. V. Nossal. 1975. Induction of B lymphocyte tolerance by an oligovalent antigen. I. Influence of the read-out system. *Cell. Immunol.* In press.

28. Coutinho, A., and G. Möller. 1975. Thymus-independent B cell induction and paralysis. *Adv. Immunol.* In press.

29. Schrader, J. W. 1974. The induction of immunological tolerance to a thymus-dependent antigen in the absence of thymus-derived cells. *J. Exp. Med.* **139**:1303.

30. Mitchell, G. F. 1974. T cell modification of B cell responses to antigen in mice. *Contemp. Top. Immunobiol.* **3**:97.

PERSISTENCE OF ANTIGEN-BINDING CELLS WITH SURFACE TOLEROGEN: ISOLOGOUS *VERSUS* HETEROLOGOUS IMMUNOGLOBULIN CARRIERS[1]

M. VENKATARAMAN, MARLENE ALDO-BENSON,[2] YVES BOREL, AND DAVID W. SCOTT[3]

From the Division of Immunology, Department of Microbiology and Immunology, Duke University Medical Center, Durham, North Carolina, 27710 and Division of Immunology, Department of Pediatrics, The Children's Hospital Medical Center, Harvard Medical School, Boston, Massachusetts, 02115

Adult mice injected intravenously with fluorescein isothiocyanate (FL) conjugated to an isologous carrier (mouse IgG2a) (FL-MGG) or to a heterologous carrier (sheep γ-globulins) (FL-SGG) are rendered specifically unresponsive to the FL-hapten. Tolerance induced by either tolerogen persists upon adoptive transfer. Since cells with tolerogen on their surface are detectable by immunofluorescence in both experimental situations, we compared the persistence of such cells in the spleens of mice injected with either tolerogen or with an immunogen (FL-KLH). FL-binding cells were seen during the first 5 days after the injection of FL-SGG, but were undetectable by day 7 and thereafter. In contrast, fluorescent cells were detectable up to 20 to 30 days in some animals after FL-MGG, although the number decreased after day 11. FL-binding cells decreased significantly after day 1 in FL-KLH-injected mice. When the number of FL-binding cells in tolerant animals had reached baseline levels, the animals were *reinjected* with their respective tolerogen. Normal or slightly increased numbers of FL-binding cells were found in the spleens of FL-SGG injected mice, whereas decreased numbers were observed in the spleens of FL-MGG injected mice. This suggests that in tolerance induced by heterologous carrier, specific lymphocytes possess free receptors but cannot respond. On the other hand, with the isologous carrier-induced tolerance, free receptors are either not available (modulated?) or blocked with undetectable amounts of tolerogen. Whether this difference in the kinetics represents a difference in the mechanism of tolerance is unknown.

Specific tolerance to haptens may be induced with the hapten bound to either isologous or heterologous immunoglobulin carriers (1–3). When mice are made tolerant by DNP-isologous IgG, the hapten DNP persists on the surface receptors of

splenic lymphocytes (4) and the presence of this hapten-bearing cell directly correlates with tolerance both *in vitro* (5) and *in vivo* (6). In rats made tolerant to fluorescein (FL) by injection of that hapten bound to sheep γ-globulin (FL-SGG), the FL has been shown to persist on lymphocytes up to 7 days after injection (2). Although it appears that these hapten-bearing cells play a role in the mechanism of tolerance induction by both isologous and heterologous immunoglobulin IgG carriers, it is possible that the function and fate of such cells may be different. The present study compares the behavior of antigen-binding cells (ABC) and the hapten-bearing cells in hapten specific tolerance by using both isologous mouse IgG2a (MGG) and heterologous SGG carriers. In both forms of tolerance the hapten persists on the surface of specific lymphocytes whereas the antigen FL-keyhole limpet hemocyanin (KLH) does not. Although both carriers were equally effective at inducting tolerance which could be adoptively transferred with spleen cells, the FL-MGG persists on splenic lymphocytes for 10 to 20 days whereas FL-SGG is present for only 5 days. In addition, the number of cells with free anti-FL receptors after hapten-bearing cells were no longer detectable was significantly less in FL-MGG-tolerant mice than in FL-SGG-injected animals. The implications of these findings in terms of possible differences in tolerance mechanisms are discussed.

MATERIALS AND METHODS

Animals. Six-week-old male (C57BL/6 × DBA/2) F_1(B6D2F$_1$) or 6- to 8-week-old male (C3H/He × DBA/2) F_1(c3D2F$_1$) mice (Jackson Laboratories, Bar Harbor, Maine) were used in all experiments.

Preparation of carrier proteins. SGG (Miles-Pentex, Fr II, Kankakee, Ill.) was further purified with sodium sulfate as previously described (2) to remove aggregated material. Murine IgG2a was obtained by starch block electrophoresis from serum of BALB/c mice bearing the tumor RPC5, as previously described (7). Keyhole limpet hemocyanin (KLH) (Pacific Biomarine Lab. Venice, Calif.) was prepared according to the method of Campbell *et al.* (8).

Fluorescein conjugates. FL-SGG conjugates were prepared as described elsewhere (2). After conjugation, the FL to protein ratio and protein concentration were determined spectrophotometrically at 276 and 493 nM by using the nomograph adopted by Goldman (9). All FL-SGG preparations contained six to seven FL groups per molecule of protein. To remove any xeno-antibody to mouse tissues, FL-SGG preparations were routinely absorbed with mouse erythrocytes, thymocytes, and spleen cells. FL-SGG preparations were ultracentrifuged for 60 min at 100,000 × G after which the upper one-third of the

Received for publication May 2, 1977.

Accepted for publication June 16, 1977.

The costs of publication of this article were defrayed in part by the payment of page charges. This article must therefore be hereby marked *advertisement* in accordance with 18 U.S.C. Section 1734 solely to indicate this fact.

[1] This work was supported by United States Public Health Service Grants AI-10716 (D.W.S.) and NOI-CB-4390, AI-11980 (Y.B.)

[2] Fellow of the Arthritis Foundation. Present address: Rheumatology Division, Dept. of Medicine, Indiana Univ. School of Medicine, 1100 W. Michigan St., Indianapolis, Indiana 48202.

[3] Research Career Development Awardee No. A1-00093.

[4] Abbreviations used in this paper: FL, fluorescein isothiocyanate; IgG, immunoglobulin; MGG, mouse IgG2a; SGG, sheep γ-globulins; ABC, antigen-binding cells.

supernatant was removed and used within 1 hr as tolerogen. FL_6-MGG and FL_6-KLH were prepared as above or by the method of Rinderknecht (10) by using 1 mg of fluorescein Celite per 1 mg carrier protein. Carrier protein concentration was determined by the Folin method and FL⁻ concentration was determined spectrophotometrically at 493 nM. FL to protein ratios were determined by using an extinction coefficient of 9.2×10^4 for fluorescein. For antigen-binding studies, FL-MGG and FG-SGG were ultracentrifuged as described above. In adoptive transfer experiments, only FL-SGG was centrifuged.

Adoptive transfer. Recipients for adoptive transfer experiments received 900 R whole body irradiation 3 to 5 hr before cell reconstitution, in a Gamma Cell 40 small animal irradiator (Atomic Energy Canada, Ltd. Ottawa, Canada). Irradiated mice received 40×10^6 syngeneic spleen cells intravenously from donors injected either with 1 mg FL-SGG (7 days) or 1 mg FL-MGG (10 days) before transfer. Preliminary experiments established that stable tolerance could be adoptively transferred as early as 48 hr after tolerogen injection. Control mice were reconstituted with 40×10^6 spleen cells from normal syngeneic donors. The recipients were then immunized with 400 μg of either FL_{6-17}-KLH, DNP_{87}-KLH, or TNP_{42}-KLH given intraperitoneally in CFA immediately after reconstitution.

The anti-FL response was assayed by a modification of the Jerne hemolytic PFC assay, by using FL-coated sheep RBC or goat RBC as target cells. FL-RBC were prepared by mixing 5 mg of FL in 10 ml 0.05 M carbonate buffer, pH 9.2, with 10 ml of 20% RBC in carbonate buffer for 30 to 40 min at room temperature (2, 11) and then washing with PBS. The PFC to TNP were assayed as previously described (1). Results are expressed as the geometric mean of PFC/spleen ($\pm$ S.E.).

Immunofluorescence and assay of ABC. At various times after intravenous injection of either 1 mg FL_7-SGG or 1 mg FL_6-MGG, the mice were sacrificed by CO_2 narcosis, and their spleen cells were suspended in cold PBS with 0.1% azide and 1 mg/ml bovine serum albumin (BSA) and washed three times with the same medium. Slides were made of the final cell suspensions and examined with a Leitz Ortholux II fluorescent microscope. More than 20,000 cells from individual mice were scanned per slide and the number of FL lymphocytes counted. Dead cells, cells in clumps, and FL cells other than lymphocytes were not counted. Determinations were made on three to four mice on separate occasions to insure reproducibility. Results are expressed as the mean ABC ($\pm$S.E.) for each group. The term FL-ABC refers to all *fluorescent* ABC and is not meant to connote hapten-specificity. Although we know that some FL-ABC are carrier-specific (2), FL-ABC will be called "hapten-binding cells" for simplicity.

Statistical analysis. Statistical analysis of both PFC assays and numbers of FL cells was done according to the Student's t-test.

RESULTS

Tolerance induction. Preliminary studies were done to confirm that FL_7-SGG and FL_6-MGG were indeed tolerogenic in mice. The results (Table I and II) demonstrate that both tolerogens produce a state of tolerance which is not reversed by adoptive transfer. This tolerance is hapten specific since recipients of tolerant cells responded normally to DNP or TNP.

Detection of FL-bearing cells in vivo. Having demonstrated the tolerogenicity of both FL-MGG and FL-SGG, we then compared the number, and kinetics of appearance of fluorescent lymphocytes in animals injected with these two tolero-

TABLE I

Adoptive transfer of mouse IgG2a carrier-determined tolerance to FL

No. of Animals	Donor Spleen Cells[a]	Immunogen	PFC(S.E.)/Spleen vs	
			FL	TNP
6	Tolerant	FL-KLH	27,386(3750)[b]	
6	Normal	FL-KLH	100,669(12812)[b]	
6	Tolerant	DNP-KLH		45231(18206)[c]
6	Normal	DNP-KLH		49657(1224)[c]

[a] 40×10^6 normal or tolerant (obtained from donor BDF_1 mice injected 10 days earlier with 1 mg FL_6 mouse IgG2a) were transferred to 900 R lethally irradiated syngeneic BDF_1 mice. Immediately after irradiation and reconstitution, mice were immunized i.p. with 400 μg of either FL_{17}-KLH or DNP_{87}-KLH in complete Freund's adjuvant (CFA). PFC assay with FL-SRBC or TNP-SRBC as targets was done 8 days later.
[b] p value < 0.001.
[c] Not significantly different.

TABLE II

Adoptive transfer of SGG carrier-determined tolerance to fluorescein

No. of Animals	Donor Spleen Cells[a]	Immunogen	PFC(S.E.)/Spleen vs	
			FL	TNP
7	Tolerant	FL-KLH	7225(1508)[b]	
6	Normal	FL-KLH	90,450(10406)[b]	
7	Tolerant	TNP-KLH		10,014(2512)[c]
6	Normal	TNP-KLH		8,140(1956)[c]

[a] 40×10^6 normal or tolerant (obtained from donor $C3D2F_1$ mice injected 7 days earlier with 1 mg FL_6-SGG) were transferred to 900 R lethally irradiated syngeneic $C3D2 F_1$ mice. Immediately after irradiation and reconstitution, mice were immunized i.p. with 400 μg of either FL_6-KLH or TNP_{42}-KLH in complete Freund's adjuvant (CFA). PFC assay with FL-GRBC or TNP-GRBC as targets was done 7 days later.
[b] p value < 0.001.
[c] Not significantly different.

gens. Animals injected with the antigen FL_6-KLH were also included as a control to insure that the presence of FL-bearing cells correlates only with tolerance. C3D2 F_1 mice injected with 1 mg of FL-MGG, FL-SGG, and FL-KLH were sacrificed at various times after injection and spleen cells inspected for FL on the surface as previously described (2). Positive cells had typical green membrane fluorescence and morphologically appeared to be lymphocytes. Figure 1 shows the number of FL-bearing cells (ABC) in the spleens taken from mice injected with FL-MGG, FL-SGG, or FL-KLH at various times before sacrifice. In all three groups of mice many FL cells were present within the 1st hour after injection. However, their persistence varied with the compounds injected. The mean number of fluorescent cells 1 day after injection of FL-MGG, FL-SGG, and FL-KLH was 142 $\pm$ 26, 192 $\pm$ 34, and 44 $\pm$ 9 per 10^5 spleen cells, respectively. In animals injected with FL-MGG, FL was detected on the lymphocytes up to 11 days after injection. In a few cases FL-bearing cells were detectable at 20 to 30 days after injection. However, in the FL-SGG injected animals, the FL-bearing cells were present at 5 days, but could not be detected at 7 days after injection. When animals were injected with the antigenic compound FL-KLH, the hhapten was detectable in a high percentage of cells 1 hr after injection, but the number of fluorescent cells decreased rapidly during the next 24 hr. The number of FL-ABC at 1 day in the spleens of mice injected with FL-KLH was significantly lower than the number of ABC in animals injected with either tolerogenic compound (p < 0.01). At 5 days postinjection, only one FL-KLH injected animal showed low numbers of FL-bearing cells. Thus all the three FL conjugates are bound by lymphocytes, but only the tolerogenic compounds (FL-MGG and FL-SGG) result in persistent hapten-bearing cells, with FL-MGG persisting longer than FL-SGG.

We then investigated whether any cells with free receptors for FL were detectable in tolerant mice at a time after hapten-bearing cells were no longer directly detectable. Mice which were injected with 1 mg FL-SGG 7 days before were reinjected with 1 mg FL-SGG, and their spleens assayed for fluorescent lymphocytes 20 hr later. Similarly, mice injected with FL-MGG were reinjected on day 20 and FL-ABC were assayed. The results (Table III) indicate that mice made tolerant with FL-MGG had fewer numbers of splenic lymphocytes with receptors free to bind the same tolerogen at day 20 (62 ± 10) than on day 1. However, mice made tolerant with FL-SGG had the same number of cells in their spleens capable of binding a

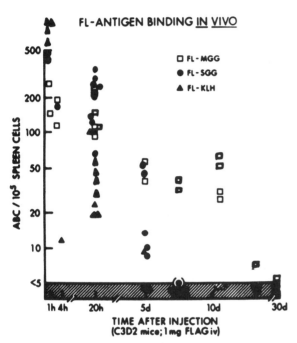

Figure 1. Detection of FL-ABC in C3D2 mouse spleens at various times after the injection of FL-mouse IgG(FL-MGG) (□), FL-sheep IgG(FL-SGG) (○), or FL-hemocyanin (FL-KLH) (▲). See *Materials and Methods* for details. FLAG = fluorescent antigen.

TABLE III

Number of FL- antigen-binding cells 1 day after tolerogen injection and reinjection

Tolerogen	ABC/10⁵ Day 1[a]	Day of Reinjection	ABC/10⁵ Day 1[a] after Reinjection
FL-SGG	192(34)[b]	7	247(17)[c]
FL-MGG	142(26)[b]	20	62(10)[c]

[a] Data are presented as the mean number (S.E.) of ABC per 10⁵ nucleated spleen cells from groups of 5 to 9 mice injected with 1 mg FL₆₋₇-SGG or 1 mg FL₅₋₇-MGG i.v. and examined 20 hr after injection or reinjection. Day 1 values are from normal mice injected with tolerogen at the same time that tolerant mice were reinjected.

[b] Data from Figure 1 for reference. Values for ABC with these tolerogens are not significantly different from each other.

[c] Number of FL-ABC are significantly lower in reinjected FL-MGG-tolerant mice compared either to reinjected FL-SGG tolerant mice (p < 0.001) or compared to FL-MGG-injected animals at day 1 (p < 0.05). Numbers of FL-ABC in FL-SGG-injected mice are not significantly different after injection *vs* reinjection (p > 0.05).

second dose of the same tolerogen (247 ± 17) as normal control mice. This is not simply a difference in timing since FL-SGG tolerant mice, which were reinjected on day 20, still possess normal numbers of ABC. (One day after injection 138 ± 24, *vs* reinjected after 20 days 140 ± 26) (M. Venkataraman, unpublished results).

DISCUSSION

The number and fate of specific ABC in tolerant animals should provide evidence concerning the mechanism of tolerance, but published studies in this area have led to contradictory results. For example, some workers have found decreased numbers of ABC in tolerant hosts (2, 3, 12) whereas others have found normal or even increased numbers (13, 14). Moreover, the finding of reduced numbers of ABC has been ascribed to blockade of cellular receptors by surface tolerogen (4, 12) in some cases rather than clonal deletion or receptor modulation. In this paper, a different approach is described. Fluoresceinated tolerogens were used to trace directly the persistence of antigen-bearing ABC during tolerance. Moreover, the ability of these tolerant cell populations to rebind further antigen *in vivo*, as previously reported (2) was studied. We were primarily interested in the differences between heterologous and isologous Ig carrier-determined tolerance to the FL hapten. The adoptive transfer of carrier-determined tolerance as well as the persistence of the hapten-bearing cells with both DNP-MGG- and FL-SGG-induced tolerance is well known (1-4). However, the comparison of hapten-bearing cells in both forms of tolerance required the demonstration of tolerance with FL-MGG in mice and that FL-bearing cells are detectable in animals tolerant to FL-MGG in a manner similar to that used to detect FL-SGG. The persistence of tolerance in adoptive transfer of both FL-MGG- and FL-SGG-tolerant cells demonstrates that this tolerance is not functionally different from that induced by DNP-IgG or from each other. As expected, FL is seen on lymphocytes of animals receiving both tolergenic and immunogenic forms of the hapten hours after injection. However, the number of cells with FL-KLH on their surface falls to low levels by day 1 and is virtually absent by day 5, whereas both tolerogenic forms of the hapten persist on the splenic lymphcoytes. This is an agreement with previous findings in DNP-IgG induced tolerance in mice (4) and FL-SGG-induced tolerance in rats (2). Although the rapid disappearance of an immunogenic molecule like KLH may be related to its large size and clearance by the reticuloendothelial system, it has been shown that nontolergenic haptenated IgG subclasses also do not persist on cells (16; see below).

Two major differences are noted in the behavior of ABC in tolerance with ultracentrifuged FL-MGG and FL-SGG. The FL-bearing cells in FL-MGG-injected mice persist for longer periods than those in FL-SGG-injected mice. This may reflect a longer half-life and more persistent tolerant state induced with isologous *vs* heterologous Ig (2, 4, 6). Consistent with this interpretation is the observation that, in rabbits, only FL-autologous IgG is tolerogenic whether FL-isologous IgG is not (15). However, the persistence in the circulation does not always correlate with the tolerogenicity of a compound (16).

A second difference between isologous and heterologous carriers is that the number of cells with free receptors for antigen returns to normal upon reinjection of mice made tolerant with FL-SGG. However, the number of cells with free receptors is reduced in FL-MGG-injected animals, when hapten-bearing cells are no longer detectable. This suggests a difference in behavior of hapten-binding cells in these two forms of tolerance. It is possible that in tolerance to FL-SGG the tolerogen is removed from the lymphocyte receptor after 5 days or the receptor-tolerogen complex is shed leaving it free to bind newly injected FL-SGG. This does not appear to be due to the presence of cytophilic antibodies in FL-SGG-tolerant mice because few anti-FL PFC are detectable in such animals (M. Venkataraman, unpublished results). Alternatively, some of the FL cells noted after reinjection of FL-SGG may represent SGG-binding cells. This cannot be excluded and is currently being tested by reinjecting FL-SGG-tolerant mice with FL-MGG. Since the disappearance of FL-MGG is not accompanied by an increase of ABC, it is unlikely that FL-MGG is removed from cell receptors. Therefore, either FL-MGG-binding cells are eliminated, FL remains on the receptor but is not longer detectable, or the receptors are modulated off and not resynthesised. The first possibility seems unlikely since reversal of tolerance to DNP-MGG by *in vitro* incubation results in a loss of hapten-bearing cells and return to normal numbers of cells with free DNP receptors (5). Also, DNP has been shown indirectly by autoradiography to remain on the cell surface for 6 weeks in tolerant mice (6). Thus it is possible that the FL-MGG is still on the lymphocyte surface, but the FL is no longer visible due to partial degradation of the molecule, or quenching of fluorescence or both. Some precedent exists for the modulation hypothesis, especially with immature B cells (17; Venkataraman and Scott, in preparation) although the exact mechanism is unknown. At present, it is impossible to distinguish between persistent tolerogen blockade and receptor modulation. Whether the differences in kinetics of ABC and hapten-bearing cells in tolerance with heterologous and isologous IgG carriers represents a difference in mechanisms of tolerance is unknown. This difference is also consistent with the view that tolerogenicity is directly related to receptor blockade. To resolve these issues, the demonstration that hapten-bearing cells mediate tolerance and the isolation of specific FL-ABC will be required. Methods are now available to isolate these specific FL-binding cells by use of anti-FL columns (18), FL-gelatin (19) or the fluorescence-activated cell sorter (20). Further studies of these hapten-bearing cells isolated from animals tolerized by these two carriers are in progress.

REFERENCES

1. Borel, Y. 1971. Induction of immunological tolerance by hapten (DNP) bound to nonimmunogenic protein carrier. Nature (New Biol) 230:180.
2. Scott, D. W. 1976. Cellular events in tolerance. V. Detection, isolation and fate of lymphoid cells which bind fluoresceinated antigen *in vivo*. Cell. Immunol. 22:311.
3. Katz, D. H., J. M. Davie, W. E. Paul, and B. Benacerraf. 1971. Carrier function in anti-hapten response: IV. Experimental conditions for the induction of anti-hapten anamnestic responses by non-immunogenic hapten polypeptide conjugates. J. Exp. Med. 134:201.
4. Aldo-Benson, M., and Y. Borel. 1974. Tolerant cell: Direct evidence for receptor blockade by tolerogen. J. Immunol. 112:1793.
5. Aldo-Benson, M., and Y. Borel. 1976. Loss of carrier-determined tolerance *in vitro* with loss of receptor blockade. J. Immunol. 116:223.
6. Aldo-Benson, M., and Y. Borel. 1977. Hapten bearing cell in carrier determined tolerance. Eur. J. Immunol. 7:175.
7. Borel, Y., D. T. Golan, L. Kilham, and H. Borel. 1976. Carrier determined tolerance with various subclasses of murine myeloma IgG. J. Immunol. 116:854.
8. Campbell, D. H., J. S. Garvey, N. E. Cremer, and D. H. Susdorf. 1963. Methods in Immunology. W. A. Benjamin, Inc., New York. P. 116.
9. Goldman, M. 1968. Fluorescent Antibody Methods. Academic Press, New York. P. 125.
10. Rinderknecht, H. 1962. Ultra-rapid fluorescent labelling of proteins. Nature 193:167.
11. Calderon, J., J. M. Kiely, J. L. Lefko, and E. R. Unanue. 1975. The modulation of lymphocyte functions by molecules secreted by macrophages. J. Exp. Med. 142:151.
12. Louis, J., J. M. Chiller, and W. O. Weigle. 1973. Fate of antigen binding cells in unresponsive and immune mice. J. Exp. Med. 137:461.
13. Möller, E., and O Sjöberg. 1972. Antigen binding cells in immune and tolerant animals. Transplant. Rev. 8:26.
14. Cooper, M. G., G. L. Ada, and R. W. Langman. 1972. The incidence of heomcyanin-binding cells in hemocyanin-tolerant rats. Cell. Immunol. 4:289.
15. Gollogly, J. R., R. E. Cathou, and Y. Borel. 1976. Carrier determined suppression of anti-fluorescein antibody in the rabbit. Proc. Soc. Exp. Med. 152:508.
16. Paley, R. S., S. Leskowitz, and Y. Borel. 1975. Effect of tolerance induction of the mode of attachment of the hapten to the carrier. J. Immunol. 115:1409.
17. Raff, M. C., J. J. T. Owen, M. D. Cooper, A. R. Lawton, M. Megson, and W. E. Gathings. 1975. Differences in susceptibility of mature and immature mouse B lymphocytes to anti-immunoglobulin-induced immunoglobulin suppression *in vitro*. J. Exp. Med. 142:1052.
18. Scott, D. W. 1976. Antifluorescein affinity columns. Isolation and immunocompetence of lymphocytes that bind fluoresceinated antigens *in vivo* or *in vitro*. J. Exp. Med. 144:69.
19. Haas, W., and J. E. Layton. 1975. Separation of antigen-specific lymphocytes. I. Enrichment of antigen-binding cells. J. Exp. Med. 141:1014.
20. Marx, J. L. 1975. Lasers in bio-medicine: Analyzing and sorting cells. Science. 188:821.

Receptor-mediated inactivation of early B lymphocytes

THE interaction of surface immunoglobulin (sIg), the antigen receptor of B lymphocytes, with ligands—whether antigen or anti-Ig antibodies— initiates a series of surface and cytoplasmic events which, depending on the nature of the ligand and the presence or absence of cooperative signals from other cells, has various outcomes (reviewed in ref. 1). Immediately on the binding of a polyvalent ligand to sIg, the complexes redistribute on the cell surface, a contractile event occurs, the complexes are endocytosed and shed from the membrane and eventually new receptors appear on the cell surface[1]. This *in vitro* cycle, in the absence of cooperative interactions, does not lead to B cell differentiation into secreting plasma cells, whereas it does so when induced in the proper conditions involving helper cells[1,2].

A crucial step in this ligand-induced cycle is the re-expression of new receptors on the cell surface, a process involving protein synthesis and requiring the activity of microtubules[1]. It is expected that an interruption at this step, leading to a lack of receptors on the cell surface, would be tantamount to an inactive B cell unable to respond to antigen. We report here the failure of B cells from young mice to re-express sIg after relatively brief exposure to anti-Ig antibodies. This failure denotes a peculiar sensitivity of early B cells to ligand–receptor interaction in the absence of cooperative effects and could readily explain several of the phenomena of unresponsiveness seen during the neonatal period. It should be noted that B cells from young mice (until about 2 weeks of age) have characteristics different from most adult B cells in that they lack C3 receptors[3,4], cap poorly[4] and bind more labelled anti-Ig antibodies[4].

The basic experiments consisted of exposing spleen lymphocytes from young (up to 2 weeks after birth) or adult (8–16 weeks) mice to rabbit anti-mouse Ig antibodies (RAMG) for 1 or 24 h, after which the cells were cultured for several days in the absence of RAMG. Surface Ig was detected by immunofluorescence. When RAMG bound to an adult B lymphocyte, sIg–anti-Ig complexes were eliminated within minutes, leaving the cell membrane cleared of sIg. Then, within a few hours of culture in the absence of anti-Ig, the cell restored its sIg, completing the process by 8–24 h later. The B lymphocyte from a young mouse spleen however, did not behave in this way. Depending on the time the cell was exposed to RAMG, the early B lymphocyte restored its sIg poorly or not at all. As Fig. 1 shows, within 1 d of a 1-h or a 24-h RAMG incubation (both of which completely cleared the cell of sIg), most of the original adult B lymphocytes restored their sIg. In contrast, only about half of the early B lymphocytes re-expressed sIg after a 1-h RAMG treatment, and almost none re-expressed it after 24 h of treatment with RAMG.

Since about half of the early B lymphocytes restored their sIg after a single cycle of clearance, it seems unlikely that the sIg initially on the cell surface represented cytophilic Ig. To strengthen this conclusion and to test whether early B lymphocytes can ever completely regenerate their sIg, cells of both ages were treated with the proteolytic enzyme Pronase at doses known to eliminate surface Ig, and then observed for the re-expression of sIg. Figure 2 shows that both early and adult B lymphocytes completely regenerated their sIg after with Pronase clearance. Note also for the early cells that a portion treated with Pronase after 24 h of incubation with RAMG did not re-express sIg. This shows that Pronase did not represent some stimulatory activity which might reverse the

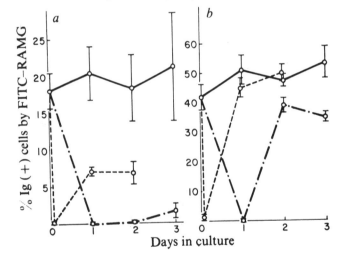

Fig. 1 Re-expression of sIg after treatment with RAMG (the RAMG used was a poly-specific rabbit anti-mouse Ig antibody previously shown to bind only to B lymphocytes in the mouse spleen). Young (6–9 d, *a*) or adult (2–4 months, *b*) C57BL/6 mouse spleens were teased into Hanks' balanced salt solution (HBSS) plus 1% HEPES plus 5% foetal calf serum and washed three times. They were then exposed at a concentration of 10^7 cells per ml to Mishell–Dutton medium[15] alone or containing RAMG (200 µg ml^{-1}) for 1 h or 24 h at 37 °C in a 5% CO_2 atmosphere. The dose of RAMG was one that totally cleared the cells of sIg. After the RAMG treatments, cells were washed five times, resuspended at 10^7 cells per ml in fresh medium and cultured for various times. For immunofluorescence the cells were washed once, pelleted, resuspended in 50 µl of medium and 50 µl of fluoresceinated RAMG (FITC-RAMG) (500 µg ml^{-1}, F:P ratio approximately 6:3). The cells were incubated for 30 min at 4 °C, washed twice, resuspended 10^7 per ml, and live cells were scored in a Leitz fluorescence microscope. ——, Control; – – –, 1-h RAMG; – . –, 24-h RAMG.

Table 1 Effect of RAMG treatment on LPS mitogenesis

Incubation (h)	Incubation medium	LPS in culture	Experiment 165 c.p.m./10⁶ cells	E−C	E/C	Experiment 167 c.p.m./10⁶ cells	E−C	E/C	Experiment 169 c.p.m./10⁶ cells	E−C	E/C	Averages E−C	Averages E/C
Young cells													
1	Medium	−	5,319± 97	15,343	3.88	7,605± 276	38,850	6.0	4,842± 407	47,759	10.86	33,964	6.95
1	Medium	+	20,662±1,047			46,455±1,648			52,601± 808				
1	RAMG	−	4,854± 138	933	1.19	6,770± 381	10,524	2.55	3,973± 338	13,433	4.38	8,297	2.71
1	RAMG	+	5,787± 671			17,294± 271			17,406± 813				
24	Medium	−	3,016± 362	7,724	3.56	5,339± 122	9,746	2.83	3,589± 189	14,595	5.07	10,688	3.82
24	Medium	+	10,739± 278			15,085± 245			18,184± 696				
24	RAMG	−	3,848± 157	−1,765	0.54	7,393± 314	−2,937	.60	4,910± 189	−250	.94	−1,650	.69
24	RAMG	+	2,083± 146			4,456± 195			3,660± 225				
Adult cells													
1	Medium	−	10,009± 654	61,930	7.19	10,979± 430	66,023	7.01	7,447± 220	67,243	10.03	66,065	8.08
1	Medium	+	71,939±2,033			77,002±1,393			74,690±1,180				
1	RAMG	−	3,092± 288	32,868	11.63	3,792± 155	35,847	10.45	5,016± 56	57,883	12.54	42,199	11.54
1	RAMG	+	35,960± 268			39,629± 561			62,899±2,185				
24	Medium	−	4,582± 117	55,723	13.16	6,972± 375	52,339	8.51	4,540± 300	57,827	13.74	55,296	11.80
24	Medium	+	60,305± 412			59,311± 321			62,367±1,027				
24	RAMG	−	3,362± 692	57,697	18.16	2,180± 54	35,786	17.42	1,909± 68	53,483	29.02	48,989	21.53
24	RAMG	+	61,059±3,344			37,966±1,660			55,392±1,448				

Young and adult C57BL/6 cells were incubated for 1 or 24 h in Mishell–Dutton medium containing RAMG (200 µg ml⁻¹) or no RAMG. They were then washed five times in HBSS plus HEPES plus FCS and resuspended to 10⁶ per ml in RPMI-1640 with penicillin and streptomycin and 5% FCS and 2 mM L-glutamine alone or with LPS (10 µg ml⁻¹). Each point was done in triplicate. After 24 h of culture in 5% CO_2, each tube was pulsed with 1 µCi in 50 µl of ³H-thymidine (2 Ci mmol⁻¹). The incorporation of the labelled thymidine into tricloroacetic acid-insoluble material was determined 24 h later. The results are expressed as the arithmetic mean of triplicate cultures ± s.e.m. E−C, Experimental mean—control mean; E/C, experimental mean/control mean.

effects of the 24-h treatment with RAMG. The main conclusion from the Pronase experiments is, therefore, that early B lymphocytes can re-express their sIg as completely and as rapidly as adult B lymphocytes but are specifically inactivated by interaction with RAMG.

The next question asked was whether these early B lymphocytes were inactivated or totally deleted from the cultures (that is, killed). To determine this, the spleen cells were examined after exposures to RAMG for the presence of another[5],

Fig. 2 Re-expression of surface Ig after treatment with Pronase. Young (6–8 d, _a_) and adult (2–4 months, _b_) C57BL/6 cells were treated with Pronase (2 mg ml⁻¹) (_Streptococcus griseus_ protease) and DNase II (20 µg ml⁻¹) for 1 h at 37 °C in HBSS plus HEPES, washed five times and then cultured and assayed as in Fig. 1. Treatment with Pronase was either immediately after cell collection or after 24-h culture in Mishell–Dutton medium containing RAMG (200 µg ml⁻¹) and three washes. For comparison, re-expression after incubations of 1 and 24 h in Mishell–Dutton medium containing RAMG (200 µg per ml) are also shown. ○--○, 1 h RAMG; ○—○, 24 h RAMG; ●--●, 1 h Pronase; ●—●, 24 h RAMG, 1 h Pronase.

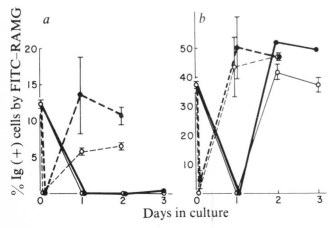

Days in culture

independent[6] B lymphocyte marker, the I region-associated antigens (Ia). As all panels of Fig. 3 show, the surface distribution of Ia was unaffected immediately after the clearance of sIg by RAMG in early and adult B lymphocytes. One day after incubation with RAMG, adult B lymphocytes still showed their Ia and had restored their sIg. In contrast, early B lymphocytes also retained their Ia but had restored their sIg poorly or not at all. We concluded, therefore, that these early B cells were still present in the culture and alive at this time but had failed to re-express their sIg. Two days after the end of a 24-h treatment with RAMG the situation seemed to be changing, however. The percentage of early Ia-positive cells was declining, implying that, as a result of the prolonged ligand–receptor interaction, the original early B lymphocytes were either dying or were changing other cellular characteristics.

In addition to surface markers, the functional property of lipopolysaccharide (LPS)-induced mitogenesis was also studied after receptor clearance. Table 1 shows that the LPS mitogenic response of adult B lymphocytes was affected minimally, if at all, after treatment with RAMG. In striking contrast, early B lymphocytes had a marked reduction in LPS-induced mitogenesis after the same procedures. Thus, the failure to re-express receptors after RAMG clearance was accompanied by the loss of responsiveness to the potent B cell mitogen LPS.

Although the mechanism of this receptor-mediated inactivation of early B lymphocytes is not clear, certain conclusions can be drawn.

First, the inactivation seems to be a direct consequence of the interaction of RAMG with sIg and does not involve suppressor-type T-cell effects. In this system the RAMG only interacts with B lymphocytes. Furthermore, in experiments involving mixtures of adult and young spleen cells, each population reacts independently, neither one influencing the other in any direction. Second, this inactivation was not associated with proliferation of B cells after treatment with RAMG. We have failed repeatedly to detect any mitogenic

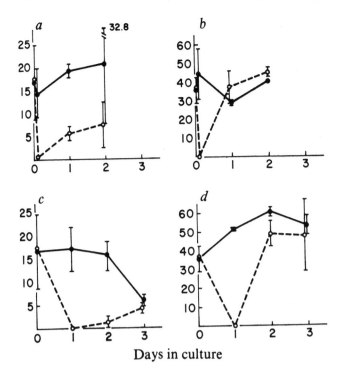

Fig. 3 Re-expression of surface Ig in relation to surface Ia. The four panels show young and adult AKR cells treated for 1 and 24 h with RAMG as in Fig. 1. For each assay point, a tube of cultured cells was washed once, pelleted, resuspended in 50 μl rhodamine-conjugated RAMG (R-RAMG), (1.3 mg ml⁻¹), incubated for 30 min at 4 °C, washed three times, and resuspended in 20 μl normal mouse serum. Twenty microlitres fluorescein-conjugated ATH anti-ATL spleen cell serum (F anti-Ia) were added, and the cells were incubated for 30 min at 4 °C, washed twice and resuspended to 10⁷ per ml. The cells were then scored for the presence of Ig (rhodamine stain) and Ia (fluorescein stain). *a*, Young cells, 1 h RAMG; *b*, adult cells, 1 h RAMG; *c*, young cells, 24 h RAMG; *d*, adult cells, 24 h RAMG. ●, % cells Ia (+) by FITC-a Ia; ○, % cells Ig (+) by R-RAMG.

effect of RAMG on early or adult murine B lymphocytes. Finally, this receptor-mediated inactivation seems to be controlled by some intracellular mechanism independent of the presence or absence of ligand–receptor complexes on the membrane. Ault *et al.*[7] showed that after binding a highly tolerogenic polypeptide of *d*-amino acids, primed adult mouse spleen cells did not regain antigen-binding capability. Also, the complexes were not totally cleared from the cell surface. One could explain these results[7] by postulating a relationship between clearance of complexes and expression of new receptors. This

explanation does not apply to our experiments, however, since the early B cell does clear its surface of complexes. Also, even if some long lasting complexes were present, the Pronase should have removed them and released the cell.

These results show that the interaction of the antigen receptors of an early B lymphocyte with a specific ligand leads to the inactivation of two fundamental B cell processes—receptor re-expression and LPS-induced mitogenesis. This phenomenon may be the basis for several situations of immunological unresponsiveness. The inactivation reported here could explain, in part, the phenomenon of allotype suppression, as well as the analogous cases of suppression of specifically committed cells by anti-receptor antibody[8,9] and of cells bearing a particular Ig class by class-specific antibody[10]. Further, young animals are notoriously easy to render specifically tolerant towards antigen. Although other cells and mechanisms are probably also operative[11], a specific antigen may directly turn off the early B lymphocytes by the process discussed here. Studies are in progress evaluating this point directly on antigen-binding lymphocytes. Finally, the inactivation of early type B lymphocytes could also have a role in adult tolerance. There is evidence that adult mouse spleen[4] and bone marrow[12] contain some B lymphocytes of the type constituting the young spleen. These may be precursors which constantly repopulate the pool of competent adult B lymphocytes[13]. It is possible that the state of unresponsiveness in the adult, although complex, could be prolonged by the ligand-induced inactivation described here, having as its target the early type of B cell found in bone marrow. In this respect, it has been shown that adult bone marrow lymphocytes are very easily rendered tolerant by antigen *in vitro*[14].

CHARLES L. SIDMAN
EMIL R. UNANUE

Department of Pathology,
Harvard Medical School,
Boston, Massachusetts 02115

Received June 16; accepted July 29, 1975.

1 Unanue, E. R., *Am. J. Path.*, **77**, 1–20 (1974).
2 Kishimoto, T., and Ishizaka, K., *J. Immun.*, **114**, 585–591 (1975).
3 Gelfand, M. C., Elfenbein, G. J., and Paul, W. E., *J. exp. Med.*, **139**, 1125–1141 (1974).
4 Sidman, C. L., and Unanue, E. R., *J. Immun.*, **114**, 1730–1735 (1975).
5 Shreffler, D. C., and David, C. S., *Adv. Immun.*, **20**, 125–195 (1975).
6 Unanue, E. R., Dorf, M. E., David, C. S., and Benacerraf, B., *Proc. natn. Acad. Sci. U.S.A.*, **71**, 5014–5016 (1974).
7 Ault, K. A., Unanue, E. R., Katz, D. H., and Benacerraf, B., *Proc. natn. Acad. Sci. U.S.A.*, **71**, 3111–3114 (1974).
8 Strayer, D. S., Cosenza, H., Lee, W. M. F., Rowley, D. A., and Köhler, H., *Science*, **186**, 640–643 (1974).
9 Köhler, H., Kaplan, D. R., and Strayer, D. S., *Science*, **186**, 643–644 (1974).
10 Lawton, A. R., Asofsky, R., Hylton, M., B., and Cooper, M. D., *J. exp. Med.*, **135**, 277–297 (1972).
11 Mosier, D. E., and Johnson, B. M., *J. exp. Med.*, **141**, 216–226 (1975).
12 Ryser, J. E., and Vassali, P., *J. Immun.*, **113**, 719–728 (1974).
13 Osmond, D. G., *J. reticuloendothel. Soc.*, **17**, 99–114 (1975).
14 Nossal, G. J. V., and Pike, B., *J. exp. Med.*, **141**, 904–917 (1975).
15 Mishell, R. I., and Dutton, R. W., *J. exp. Med.*, **126**, 423–442 (1967).

B-CELL TOLERANCE

IV. Differential Role of Surface IgM and IgD in Determining Tolerance Susceptibility of Murine B Cells*

By E. S. VITETTA, J. C. CAMBIER,‡ F. S. LIGLER, J. R. KETTMAN, AND J. W. UHR

(From the Department of Microbiology, University of Texas Southwestern Medical School, Dallas, Texas 75235)

During ontogeny IgD appears later than IgM on splenocytes of neonatal mice (1) and at a time when mice develop a markedly increased immune responsiveness (2). Based on these observations, it was suggested that IgD serves as a "triggering" isotype for induction of immune responses, whereas surface IgM functions as a tolerizing receptor (3). To test this hypothesis, the susceptibility of adult splenocytes (which are predominantly $\mu^+\delta^+$ [4–6]) and neonatal splenocytes (which bear predominantly IgM [μp^+; 1, 4–6]) to tolerance induction were compared. The results indicate that neonatal splenic B cells responsive to thymus dependent (TD) antigens are exquisitely susceptible to tolerance induction compared with those from adult mice (7–9). However, cells from *both* adult and neonatal mice were highly susceptible to tolerance induction when thymus independent (TI) antigen was used as immunogen (8). These results suggest that the major precursor for the TD response is a $\mu^+\delta^+$-cell which appears late in ontogeny and is resistant to tolerance induction and that the μp^+-cell is the major precursor for the TI response and is highly susceptible to tolerance induction. Other differences between responders for TI and TD antigens have been described previously (10–12). To test this concept, adult splenocytes were treated with papain under conditions in which IgD, but not five other surface molecules, was removed (13). Such treated splenocytes were shown to be markedly susceptible to tolerance induction, resembling TD responders from neonatal animals. This experiment was interpreted as indicating that IgD confers resistance to tolerance induction on $\mu^+\delta^+$-cells. To prove this interpretation, it is necessary to show that specific removal of IgD with anti-δ also results in increased susceptibility to tolerance induction and that treatment with anti-μ does not have a similar effect.

In the present studies, we have removed surface IgM or IgD by antibody-induced capping and assessed the tolerance susceptibility of the treated cells. Our results demonstrate that removal of IgD, but not IgM, from TD responders increases their susceptibility to tolerance induction.

Materials and Methods

Experimental Plan. BDF$_1$ (C57BL/6 × DBA/2 F$_1$) splenocytes were treated with antibrain-

* Supported by National Institutes of Health grants AI-10967, AI-11851, AI-12789, and AI-11893.

‡ Supported by National Institutes of Health postdoctoral grant AI-05021.

associated Thy-1 (BAθ) and complement (C′) to remove T cells (7). The B cells were then treated with normal rabbit Ig or antibody to "μ" or "δ" under capping conditions; the capping was assessed by immunofluorescence; treated cells were exposed to tolerogen trinitrophenyl-human immunoglobulin (TNP$_{17}$HGG) for 24 h (7) in the presence of the same antibody used for capping; cells were washed and incubated for 4 days with (TNP) on either a TI (*Brucella*) or a TD (sheep red blood cell [SRBC]) immunogen (8); direct plaque-forming cell responses to TNP and SRBC were assessed at the end of the incubation period (7).

Antisera. The preparation and specificity of rabbit antimouse Ig, rabbit anti-μ, rabbit anti-BAθ, and goat antirabbit Ig (GARIg) have been described previously (14). Rabbit antimouse-δ (15, 16) was absorbed with thymocytes and was judged to be monospecific by criteria described previously (16). These include: (*a*) a single peak on sodium dodecyl sulfate polyacrylamide gel electrophoresis after reaction with a lysate of iodinated splenocytes; (*b*) immunofluorescent staining of the predicted numbers of cells from various lymphoid tissues; (*c*) inability to stain splenocytes after their treatment with either anti-κ or allotypic anti-δ (6); (*d*) independent capping of surface molecules on splenocytes with anti-δ and anti-μ. Chromatographically purified fractions from the antisera or a 30% (NH$_4$)$_2$SO$_4$ precipitate of anti-δ were dissolved in phosphate-buffered saline pH 7.3 (PBS). The IgG fraction from GARIg was conjugated with fluorescein isothiocyanate (FITC; Sigma Chemical Co., St. Louis, Mo.) and chromatographed on DEAE-cellulose (DE52) to give a fraction with molar fluorescein-protein ratio of 2.6.

Capping and Blocking of Cell Surface Immunoglobulin. B cells were incubated for 30 min at 4°C with normal rabbit serum (NRS) or the rabbit antisera described above. Cells were washed and exposed to GARIg (17) for 90 min at 37°C in complete medium to achieve capping. After further washing, cells were cultured for 24 h in the presence of Ig fractions of the same antiserum used to induce capping (400 μg Ig/ml) and were simultaneously exposed to varying concentrations of tolerogen (TNP$_{17}$HGG).

Immunofluorescent Staining. Viable cells were prepared and treated with NRS or antisera specific for "μ," "δ," or Ig as described previously (17). After washing with balanced salt solution (BSS)-azide, the cells were resuspended at 3×10^7 cells/ml in BSS-azide containing FITC-GARIg (0.1 mg/ml). After 10 min at 4°C, the cells were washed in BSS-azide and fixed in 1% paraformaldehyde in PBS. The cells were examined at ×1,000 with a Leitz Ortholux no. 2 fluorescence microscope. For each sample, 100–200 cells were scored for fluorescence without knowledge of the identity of the sample.

Results

Before studies of function, the optimal conditions for modulating surface IgM and IgD on B cells from mouse spleens were determined by immunofluorescence studies. As can be seen in Table I, treatment with antibody was highly effective in removing the majority of the surface Ig to which the antibody was directed. It is presumed that the surface Ig remained modulated during tolerance induction because the concentrations of antibody in the incubation medium were similar to those that induced capping. These concentrations of antibody were chosen because they were also highly effective at blocking a primary immune response in vitro (18).

Fig. 1 shows a representative experiment designed to determine the effect of treating splenic B cells with antiserum on their susceptibility to tolerance induction. As seen in the left panel, only anti-δ increased the susceptibility of the TD responders to tolerance induction. Addition of IgM and IgG to the anti-δ serum did not abrogate this effect. In contrast, anti-δ had no effect on the tolerance susceptibility of TI responders (right panel). The dose-response curve of TD responsive precursors to tolerogen after treatment with anti-δ resembled the dose-response curve for untreated TI responders.

Treatment with anti-μ markedly *decreased* the tolerance susceptibility of TI

TABLE I

Removal of Surface Ig from B Cells by Treatment with Anti-μ or Anti-δ*

Capping	Antiserum used for			
	Staining			
	$\alpha\mu$	$\alpha\delta$	αIg	NRS
	% positive			
None	80	90	96	8
$\alpha\mu$	6	80	92	—
$\alpha\delta$	89	1	94	4
$\alpha\delta$ Absorbed with IgM and IgG	79	4	96	—
NRS	90	83	94	3

* Spleen cells were treated with Baθ + c′. The dead cells were removed before staining.

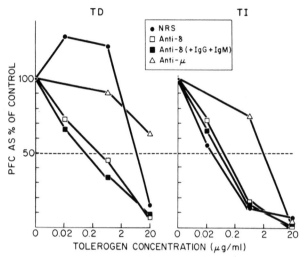

FIG. 1. The effect of removing IgM or IgD receptors on the ability to tolerize splenic B cells responding to TD (left panel) and TI (right panel) forms of TNP immunogen. The tolerogen was TNP$_{17}$HGG (7). The control SRBC plaque-forming cells (PFC) were unaffected by varying the dose of tolerogen as described previously (7, 8). The TNP responses are expressed as direct PFC/10^6 viable recovered cells. TD control responses were: NRS, 810; $\alpha\mu$, 1,098; $\alpha\delta$, 1,149; $\alpha\delta$(+IgG+IgM), 1,286; TI control responses were NRS, 412; $\alpha\mu$, 440; $\alpha\delta$, 517; $\alpha\delta$(+IgG+IgM), 435. This is a representative experiment of three that were done. Each point represents the average of the responses of duplicate cultures.

responders and also decreased slightly the tolerance susceptibility of TD responders (as judged by the 50% tolerance level).

Discussion

The present studies provide further support for our earlier hypotheses (3, 8) cited above. Thus, removal of IgD with anti-δ was shown to increase susceptibility to tolerance induction of treated-TD responders but had no effect on TI responders, confirming earlier results with papain (13). In addition, it was shown that treatment with anti-μ did not similarly increase the tolerizability of TD responders. This is a critical point because it could be argued that diminishing the concentration of *either* isotype would result in an increased susceptibility to tolerance induction. The present studies indicate that different reactivities are conferred on a $\mu^+ \delta^+$-cell by IgM and IgD with regard to tolerance induction, i.e., these two isotypes can be responsible for conveying different signals to the same cell. The absence of a similar effect of anti-δ on TI responders argues that they lack IgD or that, if it is present, it does not determine tolerance susceptibility.

It was also observed that treatment of B cells with anti-μ decreased the susceptibility of TI responders to tolerance induction. This finding implies that interaction of tolerogen with IgM receptors is a necessary event for tolerance induction in TI responders and that removing such receptors by stripping and blocking with anti-μ prevents this interaction. It is provocative that the stripping induced with anti-μ antibody itself does not give a tolerogenic signal.

In contrast to TI responders on which there may be only one functioning isotype, TD responders appear to require *both* IgM and IgD receptors for induction of an antibody response (18). Thus, our working hypothesis is that IgM receptors can deliver either a tolerogenic or triggering signal whereas IgD receptors may be able to deliver only a triggering signal. The events that determine which signal will be conveyed by IgM receptors remain to be elucidated.

Note Added in Proof. Experiments performed by Scott et al. (*J. Exp. Med.* 1977. 146:1473) using allotypic anti-δ to inhibit tolerance induction are in complete agreement with our studies using heterologous anti-δ.

We are grateful to Ms. G. Sloan, Mr. S. Lin, Ms. M. Bagby, Mr. Y. Chinn, Ms. S. Diase, and Ms. M. Neale for expert technical assistance, and Ms. J. Hahn for secretarial assistance.

Received for publication 8 August 1977.

References

1. Vitetta, E. S., U. Melcher, M. McWilliams, J. Phillips-Quagliata, M. Lamm, and J. W. Uhr. 1975. Cell surface immunoglobulin. XI. The appearance of an IgD-like molecule on murine lymphoid cells during ontogeny. *J. Exp. Med.* 141:206.
2. Spear, P. G., and G. M. Edelman. 1974. Maturation of the humoral immune response in mice. *J. Exp. Med.* 139:249.
3. Vitetta, E. S., and J. W. Uhr. 1975. Immunoglobulin-receptors revisited. *Science (Wash. D.C.).* 189:964.
4. Vitetta, E. S., and J. W. Uhr. 1976. Cell surface immunoglobulin. XV. The presence of IgM- and IgD-like molecules on the same cell in murine lymphoid tissue. *Eur. J. Immunol.* 6:140.
5. Cooper, M. D., J. F. Kearney, A. R. Lawton, E. R. Abney, R. M. E. Parkhouse, J. L. Preud'homme, and M. Seligmann. 1976. Generation of immunoglobulin class diversity in B cells: a discussion with emphasis on IgD development. *Ann. Immunol.* 127:573.
6. Goding, J. 1977. Allotypes of IgM and IgD receptors in the mouse: a profile for lymphocyte differentiation. *Contemp. Top. Immunobiol.* 8: In press.
7. Cambier, J. C., J. R. Kettman, E. S. Vitetta, and J. W. Uhr. 1976. Differential

susceptibility of neonatal and adult murine spleen cells to *in vitro* induction of B cell tolerance. *J. Exp. Med.* 144:293.

8. Cambier, J. C., E. S. Vitetta, J. W. Uhr, and J. R. Kettman, Jr. 1977. B Cell tolerance. II. TNP₁₇HgG induced tolerance in adult and neonatal murine B cells responsive to T dependent and independent forms of the same hapten. *J. Exp. Med.* 145:778.

9. Metcalf, E. S., and N. R. Klinman. 1976. *In vitro* induction of neonatal murine B cells. *J. Exp. Med.* 143:1327.

10. Playfair, J. H. L., and E. C. Purves. 1971. Separate thymus dependent and thymus independent antibody forming cell precursors. *Nat. New Biol.* 231:149.

11. Gorczynski, R. M., and M. Feldmann. 1975. B cell heterogeneity—difference in the size of B lymphocytes responding to T dependent and T independent antigens. *Cell. Immunol.* 18:88.

12. Quintáns, J., and H. Cosenza. 1976. Antibody response to phosphorylcholine *in vitro*. II. Analysis of T-dependent and T-independent responses. *Eur. J. Immunol.* 5:399.

13. Cambier, J. C., E. S. Vitetta, J. R. Kettman, G. M. Wetzel, and J. W. Uhr. 1977. B cell tolerance. III. Effect of papain-mediated cleavage of cell surface IgD on tolerance susceptibility of murine B cells. *J. Exp. Med.* 146:107.

14. Vitetta, E. S., J. Forman, and J. R. Kettman. 1976. Cell surface immunoglobulin. XVIII. Functional differences of B lymphocytes bearing different cell surface immunoglobulin isotypes. *J. Exp. Med.* 143:1055.

15. Abney, E., I. R. Hunter, and R. M. E. Parkhouse. 1976. Preparation and characterization of an antiserum to the mouse candidate for immunoglobulin D. *Nature (Lond.).* 259:404.

16. Zan-Bar, I., S. Strober, and E. S. Vitetta. 1977. The relationship between surface Ig isotype and immune function of murine B lymphocytes. I. Surface Ig isotypes on primed B cells in the spleen. *J. Exp. Med.* 145:1188.

17. Ligler, F., E. S. Vitetta, and J. W. Uhr. 1977. Cell surface immunoglobulin. XXII. Reappearance of surface IgM and IgD on murine splenocytes after removal by capping. *J. Immunol.* In press.

18. Cambier, J. C., F. S. Ligler, J. W. Uhr, J. R. Kettman, and E. S. Vitetta. Blocking of primary *in vitro* antibody responses to TNP with antisera specific for μ and δ. *Proc. Natl. Acad. Sci. U. S. A.* In press.

Chapter **VI**

Macrophages and the Issue of Antigen Presentation

Chapter VI

Macrophages and the Issue of Antigen Presentation

Although the importance of the macrophage in contributing to immune responses is now well established, the precise role of this cell in the immune response has not yet been elucidated. Initial studies centered around the observations that immune responses were enhanced when antigen was taken up by phagocytic cells (Unanue and Askonas, 1968) but depressed when such phagocytic cells were removed from in vitro test systems (Mosier, 1967). These results led to the theory that antigen must be "processed" by macrophages (or antigen-presenting cells) in order to be effective in stimulating lymphocytes.

The term "macrophage" has been used rather loosely to describe what is, in fact, a rather wide variety of phagocytic cells present in a number of different tissues. Macrophages in different tissues are referred to by an array of different names, although all of these cells contain abundant lysosomes and surface receptors for both the F_C portion of immunoglobulin and the third component of complement. For example, phagocytic cells in liver are designated Kuppfer cells, while those in skin are called Langerhans cells. Despite their ultimate localization, all phagocytic cells derive from a common hematopoietic precursor in adult bone marrow and undergo most of their differentiation in this blood-forming tissue. It is not known at this time whether all phagocytic cells are equally instrumental in presenting antigen to reactive lymphocytes, but the terms "macrophage" and "antigen-presenting cell" are currently used to describe this cellular participant in the immune response.

The macrophages used in the studies described in this chapter are usually isolated from the peritoneal cavity, spleen, or thymus of a mouse. They can be separated from the lymphoid cells with which they are found by virtue of their adherence to glass, plastic, or polysaccharide surfaces, or by their ingestion of iron filings. (Iron-filled macrophages can be separated from other cells with a magnet.)

In this chapter we will trace the development of the concept that the role of macrophages in the immune response is to take up antigen and present it in an immunogenic form to T cells by "associating" the antigen with surface Ia molecules. As is true for the cytolytic T cell, the helper T cell can apparently only recognize antigen in association with MHC gene products. While cytolytic T cells have receptors that only recognize antigen when it is presented in association with H-2K or H-2D molecules (see Chapter VIII), helper cells bear receptors that recognize antigen when it is presented in association with H-2I molecules. Thus, these two functionally distinct T cells both use gene products of the MHC to facilitate antigen recognition.[1] The macrophage is therefore a critical intermediary between free antigen and helper T cells; its function as an antigen "presenter" must involve the exposure of antigen plus Ia molecules in a manner that can be recognized by helper T cells.

It is obvious that this sort of model places the expression of Ia antigens in a key position. If MHC products are to be the elements by which helper T cells choose to recognize and respond to particular antigens, then any successful antigen-presenting cell must provide the necessary association between foreign antigen and Ia antigen. An unsuccessful or inappropriate association could fail to trigger an immune response. The papers included in this chapter concern themselves with the heart of this model: how must a macropahge present an antigen in order for that antigen to stimulate a helper T cell? The papers focus on the role of Ia antigens in this process and also raise the possibility that immune responsiveness can be controlled at the level of the macrophage.

A critical observation in the development of the role of the macrophage in antigen presentation is reported in the paper by Rosenthal and Shevach. The system that was used to analyze antigen presentation was the proliferative response of T cells derived from guinea pigs immunized in vivo with antigen and then again reexposed in vitro to the same antigen. Effective antigen recognition by the immune T cells results in their proliferation which can be measured by the incorporation of radioactive precursors into DNA.[2] Rosenthal and Shevach demonstrate that free antigen without macrophages or macrophages alone are not capable of activating immune T cells; however, if macrophages are exposed to antigen and then washed to remove free antigen (antigen-pulsed macrophages) they can cause T cell proliferation. This demonstrates that macrophages can present antigen in a stimulatory form to T cells. Even more interesting however is the fact that the antigen-pulsed macrophages had to be histocompatible with the immune T cells in order for the latter to be activated. This result implicates the histocompatibility antigens in the activation process.[3] In support of this notion was the further demonstration by Rosenthal and Shevach that alloantisera directed against the histocompatibility antigens prevented T cell activation.

In their accompanying report (Shevach and Rosenthal, 1973) (not presented here) these authors provide evidence for the fact that animals incapable of responding to a particular antigen (non-responders) are defective in the ability of their macrophages to "present" that antigen to T lymphocytes in an effective

---

[1]The I region (encoding I-associated antigens) is sub-divided into several regions, I-A, I-B, I-C, and I-E. Each segment encodes different surface molecules any one of which may serve as the relevant H-2 I molecule for the presentation of a particular antigen. Depending on the alleles which an inbred mouse strain carries at the I subregions, that strain is rendered a responder or a non-responder to specific antigens. Different subregions control the immune response to different antigens, and the genetic evidence for defining the I-A, I-B, I-C and I-E subregions comes from recombination studies among strains expressing different (responder vs. non-responder) phenotypes for various antigens.

[2]There is a strong correlation between the T cells which proliferate in this assay and helper T cells which are necessary for antibody synthesis, but it is not absolutely clear that the only cells that proliferate in response to antigen on macrophages are helper T cells.

[3]The difference between strain 2 and strain 13 guinea pigs used in these studies resides in the guinea pig equivalent of the mouse H-2 I region.

P₂ antigens they are histocompatible with P₂ macrophages even though they do not respond. This establishes that the histocompatibility restriction demonstrated by Rosenthal and Shevach was only an apparent one; the sharing of histocompatibility of antigens is not sufficient to allow interaction. This result further suggests that the receptor(s) on the helper T cell recognizes antigen in association with an I region product and that such an I region product need not be expressed by both the macrophage and the T cell. The paper by Pierce, Kapp and Benacerraf extends this idea. Working with mice[5] they demonstrate that the macrophage and the T cell can be completely histoincompatible and yet the T cell can be primed by and proliferate in response to an antigen presented by such a macrophage. These experimenst suggest that the repertoire of helper T cell receptors is independent of the genotype of the T cell at the MHC. They lead us to think of the helper T cell receptor in the same way that we think of the B cell receptor (antibody molecule). The latter recognizes the universe of antigens as free antigens while the former recognizes the universe of antigens as modified by a particular I region product. (The members of both universes may in fact be identical.)

Because of the availability of highly inbred recombinant strains, the mouse has enabled a more detailed dissection of macrophage-T cell interactions than the guinea pig. Using antisera (both conventional and monoclonal) specific for gene products of the various subregions of the I region, it has been possible to explore more precisely the role of Ia antigens in macrophage function. The paper by Sprent describes experiments which explore the in vivo role of I region products. Sprent demonstrates that antiserum directed against I region products can prevent T cell priming when administered in vivo. By using a monoclonal anti I-A reagent, he is able to show that products of the I-A and I-E subregions interact to produce a novel restricting element which differs from the restricting element encoded entirely within the I-A subregion.

This paper by Sprent also includes data that clearly demonstrate I-region restriction both for T-cell and macrophage interactions and T cell-B cell interactions. Helper T cells which have receptors specific for antigen in association with the I region product on the appropriate macrophages will only activate ("help") B cells bearing that same I region product. In this study, helper T cells are primed in vivo and then assayed for activity by their ability to help primed B cells secrete antibody when stimulated with antigen. In contrast to the experiments of Shevach and Rosenthal, these experiments are done in vivo, and measure antibody secretion (i.e. activation of B cells) as opposed to T cell proliferation.

The combined results of Thomas and Shevach studying guinea pigs and Sprent studying mice suggest that a particular helper T cell recognizes the same antigen-Ia complex on macrophages and B cells. The following scenario may provide a description of macrophage-T cell and T cell-B cell interactions. First, macrophages bearing appropriate Ia antigens capture and process some antigen, X. T cells specific for this particular complex of X and Ia are stimulated by these activated, antigen-presenting cells. These stimulated T cells in turn seek out and stimulate B cells expressing the same complex of Ia and X. Thus B cells and macrophages present similar images to a reactive T cell, although the acquisition of the antigen X has been specific in one case (the B cell) and non-specific in the other (the macrophage).

Activated macrophages release a variety of factors, some of which are described in the paper by Gery and Waksman. The first paper in the series (not included here--Gery et al., 1972) reports that a "factor" designated Lymphocyte

---

[5]The basic conclusions reached by Shevach and collaborators have also been demonstrated and extended in the mouse (see Yano, et al., 1976; Erb and Feldman, 1975; and Kappler and Marrack, 1976).

[6]Similar results are contained in the paper by Sprent (see Chapter VIII).

way. Thus, this non-responder phenotype may be attributable to a deficit in the macrophage population.[4]

   Because the non-responder phenotype is recessive (responder X non-responder) F_1 animals are responders to the antigen in question. Accordingly, immune T cells derived from such F_1 animals and placed in culture do proliferate when antigen is presented on F_1 macrophages. They fail to proliferate when antigen is presented via non-responder parental macrophages but do, of course, respond when antigen is presented by macrophages from the responder parent. The picture that emerges from this result is that antigen cannot effectively be presented by macrophages bearing the non-responder I region gene products. F_1 macrophages are functional for presentation because they express the I region products of both the responder and non-responder alleles.   This conclusion is strengthened by the result that presentation by F_1 macrophages can be totally inhibited by the presence of antiserum directed against the responder (but not the non-responder) I region gene product.

   The cell interactions described by Rosenthal and Shevach appear to suggest that the T cell and the macrophage must be histocompatible at the I region in order for effective activation of the T cell to take place. An important new view of this picture came from studies reported in the paper of Thomas and Shevach. These authors point out that the histocompatibility requirement may only be a product of the experimental protocol used by Rosenthal and Shevach. More specifically, they showed that if T cells are primed <u>in vivo</u> to a particular antigen so that the antigen was presented in association with the I region product of the host animal, then these primed T cells will only respond when they see antigen again in association with the same I region product. Since the macrophage and T cell of the conventional animal are by definition histocompatible, the priming of the T cell is obviously occurring in a histocompatible environment and will necessarily lead to an <u>in vitro</u> response that also demands histocompatibility. This does not necessarily imply that the T cell and the macrophage must express the same I region alleles; it only necessitates that the T cell be re-presented <u>in vitro</u> with the same antigen-I region complex with which it was originally primed. These authors set out to discriminate between effects attributable to the genotype of the T cell and those attributable to the phenotype of its priming environment.

   Thomas and Shevach show that $(P_1 X P_2)F_1$ T cells primed for the first time <u>in vitro</u> with P_1 macrophages will only respond to antigen when presented for a second time on P_1 macrophages. Such primed T cells will not respond to antigen presented on P_2 macrophages. Since the F_1 T cells bear on their surface P_1 and

---

[4]This paper raises the issue of immune response (Ir) gene control, a vital issue which is not considered in any detail in this book. Interested readers are referred to Katz (1977) and Golub (1981) for an overview of Ir genes and to Levine et al. (1963) and McDevitt and Sela (1965) for the initial reports of Ir genes. For the sake of this discussion, it is important to point out that the ability to mount an antibody response to antigen X can be controlled by two completely independent sets of genes: the Ig genes and the Ir genes. If, for some reason, an individual does not express the V_H and V_L genes required to make any combining site specific for X, then the individual could not make an anti-X antibody. The Ir genes exert control over immune responsiveness not by affecting the Ig gene products but rather by affecting either the expression or recognition of the Ir-encoded antigens. Although the mechanism of this regulation is not understood, the phenotypic result of this Ir gene control is the existence of strains of inbred animals which are either responders or non-responders to particular antigens. Rosenthal and Shevach consider the possibility that inappropriate or ineffective association of antigen and Ir antigens by macrophages of a non-responder animal are responsible for the immune response phenotype of that animal.

Activating Factor (LAF)[8] is released into the supernatant of cultures of mitogen-stimulated spleen cells. LAF is capable of stimulating T cell but not B cell growth. The paper included here describes the assay for LAF and the identification of a cell responsible for its production. Macrophages release the LAF (IL-I) when they are stimulated by substances like lipopolysaccharide (LPS) or phytohemagglutinin (PHA). Resting T cells exposed to LAF, (IL-I), demonstrate an enhanced responsiveness to conventional T cell mitogens like PHA or concanavalin A. This satisfyingly simple picture is complicated by the additional observation by Gery and Waksman that macrophages can themselves be activated to produce LAF by factors apparently produced by stimulated T cells. How this circular interaction of cells and factors contributes to immune responsiveness has not yet been determined.

The failure of macrophages to process and present antigen is, as previously described, associated with the inability of the organism to mount an immune response against that antigen. As noted in the paper by Pierres and Germain, this failure of macrophages to present antigen to the immune system can result in the appearance of suppressor T cells. These workers show that the addition of antigen to a culture of spleen cells depleted of macrophages results in the generation of suppressor T cells. Such suppressor T cells are specific for the antigen and can prevent any further response to that antigen even if macrophages are added back to the cultures.

As noted in Chapter VIII, suppressor T cells can bind free antigens whereas helper T cells cannot. Thus, it might be argued that the more efficiently macrophages take up and process antigen, the more likely the organism is to respond. The failure of macrophages to take up antigen results first in an excess of free antigen and subsequently in a state of antigen-specific unresponsiveness mediated by suppressor T cells. This notion has some experimental support. Suppressor T cells are readily generated either when antigen is administered in a form inimical to macrophage uptake, or when antigen is administered intravenously in high concentrations. In contrast, good immune responses are seen if insoluble complexes of antigen, readily assimilated by macrophages, are administered. The presence of adjuvants (agents capable of "activating" macrophages) further enhances the immune response.

Although the macrophage appears to play a key role in the control of immune responsiveness, immunologists are only beginning to understand the precise cellular and biochemical events involved in macrophage "processing" and "presenting" of antigens. Although these experiments support the idea that some kind of association between antigen and Ia region gene products is important in antigen presentation, the chemical basis for this phenomenon is as yet obscure. We do not know how macrophages process antigen or even what "processing" means in chemical terms. Similarly, the "association" of antigen and I region gene products may be anything from actual biochemical association of these molecules to merely proximal expression at the cell surface.

Further experiments directed at analyzing the membrane biochemistry of antigen presenting cells should be instrumental in resolving some of these issues.

[8]This factor has been renamed Interleukin I (IL-I); see Aarden et al., 1979.

Papers Included in this Chapter

1. Rosenthal, A.S. and Shevach, E.M. (1973) Function of macrophages in antigen recognition by guinea pig T lymphocytes. I. Requirement for histocompatible macrophages and lymphocytes. J. Exp. Med. 138, 1194.

2. Thomas, D.W. and Shevach, E.M. (1976) Nature of the antigenic complex recognized by T lymphocytes. I. Analysis with an in vitro primary response to soluble protein antigens. J. Exp. Med. 144, 1263.

3. Pierce, C.W., Kapp, J.A., and Benacerraf, B. (1976) Regulation by the H-2 gene complex of macrophage-lymphoid cell interactions in secondary antibody responses in vitro. J. Exp. Med. 144, 371.

4. Sprent, J. (1980) Effects of blocking helper T cell induction in vivo with anti-Ia antibodies. Possible role of I-A/E hybrid molecules as restriction elements. J. Exp. Med. 152, 996.

5. Gery, I. and Waksman, B.H. (1972) Potentiation of the T-lymphocyte response to mitogens. II. The cellular source of potentiating mediator(s). J. Exp. Med. 136, 143.

6. Pierres, M. and Germain, R.N. (1978) Antigen-specific T cell-mediated suppression. IV. Role of macrophages in generation of L-glutamic acid^{60}-L-alanine ^{30}-L-tyrosine^{10} (GAT)-specific suppressor T cells in responder mouse strains. J. Immunol. 121, 1306.

References Cited in this Chapter

Aarden, L.A., et al. (1979) Letter to the Editor: Revised nomenclature for antigen-nonspecific T cell proliferation and helper factors. J. Immunol. 123, 2928.

Erb, P. and Feldmann, M. (1975) Role of macrophages in in vitro induction of T-helper cells. Nature 254, 352.

Gery, I., Gershon, R.K., and Waksman, B.H. (1972) Potentiation of the T-lymphocyte response to mitogens. I. The responding cell. J. Exp. Med. 136, 128.

Golub, E.S. (1981) The Cellular Basis of the Immune Response (Second Edition). Sinauer Associates, Inc., Sunderland, MA.

Kappler, J.W. and Marrack, P.C. (1976) Helper T cells recognize antigen and macrophage surface components simultaneously. Nature 262, 797.

Katz, D.H. (1977) Lymphocyte Differentiation, Recognition, and Regulation. Academic Press, New York.

Levine, B., Ojeda, A., and Benacerraf, B. (1963) Studies on artificial antigens. III. The genetic control of the immune response to hapten-poly-L-lysine conjugates in guinea pigs. J. Exp. Med. 118, 953.

McDevitt, H.O. and Sela, M. (1965) Genetic control of the antibody response. I. Demonstration of determinant-specific differences in response to synthetic polypeptide antigens in two strains of inbred mice. J. Exp. Med. 122, 517.

Mosier, D. (1967) A requirement of two cell types for antibody formation _in vitro_.
    Science 158, 1573.

Shevach, E.M. and Rosenthal, A.S. (1973) Function of macrophages in antigen recog-
    nition by guinea pig T lymphocytes. II. Role of the macrophage in the regula-
    tion of genetic control of the immune response. J. Exp. Med. 138, 1213.

Unanue, E.R. and Askonas, B.A. (1968) The immune response of mice to antigen in
    macrophages. Immunology 15, 287.

Yano, A., Schwartz, R.H., and Paul, W.E. (1978) Antigen presentation in the murine
    T lymphocyte proliferative response. II. Ir-GAT-controlled T lymphocyte
    responses require antigen-presenting cells from a high responder donor. Eur.
    J. Immunol. 8, 344.

FUNCTION OF MACROPHAGES IN ANTIGEN RECOGNITION BY GUINEA PIG T LYMPHOCYTES

I. Requirement for Histocompatible Macrophages and Lymphocytes

By ALAN S. ROSENTHAL and ETHAN M. SHEVACH

(From the Laboratory of Clinical Investigation and the Laboratory of Immunology, National Institute of Allergy and Infectious Diseases, National Institutes of Health, Bethesda, Maryland 20014)

(Received for publication 12 July 1973)

The activation of immunocompetent lymphocytes by antigen is dependent upon interaction between antigen and a specific lymphocyte recognition structure. For the bone marrow-derived or B lymphocyte, a number of studies have demonstrated easily detectable membrane immunoglobulin receptors that are capable of binding the antigen for which the cell is specific (1, 2). Upon interaction with antigen, B lymphocytes proliferate and/or differentiate into plasma cells that synthesize and secrete immunoglobulin molecules with binding properties identical with that of the receptors of their precursor B lymphocyte. The mechanism by which thymus-derived or T lymphocytes recognize antigen is a matter of considerable controversy. Although some workers have observed immunoglobulin on the membrane of T lymphocytes (3, 4), others have failed to detect it in significant quantities (5, 6). Specific antigen binding to T lymphocytes has also been quite difficult to demonstrate (7, 8). Antigen binding to T cells has been measured indirectly by exposing cells in vitro to [^{125}I]-antigen of high specific activity with subsequent killing or "suicide" of the specific antigen-binding T lymphocytes (9). More recently, theta-positive, surface immunoglobulin-negative antigen-binding cells have been visualized directly after incubation of primed mouse spleen cells with relatively high concentrations of ^{125}I-labeled antigen (10) and by the cytoadherence techniques (11).

Regardless of whether or not T cells efficiently bind soluble antigens, little or no data is available on the mechanism by which these cells are activated or triggered by antigen. Indeed, a number of studies suggest that T-cell activation, as measured by lymphocyte proliferation, involves the cooperation of macrophages and lymphocytes (12–15). In the guinea pig recent studies have shown that stimulation of T-lymphocyte proliferation in vitro involves an initial uptake of soluble protein antigens by macrophages (16). This uptake is maximally acquired after only brief exposure of the macrophages to antigen (30–60 min at 37°C) (15, 17) and requires the expenditure of metabolic energy. Thus, the uptake of antigen by macrophages appears to represent more than simple surface binding (13, 15, 17).

Other studies on the genetic control of specific immune responses have raised the possibility that molecules other than immunoglobulin may play a role in antigen recognition by the T lymphocyte (18). Thus, alloantisera can specifically block the activation of T lymphocytes by antigens, the response to which is linked to the presence of histocompatibility specificities against which the alloantisera are directed (19). It was concluded from these observations that the immune response genes produce a cell surface-associated product and that this product plays a role in the mechanism of antigen recognition by the T lymphocyte.

It seemed reasonable, therefore, to evaluate the importance of macrophage and T-lymphocyte histocompatibility determinants in the expression of antigen recognition. We will demonstrate in this report that the recognition of soluble protein antigens by guinea pig T lymphocytes requires the presentation of antigen on histocompatible macrophages and that this interaction between macrophage and T lymphocyte can be blocked by alloantisera. These data are interpreted as supporting the existence of a specific associative event between macrophage-bound antigen and T lymphocyte that is mediated by histocompatibility determinants themselves or by membrane surface products of genes linked to the major histocompatibility region.

Materials and Methods

Animals and Immunization.—Strain 2 and strain 13 guinea pigs were obtained from the Division of Research Services, National Institutes of Health, Bethesda, Md. $(2 \times 13)F_1$ animals were obtained by mating strain 2 with strain 13 ainmals in our own colony. Outbred Hartley strain guinea pigs were purchased from Camm Research Institute, Inc., Wayne, N.J.

Guinea pigs were immunized with an emulsion of antigen or saline in complete Freund's adjuvant (containing 0.4 mg/ml *Mycobacterium tuberculosis* $H_{37}RA$; Difco Laboratories, Detroit, Mich.). Each animal received 0.1 ml emulsion in each footpad for a total dose of antigen of 100 μg per guinea pig.

Reagents.—Guinea pig albumin (GPA)[1] was purchased from Pentex, Biochemical, Kankakee, Ill. 2,4-Dinitrophenyl (DNP)-GPA was prepared as previously described (1). Purified protein derivative of tuberculin (PPD) was obtained from Connaught Medical Research Laboratories, Willowdale, Ontario, Canada. The strain 2 anti-strain 13 serum and the strain 13 anti-strain 2 serum were prepared and assayed as previously described (19). These sera were also used to histocompatibility type lymph node lymphocytes from outbred animals as described (20).

Cell Collection and Purification.—The techniques for collection of peritoneal exudate cells (PEC) have been described in detail previously (21). In brief, animals were injected intraperitoneally with 25 ml of sterile mineral oil (Marcol 52; Humble Oil and Refining Co., Houston, Tex.). 3–4 days later the exudate cells were harvested by lavaging the peritoneal cavity with 150 ml of Hanks' balanced salt solution. Erythrocytes were removed from the pooled exudate cells by treatment with buffered NH_4Cl (22) at 4°C and washed four times before further manipulation. The PEC population consisted of about 75% monocytes-macrophages, 10% neutrophils, and 15% lymphocytes. Peritoneal exudate lymphocytes (PELs), a population of highly enriched antigen-reactive T lymphocytes (21, 23), were obtained by purification of the PEC over adherence columns (7). After column purification the

---

[1] *Abbreviations used in this paper:* DNP, 2,4-dinitrophenyl; FCS, fetal calf serum; GPA, guinea pig albumin; H, histocompatibility; LNLs, lymph node lymphocytes; MLR, mixed leukocyte reaction; Mϕ, macrophage; NGPS, normal guinea pig serum; PECs, peritoneal exudate cells; PELs, peritoneal exudate lymphocytes; PHA, phytohemagglutinin; PPD, purified protein derivative of tuberculin; [^{3}H]TdR, tritiated thymidine.

PEL population contains about 90% lymphocytes, 2–5% neutrophils, and 5–8% monocytes or macrophages.

Lymph node cell suspensions were prepared from trimmed lymph nodes by teasing with a needle and forceps and depleted of macrophages and B lymphocytes by passage over adherence columns as previously described (7). Lymph node lymphocytes (LNLs) contained about 0.5% macrophages when tested by latex bead phagocytosis and were greater than 98% viable by trypan blue exclusion.

"Purified" macrophages were prepared from the PEC population by allowing the cells to adhere to the surface of glass Petri dishes in the presence of 10% fetal calf serum (Gray Industries, Inc., Fort Lauderdale, Fla.) at 37°C. After 3 h, the cells were washed and overlaid with fresh media containing 10% fetal calf serum (FCS) and then cultured for an additional 24 h. Nonadherent cells were again washed away; the medium was replaced with several washes of iced saline; and the dishes were cooled at 4°C. After 30 min the cells were gently scraped off the glass with a rubber policeman and washed. This cell preparation contained greater than 98% macrophages with a viability of greater than 90%.

Technique of Brief Antigen Exposure.—PECs or monolayer-purified macrophages at a concentration of 15×10^6/ml in the presence of 30 μg/ml mitomycin C (Nutritional Biochemicals Corp., Cleveland, Ohio) were allowed to equilibrate at 37°C. The appropriate concentration of antigen was then added and the cell mixtures were maintained at 37°C for 60 min (15). At the end of the exposure period, the cell suspensions were washed four times with media. Residual antigen after this washing procedure was determined using ^{125}I-labeled PPD and found to be consistently less than 0.1 μg of PPD per 10^6 macrophages after an initial 100 μg PPD/ml exposure.

In Vitro Assay of Antigen-Induced DNA Synthesis.—Antigen-pulsed macrophages at a concentration of 1×10^6/ml were mixed with immune PELs or LNLs at a concentration of 2×10^6/ml in Eagle's minimal essential media, Spinner modification (S-MEM; Microbiological Associates, Bethesda, Md.) supplemented with glutamine (300 μg/ml), penicillin 100 U/ml, streptomycin 100 μg/ml, and 10% FCS or 10% normal guinea pig serum (NGPS). Aliquots (0.2 ml) of these mixtures were added to each of four wells in a sterile round bottom microtiter plate (Cooke Engineering Co., Alexandria, Va.) covered and incubated at 37°C in a humidified atmosphere of 95% air, 5% CO_2. After 24 or 48 h, 1 μCi of tritiated thymidine ([^{3}H]thymidine, 6.7 Ci/mM; New England Nuclear, Boston, Mass.) was added to each well. 18 h later the amount of [^{3}H]thymidine incorporated into cellular DNA was measured with the aid of a semiautomated microharvesting device (24). Radioactivity was counted in a Beckman liquid scintillation counter (Beckman Instruments, Inc., Fullerton, Calif.) and expressed as either total counts per minute (cpm) per culture or difference between control and antigen-stimulated culture (Δcpm per culture). The overall scheme of the experiments is shown in Fig. 1.

RESULTS

Requirement for Syngeneic Macrophages for Activation of T Lymphocyte Proliferation.—We have shown previously that a log linear relationship exists between the numbers of syngeneic PPD-pulsed macrophages added to immune lymphocytes and the resultant DNA synthesis at ratios of macrophages to lymphocytes less than 1:1.[2] Moreover [^{3}H]TdR incorporation into new DNA was proportional to the log of the concentration of antigen used to

---

[2] Rosenstneich, D. L., and A. S. Rosenthal. Peritoneal exudate lymphocyte. III. Dissociation of antigen-reactive lymphocytes from antigen binding cells in T-lymphocyte-enriched populations in the guinea pig. Manuscript submitted for publication.

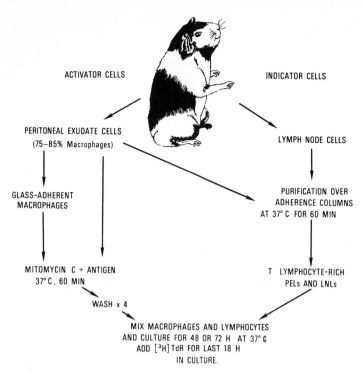

ACTIVATOR CELLS   INDICATOR CELLS

FIG. 1. Schematic diagram of the assay of macrophage-associated antigen activation of guinea pig T-DNA lymphocyte synthesis.

pulse the macrophages up to a concentration of 100 µg/ml. Subsequent experiments, therefore, routinely employed 0.1×10^6 macrophages pulsed with 100 µg/ml PPD for 60 min at 37°C and mixed with 0.2×10^6 lymphocytes.

The initial studies examined the ability of strain 2 and 13 macrophages to activate immune strain 2 and 13 PELs. The antigen-dependent incorporation of [³H]TdR into new DNA was assessed at 72 h and expressed as cpm $\times 10^{-3}$ (Table I). Strain 2 macrophage-associated antigen ($M\phi_2$) initiated immune strain 2 PEL DNA synthesis quite effectively ($M\phi_2$-PPD 180,390 cpm vs. $M\phi_2$-control 1,250 cpm), while the same macrophages mixed with immune strain 13 PELs stimulated little DNA synthesis ($M\phi_2$-PPD 13,540 vs. $M\phi_2$-control 7,400). Similarly, antigen-pulsed strain 13 macrophages induced little DNA synthesis when mixed with immune strain 2 PELs ($M\phi_{13}$-PPD 16,790 cpm vs. $M\phi_{13}$-control 6,360 cpm) but considerable DNA synthesis in immune strain 13 lymphocytes ($M\phi_{13}$-PPD 58,700 vs. $M\phi_{13}$-control 550 cpm). Also of note was the small but significant mixed leukocyte reaction (MLR) seen in the absence of antigen in mixtures of allogeneic macrophage-rich PECs and PELs.

The experimental approach shown in Table I was repeated, but in this series of experiments column-purified lymph node cells (LNLs) were used as the reactive population. The ability of $(2 \times 13)F_1$ LNL to respond to parental macrophages and the ability of F_1 macrophages to activate parental LNL were also examined. The data presented in Table II using LNLs confirm the observa-

TABLE I

The Requirement for Histocompatible Macrophages in Antigen-Mediated DNA Synthesis in Immune Guinea Pig Peritoneal Exudate Lymphocytes

Macrophage*		Lymphocyte DNA synthesis	
Strain	Antigen pulse	Strain 2	Strain 13
		$[^3H]TdR$ incorporation (cpm $\times 10^{-3}$)	
2	0	1.25 ± 0.88	7.40 ± 3.04
2	+	180.39 ± 21.79	13.54 ± 4.60
13	0	6.36 ± 2.01	0.55 ± 0.28
13	+	16.79 ± 1.64	58.70 ± 11.41

* Macrophage-rich PEC from nonimmunized strain 2 and 13 guinea pigs were incubated with mitomycin C and/or PPD for 60 min at 37°C, washed, and each mixed with both immune strain 2 and 13 PELs as described in Materials and Methods. The $[^3H]TdR$ incorporation is expressed as mean cpm $\times 10^{-3} \pm$ SE of three experiments.

TABLE II

The Requirement for Histocompatible Macrophages in Antigen-Mediated DNA Synthesis in Immune Guinea Pig Lymph Node Lymphocytes

Macrophage*		Lymphocyte DNA synthesis		
Strain	Antigen pulse	Strain 2	Strain 13	$(2 \times 13)F_1$
		$[^3H]TdR$ incorporation (cpm $\times 10^{-3}$)		
2	0	0.92 ± 0.28	5.68 ± 1.08	1.60 ± 0.30
2	+	26.38 ± 8.27	8.61 ± 2.08	6.98 ± 0.80
13	0	4.63 ± 1.86	1.66 ± 0.37	1.78 ± 0.47
13	+	3.12 ± 0.67	19.89 ± 4.47	7.81 ± 1.75
$(2 \times 13)F_1$	0	1.91 ± 0.99	4.27 ± 0.34	1.66 ± 0.53
$(2 \times 13)F_1$	+	12.42 ± 3.19	11.81 ± 1.98	12.57 ± 2.33

* Experimental design identical to Table I except that $(2 \times 13)F_1$ macrophages and lymphocytes have been added and column-purified lymph node cells have been used as indicators instead of PELs. $[^3H]TdR$ incorporation is expressed as mean cpm $\times 10^{-3} \pm$ SE of six experiments.

tion that macrophage-rich PEC pulsed with antigen efficiently activate only syngeneic lymphocytes. Thus, the requirement for syngeneic macrophages and lymphocytes for T-lymphocyte recognition is not a peculiarity of the source of the T lymphocytes. In addition, these data show that F_1 macrophage-associated antigen activates parental lymphocytes only about 50% as well as parental macrophages ($M\phi F_1 \rightarrow LNL_2$ 12,420 cpm vs. $M\phi_2 \rightarrow LNL_2$ 26,380 and $M\phi F_1 \rightarrow LNL_{13}$ 11,810 cpm vs. $M\phi_{13} \rightarrow LNL_{13}$ 19,890 cpm). Moreover, parental 2 and 13 macrophage-associated antigen activated immune F_1 lymphocyte DNA synthesis only about 50% as well as F_1 macrophage-associated antigen ($M\phi_2 \rightarrow LNL_{F_1}$ 6,980 cpm; $M\phi_{13} \rightarrow LNL_{F_1}$ 7,810 cpm vs. $M\phi_{F_1} \rightarrow$ LNL_{F_1} 12,570 cpm).

Purified Allogeneic Macrophages Fail to Activate T-Lymphocyte Proliferation.—The experiments presented thus far have used the PEC population as the source of macrophages. Inasmuch as this population of cells is composed of 75–85% macrophages, it seemed appropriate to evaluate the stimulation of immune T lymphocytes by highly purified allogeneic macrophages. Strain 13 macrophages were purified by adherence plating, washed, and pulsed with either 100 μg/ml DNP-GPA, or 10 μg/ml PPD; they were then mixed with either strain 2 or strain 13 lymphocytes (Table III). The results demonstrate again that antigen on macrophages stimulate only syngeneic lymphocyte proliferation. In addition, a similar requirement for syngeneic macrophages was shown for another antigen DNP-GPA; this demonstrates that the requirement for syngeneic macrophages is not a peculiarity of the antigen PPD. Of note is the absence of a mixed leukocyte reaction in combinations of allogeneic-purified macrophages and lymphocytes. The data show that macrophage-rich PEC and the more highly purified macrophages when pulsed with antigen consistently gave similar results when assessed for their ability to initiate allogeneic and syngeneic lymphocyte proliferation. Because of the difficulty in purifying macrophages, in sufficient numbers, the macrophages used in the remainder of this paper are the PEC population.

Absence of T-Lymphocyte Activation by Allogeneic Macrophage-Associated Antigen is Not the Result of Differences in the Kinetics of Activation.—Since DNA synthesis was measured at only a single time, the possibility remained that activation of lymphocytes by antigen on allogeneic macrophages might occur either before or after that time. The kinetics of activation of strain 2

TABLE III

Activation of Strain 2 and 13 PELs by Purified Strain 13 Macrophages

Strain 13 macrophages*	Lymphocyte DNA synthesis	
	Strain 2	Strain 13
	$[^3H]TdR$ incorporation ($cpm \times 10^{-3}$)	
Control	0.37 ± 0.10	0.55 ± 0.07
DNP-GPA pulsed	0.30 ± 0.08	18.03 ± 4.91
PPD pulsed	0.63 ± 0.14	6.01 ± 0.53

* Monolayer purified macrophages from nonimmunized strain 13 guinea pigs were incubated at 37°C for 60 min in the presence of media alone, 100 μg/ml DNP-GPA, or 100 μg/ml PPD. The cells were then washed and mixed 1:2 with immune strain 2 and 13 PELs. The data presented are the mean cpm $\times 10^{-3} \pm$ SE of three experiments.

lymphocytes by strain 2 and 13 macrophage-associated antigen was examined. As shown in Fig. 2, at neither 24, 48, nor 72 h was significant activation of strain 2 PELs by strain 13 macrophages observed.

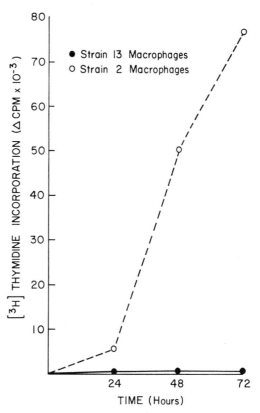

FIG. 2. Kinetics of macrophage-associated antigen activation of immune lymphocyte DNA synthesis. Strain 2 and 13 macrophages from a nonimmunized guinea pig were pulsed for 60 min at 37°C with mitomycin C alone or mitomycin C plus 100 μg/ml PPD, washed, and mixed with immune strain 2 PELs. Mixtures of syngeneic and allogeneic macrophages and lymphocytes were cultured 24, 48, and 72 h and [^{3}H]TdR incorporation expressed as Δcpm $\times$ 10^{-3}. Note that nonsyngeneic macrophages do not activate significant PEL DNA synthesis at any time in culture.

The Mixed Leukocyte Reaction Does Not Inhibit Macrophage-Associated Antigen Induction of Lymphocyte Proliferation.—To evaluate whether or not an inhibitor of blastogenesis might be released in mixtures of histoincompatible cells and thus account for the lack of stimulation in allogeneic mixtures of macrophages and lymphocytes, a number of experiments was carried out. First, strain 2 and strain 13 macrophages were combined, pulsed with antigen, and then added to either strain 2 or strain 13 lymphocytes; the resulting DNA synthesis was compared with that of individual syngeneic and allogeneic combinations (Table IV). The results demonstrate that the presence of allogeneic macrophages does not inhibit the induction of proliferation of strain 2 or strain 13 lymphocytes by syngeneic macrophages. Second, strain 2 or strain 13 macrophages were pulsed with antigen and then added to either strain 2, strain 13, or an equal mixture of 2 and 13 lymphocytes, and the resultant DNA synthesis was assessed. As can be seen from Table IV, the presence of allogeneic lymphocytes does not inhibit the activation of strain 2 or strain 13 lymphocytes by syngeneic macrophages.

TABLE IV

The Effect of the Mixed Lymphocyte Reaction on Macrophage-Associated Antigen Activation of DNA Synthesis by Immune Guinea Pig Lymphocytes

Macrophages		Lymph node lymphocyte strain		
Strain	Antigen pulse	2	13	2 + 13*
		$[^3H]TdR$ incorporation (cpm $\times$ 10^{-3})		
2	0	1.25 ± 0.25	5.47 ± 0.79	5.99 ± 1.01
2	+	39.90 ± 4.87	6.43 ± 0.47	59.92 ± 1.97
13	0	3.68 ± 0.55	2.71 ± 0.27	4.05 ± 0.31
13	+	3.83 ± 0.60	23.55 ± 1.22	15.48 ± 0.58
2 + 13‡	0	3.67 ± 0.18	6.03 ± 0.06	N.T.§
2 + 13	+	26.21 ± 0.38	21.96 ± 1.57	N.T.

Experimental conditions were identical with those in Table I. $[^3H]TdR$ incorporation is expressed as mean cpm $\times$ 10^{-3} ± SE of three experiments.

* A mixture containing 1.0×10^6/ml each of 2 and 13 lymphocytes.

‡ Mixtures containing 0.5×10^6/ml each of 2 and 13 macrophages.

§ Not tested.

An alternative approach to rule out the presence of an inhibitor of blastogenesis in mixtures of allogeneic macrophages and lymphocytes is to examine whether allogeneic macrophages will inhibit the response of lymphocytes to PPD (100 µg/ml) present for the duration of the culture or to phytohemagglutinin (PHA) (1 µg/ml). Although LNLs are obtained by column purification, they still contain sufficient (~0.5%) syngeneic macrophages to respond partially to continuous antigen or PHA. Strain 2 or strain 13 macrophages were pulsed with PPD, washed, and then added to strain 13 LNL. Again activation of significant DNA synthesis was seen only in the syngeneic combination (Table V, column 1). When PPD or PHA were then added, the strain 13 LNLs respond irrespective of the presence of strain 2 or strain 13 macrophages. The response to added PPD is greater in the syngeneic mixture and this is probably due to the fact that LNLs are relatively deficient in endogenous macrophages, and the addition of syngeneic macrophages either with or without previous antigen exposure enhances lymphocyte proliferation. These and

TABLE V

Allogeneic Macrophages Do Not Inhibit Lymphocyte DNA Synthesis on Continuous Antigen or Mitogen Exposure

Macrophage		Additions to cultures of macrophages and strain 13 lymphocytes*		
Strain	Antigen pulse	None	PPD	PHA
		$[^3H]TdR$ incorporation $(cpm \times 10^{-3})$		
2	0	5.66 ± 0.66	13.36 ± 1.67	217.01 ± 41.19
2	+	7.63 ± 1.58	13.43 ± 2.60	331.64 ± 32.17
13	0	1.78 ± 0.45	21.79 ± 0.95	357.29 ± 25.32
13	+	18.47 ± 3.77	27.29 ± 2.75	407.94 ± 47.31

* Mixture of macrophages and lymphocytes were cultured for 48 h without further addition or with addition of 100 μg/ml PPD or 1 μg/ml PHA continuously. Experimental conditions identical to Table I except that strain 2 and 13 macrophages were assayed only on strain 13 LNLs. $[^3H]TdR$ incorporation is expressed as mean cpm $\times 10^{-3} \pm$ SE of six experiments.

the previous experimental series are interpreted as indicating that no inhibitor of blastogenesis was released sufficient to account for the almost total absence of stimulation of lymphocyte DNA synthesis by allogeneic macrophage-associated antigen.

Studies Using Outbred Animals as the Source of Macrophages.—Random-bred guinea pigs have previously been shown to possess strain 2 and 13 histocompatibility determinants (25, 26). In order to evaluate whether the failure of strain 2 or strain 13 macrophages to activate allogeneic lymphocytes is related to the major serologically defined histocompatibility antigens, we studied the capacity of macrophages from outbred animals of serologically defined histocompatibility type to stimulate strain 2 and strain 13 lymphocytes. Outbred Hartley guinea pigs were individually tested with 13 anti-2 and 2 anti-13 sera for the presence of strain 2 and strain 13 determinants and grouped into three groups, [(2+, 13+), (2+, 13−), (2−, 13+)]. No (2−, 13−) guinea pigs have been identified in our colonies. Macrophages from these guinea pigs were pulsed with PPD and mixed with immune strain 2 and 13 lymphocytes. The data are given as the arithmetic mean and with individual points reflecting the data from single animals (Fig. 3). A general correlation exists between the sharing of strain 2 or 13 histocompatibility determinants and the ability of macrophages from outbred guinea pigs to stimulate the immune lymphocytes. Thus, 2+ macrophages, whether 13− or 13+, are superior to 2− macrophages in activation of strain 2 lymphocytes. Similarly, 13+ macrophages, whether 2+ or 2−, are superior to 13− macrophages in stimulation of 13 lymphocytes.

These findings establish that macrophage-lymphocyte interaction is regulated by histocompatibility (H) antigen itself or by the product of a gene that is closely linked to the H-antigen gene. Of additional interest is that 2− outbred

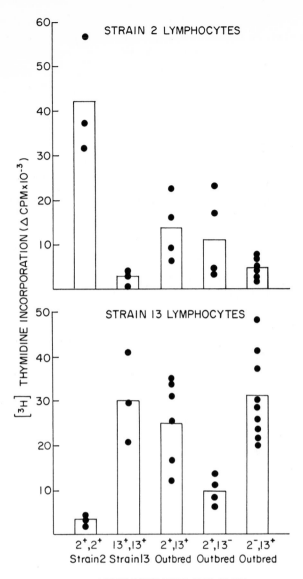

FIG. 3. The correlation of the histocompatibility type of outbred guinea pigs and the ability of their macrophages to stimulate DNA synthesis in immune strain 2 and 13 guinea pig lymphocytes. Experimental conditions are identical to Table I. LNLs are used as indicator lymphocytes. Outbred guinea pigs are categorized as (2+, 13+), (2+, 13−), and (2−, 13+). The bar graphs give the arithmetic mean cpm ± SE of the difference between stimulated and control culture. The individual value for each animal is represented by points about the mean.

macrophages activate strain 2 lymphocytes to a greater extent than do 2—inbreds (strain 13). Conversely, 13— outbreds activate strain 13 lymphocytes greater than do 13— inbreds (strain 2). The latter observations suggests that factors other than those serologically identifiable by the 13 anti-2 and 2 anti-13 sera may play some role in macrophage-lymphocyte cooperation.

Inhibition of Macrophage-Lymphocyte Cooperation by Alloantisera.—If indeed the interaction between macrophages and lymphocytes is mediated via histocompatibility antigen determinants or by membrane structures closely linked to these determinants, alloantisera might block this interaction. Strain 2, strain 13, or F_1 macrophages were pulsed with PPD or PHA, washed, and then mixed with F_1 PELs; these combinations were then cultured in normal guinea pig serum, 13 anti-2 serum, or 2 anti-13 serum (Fig. 4). Both the anti-2 and the anti-13 sera inhibited by about 30–40% the activation of F_1 PELs by F_1 macrophages pulsed with PPD. However, when strain 2 macrophages that had been pulsed with PPD were added to F_1 cells in the presence of the 13 anti-2 serum, the stimulation of the F_1 lymphocytes was markedly diminished; the strain 2 anti-strain 13 serum had essentially no effect on the activation of F_1 cells by strain 2 macrophages, but this serum did completely abolish the stimulation of F_1 cells by strain 13 macrophages. In no case did the alloantisera effect the activation of F_1 lymphocytes by macrophage-associated PHA. These data are interpreted as indicating that alloantisera block macrophage-lymphocyte interaction by acting on histocompatibility antigen determinants or sites closely linked to them; these studies do not localize on which cell the alloantisera exert their inhibitory effect.

Alloantisera Do Not Inhibit Macrophage-Lymphocyte Interaction by Acting Solely on the Macrophage.—In these experiments, F_1 macrophages were pulsed with PPD and then added to strain 2 or strain 13 PELs. When F_1 macrophages were added to strain 2 PELs, marked inhibition of the PPD response (Fig. 5) was seen in the presence of the 13 anti-2 serum and only minimal depression was seen with the 2 anti-13 serum. Similar results were observed when F_1 macrophages were added to strain 13 PELs; only in the presence of the 2 anti-13 serum was depression of the PPD response noted. These data demonstrate that the alloantisera only exert their inhibitory effect if they are directed against H-antigen determinants of both the macrophage and the lymphocyte.

DISCUSSION

The mechanism(s) by which T lymphocytes recognize antigenic signals is not known. The approach to the understanding of T-lymphocyte function used in this study is the analysis of the physiologic conditions under which T lymphocytes respond to soluble protein antigens. The assay of T-lymphocyte function employed in this study, antigen-induced proliferation, is a correlate of in vivo cellular immunity (27). Moreover, in the guinea pig, this in vitro response can be analyzed in terms of the cell types involved. T-lymphocyte antigen recognition in the guinea pig, as assessed by lymphocyte proliferation, has been shown to require macrophages (12–16). Recent

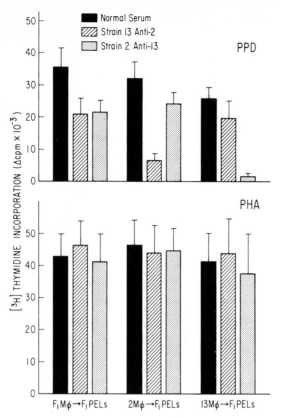

FIG. 4. The effect of alloantisera on macrophage-associated antigen activation DNA synthesis in $(2 \times 13)F_1$ PELs. Strain 2, 13, and $(2 \times 13)F_1$ macrophages obtained from a nonimmunized guinea pig were pulsed for 60 min at 37°C with mitomycin C and 100 μg/ml PPD or 10 μg/ml PHA. All cells were washed and mixed with immune $(2 \times 13)F_1$ lymphocytes and each was cultured in normal guinea pig serum, strain 13 anti-2 serum, and strain 2 anti-13 serum. After 72 h [³H]TdR incorporation was expressed as mean Δcpm $\times 10^{-3}$ $\pm$ SE. Note that both 13 anti-2 and 2 anti-13 sera suppress F_1 macrophage-associated antigen activation of F_1 lymphocytes about 40%. Strain 2 macrophage activation of F_1 lymphocyte DNA synthesis is suppressed by strain 13 anti-2 sera 80% and only 25% by strain 2 anti-13. In a reciprocal manner strain 13 macrophage activation of F_1 lymphocytes was suppressed 94% by strain 2 anti-13 and only 22% by strain 13 anti-2 antisera. The response of lymphocytes to macrophage pulsed with the nonspecific mitogen PHA is not appreciably altered by the source or specificity of the guinea pig sera used.

studies have indicated that the macrophage is the initial cell to interact with antigen (15, 16). Thus, when T-lymphocyte populations are fractionated by passage over adherence columns to deplete macrophages, exposed to high concentrations of antigens for 1–24 h, washed extensively to remove residual antigen, and then cultured for 2–3 days, no proliferation is seen (16). Indeed, no proliferation occurs even if the antigen-pulsed lymphocytes are reconstituted with macrophages before culture. However,

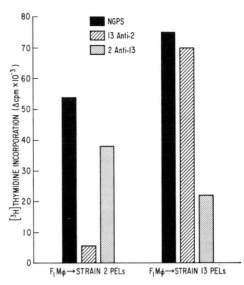

Fig. 5. The effect of alloantisera on macrophage-associated antigen activation of DNA synthesis in strain 2 or strain 13 animals. F_1 macrophages were pulsed for 60 min at 37°C with 100 μg/ml PPD in the presence of 30 μg/ml mitomycin C. The macrophages were washed and then added to either strain 2 or strain 13 PELs in the presence of NGPS, 13 anti-2, or 2 anti-13 sera. After 72 h, [³H]TdR incorporation was measured and the results are expressed as Δcpm $\times$ 10^{-3}. The inhibition of F_1 macrophage activation of parental T cell DNA synthesis is only seen when the alloantisera are directed against determinants present both on the macrophage and the T lymphocyte.

macrophages from nonimmunized animals exposed to antigen for as little as 60 min at 37°C, washed, and mixed with immune lymphocytes induce a proliferative response comparable to that of unfractionated lymphoid-rich populations (15). Lymphocytes, fibroblasts, thymocytes, guinea pig L_2C leukemia cells, or killed macrophages similarly exposed to antigen, washed, and mixed with immune lymphocytes do not induce proliferation. Furthermore, culture supernatants derived from antigen-pulsed macrophages are unable to induce immune lymphocyte proliferation (13). These data indicate that an initial and obligatory uptake of antigen by macrophages precedes immunospecific T-lymphocyte recognition and suggest that a direct physical interaction between macrophage and T lymphocyte is required for initiation of proliferation (16).

The data presented clearly establish that efficient interaction of macrophage-associated antigen with the immunospecific T lymphocyte as measured by antigen-induced lymphocyte proliferation occurs only when the macrophage and T lymphocyte are syngeneic. It is likely that this interaction is mediated by histocompatibility antigens themselves or by the products of genes closely linked to the major H-antigen complex. Combinations of semiallogeneic macrophages and T cells did lead to T-lymphocyte proliferation although this activation was somewhat less efficient than the syngeneic combinations. This indicates that the presence of a foreign H-antigen determinant does not completely block macrophage-lymphocyte cooperation. It is also unlikely that the requirement for syngeneic macrophages to initiate DNA synthesis is a peculiarity of the principal antigen used in these studies (PPD) as it was also seen with another soluble protein antigen DNP-GPA and as will be shown in the companion paper with the random copolymers of L-glutamic acid, L-lysine and L-glutamic acid, and L-tyrosine.

Although a small but significant MLR was seen in most experiments, the failure of allogeneic macrophages to activate T lymphocytes was not due to the presence of an inhibitor of blastogenesis liberated in mixtures of histo-incompatible cells. Thus, when strain 2 and strain 13 macrophages were intentionally mixed and added to lymphocytes from one of the strains, the activation of the syngeneic lymphocytes by the syngeneic macrophages was not inhibited; conversely, when the lymphocytes of the two strains were intentionally mixed and macrophages of one strain added, again no inhibition of the syngeneic macrophage-T cell interaction was observed. Furthermore, the presence of allogeneic macrophages did not inhibit the response of column-purified LNLs to high dose continuous antigen or PHA; the activation of the T lymphocytes in this situation is presumably mediated via the residual syngeneic macrophages in the LNL population. Although the activation of lymphocytes by allogeneic macrophages might exhibit a different kinetic pattern than their activation by syngeneic macrophages, no evidence of T-cell proliferation was seen when allogeneic mixtures of antigen-pulsed macrophages and lymphocytes were harvested over a 2–4 day period.

The precise role of the macrophage in a number of in vitro cell culture systems has not been defined. Indeed, it has recently been suggested that the requirement for macrophages is an in vitro "artifact" and their presence is merely required to prevent lymphocyte death (28); in the Mishell-Dutton antibody-forming system in the mouse, the plaque-forming cell capacity of nonadherent cells can be restored completely by 2-mercaptoethanol. We have already stated the evidence derived from in vitro studies of lymphocyte proliferation in the guinea pig that the macrophage is the cell that initially binds and subsequently presents antigen to the T lymphocyte. In addition, we have been unable to replace macrophage function in this system by 2-mercaptoethanol.

It has been demonstrated in a number of experimental systems that aggregation of lymphocytes occurs around macrophages and that this aggregation may be a requirement for the in vitro proliferative response (29, 30). More recent studies in this laboratory have shown that guinea pig macrophages possess an antigen-independent receptor for thymocytes as well as mature lymphocytes (36). Binding of lymphoid cells to macrophages requires active macrophage metabolism and is not mediated by lymphocyte membrane-associated immunoglobulin. The characteristics and specificities of this receptor indicate that it is distinct from other known macrophage receptors. Whether this receptor plays a functional role in macrophage-associated antigen activation or lymphocyte proliferation remains to be established.

Studies of macrophages from outbred animals strongly suggest that at least one component of the cell surface structures mediating macrophage lymphocyte interaction is the product of the major histocompatibility gene complex. Thus, a good correlation was seen between the ability of macrophages obtained from an outbred animal to activate strain 2 or 13 T cells and the presence of serologically detectable strain 2 or strain 13 determinants. However, the specificity of the interaction was not as great as that observed with inbred cells. Thus, macrophages from outbred animals that lacked strain 2 determinants did induce some lymphocyte proliferation in strain 2 animals. It should also be noted that according to the classification of guinea pig leuko-cyte antigens proposed by Sato and de Weck (31), strain 2 and strain 13 animals share a major histocompatibility antigen. Perhaps this could account for the small amount of stimulation seen when strain 2 macrophages are mixed with strain 13 T cells or vice versa; alternatively, if the major histocompatibility complex in the guinea pig is bipartite in structure and similar to that of mice and man, it is possible that strain 2 and strain 13 animals could share one major sublocus but the gene controlling macrophage-lymphocyte interaction is at or near the second sublocus. Resolution of these problems must wait further study of the guinea pig H antigens.

Although the requirement in the in vitro antigen-induced proliferative response for histocompatible macrophages and lymphocytes has not been examined in detail in another species, a number of observations suggest the possibility that a similar situation may exist in man. Cline and Swett reported that human monocytes that had been pulsed with tuberculin were capable of inducing lymphocyte proliferation only when they were derived from the same individual (12). On the other hand, Blaese et al. (32) observed that allogeneic monocytes could activate T-lymphocyte proliferation although in most cases this activation was significantly less than in the syngeneic combination. However, in this study the HL-A types of the cell donors were not determined and hence partial cooperation between allogeneic lymphocyte and macrophage may have occurred because the cells shared some HL-A specificities. Although the responding lymphocyte populations used in the study of Blaese et al. were obtained by adherence column purification and did not respond to stimulation by continuous antigen unless reconstituted with macrophages, it is possible that allogeneic macro-phages could have provided the "feeder" or nutritive function of syngeneic macro-phages and enough syngeneic macrophages remained after column purification to function as the antigen-presenting cells.

Kindred and Shreffler (33) have shown that the cooperation between T and B cells in vivo in congenitally athymic (*nu/nu*) mice requires that the participating cells be syngeneic. These observations have been confirmed and extended by Katz, Hamaoka, and Benacerraf (34) who have shown that both in vivo and in vitro T and B cell cooperation only occurs between T and B cells derived from syngeneic or semi-syngeneic donors. In contrast to the rigid requirement for syngeneic T and B cells necessary for antibody production, Katz and Unanue (35) have demonstrated that syngeneic and allogeneic antigen-pulsed macrophages activate in vitro secondary anti-DNP responses of mouse spleen cells equally well. These observations differ markedly from our results and a number of comments should be made about this discrepancy. In the studies reported by Katz and Unanue, the role of the macrophage has not been firmly established. Indeed when they attempted to assess this role by depleting macrophages on Petri dishes, the adherent cell-depleted populations were capable of mounting secondary in vitro responses in the absence of added macrophages, but the addition of macrophages that had been pulsed with antigen resulted in substantial improvement in the response particularly in the case of IgG antibody production. Furthermore, Katz and Unanue demonstrated that the antigen-presenting function of macrophages could be replaced although considerably less efficiently by antigen-pulsed fibroblasts. The differences in macrophage function in the mouse in contrast to those described in the present study in the guinea pig may represent either species differences or alternatively differences in the assay itself. With respect to the latter point, it should be noted that in the in vitro lymphocyte proliferation assay as used in our studies, a log linear relationship exists between the amount of macrophage-associated antigen and the amount of [^{3}H]TdR incorporated into new DNA. By contrast, no evidence has been presented in the mouse assays as to what the quantitative relationship was between the T helper activity and the resultant number of plaques. If T helper activity is present in excess, then significant differences in the efficiency of syngeneic vs. allogeneic macrophages might go unobserved. It is apparent that resolution of the apparent discrepancies in the function of macrophages in the mouse and guinea pig must await further studies.

Alloantisera directed against the histocompatibility antigens of the two inbred strains of guinea pigs are able to markedly inhibit macrophage-lymphocyte interaction as measured by the inhibition of antigen-mediated cell proliferation. This conclusion is derived from the experiments where parental macrophages are added to F_1 lymphocytes. In this situation inhibition of the proliferative response to PPD is only seen when the alloantisera are directed against histocompatibility antigens present on both the macrophage and T lymphocyte; when the alloantisera are directed against the determinants present only on the lymphocyte, no inhibition is seen. Conversely, when F_1 macrophages are added to parental lymphocytes, inhibition of the PPD response was again seen only when the alloantisera were directed against determinants present on both the macrophage and the lymphocyte; when the alloantiserum is directed only against specificities present on the macrophage, no inhibition is seen. This latter experiment makes it very unlikely that the inhibition of the DNA synthetic response produced by the alloantisera is secondary to their ability to kill or agglutinate macrophages, because the F_1 macrophages have both the strain 2 and the strain 13 determinants on their surface.

The conclusion to be drawn from this study is that the activation of immune lymphocytes by antigen-pulsed macrophages is dependent on the interaction of cell surface structures that are the products of the major histocompatibility complex, and this interaction can be blocked by sera that are directed against these structures or against membrane components close to these structures.

SUMMARY

Antigen activation of DNA synthesis in immune thymus-derived lympho-cytes of guinea pigs requires the cooperation of macrophages and lymphocytes. We have investigated the role of histocompatibility determinants in this macro-phage-lymphocyte interaction using cells from inbred strain 2 and 13 guinea pigs. The data demonstrate that efficient presentation of macrophage-asso-ciated antigen to the lymphocyte requires identity between macrophage and lymphocyte at some portion of the major histocompatibility complex. The failure of allogeneic macrophages to effectively initiate immune lymphocyte proliferation was not the result of the presence of an inhibitor of blastogenesis released in mixtures of allogeneic cells, peculiarities of the antigen or lymphoid cells employed, nor differing kinetics of activation by allogeneic macrophages. In addition, data were presented that demonstrated that alloantisera inhibit lymphocyte DNA synthesis by functional interference with macrophage-lymphocyte interaction.

We wish to thank Doctors I. Green and W. E. Paul for helpful discussions and critical review of the manuscript. We also wish to thank Mr. J. Thomas Blake and Mrs. Linda Lee for expert technical assistance.

REFERENCES

1. Davie, J. M., and W. E. Paul. 1971. Receptors on immunocompetent cells. II. Specificity and nature of receptors on dinitrophenylated guinea pig albumin-^{125}I-binding lymphocytes of normal guinea pigs. *J. Exp. Med.* **134**:495.

2. Davie, J. M., A. S. Rosenthal, and W. E. Paul. 1971. Receptors on immunocom-petent cells. III. Specificity and nature of receptors on dinitrophenylated guinea pig albumin-^{125}I-binding cells of immunized guinea pigs. *J. Exp. Med.* **134**:517.

3. Marchalonis, J. J., R. E. Cone, and J. L. Atwell. 1972. Isolation and partial char-acterization of lymphocyte surface immunoglobulin. *J. Exp. Med.* **135**:956.

4. Hammerling, U., and K. Rajewsky. 1971. Evidence for surface associated im-munoglobulin on T and B lymphocytes. *Eur. J. Immunol.* **1**:447.

5. Vitetta, E. S., C. Bianco, V. Nussenzweig, and J. W. Uhr. 1972. Cell surface immunoglobulin. IV. Distribution among thymocytes, bone marrow cells, and their derived populations. *J. Exp. Med.* **136**:81.

6. Perkins, W. D., M. J. Karnovsky, and E. R. Unanue. 1972. An ultrastructural study of lymphocytes with surface-bound immunoglobulin. *J. Exp. Med.* **135**: 267.

7. Rosenthal, A. S., J. M. Davie, D. L. Rosenstreich, and J. T. Blake. 1972. Depletion of antibody-forming cells and their precursors from complex lymphoid cell populations. *J. Immunol.* **108**:279.

8. Lamelin, J. P., B. Lisowska-Bernstein, A. Matter, J. E. Ryser, and P. Vassalli. 1972. Mouse thymus-independent and thymus-derived lymphoid cells. I. Im-munofluorescent and functional studies. *J. Exp. Med.* **136**:984.

9. Basten, A., J. F. A. P. Miller, N. L. Warner, and J. Pye. 1971. Specific inactiva-

tion of thymus-derived (T) and non-thymus derived (B) lymphocytes by ^{125}I-labeled antigen. *Nat. New Biol.* **231**:104.

10. Roelants, G., L. Forni, and B. Pernis. 1973. Blocking and redistribution ("capping") of antigen receptors on T and B lymphocytes by anti-immunoglobulin antibody. *J. Exp. Med.* **137**:1060.

11. Ashman, R. F., and M. C. Raff. 1973. Direct demonstration of theta-positive antigen-binding cells, with antigen-induced movement of thymus dependent cell receptors. *J. Exp. Med.* **137**:69.

12. Cline, M. J., and V. C. Swett. 1968. The interaction of human monocytes and lymphocytes. *J. Exp. Med.* **128**:1309.

13. Hersh, E. M., and J. E. Harris. 1968. Macrophage-lymphocyte interaction in antigen-induced blastogenic response of human peripheral blood leukocytes. *J. Immunol.* **100**:1184.

14. Seeger, R. C., and J. J. Oppenheim. 1970. Synergistic interaction of macrophages and lymphocytes in antigen-induced transformation of lymphocytes. *J. Exp. Med.* **132**:44.

15. Rosenstreich, D. L., and A. S. Rosenthal. 1973. Peritoneal exudate lymphocyte. II. *In vitro* lymphocyte proliferation induced by brief exposure to antigen. *J. Immunol.* **110**:934.

16. Waldron, J. A., R. G. Horn, and A. S. Rosenthal. 1973. Antigen induced proliferation of guinea pig lymphocytes *in vitro*: obligatory role of macrophages in the recognition of antigen by immune T-lymphocytes. *J. Immunol.* **111**:58.

17. Rosenthal, A. S., D. L. Rosenstreich, J. T. Blake, P. E. Lipsky, and J. W. Waldron. 1973. Mechanism of antigen recognition by T lymphocytes. *In* Proceedings Seventh Leukocyte Culture Conference. F. Daguillard, editor. Academic Press, New York. 201.

18. Benacerraf, B., and H. O. McDevitt. 1972. Histocompatibility-linked immune response genes. *Science (Wash. D.C.).* **175**:273.

19. Shevach, E. M., W. E. Paul, and I. Green. 1972. Histocompatibility-linked immune response gene function in guinea pigs. Specific inhibition of antigen-induced lymphocyte proliferation by alloantisera. *J. Exp. Med.* **136**:1207.

20. Shevach, E. M., D. L. Rosenstreich, and I. Green. 1973. The distribution of histocompatibility antigen on T and B cells in the guinea pig. *Transplantation.* **6**:126.

21. Rosenstreich, D. L., J. T. Blake, and A. S. Rosenthal. 1971. The peritoneal exudate lymphocyte. I. Differences in antigen responsiveness between peritoneal exudate and lymph node lymphocytes from immunized guinea pigs. *J. Exp. Med.* **134**:1170.

22. Roos, D., and J. A. Loos. 1970. Changes in the carbohydrate metabolism of mitogenically stimulated human peripheral lymphocytes. *Biochim. Biophys. Acta.* **222**:565.

23. Rosenstreich, D. L., E. M. Shevach, I. Green, and A. S. Rosenthal. 1972. The uropod-bearing lymphocyte of the guinea pig. Evidence for thymic origin. *J. Exp. Med.* **135**:1037.

24. Harzman, R. J., M. L. Bach, F. H. Bach, G. B. Thurman, and K. W. Sell. 1972. Precipitation of radioactively labeled samples: a semi-automatic multiple sample processor. *Cell. Immunol.* **4**:182.

25. Martin, W. J., L. Ellman, I. Green, and B. Benacerraf. 1970. Histocompatibility type and immune responsiveness in random bred Hartley strain guinea pig. *J. Exp. Med.* **132:**1259.

26. Bluestein, H. G., I. Green, and B. Benacerraf. 1971. Specific immune response genes of the guinea pig. IV. Demonstration in random-bred guinea pigs that responsiveness to a copolymer of L-glutamic acid and L-tyrosine is predicated upon the possession of a distinct strain 13 histocompatibility specificity. *J. Exp. Med.* **134:**1538.

27. Oppenheim, J. J. 1968. Relationship of in vitro lymphocyte transformation to delayed hypersensitivity in guinea pigs and man. *Fed. Proc.* **27:**21.

28. Chen, C., and J. G. Hirsch. 1972. The effects of mercaptoethanol and of peritoneal macrophages on the antibody-forming capacity of nonadherent mouse spleen cells in vitro. *J. Exp. Med.* **136:**604.

29. Mosier, D. E. 1969. Cell interactions in the primary immune response in vitro. A requirement for specific cell clusters. *J. Exp. Med.* **129:**351.

30. Hanifin, J. M., and M. J. Cline. 1970. Human monocytes and macrophages: interaction with antigen and lymphocytes. *J. Cell Biol.* **46:**97.

31. Sato, W., and A. L. de Weck. 1972. Leucocyte typing in guinea pigs. *Z. Immunitaetsforsch.* **144:**S.49.

32. Blaese, R. M., J. J. Oppenheim, R. C. Seeger, and T. A. Waldmann. 1972. Lymphocyte-macrophage interaction in antigen induced in vitro lymphocyte transformation in patients with the Wiskott-Aldrich syndrome and other diseases with anergy. *Cell. Immunol.* **4:**228.

33. Kindred, B., and D. C. Shreffler. 1972. H-2 dependence of cooperation between T and B cells *in vivo. J. Immunol.* **109:**940.

34. Katz, D. H., T. Hamaoka, and B. Benacerraf. 1973. Cell interactions between histoincompatible T and B lymphocytes. II. Failure of physiologic cooperative interactions between T and B lymphocytes from allogeneic donor strains in humoral response to hapten protein conjugates. *J. Exp. Med.* **137:**1405.

35. Katz, D. H., and E. R. Unanue. 1973. Critical role of determinant presentation in the induction of specific responses in immunocompetent lymphocytes. *J. Exp. Med.* **137:**967.

36. Lipsky, P. E., and A. S. Rosenthal. 1973. Macrophage-lymphocyte interaction. I. Characteristics of the antigen-independent binding of guinea pig thymocytes and lymphocytes to syngeneic macrophages. *J. Exp. Med.* In press.

NATURE OF THE ANTIGENIC COMPLEX RECOGNIZED
BY T LYMPHOCYTES

I. Analysis with an In Vitro Primary Response to Soluble Protein Antigens

By DAVID W. THOMAS* AND ETHAN M. SHEVACH

(From the Laboratory of Immunology, National Institute of Allergy and Infectious Diseases,
National Institutes of Health, Bethesda, Maryland 20014)

The major histocompatibility complex (MHC)[1] encodes a number of genes that control or regulate the immune response.

Several genes mapping in the immune response (*I*) region of the MHC appear to control the interaction of immunocompetent cells including the interaction of thymus-derived (T) lymphocytes with bone marrow-derived (B) lymphocytes (1) or macrophages (2, 3). Two concepts have been proposed to explain the histocompatibility restrictions in immunocompetent cell interactions. According to one proposal (the cellular interaction structure model) the *I*-region genes code for specific cellular interaction structures and that homology between these structures is necessary for effective cellular interactions. A second proposal (the complex antigenic determinant model) is based on the observations that mouse T cells sensitized to hapten- or virus-modified cells are primarily cytotoxic for similarly modified target cells which are *H-2D* or *H-2K* compatible (4, 5). These observations suggest that T cells do not recognize antigens per se, but can only be sensitized to antigen-modified membrane components or to complexes of antigen combined with certain membrane molecules.

Previous studies have demonstrated that in the guinea pig effective interaction between antigen-pulsed macrophages and immune T cells required that the macrophages and T cells be syngeneic (2, 6, 7). Because the MHC's of strain 2 and strain 13 animals differ only with respect to *I*-region antigens, it was proposed that identity between *I*-region antigens was required for effective macrophage-T-cell interaction. Identity at the B region of the guinea pig MHC, the homologue of the mouse *H-2K* and *H-2D* regions, was neither sufficient nor necessary for T-cell activation. Although these experiments are consistent with the cellular interaction structure model, it should be noted that these studies were performed with immune T cells. Another explanation for the failure of strain 13 macrophage-associated antigen to activate the immune strain 2 lymphocyte in vitro is that the strain 2 T cell has been primed in vivo only to antigen associated with strain 2 macrophages, and not strain 13 macrophages. Therefore, the failure of interaction between antigen-pulsed allogeneic macrophages and immune T cells is also consistent with the complex antigenic determinant model and in this respect can be regarded as a failure of antigen recognition rather than as a failure of cellular interaction.

---

* Supported by National Institutes of Health fellowship F32 AI 05141-01.

[1] *Abbreviations used in this paper:*   DNP-GPA, 2,4-dinitrophenyl guinea pig albumin; HSA, human serum albumin; [³H]TdR, tritiated thymidine; Ova, ovalbumin; MHC, major histocompatibility complex; MLR, mixed leukocyte reaction, NGPS, normal guinea pig serum; TNBS, trinitrobenzene sulfonate; TNP, trinitrophenol.

In order to further analyze the genetic factors involved in the regulation of macrophage-T-cell interactions, we have developed an assay for the generation of an in vitro primary response to soluble protein antigens in which nonimmune guinea pig T lymphocytes can be primed and subsequently challenged in tissue culture with antigen-pulsed macrophages. Experiments can thereby be performed in which any combination of allogeneic or syngeneic lymphocytes and macrophages can be tested. The results presented in this study are compatible with the proposal that T cells do not respond to native antigen and are primarily sensitized to an antigen-induced modification of the stimulator cell.

Materials and Methods

Animals. Inbred strain 2, strain 13, and $(2 \times 13)F_1$ guinea pigs were obtained from the Division of Research Services, National Institutes of Health, Bethesda, Md.

Preparation of Cells. Guinea pigs were injected intraperitoneally with 25 ml of sterile mineral oil (Marcol 52; Humble Oil & Refining Co., Houston, Texas) and the resulting peritoneal exudate was harvested 3–4 days later. This cell population consisting of approximately 75% macrophages, 10% neutrophils, and 15% lymphocytes was used as a source of macrophages for antigen pulsing (see below). A T-lymphocyte-enriched cell population was prepared by passing lymph node cells from animals injected in the foot pads several weeks previously with complete Freund's adjuvant (Difco Laboratories, Detroit, Mich.) over a rayon wool adherence column (8).

In Vitro Antigen Priming. Unfractionated peritoneal exudate cells $(5–10 \times 10^6/ml)$ were incubated for 1 h at 37°C in Hanks' balanced salt solution containing 25 μg/ml of mitomycin C (Sigma Chemical Co., St. Louis, Mo.) and 100 μg/ml of ovalbumin (Ova, five times crystallized; Pentex Biochemical, Kankakee, Illinois). The Ova-pulsed macrophages were washed four times to remove the unbound Ova and mitomycin C. In several experiments the macrophages were treated with trinitrobenzene sulfonate (TNBS) according to the method of Shearer et al. (5). The antigen-pulsed or TNBS-treated macrophages (1×10^6) and column-passed lymph node cells (5×10^6) were maintained in a total vol of 1.5 ml of EHAA medium (9) containing L-glutamine (300 μg/ml), penicillin (100 U/ml), 5-fluorocytosine (5 μg/ml), 2-mercaptoethanol $(5 \times 10^{-5}$ M), and 5% heat-inactivated normal guinea pig serum (NGPS). During the first culture the cells were incubated for 1 wk at 37°C in 2% CO_2 in air (Fig. 1). On the 3rd day of the first culture the medium was decanted and replaced with 1.5 ml of fresh medium.

In Vitro Assay of DNA Synthesis. The antigen-primed T cells $(1–2 \times 10^5$ per well) recovered from the first culture were restimulated in a second culture in flat-bottom microtiter plates (Cooke Laboratory Products Div., Dynatech Laboratories Inc., Arlington, Va.) with fresh Ova-pulsed peritoneal exudate cells $(1 \times 10^5$ per well) in a total vol of 0.2 ml of EHAA medium containing 5% NGPS (Fig. 1). In some experiments purified antigen-pulsed macrophages were obtained by washing away the nonadherent peritoneal exudate cells after a 1 h incubation in the microtiter wells before adding the Ova-primed T cells. After incubation for 3 days at 37°C in 2% CO_2 in air 1 μCi of tritiated thymidine ([³H]TdR, sp act 6.7 Ci/mmol; New England Nuclear, Boston, Mass.) was added to each well. The amount of radioactivity incorporated into cellular DNA was determined after an additional 18 h incubation with the aid of a semiautomated microharvesting device (10). The results of triplicate cultures are expressed as either total counts per minute per culture or as the difference between antigen-stimulated and nonstimulated control cultures (Δ counts per minute). The stimulation index refers to the counts per minute obtained from antigen-stimulated cultures divided by the counts per minute from control cultures.

Results

Kinetics and Specificity of the In Vitro Primary Response. During the course of the initial 7-day culture of strain 13 T cells and strain 13 Ova-pulsed macrophages samples of cells were harvested at daily intervals and the amount of DNA synthesis was determined (Fig. 2). The initial 7-day incubation of strain 13 T cells with Ova-pulsed strain 13 macrophages (Ova-MΦ) in tissue culture

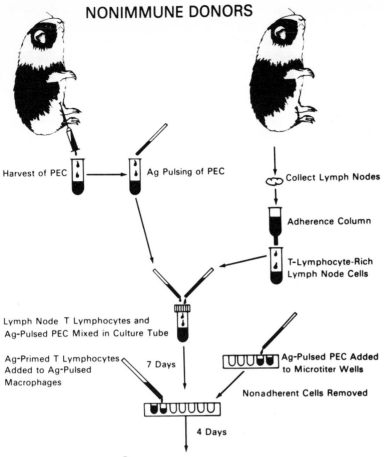

NONIMMUNE DONORS

Harvest of PEC → Ag Pulsing of PEC

Collect Lymph Nodes

Adherence Column

T-Lymphocyte-Rich Lymph Node Cells

Lymph Node T Lymphocytes and Ag-Pulsed PEC Mixed in Culture Tube

Ag-Primed T Lymphocytes Added to Ag-Pulsed Macrophages

7 Days

Ag-Pulsed PEC Added to Microtiter Wells

Nonadherent Cells Removed

4 Days

Assay [^{3}H] TdR Uptake Response to Antigen

FIG. 1. Schematic diagram for the in vitro priming of guinea pig T lymphocytes with antigen (Ag)-pulsed macrophages. The details of this procedure are described in the Materials and Methods. PEC, peritoneal exudate cells.

resulted in only a modest increase in DNA synthesis (two- to threefold) in comparison to cultures containing nonantigen-pulsed macrophages. After the first 7 days in culture approximately 20–30% of the original cell number was recovered; viability was 95% as determined by trypan blue exclusion. When these cells were transferred to microtiter wells containing fresh Ova-MΦ a substantial increase in DNA synthesis (60-fold) occurred which was maximal 4 days after transfer. It thus appeared that the T cells had become primed to the Ova and were capable of giving a substantial secondary response upon rechallenge with fresh Ova-MΦ. The in vitro priming seemed to be specific for Ova and not a nonspecific activation since Ova-primed T cells failed to respond to human serum albumin-pulsed macrophages (HSA-MΦ) in the second culture (Fig. 2). This specificity was more clearly shown in a "checkerboard" experiment in which separate aliquots of strain 13 T cells were primed with Ova or with DNP guinea pig albumin (DNP-GPA)-pulsed macrophages (Table I). The Ova-primed

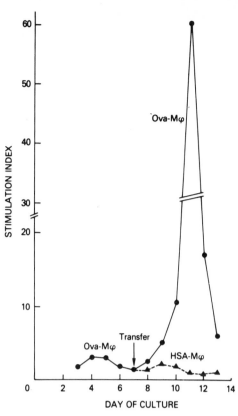

Fig. 2. Kinetics of the primary in vitro response with strain 13 T lymphocytes. Strain 13 T lymphocytes were incubated for 7 days with Ova-pulsed strain 13 macrophages (Ova-MΦ) and transferred to fresh Ova-MΦ in the microtiter wells for an additional 5-day incubation as described in the Materials and Methods. DNA synthesis was determined at daily intervals by the incorporation of [³H]TdR and the results are expressed as the stimulation index of antigen-stimulated to nonstimulated control cultures. Also shown is the response of Ova-primed T cells transferred to microtiter wells containing HSA-pulsed strain 13 macrophages (HSA-MΦ).

T cells could be restimulated with Ova-MΦ, but not with DNP-GPA-pulsed strain 13 macrophages in the second culture. In similar fashion, the DNP-GPA-primed strain 13 T cells could be restimulated in the second culture with DNP-GPA, but not Ova-pulsed macrophages.

Genetic Restriction of the In Vitro Primary Response. Because we have previously demonstrated genetic restrictions in the in vitro activation of in vivo primed T cells by antigen-pulsed macrophages (2), it was important to determine if similar genetic restrictions in T-cell-macrophage interactions occurred as a result of in vitro priming. We found that if strain 13 T cells were primed with strain 13 Ova-pulsed macrophages they could only be restimulated with Ova-pulsed strain 13, but not strain 2, macrophages in the second culture (Table II). Similarly, strain 2 T cells primed with Ova-pulsed strain 2 macrophages could only be restimulated by strain 2, but not strain 13 Ova-pulsed macrophages (Table III). This genetic restriction on the T-cell response was also obtained when TNBS-treated macrophages were used as the antigen (exp. 3,

TABLE I

Specificity of the In Vitro Primary Response with Strain 13 T Lymphocytes

First culture,* macrophage pulsed with:	Second culture,‡ [³H]TdR cpm		
	Macrophage pulsed with:		
	None	Ova	DNP-GPA
Ova	4,190	58,880	9,050
DNP-GPA	3,630	7,980	16,490

* Strain 13 T lymphocytes were incubated for 7 days with strain 13 macrophages pulsed with 100 μg of Ova or DNP-GPA, as described in the Materials and Methods.

‡ The incorporation of [³H]TdR was determined 4 days after transferring the cells recovered from the first culture to fresh antigen-pulsed strain 13 macrophages in microtiter wells as described in the Materials and Methods. Underlined values indicate cultures in which a positive response has occurred.

TABLE II

Genetic Restriction of the In Vitro Primary Response with Strain 13 T Lymphocytes

	First culture*		Antigen	Second culture,‡ [³H]TdR cpm			
				2 MΦ		13 MΦ	
	T lymphocyte	Antigen-pulsed MΦ		Unpulsed	Pulsed	Unpulsed	Pulsed
Exp. 1	13	13	Ova	3,300	1,800	2,820	103,400
	13	2		223,400	180,500	4,500	4,500
Exp. 2	13	13	Ova	9,130	11,100	10,900	147,200
	13	2		323,900	325,000	5,930	4,670
Exp. 3§	13	13	TNP	31,210	37,960	14,220	83,380
	13	2		215,700	219,900	4,220	13,160

* Strain 13 T lymphocytes were incubated for 7 days with syngeneic or allogeneic Ova-pulsed macrophages as described in the Materials and Methods.

‡ Same as in the footnote for Table I.

§ The antigen in this experiment was TNBS-treated macrophages prepared according to the method of Shearer et al. (5) instead of Ova.

TABLE III

Genetic Restriction of the In Vitro Primary Response with Strain 2 T Lymphocytes

	First culture*		Antigen	Second culture,‡ [³H]TdR cpm			
				13 MΦ		2 MΦ	
	T lymphocytes	Antigen-pulsed MΦ		Unpulsed	Pulsed	Unpulsed	Pulsed
Exp. 1	2	2	Ova	155	239	1,080	16,000
	2	13		14,240	15,190	728	719
Exp. 2	2	2	Ova	3,820	2,760	2,300	12,550
	2	13		77,800	50,870	3,490	5,700
Exp. 3§	2	2	TNP	20,520	23,750	15,530	77,620
	2	13		60,980	58,070	3,290	11,420

* Strain 2 T lymphocytes were incubated for 7 days with syngeneic or allogeneic Ova-pulsed macrophages as described in the Materials and Methods.

‡ Same as in the footnote for Table I.

§ Same as in the footnote for Table II.

Tables II and III). These results are therefore analogous to the previous findings of Rosenthal and Shevach (2) and suggest that the T cell and macrophage may require similar Ia antigens for efficient interaction. An alternate explanation may be that T cells respond only to the type of antigen-pulsed macrophage with which they were initially sensitized, irrespective of the histocompatibility type of the macrophage.

In order to test these two possibilities, strain 13 or strain 2 T cells were primed with syngeneic or allogeneic strain 2 or strain 13 antigen-pulsed macrophages and restimulated with the respective syngeneic or allogeneic macrophages in the second culture (Tables II and III). It was clear from the results, however, that this approach was not practical since a substantial secondary mixed leukocyte reaction (MLR) was generated with the allogeneic macrophages that obscured any antigen-specific stimulation that might have been present. The T cells were capable of responding to Ova or trinitrophenol (TNP), though, since in the same experiments a good response was elicited by priming with syngeneic antigen-pulsed or TNBS-treated macrophages. The relatively reduced MLR produced by strain 2 T cells responding to strain 13 macrophages, as compared to the 13 against 2 MLR, is a consistent finding as reported previously by Greineder and Rosenthal (11). In the case where the 2 against 13 MLR is relatively weak (exp. 1, Table III) it might have been expected that some antigen-specific stimulation greater than the MLR stimulation alone would have been detected. However, in several experiments the magnitude of the response of strain 2 T cells to either untreated or antigen-pulsed strain 13 macrophages was the same.

Genetic Restriction On the Response of (2 × 13)F$_1$ T Cells Primed with Antigen-Pulsed Parental Macrophages. In order to circumvent the problem of the MLR we primed (2 × 13)F$_1$ T cells and subsequently rechallenged them with the strain 2 or strain 13 Ova-pulsed macrophages (Table IV). We found that if the F$_1$ T cells were initially primed with Ova-pulsed macrophages from one parent they could be restimulated in the second culture only with the Ova-pulsed parental macrophage used for initial sensitization, and not with those of the other parent. For example, F$_1$ T cells primed with strain 13 Ova-pulsed macrophages responded only to strain 13, but not strain 2 Ova-pulsed macrophages in the second culture. Similarly, F$_1$ T cells primed with strain 2 Ova-pulsed macrophages were restimulated only by strain 2, but not strain 13, Ova-pulsed macrophages. This genetically restricted response by F$_1$ T cells was not unique to Ova since the same phenomenon occurred when fetal calf serum-pulsed or TNBS-treated parental macrophages were used as the antigen (exps. 4 and 5, Table IV). In contrast to the genetic restriction observed with parental antigen-pulsed macrophages, F$_1$ T cells primed with (2 × 13)F$_1$ antigen-pulsed macrophages could be restimulated in the second culture with either the strain 2 or strain 13 antigen-pulsed macrophages.

Discussion

The technique of an in vitro primary response to soluble protein antigens described in this study provides a method by which the mode of antigen presentation can be easily manipulated. It is thus possible to more clearly define the genetic restrictions on antigen responsiveness by controlling the histocom-

TABLE IV

In Vitro Primary Response of (2 × 13)F₁ T Lymphocytes Stimulated with Antigen-Pulsed Parental Macrophages

| | First culture* | | Antigen | Second culture,‡ [³H]TdR Δ cpm | |
| | | | | Antigen-pulsed MΦ strain: | |
	T Lymphocyte	Antigen-pulsed MΦ		2	13
Exp. 1	(2 × 13)F₁	(2 × 13)F₁	Ova	6,750	21,460
		2		2,100	0
		13		10	9,620
Exp. 2	(2 × 13)F₁	(2 × 13)F₁	Ova	27,730	95,680
		2		10,590	530
Exp. 3	(2 × 13)F₁	2	Ova	7,450	980
		13		0	8,200
Exp. 4§	(2 × 13)F₁	2	FCS	11,920	840
		13		0	12,260
Exp. 5‖	(2 × 13)F₁	(2 × 13)F₁	TNP	42,240	11,230
		2		81,220	900
		13		0	27,950

* (2 × 13)F₁ T lymphocytes were incubated for 7 days with (2 × 13)F₁, strain 2, or strain 13 antigen-pulsed macrophages as described in the Materials and Methods.

‡ Same as in the footnote for Table I except that the unstimulated control counts per minute have been subtracted and only the Δ counts per minute are shown.

§ Macrophages were pulsed with fetal calf serum (FCS) as the antigen in this experiment instead of Ova.

‖ The antigen in this experiment was TNBS-treated macrophages prepared according to the method of Shearer et al. (5) instead of Ova.

patibility type of the stimulator cell used to initially present an antigen. In agreement with the previous findings of Rosenthal and Shevach (2) using T cells immunized in vivo, we found that T cells primed with syngeneic antigen-pulsed macrophages in vitro could be restimulated only with syngeneic, but not allogeneic, antigen-pulsed macrophages. These results suggest that either identity is required between the strain 2 or strain 13 *I*-region products for efficient T-cell-macrophage interaction, or alternatively, that T cells respond only to antigen associated with the type of macrophage used for initial sensitization. A direct test of these possibilities by sensitizing T cells with antigen-pulsed allogeneic macrophages was not feasible because of the magnitude of the secondary MLR elicited by the allogeneic stimulus. However when we eliminated the MLR by using combinations of parental macrophages and (2 × 13)F₁ T cells, we found that F₁ T cells primed with antigen-pulsed macrophages from one parent could be restimulated only with the parental macrophage used for initial sensitization, and not with those of the other parent. If only *I*-region identity was required for efficient T-cell-macrophage interaction, then (2 × 13)F₁ T cells, which express both the strain 2 and strain 13 *I*-region antigens, primed with one parental antigen-pulsed macrophage would have been restimulated with anti-gen-pulsed macrophages from either parent. In contrast, our results suggest that the primary restriction on the $(2 \times 13)F_1$ T-cell response may be imposed by the type of macrophage used for initial sensitization, irrespective of the *I*-region antigen which that macrophage expressed.

Based on our results it is likely that some $(2 \times 13)F_1$ T cells are able to be sensitized to the antigenic complex containing strain 2 *I*-region products while other T cells are sensitized to the immunogenic complex containing strain 13 *I*-region products. One prediction of this proposal is that in an immunized F_1 animal separate T-cell subpopulations would exist which have been sensitized to antigen associated with either the strain 13 or strain 2 *I*-region gene products presented by the $(2 \times 13)F_1$ macrophage. In fact, using cells from immune $(2 \times 13)F_1$ donors direct evidence has been obtained for the existence of two populations of lymphocytes, one specific for antigen presented by strain 2 macrophages and the other for antigen associated with strain 13 macrophages (12). In these experiments T lymphocytes from F_1 donors immunized with Ova were positively selected by culturing them for 1 wk with either strain 2 or strain 13 Ova-pulsed macrophages. After positive selection in the primary culture the F_1 T cells showed significant responses when recultured with Ova-pulsed macrophages of the histocompatibility type used in the initial culture, but little or no response to Ova-pulsed macrophages of the other histocompatibility type.

Although the genetic restriction on the response of $(2 \times 13)F_1$ T cells stimulated with parental macrophages supports the complex antigenic determinant concept of T-cell recognition, the cellular interaction structure proposal cannot be ruled out since the F_1 T-cell and parental macrophages share Ia specificities. However, in order for our experimental results to be consistent with the cellular interaction structure model one must postulate allelic exclusion for cellular interaction structures at the T-cell level. We have been unable to demonstrate allelic exclusion for *I*-region antigens and 95–98% of $(2 \times 13)F_1$ T cells can be stained by either anti-2 or anti-13 serum using indirect immunofluorescence (E. Shevach, I. Green, and W. E. Paul, unpublished observations).

It is also difficult to reconcile the cellular interaction structure model of immunocompetent cell interaction with a number of recent experimental observations which have demonstrated cellular cooperation across an allogeneic barrier. Pfizenmaier et al. (13) have recently demonstrated that *H-2* homology between effector and stimulator cells was not obligatory. Thus, when cytotoxic T cells from F_1 radiation chimeras repopulated with lymphocytes from one parent were primed with viral-infected or hapten-modified stimulator cells of the other parental haplotype they could only kill similarly treated target cells expressing the same H-2 antigens as those present during the initial sensitization. Similarly, Pierce et al. (14) found that T-cell helper activity for antibody production from mice immunized with antigen-pulsed allogeneic macrophages could be elicited in vitro only with the same macrophage type as used for the in vivo immunization. Miller et al. (15) have analogous observations for the transfer of delayed hypersensitivity in the mouse. Although we could not directly test the ability of antigen-pulsed allogeneic macrophages to prime across the histocompatibility barrier because of the magnitude of the secondary MLR, under conditions where the MLR has been reduced by anti-Ia sera directed toward the stimulatory macrophage we could directly demonstrate that T cells can be specifically sensitized to antigens associated with allogeneic macrophages.[2] All of these results strongly suggest that T cells recognize antigen in association with MHC gene products but that MHC identity is not required for efficient immunocompetent cellular interaction.

Whereas mouse cytotoxic T cells primarily recognize antigen associated with *H-2K* or *H-2D* gene products (4, 5), our findings in the guinea pig indicate that the genetic restriction on T-cell activation as measured by proliferative assays is imposed by *I*-region antigens rather than by gene products homologous to the *H-2K* or *H-2D* regions in the mouse. Taken together, these results suggest the possibility that *H-2K*, *H-2D*, or *I*-region genes may serve a similar function in T-cell recognition. That is, either type of gene product may be included in the antigen-induced modification of self-components to create the immunogenic complex. However, T-cell subpopulations perhaps expressing different functions may exhibit a preference for the type of MHC gene product which is included in the complex. It is of interest that in our hands the proliferative response to TNP-modified macrophages or macrophages pulsed with soluble protein both appear to be exclusively under *I*-region antigen regulation. In contrast, Miller et al. (15) have recently demonstrated that *I*-region identity, but not *K* or *D* identity, was sufficient to demonstrate the transfer of delayed hypersensitivity in the mouse when fowl gamma globulin was used as an antigen, while either *I, K,* or *D* identity would permit the transfer of delayed hypersensitivity to dinitrofluorobenzene. The discrepancy in these findings may be explained by the different animal species and experimental techniques utilized.

According to the complex antigenic determinant hypothesis, T cells derived from an immunized F_1 animal should be equally sensitized with the antigen-altered Ia complex of both parental haplotypes and should be able to be stimulated in vitro by antigen-pulsed macrophages from either parent. One difficulty with this proposal is the finding by Shevach (7) that for responses controlled by *Ir* genes immune (responder × nonresponder)F_1 T cells could not be activated by antigen-pulsed nonresponder macrophages. One explanation for this finding would be to postulate that the nonresponder macrophage lacks the Ia antigen which can be altered by the antigen under study. If this is the case, then one might predict that nonresponder T cells may be able to be sensitized to antigen-pulsed responder macrophages expressing the necessary Ia antigen. Experiments are now in progress to examine this possibility.

The results we have presented in this study support the concept that T cells are incapable of being activated by antigen per se. Rather, the immunogenic complex seems to involve an antigen-induced modification of self-components which, in the guinea pig, probably includes *I*-region gene products. The exact composition of this immunogenic complex is unknown, but there are several possible roles for the function of the *I*-region products. One possibility is that antigen directly binds to the Ia molecule which modifies the antigen for presentation to the T cell. Alternatively, by binding to the *I*-region gene product the

---

[2] Thomas, D. W., and E. Shevach. Nature of the antigenic complex recognized by T lymphocytes. II. T lymphocyte activation by direct modification of macrophage histocompatibility antigens. Manuscript in preparation.

antigen may induce a modification of the Ia molecule which T cells recognize as an altered self-component. Another possibility is that the combination of antigen with *I*-region gene products creates a new complex antigen which contains determinants derived from both the Ia molecule and antigen.

Summary

In order to analyze the genetic factors involved in the regulation of macrophage-T-cell interaction we have developed an in vitro primary response to soluble protein antigens in which nonimmune guinea pig T cells can be sensitized and subsequently challenged in tissue culture with antigen-pulsed macrophages. Antigen-specific T-cell activation, as measured by increased DNA synthesis, occurred when syngeneic antigen-pulsed macrophages were used for both initial sensitization and secondary challenge. No T-cell activation occurred when allogeneic antigen-pulsed macrophages were used for secondary challenge of cells primed with syngeneic macrophages. When allogeneic antigen-pulsed macrophages were used in both primary and secondary cultures it was difficult to assess antigen-specific stimulation due to the substantial mixed leukocyte reaction. However, when T cells from F_1 animals were primed with parental antigen-pulsed macrophages they responded only to the parental macrophages used for initial sensitization but not to those of the other parent. These results are discussed with respect to T-cell recognition of a complex antigenic determinant which may include *I*-region gene products.

We thank Doctors W. E. Paul and A. S. Rosenthal for many stimulating and helpful discussions.

Received for publication 28 June 1976.

References

1. Katz, D. H., M. Graves, M. E. Dorf, H. Dimuzio, and B. Benacerraf. 1975. Cell interactions between histoincompatible T and B lymphocytes. VIII. Cooperative responses between lymphocytes are controlled by genes in the *I* region of the *H-2* complex. *J. Exp. Med.* 141:263.
2. Rosenthal, A. S., and E. Shevach. 1973. Function of macrophages in antigen recognition by guinea pig T lymphocytes. I. Requirement for histocompatible macrophages and lymphocytes. *J. Exp. Med.* 138:1194.
3. Erb, P., and M. Feldmann. 1975. The role of macrophages in the generation of T-helper cells. II. The genetic control of the macrophage-T-cell interaction for helper cell induction with soluble antigens. *J. Exp. Med.* 142:460.
4. Zinkernagel, R. M., and P. C. Doherty. 1975. *H-2* compatibility requirement for T-cell-mediated lysis of target cells infected with lymphocytic choriomeningitis virus. Different cytotoxic T-cell specificities are associated with structures coded for in *H-2K* or *H-2D*. *J. Exp. Med.* 141:1427.
5. Shearer, G. M., T. G. Rehn, and C. A. Garbarino. 1975. Cell-mediated lympholysis of trinitrophenyl-modified autologous lymphocytes. Effector cell specificity of modified cell surface components controlled by the *H-2K* and *H-2D* serological regions of the murine major histocompatibility complex. *J. Exp. Med.* 141:1348.
6. Shevach, E. M., and A. S. Rosenthal. 1973. Function of macrophages in antigen recognition of guinea pig T lymphocytes. II. Role of the macrophage in the regulation of genetic control of the immune response. *J. Exp. Med.* 138:1213.

7. Shevach, E. M. 1976. The function of macrophages in antigen recognition by guinea pig T lymphocytes. III. Genetic analysis of the antigens mediating macrophage-T lymphocyte interaction. *J. Immunol.* **116**:1482.

8. Rosenstreich, D. L., J. T. Blake, and A. S. Rosenthal. 1971. The peritoneal exudate lymphocyte. I. Differences in antigen responsiveness between peritoneal exudate and lymph node lymphocytes from immunized guinea pigs. *J. Exp. Med.* **134**:1170.

9. Click, R. E., L. Benck, and B. J. Alter. 1972. Immune responses *in vitro*. I. Culture conditions for antibody synthesis. *Cell Immunol.* **3**:264.

10. Harrison, M. R., G. B. Thurmond, and G. M. Thomas. 1974. A simple and versatile harvesting device for processing radioactive label incorporated into and/or released from cells in microculture. *J. Immunol. Methods.* **4**:11.

11. Greineder, D. K., and A. S. Rosenthal. 1975. Macrophage activation of allogeneic lymphocyte proliferation in the guinea pig mixed leukocyte culture. *J. Immunol.* **114**:1541.

12. Paul, W. E., E. Shevach, D. W. Thomas, S. F. Pickeral, and A. S. Rosenthal. Genetic restriction in T lymphocyte activation by antigen-pulsed peritoneal exudate cells. *Cold Spring Harbor Symp. Quant. Biol.* **41**:in press.

13. Pfizenmaier, K., A. Starzinski-Powitz, H. Root, M. Röllinghoff, and H. Wagner. 1976. Virus and trinitrophenol hapten-specific T-cell-mediated cytotoxicity against H-2 incompatible target cells. *J. Exp. Med.* **143**:999.

14. Pierce, C. W., J. A. Kapp, and B. Benacerraf. 1976. Stimulation of antibody responses *in vitro* by antigen-bearing syngeneic and allogeneic macrophages. *In* The Role of Products of the Histocompatibility Gene Complex in Immune Responses. D. H. Katz and B. Benacerraf, editors. Academic Press, Inc., New York. 391.

15. Miller, J. F. A. P., M. A. Vadas, A. Whitelaw, and J. Gamble. 1976. Role of major histocompatibility complex gene products in delayed type hypersensitivity. *Proc. Natl. Acad. Sci. U. S. A.* **73**:2486.

REGULATION BY THE *H-2* GENE COMPLEX OF MACROPHAGE-LYMPHOID CELL INTERACTIONS IN SECONDARY ANTIBODY RESPONSES IN VITRO*

By CARL W. PIERCE,‡ JUDITH A. KAPP, AND BARUJ BENACERRAF

(From the Department of Pathology, Harvard Medical School, Boston, Massachusetts 02115)

The development of optimal antibody responses to T-cell-dependent antigens requires the participation of and interactions among at least three distinct types of cells in the immune system: macrophages (Mϕ)[1]; thymus-derived cells (T cells); and precursors of antibody-producing cells (B cells) (1). Mϕ have the critical function of presenting antigen in a highly immunogenic form to T and B cells to initiate the immune response (1, 2). Efficient physiologic interactions among antigen-specific murine T and B cells in the development of secondary IgG antibody responses appear to require that these cells share specificities encoded by the *I* region of the *H-2* complex (3, 4). Similarly, the generation of carrier-specific helper T cells in vitro has been reported to require that Mϕ and T cells also share specificities encoded by the *I* region of the *H-2* complex (5). In another experimental system, combinations of immune guinea pig lymphocytes and Mϕ must be syngeneic for successful development of DNA synthetic responses to the immunizing antigen in vitro (6, 7). By contrast, genetic restrictions have not been observed in interactions among murine Mϕ and lymphoid cells (T cells and B cell) necessary for the development of primary plaque-forming cell (PFC) responses in vitro; lymphoid cells develop comparable primary PFC responses when incubated with antigen and syngeneic or allogeneic Mϕ (references 1, 2, and 8–14; and footnote 2).

In the course of the experiments investigating the lack of genetic restrictions in primary PFC responses, it became obvious that one significant difference between our experiments and those demonstrating genetic restrictions in Mϕ-lymphocyte interactions was that the latter used immune lymphocytes. In this communication, we report experiments investigating the ability of syngeneic and allogeneic antigen-bearing Mϕ to stimulate secondary PFC responses by immune lymphoid cells in vitro. The results demonstrate that immune lymph-

* This investigation was supported by U. S. Public Health Service Research Grants AI-09897 and AI-09920 from the National Institute of Allergy and Infectious Diseases.

‡ Recipient of U. S. Public Health Service Research Career Development Award 5KO4-AI-70173 from the National Institute of Allergy and Infectious Diseases.

[1] Abbreviations used in this paper: GAT, terpolymer of L-glutamic acid^{60}-L-alanine^{30}-L-tyrosine^{10}; HBSS, Hanks' balanced salt solution lacking sodium bicarbonate; IgG, refers to IgG$_1$ and IgG$_2$ in aggregate; Mϕ, macrophage(s); PFC, plaque-forming cell(s).

[2] Pierce, C. W., and J. A. Kapp. Lack of genetic restrictions regulating macrophage-lymphoid cell interactions in primary antibody responses *in vitro*. Manuscript in preparation.

oid cells develop secondary PFC responses preferentially when stimulated by antigen-bearing Mϕ syngeneic to the Mϕ used to immunize the lymphoid cells and that these genetic restrictions are controlled by the *H-2* complex.

Materials and Methods

Mice. Male C57BL/6 (*H-2^b^*), C57BL/10 (*H-2^b^*), P/J (*H-2^p^*), DBA/1 (*H-2^q^*), B10.G (*H-2^q^*), D1.LP (*H-2^b^*), B10.A (*H-2^a^*), A/J (*H-2^a^*), and (C57BL/6 × DBA/1)F₁ (*H-2^b/q^*) mice were obtained from The Jackson Laboratory, Bar Harbor, Maine, or the Department of Pathology Animal Facility at Harvard Medical School, Boston, Mass. The mice were maintained on laboratory chow and acidified-chlorinated water and used when 2–6-mo old.

Antigens. Sheep erythrocytes (SRBC) (Grand Island Biological Co., Grand Island, N. Y.) were washed three times with Hanks' balanced salt solution (HBSS) before use as indicator cells in the PFC assay or as antigen in culture (15). The synthetic linear random terpolymer of L-glutamic acid⁶⁰-L-alanine³⁰-L-tyrosine¹⁰ (GAT) (Miles Laboratories, Inc., Miles Research Div., Kankakee, Ill.), mol wt approximately 45,000, was prepared for use as antigen in culture (16), for preparing GAT-Mϕ (13, 17), and for coupling to SRBC for use as indicator cells in the PFC assay (16), as described previously.

Culture System and Hemolytic Plaque Assay. Spleen cells or splenic lymphoid cells, depleted of Mϕ by adherence techniques (18), at 10⁷ cells/ml in completely supplemented Eagle's minimal essential medium (MEM) containing 10% fetal calf serum (lot M26302; Reheis Chemical Co., Kankakee, Ill.), were incubated with graded numbers of GAT-Mϕ, or Mϕ and 10⁷ SRBC, or 5 µg GAT for 5 days under modified Mishell-Dutton conditions (15). Details of specific experiments are in the text. IgM and IgG PFC responses to SRBC and IgG GAT-specific PFC responses were assayed on SRBC and GAT-SRBC indicator cells, respectively, using the slide modification of the Jerne hemolytic plaque assay (15, 16). Preparation of GAT-SRBC and specificity controls for the GAT PFC assay have been described (16).

Preparation of GAT-Bearing Macrophages. Peritoneal exudate cells, as a source of Mϕ, were collected from mice injected intraperitoneally (i.p.) with 1 ml sterile 10% proteose peptone broth (Difco Laboratories, Detroit, Mich.) 3 days previously. This procedure routinely yielded 6–8 × 10⁶ cells per mouse, approximately 85% of which were morphologically Mϕ. These cells were washed three times with HBSS, adjusted to 2 × 10⁶ cells/ml in HBSS, and reacted with 100 µg/ml GAT (containing 1% [¹²⁵I]GAT) at 4°C for 45–60 min (13, 17). The cells were washed three times with 50 volumes of HBSS and adjusted to the desired density in HBSS. GAT bound to Mϕ was quantitated by counting a portion of the cells in a Packard Autogamma Spectrometer (Packard Instrument Co., Inc., Downers Grove, Ill.) and is expressed as nanograms GAT/10⁵ or 10⁶ cells in the tables. GAT-Mϕ were added in graded numbers to cultures of 10⁷ spleen or splenic lymphoid cells immediately after preparation. In one experiment (Table I) peritoneal exudate Mϕ, not reacted with GAT, were added in graded numbers to cultures of 10⁷ splenic lymphoid cells and 10⁷ SRBC.

Immunization of Mice. Mice were immunized by i.p. injection of 3–4 × 10⁶ GAT-Mϕ (bearing 20–30 ng GAT/10⁶ cells) or 10 µg GAT in a mixture of magnesium-aluminum hydroxide gel (Maalox; Wm. H. Rorer, Inc., Fort Washington, Pa.) and pertussis vaccine (Eli Lilly & Co., Indianapolis, Ind.) (16). At various intervals after immunization, spleen cells from these mice were assayed for secondary PFC responses to GAT in vitro.

Results

PFC Responses of GAT-Primed Lymphoid Cells Stimulated by Syngeneic and Allogeneic Mϕ. The lack of genetic restrictions regulating efficient Mϕ-lymphoid cell interactions in primary PFC responses will be reported in detail separately.² However, to present the data on genetic restrictions in Mϕ-lymphoid cell interactions in secondary PFC responses in the proper context, the relevant data on the lack of such restrictions in the primary response should be summarized. First, syngeneic and allogeneic Mϕ supported development of comparable primary PFC responses to SRBC and GAT in vitro. Second, when

GAT-Mϕ were incubated for 24 h alone before addition to lymphoid cell cultures, approximately 90% of the GAT initially associated with Mϕ was released into the culture medium. Since the PFC responses stimulated by these aged syngeneic and allogeneic GAT-Mϕ were comparable and not significantly less than responses stimulated by freshly prepared GAT-Mϕ, the possibility that GAT is transferred from allogeneic Mϕ to syngeneic Mϕ contaminating the splenic lymphoid cells and that this GAT actually stimulates the PFC response is unlikely. Third, since allogeneic Mϕ may stimulate a mixed lymphocyte response with resultant release of factors which stimulate antibody responses, DNA synthetic responses have been measured in the same cultures used to assess PFC responses; 7×10^4 allogeneic Mϕ have never stimulated significant DNA synthetic responses in cultures of 10^7 C57BL/6 lymphoid or spleen cells ($E/C < 1.5$). Further, factors which stimulate or enhance PFC responses have not been demonstrated in the medium from such cultures.

In other experiments, we observed that splenic lymphoid cells from mice immunized with GAT developed secondary PFC responses to GAT preferentially when incubated with syngeneic GAT-Mϕ. To probe this phenomenon further, C57BL/6 mice were immunized with GAT in Maalox-pertussis and their splenic lymphoid cells were incubated with graded numbers of syngeneic or allogeneic (DBA/1 $H\text{-}2^q$) GAT-Mϕ or normal Mϕ and SRBC (Table I). As a control, primary responses to GAT by virgin C57BL/6 lymphoid cells stimulated by the same GAT-Mϕ were determined. Syngeneic and allogeneic GAT-Mϕ stimulated comparable primary PFC responses to GAT. By contrast, GAT-primed C57BL/6 lymphoid cells preferentially developed secondary PFC responses to GAT when stimulated by syngeneic GAT-Mϕ, the same Mϕ that were involved in the immunization process of the lymphoid cells in vivo. These same GAT-primed lymphoid cells, however, developed comparable primary PFC responses to SRBC when incubated with syngeneic or allogeneic Mϕ, indicating that the restrictions observed in secondary responses are antigen-specific and not a general result of immunization.

In further experiments, C57BL/6 mice were immunized by i.p. injection of syngeneic or allogeneic (DBA/1) GAT-Mϕ or normal Mϕ as indicated in Table II. 28 days later, spleen cells from these mice were incubated with graded numbers of C57BL/6 or DBA/1 GAT-Mϕ or 5 μg soluble GAT. Normal DBA/1 Mϕ were added to some cultures for control purposes. Spleen cells from mice immunized with C57BL/6 GAT-Mϕ developed secondary PFC responses to GAT when incubated with soluble GAT or C57BL/6 GAT-Mϕ, but not when incubated with allogeneic DBA/1 GAT-Mϕ (A). By contrast, spleen cells from mice immunized with DBA/1 GAT-Mϕ developed secondary PFC responses to GAT when incubated with DBA/1 GAT-Mϕ, but not when incubated with soluble GAT or C57BL/6 GAT-Mϕ (B). Spleen cells from mice immunized with both C57BL/6 and DBA/1 GAT-Mϕ developed comparable secondary PFC responses when stimulated with soluble GAT, C57BL/6, or DBA/1 GAT-Mϕ (C). As a control, unprimed C57BL/6 spleen cells developed comparable primary PFC responses when stimulated with soluble GAT, C57BL/6, or DBA/1 GAT-Mϕ (E).

To evaluate possible effects due to sensitization of C57BL/6 lymphocytes by DBA/1 Mϕ, C57BL/6 mice were immunized with C57BL/6 GAT-Mϕ plus normal

TABLE I

PFC Responses of GAT-Primed Lymphoid Cells Stimulated by Syngeneic and Allogeneic Mφ

| GAT-Mφ/culture | Day 5 IgG GAT-specific PFC/culture | | | |
| | 10^7 virgin lymphoid cells — C57BL/6 (H-2^b) | | 10^7 GAT-primed lymphoid cells* — C57BL/6 (H-2^b) | |
	C57BL/6 Mφ‡ (H-2^b)	DBA/1 Mφ‡ (H-2^q)	C57BL/6 Mφ‡ (H-2^b)	DBA/1 Mφ‡ (H-2^q)
7×10^4	2,950	2,445	3,980	1,260
3.5×10^4	530	1,195	3,310	710
1×10^4	<10	105	1,130	105
No Mφ—5 μg GAT	<10		205	

Mφ/culture + 10^7 SRBC	Day 5 anti-SRBC PFC/culture			
	10^7 GAT-primed lymphoid cells* — C57BL/6 (H-2^b)			
	C57BL/6 Mφ§		DBA/1 Mφ§	
	IgM	IgG	IgM	IgG
7×10^4	1,535	490	1,340	490
3.5×10^4	1,095	·245	675	150
1×10^4	270	60	405	110
No Mφ + 10^7 SRBC	195	85	—	—

* Lymphoid cells were prepared from spleen cell suspensions from C57BL/6 mice immunized 53 days previously by i.p. injection of 10 μg GAT in Maalox-pertussis. At culture initiation, these cells had 25 IgG GAT-specific PFC per 10^7 cells.

‡ C57BL/6 Mφ, 1.20 ng GAT/10^5 cells; DBA/1 Mφ, 1.25 ng GAT/10^5 cells.

§ Normal, non-GAT-bearing Mφ.

DBA/1 Mφ (D). Further, cultures with both C57BL/6 GAT-Mφ and normal DBA/1 Mφ provide critical control information. Normal DBA/1 Mφ had no significant effect on responses stimulated by C57BL/6 GAT-Mφ in cultures of spleen cells from mice immunized with C57BL/6 GAT-Mφ (A), C57BL/6 GAT-Mφ and DBA/1 GAT-Mφ (C), C57BL/6 GAT-Mφ and normal DBA/1 Mφ (D), or unprimed spleen cells (E). These observations indicated that DBA/1 Mφ did not stimulate a suppressive effect on secondary responses of appropriately immunized spleen cells to C57BL/6 GAT-Mφ. Further, DBA/1 Mφ added to cultures of spleen cells from mice immunized with DBA/1 GAT-Mφ did not enhance the responses in cultures containing C57BL/6 GAT-Mφ (B). This suggested that allogeneic Mφ did not stimulate a nonspecific enhancing effect and, more importantly, that any transfer of GAT from C57BL/6 to DBA/1 Mφ was insufficient to stimulate a secondary PFC response to GAT.

Comparable results have been obtained when P/J (H-2^p) or BALB/c (H-2^d) Mφ were substituted for DBA/1 Mφ throughout the experiment. Further, significant DNA synthetic responses ($E/C < 1.5$) were not detected in cultures of spleen cells from mice immunized with allogeneic Mφ and H-2-identical alloge-

TABLE II

Secondary PFC Responses to GAT Stimulated by Syngeneic and Allogeneic Mφ in Cultures of Spleen Cells Primed with GAT-Mφ

GAT-Mφ used to prime spleen cells*	GAT-Mφ/culture	Day 5 IgG GAT-specific PFC/culture 10^7 spleen cells — C57BL/6 (H-2^b)		
		C57BL/6 Mφ‡ (H-2^b)	DBA/1 Mφ‡ (H-2^q)	5 µg GAT
A C57BL/6 (H-2^b)	5 × 10^4	425	<10	
	2.5 × 10^4	265	<10	680
	2.5 × 10^4§	420	—	
B DBA/1 (H-2^q)	5 × 10^4	20	1,020	
	2.5 × 10^4	30	760	<10
	2.5 × 10^4§	30	—	
C C57BL/6 + DBA/1	5 × 10^4	1,510	850	
	2.5 × 10^4	1,040	910	1,270
	2.5 × 10^4§	1,020	—	
D C57BL/6 + normal DBA/1	5 × 10^4	355	<10	
	2.5 × 10^4	840	<10	1,020
	2.5 × 10^4§	350	—	
E None (unprimed spleen cells)	5 × 10^4	1,110	870	
	2.5 × 10^4	410	520	1,255
	2.5 × 10^4§	1,520	—	

* C57BL/6 mice were immunized 28 days previously by i.p. injection of 4×10^6 of the indicated Mφ. C57BL/6 Mφ, 33.5 ng GAT/10^6 cells; DBA/1 Mφ, 27.0 ng GAT/10^6 cells. At culture initiation, these spleen cells had <25 IgG GAT-specific PFC per 10^7 cells.

‡ C57BL/6 Mφ, 2.78 ng GAT/10^5 cells; DBA/1 Mφ, 3.15 ng GAT/10^5 cells.

§ 2.5×10^4 non-GAT-bearing DBA/1 Mφ were added to cultures containing 2.5×10^4 C57BL/6 GAT-Mφ.

neic Mφ. These genetic restrictions regulating efficient Mφ-lymphoid cell interactions in secondary antibody responses cannot be demonstrated before 2 wk and disappear gradually 8 wk after a single immunization with GAT-Mφ.

Localization of Genetic Restrictions Regulating Macrophage-Lymphoid Cell Interactions in Secondary PFC Responses to the H-2 Complex. The next experiments were designed to determine if the restrictions regulating Mφ-lymphoid cell interactions in secondary PFC responses were controlled by the *H-2* gene complex. C57BL/6 mice were immunized with syngeneic or allogeneic (DBA/1) GAT-Mφ and 24 days later spleen cells from these mice were stimulated with GAT-Mφ from the five strains of mice indicated in Table III. Primary PFC responses to GAT by virgin spleen cells demonstrated that all the GAT-Mφ were capable of stimulating antibody responses (A). Spleen cells from mice immunized with C57BL/6 GAT-Mφ developed secondary PFC responses to GAT preferentially when stimulated with C57BL/6 (H-2^b), D1.LP (H-2^b on DBA/1 background), and (C57BL/6 × DBA/1)F$_1$ (H-$2^{b/q}$) GAT-Mφ (B). Further, spleen cells from mice immunized with allogeneic DBA/1 GAT-Mφ developed secondary PFC responses preferentially when stimulated with DBA/1 (H-2^q), B10.G

TABLE III

Secondary PFC Responses to GAT by C57BL/6 Spleen Cells Primed with GAT-Mφ
Depend on the H-2 Haplotype of the Stimulating Macrophages

		Day 5 IgG GAT-specific PFC/culture 10^7 spleen cells — C57BL/6 (H-2^b)				
GAT-Mφ used to prime spleen cells*	GAT-Mφ/ culture	C57BL/6 Mφ‡ (H-2^b)	DBA/1 Mφ‡ (H-2^q)	B10.G Mφ‡ (H-2^q)	D1.LP Mφ‡ (H-2^b)	(B6 × D1)F₁ Mφ‡ (H-$2^{b/q}$)
A None (unprimed spleen cells)	7 × 10⁴	740	800	940	500	330
	5 × 10⁴	980	850	760	500	1,050
	2.5 × 10⁴	970	630	710	380	240
	None	<10	—	—	—	—
B C57BL/6 (H-2^b)	7 × 10⁴	210	<10	20	330	570
	5 × 10⁴	350	<10	<10	350	230
	2.5 × 10⁴	40	<10	50	240	200
	None	40	—	—	—	—
C DBA/1 (H-2^q)	7 × 10⁴	30	210	480	40	430
	5 × 10⁴	20	150	260	40	310
	2.5 × 10⁴	90	220	320	40	<10
	None	30	—	—	—	—

* C57BL/6 mice were immunized 24 days previously by i.p. injection of 3×10^6 of the indicated Mφ. C57BL/6 Mφ, 33.5 ng/10^6 cells; DBA/1 Mφ, 25.0 ng/10^6 cells. At culture initiation these spleen cells had <30 IgG GAT-specific PFC per 10^7 cells.

‡ C57BL/6 Mφ, 2.25 ng GAT/10^5 cells; DBA/1 Mφ, 1.75 ng GAT/10^5 cells; B10.G Mφ, 2.00 ng GAT/10^5 cells; D1.LP Mφ, 2.00 ng GAT/10^5 cells; (C57BL/6 × DBA/1)F₁, (B6 × D1)F₁ Mφ, 1.85 ng GAT/10^5 cells.

(H-2^q on C57BL/10 background), and (C57BL/6 × DBA/1)F₁ (H-$2^{b/q}$) GAT-Mφ (C).

Thus, secondary PFC responses to GAT are elicited preferentially by Mφ syngeneic or semisyngeneic at the *H-2* complex with the Mφ which presented GAT to the lymphocytes during the in vivo immunization process. The experiment in Table IV demonstrates the involvement of the *H-2* complex in this phenomenon again, using spleen cells from B10.A (H-2^a) mice immunized with B10.A GAT-Mφ. These spleen cells developed secondary PFC responses when stimulated with B10.A or A/J (H-2^a) GAT-Mφ, but not when stimulated with allogeneic B10 (H-2^b) or P/J (H-2^p) GAT-Mφ. It should be noted in both Tables III and IV that GAT-Mφ from strains with the same genetic background and differing only at the *H-2* complex from Mφ used in the immunization process, failed to stimulate significant secondary PFC responses to GAT in vitro.

Discussion

These experiments illustrate some major points concerning genetic restrictions regulating efficient Mφ-lymphoid cell interactions in the development of primary and secondary antibody responses in vitro. First, in primary antibody responses, no genetic restrictions have been demonstrated; syngeneic and allogeneic Mφ support development of comparable responses. Second, in the second-

TABLE IV

Secondary PFC Responses to GAT by B10.A Spleen Cells Primed with GAT-Mϕ Depend on the H-2 Haplotype of Stimulating Macrophages

GAT-Mϕ used to prime spleen cells*	GAT-Mϕ/ culture	Day 5 IgG GAT-specific PFC/culture 10^7 spleen cells — B10.A (H-2^a)			
		B10.A Mϕ‡ (H-2^a)	A/J Mϕ‡ (H-2^a)	B10 Mϕ‡ (H-2^b)	P/J Mϕ‡ (H-2^p)
None (unprimed spleen cells)	7×10^4	590	450	580	470
	5×10^4	1,110	400	450	340
	None	<10	–	–	–
B10.A (H-2^a)	7×10^4	760	670	<10	<10
	5×10^4	660	780	60	60
	None	70	–	–	–

* B10.A mice were immunized 15 days previously by i.p. injection of 3×10^6 B10.A Mϕ bearing 17.0 ng GAT/10^6 cells. At culture initiation, these cells had 30 IgG GAT-specific PFC per 10^7 cells.

‡ B10.A Mϕ, 2.00 ng GAT/10^5 cells; A/J Mϕ, 2.50 ng GAT/10^5 cells; B10 Mϕ, 2.00 ng GAT/10^5 cells; P/J Mϕ, 1.80 ng GAT/10^5 cells.

ary antibody response, genetic restrictions are operative; immunized spleen cells develop antibody responses preferentially when stimulated with antigen-bearing Mϕ syngeneic with the Mϕ which presented antigen to the virgin lymphocytes during the immunization process in vivo. Third, these genetic restrictions are controlled by the *H-2* complex. Fourth, they can be demonstrated only during a limited period, from 2 to 8 wk, after a single immunization with limiting quantities of antigen on Mϕ. Fifth, these restrictions in secondary antibody responses are antigen specific; responses of the same primed lymphoid cells to another antigen do not exhibit genetic restrictions. Sixth, these genetic restrictions appear to operate at the level of the immune T cell, although restrictions at the level of the B cell have not yet been rigorously excluded (19).

The antigen used in these experiments was the synthetic linear random terpolymer of GAT. Antibody responses to GAT are under *H-2*-linked, immune response gene control; mice of the H-$2^{a,b,d,f,j,k,r,u,v}$ haplotypes are "responders" (20); spleen cells from these mice develop IgG GAT-specific PFC responses when stimulated with soluble GAT or GAT-Mϕ in Mishell-Dutton cultures. Spleen cells from mice of the H-$2^{n,p,q,s}$ haplotypes fail to respond to GAT and are "nonresponders" (16, 20). Mϕ and helper T cells are required for development of antibody responses to GAT by responder B cells in vitro, and the defect in nonresponder mice is not in Mϕ (13). Mϕ from peptone-induced peritoneal exudates of responder and nonresponder mice bind approximately equal quantities of GAT, and these Mϕ bearing nanogram quantities of GAT stimulate comparable PFC responses by responder lymphoid cells (13, 14, 17).

The genetic restrictions regulating efficient Mϕ-lymphoid cell interactions in secondary antibody responses have been investigated, to date, only with GAT. However, these restrictions have been demonstrated with three types of responder spleen or lymphoid cells: C57BL/6 (H-2^b), BALB/c (H-2^d), and B10.A (H-2^a); and with Mϕ from several murine strains, which are both responders and nonresponders to GAT and which may or may not share specificities encoded by the *H-2* complex with the responder lymphoid or spleen cells. Thus, the observations appear to reflect a general phenomenon in secondary responses to this antigen and are not restricted to limited combinations of Mϕ and lymphoid cells.

It should be pointed out, however, that the observed genetic restrictions in secondary antibody responses are the only criteria of successful in vivo immunization in these experiments available at this time. IgG GAT-specific PFC responses in spleens of mice immunized with GAT-Mϕ are insignificant compared to those elicited by 10 μg of GAT. The secondary PFC responses to GAT in vitro have the same kinetics of development and are of the same general magnitude as primary PFC responses. The possibility that immunization with GAT-Mϕ may lead to a significant shift in the Ig class of the PFC from IgG$_1$ to IgG$_2$ in the secondary response is currently under investigation.

Another observation which deserves comment at this juncture is that spleen cells from mice immunized with allogeneic GAT-Mϕ, although developing secondary PFC responses when stimulated with the same allogeneic GAT-Mϕ, do not develop significant primary PFC responses when stimulated with soluble GAT or syngeneic GAT-Mϕ. In part, this phenomenon may be related to the problem of criteria of immunization discussed above. Since these spleen cells do respond to the GAT-Mϕ used for immunization, it is clear that the immune apparatus is intact, i.e., functional specific T cells and B cells are present. Since other studies have indicated that the genetic restrictions operate at the level of the T cell and not the B cell (19), this observation implies that virgin T cells were either not present in spleens of mice immunized with GAT-Mϕ, or that, if present, their function was either pre-empted or suppressed by the primed T cells. At this time, we have no data to permit a choice among these possibilities.

The observations that secondary antibody responses by immunized lymphoid cells develop preferentially with Mϕ syngeneic with those used for immunization and that these genetic restrictions appear to operate at the level of the immune T cells, provide an explanation for the genetic restrictions in Mϕ-lymphocyte interactions observed in DNA synthetic responses to antigen by guinea pig cells (6, 7). The lymphocytes in these experiments were from animals immunized in the presence of syngeneic Mϕ. The fact that these lymphocytes develop DNA synthetic responses preferentially in the presence of syngeneic Mϕ is, therefore, entirely consistent with the present findings. More recent observations involving transfer of delayed hypersensitivity in mice indicate that sensitized T cells elicit these responses only when transferred to hosts whose Mϕ are syngeneic at the *I* region of the *H-2* complex with those present during the sensitization of the T cells (21, 22). These observations emphasize the critical nature of the Mϕ membrane-antigen complex in the sensitization of T cells and in eliciting subsequent responses by immune T cells.

The present findings appear to be in conflict with recent observations that helper T cells develop in vitro only in the presence of Mϕ sharing specificities encoded by the *I* region of the *H-2* complex with the T cells (5). However, in the assay of helper T-cell activity, Mϕ syngeneic with the T cells, but not syngeneic with Mϕ used to generate the helper T cells, have been employed. On the basis of the present findings, it could be predicted that, if Mϕ syngeneic with those used to generate helper T cells were used in the assay, the helper T cells would function preferentially with these Mφ. If the genetic restrictions involved in the generation of helper T cells are indeed explained on the basis of the secondary response restrictions demonstrated in the present experiments, both phenomena should be attributable to *I*-region gene products expressed on Mφ. Studies mapping the genetic restrictions in the present system are in progress.

The observations that specifically sensitized T cells respond preferentially when confronted with the same Mφ-antigen complex used for immunization have considerable implications for the understanding of fundamental mechanisms in immune responses to T-cell-dependent antigens. First, these observations provide evidence for a more actively specific role for Mφ in immunizing T cells and eliciting subsequent specific responses by T cells than just presentation of antigen in a highly immunogenic form. Moreover, a role for antigens encoded by the *H-2* complex in this process is demonstrated. Thus, immune T cells appear to recognize and respond to specific antigen selectively or preferentially when confronted with that antigen in the context of Mφ-membrane molecules encoded by the *H-2* complex. Whether Mφ-membrane molecules are modified in some way when presenting antigen, i.e. "altered-self concept," or whether T cells recognize antigen only in the context of specific, unaltered membrane molecules, i.e. "linked-recognition concept," is not known at this time (23).

Nevertheless, the Mφ membrane-antigen complex involved in these phenomena may be analogous to the hapten-carrier complex involved in stimulating B cells, i.e., Mφ membrane molecules may function as a "carrier" for T-cell antigen recognition. Since the genetic restrictions regulating Mφ-lymphoid cell interactions in secondary antibody responses are determined by the *H-2* complex, these phenomena may represent an expanded capacity of T cells to recognize antigens encoded by the *I* region of the *H-2* complex, analogous to the mixed lymphocyte reaction. Recent observations indicate that T cells participating in the mixed lymphocyte response can also participate in responses to nonhistocompatibility antigens (24). From an evolutionary point of view, recognition of histocompatibility antigens is a primitive mechanism. It is possible that during evolution T cells have expanded the library of antigens which they recognize and respond to by "seeing" these antigens displayed on Mφ membranes in the context of histocompatibility antigens. This ability to recognize new antigens would provide obvious survival advantages for the species.

Summary

The ability of antigen-bearing syngeneic and allogeneic peptone-induced peritoneal exudate macrophages to support development of primary and secondary antibody responses by murine lymphoid or spleen cells in vitro has been investigated. The antigen used was the terpolymer of L-glutamic acid^{60}-L-alanine^{30}-L-tyrosine^{10} (GAT). Syngeneic and allogeneic macrophages supported development of comparable primary antibody responses to GAT, indicating that genetic restrictions do not limit efficient macrophage-lymphocyte interactions in primary responses. By contrast, immunized spleen or lymphoid cells developed secondary antibody responses preferentially when stimulated in vitro with GAT on macrophages syngeneic to the macrophages used to present GAT during in vivo immunization. Thus, genetic restrictions regulate efficient macrophage-lymphocyte interactions in secondary antibody responses. These restrictions have been demonstrated from 2 to 8 wk after a single immunization with limiting quantities of GAT and are controlled by the *H-2* gene complex. The implications that immune lymphocytes selectively recognize and respond to antigen presented in the context of the macrophage membrane-antigen complex which sensitized the lymphocytes initially are considered.

We thank Ms. Barbara Teixeira and Sharon Smith for secretarial assistance in preparation of the manuscript and Ms. Fern De La Croix and Cheryl Petrell for their excellent technical assistance.

Received for publication 8 April 1976.

References

1. Pierce, C. W., and J. A. Kapp. 1976. The role of macrophages in antibody responses *in vitro. In* Immunobiology of the Macrophage. D. S. Nelson, editor. Academic Press, Inc., New York. 1–33.
2. Unanue, E. R. 1972. The regulatory role of macrophages in antigenic stimulation. *Adv. Immunol.* 15:95.
3. Katz, D. H., M. Graves, M. E. Dorf, H. Dimuzio, and B. Benacerraf. 1975. Cell interactions between histoincompatible T and B lymphocytes. VIII. Cooperative responses between lymphocytes are controlled by genes in the *I* region of the *H-2* complex. *J. Exp. Med.* 141:263.
4. Pierce, S. K., and N. R. Klinman. 1975. The allogeneic bisection of carrier-specific enhancement of monoclonal B-cell responses. *J. Exp. Med.* 142:1165.
5. Erb, P., and M. Feldmann. 1975. The role of macrophages in the generation of T-helper cells. II. The genetic control of the macrophage-T-cell interaction for helper cell induction with soluble antigens. *J. Exp. Med.* 142:460.
6. Rosenthal, A. S., and E. M. Shevach. 1973. The function of macrophages in antigen recognition by guinea pig T lymphocytes. I. Requirement for histocompatible macrophages and lymphocytes. *J. Exp. Med.* 138:1194.
7. Shevach, E. M., and A. S. Rosenthal. 1973. The function of macrophages in antigen recognition by guinea pig T lymphocytes. II. Role of the macrophages in the regulation of genetic control of the immune response. *J. Exp. Med.* 138:1213.
8. Hartmann, K.-U., R. W. Dutton, M. McCarthy, and R. I. Mishell. 1970. Cell components in the immune response. II. Cell attachment separation of immune cells. *Cell. Immunol.* 1:182.
9. Haskill, J. S., P. Byrt, and J. Marbrook. 1970. In vitro and in vivo studies of the immune response to sheep erythrocytes using partially purified cell preparations. *J. Exp. Med.* 131:57.
10. Shortman, K., and J. Palmer. 1971. The requirement for macrophages in the *in vitro* immune response. *Cell. Immunol.* 2:399.
11. Cosenza, H., and L. D. Leserman. 1972. Cell interactions *in vitro*. I. Role of the third cell in the *in vitro* response of spleen cells to erythrocyte antigens. *J. Immunol.* 108:418.
12. Katz, D. H., and E. R. Unanue. 1973. Critical role of determinant presentation in the induction of specific responses in immunocompetent cells. *J. Exp. Med.* 137:967.
13. Kapp, J. A., C. W. Pierce, and B. Benacerraf. 1973. Genetic control of immune responses in vitro. II. Cellular requirements for the development of primary plaque-forming cell responses to the random terpolymer L-glutamic acid^{60}-L-alanine^{30}-L-tyrosine^{10} (GAT) by mouse spleen cells in vitro. *J. Exp. Med.* 138:1121.

14. Pierce, C. W., J. A. Kapp, S. M. Solliday, M. E. Dorf, and B. Benacerraf. 1974. Immune responses in vitro. XI. Suppression of primary IgM and IgG plaque-forming cell responses in vitro by alloantisera against leukocyte alloantisera. *J. Exp. Med.* 140:921.

15. Pierce, C. W., B. M. Johnson, H. E. Gershon, and R. Asofsky. 1970. Immune responses in vitro. III. Development of primary γM, γG, and γA plaque-forming cell responses in mouse spleen cell cultures stimulated with heterologous erythrocytes. *J. Exp. Med.* 134:395.

16. Kapp, J. A., C. W. Pierce, and B. Benacerraf. 1973. Genetic control of immune responses in vitro. I. Development of primary and secondary plaque-forming cell responses to the random terpolymer L-glutamic acid^{60}-L-alanine^{30}-L-tyrosine^{10} (GAT) by mouse spleen cells in vitro. *J. Exp. Med.* 138:1107.

17. Pierce, C. W., J. A. Kapp, D. D. Wood, and B. Benacerraf. 1974. Immune responses *in vitro*. X. Functions of macrophages. *J. Immunol.* 112:1181.

18. Pierce, C. W. 1973. Immune responses *in vitro*. VI. Cell interactions in the development of primary IgM, IgG and IgA plaque-forming cell responses *in vitro*. *Cell. Immunol.* 9:453.

19. Pierce, C. W., J. A. Kapp, and B. Benacerraf. 1976. Stimulation of antibody responses *in vitro* by antigen-bearing syngeneic and allogeneic macrophages. *In* The Role of Products of the Histocompatibility Gene Complex in Immune Responses. D. H. Katz and B. Benacerraf, editors. Academic Press, Inc., New York. 391–401.

20. Benacerraf, B., and D. H. Katz. 1975. The histocompatibility-linked immune response genes. *Adv. Cancer Res.* 21:121.

21. Miller, J. F. A. P., M. A. Vadas, A. Whitelaw, and J. Gamble. 1975. H-2 gene complex restricts transfer of delayed-type hypersensitivity. *Proc. Natl. Acad Sci. U. S. A.* 72:5095.

22. Miller, J. F. A. P., M. A. Vadas, A. Whitelaw, and J. Gamble. 1975. H-2 linked Ir gene regulation of delayed-type hypersensitivity in mice. *In* The Role of Products of the Histocompatibility Gene Complex in Immune Responses. D. H. Katz and B. Benacerraf, editors. Academic Press, Inc., New York. 403–415.

23. Rosenthal, A. S., and E. M. Shevach. 1976. Macrophage-T lymphocyte interaction: the cellular basis for genetic control of antigen recognition. *In* The Role of Products of the Histocompatibility Gene Complex in Immune Responses. D. H. Katz and B. Benacerraf, editors. Academic Press, Inc., New York. 335–350.

24. Heber-Katz, E., and D. B. Wilson. 1976. Sheep red blood cell-specific helper activity in rat thoracic duct lymphocyte populations positively selected for reactivity to specific strong histocompatibility alloantigens. *J. Exp. Med.* 143:701.

EFFECTS OF BLOCKING HELPER T CELL
INDUCTION IN VIVO WITH ANTI-Ia ANTIBODIES
Possible Role of I-A/E Hybrid
Molecules as Restriction Elements*

BY J. SPRENT

*From the Division of Research Immunology, Department of Pathology, School of Medicine, University of
Pennsylvania, Philadelphia, Pennsylvania 19104*

A large body of evidence indicates that the specificity of most T lymphocytes is not directed to antigen per se but to antigen associated with gene products of the major histocompatibility complex (MHC)[1] (1). In the case of T proliferative responses and helper T cells involved in T-B collaboration, antigen is presented to T cells by a class of macrophagelike accessory cells (henceforth termed macrophages) that express Ia determinants (2–8). In mice, these determinants are encoded by genes situated in the *I*-region of the *H-2* complex. T cells respond to the association of antigen plus Ia determinants in an H-2-restricted fashion. Thus, in the case of normal homozygous mice, T cells have specificity for antigen plus self Ia determinants but generally (9–15), though not invariably (16, 17), fail to recognize antigen in association with allo-Ia antigens. Unlike homozygous mice, H-2-heterozygous mice contain two distinct subgroups of T cells, each of which is restricted by the H-2 (Ia) determinants of only one of the two parental strains (3, 7, 12–15, 18–21).

Studies in vitro have shown that T-macrophage interactions can be blocked by the addition of anti-Ia antibody during culture (15, 22–24). This paper examines the effects of stimulating T cells to antigen in the presence of anti-Ia antibody in vivo.

Materials and Methods

Mice. (CBA × C57BL/6[B6])F$_1$ mice were purchased from Cumberland View Farms, Clinton, Tenn., and B10.A(4R) mice were a gift from Dr. Peter Doherty, Wistar Institute of Anatomy and Biology, Philadelphia, Pa. All other mice were obtained from The Jackson Laboratory, Bar Harbor, Maine.

Cell Suspensions. Purified populations of unprimed (CBA × B6)F$_1$ T cells were prepared by passing pooled mesenteric, axillary, inguinal, and cervical lymph node (LN) cells over nylon wool columns (8).

Positive Selection. Doses of $\simeq 7 \times 10^7$ purified (CBA × B6)F$_1$ T cells plus 0.5 ml of 25% sheep erythrocytes (SRC) were transferred intravenously into (CBA × B6)F$_1$ or B10.A(4R) mice exposed to 900 rad 2 d before. Anti-Ia^k antibody (A.TH anti-A.TL antiserum or monoclonal anti-I-A^k antibody [see below]) was injected intraperitoneally in a total dose of 2 ml/mouse, 1 ml being given 4–6 h before the injection of T cells plus SRC and another 1 ml injected the next day. At 5 d after T cell transfer, the donor T cells were recovered from the spleen plus

* Supported by U. S. Public Health Service grants Ca-15822, AI-10961, AI-15393, and CA-09140.

[1] *Abbreviations used in this paper:* LN, lymph nodes; MHC, major histocompatibility complex; MLN, mesenteric lymph nodes; PFC, plaque-forming cells; SRC, sheep erythrocytes.

mesenteric lymph nodes (MLN) of the recipients for use as helper T cells. The mean percent recoveries of the injected T cells obtained from the spleen plus MLN of the host in a total of 12 experiments (7 for F_1 selection hosts and 5 for B10.A[4R] hosts) were: 33% for transfer of T cells plus SRC, 17% for T cells without SRC, 20% for T cells plus SRC plus A.TH anti-A.TL antiserum, and 17% for T cells plus SRC plus monoclonal anti-I-A^k antibody (percentages expressed with respect to number of viable cells recovered). Serum taken from the anti-Ia^k-injected selection hosts at the time of harvesting the donor T cells had no demonstrable cytotoxic activity.

Anti-Ia^k Reagents. A.TH anti-A.TL (anti-Ia^k) serum was obtained commercially from Cedarlane Laboratories Ltd., Hicksville, N. Y. Monoclonal anti-I-A^k antibody was obtained from ascites fluid of (B6 × DBA/2)F_1 mice that sustain the growth of the mouse-mouse hybridoma 10–3.6 (IgG$_{2a}$). This reagent detects a public I-A specificity (Ia.17) expressed by k, f, r, and s haplotypes but not by b, d, p, or q haplotypes (25). The hybridoma was kindly made available by the Herzenberg group, Stanford University School of Medicine, Stanford, Calif., and was forwarded to this laboratory by Dr. P. Marrack and Dr. J. Kappler, Rochester School of Medicine and Dentistry, Rochester, N. Y., after further cloning. Approximate titers of the two anti-Ia^k reagents tested in the presence of guinea pig complement on spleen cells from CBA ($K^k I-A^k I-B^k I-J^k I-E^k I-C^k S^k D^k$) (*kkkkkkkk*), B10.A(4R) (*kkbbbbbb*), B10.A(5R) (*bbbkkddd*), and B10 (*bbbbbbbb*) mice were: A.TH anti-A.TL: CBA 3^6, 4R 3^5, 5R 3^5, B10 3^3 (indicating high activity for both I-A^k and I-J/E^k but only minimal activity for I-A^b); hybridoma 10-3.6: CBA 3^9, 4R 3^9, 5R 3^0; B10 3^0. Cytotoxic indices of these anti-Ia^k reagents tested on various Ia^k-homozygous cell populations were: normal spleen 50–60%, anti-Thy-1.2-treated spleen 80–90%, normal LN 30–45%; for obscure reasons, the monoclonal anti-I-A^k reagent showed plateau lysis of 20–30% on I-A^k-heterozygous (CBA × B6)F_1 spleen cells (compared with 50–60% for A.TH anti-A.TL antiserum). With respect to T cells, both reagents had only minimal activity (<5% specific lysis) on nylon-wool-passed LN T cells. When tested on unfractionated LN, the sum of cytotoxic indices with the anti-Ia^k reagents (30–45%) plus cytotoxic indices with anti-Thy-1.2 antibody (40–60%) was generally <100%; the addition of both anti-Ia^k and anti-Thy-1.2 antibody produced 90–95% lysis. The activity of the anti-Ia^k reagents for unfractionated LN T cells thus appeared to be minimal.

T-B Collaboration. As detailed elsewhere (8), small doses (0.4–0.8 × 10^6) of the activated T cells recovered from the spleen plus MLN of the selection hosts were washed twice and transferred with B cells (SRC-primed spleen cells treated with anti-Thy-1.2 antibody and complement) and SRC into irradiated (700 rad) (CBA × B6)F_1 mice. IgM (direct) and IgG (indirect) plaque-forming cells (PFC) were measured in the spleen on day 7 posttransfer.

Results

Experimental Protocol. The general approach was to activate purified (CBA × B6)F_1 (Ia^k × Ia^b) T cells to SRC in syngeneic irradiated F_1 mice in the presence of anti-Ia^k antibody, harvest the activated T cells 5 d later, and then test their helper T function for parental strain B cells on further transfer. In initial experiments, A.TH anti-A.TL antiserum was used as a source of anti-Ia^k antibodies. On the basis of cytotoxic testing with cells from various *H-2*-recombinant strains (Materials and Methods), this antiserum had high titered (3^5-3^6) activity ($\simeq$90% lysis) for T cell-depleted spleen cells that express either I-A^k or I-E^k subregion determinants, low activity on B10 spleen (3^3), and little, if any, activity against purified (CBA × B6)F_1 T cells ($\leq$5% lysis on nylon-wool-passed LN).

Experiments with A.TH anti-A.TL antiserum were designed arbitrarily as follows: (CBA × B6)F_1 mice were exposed to lethal irradiation (900 rad) and left for a period of 2 d, i.e., to allow time for clearance of radiosensitive Ia-positive cells such as B lymphocytes. On day 0, the mice (one single mouse/experiment) received 1 ml of unabsorbed A.TH anti-A.TL antiserum intraperitoneally and then, 4–6 h later, were injected intravenously with an inoculum of $\simeq$7 × 10^7 nylon-wool purified (CBA ×

B6)F$_1$ LN T cells plus 0.5 ml of 25% SRC. 1 d later (day 1) the mice received another 1-ml intraperitoneal injection of A.TH anti-A.TL antiserum; control mice received no antiserum. On day 5, spleen and MLN cells were removed from the recipients (for number of T cells recovered, see Materials and Methods) to assess helper T function. Helper T function was measured by transferring small numbers (0.4–0.8 × 10^6) of the activated T cells plus SRC and parental strain or F$_1$ B cells (SRC-primed anti-Thy-1.2-treated spleen) into irradiated (CBA × B6)F$_1$ mice for measurement of anti-SRC PFC.

T Cell Selection in Presence of A.TH Anti-A.TL Antiserum. The helper specificity of (CBA × B6)F$_1$ T cells positively selected to SRC in an irradiated F$_1$ mouse given 2 ml of A.TH anti-A.TL (anti-Ia^k) antiserum is shown in Table I. It is evident that the capacity of the activated F$_1$ T cells to stimulate anti-SRC PFC responses by CBA (Ia^k) B cells was markedly reduced, both for IgM and IgG PFC production; these low responses were no higher than the responses given by unprimed T cells, i.e., by F$_1$ T cells passaged through irradiated F$_1$ mice in the absence of SRC. T cell help for B6 (Ia^b) B cells, by contrast, was unaffected.

TABLE I

Positive Selection of (CBA × B6)F$_1$ T Cells to SRC in Irradiated F$_1$ Mice in Presence of A.TH Anti-A.TL (Anti-Ia^k) Antiserum*: Selective Inhibition of Helper T Function for CBA B Cells

Irradiated hosts used for positive selection of (CBA × B6)F$_1$ T cells to SRC	Antiserum added during selection	SRC added during selection	Dose of helper T cells	B cells for measuring helper T function‡	Anti-SRC PFC/spleen in irradiated (CBA × B6)F$_1$ mice	
					IgM	IgG
			× 10^6			
(CBA × B6)F$_1$	None	+	0.4	CBA	3,410 (1.14)§	13,690 (1.09)
			0.8	CBA	6,020 (1.40)	27,980 (1.16)
			0.4	B6	2,760 (1.64)	4,660 (1.25)
			0.8	B6	13,160 (1.09)	10,440 (1.39)
(CBA × B6)F$_1$	A.TH anti-A.TL (anti-Ia^k)	+	0.4	CBA	380 (2.32)	50 (2.20)
			0.8	CBA	780 (1.18)	1,230 (1.19)
			0.4	B6	3,480 (1.28)	6,750 (1.27)
			0.8	B6	4,750 (1.38)	10,520 (1.13)
(CBA × B6)F$_1$	None	−	0.4	CBA	60 (1.26)	290 (1.20)
			0.8	CBA	180 (1.41)	1,710 (1.28)
			0.4	B6	110 (1.26)	0
			0.8	B6	90 (1.41)	570 (1.28)

* As described in text, irradiated (CBA × B6)F$_1$ mice exposed to 900 rad 2 d before were given 7 × 10^7 nylon-wool-passed (CBA × B6)F$_1$ LN T cells (93% Thy-1.2-positive) intravenously plus 0.5 ml of 25% SRC. The single recipient of A.TH anti-A.TL antiserum was given the serum in divided doses intraperitoneally, 1 ml 4 h before the injection of T cells plus SRC and 1 ml on the next day. Activated helper T cells were recovered from spleen plus MLN on day 5 after T cell transfer. As controls, irradiated F$_1$ mice (2 mice/group) received T cells plus SRC but not antiserum, or T cells alone.

‡ Helper T cells plus SRC (0.1 ml of 5%) were transferred together with CBA or B6 B cells (anti-Thy-1.2-treated SRC-primed spleen) into irradiated (CBA × B6)F$_1$ mice for measurement of splenic PFC on day 7.

§ Geometric mean of data from 3–4 mice/group. Number in parenthesis refers to value by which mean is multiplied or divided to give upper and lower limits, respectively, of SE. Background values given by B cells transferred without T cells have been subtracted. These values ranged between 0 and 470 PFC/spleen. PFC for T cells transferred without B cells were <100 PFC/spleen.

To map the restriction in helper T function, T-B collaboration was examined with B cells from the B10 congenic lines. As shown in Table II, (CBA × B6)F$_1$ T cells selected to SRC in the presence of A.TH anti-A.TL serum (A.TH anti-A.TL F$_1$ T cells) gave only minimal PFC responses with B10.BR (*kkkkkkkk*) and B10.A(4R) (*kkbbbbbb*) B cells but collaborated well with B10 (*bbbbbbbb*) B cells. Suppression seemed to be an unlikely explanation for the restriction because good responses were observed with (B10 × B10.BR)F$_1$ B cells. Likewise, addition of the A.TH anti-A.TL F$_1$ T cells to the control F$_1$ T cells did not cause demonstrable suppresion. Finally, five-fold higher doses of A.TH anti-A.TL F$_1$ T cells did give significant responses with B10.BR B cells. Thus, these findings are against the possibility that carry-over of anti-Ia^k antibody with the helper T cells was responsible for the restriction in helper function. In this respect it should be mentioned that, after harvest, the activated T cells were washed twice before transfer with B cells. Moreover, at the time the activated T cells were removed, the serum of the selection hosts had no detectable anit-Ia^k activity (Materials and Methods).

The data shown in Tables I and II are representative of six experiments. In all experiments the reduction in the IgG response with homozygous Ia^k-bearing B cells was >85%. For IgM responses the reduction was less marked but was usually >70%. There was generally no change in the response with B10 cells: a mild reduction was observed on occasions (Table III) but this was countered by increased responses in other experiments (Table II).

As a working hypothesis, the above findings were taken to imply that, on transfer to irradiated (CBA × B6)F$_1$ mice, antibodies in the A.TH anti-A.TL antiserum bound to the Ia^k determinants on the host macrophages. This binding blocked the capacity of the macrophages to present antigen to the Ia^k-restricted subgroup of F$_1$ T cells, but did not interfere demonstrably with presentation to the Ia^b-restricted subgroup.

TABLE II

Helper Function of (CBA × B6)F$_1$ T Cells Activated to SRC in Irradiated F$_1$ Mice in Presence of A.TH Anti-A.TL Antiserum*: T-B Collaboration with B10.BR, B10.A(4R), and B10 B Cells

Group	Irradiated hosts used for positive selection of (CBA × B6)F$_1$ T cells to SRC	Antiserum added during selection	Dose of helper T cells	B cells for measuring helper T function*	H-2 haplotype of B cells $K \frac{I}{A\,B\,J\,E\,C} S\,D$	Anti-SRC PFC/spleen in irradiated (CBA × B6)F$_1$ mice IgM	IgG
			× 10^6				
A	(CBA × B6)F$_1$	None	0.7	B10.BR	*k k k k k k k k*	10,300 (1.22)*	41,190 (1.11)
			0.7	B10.A(4R)	*k k b b b b b b*	9,380 (1.14)	35,610 (1.17)
			0.7	B10	*b b b b b b b b*	12,090 (1.16)	20,730 (1.11)
			0.7	(B10 × B10.BR)F$_1$		8,800 (1.16)	28,260 (1.32)
B	(CBA × B6)F$_1$	A.TH anti-A.TL (anti-Ia^k)	0.7	B10.BR	*k k k k k k k k*	2,760 (1.36)	4,140 (1.24)
			0.7	B10.A(4R)	*k k b b b b b b*	840 (1.15)	560 (1.91)
			0.7	B10	*b b b b b b b b*	25,780 (1.24)	55.390 (1.20)
			0.7	(B10 × B10.BR)F$_1$		6,670 (1.17)	33,440 (1.11)
	Group A + group B T cells		0.7 of each	B10.BR	*k k k k k k k k*	8,200 (1.21)	39,470 (1.27)
	Group B T cells		3.5	B10.BR	*k k k k k k k k*	4,180 (1.13)	28,570 (1.12)

* As for Table I. Subtracted background values for B cells transferred without T cells ranged from 360 to 1,280 PFC/spleen. T cells alone gave < 100 PFC/spleen.

TABLE III

Apparent Failure of Monoclonal Anti-I-A^k Antibodies to Block Positive Selection of (CBA × B6)F$_1$ T Cells*

Irradiated hosts used for positive selection of (CBA × B6)F$_1$ T cells to SRC	Antibodies added during selection	SRC added during selection	B cells for measuring helper T function*	Anti-SRC PFC/spleen in irradiated (CBA × B6)F$_1$ mice	
				IgM	IgG
(CBA × B6)F$_1$	None	+	B10.BR	8,930 (1.34)*	30,380 (1.36)
			B10	19,850 (1.41)	44,790 (1.32)
(CBA × B6)F$_1$	None	−	B10.BR	310 (1.39)	1,100 (1.38)
			B10	780 (1.91)	0
(CBA × B6)F$_1$	A.TH anti-A.TL (anti-Ia^k antiserum)	+	B10.BR	2,850 (1.33)	2,830 (1.52)
			B10	11,780 (1.67)	23,790 (1.69)
(CBA × B6)F$_1$	Monoclonal anti-I-A^k (10–3.6 ascites fluid)	+	B10.BR	22,980 (1.07)	32,820 (1.12)
			B10	25,310 (1.57)	37,830 (1.44)

* As for Table I and II. Helper T cells transferred in dose of 0.7×10^6 cells. Subtracted background values for B cells transferred without T cells ranged from 50 to 1,000 PFC/spleen, T cells alone gave <400 PFC/spleen.

Because restrictions affecting T-macrophage interactions and T-B collaboration have been mapped to the *K/I-A* subregion of the *H-2* complex (Discussion), it was reasoned that the blocking activity of the A.TH anti-A.TL antiserum was mediated by antibodies with specificity for I-A^k determinants rather than for I-E^k determinants. If so, purified anti-I-A^k antibodies would be expected to have comparable blocking activity.

T Cell Selection in Presence of Monoclonal Anti-I-A^k Antibody. Analogous experiments in which (CBA × B6)F$_1$ T cells were selected to SRC in irradiated F$_1$ mice in the presence of monoclonal anti-I-A^k antibody (2 ml of ascites fluid from mice that bear the hybridoma 10–3.6) are shown in Table III. This antibody (IgG$_{2a}$) had high cytotoxic activity for homozygous I-A^k-bearing B cells but no demonstrable activity for B10 B cells or (CBA × B6)F$_1$ T cells (Materials and Methods). It is evident from Table III that, in contrast to A.TH anti-A.TL antiserum, selection in the presence of the anti-I-A^k antibody failed to inhibit the helper T response for B10.BR B cells; this applied to both IgM and IgG responses and was observed in six separate experiments.

Various trivial explanations could be invoked to account for the apparent failure of the monoclonal anti-I-A^k antibodies to block selection: for example, the antibodies might have had poor homing capacity, a short survival time, or inadequate binding avidity, etc. For completeness, however, it was decided to repeat the experiments and include both B10.A(4R) and B10.BR B cells for measuring helper T function. The surprising finding (Table IV) was that, despite the high levels of help provided for B10.BR B cells, F$_1$ T cells selected in the presence of monoclonal anti-I-A^k antibody gave only minimal responses with B10.A(4R) B cells; three other experiments gave similar data.

T Cell Selection in B10.A(4R) Mice. The above findings are clearly difficult to reconcile with the generally accepted view that helper T cells are restricted entirely by the *I-A* subregion (Discussion). Nevertheless, on a priori grounds, the data in Table

TABLE IV

Helper Function of (CBA × B6)F₁ T Cells Activated to SRC in Irradiated F₁ Mice in Presence of Monoclonal Anti-I-A^k Antibody*: Reduction of Helper T Function for B10.A(4R) B Cells

Irradiated hosts used for positive selection of (CBA × B6)F₁ T cells to SRC	Antibodies added during selection	Dose of helper T cells	B cells*	*H-2* haplotype $K\frac{I}{A\,B\,J\,E\,C}S\,D$	Anti-SRC PFC/spleen in irradiated (CBA × B6)F₁ mice	
					IgM	IgG
		$\times 10^6$				
(CBA × B6)F₁	None	0.35	B10.BR	*k k k k k k k k*	12,100 (1.28)*	32,280 (1.03)
		0.70	B10.BR	*k k k k k k k k*	17,500 (1.09)	58,150 (1.11)
		0.35	B10.A(4R)	*k k b b b b b b*	15,820 (1.20)	52,480 (1.04)
		0.70	B10.A(4R)	*k k b b b b b b*	24,020 (1.04)	57,680 (1.07)
		0.70	B10	*b b b b b b b b*	4,700 (1.16)	15,650 (1.12)
(CBA × B6)F₁	Monoclonal Anti-I-A^k	0.35	B10.BR	*k k k k k k k k*	9,120 (1.31)	37,470 (1.09)
		0.70	B10.BR	*k k k k k k k k*	13,450 (1.22)	52,550 (1.15)
		0.35	B10.A(4R)	*k k b b b b b b*	1,020 (1.40)	1,870 (1.27)
		0.70	B10.A(4R)	*k k b b b b b b*	2,660 (1.72)	2,490 (1.64)
		0.70	B10	*b b b b b b b b*	4,830 (1.16)	17,370 (1.04)

* As for Tables I–III. Subtracted background values for B cells transferred without T cells ranged from 330 to 1,980 PFC/spleen. T cells alone gave <300 PFC/spleen.

IV might be taken to imply that, in addition to the I-A^k-restricted T cells, there also exist T cells restricted to Ia^k molecules encoded wholly or in part by genes mapping to the right of *I-A*, e.g., in the *I-E* subregion. T cells restricted to these molecules would be able to interact with B10.BR (*I-E^k*) B cells but not collaborate with B10.A(4R) (*I-E^b*) B cells. By the same token, positive selection of these E^k-restricted T cells would occur in (CBA × B6)F₁ (*I-E^{k/b}*) mice but not in B10.A(4R) mice. According to this interpretation, the capacity of (CBA × B6)F₁ T cells to collaborate with B10.BR or B10.A(4R) B cells after selection to SRC in B10.A(4R) mice would be mediated entirely by the I-A^k-restricted subgroup of T cells. If so, adding monoclonal anti-I-A^k antibody during selection in B10.A(4R) mice should block helper function for both B10.BR and B10.A(4R) B cells.

To test this prediction, (CBA × B6)F₁ T cells were positively selected to SRC in irradiated B10.A(4R) mice in the presence of either A.TH anti-A.TL antiserum, monoclonal anti-I-A^k antibody, or no antibody (Table V). As reported previously (14), the control F₁ T cells activated in B10.A(4R) mice in the absence of antibody collaborated well with I-A^k-bearing B10.BR, B10.A(4R), and (B10 × B10.BR)F₁ B cells but did not stimulate B10 (I-A^b) B cells. By contrast, F₁ T cells activated in the presence of either the broad spectrum anti-Ia^k reagent (A.TH anti-A.TL) or the monoclonal anti-I-A^k antibody failed to stimulate any of the four groups of B cells. To exclude the possiblity that the T cells in this situation had been inactivated nonspecifically, selection was studied in irradiated B10.A(4R) mice injected with T cell-depleted irradiated B10 spleen cells, i.e., a source of I-A^b-bearing antigen-presenting cells; these cells were given 1 h before the injection of F₁ T cells plus SRC (Table VI, footnote). As shown in Table VI, F₁ T cells selected to SRC in B10-spleen-injected B10.A(4R) mice in the presence of anti-I-A^k antibody provided effective help for B10 B cells while remaining unresponsive for both B10.BR and B10.A(4R) B cells.

Table V

Helper Function of (CBA × B6)F₁ T Cells Activated to SRC in Irradiated B10.A(4R) Mice in the Presence of Monoclonal Anti-I-Aᵏ Antibody*

Irradiated hosts used for positive selection of (CBA × B6)F₁ T cells to SRC	Antibodies added during selection	B cells	H-2 haplotype* $K \frac{I}{A\ BJ\ E\ C} S D$	Anti-SRC PFC/spleen in irradiated (CBA × B6)F₁ mice IgM	IgG
B10.A(4R)	None	B10.BR	k k k k k k k k	6,600 (1.30)*	22,600 (1.18)
		B10.A(4R)	k k b b b b b b	17,650 (1.71)	15,640 (1.09)
		B10	b b b b b b b b	310 (1.81)	0‡
		(B10 × B10.BR)F₁		5,220 (1.27)	21,640 (1.12)
B10.A(4R)	Monoclonal anti-I-Aᵏ	B10.BR	k k k k k k k k	300 (1.04)	110 (1.30)
		B10.A(4R)	k k b b b b b b	300 (1.30)	0
		B10	b b b b b b b b	420 (1.82)	570 (1.57)‡
		(B10 × B10.BR)F₁		320 (1.46)	280 (1.23)
B10.A(4R)	A.TH anti-A.TL (anti-Iaᵏ)	B10.BR	k k k k k k k k	920 (1.25)	1,030 (1.19)
		B10.A(4R)	k k b b b b b b	760 (1.42)	490 (1.30)
		B10	b b b b b b b b	500 (1.11)	240 (1.25)‡

* As for Table IV, except that irradiated B10.A(4R) mice were used for positive selection. Helper T cells transferred in a dose of 0.7×10^6 cells. Subtracted background values for B cells transferred without T cells were in the range of 50–250 PFC/spleen. T cells alone gave <100 PFC/spleen.

‡ These B cells gave high responses (>10,000 PFC/spleen) when transferred with 4×10^6 nylon-wool-passed SRC-primed (CBA × B6)F₁ T cells.

Role of the I-Eᵏ Region in Restricting Helper T Function. To seek direct information on whether I-Eᵏ molecules can act as restriction sites, (A/J (*kkkkkddd*) × B6)F₁ T cells were positively selected to SRC in irradiated B10.A (*kkkkkddd*) or B10.A(5R) (*bbbkkddd*) mice and then tested for helper T function with B cells from these two strains and also with B10 B Cells. As shown in Table VII, F₁ T cells selected to SRC in B10.A (*I-Aᵏ, I-Eᵏ*) mice collaborated well with B10.A B cells but did not give demonstrable helper activity for B10.A(5R) (*I-Aᵇ, I-Eᵏ*) or B10 B cells. Likewise, selection in B10.A(5R) mice failed to provide help for B10.A B cells but generated high levels of help for both B10.A(5R) and B10 B cells. These findings thus corroborate previous findings of other workers (Discussion) that I-E (or I-J) determinants per se do not act as restriction sites.

Discussion

Before attempting to interpret the confusing pattern of results considered above, it may be useful to briefly review the evidence that the restrictions affecting helper T cells map in the *I-A* subregion. The importance of this subregion is apparent from the repeated finding that, although homozygous T cells generally fail to interact with macrophages or B cells across whole *H-2* haplotype barriers, H-2 compatibility (or H-2 sharing) limited only to the *I-A* (or *K/I-A*) subregion leads to effective cell interaction (11, 13, 14, 26, 27). By contrast, total mismatching at the *I-A* subregion, but with a sharing of determinants encoded by the *I-B* through *I-C* subregions or parts thereof,

Table VI

Helper Function of (CBA × B6)F₁ T Cells Activated to SRC in the Presence of Monoclonal Anti-I-A^k Antibody in B10.A(4R) Mice Supplemented with Irradiated B10 Spleen*

Group	Irradiated hosts used for positive selection of (CBA × B6)F₁ T cells to SRC	Antibodies added during selection	B cells*	Anti-SRC PFC/spleen in irradiated (CBA × B6)F₁ mice	
				IgM	IgG
A	B10.A(4R)	None	B10.BR	17,380 (1.21)*	35,470 (1.17)
			B10.A(4R)	22,460 (1.18)	38,990 (1.20)
			B10	380 (1.49)	700 (1.36)
B	B10.A(4R) injected with 1,200 rad anti-θ-B10 spleen‡	None	B10.BR	12,640 (1.12)	29,830 (1.13)
			B10.A(4R)	16,840 (1.10)	25,760 (1.20)
			B10	2,890 (1.13)	9,630 (1.01)
C	B10.A(4R) injected with 1,200 rad anti-θ-B10 spleen‡	Monoclonal anti-I-A^k	B10.BR	500 (1.88)	230 (1.35)
			B10.A(4R)	2,170 (1.39)	940 (1.46)
			B10	7,360 (1.13)	10,350 (1.13)
	Group B + Group C T cells (0.6 × 10⁶ of each)		B10.BR	15,120 (1.22)	33,600 (1.17)

* As for Table IV. Helper T cells transferred in a dose of 0.6 × 10⁶ cells. Subtracted background values for B cells transferred without T cells ranged from 50 to 360 PFC/spleen.

‡ 5 × 10⁷ viable B10 spleen cells pretreated with anti-Thy-1.2 serum plus complement (anti-θ-spleen) and then exposed to 1,200 rad before intravenous injection into irradiated B10.A(4R) mice plus SRC; F₁ T cells given 1 h later.

does not lead to successful interaction. In the case of heterozygous T cells, this is exemplified by the helper specificity of (CBA × B6)F₁ T cells after positive selection to SRC in irradiated B10.A(4R) (*kkbbbbbb*) mice. F₁ T cells activated in these mice give high levels of help for both B10.BR and B10.A(4R) B cells but, significantly, do not interact with B10 B cells (Table V); likewise, F₁ T cells selected to SRC in B10 mice do not have demonstrable helper activity for B10.A(4R) B cells (14). In this situation, therefore, the *I-B* through *H-2D* region of the *H-2* complex does not play a discernible role in controlling T cell function, either at the level of T cell selection or during T-B collaboration.

It is important to point out that B10 and B10.A(4R) mice do not express a detectable I-E product on the cell surface (28). Hence, to examine the possible role of the *I-E* subregion in controlling cell interactions one must turn to a haplotype where an I-E product is expressed, e.g., to the *H-2^k* or *H-2^a* (*kkkkkddd*) (A/J, B10.A) haplotype. In this respect, it is significant that activation of normal (A/J × B6)F₁ T cells to SRC in irradiated B10.A mice generated T cell help for B10.A (*kkkkkddd*) B cells but not for I-E-compatible B10.A(5R) (*bbbkkddd*) B cells (Table VII). Thus, this finding argues that I-E^k determinants per se do not act as restriction elements. A number of other groups have reached a similar conclusion (11, 13, 26).[2]

Collectively, the above findings would seem to constitute strong evidence that determinants mapping to the right of the *I-A* subregion do not play a discernable role in controlling cell interactions. How, then, does one account for the finding that

_____

[2] Low (10–25% of normal) collaborative responses with H-2 compatibility limited to the *I-E/C* subregion has been reported by Martinez et al. (29) for a system in which hapten-specific T cells stimulate hapten-modified B cells polyclonally.

TABLE VII

Helper T Specificity of (A/J × B6)F₁ T Cells Positively Selected to SRC in Irradiated B10.A,
B10.A(5R), or B10 Mice

Irradiated hosts used for positive selection of (A/J × B6)F₁ T cells to SRC*	SRC added during selection	B cells‡	H-2 haplotype $K \frac{I}{A\,B\,J\,E\,C} S\,D$	Anti-SRC PFC/spleen in irradiated (A/J × B6)F₁ mice	
				IgM	IgG
B10.A	+	B10.A	k k k k k d d d	9,860 (1.43)§	41,650 (1.15)
		B10.A(5R)	b b b k k d d d	90 (1.22)	100 (1.44)
		B10	b b b b b b b b	0	60 (1.26)
B10.A	−	B10.A	k k k k k d d d	360 (1.47)	0
		B10.A(5R)	b b b k k d d d	50 (1.49)	140 (1.65)
		B10	b b b b b b b b	0	170 (1.12)
B10.A(5R)	+	B10.A	k k k k k d d d	840 (1.22)	100 (2.05)
		B10.A(5R)	b b b k k d d d	8,760 (1.23)	41,120 (1.05)
		B10	b b b b b b b b	3,780 (1.28)	21,070 (1.36)
B10	+	B10.A	k k k k k d d d	160 (1.44)	170 (1.43)
		B10.A(5R)	b b b k k d d d	10,470 (1.19)	29,600 (1.42)
		B10	b b b b b b b b	5,740 (1.28)	21,270 (1.13)

* 65×10^6 purified (A/J × B6)F₁ T cells ± SRC given to irradiated (850 rad 1 d before) selection hosts of various strains. Donor cells recovered from spleen plus MLN on day 5 and used as helper T cells in doses of 0.7×10^6 viable cells.

‡ SRC-primed B cells transferred in doses of 10^7 for B10.A, 8×10^6 for B10.A(5R), and 12×10^6 for B10.

§ Subtracted background values for B cells transferred without T cells ranged from 0 to 270 PFC/spleen. T cells alone gave <50 PFC/spleen.

selection of (CBA × B6)F₁ T cells to SRC in irradiated F₁ mice in the presence of anti-I-A^k antibody suppressed the generation of help for B10.A(4R) B cells but not for B10.BR B cells? Such a finding implies that *I*-region genes telomeric to the *I-A* subregion do influence cell interactions. In speculating on the possible function of these genes, the recent evidence on the biochemistry of *I*-region gene products is particularly germane. It is well accepted that classic I-A molecules consist of two chains, α and β, both of which are encoded by genes situated in the *I-A* subregion (28, 30). In addition, Jones et al. (31) and others (32, 33) have demonstrated the existence of a different set of 2-chain molecules encoded by two separated genes: one gene (E_α) maps in the *I-E* subregion and codes for the relatively nonpolymorphic α-chain, whereas the other gene (*Ae*) is situated in the *I-A* subregion and encodes the highly polymorphic β-chain (E_β); these A/E hybrid molecules can be formed by either *cis* or *trans* chain association.

In the case of antigens not under demonstrable *Ir* gene control, there is no direct evidence that E_β–E_α A/E hybrid molecules can act as restricting elements for cell interactions. Nevertheless, Schwartz et al. (34, 35) argue persuasively that such hybrid molecules are of critical importance in controlling T cell responses to antigens under double *Ir* gene control such as GLφ (36); responsiveness to these antigens is controlled by two distinct genes that map in the *I-A* and *I-E* subregions, respectively. It is also of interest that hybrid molecules can act as alloantigens, both for the mixed-lymphocyte reaction (37) and cell-mediated lympholysis (K. Fischer-Lindahl and B. Hausman. Personal communication.).

Based on the supposition that E_β–E_α hybrid molecules can act as restriction elements for antigens not under *Ir* gene control, the present data can be explained in terms of the following assumptions:

(*a*) (CBA × B6)F$_1$ T cells comprise at least four subgroups of T cells. Two subgroups are restricted by the *I-A*-encoded determinants of CBA (A_β^k-A_α^k) and B6 (A_β^b-A_α^b), respectively. (For simplicity, the possible existence of two additional subgroups of cells restricted by *trans*-associated A_β^b-A_α^k and A_β^k-A_α^b molecules will not be considered.) The other two T cell subgroups are restricted by two sets of A/E hybrid molecules, i.e., E_β^k-E_α^k and E_β^b-E_α^k. (It will be presumed that the apparent nonexistence of the E_α^b chain precludes the possible existence of another two subgroups of T cells restricted by E_β^k-E_α^b and E_β^b-E_α^b, respectively.)

(*b*) The activity of broad spectrum anti-Ia^k (A.TH anti-A.TL) antiserum has specificity for two sets of molecules, i.e., A_β^k-A_α^k (I-A^k) and E_β^k-E_α^k (I-A/E^k); the activity for I-A/E^k is presumably directed to the E_α^k chain because A.TH anti-A.TL antiserum has high activity for B10.A(5R) (*bbbkkddd*) B cells (Materials and Methods).[3] When injected in vivo, the anti-I-A^k and anti-I-A/E^k antibodies bind to the corresponding molecules on macrophages and block the activation of T cells that recognize antigen in association with these molecules. Thus, A.TH anti-A.TL antiserum blocks selection of (*i*) the I-A^k-restricted T cells (which can interact with either B10.BR or B10.A[4R] B cells) and (*ii*) the I-A/E^k-restricted T cells (which interact only with B10.BR B cells and not with B10.A[4R] B cells). The minimal activity of A.TH anti-A.TL antiserum for I-A^b (B10) B cells does not demonstrably impair activation of the I-A^b-restricted T cells, i.e., cells that collaborate with B10 B cells. In contrast to A.TH anti-A.TL antiserum, monoclonal anti-I-A^k antibodies block activation of only the I-A^k-restricted T cell group and not the I-A/E^k-restricted T cells.

Based on these assumptions, the data can be explained as follows (Table VIII):

TABLE VIII

Interpretation of Results in Terms of T Cell Restriction by A/E Hybrid Molecules

Antibody added during selection	Host for selection	Specificity of selected T cells				Response with B cells		
		I-A^k	I-A^b	E_β^k-E_α^k	$(E_\beta^b$-$E_\alpha^k)$*	B10.A(4R) (*kb*)‡ 1. I-A^k	B10.BR (*kk*) 1. I-A^k 2. E_β^k-E_α^k	B10(*bb*) 1. I-A^b
No serum	(CBA × B6)F$_1$	+	+	+	(+)	+	+	+
A.TH anti-A.TL (anti-I-A^{k}E^k)§	(CBA × B6)F$_1$	−	+	−	(−)	−	−	+
Anti-I-A^k	(CBA × B6)F$_1$	−	+	+	(+)	−	+	+
No serum	B10.A(4R)	+	−	−	(−)	+	+	−
Anti-I-A^k	B10.A(4R)	−	−	−	(−)	−	−	−

* Hypothetical only.
‡ I-A, I-E.
§ This antiserum could also have specificity for other *I*-region determinants.

---

[3] If A.TH anti-A.TL antiserum does have specificity for the E_α^k chain, this antiserum presumably would also have specificity for E_β^b-E_α^k molecules.

Selection in (CBA × B6)F₁ Mice. With selection of (CBA × B6)F₁ T cells to SRC in irradiated F₁ mice in the presence of A.TH anti-A.TL antiserum, inhibition of selection of both the I-A^k- and I-A/E^k-restricted T cells results in the expression of only low (unprimed) help for both B10.BR and B10.A(4R) B cells. Selection of the I-A^b-restricted T cells is not inhibited and these cells give high levels of help for B10 B cells. With selection in the presence of monoclonal anti-I-A^k antibody, blockade of the I-A^k-restricted T cells results in a marked reduction in help for B10.A(4R) B cells. Activation of the I-A/E^k-restricted T cells is not blocked, however, and these cells give good collaborative responses with I-A/E^k-bearing B10.BR B cells but do not interact with B10.A(4R) B cells which lack I-A/E^k.

Selection in B10.A(4R) Mice. Unlike (CBA × B6)F₁ mice, the antigen-presenting cells in B10.A(4R) mice express only a single set of restricting Ia molecules, i.e., I-A^k. Hence, positive selection of F₁ T cells to SRC in B10.A(4R) mice is limited to the I-A^k-restricted T cell subgroup; after selection, these cells provide help for either B10.BR or B10.A(4R) B cells but not for B10 B cells. Blocking the selection of these T cells with anti-I-A^k antibody (either A.TH anti-A.TL antiserum or monoclonal anti-I-A^k antibody) thus inhibits help for both B10.BR and B10.A(4R) B cells.

This explanation is strictly a working hypothesis, and it is worth emphasizing that the data do not rule out the possible involvement of other *I*-region genes, e.g., those situated in the *I-B* or *I-C* subregions. To test the above hypothesis, the simplest approach would be to examine the blocking activity of monoclonal anti-E^k antibodies;[4] as yet, we do not have access to such a reagent. As an alternative to using monoclonal antibodies, the technique of inducing negative selection of T cells to SRC in allogeneic irradiated mice (38) provides a useful approach. For example, in the case of CBA (*I-A^k, I-E^k*) T cells, one would expect that removal of the I-A/E^k-restricted subgroup of cells would not occur in selection hosts expressing either I-A^k or I-E^k alone (B10.A[4R] and B10.A[5R], respectively) but would occur in hosts expressing both determinants (CBA [*cis*] or [4R × 5R]F₁ [*trans*]). Preliminary experiments suggest that this is indeed the case (J. Sprent and B. Alpert. Unpublished data.).

Certain aspects of the present data require comment. First, it might seem surprising that positive selection of T cells in F₁ mice in the presence of anti-I-A^k antibody did not cause a detectable reduction in T cell help for B10.BR B cells, even when limiting-doses of T cells were used (Table IV). Thus, if the control response with B10.BR B cells reflected the combined action of both the I-A^k- and I-A/E^k-restricted T cell subgroups, inhibiting the activation of the I-A^k-restricted T cells would be expected to cause an appreciable reduction in the B10.BR B cell response. The fact that there was no trace of a reduction could be explained in terms of a finite availability of T cell space. Thus, inhibiting the activation of one subgroup of T cells might increase the space available for positive selection (clonal expansion) of the other T cell subgroups. Testing this notion would be difficult.

Recently, Perry et al. (39) have reported that injection of microliter quantities of anti-Ia antibodies can inhibit the induction and expression of delayed-type hypersensitivity. Inhibition of T cell activation in the present system, by contrast, required prodigious quantities of anti-Ia antibody. In the case of A.TH anti-A.TL antiserum,

---

[4] An alternative approach would be to examine the blocking effects of A.TH anti-A.TL antiserum after appropriate absorption to remove activity to either I-A or I-E determinants. The prohibitive cost of this reagent has discouraged such an approach.

doses of <2 ml had minimal effects (whether smaller doses of monoclonal anti-I-A^k antibody would have been effective was not studied). If the antibody does indeed act by binding to macrophages and thereby blocking T-macrophage interaction, it is perhaps not surprising that massive quantities were needed, particularly because much of the antibody was probably absorbed by other Ia-positive cells.

Summary

To examine the role of Ia antigens in controlling T cell activation in vivo, unprimed (CBA × B6)F$_1$ (H-2^k × H-2^b) T cells were positively selected to sheep erythrocytes (SRC) for 5 d in irradiated F$_1$ mice in the presence of large doses of anti-Ia^k antibody. With selection in the presence of broad-spectrum anti-Ia^k antibody (A.TH anti-A.TL antiserum), the activated T cells were markedly reduced in their capacity to collaborate with either B10.BR (I-A^k I-B^k I-J^k I-E^k I-C^k) (*kkkkk*) or B10.A(4R) (*kbbbb*) B cells but gave good helper responses with B10 (*bbbbb*) and (B10 × B10.BR)F$_1$ B cells. Because there was no evidence for suppression, these findings were taken to imply that the anti-Ia^k antibody bound to Ia determinants on radioresistant macrophagelike cells of F$_1$ host origin and blocked the activation of the Ia^k-restricted subgroup of F$_1$ T cells but did not affect activation of the Ia^b-restricted T cell subgroup.

Analogous experiments in which F$_1$ T cells were selected to SRC in F$_1$ mice in the presence of monoclonal anti-I-A^k antibody gave different results. In this situation, the reduction in T cell help for Ia^k-bearing B cells applied to B10.A(4R) B cells but not to B10.BR B cells. With selection of F$_1$ T cells in B10.A(4R) mice, by contrast, anti-I-A^k antibody blocked T cell help for both B10.A(4R) and B10.BR B cells. These data suggested that genes telomeric to the I-A subregion were involved in controlling T cell activation and T-B collaboration. Because no evidence could be found that I-B through I-C determinants per se could act as restrictions elements, the working hypothesis for the data is that Ia^k-restricted T cells consist of two subgroups of cells: one subgroup is restricted by I-A-encoded molecules, whereas the other is restricted by I-A/E hybrid molecules encoded by two separated genes situated in the I-A and I-E subregions, respectively. The notion that A/E hybrid molecules serve as restriction elements is in line with the findings of other workers that these molecules can act as alloantigens and control responses to certain antigens under double Ir gene control.

The skillful technical assistance of Ms. K. Norris and the typing expertise of Ms. K. King and Ms. T. McKearn are gratefully acknowledged. I thank Mr. F. Symington for passaging the hybridoma line.

Received for publication 13 May 1980 and in revised form 19 June 1980.

References

1. Zinkernagel, R. M., and P. C. Doherty. 1979. MHC-restricted cytotoxic T-cells: studies on the biological role of polymorphic major transplantation antigens determining T-cell restriction-specificity, function, and responsiveness. *Adv. Immunol.* **27:**51.
2. Katz, D. H., and B. Benacerraf. 1975. The function and interrelationship of T cell receptors, Ir genes, and other histocompatibility gene products. *Transplant. Rev.* **22:**175.
3. Swierkosz, J. E., K. Rock, P. Marrack, and J. W. Kapler. 1978. The role of *H-2*-linked genes in helper T-cell function. II. Isolation on antigen-pulsed macrophages of two separate

populations of F_1 helper T cells each specific for antigen and one set of parental *H-2* products. *J. Exp. Med.* **147**:554.

4. Thomas, D. W., U. Yamashita, and E. M. Shevach. 1977. The role of Ia antigens in T cell activation. *Immunol. Rev.* **35**:97.

5. Rosenthal, A. S. 1978. Determinant selection and macrophage function in genetic control of the immune response. *Immunol. Rev.* **40**:136.

6. Waldmann, H. 1978. The influence of the major histocompatibility complex on the function of T-helper cells in antibody formation. *Immunol. Rev.* **42**:202.

7. Schwartz, R. H., A. Yano, and W. E. Paul. 1978. Interaction between antigen-presenting cells and primed T lymphocytes. *Immunol. Rev.* **40**:153.

8. Sprent, J. 1978. Restricted helper function of F_1 hybrid T cells positively selected to heterologous erythrocytes in irradiated parental strain mice. I. Failure to collaborate with B cells of the opposite parental strain not associated with active suppression. *J. Exp. Med.* **147**:1142.

9. Rosenthal, A. S., and E. M. Shevach. 1973. Function of macrophages in antigen recognition by guinea pig T lymphocytes. I. Requirement for histocompatible macrophages and lymphocytes. *J. Exp. Med.* **138**:1194.

10. Katz, D. H., T. Hamaoka, and B. Benacerraf. 1973. Cell interactions between histoincompatible T and B lymphocytes. II. Failure of physiologic cooperative interactions between T and B lymphocytes from allogeneic donor strains in humoral response to hapten-protein conjugates. *J. Exp. Med.* **137**:1405.

11. Erb, P., and M. Feldmann. 1975. The role of macrophages in the generation of T-helper cells. II. The genetic control of the macrophage-T-cell interaction for helper cell induction with soluble antigens. *J. Exp. Med.* **142**:460.

12. Miller, J. F. A. P., M. A. Vadas, A. Whitelaw, and J. Gamble. 1976. Role of major histocompatibility complex gene products in delayed-type hypersensitivity. *Proc. Natl. Acad. Sci. U. S. A.* **73**:2486.

13. Yano, A., R. H. Schwartz, and W. E. Paul. 1977. Antigen presentation in the murine T-lymphocyte proliferative response. I. Requirement for genetic identity at the major histocompatibility complex. *J. Exp. Med.* **146**:828.

14. Sprent, J. 1978. Role of H-2 gene products in the function of T helper cells from normal and chimeric mice measured in vivo. *Immunol. Rev.* **42**:108.

15. Ziegler, K., and E. R. Unanue. 1979. The specific binding of *Listeria monocytogenes*-immune T lymphocytes to macrophages. I. Quantitation and role of H-2 gene products. *J. Exp. Med.* **150**:1143.

16. Pierce, C. W., J. A. Kapp, and B. Benacerraf. 1976. Regulation by the *H-2* gene complex of macrophage-lymphoid cell interactions in secondary antibody responses in vitro. *J. Exp. Med.* **144**:371.

17. Thomas, D. W., and E. M. Shevach. 1977. Nature of the antigenic complex recognized by T lymphocytes. III. Specific sensitization by antigens associated with allogenic macrophages. *Proc. Natl. Acad. Sci. U. S. A.* **74**:2104.

18. Paul, W. E., E. M. Shevach, S. Pickeral, D. W. Thomas, and A. S. Rosenthal. 1977. Independent populations of primed F_1 guinea pig T lymphocytes respond to antigen-pulsed parental peritoneal exudate cells. *J. Exp. Med.* **145**:618.

19. Thomas, D. W., and E. M. Shevach. 1978. Nature of the antigenic complex recognized by T lymphocytes. V. Genetic predisposition of independent F_1 T cell subpopulations responsive to antigen-pulsed parental macrophages. *J. Immunol.* **120**:638.

20. McDougal, J. S., and S. P. Cort. 1978. Generation of T helper cells in vitro. IV. F_1 T helper cells primed with antigen-pulsed parental macrophages are genetically restricted in their antigen-specific helper activity. *J. Immunol.* **120**:445.

21. Erb, P., B. Meier, D. Kraus, H. von Boehmer, and M. Feldmann. 1978. Nature of T cell-

macrophage interactions in vitro. I. Evidence for genetic restrictions of T-macrophage interactions prior to T cell priming. *Eur. J. Immunol.* **8:**786.

22. Shevach, E. M., W. E. Paul, and I. Green. 1972. Histocompatibility-linked immune response gene function in guinea pigs. Specific inhibition of antigen-induced lymphocyte proliferation by alloantisera. *J. Exp. Med.* **136:**1207.

23. Lyons, C. R., T. F. Tucker, and J. W. Uhr. 1977. Specific binding of T lymphocytes to macrophages. V. The role of Ia antigens on macrophages in the binding. *J. Immunol.* **122:** 1598.

24. Schwartz, R. H., C. S. David, M. E. Dorf, B. Benacerraf, and W. E. Paul. 1978. Inhibition of dual Ir gene-controlled T-lymphocyte proliferative response to poly (Glu^{56} Lys^{35} Phe^{9}) with anti-Ia antisera directed against products of either I-A or I-C subregion. *Proc. Natl. Acad. Sci. U. S. A.* **75:** 2387.

25. Oi, V. T., P. P. Jones, J. W. Goding, L. A. Herzenberg, and L. A. Herzenberg. 1978. Properties of monoclonal antibodies to mouse Ig allotypes, H-2 and Ia antigens. *Curr. Top. Microbiol. Immunol.* **81:**115.

26. Kappler, J. W., and P. Marrack. 1978. The role of *H-2* linked genes in helper T-cell function. IV. Importance of T-cell genotype and host environment in *I*-region and *Ir* gene expression. *J. Exp. Med.* **148:**1510.

27. Vadas, M. A., J. F. A. P. Miller, A. M. Whitelaw, and J. R. Gamble. 1977. Regulation by the H-2 gene complex of delayed-type hypersensitivity. *Immunogenetics.* **4:**137.

28. David, C. S. 1976. Serologic and genetic aspects of murine Ia antigens. *Transplant. Rev.* **30:** 299.

29. Martinez, A.-C., A. Coutinho, and R. R. Bernabe. 1980. Hapten-specific helper T cells restricted by the I-E(C) subregion of the MHC. *Immunogenetics.* **10:**299.

30. Jones, P. P., D. B. Murphy, and H. O. McDevitt. 1978. Identification of separate *I* region products by two-dimensional electrophoresis. *In* Ir Genes and Ia Antigens. H. O. McDevitt, editor. Academic Press, Inc., New York. 203.

31. Jones, P. P., D. B. Murphy, and H. O. McDevitt. 1978. Two-gene control of the expression of a murine Ia antigen. *J. Exp. Med.* **148:**925.

32. Uhr, J. W., J. D. Capra, E. S. Vitetta, and R. G. Cook. 1979. Organization of the immune response genes: both units of murine I-A and I-E/C molecules are encoded within the *I* region. *Science (Wash. D.C.).* **206:**292.

33. Silver, J., and W. A. Russel. 1979. Genetic mapping of the component chains of Ia antigens. *Immunogenetics.* **8:**339.

34. Schwartz, R. H., A. Yano, J. H. Stimpfling, and W. E. Paul. 1979. Gene complementation in the T-lymphocyte proliferative response to poly(Glu^{55} Lys^{36} Phe^{9})n. A demonstration that both immune response gene products must be expressed in the same antigen-presenting cell. *J. Exp. Med.* **149:**40.

35. Longo, D. L., and R. H. Schwartz. 1980. Gene complementation. Neither *Ir-GLØ* gene need be present in the proliferative T cell to generate an immune response to poly(Glu^{55}-Lys^{36}Phe^{9})n. *J. Exp. Med.* **151:**1452.

36. Katz, D. H., M. E. Dorf, and B. Benacerraf. 1976. Control of T-lymphocyte and B-lymphocyte activation by two complementing *Ir-GLØ* immune response genes. *J. Exp. Med.* **143:**906.

37. Fathman, C. G., and H. Hengartner. 1979. Crossreactive mixed lymphocyte reaction determinants recognized by cloned alloreactive T cells. *Proc. Natl. Acad. Sci. U. S. A.* **76:** 5863.

38. Sprent, J. 1978. Role of the *H-2* complex in induction of T helper cells in vivo. I. Antigen-specific selection of donor T cells to sheep erythrocytes in irradiated mice dependent upon sharing of *H-2* determinants between donor and host. *J. Exp. Med.* **148:**478.

39. Perry, L. L., M. E. Dorf, B. A. Bach, B. Benacerraf, and M. I. Greene. 1980. Mechanisms of regulation of cell-mediated immunity: anti-I-A alloantisera interfere with induction and expression of T-cell-mediated immunity to cell-bound antigen in vivo. *Clin. Immunol. Immunopathol.* **15:**299.

POTENTIATION OF THE T-LYMPHOCYTE RESPONSE TO MITOGENS

II. The Cellular Source of Potentiating Mediator(s)*

By IGAL GERY‡ AND BYRON H. WAKSMAN

(From the Department of Microbiology, Yale University School of Medicine, New Haven, Connecticut 06510)

(Received for publication 25 February 1972)

The first papers of the present series establish that mouse thymocytes and peripheral thymus-processed (T)[1] lymphocytes are stimulated to mitosis and their response to such agents as phytohemagglutinin greatly potentiated by factor(s) designated "lymphocyte-activating factor" (LAF), which are produced in cultures of syngeneic or xenogeneic lymphoid cells (1, 2). In this paper, we present data showing that adherent cells, probably macrophages, are the principal source of LAF and that its production is increased by agents which stimulate these cells.

Materials and Methods

Animals.—Male or female CBA/J mice, 6–12 wk of age, were used without treatment or after irradiation (850 R) and reconstitution with 5×10^6 syngeneic bone marrow cells (XBM), 5–8×10^7 thymocytes (XT), or both bone marrow and thymus cells (XBMT). Young adult New Zealand albino rabbits of both sexes were purchased from a local dealer and used without treatment.

Cell Preparation and Culture.—All the materials and techniques employed for cell preparation and culture are fully described in our previous papers (1–3). The separation of adherent from nonadherent cells was carried out both on plastic (4, 5) and by the use of nylon columns (6, 7). While 8% pooled normal human serum was routinely used for lymphocyte culture, the adherence technique required higher concentrations, 10% human serum being used routinely and 10% fetal calf serum (Grand Island Biological Company, Grand Island, N.Y.) in one experiment.

RESULTS

Release of Potentiating Factors by Cells from Different Sources.—Factors stimulatory to normal CBA thymocytes are released by normal human, rabbit,

---

* Supported by US Public Health Service Grants AI-06112 and AI-06455.

‡ On leave of absence from the Hebrew University-Hadassah Medical School, Department of Medical Ecology, Jerusalem, Israel.

[1] *Abbreviations used in this paper:* B, nonthymus processed; Con A, concanavalin A; LAF, lymphocyte-activating factor; LPS, lipopolysaccharide; PBL, peripheral blood lymphocytes; PHA, phytohemagglutinin; SUP, supernatant; T, thymus processed; T-^{3}H, thymidine-^{3}H; XBM, irradiation plus syngeneic bone marrow cells; XBMT, irradiation plus syngeneic bone marrow cells and thymocytes; XT, irradiation plus syngeneic thymocytes.

or syngeneic mouse lymphoid cells into the supernatant (SUP) (Table I). Human peripheral blood lymphocytes (PBL), acted on by phytohemagglutinin (PHA) or lipopolysaccharide (LPS), produce SUP which is mitogenic alone and which greatly potentiates the thymocyte response to PHA. The production of SUP activity bears no relation to DNA synthesis in the donor culture, which is marked with PHA and negligible with respect to LPS. Concanavalin A (Con A), which was mitogenic for the donor culture, gave no SUP activity in this experiment.

With syngeneic spleen cells, all three mitogens stimulate mitosis and production of SUP activity but at a lesser level than with human cells. Only Con

TABLE I

Response to Stimulants and Production of Potentiating Factors by Cells of Different Species

Experiment	Donor culture			Recipient culture* T-^{3}H uptake	
	Cells	Stimulant	T-^{3}H uptake	SUP alone	SUP + PHA
I	None	—	—	29	242
	Human blood leuko-cytes	None	63	28	488
		PHA	64,492	14,880	38,418
		LPS	747	13,426	53,898
		Con A	2,426	73	368
	Mouse spleen cells	None	355	42	417
		PHA	17,003	109	1,224
		LPS	55,697	147	14,811
		Con A	85,200	1,091	9,090
II	None	—	—	147	665
	Rabbit spleen cells	None	1,391	265	1,305
		PHA	6,154	519	1,551
		LPS	1,856	520	4,637

* Mouse thymocytes.

A–SUP stimulates DNA synthesis in recipient thymocytes without added PHA. Rabbit spleen cells give similar findings but a still lower level of SUP production. However, rabbit cells were agglutinated to some extent by human serum in the medium, and SUP production may have been affected by this. LPS was clearly the best stimulant of SUP production with all the cell types tested.

When mouse organs are compared (Table II), bone marrow, spleen, and thymus, in that order, are found to contain cells active in producing SUP. Some is released by unstimulated cells, more by cells exposed to PHA or Con A, and the most by cells treated with LPS. Only LPS was effective in stimulating marrow to produce an active SUP. While all the supernatants tested acted synergistically with PHA, only Con A–SUP of spleen and LPS-SUP from marrow were mitogenic alone, in each case more so than PHA. Again there was no correlation between donor cell mitosis and production of SUP.

Characterization of Cells which Produce Potentiating Factors.—Lethal irradiation of mouse donors and reconstitution with bone marrow alone did not significantly affect the ability of their spleen cells to produce SUP, when incubated without stimulant or with LPS (Table III). With PHA and Con A, on the other hand, there was some loss of ability to form SUP, and this was restored by the additional injection of thymocytes. These data and those of

TABLE II

Response to Stimulants and Production of Potentiating Factors by Cells of Different Mouse (Syngeneic) Organs

Experiment		Donor culture			Recipient culture T-³H uptake	
	Cell source	Stimulant	T-³H uptake		SUP alone	SUP + PHA
I	—	—	—		55	557
	Spleen	None	1,763		80	1,367
		PHA	42,061		183	3,919
		LPS	60,635		419	13,568
		Con A	112,606		5040	12,984
	Thymus	None	35		76	803
		PHA	456		146	1,141
		LPS	337		247	3,892
		Con A	7,108		199	886
II	—	—	—		69	1,693
	Spleen	None	231		56	3,046
		PHA	35,780		364	4,285
		LPS	61,310		632	23,529
		Con A	158,447		7125	25,406
	Bone marrow	None	3,336		45	2,741
		PHA	11,143		209	4,073
		LPS	10,663		5008	33,676
		Con A	11,328		369	4,883

Table II imply that T lymphocytes may contribute to the formation of active SUP with PHA and Con A but not with LPS.

When mouse spleen cells were separated into adherent and nonadherent populations by incubation in plastic Petri dishes, the ability to produce mitogenic and potentiating SUP, after stimulation with Con A and especially with LPS, was found to reside almost entirely in the adherent cells (Table IV). When the adherent cells were cultured for 2 days and washed repeatedly to remove cells other than macrophages, they remained active as sources of SUP. At this time contamination with granulocytes and lymphocytes is minimal.

Human PBL were similarly freed of adherent cells by passage through nylon columns. These cells retained their reactivity with PHA, as shown by thymidine-^{3}H (T-^{3}H) uptake after 3 days in culture, yet lost to a considerable degree their ability to produce mitogenic or potentiating factors when stimulated with either PHA or LPS (Table V). Titrations indicated that column-purified

TABLE III

Response to Stimulants and Production of Potentiating Factors by Syngeneic Normal, XBM, and XBMT Spleen Cells

Donor culture			Recipient culture T-^{3}H uptake	
Treatment of mouse	Stimulant	T-^{3}H uptake	SUP alone	SUP + PHA
—	—	—	45	879
Normal	None	221	23	1,015
	PHA	22,025	136	2,132
	LPS	38,201	125	8,432
	Con A	115,712	3413	14,364
XBM	None	1,151	44	1,608
	PHA	6,484	95	2,458
	LPS	1,582	179	11,829
	Con A	10,653	641	5,137
XBMT	None	2,552	38	2,746
	PHA	21,705	233	6,227
	LPS	4,872	1048	13,832
	Con A	50,981	3028	13,702

TABLE IV

Production of Potentiating Mediators by Adherent and Nonadherent Mouse Spleen Cells

Stimulant of donor culture	PHA in recipient culture	Mouse spleen fraction			
		Original	Nonadherent	Adherent (1 hr)	Adherent (2 days)
None	—	30*	48	37	ND
	+	238	157	789	ND
LPS	—	76	42	3,112	173
	+	2258	759	15,326	7992
Con A	—	401	92	541	ND
	+	1662	258	3,665	ND

* Values for T-^{3}H uptake (counts per minute) in recipient thymocyte cultures without added SUP were unstimulated, 55, and PHA-stimulated, 293.

cells produced less than 4% of the SUP activity released by the original unfractionated PBL. The residual activity did not disappear on dilution and may differ qualitatively from that obtained with unpurified cells.

The Indirect Action of Mitogens.—Removal of adherent cells from a normal CBA thymocyte population reduces the response of these cells in culture (T-^{3}H uptake) to PHA and Con A to less than one-half of control values and virtually eliminates the response to LPS (Table VI). The supernatants after treatment of control cultures with any of the three mitogens show SUP activity, and comparable activity is much reduced in the supernatants of nonadherent cells in culture (data not shown).

Quantitative Comparison of Different Supernatants.—A comparative titration was carried out on two active SUP obtained from human PBL stimulated with PHA and LPS (Fig. 1). Both were mitogenic alone for mouse thymocytes, and both enhanced the response of these cells to PHA. Enhancement was signifi-

TABLE V

Effect of Nylon Column Purification on Ability of Human Blood Leukocytes to Produce Potentiating Factors

Experiment	Donor culture			SUP dilution	Recipient culture T-^{3}H uptake	
	Cells	Stimulant	T-^{3}H uptake		SUP alone	SUP + PHA
I	None	—	—	—	64	1,032
	Original	None	29	Undiluted	3,748	30,139
				1:5	202	12,853
		PHA	15,869	Undiluted	19,942	51,858
				1:5	1,214	20,073
				1:25	155	6,148
		LPS	25	Undiluted	13,656	56,126
				1:5	6,071	39,127
				1:25	1,511	21,831
	Purified	None	38	—	ND	ND
		PHA	7,859	Undiluted	125	1,900
				1:5	57	1,718
		LPS	21	Undiluted	122	3,655
				1:5	100	3,084
II	None	—	—	—	53	525
	Original	None	60	Undiluted	87	833
		PHA	45,855	Undiluted	27,529	31,518
	Purified	None		—	ND	ND
		PHA	26,210	Undiluted	200	1,409

cant at concentrations of SUP below the mitogenic threshhold. The maximal level of enhanced stimulation attained was only slightly greater in each case than that achieved with the optimal concentration of SUP alone. The titration curves, both for SUP alone and for SUP + PHA, were closely similar and suggested that the same active agent was present in the two preparations.

DISCUSSION

In the previous paper (2), evidence was presented that the target cell for LAF is the central or peripheral thymus-processed (T) lymphocyte. The data of the present paper show that LAF is probably produced by macrophages in both human and mouse cultures. Active SUP were obtained from adherent cells, which could be maintained for 2 days attached to plastic and which withstood vigorous washing. Such preparations do not include significant numbers of T cells, nor do spleen cell suspensions from XBM mice, which also

TABLE VI

Stimulation of Mouse Thymus Cells by LPS: the Need for Adherent Cells

Stimulant	T-³H uptake by thymus cells		Ratio original/nonadherent
	Original	Nonadherent*	
None	130	20	6.5
LPS	343	81	4.2
PHA	441	178	2.5
LPS + PHA	1527	274	5.6
Con A	5844	2606	2.2

* Collected after two incubations in Petri dishes.

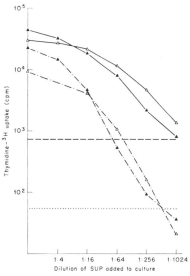

FIG. 1. Titration of mitogenic and potentiating activity of SUP prepared by incubating PBL with PHA (▲) or LPS (△). T-³H-incorporation of mouse thymocytes without additives ....., with PHA – – – –, with SUP –·–·–, or with SUP + PHA ——.

produced SUP as active as that from normal spleen. T cells may play an indirect role, nevertheless, when PHA or Con A is used to stimulate LAF production by macrophages (see below), and this may account in part for differences in LAF production by different organs stimulated with the various agents. Nonthymus processed (B) cells appear to be ruled out as a source of LAF since active SUP could be obtained from thymus cell suspensions and from

XBM spleen at a time when no B cells reactive to LPS are present (3). The most active SUP were obtained with LPS, a well-recognized stimulant of macrophages (8–11).

There is no evidence that LAF produced by "unstimulated" adherent cells differs qualitatively from that produced after stimulation with PHA or LPS or that the latter are different from each other. All are mitogenic alone and show striking synergy with PHA or with Con A used at concentrations below those giving a maximal response. Their titration curves, alone or in the presence of PHA, are perfectly parallel (Fig. 1). Apparent qualitative differences in SUP from different animal species and different organs are most readily explained as depending on differences in LAF concentration. Also, as noted, differences in the number of T cells influence LAF production by PHA or Con A as compared with LPS. There may nevertheless be some heterogeneity of factors which act on lymphocytes. We have seen (Table V) that nylon column–purified PBL can be stimulated by LPS to produce weakly active SUP whose activity resists dilution. Tridente et al. (12, 13) and Winkelstein (14) have described synergistic responses of thymocytes with other cells having different kinetic properties from those under consideration here, and the former group reports that the adherent cells from bone marrow are less effective in stimulating thymocytes than unfractionated marrow cells.

The present observations suggest that LAF may play an essential role in many or all the immunologic responses in which T cells participate, certainly in each instance where a requirement for macrophages or "adherent cells" has been demonstrated. Such a requirement in the reaction of the sensitized cells of delayed hypersensitivity with eliciting antigen is well recognized, whether this be measured by proliferation and blast transformation (6) or by production of secondary mediators such as lymphotoxin.[2] Equally important is the primary response of unimmunized T cells, as in the graft-*versus*-host reaction and its in vitro equivalent, the mixed lymphocyte reaction (15, 16), or the response to an immunizing dose of soluble antigen with adjuvant (17).[3] In such primary responses, a further cooperation between T cells of different biologic types or specificities may be essential (18, 19). Finally, the cooperation of T and B cells (20, 21) in stimulating the latter to plasma cell transformation and antibody production requires macrophages both in vivo (22) and in vitro (4, 5, 23–27). In several of these instances involving T cells alone (16, 28) or T-B cell cooperation (24, 25, 29) it has already been recognized that soluble factors comparable to LAF serve as mediator. This conclusion does not necessarily conflict with other suggested mechanisms of macrophage action in the immune response. Supernatant factors described by other investigators appear in some instances to represent altered antigen (25, 29). There is per-

---

[2] Yoshinaga, M. Unpublished data.

[3] Spiesel, S. Z., R. K. Gershon, and B. H. Waksman. Adjuvant effects on mouse thymus-derived cells. I. A survey of various classes of adjuvants. Manuscript in preparation.

suasive evidence that antigen attached in or near the cell surface of macrophages may act as "superantigen" (30–33) or that antigen linked to a special RNA provided by the macrophage may play this role (34–38).

One of the most important implications of the present observations concerns the role of macrophages in the action of adjuvants, whether to increase delayed sensitization or to enhance antibody production. In the intact animal (10, 39) some adjuvant materials, such as mineral oil, simply promote wide dissemination and long persistence of the antigenic stimulus. Others increase the flux of lymphocytes into thymus-dependent areas of stimulated peripheral lymph nodes (40), a localization which may depend on specific as well as nonspecific elements in the local response to antigen and adjuvant and a release of mediators (41, 42). Adjuvant effects are also observed in vitro however (43), and clearly must be mediated by more direct cellular events. Enhanced antigen uptake by macrophages (44–46), whether as a result of the particulate character of the antigen (47, 48) or of stimulation directed to the macrophage (49, 50), was long felt to account for such effects. This mechanism accords well with the suggested role of the macrophage in providing superantigens to the host (30, 31, 33–37). More recently, however, the suggestion was made that adjuvants act by labilizing macrophage lysosomes (51–53), and an increased production of lysosomal enzymes does in fact result from the action of LPS on macrophage monolayers (11). A target for the released lysosomal contents has not been identified; it is possible that LAF may be a lysosomal factor the target of which is the T lymphocyte. T cells in XT mice synthesize DNA rapidly in response to in vivo stimulation with a variety of adjuvant materials, among them pertussis vaccine and purified LPS.[3] Our findings, presented in this and the preceding paper, establish that the stimulus to these cells must be mediated in part by macrophages and LAF production. These findings assume greater significance in the light of the recent demonstration that the adjuvant effect of pertussis, LPS, Freund's adjuvant, or retinol on antibody formation requires the participation of T cells (17, 42) and the in vitro demonstration that phagocytic cells and T lymphocytes together are required to "help" nonmitotic B cells respond to antigens such as sheep erythrocytes (24, 29, 54).

Certain stimulatory effects observed in the intact animal and in vitro appear to be indirect. PHA and Con A produce less effective SUP from bone marrow than spleen, and its production in the latter is lessened in XBM animals as contrasted with XBMT animals (see Tables II and III). This implies that these mitogens, unlike LPS which is highly effective in producing LAF activity from bone marrow and in the presumed absence of T cells, act in part by stimulating T lymphocytes which then activate macrophages with their mediators and these in turn produce LAF, a striking circular mechanism. PHA and Con A act only on T cells (55–59). That stimulation of T cells may produce activation of macrophages indirectly by way of soluble mediators has been demonstrated repeatedly (60–62). Thus the postulated circular mechanism may come into play in any situation where T cells are activated, for example in the interesting "allogeneic effect" described by Katz et al. (63). The fact that mouse spleen reacts maximally to a dose of Con A 5-10-fold lower than required for maximal stimulation of mouse thymus[4] may simply reflect the relative paucity of macrophages in the thymus and consequent weakness of the suggested circular mechanism. Removal of macrophages and other large cells from reactive thymocytes on bovine serum albumin gradients has been shown (2, 64) to diminish their reactivity to PHA. PHA and Con A, however, may also have direct effects on macrophages, and we have shown production of LAF from adherent cells by the former (59). In consequence, as one would predict, PHA acts as an adjuvant if administered at the correct time in relation to antigen (see references 65 and 66 for reviews).

The response to LPS also illustrates a second type of indirect effect. It is well recognized (8–10) that LPS is highly stimulatory to macrophages both in vivo (10) or in vitro (11). In the mouse, LPS also stimulates B lymphocytes but has no direct action on T cells (3, 59, 67). In accord with these findings, the stimulation of thymocytes by LPS is shown in the present paper to depend entirely on the presence of adherent cells, presumably macrophages. One must suppose that the vigorous proliferative response of peripheral T cells when LPS is given to XT animals[3] may also be dependent on production of small amounts of LAF which stimulate these cells.

SUMMARY

Effective supernatants (SUP), which potentiate mouse T-cell responses to phytohemagglutin (PHA), are obtained from cells of several species (human, rabbit, rat, mouse) and indeed from syngeneic spleen, thymus, or bone marrow cells. Unstimulated cells release some SUP activity but more is produced after stimulation. Lipopolysaccharide (LPS) produced very active SUP in all cultures tested. PHA was similarly active on human leukocytes only, whereas concanavalin A (Con A) gave highly efficient SUP only with mouse spleen cells. SUP production is not correlated with a mitotic response of the donor cells and is observed in cultures unable to respond mitotically to the stimulant. Adherent mouse spleen cell populations, consisting largely or entirely of macrophages, produce active SUP, while nonadherent cells do not. Similarly, purification of human peripheral leukocytes on nylon columns, with removal of macrophages and other adherent cells, destroys their ability to produce SUP. The importance of indirect effects in stimulating mitotic responses of T cells is emphasized by the fact that the mitotic response of mouse thymocytes to LPS and its ability to potentiate the response of these cells to PHA disappears with removal of adherent cells from the thymocyte population. Con-

---

[4] Gery, I. Unpublished data.

versely the production of SUP from spleen cells stimulated by Con A requires the presence of T cells.

We wish to thank Dr. Richard K. Gershon for his helpful and stimulating discussions and Wendy C. Brown for technical assistance.

BIBLIOGRAPHY

1. Gery, I., R. K. Gershon, and B. H. Waksman. 1971. Potentiation of cultured mouse thymocyte responses by factors released by peripheral leukocytes. *J. Immunol.* **107:**1778.

2. Gery, I., R. K. Gershon, and B. H. Waksman. 1972. Potentiation of the T-lymphocyte response to mitogens. I. The responding cell. *J. Exp. Med.* **136:**128.

3. Gery, I., J. Krüger, and S. Z. Spiesel. 1972. Stimulation of B-lymphocytes by endotoxin. Reactions of thymus-deprived mice and karyotypic analysis of dividing cells in mice bearing T6T6 thymus grafts. *J. Immunol.* **108:** 1088.

4. Mosier, D. E. 1967. A requirement for two cell types for antibody formation *in vitro. Science (Wash. D.C.).* **158:**1573.

5. Tan, T., and J. Gordon. 1971. Participation of three cell types in the anti-sheep red blood cell response in vitro. Separation of antigen-reactive cells from the precursors of antibody-forming cells. *J. Exp. Med.* **133:**520.

6. Oppenheim, J. J., B. G. Leventhal, and E. M. Hersh. 1968. The transformation of column-purified lymphocytes with nonspecific and specific antigenic stimuli. *J. Immunol.* **101:**262.

7. Amos, D. B. 1969. Cytotoxicity testing. *In* Manual of Tissue Typing Techniques. Compiled by Transplantation Immunology Branch Collaborative Research, NIAID, NIH. 1.

8. Bennett, I. L., Jr., and L. E. Cluff. 1957. Bacterial pyrogens. *Pharmacol. Rev.* **9:**427.

9. Nowotny, A. 1969. Molecular aspects of endotoxic reactions. *Bacteriol. Rev.* **33:**72.

10. Neter, E. 1969. Endotoxins and the immune response. *Curr. Top Microbiol. Immunol.* **47:**82.

11. Wiener, E., and D. Levanon. 1968. The *in vitro* interaction between bacterial lipopolysaccharide and differentiating monocytes. *Lab. Invest.* **19:**584.

12. Tridente, G., G. Biasi, L. Chieco-Bianchi, and L. Fiore-Donati. 1971. Thymusmarrow cell interaction *in vitro.* Enhancing effect of different bone marrow constituents on PHA response of thymocytes. *In* Morphological and Functional Aspects of Immunity. K. Lindahl-Kiessling, G. Alm, and M. G. Hanna, Jr., editors. Plenum Publishing Corporation, New York. 633.

13. Tridente, G., G. Biasi, L. Chieco-Bianchi, and L. Fiore-Donati. 1971. Thymusmarrow cell interaction evaluated by PHA stimulation and graft-versus-host activity. *Nature (Lond.).* **234:**105.

14. Winkelstein, A. 1971. Augmentation of PHA responsiveness in mixed thymusmarrow cultures. *J. Immunol.* **107:**195.

15. Alter, B. J., and F. H. Bach. 1970. Lymphocyte reactivity *in vitro.* I. Cellular reconstitution of purified lymphocyte response. *Cell. Immunol.* **1:**207.

16. Bach, F. H., B. J. Alter, S. Solliday, D. C. Zoschke, and M. Janis. 1970. Lymphocyte reactivity *in vitro.* II. Soluble reconstituting factor permitting response of purified lymphocytes. *Cell. Immunol.* **1:**219.

17. Allison, A. C., and A. J. S. Davies. 1971. Requirement of thymus-dependent lymphocytes for potentiation by adjuvants of antibody formation. *Nature (Lond.)* **233**:330.

18. Cantor, H., R. Asofsky, and N. Talal. 1970. Synergy among lymphoid cells mediating the graft-*versus*-host response. II. Synergy in graft-*versus*-host reactions produced by BALB/c lymphoid cells of differing anatomic origin. *J. Exp. Med.* **131**:235.

19. Raff, M. C., and H. I. Cantor. 1971. Subpopulations of thymus cells and thymus-derived lymphocytes. *In* Progress in Immunology. D. B. Amos, editor. Academic Press, Inc., New York. 83.

20. Möller, G., editor. 1969. Antibody-sensitive cells. *Transplant. Rev.* **1**:3.

21. Playfair, J. H. L. 1971. Cell cooperation in the immune response. *Clin. Exp. Immunol.* **8**:839.

22. Talmage, D. W., J. Radovich, and H. Hemmingsen. 1969. Cell interaction in antibody synthesis. *J. Allergy.* **43**:323.

23. Hartmann, K., R. W. Dutton, M. M. McCarthy, and R. I. Mishell. 1970. Cell components in the immune response. II. Cell attachment separation of immune cells. *Cell. Immunol.* **1**:182.

24. Dutton, R. W., M. M. McCarthy, R. I. Mishell, and D. J. Raidt. 1970. Cell components in the immune response. IV. Relationships and possible interactions. *Cell. Immunol.* **1**:196.

25. Shortman, K., and J. Palmer. 1971. The requirements for macrophages in the *in vitro* immune response. *Cell. Immunol.* **2**:399.

26. Aaskov, J. G., and W. J. Halliday. 1971. Requirement for lymphocyte-macrophage interaction in the response of mouse spleen cultures to pneumococcal polysaccharide. *Cell. Immunol.* **2**:335.

27. Shortman, K., E. Diener, P. Russell, and W. D. Armstrong. 1970. The role of non-lymphoid accessory cells in the immune response to different antigens. *J. Exp. Med.* **131**:461.

28. Gordon, J., and L. D. MacLean. 1965. A lymphocyte-stimulating factor produced *in vitro*. *Nature (Lond.).* **208**:795.

29. Hoffmann, M., and R. W. Dutton. 1971. Immune response restoration with macrophage culture supernatants. *Science (Wash. D.C.).* **172**:1047.

30. Cruchaud, A., J. P. Despont, J. P. Girard, and B. Mach. 1970. Immunogenicity of human γ-globulin bound to macrophages after inhibition of RNA synthesis. *J. Immunol.* **104**:1256.

31. Roelants, G. E., and J. W. Goodman. 1969. The chemical nature of macrophage RNA-antigen complexes and their relevance to immune induction. *J. Exp. Med.* **130**:557.

32. Unanue, E. R., B. A. Askonas, and A. C. Allison. 1969. A role of macrophages in the stimulation of immune responses by adjuvants. *J. Immunol.* **103**:71.

33. Schmidtke, J., and E. R. Unanue. 1971. Interaction of macrophages and lymphocytes with surface immunoglobulin. *Nature (New Biol.) (Lond.).* **233**:84.

34. Campbell, D. A., and J. S. Garvey. 1961. The fate of foreign antigen and speculations as to its role in immune mechanisms. *Lab. Invest.* **10**:1126.

35. Rittenberg, M. B., and E. L. Nelson. 1960. Macrophages, nucleic acids, and the induction of antibody formation. *Am. Nat.* **94**:321.

36. Askonas, B. A., and J. M. Rhodes. 1965. Immunogenicity of antigen-containing ribonucleic acid preparations from macrophages. *Nature (Lond.).* **205**:470.

37. Fishman, M. 1961. Antibody formation in vitro. *J. Exp. Med.* **114**:837.

38. Bishop, D. C., and A. A. Gottlieb. 1970. Macrophages, RNAs and the immune response. *Curr. Top. Microbiol. Immunol.* **51**:1.

39. Paraf, A. 1970. Mécanisme d'action des adjuvants de l'immunité. *Ann. Inst. Pasteur. (Paris).* **118**:419.

40. Dresser, D. W., R. N. Taub, and A. R. Krantz. 1970. The effect of localized injection of adjuvant material on the draining lymph node. *Immunology.* **18**:663.

41. Zatz, M. M., and E. M. Lance. 1971. The distribution of ^{51}Cr-labeled lymphocytes into antigen-stimulated mice. Lymphocyte trapping. *J. Exp. Med.* **134**:224.

42. Taub, R. N., and R. K. Gershon. 1972. The effect of localized injection of adjuvant material on the draining lymph node. III. Thymus dependence. *J. Immunol.* **108**:377.

43. Ortiz-Ortiz, L., and B. N. Jaroslow. 1970. Enhancement by the adjuvant, endotoxin, of an immune response induced *in vitro. Immunology.* **19**:387.

44. Thorbecke, G. J., and B. Benacerraf. 1962. The reticulo-endothelial system and immunological phenomena. *Prog. Allergy.* **6**:559.

45. Schwartz, R. S., R. J. W. Ryder, and A. A. Gottlieb. 1970. Macrophages and antibody synthesis. *Prog. Allergy.* **14**:81.

46. Abdou, N. I., and M. Richter. 1970. The role of bone marrow in the immune response. *Adv. Immunol.* **12**:201.

47. Frei, P. C., B. Benacerraf, and G. J. Thorbecke. 1965. Phagocytosis of the antigen a crucial step in the induction of the primary response. *Proc. Natl. Acad. Sci. U.S.A.* **53**:20.

48. Schmidtke, J. R., and E. R. Unanue. 1971. Macrophage-antigen interaction. Uptake, metabolism and immunogenicity of foreign albumin. *J. Immunol.* **107**:331.

49. Galliley, R., and M. Feldman. 1966. The induction of antibody production in X-irradiated animals by macrophages that interacted with antigen. *Isr. J. Med. Sci.* **2**:358.

50. Pinckard, R. N., D. M. Weir, and W. H. McBride. 1967. Factors influencing the immune response. I. Effects of the physical state of the antigen and of lympho-reticular cell proliferation on the response to intravenous injection of bovine serum albumin in rabbits. *Clin. Exp. Immunol.* **2**:331.

51. Dresser, D. W. 1968. An assay for adjuvanticity. *Clin. Exp. Immunol.* **3**:877.

52. Spitznagel, J. K., and A. C. Allison. 1970. Mode of action of adjuvants. Retinol and other lysosome-labilizing agents as adjuvants. *J. Immunol.* **104**:119.

53. Weissmann, G., and P. Dukor. 1970. The role of lysosomes in immune responses. *Adv. Immunol.* **12**:283.

54. Haskill, J. S., J. Marbrook, and B. E. Elliott. 1971. Thymus-independent step in the immune response to sheep erythrocytes. *Nature (New Biol.) (Lond.).* **233**:237.

55. Martial-Lasfargues, C., M. Liacopoulos-Briot, and B. N. Halpern. 1967. Culture des leucocytes sanguins de souris in vitro. Etude de l'action de la phytohémagglutinine (PHA) sur les lymphocytes de souris normales et thymectomisées *C. R. Soc. Biol.* **160**:2013.

56. Greaves, M. F., I. M. Roitt, and M. E. Rose. 1968. Effect of bursectomy and thymectomy on the responses of chicken peripheral blood lymphocytes to phytohaemagglutinin. *Nature* (*Lond.*). **220**:293.

57. Doenhoff, M. J., A. J. S. Davies, E. Leuchars, and V. Wallis. 1970. The thymus and circulating lymphocytes of mice. *Proc. Roy. Soc. Lond. B Biol. Sci.* **176**:69.

58. Stobo, J. D., A. S. Rosenthal, and W. E. Paul. 1972. Functional heterogeneity of murine lymphoid cells. I. Responsiveness to and surface binding of concanavalin A and phytohemagglutinin. *J. Immunol.* **108**:1.

59. Andersson, J., G. Möller, and O. Sjöberg. 1972. Selective induction of DNA synthesis in T and B lymphocytes. *Cell. Immunol.* In press.

60. Mooney, J. J., and B. H. Waksman. 1970. Activation of normal rabbit macrophage monolayers by supernatants of antigen-stimulated lymphocytes. *J. Immunol.* **105**:1138.

61. Junge, U., J. Hoekstra, and F. Deinhardt. 1971. Stimulation of peripheral lymphocytes by allogenic and autochthonous mononucleosis lymphocyte cell lines. *J. Immunol.* **106**:1306.

62. Nathan, C. F., M. L. Karnovsky, and J. R. David. 1971. Alterations of macrophage functions by mediators from lymphocytes. *J. Exp. Med.* **133**:1356.

63. Katz, D. H., W. E. Paul, E. A. Goidl, and B. Benacerraf. 1971. Carrier function in anti-hapten antibody responses. III. Stimulation of antibody synthesis and facilitation of hapten-specific secondary antibody responses by graft-*versus*-host reactions. *J. Exp. Med.* **133**:169.

64. Moore, R. D., and M. D. Schoenberg. 1971. Synthesis of antibody by lymphocytes restimulated *in vitro* with antigen. *J. Reticuloendothel. Soc.* **9**:254.

65. Landy, M., and L. N. Chessin. 1969. The effect of plant mitogens on humoral and cellular immune responses. *Antibiot. Chemother.* **15**:199.

66. Naspitz, C. K., and M. Richter. 1968. The action of phytohemagglutinin *in vivo* and *in vitro*, a review. *Prog. Allergy.* **12**:1.

67. Peavey, D. L., W. H. Adler, and R. T. Smith. 1970. The mitogenic effects of endotoxin and staphylococcal enterotoxin B on mouse spleen cells and human peripheral lymphocytes. *J. Immunol.* **105**:1453.

ANTIGEN-SPECIFIC T CELL-MEDIATED SUPPRESSION

IV. Role of Macrophages in Generation of L-Glutamic Acid^{60}-L-Alanine^{30}-L-Tyrosine^{10} (GAT)-Specific Suppressor T Cells in Responder Mouse Strains[1]

MICHEL PIERRES[2] AND RONALD N. GERMAIN

From the Department of Pathology, Harvard Medical School, Boston, Massachusetts 02115

The role of adherent cells (MØ) in 1) primary *in vitro* PFC responses and 2) *in vitro* generation of suppressor T cells (T$_s$) to the antigen L-glutamic acid^{60}-L-alanine^{30}-L-tyrosine^{10} in responder strain mice has been investigated. Removal of MØ from spleen cell suspensions by either serial culture on plastic surfaces or passage over Sephadex G-10 beads yielded lymphocytes unable to give primary *in vitro* PFC responses under modified Mishell-Dutton conditions to either soluble GAT or the particulate antigens sheep red blood cells (SRBC) or GAT-complexed electrostatically to methylated bovine serum albumin (GAT-MBSA). However, the viability of such MØ-depleted cultures was significantly lower than control cultures of unseparated spleen cells. Addition of 5×10^{-5} M 2-mercaptoethanol (2-ME) to the culture medium restored viability and permitted nonadherent cells to respond to SRBC and GAT-MBSA, but not to soluble GAT. The response of nonadherent cells to soluble GAT could be restored in the absence of 2-ME by the addition of 1 to 3×10^5 x-irradiated, anti-Thy 1.2 + C-treated peritoneal exudate cells (MØ), or in the presence of 2-ME by 1 to 2 $\times 10^4$ MØ. This low number of MØ was identical to the number of GAT-bearing MØ necessary in the presence of 2-ME to give optimal GAT PFC responses. Taken together, these data indicate a crucial antigen-presenting role for MØ in primary *in vitro* anti-GAT responses. Removal of such antigen-presenting cells from responder spleen cells left a lymphocyte population, which, when cultured for 2 days with normally immunogenic (10 μg/ml) concentrations of GAT, gave rise to highly active Thy 1.2$^+$ GAT-specific suppressor cells. This T$_s$ induction was demonstrated in several responder strains. The GAT-T$_s$ were unable to suppress the *in vitro* responses to SRBC or the copolymer L-glutamic acid^{50}-L-tyrosine^{50} (GT) as GT-MBSA. Increasing the concentration of GAT to $\geq$ 30 μg/ml permitted the induction of GAT-T$_s$ in unseparated spleen cells. Thus, in responder mice, MØ able to present antigen in an immunogenic form play a central role in regulating the balance of activated T$_H$ and T$_s$, suggesting that an Ir gene defect in nonresponder mice at the MØ level for this presentation function may account for the predominant suppressor T cell responses of such mice.

L-glutamic acid^{60}-L-alanine^{30}-L-tyrosine^{10} (GAT)[3] is a synthetic antigen under dominant H-2-linked Ir gene control, which, when administered to nonresponder mice bearing the H-2$^{p, q, s}$ haplotypes, elicits a characteristic suppressor T cell (T$_s$) response. This T$_s$ response has been investigated in detail by this laboratory (1). These suppressor cells, which are Lyt 23$^+$ and antigen specific, prevent the development of primary PFC responses to GAT complexed to the immunogenic carrier, methylated bovine serum albumin (GAT-MBSA), in syngeneic nonresponder mice (2, 3). Extracts of spleens and thymuses from GAT-primed nonresponder mice contain a suppressor factor(s) (T$_s$F) that is antigen specific, 40 to 50,000 m.w., bears I region determinants, and functions at least in large part by inducing virgin splenic T cells to become a new population of GAT-T$_s$ (T$_{s_2}$) (4–7). More recent studies have revealed that this same GAT- T$_s$F can both act directly across H-2 barriers in other nonresponder strains (8), and also induce T$_{s_2}$ in responder mice (9).

It has not yet been possible to directly demonstrate specific GAT-T$_s$ induced by antigen *in vivo* in responder mice. However, the ability of GAT-T$_s$F from nonresponder mice to induce GAT-T$_{s_2}$ in responder spleen cells suggest that at least part of the GAT-specific suppressor mechanism is intact in those responder strains tested (BALB/c, A/J) (9). The question remains as to whether or not such Ir responder mice could, under any circumstances, generate GAT-T$_s$ from antigen directly, and if so, what was the underlying physiologic basis for the differences in T$_s$ generation between nonresponder and responder

Received for publication May 11, 1978.

Accepted for publication June 28, 1978.

The costs of publication of this article were defrayed in part by the payment of page charges. This article must therefore be hereby marked *advertisement* in accordance with 18 U.S.C. Section 1734 solely to indicate this fact.

[1] This work was supported by Grants AI-09920 and AI-00152 from the National Institute of Allergy and Infectious Diseases, and Grant CA-09130 from the National Cancer Institute, Department of Health, Education and Welfare.

[2] Recipient of Public Health Service International Research Fellowship of the National Institutes of Health no. F05TW 2381-02.

[3] Abbreviations used in this paper: GAT, random terpolymer of L-glutamic^{60}-L-alanine^{30}-L-tyrosine^{10}; GT, random copolymer of L-glutamic acid^{50}-L-tyrosine^{50}; H-2, murine major histocompatibility complex; HBSS, Hanks' balanced salt solution; Ir, immune response; MØ, macrophage(s); MBSA, methylated bovine serum albumin; MD$_H$, modified Mishell-Dutton medium with 10 mM HEPES; 2-ME, 2-mercaptoethanol; MEM-5, MEM-10, Eagle's minimum essential medium with 5 or 10% FCS; OVA, ovalbumin; PEC, peritoneal exudate cells; T$_H$, helper T cell; T$_s$, suppressor T cell; T$_s$F, T cell-derived suppressor factor.

mice challenged identically with GAT. Kapp *et al.* (10) and Pierce *et al.* (11) have demonstrated that GAT presented on MØ has special properties with respect to T helper cell priming in both nonresponder and responder mice, and it was recently shown that (responder × nonresponder) F₁ mice primed to soluble GAT give secondary *in vitro* responses only to GAT presented on responder but not nonresponder MØ (12). Furthermore, this differential responsiveness to MØ-bound GAT is associated with suppressor T cell function.[4] These data suggested that MØ might play a crucial role in determining the Ir gene-controlled balance between GAT-specific T_H and T_s activation. The current study therefore first re-examined carefully the role of MØ in *in vitro* primary PFC responses to GAT, then approached the question of whether the absence of antigen presentation by responder MØ might favor T_s induction by GAT, with responder spleen cells. The data presented below document that MØ have both nonspecific culture-supporting and specific antigen presentation roles in primary GAT PFC responses *in vitro* and also demonstrate that nonadherent (MØ depleted) responder spleen cells preferentially react to GAT exposure with T_s induction, i.e., behave phenotypically like unseparated nonresponder cells. The implications of these data for understanding GAT-Ir gene function are discussed.

MATERIALS AND METHODS

Mice. Male or female (C57BL/6 × DBA/2)F₁, [(B6D2)F₁], (H-2^{b/d}), BALB/c (H-2^d), B10.BR (H-2^k), and CBA (H-2^k) mice were purchased from the Jackson Laboratories, Bar Harbor, Maine, or Health Research Incorporated, West Seneca, N. Y. They were maintained in our animal facilities on standard laboratory chow and chlorinated water *ad libitum,* and used at 8 to 20 weeks of age.

Antigens. Sheep blood in Alsever's solution was purchased from GIBCO, Madison, Wis. Before use as antigen or indicator cells for the PFC assay, the SRBC were washed three times in sterile Hanks' balanced salt solution (HBSS), the buffy coat being removed after the first centrifugation. The random copolymers L-glutamic acid^{60}-L-alanine^{30}-L-tyrosine^{10} (GAT), average m.w. 38,500 (lot VI), and L-glutamic acid^{50}-L-tyrosine^{50} (GT), average m.w. 90,000 (lot IX), were purchased from Miles Laboratories, Miles Research Division, Elkhart, Ind. Sterile solutions for culture were prepared in saline containing 1% Na_2CO_3 and adjusted to pH 7.0. Methylated bovine serum albumin (MBSA) was purchased from Calbiochem, San Diego, Calif., and the copolymers GAT and GT were complexed to MBSA as previously described (13). Chicken ovalbumin (OVA) was purchased from Miles Laboratories, Inc., Kankakee, Ill.

Culture conditions. Single-cell suspensions were obtained by teasing spleens in cold HBSS or in Eagle's minimum essential medium containing sodium bicarbonate, 10 mM HEPES (Microbiological Associates, Bethesda, Md.), and 5 or 10% fetal calf serum (Irvine Scientific, Irvine, Calif. (lot 60138) (MEM-5 or MEM-10). After settling of coarse particules, spleen cells were centrifuged and resuspended in Mishell-Dutton medium (14) containing 10% FCS and 10 mM HEPES (MD_H). Where appropriate, 2-mercaptoethanol (2 ME) (Eastman-Kodak, Rochester, N. Y.) was added to the medium at a final concentration of 5 × 10^{-5} M. Viable spleen cells (7.5 × 10^6) were cultured in a volume of 1 ml of MD_H in Linbro plates (FB 16-24TC, Linbro Chemical Co., New Haven, Conn.), on a rocking platform at

37°C in humidified 10% CO_2, 7% O_2, 83% N_2 atmosphere. Cultures were fed daily with 70 μl of a feeding mixture composed of 50% of nutritional cocktail (14) and 50% FCS. On day 5 duplicate cultures were harvested, pooled, and the viable cell recovery was determined by trypan blue dye exclusion. The cells were then washed and assayed for PFC responses.

PFC assay. The antibody responses to SRBC, GAT, GAT-MBSA, and GT-MBSA were assayed by a slide modification of the Jerne hemolytic plaque assay described in detail elsewhere (13).

In vitro induction of suppressor cells. Viable nonadherent or unseparated spleen cells (30 × 10^6) were cultured in 3 ml of MD_H in 60-mm Falcon tissue culture Petri dishes (No. 3002) for 2 or 3 days. The cultures contained either no additional antigen, 10 μg/ml ovalbumin (OVA), 1 to 100 μg/ml GAT, or 10 μg/ml GT, and were fed daily with 200 ml of feeding mixture. On day 2 or 3, the cells were harvested, washed three times in MEM-5, then suspended in MD_H, and the viable cells were enumerated. Usually 20 to 30% viable cells were recovered depending on the duration of the culture and the nature of the cells cultured. Graded numbers of viable educated cells (9 × 10^5 to 10^3) were then tested for their ability to suppress the primary *in vitro* response of 7.5 × 10^6 viable normal spleen cells to GAT, GT-MBSA, or SRBC under modified Mishell-Dutton culture conditions.

Macrophages

Peritoneal exudate cells (PEC). Mice were injected i.p. with 1.5 ml of sterile proteose peptone (Difco, Detroit, Mich.). Three days later the peritoneal cavity was washed with sterile HBSS. Usually 8 to 15 × 10^6 PEC were obtained per mouse, and 80% of these cells were MØ-like as judged by morphologic and adherence criteria. In some experiments the PEC were irradiated at 4°C (1400 R, 100 R/min, 0.2 mm A1, 1 mm Cu filtration, Maximar III, General Electric, Waterford, N. Y.), then washed and treated in two steps by an anti-Thy 1.2 antiserum and a nontoxic rabbit complement (C) as previously described (7).

Adherent spleen cells. Viable spleen cells (5 × 10^6) suspended in 1 ml of MEM-10 were incubated in Linbro wells for 2 hr at 37°C in a 5% CO_2 atmosphere in air. The nonadherent cells were removed and the remaining firmly adherent cells were washed three times in HBSS.

GAT-MØ. The preparation of GAT bound to peritoneal macrophages (GAT-MØ) has been described previously (15).

Macrophage depletion. Two techniques were used to achieve macrophage depletion, with similar results. The first method was adherence on plastic surfaces as described by Mosier (16). Viable spleen cells (100 × 10^6) in 10 ml of MEM-10 were incubated in 100 mm × 18 mm tissue culture Petri dishes (Falcon No. 3003) for 90 min at 37°C in a 5% CO_2 atmosphere. The nonadherent spleen cells were then resuspended by gentle swirling and transferred into another dish for a second incubation of 60 min. The same operation was repeated once more. This serial passage on plastic surface usually yields 30 to 40% of the original spleen cells, containing less than 0.1% macrophages as determined by latex particle (Dow Chemical Co., Indianapolis, Ind.) uptake. The second technique of macrophage depletion was Sephadex G10 column separation by the method described by Ly and Mishell (17) with the following slight modifications. The columns were equilibrated in MEM-10 before cell addition, then incubated 45 min at 37°C before elution of cells with the same medium at 37°C. Fifty percent of the original cells (containing less than 0.1% macrophages) were recovered in most of the experiments.

---

[4] Germain, R. N. and B. Benacerraf. 1978. Involvement of suppressor T cells in Ir gene regulation of secondary antibody responses of primed (responder × nonresponder) F₁ mice to macrophage bound L-glutamic acid^{60}-L-alanine^{30}-L-tyrosine^{10} (GAT). J. Exp. Med. In press.

T cell enrichment. T-enriched spleen cells were obtained by filtration through nylon wool columns according to Julius *et al.* (18). Twenty to 30% of the original cells were recovered and 1 to 5% of the eluted cells would be stained with a fluorescent rabbit anti-mouse Ig antibody (the gift of Dr. K. A. Ault).

RESULTS

Effect of macrophage depletion on primary in vitro responses to GAT and SRBC. Depletion of splenic adherent cells is known to decrease primary *in vitro* PFC responses to SRBC and GAT (16, 19). As a first step in further analyzing the requirements for MØ in GAT responses, macrophage depletion was carried out with two different techniques, and the effects of such procedures on *in vitro* PFC responses to GAT and SRBC under the current culture conditions were examined. Spleen cells were cultured serially on plastic dishes (16) in order to remove adherent cells or passed over Sephadex G-10 columns (17). Figure 1 shows data pooled from several experiments employing the plastic adherence procedure. Macrophage-depleted spleen cells gave background PFC responses (10 to 15% of control) under these conditions to both GAT and SRBC. Similar results were obtained with Sephadex G-10 column filtration (Tables I and II). It was noted, however, that MØ-depleted cultures were more alkaline than cultures of unseparated spleen cells and had a significantly lower cell recovery (7 to 8%) than even nonantigen-stimulated whole spleen cell cultures (15 to 20%). The ability of 2-ME (known to enhance *in vitro* PFC responses and promote lymphocyte viability [20]) to reconstitute the primary PFC responses of MØ-depleted cells was therefore assessed. As shown in Figure 1, addition of 2-ME to a final concentration of 5×10^{-5} M restored both the IgM and IgG PFC responses to SRBC, but not the IgG response to GAT. However, addition of the appropriate number (3×10^5) of PEC did restore both the SRBC and GAT responses. This differential reconstitution occurred despite the ability of 2-ME to restore cell recoveries to the usual level, and appears to indicate a differential MØ requirement for these two antigens under these culture conditions.

Comparison of the macrophage dependency of primary in vitro PFC responses to soluble GAT and the insoluble complex GAT-MBSA. The above observation that SRBC but not GAT responses of macrophage-depleted spleen cell cultures could be reconstituted with 2-ME suggests, as previously reported (21), that the physical form of the antigen might affect the macrophage requirement of the corresponding response. The terpolymer GAT offered a way to investigate this parameter directly since it is immunogenic *in vitro* as soluble GAT or as the insoluble complex GAT-MBSA. Table I shows the results of two representative experiments which indicate that: 1) macrophage-depleted spleen cell cultures did not respond to soluble GAT or to GAT-MBSA. 2) 2-ME restored the response to the latter particulate antigen but not to soluble GAT. These results taken together with the reconstitution of the SRBC responses by 2-ME suggest that such particulate antigens have a less stringent macrophage requirement than soluble antigens for *in vitro* PFC responses.

Number of antigen-free macrophages or GAT-bearing macrophages required to reconstitute the in vitro response to GAT. Adherence to plastic or filtration through a Sephadex G10 column removes not only most of the macrophages in spleen cell suspensions but also a population of adherent lymphocytes. Experiments were therefore designed to establish that the cell type responsible for the lack of response in such depleted cell cultures was macrophage-like. In preliminary experiments (data not shown) it was verified that both firmly adherent spleen cells (representing about 1% of 5×10^6 spleen cells) and also 3×10^5 PEC were able to similarly reconstitute MØ-depleted spleen cell cultures. Furthermore, the same number of irradiated and anti-Thy 1.2 antiserum and C-treated PEC restored most of the PFC response to soluble GAT, indicating therefore that a non-T-derived, radioresistant accessory cell population is required for *in vitro* PFC response to GAT. Since 2-ME could not restore these GAT responses in adherent cell-depleted populations, experiments were designed to investigate the macrophage requirement of such cultures in the presence of 2-ME. Table II indicates that as few as 3 to 5×10^4 PEC could achieve such a reconstitution, suggesting that when the nonspecific culture-supporting effects of macrophages are replaced by 2-ME, a small number of PEC are nevertheless necessary and sufficient for the development of anti-GAT PFC response *in vitro*. Since GAT can be presented in highly immunogenic form in culture as GAT-bearing macrophages, the ability of irradiated and anti-Thy 1.2 + C-treated antigen-free or GAT-bearing peritoneal macrophages to reconstitute the *in vitro* response to GAT in the presence or absence of 2-ME was studied. Table II

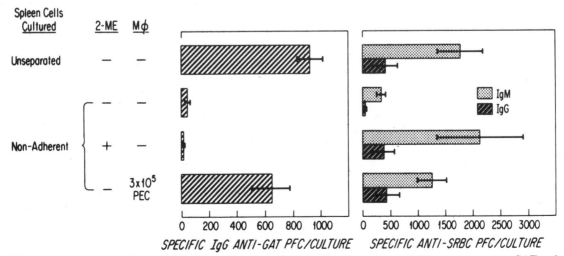

Figure 1. Effects of 2-ME (5×10^{-5} M) and MØ replacement (3×10^5 PEC) on *in vitro* primary PFC responses to 1 μg GAT or 2×10^6 SRBC of modified Mishell-Dutton cultures containing adherent cell-depleted splenocytes. The data are from several replicate experiments, with the arithmetic mean PFC/culture ± S.E. shown.

presents two similar experiments carried out with (B6D2)F$_1$ and BALB/c mice which indicate that: 1) antigen-free and GAT-bearing macrophages exert comparable replacing functions; 2) 10-fold fewer adherent cells are required to reconstitute the GAT response in the presence of 2-ME. It can be concluded from these data that a critical role for macrophage antigen presentation appears to exist for the *in vitro* response to soluble GAT in responder strains.

In vitro induction of GAT-specific suppressor cells in macrophage-depleted responder spleen cell cultures. The same antigen concentrations of GAT (1 to 10 μg/ml) known to induce optimal *in vitro* primary PFC responses with responder spleen cells lead to specific GAT suppressor T cell induction with nonresponder cells (7). Since both the current and previous

data (19) suggested a critical role for MØ antigen presentation in GAT T helper cell function in responder mice, it was possible that MØ depletion might permit suppressor cell development in the relative absence of a helper response in responder spleen cells at normally immunogenic concentrations of GAT. This was first investigated by comparing the ability of unseparated *vs* MØ-depleted responder spleen cells to develop specific GAT suppressor cells when cultured with GAT. Figure 2 depicts a representative experiment in which 30×10^6 viable unseparated or MØ-depleted (B6D2)F$_1$ spleen cells were incubated for 2 days under Mishell-Dutton conditions with 10 μg GAT or 10 μg OVA and, after washing, were tested for their ability to suppress the primary *in vitro* GAT PFC response of normal syngeneic spleen cells. The data reveal that even 10^5 unseparated spleen cells recovered from such first cultures could not affect the response to GAT, whereas both 10^5 and 3×10^4 cells from nonadherent spleen populations cultured for 2 days with 10 μg/ml GAT significantly suppressed the GAT PFC response of syngeneic cells. Nonadherent cells educated with OVA did not show this suppressive activity, indicating that GAT was required in the first culture for suppressor cell generation. It is clear, therefore, that upon stimulation with normally immunogenic concentrations of GAT, MØ-depleted, but not unseparated, spleen cells can be stimulated to develop detectable suppressor cells to GAT. The pooled data from a series of similar experiments are presented in Table III, which demonstrate that at equal cell numbers (10^5), nonadherent cells cultured without antigen have no suppressive activity, whereas GAT-exposed MØ-depleted spleen cells are completely suppressive. Furthermore, even 3×10^5 of such GAT-primed cells failed to suppress primary IgM (or, although not shown, IgG) anti-SRBC responses, indicating the specificity of the suppression.

In vitro induction of GAT-specific suppressor cells in var-

TABLE I

Effects of macrophage depletion and 2-ME on primary in vitro responses to GAT and GAT-MBSA

Experiment	Responder strain	Spleen Cells Cultures	Specific IgG Anti-GAT PFC Response per Culture to	
			1 μg GAT	1 μg GAT-MBSA
I	(B6D2)F$_1$	Unseparated	920	490
		Nonadherent[a]	70	70
		Nonadherent + 2-ME[b]	<20	670
II	BALB/c	Unseparated	1342	1297
		Sephadex G-10 filtered[c]	112	<15
		Sephadex G-10 filtered + 2-ME	<15	872

[a] Nonadherent spleen cells obtained by serial passage on plastic surface.

[b] 5×10^{-5} M final concentration.

[c] See *Materials and Methods.*

TABLE II

Critical role of macrophage presentation for GAT IgG PFC responses in vitro

Experiment	Strain	Spleen Cells Cultured[a]	2-ME[b]	Macrophages Added[c]		GAT	Specific IgG Anti-GAT PFC per Culture
				Unpulsed	GAT-pulsed		
						μg	
I.	(B6D2)F$_1$	Unseparated	−	−	−	1	1360
		Unseparated	−	−	3×10^4	−	1220
		Nonadherent[d]	−	−	−	1	50
		Nonadherent[d]	−	3×10^4	−	1	90
		Nonadherent[d]	−	10^5	−	1	350
		Nonadherent[d]	−	3×10^5	−	1	740
		Nonadherent[d]	−	−	3×10^4	−	20
		Nonadherent[d]	−	−	10^5	−	180
		Nonadherent[d]	−	−	3×10^5	−	980
		Nonadherent[d]	+	−	−	1	20
		Nonadherent[d]	+	3×10^4	−	1	820
		Nonadherent[d]	+	−	3×10^4	−	460
II.	BALB/c	Unseparated	−	−	−	5	1065
		Unseparated	−	−	3×10^4	−	1080
		Sephadex G-10 filtered[e]	−	−	−	5	112
		Sephadex G-10 filtered[e]	−	3×10^5	−	5	292
		Sephadex G-10 filtered[e]	−	6×10^5	−	5	652
		Sephadex G-10 filtered[e]	+	−	−	5	15
		Sephadex G-10 filtered[e]	+	3×10^4	−	5	765

[a] 7.5×10^6 spleen cells cultured under modified Mishell-Dutton conditions.

[b] 5×10^{-5} M.

[c] PEC obtained by washing peritoneal cavity of mice injected 3 days before with proteose peptone i.p., were irradiated (1400 R) and treated by an anti-Thy 1.2 antiserum and rabbit complement. After washing, one aliquot of cells was pulsed *in vitro* with GAT. Botn antigen-free macrophages and GAT-macrophages were then resuspended in MD$_H$, and added in various numbers to spleen cell cultures.

[d] Separated by serial passages on plastic surface.

[e] See *Materials and Methods.*

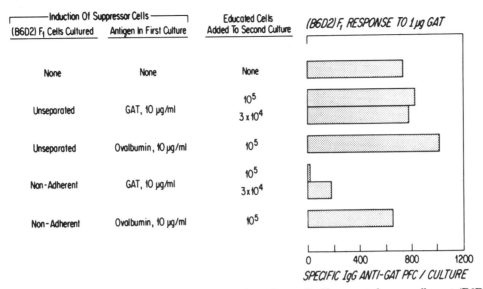

Figure 2. Induction of GAT T$_s$ with the use of nonadherent responder spleen cells. Unseparated or nonadherent (B6D2)F$_1$ spleen cells were cultured for 2 days with 10 μg/ml OVA or 10 μg/ml GAT, then washed and added to fresh modified Mishell-Dutton cultures of 7.5×10^6 syngeneic F$_1$ spleen cells stimulated with 1 μg GAT. Specific IgG GAT PFC were determined on day 5 of culture by using GAT-coated SRBC.

TABLE III

In vitro induction of GAT-specific suppressor cells with macrophage-depleted spleen cells of (B6D2)F$_1$ responder mice

Induction of Suppressor Cells (First Culture)		Educated Cells Added to Second Culture	Specific PFC per Culture[a]	
Cells cultured	Antigen		IgG anti-GAT[b]	IgM anti-SRBC[c]
Nonadherent spleen cells	None	None	1055 ± 167	2258 ± 444
Nonadherent spleen cells	None	3×10^5	1286 ± 401	2836 ± 566
Nonadherent spleen cells	None	10^5	975 ± 405	2795 ± 1375
Nonadherent spleen cells	None	3×10^4	1276 ± 460	2496 ± 502
Nonadherent spleen cells	GAT 30–50 μg/ml	3×10^5	<20	2135 ± 1035
Nonadherent spleen cells	GAT 30–50 μg/ml	10^5	<20	2293 ± 579
Nonadherent spleen cells	GAT 30–50 μg/ml	3×10^4	1025 ± 705	2485 ± 655
Nonadherent spleen cells	GAT 30–50 μg/ml	10^4	750 ± 415	2673 ± 599

[a] Arithmetic mean ± standard errors from four independent experiments.

[b] Cultures were stimulated with 1 μg GAT.

[c] Cultures were stimulated with 2×10^6 SRBC.

ious responder strains. The demonstration of GAT suppressor cell induction in MØ-depleted spleen cells of responder (B6D2)F$_1$ mice suggested studying other responder strains to determine whether or not this phenomenon could be generalized. A limited series of experiments indicate that BALB/c (H-2^d), CBA (H-2^k), and B10.BR (H-2^k), as well as (B6D2)F$_1$ (H-2$^{b/d}$), all yield T$_s$ after 2 days of culture of MØ-depleted spleen cells with 30 to 50 μg/ml GAT that are able to specifically inhibit GAT, but not SRBC, primary PFC responses *in vitro* (data not shown).

In vitro-generated suppressor cells from responder mice are thymus derived. Two experimental approaches were used to demonstrate directly that the suppressor cells in this model are T cells. In the first approach, nylon wool-passed cells, <5% Ig$^+$, and largely MØ depleted, were cultured for 2 days with 30

μg/ml GAT and tested for suppressive activity. As few as 3×10^4 such cells completely and specifically suppressed GAT PFC responses (Table IV). In the second approach, suppressor cells were obtained after 2 days of culture of MØ-depleted spleen cells with 30 μg/ml GAT, and treated with anti-Thy 1.2 plus C before assay. This treatment, but not control treatment, removed suppressive activity from the precultured cells. These experiments clearly indicate that the suppressor cells generated in responder mice under these conditions are T cells.

Fine specificity of in vitro-induced suppressor cells. Although the above experiments indicated that the cells able to suppress GAT-PFC responses were unable to inhibit anti-SRBC responses, it would be argued that the sensitivity of these two antigens to nonspecific suppression is sufficiently different to account for the apparent specificity of the suppressive effects, and that the requirement for GAT for such suppressor induction is not reflected in the fine specificity of the actual suppressor cells. To better define the antigen specificity of the GAT-induced T$_s$, the closely related polymer L-glutamic acid^{50}L-tyrosine^{50} as GT-MBSA was employed. Nonadherent spleen cells from a GT nonresponder, suppressor strain (BALB/c) that is also a GAT responder were cultured for 2 days with either no antigen, 10 μg/ml GT, or 30 μg/ml GAT. These cells were then tested for their ability to inhibit primary PFC responses to either GT-MBSA or GAT. As illustrated in Figure 3, cells educated with GT suppress only GT-MBSA responses, and cells cultured with GAT suppress only GAT responses, even at the very high cell number of 6×10^5. These data indicate that the suppression in this *in vitro* model is highly specific, and can clearly distinguish between polymers that are extensively cross-reactive at the serologic level.

Differential antigen requirements for suppressor cell induction by GAT in unseparated and MØ-depleted responder spleen cell cultures. The above series of experiments indicated that GAT-T$_s$ can be induced in responder cell cultures by immunogenic or supraimmunogenic doses of GAT. At immunogenic concentrations, MØ removal was crucial to effective T$_s$ generation. Table V documents that when the concentration of GAT is increased to a sufficiently high level, the need for MØ depletion for T$_s$ induction can be partially circumvented. Thus,

TABLE IV

In vitro generated (B6D2)F₁ GAT-specific suppressor cells are T derived

Expt. No.	Induction of Suppressor Cells (First Culture)		Educated Cells Added to Second Culture	Specific PFC per Culture[a]	
	Cells Cultured	Antigen		IgG anti-GAT[b]	IgM anti-SRBC[b]
I.			None	520	3440
	Nylon-wool filtered spleen cells	None	6×10^5	640	2010
	Nylon-wool filtered spleen cells	None	3×10^5	540	2460
	Nylon-wool filtered spleen cells	GAT 30 µg/ml	6×10^5	<20	2680
	Nylon-wool filtered spleen cells	GAT 30 µg/ml	3×10^5	100	2630
	Nylon-wool filtered spleen cells	GAT 30 µg/ml	3×10^4	<20	2170
	Nylon-wool filtered spleen cells	GAT 30 µg/ml	10^4	710	3640
II.			None	450	1300
	Nonadherent spleen cells	None	3×10^5	400	1750
	Nonadherent spleen cells	GAT 30 µg/ml	3×10^5	<20	1100
	Nonadherent spleen cells	GAT 30 µg/ml	3×10^5 treated with anti-Thy 1.2 and complement	920	1530
	Nonadherent spleen cells	GAT 30 µg/ml	3×10^5 treated with complement alone	<20	860

[a] 7.5×10^6 spleen cells were cultured in modified Mishell-Dutton conditions.

[b] Cultures stimulated with 1 µg GAT or 2×10^6 SRBC.

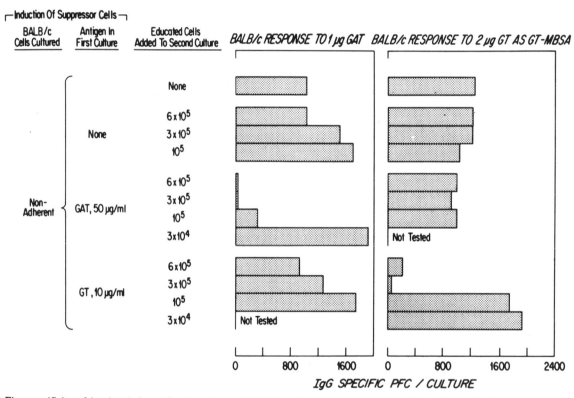

Figure 3. Fine specificity of *in vitro* induced T$_s$. Nonadherent BALB/c spleen cells were cultured for 2 days with either no antigen, 10 µg/ml GT, or 50 µg/ml GAT. After this culture period, the remaining cells were washed thoroughly and varying numbers added to fresh modified Mishell-Dutton cultures of 7.5×10^6 BALB/c cells, which were then stimulated by either 1 µg GAT or 2 µg GT as GT-MBSA. The IgG specific PFC were determined on day 5 of culture by using GAT-coated SRBC.

50 µg/ml GAT permitted detection of GAT-T$_s$ with unseparated spleen cells when 10^5 but not 3×10^4 cells were transferred to assay cultures, whereas as few as 3×10^4 cells still were suppressive if MØ-depleted cells were cultured with this concentration of antigen. Figure 4 illustrates a dose-response titration for the induction of GAT-T$_s$ with whole or nonadherent spleen cells and varying amounts of GAT (3 to 100 µg/ml). The data for 1.5×10^5 transferred cells only are shown, but similar results were obtained with 3 and 6×10^5 cells. In all cases, a 3- to 4-fold difference in the concentration of GAT necessary to induce 50% suppression at a fixed cell number was observed when comparing unseparated *vs* MØ-depleted cells. In other experiments, MØ-depleted cells required as little as ⅙ to ⅒ the concentration of GAT as unseparated cells to yield equivalent suppression. These data are all consistent with the conclusion that removal of MØ from spleen cell populations leads to more easily detected T$_s$ induction, and a decreased antigen requirement for such T$_s$ generation, but that GAT-T$_s$ can nonetheless be produced in the presence of antigen-presenting MØ, provided the concentration of GAT is sufficiently great.

DISCUSSION

The data presented above deal with two aspects of macrophage function in the *in vitro* immune response to the synthetic antigen GAT. First, a series of experiments demonstrates that MØ-depleted responder spleen cells fail to respond to soluble GAT, to SRBC, or to GAT-MBSA, and that when 2-ME is added to such cultures to enhance viability, only the responses to the particulate antigens SRBC and GAT-MBSA are reconstituted. A small number (1 to 3×10^4) of macrophages are necessary and, in the presence of 2-ME, sufficient to restore the PFC response to soluble GAT. These data, on the whole, agree with previous studies on adherent cell function in primary *in vitro* responses (16, 19). Several groups have shown that 2-ME, other reducing reagents, or macrophage or fibroblast supernatants could restore the responses of nonadherent spleen cells to particulate antigens such as SRBC (20, 22, 23). These agents appear to act in an antigen nonspecific fashion to promote lymphocyte viability and possibly to enhance blastogenesis (24, 25). However, this nonspecific supportive function of macrophages is not sufficient for all immune responses, since even in the 2-ME-containing cultures whose viability equalled control cultures, MØ-depleted spleen cells failed to respond to soluble GAT. These data are different from those obtained by Pierce *et al.* in a similar system (19). They found that 2-ME could reconstitute the GAT-PFC response of C57BL/6 nonadherent spleen cells. It is likely that variations in cell separation techniques, FCS, batch of GAT, and actual culture conditions account for this discrepancy. Nevertheless, Pierce *et al.* (19, 26) concluded that MØ played an important role in antigen presentation, as well as general culture conditioning, for GAT-PFC responses. This conclusion was based in part on detailed studies of the enhanced immunogenicity of nanogram quantities of GAT bound to macrophage surfaces (15). The data presented above clearly agree with such an interpretation. Addition of small numbers of either antigen-free or GAT-bearing MØ to nonadherent spleen cells is necessary for GAT-PFC responses in the presence of 2-ME, directly demonstrating the presentation role of these cells, and the immunogenicity of MØ-bound GAT.

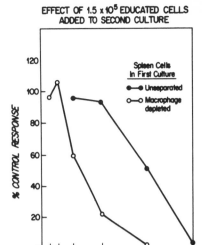

Figure 4. Differential antigen requirement for T_s induction by GAT in unseparated and macrophage-depleted BALB/c spleen cell cultures. Nonadherent spleen cells were cultured in modified Mishell-Dutton conditions in the presence of various concentrations of GAT (0 to 100 μg/ml). After 2 days 1.5×10^5 viable washed educated cells were tested for their ability to suppress the primary *in vitro* response to 1 μg GAT of 7.5×10^6 fresh BALB/c spleen cells. The IgG specific anti-GAT PFC were determined on day 5 of culture by using GAT-coated SRBC.

Second, the role of macrophages in regulating suppressor cell generation in responder mice was investigated. Previous studies from this laboratory revealed that although no difference in initiation of primary immune responses to GAT could be detected by using responder spleen cells with either GAT-pulsed responder or nonresponder macrophages (15), secondary *in vitro* responses of GAT-primed (responder × nonresponder)F_1 mice only occurred with GAT presented on the responder parental MØ (12). It has recently been found that this restriction is in part regulated by GAT-specific suppressor T cells in these primed F_1 mice.[4] Further, it has also been shown that GAT responder spleen cells could be induced by nonresponder-derived GAT-T_sF to develop GAT-specific T_s (9). Taken together these data suggested that responder mice could make suppressor T cells to GAT, and that a possible difference between responder mice and nonresponder mice lay in the ability of their MØ to present GAT efficiently to certain helper T cells. Lack of this function might account for the predominant induction of T_s in nonresponders. This led to the prediction that MØ depletion of responder spleen cells would allow preferential T_s induction upon exposure to normally immunogenic doses of soluble GAT *in vitro*. This paper documents that this prediction is correct. Incubation of MØ-depleted responder spleen cells under Mishell-Dutton conditions for 2 days with ≥10 mg/ml of GAT induces potent GAT-T_s, able to specifically inhibit the primary PFC response of syngeneic spleen cells to soluble GAT. Similar T_s can also be induced with unseparated spleen cells and 5 to 10 times higher concentrations of GAT. Such T_s fail to suppress the primary PFC response to the related polypeptide antigen GT (as GT-MBSA). This demonstration of *in vitro* GAT-T_s induction is in general agreement with the work of Feldmann and coworkers concerning both soluble protein antigens (27) and synthetic polypeptides (28). They found that high concentrations of antigen led to efficient T_s induction *in vitro* in both responders and nonresponders for GAT (28). However, our data indicate that low doses of antigen (<10 μg/ml) still induce potent T_s in nonresponders, although

TABLE V

In vitro induction of GAT-specific suppressor cells from unseparated spleen cells requires high concentrations of antigen

Induction of Suppressor Cells		Educated Cells Added to Second Culture	Specific Ig Anti-GAT PFC per culture[a]	% Suppression[b]
(B6D2)F_1 cells cultured	Antigen in first culture			
None	None	None	730	
Unseparated	Ovalbumin 10 μg/ml	10^5	1025	0
	GAT 10 μg/ml	10^5	835	0
		3×10^4	775	0
		10^4	915	0
	GAT 50 μg/ml	10^5	20	>98
		3×10^4	775	0
		10^4	485	35
Macrophage depleted	Ovalbumin 10 μg/ml	10^5	635	14
	GAT 10 μg/ml	10^5	20	>98
		3×10^4	185	75
		10^4	785	0
	GAT 50 μg/ml	10^5	20	>98
		3×10^4	65	82
		10^4	685	7

[a] 7.5×10^6 cells cultured in modified Mishell-Dutton conditions, stimulated with 1 μg GAT.

[b] % suppression of control response.

failing to do so in responders, and further point to the macrophage as critical in determining the ability of GAT to lead to T_s formation. Similar results concerning T_s induction after MØ depletion have been obtained by Ishizaka and Adachi (29) with OVA as an antigen.

These results raise several important issues about MØ function and suppressor cell induction. It is now widely believed that certain Ir genes in fact are equivalent to Ia antigens on macrophages and that the lack of the appropriate Ia prevents antigen presentation and the stimulation of "positive" effector T cells such as helper (2, 3), proliferating (30), or DTH-inducing cells (31). The GAT model is not entirely consistent with an absolute defect in MØ antigen presentation by Ir nonresponder-derived MØ, since 1) cyclophosphamide- and anti-I-J antiserum (32)-treated nonresponder mice give primary GAT responses *in vivo*; 2) nonresponder MØ present GAT for primary *in vitro* responses by responder lymphocytes. Thus, for GAT, a reduced ability of nonresponder MØ to trigger appropriate T_H may result in the predominance of a T_s response. This latter may occur due to direct recognition of soluble antigen by T_s precursors or by GAT presentation on cell membranes with or without the involvement of major histocompability complex (MHC) gene products. Either mode of presentation is compatible with the ability of MØ-depleted responder spleen cells to generate efficient T_s, and with the induction of T_s in whole spleen cells by high concentrations of antigen, since in each case, the relative proportion of antigen presented in the context of responder Ia on MØ *vs* that presented free or on other cell surfaces is reduced. It should be noted that the ability of GAT-MBSA to induce GAT PFC responses in nonresponder mice may relate not only to MBSA carrier-specific T_H function, as generally believed, but perhaps also to an alteration in MØ presentation of GAT itself, leading to preferential GAT-specific T_H *vs* T_s activation.

The precise T cell (Lyt) subset to which the *in vitro*-induced GAT-T_s belong has not yet been identified nor is the pathway by which these cells are activated known. The two major possibilities involve a) direct triggering of precursors of Lyt 23^+ T_s or b) the recently described "feedback" suppression loop involving Lyt 1^+ "inducer" cells triggering Lyt 123^+ Qa 1^+ cells to become T_s (33). If such Lyt 1^+ inducers are distinct from Lyt 1^+ helper cells and also differ in their MØ requirements, they could be involved in the T_s system described here. Careful Ly phenotyping of all the cells involved in generation of GAT T_s *in vitro* from responder spleen cells will hopefully resolve this question.

Finally, it is apparent from this investigation that Ir gene expression at the MØ level probably involves differential activation of the T_H *vs* T_s precursors in such a way that T_H predominate in responder mice whereas T_s predominate in at least certain types of nonresponders. These considerations suggest further that the role of Ia antigens is probably different in the triggering processes of these two T cell subsets. It will be important to compare the specificity of helper and suppressor factors for GAT in the context of Ia *vs* K, D antigens to determine if such cell products reflect in their bindings a predilection for one or the other MHC product together with antigen. In addition, the possibility that Ia^+ and Ia^- MØ differ in their adherence properties and frequency, and that such distinct types of MØ play different roles in T_H and T_s triggering must be considered. Lastly, the effect different Ia genotypes have during ontogeny on the ability of macrophages bearing responder *vs* nonresponder suface antigens to stimulate responses to Ir-regulated antigens must be determined to understand better the role of Ir genes in immune regulation at the MØ-T cell level.

Acknowledgment. We wish to thank Professor Baruj Benacerraf for his helpful discussions, suggestions, and support during this study and the preparation of this manuscript. The secretarial assistance of Ms. Diane Cannata is appreciated.

REFERENCES

1. Benacerraf, B., J. A. Kapp, P. Debré, C. W. Pierce, and F. de la Croix, 1975. The stimulation of specific suppressor T cells in genetic nonresponder mice by linear random copolymers of L-amino acids. Transplant. Rev. 26:21.
2. Kapp, J. A., C. W. Pierce, S. Schlossmann, and B. Benacerraf. 1974. Genetic control of immune responses *in vitro*. V. Stimulation of suppressor T cells in nonresponder mice by the terpolymer L-glutamic acid^{60}-L-alanine^{30}-L-tyrosine^{10} (GAT). J. Exp. Med. 140:648.
3. Kapp, J. A., H. Cantor, C. W. Pierce, and B. Benacerraf. 1977. Genetic control of antibody responses to GAT. Fed. Proc. 36:1224.
4. Kapp, J. A., C. W. Pierce, F. de la Croix, and B. Benacerraf. 1976. Immunosuppressive factor(s) extracted from lymphoid cells of nonresponder mice primed with L-glutamic acid^{60}-L-alanine^{30}-L-tyrosine^{10} (GAT). I. Activity and antigenic specificity. J. Immunol. 116:305.
5. Kapp, J. A., C. W. Pierce, and B. Benacerraf. 1977. Immunosuppressive factor(s) extracted from lymphoid cells of nonresponder mice primed with L-glutamic acid^{60}-L-alanine^{30}-L-tyrosine^{10} (GAT). II. Cellular source and effect on responder and nonresponder mice. J. Exp. Med. 145:828.
6. Theze, J., J. A. Kapp, and B. Benacerraf. 1977. Immunosuppressive factor(s) extracted from lymphoid cells of nonresponder mice primed with L-glutamic acid^{60}-L-alanine^{30}-L-tyrosine^{10} (GAT). III. Immunochemical properties of the GAT-specific suppressive factor. J. Exp. Med. 145:839.
7. Germain, R. N., J. Theze, J. A. Kapp, and B. Benacerraf. 1978. Antigen-specific T cell mediated suppression. I. Induction of L-glutamic acid^{60}-L-alanine^{30}-L-tyrosine^{10} specific suppressor T cells *in vitro* requires both antigen specific T cell suppressor factor and antigen. J. Exp. Med. 147:123.
8. Kapp, J. A. 1978. Immunosuppressive factors from lymphoid cells of nonresponder mice primed with L-glutamic acid^{60}-L-alanine^{30}-L-tyrosine^{10}. IV. Lack of strain restrictions among allogeneic, nonresponder donors and recipients. J. Exp. Med. 147:997.
9. Germain, R. N., and B. Benacerraf. 1978. Antigen specific T-cell mediated suppression. III. Induction of antigen specific suppressor T-cells (T_{s_2}) in L-glutamic acid^{60}-L-alanine^{30}-L-tyrosine^{10} (GAT) responder mice by nonresponder derived GAT suppressor factor (GAT-T_sF). J. Immunol. 121:602.
10. Kapp, J. A., C. W. Pierce, and B. Benacerraf. 1975. Genetic control of immune responses *in vitro*. VI. Experimental conditions for the development of helper T-cell activity specific for the terpolymer L-glutamic acid^{60}-L-alanine^{30}-L-tyrosine^{10} (GAT) in nonresponder mice. J. Exp. Med. 142:50.
11. Pierce, C. W., J. A. Kapp, and B. Benacerraf. 1976. Regulation by the H-2 gene complex of macrophage-lymphoid cell interactions in secondary antibody response *in vitro*. J. Exp. Med. 144:371.
12. Pierce, C. W., R. N. Germain, J. A. Kapp, and B. Benacerraf. 1977. Secondary antibody responses *in vitro* to L-glutamic acid^{60}-L-alanine^{30}-L-tyrosine^{10} (GAT) by (responder x nonresponder)F$_1$ spleen cells stimulated by parental GAT-macrophages. J. Exp. Med. 146:1827.
13. Kapp, J. A., C. W. Pierce, and B. Benacerraf. 1973. Genetic control of immune response *in vitro*. I. Development of primary and secondary plaque forming cell response to the random terpolymer L-glutamic acid^{60}-L-alanine^{30}-L-tyrosine^{10} (GAT) by mouse spleen cells *in vitro*. J. Exp. Med. 138:1107.
14. Mishell, R. I., and R. W. Dutton. 1967. Immunization of dissociated spleen cells cultures from normal mice. J. Exp. Med. 126:423.

15. Kapp, J. A., C. W. Pierce, and B. Benacerraf. 1973. Genetic control of immune responses *in vitro*. II. Cellular requirement for the development of primary plaque forming cell responses to the random terpolymer L-glutamic acid^{60}-L-alanine^{30}-L-tyrosine^{10} (GAT) by mouse spleen cells *in vitro*. J. Exp. Med. 138:1121.

16. Mosier, D. E. 1967. A requirement for two cell types for antibody formation *in vitro*. Science. 158:1575.

17. Ly, I. A., and R. I. Mishell, 1974. Separation of mouse spleen cells by passage through columns of Sephadex G-10. J. Immunol. Meth. 5:239.

18. Julius, M. F., E. Simpson and L. A. Herzenberg. 1973. A rapid method for the isolation of functional thymus-derived murine lymphocytes. Eur. J. Immunol. 3:645.

19. Pierce, C. W., J. A. Kapp, D. D. Wood, and B. Benacerraf. 1974. Immune responses *in vitro*. X. Function of macrophages. J. Immunol. 112:1181.

20. Cheng, C., and J. C. Hirsch. 1972. The effect of mercaptoethanol and of peritoneal macrophages on the antibody forming capacity of non-adherent mouse spleen cells *in vitro*. J. Exp. Med. 136:604.

21. Erb, P., and M. Feldmann. 1975. The role of macrophages in the generation of T helper cells. I. The requirement for macrophages in helper cell induction and characteristics of the macrophage T cell interaction. Cell. Immunol. 19:356.

22. Hoffman, M. and R. W. Dutton. 1971. Immune response restoration with macrophage culture supernatant. Science 172:1047.

23. Moller, G., H. Lemke, and H.-G. Opitz. 1977. The role of adherent cells in the immune response. Fibroblasts and products released by fibroblasts and peritoneal cells can substitute for adherent cells. Scand. J. Immunol. 5:269.

24. Opitz, H.-G., O. Opitz, H. Lemke, G. Hewlett, W. Schreml, and H.-D. Flad. 1977. The role of fetal calf serum in the primary immune response *in vitro*. J. Exp. Med. 145:1029.

25. Goodmann, M. G., and W. O. Weigle. 1977. Nonspecific activation of murine lymphocytes. I. Proliferation and polyclonal activation induced by 2-mercaptoethanol and α-thioglycerol. J. Exp. Med. 145:473.

26. Pierce, C. W., and J. A. Kapp. 1976. The role of macrophages in antibody responses *in vitro*. *In* Immunobiology of the Macrophage, Edited by D. N. Nelson, Academic Press, New York. Pp. 2–34.

27. Kontiainen, S., and M. Feldmann. 1976. Suppressor cell induction *in vitro*. I. Kinetics of induction of antigen specific suppressor cells. Eur. J. Immunol. 6:296.

28. Howie, S. 1977. *In vitro* studies on H-2 linked unresponsiveness to synthetic polypeptide antigens. II. Induction of suppressor cells in both responsive and unresponsive mice to (T,G)-A--L and GAT^{10}. Immunology 32:301.

29. Ishizaka, K., and T. Adachi. 1976. Generation of specific helper cells and suppressor cells *in vitro* for the IgE and IgG antibody responses. J. Immunol. 117:40.

30. Schwartz, R. H., and W. E. Paul. 1976. T-lymphocyte-enriched murine peritoneal exudate cells. II. Genetic control of antigen-induced T-lymphocyte proliferation. J. Exp. Med. 143:529.

31. Miller, J. F. Ap., M. S. Vadas, A. Whitelaw, J. Gamble, and C. Bernard. 1977. Histocompatibility linked immune responsiveness and restrictions imposed on sensitized lymphocytes. J. Exp. Med. 145:1613.

32. Pierres, M., R. N. Germain, M. E. Dorf, and B. Benacerraf. 1978. *In vivo* effects of anti-Ia alloantisera. I. Elimination of specific suppression by *in vivo* administration of antisera specific for I-J controlled determinants. J. Exp. Med. 147:656.

33. Eardley, D. D., J. Hugenberg, L. McVay-Boudreau, F. W. Shen, R. K. Gershon, and H. Cantor. 1978. Immunoregulatory circuits among T-cell sets. I. T-helper cells induce other T-cell sets to exert feedback inhibition. J. Exp. Med. 147:1106.

Chapter **VII**

Antigen Recognition by T Cells

Chapter **VII**

Antigen Recognition by T Cells

T cell recognition of antigen has been one of immunology's more elusive phenomena. That T cells recognize antigen has been well established for some time, as
has been discussed in Chapter I. Each T cell recognizes a specific antigen and is
clonally stimulated and expanded in response to that antigen. This is true for
functionally distinct T cells (helpers, killers, suppressors), just as it is for
B cells. To get beyond these generalities, however, has been difficult, and the
precise characterization of both the T cell receptor for antigen and the nature of
what the T cell considers antigenic have been difficult crosses for the immunologist to bear.

The readings in this chapter address these two issues: what can be learned
about the nature of the T cell receptor; what can be learned about how the T cell
recognizes antigen.

The search for the T cell receptor has been founded on the notion that T cells,
like B cells, might use immunoglobulin as an antigen receptor. (This notion stems
from the scientist's reliance on biological precedents to serve as guides in solv-
ing new problems.) That this is not true in its purest form is evident from the
fact that labelled anti-immunoglobulin reagents, which bind to B cells, do not bind
to T cells. However, immunoglobulin, already specialized to accomodate the requi-
site diversity, has been an appealing and persistent candidate in the search for
the T cell receptor. Binz and Wigzell, and Eichmann and Rajewsky, whose papers
follow, attempted to achieve two goals: 1) the preparation of an antibody against
the T cell receptor and 2) the determination of whether or not such a receptor was
immunoglobulin. These papers are followed by two papers (Black, et al., and
Krawinkel, et al.) which expand on these studies directed at elucidating the
nature of the T cell receptor.

The second part of the chapter deals with an analysis of T cells and their
receptors which demonstrates that the Ig or Ig-like nature of the receptor is too
simple a picture to account for the properties of T cells. It is shown that most
T cells recognize antigens not as freely soluble molecules, the way Ig molecules
do, but rather in association with other cell surface components. These experiments suggest that T cells recognize antigen either through the cooperative
efforts of two distinct receptors or through a single receptor considerably more
complex than the antigen-binding portion of immunoglobulin.

Binz and Wigzell, in one of the more insightful experiments of recent immunology, prepared an antiserum against a T cell receptor specific for a rat alloantigen. They then demonstrated that such an antiserum also reacts with immunoglobulin molecules which recognize the same alloantigen. In doing this experiment
they considered the fact that two histoincompatible rats, A and B, would each have
T cells reactive to the other's Ag-B (MHC) antigens. More explicitly, an A rat

would have T cells reactive to the alloantigens of the strain B. Such A anti-B
T cells would express receptors for B and would be the cells responsible for rejection of B tissue. T cells of such specificity would also be conspicuously
absent from an (A x B)F_1 animal which is tolerant of B antigens. Thus, it is
possible for an F_1 animal to recognize at least the antigen binding portion of
such a receptor as foreign. So, injection of A cells into an (A x B)F_1 would
lead first to the proliferation of A anti-B T cells and then to the F_1 antibody
response against the immunogenic determinants of such a receptor. Binz and
Wigzell demonstrate the specificity and activity of this antibody by antigen-
specific inhibition of T cell function. They further show that such an antibody,
made against the T cell receptor, also recognizes specific determinants on
immunoglobulin molecules of the same antigen binding specificity derived from
the same inbred animal. Thus, strain A, when stimulated to react against strain
B either with T cells or B cell-derived antibodies, uses serologically related
receptor material to do so. These data support the idea that the T cell receptor
resembles immunoglobulin. Binz and Wigzell further demonstrate that resting T
cells as well as T cells involved in the mixed lymphocyte response (MLR) of A
against B and in the graft versus host (GVH) reactivity of A against B express
receptors that are similar to the antibody produced by strain A against the
alloantigens of strain B.

Unlike Binz and Wigzell, Eichmann and Rajewsky begin with an antiserum directed against idiotypic determinants of the B cell's immunoglobulin. They reason
that while the T cell may not express conventional immunoglobulin, it might still
use the antigen-binding portion of such a molecule, perhaps in conjunction with
another kind of constant region. Thus, antibody directed specifically at the
idiotypic specificities of a particular immunoglobulin might recognize a T cell
receptor with identical antigen binding specificity. Eichmann and Rajewsky make
such a reagent against the restricted class of antibodies produced by A strain
mice against the A-CHO determinant of streptococcal polysaccharide. They then
assay the specificity and activity of their antibody, not by inhibition of T cell
function but rather by induction of T cell function; they use their anti-idiotypic
antibody reagent to mimic antigen and prime T cells and B cells specific for the
A-CHO determinant of streptococcal polysaccharide. The ability of anti-idiotypic
antibodies to mimic antigen is most likely due to the capacity of both these
molecules to recognize and interact with the specific antigen receptors on T and
B cells. As in the case of Binz and Wigzell, the same anti-idiotypic antibodies
detect both the T and B cell receptors.

There is an interesting point to notice in analyzing the data. Both T and B
cells expressing the A5A idiotype are primed by anti-idiotypic antibody; the B
cells are assayed via the response to Strep A and the T cells via the response to
NIP-Strep A. It is therefore possible to determine for the B cell case but not
for the T cell case whether the Strep A antigen and the anti-idiotypic antibody
stimulate completely overlapping sets of cells. That is, one can determine, as
Eichmann and Rajewsky did, what fraction of the anti-Strep A antibody response
bears the A5A idiotype after stimulation with either Strep A or anti-id. It is
not possible to make this determination for the T cells, since one is not directly assaying the population of the primed helper population but rather assaying
the activity of such a population through its capacity to help NIP-reactive B
cells.

These experiments also demonstrate that antibodies to the T cell receptor can
"tickle" the cell sufficiently to generate a memory population. It thus appears
that binding to the T cell receptor(s) is sufficient to promote certain levels of
activity. This observation may have profound consequences for intervention in the
immune system without using antigen--this topic will be discussed at greater
length elsewhere.

The experiments of Black and coworkers extend some of Rajewsky's observations
on the ability to stimulate antigen-reactive, idiotype-positive B and T cells

with anti-idiotype antibody. They test the premise that only strains which are genetically capable of making a particular idiotypic response (call it id 1) to antigen X will respond when stimulated with anti-id 1 anti-idiotypic antibody. Another strain, responding to antigen X with another idiotype (id 2), will be impervious to stimulation with anti-id 1 but sensitive to anti-id 2. More specifically, these authors compare the immune responses of Balb/c and A/J mice to the A-CHO polysaccharide of streptococcus. Both these strains make anti-A-CHO antibody but the major antibody products produced by the two strains are different: A/J responds primarily with antibody of the A5A idiotype while Balb/c responds predominantly with antibody of the S117 idiotype. A/J makes no S117 and Balb/c makes no A5A. Subsequent injection of A/J mice with anti-A5A will prime both T cells and B cells specific for A-CHO, while immunization with anti-S117 is ineffective. The reverse is true for Balb/c, which responds to anti-S117 but not to anti-A5A. Since these experiments test both T and B cell sensitivity to the same set of anti-idiotypic antibodies, the authors suggest that the idiotypic repertoire of both lymphocyte types use the same genetic information for generating at least a portion of their respective receptors.

The first set of papers in this section addresses itself to the premise that T and B lymphocytes share a set of idiotypic determinants. This, if true, is a fundamentally critical point in understanding the nature of the T cell receptor, for it makes it highly likely that at least a portion of the immunoglobulin molecule is used by the T cell to recognize the antigenic universe. This makes it reasonable to imagine that the set of V_H genes encoding the variable regions of immunoglobulin heavy chains are available for T cell use, perhaps joined to genes encoding a T cell specific "constant region" useful for T cell effector function. Alternatively, the construction of the T cell receptor could occur at the RNA or protein level, assembling V_H mRNA or peptides with T cell-associated constant region mRNA or peptides. Another structural consideration is the possible need for a T cell-specific light chain. The idiotypes discussed in these papers appear to require both V_H and V_L, but no direct evidence exists for either κ or λ chains on T cells. It is conceivable, therefore, that a particular V_H would require the association of some other chain to stabilize the antigen binding site. Alternatively, conventional light chains may be used by the T cell but in such a way as to be inaccessible to antibody without prior disruption of the receptor structure.

It is wise to remember that the data supporting the immunoglobulin nature of the T cell receptor is based on serologic cross-reaction of idiotypic determinants. It is therefore possible that we are being misled by these data, and are misreading a situation in which two receptors which bind the same substrate would, in all likelihood, have some structural similarities without being, to any real extent, identical. It would, therefore, be useful to have independent evidence, possibly of a biochemical nature, which supports the notion that some portion of the T cell receptor is indeed immunoglobulin.

Although the papers by Krawinkel and coworkers approach these biochemical and genetic questions, there   are not yet definitive answers to the ticklish T cell question. These experiments demonstrate that receptor material, derived from T cells, expresses antigenic determinants recognized by antibodies specific for immunoglobulin V_H but not V_L regions. Consistent with previous studies, these receptors do not express determinants detectable by conventional anti-immunoglobulin antibodies. On the other hand, this T cell-associated material reflects the same fine specificity (heteroclicity)[1] as the B cell immunoglobulins specific for the same antigen. This receptor material is isolated from hapten-sensitized

---

[1]Heteroclicity is the capacity of an antibody combining site to bind a non-identical but closely related antigen more tightly than the initial immunogen; it is a property inherent to the molecular structure of the antibody. Both the immunoglobulin and the T cell receptor material which Krawinkel studies bind the hapten NP more tightly than the immunizing hapten NIP.

lymphocytes by adsorption onto hapten-coated nylon fibers, but the function of
the T lymphocytes expressing and using this receptor is as yet unknown. We can
expect, however, that these scientists will provide important new confirmation
on the biochemistry of this receptor material in the near future.

Actually, Krawinkel's ability to isolate antigen-binding material from T
cells by direct interaction with antigen suggests that the donor T cell population is suppressive in function (see Chapter IV). Attempts to isolate helper
or killer T cells by direct labelling with free carrier, or by rosetting on
antigen-coated red cells have never been successful, indicating that these T
cells may have more complex requirements for antigen binding.

A pivotal step forward in the elucidation of T cell recognition of antigen
occurred in 1975 with the demonstration by Zinkernagel and Doherty that cytotoxic T cells recognize viral antigens in association with cell surface antigens
encoded by the major histocompatibility complex. This phenomenon, termed "H-2
restriction," indicates that an individual's killer T cells only lyse target cells
which bear both the foreign (often viral) antigens and the histocompatibility
antigens expressed by the cells which initially elicited the killing activity.
In contrast, normal cells, uninfected by virus, or histoincompatible cells infected with virus, are not lysed by the cytotoxic T cells reactive to "anti-virus
plus self." In physiological situations, the MHC antigens on the immunizing cells
would be self antigens, so that the specificity of the responsive T cells must be
for foreign antigens in the context of self. Thus, one's T cell population can
distinguish between self and so-called "altered self," mobilizing for activity
only upon confrontation with the latter.

Genetic dissection of this phenomenon using inbred strains of mice recombinant at the H-2 complex revealed that cytotoxic T cells are sensitive to the gene
products of the K and D loci, not those of the I region. In addition, F_1 animals
are capable of recognizing both parental haplotypes as self, so that an (A x B)F_1
infected with virus X will generate T cells capable of lysing X-infected cells of
the A, B, and F_1 haplotypes. Experiments using the technique of cold target inhibition demonstrate that the F_1 T cells responsible for the lysis of infected A-
type cells are different from those T cells responsible for lysis of infected B-
type cells. Thus, the F_1 T cell is clonally restricted in its ability to
recognize parental H-2 antigens, although all F_1 cells express both sets of
parental alloantigens on their surface. This finding makes untenable the proposal that H-2 restriction is the result of a "like with like" interaction of
two cells expressing identical MHC products.

The generality of H-2 restriction is exemplified by Sprent's elegant in vivo
experiments in which he demonstrates that helper T cells also recognize foreign
antigens in the context of an MHC gene product. In this case, the relevant gene
product(s) is encoded in the I region and the restriction applies to T macrophage
and possible T-B interactions. Like cytotoxic T cells, F_1 helper T cells are
clonally restricted in their ability to recognize both parental haplotypes
(see also the paper by Sprent in Chapter VI).

Bevan's work addresses what may be a fascinating consequence of the T cell's
need to recognize "self plus X," or foreign antigens in the context of MHC gene
products. He proposes, then tests, the idea that an individual's ability to
recognize allogeneic MHC determinants is a result of certain degeneracy in the
T cell receptor(s). That is, to a T cell, certain combinations of a particular
foreign determinant with a particular haplotype may look sufficiently like
another distinct haplotype to stimulate the rejection of infected host tissue on
the one hand, and normal foreign tissue on the other. These ingenious experiments exploit the fact that antigen stimulation is the most efficient way to
expand and select a specific population of lymphocytes. By repeatedly stimulating a heterogeneous population of lymphocytes with a specific antigen, one
selects for and expands specifically the reactive clone of cells until progeny
of that clone represent the majority of the surviving cell population. These
"enriched" cells can then be tested for cross-reaction on other kinds of target

cells. The results of Bevan's experiments suggest that killer T cells, selected for reactivity against a particular "self plus X" combination also react against unaltered cells of a different (allogeneic) MHC haplotype. Isolation of a cloned population of T cells would be necessary to prove this point, but the current results are certainly suggestive.

What remains is the question of how many receptors for antigen the T cell expresses. Although experiments discussed in Chapters V and VIII indicate that two receptors (one for MHC products, one for conventional antigen) may be more likely than a single complex one, a definitive answer will depend upon sufficient improvements in cell culture technology to enable biochemical analysis.

The next few years should see the characterization not only of the T cell receptor but also of the genes which encode it. Studies in the biochemistry of receptor material are being conducted in several laboratories, and we expect that such investigations will reveal a) whether all T cells have the same kind of receptors and b) how many receptors any given T cell uses.

The production of monospecific T cell lines, either by fusion or improved antigen selection and cell culture methods, will be a key step forward in the study of T cell receptors and T cell receptor genes. Much as monospecific myelomas have facilitated the characterization of immunoglobulin, so will such antigenically restricted T cell lines allow the analysis of T cell receptors. The creation of cloned T cell lines expressing those idiotypic determinants shared by the B cell immunoglobulin will be particularly useful in determining the role of immunoglobulin genes in encoding the T cell receptor(s).

Papers Included in this Chapter

1. Binz, H. and Wigzell, H. (1975) Shared idiotypic determinants on B and T
    lymphocytes reactive against the same antigenic determinants. I. Demonstra-
    tion of similar or identical idiotypes on IgG molecules and T cell receptors
    with specificity for the same alloantigens. J. Exp. Med. 142, 197.

2. Eichmann, K. and K. Rajewsky. (1975) Induction of T and B cell immunity by
    anti-idiotypic antibody. Eur. J. Immunol. 5, 661.

3. Black, S.J., Hämmerling, G.J., Berek, C., Rajewsky, K., and K. Eichmann (1976)
    Idiotypic analysis of lymphocytes in vitro. I. Specificity and heterogeneity
    of B and T lymphocytes reactive with anti-idiotypic antibody. J. Exp. Med.
    143, 846.

4. Krawinkel, U., Cramer, M., Imanishi-Kari, T., Jack, R.S., Rajewsky, K. and
    Mäkelä, O. (1977) Isolated hapten-binding receptors of sensitized lymphocytes.
    I. Receptors from nylon-wool enriched mouse T lymphocytes lack serological
    markers of immunoglobulin constant domains but express heavy chain variable
    portions. Eur. J. Immunol. 8, 566.

5. Zinkernagel, R.M. and Doherty, P.C. (1975) H-2 compatibility requirement for
    T cell mediated lysis of target cells infected with lymphocytic choriomenin-
    gitis virus. Different cytotoxic T cell specificities are associated with
    structures coded for in H-2K or H-2D. J. Exp. Med. 141, 1427.

6. Sprent, J. (1978) Role of the H-2 complex in induction of T helper cells in
    vivo. I. Antigen-specific selection of donor T cells to sheep erythrocytes
    in irradiated mice dependent upon sharing of H-2 determinants between donor
    and host. J. Exp. Med. 148, 478.

7. Sprent, J. (1978) Restricted helper function of F_1 hybrid T cells positively
    selected to heterologous erythrocytes in irradiated parental strain mice.
    I. Failures to collaborate with B cells of the opposite parental strain not
    associated with active suppression. J. Exp. Med. 147, 1142.

8. Bevan, M.J. (1977) Killer cells reactive to altered-self antigens can also
    be alloreactive. Proc. Nat. Acad. Sci. USA 74, 2094.

SHARED IDIOTYPIC DETERMINANTS ON
B AND T LYMPHOCYTES REACTIVE AGAINST THE SAME
ANTIGENIC DETERMINANTS

I. Demonstration of Similar or Identical Idiotypes on IgG Molecules and
T-Cell Receptors with Specificity for the Same Alloantigens*

By HANS BINZ and HANS WIGZELL

(*From the Department of Immunology, Uppsala University, Uppsala, Sweden*)

The immunocompetent cells that can recognize foreign structures by their own actively synthesized antigen-binding receptors are the lymphocytes (1). Two major groups of lymphocytes exist, with separate functions and without ability to transform into each other (2, 3). One group is called T lymphocytes, due to their dependence on the presence of the thymus for development (2). The second group of cells are called B lymphocytes, after the bursa of Fabricius in birds, where maturation of these B cells do occur (3). T lymphocytes are responsible for the major part of so called cell-bound immune reactions, whereas B lymphocytes produce conventional immunoglobulin antibodies.

The chemical structure of the antigen-binding receptors on the B lymphocytes is generally considered to be that of a classical immunoglobulin, with minor modifications as to make the molecule surface-attached (4). With regard to the antigen-binding receptors on T lymphocytes opinions do vary considerably as to the nature of the receptor, and classical immunoglobulins (5, 6) as well as molecules, determined by genes associated with the loci determining the major histocompatibility antigens of the species (7, 8), have been implied. Most workers have failed, by classical means, to detect significant amounts of immunoglobulin on the surface of T lymphocytes (9–11).

One approach to the study of the molecular basis of the receptors on T lymphocytes would be to use anti-idiotypic antibodies directed against the products of B lymphocytes, that is against antibodies specific for a particular antigen determinant. If such antiantibodies would have a detectable significant impact on T lymphocytes capable of reacting with the same antigen, this would strongly suggest that B and T lymphocytes use the same variable set of genes for the creation of antigen-binding sites on their receptors. Indications that such anti-idiotypic antibodies will have the expected impact on the relevant T lymphocytes have come from work in systems using an F_1 hybrid to parent combination taking advantage of the immunogenetical filter in the production of such specific antiantibodies (12, 13). Using this system substantial amounts of data have been accumulated to suggest that B and T lymphocytes directed against the same alloantigenic determinants may share idiotypic determinants (14–17). Final proof for such an identity is still lacking, however, as in most tests either the lymphocyte populations or the antiserum

* This work has been supported by the Swiss Foundation for Scientific Research, the Swedish Cancer Society, and NIH contract NO1-CB-33859.

used for immunization might have been contaminated with products from the other lymphocyte group.

In a previous set of articles we accumulated sizeable amount of results, indicating that anti-idiotypic antibodies produced in an F_1 hybrid rat against antibodies produced in one parental strain against the alloantigens of the other strain, will inhibit the T lymphocytes of the genotype of the antiserum donor to function in the relevant, cell-bound immune reactions (15–17). In radioimmunoassays, however, we were only able to show that such F_1 antibodies would react in a detectable manner with B lymphocytes of the relevant strain (16, 17). As we considered this to be a matter of sensitivity of the test or quality of antibodies induced we have now tried and found it possible to obtain extremely high-titered antibodies produced in F_1 hybrid rats against T lymphocytes of one of the parental strains. Such antisera were found to react with highly purified IgG alloantibodies, produced in the T donor strain against alloantigens of the other parental strain. They were also found capable of reaction with the relevant immunocompetent normal parental T lymphocytes. Absorption experiments were then performed that showed that highly purified IgG alloantibodies could remove the reactivity of the F_1 anti-idiotypic antibodies against the relevant "pure" parental T lymphocytes and vice versa.

Thus the results obtained all directly demonstrate that B and T lymphocytes directed against the same antigenic determinants have antigen-binding receptors, which are similar or identical with regard to their antigen-binding sites as judged by their shared idiotypes.

Material and Methods

Animals. Rats of the inbred strains Lewis (L) (Ag-b¹), DA (Ag-B⁴), and BN (Ag-B³) as well as F_1 hybrids between these strains were domestically raised and maintained. Either adult male or female rats were used within an experiment.

Rat Lymphocyte Preparations

SINGLE LYMPHOID CELL SUSPENSION. These were done from spleens and lymph nodes according to a standard procedure (18). Red cells were lysed by incubation in 0.84% NH_4Cl for 8 min at room temperature. Lymphoid cells were washed twice with Eagle's basal medium (BME)¹ containing 5% heat-inactivated fetal calf serum (FCS). The cell viability was judged by trypan-blue exclusion.

PREPARATION OF RAT T LYMPHOCYTES. This was done as already described by using Degalan V-26 beads coated with rat gamma globulin and rabbit antirat gamma globulin (16). Such T-cell preparations showed normally less than 2% B-cell contamination as judged by autoradiography using ¹²⁵iodine-labeled rabbit antirat gamma globulin or by immunofluorescence technique using rabbit antirat gamma globulin and FITC-coupled goat IgG antirabbit IgG.

PURIFICATION OF RAT B LYMPHOCYTES. For this a rabbit antirat T-lymphocyte serum was used in a cytolytic assay. A rabbit was inoculated with 5×10^8 peripheral purified rat T lymphocytes twice at 2-wk intervals and bled 1 wk after last injection. The serum was heat-inactivated and then absorbed extensively with rat liver cells until a serum was produced that killed 30–40% of rat spleen cells and 60–70% of rat lymph node cells in the presence of guinea pig complement. The killing was at the same plateau level over several twofold dilutions up to 1:500 serum dilution. For purification of spleen B cells, cells were incubated with the antiserum at 4°C for 1 h whereafter guinea pig complement was added and the suspension was incubated at 37°C for 40 min (using dilutions of antisera and complement as to achieve the above indicated percentual lysis of spleen cells). Cells were then filtered through sterile gauze to remove debris and dead cells and washed three times. Such cell suspensions were enriched for Ig surface-positive cells to more than 90% positive cells.

_____

¹ *Abbreviations used in this paper:* BME, Eagle's basal medium; FCS, fetal calf serum; MLC, mixed leukocyte culture; PBS, phosphate-buffered saline.

Production of Rat Antisera

PRODUCTION OF ALLOANTISERA. Rat alloantisera of specificity Lewis anti-DA and DA anti-Lewis were raised by full thickness skin grafts from one strain to another (18). 3 wk after a single skin graft animals were bled and the sera heat-inactivated at 56°C for 30 min, Millipore filtered, and stored at − 20°C. Such alloantisera showed titers between 1:2,048 and 1:8,192 as measured in a hemagglutination assay described elsewhere (19).

PRODUCTION OF ANTI-IDIOTYPIC ANTISERA. See Results.

Purification of Rat IgG. This was done as described previously (13). Briefly, the gamma globulin fraction was precipitated with saturated $(NH_4)_2SO_4$ or by dialysis against 18% and 15% Na_2SO_4. The IgG fraction was then purified on DEAE-cellulose. Such preparations showed a single IgG peak on a calibrated G-200 column. In immunoelectrophoresis the preparations showed a single IgG precipitation line against a rabbit antiserum against whole rat serum.

Graft-vs.-Host Reactions. These were done in 5- to 8-wk old F_1 rats as described previously (15). 4×10^6 viable parental lymphocytes in 0.1 ml BME were injected into each foot pad. Animals were killed 7 days later and the lymph nodes were excised and weighed (20).

Mixed Leukocyte Culture (MLC). Rats were anesthesized with ether and then killed by cervical dislocation and their spleens and lymph nodes aseptically teased with forceps in petri dishes containing RPMI 1640 (Flow Laboratories, Inc., Rockville, Md.) complemented with glutamin (2.5 mg/ml), 100 IU of penicillin, 100 μg of streptomycin, and 5% FCS (Flow Laboratories). The undissociated cells were allowed to settle for 5 min at 4°C. The single cell suspension was washed twice with the same medium. Stimulator cells were adjusted to 1×10^8 cells per ml and were irradiated with 2,000 rad from an X-ray machine. The cells were washed again and adjusted to 7.5×10^6 per ml. Untreated responder cells were adjusted to 1.5×10^7 cells per ml. 5×10^7 pelleted responder cells were incubated with 100 μl of 1:2 diluted anti-idiotypic or normal F_1 serum for 60 min at 4°C after which guinea pig complement (absorbed with agarose, 80 mg/ml serum) was added to a final dilution of 1:3. After a further incubation for 40 min at room temperature, cells were washed twice and finally adjusted to 1.5×10^7 viable cells per ml. Both the F_1 anti-idiotypic and the normal F_1 serum were previously absorbed with 1×10^7 (Lewis × DA)F_1 lymphoid cells per ml serum.

The one-way mixed leuckoyte culture (MLC) was performed in Cooke round bottom microtiter plates (Cooke Laboratory Products, Div. Dynatech Laboratories, Alexandria Va.) using 1.5×10^6 responder cells and 0.75×10^6 stimulator cells in a total vol of 0.2 ml. The culture medium was RPMI 1640 (Flow Laboratories) complemented with glutamin, penicillin and streptomycin (see above) and 5% (Lewis × DA)F_1 normal rat serum previously absorbed with 1×10^7 (Lewis × DA)F_1 lymphoid cells per ml serum. Control cultures contained responder cells or stimulator cells alone in concentrations as in the one-way MLC. All cultures were pulsed for 16 h on day 3 with 2 μCi of [^{3}H-]thymidine (Radiochemical Centre, Amersham) with a spec act of 5 Ci per mmol in 20 μl culture medium without normal rat serum. On day 4 the cells were harvested using a Skatron collector and washed on glassfiber filters. The filters were dried and placed into plastic scintillation vials. All samples were counted in 3 ml scintillation fluid in an Intertechnique SL 30 liquid scintillation spectrometer.

Immunabsorbens Technique. IgG preparations from Lewis anti-DA alloantibodies, Lewis, and (Lewis × DA)F_1 normal sera were coupled to CnBr-activated Sepharose 4 B (Pharmacia Fine Chemicals, Uppsala, Sweden) using carbonate buffer pH 9.0. After incubation for 4 h at room temperature the beads were washed once with coupling buffer followed by an incubation with 1 M ethanolamin pH 8.0 for 2 h at room temperature. Beads were then washed with isotonic glycin-HCl buffer pH 2.8 containing 2 M NaCl followed by 0.1 M carbonate buffer pH 8.0 containing 1 M NaCl. This was done three times in a cyclic way. The beads were finally washed with phosphate-buffered saline (PBS) and filled into plastic syringes and washed again with excess PBS. Serum to be analyzed was passed over the column and washed out with PBS. The eluate was collected and passed again through the same column and finally sterile filtered. Columns were washed with excess PBS until the washing fluid contained less than 10 μg per ml of protein. Absorbed material was then eluted with isotonic glycin-HCl buffer pH 2.8 containing 2 M NaCl. The eluate was neutralized immediately, dialyzed against PBS and concentrated by vacuum dialysis and finally sterile filtered.

Indirect Radioimmunoassay

USING ^{125}I-RADIOLABELED RABBIT ANTI-RAT IG. Normally, 1×10^7 purified rat T lymphocytes were distributed into two or three parallel 2-ml plastic tubes. To the cell pellet 50 μl of 1:10 diluted (or as indicated in the tables) serum was added. The mixture was incubated for 1 h at 4°C. Cells were then washed three times with BME containing 5% FCS. 0.1 ml of ^{125}I-radiolabeled rabbit IgG antirat Ig (spec act 1 mCi per mg protein) corresponding to about $1-2 \times 10^6$ cpm of ^{125}I were added to the cell suspension and incubated for 1 h at 4°C after which the cells were washed five times. After the final wash the cells were transferred into new plastic tubes and counted for 1 min in a Intertechnique CG 30 automatic gamma spectrometer.

For quantitation of many samples the test was modified in the following manner (22). 1×10^7 lymphoid cells in 25 μl were distributed in round bottom microtiter plates (Cooke Laboratory), 25 μl of 1:10 diluted (or as indicated) serum was added into each well and the plates gently shaked and incubated for 1 h at 4°C. The plates were then centrifuged for 10 min at 1,500 rpm in an X-3 Wifug (Winkelcentrifug, Stockholm) centrifuge in special holders fitting the microtiter plates. The supernates were smashed out into a sink and the cells resuspended with BME; cells were washed three times. 25 μl of ^{125}I-labeled rabbit IgG antirat Ig (corresponding to about $2.5-5 \times 10^5$ cpm) were then added into each well. After incubation for 1 h at 4°C the cells were washed again three times as described above. Finally the cells were transferred into plastic tubes for gamma counting.

USING ^{125}I-LABELED PROTEIN A FROM STAPHYLOCOCCUS AUREUS. Protein A from Staphylococcus aureus is a molecule with specificity for Fc of most mammalian IgG classes (23). It can be iodinated and used as a sensitive probe for cell-bound IgG molecules (21). The procedure was identical to the second one described previously for radiolabeled rabbit antirat Ig, and has been described previously in detail (21).

Passive Hemagglutination Assay. This was done as described elsewhere (24). 300 μg IgG from relevant rat serum was dissolved in 200 μl saline. 10 μl of 0.07% CrCl$_3$ solution and 100 μl of 50%, five times washed sheep red blood cells (SRBC) were added and the suspension well mixed. The CrCl$_3$ was dissolved in saline and was kept half an hour at room temperature before used. After incubation of The elegant technical assistance of Berit Olsson, Ann Sjölund, and Lisbeth Ostberg is gratefully heat-inactivated FCS. After the final wash the coated SRBC were adjusted to a final concentration of 0.2% with saline containing 5% FCS. 25 μl of 0.2% coated SRBC were incubated with 25 μl of anti-idiotypic or normal sera dilutions for 1 h at room temperature followed by 2 h at 4°C or overnight at 4°C in V-bottom micro titer trays (BIO-cult, Linbor, IS MVC 96). Before used in this assay the anti-idiotypic sera were absorbed once with one tenth (vol/vol) of packed SRBC. As controls, uncoated SRBC or SRBC-treated with CrCl$_3$ only were used.

Results

(DA $\times$ Lewis)F$_1$ Hybrid Rats Inoculated with Parental T Lymphocytes Produce Antibodies Against Idiotypic Determinants on Relevant Parental Alloantibodies. Adult F$_1$ hybrid rats were inoculated intraperitoneally with graded doses of normal parental T plus B lymphocytes (spleen and lymph node cells) or normal T lymphocytes (spleen and lymph node cells that had been filtered through an anti-Ig column). They would then be expected to produce antibodies against cell-bound receptors, present on the inoculated cells and with specificity for alloantigens of the other parental strain (12). At various times thereafter the animals were bled and their sera analyzed for possible content of antialloantibodies in the indirect hemagglutination assay. Representative results of such experiments are shown in Table I and indicate that the F$_1$ hybrid rats will make such antibodies [as shown before (13, 16)]. The best way to obtain high-titered antialloantibodies was to inoculate column-purified parental T lymphocytes, whereas the presence of parental B lymphocytes, if anything, seemed to have an inhibitory effect. The present results would seem to indicate that either T and B lymphocytes directed against the same alloantigens do share idiotypic determinants or, alternatively a few contaminating B cells amongst the inoculated parental T cells do carry the antigen-binding receptors responsible for the induction of these anti-idiotypic determinants.

TABLE I

Induction of Antialloantibodies Reactive with Alloantibody IgG of Relevant Specificity in (DA × Lewis)F₁ Rats Inoculated with DA or Lewis Lymphocytes

Exp.	Immunizing cells	Target cells	Days postimmunization				
			4	5	7	10	15
1	DA, T + B,5 × 10⁷	SRBC, DA anti-Lewis IgG	3.5±0.3		5.8±0.5	6.5±1.0	6.3±1.0
2	DA, T + B, 2.5 × 10⁷	SRBC, DA anti-Lewis IgG		2.0±0.0			
	DA, T 2.5 × 10⁷	"		3.0±0.4			
3	Lewis, T + B, 5 × 10⁷	SRBC, Lewis Anti-DA IgG	3.0±1.0		2.7±0.9	4.0±1.5	3.5±0.5
4	Lewis, T + B, 2.5 × 10⁷	SRBC, Lewis Anti-DA IgG		2.4±0.5			
	Lewis, T 2.5 × 10⁷	"		7.0±1.1			

All sera tested against SRBC coated with IgG from DA, Lewis or (DA × Lewis)F₁ normal sera as well as against SRBC coated with IgG from DA anti-Lewis or Lewis anti-DA immune sera. All sera were preadsorbed with SRBC before testing. Only groups where agglutination occurred are indicated in the table. Starting size of experimental groups normally five animals per group.

The rest of the present article we will use to try to prove that in fact T and B lymphocytes directed against the same alloantigens do have shared (or identical) antigen-binding receptors as judged by the use of anti-idiotypic antisera. For this purpose we inoculated a large number of F₁ hybrid rats intraperitoneally with 2×10^7 DA or Lewis column-purified T lymphocytes once every week and tested the antisera obtained at various time intervals for anti-idiotypic antibodies in the agglutination assay. A pool of unusually high-titered anti-idiotypic antibodies produced against Lewis T lymphocytes was obtained by such a procedure. Most of the present results to be described were carried out with this single, large pool of well-defined antibodies. This serum pool will be called 1003, and was obtained from a group of five (DA × Lewis)F₁ rats bled after inoculation 4–7 times with 25×10^6 Lewis T lymphocytes intraperitoneally. The pool was heat-inactivated at 56°C for 30 min, and was then adsorbed with normal (DA × Lewis)F₁ hybrid spleen and lymph node cells to remove possible autoimmune antibodies (25).

Characteristics of the (DA × Lewis)F₁ Serum 1003 Obtained by Immunization with Lewis T Lymphocytes

BINDING TO LEWIS-ANTI-DA IgG ALLOANTIBODIES. As stated before the 1003 serum was selected for its unusually high titer of anti-(Lewis anti-DA) antibodies. The titers and demonstration of the anti-idiotypic nature of the 1003 serum is shown in Table II. As seen, not only had the 1003 serum an extremely high titer against sheep erythrocytes coated with IgG from Lewis anti-DA alloantiserum (2^{18} in titer), but it was also precipitating in gel against such alloantisera. A similar precipitating capacity of antialloantibodies has been reported before (26). As the 1003 serum in the hemagglutination or gel-diffusion tests with IgG did not react with any other normal or immune serum than the Lewis anti-DA serum (e.g. no reaction with Lewis anti-BN alloantisera) we conclude that this serum must contain a high concentration of true anti-idiotypic antibodies with specificity for Lewis antibodies directed against DA alloantigens.

BINDING TO NORMAL LEWIS T AND B LYMPHOCYTES. The 1003 serum was now tested for its capacity to react with normal lymphocytes from various rat strains or F_1-hybrids using the indirect radioimmunoassays as described in the methods section. Of the tested strains, only Lewis normal lymphocytes could be shown to bind 1003 antibodies as seen in Table III. When the distribution of binding Lewis lymphocytes was analyzed as to T or B type or to organ distribution sizeable,

TABLE II

Presence of Anti-Idiotypic Antibodies with Specificity for Lewis Anti-DA Alloantibodies in 1003 Serum

Indirect hemagglutination		Gel diffusion	
Target	Titer	Target serum	Line
SRBC-Lewis normal IgG	Neg	Lewis normal	−
SRBC-Lewis anti-DA IgG	2^{16}–2^{18}	Lewis anti-DA	+
SRBC-Lewis anti-BN IgG	Neg	Lewis anti-BN	−
SRBC-DA normal IgG	Neg	DA normal	−
SRBC-DA anti-Lewis IgG	Neg	DA anti-Lewis	−
SRBC-(DA × L)F_1 normal IgG	Neg	(DA × L)F normal	−

TABLE III

Uptake of Serum 1003 by Lewis Spleen Cells and Lewis Purified T Cells as Measured by ^{125}I-labeled Rabbit Antirat Ig

Lymphoid cells*	Incubated with:	Uptake of [^{125}I]rabbit antirat Ig‡ (Mean cpm of quadruplicates ± SE)
LS	Serum 1003	16.878 ± 1.274
LS	N rat S	7.712 ± 1.018
LT	Serum 1003	12.756 ± 700
LT	N rat S	6.869 ± 534
L Th P	Serum 1003	1.269 ± 106
L Th P	N rat S	1.105 ± 44
DA S	Serum 1003	9.119 ± 1.513
DA S	N rat S	8.578 ± 489
DA T	Serum 1003	6.350 ± 457
DA T	N rat S	5.580 ± 552
DA Th p	Serum 1003	566 ± 53
DA Th p	N rat S	498 ± 55
F_1 S	Serum 1003	4.622 ± 306
F_1 S	N rat S	4.818 ± 112
F_1 T	Serum 1003	3.346 ± 490
F_1 T	N rat S	3.927 ± 523
F_1 Th p	Serum 1003	528 ± 41
F_1 Th p	N rat S	415 ± 33
BN pT	Serum 1003	4.196 ± 58
BN T	N rat S	5.393 ± 78
BN Th p	Serum 1003	480 ± 42
BN Th p	N rat S	342 ± 28

* S denotes a mixture of spleen and lymph node cells; T denotes T lymphocytes prepared by fractionation of spleen and lymph node cells over anti-Ig bead columns; Thp denotes thymocytes which have been passed over anti-Ig columns.

‡ Input per tube: 5×10^5 cpm of [^{125}I]rabbit antirat Ig.

specific binding to both T and B lymphocytes was observed as shown in Tables III and IV. The binding to normal Lewis thymocytes, on the other hand, was insignificant. It should be noted that some of the tests in Table IV were performed using radiolabeled protein A from Staphylococcus, demonstrating that the 1003 serum contained IgG anti-idiotypic T-cell-reactive antibodies. The fact that the indirect radioimmunoassay even did function when using a mixture of T and B lymphocytes as targets (thus having a highly significant background noise due to the binding to surface Ig on normal B lymphocytes) would indicate a

TABLE IV

Capacity of Lewis T or B Lymphocytes to Bind (DA × L)F$_1$ anti-Lewis T Antibodies

Cells*	Antiserum‡	[^{125}I]pA§	Mean ± SE‖
Lewis T	[^{125}I]F$_1$ anti-Lewis T	−	*6.399 ± 256*
Lewis T	F$_1$ anti-Lewis T	+	*1.074 ± 78*
Lewis T	F$_1$ normal	+	222 ± 15
Lewis B	[^{125}I]F$_1$ anti-Lewis T	−	*11.826 ± 650*
Lewis B	F$_1$ anti-Lewis T	+	*26.325 ± 1.161*
Lewis B	F$_1$ normal	+	12.445 ± 981
DA T	[^{125}I]F$_1$ anti-Lewis T	−	2.098 ± 283
DA T	F$_1$ anti-Lewis T	+	280 ± 26
DA T	F$_1$ normal	+	342 ± 22
DA B	[^{125}I]F$_1$ anti-Lewis T	−	1.870 ± 261
DA B	F$_1$ anti-Lewis T	+	12.075 ± 639
DA B	F$_1$ normal	+	14.349 ± 137

* Lewis T cells were prepared by fractionation of spleen and lymph node cells over anti-Ig bead columns. B cells were produced by cytolysis of T cells by a rabbit antirat T-cell serum in presence of complement. Purity with regard to T and B respectively exceeded 95%.

‡ Directly ^{125}I-labeled or unlabeled F$_1$ anti-Lewis T serum was used. When ^{125}I-labeled, the IgG fraction of the antiserum was used. Input per well: 3×10^5 cpm.

§ + denotes secondary addition of [^{125}I]protein A to the test. Input per well: 5×10^4 cpm.

‖ Mean ± SEM. Triplicates. Italicized figures denote significantly increased specific binding compared to DA cell controls.

high number of idiotype-positive cells in the present system as subsequently found to be true.[2]

PROOF THAT THE BINDING OF THE 1003 ANTIBODIES OCCURS WITH RECEPTORS PRESENT ON NORMAL LEWIS T LYMPHOCYTES AND WITH SPECIFICITY FOR DA ALLO-ANTIGENS AND NOT WITH ALL LEWIS T CELLS. We have previously found that F$_1$ hybrid antisera produced against parental alloantisera in rats, in the presence of complement, will inhibit the capacity of the relevant parental T lymphocytes to function in a given graft-vs.-host (GVH) reaction (17). We now first tested if the

_____

[2] Binz, H., and H. Wigzell. Manuscript in preparation.

1003 serum could selectively inactivate Lewis T lymphocytes from participating in immune reactions against DA alloantigens, whilst sparing the reactivity against alloantigens of other strains intact. Two test systems were used: GVH reactions as measured by the local popliteal lymph node assay and the MLC system. Examples of the results obtained in the two systems are shown in Table V (see also Table VII) and demonstrate clearly that the 1003 serum contains antibodies specific for receptors on those Lewis T lymphocytes that can react with DA alloantigens. Thus, a close to complete inhibition of the reactivity against DA alloantigens was observed in both systems, whereas reactivity against BN was left largely untouched.

TABLE V

Serum 1003 Specifically Eliminates Lewis Anti-DA GVH-Reactive T Cells

Cells injected*	Host	Injected cells treated with:	Mean of "Lewis" lymph node weights‡	Mean of "DA" lymph node weights‡	Mean of "BN" lymph node weights‡	Mean log ratio ± SE
L	$(L \times DA)F_1$	$(L \times DA)F_1 NS + C'$	70.7 ± 8.6			
DA		$(L \times DA)F_1 NS + C'$		70.2 ± 7.6		-0.01 ± 0.02§
L	$(L \times DA)F_1$	Serum $1003 + C'$	11.2 ± 0.9			
DA		Serum $1003 + C'$		73.2 ± 5.03		$0.81 \pm 0.02 \|$
L	$(L \times BN)F_1$	$(L \times DA)F_1 NS + C'$	33.7 ± 5.3			
BN		$(L \times DA)F_1 NS + C'$			32.2 ± 2.5	-0.01 ± 0.05§
L	$(L \times BN)F_1$	Serum $1003 + C'$	31.5 ± 3.7			
BN		Serum $1003 + C'$			31.9 ± 4.3	-0.00 ± 0.07§

* 5×10^6 spleen and lymph node cells were injected into each foot pad.
‡ Mean weights of four nodes (mg).
§ Not significantly different from 0.
‖ Significantly different from 0 ($P < 0.01$).

PROOF THAT THE 1003 ANTIBODIES DIRECTED AGAINST IDIOTYPIC DETERMINANTS ON T LYMPHOCYTES AND THOSE DIRECTED AGAINST THE IDIOTYPE-POSITIVE LEWIS-ANTI-DA IgG ALLOANTIBODIES ARE THE VERY SAME ANTIBODY MOLECULES. The antibodies in the 1003 that are capable of reaction with those Lewis T lymphocytes that are immunocompetent to react against DA alloantigens and those 1003 antibodies in the 1003 that are capable of reaction with those Lewis T lymphocytes that are immunocompetent to react against DA alloantigens and those 1003 antibodies that react with Lewis anti-DA IgG antibodies had now both been shown to be idiotypic. It now remained to finally demonstrate that the normal Lewis T lymphocytes with receptors for DA alloantigen did share idiotypic determinants with the Lewis anti-DA alloantibodies produced by B cells; that is, to prove identity or similarity between the antigen-binding sites of the two groups of reactive molecules. Two approaches were used. In the first, highly purified IgG was produced from Lewis anti-DA immunserum (see Methods) and covalently linked to Sepharose beads. Such immunosorbant columns (and controls) were now analyzed for their capacity to remove from the 1003 serum the capacity to react with Lewis T lymphocytes as measured in radioimmunoassays or GVH or MLC reactions. The tests were performed testing both the immunosorbent column passed 1003 serum as well as the adsorbed and at low pH eluted 1003 material for inhibitory or binding capacity. Results of such experiments are shown in Tables VI, VII and VIII, and demonstrate the selective capacity of highly purified Lewis anti-DA IgG columns to remove in a recoverable form the 1003 antibodies with capacity to react with normal Lewis anti-DA-reactive T

TABLE VI

Capacity of Lewis Anti-DA Immunoabsorbens to Remove (Lewis × DA)F$_1$ Anti-Lewis T Anti-Idiotypic Antibodies

Lymphoid cells		Uptake of [^{125}I] rabbit antirat Ig Mean cpm of triplicates ± SE*
LT	Serum 1003	70.074 ± 2.119
LT	N rat S	22.367 ± 2.026
LT	Serum 1003 absorbed on L anti-DA IgG	24.505 ± 787
LT	Serum 1003 absorbed on L normal IgG	51.545 ± 6.648
LT	Serum 1003 absorbed on (L × DA)F$_1$ normal IgG	56.551 ± 3.355
LT	Serum 1003 bound and eluted from L anti-DA IgG	64.931 ± 9.626
LT	Serum 1003 bound and eluted from L normal IgG	19.542 ± 469
LT	Serum 1003 bound and eluted from (L × DA)F$_1$ normal IgG	21.618 ± 2.401
DA T	Serum 1003	29.271 ± 2.819
DA T	N rat S	25.444 ± 384
DA T	Serum 1003 absorbed on L anti-DA IgG	32.788 ± 2.350
DA T	Serum 1003 absorbed on L normal IgG	28.713 ± 1.164
DA T	Serum 1003 absorbed on (L × DA)F$_1$ normal IgG	27.542 ± 1.037
DA T	Serum 1003 bound and eluted from L anti-DA IgG	25.659 ± 304
DA T	Serum 1003 bound and eluted from L normal IgG	28.245 ± 2.537
DA T	Serum 1003 bound and eluted from (L × DA)F$_1$ normal IgG	25.786 ± 2.829

* Input per well: 2×10^6 cpm of [^{125}I]rabbit antirat Ig.

cells. The included controls demonstrated the Ig nature of the inhibitory capacity of the 1003 serum. Furthermore, the specificity of 1003 removing capacity of the IgG Lewis anti-DA column was further proven by the failure of such columns to remove DA antisheep red blood cell antibodies, thus also excluding the possibility that any anti-DA allotype-specific antibodies might have been on the column. Thus, these data strongly suggest the presence of identical idiotypes on IgG molecules produced in Lewis B cells against DA alloantigens and on the antigen-binding surface receptor on normal Lewis T cells with capacity to react against DA alloantigens.

To further prove this point normal Lewis T or T + B cells were used as specific immunosorbents to remove from the 1003 serum the antibodies, capable of agglutinating in the anti-idiotypic agglutination test, erythrocytes coated with

TABLE VII

Effect of Serum 1003 on the MLC Reaction. Selective Removal of Inhibitory Capacity with L anti-DA Alloantibody IgG Immunosorbant

Responder cells	Target cells	Responder cells treated with	[³H]Thymidine incorporation of mixture. Mean cpm of quadruplicates ± SE	[³H]Thymidine incorporation of responder cells alone. Mean cpm of triplicates ± SE	[³H]Thymidine incorporation of target cells alone. Mean cpm of duplicates ± SE
L	$(L \times DA)F_1$	—	32.933 ± 245	1.017 ± 84	167 ± 11
L	"	$(L \times DA)F_1$ NS + C'	32.112 ± 145	1.566 ± 352	187 ± 26
L	"	Serum 1003 + C'	1.938 ± 29	1.133 ± 43	199 ± 2
L	"	Serum 1003 abs L anti-DA IgG + C'	30.929 ± 292	1.006 ± 48	179 ± 23
L	"	Serum 1003 abs LNS IgG + C'	1.680 ± 419	1.040 ± 62	223 ± 41
L	"	Serum 1003 abs $(L \times DA)F_1$ IgG + C'	1.609 ± 113	1.015 ± 57	244 ± 49
L	"	Serum 1003 bound and eluted from L anti-DA IgG + C'	3.013 ± 273	1.087 ± 72	214 ± 102
L	"	Serum 1003 bound and eluted from LNS IgG + C'	31.525 ± 743	2.037 ± 741	249 ± 27
L	"	Serum 1003 bound and eluted from $(L \times DA)F_1$ IgG + C'	30.987 ± 1.155	1.777 ± 373	296 ± 65
L	$(L \times BN)F_1$	$(L \times DA)F_1$ NS + C'	22.857 ± 891	1.814 ± 799	252 ± 111
L	$(L \times BN)F_1$	Serum 1003 + C'[1]	21.844 ± 1.091	1.007 ± 154	366 ± 220
DA	$(L \times DA)F_1$	$(L \times DA)F_1$ NS + C'	30.294 ± 528	1.134 ± 61	195 ± 72
DA	$(L \times DA)F_1$	Serum 1003 + C'	29.915 ± 257	1.227 ± 81	214 ± 35

TABLE VIII

Capacity of L Anti-DA Alloantibody IgG to Remove Anti-Idiotypic Antibodies from 1003 Serum as Measured by Inhibition of GVH-Reaction

Cells injected*	Host	Injected cells treated with:	Mean of "Lewis" lymph node weights*	Mean of "DA" lymph node weights*	Mean log Ratio ± SE
L, DA	$(L \times DA)F_1$	$(L \times DA)F_1$ N S + C'	28.5±3.3	26.4±2.9	−0.02±0.02
L, DA	"	Serum 1003 + C'	6.1±1	32.6±5.6	0.73±0.06
L, DA	"	Serum 1003 abs L anti-DA IgG + C'	27.7±1.6	25.0±0.3	−0.04±0.02
L, DA	"	Serum 1003 abs L N S IgG + C'	6.2±0.6	33.6±3.5	0.70±0.07
L, DA	"	Serum 1003 $(L \times DA)F_1$ N S IgG + C'	4.1±0.4	33.0±2.2	0.89±0.01
L, DA	"	Serum 1003 bound and eluted from L anti-DA IgG + C'	5.1±1.0	29.7±2.7	0.78±0.09
L, DA	"	Serum 1003 bound and eluted from L N S IgG + C'	21.8±1.0	22.9±0.7	0.02±0.01
L, DA	"	Serum 1003 bound and eluted from F_1 N S IgG + C'	26.2±1.2	23.5±1.4	0.05±0.05

* 3×10^6 spleen cells and lymph node cells were injected into each foot pad.

IgG Lewis anti-DA alloantibodies. As seen in Table IX purified Lewis T lymphocytes were, if anything, better than T+ B cell mixtures in removing such specific anti-idiotypic antibodies. We could thus conclude that, by all measurements, Lewis normal T lymphocytes, with the capacity to react with the DA

TABLE IX

Capacity of Normal Lewis T Lymphocytes to Adsorb away Serum 1003 Antibodies as Subsequently Measured in the Indirect Hemagglutination Assay

Exp.	Inhibitory cells*	Agglutination titers‡
1	Lewis T	2^5
	(Lewis × DA)F$_1$	2^{15}
	0	2^{16}
2	Lewis, T + B	2^7
	Lewis, T	2^5
	(Lewis × DA)F$_1$, T + B	2^{11}
	(Lewis × DA)F$_1$, T	2^{12}
	0	2^{13}

* 10^7 cells of indicated type was added to each hole in the microplate, whereafter twofold dilutions of 1003 serum was added. Purity of T cells with regard to B-cell contamination = 98% or more. B-cell contamination in T + B cell populations = 50–60%. T + B = spleen and lymph node cells mixture. T = T cells purified out of the T + B population via passage through anti-Ig columns.

‡ Agglutination as assessed against SRBC coated with Lewis-anti-DA IgG antibodies. Controls including SRBC coated with IgG from normal Lewis, DA or F$_1$ sera as well as IgG from DA-anti-Lewis immune sera were negative.

strain alloantigens, have receptors with identical idiotypes to the IgG antibodies that Lewis B cells will produce against the same alloantigens.

Discussion

The aim of the present article is to demonstrate that the antigen-binding receptors on B and T lymphocytes reactive against the same antigenic determinants do share idiotypic determinants. F$_1$ hybrid animals can be immunized against parental lymphocytes or alloantibodies produced in one parental strain against alloantigens of the other parental strain. Such antireceptor antisera made in F$_1$ hybrid mice or rats have been found capable of selective interference with relevant parental lymphocytes as measured by in vitro (12–14, 16, 27) or in vivo (15, 17, 26–28) tests.

Using radiolabeled F$_1$ antialloantibody IgG molecules it was previously found possible to label parental lymphocytes of the relevant strain (13, 16). Here, the presence of idiotypic receptors on normal parental lymphocytes was parallelled by the selective ability of such lymphocytes to bind in vitro to the relevant allogeneic monolayer of cells (16, 17). Although such antialloantibody serum was found able to selectively interfere with the specific binding to the monolayer of both B and T lymphocytes (16, 17), in the previous assays only B lymphocytes could be shown to bind enough radiolabeled anti-idiotypic alloantibody IgG molecules to allow detection. However, we then favored the interpretation that T cells as well as B cells should bind in such anti-idiotypic assays, provided the radioimmunoassay became sensitive enough or the anti-idiotypic antibodies had the relevant specificity. The present results have indeed shown that when F_1 hybrids make high-titered antisera against parental T-cell receptors with specificity for alloantigen of the other parental strain, such sera will now also be capable of showing directly demonstrable binding to parental immunocompetent T lymphocytes carrying such idiotypic, antigen-binding receptors.

A great help in the analysis of anti-idiotypic antibodies produced in F_1 hybrids against inoculated alloantiserum, produced in one parental strain against the alloantigens of the other parental, came with the development of a sensitive indirect hemagglutination assay to detect such antibodies (24, 26). Taking advantage of these findings we have been able to confirm and extend the results previously obtained in the radioimmunoassays and the cellular immune reactions (15–17), and as we believe, to finally prove the existence of shared idiotypic determinants on antigen-binding receptors on T or B lymphocytes reactive against the same antigenic determinants. The crucial issue to consider in such an approach is the purity of the reagents or cells used, as "pure" T lymphocytes could always be contaminated by a certain low but still decisive number of B cells (or cytophilic B-cell products) whereas on the other hand, alloantiserum used in the previous assays might have contained soluble T-cell receptors (12, 14, 16, 17). Our approach in the present article was to immunize F_1 hybrids with "pure" parental T lymphocytes, trying to make a pool of high-titered antiserum, reactive with idiotypic receptors or antibodies with the relevant antigen-binding specificity. T lymphocytes were used as "antigen" as we had previously had problems to get high enough titered antisera by the use of injection of alloantibodies only (16, 17). There are reasons to believe that receptors on immunocompetent T cells would be most immunogeneic in this system. We succeeded in producing such high-titered F_1 antisera using as a screening assay sheep erythrocytes coated with IgG from relevant immune parental alloantiserum; that is, looking for antibodies directed against idiotypic determinants present on B-cell products, the IgG molecules. Subsequently we analyzed these high-titered F_1 hybrid antisera for their capacity to react in a radioimmunoassay with normal, parental "pure" T lymphocytes. We could find a nice positive correlation between anti-idiotypic titer as measured by agglutination of the alloantibody IgG-coated erythrocytes and the capacity to react with the relevant, parental T lymphocytes.[2]

We then demonstrated that the binding of F_1 antibodies as measured in the radioimmunoassay onto the normal parental T lymphocytes was indeed directed against idiotypic antigen-binding receptors on a minority of these cells. This was demonstrated by the capacity of such anti-idiotypic antisera to, in the presence of complement, selectively abolish the capacity of those parental T lymphocytes to react against the relevant allogeneic cells as measured in local GVH-reaction or in vitro MLC, whilst sparing the reactivity of the cell population towards other, third party allogeneic cells. Furthermore, the strength of the reaction allowed a direct enumeration of the idiotype-positive T and B cells.[2]

The final proof demonstrating identity between the anti-idiotypic antibodies reactive with the B-cell product, the alloantibodies of IgG type, and those that react with the idiotypic receptors on the parental T cells came from a series of experiments, where reciprocal adsorptions using highly purified T cells or IgG-immunosorbents were used to selectively remove the anti-idiotypic antibodies reactive with the "other" idiotype group. Highly purified IgG from the relevant parental alloantiserum coupled to Sepharose was shown capable of selective removal of the F_1 anti-idiotypic antibodies reactive with the relevant T-cell receptors. As measurements were taken to ensure highest possible purity of the IgG alloantibodies before coupling to Sepharose, and contamination with soluble T-cell receptors for antigen would seem unlikely, unless such receptors are indeed physically unseparable from conventional IgG molecules. Reciprocally, purified parental T cells were capable of selective removal of F_1 antialloantibody antibodies as subsequently tested in the hemagglutination assay. Here the fact that normal purified parental T lymphocytes were seemingly more efficient than a mixture of T and B cells in removing specific anti-idiotypic antibodies as subsequently measured in the hemagglutination assay would strongly argue against possible B-cell contamination as a cause for the capacity of T cells to adsorb away these antibodies.

In conclusion, our results demonstrate that T and B lymphocytes directed against the same alloantigens are using antigen-binding receptors or antibodies with shared idiotypic determinants. The simplest interpretation of these findings would be that T and B lymphocytes are indeed using the same set (or subset) of variable genes to create the antigen-binding area of their respective receptors for antigen.

Summary

Antigen-binding receptors on T lymphocytes and IgG antibodies with the same antigen-binding specificity as the T-cell receptors display shared or identical idiotypes. This was shown using a system where adult F_1 hybrid rats between two inbred strains were inoculated with T lymphocytes from one parental strain. Such F_1 hybrid rats produce antibodies directed against idiotypic determinants present on IgG alloantibodies, produced in the T donor genotype strain and with specificity for the alloantigens of the other parental strain. The idiotypic nature of the F_1 antialloantibody serum against the parental alloantibodies was demonstrated both by indirect hemagglutination tests or by gel diffusion using alloantisera with different specificity as targets. Furthermore, the F_1 anti-T-lymphocyte sera could be shown to contain antibodies against idiotypic parental T lymphocytes as well. This was shown by the capacity of the antisera, in the presence of complement, to wipe out the relevant parental T-cell reactivity against the other parental strain (as measured in MLC or GVH) whilst leaving the T-lymphocyte reactivity against a third, unrelated allogeneic strain intact.

These findings demonstrate that F_1 hybrid rats inoculated with parental T lymphocytes make anti-idiotypic antibodies directed against both the T cell receptors and IgG alloantibodies of that parental strain with specificity for alloantigens of the other parental strain. In order to prove identity between the anti-idiotypic antibodies against the B and T-cell antigen-binding molecules the following experiments were carried out: highly purified IgG from relevant alloantibody-containing serum in immunosorbent form could be shown to selectively remove both anti-idiotypic activities from the F_1 antiserum. Furthermore, parental normal T lymphocytes could be shown capable of removing from the anti-idiotypic antisera all those antibodies that would cause agglutination of the relevant alloantibody-coated erythrocytes in the indirect agglutination assay. We would thus conclude that T and B lymphocytes reactive against a given antigenic determinant use receptors with antigen-binding areas coded for by the same variable gene subset(s).

The elegant technical assistance of Berit Olsson, Ann Sjölund, and Lisbeth Östberg is gratefully, acknowledged.

Received for publication 5 March 1975

Note added in proof. Anti-idiotypic antibodies directed against a mouse-antistreptococcal determinant can, if passively administered to normal mice, prime these animals for helper activity against the streptococcal antigens (K. Eichmann, and K. Rajewsky, personal communication). It is possible that these findings have the same meaning as those results in the present article, that is, that T and B lymphocytes reactive against the same antigenic determinant may share idiotypic determinants.

References

1. Gowans, J. L., and D. D. McGregor. 1965. The immunological activities of lymphocytes. *Progr. Allergy.* **9:**1.
2. Miller, J. F. A. P., and G. F. Mitchell. 1969. Thymus and antigen reactive cells. *Transplant. Rev.* **1:**3.
3. Cooper, M. D., R. D. A. Peterson, M. A. South, and R. A. Good. 1966. The functions of the thymus system and the bursa system in the chicken. *J. Exp. Med.* **123:**75.
4. Wigzell, H. 1973. On the relationship between cellular and humoral antibodies. *Contemp. Top. Immunobiol.* **3:**77.
5. Marchalonis, J. J., R. E. Cone, and J. L. Atwell. 1972. Isolation and partial characterization of lymphocyte surface immunoglobulins. *J. Exp. Med.* **135:**956.
6. Feldmann, M. 1974. T cell suppression in vitro. II. Nature of the suppressive specific factor. *Eur. J. Immuno.* **4:**667.
7. Benacerraf, B., and H. O. McDevitt. 1972. Histocompatibility-linked immune response genes. *Science (Wash. D. C.).* **175:**273.
8. Shevach, E. M., I. Green, and W. E. Paul. 1974. Alloantiserum-induced inhibition of immune response gene product function. II. Genetic analysis of target antigens. *J. Exp. Med.* **139:**679.
9. Fröland, S. S., and J. Natvig. 1971. Effect of poly-specific rabbit anti-immunoglobulin-antisera on the response of human lymphocytes in mixed lymphocyte culture. *Int. Arch. All.* **41:**248.
10. Crone, M., C. Koch, and M. Simonsen. 1972. The elusive T cell receptor. *Transplant. Rev.* **10:**36.
11. Mond, J. J., A. L. Luzzati, and G. J. Thorbecke. 1972. Surface antigens of immunocompetent cells. IV. Effect of anti-G5 allotype on immunologic responses of homozygous b5 rabbit cells in vitro. *J. Immunol.* **108:**566.
12. Ramseier, H., and J. Lindenmann. 1971. Cellular receptors. Effect of antialloantibodies on the recognition of transplantation antigens. *J. Exp. Med.* **134:**1083.
13. Binz, H., and J. Lindenmann. 1972. Cellular receptors. Binding of radioactively labeled antialloantiserum. *J. Exp. Med.* **136:**872.
14. Ramseier, H., and J. Lindenmann. 1972. Similarity of cellular recognition structures for histocompatibility antigens and of combining sites of corresponding alloantibodies. *Eur J. Immunol.* **2:**109.

15. Binz, H., J. Lindenmann, and H. Wigzell. 1973. Inhibition of local graft-versus-host reaction by antialloantibodies. *Nature (Lond.)* **246:**146.

16. Binz, H., J. Lindenmann, and H. Wigzell. 1974. Cell-bound receptors for alloantigen on normal lymphocytes. I. Characterization of receptor-carrying cells by the use of antibodies to alloantibodies. *J. Exp. Med.* **4:**877.

17. Binz, H., J. Lindenmann, and H. Wigzell. 1974. Cell-bound receptors for alloantigen on normal lymphocytes. II. Antialloantibody serum contains specific factors reacting with relevant immunocompetent T lymphocytes. *J. Exp. Med.* **140:**731.

18. Billingham, R. E., and W. K. Silvers, editors. 1961. *In* Transplantation of Tissues on Cells. The Wistar Institute Press, Philadelphia.

19. Stimpfling, S. H. 1964. Methods for detection of hemagglutinins in mouse isoantisera. *Methods Med. Res.* **10:**27.

20. Ford, W. L., W. Burr, and M. M. Simonsen. 1970. A lymph node weight assay for the graft-versus-host activity of rat lymphoid cells. *Transplantation.* **10:**258.

21. Dorval, G., K. I. Welsh, and H. Wigzell. 1974. Labeled Staphylococcal protein A as an immunological probe in the analysis of cell surface markers. *Scand. J. Immunol.* **3:**405.

22. Dorval, G., K. I. Welsh, and H. Wigzell. 1975. A new radio immunoassay of cellular surface antigens on living cells. I. Use of iodinated soluble protein A from Staphylococcus aureus. *J. Immunol. Meth.* In press.

23. Forsgren, A., and J. Sjöqvist. 1966. "Protein A" from Staphylococcus aureus. I. Pseudo-immune reaction with human gammaglobulin. *J. Immunol.* **97:**822.

24. Wells, J. V., A.-C. Wang, Ch. T. Thornsvard, M. S. Schanfield, J. F. Bleumers, and H. H. Fudenberg. 1973. Detection of idiotypic determinants on human monoclonal proteins with specific rabbit antisera and hemagglutination-inhibition. *J. Immunol.* **111:**841.

25. Lindholm, L., L. Rydberg, and Ö Strannegård. 1973. Development of host plasma cells during graft-versus-host reactions in mice. *Eur. J. Immunol.* **3:**511.

26. McKearn, T. J., F. P. Stuart, and F. W. Fitch. 1974. Anti-idiotypic antibody in rat transplantation immunity. I. Production of anti-idiotypic antibody in animals repeatedly immunized with alloantigens. *J. Immunol.* **113:**1876.

27. Binz, H. 1975. Local graft-versus-host reaction in mice specifically inhibited by anti-receptor antibodies. *Scand. J. Immunol.* **4:**79.

28. McKearn, T. J. 1974. Antireceptor antiserum causes specific inhibition of reaction to rat histocompatibility antigens. *Science (Wash. D. C.).* **183:**94.

K. Eichmann and K. Rajewsky

Institute for Genetics, University of Cologne, Cologne

Induction of T and B cell immunity by anti-idiotypic antibody*

A small dose of the IgG_1 fraction of anti-idiotypic antibody ($\overline{a}Id1$) raised in guinea pigs against a strain A/J antibody specific for streptococcal Group A carbohydrate sensitizes A/J mice against Group A streptococci. This is opposed to the previously established suppressive function of anti-idiotypic antibody of the IgG_2 class ($\overline{a}Id2$). Correspondingly, $\overline{a}Id1$ but not $\overline{a}Id2$ is eliminated from the circulation in the way typical of an immunogenic molecule. However, the stimulatory component in the IgG_1 fraction is not necessarily itself IgG_1 antibody.

Sensitization occurs in both B and helper T lymphocytes and is specific for Group A streptococci. In the B cell compartment sensitization is restricted to precursor cells expressing the idiotype. The concomitant activation of T helper cells therefore suggests that these cells make use of receptors with a similar or identical idiotype.

Efficient sensitization by $\overline{a}Id1$ of both T and B cells is also demonstrated in strain C57L/J mice which upon immunization with Group A streptococci express a partially cross-reacting idiotype as a minor component. When such animals were primed with $\overline{a}Id1$, essentially all of the anti-carbohydrate antibody carried the partially cross-reacting idiotype.

1. Introduction

The binding specificity of antibody molecules is determined by their variable portions. However, in addition to providing a particular binding site for antigen, each variable portion possesses a characteristic set of antigenic determinants, so-called idiotypic determinants or idiotopes, which distinguish it from the variable part of other antibody molecules and against which specific anti-idiotypic antibody can be raised. The variable portion of a given antibody will thus specifically combine with a particular antigen as well as with a particular kind of anti-idiotypic antibody.

Since at least in the case of B lymphocytes antibody molecules serve as surface receptors which enable the cells to bind antigen, one might envisage that anti-idiotypic antibody could substitute for antigen in the induction of an immune response.

In the case of immunological paralysis such a functional equivalence of antigen and anti-idiotypic antibody has already been demonstrated. Administration of anti-idiotypic antibody usually leads to specific idiotype suppression [1–3] and potentially can establish unresponsiveness to the corresponding antigen [2, 4]. However, recent studies in this laboratory [3] and by Trenkner and Riblet [5] suggest that under certain conditions anti-idiotypic antibody can also enhance the immune response.

Here we demonstrate efficient priming of both T and B lymphocytes by anti-idiotypic antibodies for a secondary immune response to Group A streptococcal carbohydrate (A-CHO).

[I 1130]

* We dedicate this paper to our late colleague and friend Dr. Hermann Seiler, who first had the idea to attempt lymphocyte stimulation by anti-idiotypic antibody. His death in 1972 made it impossible for him to pursue this project.

Correspondence: Klaus Rajewsky and Klaus Eichmann, Institut für Genetik der Universität Köln, D-5 Köln 41, Weyertal 121, Fed. Rep. Germany

Abbreviations: $\overline{a}Id$: IgG fraction from guinea pig anti-A5A antiserum $\overline{a}Id1$ and $\overline{a}Id2$: IgG_1 and IgG_2 fractions, respectively, from same antiserum **GPIgG:** Guinea pig IgG **Strep.A:** Group A streptococcal vaccine **A-CHO:** Purified Group A streptococcal carbohydrate **NIP:** (4-Hydroxy-5-iodo-3-nitrophenyl)acetyl **CG:** Chicken IgG **RAG:** Rabbit anti-guinea pig IgG antiserum **IBC:** Idiotype binding capacity

2. Materials and methods

2.1. Animals

Six to eight wk old A/J and C57L/J mice were obtained from the Jackson Laboratory, Bar Harbor. The animals were vaccinated against ectromelia and given a rest of at least 4 weeks before they entered the experiments. Guinea pigs were obtained from a local breeder.

2.2. Previously described methods

These include: The preparation of NIP-protein conjugates and of streptococcal vaccines [6, 7]; the purification of normal mouse and guinea pig immunoglobulin [3, 8, 9] and of antibodies to streptococcal carbohydrates [9, 10]; the preparation of anti-idiotypic antibody in IgG-tolerant guinea pigs [11]; the radioiodination of proteins and carbohydrates [9, 12]; the quantitative determination of antibodies to NIP [13, 14] and A-CHO [9] and of idiotypic and anti-idiotypic antibody in radioactive binding assays [15]; the isolation and propagation of clone A5A by *in vivo* spleen cell cloning [9]; the separation of guinea pig IgG_1 and IgG_2 by preparative agarose electrophoresis [3]; the elimination of T lymphocytes from spleen cell populations by treatment of the cells with antiserum to Thy-1.2 and complement [16–18]; the elimination of B lymphocytes from spleen cell populations by passage of the cells over Degalan columns coated with complexes of mouse immunoglobulin and rabbit antibodies to mouse immunoglobulin [19, 20]; the induction of an adoptive secondary response by transfer of hapten and carrier-primed spleen cells into sublethally irradiated syngeneic hosts and subsequent stimulation by hapten-carrier conjugates [13, 20, 21]. In the cell transfer experiments, irradiation of the hosts was 350 r, and $6 \times 10^6 - 40 \times 10^6$ cells of each type were injected per host.

The reagents which are required in the various technical procedures have either been characterized in our previous papers (see references above) or are described below. Group A streptococci and purified A-CHO were a gift from Dr. R.M. Krause, The Rockefeller University, New York.

2.3. Immunization

2.3.1. Priming

Priming with anti-idiotypic antibody against A5A consisted of a single intravenous injection of purified IgG_1 from anti-idiotypic serum ($\bar{a}Id1$). The dose corresponded to 0.1 μg idiotype binding capacity (IBC), which is equivalent to $3-5$ μg guinea pig IgG (GPIgG). Before injection, the $\bar{a}Id$ was centrifuged for 30 min at 48 000 RCF. Priming with Group A streptotoccal vaccine (Strep.A) consisted of a single intraperitoneal injection of 1×10^9 particles. The sensitization of mice with (4-hydroxy-5-iodo-3-nitrophenyl)acetyl coupled chicken IgG (NIP-CG) has been described [13, 20]. Our conjugates contained $6-15$ NIP groups per 150 000 daltons. Cells from sensitized mice were adoptively transferred $6-8$ weeks after injection of $\bar{a}Id1$, Strep.A, or NIP-CG.

2.3.2. Challenge

Secondary injections of antigen were given intraperitoneally within the first 24 hours after cell transfer. Standard doses were 0.2 μg $NIP_{10}CG$, 20 μg NIP_4GPIgG and 1×10^9 Strep.A particles conjugated with NIP (NIP-Strep.A). Coupling of NIP groups to streptococci was achieved by addition of 1 mg NIP azide (a vast excess) in 25 μl dimethylformamide to a suspension of 6×10^{10} particles in 2 ml 0.2 M sodium bicarbonate. The particles were then washed $3-4$ times with saline, suspended in 6 ml saline and briefly sonicated in order to disrupt aggregates.

2.4. Elimination experiments

2.4.1. Elimination of IgG_1 and IgG_2 anti-idiotypic antibodies

IgG_1 and IgG_2 fractions from anti-idiotypic antisera were centrifuged for 30 min at 48 000 RCF and injected intravenously into groups of 6 mice in amounts equivalent to 10 μg IBC (between 300 and 500 μg GPIgG). The first bleeding was taken 30 min after injection and subsequent bleedings were drawn every 2 to 3 days. Circulating $\bar{a}Id$ was determined by measuring the IBC of the sera at 10^{-9} M ^{125}I-labeled A5A antibody, using a rabbit anti-guinea pig IgG antiserum (RAG) for precipitation.

2.4.2. Elimination of IgG_1 and IgG_2 immunoglobulins

The same IgG fractions that were used to determine the elimination of anti-idiotypic antibodies were labeled with ^{125}I and mixed with cold IgG to a specific activity of 4×10^6 cpm/500 μg IgG. This amount was injected intravenously in groups of 6 mice and bleedings were drawn as described under 2.4.1. The radioactivity was determined using a few drops of whole blood and was calculated to cmp/100 μg of blood. In weekly intervals, serum was prepared and the radioactivity was precipited with RAG. In all determinations, between 92 % and 81 % of the radioactivity was precipitable.

3. Results

3.1. The idiotypic system

A major portion of the antibodies of strain A/J mice to Group A streptococcal carbohydrate (A-CHO) represents a single antibody species which is defined by isoelectric focusing spectrum [3] and by idiotype [9, 15]. The idiotype is detected by guinea pig anti-idiotypic antisera raised against the anti-A-CHO antibody produced by the A/J lymphocyte clone A5A [9]. The expression of the A5A idiotype in inbred mice is controlled by a single locus which is linked to the heavy chain constant region allotype [15, 22, 23].

Anti-idiotypic antibody against the A5A idiotype can be fractionated into the IgG_1 ($\bar{a}Id1$) and IgG_2 ($\bar{a}Id2$) classes of GPIgG [3, 24]. Previous work has shown that pretreatment of A/J mice with $\bar{a}Id2$ resulted in an idiotype-negative response to subsequent challenge with Strep.A suggesting a specific unresponsiveness of idiotype positive precursor cells [3, 25]. In contrast, pretreatment with $\bar{a}Id1$ resulted in an enhanced proportion of idiotype-positive antibodies upon subsequent challenge with Strep.A, suggesting an enhanced responsiveness of idiotype positive precursors [3]. The experiments reported in the following sections aim at defining conditions in which IgG_1 anti-idiotypic antibody successfully induces the expression of the A5A idiotype and hence immunity to Group A streptococci in the A/J strain. In addition, an attempt is made to similarly induce immunity by injecting the anti-A5A idiotype into C57L/J mice which, in contrast to A/J mice, express a partially cross-reactive idiotype as a minor component of their anti-A-CHO antibodies [15].

3.2. Sensitization of B and helper T lymphocytes by anti-idiotypic antibody

3.2.1. Priming of A/J mice with anti-A5A idiotype

In a large series of unsuccessful experiments we tried to induce an adoptive secondary anti-A-CHO response by administration of anti-idiotypic antibody instead of antigen. This was done by transferring spleen cells from A/J mice, preimmunized with Strep.A, into sublethally irradiated syngeneic hosts which were then boosted with anti-idiotypic antibodies. Boosting doses ranged from 10 ng to 10 μg IBC in these experiments, specifically purified $\bar{a}Id$ and also, when it became available, $\bar{a}Id1$ and $\bar{a}Id2$, were used, and in some cases the $\bar{a}Id$ was administered in combination with GPIgG-primed helper cells. Ten days after challenge the animals were bled and the sera were analysed for the presence of antibody to A-CHO and for A5A idiotype. Although the animals produced high antibody titers upon challenge with Strep.A, we could in no instance detect a response in animals that had received $\bar{a}Id$ instead.

We therefore decided to reverse the experiment and to ask the question whether anti-idiotypic antibody would prime the animals for a secondary response to streptococci.

The experiments were designed in such a way that priming in both the B and the T cell compartment could be detected. Spleen cells from animals that had received either Strep.A or $\bar{a}Id$ $6-8$ weeks previously were transferred into sublethally irradiated syngeneic hosts together with spleen cells from syngeneic animals primed with NIP-CG. The animals were then boosted with NIP-conjugated streptococci (NIP-Strep.A), bled $8-10$ days later and the sera were titrated for antibodies against NIP and A-CHO, respectively, and for A5A idiotype. It was established and is again born out by the present data that there is no significant serological cross-reaction between the NIP hapten and A-CHO. The induction of a secondary response in such a system relies on the cooperation of antibody-forming cell precursors and T helper cells [26, 17]. The anti-hapten (NIP) response thus depends, in the presence of hapten-primed precursor cells, entirely on carrier-specific helpers and can accordingly be used to measure priming to streptococcal antigens in the T cell compartment. On the other hand, the combined determination of anti-A-CHO and idiotype titers reveals the extent of specific priming in the B cell compartment.

The results of a typical experiment are compiled in Table 1. It is apparent that high anti-A-CHO and idiotype responses are obtained when cells from animals primed with either streptococci or āId1 are transferred, whereas cells from animals injected with NIP-CG do not exhibit A-CHO specific B cell memory. It is particularly significant that the āId1-primed cells produce anti-A-CHO and idiotype titers of equal magnitude, whereas priming with Strep.A results in an anti-A-CHO response in which only about 25 % of the antibodies carry the A5A idiotype. Thus, in the B cell population, sensitization by āId1 selectively affects idiotype-positive precursor cells. This finding is most easily interpreted by assuming that B cell priming has occurred upon injection of āId1.

Table 1. ā Id1 induces B cell priming and specific helper function

Cells analyzed for[a] Strep.A sensitivity	Antigen	Anti-NIP[b] response	Anti-A-CHO[c] response	Idiotype[d] response
–	NIP-CG	157 (1.2)	< 2	< 0.2
–	NIP-Strep.A	2.5 (1.2)	5.2 (1.5)	3.8 (1.4)
1° āId1	NIP-Strep.A	152 (1.1)	180 (1.1)	174 (1.1)
1° Strep.A	NIP-Strep.A	36 (1.1)	413 (1.1)	98 (1.1)

a) Hosts received 30 x 10^6 cells primed with NIP-CG and additional cells (40 x 10^6/host) as indicated in the table. Groups of 6 animals.
b) Geometric means of molar serum binding capacities x 10^8 with standard error, determined at 10^{-8} M hapten.
c) Geometric means of anti-A-CHO serum concentrations (μg/ml) with standard error.
d) Geometric means of A5A serum concentrations (μg/ml) with standard error.

The anti-hapten responses strikingly demonstrate the presence of specific helper function in both Strep.A and āId1-primed spleen cells. In order to verify that helper function was indeed exerted by T lymphocytes, a similar but more detailed experiment was carried out, in which the spleen cell populations were either treated with anti-Thy-1.2 serum in order to eliminate T cells [17] or were passed over Degalan columns coated with complexes of mouse Ig and rabbit antibodies to mouse Ig [19] in order to eliminate B cells (Table 2). Again, efficient priming of both B and T helper cells was achieved upon injection of either Strep.A or āId1. The helper cells induced by āId1 exhibited specificity for streptococci, since they did not help in the response to a conjugate of the hapten with the unrelated carrier guinea pig immunoglobulin. Irrespective of whether helper function had been induced by āId1 or Strep A, the elimination of T cells from the helper cell population led to the elimination of help for the anti-NIP response. The response to A-CHO, however, was only slightly reduced, demonstrating that our antiserum to Thy-1.2 was not toxic for B lymphocytes. We think that in this situation A-CHO-specific precursor cells can cooperate with NIP-specific helper cells present in the NIP-CG-primed cell population. This interpretation is supported by our unpublished finding that in animals which had received āId1- and NIP-CG-primed cells, NIP-Strep.A induced a much higher response to A-CHO than uncoupled streptococci.

Table 2. ā Id1-induced helper function is exerted by T cells

Cells analyzed for[a] Strep.A sensitivity	Antigen	Anti-NIP[b] response	Anti-A-CHO[c] response	Idiotype[d] response
–	NIP-CG	74 (1.0)	< 2	n.d.[e]
–	NIP-Strep.A	5.6 (1.4)	4.1 (1.6)	n.d.
–	NIP-GPIgG	0.4	< 2	n.d.
1° āId1	NIP-GPIgG	1.6 (1.3)	< 2	n.d.
1° āId1	NIP-Strep.A	54.5 (1.3)	89 (1.2)	74 (1.2)
1° āId1, "B cells"	NIP-Strep.A	1.1 (1.2)	36 (1.2)	38 (1.4)
1° āId1, "T cells"	NIP-Strep.A	42.8 (1.4)	6.3 (1.6)	n.d.
1° Strep.A	NIP-Strep.A	32.8 (1.5)	172 (1.1)	48 (1.2)
1° Strep.A, "B cells"	NIP-Strep.A	2.9 (1.3)	85 (1.3)	18 (1.4)
1° Strep.A, "T cells"	NIP-Strep.A	44.4 (1.8)	10.6 (2.3)	n.d.

a) Hosts received 15 x 10^6 cells primed with NIP-CG and additional cells as indicated in the table. "B cells" stands for spleen cells treated with antiserum to Thy-1.2 and complement. "T cells" stands for spleen cells passaged through Degalan columns coated with MIg-anti-MIg complexes (cell recovery was 10 %). In the case of āId1-primed cells, each host received 25 x 10^6 untreated cells or 9 x 10^6 "B cells" or 10 x 10^6 "T cells". In the case of cells primed with A streptococci the corresponding cell numbers were 20 x 10^6 untreated cells, 6 x 10^6 "B cells" and 7.5 x 10^6 "T cells".
b) – d) see Table 1.
e) n.d. = Not done.

In accord with these data are the results obtained with column-passaged cells. The elimination of B cells, verified by the drastic reduction of anti-A-CHO titers, left helper activity untouched.

Helper function in this system is thus mediated by T cells and we accordingly conclude that the administration of āId1 confers specific immunity to both the T and B cell system in A/J mice.

3.2.2. Priming of C57L mice with anti-A5A idiotype

Can we achieve sensitization by anti-idiotypic antibody also in animals in which the idiotype usually represents only a minor fraction of the antigen-induced antibody population? For the investigation of this question we used strain C57L whose anti-A-CHO antibodies contain a minor component that carries an idiotype cross reactive with A5A ([15], see also below, Fig. 1). The results collected in Table 3 show that also in this mouse strain immunity to A streptococci can be induced by injection of āId1. Although the anti-NIP titers in this experiment were considerably lower than in the previous ones, we clearly detect priming of both T helper and B precursor cells in the adoptive transfer system, and again the sensitization of helper T cells with anti-idiotypic antibody appears to be at least as efficient as that with streptococcal vaccine.

The highly selective B cell priming by āId1 is demonstrated by the inhibition test depicted in Fig. 1, which compares the proportions of idiotype-bearing molecules in anti-A-CHO antibodies produced by C57L cells primed with āId1 (group 5/6 in table 3) and with Strep.A (group 7/8 in Table 3), respectively. This proportion is revealed by the distance of each inhibition curve to the standard inhibition curve by the homologous A5A antibody [3, 15]. Only 5 out of the 12 inhibition curves for each cell type are shown because many of them were practically superimposable. For comparison, high titer antisera from 5 C57L mice hyperimmunized with Strep.A were included in the experiment.

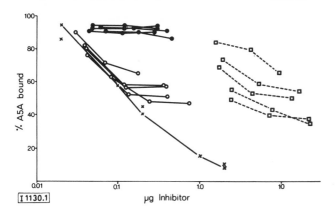

Figure 1. Idiotypic analysis of C57L antibodies to A streptococci in the binding inhibition assay. The curves represent inhibition of binding of radioactively labeled A5A to anti-idiotypic antibody (ordinate) by increasing amounts of unlabeled anti-streptococcal antibodies (abscissa). X——X Standard inhibition by the homologous A5A idiotype; ●——● C57L antibodies produced by cells which were primed with Strep.A (Table 3, group 7/8); o——o C57L antibodies produced by āId1-primed cells (Table 3 group 6/7); □———□ antibodies of C57L mice hyperimmunized with Strep.A.

Table 3. Sensitization of C57L mice with āId1

Group no.[a]	Cells analyzed for[b] Strep.A sensitivity	Antigen	Anti-NIP[c] response	Anti-A-CHO[d] response
1	–	NIP-CG	21.2 (1.1)	< 2
2	–	NIP-Strep.A	0.7 (2.0)	< 2
3	non-sensitized cells	NIP-Strep.A	1.0 –	< 2
4	non-sensitized cells	NIP-Strep.A	2.5 (1.5)	3.1 (1.6)
5	1° āId1	NIP-Strep.A	8.8 (1.5)	48 (1.3)
6	1° āId1	NIP-Strep.A	8.8 (1.4)	37 (1.4)
7	1° Strep.A	NIP-Strep.A	6.5 (1.3)	123 (1.2)
8	1° Strep.A	NIP-Strep.A	6.8 (1.3)	141 (1.1)

a) Groups of 6 animals.
b) Hosts received 20×10^6 cells primed with NIP-CG and additional cells (30×10^6) as indicated in the table. Groups receiving the same type of helper cells received these from different donors.
c) – d) see Table 1; (b–c) for results of idiotypic analysis see Fig. 1.

The data show that in the sera from hyperimmune animals a small proportion of the antibodies (1–5 %) carry an idiotype which partially cross-reacts with A5A. Because of its presence in low proportions we cannot expect, and in fact fail, to detect this idiotype in the relatively small quantity of antibody produced upon adoptive transfer of cells primed with Strep.A. In contrast, when āId1-primed cells are adoptively transferred, the vast majority if not all of the anti-A-CHO antibody molecules appear to carry A5A idiotypic determinants. Again, however, and very significantly, the antibodies only partially cross-react with the A5A idiotype.

3.3. Elimination patterns of āId1 and āId2

Induction of immunity to a circulating antigen results in its rapid clearance from the circulation. Likewise, if anti-idiotypic antibody induces the expression of idiotype in the animals, this process should be reflected in the elimination pattern of the anti-idiotypic antibody. This is indeed what we find. Fig. 2 shows the elimination curves for IgG_1 and IgG_2 immunoglobulins and anti-idiotypic antibodies in A/J mice.

āId1 is eliminated slowly during the first 10 days, but then rapidly to almost complete disappearence from the serum on

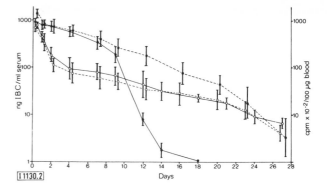

Figure 2. Elimination of guinea pig IgG_1 (closed circles) and IgG_2 (open circles). The IBC represents the anti-idiotypic antibody (solid lines) and the radioactivity represents total immunoglobulin (dotted lines). Radioactivity was corrected for decay of ^{125}I. In all cases geometric means and standard deviations of groups of 6 mice are given.

day 14. This sharp drop which is typical for immune elimination is not observed with the bulk of the IgG_1 fraction and therefore almost necessarily reflects induction and subsequent binding of idiotype. It is puzzling that immune elimination of the total IgG_1 fraction does not likewise occur. The reason for this may be trivial in that the total amount of IgG_1 which we inject may be sufficient to induce paralysis to those determinants which are shared by all Ig molecules in the mixture. However, the selective clearance of anti-idiotypic activity could also indicate that the sensitizing anti-idiotypic component in āId1 is itself not IgG_1 antibody. This interesting possibility is presently being investigated.

The situation is entirely different in the case of IgG_2, where both anti-idiotype and total immunoglobulin show similar elimination curves with rapid initial elimination of approximately 75 % of the material during the first two days, and subsequent slow clearance with a half life of approximately 12 days. Since previous work has indicated a suppressive rather than stimulatory effect of āId2 [3], we reach the satisfying conclusion that of the two classes of antiidiotypic antibody, āId1 contains an anti-idiotypic activity that induces specific immune clearance as would be expected from its ability to specifically sensitize the immune system.

4. Discussion

The present data establish that anti-idiotypic antibody can presensitize an animal for a secondary response to the antigen which is recognized by the corresponding idiotype. With respect to both T helper and B precursor function, cells from animals injected with āId1 behave exactly like cells from animals primed with the streptococcal antigen in the adoptive transfer system, with one important exception: In the case of priming with āId1 exclusively idiotypic antibody appears in the secondary response. Priming with antiidiotypic antibody is thus limited, at least in the B cell compartment, to precursor cells which express the idiotype. This clearly demonstrates the specificity of sensitization by āId1 and strongly argues against a contamination of our preparations of anti-idiotypic antibody with antigen. However, since this argument applies only to the situation in the B cell compartment, it should be pointed out that contamination by antigen is extremely unlikely for two additional reasons. Firstly, antigen contamination would have to be invoked selectively for āId1, since āId2, prepared from the same serum, showed suppressive rather than stimulatory activity [3, 25]. Secondly, the preparation of āId involves two ammonium sulfate precipitations and preparative agarose electrophoresis with several dialysis steps in between. Any residual contamination by the N-acetyl-glucosamine monosaccharide, which is used to elute the A5A antibody from the affinity column [7, 9] and which may have been injected into the guinea pig together with A5A, should be eliminated by these procedures. Furthermore, IgG$_1$ fractions from all of three anti-A5A antisera thus far studied were equally efficient.

The mechanism by which āId1 sensitizes B and T lymphocytes is at this stage a matter of speculation, but there is nothing to suggest that it should differ fundamentally from that of antigen-mediated sensitization. Even the elimination pattern of āId1 strikingly resembles classical immune elimination of antigen, and experiments in progress indicate that despite our previous failure (see Section 3.2.1.) and in accord with Trenkner and Riblet's data [5], anti-idiotypic antibody can under suitable conditions drive precursor B cells through the entire pathway of differentiation to the stage of antibody secretion.

The functional analogy between antigen and anti-idiotypic antibody can be extended even further. Like in the world of classical antigens, we encounter upon introduction of antiidiotypes the immune system's dual response pattern in that āId1 enhances and āId2 suppresses the immune response. It should be pointed out that āId2 has not yet been tested for its effect on T cells, but B cell immunity was clearly shown to be suppressed either directly or via the induction of suppressor cells [3, 25]. Attention has recently been drawn by Jerne to the possibility that idiotype-anti-idiotype interactions may be essential for the regulation of the immune system, which in the frame of his hypothesis is considered a network of interacting variable domains of immunoglobulin molecules [27]. It is intriguing to speculate that many of the various regulatory phenomena which have been described in antigen-mediated immunity may be rediscovered in the idiotypic network and that some of them will only there reveal their true physiological significance and nature. Strong support for this notion would be at hand if one would be able to assign enhancing and suppressive functions also to certain classes of *autogeneous* anti-idiotypic antibody, which is already known to arise under physiological conditions following idiotype induction [28, 29]. It is important to note in this context that, as pointed out in Section 3.3., the molecular identity of the sensitizing anti-idiotypic component in āId1 is not as yet established.

Since the sensitizing capacity of āId1 must almost necessarily relate to its ability to bind the idiotypic receptors on the lymphocyte surface, it is clear that the idiotype stimulation phenomenon offers a unique and powerful approach to the analysis, identification and, hopefully, isolation of receptor molecules in the immune system. Thus, the data presented here and those of Trenkner and Riblet [5] leave hardly any doubt that the precursors of idiotype-secreting B lymphocytes express the idiotype as functional receptor molecules. This is close to formally proving the basic assumption of selective theories of antibody formation, namely of those antibody molecules which they eventually secrete.

Receptor analysis by idiotype stimulation has a unique advantage in that it allows one to approach the question of whether a particular idiotypic receptor is represented in an animal's receptor repertoire, irrespective of whether the corresponding receptor-bearing precursor cells are ever sufficiently activated and expanded by antigenic stimulation alone in order to become detectable. This is exemplified in our experiment in the C57L strain, where aId1 drastically expands minor B cell clones expressing an idiotype which under similar experimental conditions remains undetectable in cells primed with antigen instead of āId1. We are now in the position, therefore, to investigate whether rarely occurring idiotypes are restricted to certain individuals of a given strain of mice or occur in all its members but on a minor fraction of the lymphocyte population. The answer to this question will be relevant for the problem of how antibody diversity is generated. Our results in the C57L strain show that these animals possess precursor cells which are precomitted to the synthesis of an idiotype *cross-reactive* with A5A, but lack precursors expressing A5A itself. In addition, in unpublished experiments we failed to induce A5A expression by āId1 in CBA, Balb/c and C57Bl/10 mice, which have been previously shown not to produce detectable A5A in the response to Group A streptococci. So far, therefore, our data support the view that the potential receptor repertoire of the immune system is strictly germ line controlled.

Most interesting, of course, is the finding that injection of āId1 leads to sensitization not only of antibody-forming B cell precursors, but also of T helper cells. The most straightforward interpretation of this result is to attribute idiotype-bearing receptor molecules also to T helper cells. If this turns out to be true, it would represent a powerful argument in favour of the view [20, 30] that T lymphocytes or at least T helper cells make use of the same recognition system as B cells, namely the products of the immunoglobulin variable genes. The results of Binz et al. point in the same direction in showing that anti-recognition site antibody reacts with humoral allo-antibody as well as with cells mediating graft versus host reaction [31, 32] and mixed lymphocyte reactions (personal communication).

Before such a conclusion can be reached, however, a number of alternative interpretations have to be excluded. Thus, one might argue that our A5A idiotype preparations may be contaminated with T cell-derived carbohydrate-binding receptor molecules and hence our anti-idiotypic sera with antiidiotypic- antibodies against these contaminating receptors. This rather remote possibility (and other even more remote ones) can be ruled out by establishing that the induction of T cell immunity by anti-idiotype is restricted to mouse strains which express the corresponding idiotype. So far our experiments support this notion. Anti-A5A idiotype efficiently sensitized C57L mice (Table 3), but was ineffective in CBA and Balb/c mice, which do not express the A5A idiotype [15]. Conversely, anti-idiotypic antibody against the BALB/c myeloma S117 (gift from Dr. M. Cohn), which specifically binds A-CHO [33], induced T and B cell immunity in BALB/c but not A/J mice, which in contrast to BALB/c do not express the S117 idiotype in their antibodies to A-CHO (Berek, Eichmann and Rajewsky, to be published). The experiments with the S117 idiotype are particularly relevant here, since the myeloma protein would hardly be expected to contain any T-cell-derived contamination. In addition, the strain-specificity of stimulation again virtually excludes contamination of āId1 with antigen.

More serious is another objection, namely that T cell sensitization may be indirect and mediated through B cell induction by anti-idiotypic antibody. There are essentially two alternative ways in which this could happen. One is that B cell immunity would render an internal antigen immunogenic, in analogy to the enhancement of the immune response by 19 S antibody [34, 35]. Against this interpretation are preliminary experiments in which (i) helper cell sensitization by anti-idiotypic antibody was achieved within 5 days *in vivo* and (ii) T helper cells primed with āId1 were efficiently and specifically inhibited by anti-idiotypic antibody and also by purified A-CHO *in vitro* (S. Black, K. Eichmann, G. Hämmerling and K. Rajewsky, unpublished).

The alternative view, namely that T cells pick up and make use of B cell-derived receptor molecules in the ways suggested by Crone et al. [36] and by Playfair [37], stands up against all our experimental evidence and in fact our data have no bearing on the question whether idiotypic T cell receptors are produced by the T cells themselves or passively acquired. It appears to us, however, that this question as well as the problem of the molecular nature of the T cell receptor can now be approached by direct experimentation.

Finally, we would like to mention the potential importance of the present system for applied immunology. We achieve efficient sensitization of our animals by a single injection of a minute quantity of anti-idiotypic antibody. The animals do not receive any antigen. Sensitization extends to both the T and B cell compartment, and we are presently exploring ways in which it could be limited to either to the two. Sensitization is also efficient in cases where the anti-idiotypic antibody is specific for an idiotype which occurs in low frequency. In certain instances anti-idiotypic antibody may thus serve as a new type of vaccine.

We are grateful to Dr. Rüdiger Mohr for carrying out T and B cell separation and for a gift of NIP-GPIgG and to Ms. Gertrud von Hensberg and Ms. Hannelore Wirges-Koch for able technical help. This work was supported by the Deutsche Forschungsgemeinschaft through Sonderforschungsbereich 74.

Received May 12, 1975.

5. References

1 Hart, D.A., Wang, A.L., Pawlak, L.L. and Nisonoff, A., *J. Exp. Med.* 1972. *135*: 1293.

2 Cosenza, H. and Köhler, H., *Proc. Nat. Acad. Sci. US* 1972. *69*: 2701.

3 Eichmann, K., *Eur. J. Immunol.* 1974. *4*: 296.

4 Strayer, S.D., Cosenza, H., Lee, W.M.F., Rowley, D.A. and Köhler, H., *Science,* in press.

5 Trenkner, E. and Riblet, R., *J. Exp. Med.*, submitted for publication.

6 Brownstone, A., Mitchison, N.A. and Pitt-Rivers, R., *Immunology* 1966. *10*: 481.

7 Krause, R.M., *Advan. Immunol.* 1970. *12*: 1.

8 Levy, H.B. and Sober, H.A., *Proc. Soc. Exp. Biol. Med.* 1960. *103*: 250.

9 Eichmann, K., *Eur. J. Immunol.* 1972. *2*: 301.

10 Eichmann, K. and Greenblatt, J., *J. Exp. Med.* 1971. *133*: 424.

11 Eichmann, K. and Kindt, T.J., *J. Exp. Med.* 1971. *134*: 532.

12 Greenwood, F.C., Hunter, W.M. and Glover, J.S., *Biochem. J.* 1963. *89*: 114.

13 Mitchison, N.A., *Eur. J. Immunol.* 1971. *1*: 10.

14 Rajewsky, K. and Pohlit, H., in Amos, B. (Ed.), *Progress in Immunology,* Academic Press, N.Y. and London 1971, p. 337.

15 Eichmann, K., *J. Exp. Med.* 1973. *137*: 603.

16 Reif, A.E. and Allen, J.M.V., *Nature* 1966. *209*: 521.

17 Raff, M.C., *Nature* 1970. *226*: 1257.

18 Rajewsky, K., Roelants, G.E. and Askonas, B.A., *Eur. J. Immunol.* 1972. *2*: 592.

19 Wigzell, H., Sundquist, K.G. and Yoshida, T.O., *Scand. J. Immunol.* 1972. *1*: 75.

20 Rajewsky, K. and Mohr, R., *Eur. J. Immunol.* 1974. *4*: 111.

21 Mitchison, N.A., *Eur. J. Immunol.* 1971. *1*: 18.

22 Eichmann, K. and Berek, C., *Eur. J. Immunol.* 1973. *3*: 599.

23 Eichmann, K., Tung, A.S. and Nisonoff, A., *Nature* 1974. *250*: 509.

24 Benacerraf, B., Ovary, Z., Bloch, K.J. and Franklin, E.C., *J. Exp. Med.* 1963. *117*: 937.

25 Eichmann, K., *Eur. J. Immunol.*, submitted for publication.

26 Mitchison, N.A., Rajewsky, K. and Taylor, R.B., in Šterzl, J. and Říha, I. (Eds.), *Developmental Aspects of Antibody Formation and Structure,* Publishing House of the Czechoslovak Academy of Science, Prague 1970. *II*: 547.

27 Jerne, N.K., *Ann. Immunol. (Inst. Pasteur)* 1974. *125c*: 373.

28 Kluskens, L. and Köhler, H., *Proc. Nat. Acad. Sci. US*, in press.

29 McKearn, T.J., Stuart, F.P. and Fitch, F.W., *J. Immunol.* 1974. *113*: 1876.

30 Schlossmann, S.F., *Transplant. Rev.* 1972. *10*: 97.

31 Binz, H., Lindenmann, J. and Wigzell, H., *J. Exp. Med.* 1974. *139*: 877.

32 Binz, H., Lindenmann, J. and Wigzell, H., *J. Exp. Med.* 1974. *140*: 731.

33 Vicari, G., Sher, A., Cohn, M. and Kabat, E., *Immunochemistry* 1970. *7*: 829.

34 Henry, C. and Jerne, N.K., *J. Exp. Med.* 1968. *128*: 133.

35 Dennert, G., Pohlit, H. and Rajewsky, K., in Cross, A., Kosunen, T. and Mäkelä, O. (Eds.), *Cell Interactions and Receptor Antibodies in Immune Responses.* Academic Press, New York 1971, p. 3.

36 Crone, M., Koch, C. and Simonson, M., *Transplant. Rev.* 1972. *10*: 36.

37 Playfair, J.H.L., *Clin. Exp. Immunol.* 1974. *17*: 1.

IDIOTYPIC ANALYSIS OF LYMPHOCYTES IN VITRO

I. Specificity and Heterogeneity of B and T Lymphocytes Reactive with Anti-Idiotypic Antibody*

By S. J. BLACK,‡ G. J. HÄMMERLING, C. BEREK, K. RAJEWSKY, AND K. EICHMANN

(From the Institute for Genetics, University of Cologne, Cologne, Federal Republic of Germany)

Previous work from this laboratory has shown that passively administered anti-idiotypic antibody (anti-Id),[1] under certain experimental conditions, sensitizes mice to the antigen which is recognized by the corresponding idiotype (Id). This sensitization was shown to occur in T-helper as well as in B-precursor lymphocytes, suggesting that antigen receptors on both types of cells possess the same or similar idiotypes, and that the interaction of these receptors with anti-Id is functionally equivalent to that with antigen (1).

This work employed antibodies to Group A streptococcal carbohydrate (A-CHO), generated in strain A/J mice by immunization with Group A streptococcal vaccine (Strep.A) (2). A major portion of these antibodies represents a single antibody species which is defined by isoelectric-focusing spectrum (3) and by Id (2–4). The Id is detected by guinea pig anti-Id raised against the anti-A-CHO antibody produced by the A/J lymphocyte clone A5A (2–4). The expression of the A5A Id is controlled by a single gene within the Ig-1 complex of A/J mice (5). Anti-Id against the A5A Id can be fractionated into the IgG1 and IgG2 classes of guinea pig IgG (3, 6, 7). Whereas anti-Id belonging to the IgG2 class was shown to be suppressive (3, 6), stimulation of immunity was achieved with anti-Id of the IgG1 class (anti-A5A Id1) (1).

In this paper, we introduce an additional idiotypic system which employs a myeloma protein rather than an induced antibody for the production of anti-Id. The transplantable BALB/c plasmacytoma S117 secretes an IgA/k immunoglobulin with specificity for N-acetyl-glucosamine, the major antigenic determinant of A-CHO (8). Guinea pig anti-Id against this myeloma protein reacts with the antibodies of BALB/c mice immunized with Strep.A, and genetic analysis demonstrated that the expression of the S117 idiotypic marker is controlled by a gene within the Ig-1 complex of strain BALB/c (reference 9, and footnote 2). Unpublished experiments in vivo had indicated that passively administered anti-S117 Id of the IgG1 class (anti-S117 Id1) primed BALB/c mice to produce a secondary

---

* This work was supported by the Deutsche Forschungsgemeinschaft through Sonderforschungsbereich 74.

‡ Supported by a Royal Society (London) European Fellowship.

[1] *Abbreviations used in this paper:* A-CHO, Group A streptococcal carbohydrate; anti-Id, anti-idiotypic antibody; anti-Id1 and anti-Id2, anti-Id of the IgG1 and IgG2 classes of guinea pig IgG, respectively; BSS, balanced salt solution; C-CHO, Group C streptococcal carbohydrate; IBC, idiotype-binding capacity; Id, idiotype; NGPS, normal guinea pig serum; PFC, plaque-forming cells; Strep.A, Group A streptococcal vaccine; TNBS, 2,4,6-trinitrobenzene sulfonic acid; TNP, 2,4,6-trinitrophenyl; V, variable.

[2] Berek, C., and K. Eichmann. Manuscript in preparation.

anti-A-CHO response upon subsequent challenge with Strep.A, suggesting that induction of specific immunity had occurred.

Employing both these idiotypic systems for the in vitro analysis of lymphocytes primed with anti-Id, we investigated the specificity and the heterogeneity of Id-bearing T and B cells, as well as several possible objections raised against the interpretation of our previous in vivo experiments (1). The results suggest: (*a*) Antigen contamination is not responsible for the sensitization of lymphocytes with anti-Id. (*b*) Anti-Id produced against a myeloma protein sensitizes T-helper cells as effectively as does anti-Id against an induced antibody. This eliminates a possible role of antibodies to T-cell receptors in T-cell sensitization. (*c*) T and B lymphocytes primed with the same anti-Id possess specificity for the same carbohydrate determinant. (*d*) The restriction in the heterogeneity of the lymphocyte populations primed with anti-Id, as compared to that of lymphocytes primed with Strep.A, is similar for T and B cells. A brief account of these experiments has been previously published (10). Taken together, these data strongly suggest that the antigen recognition systems of B- and T-helper cells are alike with respect to heterogeneity, specificity, and idiotypic determinants. In the accompanying paper (11), we will present evidence that the same genes may participate in the coding for T- and B-cell receptor variable (V) regions.

Materials and Methods

Animals. 6- to 8-wk-old A/J mice were purchased from The Jackson Laboratory, Bar Harbor, Maine; and BALB/c mice were obtained from Zentralinstitut für Versuchstierzucht, Hannover, West Germany. All animals were vaccinated against ectromelia and rested at least 4 wk before being used in experiments. Guinea pigs were obtained from a local breeder.

Antigens. Sheep erythrocytes (SRBC) were purchased from Behringwerke, Marburg-Lahn, West Germany. Haptenation of SRBC with 2,4,6-trinitrobenzene sulfonic acid (TNBS) was performed by the method of Rittenberg and Pratt (12). The preparation of the stearoyl derivative of A-CHO and its coupling to SRBC were done by the method of Pavlovskis and Slade (13). Purified A-CHO and Group C streptococcal carbohydrate (C-CHO) were a gift from Dr. R. M. Krause, The Rockefeller University, New York.

Group A streptococci, strain J17A4, were also a gift from Dr. R. M. Krause and the preparation of the vaccine (Strep.A) was done as previously described (14). The haptenation of Group A streptococci with TNBS was achieved by a modification of the Rittenberg and Pratt technique (12). A 200 mg portion of TNBS was dissolved in 42 ml of cacodylate buffer. 10 ml of cacodylate buffer containing 1.5×10^{11} Strep.A particles were added under rapid mixing and the suspension was then stirred slowly at room temperature for 30 min. Coupled particles (TNP-Strep.A) were centrifuged for 10 min at 300 g, the supernate was decanted and replaced by 30 ml of cacodylate buffer containing 50 mg of glycyl-glycine. The particles were resuspended, centrifuged, and the supernate decanted as before. This procedure was followed by four washes in 50 ml of 0.15 M saline. The haptenation reaction was conducted sterilely in foil wrapped glass vessels.

Idiotypes and Anti-Idiotypic Antibodies. The preparation of antibody A5A from the sera of irradiated A/J mice reconstituted with cloned spleen cells has been described (2). Anti-Id antibodies to antibody A5A were prepared in guinea pigs and fractionated into their IgG1 and IgG2 classes by previously described methods (2, 3). The S117 myeloma protein was isolated from the sera or ascites of mice bearing myeloma tumor S117, using *p*-amino-phenyl-*N*-acetyl-glucosamine coupled to Sepharose 2B as an immunoadsorbent (15, 16). Anti-Id antibodies to S117 and their IgG1 and IgG2 fractions were prepared as described for antisera to A5A (2, 3), except that the anti-S117 sera were adsorbed on Sepharose 2B-coupled IgA/k myeloma proteins TEPC15 and J539. Details on the S117 idiotypic system will be published elsewhere.[2] The Id-binding capacity (IBC) of anti-Id sera and of IgG fractions purified from them was determined at 10^{-8} M radioiodinated Id as previously described (4).

Immunization. Priming was done by intraperitoneal injection of 1×10^9 Strep.A organisms or 0.1 μg IBC of anti-A5A Id1, which was found as effective as the intravenous injections used previously (1). Except for a few experiments in which two priming injections were given at a 2 wk interval, a single injection was found satisfactory for both Strep.A and anti-A5A Id1. Priming with anti-S117 Id1 was done as with anti-A5A Id1. To induce SRBC priming, 10^8 erythrocytes were injected intraperitoneally. Mice immunized with anti-A5A Id1, anti-S117 Id1, or Strep.A were rested at least 4 wk before being used in experiments. Mice immunized with SRBC were used 9 days later for culture experiments.

Cell Preparations. Mice were killed by cervical dislocation and their spleens were removed sterily and immersed in cold sterile balanced salt solution (BSS). The spleens were then disrupted by teasing and the cell suspensions were allowed to sediment on ice. After 10 min, supernates were separated from sedimented cell debris, centrifuged, and the cell pellets were resuspended and washed in BSS.

Enriched T-cell preparations (85–95% Thy 1.2 positive) were obtained by passage of spleen cells through nylon wool columns according to the method of Julius et al. (17). Spleen cells were depleted of T lymphocytes by treatment with an AKR anti-CBA anti-Thy 1.2 serum and agarose-absorbed (18) guinea pig complement. The characteristics of the anti-Thy 1.2 serum and the methods for T-cell depletion have been described (19). The viability of treated cells was examined by trypan blue exclusion to determine the efficency of specific killing. Then the cells were centrifuged and washed three times in BSS before use in culture experiments.

Lymphocyte Microculture. Lymphocyte microcultures were performed in Micro-Test II tissue culture trays (no. 3040; Falcon Plastics, Div. of BioQuest, Oxnard, Calif.) using a modification of the Mishell and Dutton technique (20). The culture medium, nutritional cocktail, and gas mixture were the same as used in the standard macroculture system (20). Individual wells of a Micro-Test II tissue culture tray were seeded as follows: Antigen was added in 50 μl of medium, followed by 50 μl of medium containing 10^6 spleen cells. For inhibition experiments the 50 μl of cells were added first and subsequently the inhibitor was added in 10 μl of medium. The mixtures were incubated at room temperature for 20 min. Then antigen, in 50 μl of medium, was added.

Inhibition experiments employed purified A-CHO and C-CHO, as well as anti-idiotypic anti-sera and normal guinea pig serum (NGPS) as control. All guinea pig sera were absorbed twice with 10% (vol/vol) normal spleen cells for 30 min at 4°C and centrifuged for 30 min at 17,000 g before use. All inhibitors were present for the entire culture period.

Cultures were maintained without rocking at 37°C in a humid atmosphere of 10% CO_2, 8% O_2, and 82% N_2, and were supplemented on each day with 10 μl of nutritional cocktail. After 4 days the Micro-Test II trays were centrifuged at 100 g for 10 min. The culture medium was removed by suction and replaced with 100 μl of BSS. Each culture well was then assayed for direct and indirect plaque-forming cells (PFC) using the local hemolysis in gel assay.

Local Hemolysis in Gel Assay. Antibody-secreting cells were detected by the plaque assay of Jerne et al. (21) as modified by Mishell and Dutton (20), using either A-CHO-SRBC or TNP-SRBC, or both. Attempts were made to detect IgG PFC according to the method of Dresser and Wortis (22) using as developing serum a rabbit antimouse Ig serum of proven ability. There were no anti-TNP IgG-secreting PFC detected in our cultures and we also consistently failed to detect indirect PFC specific for A-CHO even though such cells would have been expected to be generated in culture. This observation agrees with findings of other workers (23, 24). Thus, only the number of direct PFC are presented in this paper. The specificity of PFC was ascertained by plaquing against uncoupled SRBC or by addition of various inhibitors including TNP-bovine serum albumin, A-CHO, or anti-Id, to the agarose gel.

Results

Precursor and Helper Activity In Vitro of Cells Primed In Vivo with Strep.A or with Anti-Id. The experiments were designed in such a way that priming of both B and T lymphocytes in vivo could be detected by challenge of spleen cells with the appropriate antigen in vitro. Priming of B precursor cells was detected by challenge with Strep.A and plaquing against A-CHO-SRBC. Priming of T-helper cells was detected by challenge with TNP-Strep.A and plaquing against TNP-SRBC.

Fig. 1 shows the results of experiments performed to assess the optimum dose

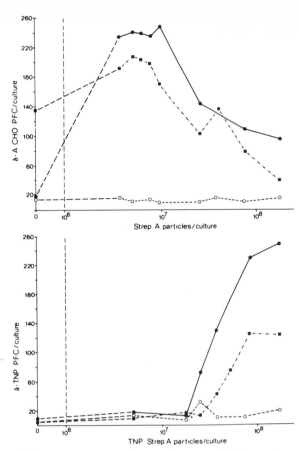

FIG. 1. In vitro PFC response to A-CHO (upper frame), to TNP (lower frame) of normal A/J spleen cells (O---O), and of spleen cells from A/J mice that had been primed with Strep.A (■---■) or anti-A5A Id1 (●——●). Cells were challenged with increasing concentrations of Strep.A particles (upper frame) or Strep.A particles coupled with TNP (lower frame).

for the in vitro challenge with Strep.A (upper frame) and TNP-Strep.A (lower frame), respectively. Spleen cells from normal A/J mice and from A/J mice that had received two intraperitoneal injections of 1×10^9 Strep.A particles or two intraperitoneal injections of 0.1 μg IBC of anti-A5A Id1, respectively, were analyzed in these experiments. As can be seen from the dose-response curves in Fig. 1, cells from mice primed with Strep.A and with anti-A5A Id1 produced substantial numbers of PFC to A-CHO as well as to TNP when challenged in vitro with the appropriate antigen. Spleen cells from unprimed mice failed to respond to either antigen at any of the antigen concentrations tested. The data clearly show that priming with anti-A5A Id1 generates precursor cells for a PFC response to A-CHO as well as helper activity for the response to a hapten that is coupled to Strep.A. Priming with anti-Id is at least as effective as priming with Strep.A. It is apparent that the optimal particle concentration for the induction of anti-TNP PFC is an order of magnitude greater than that for the induction of anti-A-CHO PFC. Similarly, cultures stimulated with TNP-Strep.A for an optimal anti-TNP PFC response produced hardly any anti-A-CHO PFC (data not shown). Therefore, precursor and helper activity were analyzed in separate experiments throughout this study.

In order to verify that helper function was indeed exerted by T lymphocytes, normal, Strep. A-, or anti-A5A Id1-primed spleen cells were treated either with normal mouse serum and complement or with anti-Thy-1.2 serum and complement before being exposed to TNP-Strep.A in culture. The results from a representative experiment are presented in Table I. Treatment with anti-Thy-1.2 serum and complement completely abolished the ability of all spleen cell preparations to generate TNP-specific PFC on challenge with TNP-Strep.A. Normal mouse serum and complement had little effect on this response. A more direct demonstration that T lymphocytes acted as helper cells in our culture system came from experiments in which nylon wool-enriched T lymphocytes were used. In order to substitute for B cells, untreated spleen cells or anti-Thy-1.2 and complement-treated spleen cells from normal mice were added to the cultures. Control cultures contained normal spleen cells, unfractionated primed spleen cells, or nylon wool-enriched T cells alone. A typical result is presented in Table II. It is evident that unfractionated primed spleen cells and nylon wool-enriched primed spleen cells were equally effective as helpers in the response to TNP-Strep.A. Pretreatment of normal spleen cells with anti-Thy-1.2 serum and complement before mixing with nylon wool-enriched T cells did not reduce the responsiveness of mixed populations to challenge with TNP-Strep.A. Taken together the experiments presented in Tables I and II show that the in vitro anti-TNP PFC response to conjugates of TNP-Strep.A is dependent on the presence of helper T cells primed to Strep.A. The results of this in vitro analysis are essentially in accord with that previously obtained from in vivo adoptive transfers (1), in showing that the helper activity of spleen cells from mice primed

TABLE I

The T-Cell Dependency of the In Vitro Anti-TNP PFC Responses Generated by Strep.A and Anti-A5A Id1-Primed A/J Spleen Cells as Shown by T-Cell Depletion

Priming antigen	Pretreatment of cells	Anti-TNP PFC/culture*
—	—	3 (1.34)
—	NMS + C'‡	3 (1.52)
—	Anti-Thy 1.2 + C'	4 (1.23)
Strep.A	—	175 (1.10)
Strep.A	NMS + C'	108 (1.14)
Strep.A	Anti-Thy 1.2 + C'	7 (1.66)
Anti-A5A Id1	—	152 (1.23)
Anti-A5A Id1	NMS + C'	123 (1.10)
Anti-A5A Id1	Anti-Thy 1.2 + C'	5 (1.28)

* Geometric mean of direct PFC per culture (with standard deviation coefficient). Each result is derived from observations on eight microcultures and is corrected for background anti-SRBC PFC. The dose of TNP-Strep.A in this experiment and in all of the following experiments was 1.5 × 10⁸ particles per culture.

‡ NMS + C', normal mouse serum plus guinea pig complement.

TABLE II

The T-Cell Dependency of the In Vitro Anti-TNP PFC Responses Generated by Strep.A and Anti-A5A Id1-Primed A/J Spleen Cells as Shown by T-Cell Enrichment

Cells in culture		Anti-TNP PFC/culture§
Primed*	Unprimed‡	
—	10^6	17 (1.37)
Strep.A, 10^6	—	282 (1.08)
Strep.A, 5×10^5	10^6	203 (1.11)
Strep.A; T cells, 5×10^5‖	10^6	192 (1.14)
Strep.A; T cells, 1.5×10^6	—	4 (1.70)
—	10^6	82 (1.12)
Anti-A5A Id1, 10^6	—	550 (1.23)
Anti-A5A Id1, 5×10^5	10^6	532 (1.17)
Anti-A5A Id1; T cells, 3.2×10^5	10^6	397 (1.15)
Anti-A5A Id1; T cells 10^6	—	1 (1.11)
—	Anti-Thy 1.2 + C', 10^6	14 (1.26)
Anti-A5A Id1; T cells, 4×10^5	Anti-Thy 1.2 + C', 6×10^5	410 (1.07)

* Spleen cells from A/J mice immunized with Strep.A or anti-A5A Id1 as indicated.
‡ Spleen cells from normal A/J mice.
§ See footnotes Table I.
‖ T-cell-enriched population obtained by passage of spleen cells over a nylon wool column.

with anti-A5A Id1 is exerted by T lymphocytes specific for streptococcal antigens.

Specificity and Heterogeneity of Primed Cell Populations. In previous experiments it had been shown that A/J spleen cells primed with anti-A5A Id1 exclusively produced A5A-Id-positive antibodies when transferred into an adoptive host and challenged with Strep.A (1). This is in contrast to cells primed with Strep.A, which produced antibodies only about 25% of which were A5A-Id positive. This result indicated that anti-A5A Id1 specifically activated precursor B cells expressing the A5A Id. In addition to providing data on B cells, the present in vitro system enabled us to assess and compare the specificities of both B- and T-lymphocyte populations.

For the study of B cells, Strep.A- and anti-A5A Id1-primed cells were challenged in vitro with Strep.A and analyzed for A-CHO-specific PFC. Each set of cultured spleen cells was divided into four portions; one was plaqued normally against the A-CHO-SRBC, the second was plaqued in the presence of 20 μg A-CHO, the third was plaqued in the presence of 0.5 μg IBC of anti-A5A serum, and the last was plaqued in the presence of an equivalent dilution of NGPS. Results from a representative experiment are presented in Table III. Addition of A-CHO inhibited the PFC of both Strep.A- and anti-A5A Id1-primed cells by about 80%. NGPS at a final dilution of 1:250 had little effect on the appearance of A-CHO-specific PFC. In contrast, anti-A5A serum at the same dilution reduced the number of PFC generated by Strep.A-primed spleen cells by approximately 40% and the number of PFC generated by anti-A5A Id1-primed cells by 94%. In other experiments Strep.A-primed spleen cells produced as few as 10% A5A-

TABLE III

The Idiotypic Specificity of Precursor Cell Responses to A-CHO Generated by Strep.A and Anti-A5A Id1-Primed A/J Spleen Cells In Vitro, as Shown by Inhibition of Plaque Formation

Priming antigen	PFC-inhibitor*	Anti-A-CHO PFC/culture‡
Strep.A	−	169 (1.04)
Strep.A	20 μg A/CHO	34 (1.11)
Strep.A	1:250 NGPS	120 (1.33)
Strep.A	0.5 μg IBC anti-A5A	100 (1:08)
Anti-A5A Id1	−	249 (1.08)
Anti-A5A Id1	20 μg A-CHO	57 (1.21)
Anti-A5A Id1	1:250 NGPS	204 (1.08)
Anti-A5A Id1	0.5 μg IBC anti-A5A	6 (1.00)

* Inhibitors were added to the agarose solution upon performing the plaque test, not to the culture medium. Figures indicate total amounts per 0.5 ml agarose.

‡ Geometric mean of direct PFC per culture (with standard deviation coefficient). Each result is derived from observations on eight cultures and is corrected for background anti-SRBC PFC.

positive PFC, whereas anti-A5A Id1-primed cells invariably generated more than 90% A5A-positive PFC. It thus seems that immunization with anti-A5A Id1 can drastically expand the A5A-positive PFC precursor pool. These data are in strict accordance with that obtained in vivo (1).

To study the specificity and heterogeneity of helper T cells, Strep.A- and anti-A5A Id1-primed spleen cells were challenged in vitro with TNP-Strep.A and analyzed for TNP-specific PFC. Inhibition experiments analogous to those performed to study B-cell specificity were done, but now the inhibitors were included into the culture medium in order to inhibit helper activity. Cultures in which inhibitions were performed were inoculated with: 0.1 or 0.01 μg of A-CHO or of non-cross-reactive C-CHO, or with 0.1, 0.05, or 0.01 μg IBC of anti-A5A serum or with equivalent dilutions of NGPS. The results from a typical experiment are presented in Table IV and are representative of those obtained in several separate trials.

The data show that neither of the primed helper cell populations was inhibited by C-CHO. In contrast, A-CHO, at both concentrations tested, had a strong inhibitory effect on the response of anti-A5A Id1-primed cells and, to a much lesser extent, on Strep.A-primed cells. A similar but more pronounced difference between Strep.A and anti-A5A Id1-primed helper cells was seen upon inhibition with anti-Id. Pronounced inhibition of anti-A5A Id1-primed helper cells is achieved with anti-A5A serum, whereas the effect of this antiserum on Strep.A-primed helper cells was approximately 10 times less. The PFC numbers of cultures that had contained anti-Id have to be compared to that of the cultures with equivalent dilutions of NGPS. As can be seen, NGPS had a slight toxic effect on the cultures, which was, however, nonspecific and could be readily distinguished from the inhibitory effect of anti-Id. Not shown here are the results of control experiments in which the anti-A5A antiserum was tested for inhibition of the anti-TNP PFC response of SRBC-primed cells. No inhibition was found in these experiments, even with 0.1 μg IBC of anti-A5A serum.

TABLE IV

The Idiotypic Specificity of Helper T-Cells Generated in A/J Mice by Immunization with Strep.A or Anti-A5A Id1, as Shown by Inhibition of Carrier Recognition In Vitro

Priming antigen	Inhibitor*	Anti-TNP PFC/culture‡
—	—	5 (1.42)
Strep.A	—	203 (1.11)
Strep.A	0.1 µg C-CHO	312 (1.21)
Strep.A	0.01 µg C-CHO	215 (1.14)
Strep.A	0.1 µg A-CHO	133 (1.22)
Strep.A	0.01 µg A-CHO	231 (1.06)
Strep.A	NGPS 1:250	108 (1.11)
Strep.A	NGPS 1:500	151 (1.03)
Strep.A	NGPS 1:2,500	185 (1.21)
Strep.A	0.1 µg IBC anti-A5A	54 (1.17)
Strep.A	0.05 µg IBC anti-A5A	133 (1.22)
Strep.A	0.01 µg IBC anti-A5A	176 (1.12)
—	—	3 (1.27)
Anti-A5A Id1	—	186 (1.13)
Anti-A5A Id1	0.1 µg C-CHO	215 (1.15)
Anti-A5A Id1	0.01 µg C-CHO	183 (1.09)
Anti-A5A Id1	0.1 µg A-CHO	39 (1.24)
Anti-A5A Id1	0.01 µg A-CHO	89 (1.18)
Anti-A5A Id1	NGPS 1:250	109 (1.12)
Anti-A5A Id1	NGPS 1:500	169 (1.21)
Anti-A5A Id1	NGPS 1:2,500	201 (1.12)
Anti-A5A Id1	0.1 µg IBC anti-A5A	11 (1.34)
Anti-A5A Id1	0.05 µg IBC anti-A5A	31 (1.40)
Anti-A5A Id1	0.01 µg IBC anti-A5A	59 (1.61)

* Purified A-CHO, C-CHO, anti-A5A serum, or NGPS added to the cultures at the concentration indicated. All figures given represent the final concentration of each inhibitor per culture. The anti-A5A serum used had an IBC of 250 µg/ml.

‡ See footnotes Table I.

These data clearly indicate that anti-A5A Id1-primed helper cells differ from Strep.A-primed helper cells in that the former consist essentially of A5A Id-positive cells, whereas the latter are heterogeneous and consist of a mixture of Id-positive and Id-negative cells. Thus, anti-A5A Id1 expands a specific A5A-positive helper T-cell population, analogous to the expansion of A5A-positive B cells shown above.

Analysis of Helper T Cells Carrying the Id of Myeloma Protein S117. To investigate the question whether the anti-Id antibodies that activate and inhibit T cells are indeed directed against immunoglobulin idiotypes, we used the BALB/c myeloma protein S117, which binds A-CHO and whose Id is expressed on induced antibodies of BALB/c mice immunized with Strep.A. In contrast to antigen-induced antibodies, a myeloma protein is not expected to contain any T-cell-derived receptor molecules, and an anti-Id produced against a purified myeloma protein can be expected to be exclusively directed to immunoglobulin idiotypes.

Cells from normal BALB/c and from BALB/c mice injected with Strep.A or with anti-S117 Id1 were analyzed in experiments analogous to those performed for the A5A Id in A/J mice. The specificity of the ensuing helper T-cell response was measured by addition of 0.01 μg IBC of anti-S117 serum to cultures containing either of the primed spleen cell populations. Control cultures received an appropriate dilution of NGPS. Cultures were challenged with TNP-Strep.A and analyzed for anti-TNP PFC.

The results in Table V show that spleen cells from unimmunized BALB/c mice failed to generate TNP-specific PFC in culture, whereas spleen cells harvested from Strep.A- or anti-S117 Id1-primed mice produced a substantial TNP-specific PFC response when challenged with TNP-Strep.A. Addition of 0.01 μg IBC of anti-S117 serum to the cultures completely abolished the response of anti-S117 Id1-primed cells, but did not reduce the response of Strep.A-primed cells any more than did NGPS in equivalent dilution. This is in agreement with results on the antibody response of BALB/c mice to Strep.A, which is characterized by a small proportion of S117 Id-positive antibodies (C. Berek and K. Eichmann, unpublished observation). Thus, as in the A5A system of strain A/J, the heterogeneity of A-CHO-binding T-helper cells in BALB/c mice closely reflects the heterogeneity of the corresponding B-cell population. Furthermore, as pointed out above, the anti-Id that react with T-helper cells in the S117 system are unequivocally directed against immunoglobulin idiotypes.

Strain Specificity of T-Cell Idiotypes. The phenotypic polymorphism observed for the A5A and S117 idiotypes in strain A/J and BALB/c antibodies to A-CHO reflects a genetic polymorphism in the genes that encode the V regions of the heavy chains of these antibodies (10). A similar polymorphism on the level of the helper T cells of the two strains would be a further indication for the molecular identity of T- and B-cell receptor V regions. Furthermore, objections such as the contamination of the anti-Id with antigen, or the possible cross-reactivity of the anti-Id with Strep.A could be dismissed.

TABLE V

The Anti-TNP PFC Responses of Normal, Strep.A, and Anti-S117 Id1 Primed BALB/c Spleen Cells Challenged In Vitro with TNP-Strep.A

Priming antigen	Inhibitor	Anti-TNP PFC/culture*
–	–	3 (1.11)
Strep.A	–	145 (1.06)
Strep.A	1:400 NGPS	64 (1.14)
Strep.A	0.01 μg IBC anti-S117‡	57 (1.12)
Anti-S117 Id1	–	97 (1.06)
Anti-S117 Id1	1:400-NGPS	61 (1.12)
Anti-S117 Id1	0.01 μg IBC anti-S117	4 (1.35)

* See footnotes Table I.

‡ The anti-S117 serum used had an IBC of 40 μg/ml.

To analyze idiotypic polymorphism on the helper T-cell level, A/J and BALB/c mice were primed in a crisscross fashion with Strep.A, anti-A5A Id1 or anti-S117 Id1. Their spleen cells were challenged in culture with TNP-Strep.A and 4 days later the number of TNP-specific PFC in each culture was measured. Some cultures of each of the homologously primed spleen cells, i.e. anti-A5A Id1-primed A/J spleen cells and anti-S117 Id1-primed BALB/c spleen cells, received anti-A5A or anti-S117 serum to examine the specificity of priming, and control cultures received NGPS. The results of these experiments are presented in Table VI.

It is apparent from the data that although helper cells were elicited in both strains by priming with Strep.A, only strain A/J produced helper cells in response to priming with anti-A5A Id1 and only strain BALB/c produced helper cells in response to priming with anti-S117 Id1. Furthermore, the response of anti-A5A Id1-primed A/J cells was inhibited by addition of anti-A5A but not anti-S117 serum to the cultures. Conversely, the response of anti-S117 Id1-primed BALB/c spleen cells could be inhibited by addition of anti-S117 but not anti-A5A serum to the cultures. It is clear from the data that helper cell priming as well as inhibition of the helper function by anti-Id reveal the same idiotypic polymorphism as has been observed for the corresponding antibodies.

TABLE VI

The Anti-TNP PFC Responses of A/J and BALB/c Spleen Cells Primed with Strep.A, anti-A5A Id1, or Anti-S117 Id1

Mouse strain	Priming antigen	Inhibitor	Anti-TNP PFC/culture*
A/J	Strep.A	—	179 (1.03)
	Anti-A5A Id1	—	171 (1.08)
	AntiS117 Id1	—	13 (1.29)
	—	—	11 (1.11)
BALB/c	Strep.A	—	94 (1.17)
	Anti-A5A Id1	—	9 (1.74)
	Anti-S117 Id1	—	84 (1.19)
	—	—	7 (1.49)
A/J	Anti-A5A Id1	—	133 (1.16)
	Anti-A5A Id1	0.01 μg IBC anti-A5A‡	28 (1.10)
	Anti-A5A Id1	1:2,500 NGPS	142 (1.15)
	Anti-A5A Id1	0.01 μg IBC anti-S117	83 (1.32)
	Anti-A5A Id1	1:400 NGPS	79 (1.41)
BALB/c	Anti-S117 Id1	—	103 (1.21)
	Anti-S117 Id1	0.01 μg IBC anti-A5A	147 (1.51)
	Anti-S117 Id1	1:2,500 NGPS	162 (1.16)
	Anti-S117 Id1	0.01 μg IBC anti-S117	9 (1.21)
	Anti-S117 Id1	1:400 NGPS	69 (1.31)

* See footnotes Table I.

‡ The anti-A5A serum had an IBC of 250 μg/ml and the anti-S117 serum used had an IBC of 40 μg/ml.

Discussion

The present data confirm our previous observations (1) in showing that anti-Id can presensitize an animal for a secondary response to the antigen which is complementary to the corresponding Id. B cells primed with anti-Id to an A-CHO-binding immunoglobulin generated a strong anti-A-CHO PFC response when challenged with Strep.A in vitro. T cells primed in the same way were able to act as helper cells in the primary immune response to TNP upon challenge in vitro with TNP-Strep.A. Since the magnitude of the responses of cells primed with anti-Id was equal to and, on occasions, exceeded that of cells primed with Strep.A, it appears that priming with anti-Id is as effective as priming with antigen.

The present data extend the previous observations (1), mainly in two ways: (a) The in vitro system afforded a detailed analysis of the helper T-cell population that has not been possible in vivo, and (b) using the S117 idiotypic system as a counterpart to the A5A idiotypic system it was possible to rule out a number of trivial explanations for the stimulation of lymphocytes with anti-Id. These explanations included the suggestion that our anti-Id preparations were contaminated with antigen, that there exists a cross-reaction between idiotypic determinants of the anti-Id and some streptococcal antigen, and that our anti-Id antisera may contain antibodies to T-cell receptor molecules that may have been copurified with the original A5A antibody and, accordingly, were injected into the guinea pigs together with it. Whereas the former had been considered as rather remote possibilities (1), the latter was taken as a true alternative to account for T-cell activation by anti-Id. However, the experiments that employed the Id of myeloma protein S117 reveal that the anti-Id that interact with T cells are indeed directed against immunoglobulin idiotypes. Since an immunoglobulin secreted by a plasmacytoma tumor is not expected to contain any T-cell-derived receptor molecules, the possibility of anti-T-cell receptor antibodies contaminating the anti-Id antisera is ruled out at least for the S117 system. Furthermore, the strain specificity of lymphocyte sensitization by anti-Id completely eliminates the possibility of antigen contamination as well as the possible cross-reaction between the anti-Id and some streptococcal antigen.

The mechanism by which anti-Id sensitizes B and T lymphocytes is at this stage a matter of speculation. The association of the stimulatory capacity with the IgG1 fraction of anti-Id has been confirmed in unpublished experiments in which IgG2 fractions, isolated from the same sera, were found nonstimulatory at any of the doses (10 ng to 1 μg IBC) at which stimulation was achieved with the IgG1 fractions. However, IgG1 fractions were found to contain small amounts (<1% of the total IBC) of IgM anti-Id, so that the stimulatory component may not be the IgG1 anti-Id itself. Thus, the requirements for lymphocyte stimulation by anti-Id are poorly understood. No doubt, however, exists that at some stage this process involves the binding of anti-Id to the antigen receptors on the lymphocyte surface. For the helper T cells, this binding is also demonstrated by the inhibition of helper function by anti-Id in vitro. The functional consequences of the binding of anti-Id to antigen receptors imply a functional role in addition to the mere physical presence of Id-bearing molecules on lymphocytes, particularly on helper T cells.

Whereas in the previously reported experiments (1) the specificity and heterogeneity of primed B cells could be studied by idiotypic analysis of antibody products, it was impossible to investigate equivalent properties of the primed helper cell population in vivo. This we have now done in vitro using classical carrier inhibition analysis to ascertain the specificity of T cells primed with antigen and with anti-Id. The results of these experiments clearly revealed that T-helper cells primed with Strep.A were heterogeneous in that only a fraction were inhibitable by A-CHO or anti-Id. This heterogeneity of Strep.A-primed helper cells is probably due to antigenic heterogeneity of the heat-killed and pepsin-treated streptococcus (14) which, in addition to the group-specific carbohydrate antigen, carries peptidoglycan and probably also residual proteins on its surface (25). As immunization with Strep.A gives rise to antibodies against all of these antigens, it is not surprising that a similar spectrum of helper cell specificities is generated, and the partial inhibition of Strep.A-primed helper cells by A-CHO supports this view. The question whether there exists idiotypic heterogeneity within the A-CHO-specific portion of Strep.A-primed helper cells could not be answered beyond doubt, although the results of some experiments appeared to suggest that not all A-CHO-specific helper cells were inhibited by anti-Id. An idiotypic heterogeneity among helper T cells specific for one antigen would indicate the clonal distribution of idiotypes on T cells.

In contrast, helper cells primed with anti-Id were readily and equally well inhibited by anti-Id and A-CHO. This suggests that, out of a heterogeneous population of T-helper cells, anti-Id can identify and expand an apparently uniform cell population. The same was found to be the case for B cells, as shown in the previous paper (1) and confirmed by the present experiments (see Table III). The crucial question to be raised in this connection is whether the process of selecting a specific T-cell population resembles clonal selection as it occurs on the B-cell level. Although this is suggested by the existence of idiotypically uniform and monospecific T-cell populations, a definite answer to this question requires the demonstration that idiotypes are indeed clonal markers not only for B cells but also for T cells.

Another surprising finding which came from these studies is that T cells may have specificity for carbohydrate antigens. The experience of many laboratories that the immune responses to soluble polysaccharide antigens are T-cell independent may have led to the conclusion that T cells are unable to recognize polysaccharide determinants in general. This generalization has been disproved by our experiments. Moreover, T cells can distinguish between carbohydrate antigens as closely related as A-CHO and C-CHO, which are distinct by terminal hexosamines but possess the same rhamnose backbone (25, 26). Thus, the discriminatory capacity of T cells with respect to these antigens is just as good as that of antibodies. This is in agreement with previous studies from this and other laboratories on helper cell specificity (27–29).

The genetic analysis of T- and B-cell idiotypes will be discussed in detail in the accompanying paper (11). In this context it may suffice to mention that both the A5A and S117 idiotypes have been assigned to the V region of the heavy chain of anti-A-CHO antibodies by means of their genetic linkage to the heavy-chain allotype (5, 6, 10). The parallel expression of the same idiotypic polymorphism in both the B- and T-cell populations suggests that B- and T-cell idiotypes are governed by the same V_H genes, and that the V region of the immunoglobulin H chain is shared between B- and T-cell antigen receptors. Our experiments have no bearing on any other part of the immunoglobulin molecule.

In addition to the sharing of idiotypes between B and T cells, the similarities in specificity and heterogeneity discussed above are further strong arguments in favor of the view that helper T cells use at least partially the same antigen recognition system as B cells, namely the products of H-chain V-region genes. Such conclusions are in agreement with recent experiments of Binz et al., which shown that antibody directed against a specific T-cell alloantigen recognition site also reacts with humoral alloantibody as well as with cells mediating graft vs. host reaction (30, 31). Similarly, McKearn has shown that antialloantibody in rats inhibits graft rejection (32). Binz and Wigzell, however, have also obtained evidence that Id-bearing molecules produced by T cells have properties that clearly distinguish them from immunoglobulin molecules (33). Thus, it appears that evidence for the sharing of idiotypes between T-cell receptors and antibodies has now been obtained in several unrelated experimental systems. Future studies will be directed towards the isolation and characterization of Id-bearing T-cell receptor molecules, using anti-Id as a probe.

Summary

Guinea pig anti-idiotypic antibodies (anti-Id) of the IgG1 class, directed to an A/J antibody to Group A streptococcal carbohydrate (A-CHO), or directed to a BALB/c myeloma protein that binds the same antigen, stimulate B-precursor cells as well as T-helper cells when injected into mice of the appropriate strain. The strain-specific induction of both precursor and helper activity was detected by in vitro secondary responses of primed spleen cells to A-CHO or to 2,4,6-trinitrophenyl (TNP) upon challenge with Group A streptococcal vaccine (Strep.A) or with TNP-Strep.A, respectively. B- and T-cell populations primed with anti-Id were uniform with respect to the binding of antigen and of anti-Id. This was in contrast to cells primed with Strep.A, which were heterogeneous. Taken together, B and T cells that possess the same antigen-binding specificity share idiotypic determinants, reveal the same idiotypic polymorphism, and may display similar degrees of heterogeneity with respect to the binding of antigen and anti-Id. Since the anti-Id used in this study detect Id determinants associated with the heavy chain of the variable region of mouse antibodies, the data suggest that this region of the immunoglobulin molecule is shared between T- and B-cell antigen receptors.

We thank Dr. Melvin Cohn for the S117 plasmocytoma and Dr. Michael Potter for the other myeloma proteins and tumors used in this study. The excellent technical assistance of Ms. Monica Schöld and Ms. Hannelore Wirges-Koch is gratefully acknowledged.

Received for publication 2 December 1975.

References

1. Eichmann, K., and K. Rajewsky. 1975. Induction of T and B cell immunity by anti-idiotypic antibody. *Eur. J. Immunol.* 5:661.

2. Eichmann, K. 1972. Idiotypic identity of antibodies to streptococcal carbohydrate in inbred mice. *Eur. J. Immunol.* 2:301.

3. Eichmann, K. 1974. Idiotype suppression. I. Influence of the dose and of the effector function of anti-idiotypic antibody on the production of an idiotype. *Eur. J. Immunol.* 4:296.

4. Eichmann, K. 1973. Idiotype expression and the inheritance of mouse antibody clones. *J. Exp. Med.* 137:603.

5. Eichmann, K., and C. Berek. 1973. Mendelian segregation of a mouse antibody idiotype. *Eur. J. Immunol.* 3:599.

6. Eichmann, K. 1975. Idiotype suppression. II. Amplification of a suppressor T cell with anti-idiotypic activity. *Eur. J. Immunol.* 5:511.

7. Benacerraf, B., Z. Ovary, K. J. Bloch, and E. C. Franklin. 1963. Properties of guinea pig 7S antibodies. I. Electrophoretic separation of two types of guinea pig 7S antibodies. *J. Exp. Med.* 117:937.

8. Vicari, G., A. Sher, M. Cohn, and E. Kabat. 1970. Immunochemical studies on a mouse myeloma protein with specificity to certain β-linked terminal residues of N-acetyl-glucosamine. *Immunochemistry.* 7:829.

9. Eichmann, K. 1975. Genetic control of antibody specificity in the mouse. *Immunogenetics.* 2:491.

10. Black, S. J., K. Eichmann, G. J. Hämmerling, and K. Rajewsky. 1975. Stimulation and inhibition of T cell helper function by anti-idiotypic antibody. *In* Membrane Receptors of Lymphocytes. M. Seligmann, J. L. Preud'homme, and F. M. Kourilsky, editors. North-Holland Publishing Co., Amsterdam and New York. 117.

11. Hämmerling, G. J., S. J. Black, C. Berek, K. Eichmann, and K. Rajewsky. 1975. Idiotypic analysis of lymphocytes in vitro. II. Genetic control of T helper cell responsiveness to anti-idiotypic antibody. *J. Exp. Med.* 143:861.

12. Rittenberg, M. B., and K. L. Pratt. 1969. Anti-TNP plaque assay — primary response of Balb/c mice to soluble and particulate immunogen. *Proc. Soc. Exp. Biol. Med.* 132:575.

13. Pavlovskis, O., and H. D. Slade. 1969. Adsorption of ³H-fatty acid esters of streptococcal groups A and E cell wall polysaccharide antigens by red blood cells. *J. Bacteriol.* 100:641.

14. Krause, R. M., 1970. The search for antibodies with molecular uniformity. *Adv. Immunol.* 12:1.

15. Kristiansen, T., L. Sundberg, and J. Porath. 1969. Studies on blood group substances. II. Coupling of blood group substance A to hydroxyl-containing matrixes, including aminoethyl cellulose and agarose. *Biochim. Biophys. Acta.* 184:93.

16. Eichmann, K., and J. Greenblatt. 1971. Relationships between relative binding affinity and electrophoretic behavior of rabbit antibodies to streptococcal carbohydrates. *J. Exp. Med.* 133:424.

17. Julius, M. H., E. Simson, and L. A. Herzenberg. 1973. A rapid method for the isolation of functional thymus-derived murine lymphocytes. *Eur. J. Immunol.* 3:645.

18. Douglas, T. L. 1972. Occurrence of theta-like antigen in rats. *J. Exp. Med.* 136:1054.

19. Rajewsky, K., G. E. Roelants, and B. A. Askonas. 1972. Carrier specificity and the allogeneic effect in mice. *Eur. J. Immunol.* 2:592.

20. Mishell, R. I., and R. W. Dutton. 1967. Immunization of dissociated spleen cell cultures from normal mice. *J. Exp. Med.* 126:423.

21. Jerne, N. K., A. A. Nordin, and C. Henry. 1963. The agar plate technique for recognizing antibody producing cells. *In* Cell-Bound Antibodies. The Wistar Institute Press, Philadelphia, Pa. 109.

22. Dresser, D. W., and H. H. Wortis. 1967. Localized haemolysis in gel. *In* Handbook of

Experimental Immunology. D. W. Weir, editor. Blackwell Scientific Publishing, Ltd., Oxford, England. 1054.

23. Briles, D., and J. Davie. 1975. Clonal dominance. I. Restricted nature of the IgM antibody response to Group A streptococcal carbohydrate in mice. *J. Exp. Med.* **141**:1291.

24. Read, S. E., and D. G. Braun. 1974. In vitro antibody response of primed rabbit peripheral blood lymphocytes to Group A variant streptococcal polysaccharide. *Eur. J. Immunol.* **4**:422.

25. McCarty, M. 1969. The streptococcal cell wall. *Harvey Lect.* **65**:73.

26. Coligan, J. E., W. C. Schnute, and T. J. Kindt. 1975. Immunochemical and chemical studies on streptococcal group-specific carbohydrates. *J. Immunol.* **114**:1654.

27. Rajewsky, K., and R. T. Mohr. 1974. Specificity and heterogeneity of helper T-cells in the response to serum albumins in mice. *Eur. J. Immunol.* **4**:111.

28. Schlossmann, S. F. 1972. Antigen recognition: the specificity of T-cells involved in the cellular immune response. *Transplant. Rev.* **10**:97.

29. Paul, W. E., and G. W. Siskind. 1970. Hapten specificity of cellular immune responses as compared with the specificity of serum anti-hapten antibody. *Immunology.* **18**:921.

30. Binz, H., J. Lindenmann, and H. Wigzell. 1974. Cell-bound receptors of alloantigens on normal lymphocytes. II. Antialloantibody serum contains specific factors reacting with relevant immunocompetent T lymphocytes. *J. Exp. Med.* **140**:731.

31. Binz, H., and H. Wigzell. 1975. Shared idiotypic determinants on B and T lymphocytes reactive against the same antigenic determinants. I. Demonstration of similar or identical idiotypes on IgG molecules and T-cell receptors with specificity for the same alloantigens. *J. Exp. Med.* **152**:197.

32. McKearn, T. J. 1974. Antireceptor antiserum causes specific inhibition of reaction to rat histocompatibility antigens. *Science (Wash. D. C.).* **183**:94.

33. Binz, H., and H. Wigzell. 1975. Shared idiotypic determinants on B and T lymphocytes reactive against the same antigenic determinant. IV. Isolation of two groups of naturally occurring idiotypic molecules with specific antigen binding activity in serum and urine from normal rats. *Scand. J. Immunol.* **4**:591.

U. Krawinkel[+], M. Cramer, Thereza
Imanishi-Kari[o], R.S. Jack, K. Rajewsky
and O. Mäkelä[□]

Institute for Genetics, University of
Cologne, Cologne and Department of
Bacteriology and Immunology, University
of Helsinki, Helsinki[□]

Isolated hapten-binding receptors of sensitized lymphocytes

I. Receptors from nylon wool-enriched mouse T lymphocytes lack serological markers of immunoglobulin constant domains but express heavy chain variable portions*

Hapten-binding receptor material was isolated from hapten-sensitized mouse lymphocytes, as described previously (*Eur. J. Immunol.* 1976. 6: 529; *Cold Spring Harbor Symp. Quant. Biol.* 1977. 41: 285). The material was separated into a fraction expressing immunoglobulin determinants (anti-Ig[+] fraction) and a fraction lacking determinants of the known Ig constant domains (anti-Ig[−] fraction). We present further evidence in support of the notion that the anti-Ig[+] fraction is B cell-derived, whereas the anti-Ig[−] fraction originates from T lymphocytes. Receptors derived from C57BL/6 mice of the anti-Ig[−] phenotype with specificity for the hapten (4-hydroxy-3-nitrophenyl)acetyl (NP) are found to express markers which are characteristic for the variable portion of primary anti-NP antibodies. One of these markers relates to the fine specificity of hapten binding [6, 24], the other is defined by anti-idiotypic antibodies. Genetic studies show that the expression of these markers both in antibodies and the anti-Ig[−] receptor fraction is controlled by genes in the heavy chain linkage group.

The results demonstrate that in this system, humoral antibodies and receptor molecules of both the anti-Ig[+] and anti-Ig[−] phenotype bind the hapten with strikingly similar affinity and fine specificity. More specifically, they suggest that the molecules in the anti-Ig[−] receptor fraction carry the variable region (and in fact the *entire* variable region) of the Ig heavy chain.

1. Introduction

We have recently shown that hapten-binding receptor molecules can be specifically isolated from hapten-sensitized lymphocytes [1−3]. The cells were adsorbed to hapten-coated nylon mesh [4, 5] and released from the nylon by temperature shift [5]. Hapten-binding material could subsequently be recovered from the derivatized nylon. Two fractions of receptor material could be distinguished serologically: (a) a fraction binding to anti-Ig antibody and therefore expressing class-specific Ig determinants (anti-Ig[+] fraction), and (b) a fraction which did not detectably express antigenic determinants of any of the known classes of Ig polypeptide chains (anti-Ig[−] fraction). Cell fractionation experiments indicated that the anti-Ig[+] fraction was derived from B lymphocytes,

[I 1721]

whereas the anti-Ig[−] fraction originated from T cells. In the present paper we further substantiate these points and also demonstrate that the variable portions of the molecules present in the anti-Ig[−] fraction are at least partially encoded by genes in the Ig heavy chain linkage group. The evidence is provided by studies which employ two genetic markers controlled by genes encoding variable portions of Ig heavy chains (V_H genes). The first is a fine specificity marker found in antibodies of C57BL/6 mice against the hapten (4-hydroxy-3-nitrophenyl)acetyl (NP) [6]. The second is an idiotypic marker expressed in the same antibody population [7, 8].

2. Materials and methods

2.1. Animals

C57BL/6 mice were obtained from the Zentralinstitut für Versuchstierzucht, Hannover, FRG and Gl. Bomholtgaard Ltd., Ry, Denmark, and CBA/J mice from the Jackson Laboratories, Bar Harbor, Maine; S. Ivanovas Ltd., Kisslegg, FRG, and as a gift from the Institut für Biologisch-Medizinische Forschung, Füllingsdorf, Switzerland. Male and female mice, aged 6−12 weeks, were vaccinated against ectromelia and rested for two weeks before entry into an experiment. (C57BL/6 x CBA/J)F_1, (C57BL/6 x CBA/J)F_2, and [(C57BL/6 x CBA/J)F_1 x CBA/J]B_1 hybrids were bred in Cologne. Reciprocal crosses were included. Mice of every generation were taken from 3 to 6 independent litters, vaccinated against ectromelia and rested as described above.

2.2. Determination of Ig-1 allotypes

Anti-allotypic alloantisera were raised according to the method of Dresser and Wortis [9]. An alloantiserum against Ig-1[a] (CBA/J)

* This work was supported by the Deutsche Forschungsgemeinschaft through Sonderforschungsbereich 74.

+ In partial fulfilment of doctoral thesis requirements.

o Recipient of a Long-Term Fellowship from the European Molecular Biology Organization, EMBO.

Correspondence: U. Krawinkel, Institut für Genetik, Universität zu Köln, Weyertal 121, D-5000 Köln 41, FRG

Abbreviations: BSA: Bovine serum albumin **cap:** 6-Amino-caproic acid **CG:** Chicken Ig **DNase:** Calf thymus desoxyribonuclease II **DNFB:** 1-Fluoro-2,4-dinitro-benzene **DNP:** 2,4-Dinitrophenyl **DNP-cap T4:** Bacteriophage T4 coupled with DNP-cap **FCS:** Fetal calf serum **GaMIg:** Polyspecific goat anti-mouse Ig antiserum **HPI:** Haptenated phage inactivation **HPII:** Inhibition of HPI **KLH:** Keyhole limpet hemocyanin **NAP:** 4-Azido-2-nitrophenyl **NIP:** (4-Hydroxy-5-iodo-3-nitrophenyl)acetyl **NIP-cap T4:** Bacteriophage T4 coupled with NIP-cap **NP:** (4-Hydroxy-3-nitrophenyl)acetyl **PBS:** Phosphate-buffered saline **RaMIg:** Polyspecific rabbit anti-mouse Ig antiserum was prepared in C57BL/6 mice and an alloantiserum against Ig-1^b (C57BL/6) in CBA/J mice. Allotypic specificities were determined in double diffusion in 0.9 % agarose.

2.3. Haptens, antigens and immunization

NP-caproic acid (NP-cap) and (4-hydroxy-5-iodo-3-nitro-phenyl)-acetyl (NIP)-cap were synthesized according to the methods of Brownstone et al. [10]. The other compounds used were: 1-fluoro-2,4-dinitro-benzene (DNFB, Serva, Heidelberg, FRG), 2,4-dinitro-phenyl(DNP)-cap (Sigma, München, FRG), and 2,4-dinitrophenol (DNP-OH, Riedel-De Häen, Seelze, FRG). The following hapten-protein conjugates were prepared: NP_{12}-chicken gamma-globulin (NP_{12}CG), NP_{16}CG and NIP_{10}CG were synthesized from the corresponding azides [10] as described previously [1]. DNP_{11}KLH (keyhole limpet hemocyanin) was a gift of Dr. B. Rubin. Purified KLH (Pacific Biomarine Supply Comp., Venice, CA) was coupled with DNFB to yield DNP_9-KLH [11]. Mice were immunized with these antigens as previously described [1]. 4-Azido-2-nitro-phenyl $(NAP)_{10}$CG was kindly donated by Dr. J.C.A. Knott. In this case mice were immunized in the dark with 100 μg NAP_{10}CG in complete Freund's adjuvant (Difco Laboratories, Detroit, Michigan) distributed into all four footpads.

2.4. Enrichment of splenic T and B lymphocytes

Spleen cell suspensions were prepared as described previously [1]. The medium contained $10-15$ μg/ml calf thymus desoxyribonuclease (DNase, Roth, Karlsruhe, FRG). Splenic T lymphocytes were enriched by the method of Julius et al. [12]. The procedure was slightly modified in that 10 mg/ml bovine serum albumin (BSA, Behringwerke, Marburg, FRG) was used instead of 5 % fetal calf serum (FCS) to wash the nylon wool columns before addition of cells, and serum-free medium was used in all other steps. Splenic B cells were enriched in the following way: spleen cells were incubated with the appropriate anti-Thy-1.2 antiserum [13] dilution for 30 min at 4 °C, and mixed with guinea pig complement (Behringwerke). After 45 min at 37 °C the cells were washed four times with cold DNase-containing medium before further experimentation.

2.5. Isolation of hapten-specific receptor material

The method of Kiefer [5, 14] was used to bind and release hapten-specific lymphocytes from haptenated nylon discs as described previously [1]. Elution of nylon-bound, hapten-specific material was achieved in three alternative ways: (a) by acid buffer as described previously [1], (b) by incubation of the nylon discs with 3.5 M KSCN, 0.1 M Na_2HPO_4, pH 7.0 for $10-30$ min at 4 °C, and (c) by incubation with the corresponding free hapten (5 x 10^{-4} M NP-cap in phosphate-buffered saline, pH 7.2 (PBS), 10^{-3} M DNP-cap in PBS, or 10^{-1} M DNP-OH in PBS for the elution of NAP receptors) for $2-16$ h at 4 °C. Concentration, extensive dialysis ($48-72$ h) and storage of receptor material were as previously described [1].

2.6. Antisera

The anti-idiotypic antiserum is described and characterized elsewhere [8]. In short, anti-NP antibodies raised in C57BL/6 mice were purified on NIP-BSA-Sepharose columns and injected into guinea pigs which were concomitantly tolerized

to mouse Ig. The resulting antiserum was absorbed on normal C57BL/6 Ig, and subsequently either on normal BALB/c Ig or on BALB/c serum containing MOPC 104E protein. Anti-idiotypic activity was verified by radioactive binding inhibition assays [8, 15]. Anti-idiotypic antisera were further fractionated by agarose block electrophoresis. The polyspecific rabbit anti-mouse Ig antiserum (RaMIg) was described previously [1]. Another polyspecific antiserum to mouse Ig (GaMIg) with the same characteristics was raised by the same procedure in a goat. Fluorescein-isothiocyanate-coupled rabbit antibodies against mouse Ig were obtained from Behringwerke. Washed cells in 5 % FCS were incubated with a 1:10 dilution of fluorescent antibodies for 20 min on ice, washed four times and examined under a fluorescent microscope (Leitz, Wetzlar, FRG). The AKR anti-C3H thymocyte antiserum (anti-Thy-1.2) was characterized previously [13].

2.7. Immunosorbents and adsorptions

Preparations of immunosorbents of RaMIg, GaMIg, and anti-idiotypic antiserum or its IgG fraction as well as absorptions (2 h) were carried out as described previously [1].

2.8. Haptenated phage inactivation (HPI) assays

Highly sensitive NIP-cap-and DNP-cap T4 bacteriophages were prepared as previously described [16]. HPI assays followed standard procedures [16–18]. After a rough estimation of inactivation titers, a fine titration of inactivators was performed. The fraction of surviving phages (P/P_0, logarithmic) was plotted against the concentration of inactivators (linear). The slopes of the resulting straight lines are used as a direct measure of inactivator concentrations in the absorption studies described in this publication. The phage-inactivating activity of a given receptor preparation is expressed in the following way:

1 HPI unit of receptor activity = HPI titer of the receptor preparation x volume of receptor preparation [ml].

For comparisons the yields of phage-inactivating activity of the various receptor preparations are calculated in HPI units per 10^8 input lymphocytes. Inhibition of haptenated phage inactivation (HPII) was carried out with NP-cap and NIP-cap in PBS.

3. Results

3.1. Two classes of hapten-binding receptors: the anti-Ig⁺ and the anti-Ig[−] fractions

Within the hapten-binding receptor material isolated from splenic lymphocytes by the nylon method two classes of receptors can be distinguished [1]. A major class ($65-80$ % of the hapten-binding activity) is characterized by its binding to class-specific anti-Ig antibodies. We call this class of receptor molecules the anti-Ig⁺ fraction. A minor class ($25-35$ % of the hapten-binding activity) does not detectably express antigenic determinants of the known Ig classes. This class of receptor molecules is called the anti-Ig[−] fraction.

A large variety of anti-Ig antibodies have been shown to be unreactive with the anti-Ig[−] fraction [1−3]. The protocol for these experiments was as follows. The spleen cell-derived receptor material was first absorbed with insolubilized polyspecific anti-mouse Ig serum. In a second absorption step it was then determined whether a given insolubilized anti-Ig antiserum could detectably absorb additional hapten-binding activity from the preabsorbed material. We have thereby established that the anti-Ig⁻ fraction does not detectably express determinants of murine γ, α, μ, δ, κ and λ Ig polypeptide chains as defined by antisera raised in rabbits and guinea pigs [1, 3]. This statement was recently reconfirmed for λ chains by the use of a rabbit antiserum against isolated λ chains of the mouse myeloma protein MOPC 104E (the antiserum was a gift of Dr. H.C. Morse; data not shown). The anti-Ig⁻ fraction is also devoid of determinants which chicken antibodies detect on the BALB/c myeloma proteins MOPC 104E (μ, λ) and TEPC 183 (μ, κ) [3]; (the chicken antisera were a gift of Dr. U. Hämmerling and are characterized in ref. [19]). At this stage, the anti-Ig⁻ fraction thus carries only one positive marker, namely its specificity for antigen, and is only negatively distinguished from the anti-Ig⁺ fraction.

Several points argue against the trivial hypothesis that the anti-Ig⁻ fraction consists of denatured molecules of the anti-Ig⁺ fraction. First, the molecules in the anti-Ig⁻ fraction appear functionally intact in that they bind hapten specifically and with high affinity (see also Sect. 3.3.1.). Second, the properties of the anti-Ig⁻ fraction including its relative size do not appear to depend on the details of the elution procedure, although the total yield of hapten-binding activity is optimal when the material is eluted from the nylon mesh with free hapten.

This is documented in Table 1. These data also indicate that anti-Ig⁺ and anti-Ig⁻ hapten-specific molecules can be recovered in the usual proportions from the medium into which the cells are released from the nylon by temperature shift [2], i.e. without any particular elution procedure (Table 1). Affinity titrations have shown [20] that as expected, this fraction

of receptors has a lower average affinity for hapten than the receptors sticking to (and being eluted from) the nylon mesh. The strongest argument against the denaturation hypothesis however, arises from experiments aiming at the elucidation of the cellular origin of the anti-Ig⁺ and the anti-Ig⁻ fraction (see Sect. 3.2.).

3.2. Cellular origin of the anti-Ig⁺ and the anti-Ig⁻ fraction

Previous experiments in which receptor material was isolated from cell populations enriched in T or B lymphocytes suggested that the receptor molecules belonging to the anti-Ig⁺ fraction originate from B lymphocytes, whereas the anti-Ig⁻ fraction is T cell-derived [1, 2]. In Table 2 we present a summary of our present knowledge on this point in the mouse system. In the accompanying publication we show that in the rabbit the situation is strictly analogous [21]. The data in Tables 1 and 2 demonstrate that (a) in receptor material isolated from unfractionated spleen cells the anti-Ig⁻ fraction regularly amounts to 25–35 % of the total hapten-binding activity, (b) the anti-Ig⁺ fraction predominates in receptors isolated from enriched splenic B cells or from a hybrid cell line producing monoclonal antibodies with specificity for NP ([22, 23], and M. Reth, G. Hämmerling and K. Rajewsky, in preparation), and (c) the anti-Ig⁻ fraction predominates in receptor material isolated from enriched splenic T lymphocytes. It is noteworthy that there is a good correlation between the proportion of T cells in the input cell population and the relative size of the anti-Ig⁻ fraction. It should also be mentioned that in all cases 0.5–1 % of the input cell population bound specifically to the derivatized nylon and that this binding could be specifically inhibited by free antigen to about 60 % ([1], and undocumented observations). The simplest interpretation of the overall data is to ascribe the anti-Ig⁺ fraction to T cells.

Table 1. Various elution procedures in the isolation of receptor material

Exp. no.	Receptor preparation origin and specificity	Elution from nylon discs with	Yield of receptor activity[a]	Anti-Ig⁻ fraction[b] (%)
1	C57BL/6 anti-NP	0.1 M Glycine/HCl, pH 3.2 (3 preparations)	58, 23, 23	34, 25, 25
2		3.5 M KSCN, pH 7.0	69	25
3		5 x 10⁻⁴ M NP-cap, pH 7.2	86[d]	34
4		– [c]	28[d]	34
5	C57BL/6 anti-DNP	0.1 M Glycine/HCl, pH 3.2	5	31
6		10⁻³ M DNP-cap, pH 7.2	33	31
7	CBA anti-NP	0.1 M Glycine/HCl, pH 3.2	26	36
8		5 x 10⁻⁴ M NP-cap, pH 7.2	142[e], 272	38, 33
9		– [c]	54[e]	34
10	CBA anti-DNP	0.1 M Glycine/HCl, pH 3.2	2.5	44
11	CBA anti-NAP	0.1 M DNP-OH, pH 7.2	2	47

a) Expressed in HPI-units per 10⁸ input lymphocytes as described in Sect. 2.8.
b) Fraction of antigen-specific receptor material not binding to RaMIg or GaMIg.
c) Receptor material from medium into which cells were released at room temperature after binding to hapten-coupled nylon discs.
d) These two preparations originate from the same lymphocyte population.
e) These two preparations originate from the same lymphocyte population.

Table 2. Cellular origin of receptor material

Exp. .no.	Receptor preparation Origin and specificity[a)	Treatment of cells[b)	% Cells carrying[c) surface Ig	Thy-1.2	Yield of receptor activity[d)	Anti-Ig⁻ fraction[e) (%)
1	C57BL/6 anti-NP	Nylon wool	25	n.d.[f)	8	65
2	C57BL/6 anti-NP	Nylon wool	15	n.d.	4	86
3	CBA/J anti-NIP	Nylon wool	n.d.	79	32	74
4	CBA/J anti-NAP	Nylon wool	5	n.d.	4.5	85
5	C57BL/6 anti-NP	Anti-Thy-1.2 + C'	81	n.d.	1.5	10
6	Clone S43[g) anti-NP	–	n.d.	n.d.	165	5

a) Receptor preparations were isolated from hapten-sensitized splenic lymphocytes on nylon discs coupled with the relevant hapten. Hapten-binding material was eluted with 0.1 M glycine/HCl, pH 3.2 in Expts. 1, 2 and 5 and with free hapten in Expts. 3, 4 and 6.
b) T lymphocytes were enriched by passage of spleen cells through nylon wool columns according to the method of Julius [12] before receptor isolation. B lymphocytes were enriched by treating spleen cells with anti-Thy-1.2 and complement.
c) Surface Ig was determined by staining cells with fluorescein-coupled rabbit anti-mouse Ig antibody. Thy-1.2 was determined by a cytotoxic assay with anti-Thy-1.2 serum and complement.
d) see legend of Table 1 and Sect. 2.8.
e) see legend of Table 1.
f) n.d. = Not done.
g) The method of Köhler and Milstein [22] was used to fuse spleen cells of C57BL/6 mice hyperimmunized to NP with X63Ag8 myeloma cells; the resulting hybrid clones were assayed for production of NP-specific antibodies ([22] and Reth, M., Hämmerling, G. and Rajewsky, K., manuscript in preparation). Clone S43 represents one of the hybrid cell lines thus obtained.

3.3. Presence of immunoglobulin V_H region markers on the anti-Ig⁻ fraction

The experiments reported in this section represent an attempt to identify genetic markers of the variable region of antibody polypeptide chains on the anti-Ig⁻ fraction of our receptor molecules. The analysis consists of two steps. (a) The selection of suitable markers and their detection on the receptor material, and (b) the demonstration that the expression of the variable region markers in the anti-Ig⁻ fraction and in antibody molecules is controlled by the same genetic region.

In the absence of genetic markers for the variable region of Ig light chains two different V_H markers were selected which are both expressed in a particular population of antibody molecules, namely the antibody population which C57BL/6 mice generate in the primary response against the hapten NP. These antibodies carry a strain-specific fine specificity marker in that their affinity for the cross-reacting hapten NIP is higher than for the homologous hapten NP [24]. The same antibodies can also be characterized idiotypically [7, 8]. Both the fine specificity and the idiotypic marker of C57BL/6 anti-NP antibodies are inherited in a simple codominant Mendelian fashion and can be genetically mapped to the heavy chain linkage group [6–8]. Our interpretation of this result is that both fine specificity and idiotype represent markers of the variable region of the heavy chain of C57BL/6 anti-NP antibodies.

Both markers are absent from anti-NP antibodies of CBA origin [6–8]. Our genetic experiments were therefore carried out in C57BL/6 and CBA animals. A (C57BL/6 x CBA)F_1 generation was bred, and from the F_1 animals an F_2 generation and a backcross to CBA was developed. The animals were classified according to the Ig-1 allotype and immunized with NP-CG. The analysis of the humoral anti-NP response established that both the fine specificity and the idiotypic marker showed linkage to the Ig-1 locus. This is described in detail in a separate paper [8]. For the present analysis, animals from F_1, F_2 and backcross

generations were grouped according to Ig allotype, receptor material was isolated from pooled spleen cells of these groups and analyzed for the presence or absence of the two V_H region markers.

3.3.1. Fine specificity of the anti-Ig⁻ receptor fraction

The heteroclitic property which is characteristic for C57BL/6 anti-NP antibodies is also found in NP-specific receptor material isolated from NP-CG-sensitized C57BL/6 mice. The data in Fig. 1b demonstrate that both the total receptor material and the anti-Ig⁻ fraction of this material exhibit a higher average affinity for the NIP than for the NP hapten. NP-specific receptors of CBA origin behave differently in the fine specificity analysis (Fig. 1a). No heteroclicity can be detected in the anti-Ig⁻ fraction. The total receptor material (i.e. essentially the anti-Ig⁺ fraction) is slightly heteroclitic at high but not at low hapten concentrations. Slight heteroclicity of CBA-derived anti-Ig⁺ receptors and also of CBA anti-NP antibodies has repeatedly been found in our experiments when NIP-cap T4 instead of NP-coupled T4 bacteriophage was used in the HPII assay (undocumented results; see also below, Fig. 2). The NIP-cap T4 bacteriophage which probably detects only a fraction of the NP-binding activity of CBA origin and may very well strongly select for minor heteroclitic components in CBA antibodies or anti-Ig⁺ receptor material, was nevertheless used for our studies because of its exquisite sensitivity in the HPI assay. The heteroclicity in CBA anti-NP antibodies or CBA-derived NP-specific anti-Ig⁺ receptor material was variable, expressed at certain concentrations of hapten only and never found in the anti-Ig⁻ fraction. In contrast, in C57BL/6 heteroclicity was regular, mostly expressed over the entire range of hapten concentrations and always detectable in the anti-Ig⁻ fraction.

Fig. 2 depicts the fine specificity of receptor material which was isolated from the allotypically defined groups of the F_2 and backcross generations. Again, the situation is particularly

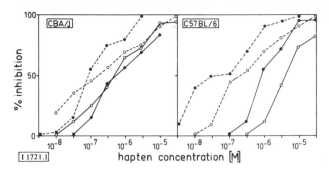

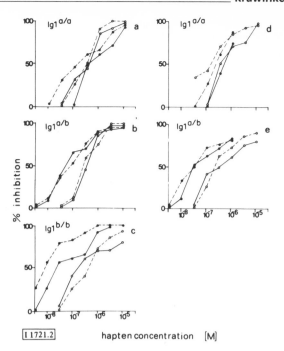

Figure 1. Fine specificity of NP-binding material from $NP_{12}CG$-sensitized spleen cells isolated on NP-cap-coupled nylon discs. (a) Receptor material from CBA/J cells, elution from nylon by washing with 0.1 M glycine/HCl buffer pH 3.2. (b) Receptor material from C57BL/6 cells, elution from nylon with 3.5 M KSCN. Inactivation of NIP-cap T4 by NP-binding material before (broken lines) and after (unbroken lines) absorption with insolubilized GaMIg inhibited by NP-cap (○) and NIP-cap (●).

Figure 2. Fine specificity of NP-binding material from $NP_{12}CG$-sensitized spleen cells eluted from NP-cap-coupled nylon with 5×10^{-4} M NP-cap. The cell donors differ for each frame and are all derived from a (C57BL/6 x CBA/J)F_1 generation. (a) Ten mice of the F_2 generation, homozygous for the Ig-1^a allotype, yield of receptor activity: 145 HPI units/10^8 input cells. (b) Twenty-one mice of the F_2 generation, heterozygous for the Ig-1^a and the Ig-1^b allotypes; yield of receptor activity: 97 HPI units/10^8 input cells. (c) Nine mice of the F_2 generation, homozygous for the Ig-1^b allotype; yield of receptor activity: 86 HPI units/10^8 input cells. (d) Nine mice of the backcross generation $[(C57BL/6 x CBA/J)F_1 x CBA/J]B_1$, homozygous for the Ig-1^a allotype; yield of receptor activity: 200 HPI units/10^8 input cells. (e) Four mice of the backcross generation $[(C57BL/6 x CBA/J)F_1 x CBA/J]B_1$, heterozygous for the Ig-1^a and the Ig-1^b allotypes; yield of receptor activity: 183 HPI units/10^8 input cells. Inactivation of NIP-cap T4 by NP-binding material before (broken lines) an after (unbroken lines) absorption with insolubilized GaMIg, inhibited by NP-cap (○) and NIP-cap (●).

clear for the anti-Ig$^-$ fraction. This fraction is strongly heteroclitic in b/b homozygous and a/b heterozygous animals and nonheteroclitic in a/a homozygous animals. Essentially the same is true for unseparated receptors with the exception of slight heteroclicity in a/a homozygous animals of the F_2 generation in the high affinity region (Fig. 2a). The extent of heteroclicity in the latter case is small, however, as compared to the striking heteroclicity found in receptor material derived from mice carrying the Ig-1^b allotype (Fig. 2b, d, e).

We conclude that in C57BL/6 mice, isolated receptor molecules and humoral antibodies with specificity for NP carry a similar, possibly the same, fine specificity marker. In the case of humoral antibody this marker is genetically linked to the Ig-1 allotype [6], and the present data suggest that this is also true for the fine specificity marker expressed in the receptor material and particularly in the anti-Ig$^-$ receptor fraction.

It should be mentioned that the data in Fig. 2 in particular suggest a rather similar affinity distribution in isolated receptors of both the anti-Ig$^+$ and the anti-Ig$^-$ type and also in humoral antibodies [1, 24]. The heteroclitic property of receptor material and humoral antibody thus appears similar not only quantitatively but also in terms of the ranges of affinities for the haptens involved. In our original study the anti-Ig$^-$ fraction of C57BL/6-derived NP-specific receptors was found to have a dramatically lower average for the hapten than the anti-Ig$^+$ fraction [1]. It would now appear that this finding represented an exception rather than the rule, or even that the discrepancy is due to technical matters. The receptors analyzed in Fig. 2 were all eluted from the nylon with free hapten, whereas previously acidic buffer was used for this purpose. It is of interest in this connection to see that the material represented in Fig. 1b (where again the anti-Ig$^-$ fraction has a distinctly lower average affinity for hapten than the anti-Ig$^+$ fraction) was eluted from the nylon with 3.5 M KSCN. Certain elution procedures may thus lead to a selective loss of affinity in the anti-Ig$^-$ fraction. If this interpretation is correct it would, however, only hold for C57BL/6 and not CBA-derived receptor material ([2], and undocumented experiments; compare also Fig. 1b which depicts results obtained with acid-eluted material of CBA origin).

Considering the overall data, there still appears to be a tendency in the anti-Ig$^-$ fraction towards slightly lower affinities for hapten as compared to the corresponding affinities of the anti-Ig$^+$ fraction and of humoral antibodies. The differences are small, however, and there can be no question that the molecules of the anti-Ig$^-$ phenotype exhibit high binding affinities for haptens like NP, NIP or DNP (for DNP see ref. [2]).

3.3.2. Idiotypic specificity of the anti-Ig$^-$ receptor fraction

The same receptor material which in the above experiments had been analyzed for its fine specificity was also subjected to idiotypic analysis. For this purpose an anti-idiotypic antiserum was used which is characterized in detail in the preceding publication [8]. The serum was prepared by immunizing a guinea pig with primary C57BL/6 anti-NP antibody purified by affinity chromatography. After suitable absorption the antiserum bound specifically more than 90 % of the antibody preparation and of other samples of anti-NP antibody from mouse strains carrying the Ig-1^b allotype, but did not react detectably with a large variety of other Ig preparations originating from C57BL/6 or other mouse strains.

The apparent idiotypic specificity of the antiserum in conventional idiotypic binding and binding inhibition analysis turned out to be insufficient for a direct idiotypic analysis of the receptor material. When the antiserum was insolubilized and various receptor and antibody preparations were absorbed with the resulting immunosorbents, it was found that up to 50 % of the hapten-binding activity of nonidiotypic material was "nonspecifically" bound to the immunosorbent. This contrasted with a removal of 80–90 % of the activity from C57BL/6 anti-NP antibody (data not shown). The high "nonspecific" absorption under these experimental conditions is not surprising if one takes into account the minute amounts of hapten-binding activity in the system together with the probably overwhelming excess of anti-idiotypic antibody on the immunosorbent.

The problem found an unexpected solution when we discovered that the "nonspecific" absorption completely disappeared when the receptor material had been preabsorbed with insolubilized polyspecific anti-Ig, *i.e.* when the anti-Ig⁻ fraction was analyzed. The situation is depicted in Fig. 3. It can be seen that the anti-idiotypic immunosorbents eliminated 80 % of the hapten-binding activity of the anti-Ig⁻ fraction of C57BL/6 anti-NP receptor material isolated either from unseparated spleen cells (Fig. 3a) or enriched T lymphocytes (Fig. 3b), whereas there was no detectable removal of activity from the anti-Ig⁻ fractions of (a) anti-NP receptors from CBA mice and (b) anti-DNP receptors from C57BL/6 mice. For the

analysis of the anti-Ig⁻ receptor fraction, the anti-idiotypic immunosorbent thus exhibits remarkable specificity and, in addition, detects idiotypic determinants on a surprisingly large fraction of anti-Ig⁻ receptors with appropriate specificity originating from idiotype-positive mice.

The data depicted in Fig. 4 represent the genetic analysis in the idiotypic system and are strictly analogous to the data on fine specificity in Fig. 2. The clear-cut results strongly support the concept that the idiotypic determinants on the anti-Ig⁻ receptor fraction are controlled by genes in the heavy chain linkage group. The presence of idiotypic receptor molecules of the anti-Ig⁻ phenotype appears to correlate strictly with the presence of the Ig-1^b allotype in the lymphocyte donors.

4. Discussion

The present data concern the analysis of the antigen-binding receptor molecules which we isolated from sensitized lymphocytes [1–3]. The way in which these molecules are specifically isolated, their specificity for antigen, the yield of receptor material in terms of hapten-binding activity, its serological analysis with anti-Ig antisera and its cellular origin have been discussed in detail in our previous reports, and we will only briefly comment on these points here.

The data in Table 1 show that the yield of hapten-binding material can be considerably improved when the elution of receptors from the haptenated nylon is carried out with free

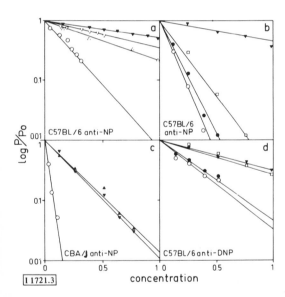

Figure 3. Idiotypic analysis of receptor material. The following receptor preparations were analyzed: receptors isolated from (a) NP$_{12}$CG-sensitized C57BL/6 spleen cells on NP-cap-coupled nylon discs (preparation 2 in Table 1); (b) NP$_{16}$CG-sensitized C57BL/6 splenic T cells enriched on nylon wool (preparation 1 in Table 2); (c) NP$_{12}$CG-sensitized CBA/J spleen cells (preparation 9 in Table 1); (d) DNP$_9$KLH-sensitized C57BL/6 spleen cells on DNP-cap-coupled nylon discs (preparation 6 in Table 1). The concentration of receptor material on the abscissa is given in arbitrary units. Inactivation of NIP-cap T4 (a–c) or DNP-cap T4 (d) by (○—○) nonabsorbed material or material absorbed with (●—●) insolubilized normal guinea pig serum, (△—△) GaMIg immunosorbent, (▽—▽) anti-idiotypic immunosorbent, (▲—▲) GaMIg immunosorbent and insolubilized normal guinea pig serum, (▼—▼) GaMIg immunosorbent and anti-idiotypic immunosorbent and (□—□) GaMIg immunosorbent in two sequential absorption steps. Anti-idiotypic immunosorbent consisted of Sepharose-bound anti-idiotypic serum (Fig. 3a, c) or the Sepharose-bound IgG fraction of the same antiserum (Fig. 3b, d).

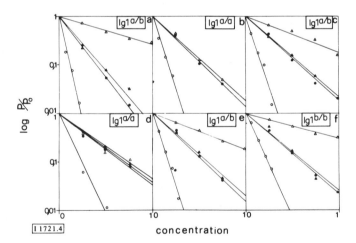

Figure 4. Genetic control of idiotypic specificity expressed in receptor material. NP-specific receptor material was isolated from spleen cells of groups of mice which belonged to a (C57BL/6 × CBA/J)F$_1$ (a) or (C57BL/6 × CBA/J)F$_2$ generation (d–f), or to the backcross [(C57BL/6 × CBA/J)F$_1$ × CBA/J]B$_1$ (b, c). The receptor material was eluted from the nylon with 5 × 10^{-4} M NP-cap. The cell donors were grouped according to their Ig-1 allotype and the groups were as follows: (a) F$_1$ generation, Ig-1$^{a/b}$, 8 animals, yield of receptor activity: 360 HPI units/10^8 input cells; (b) backcross generation, Ig-1$^{a/a}$; (c) backcross generation, Ig-1$^{a/b}$; (d) F$_2$ generation, Ig-1$^{a/a}$; (e) F$_2$ generation, Ig-1$^{a/b}$ and (f) F$_2$ generation, Ig-1$^{b/b}$. For numbers of cell donors and yields of receptor activity see legend to Fig. 2. Inactivation of NIP-cap T4 (ordinate) by (○—○) nonabsorbed material, (●—●) material absorbed with GaMIg immunosorbent, (▲—▲) material absorbed twice with GaMIg immunosorbent and (△—△) material absorbed by GaMIg immunosorbent and subsequently by anti-idiotypic immunosorbent. The immunosorbent was prepared by coupling the IgG fraction of the anti-idiotypic antiserum to activated Sepharose.

hapten. Irrespective of the elution procedure, however, we isolate hapten-specific material which can invariably be classified into two serologically defined fractions, the anti-Ig+ fraction which binds to anti-Ig antisera and the anti-Ig− fraction which on the basis of immunosorbent analysis, appears to lack any determinant of the known constant regions of Ig polypeptide chains. In a series of experiments it was found that the relative sizes of the anti-Ig+ and anti-Ig− fractions were solely dependent on the cellular composition of the input lymphocyte population. There was a striking correlation between anti-Ig+ fraction and B cell input on the one hand and anti-Ig− fraction and T cell input on the other. These results suggest that the two fractions of receptor material originate from different cell populations. The simplest and most straightforward hypothesis ascribes the anti-Ig− fraction to T cells [1−3]. We should like to stress, however, that so far we cannot formally prove that this is the correct interpretation of the data, and indeed, the formal proof of this point is a difficult goal to reach. In addition, we do not yet know whether the anti-Ig− molecules are a product of the cells from which they are isolated. One way to answer these questions may be the use of chimeric mice whose B cells come from a strain which produces heteroclitic and idiotypically defined antibodies whose T cells, however, come from a different strain. Finally, although we consider it likely that both the anti-Ig+ and the anti-Ig− fractions represent receptor material from the cell surface, the localization of the material in or on the cell is at present a matter of speculation.

The main finding of the present study is the discovery of two markers of the variable region of Ig heavy chains on the hapten-binding molecules of the anti-Ig− phenotype. Upon primary immunization with an NP-carrier conjugate C57BL/6 mice produce anti-NP antibodies which are heteroclitic in that they exhibit higher affinity for the heterologous hapten NIP than for the homologous hapten NP [24]. This heteroclicity is not found as a predominant feature of the anti-NP response in other strains of mice and is transmitted in genetic crosses as a genetic marker which follows simple Mendelian codominant inheritance and can be mapped to the Ig-1 locus [6]. It was recently established that most if not all of these heteroclitic antibodies carry λ chains [8, 23]. The fine specificity marker of C57BL/6 anti-NP antibodies is by definition closely associated with the antibody-combining site and therefore probably with amino acid sequences in or around the hypervariable regions of the heavy chains. As discussed in a separate publication [23], the λ chain may, however, contribute to the expression of the heteroclitic property.

The same antibody population in C57BL/6 mice can also be characterized by anti-idiotypic antisera [7, 8]. The idiotypic marker defined in this way is again strain-specific and transmitted in genetic crosses together with heteroclicity and the Ig heavy chain allotype. The reaction of the particular anti-idiotypic antiserum employed in the present study with the idiotype could neither be inhibited by free NIP hapten nor by NIP8 BSA conjugate even when the latter was present in very large excess [8]. The antiserum therefore appears to detect determinants of the variable region of the heavy chain which are distant from the antigen-binding site and therefore may represent framework residues of the V_H region (for a detailed discussion see the preceding paper [8]). The combined determination of the two V_H region markers would thus relate to both framework and hypervariable positions of the variable region of the Ig heavy chain.

Our analysis demonstrates that both the fine specificity and the idiotypic marker can be detected in NP-specific C57BL/6 receptor molecules of the anti-Ig− phenotype. The fine specificity marker is of course only indirectly related to the coding sequence which controls the hapten-binding site. It is reassuring, however, to find the heteroclitic property in humoral antibodies and the anti-Ig− receptor material at similar levels of affinities for hapten if the receptor material is prepared under mild conditions (Fig. 2). This is the case although at least the constant portions of λ chains are not present in the anti-Ig− fraction. The idiotypic marker is defined by an anti-idiotypic reagent which exhibits the same remarkable specificity at the level of humoral antibody [8] and of our isolated receptor material (Fig. 3). It has been verified in undocumented experiments that the anti-idiotypic immunosorbents react with the anti-Ig− fraction via determinants *not* associated with the constant region of λ chains. The idiotypic marker appears to be expressed on most of the hapten-specific molecules in the anti-Ig− receptor fraction. The same is true for anti-NP antibodies in the primary, but not the secondary anti-NP response of C57BL/6 mice. In the latter, only 20 % of the antibody population are idiotype-positive when analyzed with the anti-idiotypic serum used in this study. One wonders whether the anti-Ig− fraction (T cell receptors, as we think) might be more restricted to the expression of major (germ line) idiotypes than the anti-Ig+ fraction (representing B cell receptors). This point will be established as soon as we are able to determine the idiotype frequency in the anti-Ig+ receptor fraction.

A genetic analysis strongly suggests that the expression of both markers in the anti-Ig− fraction is controlled by genes in the heavy chain linkage group, as has been found to be the case for humoral antibody [6−8]. The genetic aspect is of fundamental importance for the interpretation of the results, since we have so far no way of proving that the two markers correspond to identical molecular structures on humoral antibody and the anti-Ig− receptor fraction. If they do, the genetic data suggest that the *same* V_H genes code for variable portions of Ig heavy chains and of molecules in the anti-Ig− fraction, and the lack of "nonspecific" adsorption of anti-Ig− molecules to insolubilized anti-idiotype (*c.f.* Sect. 3.3.2.) is a further strong argument for the absence of constant parts of conventional Ig polypeptide chains from the anti-Ig− receptor fraction. If they do not (and, indeed, the lack of "nonspecific" adsorption mentioned above could serve as an argument here), it would still appear that genes in the heavy chain linkage group code for variable portions of receptor molecules in the anti-Ig− fraction.

Despite these possible complications it seems clear that the experimental evidence supports the simple view that the molecules in the anti-Ig− receptor fraction express variable portions with framework and hypervariable residues as products of V_H genes. This view is reinforced by our results presented in another paper [21] in which we show that anti-Ig− receptor molecules can also be isolated from rabbit lymphocytes and that these molecules again appear to be T cell-derived, to lack class-specific determinants of Ig polypeptide chains, but to express a-locus allotypic determinants.

We are thus faced with a class of antigen-binding receptor molecules which express V_H regions but none of the known Ig constant domains, and which appear to be T cell-derived. Binz et al., in their experimental system of alloreactivity, have found very similar receptor molecules in the rat [25]. The molecule which they describe appears to have a molecular weight of 150 000 Daltons and to consist of two polypeptide chains of identical size [26]. Our preliminary biochemical characterization of the anti-Ig⁻ fraction is in accord with these results [1, 2]. In particular, the bulk of the active molecules in our receptor preparations appears to possess a molecular weight above 100 000 Daltons, *i. e.* these molecules appear to differ in terms of size from the antigen-specific Ia-bearing T cell-derived factors described by other authors [27−30]. Also, we have so far been unable, as have been Binz and Wigzell in their analogous system [25], to demonstrate determinants coded by the major histocompatibility complex on the molecules in the anti-Ig⁻ fraction [3].

The present evidence is in accord not only with that of Binz and Wigzell, but also with our own previous experiments in which it was shown that functional antigen receptors on T helper cells express idiotypic determinants of Ig heavy chains [2]. Again, genetic data suggested that V_H genes code for variable portions not only of antibody molecules, but also of T cell receptors.

Detailed discussions of how the molecular structure of the T cell receptor for antigen can be visualized on the basis of the available experimental evidence have been given recently [31, 32] and will not be repeated here. The constant part of the polypeptide chains carrying V_H regions in T cell receptors remains unknown at present. It may simply define a new Ig class. Most important, however, appears to be the unresolved problem of whether the V_H region is the only component of the antigen-binding site of T cell receptors. Three observations may be relevant in this context. (a) The molecules of the anti-Ig⁻ fraction show a surprisingly high apparent average affinity for hapten, and (b) are clearly heteroclitic, a property which in antibodies is correlated with the presence of a λ chain [23]; these observations could suggest the presence of light chain variable, but not constant, regions on the T cell receptor. On the other hand, (c) indirect evidence on idiotypic properties of T helper cell receptors suggests that light chain variable regions are *not* expressed in T cell receptors [2]. If the latter turned out to be true, one would wonder whether both specificity and affinity of these receptors are provided by V_H alone or whether we still have to discover in the isolated receptor molecules a new type of polypeptide chain with a new type of variable region which might be under the control of the major histocompatibility complex (see [31]). These questions will have to be answered by straightforward serological and chemical analysis.

We wish to thank the Institut für Biologisch-Medizinische Forschung, Füllingsdorf, for the generous supply of CBA/J mice, and are grateful to Drs. J.C.A. Knott and B. Rubin for the gift of valuable reagents and to Drs. U. Hämmerling and H.C. Morse for the gift of antisera.

Received May 2, 1977.

5. References

1 Krawinkel, U. and Rajewsky, K., *Eur. J. Immunol.* 1976. *6*: 529.

2 Krawinkel, U., Cramer, M., Berek, C., Hämmerling, G., Black, S.J., Rajewsky, K. and Eichmann, K., *Cold Spring Harbor Symp. Quant. Biol.* 1976. *41*: 285.

3 Cramer, M., Krawinkel, U., Hämmerling, G., Black, S.J., Berek, C., Eichmann, K. and Rajewsky, K., in McDevitt, H.O., Paul, W.E. and Benacerraf, B. (Eds.), *Proceedings of the third Ir gene workshop*, Academic Press, New York 1977, in press.

4 Rutishauser, U. and Edelman, G.M., *Proc. Nat. Acad. Sci. US* 1972. *69*: 3774.

5 Kiefer, H., *Eur. J. Immunol.* 1973. *3*: 181.

6 Imanishi, T. and Mäkelä, O., *J. Exp. Med.* 1974. *140*: 1498.

7 Mäkelä, O. and Karjalainen, K., *Immunol. Rev.* 1977, in press.

8 Jack, R.S., Imanishi-Kari, T. and Rajewsky, K., *Eur. J. Immunol.* 1977. *7*: 559.

9 Dresser, D.W. and Wortis, H.H., in Weir, D.M. (Ed.), *Handbook of Experimental Immunology*, Blackwell Scientific Publications Ltd., Oxford 1967, p. 1054.

10 Brownstone, A., Mitchison, N.A. and Pitt-Rivers, R., *Immunology* 1966. *10*: 481.

11 Feldmann, M., *Nature-New Biol.* 1971. *231*: 21.

12 Julius, M.H., Simpson, E. and Herzenberg, L.A., *Eur. J. Immunol.* 1973. *3*: 645.

13 Rajewsky, K., Roelants, G.E. and Askonas, B.A., *Eur. J. Immunol.* 1972. *2*: 592.

14 Kiefer, H., *Eur. J. Immunol.* 1975. *5*: 624.

15 Eichmann, K., *Eur. J. Immunol.* 1972. *2*: 301.

16 Becker, M. and Mäkelä, O., *Immunochemistry* 1975. *12*: 329.

17 Mäkelä, O., *Immunology* 1966. *10*: 81.

18 Haimovich, J. and Sela, M., *J. Immunol.* 1966. *97*: 338.

19 Hämmerling, U., Mack, C. and Pickel, H.-G., *Immunochemistry* 1976. *13*: 525.

20 Krawinkel, U., *Spezifische Anreicherung und Charakterisierung hapten-bindender Rezeptoren von T und B Lymphozyten aus Mäusen und Kaninchen*, Ph. D. Thesis, University of Cologne 1977.

21 Krawinkel, U., Cramer, M., Mage, R.G., Kelus, A.S. and Rajewsky, K., *J. Exp. Med.* 1977. *146*, in press.

22 Köhler, G. and Milstein, C., *Nature* 1975. *256*: 495.

23 Reth, M., Imanishi-Kari, T., Jack, R.S., Cramer, M., Krawinkel, U., Hämmerling, G.J. and Rajewsky, K., in Sercarz, E., Herzenberg, L. and Fox, C.F. (Eds.), *The Immune System: Genes and the Cells in Which They Function. ICN-UCLA Symposia on Molecular and Cellular Biology*, vol. *8*, Academic Press, New York, N.Y. 1977, in press.

24 Imanishi, T. and Mäkelä, O., *Eur. J. Immunol.* 1973. *3*: 323.

25 Binz, H., Wigzell, H. and Bazin, H., *Nature* 1976. *264*: 639.

26 Binz, H. and Wigzell, H., *Scand. J. Immunol.* 1976. *5*: 559.

27 Munro, A.J. and Taussig, M.J., *Nature* 1975. *256*: 103.

28 Mozes, E., in Katz, D.H. and Benacerraf, B. (Eds.), *The Role of Products of the Histocompatibility Gene Complex in Immune Responses*, Academic Press, New York 1976, p. 485.

29 Tada, T., Taniguchi, M. and Takemori, T., *Transplant. Rev.* 1975. *26*: 106.

30 Takemori, T. and Tada, T., *J. Exp. Med.* 1975. *142*: 1241.

31 Rajewsky, K. and Eichmann, K., *Contemp. Top. Immunbiol.* 1977. *7*: 69.

32 Janeway, C.A., Jr., Wigzell, H. and Binz, H., *Scand. J. Immunol.* 1976. *5*: 993.

H-2 COMPATIBILITY REQUIREMENT FOR
T-CELL-MEDIATED LYSIS OF TARGET CELLS INFECTED
WITH LYMPHOCYTIC CHORIOMENINGITIS VIRUS

Different Cytotoxic T-Cell Specificities are Associated with Structures Coded for in *H-2K* or *H-2D*

By ROLF M. ZINKERNAGEL AND PETER C. DOHERTY

…ent of Microbiology, The John Curtin School of Medical Research, Australian …ational University, Canberra City, A. C. T. 2601, Australia)

…y at the *H-2*-gene complex is essential for interaction of sensitized … lymphocytes (T cells) and target cells infected with either …horiomeningitis (LCM)[1] virus or with ectromelia virus. This …been demonstrated both in vitro (1–4), by the use of ^{51}Cr-release …vivo in adoptive transfer experiments (5, 6). Similar restrictions …n described for T-cell-mediated lysis of trinitrophenyl-modified …7), for the T-cell helper effect (8–10), for the transfer of cell-…unity to *Listeria monocytogenes* (11) and for the proliferation of …nea pig lymphocytes exposed to antigen-pulsed macrophages (12). …in vitro cell-mediated lysis of Rous sarcoma virus-transformed cells … restricted in the same way (13).

…ly exclusive hypotheses have been proposed to explain these phenomona (1, …that genes in the *H-2*-gene complex, perhaps located in the *Ir* region, code …nvolved in recognition of other somatic cells by T cells. This implies the …dual recognition system, requiring both presence of structures involved in …teraction and an antigen-specific T-cell receptor. The second possibility is …ce components specified in the *H-2K* or *H-2D* regions are involved in antigen …T cells are sensitized to "altered-self," either modified *H-2* antigens or …ded for by the *H-2*-gene complex that are not normally expressed on the cell … some complex of viral and *H-2* antigens.

…presented here supports the second hypothesis, by establishing that …totoxic T-cell specificities are associated with each parental haplo-…mice, or with either the *H-2K* or *H-2D* region as demonstrated with …binant strains.

Materials and Methods

Mice. CBA/H, BALB/c, CBA/H × BALB/c F$_1$, and A/J mice were from colonies at the Australian National University, Canberra City, Australia. The C3H.OH mice were originally obtained from

[1]*Abbreviations used in this paper:* FCS, Fetal calf serum; L cells, L929 fibroblasts (*H-2*^{k}); LCM, lymphocytic choriomeningitis; LU, lytic units (19); P-815, DBA/2 mastocytoma cells (*H-2*^{d}).

Doctors D. C. Shreffler and C. S. David, Department of Human Genetics, the University of Michigan Medical School, Ann Arbor, Mich., and were then bred locally. The *H-2* genotypes (15) of the mouse strains used throughout these experiments are shown in Table I.

Immunization. 7- to 10-wk old mice were injected intracerebrally (i.c.) with 300 LD_{50}, or intravenously (i.v.) with a mean lethal dose (LD_{50}) of the WE3 strain of LCM virus.

Target cells. C3H (*H-2^k*) mouse L929 fibroblasts (L cells) were grown in Eagle's minimal essential medium (F-15; Grand Island Biological Co., Grand Island, N. Y.) plus 10% fetal calf serum (FCS) in 250-ml Falcon plastic tissue culture flasks (Falcon Plastics, Div. of BioQuest, Oxnard, Calif.). When confluent, the monolayer was infected with supernates of a 10% suspension of LCM-infected guinea pig lung for 1 h at 37°C. After one wash with 20 ml of prewarmed medium the cells were incubated for 24 h at 37°C. After trypsinization, normal or infected L cells were transferred into spinner flasks, at a density of $2-5 \times 10^5$/ml, and maintained as spinner cultures for a further 24 h before being used as ^{51}Cr-labeled targets, or as unlabeled competitor cells.

Mastocytoma cells of DBA/2 (*H-2^d*) (P-815) origin were cultured in Dulbecco's modified Eagle's medium (H-16, GIBCO) plus 10% FCS. LCM-infected P-815 cells were maintained as a continuously infected line.

Peritoneal Macrophages. Peritoneal macrophages were obtained from mice killed by cervical

TABLE I
H-2 Genotypes of the Mouse Strains Used (15)

Strain	H-2 genotype					
	K	IA	IB	IC	Ss-Slp	D
BALB/c	d	d	d	d	d	d
	d	d	d	d	d	d
CBA/H	k	k	k	k	k	k
	k	k	k	k	k	k
CBA/H × BALB/c	k	k	k	k	k	k
	d	d	d	d	d	d
A/J	k	k	k	d	d	d
	k	k	k	d	d	d
C3H.OH	d	d	d	d	d	k
	d	d	d	d	d	k

dislocation. The abdominal skin was reflected and 10–15 ml of cold (4°C) Puck's saline A (16) were injected vigorously into the abdominal cavity, aspirated, and reinjected twice more. Between 2×10^6 to 2×10^7 viable large nucleated cells were recovered from each mouse. These cells were plated at a density of 2×10^5 large cells per well of 96-hole tissue culture trays (Linbro Chemical Co., New Haven, Conn.). Cell losses during infection, ^{51}Cr labeling, and washings left about 5×10^4 cells per well.

Cytotoxicity Assay. The assay for measuring T-cell-mediated virus-specific ^{51}Cr release was performed, with some modification, as described previously (17). LCM virus-infected or normal target cells were labeled with ^{51}Cr and washed. Target cells in suspension were dispensed in 50-μl aliquots, containing 5×10^4 cells, into individual flat-bottomed wells of 96-hole plastic trays.

Targets were overlayed at a ratio of 20:1, except where otherwise stated. After incubation for 8–10 h at 37°C, the supernates were pipetted and the remaining ^{51}Cr was recovered by water lysis. Results are expressed as the mean ± SEM percent ^{51}Cr release for four replicates (17). The cytotoxic activity per spleen was calculated by determining the total amount of arbitrary lytic units (LU), defined as numbers of nucleated spleen cells necessary to specifically lyse 33% of the target cells in one well (19).

Competing unlabeled target cells were added, at appropriate concentrations, in 50-μl vol to wells containing ^{51}Cr-labeled targets. Spleen cells were then added in a further 200 μl, using a tuberculin syringe fitted with a 26 gauge needle to achieve good mixing.

Results

Assay on F_1 and H-2 Recombinant Target Cells. Compatibility of immune spleen cells and target cells at one *H-2* haplotype, or at *H-2K* or *H-2D*, was sufficient for the lytic interaction to occur (Table II). Providing that this degree of identity was achieved, presence of foreign *H-2* specificities on T cells or target cells did not cause impairment of cytoxicity, establishing that the requirement for *H-2* compatibility in no way reflects allogeneic inhibition (18).

Relative Cytotoxic Activity Generated by F_1 and H-2 Recombinant Mice. The capacity of immune F_1 and *H-2* recombinant spleen cells to lyse H-2^k or H-2^d virus-infected targets was determined firstly as LU (19), secondly as LU per spleen, and thirdly as percent of cytotoxic activity for syngeneic H-2^k or H-2^d combinations (Table III). Comparisons of absolute LU values can only be made

TABLE II

LCM-Specific Cytotoxic Activity of Spleen Cells Assayed on F_1 and H-2 Recombinant Macrophage* Target Cells

Mouse strain	Spleen cells §	^{51}Cr release‡ from target macrophages							
		CBA/H		BALB/c		CBA/H × BALB/c F$_1$		A/J	
		LCM	Normal	LCM	Normal	LCM	Normal	LCM	Normal
		%	%	%	%	%	%	%	%
CBA/H	Immune	61.0±1.3‖	28.3±1.1	34.6±2.4	36.4±1.3	55.1±0.8‖	31.9±0.9	57.5±1.2‖	37.0±2.6
	Normal	31.4±2.1	27.5±0.6	38.4±3.3	34.6±0.7	34.7±2.8	31.3±2.5	36.3±0.3	38.0±2.2
BALB/c	Immune	35.2±1.4	32.2±1.3	68.0±2.8‖	35.2±1.9	71.7±2.0‖	29.3±0.4	69.3±3.9‖	38.2±1.2
	Normal	34.0±1.5	31.9±1.3	36.6±2.6	38.9±1.6	36.9±3.3	31.3±2.5	33.6±4.2	38.4±3.4
CBA/H × BALB/cF$_1$	Immune	56.4±1.5‖	29.2±2.0	69.9±2.7‖	39.6±2.3	73.1±1.2‖	29.3±0.4	NT	NT
	Normal	33.4±4.3	28.7±2.1	33.3±1.3	37.5±2.5	35.8±2.6	31.6±1.7	NT	NT
A/J	Immune	50.0±2.8‖	32.4±2.2	67.8±0.2‖	32.5±0.6	63.0±1.0‖	31.4±0.7	68.7±4.8‖	38.7±2.3
	Normal	28.5±0.7	31.2±0.4	43.8±6.4	39.8±3.7	36.8±1.5	29.3±2.3	31.4±4.8	40.1±1.7

* Peritoneal macrophages were infected 12 h after plating and assayed 24 h later.
‡ Spleen cells were overlayed at 30:1 for 9 h at 37°C. Mean ± SEM of four replicates.
§ Donor mice were infected i.v. with 2,000 LD$_{50}$ WE3 9 days previously.
‖ Significantly greater than control values ($P < 0.001$).

within the same target system, as the P-815 are apparently more readily lysed than are the L cells. In both systems T cells from F_1 mice showed from 70 to 100% of the activity of parental strain lymphocytes, depending on the method of comparison. Recombinant T cells were relatively more effective when there was compatibility at *H-2D* (Tables I and III), an observation indicating that there may be at least two specificities of cytotoxic T cells, those associated with the *D* end being the more potent.

Cold Target Competition In Vitro. Further evidence for more than one specificity of cytotoxic T cells in LCM-immune mice was found from cold target competitive inhibition experiments in vitro (20). The cytotoxic activity of immune T cells for ^{51}Cr-labeled, virus-infected target cells could only be inhibited specifically by mixture with unlabeled, syngeneic virus-infected cells (Fig. 1). Syngeneic normal cells, or virus-infected allogeneic cells, had no effect.

TABLE III

Relative Activity of Immune Spleen* Cells from Parent, F₁, and H-2 Recombinant Mice

Mouse strain	Total N cells per spleen	H-2^k LCM target			H-2^d LCM target		
		LU‡	LU/ spleen	CBA/H (H-2^k) activity	LU	LU/ spleen	BALB/c (H-2^d) activity
				%			%
CBA/H	4.2×10^7	5.5	77	100	$>10^7$	<5	<2
BALB/c	6.8×10^7	$>10^7$	<7	<9	2.0×10^5	340	100
CBA/H × BALB/c F₁	3.0×10^7	5.5×10^5	55	72	1.1×10^5	268	79
A/J	3.7×10^7	1.1×10^6	35	46	1.6×10^5	237	70
C3H.OH	6.1×10^7	1.0×10^6	59	77	4.0×10^5	153	45

* All mice were sampled on day 7. The kinetics of the cytotoxic T-cell response is comparable in different mouse strains (3).

‡ LU, i.e., number of nucleated (N) cells necessary to specifically lyse 33% of the LCM-infected targets (19).

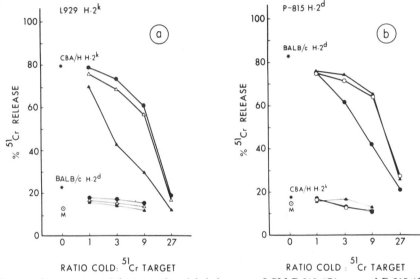

FIG. 1. Competitive inhibition with unlabeled targets, LCM-P-815 (●), normal P-815 (○), LCM-L cells (▲), or normal L cells (△), added at various ratios of the LCM-specific ^{51}Cr release (*) by 7 day immune CBA/H (H-2^k) and BALB/c (H-2^d) spleen cells assayed on (a) LCM-L cells (H-2^k) and (b) LCM-P-815 (H-2^d). Spontaneous ^{51}Cr (⊙) release was not significantly different from that observed for normal target cells.

Competition was most apparent when cold targets were at from three to nine times in excess. Use of higher concentrations resulted in nonspecific steric hindrance.

Cytolytic capacity of immune spleen cells from H-2$^{k/d}$ F₁ mice was inhibited only when the competing virus-infected cells were syngeneic with the target cells (Fig. 2). Presence of excess H-2$^{d/d}$ virus-infected mastocytoma cells caused no specific decrease in killing of H-2$^{k/k}$ virus-infected L-cells, and the converse was also true. This finding is most readily interpreted as indicating that there are at least two specificities of LCM-immune cytotoxic T cells in H-2$^{k/d}$ F₁ mice, each associated with altered-self characteristics of one parental H-2 type (14).

Essentially similar results were recorded for recombinant mice (Table I and Fig. 3). Cytotoxic activity of immune lymphocytes for *H-2K* or *H-2D* compatible virus-infected targets was specifically inhibited only when there was *H-2* identity between target and competing cells, and was not diminished by presence of

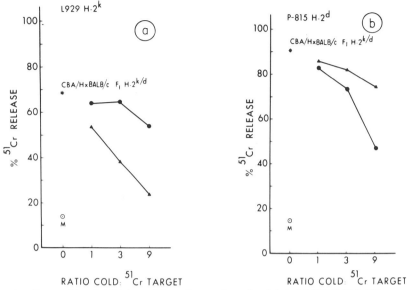

FIG. 2. Competitive inhibition with various ratios of unlabeled LCM-P-815 (●), or with LCM-L cells (▲), of the LCM-specific ^{51}Cr release by 7 day immune CBA/H × BALB/c F$_1$ (*H-2$^{k/d}$*) spleen cells (*) assayed on LCM-L cells (*H-2^k*) or LCM-P-815 (*H-2^d*) target cells.

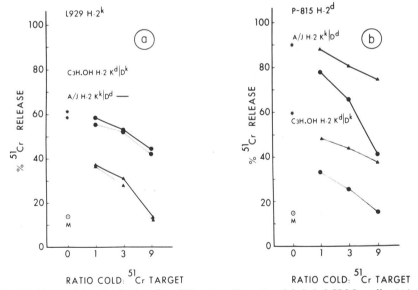

FIG. 3. Competitive inhibition by different ratios of unlabeled LCM-L cells (▲) or LCM-P-815 (●) of the LCM-specific ^{51}Cr release from labeled LCM-L cells or LCM-P-815 by 7 day immune A/J (*H-2K^k/D^d*) or C3H.OH (*H-2K^k/D^k*) (*) spleen cells.

virus-infected cells compatible with the T cell at the irrelevant *H-2* locus. Again, these experiments indicate that there are at least two specificities of cytotoxic T cells in recombinant mice, one associated with *H-2K* and the other with *H-2D*.

Selective Proliferation In Vivo. Evidence for cytotoxic T cells of more than one specificity was also found from in vivo experiments. The system used has been described previously (14). Briefly, donors are killed at 7 days after i.v. inoculation with 2,000 LD_{50} of WE3 LCM virus. Recipients are irradiated (850 R) 24 h before adoptive transfer, dosed with 10^6 LD_{50} of WE3 LCM virus 18 h later, inoculated with 10^8 immune donor lymphoid cells, and cytotoxic activity in pooled spleen and lymph nodes determined after a further 3 days. Donor cells (14) proliferate in the recipient throughout this interval, as shown by failure of irradiated (850 R) spleen cells to multiply further (Table IV). Replication depends on continued exposure to histocompatible, virus-infected recipient cells

TABLE IV

Failure of Irradiated (850 R) LCM-Immune CBA/H × C57BL F_1 Immune T Cells to Show Cytotoxic Activity at 72 h after Transfer to Irradiated, Virus-Infected CBA/H Recipients

Cell population	Treatment	^{51}Cr release from L cells	
		Infected	Normal
		%	%
Before transfer*	None	53.9 ± 3.2	15.3 ± 1.2
	850 R	54.6 ± 4.2	14.7 ± 2.1
72 h after transfer‡	None	98.2 ± 1.9	18.8 ± 1.9
	850 R	19.1 ± 0.5	16.9 ± 3.1
Control LCM immune‡		93.9 ± 2.3	18.4 ± 0.9
Control normal spleen		18.9 ± 1.4	17.7 ± 1.7

* Assayed for 4 h at 37°C.
‡ Assayed for 15 h at 37°C.

(5, 14). Presence of alloantigens, in an F_1 donor or recipient, has no inhibitory effect (5, 14), providing one haplotype is shared.

T cells from 7 day immune F_1 ($H-2^{k/d}$) mice readily lyse virus-infected targets of both parental *H-2* types (Table II). However passage through $H-2^k$ recipients results in selective proliferation of lymphocytes with significant cytotoxic activity only for $H-2^k$ virus-infected fibroblasts (Table V). The converse applies after transfer through $H-2^d$ recipients. These results are best explained as representing preferential multiplication of cytotoxic T cells of two distinct specificities, each associated with one parental *H-2* type (14).

Similar observations have also been made for T cells from recombinants (Table I). Spleen cells from immune A/J or C3H.OH mice lyse both $H-2^k$ and $H-2^d$ virus-infected targets (Fig. 3). Passage through $H-2^k$ recipients selects a population(s) of T cells which preferentially kill $H-2^k$ targets, i.e., specific for altered *H-2K* in the A/J mice or altered *H-2D* in the C3H.OH mice (Table V).

TABLE V

Cytotoxic T-Cell Activity in Lymphoid Tissue of Irradiated LCM-Infected Recipients Dosed with LCM-Immune Spleen Cells 3 days Previously*

		^{51}Cr release‡			
		L-929		P-815	
	Recipients	Infected	Normal	Infected	Normal
		%	%	%	%
Donors					
CBA/H × BALB/c F$_1$	BALB/c	21.2 ± 1.2	18.5 ± 1.2	78.0 ± 1.2	26.9 ± 3.1
CBA/H × BALB/c F$_1$	CBA/H	84.9 ± 0.9	25.5 ± 2.1	29.8 ± 0.8	22.7 ± 2.1
C3H.OH	CBA/H	48.3 ± 0.3	21.9 ± 1.2	25.8 ± 1.5	23.8 ± 3.1
A/J	CBA/H	63.3 ± 1.1	21.0 ± 1.1	32.9 ± 1.5	28.9 ± 3.3
Controls					
BALB/c immune		24.9 ± 1.0	18.6 ± 0.6	70.2 ± 3.9	20.5 ± 1.0
BALB/c normal		18.6 ± 1.5	15.6 ± 1.6	19.9 ± 1.8	23.6 ± 0.6
CBA/H immune		85.7 ± 1.6	22.7 ± 2.9	23.6 ± 0.6	23.4 ± 0.7
CBA/H normal		18.3 ± 0.5	23.8 ± 1.4	22.1 ± 3.2	21.6 ± 1.1
Medium		18.6 ± 0.5	17.8 ± 0.7	18.7 ± 0.3	22.4 ± 2.7

* Donor mice were injected i.v. with 2,000 LD$_{50}$ of WE3 LCM virus and killed after 7 days. Recipient mice were irradiated (850 R) 24 h previously and dosed with 10^6 LD$_{50}$ of WE3 LCM 6 h before cell transfer. Activity of donor cells in Table III and Figs. 1–3.

‡ Target cells were overlayed at a ratio of 30:1 and incubated for 10 h at 37°C. Means ± SEM of four replicates.

Discussion

LCM-immune mice apparently possess cytotoxic T cells of at least two broad specificities, associated with either *H-2K* or *H-2D*. The evidence for this is derived from a variety of separate approaches. Genetic mapping studies, using recombinant mice, have rigorously shown that T cells and virus-infected targets need be compatible only in the region of either the *H-2K* or the *H-2D* locus.[2] The same is true for the ectromelia model.[2] Identity at *Ir*-Ss/Slp is neither essential, nor is it sufficient, for successful interaction to occur.[2] It must be stressed that there is no indication that this *H-2* compatibility requirement for cytotoxic T-cell killing of virus-infected cells reflects presence of a physiological interaction mechanism coded for by genes in the *Ir* region (9, 10, 12, 21).

Even so, existence of two separate self-recognition systems, specified by genes in either the *H-2K* or *H-2D* regions, may still be invoked, e.g., in this case the observation that compatibility at *H-2D* results in greater killing could indicate that the physiological interaction mechanism coded for at the *D* end is more effective. The apparent separation of two distinct T-cell specificities in F$_1$ mice, one associated with either parental haplotype, might be explained by allelic exclusion in this hypothetical physiological interaction mechanism (22,

---

[2] Blanden, R. V., P. C. Doherty, M. B. C. Dunlop, I. D. Gardner, R. M. Zinkernagel, and C. S. David. 1975. Genes required for T-cell-mediated cytotoxicity against virus-infected target cells are in the *K* and *D* regions of the *H-2*-gene complex. Manuscript submitted for publication.

23). Such an argument cannot, however, be applied to the comparable results obtained for *H-2* recombinant mice, as they are homozygous. It would thus be necessary to postulate that only one, or other, of the self-recognition structures coded for at *H-2K* or *H-2D* could be expressed in any one T cell, i.e., some form of *H-2K* or *H-2D* receptor exclusion. Furthermore, the results of the selective proliferation experiments indicate that any such exclusion is clonal.

This complex mechanism is, to our knowledge, without precedent. We therefore favor the simpler postulate that the T cell is sensitized to altered self (1–3, 14), the self antigens involved being coded for at, or near to, either *H-2K* or *H-2D*. Altered self is interpreted broadly as reflecting either short- or long-range virus-induced modification of *H-2* antigens, perhaps associated with the *H-2* private specificities or products of genes linked closely to them. Such a concept accommodates recognition of alloantigens, or of changes induced by viruses, intracellular bacteria or chemicals, within the same model (14). Furthermore, it is in general agreement with Bodmer's modification (24) of Jerne's argument for generation of immunological diversity (25). In the present model all T-cell specificities may be considered to have arisen by somatic mutation from genes originally coding for recognition of histocompatibility antigens.

The altered-self concept is also in accord with Burnet's immunological surveillance theory, broadened somewhat to include response to infectious agents as an essential part (26). Viruses and bacteria may, in evolutionary terms, be considered as potent selective forces, affecting young animals particularly. The T-cell-mediated response to such agents is, in the great majority of cases, protective. The realization that T cells in virus-infected mice are sensitized to either altered *H-2K* or *H-2D* thus provides a reason for gene duplication in the *H-2*-gene complex (27). The overall T-cell response is, as numerically analyzed here, considerably augmented by the presence of two distinct cytotoxic T-cell specificities. An identical argument may be used to explain selective advantage of heterozygosity of the *H-2*-gene complex, four specificities are more effective than two. The extreme polymorphism of histocompatibility antigen systems (24) may thus, on a purely mechanistic basis, be considered to have developed in response to selective pressure exerted by infectious agents.

Summary

Use of syngeneic, allogeneic, F_1, and *H-2* recombinant mice has shown that animals injected with lymphocytic choriomeningitis (LCM) virus generate T cells which are cytotoxic for *H-2K* or *H-2D* compatible, but not *H-2* different, virus-infected target cells. Three separate lines of evidence are presented which indicate that these immune T cells are sensitized to "altered-self," the self antigens involved being coded for in the *H-2K* or *H-2D* regions. Firstly, cytotoxic activity associated with mutuality at *H-2D* is greater than that observed when T cells and targets are identical at *H-2K*. Secondly, lysis mediated by immune T cells from F_1 or *H-2* recombinant mice is specifically inhibited only by presence of unlabeled, virus-infected cells that are *H-2* compatible with the targets. Thirdly, LCM-immune F_1 and *H-2* recombinant T cells inoculated into irradiated, virus-infected recipients proliferate only to kill target cells that are *H-2* compatible with both the donor and the recipient.

All of these experiments establish that there is a dissociation of T-cell activities between parental haplotypes in F_1 mice, and between *H-2K* and *H-2D* in recombinants. It would thus seem that there are at least two specificities of LCM-immune T cells in homozygotes, associated with either *H-2K* or *H-2D*, and four specificities in F_1 hybrids. The significance of these findings, with respect both to gene duplication and to the marked polymorphism in the *H-2* system, is discussed.

We wish to thank Dr. M. H. R. MacDonald for helpful discussions, Doctors D. C. Shreffler and C. S. David for the original breeding pairs of C3H.OH mice, and Miss Gail Essery for her capable technical assistance.

Received for publication 3 February 1975.

References

1. Zinkernagel, R. M., and P. C. Doherty. 1974. Restriction of *in vitro* T cell-mediated cytotoxicity in lymphocytic choriomeningitis within a syngeneic or semiallogeneic system. *Nature (Lond.).* **248:**701.

2. Doherty, P. C., and R. M. Zinkernagel. 1974. T cell-mediated immunopathology in viral infections. *Transplant. Rev.* **19:**89.

3. Doherty, P. C., and R. M. Zinkernagel. 1975. *H-2* compatibility is required for T-cell-mediated lysis of target cells infected with lymphocytic chorimeningitis virus. *J. Exp. Med.* **141:**502.

4. Gardner, I., N. A. Bowern, and R. V. Blanden. 1974. Cell-mediated cytotoxicity against ectromelia virus-infected target cells. III. Role of the H-2 gene complex. *Eur. J. Immunol.* **4:**63.

5. Doherty, P. C., and R. M. Zinkernagel. 1975. Capacity of sensitized thymus derived lymphocytes to induce fatal lymphocytic choriomeningitis is restricted by the H-2 gene complex. *J. Immunol.* **114:**30.

6. Blanden, R. V. 1975. Mechanisms of cell-mediated immunity in viral infection. *In* Proceedings of the Second International Congress of Immunology. Brighton, England. L. Brent and J. Holborow, editors. **2(4):**17.

7. Shearer, G. M. 1974. Cell-mediated cytotoxicity to trinitrophenyl-modified syngeneic lymphocytes. *Eur. J. Immunol.* **4:**527.

8. Kindred, B., and D. C. Shreffler. 1972. H-2 dependence of co-operation between T and B cells *in vivo. J. Immunol.* **109:**940.

9. Katz, D. H., T. Hamaoka, and B. Benacerraf. 1973. Cell interactions between histoincompatible T and B lymphocytes. II. Failure of physiologic cooperative interactions between T and B lymphocytes from allogeneic donor strains in humoral response to hapten-protein conjugates. *J. Exp. Med.* **137:**1405.

10. Katz, D. H., T. Hamaoka, M. E. Dorf, and B. Benacerraf. 1973. Cell interactions between histoincompatible T and B lymphocytes. The H-2 gene complex determines successful physiologic lymphocyte interactions. *Proc. Natl. Acad. Sci. U. S. A. 70:* 2624.

11. Zinkernagel, R. M. 1974. Restriction by the H-2 gene complex of the transfer of cell-mediated immunity to *Listeria monocytogenes. Nature (Lond.)* **251:**230.

12. Rosenthal, A. S., and E. M. Shevach. 1973. Function of macrophages in antigen recognition by guinea pig T lymphocytes. I. Requirement for histocompatible macrophages and lymphocytes. *J. Exp. Med.* **138:**1194.

13. Wainberg, M. A., Y. Markson, D. W. Weiss, and F. Donjanski. 1974. Cellular immunity against Rous sarcomas of chickens. Preferential reactivity against autoch-

thonous target cells as determined by lymphocyte adherence and cytotoxicity tests *in vitro*. *Proc. Natl. Acad. Sci. U. S. A.* **71**:3565.

14. Zinkernagel, R. M., and P. C. Doherty. 1974. Immunological surveillance against altered self by sensitized thymus derived lymphocytes in lymphocytic choriomeningitis. *Nature (Lond.).* **251**:547.

15. Shreffler, D. C. and C. S. David. 1974. The *H-2* major histocompatibility complex and the I immune response region: genetic variation, function, and organization. *Adva. Immunol.* In press.

16. Blanden, R. V., and R. E. Langman. 1972. Cell-mediated immunity to bacterial infection in the mouse. Thymus-derived cells as effectors of acquired resistance to *Listeria monocytogenes. Scand. J. Immunol.* **1**:379.

17. Zinkernagel, R. M., and P. C. Doherty. 1974. Characteristics of the interaction *in vitro* between cytotoxic thymus-derived lymphocytes and target monolayers infected with lymphocytic choriomeningitis virus. *Scand. J. Immunol.* **3**:287.

18. Hellström, K. E., and G. Möller. 1965. Immunological and immunogenetic aspects of tumor transplantation. *Prog. Allergy.* **9**:158.

19. Cerottini, J. C., and K. T. Brunner. 1974. Cell-mediated cytotoxicity, allograft rejection, and tumor immunity. *Adv. Immunol.* **18**:67.

20. Ortiz de Landazuri, M., and R. B. Herberman. 1972. Specificity of cellular immune reactivity to virus-induced tumors. *Nat. New Biol.* **238**:18.

21. Katz, D. H., and B. Benacerraf. 1974. The function and interrelationship of T cell receptors, Ir genes and other histocompatibility gene products. *Transplant. Rev.* In press.

22. Pernis, B., G. Chiappino, A. S. Kelus, and P. G. H. Gell. 1965. Cellular localization of immunoglobulins with different allotypic specificities in rabbit lymphoid tissue. *J. Exp. Med.* **122**:853.

23. Gell, P. G. H. 1967. Restriction on antibody production by single cells. *Cold Spring Harbor Symp. Quant. Biol.* **32**:441.

24. Bodmer, W. F. 1972. Evoluntionary significance of the HL-A system. *Nature (London).* **237**:139.

25. Jerne, N. K. 1971. The somatic generation of immune recognition. *Eur. J. Immunol.* **1**:1.

26. Burnet, M. 1970. Immunological Surveillance. Pergamon Press (Australia) Pty. Ltd., Sydney, Australia.

27. Klein, J., and D. C. Shreffler. 1971. The H-2 model for the major histocompatibility systems. *Transplant. Rev.* **6**:3.

ROLE OF THE *H-2* COMPLEX IN INDUCTION OF T HELPER CELLS IN VIVO

I. Antigen-Specific Selection of Donor T Cells to Sheep Erythrocytes in Irradiated Mice Dependent upon Sharing of *H-2* Determinants between Donor and Host*

By J. SPRENT

(From the Immunobiology Unit, Department of Pathology, University of Pennsylvania School of Medicine and the Wistar Institute for Anatomy and Biology, Philadelphia, Pennsylvania 19104)

Optimal stimulation of T cells by antigen in vitro requires the presence of macrophages (1). The precise function of these cells in antigen presentation is controversial. A number of workers contend that macrophages render antigen immunogenic by presenting T cells with a complex of antigen bound to major histocompatibility complex (MHC)[1] determinants (2–6). Certain of these groups (3, 6), though not others (5), conclude that T-cell activation depends upon macrophages and the responding T cells sharing MHC determinants. By contrast, other groups argue that at least for some antigens, the main role of macrophages in vitro is to release factors which promote cell viability (7, 8).

It is important to establish whether macrophages and MHC gene products play a role in T-cell activation in vivo. The approach adopted in the present paper to study this problem is based on the observation that T cells encountering specific antigen in vivo leave the circulation, e.g. thoracic duct lymph, for a period of 1–2 days (9–12). During this stage of negative selection the cells become selectively sequestered in regions where the antigen is concentrated, e.g. the spleen. Here the cells proliferate extensively before re-entering the circulation in expanded numbers after 4–5 days — the stage of positive selection.

This chain of events could either reflect T-cell stimulation by antigen per se or, alternatively, by antigen processed by macrophages or related cells. If the second possibility were correct and, in addition, if the macrophages presenting the antigen had to be MHC-compatible with the T cells, then activation of T cells to antigen should not occur in an MHC-different environment.

The present studies verify this prediction with the demonstration that purified T cells transferred with a particulate antigen (heterologous erythrocytes) into irradiated mice fail to undergo either negative or positive selection to the antigen unless the donor and host share *H-2* determinants.

Materials and Methods

Mice. CBA/J (CBA) (*H-2^k*), C57BL/6 (B6) (*H-2^b*), C57BL/10 (B10) (*H-2^b*), B10.D2 (*H-2^d*), BALB/c (*H-2^d*), C3H/He (H-2^k), and (B6 × DBA/2 (*H-2^d*))F₁ mice were obtained from The

* Supported by grants AI-10961 and CA-15822 from the U. S. Public Health Services.

[1] *Abbreviations used in this paper:* HRC, horse erythrocytes; LN, lymph nodes; MHC, major histocompatibility complex; PFC, plaque-forming cells; SRC, sheep erythrocytes; TDL, thoracic duct lymphocytes.

Jackson Laboratory, Bar Harbor, Maine. (CBA × B6)F$_1$ mice were obtained from Cumberland View Farms, Clinton, Tenn.

Media. RPMI-1640 (Microbiological Associates, Walkerville, Md.) supplemented with 10% fetal calf serum was used.

Injections. All suspensions of lymphoid cells, sheep erythrocytes (SRC), and horse erythrocytes (HRC) were given intravenously.

Cells. Thoracic duct lymphocytes (TDL) were obtained as described elsewhere (13). Suspensions of lymph node (LN) cells were prepared by teasing with fine forceps through an 80-mesh stainless steel sieve in cold medium.

Irradiation. Mice were exposed to ^{137}Cs γ-irradiation at a dose rate of 106 rads/min (14).

Cell Identification with Alloantisera. CBA anti-B6, B6 anti-CBA, and CBA anti-DBA/2 *H-2*-alloantisera and anti-Thy 1.2 (AKR anti-C3H thymus) antiserum were prepared as described elsewhere (14). For cell identification after negative selection, TDL were treated with alloantisera and complement by a two-step procedure (14). Cytotoxic indices were established with respect to control samples incubated with antisera without complement or with normal mouse serum plus complement. Percent lysis with the control samples was invariably <5%.

T-Cell Purification. The T cells used for negative selection were obtained from pooled mesenteric, axillary, cervical, and inguinal LN. Most Thy 1.2-negative cells were removed by passing cells over nylon-wool columns (15). The effluent T cells were >90% Thy 1.2-positive.

Preparation of B Cells. Spleen cells from mice primed with both SRC and HRC 2–4 mo previously were treated with anti-Thy 1.2 antiserum and complement as described previously (14).

Measurement of T-B Collaboration. Unless stated otherwise, the helper function of CBA T cells after negative selection was studied by transferring the cells in small doses (2×10^6) into irradiated (700 rads) CBA mice together with SRC and HRC (0.1 ml of 5% solution of each) and 5×10^6 anti-Thy 1.2-serum-treated spleen cells from SRC- and HRC-primed CBA mice as a source of B cells. Direct (IgM) and indirect (IgG) plaque-forming cells (PFC) to both SRC and HRC were then measured in the spleens of the recipients 7 days later (14).

Priming with Heterologous Erythrocytes. Unless stated otherwise, the T and B cells were both prepared from mice primed intraperitoneally with 0.2 ml of a 25% solution of a mixture of SRC and HRC 2–4 mo previously.

Statistical Analysis. Geometric means and values used to derive upper and lower limits of the SE of the mean (these values represent the anti-log of SE of the $\log_{10}$ geometric mean) were calculated from the $\log_{10}$ of the PFC counts. *P* values were determined by Students' *t* test. In the comparison of the mean of any two groups of observations a significance level of 0.05 was chosen.

Results

Experimental Design. The general plan of the experiments was to inject purified CBA T cells into irradiated syngeneic or allogeneic mice together with SRC and study the helper function of the donor-derived T cells recovered from thoracic duct lymph of the recipients 1–2 days later. The T cells were obtained from LN of mice primed to both SRC and HRC 2–4 mo previously. Before transfer, the T cells were depleted of B cells (and presumably most macrophages) by nylon wool filtration. The effluent T cells (>90% Thy 1.2-positive) were transferred intravenously in a dose of 10^8 viable cells together with SRC (0.5 ml of 50% solution) into mice that had received 900 rads 1 day before. Thoracic duct fistulas were inserted in the recipients 20 h later and TDL were collected between 24 and 40 h postinjection. Testing with anti-Thy 1.2 antiserum and complement showed that the lymph-borne cells consisted almost entirely (≥97%) of T (Thy 1.2-positive) cells. In situations where the cells were filtered through *H-2*-incompatible hosts, testing with appropriate alloantisera showed that the lymph-borne cells were invariably > 90% of donor origin. Cell viability was 99–100% and the yield of cells (compared with the numbers initially injected) was in the order of 10–15%.

TABLE I

Helper Function of Primed CBA T Cells Negatively Selected to SRC in Irradiated CBA Mice. T Cells Harvested from Lymph at 1 Day Post-Transfer

T-Cell group	Helper cells* (2.5×10^6)	SRC Added during negative selec- tion	B Cells‡	PFC/Spleen at 7 days in irradiated CBA mice			
				Anti-SRC		Anti-HRC	
				IgM	IgG	IgM	IgG
A	CBA T Cells → irradiated CBA	–	CBA	11,880(1.09)§	36,530(1.25)	6,020(1.17)	44,670(1.16)
B	CBA T Cells → irradiated CBA	+	CBA	0	0	5,730(1.44)	36,620(1.45)
	Group A + group B (2.5×10^6 of each)		CBA	11,460(1.18)	42,960(1.34)	13,580(1.12)	79,530(1.04)

* 10^8 Nylon-wool-passed LN T cells from CBA mice primed with both SRC and HRC 2–4 mo previously were transferred intravenously ± 0.5 ml of 50% SRC into CBA mice given 900 rads 1 day before. Mice were cannulated at 20 h postinjection and TDL collected between 24 and 40 h. To measure helper function, the lymph-borne T cells were transferred with B cells, SRC, and HRC into irradiated (700 rads) CBA mice.

‡ 5×10^6 Viable anti-Thy 1.2-serum-treated spleen cells from mice primed with SRC and HRC.

§ Geometric mean of four mice per group; figure in parenthesis refers to value by which mean is multiplied by or divided by to give upper and lower limits, respectively, of SE. Background values given by B cells transferred without T cells have been subtracted. These values (PFC/spleen) were; 1,850 (1.24) (IgM SRC), 2,080 (1.73) (IgG SRC), 1.140 (1.18) (IgM HRC), 1,670 (1.68) (IgG HRC). Values for T cells transferred without B cells were all < 100 PFC/spleen.

To study their helper function the lymph-borne cells were transferred in a dose of 2×10^6 cells into irradiated (700 rads) CBA mice together with SRC and HRC (0.1 ml of 5% solution of each) and syngeneic B cells (5×10^6 anti-Thy 1.2-serum-treated spleen cells from CBA mice primed with both SRC and HRC). Numbers of direct (IgM) and indirect (IgG) PFC to SRC and HRC were then measured in the spleen 7 days later.

Negative Selection in Syngeneic Mice. When CBA T cells were transferred into irradiated CBA mice and recovered from thoracic duct lymph of the recipients 1 day later, the cells provided effective help for both SRC and HRC (Table I). By contrast, CBA T cells filtered through irradiated CBA mice in the presence of SRC failed to stimulate either IgM or IgG anti-SRC PFC responses. This abolition of help for SRC was specific since good responses were observed against HRC. The phenomenon did not appear to be due to suppression since the injection of both TDL populations together gave high responses. (For simplicity, background values obtained when B cells were transferred without T cells have been subtracted from the data shown in Table I and in subsequent tables; these values are shown in the footnotes to the tables.)

The specific unresponsiveness of T cells taken from the central lymph at 1 day after cell transfer was followed by a period of hyper-reactivity where the lymph-borne cells gave markedly increased responses to the injected antigen. Thus, as illustrated in Table II, T cells collected from the lymph at 5 days post-transfer, i.e. during the stage of positive selection to antigen, gave high IgM and IgG anti-SRC responses with as few as 0.5×10^6 T cells. These responses were over 40-fold higher than with the same dose of cells taken from irradiated mice given T cells but not antigen 5 days before. Responses against HRC were low with both T-cell populations unless higher doses of helper cells (2.5×10^6) were used.

Negative Selection in Allogeneic Mice. The negative selection to SRC observed when CBA (H-2^k) T cells were filtered from blood to lymph for 1 day

TABLE II

Helper Function of Primed CBA T Cells Positively Selected to SRC in Irradiated CBA Mice. T Cells Harvested from Lymph at 5 Days Post-Transfer

Helper cells*	SRC Added during positive selec- tion	Dose of helper cells	B cells‡	PFC/Spleen at 7 days in irradiated CBA mice			
				Anti-SRC		Anti-HRC	
				IgM	IgG	IgM	IgG
CBA T Cells → ir- radiated CBA		0.5×10^6	CBA	460 (1.40)§‖	1,120 (1.54)‖	220 (1.77)‖	2,350 (1.30)‖
		2.5×10^6	CBA	4,420 (1.26)	20,320 (1.38)	1,890 (1.26)	18,700 (1.16)
CBA T Cells → ir- radiated CBA	+	0.5×10^6	CBA	8,280 (1.15)	51,180 (1.18)	670 (1.21)	4,490 (1.22)
		2.5×10^6	CBA	16,470 (1.10)	65,860 (1.13)	1,980 (1.15)	24,480 (1.07)

* As for footnote to Table I except that mice were cannulated to obtain TDL as helper cells at 5 days after T-cell transfer.

‡ As for footnote to Table I.

§ As for footnote to Table I. Subtracted values of B cells transferred without T cells were: 1,760(1.04) (IgM SRC), 2,920 (1.43) (IgG SRC), 630(1.17) (IgM HRC), 4,020(1.22) (IgG HRC). Values for T cells transferred without B cells were < 300 PFC/spleen.

‖ Not significantly above background values of B cells transferred without T cells.

TABLE III

Negative Selection of Primed CBA T Cells to SRC in Semiallogeneic Hosts but not in Allogeneic Hosts. T Cells Harvested from Lymph at 1 Day Post-Transfer

Helper cells* (2×10^6)	SRC Added during nega- tive selec- tion	B Cells‡	PFC/Spleen at 7 days in irradiated CBA mice			
			Anti-SRC		Anti-HRC	
			IgM	IgG	IgM	IgG
CBA T Cells → irradiated B6	–	CBA	5,420(1.33)§	8,670(1.44)	2,950(1.27)	12,510(1.30)
CBA T Cells → irradiated B6	+	CBA	2,560(1.27)	8,740(1.35)	2,240(1.07)	18,700(1.25)
CBA T Cells → irradiated (CBA × B6)F₁	–	CBA	5,240(1.20)	7,550(1.07)	1,180(1.13)	9,520(1.27)
CBA T Cells → irradiated (CBA × B6)F₁	+	CBA	0	110(1.50)‖	2,210(1.40)	11,540(1.08)

*‡ As for footnote to Table I. After filtration, > 94% of the lymph-borne cells were resistant to lysis by CBA anti-B6 alloantiserum; > 97% were lysed with anti-Thy 1.2 antiserum.

§ As for footnote to Table I. Subtracted values of B cells transferred with T cells were: 440 (1.49) (IgM SRC), 130 (1.18) (IgG SRC), 220 (1.42) (IgM HRC), 680 (1.38) (IgG HRC). Values for T cells transferred without B cells were < 200 PFC/spleen.

‖ Not significantly above value of B cells transferred without T cells.

through CBA mice did not occur when the cells were filtered through allogeneic B6 (H-2^b) mice. Thus, as shown in Table III, acute recirculation of CBA T cells through irradiated B6 mice in the presence of SRC did not affect the helper function of the cells for SRC. This was observed in five separate experiments. In no experiment was the IgG anti-SRC response reduced significantly; the IgM response was reduced by approximately 50% in two experiments (significant in only one experiment—Table III) but was not reduced in three other experiments.

TABLE IV

Negative Selection of Primed (CBA × B6)F₁ T Cells to SRC in Irradiated (DBA/2 × B6)F₁ Mice. T Cells Harvested from Lymph at 1 Day Post-Transfer

| T-Cell group | Helper cells* (2 × 10⁶) | SRC Added during negative selection | B Cells‡ | PFC/Spleen at 7 days in irradiated (CBA × B6)F₁ mice | | |
| | | | | Anti-SRC | | Anti-HRC |
				IgM	IgG	IgG
A	(CBA×B6)F₁ T Cells → irradiated (DBA/2 × B6)F₁	–	B6	4,520(1.23)§	19,740(1.15)	14,460(1.10)
B	(CBA × B6)F₁ T cells → irradiated (DBA/2 × B6)F₁	+	B6	0	1,640(1.07)‖	17,190(1.19)
	Group A + group B		B6	5,760(1.13)	26,370(1.20)	39,140(1.31)

*‡ As for footnote to Table I. After filtration, > 90% of the lymph-borne cells were lysed by B6 anti-CBA alloantiserum which had been absorbed with DBA/2 spleen (this antisera lysed 100% of (CBA × B6)F₁ LN but < 10% of (DBA/2 × B6)F₁ LN); > 98% of the cells were lysed by anti-Thy 1.2 antiserum.

§ As for footnote for Table I. Subtracted values of B cells transferred without T cells were: 990(1.34) (IgM SRC), 3,200(1.37) (IgG SRC), 1,100(1.19) (IgG HRC). Values for T cells transferred without B cells were <400 PFC/spleen.

‖ Not significantly above background of B cells transferred without T cells.

Three trivial explanations might account for the failure to observe negative selection to SRC in allogeneic intermediate hosts. The first possibility is that the concomitant induction of a graft-versus-host reaction inhibited selection. The fact that excellent selection to SRC occurred in semiallogeneic (CBA × B6)-F₁ mice (Table III) rules out this possibility.

A second explanation is that a host-versus-graft reaction impaired selection. This was investigated by filtering (CBA × B6)F₁ T cells through irradiated (DBA/2 (*H-2ᵈ*) × B6)F₁ mice in the presence of SRC and studying help provided for B6 Bcells. As shown in Table IV, effective selection to SRC occurred in this situation where both a graft-versus-host reaction and a host-versus-graft reaction could occur.

The third possibility is that B6 mice are nonspecifically less effective than CBA mice at being able to present antigen in an immunogeneic form to T cells. If so, syngeneic T cells would not undergo selection to antigen in B6 mice. The data shown in Table V rules out this possibility. Here it can be seen that effective selection occurred when B6 T cells were filtered with SRC through irradiated B6 mice and tested for their capacity to help B6 B cells.

Role of H-2 Complex in Negative Selection.   In the experiments considered above, it is apparent that negative selection to the injected antigen occurred whenever *H-2*-determinants were shared between the donor T cells and the host presenting antigen to the T cells. To determine whether the phenomenon was indeed *H-2*-linked, negative selection of CBA (*H-2ᵏ*) T cells was studied in irradiated B10.Br (*H-2ᵏ*), B10 (*H-2ᵇ*), and B10.D2 (*H-2ᵈ*) mice (these congenic resistant strains have identical genetic backgrounds and differ only at the *H-2* complex). As shown in Table VI, effective selection to SRC occurred in *H-2-*

TABLE V

Negative Selection of Primed B6 T Cells to SRC in Irradiated B6 Mice. T Cells Harvested from Lymph at 1 Day Post-Transfer

T Cell group	Helper cells* (2 × 10⁶)	SRC Added during negative selection	B Cells‡	Anti-SRC		Anti-HRC
				IgM	IgG	IgG
A	B6 T Cells → irradiated B6	−	B6	9,790(1.22)§	12,910(1.21)	27,220(1.31)
B	B6 T Cells → irradiated B6	+	B6	950(1.22)‖	1,070(1.19)‖	24,220(1.19)
	Group A + group B		B6	14,690(1.18)	15,050(1.19)	52,960(1.13)

*‡ As for footnote to Table I except that B6 T and B cells were used.

§ As for footnote to Table I. Background values of B cells transferred without T cells were: 1,230(1.54) (IgM SRC), 1,770(1.49) (IgG SRC), 880(1.13) (IgG HRC).

‖ Not significantly above values of B cells transferred without T cells.

TABLE VI

Negative Selection of Primed CBA T Cells to SRC in Irradiated B10.Br Mice but not in B10 or B10.D2 Mice. T Cells Harvested from Lymph at 1 Day Post-Transfer

Helper cells* (2 × 10⁶)	*H-2* Region of filtration host	SRC Added during negative selection	B Cells‡	Anti-SRC		Anti-HRC
				IgM	IgG	IgG
CBA T Cells → irradiated B10.Br	*H-2*ᵏ	−	CBA	8,930(1.28)§	45,300(1.11)	42,910(1.22)
CBA T Cells → irradiated B10.Br	*H-2*ᵏ	+	CBA	650(1.37)‖	320(1.56)‖	47,130(1.19)
CBA T Cells → irradiated B10	*H-2*ᵇ	+	CBA	7,840(1.20)	39,170(1.18)	37,070(1.15)
CBA T Cells → irradiated B10.D2	*H-2*ᵈ	+	CBA	9,150(1.11)	48,100(1.11)	42,120(1.17)

*‡ As for footnote to Table I. With cells filtered through B10 and B10.D2 mice, > 90% of the cells were resistant to lysis by CBA anti-B6 and CBA anti-DBA/2 alloantisera, respectively. All lymph-borne T cells were > 96% Thy 1.2 positive.

§ As for footnote Table I. Subtracted values of B cells transferred without T cells were: 950(1.08) (IgM SRC), 4,600(1.27) (IgG SRC), 880(1.14) (IgG HRC). Values for T cells transferred without B cells were < 400 PFC/spleen.

‖ Not significantly above values of B cells transferred without T cells.

compatible B10.Br mice but not in *H-2*-incompatible B10 or B10.D2 mice.

Negative Selection with Unprimed T Cells. All of the experiments considered above employed T cells taken from mice primed with both SRC and HRC. The experiment illustrated in Table VII shows that similar results occurred when the T cells were taken from unprimed mice. Thus, it is evident that unprimed CBA T cells underwent negative selection to SRC in *H-2*-compatible C3H (*H-2*ᵏ) mice but not in *H-2*-incompatible BALB/c (*H-2*ᵈ) mice. (To ensure good helper responses in this experiment, the T and B cells were transferred in higher doses (4.5×10^6 and 7×10^6, respectively) than in the preceding experiments with primed T cells.)

Positive Selection in Allogeneic Mice. If negative selection is a prerequisite for positive selection (which seems very likely), the failure to observe negative selection to SRC in *H-2*-different hosts should be associated with an inability to induce T helper cell induction in this situation, i.e., an inability to induce positive selection to SRC. A priori, this might be tested by transferring T cells plus SRC to irradiated allogeneic mice and studying the helper function of cells harvested from the recipients 5 days later. This would be an unsatisfactory approach because at this stage most of the cells in the lymphoid tissues and the central lymph would be alloaggressive blast cells stimulated by the host alloantigens (16). This problem can be avoided by using T cells which have been depleted of specific alloreactive lymphocytes, e.g. by prior acute recirculation (without SRC) through irradiated mice of the strain to be used for activation (17).

Accordingly, LN cells from SRC-primed CBA mice were injected intravenously without SRC into irradiated (CBA $\times$ B6)F$_1$ mice and recovered from thoracic duct lymph of the recipients 20–40 h later (footnote, Table VIII). Of the cells harvested from the recipients, 95% were of donor origin (were resistant to lysis by CBA anti-B6 alloantiserum and complement) and >99% were T (Thy 1.2-positive) cells. These cells were transferred intravenously in a dose of 2×10^7 cells with or without SRC (0.1 ml of 50% solution) into (CBA $\times$ B6)F$_1$ and B6 mice that had received 850 rads 1 day previously. The recipients were cannulated 5 days later and TDL collected overnight. The lymph-borne cells (>90% T cells of donor CBA origin) were transferred with CBA B cells and SRC into irradiated CBA mice to measure their helper function.

TABLE VII

Negative Selection of Unprimed CBA T Cells to SRC in Irradiated C3H Mice but not in BALB/c Mice. T Cells Harvested from Lymph at 1 Day Post-Transfer

Helper cells* (4.5×10^6)	SRC Added during negative selection	B Cells‡	PFC/Spleen at 7 days in irradiated CBA mice		
			Anti-SRC		Anti-HRC
			IgM	IgG	IgG
CBA T Cells → irradiated C3H	−	CBA	20,760(1.23)§	100,560(1.35)	48,700(1.17)
CBA T Cells → irradiated C3H	+	CBA	2,790(1.23)	9,750(1.10)‖	53,190(1.14)
CBA T Cells → irradiated BALB/c	+	CBA	24,200(1.04)	102,710(1.20)	58,350(1.17)

*‡ As for Table I except that the purified T cells were prepared from unprimed mice. The B cells (taken from primed mice) were transferred in a dose of 7×10^6 viable cells.

§ As for footnote to Table I. Subtracted values of B cells transferred without T cells were: 980(1.43) (IgM SRC), 7,120(2.25) (IgG SRC), 1,880(1.98) (IgG HRC). Values for T cells transferred without B cells were < 200 PFC/spleen.

‖ Not significantly above background value of B cells transferred without T cells.

TABLE VIII

Positive Selection of Primed CBA T Cells to SRC in Irradiated (CBA × B6)F₁ Mice but not in B6 Mice. T Cells Rendered Unresponsive to B6 Alloantigens by Prior Filtration through Irradiated (CBA × B6)F₁ Mice. T Cells Recovered from Lymph of Secondary Recipients at 5 Days Post-Transfer

T-Cell group	Helper cells* (prefiltered through irradiated (CBA × B6)F₁ mice)	SRC Added during positive selection	Dose of helper cells	B Cells‡	Anti-SRC PFC/spleen in irradiated CBA mice	
					IgM	IgG
A	CBA T Cells → irradiated (CBA × B6)F₁	+	1×10^5	CBA	5,970(1.59)§	35,340(1.08)
			8×10^5	CBA	6,030(1.09)	83,040(1.23)
B	CBA T Cells → irradiated B6	+	8×10^5	CBA	950(1.22)	4,200(1.16)
C	CBA T Cells → irradiated B6	−	8×10^5	CBA	950(1.21)	4,500(1.19)
	Group A + Group B		8×10^5 of each	CBA	6,240(1.29)	79,560(1.10)

* Unfractionated LN cells from SRC-primed CBA mice were transferred intravenously in a dose of 1.5×10^8 cells without SRC into (CBA × B6)F₁ mice given 900 rads 8 h previously. Recipients were cannulated 18 h later and TDL collected between 20 and 40 h postinjection. The lymph-borne CBA T cells (99% Thy 1.2-positive and 95% resistant to lysis by CBA anti-B6 alloantiserum) were transferred intravenously in a dose of 2×10^7 cells ± 0.1 ml of 50% SRC into (CBA × B6)F₁ or B6 mice given 850 rads 1 day before (two mice per group). These mice were cannulated 5 days later and TDL collected overnight; for each group the TDL were > 97% Thy 1.2 positive and >90% resistant to lysis by CBA anti-B6 alloantiserum. Helper function was measured by transferring the lymph-borne cells with 0.1 ml of 5% SRC plus SRC-primed B cells into irradiated CBA mice as described in footnote to Table I.

‡ As for footnote to Table I.

§ As for footnote to Table I. Subtracted values of B cells transferred without T cells were: 100(1.50) (IgM SRC), 690(1.55) (IgG SRC). All values shown in Table are significantly above background values of B cells transferred without T cells. Values for T cells transferred without B cells were < 100 PFC/spleen.

As shown in Table VIII, CBA T cells positively selected to SRC in irradiated (CBA × B6)F₁ mice (group A) gave high anti-SRC responses with cell doses as low as 10^5. CBA T cells selected to SRC in B6 mice (group B), by contrast, gave only low responses, even with doses as high as 8×10^5. Though significant, these responses were no higher than the responses given by equivalent numbers of CBA T cells harvested from B6 mice not given SRC (group C). The fact that T cells from group B did not inhibit the helper function of group A T cells suggests that the failure to observe positive selection to SRC in irradiated B6 mice was not the result of suppression.

Discussion

The present studies demonstrate that marked negative selection to SRC occurs when purified T cells are acutely recirculated in the presence of SRC through irradiated mice which share *H-2* determinants with the T cells. Thus, when the data shown in Tables I, III, and VI are pooled, anti-SRC helper responses by primed CBA (*H-2ᵏ*) T cells were reduced by a mean of 96% (95%

IgM, 97% IgG) when the cells were filtered in the presence of SRC through irradiated *H-2^k* (CBA or B10.Br) mice or through (*H-2^k* × *H-2^b*)F₁ ((CBA × B6)F₁) mice. The response to a third-party antigen (HRC) was not affected (1% reduction of IgG response). Negative selection to SRC did not occur, however, when CBA T cells were filtered through *H-2*-different mice, i.e. through *H-2^b* (B6 or B10) or *H-2^d* (B10.D2) mice. In this situation, the capacity of the filtered cells to evoke IgG anti-SRC responses was reduced by a mean of only 1% (data pooled from Tables III and VI and from three other unpublished experiments). IgM responses were reduced significantly in one experiment (Table III) but not in four others (the mean reduction of IgM responses in five experiments was 15%). Similar results were found with unprimed T cells (Table VII).

No evidence was found that the failure to induce negative selection to SRC in *H-2*-different hosts was due to such factors as the concomitant onset of a graft-versus-host or host-versus-graft reaction (Tables III and IV). It would seem reasonable to conclude, therefore, that the T cells were unable to recognize the antigen in an *H-2*-different environment, even though the antigen was injected in a massive dose, i.e. 0.5 ml of 50% solution of SRC/10⁸ lymphocytes (approximately 25 erythrocytes/lymphocyte). The reciprocal failure to detect positive selection (clonal expansion) of T cells in *H-2*-different hosts (Table VIII) strengthens this viewpoint.

The simplest interpretation of the data is that T cells recognize antigen in vivo only when it is presented in association with a cell type which shares *H-2* determinants with the T cells. The identity of the cell(s) presenting antigen in vivo is not known. In the present system radioresistant macrophages (or related cells) of the transfer hosts are the logical choice since macrophages are known to be radioresistant (18) and, at least for certain antigens, these cells play an important role in antigen presentation in vitro (Introduction). Currently we are attempting to identify the antigen-presenting cell by determining whether negative selection can be induced in *H-2*-different hosts by supplementing the donor T cells with other cell types. Preliminary results suggest that addition of macrophage-enriched populations, e.g. peritoneal exudate cells, will indeed lead to selection in this situation. Addition of small B lymphocytes from thoracic duct lymph has not been successful.

Precisely which genes are responsible for selection to antigen in vivo has yet to be established. Preliminary studies with recombinant mice as filtration hosts suggest that the *I* region per se is important, although which subregion is involved has not been studied.

Can one conclude by extrapolation from the present data that T cells are unable to recognize free antigen in vivo? This would be consistent with the evidence that T helper cells (or T cells with the Ly 1⁺ 2⁻ 3⁻ phenotype) fail to bind antigen in vitro (19–22). While the data are in line with this viewpoint, an alternative argument is that T cells do recognize free antigen in vivo but that in an *H-2*-different environment the antigen is rapidly degraded by the allogeneic macrophages and thereby rendered nonimmunogenic. A critical question therefore is whether selection to antigen would occur in a syngeneic environment after blockade of the reticuloendothelial system. This is currently being investigated. Predicting the outcome is difficult since it hinges on the central question of whether T cells have receptors for free antigen or only for MHC-associated antigen (23–25).

Perhaps the most surprising aspect of the present data is that the restriction observed applied to a crude particulate antigen (heterologous erythrocytes). In this respect, *H-2* gene control of T-macrophage interactions in vitro is reported not to apply to particulate antigens (26). Indeed, certain groups have concluded that the immune response to heterologous erythrocytes does not require macrophages (7, 8). The answer here is possibly that many antigens which remain in a particulate form in vitro are rapidly degraded to a soluble form in vivo. Consequently, one might expect *H-2* gene control of T helper cell induction in vivo to apply to a wide range of antigens. This remains to be proved.

Finally, it should be emphasized that the B cells used in this study were always syngeneic with the T cells. Hence, it might be asked whether negative selection would be apparent in *H-2*-different hosts if allogeneic B cells of this strain were used to monitor selection. Here it should be emphasized that in our hands T cells from normal (nonchimeric) mice do not stimulate *H-2*-different B cells in vivo (17). This applies even with unprimed parental strain T cells activated to antigen (SRC) in irradiated F_1 mice (J. Sprent, unpublished data).

Summary

When purified CBA lymph node T cells were mixed with sheep erythrocytes (SRC) and filtered from blood to lymph through irradiated syngeneic mice for 1–2 days, the donor cells lost their capacity to stimulate anti-SRC responses by CBA B cells; the response to a third-party antigen (horse erythrocytes) was unaffected and active suppression was not involved. This process of specific negative selection to SRC also occurred when semiallogeneic mice were used as filtration hosts. By contrast, when allogeneic hosts were used the helper function of the donor cells was not reduced; this applied to both primed and unprimed T cells. Studied with congeneic resistant strains indicated that negative selection to SRC occurred only when the donor and host shared *H-2* determinants.

Studies with T cells depleted of alloreactive lymphocytes showed that negative selection to SRC in irradiated F_1 hybrid mice was followed by a stage of positive selection where the donor cells gave greatly increased responses to the injected antigen. Positive selection did not occur in *H-2*-different mice, however, and the helper function of the donor cells remained unchanged.

By these parameters it was concluded that homozygous T helper cells have no detectable capacity to recognize antigen in an *H-2*-different environment.

Stimulating discussion with D. B. Wilson and the typing skill of Miss Karen Nowell is gratefully acknowledged.

Received for publication 3 May 1978.

References

1. Immunobiology of the Macrophage. 1978. D. S. Nelson, editor. Academic Press, Inc. New York. 1–633.
2. Rosenthal, A. S., and E. M. Shevach. 1973. Function of macrophages in antigen recognition by guinea pig T lymphocytes. I. Requirement for histocompatible macrophages and lymphocytes. *J. Exp. Med.* 138:1194.
3. Erb, P., and M. Feldmann. 1975. The role of macrophages in the generation of T-helper cells. II. The genetic control of the macrophage-T cell interaction for helper cell induction with soluble antigens. *J. Exp. Med.* 142:460.
4. Kappler, J. W., and P. C. Marrack. 1976. Helper T cells recognize antigen and macrophage components simultaneously. *Nature (Lond.).* 262:797.
5. Pierce, C. W., and J. A. Kapp. 1978. *H-2* gene complex regulation of macrophage-lymphocyte interactions *in vitro*. In Ir Genes and Ia Antigens. H. O. McDevitt, editor. Academic Press, Inc., New York. p. 357.

6. Yano, A., R. H. Schwartz, and W. E. Paul. 1977. Antigen presentation in the murine T-lymphocyte proliferative response. I. Requirements for genetic identity at the major histocompatibility complex. *J. Exp. Med.* 146:828.

7. Chen, C., and J. G. Hirsch. 1972. The effect of mercaptoethanol and of peritoneal macrophages on the antibody-forming capacity of non-adherent mouse spleen cells in vitro. *J. Exp. Med.* 136:604.

8. Heber-Katz, E., and D. B. Wilson. 1975. Collaboration of allogeneic T and B lymphocytes in the primary antibody response to sheep erythrocytes in vitro. *J. Exp. Med.* 142:928.

9. Sprent, J., J. F. A. P. Miller, and G. F. Mitchell. 1971. Antigen-induced selective recruitment of circulating lymphocytes. *Cell. Immunol.* 2:171.

10. Rowley, D. A., J. L. Gowans, R. C. Atkins, W. L. Ford, and M. E. Smith. 1972. The specific selection of recirculating lymphocytes by antigen in normal and preimmunized rats. *J. Exp. Med.* 136:499.

11. Sprent, J., and I. Lefkovits. 1976. Effect of recent antigen priming on adoptive immune responses. IV. Antigen-induced selective recruitment of recirculating lymphocytes to the spleen demonstrable with a microculture system. *J. Exp. Med.* 143:1289.

12. Ford, W. L., S. J. Simmonds, and R. C. Atkins. 1975. Early events in systemic graft-vs-host reaction. II. Autoradiographic estimates of the frequency of donor lymphocytes which respond to each Ag-B-determined antigenic complex. *J. Exp. Med.* 141:681.

13. Sprent, J. 1973. Circulating T and B lymphocytes of the mouse. I. Migratory properties. *Cell. Immunol.* 7:10.

14. Sprent, J. 1978. Restricted helper function of F_1 hybrid T cells positively selected to heterologous erythrocytes in irradiated parental strain mice. I. Failure to collaborate with B cells of the opposite parental strain not associated with active suppression. *J. Exp. Med.* 147:1142.

15. Julius, M. H., E. Simpson, and L. A. Herzenberg. 1973. A rapid method for the isolation of thymus-derived murine lymphocytes. *Eur. J. Immunol.* 3:645.

16. Sprent, J., and J. F. A. P. Miller. 1972. The interaction of thymus lymphocytes with histoincompatible cells. II. Recirculating lymphocytes derived from antigen-activated thymus cells. *Cell. Immunol.* 3:385.

17. Sprent, J., and H. von Boehmer. 1976. Helper function of T cells depleted of alloantigen-reactive lymphocytes by filtration through irradiated F_1 hybrids. I. Failure to collaborate with allogeneic B cells in a secondary response to sheep erythrocytes measured in vivo. *J. Exp. Med.* 144:617.

18. Anderson, R. E., and N. L. Warner. 1976. Ionizing radiation and the immune response. *Adv. Immunol.* 24:215.

19. Elliot, B. E., J. S. Haskill, and M. A. Axelrad. 1973. Thymus-derived rosettes are not helper cells. *J. Exp. Med.* 138:1133.

20. Okumura, K., T. Takemori, T. Tokuhisa, and T. Tada. 1977. Specific enrichment of the suppressor T cell bearing I-J determinants. *J. Exp. Med.* 146:1234.

21. Taniguchi, M., and J. F. A. P. Miller. 1977. Enrichment of specific suppressor T cells and characterization of their surface markers. *J. Exp. Med.* 146:1450.

22. Cone, R. E., R. K. Gershon, and P. W. Askenase. 1977. Nylon adherent antigen-specific rosette-forming T cells. *J. Exp.Med.* 146:1390.

23. Doherty, P. C., R. V. Blanden, and R. M. Zinkernagel. 1976. Specificity of virus-immune effector T cells for H-2K and H-2D compatible interactions: implications for H-antigen diversity. *Transplant. Rev.* 29:89.

24. Miller, J. F. A. P., M. A. Vadas, A. Whitelaw, and J. Gamble. 1976. Role of major histocompatibility complex gene products in delayed-type hypersensitivity. *Proc. Natl. Acad. Sci. U.S.A.* 73:2486.

25. Paul, W. E., E. M. Shevach, S. Pickeral, D. W. Thomas, and A. S. Rosenthal. 1977. Independent populations of primed F_1 guinea pig T lymphocytes respond to antigen-pulsed parental peritoneal exudate cells. *J. Exp. Med.* 145:618.

26. Erb, P., and M. Feldmann. 1975. The role of macrophages in the generation of T helper cells. I. The requirements for macrophages in helper cell induction and characteristics of the macrophage-T cell interaction *Cell. Immunol.* 19:356.

RESTRICTED HELPER FUNCTION OF F₁ HYBRID T CELLS POSITIVELY SELECTED TO HETEROLOGOUS ERYTHROCYTES IN IRRADIATED PARENTAL STRAIN MICE

I. Failure to Collaborate with B Cells of the Opposite Parental Strain Not Associated with Active Suppression*

By J. SPRENT

(From the Immunobiology Research Unit, Department of Pathology, University of Pennsylvania
School of Medicine, and the Wistar Institute of Anatomy and Biology,
Philadelphia, Pennsylvania 19104)

Several studies have shown that F_1 hybrids derived from parental strains differing at the major histocompatibility complex (MHC)[1] contain two distinct subpopulations of T cells, each reactive to antigen presented in the context of one of the two parental strains. This has been demonstrated with respect to a variety of T-cell functions, e.g., measurement of proliferative responses to antigen-pulsed macrophages in vitro (1, 2), elicitation of delayed type hypersensitivity (DTH) on adoptive transfer (3), and expression of cell-mediated lympholysis (CML) directed against viruses (4), haptens (5), and minor histocompatibility determinants (6). With respect to two of these functions—T-cell proliferation and expression of DTH—the response involves T cells with the Ly $1^+ 2^- 3^-$ phenotype (7, and R. H. Schwartz, personal communication). This phenotype is also expressed by helper T cells involved in T-B collaboration (8).

Recent evidence suggests that macrophages play a critical role in presenting antigen in an immunogenic form to Ly $1^+ 2^- 3^-$ T cells. The precise contribution of macrophages in antigen presentation is not clear, although with certain antigens it has been observed that T-cell activation depends upon the responding T cells and macrophages sharing MHC determinants. In this respect, studies of Erb and Feldmann (9–11) suggest that, in vitro, mouse macrophages process antigen and complex it to small particles coded for by the *I-A* subregion of the *H-2* complex. These complexes stimulated T cells to express helper activity, but only when the T cells and the macrophages forming the complex were I-A-compatible. It is perhaps tempting to conclude from these data that all helper T cells, and perhaps Ly $1^+ 2^- 3^-$ cells in general, are unable to recognize antigen unless it has been processed by MHC-compatible macrophages. At present, this generalization seems premature because it fails to explain why certain antigens, e.g., particulate antigens such as heterologous erythrocytes or keyhole limpet hemocyanin cross-linked to Sepharose beads, are able to induce helper function in vitro either in the virtual absence of macrophages (12) or in the presence of MHC-incompatible macrophages (9).

The purpose of the studies presented in these two papers was to determine whether or

---

* Supported by grants AI-10961 and CA-15822 from the U. S. Public Health Service.

[1] *Abbreviations used in this paper*: CML, cell-mediated lympholysis; DTH, delayed-type hypersensitivity; HRC, horse red blood cells; LN, lymph nodes; MHC, major histocompatibility complex; PE, peritoneal exudate; PFC, plaque-forming cells; SRC, sheep red blood cells; TDL, thoracic duct lymphocytes.

not the reported dichotomy in the reactivity of F_1 hybrid T cells involved in such functions as CML and DTH also applies to helper T cells involved in T-B collaboration. This question was approached by activating purified F_1 T cells to heterologous erythrocytes in irradiated mice of one parental strain, and then harvesting the cells at a time (5 days) when the antigen-reactive precursors were presumed to have undergone clonal expansion (positive selection) after proliferating in response to the antigen in such organs as the spleen (13). The helper function of the activated T cells was then assessed in terms of their capacity to collaborate with parental strain B cells. The results show that despite the particulate nature of the antigen used, F_1 T cells exposed to sheep or horse erythrocytes in irradiated mice of one parental strain developed excellent helper activity for B cells derived from this strain, but gave only minimal help for B cells of the opposite strain. In contrast to the results of an analogous study by Skidmore and Katz (14), good collaboration was observed with F_1 B cells, and there was no evidence that suppressor cells accounted for the phenomenon.

Materials and Methods

Mice. For most experiments CBA/Cum (CBA) ($H-2^k$), C57BL/6 Cum (B6) ($H-2^b$), and (CBA × B6)F_1 hybrids were used. These mice were obtained from Cumberland View Farms, Clinton, Tenn. CBA/J mice obtained from The Jackson Laboratory, Bar Harbor, Maine were also used in some experiments. CBA/J and CBA/Cum failed to respond to each other in mixed lymphocyte culture, and were therefore assumed to be H-2- and *Mls*-identical (*Mls*^{d}). Results obtained with B cells derived from CBA/J compared with CBA/Cum mice were not discernibly different. DBA/2 ($H-2^d$) and (DBA/2 × B6)F_1 mice, also obtained from Cumberland View Farms, were used in one experiment.

Media. RPMI-1640 (Microbiological Associates, Walkersville, Md.) supplemented with 10% fetal calf serum (Microbiological Associates) was used.

Injections. All suspensions of lymphoid cells and sheep erythrocytes (SRC) and horse erythrocytes (HRC) were given intravenously unless stated otherwise.

Cells. Thoracic duct lymphocytes (TDL) were obtained as described elsewhere (15). Suspensions of spleen and lymph node (LN) cells were prepared by teasing the organs with fine forceps through an 80-mesh stainless steel sieve in cold medium.

Irradiation. Mice were exposed to ^{137}Cs γ-irradiation at a dose rate of 106 rads/min.

Cell Identification with Alloantisera. CBA anti-B6 and B6 anti-CBA *H-2* alloantisera and anti-thy 1.2 (AKR anti-C3H thymus) antiserum were prepared as described previously (16). These antisera, all of which were shown to be specific before use, were employed to establish the identity of the cells in the central lymph of irradiated B6 or CBA mice given (CBA × B6)F_1 T cells plus SRC 5 days before. In the experiment in which (DBA/2 × B6)F_1 T cells were activated in irradiated DBA/2 and B6 mice, DBA/2 anti-B6 and B6 anti-DBA/2 alloantisera were used; these sera were prepared as referred to above. 50 μl of the test lymphocytes (5 × 10^6/ml) were incubated at 4°C for 30 min with 25 μl of antisera in small plastic tubes and washed once by centrifugation. The cell pellet was resuspended in 50 μl of complement (1:5 dilution of guinea pig serum) for 30 min at 37°C. Cytotoxic indices were then calculated by dye exclusion with respect to controls incubated with normal mouse serum plus complement or with antiserum without complement. Percent lysis with the control samples was invariably <5%. The anti-thy 1.2 serum appeared to be specific for T cells, since it lysed 97–100% of LN cells filtered from blood to lymph through irradiated mice (16), 20–30% of spleen cells, and <5% of bone marrow cells.

Preparation of T Cells for Positive Selection. The T cells used for positive selection were obtained from pooled mesenteric, inguinal, axillary, and cervical LN of unprimed (CBA × B6)F_1 or (DBA × B6)F_1 mice. LN cells were depleted of most thy 1.2-negative cells by passage over nylon-wool columns as described by Julius et al (17). The majority of the effluent cells recovered from the columns were T cells, since 86–95% (mean – 91%, 15 experiments) were susceptible to lysis by anti-thy 1.2 antiserum in the presence of complement. The recovery of T cells after nylon-wool passage was usually on the order of 60–80% (established with respect to the number of thy 1.2-positive cells in the LN cell preparation before passage [50–70%]). The viability of the cells recovered from the column was usually >85%.

Positive Selection to Heterologous Erythrocytes in Irradiated Mice. In most experiments, 5×10^7 viable, unprimed, nylon-wool-passed (CBA $\times$ B6)F_1 T cells in a 0.5-ml volume were mixed with 0.5 ml of 25% SRC (or HRC) and injected intravenously into CBA, B6, or (CBA $\times$ B6)F_1 mice (2-5 mice per group) given 800 rads 1 day previously. (DBA/2 $\times$ B6)F_1 T cells were used in one experiment, and these cells were transferred with SRC into irradiated B6 or DBA/2 mice. Thoracic duct fistulas were inserted in the mice 5 days later. TDL were collected overnight and used as helper T cells after establishing their identity with appropriate alloantisera.

Preparation of B Cells. Spleen cells from mice primed with SRC or with both SRC and HRC were resuspended in undiluted anti-thy 1.2 antiserum (0.1 ml of antisera/2×10^7 spleen cells) and kept at 4°C for 30 min. After washing twice by centrifugation, the cells were incubated at 37°C for 30 min with guinea pig complement (1:6 dilution, 2×10^7 cells/ml). The cells were then washed once and resuspended to a volume suitable for injection. This treatment lysed 20-45% of the spleen cells.

Measurement of T-B Collaboration. In most experiments, helper T cells (usually 0.8×10^6 TDL from irradiated mice given F_1 T cells plus SRC 5 days before) were transferred with 0.1 ml of 5% SRC into irradiated (750 rads 1 day before) F_1 hybrid mice with B cells (anti-thy 1.2-treated spleen cells) prepared from SRC-primed F_1 hybrid or parental strain mice. With one exception, B cells were transferred in a dose of 5×10^6 viable cells. The exception occurred when B6 B cells were transferred to irradiated (CBA $\times$ B6)F_1 recipients; here, presumably because of the strong *Hh* barrier existing in this situation (18), a higher dose of cells ($8-10 \times 10^6$) had to be transferred to get responses equivalent to those given by the other B-cell populations. In experiments in which responses to HRC were measured, both SRC and HRC (0.1 ml of 5% of each) were added to the injected mixture of T and B cells; in this situation, the B cells were prepared from mice primed to both SRC and HRC.

Plaque-Forming Cell (PFC) Assays. Direct (19s, IgM) PFC were detected by the method of Cunningham and Szenberg (19). Numbers of indirect (7s, IgG) PFC were measured by adding a polyvalent rabbit anti-mouse immunoglobulin reagent to the reactive mixture in the presence of specific goat anti-mouse μ-chain serum. As described elsewhere (16), the anti-μ serum (kindly provided by Dr. M. Feldmann, University College, London, England) suppressed IgM but not IgG PFC.

Serum Hemagglutinin Assays. These were performed by the method of Dietrich (20).

Priming with Heterologous Erythrocytes. SRC and HRC (obtained from Gibco Diagnostics, The Mogul Corp., Chagrin Falls, Ohio) were stored in Alsever's solution and washing three times before use. Mice were primed intraperitoneally with 0.1 ml of 25% SRC (or also with a similar dose of HRC) and used as B-cell donors 2-4 mo later.

Peritoneal Exudate (PE) Cells. PE cells were obtained from mice given a 1-ml intraperitoneal injection of 4% thyoglycollate solution (Difco Laboratories, Detroit, Mich.) 4 days earlier. The cells were treated with anti-thy 1.2 serum and complement according to the procedure used for preparing B cells from spleen.

Statistical Analysis. Geometric means and values used to derive the upper and lower limits of the SE of the mean (these values represent the anti-log of SE of $\log_{10}$ geometric mean) were calculated from the $\log_{10}$ of the PFC counts. Arithmetic means and standard errors of the mean were calculated from the reciprocal of the $\log_2$ values for the hemagglutinin titers. *P* values were determined by Student's *t* test. In the comparison of the means of any two groups of observations, a significance level of 0.05 was chosen.

Results

Experimental Design. The general plan of the experiments was to transfer LN T cells from unprimed (CBA $\times$ B6)F_1 mice into irradiated CBA, B6, and (CBA $\times$ B6)F_1 mice together with SRC, and then to study the helper function of the SRC-activated T cells recovered from the thoracic duct lymph of the recipients 5 days later. Before transfer, the LN cells were depleted of B cells (and presumably most macrophages) by nylon-wool filtration. The effluent T cells (>90% thy-1.2-positive) were transferred intravenously in a dose of 5×10^7 cells with SRC (0.5 ml of 25% solution), also given intravenously. To limit the

TABLE I

Identity of TDL Recovered from Thoracic Duct Lymph of Irradiated CBA, B6, and (CBA × B6)F₁ Mice Injected with Unprimed Nylon-Wool-Purified (CBA × B6)F₁ LN Cells Plus SRC 5 Days Previously

Recipients of (CBA × B6)F$_1$ T cells plus SRC*	Average no. of TDL ($\times 10^{-6}$) collected from each recipient over 18 h‡	Cytotoxic index with alloanti-sera + complement§			Host origin
		B6 anti-CBA	CBA anti-B6	Anti-thy 1.2	
					%
Irrad. CBA	14.5	99	95	97	5
Irrad. B6	12.1	95	100	98	4
Irrad. (CBA × B6)F$_1$	14.3	99	99	97	—

* Nylon-wool-purified LN T cells (93% thy 1.2-positive) were transferred intravenously in a dose of 5×10^7 viable cells with 0.5 ml of 25% SRC; the cell recipients (three mice per group) were exposed to 800 rads 1 day before.

‡ Thoracic duct fistulas were inserted at 5 days after cell transfer. TDL were collected overnight, the collections commencing at about 4 h after establishing the fistulas. Lymph samples were pooled from the three mice in each group before counting.

§ See Materials and Methods. Background lysis with cells treated with normal mouse serum plus complement was <3%.

possibility of a host-versus-graft reaction, the transfer hosts received irradiation at a high dose (800 rads) 1 day before cell transfer. Thoracic duct fistulas were inserted in the cell recipients at 5 days post-transfer, i.e. during the stage of positive selection when it was presumed that the progeny of the cells responding to the injected antigen in the lymphoid tissues entered the circulation in large numbers (21, 22). The features of the lymph-borne cells collected over a 16-h period are shown in Table I; the data are from one experiment which is representative of many others. Testing with anti-thy 1.2 serum and appropriate alloantisera in the presence of complement showed that the lymph-borne cells consisted almost entirely (≥97%) of T cells, nearly all of which were of donor origin (>94%). Cell viability was 99–100%. The yield of cells (compared with the numbers initially injected) was high (≈30%).

To study their helper function, the cells were transferred intravenously in small numbers (0.8×10^6) into irradiated (750 rads) (CBA × B6)F$_1$ mice, together with SRC (0.1 ml of 5% solution), and B cells (5–8 × 10^6 anti-thy 1.2-serum-treated spleen cells from SRC-primed mice). Numbers of direct (IgM) and indirect (IgG) PFC to SRC were measured in the spleen 7 days later.

As shown in Table II, (CBA × B6)F$_1$ T cells activated to SRC in irradiated (CBA × B6)F$_1$ mice (F$_1$ T$_{+(SRC-F_1)}$), provided effective help for B cells derived from either of the two parental strains, as well as for F$_1$ B cells. By contrast, F$_1$ T cells activated to SRC in irradiated B6 mice (F$_1$ T$_{+(SRC-B6)}$) or in CBA mice (F$_1$ T$_{+(SRC-CBA)}$) cooperated well with B cells derived from only one of the two parental strains, namely the strain in which the T cells were initially activated. The failure to stimulate more than minimal responses by B cells of the opposite parental strain did not seem to be the result of active suppression, since a supplemental injection of F$_1$ T$_{+(SRC-F_1)}$ cells led to high responses. Moreover,

TABLE II

Helper Activity of T Cells Recovered from Thoracic Duct Lymph of Irradiated CBA, B6, and (CBA × B6)F_1 Mice Injected 5 Days Previously with Unprimed (CBA × B6)F_1 T Cells Plus SRC*

T-cell group	T cells (0.8 × 10^6)‡	B cells§	Anti-SRC PFC/spleen at 7 days in irradiated F_1 mice	
			IgM	IgG
A	F_1 T$_{+(SRC-B6)}$	CBA	1,150 (1.08)‖	8,610 (1.01)
		B6	40,590 (1.13)	118,870 (1.15)
		(CBA × B6)F_1	7,370 (1.28)	84,660 (1.16)
B	F_1 T$_{+(SRC-CBA)}$	CBA	30,490 (1.02)	275,400 (1.17)
		B6	3,940 (1.05)	9,360 (1.15)
		(CBA × B6)F_1	13,060 (1.13)	74,500 (1.11)
C	F_1 T$_{+(SRC-F_1)}$	CBA	21,190 (1.25)	125,690 (1.08)
		B6	13,890 (1.18)	56,180 (1.14)
		(CBA × B6)F_1	14,680 (1.16)	89,360 (1.06)
Group A + group C		CBA	19,170 (1.30)	131,390 (1.25)
Group B + group C		B6	20,380 (1.23)	64,960 (1.27)
	—	CBA	200 (1.40)	510 (1.07)
	—	B6	990 (1.39)	1,740 (1.26)
	—	(CBA × B6)F_1	<100	610 (1.22)
Group A		—	<100	<100
Group B		—	<100	<100
Group C		—	<100	<100

* Cell recipients were exposed to 800 rads 1 day before being injected intravenously with $5 × 10^7$ unprimed nylon-wool-purified (CBA × B6)F_1 T cells plus 0.5 ml of 25% SRC.

‡ Recipients of F_1 T cells and SRC were cannulated at 5 days post-transfer and the lymph-borne cells were collected overnight. The helper activity of the cells was measured by transferring 0.8 × 10^6 TDL (nearly all of which were T cells of F_1 origin; Table I) intravenously into irradiated (750 rads) (CBA × B6)F_1 mice with B cells and 0.1 ml of 5% SRC. When transferring mixtures of T cells to check for suppression, a dose of 0.8 × 10^6 of each T-cell population was used.

§ Anti-thy 1.2-treated spleen cells from mice primed with SRC 2–4 mo previously; $5 × 10^6$ (CBA and (CBA × B6)F_1) or $8 × 10^6$ (B6) viable cells transferred.

‖ Geometric mean of data from four mice per group (two mice only for T cells transferred without B cells); number in parenthesis refers to value by which mean is multiplied by or divided to give upper and lower limits, respectively, of SE. All values shown are significantly above ($P < 0.05$) the background values of B cells transferred without T cells.

effective cooperation was observed with F_1 B cells. Five other experiments gave similar results.

Unless stated otherwise, the design of the experiments to be considered below was identical to the protocol used in Table II. In all experiments, the helper cells were recovered from thoracic duct lymph at day 5–6 post-transfer, and they were pooled from at least 2 mice per group. In each experiment the identity of the lymph-borne cells was checked with anti-thy 1.2 serum and appropriate *H-2* alloantisera. By these parameters, the lymph-borne cells were invariably >90% (and usually >95%) T cells of donor F_1 origin. For simplicity, the background values obtained when B cells were transferred without T cells have been subtracted from the data shown in Tables III–VIII. These values were generally

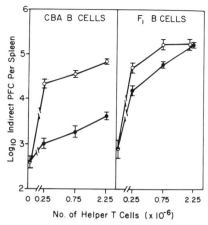

F<small>IG</small>. 1. Dose response of helper activity of (CBA × B6)F_1 T cells activated to SRC in irradiated B6 (●) or irradiated (CBA × B6)F_1 (○) mice and harvested from thoracic duct lymph of the recipients 5 days later. The fig. shows numbers of anti-SRC IgG PFC/spleen in irradiated (CBA × B6)F_1 mice injected intravenously 7 days previously with varying doses of one of the two T-cell populations plus 0.1 ml of 5% SRC and either CBA B cells or (CBA × B6)F_1 B cells. ◖, B cells transferred without T cells; PFC numbers with T cells transferred without B cells were < 100 PFC/spleen. The B cells ($5 × 10^6$ viable anti-thy 1.2-serum-treated spleen cells) were from mice primed with SRC 2 mo previously. Each point represents geometric mean of data from four mice; vertical bars show upper and lower limits of SE.

quite low (200–2,000 PFC/spleen) and are shown in the footnotes to the tables. T cells transferred without B cells gave negligible responses (<200 PFC/spleen). It should be stressed that the irradiated recipients used for measuring T-B collaboration were invariably F_1 hybrids, i.e. mice syngeneic or semi-syngeneic with the T and B cells.

Effect of Varying the Dose of Helper T Cells. Although the helper function of F_1 T cells activated to SRC in mice of one parental strain was very low for B cells of the opposite parental strain, it was significant. This is illustrated in Fig. 1 where it can be seen that F_1 $T_{+(SRC-B6)}$ cells gave a linear response when transferred in varying doses with a constant number of CBA B cells. With the highest dose of T cells ($2.25 × 10^6$), the response was 20-fold above background. It is apparent that the slope of the dose-response curve obtained in this situation paralleled the slope of the (far greater) response observed when CBA B cells were transferred with varying doses of F_1 $T_{+(SRC-F_1)}$ cells.

In the case of responses with F_1 B cells, it is of interest that the helper activity of F_1 $T_{+(SRC-B6)}$ cells was appreciably less than that of F_1 $T_{+(SRC-F1)}$ cells; this was apparent where low doses of T cells were transferred, though not with high doses. The significance of this curious finding with low T-cell doses (which is also evident in Table IV, V, and Fig. 2) will be considered in the following paper.

Effect of Transferring F_1 T Cells to Irradiated Parental Strain Mice with or without Antigen. When (CBA × B6)F_1 T cells were transferred to irradiated B6 mice without SRC, the lymph-borne cells collected at day 5 (F_1 T_{+B6}) gave significant though small anti-SRC and anti-HRC responses with CBA, B6, and

TABLE III

Helper Activity of T Cells Recovered from Thoracic Duct Lymph of Irradiated B6 Mice Injected 5 Days Previously With Unprimed (CBA × B6)F$_1$ T Cells Given with or without SRC*

T-cell group	T cells	Dose of T cells (× 10^{-6})	B cells‡	PFC/spleen at 7 days in irradiated F$_1$ mice		
				Anti-SRC		Anti-HRC
				IgM	IgG	IgM
A	F$_1$ T$_{+B6}$	0.4	CBA	1,180 (1.29)§	1,390 (1.29)‖	20 (1.65)‖
		2.0	CBA	4,210 (1.12)	10,760 (1.20)	830 (1.30)
		0.4	B6	1,300 (1.55)‖	1,930 (1.34)‖	1,250 (1.58)‖
		2.0	B6	3,560 (1.25)	5,680 (1.19)	3,690 (1.20)
		0.4	(CBA × B6)F$_1$	150 (1.09)‖	3,350 (1.15)	140 (1.46)‖
		2.0	(CBA × B6)F$_1$	4,680 (1.15)	21,870 (1.18)	4,510 (1.23)
B	F$_1$ T$_{+(SRC-B6)}$	0.4	CBA	1,620 (1.21)	1,550 (1.06)‖	0
		2.0	CBA	4,750 (1.25)	12,420 (1.25)	1,180 (1.20)
		0.4	B6	26,340 (1.20)	66,050 (1.12)	1,750 (1.23)
		2.0	B6	55,060 (1.23)	100,440 (1.13)	6,950 (1.19)
		0.4	(CBA × B6)F$_1$	8,140 (1.25)	35,900 (1.24)	640 (1.43)‖
		2.0	(CBA × B6)F$_1$	58,130 (1.36)	188,680 (1.28)	6,540 (1.53)

* As for footnote to Table II; group A. T cells were TDL from irradiated B6 mice given $5 × 10^7$ unprimed (CBA × B6)F$_1$ T cells, but not SRC, 5 days previously; group B. T cells were from irradiated B6 mice given F$_1$ T cells with SRC (0.5 ml of 25%) 5 days before.

‡ As for footnote to Table II except that the B cells were taken from mice primed with both SRC and HRC.

§ As for footnote to Table II except that background values given by B cells transferred without T cells have been subtracted. These values (PFC/spleen) were: CBA 950(1.18) (IgM SRC), 2,050(1.36) (IgG SRC), 440(1.41) (IgM HRC); B6 850(1.50) (IgM SRC), 1,890(1.70) (IgG SRC), 430(1.45) (IgM HRC); (CBA × B6)F$_1$ 1,000(1.17) (IgM SRC), 2,840(1.19) (IgG SRC), 870(1.36) (IgM HRC). Values for T cells transferred without B cells were all <100 PFC/spleen. In the table values in parentheses used to derive limits of SE (see footnote to Table II) were calculated before subtraction of the background PFC numbers.

‖ Not significantly above background values of B cells transferred alone ($P > 0.05$).

F$_1$ B cells (Table III). These responses were dose-dependent, i.e. $2 × 10^6$ T cells produced moderate numbers of PFC, whereas $0.4 × 10^6$ T cells gave responses which were generally not significantly higher than the background values observed when B cells alone were transferred.

Markedly different results occurred when SRC were added to the F$_1$ T cells transferred to the irradiated B6 mice. Three points can be made concerning the helper activity of these cells. First, even small doses of these cells (F$_1$ T$_{+(SRC-B6)}$) gave high anti-SRC responses with B6 and F$_1$ B cells, though not with CBA B cells. Second, the low response observed with CBA B cells was similar in magnitude to that given by T cells from recipients not given SRC (F$_1$ T$_{+B6}$ cells). Third, the response to a different antigen, HRC, was low with all three populations of B cells.

Specificity of Positive Selection. To obtain further information on the specificity of the F$_1$ T cells positively selected to SRC in the above experiments, F$_1$ T cells were activated in (*a*) irradiated B6 mice given SRC (F$_1$ T$_{+(SRC-B6)}$), (*b*) irradiated B6 mice given HRC (F$_1$ T$_{+(HRC-B6)}$), and (*c*) irradiated F$_1$ mice given

Table IV

Helper Activity of T Cells Recovered from Thoracic Duct Lymph of Irradiated B6 Mice Injected 5 Days Previously with (CBA × B6)F₁ T Cells Plus Either SRC or HRC*

T-cell group	T cells (0.8×10^6)‡	B cells§	IgG PFC/spleen at 7 days in irradiated F₁ mice	
			Anti-SRC	Anti-HRC
A	F₁ T₊₍SRC₋B6₎	CBA	960 (1.27)‖	1,160 (1.28)
		B6	40,770 (1.28)	5,830 (1.08)
		(CBA × B6)F₁	16,360 (1.15)	3,180 (1.16)
B	F₁ T₊₍HRC₋B6₎	CBA	460 (1.23)¶	940 (1.53)¶
		B6	1,940 (1.37)	61,420 (1.13)
		(CBA × B6)F₁	2,500 (1.13)	32,110 (1.09)
C	F₁ T₊₍SRC₊HRC₋F₁₎	CBA	35,310 (1.14)	26,550 (1.17)
		B6	12,390 (1.24)	29,400 (1.26)
		(CBA × B6)F₁	42,360 (1.13)	44,990 (1.18)

* As for footnote to Table II; T cells were taken from lymph of irradiated B6 mice given, 5 days previously, 5×10^7 unprimed (CBA × B6)F₁ T cells with 0.5 ml of 25% SRC (group A) or HRC (group B); group C. T cells were from irradiated (CBA × B6)F₁ mice given F₁ T cells together with both SRC and HRC.

‡ As for footnote to Table II except that the T and B cells were transferred with both SRC and HRC (0.1 ml of 5% of each).

§ As for footnote to Table II except that B cells were prepared from mice primed with both SRC and HRC.

‖ As for footnote to Table II except that background values given by B cells transferred without T cells have been subtracted. These values (PFC/spleen) were: CBA 350(1.45) (SRC), 470(1.33) (HRC); B6 1,000(1.26) (SRC), 800(1.19) (HRC); (CBA × B6)F₁ 950(1.36) (SRC), 290(1.07) (HRC). In the table the values in parentheses used to derive limits of SE (see footnote to Table II) were calculated before subtraction of the background PFC numbers. Numbers of PFC for T cells transferred without B cells were all <100/spleen.

¶ Not significantly above background values of B cells transferred alone ($P > 0.05$).

both SRC and HRC (F₁ T₊₍SRC₊HRC₋B6₎). The helper function of these three groups of cells is shown in Table IV. In the case of F₁ T₊₍SRC₋B6₎ cells, anti-SRC responses were high with B6 and F₁ B cells but not with CBA B cells; anti-HRC responses were low with all three B-cell populations. Reciprocal results were found with F₁ T₊₍HRC₋B6₎ cells, i.e. anti-SRC responses were all low, whereas anti-HRC responses were high with B6 and F₁ B cells, but not with CBA B cells. F₁ T₊₍SRC₊HRC₋B6₎ cells stimulated all three B-cell populations to produce high responses to both SRC and HRC.

Time of PFC Assay. In the preceding experiments, numbers of PFC were measured arbitrarily on day 7 post-transfer. Fig. 2 shows that essentially similar results occurred when PFC were assayed on days 5, 7, and 9. Thus, at no stage of assay did F₁ T₊₍SRC₋B6₎ cells give > 1,800 anti-SRC PFC/spleen with CBA B cells. By contrast, F₁ T₊₍SRC₋F1₎ cells elicited high responses with CBA B cells, PFC numbers reaching a peak on day 7 (46,730 IgG PFC), and then declining.

Serum Hemagglutinin Titers. It might be argued that despite the failure of F₁ T₊₍SRC₋B6₎ cells to stimulate CBA B cells to differentiate into PFC in the spleen, effective collaboration could have occurred in other regions. If so, this

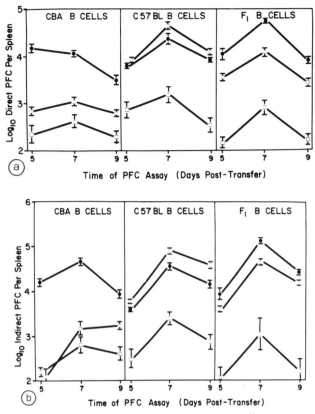

FIG. 2.   Time response of helper activity of (CBA × B6)F$_1$ T cells activated to SRC in irradiated B6 (○) or (CBA × B6)F$_1$ (●) mice and harvested from thoracic duct lymph of the recipients 5 days later. The fig. shows numbers of anti-SRC IgM (direct) (a) and IgG (indirect) (b) PFC/spleen in irradiated (CBA × B6)F$_1$ mice at various times after intra-venous injection of one of the two T-cell populations (0.8 × 10^6 cells) plus 0.1 ml of 5% SRC and B cells from CBA, B6 (C57BL), or (CBA × B6)F$_1$ mice. □, B cells transferred without T cells. The B cells (anti-thy 1.2-serum-treated spleen cells) were from SRC-primed mice and were transferred in a dose of 5 × 10^6 viable cells (8 × 10^6 for B6 B cells). Number of PFC with T cells transferred without B cells was < 100 PFC/spleen at all time points. Each point represents geometric mean of data from four mice; vertical bars show upper and lower limits of SE.

would presumably be reflected in the levels of serum hemagglutinins. To investigate this possibility, the mice used for PFC assay at day 9 in the above experiment (Fig. 2) were exsanguinated at the time of sacrifice, and these sera were measured for anti-SRC hemagglutinin activity. As shown in Table V, the serum hemagglutinin titers closely reflected the results obtained by measuring numbers of PFC in the spleen (compare with Fig. 2).

Effect of Adding PE Cells during Positive Selection.   Macrophages present in the irradiated parental strain mice used for positive selection might be responsible for the restricted helper function observed after selection. If so, the restriction would be expected not to occur if the mice used for selection were supplemented with macrophages of the opposite parental strain. To investigate

TABLE V

*Helper Activity of T Cells Recovered from Thoracic Duct Lymph of Irradiated B6 and (CBA × B6)F₁ Mice Injected 5 Days Previously with (CBA × B6)F₁ T Cells and SRC: Serum Hemagglutinin Titers after Transfer of T and B Cells to Irradiated (CBA × B6)F₁ Mice**

T cells (0.8×10^6)	B cells	Anti-SRC hemagglutinin titers $(1/\log_2)$ at 9 days after transfer to irradiated F_1 mice
F_1 $T_{+(SRC-B6)}$	CBA	1.6 (±0.8)‡
	B6	5.7 (±0.5)
	(CBA × B6)F₁	6.7 (±0.5)
F_1 $T_{+(SRC-F_1)}$	CBA	6.1 (±0.9)
	B6	4.8 (±0.8)
	(CBA × B6)F₁	9.4 (±1.1)

* For details see legend to Fig. 2. Sera tested were from mice assayed for splenic PFC at day 9 after transfer of T and B cells to irradiated (CBA × B6)F₁ mice; the numbers of PFC for these mice are shown in Fig. 2.

‡ Arithmetic mean (±SE) after subtraction of background values given by B cells transferred without T cells. These values were: CBA 1.9 (±0.5); B6 3.6 (±0.9); (CBA × B6)F₁ 2.9 (±0.7). SE were calculated before background subtraction. Titers from mice given T cells but not B cells were <1.4 mice per group. The value asterisked is not significantly above background value of B cells transferred alone; all other values are significantly above background values ($P < 0.05$).

this possibility, (CBA × B6)F₁ T cells were positively selected to SRC in irradiated B6 mice injected with 3×10^7 anti-thy 1.2-serum-treated PE cells derived from either CBA or B6 mice (groups B, C; Table VI); of these cells, >80% had features of macrophages, i.e. they were large cells and contained numerous granules. The F₁ T cells and SRC were mixed with the PE cells and injected intraperitoneally; attempts to transfer this mixture intravenously caused respiratory distress and rapid death, presumably because of clumping of the PE cells. For controls, F₁ T cells and SRC were transferred intravenously into irradiated B6 and CBA mice (groups A, D; Table VI).

The helper activity of TDL recovered from these four groups of mice at day 5 post-transfer is shown in Table VI. It is apparent that addition of CBA PE cells during positive selection in irradiated B6 mice did not lead to restriction in helper function, i.e. the TDL recovered from these mice collaborated well with either CBA or B6 B cells. By contrast, TDL recovered after positive selection in the presence of B6 PE cells led to high responses with B6 B cells, but not with CBA B cells. With respect to the helper cells recovered from Group B, it should be mentioned that >98% of the cells were susceptible to lysis with either anti-thy 1.2 serum or CBA anti-B6 alloantiserum. Thus, there was no evidence that the CBA PE cells given to these mice reached the central lymph.

Positive Selection with (DBA/2 × B6)F₁ T Cells. To determine whether or not the above findings with (CBA × B6)F₁ T cells also applied to other strain combinations, (DBA/2 × B6)F₁ T cells were activated to SRC in irradiated DBA/

TABLE VI

Helper Activity of T Cells Recovered from Thoracic Duct Lymph of Irradiated B6 Mice Injected 5 Days Previously with Unprimed (CBA × B6)F₁ T Cells Plus CBA PE Cells and SRC*

T cell group	T cells (0.8×10^6)‡	PE cells added during positive selection	B cells§	Anti-SRC PFC/spleen at 7 days in irradiated F₁	
				IgM	IgG
A	F₁ T₊₍SRC–B6₎	–	CBA	1,770 (1.12)‖	1,580 (1.52)¶
			B6	17,280 (1.14)	37,050 (1.13)
B	F₁ T₊₍SRC–B6₎	3×10^7 CBA PE	CBA	26,950 (1.16)	67,740 (1.13)
			B6	21,160 (1.19)	47,880 (1.16)
C	F₁ T₊₍SRC–B6₎	3×10^7 B6 PE	CBA	2,150 (1.34)	8,140 (1.37)
			B6	57,330 (1.24)	114,910 (1.02)
D	F₁ T₊₍SRC–CBA₎	–	CBA	19,740 (1.12)	59,400 (1.06)
			B6	1,300 (1.30)	2,160 (1.14)

* As for footnote to Table II except that in the case of groups B and C, the F₁ T cells (5×10^7 nylon-wool-passed), PE cells and SRC were all injected intraperitoneally in one injection; in groups A and D, the F₁ T cells and SRC were transferred intravenously. The PE cells were from mice given thyoglycolate 4 days before and were treated with anti-thy 1.2 antiserum and complement before injection.

§ As for footnote to Table II.

‖ As for footnote to Table II except that background values given by B cells transferred without T cells have been subtracted. These values (PFC/spleen) were: CBA 190(1.57) (IgM), 700(1.76) (IgG); B6 130(1.44) (IgM), 370(1.46) (IgG). Values for T cells transferred without B cells were all <400 PFC/spleen. In the table values in parentheses used to derive limits of SE (see footnote to Table II) were calculated before subtraction of background PFC numbers.

¶ Not significantly above background value of B cells transferred without T cells ($P > 0.05$).

2 or B6 mice. Their helper activity was then assayed in irradiated (DBA/2 × B6)F₁ mice. As shown in Table VII, F₁ T₊₍SRC–DBA/2₎ cells collaborated well with DBA/2 and F₁ B cells, but not with B6 B cells. By contrast, F₁ T₊₍SRC–B6₎ cells stimulated B6 and F₁ B cells, but not DBA/2 B cells. Injection of both populations of T cells with either of the two parental strain B-cell populations led to high responses, i.e. suppression was not observed.

Positive Selection Using Activated T Cells Recovered from the Spleen. All of the preceding experiments involved the use of activated helper cells recovered from thoracic duct lymph of the intermediate irradiated hosts. Two experiments were also performed with helper cells recovered from the spleen. Cells from this region were less satisfactory to work with because of their low viability (20–40%) and greater contamination with radioresistant host cells (15–30%). Nevertheless, the qualitative helper activity of these cells was similar to that of the cells obtained from the central lymph. Thus, it can be seen in Table VIII that (CBA × B6)F₁ T cells positively selected to SRC in irradiated B6 mice and recovered from the spleen of the recipients 5 days later collaborated well with (CBA × B6)F₁ B cells, but poorly with CBA B cells. By contrast, F₁ T cells activated in irradiated F₁ mice stimulated both CBA and F₁ B cells. Mixtures of these two T-cell populations gave high responses with CBA B cells.

Table VII

*Helper Activity of T Cells Recovered from Thoracic Duct Lymph of Irradiated B6 and DBA/2 Mice Injected 5 Days Previously with Unprimed (DBA/2 × B6)F$_1$ T Cells Plus SRC**

T-cell group	T cells (0.8 × 10⁶)‡	B cells§	Anti-SRC PFC/spleen at 7 days in irradiated (DBA/2 × B6)F$_1$ mice	
			IgM	IgG
A	F$_1$ T$_{+(SRC-B6)}$	B6	17,790 (1.13)‖	34,400 (1.26)
		DBA/2	310 (1.18)	390 (1.60)
		(DBA/2 × B6)F$_1$	7,840 (1.23)	35,860 (1.09)
B	F$_1$ T$_{+(SRC-DBA/2)}$	B6	390 (1.38)	250 (1.28)
		DBA/2	4,340 (1.05)	17,990 (1.10)
		(DBA/2 × B6)F$_1$	9,360 (1.14)	29,010 (1.02)
	Group A + group B	B6	17,980 (1.48)	32,690 (1.35)
	Group A + group B	DBA/2	3,730 (1.27)	18,180 (1.14)

* Cell recipients were exposed to 800 rads 1 day before being injected intravenously with 5×10^7 unprimed nylon-wool-purified (DBA/2 × B6)F$_1$ T cells plus 0.5 ml of 25% SRC.

‡ As for footnote to Table II except that irradiated (DBA/2 × B6)F$_1$ mice were used as the recipients of the T and B cells. When transferring a mixture of group A and group B cells to check for suppression, a dose of 0.8×10^6 of each T-cell population was used.

§ Anti-thy 1.2-treated spleen cells from mice primed with SRC 3 mo previously; 5×10^6 (DBA/2 and (DBA/2 × B6)F$_1$) or 8×10^6 (B6) viable cells transferred.

‖ As for footnote to Table II except that background values given by B cells transferred without T cells have been subtracted. These values (PFC/spleen) were: B6 100(1.37) (IgM), <100 (IgG); DBA/2 < 100 (IgM), 140(1.24) (IgG); F$_1$ 160(1.30) (IgM), 340(2.27) (IgG). PFC numbers for T cells transferred without B cells were <100/spleen. In the table values in parentheses used to derive limits of SE were calculated before subtraction of background PFC numbers. All values shown in the Table are significantly above the values of B cells transferred without T cells ($P < 0.05$).

Discussion

A priori, the restriction in helper function observed in the present studies might have been due to the induction of suppressor cells during T-cell activation. In this respect it is of particular relevance that Skidmore and Katz (14), using a hapten-carrier system, recently reported that F$_1$ T cells activated to the carrier in irradiated mice of one parental strain stimulated anti-hapten responses by B cells of this strain, but not those of the opposite strain. In marked contrast to the present studies, however, cooperation was not observed with F$_1$ B cells. The authors concluded that this failure to trigger F$_1$ B cells reflected the presence of suppressor cells generated during T-cell activation. It was not determined whether the suppressors were of host or of donor origin and whether or not they were antigen-specific. On the basis of these findings the authors argued that F$_1$ T cells contain two subpopulations of helper cells, each able to collaborate with only one of the two populations of parental strain B cells. However, it is difficult to justify this reasoning since the presence of suppressor cells in their system made it impossible to determine whether the failure to observe collaboration with B cells of the opposite parental strain reflected an innate restriction in the specificity of the helper cells rather than simple destruction of either the T cells, B cells, or both by the suppressor cells.

TABLE VIII

Helper Activity of T Cells Recovered from Spleen of Irradiated B6 and (CBA × B6)F₁
Mice Injected 5 Days Previously with Unprimed (CBA × B6)F₁ T Cells Plus SRC*

T-cell group	T cells‡ (10^6)	B cells§	Anti-SRC PFC/spleen at 7 days in irradiated F₁ mice	
			IgM	IgG
A	$F_1 T_{+(SRC–B6)}$	CBA	980 (1.22)‖	6,930 (1.17)
		(CBA × B6)F₁	11,890 (1.24)	60,760 (1.21)
B	$F_1 T_{+(SRC–F_1)}$	CBA	10,010 (1.13)	60,900 (1.18)
		(CBA × B6)F₁	38,580 (1.25)	108,380 (1.07)
	Group A + group B	CBA	19,280 (1.47)	85,140 (1.22)

* As for footnote to Table II.

‡ T cells were harvested from the spleens of four mice per group at 5 days after transfer of F₁ T
cells plus SRC. Most of the cells recovered from the spleens were (*a*) thy 1.2-positive, i.e. 93%
for group A and 90% for group B, and (*b*) of donor F₁ origin, i.e. 85% of group A cells were killed
by B6 anti-CBA alloantiserum. When transferring a mixture of group A and group B cells to
check for suppression, a dose of 10^6 viable cells of each T-cell population was used.

§ As for footnote to Table II.

‖ As for footnote to Table II except that background values given by B cells transferred without T
cells have been subtracted. These values (PFC/spleen) were: CBA 160(1.07) (IgM), 170(1.22)
(IgG); (CBA × B6)F₁ 630(1.27) (IgM), 2,490(1.27) (IgG). In the table values in parentheses used
to derive limits of SE (see footnote to Table II) were calculated before subtraction of background
PFC numbers. Numbers of PFC for T cells transferred without B cells were all <200/spleen. All
values shown in the table are significantly above the background values of B cells transferred
without T cells ($P < 0.05$).

In the present studies, three lines of evidence suggest that the restriction in
helper function was not due to the presence of suppressor cells. First, the failure
of F₁ T cells activated to antigen in one parental strain to collaborate with B
cells of the opposite strain was not infectious, i.e., high responses were
invariably observed when these populations of T and B cells were transferred in
the presence of F₁ T cells activated either in mice of the other parental strain or
in F₁ mice (Tables II, VII, VIII). Second, F₁ T cells activated in parental strain
mice cooperated well with F₁ B cells. Third, the poor responses found with B
cells of the opposite parental strain, although very low in magnitude, were not
insignificant. With small numbers of helper cells these responses were close to
background levels but rose in a linear fashion to appreciable levels as the dose
of T cells was increased (see Fig. 1). Moreover, the magnitude of these responses
was comparable to the responses observed with similar doses of unprimed F₁ T
cells (unpublished data) or with unprimed F₁ T cells passaged through irradiated parental strain mice without antigen (Table III).

In view of the evidence mentioned earlier that F₁ T cells contain two discrete
populations of antigen-reactive cells, it would seem reasonable that such a
dichotomy accounted for the present findings. If so, can the data be taken to
indicate that F₁ T cells do consist of two subpopulations of helper cells, each
able to collaborate with B cells derived from only one of the two parental
strains? Before approaching this question it is first necessary to consider the
nature of positive selection to antigen in the irradiated parental strain

environment. Since the function of positively selected T cells was measured in terms of B-cell stimulation, one might argue that antigen presentation in the irradiated intermediate hosts was controlled by host B cells. This seems unlikely since B lymphocytes are exquisitely radiosensitive and die within a few hours of irradiation (23). Moreover, there is no evidence that B lymphocytes can present antigen to T cells. Indeed there is evidence against this notion. Thus, if antigen bound to B cells could trigger helper cell induction, removal of B cells from the inoculum of F_1 T cells used for activation should be a critical step for demonstrating restriction in helper function. Preliminary experiments suggest that this is not the case since restriction in helper function after activation in parental strain mice was no less evident when normal LN (which contain 30–40% B cells but few macrophages) rather than nylon-wool-passed T cells were used for activation (unpublished data). Macrophages (or related cells) are more likely candidates for antigen presentation in the present system, not only because of their well-accepted role in T-cell triggering, but also because they are highly radioresistant (23). The fact that the restriction in helper function could be overcome by the addition of PE cells (>80% macrophages) of the opposite parental strain (Table VI) supports this view. However, the precise identity of the cell(s) presenting antigen has yet to be established.

Whatever the nature of the cell presenting antigen to the T cell, it is clearly necessary to assume that despite the particulate nature of the antigen used (heterologous erythrocytes), the T cells were unable to recognize the antigen in terms of helper cell induction unless the antigen was first processed in a specific way in the irradiated environment. Thus, of the two putative T-cell subgroups in the F_1 T cell inoculum, only one apparently underwent clonal expansion when exposed to antigen in irradiated mice of one parental strain; the other subgroup failed to recognize the antigen and therefore remained in an unprimed (nonexpanded) state. Although this interpretation is in line with the generally accepted view that T-macrophage interactions are under MHC control, it is less easy to understand why, in the present system, such interactions were associated with apparent restrictions at the level of T-B collaboration.

One explanation for this paradox is that the data reflect restrictions acting at both the level of T-macrophage interactions and during T-B collaboration. Accordingly, one could argue that each subgroup of F_1 T cells is reactive to MHC-associated antigen presented first on host macrophages during helper cell induction in one of the two parental strain environments, and second, on the corresponding population of parental strain B cells during T-B collaboration. T-cell activation in one parental strain would thus induce clonal expansion of the subgroup of F_1 T cells able to collaborate only with B cells of that parental strain (or with F_1 B cells). The other T-cell subgroup would not be stimulated and would consequently give only a low primary response with its corresponding B-cell population, i.e. B cells of the opposite parental strain. The magnitude of this response would be equivalent to that of unprimed F_1 T cells.

The data, in toto, are consistent with this viewpoint. Nevertheless, it is clearly difficult to assess this interpretation without information on the genetic control of the restriction(s) observed. It is also necessary to consider another possibility; that the failure of F_1 T cells positively selected in one parent to collaborate with B cells of the opposite parent simply reflected a lack of appropriate macrophages during the stage of T-B collaboration. In other words, the cells failed to manifest their helper function because on subsequent transfer with B cells of the strain opposite to that used for activation, the macrophages presenting antigen to the helper cells were now different from those encountered during the initial activation. Although this possibility might seem unlikely since F_1 mice were always used for measurement of T-B collaboration, it needs to be excluded.

The following paper is addressed to this question and shows that the restriction(s) observed in the present paper cannot be explained in terms of a lack of appropriate macrophages during the stage of T-B collaboration. The data are interpreted as indicating that restrictions do exist at the level of both T helper cell induction and T-B collaboration, each restriction mapping to the K-end of the H-2 complex.

Summary

Unprimed (CBA $\times$ C57BL/6)F_1 lymph node T cells were transferred with sheep erythrocytes (SRC) into heavily irradiated F_1 or parental strain mice and recovered from thoracic duct lymph or spleens of the recipients 5 days later. To study their helper function, the harvested F_1 T cells were transferred with antigen into irradiated F_1 mice plus B cells from either the two parental strains or from F_1 mice. F_1 T cells activated in F_1 mice gave high IgM and IgG anti-SRC responses with all three populations of B cells. By contrast, F_1 T cells activated in mice of one parental strain collaborated well with B cells of this strain, but poorly with B cells of the opposite strain. Active suppression was considered an unlikely explanation for this result since (a) good responses were found with F_1 B cells, and (b) addition experiments showed that the poor response with B cells of the opposite parental strain (which was equivalent to that produced by unprimed F_1 T cells) could be converted to a high response by a supplemental injection of F_1 T cells activated in F_1 mice. The phenomenon (a) was specific for the antigen used for activation (criss-cross experiments were performed with horse erythrocytes), (b) was reflected in levels of serum hemagglutinins as well as in numbers of splenic plaque-forming cells, (c) applied also to comparable activation of (DBA/2 $\times$ C57BL/6)F_1 T cells, and (d) could be prevented by activating F_1 T cells in mice of one parental strain in the presence of peritoneal exudate cells of the opposite parental strain.

The hypothesis was advanced that F_1 T cells contain two discrete subpopulations of antigen-reactive cells, each subject to restrictions acting at two different levels: (a) during T-macrophage interactions and (b) during T-B collaboration. It was suggested that when F_1 T cells are activated to antigen in a parental strain environment, radioresistant macrophages activate only one of the two subgroups of T cells, and this subgroup is able to collaborate with B cells of the strain used for activation (and with F_1 B cells) but not with B cells of the opposite parental strain. The other subgroup of T cells remains in an unprimed (nonactivated) state.

Stimulating discussion with D. B. Wilson, the technical assistance of Ms. L. Collins, and the skillful typing of Miss K. D. Nowell are gratefully acknowledged.

Received for publication 22 December 1977.

References

1. Thomas, D. W., and E. M. Shevach. 1976. Nature of the antigenic complex recognized by T lymphocytes. I. Analysis with an in vitro primary response to soluble protein antigens. *J. Exp. Med.* **144**:1263.
2. Paul, W. E., E. M. Shevach, S. Pickeral, D. W. Thomas, and A. S. Rosenthal. 1977. Independent populations of primed F_1 guinea pig T lymphocytes respond to antigen-pulsed parental peritoneal exudate cells. *J. Exp. Med.* **145**:618.
3. Miller, J. F. A. P., M. A. Vadas, A. Whitelaw, and J. Gamble. 1976. Role of major histocompatibility complex gene products in delayed-type hypersensitivity. *Proc. Natl. Acad. Sci. U. S. A.* **73**:2486.
4. Zinkernagel, R. M., and P. C. Doherty. 1975. *H-2* compatibility requirement for T-cell-mediated lysis of target cells infected with lymphocytic choriomeningitis virus. *J. Exp. Med.* **141**:1427.
5. Shearer, G. M., T. G. Rehn, and C. A. Garbarino. 1975. Cell-mediated lympholysis of trinitrophenyl-modified autologous lymphocytes. Effector cell specificity to modified cell surface components controlled by the *H-2K* and *H-2D* serological regions of the murine major histocompatibility complex. *J. Exp. Med.* **141**:1348.
6. Matzinger, P., and M. J. Bevan. 1977. Induction of H-2-restricted cytotoxic T cells: in vivo induction has the appreance of being unrestricted. *Cell. Immunol.* **33**:92.
7. Vadas, M. A., J. F. A. P. Miller, I. F. C. McKenzie, S. E. Chism, F-W. Shen, E. A. Boyse, J. R. Gamble, and A. M. Whitelaw. 1976. Ly and Ia antigen phenotypes of T cells involved in delayed-type hypersensitivity and in suppression. *J. Exp. Med.* **144**:10.
8. Feldmann, M., P. C. L. Beverley, J. Woody, and I. F. C. McKenzie. 1977. T-T interactions in the induction of suppressor and helper T cells: analysis of membrane phenotype of precursor and amplifier cells. *J. Exp. Med.* **145**:793.
9. Erb, P., and M. Feldmann. 1975. The role of macrophages in the generation of T-helper cells. I. The requirement for macrophages in helper cell induction and characteristics of the macrophage-T cell interaction. *Cell. Immunol.* **19**:356.
10. Erb, P., and M. Feldmann. 1975. The role of macrophages in the generation of T-helper cells. II. The genetic control of the macrophage-T-cell interaction for helper cell induction with soluble antigens. *J. Exp. Med.* **142**:460.
11. Erb, P., and M. Feldmann. 1975. The role of macrophages in the generation of T-helper cells. III. Influence of macrophage-derived factors in helper cell induction. *Eur. J. Immunol.* **5**:159.
12. Heber-Katz, E., and D. B. Wilson. 1975. Collaboration of allogeneic T and B lymphocytes in the primary antibody response to sheep erythrocytes in vitro. *J. Exp. Med.* **142**:928.
13. Sprent, J., J. F. A. P. Miller, and G. F. Mitchell. 1971. Antigen-induced selective recruitment of circulating lymphocytes. *Cell. Immunol.* **2**:171.
14. Skidmore, B. J., and D. H. Katz. 1977. Haplotype preference in lymphocyte differentiation. I. Development of haplotype-specific helper and suppressor activities in F_1 hybrid-activated T cell populations. *J. Immunol.* **119**:694.
15. Sprent, J. 1973. Circulating T and B lymphocytes of the mouse. I. Migratory properties. *Cell. Immunol.* **7**:10.
16. Sprent, J., and H. von Boehmer. 1976. Helper function of T cells depleted of alloantigen-reactive lymphocytes by filtration through irradiated F_1 hybrid recipients. I. Failure to collaborate with allogeneic B cells in a secondary response to sheep erythrocytes measured in vivo. *J. Exp. Med.* **144**:617.
17. Julius, M. H., E. Simpson, and L. A. Herzenberg. 1973. A rapid method for the isolation of thymus-derived murine lymphocytes. *Eur. J. Immunol.* **3**:645.
18. Cudkowicz, G., and M. Bennett. 1971. Peculiar immunobiology of bone marrow allografts. II. Rejection of parental grafts by resistant F_1 hybrid mice. *J. Exp. Med.* **134**:1513.
19. Cunningham, A. J., and A. Szenberg. 1968. Further improvements in the plaque technique for detecting single antibody-forming cells. *Immunology.* **14**:599.
20. Dietrich, F. M. 1966. The immune response to heterologous erythrocytes in mice. *Immunology.* **10**:365.
21. Sprent, J., and J. F. A. P. Miller. 1971. Activation of thymus cells by histocompatibility antigens. *Nat. New Biol.* **234**:195.
22. Ford, W. L., and R. C. Atkins. 1971. Specific unresponsiveness of recirculating lymphocytes after exposure to histocompatibility antigen in F_1 hybrid rats. *Nat. New Biol.* **234**:178.
23. Anderson, R. E., and N. L. Warner. 1976. Ionizing radiation and the immune response. *Adv. Immunol.* **24**:215.

Killer cells reactive to altered-self antigens can also be alloreactive

(cytotoxic lymphocytes/major histocompatibility antigens/minor histocompatibility antigens)

MICHAEL J. BEVAN*

The Salk Institute for Biological Studies, Post Office Box 1809, San Diego, California 92112

Communicated by George D. Snell, February 16, 1977

ABSTRACT   Murine cytotoxic thymus-derived lymphocytes immunized against cells bearing foreign minor histocompatibility antigens are specific for the immunizing minor antigens and for their own major H-2 antigens; they do not lyse target cells that bear the correct minor antigens plus a different *H-2* haplotype. These are referred to as "altered-self" or "self-plus-X" killer cells. Alloreactive killer cells are those which respond to allogeneic cells expressing a foreign (non-self) *H-2* haplotype. In this study, cytotoxic lymphocytes were immunized against minor histocompatibility differences *in vivo* and *in vitro*. These effector cells killed the immunizing altered-self target very well and showed about 1% cross-reactive lysis of an allogeneic target differing from themselves only at *H-2*. These cross-reactive clones were then selected for by repeated *in vitro* stimulation with the cells bearing foreign H-2 such that an effector population was obtained which lysed both the altered-self and the alloreactive target with the same efficiency. Cold target competition experiments established that the same killer cell could lyse either target; however, it was not determined if a killer cell uses the same receptor to respond to altered-self antigens as it does to respond to foreign H-2 antigens.

A very large fraction of thymus-derived (T) lymphocytes is committed to respond to any allogeneic cell that differs genetically from the responder at the major histocompatibility complex (MHC, the H-2 complex in mice, refs. 1–4). Between 1 and 10% of T cells respond to any foreign MHC haplotype and, because this region is polymorphic, it is likely that most T cells are alloreactive to one or another of the non-self *H-2* haplotypes. A large fraction of cytotoxic T lymphocytes (CTL), a subfraction of T cells, can be induced to kill cells bearing any foreign allele of $H-2^k$ or *H-2D* (5–7). In the mouse, then, *H-2* differences are the strongest barriers to tissue transplantation; but it is probably not the normal physiological function of T cells to respond to allogeneic cells.

The most important recent discovery relating the MHC to the physiological function of T cells is that of *H-2* restriction. It was found that CTL of $H-2^a$ haplotype immunized against autologous cells that had been altered chemically or by virus infection (X) could lyse X-altered-$H-2^a$ targets but not X-altered-$H-2^b$ or X-altered-$H-2^c$ targets (8, 9). The cytotoxic response to minor H antigens [i.e., when responder and stimulator cells bear the same *H-2* but differ in non-*H-2* coded minor H loci (10)] is similarly *H-2* restricted (11, 12). *H-2* coded structures of the target are therefore involved in the lytic interaction even when the CTL and target carry the same *H-2* genes—i.e., in the response to "altered-self" antigens.

According to the altered-self hypothesis (8, 9, 13) the explanation of *H-2* restriction is that the CTL has one receptor which

binds two components of the target cell, X-plus-H-2. The dual-recognition hypothesis (deriving from previous work on interactions between T cells and thymus-independent lymphocytes and between T cells and macrophages) postulates that $H-2^k$ and D serve as self-markers for CTL, which therefore has two separate receptors, one for self H-2 and one for X (14, 15).

In understanding the function of T cells and their antigen binding requirements, it is important that we determine the relationship between reactions to altered-self targets and to targets that bear foreign H-2 antigens (alloreactions). We need to determine: (*i*) whether alloreactive T cells can also respond to altered-self antigens, and (*ii*) if they do, whether the receptor systems used are the same in both cases. With regard to the first point, although it has been established that most of the effector and precursor CTL involved in anti-H-2 reactions are Ly-1$^-$, -2$^+$, -3$^+$, (16, 17), there is recent evidence that suggests that precursor or effector CTL for altered-self reactions are Ly-1$^+$. The precursor CTL for TNP-altered-self may be Ly-1$^+$, -2$^+$, -3$^+$ (18), the effector CTL in response to a syngeneic tumor may be Ly-1$^+$, -2$^+$, -3$^+$ (19), and effector CTL to virus-infected cells after primary *in vivo* immunization may also be Ly-1$^+$ (20). This implies that either alloreactive T cells and altered-self reactive T cells belong to two separate subsets or that the antigenic stimulus delivered to T cells by these antigens is different. On the other hand, Heber-Katz and Wilson (21) presented evidence that alloreactive cells include those T cells responsive to conventional antigens. They positively selected T cells responsive to one MHC haplotype in mixed lymphocyte culture, parked them in animals that lacked T cells, and later demonstrated that the parked cells contained helper T cell activity for the antibody response to sheep erythrocytes. In this work there could be no definitive proof that the same T cell could respond to both antigens because the assay systems for the responses were so different. The evidence rested on the completeness of the negative selection against bystander T cells in the mixed lymphocyte culture.

Here I report studies that show conclusively that an altered-self reactive CTL can also respond to and lyse cells that differ at H-2. F$_1$ (Balb/c × Balb.B) ($H-2^{d/b}$) CTL reactive to the minor H differences of B10.D2($H-2^d$) were positively selected for by *in vivo* and *in vitro* immunization with B10.D2. These effectors showed a small degree of cross-reactive lysis of Balb.K($H-2^k$) targets. The cells lysing Balb.K were further selected for by repeated stimulation in long-term mixed lymphocyte cultures with Balb.K. A highly selected CTL population capable of lysing B10.D2 and Balb.K targets almost equally was eventually obtained. It was established that the same killer cell could lyse either target by showing that the rate of lysis of labeled B10.D2 targets was specifically inhibited by addition of an excess of unlabeled Balb.K targets (cold target competition

Abbreviations: T lymphocyte, thymus-derived lymphocyte; MHC, major histocompatibility complex; CTL, cytotoxic T lymphocyte; X, foreign minor antigen.

* Present address: Center for Cancer Research, Massachusetts Institute of Technology, Cambridge, MA 02139.

experiments). Similarly, lysis of labeled Balb.K targets was inhibited in the presence of B10.D2 targets. How this finding may be related to the number of receptors on T cells as proposed by the altered-self and dual-recognition hypotheses is discussed.

MATERIALS AND METHODS

Mice. Balb.K($H-2^k$) and F_1(Balb/c × Balb.B) ($H-2^{d/b}$) mice were bred at The Salk Institute. Balb.K and Balb.B stocks were originally obtained from F. Lilly. C57BL/10Sn (B10, $H-2^b$), B10.D/2nSn($H-2^d$), and B10.BR/SgSn($H-2^k$) mice were purchased from the Jackson Laboratory, Bar Harbor, ME.

In Vivo **Immunization.** Young adult female F_1(Balb/c × Balb.B) mice were injected intraperitoneally with $2 × 10^7$ viable B10.D2 spleen cells suspended in Hanks' balanced salt solution.

Mixed Lymphocyte Cultures. The culture medium was RPMI 1640 medium (Microbiological Associates, Bethesda, MD) supplemented with 50 μM 2-mercaptoethanol and 5% (vol/vol) fetal calf serum. For the first *in vitro* immunization, responder spleen cells at $4 × 10^6$ viable cells per ml were stimulated with an equal number of irradiated (1000 rads from a ^{60}Co source) stimulator cells in 35-mm tissue culture dishes (Falcon Plastics, Bioquest, Oxnard, CA). Dishes were placed in a box, gassed with a mixture of 10% CO_2/7% O_2/83% N_2, and placed on a rocker platform at 37°. After 5 days, some dishes were used to assay for cytotoxic activity and the remainder were harvested by pipetting and the cells were placed in tissue culture flasks (Falcon, no. 3012), 10–25 ml per flask. The flasks were loosely stoppered and incubated upright in a 10% CO_2/air incubator at 37°. When the cells had settled, about half of the supernatant medium was replaced with fresh medium.

For subsequent *in vitro* stimulations the cells were removed from flasks and centrifuged, and the viable cells were counted. They were restimulated with irradiated cells in dishes on a rocking platform again, but this time at a density of 1 to $2 × 10^6$ cells per ml. Secondary stimulation was for 4 days. Then some cells were used in cytotoxicity assays and the remainder were placed in new flasks.

^{51}**Cr-Release Cytotoxicity Assay.** This was performed exactly as described previously (12). Target cells were spleen cells cultured for 2 days with lipopolysaccharide (thymus-independent cell blasts) or concanavalin A (T cell blasts). Targets were labeled with sodium [^{51}Cr]-chromate (Amersham/Searle, Arlington Heights, IL) and washed twice before use. Serial 3-fold dilutions of the killer cells were titrated against 3 to 4 × 10^4 ^{51}Cr-labeled targets for 3–4 hr in 1 ml of medium in 35-mm petri dishes on a rocking platform. At the end of the assay the contents of the dishes were transferred to tubes and centrifuged, and 0.5 ml of the supernatant was removed for counting. Percent specific lysis was calculated as follows:

$$\frac{\begin{array}{c}\text{cpm released}\\\text{in presence}\\\text{of killers}\end{array} - \begin{array}{c}\text{cpm released}\\\text{spontaneously}\end{array}}{\begin{array}{c}\text{total cpm originally}\\\text{in targets}\end{array} - \begin{array}{c}\text{cpm released}\\\text{spontaneously}\end{array}} × 100.$$

RESULTS

Long-Term Mixed Lymphocyte Culture. The CTL response of F_1(Balb/c × Balb.B) ($H-2^{d/b}$) mice to the minor H antigens of B10.D2($H-2^d$) is $H-2$ restricted and the effectors do not lyse B10($H-2^b$) targets (12). They do crossreact to a slight extent on B10.BR($H-2^k$) targets; this is not due to non-$H-2^d$ restricted

recognition of B10 minors because the same degree of cross-reaction is observed on Balb.K($H-2^k$) targets. The F_1 responder and Balb.K mice are congenics and differ only at the $H-2$ region. The crossreactive lysis of B10.BR and Balb.K targets is mediated by the same fraction of effector cells as demonstrated in cold target competition experiments (unpublished data).

Spleen cells from F_1 mice that had been primed *in vivo* with B10.D2 cells were stimulated in mixed lymphocyte culture for 5 days with B10.D2. On day 5, the CTL lysed B10.D2 targets efficiently, caused no lysis of syngeneic targets, and showed a slight crossreaction (probably about 1%) on Balb.K targets (Fig. 1). The cells were parked in flasks until day 14 when they were restimulated with either Balb.K or B10.D2 cells and then assayed for cytotoxicity on day 18. CTL rebooted with B10.D2 showed the same pattern of lysis of B10.D2 and Balb.K targets as on day 5 although this time lysis of B10.D2 was about 3-fold more efficient on a per cell basis (on day 5, 40% lysis of B10.D2 at a ratio of 3.3:1; on day 18, 40% lysis at 1:1). CTL boosted on day 14 with the alloreactive Balb.K stimulus were about 13-fold less active in lysing B10.D2 (40% lysis at 13:1) than the cells boosted with B10.D2 on day 14. Therefore, the recall of cytotoxic activity is antigen-specific to a large degree. These Balb.K-boosted CTL lysed Balb.K targets quite effectively, approximately one-seventh as well as they lysed B10.D2.

The cells that had been boosted with Balb.K on day 14 were parked once more in flasks until day 25 when they were cultured with irradiated Balb.K again or with no irradiated cells. Cytotoxicity assays were performed on day 29. The unstimulated cells had little activity against any target but those boosted with irradiated Balb.K cells lysed B10.D2 and Balb.K targets almost equally effectively (less than a 2-fold difference in the rates of lysis).

The data so far have shown that recall of CTL effector function in these long-term cultures is largely antigen-specific. The cold target competition experiments presented below show that most of the CTL clones selected by stimulation with Balb.K can lyse both Balb.K and the original altered-self B10.D2 target. These two points together indicate that the crossreactive clones can also be *induced* to become effectors with either antigen.

Cold Target Competition Experiments. The first experiment with these long-term cultured F_1 cells was done on day 29. Lysis of labeled B10.D2 and Balb.K blasts by the same highly selected population of CTL was studied in the presence of an excess of unlabeled F_1(Balb/c × Balb.B), B10.D2, or Balb.K blasts (Fig. 2). The ratio between unlabeled and labeled blasts in the assay was 75:1. Lysis of either target was inhibited about 3-fold by F_1 blasts which are syngeneic to the CTL. This syngeneic inhibition is always seen when relatively large numbers of unlabeled cells are added (12, 22). Syngeneic inhibition is taken here as "nonspecific" and is the base line from which specific blocking is considered. Fig. 2 *left* shows that unlabeled alloreactive Balb.K blasts specifically inhibited the lysis of the altered-self target, ^{51}Cr-labeled B10.D2, 3-fold; unlabeled B10.D2 inhibited 20-fold. With ^{51}Cr-labeled Balb.K as indicator, unlabeled B10.D2 caused 4-fold inhibition and unlabeled Balb.K caused 11-fold inhibition (Fig. 2 *right*).

A second cold target competition experiment was done with anti-B10.D2 CTL that had been boosted with irradiated Balb.K on day 25, parked in culture flasks until day 36, and then boosted once more with Balb.K. The assay was done on day 40 of culture. Lysis of the original altered-self target, B10.D2, was not significantly different from lysis of $H-2^k$ targets (either B10.BR or Balb.K) at this time. B10.D2 and B10.BR blasts were used as labeled indicator cells in this experiment, with a 60-fold excess of unlabeled B10($H-2^b$), B10.D2, and B10.BR as blockers.

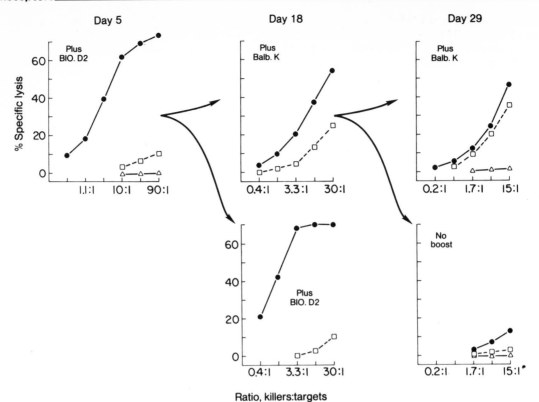

Ratio, killers:targets

FIG. 1.   Cytotoxic activity of responding F_1(Balb/c × Balb.B) (H-$2^{d/b}$) spleen cells in long-term mixed lymphocyte culture. Spleen cells were removed from F_1 mice injected 18 days previously with 2×10^7 B10.D2(H-2^d) cells and cultured on day 0 with irradiated B10.D2 spleen cells. Some of the cells were assayed for cytotoxicity on day 5, and the remainder were parked in culture flasks. Cells were recultured with irradiated Balb.K(H-2^k) or B10.D2(H-2^d) cells on day 14 and assayed on day 18. On day 25 of culture, the cells stimulated with Balb.K on day 14 were recultured with irradiated Balb.K or with no stimulating cells and assayed on day 29. Viable cell recovery from days 0–5 was 68%; from days 14–18 it was 130% after boosting with Balb.K and 213% after boosting with B10.D2; from days 25–29 it was 164% after boosting with Balb.K and 71% with no added stimulus. Targets used were ^{51}Cr-labeled blasts of F_1(Balb/c × Balb.B) (△), B10.D2 (●), and Balb.K (□). Spontaneous release of ^{51}Cr varied from 10.8 to 21.1%.

Lysis of labeled B10.D2 targets was specifically inhibited 7-fold by unlabeled B10.BR, and 27-fold by B10.D2 (Fig. 3 *left*). Lysis of labeled B10.BR targets was inhibited 4-fold by B10.D2, and 20-fold by B10.BR (Fig. 3 *right*).

This highly selected population of CTL, immunized first *in vivo* and on day 0 of culture with B10.D2 and on days 14, 25,

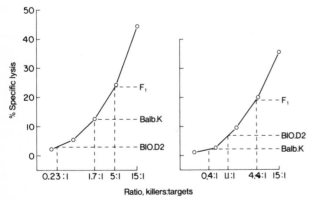

FIG. 2.   Crossinhibition of lysis of ^{51}Cr-labeled B10.D2(H-2^d) (*Left*) and Balb.K(H-2^k) (*Right*) targets by either cold target. Solid lines show lysis of 3×10^4 target cells by serial 3-fold dilutions of the F_1(Balb/c × Balb.B)(H-$2^{d/b}$) CTL population from day 29 of long-term mixed lymphocyte culture (Fig. 1) in the absence of unlabeled inhibitor cells. Horizontal broken lines indicate the lysis observed in the presence of 2.2×10^6 unlabeled blasts as marked, at the killer:target ratio of 15:1. The inhibited level of lysis is extrapolated from the standard uninhibited curve to give the equivalent killer:target ratio. Spontaneous release of ^{51}Cr was 10.8% and 19.5%.

and 36 with Balb.K, appeared to consist largely of clones that could lyse B10.D2 and Balb.K. The cold target competition experiments indicate that lysis of the original altered-self target, B10.D2, is inhibited 75–87% by an excess of Balb.K or B10.BR, and lysis of the alloreactive H-2^k targets is specifically inhibited 80% by B10.D2.

In order to show that such cross-inhibition does not normally occur with F_1 anti-B10.D2 and F_1 anti-Balb.K CTL, the following control was performed. Both populations of CTL were generated separately—F_1 anti-B10.D2 by *in vivo* priming followed by 5 days of mixed lymphocyte culture, and F_1 anti-Balb.K by a 5-day primary mixed lymphocyte culture immunization. The effectors were mixed to give approximately equal lysis of B10.D2 and Balb.K targets. Inhibition of lysis of labeled B10.D2 and Balb.K targets was then studied with a 65-fold excess of unlabeled F_1(Balb/c × Balb.B), B10.D2, or Balb.K blasts. Table 1 shows that under these conditions, when separate effectors were lysing either target, no crossinhibition greater than syngeneic inhibition could be detected. For example, lysis of ^{51}Cr-labeled B10.D2 was inhibited well by unlabeled B10.D2, but unlabeled Balb.K showed no more inhibition than was caused by unlabeled F_1 blasts.

DISCUSSION

The cold target competition experiments reported here establish that one F_1(Balb/c × Balb.B)(H-$2^{d/b}$) effector CTL can lyse either an altered-self target cell, in this case B10.D2(H-2^d), or a target cell that differs from itself at H-2, in this case Balb.K (H-2^k). The data also establish something of the nature of CTL

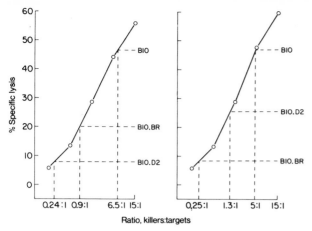

FIG. 3. Cold target competition experiment with F_1(Balb/c × Balb.B) (H-$2^{d/b}$) CTL selected to lyse both altered-self B10.D2 (H-2^d) ^{51}Cr-labeled targets (*Left*) and alloreactive (H-2^k) ^{51}Cr-labeled targets, Balb.K and B10.BR (*Right*). CTL were immunized originally against B10.D2 and then against Balb.K on days 14, 25, and 36 of long-term mixed lymphocyte culture. On day 40, serial 3-fold dilutions were assayed for lysis of $3.5 × 10^4$ ^{51}Cr-labeled blasts of B10.D2 (*Left*) and B10.BR (*Right*) (solid lines). At the 15:1 killer:target ratio, lysis was inhibited by the addition of $2.1 × 10^6$ unlabeled blasts of B10(H-2^b), B10.D2, or B10.BR (horizontal broken lines). Spontaneous release of ^{51}Cr was 10.5% and 11.1%. Labeled Balb.K targets were lysed as well as B10.BR; B10 targets were lysed very poorly.

crossreactivity. Killers immunized against cells bearing foreign *H-2* often crossreact quite strongly on third-party cells bearing a different *H-2* type (23, 24), and anti-minor H CTL may also crossreact detectably on *H-2* dissimilar cells (Fig. 1; ref. 25). Theoretically, crossreactive lysis could have occurred if 100% of the clones lysed the third-party targets with 1% efficiency (compared to the lysis of the immunizing targets). Alternatively, crossreactive lysis could have been due to only 1% of the clones lysing both targets efficiently, while 99% of them lysed only the specific target. Because the crossreaction can be selected for by immunization with the third-party cells, until lysis of both targets becomes approximately equal, the latter explanation is probably correct.

The selection procedure which I placed on these cytotoxic T cells in culture is probably analogous to the experiments with thymus-independent cells performed by Richards *et al.* (26). They immunized rabbits with hapten A and showed by iso-electric focusing that a small, specific fraction of the serum antibody to A could also bind hapten B. Secondary immunization with B induced these crossreactive clones without inducing the noncrossreactive anti-A clones.

The first immunization of F_1(Balb/c × Balb.B) cells in culture with B10.D2 selects for (positive selection) the survival of anti-B10.D2 lymphocytes and selects against (negative selection) cells that do not react with B10.D2. Cells selected against include most of the F_1 cells which would normally respond to the H-2^k coded differences of Balb.K. Secondary immunization with Balb.K is therefore selective for the clones that react with B10.D2 and Balb.K. The cold target competition experiments (Figs. 2 and 3) showed that cross-inhibition of lysis (e.g., of ^{51}Cr-labeled Balb.K by unlabeled B10.D2) was not quite as effective as homologous inhibition. This is probably because selection against anti-Balb.K-only cells in the early immunizations with B10.D2 was not complete. Thus, a small number of cells that could lyse Balb.K but not B10.D2 were left at day 14 when the first immunization with Balb.K was done. Because of the incompleteness of negative selection, the cold target

Table 1. Cold target competition experiment with artificially mixed population of altered-self and alloreactive CTL*

Ratio, CTL: ^{51}Cr-blasts	Unlabeled blasts†	% Specific lysis	
		^{51}Cr-B10.D2	^{51}Cr-Balb.K
28:1	None	52.8	58.5
9.3:1	None	37.6	40.2
3.1:1	None	17.5	19.0
1:1	None	6.5	7.6
0.3:1	None	1.9	3.0
28:1	F_1	41.8	43.0
28:1	B10.D2	1.2	42.1
28:1	Balb.K	40.1	3.9

* F_1(Balb/c × Balb.B) anti-B10.D2 and anti-Balb.K effector CTL were mixed in 1.3:1 ratio to give almost equal rates of lysis of either target.
† A 65-fold excess of unlabeled blasts over ^{51}Cr-labeled blasts was used. $F_1 = F_1$(Balb/c × Balb.B).

competition experiments had to be performed to show that most of the CTL lysed both targets.

Recall of cytotoxic effector function in these cultures appeared to be quite antigen-specific. Thus, on day 14, when most of the memory CTL were specific for B10.D2 only, addition of irradiated B10.D2 cells recalled 13-fold more anti-B10.D2 CTL activity compared to addition of irradiated Balb.K cells (Fig. 1). If a helper T cell response is also required to re-induce CTL and if recall of helper activity was antigen-specific in these cultures, to explain the fact that Balb.K can specifically re-induce cells that were first induced by B10.D2 one might postulate that Balb.K and B10.D2 share helper determinants as well as cytotoxic determinants. Since it is not established that helper T cells were completely selected out or that there was not sufficient helper activity remaining in the cultures even in the absence of added B10.D2 antigen, this postulate can only be made tentatively. Furthermore, it is not established that helper T cells are required to induce CTL.

I do not know how the results presented here can be reconciled with the findings that alloreactive CTL and altered-self reactive CTL may differ in their Ly phenotypes (18–20). I do not believe that there are two separate, nonoverlapping populations of CTL. Differences in Ly phenotype might be correlated with the stage of T cell development which in turn may depend on the receptor specificity as defined by "closeness to self antigens." Foreign H-2 antigens might select cells from one end of this spectrum of receptors while the self-plus-X antigens so far studied select from the other end.

These results show that one killer cell can respond to foreign cells that are antigenic due to differences either at minor H loci or at major H loci. There are several possible explanations for this at the level of surface receptors on T cells. First, in analogy with what has been suggested for immunoglobulin, one CTL receptor molecule might have separate antigen binding sites for different determinants expressed on the two antigens (26). Second, one T cell may have more than one species of antigen-binding receptor on its surface. Because so many lymphocytes respond to major H antigens, thereby threatening to overoccupy the T cell repertoire with alloreactivity, many investigators have proposed this model (1, 2). Third, the dual-recognition model would say that H-2 antigens differ qualitatively from all other cell-surface antigens. To respond to minor H antigens requires, in addition to the antigen-specific anti-X receptor binding of the minor antigen, that the anti-self H-2 receptor also binds to the target. In the case of a foreign H-2 antigen, reaction of one receptor with foreign H-2 is sufficient. In this model, H-2 molecules of the stimulating or target cell are given special intrinsic properties such as being the only site at which CTL killing signals can be delivered. Contrary to this notion is the finding that target cells that lack H-2 can still be lysed by CTL when agglutinated with phytohemagglutinin or concanavalin A (27, 28). If the killing acceptor site is not part of H-2 but only normally associated on the membrane with H-2, then this objection could be overruled (R. Langman, personal communication).

Finally, T cells may have one receptor, and the combining site that binds self H-2^d-plus-foreign B10 minor antigen is the same as that binding foreign H-2^k-plus-self Balb minors. That is, if H-2 influences the immunogenicity of all other membrane components (as it does in altered-self reactions), then allo-reactivity to cells bearing foreign H-2 might not be to H-2 seen in isolation but to H-2 seen in combination with many other surface components (29). This model, based on the altered-self interpretation of *H-2* restriction, was proposed to explain the high frequency of T cells responsive to cells bearing any foreign *H-2* haplotype. According to this, the special stimulatory property of H-2 is due to the specificity range of T cell receptors which is somehow designed to bind antigens only when they are associated with H-2.

I thank Polly Matzinger and Mel Cohn for helpful discussion. This work was supported by National Institutes of Health Grants CA 19754, AI 100430, and AI 05875.

1. Simonsen, M. (1967) *Cold Spring Harbor Symp. Quant. Biol.* **32**, 517–523.
2. Wilson, D. B. (1974) in *Progress in Immunology II*, eds. Brent, L. & Holborow, J. (North Holland, Amsterdam), Vol. 2, pp. 145–156.
3. Ford, W. L., Simmonds, S. J. & Atkins, R. C. (1975) *J. Exp. Med.* **141**, 681–696.
4. Binz, H. & Wigzell, H. (1975) *J. Exp. Med.* **142**, 1218–1230.
5. Bevan, M. J., Langman, R. E. & Cohn, M. (1976) *Eur. J. Immunol.* **6**, 150–156.
6. Skinner, M. A. & Marbrook, J. (1976) *J. Exp. Med.* **143**, 1562–1567.
7. Lindahl, K. F. & Wilson, D. B. (1976) *J. Exp. Med.* **145**, 508–522.
8. Doherty, P. C., Blanden, R. V. & Zinkernagel, R. M. (1976) *Transplant. Rev.* **29**, 89–124.
9. Shearer, G. M., Rehn, T. G. & Schmitt-Verhulst, A. M. (1976) *Transplant. Rev.* **29**, 222–248.
10. Snell, G. D., Dausset, J. & Nathenson, S. G. (1976) *Histocompatibility* (Academic Press, New York).
11. Gordon, R. D., Simpson, E. & Samelson, L. E. (1975) *J. Exp. Med.* **142**, 1108–1120.
12. Bevan, M. J. (1975) *J. Exp. Med.* **142**, 1349–1364.
13. Bevan, M. J. (1976) *Cold Spring Harbor Symp. Quant. Biol.* **41**, 519–527.
14. Katz, D. H. & Benacerraf, B. (1975) *Transplant. Rev.* **22**, 175–195.
15. Shevach, E. M. & Rosenthal, A. S. (1973) *J. Exp. Med.* **138**, 1213–1229.
16. Canter, H. & Boyse, E. A. (1975) *J. Exp. Med.* **141**, 1376–1389.
17. Kisielow, P., Hirst, J., Shiku, H., Beverly, P. C. L., Hoffman, M. K., Boyse, E. A. & Oettgen, H. F. (1975) *Nature* **253**, 219–221.
18. Cantor, H. & Boyse, E. A. (1976) *Cold Spring Harbor Symp. Quant. Biol.* **41**, 23–32.
19. Shiku, H., Takahashi, T., Bean, M. A., Old, L. J. & Oettgen, H. F. (1976) *J. Exp. Med.* **144**, 1116–1120.
20. Pang, T., McKenzie, I. F. C. & Blanden, R. V. (1976) *Cell. Immunol.* **26**, 153–159.
21. Heber-Katz, E. & Wilson, D. B. (1976) *J. Exp. Med.* **143**, 701–706.
22. Bevan, M. J. (1975) *J. Immunol.* **114**, 316–319.
23. Peavy, D. & Pierce, C. W. (1975) *J. Immunol.* **115**, 1515–1520.
24. Lindahl, K. F., Peck, A. B. & Bach, F. H. (1975) *Scand. J. Immunol.* **4**, 544–553.
25. Bevan, M. J. (1976) *Immunogenetics* **3**, 177–184.
26. Richards, F. F., Konigsberg, W. H., Rosenstein, R. W. & Varga, J. M. (1975) *Science* **187**, 130–137.
27. Golstein, P., Kelly, F., Avner, P. & Gachelin, G. (1976) *Nature* **262**, 693–695.
28. Bevan, M. J. & Hyman, R. (1977) *Immunogenetics* **4**, 7–16.
29. Matzinger, P. & Bevan, M. J. (1977) *Cell. Immunol.*, **29**, 1–5.

Chapter VIII

Thymic Education

Chapter VIII

Thymic Education

As discussed in a previous chapter, two functionally distinct sets of T lymphocytes are able to recognize antigen together with gene products of the major histocompatibility complex. In reviewing the experiments of Zinkernagel and Doherty in cytotoxic T cells, it was apparent that a given animal, when immunized with a virus, generated T cells which recognized both the relevant viral antigen and the appropriate (K, D) gene product of its own MHC complex. Under the specified conditions of immunization, the investigators could not demonstrate significant levels of reactivity against the appropriate viral antigens in association with non-self (or allogeneic) MHC gene products.

These results suggest that an individual can generate a T cell population in which the pool of lymphocytes is better equipped to recognize self-MHC antigens than foreign, or allogeneic, MHC antigens. The ability of these cells to make an immune response is coupled to their ability to recognize both a foreign determinant and a self MHC antigen. The question to be addressed in this chapter deals with how the population of T cells acquired this capacity to preferentially recognize self MHC antigens.

Early indications that a T cell's notion of "self" could be affected by factors within the host came from experiments aimed at determining whether T and B cells need to be histocompatible before they could cooperate successfully. Two groups working on this problem arrived at dramatically opposed results (Katz, et al., 1973; Bechtol, et al., 1974a,b). As described below, attempts to resolve these conflicting data eventually led to our current understanding of the role of histocompatibility antigen in T cell-B cell collaboration.

Before considering the details of those and subsequent experiments, it is worth thinking about the nature of the underlying problem. The fundamental question at issue here is the immunological notion of self. As discussed in the preface to Chapter I, the immune system is devoted to distinguishing self from non-self. The ability to make this distinction is based, at least formally, on two factors: the expression of self and the recognition of self. The expression of self is achieved, at least partially, through the expression of appropriate cell surface antigens encoded by genes in the MHC. The recognition of self is quite another matter. On the one hand, expression and recognition could both be achieved by the same products of the MHC. Katz, the proponent of such a "cell-interaction" molecule, puts forth the simple view that T and B cells of the same individual express the same MHC antigens. These two identical molecules on the two different cells provide a like-with-like interaction that signals, or allows, cooperation to proceed:

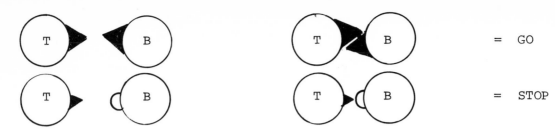

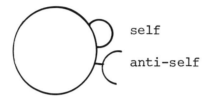

(Such like-with-like interactions are reminiscent of certain multimeric enzymes which are composed of identical subunits, all of which are required for enzymatic activity.) Here, expression of self is inextricably linked to recognition of self.

The alternative model separates the act of expression from the act of recognition. In this model, one set of gene products is used to express self while another is used to recognize self:

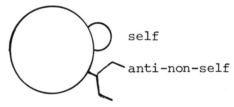

In this case, recognition of MHC gene products is not limited to recognition of self MHC antigens:

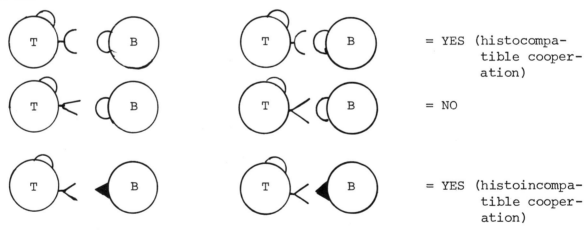

This sort of situation would also result in different possibilities for cell cooperation:

With these alternatives in mind, one can consider the experiments of Katz and Bechtol. Katz and coworkers (1973) could not demonstrate a humoral response using primed histoincompatible T and B cells in adoptive transfer, while Bechtol et al. (1974a,b) could demonstrate cooperation between histoincompatible T and B cells which had matured together in an allophenic, or chimeric, environment. Katz argues that his inability to demonstrate histoincompatible collaboration is the obvious consequence of the T and B cells expressing different cell-interaction molecules. Bechtol's work suggests that this is an insufficient explanation, since the T and B cells of the chimera express different histocompatibility antigens but still cooperate. In addition, this work raises the possibility that allogeneic T and B cell precursors which mature together become able to cooperate regardless of the fact that they do not share MHC surface antigens. Thus, expression of self MHC antigens can be separated from recognition of self MHC antigens; the T cell's notion of "self", in this case, has been broadened to include the MHC antigens of the B cell donor.

Paper 1, by von Boehmer and colleagues, supports the conclusions of the Bechtol work. It reports that T cells of the H-2^k haplotype, which have matured in an environment which includes cells of the H-2^d haplotype, can cooperate with B cells of the H-2^d haplotype. By contrast, T cells of the H-2^k haplotype which have matured in a pure H-2^k environment do not collaborate with H-2^d B cells. This is inconsistent with the model proposing like-with-like cell interaction molecules and demonstrates that the expression of self MHC molecules is distinct from the recognition of those MHC molecules.

The first indication that cytolytic T cells, restricted to killing target cells bearing self-MHC antigens plus exogenous antigenic determinants (self + X), could also be manipulated to alter their definition of "self" came from an experiment reported by Bevan (paper 2). Bevan employed the methodology described in paper 1 to construct bone marrow chimeras using bone marrow cells from the F_1 progeny of two histoincompatible parents, symbolically designated P_1 and P_2. These bone marrow cells (containing immature lymphocyte precursors) were injected into lethally irradiated recipients of either the P_1 or P_2 haplotype where they were allowed to mature. Subsequently, they were challenged with an array of target cells chosen to allow one to determine what MHC antigens the cytolytic T cells, derived from the $(P_1 \times P_2)F_1$ bone marrow cells, recognized as "self." This part of the experiment is similar in design to that used by Zinkernagel and Doherty and discussed in Chapter VII. Instead of using viruses as the source of exogenous antigen (X), Bevan exploited the family of minor histocompatibility antigens which are expressed by cells along with the antigens of the major histocompatibility complex. Thus, the general scheme of "self + X" becomes especially "MHC antigens + minor antigens". What Bevan found was that F_1 T cells allowed to mature in a parent of the P_1 haplotype preferentially recognized P_1 MHC antigens as "self". This preference was reflected in a preferential lysis of target cells bearing the appropriate parental MHC antigens. The control for the experiment is, of course, that F_1 T cells that mature in an F_1 environment recognize target cells from either parent with equal ease. The implication is that the T cells from the F_1 animal, whose normal notion of self encompasses both parental haplotypes, can be affected by their environment to confine and alter this definition of self.

Zinkernagel, making similar observations at about the same time, attempted to determine which factors in the maturational environment of the T cell were responsible for inculcating this concept of self. As discussed in paper 3, Zinkernagel, like Bevan, found that bone marrow cells of a $(P_1 \times P_2)F_1$ animal allowed to mature in a P_1 host, preferentially recognized P_1 MHC antigens as self. In contrast, splenic T cells from a similar F_1 animal did not alter their definition of self when allowed to proliferate in an irradiated P_1 host; these T cells still recognized both parental haplotypes as self. One interpretation of these results is that immature lymphocytes from donor bone marrow need to undergo differentiation in the thymus before they can become immunocompetent T cells, while splenic T cells are already fully competent before transfer into the irradiated host. Thus, the thymus could exert the influence necessary to "teach" maturing T cells about self.

This hypothesis of "thymic education" was tested by Zinkernagel directly. In Table V of paper 3 he presents the data from a slightly different sort of reconstitution experiment. Once again the donor bone marrow cells were from $(P_1 \times P_2)F_1$ animals but this time the recipients were also of the F_1 haplotype. These irradiated recipients had been previously thymectomized and grafted with a thymus of either P_1 or P_2 origin. After the donor bone marrow cells had matured and migrated to the spleen, they were assayed for cytolytic activity against virally infected targets of various haplotypes. The data indicate that

F_1 cells which matured in an F_1 recipient with a P_1 thymus were more efficient at lysing virally infected cells of the P_1 haplotype than of the P_2 haplotype. The converse was true for cells which had come under the influence of the P_2 thymus. The cytolytic T cells being of F_1 origin expressed both P_1 and P_2 MHC antigens. These results suggest that the thymus alone was important in influencing which MHC antigens are recognized as "self" by cytolytic T cells.

It is important to keep in mind that the assay systems for measuring "self recognition" are based on the requirement of T cells to recognize conventional antigen plus MHC antigens. Thus, the assessment of self-preference is really a determination of which MHC antigens the test population of T cells prefers to see: T cells from an $H-2^b$ animal injected with virus X lyse target cells (also infected with X) of the $H-2^b$ haplotype more effectively than infected target cells of the $H-2^k$ haplotype. Does this really mean that the population of naive T cells from the $H-2^b$ animal is unable to recognize $H-2^k$ + X, or does it merely reflect a preference of the immune population to recognize $H-2^b$ + X? The latter point of view would predict that the non-immune T cell population has the capacity to recognize X in association with any MHC haplotype. Immunization then effects a selection for those cells specific for X in association with self antigens, because X is only accessible for presentation by host cells.

For example, in the test systems described in papers 2 and 3, the donor bone marrow cells (of the F_1 haplotype) mature and encounter antigen in a P_1 environment. Thus, these developing F_1 T cells are in effect being primed by antigen X plus MHC antigens of the P_1 haplotype. Before assessment of cytolytic activity these cells are boosted in vitro with another dose of X on P_1, P_2 or F_1 cells. Preferential lysis of X-infected P_1 cells is considered evidence of education to the P_1 haplotype.

One interpretation of this kind of result is that the preferential lysis of P_1 + X target cells reflects the preferential expansion of T cell clones specific for the priming antigen, P_1 + X, during the initial in vivo exposure to antigen. This expanded population is then evoked by the challenging exposure to P_1 + X in vitro. This interpretation predicts that if developing P_1 T cells could be primed to antigen on histoincompatible P_2 cells, these P_1 T cells would demonstrate preferential lysis for X-infected P_2 cells. That such might be the case is demonstrated by data presented in paper 4. Matzinger extends the work of Bevan and shows that T cells derived from donor bone marrow can respond to minor histocompatibility antigens in association with MHC antigens of either donor or host type; the preference is determined by the haplotype of the priming antigen.

These data argue for the point of view thta the T cells of a given individual have the potential to recognize antigen in association with non-self MHC gene products, although realization of this potential would necessitate the creation of a non-physiological situation in which antigen priming is effected by allogeneic cells.

What, then, of the thymus?[1] There is little doubt that the thymus can bias what the cytolytic T cell population chooses to see as "self". Does it exert this influence by some yet-to-be-discovered unique mechanism, or could it possibly be working by being the first source of efficient, radioresistant antigen-presenting cells which a developing T cell encounters? Further experiments are required to resolve these issues.

The experiments of Zinkernagel and Bevan essentially propose that the thymus determines the repertoire of T cell receptors for antigen by "educating", in some

---

[1]Katz and coworkers have been arguing for some time that the thymus has little if anything to do with determining the specificity of T cells (1979), and an extensive review of many of the salient features of his model of adaptive differentiation has recently been published (1980). Unfortunately, this review is too long to reprint here but the interested reader is referred to it for an alternative point of view about the role of the thymus.

unknown way, the T cell population to recognize a limited set of MHC antigens. In the last paper of this chapter, von Boehmer, Haas, and Jerne extend this model and propose that events in the thymus direct the repertoire of T cell receptors for both MHC and conventional antigens. They suggest that determination of the pool of receptors to be used by the T cell population occurs within the thymus, and that the range of T cell specificities which ultimately emerges from this selective process is related to the MHC haplotype of the T cells themselves. In other words, what comes out of the thymus, in the way of immunocompetent T cells, is directly related to the genotype of the cells that went into the thymus.

The model is based on results obtained from experiments in which von Boehmer and colleagues study the immune response of female mice to an antigen, H-Y, present on the surfaces of otherwise histocompatible male cells. The cytolytic T cell response to H-Y, like similar responses to other minor histocompatibility antigens, is H-2 restricted. In addition, immune responsiveness to H-Y is controlled by genes in the H-2 complex, so that female mice of some H-2 haplotypes are high responders to H-Y, while females of other haplotypes are low responders.

Von Boehmer et al. attempted to correct the immunological defect of low responder mice by allowing them to mature in the thymic environment of a high-responder strain. They anticipated that altering the education process of low responder cells might alter their response phenotype. The authors found that bone marrow cells of one low responder H-2 haplotype, $H-2^{i5}$ (K^b $I-A^b$ D^d), can give rise to cytotoxic T cells which respond to H-Y by transplanting them into an irradiated ($H-2^k$ x $H-2^b$)F_1 responder mouse and allowing them to differentiate there. In contrast, the bone marrow cells of another low responder H-2 haplotype, $H-2^k$ (K^k $I-A^k$ D^k), do not give rise to responding cytotoxic cells when placed into an ($H-2^k$ x $H-2^b$)F_1 animal to mature. In addition, T cells derived from ($H-2^k$ x $H-2^b$)F_1 (responder phenotype) bone marrow will not lyse syngeneic H-Y bearing cells if the bone marrow matured in an $H-2^k$ environment. These experiments imply that the origin of nonresponsiveness in these two nonresponder haplotypes might be different. In one case, the H-2 haplotypes of the educating environment can alter the response pattern of the T cells. In the other case, the H-2 haplotype of the T cell itself seems to affect its ability to recognize and respond to certain antigens.

The authors use the data presented in this paper as the basis for a theory about the nature of T cell receptors for antigen and the method by which the diversity of the T cell repertoire for antigen recognition is generated.

The theory makes one very important and unique point: the breadth of the antigenic universe recognized by an individual's T cell population is a direct consequence of that individual's MHC haplotype. The authors argue that all immature T cells in an individual begin with two receptors which recognize self MHC antigens. As these immature T cells encounter self antigens (in the thymus, for example) they are induced to proliferate and the genes encoding one of the anti-self receptors undergoes mutation. These mutations alter the specificity of the receptor so that it becomes specific for some determinant related to but distinct from self. The number of receptors with different specificities which ultimately can be generated is related to the number of mutations which the gene for the T cell receptor can accumulate. Thus the immature "pre-T" cells of an animal of the $H-2^k$ haplotype has two receptors specific for $H-2^k$ antigens (either K/D, or I antigens, depending on whether the pre-T cell will become a helper T cell or a cytotoxic T cell). One of these (R_1) can diversify to recognize a whole range of foreign antigens. The other (R_0) retains its specificity for $H-2^k$, or "self". Both receptors are used by the T cell to recognize antigen, resulting in the phenomenon of H-2 restriction. Mice of different haplotypes will have somewhat different T cell repertoires because their R_1 receptors will be the results of mutations "away" from different anti-self receptors.

Whether any aspect of this model should prove true, it deserves credit for its effort to account for the phenomena of H-2 restriction, thymic education and H-2 linked immune responsiveness. While collecting all of one's unsolved problems into a single difficult-to-test theory may not be the most profitable way to proceed, it does provide some food for thought. Certainly we are only beginning to understand the role of the thymus in influencing our immunological notion of self.

488   VIII   Thymic Education _____ VIII.Introduction

into a single difficult-to-test theory may not be the most profitable way to proceed, it does provide some food for thought. Certainly we are only beginning to understand the role of the thymus in influencing our immunological notion of self.

Papers Included in this Chapter

1. von Boehmer, H., Hudson, L., and Sprent, J. (1975) Collaboration of histo-
    incompatible T and B lymphocytes using cells from tetraparental bone marrow
    chimeras. J. Exp. Med. <u>142</u>, 989.

2. Bevan, M.J. (1977) In a radiation chimaera, host H-2 antigens determine immune
    responsiveness of donor cytotoxic cells. Nature <u>269</u>, 417.

3. Zinkernagel, R.M., Callahan, G.N., Althage, A., Cooper, S., Klein, P.A., and
    Klein, J. (1978) On the thymus in the differentiation of "H-2 self-recogni-
    tion" by T cells: Evidence for dual recognition? J. Exp. Med. <u>147</u>, 882.

4. Matzinger, P. and Mirkwood, G. (1978) In a fully H-2 incompatible chimera,
    T cells of donor origin can respond to minor histocompatibility antigens
    in association with either donor or host H-2 type. J. Exp. Med. <u>147</u>, 84.

5. von Boehmer, H., Haas, W., and Jerne, N.K. (1978) Major histocompatibility
    complex-linked immune responsiveness is acquired by lymphocytes of low-
    responder mice differentiating in thymus of high-responder mice. Proc. Natl.
    Acad. Sci. USA <u>75</u>, 2439.

References Cited in this Chapter

Katz, D.H., Katz, L.R., Bogowitz, C.A., and Skidmore, B.J. (1979) Adaptive
    differentiation of murine lymphocytes. II. The thymic microenvironment
    does not restrict the cooperative partner cell preference of helper T
    cells differentiating in F_1--F_1 thymic chimeras. J. Exp. Med. <u>149</u>, 1360.

Katz, D.H. (1980) Adaptive differentiation of lymphocytes: Theoretical
    implications for mechanisms of cell-cell recognition and regulation of
    immune responses. Adv. Immunol. <u>29</u>, 137.

COLLABORATION OF HISTOINCOMPATIBLE T
AND B LYMPHOCYTES USING CELLS FROM TETRAPARENTAL
BONE MARROW CHIMERAS

By H. von BOEHMER, L. HUDSON, and J. SPRENT*

(From the Basel Institute for Immunology, CH-4058 Basel, Switzerland)

Experiments have been reported suggesting that under conditions where an allogeneic effect could not be demonstrated, primed histoincompatible T and B cells did not cooperate in a humoral immune response (1, 2). By contrast semiallogeneic T and B cells were found to cooperate. The authors therefore concluded that in order to obtain efficient T-B collaboration the cells concerned must share determinants coded for by the *H-2*-gene complex, most likely the *I* region (3).

In different experiments Bechthol et al. (4, 5) demonstrated that allophenic mice, produced by the fusion of embryos of strains giving a high and low response to the synthetic antigen poly-L(Tyr,Glu)-poly-D,L-Ala--poly-L-Lys (TGAL),[1] produced TGAL-specific antibody of low responder allotype. The authors interpreted this finding as an indication of collaboration between histoincompatible high responder T cells and low responder B cells. While this interpretation may be correct, direct evidence has not been presented to support the notion. It could be argued, for example, that in allophenic mice low responder T cells show a heightened response to TGAL. While the mechanism of such a conversion is unknown the possibility should be considered until disproven by experimental data.

It is obviously important to know whether the presence of certain determinants on allogeneic T and B cells prevent T-B collaboration under all circumstances, or whether this is only so when the cells are not mutually tolerant. We have studied the collaboration of histoincompatible T and B lymphocytes in response to sheep erythrocytes (SRBC) in an adoptive transfer system using primed T cells from tetraparental mice and primed B cells from both normal and chimeric mice. In previous reports we have shown that T cells from tetraparental bone marrow chimeras (TBMC) are specifically unresponsive to the determinants of the *H-2* complex of the host in the mixed leukocyte reaction (MLR) as well as in cell-mediated lympholysis (CML) (6, 7). The data presented here suggests that primed T cells are able to cooperate with allogeneic B cells to which they have been tolerized in a chimeric environment.

---

* Present address: Department of Zoology, University College, London, England.

[1] *Abbreviations used in this paper:* CML, cell-mediated lympholysis; LN, lymph node; MLR, mixed leukocyte reaction; PFC, plaque-forming cell; TBMC, tetraparental bone marrow chimeras; TDL, thoracic duct lymphocytes; TGAL, poly-L(Tyr,Glu)-poly-D,L-Ala--poly-L-Lys.

Materials and Methods

Mice. (CBA/J × DBA-2/J)F₁ and (C3H/J × SJL)F₁ hybrids, as well as the corresponding parental strains were used.

Cells. Bone marrow, spleen, lymph node (LN), and thoracic duct lymphocytes (TDL) were prepared as described previously (6).

Immunization. Mice were given a single intraperitoneal injection of 0.1 ml of 25% SRBC in saline and used 1–3 mo later.

Anti-H-2 and Anti-θ (Thy 1ᵇ) Sera. The sera were prepared by cross-immunizing mice as described previously (references 6 and 8, respectively).

TBMC. TBMC were obtained by injecting lethally irradiated (950 R) F₁ hybrids with 10⁷ anti-θ-treated bone marrow cells from both parental strains (described in detail in reference 6).

Preparation of CBA Helper T Cells from TBMC. TBMC were given an intraperitoneal injection of 0.1 ml of 25% SRBC suspension 4 mo after irradiation and injection with bone marrow cells. 1–2 mo after immunization suspensions of TDL, spleen, and LN cells were prepared and Ig-positive lymphocytes removed either by passage through a column of Sephadex G100 to which anti-immunoglobulin antibodies had been covalently bound (modified from Schlossman and Hudson, reference 9) or by passage through nylon wool columns (10). In some experiments the filtered suspension containing lymphocytes of both parental strains was treated with anti-H-2 sera to obtain helper T cells of a single *H-2* type. The damaged cells were removed by a simple filtration method (11) and the proportion of the remaining viable cells carrying alloantigens of the other parental strain was determined by a cytotoxicity test.

Preparation of B Cells Primed to SRBC. 1–4 mo after priming with SRBC, spleens were removed and cell suspensions prepared. The suspensions were treated with anti-θ serum plus complement (C) as described elsewhere (8).

Adoptive Transfer System for a Secondary Plaque-Forming Cell (PFC) Response against SRBC. F₁ recipient mice were X irradiated (850 R). The mice were injected intravenously with a mixture of primed B cells and varying numbers of primed T cells from TBMC, and other recipients were injected with the separate cell populations. In control experiments X-irradiated parental strain mice received unfractionated spleen cells from mice which served as donors for the B cells. At the same time, each mouse in the experimental and control groups received an intravenous injection of 5 × 10⁸ SRBC. 7 days after cell transfer, the spleens were taken out and the number of direct (19S) and indirect (7S) PFC determined.

Assay for PFC. PFC were detected using the technique described by Cunningham and Szenberg (12). Indirect (7S) PFC were developed by the addition of rabbit antimouse-Ig serum to the cell mixture. This antiserum, at the concentration used, did not suppress the PFC response (mainly IgM) of mice primed for 4 days with SRBC. Indirect PFC were calculated by subtracting the number of PFC obtained in the absence of developing serum from the numbers obtained after addition of the serum.

Treatment of PFC with Anti-H-2 Sera. Spleen cell suspensions were incubated with anti-H-2 sera or normal mouse serum for 30 min at 4°C, washed once, and then incubated with C for 30 min at 37°C. After further washing, the suspension was tested in the plaque assay.

Results

Lymphoid Cell Chimerism of TBMC used as Donors of Helper T Cells. For the first set of experiments a total of 8 TBMC were used; these were prepared by injecting (CBA/J × DBA-2/J)F₁ hybrids with bone marrow cells from both parental strains. Each mouse was tested for lymphoid cell chimerism using either TDL or LN cells from which the damaged cells had been removed. Each mouse showed a lymphoid cell chimerism corresponding to roughly equal proportions of CBA and DBA/2 lymphocytes (Table I). As with previous experiments (6) no significant number of lymphocytes of host origin were detected.

Purity of CBA T Cells Prepared from TBMC. Mixtures of TDL and LN cells prepared from TBMC originally contained approximately 30% of immunoglobu-

TABLE I

Lymphoid Cell Chimerism in TBMC used as Donors for Helper T Cells

Cells	Dead cells after incubation with:			
	Anti-CBA + C	Anti-DBA + C	Saline + C	Anti-CBA + anti-DBA + C
	%	%	%	%
CBA LN	100	11	15	99
DBA LN	12	99	14	95
F$_1$ (CBA × DBA) LN	97	99	11	98
Chimera 1 TDL	55	50	0	99
Chimera 2 LN	49	56	15	99
Chimera 3 LN	70	36	20	100
Chimera 4 LN	67	40	22	98
Chimera 5 LN	49	56	11	ND*
Chimera 6 LN	49	63	14	ND
Chimera 7 LN	62	50	18	ND
Chimera 8 TDL	54	48	9	ND

* ND, not determined.

TABLE II

Purity of CBA T Cells Prepared from TBMC

Exp.	Cells	Ig-positive cells		Dead* CBA cells after incubation with:		
		Original population	Anti-Ig column effluent population	Saline + C'	Anti DBA + C'	Anti CBA + C'
		%	%	%	%	%
1	LN + TDL	31	0.8	12	11	100
2		27	0.4	14	15	100

* After anti-Ig column filtration the CBA/DBA/2 cell mixture was incubated with anti-DBA/2 plus C and damaged cells were removed. The number of CBA cells in the remaining suspension was then estimated as shown.

lin-bearing (B) lymphocytes. This proportion was reduced to less than 1% after anti-Ig column filtration in two separate experiments (Table II). The column-filtered cells were then treated with an anti-DBA/2 serum and C. After filtration through cotton wool under special conditions to remove the damaged cells (11), up to 100% of the remaining cells could be killed with an anti-CBA serum plus C, showing that a suspension of T cells had been obtained, the vast majority of which carried CBA alloantigens.

The Helper Effect of Primed CBA T Cells from TBMC for Primed Syngeneic or Allogeneic B Cells. In a preliminary experiment, 10^7 anti-θ-treated CBA spleen cells primed to SRBC were transferred together with varying numbers of primed anti-Ig column-passed LN cells from CBA into lethally irradiated syngeneic recipients. At the same time the mice were given an intraperitoneal injection of SRBC and the PFC response measured 7 days after cell transfer.

This experiment was carried out to determine the minimum number of T cells required to give a detectable helper effect. As shown in Fig. 1, 5×10^5 helper cells increased the number of direct and indirect PFC, and a 10-fold increase in PFC was obtained with $1-2 \times 10^6$ helper cells. Transfer of 2×10^6 CBA T cells alone, produced no increase in the background PFC observed in X-irradiated mice not receiving B cells (<150 PFC/spleen). Transfer of 10^7 primed unfractionated CBA spleen cells produced 40,000 IgG PFC, suggesting that in the titration experiments T cells were the limiting cell population.

SRBC-primed CBA T cells derived from TBMC, and therefore tolerant to DBA/2 alloantigens (6), were transferred with primed CBA or DBA/2 anti-θ-plus C-treated spleen cells into lethally irradiated (CBA $\times$ DBA/2)F$_1$ hybrids. These CBA T cells augmented the IgG PFC response of the primed syngeneic CBA or allogeneic DBA/2 B cells: with 2×10^6 helper CBA T cells a 10-fold increase in IgG PFC was observed with both syngeneic and allogeneic anti-θ-treated primed spleen cells (Fig. 2).

When 10^7 primed CBA or DBA/2 spleen cells from the donors used for the T-B collaboration experiments were transferred to syngeneic CBA or DBA/2 recipients, a response of 35,000 IgG PFC was observed with CBA spleen cells and a response of 60,000 IgG PFC with DBA/2 spleen cells. The titration experiment using CBA T cells from TBMC therefore suggests that those helper cells, tolerant to *H-2* determinants of DBA/2, could effectively augment the secondary response of primed syngeneic CBA or allogeneic DBA/2 cells. Moreover, the data suggest that a similar number of helper T cells was required to achieve a 10-fold "background" response with either syngeneic or allogeneic anti-θ-treated primed cells.

This finding was confirmed in a further experiment where only one dose of CBA helper T cells from TBMC was tested (Table III). In this experiment, CBA anti-θ-treated SRBC-primed spleen plus 2×10^6 T cells gave a higher absolute response than DBA/2 cells plus T cells in contrast to the preceding experiment.

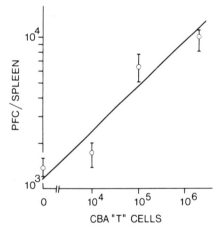

FIG. 1.   The helper effect of SRBC-primed CBA T cells from LN with 10^7 syngeneic SRBC-primed anti-θ-treated spleen cells in an adoptive transfer system. The geometric mean and upper and lower limits of PFC found in five recipients per experimental group is given.

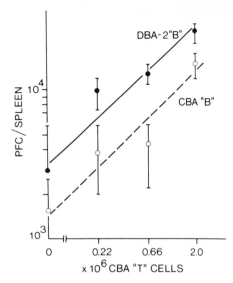

FIG. 2. The helper effect of SRBC-primed CBA T cells obtained from TDL and LN of TBMC with 10^7 syngeneic (○) or allogeneic (●) SRBC-primed anti-θ-treated spleen cells transferred into X-irradiated (CBA x DBA)F$_1$ recipients. The geometric mean and upper and lower limits of IgG PFC found in four to five recipients per experimental group is given.

TABLE III

Helper Effect of SRBC-Primed CBA T Cells from TBMC for Syngeneic and Allogeneic-Primed B Cells

Cells transferred	IgG plaques
2×10^6 CBA T cells from TBMC	964 (1,292–719)*
10^7 CBA B cells	2,668 (3,941–1,8096)
10^7 DBA B cells	907 (1,146–717)
2×10^6 CBA T$_{TBMC}$ + 10^7 CBA B	35,107 (44,679–27,585)
2×10^6 CBA T$_{TBMC}$ + 10^7 DBA B	17,763 (18,977–16,627)

* Geometric mean; upper and lower limits are given in parentheses.

Normal primed spleen cells were not tested in this experiment. It is again apparent that CBA helper T cells from TBMC significantly increased the PFC response of both syngeneic and allogeneic anti-θ-treated spleen cells.

The Helper Effect of T Cells from TBMC on a Mixed Population of B Cells from TBMC. To test whether T cells when transferred with a mixture of syngeneic and allogeneic B cells would preferentially help syngeneic B cells the following experiments were conducted: TBMC were prepared by injecting (C3H × SJL)F$_1$ hybrids with bone marrow cells from both parental strains. After 2 mo the mice were primed with SRBC and 1 mo later T and B cells were prepared. One group of X-irradiated (750 R) (C3H × SJL)F$_1$ recipients received 2×10^6 T cells from TBMC together with 6×10^6 B cells from TBMC. (Control groups received either T or B cells alone.) A second group of recipients received 2×10^6 C3H T cells from TBMC, prepared by treating the TBMC with anti-SJL serum

(45% lysis), together with 6×10^6 C3H and SJL B cells from TBMC. 7 days after priming with SRBC, PFC were determined and their origin was tested with anti-C3H and anti-SJL sera as described in the Materials and Methods. It would be expected that if the C3H T cells cooperated more efficiently with syngeneic than with allogeneic B cells then the proportion of C3H PFC would be higher in the recipients receiving C3H helper T cells compared with those recipients receiving a mixed population of C3H and SJL helper T cells. As shown in Table IV this was not observed. 2×10^6 T cells, consisting of approximately equal proportions of C3H and SJL T cells, had a similar helper effect as 2×10^6 C3H T cells, obtained after treatment of the cells with anti-SJL serum. In addition, irrespective of whether a mixture of C3H and SJL T cells or C3H T cells alone were injected, approximately 30% of the PFC were C3H and 70% of SJL origin. This suggests that under conditions where mutually tolerant populations of T and B cells are used the T cells can help the allogeneic B cells as efficiently as syngeneic B cells.

Discussion

TBMC seem to provide an experimental model which allows a detailed study of the regulation of the immune response. For this purpose it is important to know whether T and B cells from different strains will cooperate in TBMC since regulatory influences of T cells may alter the immune response pattern of B cells. The design of the experiments reported here is in many aspects different from that used by Katz and his colleagues (1–3). We have used not only a different protocol for the preparation of tolerant T cells but also different mouse strains and antigens. In our system T cells were generated in vivo in a histoin-compatible environment, and so were specifically tolerant to H-2 determinants of the allogeneic B cells with which they were transferred in the adoptive

TABLE IV

Helper Effect of SRBC-Primed T Cells from TBMC for a Mixture of Allogeneic and Syngeneic B Cells from TBMC

Cells injected		Proportion of T cells killed by:		PFC IgG	Proportion of:	
T cells	B cells	αC3H serum	αSJL serum		C3H-PFC	SJL-PFC
		%	%		%	%
2×10^6	6×10^6	58	45	3,124 (3,845–2,539)	29.8	64.0
2×10^6	—	58	45	36 (54–24)	ND	ND
—	6×10^6	—	—	305 (373–244)	ND	ND
2×10^6 anti-SJL treated	6×10^6	98	—	3,044 (3,939–2,352)	30.4	70.1

All cells were from TBMC. The proportion of C3H and SJL PFC was determined as described in the Materials and Methods: 4 samples of a suspension containing PFC were treated either with normal mouse serum, anti-C3H, anti-SJL, or both antisera together and C. Numbers obtained after treatment with normal mouse serum and C were 100%. Both antisera together killed more than 95% of all PFC.

transfer experiments. In some experiments the B cells were also derived from tolerant donors. In addition, the T cells had been primed to SRBC in the presence of allogeneic T and B cells.

There are other experiments by Toivanen and Toivanen (13) and by Kindred (14) demonstrating the failure of effective T-B collaboration in either chickens reconstituted with allogeneic bursa cells or mice grafted with allogeneic thymus cells. These experiments differ from the experiments described here since the reconstitution was in both instances performed in animals in which the development of the hemopoietic system was well advanced. In the present experiments stem cell-rich allogeneic bone marrow cells were transferred into lethally irradiated hosts, thus allowing the stem cells of both populations to differentiate into mature lymphocytes at the same time. It might be argued that T cells can only cooperate with allogeneic B cells if they are fully tolerant to those B cells, or when the T cells recognizing determinants on allogeneic B cells are absent.

While the effect of the circumstances mentioned above on the generation of helper T cells is unknown, the results presented in this communication are compatible with conclusions drawn by Bechthol et al. (4, 5) from experiments in allophenic mice. These authors concluded that collaboration between high responder T cells and histoincompatible low responder B cells could occur in these mice.

It might be argued that the collaboration across *H-2* barriers reported here and postulated by Bechthol et al. (4, 5) in allophenic mice is due to an allogeneic effect. However, this explanation seems unlikely for three reasons: (*a*) Bechthol et al. (4, 5) could not demonstrate an allogeneic effect in allophenic mice produced by embryo fusion of two histoincompatible low responder strains, i.e., in this situation no elevated levels of TGAL antibody compared to the low responder strains could be demonstrated (5); (*b*) in contrast to a report by Phillips et al. (15), we could not demonstrate elevated "background" of thymidine incorporation by thymus and LN cells in vitro as an indication of allogeneic stimulation in vivo. To the contrary, our results showed a specific deletion of MLR as well as CML-reactive cells in TBMC and no evidence was obtained for either suppressor cells or soluble blocking factors (6), and (*c*) the same number of helper T cells were required for a significant helper effect with syngeneic as well as allogeneic B cells.

Although the data presented here clearly indicate that allogeneic T-B collaboration can occur as readily as syngeneic collaboration in this system, we cannot, as yet, explain the conflict of our findings with those published by Katz et al. (1–3). One could speculate that either allogeneic T or anti-θ-treated spleen cell populations may recognize each other and as a consequence produce some factors interfering with T-B collaboration. This may depend on the cell numbers used, the degree of removal of T cells by anti-θ serum, and the method used to inactivate the T cells in their response to the allogeneic B cells. It seems that all these complicating factors are eliminated in an experimental situation where both allogeneic T and B cells are mutually tolerant. In this situation T helper cells do not seem to have a preference for syngeneic B cells as compared to allogeneic B cells. Since allogeneic T and B cells in TBMC retain their surface H-2 antigens and their ability to stimulate in MLR, it is possible to justify the argument that the differing surface antigens on allogeneic T and B cells do not alone inhibit T-B collaboration. If our speculation is correct one can postulate that some other determinants involved in T-B collaboration must be easily recognized by allogeneic nontolerant cells.

Summary

T-B collaboration has been studied in a secondary response to sheep erythrocytes using either syngeneic or allogeneic T- and B-cell combinations. T cells prepared from tetraparental bone marrow chimeras (TBMC), carrying H-2 determinants of one parental strain only, cooperated with syngeneic, as well as with allogeneic B cells carrying the alloantigens to which the T cells had been tolerized in the chimeric environment. When TBMC-derived cells of a single *H-2* specificity were transferred with a mixture of TBMC-derived B cells of both *H-2* types of the parental strains, no preference for syngeneic cooperation was found.

The data therefore suggest that the presence of differing *H-2*-complex determinants on the allogeneic T- and B-cell populations of the two different strain combinations tested do not interfere with T-B collaboration when the cell populations studied are mutually tolerant.

Received for publication 18 June 1975.

References

1. Katz, D. H., T. Homaoka, and B. Benacerraf. 1973. Cell interactions between histoincompatible T and B lymphocytes. II. Failure of physiologic cooperative interactions between T and B lymphocytes from allogeneic donor strains in humoral response to hapten-protein conjugates. *J. Exp. Med.* **137**:1405.

2. Katz, D. H., M. E. Dorf, and B. Benacerraf. 1974. Cell interactions between histoincompatible T and B lymphocytes. VI. Cooperative responses between lymphocytes derived from mouse strains differing at genes in the S and D region of the *H-2* complex. *J. Exp. Med.* **140**:290.

3. Dorf, M. E., D. H. Katz, M. Graves, H. Dimuzio, and B. Benacerraf. 1975. Cell interactions between histoincompatible T and B lymphocytes. VIII. *In vitro* cooperative responses between lymphocytes are controlled by genes in the K-end of the H-2 complex. *J. Immunol.* **114**:1717.

4. Bechthol, R. B., T. G. Wegmann, J. H. Freed, F. C. Grument, B. W. Chesebro, L. A. Herzenberg, and H. O. McDevitt. 1974. Genetic control of the immune response to TGAL in C3H—C57 tetraparental mice. *Cell. Immunol.* **13**:264.

5. Bechthol, R. B., J. H. Freed, L. A. Herzenberg, and H. O. McDevitt. 1974. Genetic control of the antibody response to poly-L(Tyr,Glu)-poly-D,L-Ala--poly-L-Lys in C3H—CWB tetraparental mice. *J. Exp. Med.* **140**:1660.

6. Von Boehmer, H., J. Sprent, and M. Nabholz. 1975. Tolerance to histocompatibility determinants in tetraparental bone marrow chimeras. *J. Exp. Med.* **141**:332.

7. Von Boehmer, H., J. Sprent, M. Nabholz, and N. Ehrhardt. 1975. Tolerance induction towards determinants of the MHC: evidence for deletion of T cell subsets. In Proceedings of Ninth Leucocyte Culture Conference. A. Rosenthal, editor. Academic Press, Inc., New York. 805.

8. Miller, J. F. A. P., and J. Sprent. 1971. Cell to cell interaction in the immune response. VI. Contribution of thymus-derived cells and antibody-forming cell precursors to immunological memory. *J. Exp. Med.* **134**:66.

9. Schlossman, S. F. and L. Hudson. 1973. Specific purification of lymphocyte populations on a digestible immunoadsorbent. *J. Immunol.* **110**:313.

10. Julius, M. H., E. Simpson, and L. A. Herzenberg. 1973. A rapid method for the isolation of functional thymus-derived murine lymphocytes. *Eur. J. Immunol.* **3**:645.

11. Von Boehmer, H., and R. Shortman. 1973. The separation of different cell classes from lymphoid organs. IX. A simple and rapid method for removal of damaged cells from lymphoid cell suspensions. *J. Immunol. Methods.* **2**:293.

12. Cunningham, A. J., and A. Szenberg. 1968. Further improvements in the plaque technique for detecting single antibody forming cells. *Immunology.* **14**:599.

13. Toivanen, A., and P. Toivanen. 1975. Histocompatibility requirement for cooperation between donor and host cells in transplantation of bursal stem cells into immunodeficient chicks. *Transplant. Proc.* In press.

14. Kindred, B. 1975. The inception of the response to SRBC in nude mice injected with various doses of congeneic or allogeneic thymus cells. *Cell. Immunol.* **17**:277.

15. Phillips, S. M. and T. G. Wegmann. 1973. Active suppression as a possible mechanism of tolerance in tetraparental mice. *J. Exp. Med.* **137**:291.

In a radiation chimaera, host H–2 antigens determine immune responsiveness of donor cytotoxic cells

CELL membrane structures controlled by genes in the major histocompatibility complex (H–2 in mice) are involved in most immune interactions between T lymphocytes and other cells[1]. Cytotoxic T lymphocytes (CTL) immunised against viruses[2], haptens[3], minor histocompatibility antigens[4] or tumour antigens[5], are specific for self H–2 antigens as well as for the foreign antigen. But CTL are not restricted to recognising antigens in combination with only self H–2. H–2^d homozygous CTL which have matured in an irradiated H–2^d/H–2^k host can respond to antigen plus H–2^k in addition to antigen plus H–2^d (refs 6–8). It is not known whether the H–2 environment in which T cells mature influences their range of specificity, that is, whether CTL from a normal mouse can respond quantitatively as well to antigen plus foreign H–2 as they do to antigen plus self H–2. These experiments were designed to test this influence. The results suggest that host H–2 antigens do exert an effect on the specificity of T-cell responses.

A single suspension of bone marrow cells from F_1(BALB/c × BALB.B) (F_1(C × C.B), H–2^d/H–2^b) mice was used to reconstitute groups of lethally irradiated parental mice, C(H–2^d) ([$F_1 \rightarrow$C] chimaeras) and C.B(H–2^b) ([$F_1 \rightarrow$C.B] chimaeras). Eight weeks later these chimaeric mice and normal F_1(C × C.B) mice were primed against minor H antigens by injecting 8×10^6 F_1(B10 × B10.D2) (H–2^b/H–2^d) spleen cells. The B10 background offers more than 30 minor histocompatibility antigenic differences that can be recognised by BALB mice[9,10]. Some weeks later the primed spleen cells were boosted in culture with irradiated F_1(B10 × B10.D2) stimulator cells and assayed for cytotoxicity 5 d later.

Following this immunisation procedure, cells from normal F_1(C × C.B) mice lysed B10 targets and B10.D2 targets almost equally (Table 1). (The two activities are mediated by separate pools of CTL[2–4]) The chimaeras responded differently. In the same conditions of immunisation with F_1(B10 × B10.D2) cells, they responded preferentially to the minor antigens in association with the H–2 antigens of the host. CTL from the [$F_1 \rightarrow$C] chimaeras killed B10.D2 targets better than B10 targets, whereas CTL from [$F_1 \rightarrow$C.B] chimaeras lysed B10 targets better than B10.D2 targets (Table 1).

Spleen cells from five chimaeras were assayed for their content of host and donor cells at time of killing. Complement-mediated lysis with H–2^b anti-H–2^d serum and with H–2^d anti-H–2^b serum indicated that in all cases at least 85% of the cells were of F_1 (donor) origin. The cytotoxic effector cells were also lysed with anti-H–2 serum and complement just before the ^{51}Cr-release assay (Table 2). Here, the [$F_1 \rightarrow$C] chimaera cells lysed B10.D2 targets ninefold more efficiently than they lysed B10 targets (data not shown). The killer cells were treated with antiserum plus complement, washed, and assayed for lysis of labelled B10.D2. Table 2 shows, most importantly, that BALB/c anti-C57BL/6 (anti-H–2^b) serum reduced the cytotoxic activity 86% compared with controls. This antiserum does not lyse BALB/c effector cells. Therefore, at least 86% of the CTL were of F_1 bone marrow origin.

These experiments show that H–2^d/H–2^b cytotoxic cells which mature in an irradiated H–2^d host respond preferentially to antigens plus H–2^d, whereas H–2^d/H–2^b cells which mature in an irradiated H–2^b host respond preferentially to the same antigens in conjunction with H–2^b gene products. The experiments were designed to test the 1971 Jerne hypothesis[14], or a modified version of it[4,15]. The hypothesis accepts that a somatic theory of generation of receptor diversity is correct and proposes that self-H–2 antigens drive the diversity. Immature T cells first express an anti-self-H–2 receptor, leading to proliferation and to accumulation of V gene mutations until there is no significant reaction with self-H–2. According to this hypothesis, the receptor repertoire of A strain T cells which had matured in an A environment would be quite different from that of A strain T cells which had matured in a B environment. The results presented here are compatible with this hypothesis and with another theory of 'adaptive differentiation'[1].

There is an alternative explanation of the results. It may be that

Table 1 Specificity of H–2^d/H–2^b cytotoxic cells from normal and chimaeric mice

Responder*	Immunised with†	B10 H–2^b	% Specific lysis of targets‡ B10.D2 H–2^d	B10.BR H–2^k	F_1(C × C.B) H–2^d/H–2^b	Ratio of lytic activity on§ B10/B10.D2
Experiment 1						
Normal F_1 (C × C.B)(H–2^d/H–2^b)	F_1(B10 × B10.D2)(H–2^b/H–2^d)	56.2	62.2	1.5	ND	0.7
Chimaera [$F_1 \rightarrow$C]	F_1(B10 × B10.D2)(H–2^b/H–2^d)	19.3	72.7	2.9	ND	0.02
Chimaera [$F_1 \rightarrow$C.B]	F_1(B10 × B10.D2)(H–2^b/H–2^d)	51.8	29.1	0.1	ND	5.2
Normal F_1(C × C.B)	C3H(H–2^k)	4.5	3.8	71.5	ND	—
Experiment 2						
Normal F_1(C × C.B)	F_1(B10 × B10.D2)	54.8	68.4	ND	0.9	0.5
Chimaera [$F_1 \rightarrow$C]	F_1(B10 × B10.D2)	4.8	80.5	ND	1.5	<0.02
Chimaera [$F_1 \rightarrow$C.B]	F_1(B10 × B10.D2)	61.3	8.8	ND	0.1	>43.0

*Chimaeras were prepared as follows: BALB/c(C(H–2^d)) and BALB.B(C.B(H–2^b)) mice were irradiated with 850 R and reconstituted on the same day with 13.4×10^6 anti-Thy-1 plus complement (C')-treated F_1(C × C.B) bone marrow cells. Eight weeks later they were immunised.

†Primed against minor H antigens by injecting 8×10^6 viable F_1(B10 × B10.D2) spleen cells intraperitoneally. Spleen cell suspensions were prepared 4 weeks later (experiment 1) or 6 weeks later (experiment 2), and boosted for 5 d in culture with an equal number of 1,000 R irradiated F_1(B10 × B10.D2) or C3HebFe/J spleen cells as described previously[11].

‡2-d con A (conavalin A) blasts were labelled with ^{51}Cr-sodium chromate and used as targets as described previously[11]. Serial dilutions of the killers were assayed against a constant number of targets, and the figures for specific lysis presented here are for a killer:target ratio of 100:1. Spontaneous release of ^{51}Cr varied from 18.1–22% in experiment 1 and from 9.6–13.7% in experiment 2.

§Calculated from the titrations of killers:targets[12,13]. For example, in experiment 1, the [$F_1 \rightarrow$C] cells lysed B10.D2 targets 50 times better than they lysed B10 since a 100:1 ratio caused 19.3% specific release from ^{51}Cr-B10, whereas the same amount of specific lysis of ^{51}Cr-B10.D2 was obtained at a ratio of 2:1.

ND, Not determined.

Table 2 Sensitivity of chimaera cytotoxic cells to anti-H–2 serum plus C'

Cytotoxic cells*	Treatment of effector cells†	% Specific lysis of ^{51}Cr-B10.D2‡	% Reduction in lytic activity§
[F$_1$→C] anti-F$_1$(B10 × B10.D2)	Medium	34.0	—
	Normal B10 serum + C'	34.3	0
	Anti-H–2^b + C'	9.8	86
	Anti-H–2^d + C'	5.1	94

*[F$_1$→C] chimaeras were primed *in vivo* 8 weeks after reconstitution, their spleen cells boosted in culture 20 weeks later and assayed for cytotoxicity on day 5 of culture.

†BALB/c anti-C57BL/6 and C57BL/6 anti-BALB/c sera were prepared by hyperimmunisation with spleen cells. Effector cells were incubated with mouse sera 1:2, washed, and incubated in guinea pig serum 1:9 as a source of complement (C'). Cells were re-suspended to the same volume and assayed.

‡Con A blasts from B10.D2 mice were labelled with ^{51}Cr and used as target. Data presented are for original number of responder spleen cells:target cells of 7:1. Spontaneous release of ^{51}Cr was 20.6%.

§Calculated from the titrations of killer:targets as in Table 1.

the host haplotype preference seen at the level of effector CTL does not reflect a bias in specificity at the level of precursor CTL. The H–2^d/H–2^b precursor CTL in the H–2^d host may have exactly the same range of reactivity as those in the H–2^b host. The haplotype preference of the effector CTL would then be due to the way

antigen is presented to CTL precursors. Even though the immunogen (B10 minor antigens) was introduced on H–2 heterozygous cells, the antigen which was responsible for priming CTL may have been processed antigen presented on radiation resistant host cells[4]. In the [F$_1$→C] chimaera such radiation-resistant antigen-presenting cells would be homozygous H–2^d and would naturally stimulate only anti-B10.D2(H–2^d) CTL, not anti-B10(H–2^b) CTL. Experiments to decide between these interpretations are in progress.

This work was supported by USPHS (grants AI14269 and CA14051).

MICHAEL J. BEVAN

*Center for Cancer Research and Department of Biology,
Massachusetts Institute of Technology,
Cambridge, Massachusetts 02139*

Received 6 June; accepted 22 July 1977.

1. Katz. D. H. *Lymphocyte Differentiation. Recognition and Function* (Academic. New York. 1977).
2. Doherty. P. C.. Blanden. R. V. & Zinkernagel. R. M. *Transplant. Rev.* **29**, 89–124 (1976).
3. Shearer. G. M.. Rehn. T. G. & Schmitt-Verhulst. A. M. *Transplant. Rev.* **29**, 222–248 (1976).
4. Bevan. M. J. *Cold Spring Harb. Symp. quant. Biol.* **41**, 519–527 (1976).
5. Blank. K. J.. Freedman. H. A. & Lilly. F. *Nature* **260**, 250–252 (1976).
6. Pfizenmaier. K. *et al. J. exp. Med.* **143**, 999–1004 (1976).
7. Zinkernagel. R. M. *Nature* **261**, 139–141 (1976).
8. von Boehmer. H. & Haas. W. *Nature* **261**, 141–142 (1976).
9. Bailey. D. W. *Immunogenetics* **2**, 249–256 (1975).
10. Staats. J. *Cancer Res.* **36**, 4333–4377 (1976).
11. Bevan. M. J. *J. exp. Med.* **142**, 1349–1365 (1975).
12. MacDonald. H. R.. Philipps. R. A. & Miller. R. G. *J. Immun.* **111**, 565–574 (1973).
13. Bevan. M. J. *J. Immun.* **114**, 316–319 (1975).
14. Jerne. N. K. *Eur. J. Immun.* **1**, 1–9 (1973).
15. Nabholz. M. & Miggiano. V. in *B and T cells in Immune Recognition* (eds Loor. F. & Roelants. G. E.) (Wiley. Chichester. UK. 1977).

ON THE THYMUS IN THE DIFFERENTIATION OF
"H-2 SELF-RECOGNITION" BY T CELLS:
EVIDENCE FOR DUAL RECOGNITION?

By R. M. ZINKERNAGEL,* G. N. CALLAHAN, A. ALTHAGE, S. COOPER, P. A. KLEIN,
AND J. KLEIN

*(From the Departments of Cellular and Developmental Immunology and Molecular Immunology
of the Research Institute of Scripps Clinic, La Jolla, California 92037, Tumor Biology Unit, the
Department of Pathology, University of Florida, Gainesville, Florida 32610, and the Department
of Microbiology, Southwestern Medical School, Dallas, Texas 75235)*

The dual specificity of thymus-derived lymphocytes (T cells) for structures coded by the major histocompatibility gene complex (MHC)[1] and for foreign antigens (X) appears to be a general finding in mice and is likely to be a more universal phenomenon in higher vertebrates. Thus, all T-cell functions that have been tested in mice i.e., T cells involved in nonlytic helper, delayed type hypersensitivity, and macrophage activation functions are specific for the murine MHC (*H-2*) coded structure mapping to the *I* region (1–5), whereas cytotoxic T-cell activity is specific for *H-2K* or *D* (6–11). Similarly, cytotoxicity appears to be MHC-restricted in the rat (12), in humans (13), and in chickens (14). This restriction of T cells by MHC determinants contrasts with the apparently complete absence of H-2 restrictedness of B-cell functions or of their antibody products.

Although the phenomenology is clear and well accepted, the explanation of it is controversial and therefore many hypotheses and speculations have attempted to catch the elusive nature of the T-cell receptor; because it is generally felt that this dual specificity of T cells reflects and may therefore reveal unique properties of the T-cell receptor(s). Two models of associative T-cell recognition have been proposed: first the dual recognition model (1–9, 14, 15) where T cells recognize two distinct antigenic entities i.e., self-H-2 structures and foreign antigen X with two separate receptors. Second, the single receptor model, which proposes that T cells possess one single receptor specificity that recognizes a neoantigenic determinant (NAD) formed either by a complex of self and X, by an antigen-specific modification of the self-H-2 structure, by host-specific, or by

* Publication no. 45 from the Department of Cellular and Developmental Immunology, publication no. 1398 from the Immunology Departments, Scripps Clinic and Research Foundation, La Jolla, Calif. 92037, and publication no. 132 from the Tumor Biology Unit, Department of Pathology, University of Florida.
[1] Abbreviations used in this paper: ATxBM, adult thymectomized irradiated and bone marrow reconstituted mice; C, complement; GVHR, graft-versus-host reaction; H-2, murine MHC; HLA i.v., human MHC; MHC, major histocompatibility gene complex; NAD, neoantigenic determinant; PFU, plaque-forming unit.

virus-specific modification of the foreign antigen X (6–11).

Experiments that apparently contradicted the rule of *H-2* restriction with chimeric mice of various combinations demonstrated that *H-2* incompatible T and B cells could cooperate (16, 17). Katz and Benacerraf have interpreted these results (18, 19) to indicate that in chimeras lymphocytes "learn to get along with each other." This proposition of lymphocytes differentiating cell interaction molecules for the *H-2* type of the cells with which they cohabitate during immunological reconstitution was called "adaptive differentiation"; our results support this as a general idea.

This paper and the following report aims at analyzing the question whether virus-specific cytotoxic T cells act via one or two receptors. The approach we chose was to determine whether *mere tolerance to alloantigen* was sufficient to allow generation of killer T cells that were specific for the tolerated alloantigen and virus or whether *additional* and *what kind* of conditions must be fulfilled for chimeras to produce *H-2*-restricted T cells that were specific for infected tolerated allogeneic targets. If T cells possess two receptor qualities, then the prediction is that the receptors for the anti-self-H-2 structures differentiate and are selected for independently for recognition of a foreign antigen. In contrast if T cells have one receptor quality then the specificity for the self-H-2 structure and for foreign antigen must be selected for simultaneously and restriction is imposed solely by the stimulating antigenic entity.

In designing experiments to probe this puzzle we started from two contrasting sets of results. Parent A → (A × B)F_1 chimeric T cells lyse trinitrophenol-modified targets of both parental *H-2* types and virus-infected parent A → (A × B)F_1 chimeras generated cytotoxic T cells that lyse virus-infected targets of the parent A type and/or (less efficiently) of the other incompatible parental B *H-2* type (20–22). In contrast mice made tolerant to *B* during neonatal life that were infected with virus generated virus-specific cytotoxic T cells exclusively for their own *H-2* type, but failed to react to virus-infected targets of the tolerated *H-2* (23). One could explain that in neonatally tolerant mice, unlike the chimeras, the tolerated haplotype is present in such trace quantities that virus plus tolerated *H-2* will be a very minor antigenic determinant that fails to trigger measurable cytotoxic activity; or that the mechanism of tolerance in parent → F_1 and neonatally tolerant mice are simply different. Recently, Bevan demonstrated that in F_1 → parent irradiation bone marrow chimeras H-2-restricted cytotoxic T-cell specificity was preferentially associated with the recipient H-2 type, thus the recipient determined the specificity of H-2 restriction (24). Independently we used a similar experimental approach.

In this report we analyze the specificity for self-H-2 plus virus of cytotoxic T cells generated in various types of irradiation bone marrow chimeras of F_1 → parent, parent → F_1, partially H-2 incompatible F_1 → F_1 or recombinant → recombinant and completely H-2 incompatible chimeras; in addition adult thymectomized irradiated and bone marrow reconstituted mice reconstituted with a semiallogeneic thymus were tested. We demonstrate that lymphopoetic stem cells differentiate in the thymus to mature T cells and during this process acquire a recognition structure with specificity for H-2 as it is expressed on the thymic epithelium; self-recognition is thus selected for thymus H-2, is independent of the lymphocytes' own H-2, and differentiates independently (with respect to time, i.e., ontogenetically, and with respect to site) of recognition of a foreign antigen.

Materials and Methods

Animals. C57BL/10, B10.D2, B10.BR, B10.A, B10.A(4R), B10.A(5R) (C57BL/6 × A)F_1, and (BALB/c × A)F_1 mice were purchased from The Jackson Laboratory, Bar Harbor, Maine. BALB/c, C3H, A and (BALB/c × C57BL/6)F_1 mice came from the Strong Foundation, San Diego, Calif. Some C3H and C3H × C57BL/6 mice were purchased from the Cumberland Farm, Clinnton, Tenn. 37761. Mice were usually 6–14 wk of age when used and whenever possible were matched for sex and age in any given experiment.

Chimeras. Chimeras were produced using the general protocol of Sprent et al. (25). The mice were irradiated with a supralethal dose of 900–950 rads from a cesium source and on the same day were reconstituted with 1.5–2.5×10^7 anti-θ plus C-treated (in part a gift from Doctors D. H. Katz and B. S. Skidmore, Scripps Clinic, La Jolla, Calif.) bone marrow cells or fetal (14–16 day) liver cells injected intravenously (i.v.). All chimeras were analyzed individually 6–12 wk later.

Thymic chimeras were made as follows. Several days to 2 wk after adult mice were thymectomized, they were irradiated with 900–950 rads and reconstituted with anti-θ plus C-treated bone marrow cells (ATxBM). The same day thymus lobes (six per recipient) were prepared from 6- to 8-wk-old mouse donors, irradiated with 875 rads, and transferred to the thymectomized and irradiated recipients subcutaneously. Thymuses were irradiated to prevent proliferating thymus cells from emigrating into the periphery, as several investigators have described (26–27). Thymic chimeras were tested for thymic function 2–3 mo after reconstitution and controlled to ensure that no rests of the original thymus were functional.

Virus and Immunization. Mice were infected i.v. with about 10^7 plaque-forming units (PFU) of vaccinia virus and sacrificed 6 days later as described elsewhere (7, 22, 29).

Cell Preparations and H-2 Typing. All cells were prepared and used in minimal essential medium supplemented with nonessential amino acids, pyruvate, bicarbonate, antibiotics, and 10% heat inactivated fetal calf serum. All these ingradients were from Flow Laboratories, Inc. (Inglewood, Calif. 90025). Spleen cells from the virus-infected mice were processed and typed for H-2 as described previously (7, 22, 23, 29). The antisera used for typing were from the NIH collection (prefix D) or from Dr. J. Klein (prefix K). In each typing assay, positive and negative control cells were included to confirm the specificity of the antisera treatment and for C activity (28).

Target Cells and Cytotoxicity Assay. The methods used and the target cells we used have been described (22, 28). L929 (*H-2K*) originate from C3H mice, MC57G (*H-2^b*) from C57BL/6, mice, and D2 (*H-2^d*), a methylcholanthrene-induced cell line is of B10.D2 origin. The D2-line spontaneously released the most ^{51}Cr (20–30% for 6 h, 30–45% for 16 h), whereas MC57G released 8–12% during a 6 h test and 12–25% over 16 h, and L929 15–25% and 20–40%, respectively. The percent of ^{51}Cr release was calculated as percent of water-release (100%) and was not corrected for medium release. Water usually released about 80–85% of the ^{51}Cr.

Statistical Methods. Means and SEM of triplicates were determined and compared by Student's *t* test. SEM was usually smaller than 3%, and always smaller than 5%.

Results

Specificity for Self-H-2 of Cytotoxic T Cells from $F_1 \rightarrow P$ Bone Marrow Chimeras. Various combinations of fully reconstituted $F_1 \rightarrow$ parent chimeras were tested for virus-specific cytotoxicity against vaccinia-infected targets (Table I). Reconstitution was assessed in each mouse individually and was complete within the limits of the serological H-2 typing method which leaves an uncertainty factor of about 3–5%. The results show that $F_1 \rightarrow$ parent irradiation *bone marrow* chimeras generated virus-specific activity associated with the donor compatible parental H-2 type, but not with the parental H-2 type that was incompatible (Table I, exp. 1).

TABLE I

Virus-Specific Cytotoxic Activity in $F_1 \rightarrow$ Parent Irradiation Chimeras*

Donor $\longrightarrow$ Recipient (H-2 types)‡		Spleen cell to target cell ratio	% ^{51}Cr Release from vaccinia infected target cells§		
			D2(d)	MC57G(b)	L(k)
Experiment 1—bone marrow chimeras:					
1. BALB/c × C57BL/6 → BALB/c		40:1	71 ‖	9	12
(d × b) (d)		13:1	66	9	14
		4:1	44	8	12
2. B10.D2 × C57BL/10 → C57BL/6		40:1	28	36	13
(d × b) (b)		13:1	27	27	13
		4:1	24	11	11
BALB/c × C57BL/6		40:1	92	43	17
(d × b)					
C3H (k)		40:1	28	10	90
Medium			30	11	18
Experiment 2—adult spleen cell chimeras:					
3. BALB/c × C3H → C3H (36 wk)		40:1	88	23	96
(d × k) (k)		13:1	79	21	63
4. BALB/c × C57BL/6 → BALB/c		40:1	85	50	45
(d × b) (d)		13:1	80	30	43
Medium			36	22	46

* Mice of 6–10 wk of age were irradiated with 925 rads and reconstituted with 2×10^7 anti-θ + C-treated bone marrow cells (exp. 1) or untreated adult spleen cells (exp. 2); chimeras were infected with about 10^7 PFU of vaccinia virus 10 or 14 wk (exp. 1), or 36 wk (exp. 2) after reconstitution.

‡ H-2 typing was performed as described in Materials and Methods. Positive and negative controls were included for each antiserum to control specificity. Typing results were for:

Chimera 1		Chimera 2	
Anti-K^d (D-31)	>95%	Anti-K^d (D-31)	>90%
Anti-K^b (D-33)	>95%	Anti-K^b (D-33)	>95%
Control	<5%	Control	<5%

Similarly chimeras 3 and 4 were fully reconstituted by F_1 cells.

§ Uncorrected values are given as means of triplicate determination; SEM were all smaller than 3%. Duration of test was 6 h (exp. 1) and 16 h (exp. 2). The release from uninfected targets was tested for all cell populations and was in no case significantly different from medium controls.

‖ Statistically significant values ($P < 0.01$) are boxed.

Thus, these results show that the presence of tolerated H-2 antigens in the reconstituted lymphoreticular compartment was insufficient for this *H-2* type to be treated as self-marker. $F_1 \rightarrow$ parent chimeras obviously differed in at least two ways from normal F_1 animals that generated virus-specific activity always associated with both parental haplotypes: (*a*) $F_1 \rightarrow$ parent chimeras have >95% of the somatic cells of the recipient parental *H-2* type and (*b*) $F_1 \rightarrow$ parent chimeras possess the thymus and the stroma of the lymphatic organs of recipient parental H-2 type exclusively.

Irradiation Chimeras Reconstituted with F_1 Adult Spleen Cells. In Table I,

TABLE II

Vaccinia-Specific Cytotoxicity Generated in Semisyngeneic-Semiallogeneic Chimeras

Donor ———→ Recipient	Spleen cell to target cell ratio	^{51}Cr Release from vaccinia infected target cells‡		
		D2(d)	L(k)	MC57G(b)
BALB/c × C57BL/6 → BALB/c × C3H	40:1	67 ‖	20	9
($d × b$)　　　　　　($d × k$)§	13:1	59	17	10
	4:1	37	16	10
BALB/c	40:1	76	14	8
C3H	40:1	26	90	11
C57BL/6	40:1	28	17	74
Medium		28	15	10

* Recipient mice were irradiated with 925 rads and reconstituted with $2 × 10^7$ anti-θ + C-treated bone marrow cells 24 h later. Chimeras were infected 6 wk later and killed 6 days after infection. Spleen cells were tested for *H-2* type and for virus-specific cytotoxicity.

‡ Results are uncorrected means of triplicate determination; the SEM was below 3.5%. The test duration was 6 h. Release from uninfected target cells was not significant.

§ The antiserum treatment was controlled for specificity by included positive and negative control cells. The typing results were: anti-K^k (K-603) <10%, anti-K^b (D-33) >90%, controls <10%.

‖ The results were compared statistically with the highest values of medium release or release by immune spleen cells from H-2 incompatible donor mice. Significant values ($P < 0.01$) are boxed.

exp. 2, parental A or B recipient mice were reconstituted with adult spleen cells from F_1 donors instead of anti-θ plus C-treated bone marrow cells or fetal liver cells. Quite different from the previous experiments, infected chimeras generated cytotoxic activity that was associated with both parental haplotypes. Furthermore, the cytotoxic activity could be elicited in these chimeras very rapidly after transfer, i.e., within 2–3 wk (data not shown); this contrasts with the irradiation bone marrow chimeras in which immunocompetence appeared only 4 wk after reconstitution.

These results indicate, that if mature, differentiated T cells are transferred under conditions in which no dramatic graft-versus-host reactions occur (such as in $F_1 →$ parent chimeras) these T cells express their immunocompetence without further differentiation in the thymus.

Self-Specificity of Virus-Specific Cytotoxic T Cells from Reciprocally Semiallogeneic (A × B) → (A × C) Chimeras.　Chimeras that were reciprocally semisyngeneic were made by transfusing irradiated (BALB/c × C3H)F_1 (*H-2^d* × *H-2^k*) with anti-θ plus C-treated bone marrow cells from (BABL/c × C57BL/6)F_1 (*H-2^d* × *H-2^b*) mice (Table II). When infected with vaccinia virus these chimeras generated virus-specific cytotoxic T cells that lysed infected *H-2^d* targets only. This occurred despite the fact that the chimeras were reconstituted fully with (BALB/c × C57BL/6)F_1 cells and that the host's nonlymphoid somatic cells expressed *H-2^k*.

Virus-Specific Cytotoxicity Generated in K, I Compatible, K, I Incompatible, or H-2 Incompatible Chimeras.　Chimeras were made by combining recombinant inbred strains of mice that were compatible for the *K, I* or *S, D* regions of

H-2. The construction of some of these chimeras was difficult, particularly those with I incompatibility. The survival rate of K, I compatible combinations was about 70%, which is comparable with the survival of parent $\rightarrow$ F$_1$ or F$_1$ $\rightarrow$ parent chimeras (50–90%); I incompatible chimera survived to about 20%.

The three kinds of chimeras in Exp. I in Table III were made on the same day under identical conditions. The (K, I, S, D) compatible semisyngeneic B10.D2 $\rightarrow$ B10.BR $\times$ B10.D2 chimera and the (K, IA) compatible B10.A(4R) $\rightarrow$ B10.A chimera both generated measurable cytotoxic activity when injected with virus (Table III, exp. 1). However, in parent $\rightarrow$ F$_1$ chimeras activity was specific for the reconstituting parental haplotype; in (K, IA) compatible chimeras activity was generated against infected targets of the K haplotype only, but neither against target of donor nor of the recipient D haplotype. In contrast the (IC, S, D) compatible B10.D2 $\rightarrow$ B10.A chimeras failed to generate a measurable response. Since the other two types of chimeras responded well, the 10 wk time between irradiation plus stem cell reconstitution and infection were sufficient for the differentiation of immunocompetence.

The capacity of (K, IA) and (D) compatible chimeras to generate virus-specific cytotoxic T cells is given in exp. 2 in Table III. Included is one example (C57BL/10 $\rightarrow$ B10.A(2R) in which the D region is the only one of compatibility. Again no measurable activity was associated with either compatible D^b, with donor K^b nor with recipient K^k. Similarly, completely H-2 incompatible chimeras of the C3H $(H\text{-}2^k) \rightarrow$ C57BL/10 $(H\text{-}2^b)$ did not generate measurable virus-specific cytotoxic activity (Table IV) associated with either $H\text{-}2^k$ or $H\text{-}2^b$ 3 mo after reconstitution. The experiments in Tables I–III demonstrate that in chimeras specific cytotoxicity was generated against virus and only the $H\text{-}2$ haplotype or the K haplotype that was shared between recipient and donor. No measurable activity was found when K or D specificities were expressed only on the donor or the recipient cells alone. In these examples (A $\times$ B)F$_1$ $\rightarrow$ A, (A $\times$ B)F$_1$ $\rightarrow$ (A $\times$ C)F$_1$, recombinant A|B $\rightarrow$ recombinant A|C or recombinant B|A $\rightarrow$ recombinant C|A there was plenty of B-haplotype expressed by the lymphoid cells in the chimeras besides the abundance of a haplotype C expressed by the somatic cells of the recipient, therefore it is not simply lack of expression of the K or D haplotypes that explained unresponsiveness. However, the significant result is that whenever donor stem cells and recipient shared the I region plus K and/or D region(s) virus-specific activity was always generated. Since (A $\times$ B)F$_1$ $\rightarrow$ A chimeras generated virus-specific T cells against infected A targets only these results suggested that the H-2 of the host was probably determining the specificity for self of T cells. Since no activity was generated in irradiated thymectomized recipients that were reconstituted with stem cells under otherwise comparable conditions, these results and those from irradiated F$_1$ $\rightarrow$ P spleen cell chimeras suggested that it may be the thymus of the recipient that not only promotes differentiation of precursor T cells to mature T cells but that the thymus determines also what $K, I,$ or D region could be eventually recognized together with a foreign viral antigen.

The H-2 Type of Transplanted Thymuses Determine the Self-Specificity of Virus-Immune Cytotoxic T Cells. In view of our results we analyzed how the H-2 haplotype of the thymus influences the differentiation of immunocompetent

TABLE III

*Virus-Specific Cytotoxicity Generated in K, I, or S, D Compatible Irradiation Bone Marrow Chimeras**

Donor ⟶ Recipient	H-2 Compatibility	Spleen cell to target cell ratio	⁵¹Cr Release from vaccinia infected target cells‡		
			D2(d)	L(k)	MC57G(b)
Experiment 1:					
1. B10.D2 → B10.A	IC,S,D	40:1	38	34	23
(d)　　($k\|d$)§		13:1	35	35	23
2. B10.A(4R) → B10.A	K,IA	40:1	29	76 ‖	22
($k\|b$)　　($k\|d$)		13:1	29	41	23
		4:1	28	37	20
3. B10.D2 → B10.BR × B10.D2	K,I,S,D	40:1	73	35	24
(d)　　　($k × d$)		13:1	53	30	22
		4:1	37	31	23
Controls:					
B10.D2		40:1	80	38	32
B10.A(4R)		40:1	33	73	60
Experiment 2:					
4. B10.A(4R) → B10.A	K,IA	40:1	28	61	15
($k\|b$)　　($k\|d$)		13:1	23	43	13
		4:1	23	26	11
5. C57/BL.10 → B10.A(2R)		40:1	23	12	14
(b)　　　($k\|b$)		13:1	22	13	12
Controls:					
B10.D2(d)		40:1	97	13	15
B10.BR × C57/BL.10		40:1	26	82	57
($k × b$)					

* Recipient mice were irradiated with 925 rads 1 day before reconstitution with anti-θ + C-treated bone marrow cells from donors that were compatible for various parts of H-2. The recipients of exp. 1 were all reconstituted on the same day and infected with virus 7 wk later. Recipients were infected at the following times after reconstitution in exp. 1: 10 wk, exp. 2: 12 wk. 6 days after infection, mice were sacrificed and spleen cells tested for H-2 types and for cytotoxicity.

‡ Results are expressed as uncorrected means of triplicate determinations; SEM was smaller than 2.7%. The test duration was 6 h for exp. 1; 16 h for exp. 2. ⁵¹Cr release from respective uninfected targets was measured for all lymphocyte populations but no significant lysis was found.

§ Antisera treatment were controlled by positive and negative cell preparations; the anti-D^b (D-2) antiserum was also low (>80) on positive H-2^b controls in exp. 1. The typing results were for:

Chimera 1	*Chimera 2*	*Chimera 3*
Anti-K^k (K-603) <5%	Anti-D^d (D-4) <10%	Anti-K^k (K-603) >95%
Anti-K^d (D-31) >90%	Anti-D^b (D-2) >75%	Anti-K^d (D-31) <5%
Control <5%	Control <10%	Control <5%

Chimera 4	*Chimera 5*
AntiD^b (D-2) >95%	Anti-K^b (D-33) >95%
Anti-K^k (K-603) >95%	Anti-K^k (K-603) <5%
Anti-D^d (D-4) <5%	Control <5%

‖ Results were compared statistically with the highest value of medium control, immune cells on H-2 incompatible infected targets and immune cells on uninfected targets; statistically significant values ($P < 0.01$) are boxed.

TABLE IV
Failure of 3 Mo Old H-2 Incompatible Chimeras to Generate Measurable Virus-Specific Cytotoxic T Cells*

Donor	Recipient	Spleen cell to target cell ratio	^{51}Cr Release from vaccinia infected target cells§	
(H-2 typing)‡ →			L(k)	MC57G b)
C3H	C57BL/10	40:1	20	9
(k)	(b)	13:1	11	8
		4:1	11	8
C3H		40:1	69	13
(k)		13:1	46	12
C57BL/6		40:1	16	68
(b)		13:1	13	50
Medium			14	10

* Recipient mice were irradiated with 950 rads and reconstituted with 2×10^7 anti-θ + C-treated bone marrow cells. The chimera was infected 3 mo after reconstitution with about 10^7 PFU of vaccinia virus and its spleen cells tested for cytotoxicity 6 days later.

‡ See Materials and Methods. Positive and negative controls were included for each antiserum. The type results were: Anti-K^b (D-33) <20%, Anti-K^k (K-603) >95%, Control <20%.

§ Uncorrected means of triplicate determinations, the SEM was smaller than 5%; statistically significant results ($P < 0.01$) are boxed. Cytotoxicity was also tested on uninfected target cells; no significant lysis occurred.

T cells and their specificity spectrum with respect to self-H-2. (BALB/c × A)F$_1$ and (C57BL/6 × A)F$_1$ animals were thymectomized at the age of about 8 wk. 2 wk later they were irradiated and reconstituted with anti-θ-treated bone marrow cells or fetal liver cells from appropriate F$_1$ donors. These ATxBM mice were implanted six irradiated (875 rads) adult thymus lobes from BALB/c, C57BL/6, or alternatively from A donors. Of 35 such thymic chimeras only about 20% were able to generate virus-specific cytotoxic T cells 3–5 mo later (Table V). All these responder animals were the only ones that possessed functional transplanted thymus tissue as assessed macroscopically and microscopically and had no remaining functional original thymus tissue. Reconstitution of (A × B)F$_1$ ATxBM mice with irradiated thymuses of parent A generated virus specific cytotoxic T cells against infected A targets only, whereas B thymus reconstituted animals responded to B plus virus only. These animals differed from normal F$_1$ mice only in that their thymus bore just one parental *H-2* type, not both. Therefore, the results are compatible with the interpretation that the *H-2* type of the radioresistant portion of the thymus determines the differentiation of self-recognition in maturing T cells.

Discussion

Our results establishing the role of a host and its thymus in determining the H-2-restricted specificity of cytotoxic T cells are summarized in Table VI and as follows: (*a*) Chimeras in which the F$_1$ cells are used to reconstitute an irradiated recipient of the same haplotype as one of the hybrid's parents (F$_1$ → P) generate virus-specific cytotoxic T cells that kill only those targets which carry the recipient parental H-2 type. This is the case when F$_1$ stem cells are used,

TABLE V

Influence of Transplanted Irradiated Parental Thymus on Virus-Specific Cytotoxicity Generated in Adult Thymectomized with Syngeneic Bone Marrow Reconstituted F_1 Recipients*

Fetal liver or bone marrow donor	Thymus donor	Thymecto-mized recipient	Spleen cell to target cell ratio	^{51}Cr Release from vaccinia infected target cells‡		
				D2(d)	L(k)	MC57G(b)
Experiment 1:						
BALB/c × A ($d \times k\|d$) (Fetal liver)	None	BALB/c × A ($d \times k\|d$)	40:1	35	39	21
			13:1	33	38	17
BALB/c × A ($d \times k\|d$)	BALB/c (d)	BALB/c × A ($d \times k\|d$)	40:1	85 §	40	18
			13:1	76	39	16
			4:1	63	39	15
C57BL/6 × A ($b \times k\|d$) (Bone marrow)	A ($k\|d$)	C57BL/6 × A ($b \times k\|d$)	40:1	75	80	17
			13:1	79	67	18
			4:1	37	54	16
	Control:‡	C57BL/10	40:1	39	40	79
			13:1	33	38	87
Experiment 2:						
BALB/c × A ($d \times k\|d$) (Fetal liver)	BALB/c (d)	BALB/c × A ($d \times k\|d$)	40:1	74 §	30	34
			13:1	67	27	29
			4:1	44	26	28
C57BL/6 × A ($b \times k\|d$) (Bone Marrow)	C57BL/6 (b)	C57BL/6 × A ($b \times k\|d$)	40:1	39	33	100
			13:1	39	32	66
			4:1	32	29	45
	Controls:‡	BALB/c × A	40:1	72	59	33
			13:1	57	46	29
		C57BL/6 × A	40:1	74	72	100
			13:1	62	59	93
	Medium			35	30	32

* Recipient mice were thymectomized 1 to 2 wk before irradiation (BALB/c × A: 875 rads; C57BL/6 × A: 925 rads and transfused with 1.5×10^7 anti-θ + C-treated bone marrow cells. The mice were transplanted subcutaneously with the equivalent of three (875 rads) thymuses of one parental haplotype. These recipients were infected with about 10^7 PFU of vaccinia virus 10 wk (exp. 1) or 13 wk (exp 2) later. The mice were killed 6 days after infection and their spleen cells tested for *H-2* type and cytotoxicity. The test duration was for exp. 1: 16 h; exp. 2: 6 h. All spleen cells carried both parental *H-2* haplotypes. The thymectomy was controlled macroscopically and microscopically and the transplanted thymuses were examined by histology.

‡ Results are uncorrected means of triplicate determinations; the SEM was smaller than 3%. The results were compared statistically with the higher of the values for medium release or release by immune *H-2* incompatible spleen cells. The lymphocytes were also tested on uninfected target cells; no significant lysis was detected; control mice were normal mice immunized with vaccinia virus.

§ Boxed results are statistically significantly different from controls ($P < 0.01$).

TABLE VI

Summary of the Experiments Described in Tables I through V

Table	Donor cells	Recipient	Thymus of chimera	Lysis of virus-infected target cells A, B, C
I	A × B (BM)	A	A	A
I	A × B (BM)	B	B	B
I	A × B (Adult)	A	A	A,B
I	A × B (Adult)	B	B	A,B
II	A × C (BM)	A × B	A × B	A
III	$K^A I^A | D^C$ (BM)	$K^A I^A | D^B$	$K^A I^A | D^B$	A
III	$K^C I^C | D^B$ (BM)	$K^A I^A | D^B$	$K^A I^A | D^B$	None
IV	B (BM)	A	A	None
V	A × B (FL)	A × B	A	A
V	A × B (FL)	A × B	B	B

BM, anti-θ plus C-treated bone marrow cells; adult: adult spleen cells; FL: fetal liver from 14- to 17-day-old fetal mice.

however, if the stem cells are contaminated with mature F_1 T cells or adult F_1 spleen cells are used for reconstitution, the chimera generates T cells that kill targets of both parental H-2 types. The F_1 spleen cell → parent chimeras were immunologically active at 2 wk (the earliest time tested) whereas F_1 stem cell → parent chimeras took 4–6 wk to become immunocompetent. (*b*) Chimeras analogous to the F_1 → P type can also be generated by grafting a parental thymus into an F_1 that has been thymectomized, irradiated, and protected with F_1 stem cells. (*c*) Chimeras of the type F_1 (A × B) → F_1 (A × C) generate T killer cells capable of lysing targets that carry H-2 antigens of the A type only. (*d*) Chimeras made from H-2 recombinants instead of F_1 mice were of the general type recombinant (A|C) → recombinant (A|B). *H-2K* and *H-2I* region compatible chimeras generated killer T cells restricted for the shared H-2K antigens only. However, when the *H-2D* region was kept constant and the *H-2K* and *H-2I* regions were varied, i.e. ($K^C I^C D^B$) → ($K^A I^A D^B$) chimeras, no cytotoxic activity was detectable on any of the targets tested. Similarly, the fully H-2 incompatible chimera A → B did not generate killer cells either.

The general rule emerges from these results that the radio-resistant portion of the thymus determines which H-2 antigens will participate in restricting virus-specific killer T-cell precursors, and the H-2 antigens carried by cells of lympho-hemopoietic origin in the chimera determine which specificities of H-2 restricted precursors will be expressed by effector T killer cells (Zinkernagel et al. *J. Exp. Med.* 147:897).

Two further important conclusions can be drawn from the data. First, if T cells are selected in the thymus for which H-2 structures are to be treated as "self", then differentiation of restriction of cytotoxic T-cell activity to certain self-H-2 antigens must be unrelated to confrontation with viral or other non-self antigens. Thus, recognition of antigens by the T-cell receptor complex may involve two independent recognition sites, which may be on either one or two membrane molecules. In addition, as demonstrated earlier and here, both recognition structures are clonally distributed (9, 22, 29, 31). Furthermore, recognition of self-H-2 markers is most likely via an antibody-antigen-like and not a self-self-like interaction. Second, *H-2I* region compatibility must exist between the antigens of the lymphoid cells donated to the chimeric host and the antigens on that host's thymic epithelium, in addition to K or D compatibility.

Two experimental factors were critical for constructing chimeras that produced clearcut results: the use of a supralethal dose of irradiation to guarantee elimination of host cells and full reconstitution by donor stem cells and the use of fetal liver cells or bone marrow cells twice treated with anti-θ to eliminate all immunocompetent T cell in the reconstituting inoculum. The difference between the results here, e.g., that A $\times$ B $\rightarrow$ A $\times$ C chimeras generate A-specific activity only, and our and other's earlier results (20, 22) showing that P $\rightarrow$ F$_1$ chimeras were able to generate cytotoxic T cells mainly against the compatible but also against the incompatible parental targets are probably best explained by the use of relatively, with respect to the mouse strain used, lower doses of irradiation in the earlier experiments which may have allowed enough stimulating cells of host origin to survive. This will be analyzed further in the subsequent paper.

Vaccinia virus was chosen because it is not pathogenic for mice, and its proliferative capacity in mice seems to be limited. Otherwise, these chimeras might have died from virus infection. A noteworthy exception were the *H-2,K, I,* or *H-2* incompatible chimeras; of which some died from the vaccinia infection. This is compatible with the finding that, in contrast to all others, these chimeras did not generate measurable anti-viral cytotoxicity.

Experimentally, it is not yet possible to rule out alternative explanations for the lack of *H-2^B* type restriction seen in (A $\times$ B)F$_1$ $\rightarrow$ A chimeras. Theoretically, anti-idiotypic antibodies could arise as described in the classical Ramseier-Lindenmann experiment where (A $\times$ B)F$_1$ lymphocytes are sensitized against anti-B receptors present in irradiated A recipients (30). However, this interpretation also requires that H-2 restricted T-cell specificity functions through a dual recognition mechanism. Furthermore, other H-2-specific suppressor mechanisms cannot be fully excluded as yet, but are under investigation.

Which is the best interpretation: T cells possess one or two recognition sites? Our and Bevan's (24) experiments might be explained along the lines of a single receptor model for an NAD. Accordingly, precursor T cells may gain their maturity by differentiating receptors for foreign antigens (NAD) in the thymus as proposed by Jerne (31). Thus in a mouse of *H-2^A* type, T cells with specificity for A will, during ontogeny, proliferate in the thymus upon contact with self-H-2^A structures; some of these proliferating T lymphocytes will mutate to express a receptor not any longer specific for A but for "slightly different from A". Through a "filter mechanism", T cells with specificity for A will not be able to leave the thymus and would be pushed to proliferate further, or alternatively are inactivated. Only T cells having a mutated recognition specificity for "different from A" will be released from the thymus. This could explain the preference of T cells of (A $\times$ B)F$_1$ $\rightarrow$ A chimeras for "altered A". Since these chimeric T cells cannot react to "altered B" we must introduce the rule that the specificity spectrum anti-foreign antigens X that can be generated in a thymus A cannot overlap with that generated in a thymus B. When applied to a model of single receptor specificity for NAD this rule is unpredicted and seems unlikely, but cannot be formally excluded as yet. In the light of the data presented here, which requires the T cell's receptor quality for anti-self-H-2 structures to be specifically learned, i.e., selected for and expressed independently of the receptor quality for anti-foreign antigen X, the dual recognition model provides a simple interpretation. One can reduce the constraints these results impose on a one receptor model for the T-cell receptor to requiring that the receptor specifically recognizes self-H-2 structures, as defined by the H-2 antigens present on thymic epithelium, independent of recognition of foreign antigen X. However, to account for specificity against foreign antigen X in the lytic effector reactions, the receptor must recognize this antigen also. To argue that one receptor has two independent recognition sites is tantamount to accepting two independent recognition sites i.e., dual recognition. The distinction between models with one and two receptors may, therefore, be reduced to deciding whether signals generating by antigen binding to the anti-self-H-2 structure and anti-foreign antigen X sites are transmitted intracellularly via one or two molecules (7, 18, 32–35).

What are the Practical Implications of these Results for Reconstituting Humans with Immunodeficiency Diseases? Thymus or lymphoid stem cell deficiencies alone or combined have been treated by transplanting such patients with thymus and/or stem cell-grafts. To avoid graft versus-host reactions (GVHR), stem cell grafts usually consist of optimally HLA matched and mixed lymphocyte reaction negative bone marrow cells, or, alternatively, of liver cells from 6- to 10-wk-old fetuses. Thymic-deficient patients are usually transplanted with fetal thymus tissue. Since transplantation of these immunoincompetent fetal tissues does not cause a GVHR, these grafts are not usually HLA matched (36–38). However, many such attempts at reconstitution fail; this may perhaps be explained in part by our results. If a patient, A, with a thymic epithelial defect, is transplanted with a fetal thymus of incompatible $HLA\text{-}A^X\text{-}B^X\text{-}D^X$ (the $HLA\text{-}D$ region corresponds to the murine $H\text{-}2I$ region, which was shown both in this and the companion report to be crucial for reconstitution of cellular immunocompetence) then the differentiating stem cells learn to recognize the donor's $HLA\text{-}A^X\text{-}B^X\text{-}D^X$ markers as self. (but not his own $HLA\text{-}A^A\text{-}B^AD^A$ markers). Therefore, these donor-derived T cells will be unable to interact with their own cell surface structures of A type to provide help for the lymphocytes or to get involved in any of the complex cellular interaction or communication processes. A similar paradox arises when a patient with a stem cell defect but a potentially functional thymus is transplanted with fetal liver cells that are $HLA\text{-}A\text{-}B\text{-}D$ incompatible with the host's HLA markers expressed by thymic epithelium.

Obviously these examples are extremes and assume that, the thymic epithelium of the first host, or the stem cells of the second recipient, do not function at all. However, we propose that in addition to the well recognized need to transplant immunoincompetent tissues to avoid GVHR, the following minimal requirements should be fulfilled to reconstitute immunocompetence: (*a*) thymus epithelium and lymphoreticular stem cells must share one $HLA\text{-}D$ haplotype, (*b*) thymus epithelium, stem cells, and the rest of the somatic cells must share at least one of the possible four $HLA\text{-}A$ or $HLA\text{-}B$ antigens. For reconstitution of pure thymus hormone deficiency these rules may not apply necessarily.

Furthermore, the experimental results and the interpretations presented here suggest that to assess cellular immunocompetence of reconstituted patients or mice the measurement of only alloreactivity or activation by T-cell mitogens may *not be meaningful*; these tests probably do not reflect the capacity of T cells to recognize and interact with HLA-self structures. Therefore, to assess syngeneic immunologic competence (i.e., the one which is crucial for the host's immune protection) after reconstitution it is necessary to assess the potential for antigen-specific and HLA-restricted T-cell functions.

Summary

In the thymus, precursor T cells differentiate recognition structures for self that are specific for the *H-2K, D,* and *I* markers expressed by the thymic epithelium. Thus recognition of self-H-2 differentiates independently of the T cells H-2 type and independently of recognition of nonself antigen X. This is readily compatible with dual recognition by T cells but does not formally exclude a single recognition model. These conclusions derive from experiments with bone marrow and thymic chimeras. Irradiated mice reconstituted with bone marrow to form chimeras of $(A \times B)F_1 \rightarrow A$ type generate virus-specific cytotoxic T cells for infected targets A only. Therefore, the H-2 type of the host determines the H-2-restricted activity of killer T cells alone. In contrast, chimeras made by reconstituting irradiated A mice with adult spleen cells of $(A \times B)F_1$ origin generate virus-specific cytotoxic activity for infected A and B targets, suggesting that mature T cells do not change their self-specificity readily. $(A \times B)F_1 \rightarrow (A \times C)F_1$ and $(K^AI^A|D^C) \rightarrow (K^AI^A|D^B)$ irradiation bone marrow chimeras responded against infected A but not B or C targets. This suggests that cytotoxicity is not generated against D^C because it is absent from the host's thymus epithelium and not against D^B because it is not expressed by the reconstituting lymphoreticular system. $(K^BI^B|D^A) \rightarrow (K^CI^C|D^A)$ K, I incompatible, or completely H-2 incompatible A $\rightarrow$ B chimeras fail to generate any measurable virus specific cytotoxicity, indicating the necessity for *I*-specific helper T cells for the generation of killer T cells. Finally adult thymectomized, irradiated and bone marrow reconstituted $(A \times B)F_1$ mice, transplanted with an irradiated thymus of A origin, generate virus-specific cytotoxic T cells specific for infected A targets but not for B targets; this result formally demonstrates the crucial role of thymic epithelial cells in the differentiation of anti-self-H-2 specificities of T cells.

Received for publication 14 September 1977.

References

1. Katz, D. H., T. Hamaoka, and B. Benacerraf. 1973. Cell interactions between histoincompatible T and B lymphocytes. II. Failure of physiologic cooperative interactions between T and B lymphocytes from allogeneic donor strains in humoral response to hapten-protein conjugates. *J. Exp. Med.* 137:1405.
2. Katz, D. H., and B. Benacerraf. 1975. The function and interrelationship of T cell receptors, Ir genes and other histocompatibility gene products. *Transplant. Rev.* 22: 175.
3. Miller, J. F. A. P., M. A. Vadas, A. Whitelaw, and J. Gamble. 1975. H-2 gene

complex restricts transfer of delayed-type hypersensitivity in mice. *Proc. Natl. Acad. Sci. U. S. A.* **72**:5095.

4. Rosenthal, A. S., and E. M. Shevach. 1973. Function of macrophages in antigen recognition by guinea pig T lymphocytes. I. Requirement for histocompatible macrophages and lymphocytes. *J. Exp. Med.* **138**:1194.

5. Paul, W. E., E. M. Shevach, D. W. Thomas, S. F. Pickeral, and A. S. Rosenthal. 1977. Genetic restriction in T-lymphocyte activation by antigen-pulsed peritoneal exudate cells. *In* Origins of Lymphocyte Diversity. *Cold Spring Harbor Symp. Quant. Biol.* **XLI**:571.

6. Zinkernagel, R. M., and P. C. Doherty. 1974. Activity of sensitized thymus derived lymphocytes in lymphocytic choriomeningitis reflects immunological surveillance against altered self components. *Nature (Lond.).* **251**:547.

7. Doherty, P. C., R. V. Blanden, and R. M. Zinkernagel. 1976. Specificity of virus-immune effector T cells for H-2K or H-2D compatible interactions: implications for H-antigen diversity. *Transplant. Rev.* **29**:89.

8. Shearer, G. M. 1974. Cell-mediated cytotoxicity to trinitrophenyl-modified syngeneic lymphocytes. *Eur. J. Immunol.* **4**:257.

9. Shearer, G. M., T. G. Rehn, and A. M. Schmitt-Verhulst. 1976. Role of the murine major histocompatibility complex in the specificity of *in vitro* T cell mediated lympholysis against chemically-modified autologous lymphocytes. *Transplant. Rev.* **29**:222.

10. Bevan, M. J. 1975. The major histocompatibility complex determines susceptibility to cytotoxic T cells directed against minor histocompatibility antigens. *J. Exp. Med.* **142**:1349.

11. Gordon, R. D., E. Simpson, and L. E. Samelson. 1975. In vitro cell-mediated immune responses to the male specific (H-Y) antigen in mice. *J. Exp. Med.* **142**:1108.

12. Zinkernagel, R. M., A. Althage, and F. C. Jensen. 1977. Cell-mediated immune response to lymphocytic choriomeningitis and vaccinia virus in rats. *J. Immunol.* **119**:1242.

13. Goulmy, E., A. Termijtelen, B. A. Bradley, and J. J. Van Rood. 1976. Y-antigen killing by T cells of women is restricted by HLA. *Nature (Lond.).* **266**:544.

14. Toivanen, P., A. Toivanen, and O. Vainio. 1974. Complete restoration of bursa-dependent immune system after transplantation of semiallogeneic stem cells into immunodeficient chicks. *J. Exp. Med.* **139**:1344.

15. Shevach, E. M., and A. S. Rosenthal. 1973. Function of macrophages in antigen recognition by guinea pig T lymphocytes. II. Role of the macrophage in the regulation of genetic control of the immune response. *J. Exp. Med.* **138**:1213.

16. Bechtol, K. B., J. H. Freed, L. A. Herzenberg, and H. O. McDevitt. 1974. Genetic control of the antibody response to POLY-L(TYR, GLU)-POLY-D,L-ALA-POLY-L-LYS in C3H ↔ CWB tetraparental mice. *J. Exp. Med.* **140**:1660.

17. von Boehmer, H., L. Hudson, and J. Sprent. 1975. Collaboration of histoincompatible T and B lymphocytes using cells from tetraparental bone marrow chimeras. *J. Exp. Med.* **142**:989.

18. Katz, D. H., and B. Benacerraf. 1976. Genetic control of lymphocyte interactions and differentiation. *In* The Role of Products of the Histocompatibility Gene Complex in Immune Responses. D. H. Katz and B. Benacerraf, editors, Acacemic Press, Inc. New York. p. 335.

19. Katz, D. H. 1976. The role of the histocompatibility gene complex in lymphocyte differentiation. *Cold Spring Harbor Symp. Quant. Biol.* **41**:611.

20. Pfizenmaier, K., A. Starzinski-Powitz, H. Rodt, M. Rollinghoff, and H. Wagner. 1976. Virus and trinitrophenol-hapten-specific T-cell-mediated cytotoxicity against H-2 incompatible target cells. *J. Exp. Med.* **143**:999.

21. von Boehmer, H., and W. Haas. 1976. Cytotoxic T lymphocytes recognize allogeneic tolerated TNP-conjugated cells. *Nature (Lond.)*. **261**:139.

22. Zinkernagel, R. M. 1976. *H-2* restriction of virus-specific cytotoxicity across the *H-2* barrier. Separate effector T-cell specificities are associated with self-H-2 and with the tolerated allogeneic H-2 in chimeras. *J. Exp. Med.* **144**:933.

23. Zinkernagel, R. M., G. N. Callahan, J. W. Streilein, and J. Klein. 1977. Neonatally tolerant mice fail to react against virus-infected tolerated cells. *Nature (Lond.)*. **266**:837.

24. Bevan, M. J. 1977. In a radiation chimeras host H-2 antigens determine the immune responsiveness of donor cytotoxic cells. *Nature (Lond.)*. **269**:417.

25. Sprent, J., H. von Boehmer, and M. Nabholz. 1975. Association of immunity and tolerance to host *H-2* determinants in irradiated F₁ hybrid mice reconstituted with bone marrow cells from one parental strain. *J. Exp. Med.* **142**:321.

26. Miller, J. F. A. P., and D. Osoba. 1967. Current concepts of the immunological function of the thymus. *Physiol. Rev.* **47**:437.

27. Davies, A. J. S. 1969. The thymus and the cellular basis of immunity. *Transplant. Rev.* **1**:43.

28. Callahan, G. N., S. Ferrone, M. D. Poulik, R. A. Reisfeld, and J. Klein. 1976. *J. Immunol.* **117**:1351.

29. Zinkernagel, R. M., and P. C. Doherty. 1975. *H-2* compatibility requirement for T-cell-mediated lysis of targets infected with lymphocytic choriomeningitis virus. Different cytotoxic T-cell specificities are associated with structures coded for in *H-2K* or *H-2D*. *J. Exp. Med.* **141**:1427.

30. Ramseier, H., M. Aguet, and J. Lindenmann. 1977. Similarity of idiotypic determinants of T and B lymphocyte receptors for alloantigens. *Immunol. Rev.* **34**:50.

31. Jerne, N. K. 1971. The somatic generation of immune recognition. *Eur. J. Immunol.* **1**:1.

32. Langman, R. E. 1977. The role of the major histocompatibility complex in immunity: a new concept in the functioning of a cell-mediated immune system. *Rev. Physiol. Biochem. Pharmacol.* In press.

33. Doherty, P. C., D. Gotze, G. Trinchieri, and R. M. Zinkernagel. 1976. Models for regulation of virally modified cells by immune thymus derived lymphocytes. *Immunogenetics*. **3**:517.

34. Zinkernagel, R. M., and P. C. Doherty. 1977. The concept that surveillance of self is mediated via the same set of genes that determines recognition of allogenic cells. *Cold Spring Harbor Symp. Quant. Biol.* **XLI**:505.

35. Janeway, C. A., Jr., H. Wigzell, and H. Binz. 1976. Two different V_H gene products make up the T cell receptors. *Scand. J. Immunol.* **5**:993.

36. Githens, J. H. 1976. Immunologic reconstitution with fetal tissue. *N. Engl. J. Med.* **294**:1116.

37. Cooper, M. D. 1975. Defective thymus development: a cause of combined immunodeficiency. *N. Engl. J. Med.* **293**:450.

38. Pahwa, R. S., R. A. Good, G. S. Incefy, and R. J. O'Reilly. 1977. Rationale for combined use of fetal liver and thymus for immunological reconstitution in patients with variants of severe combined immunodeficiency. *Proc. Natl. Acad. Sci. U. S. A.* **74**:3002.

IN A FULLY H-2 INCOMPATIBLE CHIMERA, T CELLS OF DONOR ORIGIN CAN RESPOND TO MINOR HISTOCOMPATIBILITY ANTIGENS IN ASSOCIATION WITH EITHER DONOR OR HOST H-2 TYPE*

By POLLY MATZINGER AND GALADRIEL MIRKWOOD

(From the Department of Biology, University of California San Diego, La Jolla, California 92093)

Despite much recent interest and effort, the role played by major histocompatibility complex products in the regulation of T-cell responses remains perplexing. In 1972 it was observed that mouse T and B cells would only cooperate in an antibody reponse if they shared certain regions of H-2 (1). Subsequently, H-2 gene products were also found to be involved in cytotoxic T-cell reactions, and it was postulated that the killer T cell must bear H-2 molecules in common with those of its target in order to effect lysis (2–6). Later studies with radiation chimeras showed that this is not the case, but that the H-2 region must be shared between the cells used to stimulate the response and the targets; a killer T cell that was itself H-2 type A, after having grown up in an $(A \times B)F_1$ could be stimulated to lyse H-2 type B virus-infected or trinitrophenyl-modified targets (7–9). Such chimeras were also found to contain A type helper T cells which can cooperate with B type B cells (10). It was then postulated that T-cell precursors "learn" to recognize the H-2 type of the host as self (11). Recent evidence shows that the host H-2 type of a chimera does distinctly influence the specificity of the responding T-cell population (12, 13) and that it is the H-2 type of the thymus that is important (13). Most of this work has been done with semiallogeneic chimeras (e.g., "A" bone marrow into an irradiated $[A \times B]F_1$, or $[A \times B]F_1$ bone marrow into an "A" or $[A \times C]F_1$) where the responses were very strongly restricted by the H-2 type of the host. A small number of completely allogeneic chimeras was tested (e.g., "A" bone marrow into "B") and appeared to be immunoincompetent. The virtually absolute restriction of the semiallogeneic chimeras as well as the immunoincompetence of the fully allogeneic chimeras has led to much speculation and has been quoted as suggestive evidence for the dual recognition model of T-cell receptors (13).

We report here that in contrast to the results with virus-infected mice, fully allogeneic chimeras made by repopulating irradiated $BALB/c(H-2^d)$ mice with $BALB.B(H-2^b)$ bone marrow are well able to respond to minor histocompatibility

* Supported by U. S. Public Health Service grants CA 09174 and AI 08795.
[1] *Abbreviations used in this paper*: B10, C57BL/10Sn; C, BALB/c; C.B, BALB.B; C.K, BALB.K; Con A, concanavalin A; CTL, cytotoxic T lymphocyte; H antigen, histocompatibility antigen.

(H)[2] antigens, and that the killer T cells that are themselves H-2[b] can recognize minor H antigens on either H-2[b] or H-2[d] targets.

Materials and Methods

Mice. C57BL/10Sn(H-2[b]) (B10), B10.D2/nSn(H-2[d]), B10.G(H-2[q]), B10.BR/SgSn(H-2[k]), and D1.C(H-2[d]) mice were purchased from The Jackson Laboratory, Bar Harbor, Maine. The BALB/cKe(H-2[d]) (C), and BALB.B(H-2[b]) (C.B.) mice used for making the chimeras were bred at the Salk Institute, San Diego, Calif. The C, C.B, BALB.K(H-2[k]) (C.K), and (BALB/c × BALB.B) (C × C.B) used for target and stimulator cells were bred at the University of California San Diego, from breeding pairs generously given to us by Dr. Frank Lilly (Albert Einstein Medical College, N.Y.).

Relationship of B10 Series, BALB Series, and DBA/1 Series Mice. B10, B10.D2, B10.BR, and B10.G differ genetically by a small segment of chromosome 17 which carries the H-2 gene complex and Tla, and are otherwise probably identical (14, 15). D1.C is a congenic line made by putting the H-2[d] of BALB/c on the DBA/1 background. D1.C has some minor H antigens in common with the B10 series mice that are not expressed by BALB/c, and was sometimes used as a substitute for B10.D2 which were in short supply.

C, C.B, and C.K also differ only at the H-2 gene complex. The C mice from Salk Institute differ from those of Lilly in that they possess some minor H antigen differences which were discovered when some Salk C mice immunized with B10.D2 showed slight activity on C mice from Lilly. In general, such differences were not a problem although every so often we came across a chimera which had slight cytotoxic activity on Lilly C, C.B, or (C × C.B)F₁ targets.

Chimeras. 20 C ♂ mice were given 900 rads from a Cobalt 60 source and immediately given an i.v. injection of 10[7] C.B pooled ♀ + ♂ viable bone marrow cells. T cells were removed from the marrow by treatment with AKR anti-C3H (anti-Thy 1.2) serum (15 min, 4°C) followed by agarose-absorbed guinea pig complement (45 min, 37°C). 11 of these chimeras were used between 5 wk and 13 mo after reconstitution. A total of 17 chimeras survived for more than 4 mo.

Testing Chimerism. The presence of host cells in chimeric spleen cell suspensions was assayed before in vitro culture with a cytotoxic antiserum (C57BL/6 anti-P815 (H-2[d]) (kindly given us by Joseph Coha, University of California San Diego). Spleen cells were incubated with antiserum and selected nontoxic rabbit complement for 45 min at 37°C. The ratio of viable to dead cells was determined using trypan blue. On each test, positive and negative control cells were included to ensure the specificity and activity of the antiserum.

The H-2 type of the cytotoxic effector cells generated in vitro was tested just before the ⁵¹Cr release assay, using antisera produced, absorbed, and extensively tested in functional assays by Dr. Michael Bevan, Center for Cancer Research, Massachusetts Institute of Technology, Cambridge, Mass. We used his C.B anti-B10.D2(αH-2[d]) and C3H × DBA/2 anti-C.B(αH-2[b]). The test cells were washed, resuspended to 12 × 10[6] viable cells/ml, incubated with antiserum at 4°C for 30 min, centrifuged, and resuspended in guinea pig C for 45 min at 37°C. They were then washed twice, resuspended in equal volumes, and titrated against ⁵¹Cr-labeled targets.

Immunizations. Chimeras were primed by injection of 10[7] viable spleen cells i.p. in Hanks' balanced salt solution and tested in culture from 3 wk to 12 mo later without apparent differences in response.

Cytotoxic Assay. Spleen cells from primed chimeras were cultured for 5 days at a ratio of 4:1 with mitomycin C-treated stimulator spleen cells as described (4), and were then washed once before use in the ⁵¹Cr release assay.

Serial dilutions of such in vitro-boosted cells were then titrated against 4 × 10[4] ⁵¹Cr-labeled concanavalin A (Con A) blasts (4, 16). The percentage of specific release after 3 to 5 h of incubation was calculated as:

$$\% \text{ specific release} = \frac{\text{experimental release (cpm)} - \text{spontaneous release (cpm)}}{\text{total incorporated (cpm)} - \text{spontaneous release (cpm)}} \times 100.$$

Spontaneous release varied from 9.6 to 35.6% of total in different experiments. The greatest differences we saw in spontaneous release for different targets within a single experiment was 8%. When effector:target ratio is given, the effector number is based on the number of primed

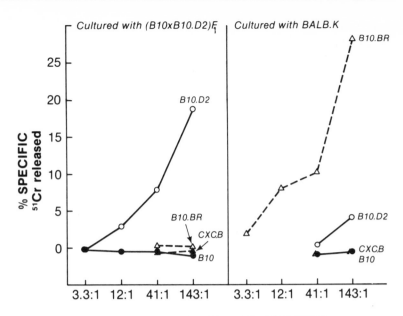

Fig. 1. Cytotoxic activity of C.B → C chimeric spleen cells primed in vivo with D1.C, cultured against (B10 × B10.D2)F₁ or BALB.K (C.K) stimulator spleen cells, and tested on ⁵¹Cr-labeled B10.D2 (○), B10 (●), B10.BR (△), or (C × C.B)F₁ (▲) Con A blasts.

cells originally put into the boosting culture, not viable cells recovered. Recovery varied from 20 to 60% in different experiments.

Results

Can Fully Allogeneic Chimeras Respond to Minor H Antigens?   Spleen cells from a C.B → C chimera, primed with 15×10^6 D1.C spleen cells 5 wk after reconstitution, were checked for the presence of residual host H-2ᵈ cells and were found to contain <1%. The cells were then cultured with (B10 × B10.D2)F₁ or C.K stimulatory cells for 5 days before testing on B10, B10.D2, B10.BR, or (C × C.B)F₁ targets. As illustrated in Fig. 1, the cytotoxic T lymphocytes (CTL) cultured with (B10 × B10.D2)F₁ stimulators lysed B10.D2 targets very well and had no activity on (C × C.B)F₁, B10, or B10.BR targets. Thus these CTL are specific for the combination of minor H + H-2ᵈ. The CTL cultured with C.K lysed B10.BR specifically, showed a slight cross reaction on B10.D2, and had no activity on B10 or (C × C.B)F₁. A similar cross reaction has been noticed before in normal C mice primed against B10.D2 and tested on B10.BR or C.K targets (16), and has been extensively studied elsewhere (17). It appears then that a fully allogeneic chimera can respond to minor H antigens and is tolerant to both marrow donor and host type cells.

Can These Chimeras Respond to Minor H Antigens in Combination with Either H-2ᵇ or H-2ᵈ?   In the previous experiment, cross priming should have led to priming of anti-B10 CTL, which should then have been boosted in the culture with (B10 × B10.D2)F₁ stimulators (16); however, no killing was seen on B10 targets. This could have been explained in three ways. Either (*a*) there really were no CTL precursors capable of responding to B10, (*b*) anti-B10 CTL

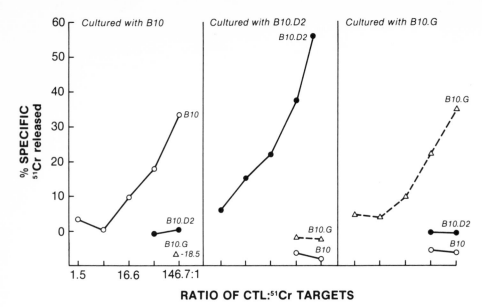

FIG. 2. Cytotoxic activity of C.B → C chimeric spleen cells primed in vivo with D1.C, cultured with B10, B10.D2, or B10.G spleen cell stimulators, and tested on ^{51}Cr-labeled B10 (○), B10.D2 (●), or B10.G (△) Con A blasts.

precursors did exist but cross priming had not occurred, or (c) such CTL did exist and were primed but were such a minority population that they were not boosted well by the (B10 × B10.D2)F₁ stimulators. We therefore decided to boost with B10 and B10.D2 separately to see whether we could raise a population of effector cells against B10.

Spleen cells from a chimera primed with D1.C were checked for the presence of H-2^d positive cells, found to be <1% positive, and then cultured with B10, B10.D2, or B10.G stimulators for 5 days before testing on B10, B10.D2, and B10.G targets. Fig. 2 shows that the CTL cultured with B10 stimulators lysed B10 cells, CTL cultured with B10.D2 lysed B10.D2 cells, and those cultured with B10.G lysed B10.G. The activity seen on B10, although sixfold lower than that on B10.D2 targets, is quite good. It seemed that we had two responding populations in the chimeras primed with D1.C, a large (or very active) set of CTL precursors against B10.D2 and a smaller (or less active) set against B10. This shows that C.B → C chimeric T cells that have grown up under the influence of an H-2^d thymus can nevertheless react against minor H antigens plus H-2^b. It also indicates that the type of boost given may be important. When the population of T cells one is looking for is likely to be small, it may be best to immunize in a way that will expand that population preferentially over other possible responders.

Table I is a summary of results from eight chimeras primed against B10 minor H antigens and boosted in vitro with either (B10 × B10.D2)F₁ or B10 and B10.D2 separately. In lines 1–4 there is virtually no activity on B10 targets whereas B10.D2 targets are lysed very well. From such data we could conclude that these chimeras are only able to respond to minors presented with H-2^d (even though cross priming should have led to activity on B10). However, lines

TABLE I

The H-2 Type of Cells Used to Prime and Boost Influences the Cytotoxic
*Responses Detected in H-2 Incompatible C.B → C Chimeras**

Experiment‡	Primed with	Boosted with	Aggressor: target ratio	Specific ^{51}Cr release	
				B10	B10.D2
75	B10.D2	(B10 × B10.D2)	8:1	0.6	63
75a	B10.D2	"	6:1	3.3	60.4
47	D1.C	"	142:1	−2.0	18.5
67	D1.C	"	50:1	3.6	76.0
62	D1.C	B10	147:1	33.2	0.4
		B10.D2	147:1	−8.0	56.0
59	D1.C	B10	97:1	23.0	−2.4
		B10.D2	49:1	1.5	65.2
67a	B10	(B10 × B10.D2)	50:1	31.6	12.8
62a	B10	B10	147:1	15.0	−1.3
		B10.D2	147:1	−4.6	7.9

* Chimeras were primed in vivo and boosted in vitro (for 5 days) with spleen cells and tested in a 4-h ^{51}Cr release assay. Although each assay was done as a titration, we report only one ratio here for convenience. No significant activity was seen on C, C.B, or (C × C.B)F_1 targets.

‡ Each number represents an individual mouse.

5–11 negate such a conclusion and show that the antigen used to prime or boost can definitely influence the response. Mice primed with minor + H-2^d and boosted with B10 definitely respond to B10, as do mice primed with B10 and boosted with the F_1(B10 × B10.D2). It appears that CTL precursors reactive to B10 do exist and can be seen if their numbers are expanded preferentially either in the priming or boosting immunizations.

Are the Effector Cells Entirely of Donor Origin? Even though we could not detect any residual host cells in the chimeric spleens, the possibility remained that they did exist in small numbers, were expanded during culture, and were responsible for at least some of the activity we saw after 5 days. We therefore killed the effector cells just before the ^{51}Cr release assay with antiserum against H-2^b or H-2^d + C to see whether any activity was due to H-2^d positive cells.

The antisera were previously tested on C anti-C.B and C.B anti-C CTL using exactly the same protocol and reagents used here and were found to eliminate the activity of the appropriate effector cells completely (M. Bevan, personal communication).

Spleen cells from two chimeras that had been primed with B10.D2 7 mo after reconstitution (ample time for the host to regenerate if it were going to) were cultured 6 mo later with (B10 × B10.D2)F_1 or B10.G stimulators. After 5 days, a sample of each culture was assayed on B10, B10.D2, B10.G, C, and C.B targets and both chimeras were found to have some activity specific for B10, excellent activity on B10.D2 and B10.G targets, and none on C or C.B. The remaining B10.D2-boosted CTL were then treated with anti-H-2^b or anti-H-2^d plus C and titrated on B10.D2 targets.

Fig. 3 shows that CTL treated with anti-H-2^d + C were completely active

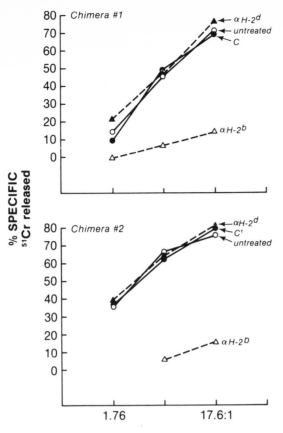

Fig. 3. Effect of treatment with anti-H-2^b or anti-H-2^d serum + C on the ability of C.B → C chimeric CTL to mediate lysis. These CTL were primed in vivo with B10.D2, cultured with B10.D2 stimulators for 5 days, treated with antiserum + C, C alone, or left untreated and then tested for their ability to lyse ^{51}Cr-labeled B10.D2 targets.

whereas those treated with anti-H-2^b + C were no longer able to function. Thus there is no detectable contribution by the irradiated host to the CTL activity we see. All the activity against B10.D2 must be due to H-2^b T cells.

Discussion

We began our experiments to ask whether H-2^b T cells could respond to minor histocompatibility antigens associated with H-2^d in a fully allogeneic chimera, and also to ask whether cross priming occurred in these animals. We found that these chimeras do respond to minor H antigens on both H-2 types, they give allogeneic reactions, they exhibit a type of cross-reactivity seen in normal C and (C × C.B) mice, and they can be cross primed. In every aspect but one they have the same CTL responses as a perfectly normal (C × C.B)F$_1$. The one difference is that, although they have a population of CTL precursors which can be activated against minor H + H-2^b (B10), they do show some preference for minor + H-2^d (B10.D2). This, from the results of Bevan (12) and Zinkernagel et al. (13) would be expected of T cells that have grown up in an H-2^d thymus.

The present results differ from those reported by Zinkernagel et al. in two important and related respects. Firstly, like Bevan, we find a preference in CTL activity toward antigen in association with host H-2 rather than the virtually absolute restriction reported by Zinkernagel et al. Secondly, whereas fully allogeneic chimeras were reported to be immunoincompetent (13), we find that they are completely capable of responding to minor H antigens and to foreign H-2. There are technical differences between our systems that could lead to these disparities.

(a) Zinkernagel et al. looked at an in vivo primary response whereas we assay a secondary response in vitro. Perhaps a boost is required to raise effector cell levels to those that can be detected in a ^{51}Cr release assay. Thus the population capable of responding to B10 in C.B → C chimeras would be undetectable without specific priming or boosting (Table I, Fig. 2).

(b) After inoculation with live virus, host cells must present viral antigen to CTL precursors. A host antigen-presenting system may not be necessary for direct priming against minor H antigens on spleen cells and the priming may be more efficient as a consequence. Thus, a chimera may respond detectably to minor H antigens while it remains unresponsive to viral antigens.

(c) The apparent immunoincompetence of fully allogeneic chimeras in viral systems was explained by postulating that CTL precursors require a helper T cell to be activated (13). In our case, therefore, T helpers, having grown up in an H-2^d thymus, should be restricted to seeing antigen in the context of H-2^d and should not be able to give assistance to the H-2^b repopulating CTL precursors. It is possible that an injection of B10.D2 spleen cells leads to a positive allogeneic effect against the H-2^b CTL, thereby sidestepping any requirement for T helpers. We feel this is unlikely because it was previously shown (16) that an allogeneic effect is neither necessary nor sufficient for priming against minor H antigens. In any case, B10 spleen cells should not be able to give a positive allogeneic effect against the C.B. CTL, and yet they are perfectly able to prime these chimeras.

It is impossible to exclude altogether a role for mature T cells contaminating the marrow inoculum, despite the treatment with anti-Thy-1 + C. In theory, such mature helper T cells, having grown up in an H-2^b thymus, could retain complete restriction to H-2^b and thus provide help for contaminating mature H-2^b CTL precursors which were themselves H-2^b restricted, or for CTL precursors grown up in the host thymus, completely restricted to H-2^d. This interpretation seems unlikely on quantitative grounds, but must be borne in mind when absolute restrictions are not seen.

We feel rather that the restriction imposed by the thymus is profound but not absolute, so that an H-2^d mouse does have some T-cell precursors capable of reacting to antigen + H-2^b (or H-2^q or H-2^s etc.) and that these precursors will be found in all classes of T cells. Therefore, if T helpers are necessary for CTL function, our fully allogeneic chimeras must have an adequate supply of such T helpers.

In summary we have found that congenic, fully H-2 incompatible radiation chimeras, primed in vivo and boosted in vitro with cells bearing minor H differences, produce H-2-restricted cytotoxic cells specific for the minor H antigens. The specificity of restriction can be either to donor or host H-2 and the activity observed depends on the H-2 type of the cells used to prime or boost.

Summary

Fully H-2 incompatible radiation chimeras were prepared using BALB congenic mice. Such chimeric mice were immunized in vivo against histocompatibility antigens of the C57BL/10Sn (B10) background in association with either of the parental H-2 haplotypes, and their spleen cells subsequently boosted in vitro with the same minor antigens. Strong H-2-restricted cytotoxic activity against minor antigens was detected, and the specificity of the restriction could be to the H-2 haplotype of the donor or the host depending on the cells used for priming or boosting. Cross priming could also be demonstrated in these mice. The results show that fully allogenic radiation chimeras can produce H-2-restricted T-cell responses to minor histocompatibility (H) antigens, and are discussed in relation to contrasting results recently obtained against viral antigens.

We thank Richard Dutton for his support and discussion of the data and Rolf Zinkernagel for going over the manuscript. A special thanks to Michael Bevan for his wonderful gifts of antisera, discussions, and helpful hints.

Received for publication 27 February 1978.

References

1. Katz, D. H., T. Hamaoka, and B. Benacerraf. 1973. Cell interactions between histoincompatible T and B lymphocytes. II. Failure of physiologic cooperative interactions between T and B lymphocytes from allogeneic donor strains in humoral response to hapten-protein conjugates. *J Exp. Med.* **137**:1405.
2. Zinkernagel, R. M., and P. C. Doherty. 1974. Activity of sensitized thymus derived lymphocytes in lymphocytic choriomeningitis reflects immunological surveillance against altered self components. *Nature (Lond.).* **251**:547.
3. Shearer, G. M. 1974. Cell-mediated cytotoxicity to trinitrophenyl-modified syngeneic lymphocytes. *Eur. J. Immunol.* **4**:257.
4. Bevan, M. J. 1975. The major histocompatibility complex determines susceptibility to cytotoxic T cells directed against minor histocompatibility antigens. *J. Exp. Med.* **142**:1349.
5. Gordon, R. D., E. Simpson, and L. E. Samelson. 1975. In vitro cell-mediated immune responses to the male specific (H-Y) antigen in mice. *J. Exp. Med.* **142**:1108.
6. Wainberg, M. A., Y. Markson, D. W. Weiss, and F. Doljanski. 1974. Cellular immunity against Rous sarcomas of chickens: preferential reactivity against autochthonous target cells as determined by lymphocyte adherence and cytotoxicity tests in vitro. *Proc. Natl. Acad. Sci. U. S. A.* **71**:3565.
7. Pfizenmaier, K., A. Starzinski-Powitz, H. Rodt, M. Rollinghoff, and H. Wagner. 1976. Virus and trinitrophenel hapten-specific T-cell-mediated cytotoxicity against H-2 incompatible target cells. *J. Exp. Med.* **143**:999.
8. Zinkernagel, R. M. 1976. *H-2* restriction of virus-specific cytotoxicity across the *H-2* barrier: separate effector T-cell specificities are associated with self-H-2 and with the tolerated allogeneic H-2 in chimeras. *J. Exp. Med.* **144**:933.
9. von Boehmer, H., and W. Haas. 1976. Cytotoxic T lymphocytes recognize allogeneic

tolerated TNP-conjugated cells. *Nature (Lond.).* 261:139.

10. von Boehmer, H., L. Hudson, and J. Sprent. 1975. Collaboration of histoincompatible T and B lymphocytes using cells from tetraparental bone marrow chimeras. *J. Exp. Med.* 142:989.

11. Katz, D. H. 1976. The role of the histocompatibility gene complex in lymphocyte differentiation. *Cold Spring Harbor Symp. Quant. Biol.* 41:611.

12. Bevan, M. J. 1977. In a radiation chimera host H-2 antigens determine the immune responsiveness of donor cytotoxic cells. *Nature (Lond.).* 269:417.

13. Zinkernagel, R. M., G. N. Callahan, A. Althage, S. Cooper, P. A. Klein, and J. Klein. 1978. On the thymus in the differentiation of "H-2 self-recognition" by T cells: evidence for dual recognition? *J. Exp. Med.* 3:882.

14. Klein, J. 1973. List of congenic lines of mice. *Transplantation (Baltimore).* 15:137.

15. Staats, J. 1976. Standardized nomenclature for inbred strains of mice: sixth listing. *Cancer Res.* 36:4333.

16. Matzinger, P., and M. J. Bevan. 1977. Induction of H-2 restricted cytotoxic T cells: in vivo induction has the appearance of being unrestricted. *Cell. Immunol.* 33:92.

17. Bevan, M. J. 1977. Killer cells reactive to altered-self antigens can also be alloreactive. *Proc. Natl. Acad. Sci. U. S. A.* 74:2094.

Major histocompatibility complex-linked immune-responsiveness is acquired by lymphocytes of low-responder mice differentiating in thymus of high-responder mice

(generation of diversity/chimera/T-cell receptor)

Harald von Boehmer, Werner Haas, and Niels K. Jerne

Basel Institute for Immunology, 487 Grenzacherstrasse, Postfach 4005 Basel 5, Switzerland

Contributed by Niels K. Jerne, March 7, 1978

ABSTRACT   Female murine T cells can respond to the Y antigen of male cells by generating cytotoxic T-killer lymphocytes. Responsiveness is linked to several *H-2* genes. Two types of low responders can be distinguished: the B10.A(5R) (*H-2 i5*) strain, a low responder because it lacks Y-specific precursor T cells able to differentiate into cytotoxic T-killer cells; and the CBA/J (*H-2 k*) strain, a low responder because it lacks Y-specific T-helper cells able to support differentiation of T-killer cell precursors. B10.A(5R) stem cells differentiating in an x-irradiated (CBA/J × C57BL/6) (*H-2 k × H-2 b*)F$_1$ host respond to Y antigen by generating T-killer cells whereas CBA/J stem cells do not. The results are consistent with the hypothesis that diversity of T-cell receptors is generated by somatic mutation of germ-line genes encoding specificity for self-*H-2*. A detailed account of this hypothesis is presented.

Jerne has proposed (1) that the germ cells of an animal carry a set of genes encoding the combining sites of lymphocyte receptors directed against a complete set of certain major histocompatibility complex (MHC)-encoded antigens of the species to which the animal belongs. One pair of these germ-line genes is expressed on a lymphocyte at an early stage of differentiation. Mutants that recognize foreign antigens arise in lymphocyte clones expressing germ-line genes coding for receptors directed against MHC gene products of the individual itself. The primary lymphoid organs (e.g., the thymus) are viewed as mutant breeding organs.

Recent experiments indicate that cells expressing receptors for MHC antigens are selected by these MHC antigens as expressed on nonlymphoid tissue of the thymus (2). The selected cells may acquire specificity for other antigens while retaining specificity for the selecting MHC antigens—i.e., a T lymphocyte expresses two classes of receptors, one recognizing a MHC antigen and a second recognizing a different antigen.

The MHC antigens encoded by the genome of an animal determine which of its receptor-encoding germ-line genes will be available for mutation. The selection of these mutated genes leads to the generation of a diversity of receptors of the second class that recognize many but not all antigens.

The hypothesis predicts that MHC-linked responsiveness can be acquired by low-responder lymphocytes differentiating in the thymus of high-responder strains. Results consistent with this prediction are described in this report.

MATERIAL AND METHODS

Mice. B10.A(5R) mice were obtained from the Jackson Laboratory (Bar Harbor, ME) and bred at the Basel Institute

for Immunology. CBA/J, C57BL/6, and their F$_1$ hybrids were obtained from the Institut für Biologische-Medizinische Forschung AG (Füllinsdorf, Switzerland). The MHC haplotypes of the mice are given in Table 1, according to David (3).

Chimeras. Chimeras were prepared by injecting lethally irradiated female recipients with anti-Thy 1.2-treated female bone marrow cells from various strains (4). Chimeras prepared by injecting CBA/J bone marrow cells into (CBA/J × C57BL/6)F$_1$ recipients will be referred to as *kkk → (kkk × bbb)*F$_1$ chimeras (see Table 1). In all chimeras described, more than 90% of lymphoid cells were donor-derived as shown by anti-H-2 antibody-mediated cytotoxicity. Two months after bone marrow reconstitution, the chimeras were immunized against Y antigen by intravenous injection of $2 × 10^7$ (*kkk × bbb*)F$_1$ male cells that had been x-irradiated [2000 R (5.2 C/kg)].

Cytotoxic Anti-Male Responses. Between 2 and 8 weeks after priming, spleen cells were cultured for 5 days with x-irradiated male cells and cytotoxic tests were performed on ^{51}Cr-labeled lipopolysaccharide-induced blasts as described (4, 5).

RESULTS

For simplicity, we use the notation given in Table 1 to identify the mouse strains used.

Low-Responder T-Cells Can Acquire Responsiveness in a Chimera. The interaction of murine T cells with other cells is restricted by MHC gene products: T-killer cells are restricted by *K* and *D* region gene products (6, 7) and T-helper cells, by *I* region products (8, 9). Virus-specific and hapten-specific T-killer cells from a (*bbb*) mouse lyse targets expressing either K^b or D^b antigens equally well (10, 11). T-killer cells specific for Y antigen (the test system used in our experiments) from a (*bbb*) mouse lyse only male targets expressing D^b antigens (5, 12). This implies that, on T-killer cells from *bbb* mice, the expression of receptors recognizing Y antigen is linked to the expression of receptors recognizing D^b antigen.

It has been impossible to obtain killer cells recognizing Y antigen in association with either K^b or D^d antigens in all strain combinations tested so far (5, 12). The *b*(*b*)*d* mouse expressing K^b and D^d is therefore a low responder. The (*k*)*k*(*k*) mouse is also a low responder even though T-killer cells lysing male targets expressing either K^k or D^k antigens can be obtained from appropriate hybrids such as ((*k*)*k*(*k*) × *b*(*b*)*b*)F$_1$ (5, 12). Because (*k*)*k*(*k*) mice possess permissive (encircled K^k and D^k) antigens for Y-specific T-killer cells, their low responsiveness is not due

Table 1. MHC haplotypes of mice used

| | | H-2 regions | | | | | | | | | Responder to | Notation used |
Strains	Haplotype	K	A	B	J	E	C	S	G	D	Y antigen	in text*
C57BL/6	$H\text{-}2^b$	b	b	b	b	b	b	b	b	b	+	bbb
CBA/J	$H\text{-}2^k$	k	k	k	k	k	k	k	k	k	−	kkk
B10.A(5R)	$H\text{-}2^{i5}$	b	b	b	k	k	d	d	d	d	−	bbd

* To avoid too cumbersome a notation in the text of this paper, for simplicity we refer to $H\text{-}2^b$ mice expressing K^b, IA^b, and D^b antigens as bbb, to $H\text{-}2^k$ mice as kkk, and to $H\text{-}2^{i5}$ mice as bbd mice. F_1 hybrids of $H\text{-}2^b$ and $H\text{-}2^k$ mice are referred to as $(bbb \times kkk)F_1$ mice. The generation of Y-antigen specific T-killer cell precursors requires the presence, in the thymus, of D^b, K^k, or D^k antigen. The generation of Y-antigen specific T-helper cells requires the presence, in the thymus, of IA^b antigen. In the text, these "permissive" MHC alleles have been encircled—e.g., $b\,ⓑ\,ⓑ$, $ⓚk\,ⓚ$, and $b\,ⓑ\,d$.

to the absence of Y-specific T-killer cell precursors. Recent experiments by Zinkernagel *et al.* (13) suggest that T-helper cells interact with T-killer cells or their precursors expressing appropriate *I* region products. Thus, $ⓚk\,ⓚ$ mice may be low responders because they express nonpermissive I_a^k antigens for Y-specific T-helper cells. One would therefore predict that lymphocyte stem cells from low-responder $ⓚk\,ⓚ$ and $b\,ⓑ\,d$ mice differentiating in an x-irradiated $(ⓚk\,ⓚ \times b\,ⓑ\,ⓑ)F_1$ host would behave differently in their response to Y antigen. $ⓚk\,ⓚ$

T cells from such chimeras should still be unresponsive, because of the lack of Y-specific T-helper cells able to interact with Y-specific T-killer cells expressing nonpermissive I_a^k antigens. On the other hand, $b\,ⓑ\,d$ cells should be responsive, because $b\,ⓑ\,d$ T-killer cell precursors expressing permissive I_a^b antigens for Y-specific T-helper cells and recognizing Y antigen in association with permissive D^b, K^k, and D^k antigens should be generated in the thymus of the recipient.

As shown in Fig. 1, results consistent with this prediction were obtained. Female $ⓚk\,ⓚ$ T cells derived from chimeras $ⓚk\,ⓚ \to (ⓚk\,ⓚ \times b\,ⓑ\,ⓑ)F_1$ could not be induced to lyse male targets (Fig. 1A). In contrast, female $b\,ⓑ\,d$ T cells derived from chimeras $b\,ⓑ\,d \to (ⓚk\,ⓚ \times b\,ⓑ\,ⓑ)F_1$ lysed male $ⓚk\,ⓚ$ as well as male $b\,ⓑ\,ⓑ$ targets (Fig. 1B). T cells from both chimeras responded equally well to allogeneic ddd cells. Female kkk or bbb targets were not lysed. The fact that $b\,ⓑ\,d$ T-killer cells from chimeras lysed male $ⓚk\,ⓚ$ targets better than male $b\,ⓑ\,ⓑ$ targets is in accordance with data by Simpson and Gordon (12) showing that mice immunized with male $(ⓚk\,ⓚ \times b\,ⓑ\,ⓑ)F_1$ cells usually respond better to male $ⓚk\,ⓚ$ cells than to male $b\,ⓑ\,ⓑ$ cells.

Female Responder $(ⓚk\,ⓚ \times b\,ⓑ\,ⓑ)F_1$ Lymphocytes, Differentiating in Low-Responder $ⓚk\,ⓚ$ Parental Thymus, Are Unresponsive to Male Cells. Low responsiveness in the $ⓚk\,ⓚ$ strain is apparently due to the absence of Y-antigen specific T-helper cells. T cells from a $(ⓚk\,ⓚ \times b\,ⓑ\,ⓑ)F_1$ responder $\to ⓚk\,ⓚ$ low responder chimera should only be selected for interaction with nonpermissive I_a^k antigens and consequently not possess Y-antigen specific T-helper cells. Fig. 2 shows that female T cells from such chimeras indeed failed

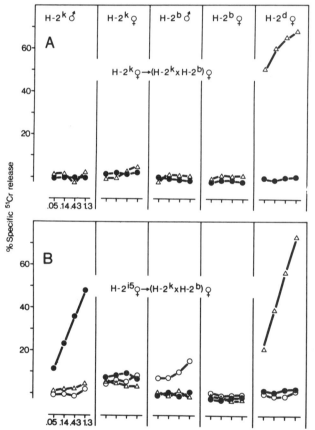

FIG. 1. Cytotoxic responses of T cells from different chimeras against male and female targets of various strains. (A) ♀$ⓚk\,ⓚ \to$ ♀$(ⓚk\,ⓚ \times b\,ⓑ\,ⓑ)F_1$ chimeras were immunized *in vivo* with ♂ $(kkk \times bbb)F_1$ cells and restimulated *in vitro* with either ♂ kkk (●—●) or ♀ ddd (△—△) cells. (B) ♀ $b\,ⓑ\,d \to$ ♀ $(ⓚk\,ⓚ \times b\,ⓑ\,ⓑ)F_1$ chimeras were immunized *in vivo* with ♂ $(kkk \times bbb)F_1$ cells and restimulated *in vitro* with ♂ kkk (●—●), ♂ bbb (○—○), or ♀ ddd (△—△) cells. Numbers on the abscissa indicate the number of responder cells ($\times 10^6$) cultured on day 0, the descendants of which are killer cells tested on 2×10^4 target cells (as indicated) on day 5. Both experiments were repeated twice with cells from other chimeras with similar results. (In this figure, the haplotype notation of Table 1 is used.)

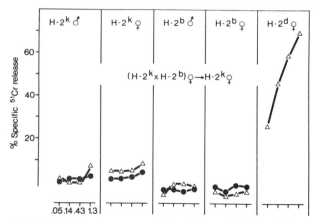

FIG. 2. Cytotoxic responses of T cells from ♀ $(ⓚk\,ⓚ \times b\,ⓑ\,ⓑ)F_1$ $\to$ ♀ $(ⓚk\,ⓚ)$ chimeras against male and female targets of various strains. The chimeras were immunized with ♂ $(kkk \times bbb)F_1$ cells and restimulated *in vitro* with either ♂ $(kkk \times bbb)F_1$ (●—●) or ♀ (ddd) (△—△) cells. For further details see Fig. 1. The experiment was repeated with cells from another chimera with similar results. (In this figure, the haplotype notation of Table 1 is used.)

to respond to male $(k)k(k)$ cells or male $b(b)b$ cells but responded well to allogeneic *ddd* cells.

DISCUSSION

The experiments described here are consistent with Jerne's hypothesis (1) that T cells are selected in the thymus according to their germ-line encoded specificity for H-2 antigens of the species and that diversity is generated by somatic mutation of germ-line genes expressing specificity for H-2 antigens of the individual itself.

For interpretation of the restriction of T-cell interactions with other cells and of other recent findings (2, 14), this hypothesis needs to be augmented by the following rules:

(*i*) A T lymphocyte expresses two classes of receptors, R_0 and R_1, recognizing two different antigens. As a rule, both R_0 and R_1 must recognize antigen displayed on the surface of another cell for cell interactions to occur.

(*ii*) At an initial stage of differentiation, R_0 and R_1 possess identical combining sites. T-helper cell precursors recognize a complete set of certain *I*-region products of the species, whereas T-killer cells recognize a complete set of certain *D*- and *K*-region products of the species.

(*iii*) MHC gene products on thymus epithelium interact with T-cell receptors that recognize these self-antigens. This leads to exhaustive proliferation and to the selection of T cells that express an unaltered R_0 receptor together with a mutant R_1 receptor.

(*iv*) Each of the selected T-helper cells is equipped with two classes of receptors: R_0 recognizing I_a and R_1 recognizing some other antigen. In this way, T-helper cell precursors recognizing a given thymic *I*-region antigen will give rise to a population of T-helper cells whose R_1 receptors recognize a set of foreign antigens. This set does not encompass all foreign antigens, because some may not be recognized by any of the occurring R_1 mutants.

(*v*) Likewise, T-killer cell precursors recognizing a given thymic *K*- or *D*-region antigen will give rise to a population of potential T-killer cells. Each of the cells of this population is equipped with two classes of receptors: R_0 recognizing K and R_1 recognizing an incomplete set of other antigens; or R_0 recognizing D and R_1 recognizing another incomplete set of antigens.

(*vi*) In normal animals, T cells and thymus epithelial cells will have identical genomes and therefore identical MHC encoded antigens. In chimeras the thymus epithelium of recipient animals may possess MHC alleles that differ from the MHC alleles of the lymphocytes arising from the bone marrow stem cells of the donor. The sets of antigens recognized by the R_0 and R_1 receptors of T cells in chimeras will be determined by the MHC of the recipient thymus, and not by the MHC of the donor T cells themselves.

(*vii*) Differentiation of executive lymphocytes (i.e., T-killer cells and antibody-secreting B cells) is supported by T-helper cells (except for allogeneic killer cells and T-independent antigens). These T-helper cells are restricted by MHC: their R_0 receptors must recognize that I_a antigen of the cell with which they interact.

(*viii*) The first requirement for obtaining T-killer cells in a chimera is, therefore, that the *I* allele expressed by the recipient thymus must be identical to the *I* allele expressed on the T-cell precursors of the donor.

(*ix*) With respect to a given foreign antigen, the second requirement is that the R_1 receptor initially recognizing this I product can mutate to a receptor recognizing the given antigen.

(*x*) And, as the third requirement, the T-killer cell precursor whose R_0 receptor recognizes a K or D product must be able, by mutation, to develop an R_1 receptor recognizing the given antigen.

Thus, in order to obtain responsiveness to Y antigen in a chimera, the recipient thymus as well as the donor lymphocytes must express I_a^b antigens. For generation of Y-antigen specific T-killer cell precursors, the thymus must express permissive K or D antigens. We were not able to detect Y-specific killing in mice unless they expressed I_a^b antigens. This is in contrast to findings of Simpson and Gordon (12) demonstrating Y-specific killing in various mice (especially F_1 hybrids) not expressing I_a^b antigens. We have reason to believe that in some of these latter experiments the generation of Y-specific killers was independent of Y-specific T-helper cells.

Our experiments on Y-antigen specific T-cells are in good agreement with data on rejection of male skin grafts: all mouse strains that rapidly reject male skin grafts express I^b antigens (15–17). Also, the $b(b)d$ strain expressing nonpermissive K and D antigens for Y-antigen specific T-killer cells rapidly rejects male skin grafts. We conclude therefore that, in mice that are unable to generate cytotoxic T lymphocytes, I^b restricted T cells reject male skin grafts. This is not surprising because *I* restricted T cells can cause delayed type hypersensitivity, an inflammatory reaction in skin (18). In all other mouse strains tested, there is either no or only late skin graft rejection (15–17), and in our experiments there was no significant induction of cytotoxic T cells. This means that Y-specific T-killer cell precursors cannot mediate rapid male skin graft rejection without Y-antigen specific T-helper cells.

The hypothesis states that, at an initial stage of differentiation, R_0 and R_1 express the same specificity, encoded by a *V* gene or a pair of *V* genes of a certain set. Only those cells in which the R_1 class of receptors mutates are selected for maturation.

T-killer cells and T-helper cells must use a different set of germ-line *V* genes. The occurrence of shared *V* gene allotypes and idiotypes on T- and B-cell receptors suggests that B cells and T cells make use of the same sets of *V* genes (19–21). The repertoire of mature B cells, however, appears to be different from that of T cells because no clear *H-2* linkage of B-cell specificity has been documented. The somatic generation of diversity in B cells may therefore require a selection mechanism different from that operating for T cells.

MHC polymorphism increases the total diversity of T-cell receptors in a population of animals of the same species. All individuals of a species therefore would not be vulnerable to the same calamity. It might be disadvantageous for every individual to express this total diversity because of certain MHC-linked diseases.

1. Jerne, N. K. (1971) *Eur. J. Immunol.* **1**, 1–9.
2. Zinkernagel, R. M., Callahan, G. N., Klein, J. & Dennert, G. (1978) *Nature* **271**, 251–253.
3. David, C. S. (1976) *Transplant. Rev.* **30**, 299–322.
4. von Boehmer, H., Sprent, J. & Nabholz, M. (1975) *J. Exp. Med.* **141**, 332–334.
5. von Boehmer, H., Fathman, C. G. & Haas, W. (1977) *Eur. J. Immunol.* **7**, 443–447.
6. Doherty, P. C., Blanden, R. V. & Zinkernagel, R. M. (1976) *Transplant. Rev.* **29**, 89–124.
7. Shearer, G. M., Rehn, T. G. & Schmitt-Verhulst, A. M. (1976) *Transplant. Rev.* **29**, 222–248.
8. Katz, D. H. & Benacerraf, B. (1975) *Transplant. Rev.* **22**, 175–1975.
9. Sprent, J. (1978) *J. Exp. Med.*, in press.

10. Zinkernagel, R. M. (1976) *J. Exp. Med.* **143**, 437–443.

11. Shearer, G. M., Rehn, T. G. & Garbarino, C. A. (1975) *J. Exp. Med.* **141**, 1348–1364.

12. Simpson, E. & Gordon, R. D. (1977) *Immunol. Rev.* **35**, 59–75.

13. Zinkernagel, R. M., Callahan, G. N., Althage, A., Cooper, S., Streilein, J. W. & Klein, J. (1978) *J. Exp. Med.*, **147**, 897–911.

14. von Boehmer, H., Haas, W. & Pohlit, H. (1978) *J. Exp. Med.*, in press.

15. Bailey, D. W. (1971) *Transplantation* **11**, 426–428.

16. Gasser, D. L. & Silvers, W. K. (1972) *Adv. Immunol.* **15**, 215–247.

17. Gasser, D. L. & Shreffler, D. C. (1974) *Immunogenetics* **1**, 133–140.

18. Miller, J. F. A. P., Vadas, M. A., Whitelaw, J. & Gamble, J. (1976) *Proc. Natl. Acad. Sci. USA* **73**, 2486–2490.

19. Rajewsky, K. & Eichmann, K. (1977) *Contemp. Top. Mol. Immunol.* **7**, 69–112.

20. Binz, H. & Wigzell, H. (1977) *Contemp. Top. Mol. Immunol.* **7**, 113–177.

21. Cosenza, H., Julius, M. H. & Augustin, A. A. (1977) *Immunol. Rev.* **34**, 3–33.

Chapter IX

Regulation of Antibody Synthesis: The Network Theory

Chapter IX

Regulation of Antibody Synthesis: The Network Theory

In considering the synthesis of antibody by the immune system it is apparent that regulation of both the quality and quantity of the antibodies made is necessary to induce effective protection for the organism. In preceeding chapters, we have discussed various aspects of cell interaction that contribute to the regulation of antibody production. Most of this discussion has centered around the functional specialization of cells involved in the immune response; for example, the ability of macrophages to present antigen, the inhibitory action of suppressor T cells, and the stimulatory function of helper T cells. This chapter will consider regulation of the immune response in another way. It will consider the proposition of the network theory that the immune system essentially uses an immune response against immunoglobulin determinants to regulate itself. That is, as certain lymphocytes proliferate in response to an antigenic stimulus, the receptors on those lymphocytes themselves act as an antigenic stimulus to other lymphocytes which then exert an inhibitory influence on the initial population.

The network theory was proposed by Niels Jerne, and this chapter begins with a paper describing the theory (paper 1). The chapter also reviews some developments in immunology which stimulated the formulation of the theory and some experiments that were performed in attempts to test the theory. To date the theory has been neither proven nor disproven, but it has remained as an important conceptual framework for a large number of experiments conducted in the past ten years.

The network theory relies on the proposition that antibodies themselves can act as antigens and provoke the synthesis of anti-antibodies in the host animal (Sirisinha and Eisen, 1971). The features of an antibody molecule that make it immunogenic either to the host or to another member of an inbred strain are exclusively in the variable portions of the heavy and light chains of the molecule (this is somewhat obvious in that the constant portions of a class of antibody molecules are identical). These individual-specific, distinguishing features are referred to as idiotypic determinants or idiotypes. A set of such related determinants present on a set of antibody molecules is collectively called the idiotope. Antibodies that recognize such determinants are designated anti-idiotypic antibodies or <u>anti-idiotypes</u>.[1]

Jerne proposes than an organism responds to an antigenic challenge by the proliferation of lymphocytes specific for the antigen. This results in an increase in the concentration of receptors (immunoglobulins, in the case of the B cell) specific for the antigen. This increase in receptor concentration makes them sufficiently immunogenic to trigger the proliferation of other lymphocytes with receptors specific for the idiotypic determinants on the first set of receptors. Ultimately, the organism will have synthesized antibodies to the antigen and anti-idiotypes to the antibodies. Since the set of anti-idiotypes can in turn stimulate anti-anti-idiotypes, an entire network of interconnected cells and molecules is perturbed.[2] The perturbation reflects both the steady state interaction of all of the idiotypes and anti-idiotypes present prior to antigenic challenge as well as those induced specifically as a consequence of a particular antigenic challenge. If we further assume that some idiotype-anti-idiotype interactions are stimulatory in nature (i.e. provoke the synthesis of yet more antibodies) and others are inhibitory in nature, the system comes to equilibrium following perturbation. An unusual feature of these presumed network interactions is that certain antibodies bearing particular idiotypes may be regulated independently of their ability to interact with the initial challenging antigen. An experimental observation that contributed to the formulation of the network theory is described below.

Oudin and Cazenave (see Oudin and Casenave, 1971) observed that certain antibodies induced by a complex antigen bearing many unrelated immunological determinants could be recognized by the same anti-idiotype. Furthermore, some of the antibodies which were induced did not bind at all to the stimulating antigen, although they did share idiotypic determinants with some portion of the induced antibodies that did bind to antigen. In other words, the elicited antibodies shared idiotypic determinants but not antigen specificity. One explanation for this observation is that the antigen non-binding molecules were stimulated to appear by anti-idiotypic molecules which were themselves stimulated by the idiotypic antibodies responding to the antigen. For this explanation to be correct it is obviously necessary to assume that expression of specific idiotypes is not absolutely correlated to particular antibody specificities.

The observations reported by Hart, Wang, Pawlak and Nisonoff in paper 2 also contributed to the idea of network regulation through idiotype-anti-idiotype interactions. The authors found that administration of anti-idiotypic antibody to an animal prior to antigenic challenge profoundly affected the nature of the subsequently induced antibodies. An unusual immune response was employed to study this phenomenon; a response in which the administration of antigens leads to the synthesis of antibodies a large fraction of which bear a common or cross-reactive idiotype. A more typical immune response is one in which many different idiotypes are synthesized such that no one is represented in large enough proportion to be detected readily. Virtually all of the studies dealing with the analysis of idiotype-anti-idiotype interactions deal with the exceptional cases in which a "dominant" idiotype results upon immunization with a particular antigen. Although we do not know why such responses occur, the resulting predominance of a particular

---

[1]Idiotypic determinants can, in principle, occur on various aspects of the variable regions of the immunoglobulin molecule: hypervariable regions, framework regions, D regions, or J regions. Different molecules may share subsets of idiotopes and different anti-idiotypic sera may detect different subsets of these determinants. Thus, not all anti-idiotypic antibodies, raised against the same population of idiotypic molecules, may be identical (Seppala and Eichmann, 1979; Marshak-Rothstein, et al., 1980).

[2]Jerne does not assign specific regulatory roles within the network to T or B cells. If T and B cells share idiotypes, then both may participate in the network.

set of antibodies facilitates certain kinds of immunological studies. Hart and coworkers found that animals immunized with the hapten azobenzene arsonate produce a significant fraction of antibodies which bear a cross-reactive idiotype. Prior administration of anti-idiotypic antibody to such animals results in the synthesis of large amounts of anti-arsonate antibodies but the usual large fraction (approximately 50%) that bear the idiotypic determinants is absent. It appears, therefore, that exogenous administration of anti-idiotype can influence the production of idiotype during an immune response. Similar observations have been made in other experimental systems in which a particular idiotype dominates the response (see Cosenza and Kohler, 1972, and Eichmann, 1974).

The proposal by Jerne put forward in 1974 has stimulated a great deal of research aimed at confirming or finding experimental support for the network theory. One line of research examines phenomena that can only be explained readily by network theory while another line of investigation deals with limited idiotype-anti-idiotype interactions, which may or may not reflect network-directed interactions. The papers by Eichmann, Coutinho, and Melchers (paper 3) and by Cosenza (paper 4) present research that fits into the former category.

Eichmann et al. report findings similar to those of Oudin and Casenave. They find that the spleen of an animal immunized with a particular antigen contains an increased number of cells which secrete antibody with no specificity for the initial antigen, as well as an increased number of cells secreting antibody which are specific for the initial antigen. These two kinds of antibodies, however, do share a common idiotype. In addition, the ratio of idiotype-bearing, non-antigen binding antibodies to idiotype-bearing, antigen-binding antibodies is the same both before and after antigenic stimulation. Antigen stimulation thus results in increased numbers of both types of molecules. This study supports the prediction of the network theory that antibodies which share idiotypic determinants but not antigen binding specificity could be coordinately regulated by a common anti-idiotype. An alternative point of view is presented by Seppala and Eichmann (1979).

The network theory also predicts that an individual animal will have the capacity to make both idiotypic and anti-idiotypic antibodies and the paper by Cosenza presenta data that makes just this point. When Balb/c mice are immunized with phosphoryl choline (PC) they make first an anti-PC response characterized by the dominant idiotype TEPC 15 and then an anti-TEPC 15, or anti-idiotype, response.

In contrast, the papers that follow employ heterologous anti-idiotypic antibodies which have been raised in rabbits or guinea pigs immunized with murine antibody molecules expressing particular idiotypic determinants. As mentioned above, the studies described in these papers deal with just one aspect of network theory: the consequences of idiotype-anti-idiotype interactions.[4]

The paper by Owen, Ju, and Nisonoff extends earlier work by Hart et al. which showed that administration of a particular anti-idiotypic antibody prior to immunization with antigen results in an immune response in which the idiotype in question does not appear. Owen et al. show that this phenomenon is due to active suppressor T cells that specifically prevent the appearance of idiotype-bearing molecules. This idiotype-specific suppression can be transferred from a suppressed mouse to a naive mouse with T cells. A similar result has been reported by Eichmann and colleagues (Eichmann, 1975).

A further analysis of the idiotype-specific suppressor cells by Owen et al. revealed that these cells could bind to immunoglobulin molecules expressing the appropriate idiotype and not to pre-immune immunoglobulin which contains no molecules expressing the idiotype. This observation is important for two reasons:

---

[4]The paper by Eichmann and Rajewsky, which is included in Chapter VIII, is relevant to this discussion. It demonstrates that anti-idiotypic antibody is capable of priming both T cells and B cells.

it demonstrates that idiotype-anti-idiotype interactions can involve both T and B cells, and that T cells capable of recognizing idiotypic determinants on immuno-globulins can be induced by antibodies capable of recognizing the same idiotypic determinants. It is curious that anti-idiotypic antibodies induce T suppressor cells with anti-idiotypic receptors. What could be the link between the two? It is possible that administration of anti-idiotypic antibodies provokes the synthesis of idiotype-bearing T or B cells which in turn provokes the synthesis of anti-idiotype-bearing suppressor T cells. The question as to why administration of anti-idiotypic antibodies leads to the induction of suppressor cells as opposed to helper T cells is not clear (see Eichmann, 1974).

Some support for this type of interaction is provided by the work of Dietz, et al. These authors report that two types of T cells exist, both of which partici-pate in the suppression of a particular idiotype. They use delayed-type hyper-sensitivity as a measure of T cell function and propose that two kinds of suppres-sor T cells exist: Ts_1 has receptors with idiotypic determinants and Ts_2 has receptors with anti-idiotypic specificity. The authors propose that anti-idiotypic antibodies induce Ts_1, which in turn induce Ts_2. The Ts_2 cell then acts in some yet to be determined way to inhibit the ultimate expression of idiotype. Further-more, if the authors trace this chain of events back far enough, they find that administration of antigen itself (azobenzene arsonate) can result in the suppres-sion of idiotype. Thus it is in principle possible to imagine a chain of events in which antigen induces idiotype which in turn induces anti-idiotype which then induces Ts_1 which finally induces Ts_2. Whether or not such a hypothetical sequence of events actually contributes to immune regulation remains to be seen, but it is interesting at this point to consider the possibilities raised by these studies.

It is obvious that Jerne's idea that idiotype-anti-idiotype interactions contribute to the regulation of the immune response has piqued the curiosity of immunologists. The numerous experiments which the network theory has inspired have certainly contributed to our knowledge about the immune system but have not really offered any definitive test of the overall theory. Like many theoretical papers, its ultimate contribution may in fact lie in the experiments which it has provoked and in its consideration of the possible relationships between the nervous and immune systems. Nonetheless, the paper by Jerne which is included in this chapter is one of the most fascinating papers to read in immunology. The reader is encouraged to reread it and to marvel at the creativity and insight of its author.

Papers Included in this Chapter

1. Jerne, N.K. (1974) Towards a network theory of the immune system. Ann. Immunol. (Inst. Pasteur) 125 C, 373.

2. Hart, D.A., Wang, A.-L., Pawlak, L.L., and Nisonoff, A. (1972) Suppression of idiotypic specificities in adult mice by adminstration of anti-idiotypic antibody. J. Exp. Med. 135, 1293.

3. Eichmann, K., Coutinho, A., and Melchers, F. (1977) Absolute frequencies of lipopolysaccharide-reactive B cells producing A5A idiotype in unprimed, streptococcal A carbohydrate-primed, anti-A5A idiotype-sensitized and anti-A5A idiotype-suppressed A/J mice. J. Exp. Med. 146, 1436.

4. Cosenza, H. (1976) Detection of anti-idiotype reactive cells in the response to phosphorylcholine. Eur. J. Immunol. 6, 114.

5. Owen, F.L., Ju, S.-T., and Nisonoff, A. (1977) Presence on idiotype-specific suppressor T cells of receptors that interact with molecules bearing the idiotype. J. Exp. Med. 145, 1559.

6. Dietz, M.H., Sy, M.-S., Benacerraf, B., Nisonoff, A., Greene, M.I., and Germain, R.N. (1981) Antigen- and receptor-driven regulatory mechanisms. VII. H-2-restricted anti-idiotypic suppressor factor from efferent suppressor T cells. J. Exp. Med. 153, 450.

References Cited in this Chapter

Cosenza, H. and Kohler, H. (1972) Specific suppression of the antibody response by antibodies to receptors. Proc. Natl. Acad. Sci. USA 69, 2701.

Eichmann, K. (1972) Idiotypic identity of antibodies to streptococcal carbohydrate in inbred mice. Eur. J. Immunol. 2, 301.

Eichmann, K. (1974) Idiotype suppression. I. Influence of the dose and of the effector functions of anti-idiotypic antibody on the production of an idiotype. Eur. J. Immunol. 4, 296.

Eichmann, K. (1975) Idiotype suppression. II. Amplification of a suppressor T cell with anti-idiotypic activity. Eur. J. Immunol. 5, 511.

Jacobson, E.B., Herzenberg, L.A., Riblet, R., and Herzenberg, L.A. (1972) Active suppression of immunoglobulin allotype synthesis. II. Transfer of suppressing factor with spleen cells. J. Exp. Med. 135, 1163.

Marshak-Rothstein, A., Benedetto, J., Kirsch, R., and Gefter, M.L. (1980) Unique determinants associated with hybridoma proteins expressing a cross-reactive idiotype. Frequency among individual sera. J. Immunol. 125, 1987.

Oudin, J. and Casenave, A. (1971) Similar idiotypic specificities in immunoglobulin fractions with different antibody functions or even without detectable antibody function. Proc. Natl. Acad. Sci. USA 68, 2616.

Seppala, I.J.T. and Eichmann, K. (1979) Induction and characterization of isogenic anti-idiotypic antibodies to Balb/c myeloma S117: lack of reactivity with major idiotypic determinants. Eur. J. Immunol. 9, 243.

Sirisinha, S. and Eisen, H.N. (1971) Autoimmune-like antibodies to the ligand-binding sites of myeloma proteins. Proc. Natl. Acad. Sci. USA 68, 3130.

TOWARDS A NETWORK THEORY
OF THE IMMUNE SYSTEM

by **N. K. Jerne**

Basel Institute for Immunology, CH-4058 Basel, Switzerland

> The essence of society is repression of the individual.
>
> Norman O. Brown
> (Life against Death)

I. — INTRODUCTION

While listening to this week's colloquium on *the genetics of immunoglobulins and of the immune response*, I have given some thought to the long road along which immunology has travelled since the days of Louis Pasteur. I should like to place my concluding remarks in this perspective, by briefly looking at the past, and at our present situation, and by attempting to predict in what direction immunological theory will develop in the next years.

The development of a science depends on the thoughts and interactions of successive generations of scientists; by their work, these scientists uncover additional phenomena and facts, expanding the territory and the number of problems to be considered. A steady rate of progression of a science therefore requires something like an exponential growth in the number of people engaged. This is what has happened in immunology over the past hundred years. I shall illustrate the growth with some numbers that are rough estimates, and starting with the year 1970, I shall go back in time by steps of twenty years.

The number of research workers in immunology in 1970 was estimated to be about 10,000. Assuming that the number of immunologists in the world has, on the average, tripled every twenty years, there were 3,300 immunologists in 1950, 1,100 in 1930, 360 in 1910, 120 in 1890, and 40 in 1870. I now quite arbitrarily introduce the notion of « outstanding immunologist » and estimate that, on the average, among 40 immunologists there will be one that is « outstanding ». This would mean that at present, or in 1970, there are 250 outstanding immunologists. This number makes sense, I think. In fact, many of these are now present in this room. Going back in time, we would expect to find about 80 outstanding immunologists in 1950, 27 in 1930, 9 in 1910, 3 in 1890 and 1 in 1870. This one, obviously, was Louis Pasteur, the father of immunology, to whom the present colloquium is dedicated.

II. — PROGRESS OF IDEAS

I shall now make an equally brief sketch (summarized in table I) of the way I see the development of ideas in immunology during these successive periods of 20 years, and shall start out with the period 1870-1890 which is the time during which Pasteur has turned his attention to immunology.

TABLE I. — **Progress of ideas.**

Aims	Periods	Pioneers Types of theory	Notions
Application	1870-1890	Pasteur Metchnikoff	Immunization Phagocytosis
Description	1890-1910	Behring Ehrlich	Antibodies Cell receptors
	1910-1930	Bordet Landsteiner	Specificity Haptens
Mechanisms	1930-1950	Sub-cellular	Antibody synthesis Antigen template
	1950-1970	Cellular	Selection Clones
System analysis	1970-1990	Multi-cellular	Network Cooperation and suppression

Practical applications of a science often precede the birth of this science itself. Gun-powder and the compass were invented and used long before chemistry and physics developed into sciences. Before Pasteur, knowledge of immunology was restricted to smallpox vaccination. The efforts of Pasteur were likewise directed towards immediate application of the phenomenon of immunity in preventive medicine. His great contribution was to demonstrate that Jennerian vaccination was only a special case that could be generalized to apply to many other microbial infections. A brilliant generalization such as this was needed and sufficient to create the science of immunology.

The period of Pasteur was followed by about forty years which I would briefly call the early descriptive period. During this time a large range of immunological phenomena were described, outlining, as it were, the territory with which this science would have to deal. Elie Metchnikoff discovered phagocytosis and emphasized cellular aspects, whereas the existence of antibodies in serum was discovered by Behring and Kitasato in 1890, and applied in therapeutic medicine that same year. A dominating figure in the period 1890-1910 was Paul Ehrlich who extended the phenomenon of specific immune responsiveness far beyond the special cases of responses to microbial antigens; he coined the term « horror autotoxicus » to describe the astonishing fact that we do not seem to make antibody to our own tissue components; and he originated the field of tumour immunology. Even the insight that transplantation problems belong to immunology originated in that period (Schöne, 1912).

There was a number of outstanding immunologists in the period 1910-1930 among whom Jules Bordet was one of the greatest. This period brought a further expansion and characterization of the notion of immunological specificity, particularly through the work of Landsteiner. Not only did antibody specificity turn out to be exquisite, but the immune system was shown capable of producing specific antibodies against artificially synthesized chemicals that had never existed in the world.

Though application and description of immunology of course continued, the time in 1930 was ripe for the next period of 40 years during which attention was turned towards discovering the mechanisms underlying the immune response, particularly antibody formation. The theoretical proposals originating during the first 20 years of that period, 1930-1950, were mainly of a sub-cellular type. They focussed on the biosynthesis of antibody molecules. It was clear that antibody molecules were made by cells. It was equally clear that a producing cell could have no prior information regarding the antigen, most obviously so for the case of artificial antigens. The conclusion seemed to be inescapable, that the antigen must bring into the cell information concerning the complementary structure of the antibody molecule.

The following 20 years, 1950-1970, saw the decline of these early antigen-template theories of antibody formation, in favour of selective theories. We might now say that the most obvious reason for this transition lies in the fact that antibody molecules are synthesized like other proteins, and that the template for their amino acid sequence is not antigen but messenger-RNA. Historically, of course, the selective theories emerged before this was known, and before antigen-template mechanisms were proven wrong. The prototype of these theories is, of course, the clonal selection theory of Burnet [4] which is a cellular theory. We are still standing, at least with one leg, in this theoretical period and we try to interpret experimental findings, including those presented at this colloquium, within this theoretical framework. I shall therefore recapitulate some essential points of this cellular hypothesis, in order to evaluate whether it continues to deal satisfactorily with present-day experimental findings.

This will then lead me to a prediction concerning the theories that will be developed in the period 1970-1990 which, I think, will be of a multicellular type of which attempts to deal with the interaction of T cells and B cells represent early beginnings. Though the search for mechanisms at the sub-cellular, cellular and inter-cellular levels will of course continue, I think that emphasis will shift to a structural analysis of the entire immune system.

III. — CLONAL SELECTION

The key postulate of the theoretical framework established in the period 1950-1970 is one of the very few general laws pertaining to the immune system. It is the law that one cell makes only one antibody. This law can be formulated with greater precision. For example, by saying that the immune system contains a population of antigen-sensitive lymphocytes each of which is already committed to the synthesis of antibody molecules of a single specificity and, upon stimulation, will transmit to all the progeny cells of its clone that same single commitment. It can be argued, of course, that the switch from IgM to IgG production within one clone of cells, and the findings reported by Herzenberg [1], that cells presenting IgM receptors can become IgG or IgA producers, and by Pernis and Rowe, that one lymphocyte can simultaneously present IgM and IgD receptors, all indicate that the law one cell → one antibody can refer only to the variable regions of the antibody molecules, or to the expression of v-genes. Furthermore, individual cells of one clone can change their inherited commitment by mutation, for which Milstein [1] brought experimental evidence. Attempts can be made to define the stage differentiation of a cell at which its single commitment becomes fixed. There must obviously be non-committed stem cells which are precursors of antigen-sensitive cells. In spite of such refinements, the law remains generally valid that all antibody molecules synthesized by a single cell and its progeny cells have identical combining sites because all cells of a clone express the same pair of v-genes for heavy and light chains. This law is now supported by many experimental findings, such as the failure of attempts to show that one cell produces antibody molecules of more than one specificity, and the success of removing responsiveness to a given antigen from a suspension of lymphocytes, by removal of a small fraction of precommitted cells that recognize this antigen. It should be noted that the latter finding was found to apply to T cells as well as to B cells. Herzenberg pointed out [1] that the positive demonstration has also been made, namely to retrieve responsiveness to the antigen in the small fraction of cells that were specifically removed.

As Pernis showed, the law one cell → one antibody leads to several predictions.

A) Allelic exclusion. This was said by Todd [1] to be one of the major contributions of allotype research.

B) All antibody-like receptors displayed by a lymphocyte should be identical or at least have identical light chains and identical heavy chain variable regions. Studies by fluorescent labelling of B cell receptors support this implication.

C) All antibodies produced by a single cell and its progeny should have the identical idiotype. The experiments on suppression of particular idiotypes reported at this colloquium by Nisonoff [8, 11, 12] support this implication.

D) Among the lymphocytes of one animal the diversity with respect to precommitment must be equal to the diversity of the antibodies this animal can produce. In other words, the repertoire of antibody specificities and idiotypes available to an animal equals the repertoire of lymphocytes expressing different pairs of v-genes.

This last point will now be conceded by almost everybody, I believe. During the last years of the period 1950-1970 this led to a controversy among immunologists concerning the origin of the v-genes expressed by lymphocytes. The proponents of the germ-line theory briefly argue as follows: all genes present in any somatic cell are copies of genes that were present in the zygote; *ergo*, the v-genes expressed by any lymphocyte are germ-line genes. The proponents of the somatic theories answer: this is right, except that genes can be modified in somatic cells, for example by mutation; they are staggered by the size of the immunological repertoire and do not believe that the number of v-genes required for encoding this repertoire could arise, nor be maintained, in evolution; they, therefore, propose that the zygote contains a small set of v-genes, and that the large set of v-genes expressed by the population of lymphocytes has arisen from the selection of somatic mutant cells. Both Milstein's and Oudin's reports at this colloquium lend some slender experimental support to these ideas [1].

This controversy dominated theoretical immunology during the last years of the 1960's, and all arguments pro et contra have been exhaustively reported in the literature [7]. I shall therefore turn my attention to other matters, except for noting that the clonal selection theory, or the law one cell → one antibody, is compatible with any view of the origin of the v-genes expressed in the lymphocyte population.

IV. — RECENT ADVANCES

Looking back at the papers presented at this colloquium, it strikes me that in 1960 practically nothing was known about the subjects discussed, and that in 1950 most of these subjects did not even exist.

Nothing was known of the primary structure of antibody molecules nor of allotypes and idiotypes; the genetic determination of specific immune responsiveness and recognition had hardly begun to be explored; intracellular antibody synthesis was misunderstood. The account that Askonas presented [1] shows the enormous advances in knowledge of antibody biosynthesis achieved in the past twenty years. What now seems most incredible is that it was hardly even suspected twenty years ago that the small lymphocytes had anything to do with the immune system, whereas we now know that they *are* the immune system, or at least 98 % of it. I still count it as a point in my own favour that I suggested this in 1955, and likewise prematurely proposed antibody diversity to be generated in the cells of the thymus [16]. Experimental proof of the involvement of lymphocytes and thymocytes came only in the 1960's through the work of Gowans, Miller, Nossal, and many others, so that we can now state quite simply that the immune system in an adult person consists of about 10^{12} lymphocytes and about 10^{20} antibody molecules. It is clear that a science that advances as rapidly as immunology requires a continuous revision of its theoretical foundations as well as a continuous expansion of the theoretical framework in order to encompass phenomena that older theories did not have to deal with.

1) Let us start with the primary structure of antibody polypeptide chains. Last year's Nobel Prizes testified to the enormous importance of this structural work to all immunological thinking. I remember the delight with which in 1963 we learned of Hilschmann's and Craig's demonstration of the sequence variability among the 107 N-terminal amino acid residues of two light chains. In one stroke, a structural basis had been provided for a potentially enormous variation in antibody specificity. Porter and Milstein [1] have pointed out the usefulness of further sequence work for establishing genetic markers that can serve in determining allelic polymorphisms in the structural c- and v-genes, for proving that two genes code for one polypeptide chain, and for clarifying the genetic origin of the multiple variable regions. These studies are pertinent to the theories of phylogenic versus ontogenic generation of antibody diversity that we have already briefly mentioned.

2) Serological determinations of allotypes and idiotypes have proven to be useful and illuminating experimental alternatives in the common situations where antibody heterogeneity prohibits the acquisition of sequence data. As Todd pointed out [1] allotype research has established genetic linkages between the structural v- and c-genes for antibodies as well as the allotypic identity of the receptor molecules on a single B cell and of the antibody molecules produced by one cell, thus lending additional support to the law one cell → one antibody. Idiotypes which are determined by the same variable regions that determine antibody combining sites are, in my opinion, of particular importance for understanding immunology, and I consider Oudin's report of the recent work of himself and his colleagues [1] among the most fascinating developments of these years. Mage, Herzenberg and Nisonoff briefly discussed allotype suppression and idiotype suppression, and I intend to return to these essential phenomena in the following sections, where I shall also attempt to reconcile with current immunological concepts the apparently paradoxal findings reported by Oudin [1] that antibodies directed against different determinants on the same antigenic molecule are apt to have similar idiotypes and that immunization results in the parallel production of unspecific immunoglobulin molecules of the same idiotypes as the specific antibodies [5, 6, 26].

3) Another apparent paradox that came to light at the end of the

1950-1970 period was the finding of a genetic determination and simple Mendelian inheritance of specific immune responsiveness. The experimental facts were presented by Biozzi, Benacerraf and McDevitt in this colloquium. So-called Ir-genes that are closely linked to the histocompatibility locus exert a dominant control of the response of T cells to specific antigens, and, through the failure of T cell cooperation, influence the antibody response by B cells to these antigens and to haptens coupled to them [1]. This seems compatible neither with a random somatic generation of antibody diversity, nor with a germ-line theory of antibody diversity, as the Ir-genes are not linked to the structural genes for antibodies. In discussing the implications of these findings, Benacerraf and McDevitt regarded inescapable the conclusion that there exists a class of molecules encoded by these Ir-genes, which are responsible for the recognition of specificity at the T cell level, and that these molecules are not immunoglobulins. A serious difficulty to be dealt with is the large repertoire of T cell specificities which would seem to require a larger number of structural genes than the relatively limited region of the genome, the major histocompatibility complex, can harbour. Other structural genes for lymphocyte surface proteins located in this genome region are the SD-genes (for serologically determined histocompatibility antigens) and LD-genes (for the antigens that determine T cell responses in mixed lymphocyte cultures). These two sets of genes are dominant in determining tolerance, *i. e.* the possession of one allele leads to tolerance of lymphocytes to the product of this allele. We could imagine that the presence of an Ir-product on the T cell surface likewise engenders tolerance towards this product. All that would then be needed is to postulate that the acquisition of tolerance to an Ir-product leads to T cell responsiveness to a set of antigens [18]. The expression of this need, it may be said, is not a scientific theory. But then neither is the expression of the need for an unidentified class of non-immunoglobulin recognition molecules. Benacerraf and McDevitt called their proposal revolutionary and heretical, respectively, and I suggest that this proposal should be resisted as long as other ways of interpretation remain possible, that do not sacrifice the idea that immunological specificity resides solely in antibody molecules.

4) Striking advances have been made in the last few years in characterizing sub-populations of lymphocytes. Gowans, Miller and Nossal presented the latest findings in this population analysis [1]. Nossal produced further evidence for the notion that the law one cell → one antibody applies equally to T- and B- cells, and for the belief that the functional receptor on T lymphocytes is an IgM-like molecule. This view is supported by accumulating experimental evidence [29]. Furthermore, Gowans [1] showed that not all functional B cells have easily demonstrable immunoglobulin receptors, whereas McDevitt [1] found that anti-IgM inhibits antigen binding by T cells. Moreover, IgM bearing T cell lymphomas are known to exist. Nossal and Benacerraf seemed willing to allow for the possibility that T cells could make use of two different recognition systems, one based on immunoglobulins and the other on the Ir-gene products. This kind of compromise, I believe, will not remain tenable. Miller claimed a key role

for the T cell in the immune response and almost relegated the B cell to a class of secondary effector cells. To me, the most striking facts remain the dichotomy of lymphocytes into T cells and B cells, and the demonstration of cooperative as well as repressive interactions between cells of these two populations.

Though the T cell/B cell dichotomy does not directly contradict the theoretical foundations that were established during the period 1950-1970, it certainly shows that this theoretical frame needs widening in order to remain a useful basis for the interpretation of experimental findings in immunology.

V. -- FORMAL NETWORK

I shall now attempt to predict the type of theoretical developments that I foresee for the period 1970-1990. By this, I do not mean to set forth a set of partial hypotheses to explain the signal or the constellation of signals to which an antigen-sensitive lymphocyte either responds or does not respond, nor to explain mechanisms of interaction among and between T cells and B cells, or of interactions between antibody molecules and cells, nor to deal with the effector mechanisms that may become operative as the results of such interactions.

The determination of mechanisms will remain important, but I think there is now a need for a novel and fundamental idea that may give a new look to immunological theory, similar to the impact that the idea of selection had on theoretical developments in the period 1950-1970. Before coming to this, I shall make a few suggestions on terminology and then start out by considering: (I) Repertoires, (II) Dualisms, and (III) Suppression.

Immunological terminology is sometimes clumsy and I now take up a proposal I made many years ago [8], namely to replace the term « antigenic determinant » by « *epitope* », and the term « antibody combining site » by « *paratope* ». An epitope is a patch on an antigen molecule, which presents a certain relief or pattern that can be recognized with various degrees of precision by complementary patterns of paratopes located on antibody molecules. Epitopes and paratopes are two essential features of the immune system enabling it to accomplish its task which is pattern recognition. Two outstanding accomplishments of the recent period are the discoveries of allotypes and idiotypes, or the finding that antibody molecules present epitopes and that these can play a functional role, as experimentally demonstrated by allotype and idiotype suppression. I draw particular attention to idiotypes which were defined by Oudin as the antigenic specificities of antibodies produced by an individual or a group of individuals in response to a given antigen [1]. An idiotype thus is a set of epitopes displayed by the variable regions of a set of antibody molecules. Each single idiotypic epitope I shall call « *idiotope* ». An idiotype then denotes a certain set of idiotopes. The patterns of idiotopes are determined by the same variable regions of antibody polypeptide chains that also determine the paratopes. In short, each Fab arm of an antibody molecule displays one paratope and a small set of idiotopes.

I a) Let us now come to the « *repertoires* », and first consider the repertoire of antibody combining sites or the total number of different paratopes in the immune system. We can here distinguish between the total number of paratopes that could potentially arise in an individual possessing a given genome, and the number of paratopes that are actually present in a given immune system [19]. The « potential » repertoire is obviously far larger than the « available » repertoire which is the number of different paratopes in the immune system of an individual at a given moment. I have previously tried to estimate the size of the available repertoire [16, 17], which may obviously vary in different species of animals and between different individuals. For the moment I shall simply assume that the available repertoire of paratopes in a given immune system is of the order of several millions.

I b) In considering the repertoire of idiotopes, I refer to the experiments of Oudin *et al.* [27, 28] and those of Kelus and Gell [20], in which the idiotypes of antibodies produced by different rabbits to the same bacterial vaccine were studied. These studies, as well as those by Eisen, Kunkel and others on the idiotypes of murine and human myeloma proteins [22, 23] suggest that the repertoire of idiotopes is enormously large. This conclusion is supported by the findings reported by Nisonoff [1] that a given idiotype is exceedingly rare in normal serum globulin. For the moment, I shall simply assume that the repertoires of paratopes and of idiotopes are both of the same order of magnitude. The studies of Askonas, Williamson, Wright and Kreth [3] on the differences in iso-electric point between antibody molecules produced by mice of one inbred strain to the same antigen clearly suggest that the potential repertoires of this mouse strain are far larger than the repertoires that are apparently made use of by a single mouse of this strain. The demonstration that animals can be stimulated to make antibodies to the idiotopes of antibody molecules produced by other animals of the same species or strain, indicates that a given animal possesses available paratopes that can recognize any idiotope occurring in the species. I therefore find it reasonable to assume that, within the immune system of one given individual, any idiotope can be recognized by a set of paratopes and that any paratope can recognize a set of idiotopes. I have tried to show in an earlier paper [19] that this recognition of antibody molecules by other antibody molecules in the same serum will result in only a negligible concentration of complexes if the product of the concentrations of reacting paratopes and idiotopes remains sufficiently small. This proviso can be satisfied by assuming sufficiently large repertoires of paratopes and idiotopes. If we accept these propositions we can conclude that, formally, the immune system is an enormous and complex network of paratopes that recognize sets of idiotopes, and of idiotopes that are recognized by sets of paratopes. I here make the additional assumption that the presence of identical paratopes on two antibody molecules does not necessarily imply that these molecules present identical idiotopes.

II) The second point I need to make is that the immune system displays a number of *dualisms*. The first of these is the occurrence of T lymphocytes and B lymphocytes with partly synergic, partly antagonistic interactions. The second dualism is the one we have just discussed, namely that antibody molecules can recognize as well as be recognized. These properties lead to the establishment of a network, and as antibody molecules occur both free and as receptor molecules on lymphocytes, this network intertwines cells and molecules. The third dualism I draw attention to is the ability of an antigen-sensitive lymphocyte to respond either positively or negatively to a recognition signal. A positive response leads to cell proliferation, cell activation and to antibody secretion, a negative response results in tolerance and suppression. We do not know the precise conditions that determine whether the recognition of an epitope by the paratope of a cell receptor will lead to a positive or a negative response. It may be that the presence of both IgM and IgD receptors on single B cells (described by Pernis and Rowe) is related to this potential dualism of response. For certain foreign epitopes the induction of high zone and low zone tolerance and of stimulation in an intermediate zone have been demonstrated. For the recognition of idiotopes by paratopes we have the findings of Iverson and Dresser [14] that the injection of plain myeloma protein into a mouse leads to tolerance to the idiotype of this immunoglobulin, whereas polymerized or modified myeloma protein can induce an anti-idiotypic antibody response. Thus, a lymphocyte can respond in two ways when the paratopes of its receptor molecules recognize an epitope, or an idiotype.

What happens to a lymphocyte in the reverse situation when the idiotopes of its receptor molecules are recognized by the paratopes of free antibody molecules or by the paratopes of the receptors of another cell ? Most experimental findings suggest that the lymphocyte is then repressed.

III) My third point is to stress the importance of *suppression*. I have become increasingly convinced that the essence of the immune system is the repression of its lymphocytes. Lymphocytes are suppressed by anti-allotypic antibody [24]. As Herzenberg and Jacobson have shown [15], this allotype suppression is maintained by T cells that recognize the allotype of B cell receptors. The situation is similar for the suppressive effect of anti-idiotypic antibody and, by analogy, T cells recognizing the idiotopes of B cell receptors may be assumed likewise to maintain B cell suppression. Conversely, we could conclude that, normally, B cells remain functional because of the absence of sufficient numbers of specific suppressive T cells, or in other words, that B cells can suppress the emergence of recognizing T cells in certain situations. In general, this seems to lead to some dynamic equilibrium or steady state among the elements of the immune system. As tolerance induction shows that B cells and T cells can also be suppressed when the paratopes of their receptors recognize a foreign epitope or an idiotope, the number of negative, suppressive situations appears to exceed the number of positive, stimulatory situations.

VI. — FUNCTIONAL NETWORK

Even if conceding the possibility of viewing the immune system as a formal network, one may question whether the concentrations of individual paratopes and idiotopes are sufficient to establish an internal functional network. Experiments have shown that the concentrations of epitopes needed to induce low-zone tolerance are often far lower than what most of us would estimate to be the average concentration of a single paratope or idiotope in the system. If, for example, we assume a repertoire size of five millions, both for paratopes and for idiotopes, then the average concentration in our blood of antibody molecules representing one member of a repertoire would be 10^{10} per ml, whereas low-zone tolerance has been achieved with epitope concentrations ranging from 10^6-10^{11} per ml [2, 25, 31].

We conclude that the immune system, even in the absence of antigens that do not belong to the system, must display an eigen-behaviour mainly resulting from paratope-idiotope interaction within the system. By its eigen-behaviour the immune system achieves a dynamic steady state as its elements interact between themselves, and as some elements decay and new ones emerge. A complete network theory of the immune system would require greater precision in stating the expected results of the various interactions within the network, and the following exposition is meant only as an illustration of certain interpretative potentialities of such a theory.

Let us start with considering an immunogenic substance which presents epitope E to the immune system. This epitope is recognized (with various degrees of precision) by a set of different antibody combining sites, or paratopes. I call this set p_1 and note that these paratopes occur on antibody and receptor molecules together with certain idiotopes, so that the set p_1 of paratopes is associated with a set i_1 of idiotopes. The symbol $p_1i_1(E)$ denotes the total set of recognizing antibody molecules and potentially responding lymphocytes with respect to epitope E (see fig. 1). Within the network of the immune system, each paratope of the set p_1 recognizes a set of idiotopes, and the entire set p_1 recognizes an even larger set of idiotopes. This set i_2 of idiotopes is, as it were, the « internal image » of epitope E because it is recognized by the same set p_1 that recognizes E. The set i_2 is associated with a set p_2 of paratopes occurring on the molecules and cell receptors of the set $p_2i_2(E)$.

Furthermore, each idiotope of the set $p_1i_1(E)$ is recognized by a set of paratopes, so that the entire set i_1 is recognized by an even larger set p_3 of paratopes which occur together with a set i_3 of idiotopes on antibodies and lymphocytes of the anti-idiotypic set $p_3i_3(E)$. We come in this way to ever larger sets that recognize or are recognized by previously defined sets within the network. As a first gross approximation to a functional network

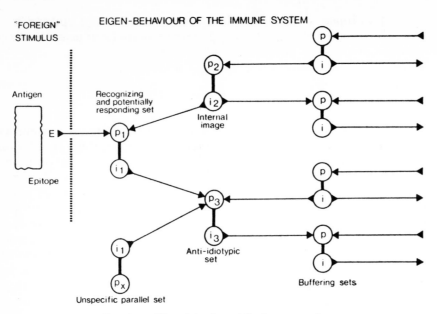

FIG. 1. — *Eigen-behaviour of the immune system.*

The immune system contains a set p_1 of combining sites (paratopes) on immunoglobulin molecules and on cell receptors which recognize a given epitope (E) of an antigen. This recognizing set includes the potentially responding lymphocytes. The molecules of set p_1 also display a set i_1 of idiotopes. Apart from recognizing the foreign epitope, the set p_1 likewise recognizes a set i_2 of idiotopes which thus constitutes, within the immune system, a kind of internal image of the foreign epitope. This set i_2 occurs in molecular association with a set p_2 of paratopes. Likewise, the set i_1 is recognized, within the immune system, by a set p_3 of paratopes which represent anti-idiotypic antibodies. Beside the recognizing set p_1i_1 there is a parallel set p_xi_1 of immunoglobulins and cell receptors which display idiotopes of the set i_1 in molecular association with combining sites that do not fit the foreign epitope. As a first approximation, the arrows indicate a stimulatory effect when idiotopes are recognized by paratopes on cell receptors and a suppressive effect when paratopes recognize idiotopes on cell receptors. Successive groups of ever larger sets encompass the entire network of the immune system.

we could assume that the elements (molecules and cells) of the « internal image » set $p_2i_2(E)$ are largely stimulatory towards the potentially responsive set $p_1i_1(E)$ of lymphocytes, whereas the elements of the anti-idiotypic set $p_3i_3(E)$ are largely inhibitory. In the eigen-behaviour of the immune system these opposing forces result in a balanced suppression which must be overcome in order to obtain an immune response to E.

I should now like to confront this first crude approach towards constructing a *network theory* with a number of phenomena and experimental findings, some of which were reported at this colloquium. The purpose is to show that a network theory might offer alternatives to current explanations.

1) *The immune response.* The elimination by the antigen of natural antibody of the set $p_1i_1(E)$ removes its inhibitory effect on the « internal image » as well as its stimulatory effect on the anti-idiotypic set. Both effects favour an escape from suppression of cells of the set $p_1i_1(E)$.

2) *Regulation.* The enhancement of $p_1i_1(E)$ will tend to reverse the above effects. The network attempts to restore its equilibrium.

3) *Low-zone tolerance.* Only a minimal fraction of natural antibody is removed, but in complex with antigen its idiotopes preferentially stimulate the suppressive anti-idiotypic set $p_3i_3(E)$.

4) *7S-inhibition.* The inhibition of the immune response by small amounts of passively given specific antibody may have to do with the recognition by this antibody of idiotopes of the internal image set $p_2i_2(E)$. The suppression of this stimulatory set may lead to non-responsiveness of $p_1i_1(E)$. This suggestion is reminiscent of the explanation offered by Rowley and Fitch [30] who have proposed that potentially antibody forming cells themselves carry an « image » of the antigen to which their antibody product is directed.

5) *Antigen-competition.* The modulations of the network required for the responses to two antigens may not be simultaneously possible.

6) *The Weigle phenomenon* [7]. Cells of the set $p_1i_1(E)$ remain suppressed when the inhibitory effect of $p_3i_3(E)$ continues to dominate over the stimulatory effect of $p_2i_2(E)$. The breakage of this tolerance could result if cross reacting antigen influences corresponding overlapping sets sufficiently.

7) *The Oudin-Cazenave enigma* [1], namely that antibodies produced against different non cross-reacting epitopes E_1 and E_2 of one antigenic molecule may have similar idiotypes. Considering the magnitudes of the paratope and idiotope repertoires, this finding seems to make no sense within current theoretical concepts, since it would seem to be highly unlikely that two different, independently selected paratopes should occur on antibody molecules of similar idiotype. In a network theory we might seek for an explanation along the following lines. The epitopes E_1 and E_2 are recognized by the two potentially responsive sets $p_1i_1(E_1)$ and $p_1i_1(E_2)$. Though non cross-reactivity implies that these two sets do not overlap with respect to p_1, they may overlap with respect to i_1. This possibility becomes more likely if we assume that a given paratope does not necessarily occur on all molecules in association with the same idiotopes, because then the set i_1 is larger than the set p_1. Furthermore, the larger suppressive sets $p_3i_3(E_1)$ and $p_3i_3(E_2)$ may well overlap with respect to p_3. If we now make the *ad hoc* assumption that such an overlap of paratopes, that these sets $p_3i_3(E_1)$ and $p_3i_3(E_2)$ have in common, will favour the escape from suppression of certain cells of the sets $p_1i_1(E_1)$ and $p_1i_1(E_2)$ which thereby become the major effectors of the antibody responses to both E_1 and E_2, we have explained the Oudin-Cazenave findings without sacrificing the basic concept that an antigen-sensitive lymphocyte is precommitted and does not recognize an entire antigen molecule but only one epitope on this molecule.

8) *Unspecific response.* The production of unspecific immunoglobulin after an immunogenic stimulus has often been observed. Our chairman

this afternoon, Dr Humphrey, has devoted several studies to this problem [13]. How can we explain the report by Oudin [1] that such unspecific immunoglobulin molecules were found to be similar in idiotype to the specific antibodies [5] ? If we consider epitope E and the potentially responding set of lymphocytes that belong to the set $p_1i_1(E)$ we may admit that certain idiotypes of this set i_1 occur also on molecules which have paratopes that do not belong to the set p_1. Let us denote this parallel set by $p_xi_1(E)$. The release of lymphocytes of set $p_1i_1(E)$ from suppression may at the same time release from suppression lymphocytes of the parallel set displaying the same idiotopes. These cells will then produce globulin molecules of this idiotype but with combining sites of set p_x which are « unspecific » as they do not fit the immunizing epitope E.

The weakness of this incipient network theory lies in its lack of precision. The formal sets p_1i_1, p_2i_2, p_3i_3, etc., refer to sets of antibody molecules that can occur both as freely circulating antibodies of all immunoglobulin classes and as functional receptors on T cells and B cells. This leaves an ambiguity in the answer to the question whether the relations between two sets is suppressive or stimulatory, or partly one and partly the other, and thus permits us to postulate interactions that suit our explanatory needs. The properties of all possible different types of interaction must therefore be clarified in order to be able to understand and eventually to manipulate the immune system. This appears to be the immense task facing immunology in the period ahead of us. Another question that needs consideration is the ontogenic development of the network in the adult immune system from a small germ-line determined, embryonic set of elements. We suspect that the initial shaping of the immune network is influenced to a large extent by the genes that cluster around the histocompatibility locus, and that this view will yield a more plausible explanation of the findings reported by Benacerraf and McDevitt [1] than postulating an entirely new second recognition system based on molecules that are not immunoglobulins. Partly because of polymorphisms in the cluster of genes at the histocompatibility locus, and partly because of a certain randomness in the ontogenic emergence of new paratopes and idiotopes, each individual will develop a different immune network. When established, the immune network must possess quite stable features. Early modulations of the network must be hard to undo later. This may be relevant to Fazekas de St Groth's concept of the original antigenic sin [9] and it suggests that immunological memory may be more firmly deposited in a persistent network modulation than in the persistence of a population of so-called « memory cells ».

I cannot pursue these matters further within the frame of my concluding remarks to this colloquium, except for saying that the clonal selection theory has regarded antigen-sensitive lymphocytes as independent cells among which selection operates in a way that is similar to the selection of particular bacterial clones in a population of bacteria. Our proposal incorporates the clonal selection theory into the wider frame of a network, without sacrificing it.

VII. — THE IMMUNE SYSTEM AND THE NERVOUS SYSTEM

In finishing, I should only like to point out that the immune system, when viewed as a functional network dominated by a mainly suppressive Eigen-behaviour, but open to stimuli from the outside, bears a striking resemblance to the nervous system. These two systems stand out among all other organs of our body by their ability to respond adequately to an enormous variety of signals. Both systems display dichotomies and dualisms. The cells of both systems can receive as well as transmit signals. In both systems the signals can be either excitatory or inhibitory. The two systems penetrate most other tissues of our body, but they seem to be kept separate from each other by the so-called blood-brain barrier. The nervous system is a network of neurons in which the axon and the dendrites of one nerve cell form synaptic connections with sets of other nerve cells. In the human body there are about 10^{12} lymphocytes as compared to 10^{10} nerve cells. Lymphocytes are thus a hundred times more numerous than nerve cells. They do not need connections by fibres in order to form a network. As lymphocytes can move about freely, they can interact either by direct encounters or through the antibody molecules they release. The network resides in the ability of these elements to recognize as well as to be recognized. Like for the nervous system, the modulation of the network by foreign signals represents its adaptation to the outside world. Early imprints leave the deepest traces. Both systems thereby learn from experience and build up a memory that is sustained by reinforcement and that is deposited in persistent network modifications, which cannot be transmitted to our offspring.

These striking phenotypic analogies between the immune system and the nervous system may result from similarities in the sets of genes that govern their expression and regulation.

KEY-WORDS: Immun system Immunoglobulin, Lymphocyte. Genetics, Immunogenesis; Network, Clonal selection, Eigen-behaviour, idiotype.

*
* *

Finally, I am sure to speak for all participants in thanking the organizors of this colloquium, and particularly Professor Jacques Oudin, for having brought us together at this memorable occasion.

REFERENCES

[1] *This Colloquium.*

[2] ADA, G. L. & PARISH, C. R., Low zone tolerance to bacterial flagellin in adult rats: a possible role for antigen localized in lymphoid follicles. *Proc. nat. Acad. Sci.* (Wash.), 1968, **61**, 556.

[3] ASKONAS, B. A., WILLIAMSON, A. R. & WRIGHT, B. E. G., Selection of a single antibody-forming cell clone and its propagation in syngeneic mice. *Proc. nat. Acad. Sci.* (Wash.), 1970, **67**, 1389.

[4] BURNET, F. M., The clonal selection theory of acquired immunity. Cambridge University Press, 1959.

[5] CAZENAVE, P.-A., L'idiotypie comparée des anticorps qui, dans le sérum d'un lapin immunisé contre la sérumalbumine humaine, sont dirigés contre des régions différentes de cet antigène. *FEBS Letters*, 1973, **31**, 348.

[6] CAZENAVE, P.-A. & OUDIN, J., L'idiotypie des anticorps qui, dans le sérum d'un lapin immunisé contre le fibrinogène humain, sont dirigés contre deux fragments distincts de cet antigène. *C. R. Acad. Sci.* (Paris) (Sér. D), 1973, **276**, 243.

[7] COHN, M., The take-home lesson. *Ann. N. Y. Acad. Sci.*, 1971, **190**, 529.

[8] COSENZA, H. & KOHLER, H., Specific suppression of the antibody response by antibodies to receptors. *Proc. nat. Acad. Sci* (Wash.), 1972, **69**, 2701.

[9] FAZEKAS de ST GROTH, S. & WEBSTER, R. G., Disquisitions on original antigenic sin. I. Evidence in man. *J. exp. Med.*, 1966, **124**, 331-345 & II. Proof in lower creatures. *Ibid.*, 346-361.

[10] HANNESTAD, K., KAO, M. S. & EISEN, H. N., Cell-bound myeloma proteins on the surface of myeloma cells: potential targets for the immune system. *Proc. nat. Acad. Sci.* (Wash.), 1972, **69**, 2295.

[11] HART, D. A., WANG, A. L., PAWLAK, L. L. & NISONOFF, A., Suppression of idiotypic specificities in adult mice by administration of antiidiotypic antibody. *J. exp. Med.*, 1972, **135**, 1293.

[12] HOPPER, J. E. & NISONOFF, A., Individual antigenic specificity of immunoglobulins. *Advanc. Immunol.*, 1971, **13**, 57.

[13] HUMPHREY, J. H., *in* « La tolérance acquise et la tolérance naturelle à l'égard de substances antigéniques définies (Colloque international du CNRS, n° 116, p. 402). Éditions du CNRS, Paris, 1963.

[14] IVERSON, G. M. & DRESSER, D. W., Immunological paralysis induced by an idiotypic antigen. *Nature* (Lond.), 1970, **227**, 274.

[15] JACOBSON, E. B., HERZENBERG, L. A., RIBLET, R. & HERZENBERG, L. A., Active suppression of immunoglobulin allotype synthesis. II. Transfer of suppressing factor with spleen cells. *J. exp. Med.*, 1972, **135**, 1163.

[16] JERNE, N. K., The natural selection theory of antibody formation. *Proc. nat. Acad. Sci.* (Wash.), 1955, **41**, 849.

[17] JERNE, N. K., Immunological speculations. *Ann. Rev. Microbiol.*, 1960, **14**, 341.

[18] JERNE, N. K., The somatic generation of immune recognition. *Europ. J. Immunol.*, 1971, **1**, 1.

[19] JERNE, N. K., What precedes clonal selection ? *in*, « Ontogeny of acquired immunity » (Ciba Foundation Symposium 1971), Associated Scientific Publ., Amsterdam, 1972.

[20] KELUS, A. S. & GELL, P. G. H., Immunological analysis of rabbit anti-antibody systems. *J. exp. Med.*, 1968, **127**, 215.

[21] KRETH, H. W. & WILLIAMSON, A. R., The extent of diversity of anti-hapten antibodies in inbred mice: anti-NIP (4 hydroxy-5-iodo-3-nitro-phenacetyl) antibodies in CBA/H mice. *Europ. J. Immunol.*, 1973, **3**, 141.

[22] KUNKEL, H. G., Individual antigenic specificity, cross specificity and diversity of human antibodies. *Fed. Proc.*, 1970, **29**, 55.

[23] LYNCH, R. G., GRAFF, R. J., SIRISINHA, S., SIMMS, E. S. & EISEN, H. N., Myeloma proteins as tumor specific transplantation antigens. *Proc. nat. Acad. Sci.* (Wash.), 1972, **69**, 1540.

[24] MAGE, R. G., Altered quantitative expression of immunoglobulin allotypes in the rabbit. *Current Top. Microbiol. Immunol.*, 1973 (sous presse).

[25] MITCHISON, N. A., Induction of immunological paralysis in two zones of dosage. *Proc. roy. Soc. B*, 1964, **161**, 275.

[26] OUDIN, J. & CAZENAVE, P.-A., Similar idiotypic specificities in immuno-globulin fractions with different antibody functions or even without detectable antibody function. *Proc. nat. Acad. Sci.* (Wash.), 1971, **68**, 2616.

[27] OUDIN, J. & MICHEL, M., Une nouvelle forme d'allotypie des globulines γ du sérum de lapin apparemment liée à la fonction et à la spécificité anticorps. *C. R. Acad. Sci.* (Paris), 1963, **257**, 805.

[28] OUDIN, J. & MICHEL, M., Sur les spécificités idiotypiques des anticorps de lapin anti-*S. typhi. C. R. Acad. Sci.* (Paris) (Sér. D), 1969, **268**, 230-233.

[29] ROELANTS, G., FORNI, L. & PERNIS, B., Bloking and redistribution (« catting ») of antigen receptors on T and B lymphocytes by anti-immunoglobulin antibody. *J. exp. Med.*, 1973, **137**, 1060.

[30] ROWLEY, D. A. & FITCH, F. W., *in* « Regulation of the antibody response » (B. Cinader) (p. 127). Charles C. Thomas, Springfield (Illinois), 1968.

[31] SHELLAM, G. R. & NOSSAL, G. J. V., Mechanism of induction of immunological tolerance. IV. The effects of ultra-low doses of flagellin. *Immunology*, 1968, **14**, 273.

SUPPRESSION OF IDIOTYPIC SPECIFICITIES IN ADULT MICE BY ADMINISTRATION OF ANTIIDIOTYPIC ANTIBODY*

By DAVID A. HART,‡ Ai-LAN WANG, LAURA L. PAWLAK, AND ALFRED NISONOFF

(*From the Department of Biological Chemistry, University of Illinois at the Medical Center, Chicago, Illinois 60612*)

(Received for publication 31 January 1972)

There is a low frequency of idiotypic cross-reactions among antibodies of the same specificity from different outbred rabbits (1–4). When such cross-reactions do occur they are generally weak and involve a small proportion of the antiidiotypic antibody population (5, 6). This generalization applies also to rabbits belonging to the same family group (2, 7). Contrasting results were reported by Eichmann and Kindt (8), who investigated a family of rabbits consisting of a brother-sister mating pair and their F_1 and F_2 descendants. The mating pair was chosen on the basis of the capacity of both rabbits to produce high titers of antistreptococcal antibody of restricted heterogeneity. Antistreptococcal antibodies elicited in several members of the family group exhibited strong idiotypic cross-reactivity.

In mice of an inbred strain, idiotypic cross-reactions are more common. They have been observed among BALB/c myeloma proteins with antibody activity directed to the phosphoryl choline hapten group (9, 10). These same myeloma proteins also share idiotypic determinants with antiphosphoryl choline antibodies induced by immunization of some BALB/c mice with pneumococci (9). Strong intrastrain idiotypic cross-reactions have also been observed among antibenzoate antibodies of all BALB/c or A/J mice tested so far and among antiphenylarsonate antibodies of A/J mice (11). Examples of both strong and weak interstrain cross-reactivity were also noted (11). One cannot, however, make the generalization that the antibodies of all mice of the same strain exhibit idiotypic cross-reactions, since we have failed to detect strong cross-reactions, with two antiidiotypic antisera prepared so far, among antiphenyl-arsonate antibodies of C57/BL mice.

The existence of extensive cross-reactions within a strain makes it possible to test the effect in vivo of antiidiotypic antiserum on the immune response of isologous mice. We report here that the formation of antiphenylarsonate antibodies of a particular idiotype in adult A/J mice, in response to the administration of keyhole limpet hemocyanin (KLH)–azophenylarsonate,[1] is greatly sup-

---

* This work was supported by grants from the National Institutes of Health (AI-06281 and AI-10220).

‡ Supported by US Public Health Service Postdoctoral Fellowship AI-49223 FSG.

[1] *Abbreviations used in this paper:* D antibody, specifically purified anti–*p*-azophenylarsonate antibody from mouse 413 (this antibody was used to elicit antiidiotypic (anti-D) antibody in a recipient rabbit); KLH, keyhole limpet hemocyanin; RIgG, rabbit IgG.

pressed by prior administration of antiidiotypic antiserum directed toward the antihapten antibody. This suggests an alternative method for selectively inhibiting the formation of a particular antibody population without generalized suppression of the immune response.

Materials and Methods

Immunization of Mice.—Methods for the preparation of antigens used for immunization and testing have been described (11). Anti–*p*-azophenylarsonate antibodies used to elicit antiidiotypic antibody were produced in mouse 413 by immunization with KLH–*p*-azophenyl-arsonate. The protocol for immunization of the mice used in suppression experiments is given below.

Antiidiotypic Antiserum Directed to Anti–p-Azophenylarsonate Antibodies of an A/J Mouse.—The antiidiotypic antiserum used in these experiments was from the same pool as that employed in previous experiments (11); it was prepared in a rabbit against anti–*p*-azophenyl-arsonate antibody of A/J mouse 413. Evidence that the rabbit antiserum, after absorption with mouse immunoglobulin, was specific for idiotypic determinants was described in detail. The mouse immunoglobulin fraction that was used for absorption of the rabbit antiserum was prepared from normal A/J serum by two precipitations with sodium sulfate at final concentrations of 18% and 14%, respectively. For brevity, the immunogen (purified antihapten antibody from mouse 413) will be designated D and the antiidiotypic antiserum, anti-D. Antiphenylarsonate antibodies from other A/J mice cross-reacted extensively with the absorbed antiserum (11).

^{125}I-labeled, specifically purified anti–*p*-azophenylarsonate antibodies from mouse 413 (^{125}I-labeled D) were prepared as described, and tested in an indirect assay system (1i), which comprised 0.01 μg of ^{125}I-labeled D, 5 μl of a 10:1 dilution of the absorbed rabbit anti-D, and 30 μl of goat antiserum specific for the Fc fragment of rabbit IgG; the goat antiserum had been absorbed with A/J mouse serum. The anti-D antibody was present in slightly less than an optimal amount; the goat anti-Fc was present in a small excess over the amount needed for maximal precipitation, as shown by preliminary experiments. Immune precipitates were washed three times after standing overnight, and the percentage of radioactivity precipitated was determined. Unlabeled inhibitory sera, when present, were mixed with the antiidiotypic antiserum before addition of the ^{125}I-labeled D.

Immunosuppression Experiments.—An IgG fraction was prepared from the absorbed rabbit anti-D antiserum by two precipitations with sodium sulfate (18 and 14%) followed by chromatography on diethylaminoethyl (DEAE)-cellulose (12). The rabbit IgG was concentrated to 10 mg/ml, dialyzed against 0.1 M NaCl-phosphate buffer, pH 7.2, and sterilized by passage through a Millipore filter (Millipore Corp., Bedford, Mass.). Rabbit IgG from pooled normal serum was prepared and sterilized by the same procedure.

Normal 8-wk-old A/J mice (Jackson Laboratory, Bar Harbor, Maine) were injected intraperitoneally with 1 mg of the IgG fraction of anti-D antiserum (group 1) or with 1 mg of nonspecific rabbit IgG (group 2). There were nine mice in each group. 3 days later the injections were repeated. After 2 wk all animals were challenged intraperitoneally with 500 μg of KLH–*p*-azophenylarsonate in complete Freund's adjuvant. 2 wk later (wk 4) a second, identical injection was given; and the animals were bled retroorbitally 1 wk later (wk 5). Additional injections of 500 μg of antigen in incomplete Freund's adjuvant were given at wk 6 and 10; a bleeding was taken 1 wk after each inoculation. Each serum was tested by Ouchterlony analysis for its reactivity with KLH (the carrier protein) and with RIgG–*p*-azophenylarsonate. The serum of each bleeding was also analyzed for its capacity to inhibit the indirect precipitation of ^{125}I-labeled D from mouse 413 with its antiidiotypic antiserum.

Each serum was also tested by Ouchterlony analysis against rabbit IgG to determine whether the 2 mg of IgG used for suppression had elicited antibodies. No precipitin lines were observed, presumably reflecting the fact that the two injections of rabbit IgG were given only 3 days apart and without adjuvant.

All of the 18 mice in the two groups survived the experiment, except for one mouse in the control group that died after the wk-7 bleeding.

RESULTS

All of the mice in both the control and suppressed groups produced precipitating antibodies against the carrier protein, KLH, and against the azophenylarsonate hapten group; the latter was shown by precipitation with RIgG-azophenylarsonate. These precipitating antibodies were present in each of the bleedings taken, i.e., at wk 5, 7, and 11 after the injection of normal or anti-idiotypic rabbit IgG. Tests of precipitation were done by the Ouchterlony

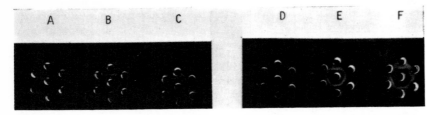

FIG. 1. Ouchterlony analysis of antisera to KLH–p-azophenylarsonate of a nonsuppressed mouse (C-2), patterns A, B, and C, and a suppressed mouse (S-1), patterns D, E, and F. Antiserum, taken 11 wk after injection of rabbit IgG (or 9 wk after the start of immunization), is in the center well; the test antigen is in the outer wells. In patterns A and D the antigen is RIgG; in B and E, RIgG–p-azophenylarsonate; in C and F, KLH (the carrier protein). Antigen concentrations, reading clockwise from the top well, are: patterns A, B, D, and E: 2, 1.5, 1, 0.5, 0.3, and 0.1 mg/ml; patterns C and F: 3, 2, 1.5, 1, 0.5, and 0.1 mg/ml.

method. A typical pattern is shown in Fig. 1. Although there were differences in the strength of the reactions among individual mice, the average antibody titer did not differ markedly among the two groups of mice. Photographs were taken of all Ouchterlony plates, and the distances of precipitin lines from the antigen and antibody wells were correlated with antigen concentration. By this rough criterion there was no significant difference between the antihapten and anti-KLH titers in the two groups of mice. Thus, there was no generalized suppression of the immune response.

The results in Table I demonstrate that the group of mice preinoculated with the antiidiotypic (anti-D) antibodies produced far less antiphenylarsonate antibody capable of displacing the ^{125}I-labeled antibody of mouse 413 from its antiidiotypic antibodies. 10 μl of antiserum from the wk-11 bleeding of each suppressed mouse caused less inhibition than 0.3 μl of antiserum from any nonsuppressed mouse. Thus the degree of idiotypic suppression exceeded 97% in each mouse. It appears, however, that suppression was not complete, since a few sera from suppressed mice had a small but significant inhibitory capacity.

This did not appear to increase as a function of time of immunization. The investigations are continuing with the surviving mice.

In previous work (11) all immune sera were shown to lose their inhibitory capacity entirely upon adsorption with Sepharose (Pharmacia Fine Chemicals Inc., Piscataway, N. J.) coupled to RIgG–p-azophenylarsonate, but not with

TABLE I

*Inhibition of Binding of ^{125}I-Labeled Antiphenylarsonate Antibody from Mouse 413 to Its Rabbit Antiidiotypic Antiserum**

Mouse No.	Pre-immune serum (10 µl)	^{125}I-labeled purified D antibody bound (% of control)											
		Wk-5 bleeding‡				Wk-7 bleeding‡				Wk-11 bleeding‡			
		µl of serum tested as inhibitor											
		0.03	0.3	3	10	0.03	0.3	3	10	0.03	0.3	3	10
Controls§													
C-1	95 (2)	87 (2)	26 (0)	23 (3)		75 (1)	13 (2)	6 (4)	8 (1)				
C-2	100 (2)	95 (2)	76 (7)	37 (0)		93 (6)	38 (0)	26 (7)	10 (1)	92 (4)	56 (0)	23 (2)	
C-3	95 (2)	77 (3)	45 (3)	22 (0)		72 (2)	26 (3)	14 (3)	10 (2)	85 (6)	25 (2)	7 (3)	
C-4	95 (2)	84 (2)	41 (0)	28 (0)		73 (10)	29 (1)	8 (3)	14 (1)	92 (1)	41 (0)	10 (3)	
C-5	102 (1)	98 (5)	79 (2)	28 (1)		86 (3)	41 (7)	20 (4)	10 (0)	95 (6)	34 (1)	17 (1)	
C-6	95 (1)	88 (0)	59 (0)	25 (0)		82 (5)	44 (2)	21 (2)	10 (2)	92 (0)	36 (4)	12 (1)	
C-7	102 (2)	84 (3)	52 (1)	28 (1)		83 (0)	38 (1)	17 (1)	14 (4)	67 (7)	22 (1)	8 (1)	
C-8	98 (1)	78 (5)	50 (5)	30 (3)		72 (5)	36 (1)	17 (1)	11 (2)	80 (3)	22 (3)	11 (3)	
C-9	98 (1)	80 (1)	48 (0)	29 (3)		75 (8)	53 (2)	28 (2)	13 (5)	99 (5)	49 (4)	12 (1)	
Suppressed‖													
S-1	91 (6)			95 (0)	87 (5)			74 (2)	75 (2)			82 (2)	70 (4)
S-2	97 (1)			95 (1)	88 (1)			89 (3)	86 (2)			100 (1)	90 (2)
S-3	102 (3)			90 (0)	80 (3)			82 (1)	75 (2)			90 (1)	84 (1)
S-4	96 (2)			101 (2)	N.D.			84 (8)	88 (2)			91 (2)	92 (1)
S-5	100 (0)			92 (2)	81 (3)			77 (2)	75 (1)			77 (1)	75 (2)
S-6	100 (0)			93 (1)	82 (3)			83 (4)	88 (3)			93 (3)	94 (6)
S-7	100 (0)			90 (1)	88 (2)			86 (6)	74 (1)			85 (0)	84 (1)
S-8	98 (1)			96 (0)	86 (2)			89 (2)	87 (2)			84 (6)	78 (1)
S-9	109 (3)			87 (0)	77 (3)			83 (2)	80 (2)			93 (2)	82 (1)

* Tests were carried out as described under Materials and Methods. In the control for a given set of experiments, the same amount of pooled normal A/J serum was substituted for the sera tested as inhibitors. In the absence of inhibitor, an average of 54.8% of the labeled D was precipitated (corrected for 5.5% precipitation, which occurred when rabbit antiovalbumin was substituted for anti-D). Experiments were done in duplicate; the average deviations are shown in parentheses.

‡ Number of weeks after the injection of normal or antiidiotypic rabbit IgG.

§ Preinjected with 2 × 1 mg of normal rabbit IgG.

‖ Preinjected with 2 × 1 mg of rabbit antiidiotypic IgG.

unconjugated Sepharose. This finding was confirmed in the present study by testing sera of the wk-11 bleeding of three mice in the control group. In each case all inhibitory capacity was lost upon treatment with the immunoadsorbent.

Sera from both the control and suppressed groups reacted strongly with monospecific goat antisera to mouse IgG1 and IgG2.

DISCUSSION

It is evident that antibodies having idiotypic specificities shared with the antiphenylarsonate antibodies of mouse 413 (D antibody) were almost com-

pletely suppressed in the sera of the group of nine A/J mice that had been
treated with 2 mg of an IgG fraction of rabbit antiidiotypic antiserum (anti-D)
2 wk before the 9-wk course of immunization with KLH–*p*-azophenylarsonate.
In contrast, hyperimmune sera of each of nine control mice of the same strain,
preinoculated with nonspecific rabbit IgG, contained antibodies that reacted
with nearly all of the antiidiotypic antibody population, as evidenced by inhibition of the indirect precipitation of ^{125}I-labeled D. These inhibitory antibodies
were present in bleedings taken 3, 5, and 9 wk after the start of immunization
(or 5, 7, and 11 wk after the injection of rabbit IgG). Immune adsorption of
three of these control sera showed that, as in previous studies (11), the inhibition was caused by antiphenylarsonate antibodies. All preimmune sera were
noninhibitory. In these and other experiments we have tested antiphenylarsonate antisera from a total of more than 20 hyperimmunized A/J mice; each
antiserum displaced at least 75 % of ^{125}I-labeled D of mouse 413 from its anti-
idiotypic antiserum.

Treatment of the experimental group with anti-D did not prevent the appearance of antibodies to the carrier protein (KLH) or to the hapten (*p*-azo-
phenylarsonate). Ouchterlony patterns suggested that the titers of these antibodies were roughly comparable in the experimental and control groups, but
the possibility of some suppression cannot be ruled out. The antiphenylarsonate
antibodies that did appear thus possessed idiotypic specificities not recognized
by the anti-D antibodies. The results are not attributable to generalized suppression of a major subgroup of IgG since sera of the suppressed animals reacted strongly with monospecific antisera to mouse IgG1 and mouse IgG2.
Also, one would not anticipate suppression of a subgroup of IgG since the
antiidiotypic antiserum injected was absorbed with excess mouse immunoglobulin and must also have been exposed in the recipient mouse to a large
excess of subgroup-specific antigenic determinants. The same arguments would
indicate that the effects observed were not attributable to suppression of a
class other than IgG.

Three major possibilities may be considered to account for the suppression
of idiotypic antibodies: (*a*) There was no suppression of biosynthesis but the
antiidiotypic antibodies injected before immunization with KLH–*p*-azophenyl-
arsonate simply absorbed the idiotypic antibodies produced; (*b*) a small amount
of idiotypic antibody was injected along with the anti-D, and the presence of
this antibody suppressed the formation of identical antibody molecules (13);
(*c*) there was central suppression, acting at the cellular level.

Possibility (*a*) can be ruled out by quantitative considerations. In each control animal, 11 wk after the start of the experiment, 0.3 μl of antiserum displaced more than 50 % of the ^{125}I-labeled D antibodies from the antiidiotypic
antibodies in the test system. Therefore, the number of anti-D molecules present
in each test was not capable of binding more than the number of D molecules
present in 0.6 μl of hyperimmune serum. Actually, the binding capacity of the
anti-D molecules must have been smaller than this; because of competition with

labeled D, some unlabeled D molecules must have remained unbound. Thus the 0.5 μl of anti-D antiserum used in each test was capable of binding the D antibodies in less than 0.6 μl of control hyperimmune sera. The total quantity of rabbit anti-D IgG injected into suppressed animals was 2 mg, which is equivalent to about 200 μl of antiserum. The total binding capacity of the injected rabbit anti-D, therefore, could not exceed the amount of D antibodies present in $200 \times 0.6/0.5$ or 240 μl of a hyperimmune mouse serum. Thus at the time of the 11-wk bleeding, the serum of a hyperimmune control mouse contained far more D molecules than necessary to saturate all of the rabbit anti-D injected into a suppressed animal. In addition, idiotypic antibodies were present in comparable titer throughout a 6-wk period preceding this bleeding (Table I). Since the half-life of mouse IgG is about 5 days (14), the control mice must have synthesized many times as much idiotypic antibody as necessary to saturate the anti-D antibodies during the 6-wk interval. This calculation is based on the assumption that *none* of the rabbit anti-D IgG was eliminated from the mouse during the 10-wk interval; actually, most of it was probably cleared rapidly since it was injected without adjuvant. The differences between the control and the experimental groups therefore cannot be attributed to simple absorption of D antibodies by rabbit anti-D.

It is not possible to rule out theoretically suppression of D-antibody formation by traces of D antibodies that might have been present in the IgG fraction of rabbit anti-D. However, any such contamination must have been minute. The rabbit was challenged with a total of 3 mg of mouse antiphenylarsonate antibody. Even if none of this mouse antibody was cleared, the maximum degree of contamination by mouse antibody of the IgG in a 3-kg rabbit would have been less than 3 parts in 1000, or 6 μg in the 2 mg used for suppression. The actual contamination was undoubtedly much lower. Furthermore, the weight of KLH–p-azophenylarsonate injected into mice during the course of immunization was 2 mg. If suppression of antibody biosynthesis by antibody is due to sequestration of antigen (13), the effect of trace contamination of rabbit IgG by mouse antiphenylarsonate antibodies must have been completely negligible.

It seems very probable, then, that antiidiotypic antibodies act at the cellular level to suppress the formation of antibodies with the corresponding idiotype. This appears quite analogous to the suppression of allotypic specificities by anti-allotypic antisera (15–17). Allotype suppression, however, must be initiated in the newborn animal to be successful. Suppression of idiotype in the adult is probably feasible because only a very limited number of lymphocytes bear receptors with a given idiotypic specificity.

In all probability the antiidiotypic antibodies interact with immunoglobulin receptors bearing the idiotypic determinants which are present on the surfaces of lymphocytes. If the receptor corresponds in structure to the antibody to be produced by descendants of the lymphocyte after antigenic stimulation, inactivation of the cell through interaction with antiidiotypic antiserum would account for the selective suppression observed. The cell type involved is probably the thymus-independent B lymphocyte.

Suppression of idiotype should provide a useful new approach for exploration of the mechanism of antibody biosynthesis. In addition, it represents another technique for specifically inhibiting the formation of a selected antibody population without generalized suppression of the immune response. Possible applications may eventually be found in the areas of transplantation and allergy.

SUMMARY

It has previously been shown that there are extensive idiotypic cross-reactions among antiphenylarsonate antibodies of A/J mice. The present work indicates that administration, into normal, adult A/J mice, of rabbit antiidiotypic antibody directed to A/J antiphenylarsonate antibody suppresses almost completely the subsequent production of antibody of the corresponding idiotype. No effect was noted on the formation of antibodies to the protein carrier or of antiphenylarsonate antibody of a different idiotype. The data are consistent with central suppression of production of the idiotypic antibody mediated through interaction with immunoglobulin receptors on lymphocytes.

Note Added in Proof.—Two control and three suppressed mice were tested by allowing them to rest for 10 wk after the wk-11 bleeding, then challenging with the KLH–*p*-azophenylarsonate antigen. A week later (wk 22) the animals were bled. It was found that 3 μl of each control serum caused nearly complete inhibition of indirect precipitation, whereas 10 μl of the serum from each suppressed mouse caused less than 12% inhibition.

REFERENCES

1. Oudin, J., and M. Michel. 1963. Une nouvelle forme d'allotypie des globulines γ du serum de lapin apparemment liée à la jonction et à la specificité anticorps. *C. R. Seances Acad. Agr. Fr.* **257**:805.
2. Kelus, A. S., and P. G. Gell. 1968. Immunological analysis of rabbit anti-antibody systems. *J. Exp. Med.* **127**:215.
3. Daugharty, H., J. E. Hopper, A. B. MacDonald, and A. Nisonoff. 1969. Quantitative investigations of idiotypic antibodies. I. Analysis of precipitating antibody populations. *J. Exp. Med.* **130**:1047.
4. Braun, D. G., and R. M. Krause. 1968. The individual antigenic specificity of antibodies to the streptococcal carbohydrate. *J. Exp. Med.* **128**:969.
5. Hopper, J. E., A. B. MacDonald, and A. Nisonoff. 1970. Quantitative investigations of idiotypic antibodies. II. Nonprecipitating antibodies. *J. Exp. Med.* **131**:41.
6. Oudin, J., and G. Bordenave. 1971. Observations on idiotypy of rabbit antibodies against *Salmonella abortus-equi. Nature (New Biol.) (London).* **231**:86.
7. Oudin, J., and M. Michel. 1969. Idiotypy of rabbit antibodies. I. Comparison of

idiotypy of antibodies against *Salmonella typhi* with that of antibodies against other bacteria in the same rabbits, or of antibodies against *Salmonella typhi* in various rabbits. *J. Exp. Med.* **130:**595.

8. Eichmann, K., and T. J. Kindt. 1971. The inheritance of individual antigenic specificities of rabbit antibodies to streptococcal carbohydrates. *J. Exp. Med.* **134:**532.

9. Cohn, M., G. Notani, and S. A. Rice. 1969. Characterization of the antibody to the C-carbohydrate produced by a transplantable mouse plasmacytoma. *Immunochemistry.* **6:**111.

10. Potter, M., and R. Liebermann. 1970. Common individual antigenic determinants in five of eight BALB/c IgA myeloma protein that bind phosphoryl choline. *J. Exp. Med.* **132:**737.

11. Kuettner, M. K., A. L. Wang, and A. Nisonoff. 1972. Quantitative investigations of idiotypic antibodies. VI. Idiotypic specificity as a potential genetic marker for the variable regions of mouse immunoglobulin polypeptide chains. *J. Exp. Med.* **135:**579.

12. Levy, H. B., and H. A. Sober. 1960. A simple chromatographic method for preparation of gamma globulin. *Proc. Soc. Exp. Biol. Med.* **103:**250.

13. Uhr, J. W., and G. Möller. 1968. Regulatory effect of antibody on the immune response. *Advan. Immunol.* **8:**81.

14. Waldman, T. A., and W. Strober. 1969. Metabolism of immunoglobulins. *Progr. Allergy.* **13:**1.

15. Mage, R., and S. Dray. 1965. Persistent altered phenotypic expression of allelic γG-immunoglobulin allotypes in heterozygous rabbits exposed to isoantibodies in fetal and neonatal life. *J. Immunol.* **95:**525.

16. Dubiski, S., 1967. Suppression of synthesis of allotypically defined immunoglobulin and compensation by another sub-class of immunoglobulin. *Nature (London).* **214:**1365.

17. Herzenberg, L. A., L. A. Herzenberg, R. C. Goodlin, and E. C. Rivera. 1967. Immunoglobulin synthesis in mice. Suppression by anti-allotype antibody. *J. Exp. Med.* **126:**701.

ABSOLUTE FREQUENCIES OF
LIPOPOLYSACCHARIDE-REACTIVE B CELLS PRODUCING
A5A IDIOTYPE IN UNPRIMED, STREPTOCOCCAL
A CARBOHYDRATE-PRIMED,
ANTI-A5A IDIOTYPE-SENSITIZED AND
ANTI-A5A IDIOTYPE-SUPPRESSED A/J MICE

By KLAUS EICHMANN, ANTONIO COUTINHO, and FRITZ MELCHERS

(From the Institut für Immunologie und Genetik, Deutsches Krebsforschungszentrum, Heidelberg, West Germany and the Basel Institute for Immunology, Basel, Switzerland)

The B-cell mitogen lipopolysaccharide (LPS) (1) stimulates every third B cell in spleen of A/J-mice to clonal growth and IgM secretion in vitro (2, 3). Each clone secretes one set of variable (v)[1] regions on Ig molecules and, therefore, Ig molecules with a given specificity for antigen. All clones of IgM-secreting, plaque-forming cells (PFC), stimulated by LPS, secrete the entire repertoire of sets of v regions expressed in the pool of LPS-reactive B cells, i.e. in one third of all splenic B cells. The repertoire can be defined in quantitative terms as the frequency of B cells which are precursors for clones secreting IgM with a given specificity. These frequencies are determined in vitro by analyses limiting the number of specific precursors in culture to one, and can be correlated to the total number of LPS-reactive B cells. By this method we have previously determined within the pool of LPS-reactive B cells, the absolute frequencies of IgM-secreting, PFC clones-specific for sheep and horse erythrocytes, trinitrophenylated and nitroiodophenylated sheep erythrocytes (4).

In this paper we determined by the same method the approximate frequency of B-lymphocyte precursors for cells secreting IgM molecules which bear the A5A idiotype. This idiotype defines the major species of antibody molecules secreted by strain A/J mice in response to immunization with group A streptococcal vaccine (Strep A) (5, 6). The A5A idiotype is detected by guinea pig antisera against a monoclonal antibody to group A streptococcal carbohydrate (A-CHO) produced by irradiated A/J mice that have been reconstituted with small numbers (~10^6) of pooled A/J spleen cells, and were subsequently immunized with Strep A (6).

---

[1] *Abbreviations used in this paper:* A5A, strain A/J, idiotypically defined, anti-A-CHO antibody; A-CHO, group A streptococcal carbohydrate; B, bone marrow derived-, bursa equivalent-; IBC, idiotype-binding capacity; Id, idiotype; Ig, immunoglobulin; LPS, bacterial lipopolysaccharides; PFC, plaque-forming cell; SRC, sheep erythrocytes; Strep A, group A streptococcal vaccine; v, variable.

The A5A antibody is also defined by its spectrotype on isoelectric focusing (7) and the combined use of this method and of anti-idiotypic antisera revealed that over 90% of all A/J mice produce the A5A antibody after immunization with Strep A (5–7). Within the total antibody produced, the A5A idiotype accounts for about 25% (7, 8). Presensitization of A/J mice with the IgG1 fraction of anti-A5A antisera before the injection of Strep A leads to an increase in this proportion to close to 100% indicating that this class of anti-idiotype antibody can expand the precursor pool of A5A-positive B cells and, correspondingly, induces A-CHO-specific memory (8, 9).

In contrast, pretreatment of A/J mice with the IgG2 class of anti-A5A antisera leads to a drastic decrease of the proportion of A5A-positive antibodies upon subsequent immunization with Strep A (7, 10), and, by using low doses of IgG2 antibodies, this suppression becomes permanent and is maintained by suppressor T cells (10).

In this system, we have determined frequencies of precursor B cells-producing A5A idiotype in vitro to ask the following questions: (*a*) What is the frequency of A5A-bearing precursor B cells in normal mice? (*b*) Does the induction of A-CHO-specific memory result in an increase of the frequency of A5A-bearing precursor B cells and is the difference in the idiotypic heterogeneity of memory cells induced by IgG1 anti-A5A and by Strep A reflected in the respective precursor frequencies? (*c*) Is the state of idiotype suppression, which is actively maintained by suppressor T cells, reflected by a decreased frequency for precursor B cell? (*d*) How many of the B-cell clones that bear the A5A idiotype possess specificity for A-CHO? We can answer each of these questions in quantitative terms, by testing the supernates of LPS-stimulated spleen cell cultures at limiting cell concentrations for the presence or absence of A5A idiotype and analyzing the data according to Poisson's distribution.

Materials and Methods

Animals. A/J-mice were obtained from the Institut für Biologisch-Medizinische Forschung AG, Füllinsdorf, Switzerland, or from Gr. Bomholtgaard, Ry, Denmark. Male or female mice, 4–10 wk old, were used in all experiments. Lewis strain rats were obtained from the Institut für Biologisch-Medizinische Forschung.

Anti-Idiotype Antisera. Anti-idiotype (Anti-Id) antisera to antibody A5A were prepared in guinea pigs that were rendered tolerant to mouse IgG as previously described (6, 11). IgG1 and IgG2 fractions from these antisera were obtained by agarose block electrophoresis as described (7).

Immunizations. Mice were immunized either with $1–2 \times 10^9$ Strep A organisms (5, 6), or with the IgG1 and IgG2 fractions, respectively, of an anti-A5A idiotype antiserum (8). IgG1 and IgG2 fractions were calibrated according to their binding capacity IBC and 0.1 μg IBC/mouse was injected, as previously described (8).

Cells. Spleen cells and growth-supporting thymus cells were prepared as described previously (2). Rat thymus cells were used in all experiments as growth-supporting cells.

Culture Medium and Mitogen. Spleen cells with growth-supporting thymus cells (3×10^6 cells/ml) were grown in RPMI 1640 medium (Grand Island Biological Co., Grand Island, N.Y.) containing glutamine (2 mM), penicillin, and streptomycin (5,000 IU/ml each), HEPES-buffer (10 mM), pH 7.3, 2-mercaptoethanol (5×10^{-5} M), fetal calf serum (10%; BioCult, Irvine, Scotland, batch K 255701 D), and LPS (50 μg/ml). LPS-S (EDTEN 18735 and S 435/188049) was kindly prepared for us by Doctors C. Galanos and O. Lüderitz, Max-Planck-Institut für Immunbiologie, Freiburg, W. Germany. Cultures were kept either in Costar tissue culture trays (no. 3524, Costar, Data Packaging Corp., Cambridge, Mass.), in 1-ml aliquots, or else in Falcon Microtest II-plates (Falcon Plastics, Div. of Bioquest, Oxnard, Calif. no. 3040) or in 5-ml Falcon tubes (no.

2058), in 0.2-ml aliquots, in a humidified 37°C incubator gassed with a mixture of 5% CO_2 in air. The efficiency of growth and maturation to IgM-secreting cells was assayed in each experiment at day 5, in 10 separate cultures with spleen cell concentrations limiting the number of LPS-reactive B cells developing an IgM response to one (see Results). For the analysis of A5A idiotypes, supernates were collected by centrifugation after 10–14 days of culture. The supernatant medium could be kept frozen until it was analyzed for secreted A5A idiotype.

Plaque Assay for Secreting Cells.   A modified hemolytic plaque assay with protein A-coated sheep erythrocytes (SRC) and mouse Ig-specific antisera as developing antibodies in the presence of complement was employed to detect and enumerate all Ig-secreting cells (12). IgM-secreting PFC were scored by employing specific antisera which were raised by repeated injections of 0.5–1 mg purified MOPC 104 E 19S IgM in Freund's incomplete adjuvants. The antisera detected μH- and λL-chains, but not γH or αH or κL-chains secreting cells as tested in plaque assays with corresponding myeloma tumor cells. Protein A used in coating of SRC was obtained from Dr. H. Wigzell, Biomedicum, University of Uppsala, Uppsala, Sweden. Complement, appropriately diluted with BSS, was from BioCult.

Assay for A5A Idiotype.   Determination of A5A idiotype: the quantitative determination of A5A idiotype in culture supernates employed our previously described radioimmune inhibition assay (5) in slightly modified form. 2–3 ng of radioiodinated-purified A5A antibody (15,000–20,000 cpm) was incubated with guinea pig anti-A5A antiserum at a dilution that binds about 70% of the radiolabeled A5A, in a total vol of 80 μl. To this, 100 μl of culture supernate was added and after a 4-h incubation, 20 μl of diluted rabbit anti-guinea pig antiserum was added. After overnight incubation, the radioactivity in the precipitates was determined as described (5). A standard inhibition curve was constructed by using cold A5A as inhibitor in concentrations ranging from 2 to 6,000 ng/ml. 50% inhibition is achieved at about 200 ng/ml. The concentration of A5A idiotype in each culture supernate equals that which causes the same degree of inhibition in the standard curve. Initially, a number of supernates were tested in 3.3-fold dilution series and, invariably, inhibition curves parallel to the standard curve were observed. Thereafter, a single point was considered sufficient to estimate the concentration of A5A idiotype in a culture supernate.

A supernate was considered positive for A5A idiotype when 100 μl of it inhibited more than 20% of the binding of the radioactive A5A to anti-A5A. This point in the standard curve is equivalent to 27 ng A5A/ml, i.e. 2.7 ng A5A/100 μl undiluted supernate. Thus, the limit of sensitivity for this assay is about 5–6 ng/culture, in a 0.2-ml culture. As will be seen later, these borderline concentrations were hardly ever observed, since a single clone produces around 30 ng A5A.

Detection of Anti-Strep A-Antibody in Culture Supernates.   To determine whether the A5A idiotype in culture supernates was associated with antibodies to A-CHO, 50 μl of supernate was added to 1×10^{10} packed Strep A particles and another 50 μl was added to an about equal amount of packed *Bordetella pertussis* for control. The organisms were resuspended in the supernates and the tubes were rotated at 4°C for 1 h. After centrifugation, the amounts of A5A idiotype in both aliquots were determined by the radioimmune inhibition assay (see above).

Results

Frequencies of A5A Idiotype Producing B cells.   All frequency determinations were done in vitro at cell concentrations limiting the number of clones producing either A5A idiotype or IgM to around one per culture. According to Poisson's distribution, one B-cell precursor for a clone of cells secreting a given type of IgM molecule is present in that number of cells plated in individual cultures which let 63% of all cultures appear positive in the assay.

We first screened spleen cells of unprimed, Strep A-CHO-primed, and anti-idiotype-primed mice for the approximate frequencies of A5A idiotype-producing B cells and for the frequencies of total IgM-secreting PFC clones in 3.3-fold dilutions from 2×10^5 cells to two cells per culture. 10 cultures were assayed at each concentration with the two tests. Data in Table I show that all spleen cell

TABLE I

LPS-Stimulated B-Cell Responses at Different Concentrations of Spleen Cells from Unprimed, Strep A-Primed and Anti-A5A Idiotype-Primed A/J Mice Detected as A5A Idiotype-Secreting Cultures and as IgM-Secreting, PFC-Positive Cultures

Number of spleen cells per culture	Source of spleen cells					
	Unprimed		Strep A-primed		Anti-A5A IgG1-Primed	
	Assay: IgM PFC	A5A idio-type	IgM PFC	A5A idio-type	IgM PFC	A5A idio-type
2×10^5	ND	10	ND	10	ND	10
6×10^4	ND	10	ND	10	ND	10
2×10^4	ND	10	ND	10	ND	10
6×10^3	ND	8	ND	10	ND	10
2×10^3	ND	1	ND	7	ND	10
6×10^2	10	1	10	5	10	8
2×10^2	10	0	10	0	10	2
6×10^1	10	0	10	0	10	1
2×10^1	9	0	10	0	10	0
6×10^0	6	ND	7	ND	5	ND
2×10^0	2	ND	2	ND	1	ND

Number of positive cultures in a total of 10 cultures tested. ND, not done.

preparations, either from unprimed or primed mice had, approximately the same number of LPS-reactive B cells, i.e. one in six spleen cells, as assayed by the development of IgM-secreting PFC clones. This absolute number of LPS-reactive B cells was also determined in all other experiments reported below and was found, in all cases, to be close to one in six spleen cells.

The approximate number of A5A idiotype-producing B-cell clones secreting into the culture medium idiotype detectable by the specific radioimmunoassay, were 1 in 6×10^3 spleen cells in unprimed mice, 1 in 2×10^3 in Strep A-CHO-primed mice and 1 in 6×10^2 in anti-A5A IgG1-primed mice (Table I).

In the range of cell concentrations where A5A idiotype production became fluctuating, indicating that the LPS-reactive A5A idiotype-producing B-cell precursors had become limiting, narrower differences of cell concentrations and larger numbers of cultures were tested. Table II summarizes the results from three experiments. We have pooled the data from these individual experiments for a given type of spleen cell source for the frequency determinations shown in Fig. 1, since in all experiments the number of LPS-reactive B cells stimulated to IgM-secreting PFC clones was almost the same, i.e. close to one in six spleen cells.

For all four cell types analyzed, the number of cultured spleen cells containing one A5A-positive B-cell precursor, developing into a clone of IgM-secreting cells after LPS stimulation, were extrapolated from the data in Fig. 1 as that number of cells with which 37% of all cultures did not yield a positive response. These frequency determinations are summarized in Table III.

Since we can detect all IgM-secreting clones stimulated by LPS we can correlate the frequencies of A5A idiotype-producing B cells to the total number

Table II

Number of LPS-Stimulated Spleen Cell Cultures with A5A Idiotype-Positive Supernates at Varying Numbers of Spleen Cells from A/J Mice that were either Unprimed or Pretreated with Strep A, Anti-A5A IgG1, and Anti-A5A IgG2

Number of spleen cells per culture	Unprimed			Strep A-primed			Anti-A5A IgG1-primed			Anti-A5A IgG2-primed		
	No. tested	No. positive	Percent positive	No. tested	No. positive	Percent positive	No. tested	No. positive	Percent positive	No. tested	No. positive	Percent positive
2×10^4										20	20	100
1.2×10^4	86	83	97									
1×10^4	70	63	90									
8×10^3	20	13	65									
6×10^3	80	25	39	30	29	97				20	3	15
4×10^3	178	37	21	20	18	90						
2×10^3	244	33	14	80	53	67	122	115	94	20	1	5
1.6×10^3							140	123	87			
1.2×10^3				50	11	22						
1.0×10^3							50	39	78			
6×10^2							60	20	33	20	1	5
2×10							60	4	6			
6×10										20	0	0

Table III

Frequencies of LPS-Reactive B-Cells Producing A5A Idiotype in Unprimed, Primed, and Suppressed A/J Mice

Source of spleen cells	Frequency*		
	In total spleen cells	Approximate in splenic B cells‡	Absolute within the population of LPS-reactive B cells§
Unprimed	$1:15.3 \times 10^3$	$1:7.6 \times 10^3$	$1:2.5 \times 10^3$
Strep A-primed	$1:1.7 \times 10^3$	$1:0.85 \times 10^3$	$1:2.8 \times 10^2$
Anti-A5A-IgG1 primed	$1:0.8 \times 10^3$	$1:0.4 \times 10^3$	$1:1.3 \times 10^2$
Anti-A5A-IgG2 primed	$1:40 \times 10^3$	$1:20 \times 10^3$	$1:6.7 \times 10^3$

* Extrapolated from data in Fig. 1 for the cell concentration at which 37% of all cultures did not respond.

‡ Assuming approximately 50% of all splenic lymphocytes to be B cells.

§ Determined, in the same experiments, by frequency determinations of IgM-secreting PFC clones.

of LPS-reactive B cells, and, therefore, can calculate the absolute frequencies of A5A idiotype-producing precursors within the pool of LPS-reactive B cells (one in six spleen cells, see above and [2, 3]) (Table III).

Of the four cell types analyzed, only the data obtained with cells primed with Strep A and with cells primed with anti-A5A IgG1 could be fitted to a linear degression of the logarithm of the fraction of idiotype-negative cultures, indicating that the A5A idiotype-producing B-cells were limiting in these cultures (Table II, Fig. 1). Normal unprimed cells showed a sharp drop in their degression line when the cell number exceeded 6×10^3/culture, and a similar drop was also observed with A5A-suppressed cells. We have analyzed this situation in more detail and have found that at a concentration of 3×10^4 cells/ml (6×10^3/culture), the culture supernates contain between 100 μg and 1 mg/

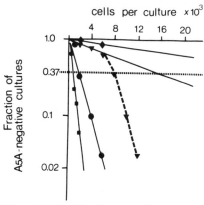

Fig. 1. Titration of LPS-reactive B cells of unprimed (▼), Strep A-primed (●), anti-A5A idiotype (IgG1)-sensitized (■) and anti-A5A idiotype (IgG2)-suppressed (◆) A/J mice. Cells were grown in the presence of 50 μg/ml LPS and 3×10^6 thymus cells/ml for 10–14 days and A5A idiotype-bearing Ig measured in the supernates of the cultures by a radioimmunoassay described in the Materials and Methods section. Each point in the Figure is based on the number of assays given in Table II. The dotted line (· · · · ·) gives the numbers of cells at which 37% of all cultures did not contain A5A idiotype, i.e. which, according to Poisson's distribution, yield that number of spleen cells in which one A5A idiotype-producing, LPS-reactive B precursor is present. The dashed line (-----) shows the actually measured curve for unprimed cells, and indicates where the assay quits to be linear, due to excess normal Ig interfering in the radioimmunoassay for A5A idiotype (see the text).

ml of non-A5A idiotype-bearing, nonspecific Ig. More than 100 μg of this Ig had to be added to the radioimmune inhibition assay (see below and Materials and Methods), to identify the small quantities of idiotype in these cultures. At this concentration of mouse Ig, the degree of precipitation between the rabbit anti-guinea pig Ig antiserum and the guinea pig anti-Id antiserum becomes incomplete (5), so that the test does no longer detect A5A idiotype only. This results in detection of false positives, consequently increasing the number of positive cultures to values higher than expected and in a rapid drop of the curves in Fig. 1. Nevertheless, the data obtained with unprimed and suppressed cells at cell concentrations lower than 6×10^3/culture (3×10^4/ml), fit a linear degression line and we have, therefore, extrapolated these degression lines in a linear fashion.

The Amount of A5A Idiotype Produced by Single Clones. The radioimmunoassay for A5A idiotype allowed a quantitative evaluation of the amounts of idiotype-bearing Ig secreted into the culture medium (see Materials and Methods). Fig. 2 shows the distribution of quantities of A5A idiotype produced in individual cultures which contained numbers of A5A idiotype-producing B-cell clones near one (see also Table II and Fig. 1). It is evident that the distributions for unprimed as well as for primed spleen cells are discontinuous. A5A idiotype-positive cultures accumulated around 25–30 ng, 55–60 ng, and 85–90 ng per culture. In most cases, the numbers of cultures under each of the observed peaks fit within the 95% confidence limits of those calculated with Poisson's distribution, with the assumption that the clonal progeny of a single A5A idiotype-bearing LPS-reactive B cell secreted 30 ng of A5A idiotype under

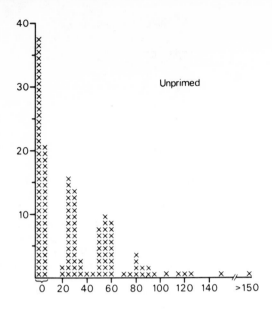

Unprimed

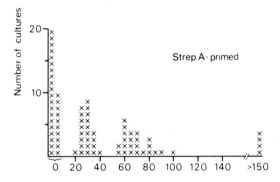

Strep A- primed

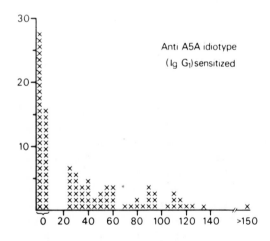

Anti A5A idiotype
(Ig G₁) sensitized

ng A5A idiotype/culture

TABLE IV

Frequency of Cultures Developing None, One (30 ng), Two (60 ng), or Three (90 ng) Clones of A5A-Secreting, LPS-Reactive B Cells, at Cell Concentrations Limiting the Number of Reactive Cells to Around One Per Culture

Number of cultures with:	Cells in culture		
	Normal	Strep A-primed	Anti-A5A IgG1-primed
No clone	59:42% (32–50:40.6%)*	30:37.5% (28–46:36.7%)*	44:44% (35–54:44.9%)*
One clone	36:25% (27–45:36.5%)	23:28% (28–46:36.7%)	22:22% (27–44:35.9%)
Two clones	29:20% (9–24:16.5%)	16:20% (13–28:18%)	14:14% (9–22:14.4%)
Three clones	10:7% (2–10:5%)	7:9% (4–16:6%)	11:11% (1–9.5:3.8%)
>Three clones	6:4% (0–6:1%)	4:5% (2–10:2%)	10:10% (0–4.6:0.7%)
Total	140:100%	80:100%	100:100%

* Expected percent and 95% confidence limits of the experimental values in Poisson's distribution.

our culture conditions. These numbers and the 95% confidence limits are summarized in Table IV.

Association of A5A Idiotype with A-CHO Specificity. Since we can enumerate from the concentration of A5A idiotype in the supernate the number of A5A-secreting clones in each culture, we can analyze, at the single clone level, how many of the clones that secrete A5A idiotype possess specificity of A-CHO. This was done by absorbing equal volumes of supernates from single, double, and triple clone cultures to group A streptococcal organisms and, as a control, to *B. pertussis*. Thereafter, the supernates were analyzed for reduction in idiotype content.

The data on cultures from unprimed cells and from cells primed with Strep A and with anti-A5A IgG1 are summarized in Table V. The number of idiotype-positive clones whose antibodies absorbed to Strep A was between 40 and 50% of the total idiotype-positive clones analyzed for each cell type.

The quantitative absorption data obtained on single, double, and triple clone cultures supported the concept that each clone produced around 30 ng idiotype, as the absorption of idiotype clearly occurred in 30-ng portions and each

FIG. 2. Distribution of the amount of A5A idiotype-bearing Ig in individual cultures of unprimed, Strep A-primed, and anti-A5A-idiotype-(IgG1)-sensitized A/J spleen cells grown in the presence of 50 μg/ml LPS and 3×10^6 thymus cells/ml under culture conditions limiting the number of A5A-producing B-cell precursors to around 1 (see Fig. 1 and Table IV). A5A idiotype-bearing Ig was quantitated in the supernates of the cultures kept for 10–14 days by a radioimmunoassay described in the Materials and Methods section.

TABLE V

The Proportions of A5A Idiotype-Secreting Clones that Possess Specificity for A-CHO in Unprimed Cells and in Cells Primed with Strep A and with Anti-A5A IgG1

	Source of spleen cells		
	Unprimed	Primed with Strep A	Primed with anti-A5A IgG1
Number of A5A-bearing clones tested	131	167	129
Number of clones specifically absorbed to Strep A	58	77	50
Percent of A-CHO-specific among A5A-bearing clones	44.3	46.1	38.8

TABLE VI

Absorption of Idiotype on Strep A Organisms in Supernates of Cultures that Contained One, Two, or Three Clones of A5A-Producing Cells*

Number of A5A-bearing clones/culture					
One clone		Two clones		Three clones	
A5A idiotype (ng/200 μl) after absorption on:					
B. pertussis	Strep A	B. pertussis	Strep A	B. pertussis	Strep A
40	0	52	28	80	60
32	0	60	28	88	60
28	0	56	28	92	60
28	0	64	32	88	56
36	0	64	36	88	64
32	0	60	32	80	32
32	0	56	0	96	28
36	0	56	0	92	36
28	0	60	0	92	28
32	0	52	0	92	0

* Only cultures which contained A5A idiotype absorbing to Strep A are shown on this Table.

portion absorbed independently of the others. 10 examples for absorptions of supernates containing A5A idiotype from one, two, or three clones are shown in Table VI.

Discussion

The B-cell mitogen LPS stimulates every third B cell of A/J-spleen to clonal growth and IgM secretion in vitro (2, 3). We have, therefore, analyzed with our frequency determinations the representation of A5A idiotype-producing cells in the LPS-reactive third of the total B-cell compartment. We cannot, at the present time, probe for the repertoires of the other two thirds of all B cells, since we have not yet identified the mitogens which stimulate them. It remains, therefore, the question as to how representative our frequency determinations are for the total pool of B cells. At the present, we have no indication that different B-cell subpopulations as defined by mitogen reactivity, selectively express different sets of v regions.

The frequency determinations of A5A idiotype-producing B cells in unprimed spleen add another number to our quantitative description of the repertoire of v gene specificities in normal murine B cells, previously done for SRC-, horse erythrocytes, trinitrophenylated-SRC- and nitroiodophenylated-SRC-specific B cells. (4). All these frequencies, including that for A5A idiotype, are surprisingly high and appear to disagree with frequencies of specific antibodies in the total pool of natural serum Ig. If natural serum Ig constitutes a representative mixture of products of all existing B cells, we would expect 1 in every 2,500 Ig molecules of normal serum to carry the A5A idiotype.

Since the detection limit of our radioimmune assay is in the order of 30 ng of A5A/ml and since the assay is limited by concentrations of nonspecific Ig in the order of 1 mg/ml, we would be able to detect A5A-bearing Ig molecules in normal serum Ig if their frequency would exceed 1 in 30,000. This, however, is not the case. As reported previously (5), the proportion of A5A-bearing Ig molecules in normal Ig is less than 1 in 30,000. This figure is more than an order of magnitude lower than that expected from the frequency of precursor cells.

The discrepancy between these figures can have a number of reasons. We can assume, for example, that the proportion of A5A-bearing molecules in the predominant class of serum immunoglobulin IgG is dramatically smaller than in serum IgM, since in our culture system predominantly IgM-producing precursor cells are estimated (W. Gerhard and F. Melchers, unpublished observations). This would result in a low proportion of A5A idiotype on the total serum Ig even if 1 in 2,500 of the IgM molecules has this idiotype. Examples for a high concentration of an antibody or idiotype in normal IgM in combination with the absence of this antibody from the normal IgG fraction are known for natural anti-phosphorylcholine antibodies (13) and for natural antibodies to haptens such as nitroiodophenyl and nitrophenyl (14, 15). If this were a general phenomenon, we would have to conclude that the IgG pool represents a selection from the IgM precursor repertoire. We should like to state that the discrepancies observed between the serum levels and the precursor frequencies remain a problem that needs further investigation.

Memory was clearly demonstrable in quantitative terms, in the B-cell compartment of Strep A and of anti-A5A idiotype IgG1 sensitized mice as a 10--20-fold increase in the frequency of A5A idiotype-producing cells. This increase is in agreement with the 10–20-fold greater idiotype production in primed mice as compared to normal mice upon Strep A challenge (5, 6, 8). It is interesting to note that the amounts of idiotype produced in primed and unprimed cultures are exactly the same on a per clone basis (discussed below). Thus, it appears that the generation of memory within the LPS-reactive B-cell compartment can be expressed as an augmentation of the number of precursor cells without any change in the capacity of the cells to mature and secrete antibody.

The data obtained on A5A-suppressed cells are somewhat less conclusive. The cells analyzed for these experiments came from mice in which suppression is mediated by suppressor T cells induced by injection of a low dose (0.1 μg) of anti-A5A IgG2 (10). In such mice, it is not clear whether the 20--50-fold reduction in their idiotype response to Strep A challenge is the result of a reduction of A5A precursor B cells or of the suppression of T-helper cells that cooperate with such B cells. From the idiotype response of suppressed mice in vivo we would expect a much greater reduction in the A5A precursor frequency than the 2.5–3-fold reduction observed here, if suppression and reduction of B precursor cells were correlated (10). It should be mentioned, however, that (*a*) the numbers of suppressed cultures tested were small (Table II) and (*b*) the point in the degression line on which the frequency estimate relies (see Fig. 1) is already in the zone at which nonspecific Ig interferes with the test system. Therefore, our frequency determination for A5A precursor B cells in mice carrying idiotype suppressor T cells is most likely an overestimate, and the reduction of such cells is probably more pronounced. Because of the limitation of the test system, we cannot, with safety, quantitatively estimate to which extent A5A-producing B cells have been reduced in suppressed mice. On the other hand, the data reveal the differential effect of IgG1 and IgG2 anti-idiotype antibodies on idiotype-bearing precursor lymphocytes: whereas IgG1 antibodies expand the precursor pool, there is no such effect associated with IgG2 antibodies (7, 8).

The present results show that memory (and perhaps suppression) in the B-cell compartment can be quantitated as frequencies of specific cells. The observed frequencies of A5A idiotype-producing B cells in normal and primed mice strongly suggests that the LPS-reactive third of all B cells in spleen is representative for all B cells to the extent that it is influenced by priming with antigen and anti-idiotype in parallel with the total pool of specific cells reacting to an antigenic challenge (6–8).

It becomes evident from our experiments with unprimed and particularly with suppressed mice that frequency determinations are not possible for idiotype-producing cells occurring with frequencies lower than one in 3×10^3 as long as the current culture conditions and the current version of the radioimmunoassay are used. Cultures with 1 A5A idiotype-producing B-cell clone contain 3,000 other LPS-reactive B-cell clones, all secreting IgM. Total IgM reach concentrations between 100 μg and 1 mg/ml in the cultures. This concentration of nonspecific IgM interferes in the radioimmunoassay, such that even without the presence of A5A idiotype-bearing antibodies inhibition is observed. Thus, cultures with more than 5×10^3 cells cannot reliably be tested for A5A idiotype. Since also the hemolytic plaque assay for any determinant coupled to SRC is limited by the frequency of SRC-specific B cells, i.e. 1 in 1,000 LPS-reactive cells (4), we have at present no assay to detect any specific B cells which occur with frequencies below 1 in 3×10^3.

Our radioimmune assay, though limited by the interference of normal Ig at high concentration, allows the quantitative determination of idiotype-bearing immunoglobulin molecules in culture supernates. It is very surprising that one clone of A5A idiotype-producing B cells within 10^2–2×10^3 clones secreting IgM with other idiotypes secretes such a constant amount of A5A idiotype (30 ng). This appears to be a property of the culture system and not due to detection limits in the radioimmunoassay (see Materials and Methods). It is furthermore supported by data obtained for parainfluenza- and influenza virus-specific antibodies in another radioimmunoassay, in which one clone of B cells-producing specific antibody, mainly of IgM class, within 10^2–10^3 other B-cell clones

clones have been found to secrete again 30 ng-specific antibody (W. Gerhard and F. Melchers, unpublished observations). The fixed amount of secreted Ig per clone of growing B cells is all the more surprising, since culture media were not collected at a fixed time, but at various times between 10 and 14 days of culture. Although we know that LPS-stimulated clones of B cells divide quite regularly every 18 h, at least within the first 5 days of culture (2), we also know that they cease to grow after 1 wk of culture and reach a plateau level of 30 ng/clone between 7 and 8 days of culture. The regulatory factor(s) which cause(s) this strict limitation in growth and secretion remain(s) to be investigated.

It should be mentioned that the figure of 30 ng/clone is essentially reasonable as it would require 1 day of secretion of 500 plasma cells each producing 2,000 molecules/s. This number of cells is generated after eight divisions of a single precursor cell within 6–7 days.

We observed a high association between the presence of the A5A idiotype and the specificity of the antibody for group A-CHO. Out of two clones producing antibody with A5A idiotype, one produces antibody which also binds to Strep A organisms and presumably has specificity to A-CHO. This high association can be explained by the specificity of the anti-A5A antiserum used in these studies. The majority of its antibodies (~90%) reacted with idiotype determinants generated by the combination of heavy and light chain, while a minority reacted with heavy chain determinants alone (16). Thus, this antiserum recognizes a particular V_H-V_L combination defining an A-CHO-binding site.

It is, however, surprising that even after priming with the antigen, Strep A-CHO, resulting in a 10-fold increase in the number of idiotype-bearing B-cell precursors, one half of all idiotype-producing clones still lack specificity for A-CHO. Such Ig molecules are comparable to the idiotype-bearing, non-antigen-reactive Ig molecules described by Oudin and Cazenave (17).

Priming of nonantigen-recognizing, but idiotype-positive B cells upon antigenic challenge is predicted by Jerne in his network hypothesis (18). It assumes that two or more Ig molecules (on two or more cells) which are structurally different in their combining site and, therefore, recognize different epitopes, may bear similar idiotypic determinants and, therefore, be under control of overlapping anti-idiotypic sets (cells). An initial increase in the number of antigen-reactive Ig molecules (i.e. numbers of cells) bearing the idiotype, effected by antigen exposure of the antigen-binding subset of idiotype-bearing molecules (cells) will result in derepression from anti-idiotypic action of the entire idiotype-bearing population and, concomitantly, in equal increases in the number of all other idiotype-bearing molecules (cells) independent of their antigen reactivity.

Summary

The absolute frequencies of B cells-producing A5A idiotype have been determined in vitro by limiting dilution analysis in a culture system in which every LPS-reactive B cell grows into a clone of IgM-secreting cells. Spleen cells from normal A/J mice contain 1 A5A-idiotype-producing B-cell precursor in 2.5 $\times 10^3$ LPS-reactive B cells. Approximately a 10–20-fold increase in frequencies of precursor cells results from antigen priming with Strep A-CHO (1 in 2.8×10^2) or from sensitization with IgG1 anti-A5A idiotype (1 in 1.3×10^2). Injection of IgG2 anti-A5A idiotype which has been shown to suppress A5A idiotype in vivo results in only a marginal and maybe insignificant decrease in precursor frequencies (1 in 6.7×10^3). On the other hand, priming does not result in a detectable qualitative difference in the specific precursor cells, since each clone of B cells secretes 30 ng of A5A-bearing Ig within 8 days of culture, regardless of being unprimed or primed.

Nearly half of all A5A idiotype-producing clones, both from unprimed as well as from primed mice, show antigen specificity in binding A-CHO. Priming by antigen, therefore, also results in a 10-fold increase in the frequency of idiotype positive B cells without antigen specificity. This result is a prediction of the network hypothesis.

We thank Ms Helen Campbell, Ms Catherine Giacometti, Ms Ingrid Falk, and Ms Christel Schenkel for able technical assistance.

This paper is dedicated to Niels Kaj Jerne.

Received for publication 2 May 1977.

References

1. Andersson, J., O. Sjöberg, and G. Möller. 1972. Induction of immunoglobulin and antibody synthesis *in vitro* by lipopolysaccharides. *Eur. J. Immunol.* 2:349.
2. Andersson, J., A. Coutinho, W. Lernhardt, and F. Melchers. 1977. Clonal growth and maturation to immunoglobulin secretion "in vitro" of every growth-inducible B-lymphocyte. *Cell.* 10:27.
3. Andersson, J., A. Coutinho, and F. Melchers. 1977. Frequencies of mitogen-reactive B cells in the mouse. I. Distribution in different lymphoid organs from different inbred strains of mice at different ages. *J. Exp. Med.* 145:1511.
4. Andersson, J., A. Coutinho, and F. Melchers. 1977. Frequencies of mitogen-reactive B cells in the mouse. II. Frequencies of B cells producing antibodies which lyse sheep or horse erythrocytes, and trinitrophenylated or nitroiodophenylated sheep erythrocytes. *J. Exp. Med.* 145:1520.
5. Eichmann, K. 1973. Idiotype expression and the inheritance of mouse antibody clones. *J. Exp. Med.* 137:603.
6. Eichmann, K. 1972. Idiotypic identity of antibodies to streptococcal carbohydrate in inbred mice. *Eur. J. Immunol.* 2:301.
7. Eichmann, K. 1974. Idiotype suppression. I. Influence of the dose and of the effector functions of anti-idiotypic antibody on the production of an idiotype. *Eur. J. Immunol.* 4:296.
8. Eichmann, K. and K. Rajewsky. 1975. Induction of T and B cell immunity by anti-idiotypic antibody. *Eur. J. Immunol.* 5:661.
9. Black, S. J., S. Hämmerling, C. Berek, K. Rajewsky, and K. Eichmann. 1976. Idiotypic analysis of lymphocytes in vitro. I. Specificity and heterogeneity of B and T lymphocytes reactive with anti-idiotypic antibody. *J. Exp. Med.* 143:846.
10. Eichmann, K. 1975. Idiotype Suppression. II. Amplification of a suppressor T cell with anti-idiotypic activity. *Eur. J. Immunol.* 5:511.
11. Eichmann, K. and T. J. Kindt. 1971. The inheritance of individual antigenic specificities of rabbit antibodies to streptococcal carbohydrates. *J. Exp. Med.* 134:532.
12. Gronowicz, E., A. Coutinho, and F. Melchers. 1976. A plaque assay for all cells secreting Ig of a given type or class. *Eur. J. Immunol.* 6:588.

13. Liebermann, R., M. Potter, E. B. Muschinsky, J. R. W. Humphrey, and S. Rudikoff. 1974. Genetics of a new IgV_H (T15 idiotype) marker in the mouse regulating natural antibody to phosphorylcholine. *J. Exp. Med.* 139:983.

14. Imanishi, T. and O. Mäkelä. 1973. Strain differences in the fine specificity of mouse anti-hapten antibodies. *Eur. J. Immunol.* 3:323.

15. Imanishi, T. and O. Mäkelä. 1974. Inheritance of antibody specificity. I. Anti-(4-hydroxy-3-nitrophenyl) acetyl of the mouse primary response. *J. Exp. Med.* 140:1498.

16. Krawinkel, U., M. Cramer, C. Berek, G. Hammerling, S. J. Black, K. Rajewsky, and K. Eichmann. 1977. *Cold Spring Harbor Symp. Quant. Biol.* 41:285.

17. Oudin, J., and P. A. Cazenave. 1971. Similar Idiotypic Specificities in immunoglobulin fractions with different antibody functions or even without detectable antibody function. *Proc. Natl. Acad. Sci. U.S.A.* 68:2616.

18. Jerne, N. K. 1974. Towards a Network Theory of the Immune System. *Ann. Immunol.* (*Inst. Pasteur*). 125C:373.

H. Cosenza

Basel Institute for Immunology, Basel

Detection of anti-idiotype reactive cells in the response to phosphorylcholine

Upon immunization with phosphorylcholine (PC), BALB/c mice produce antibodies bearing mainly one idiotype which is identical to that of the TEPC 15 PC-binding myeloma protein. By coating sheep red cells (SRC) with TEPC 15 it was possible to detect anti-idiotypic plaques during the course of an immune response to PC. These plaques are detectable only after the anti-PC response has reached a peak and begun to decline. They are specific, since addition of free TEPC 15 and not MOPC 315 prevented their detection. It is suggested that such anti-idiotypic reactive cells might regulate the response to PC.

1. Introduction

Bone marrow-derived lymphocytes (B cells) and thymus-derived lymphocytes (T cells) are known to interact in response to challenge with antigen [1]. The antibodies eventually secreted by B cells have specificity for the antigen and are extremely heterogeneous [2, 3]. Furthermore, antibodies also carry specific antigenic determinants (idiotypes) which in turn can be immunogenic in individuals of the same species [4–9] or in the same individual [10–12]. These findings have led to the postulate that interactions between idiotypes and anti-idiotypes are of central importance in the regulation of the immune response [13–14]. Therefore, it should be possible to detect idiotype-secreting cells (cells synthesizing antibody to the antigen) and anti-idiotypic secreting cells (cells producing antibody to the idiotypic determinants of the antibodies to the antigen) during the course of an antibody response. To test this possibility, the response to phosphorylcholine (PC) has been used as a model since the antibodies secreted by BALB/c mice in response to immunization with PC bear a single idiotype [15]. Taking advantage of this "idiotypic monoclonicity", the present report deals with the detection of anti-idiotypic plaque-forming cells (PFC) during the course of the immune response to PC. These plaques are specific for the TEPC 15 idiotype since they are not inhibited by other myeloma proteins. The simultaneous detection of idiotypic and anti-idiotypic cells indicates that interactions between idiotypes and anti-idiotypes might play a role in immune regulation.

2. Materials and methods

2.1. Animals

BALB/c ♀ mice, 2–6 months of age, purchased from Cumberland View Farms, Clinton, Tenn. USA, were used throughout the present study. These mice have been found to give a consistently good response to PC.

[I 1282]

Correspondence: Humberto Cosenza, Basel Institute for Immunology, 487, Grenzacherstraße, CH-4005 Basel 5, Switzerland

Abbreviations: B cell: Bone marrow-derived cell(s) **T cell:** Thymus-derived cell(s) **PC:** Phosphorylcholine **Pn:** Heat-killed vaccine of Pneumococci R36A **PnC:** C-polysaccharide from Pneumococci R R36A **SRC:** Sheep red cells **PnC-SRC:** SRC coated with PnC **TEPC 15-SRC:** SRC coated with TEPC 15 purified myeloma protein **PFC:** Plaque-forming cell(s) **PBS:** Phosphate buffered saline

2.2 Antigen and myeloma proteins

The heat-killed vaccine of rough strain *Pneumococci pneumoniae* R36A (American Type Culture, Rockville, Md., USA) was used as antigen (Pn). Each mouse was injected intravenously with 10^8 bacteria cells in 0.2 ml of saline.

TEPC 15 and MOPC 315 ascites were purchased from Litton Bionetics (Bethesda, Md., USA). The purified proteins were obtained after passage over the appropriate immunoadsorbent column, as described previously [6, 16].

2.3. Preparation of indicator sheep red cells (SRC)

To detect PFC against PC (idiotypic plaques) the SRC were coated with Pneumococcal C-polysaccharide (PnC) from the R36A strain as previously described [17]. The PnC contains PC and all the plaques detected were inhibited with the free hapten [15].

Plaques against TEPC 15 (anti-idiotypic plaques) were detected by coating SRC with the purified IgA monomer using $CrCl_3$ as coupling agent. Equal volumes of TEPC 15 (1 mg/ml), 50 % SRC and $CrCl_3$ (1 mg/ml) in saline were mixed in a test tube for 4 min at room temperature with occasional shaking. The cells were washed three times with PBS and resuspended to a 1 % concentration.

2.4. PFC technique

The spleen cell suspensions of 3 mice were pooled and resuspended up to 30 ml with Hanks' BSS supplemented with 10 % fetal calf serum and 10 mM HEPES. The same spleen cell suspension was assayed for plaques to SRC, PnC-SRC and TEPC 15-SRC by the Kennedy-Axelrad modification [18] of the hemolytic-plaque technique [19]. Briefly, to each poly-L-lysine-coated petri dish was added 2 ml of the appropriate SRC suspension to prepare the indicator red cell layers. After 1 h at room temperature the dishes were rinsed in phosphate buffered saline (PBS), 2.5 ml of the spleen cell suspension was carefully pipetted onto each petri dish containing the attached SRC layer and 0.1 ml of guinea pig complement, previously adsorbed with SRC, was quickly added. The dishes were incubated for 2 h at 37 °C in a 5 % CO_2 incubator. The spleen cell suspension was discarded, the dishes rinsed in PBS, and the attached red cell layer was fixed for 10 min with 0.25 % glutaraldehyde in PBS. Triplicate dishes were prepared against each indicator SRC preparation.

3. Results

3.1. Detection of anti-idiotypic plaques

The advantage of the Kennedy-Axelrad modification of the hemolytic plaque technique is that it allows the detection of very few PFC among a large number of bystander cells. It was thought that few anti-idiotypic PFC would be detected during the course of an immune response so that it would be necessary to screen large numbers of cells.

BALB/c mice were immunized i.v. with Pn and allowed to rest for 2 months. At this time, different groups of three mice were immunized daily with Pn so as to be able to assay for PFC on the same day and cover a span of time up to 12 days after immunization. At the time of assay, all groups were assayed for PFC against SRC, PnC-SRC and TEPC 15-SRC layers. Results from a representative experiment may be seen in Fig. 1. The background plaques to SRC have been subtracted from the plotted results. As controls, nonimmunized mice have low levels of idiotypic and undetectable levels of anti-idiotypic plaques. The idiotypic PFC increase exponentially from day 2 up to day 5 and then slowly decrease in numbers. In contrast, no anti-idiotypic PFC could be detected up to day 5 and only very few were found on day 6. They increased in number, at the time when the idiotypic plaques were decreasing, up to day 8 and then rapidly decreased to background levels.

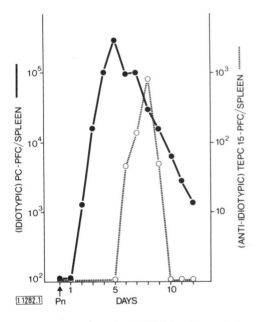

Figure 1. Detection of anti-TEPC 15 plaques during the response to PC. Different groups of 3 BALB/c mice were immunized daily with Pn. On the day of assay, the spleens from each group were pooled and each spleen cell suspension was assayed for plaques against SRC, PnC-SRC and TEPC 15-SRC as indicator cells. The background plaques to SRC have been subtracted from the results shown above.

3.2. Specific inhibition of anti-idiotypic plaques

In several experiments the peak of the anti-TEPC 15 plaques was on day 7 or 8 after immunization. Even at the peak of this response, we were able to detect a maximum of about 2000 PFC/spleen. It is therefore important to establish the idiotypic specificity of the few anti-TEPC 15 plaques. To

this end, a cell suspension was prepared from the pooled spleens of 3 BALB/c mice immunized 7 days previously with Pn and assayed for PFC against SRC, PnC-SRC and TEPC 15-SRC indicator layers. As inhibitors, purified TEPC 15 or MOPC 315 myeloma proteins (0.1 mg/ml) were added to some of the latter dishes before allowing the plaques to develop in the presence of complement. It may be seen in Table 1 that TEPC 15 completely inhibited plaque formation while MOPC 315 had no inhibitory effect. Other myeloma proteins such as MOPC 21, UPC 10 and MOPC 104E were also used as inhibitors of TEPC 15 plaques with negative results. Furthermore, when SRC were coated with TEPC 15 or MOPC 315 and then used as indicator cells, the anti-idiotypic plaques could only be detected on the TEPC 15-coated SRC.

Table 1. Specific inhibition of anti-idiotypic plaques

Indicator cells	Inhibitor	Anti-idiotypic PFC[a]
SRC	–	360 ± 52
PnC-SRC	–	4596 ± 140
TEPC 15-SRC	–	1366 ± 30
TEPC 15-SRC	TEPC 15	336 ± 46
TEPC 15-SRC	MOPC 315	1246 ± 112

a) Spleen cells assayed were obtained from the pooled spleens of 3 BALB/c mice 7 days after immunization with Pn. The results are expressed as the mean of triplicate dishes ± S.E.M. as PFC/spleen.

4. Discussion

Several recent reports have indicated that an individual, under appropriate conditions, is capable of recognizing and producing antibodies against its own immunoglobulins [10–12]. The question then arises if besides making antibody to the antigen, an individual is also capable of synthesizing anti-antibodies during the course of an immune response. The present report would indicate that in the response to PC, where the majority of the antibody bears a single idiotype, cells can also be detected which synthesize antibody to the antibodies. It is possible that after stimulation with antigen, the antibodies secreted reach a threshold concentration which in turn renders their unique V-region determinants (idiotypes) antigenic. The latter could by themselves or in combination with antigen as a carrier, stimulate anti-idiotypic cells to synthesize anti-idiotypic antibodies. The results presented here seem to support this proposition since anti-idiotypic PFC could be detected only after the idiotypic PFC had reached a peak and had actually begun to decline. It is interesting to speculate that the time of appearance of idiotypic and anti-idiotypic plaques is not just coincidental but that the latter regulates the recruitment of immunocompetent cells into the ongoing response to a particular antigen.

The response to PC is ideal for looking at the appearance of anti-idiotypic reactive cells since the antibodies secreted are of a single idiotype which is also identical with the idiotype of the TEPC 15 PC-binding myeloma [15]. However, even under these favorable conditions the detection of anti-idiotypic plaques is not an easy task and several parameters have to be well delineated. We have found coating the SRC with TEPC 15 to be of utmost importance and cell preparations coated with different concentrations of the purified myeloma protein must be tested in order to standardize the procedure.

Also, there seems to be a relationship between the appearance of anti-idiotypic plaques and the level of the response to PC, *i.e.* the plaques were more consistently detected when the anti-PC response was high. It is possible that anti-idiotypic PFC could be induced by a low anti-PC response but that they are still too few to significantly differentiate from background SRC plaques. Finally, the indicator red cell layer on the poly-L-lysine-coated dishes must be confluent and of equal density to allow for a reliable scoring of the plaques, *i.e.* if the SRC are overcoupled with the myeloma proteins a confluent layer is difficult to achieve.

Attempts to develop more anti-idiotypic plaques by adding a goat anti-mouse Ig serum at dilutions known to develop indirect plaques to SRC have met with failure. Also, in contrast to a recent report [20], we have not been able to detect anti-TEPC 15 plaques in mice rendered unresponsive to PC by a paralyzing dose of PnC. Even though no specificity controls for the TEPC 15 plaques were presented, our failure might be due to differences in the techniques used to screen for TEPC 15 plaques. It has also been reported that it is possible to detect idiotypic and anti-idiotypic antibodies in the sera of BALB/c mice hyperimmunized with Pn [21]. Besides the obvious difficulties involved in trying to rule out contamination of the serum with antigen, we have failed to detect agglutination of TEPC 15-SRC by sera from mice which were positive for splenic anti-TEPC 15 PFC. However, it could be that repeated immunization of BALB/c mice with Pn expands the anti-TEPC 15 clone and allows for the detection of anti-TEPC 15 antibodies in sera also containing large amounts of both antibodies to PC and antigen.

Other reports also seem to indicate the presence of anti-idiotypic reactive cells. It has been possible to "prime" BALB/c mice to two idiotypically noncross-reactive BALB/c myeloma proteins so that upon challenge with DNP-coupled myeloma protein, an anti-DNP response was observed only when the mice had been primed with the appropriate myeloma protein [22]. This carrier effect was sensitive to treatment with anti-Θ serum and complement, indicating that T cells recognize idiotypic determinants present on the animals' own Ig molecules. In another system [23], injection of low amounts of anti-idiotypic antibodies, directed to the A5A anti-streptococcal antibody clone, into A/J mice suppresses the expression of the A5A clone. This suppression could be repeatedly transferred by injection of spleen cells from suppressed mice into normal recipients. Thus, within the same individual there are cells which recognize the A5A idiotype and suppress its expression.

In the PC system, immunization of BALB/c mice with several idiotypically different BALB/c PC-binding myelomas led to the induction of anti-idiotypic antibodies [12]. The only exception was TEPC 15 and this was attributed to the high levels of this idiotype found in normal BALB/c mice. It is possible that this failure may be overcome by other immuni-

zation procedures or that TEPC 15 complexed with antigen is a more potent inducer of anti-TEPC 15 antibodies in BALB/c mice. To this point, we have formed complexes — in antibody excess — between Pn and TEPC 15 and after injecting these into BALB/c mice, we were also able to detect anti-TEPC 15 plaques.

In conclusion, an individual can recognize the idiotypic determinants carried on the V-regions of its own antibody molecules, and this fact opens the possibility that interaction between idiotypes and anti-idiotypes might play a role in immune regulation. However, direct functional data to substantiate this proposition still require much more intensive experimentation.

I thank Drs. A.A. Nordin and A. Augustin for reviewing the present manuscript. The technical assistance of Ms. Lyndall Poulter and Mr. John Bews is deeply appreciated.

Received October 22, 1975.

5. References

1 Claman, H.N. and Chaperon, E.A., *Transplant. Rev.* 1969. *1*: 92.

2 Hopper, J.E. and Nisonoff, A., *Advan. Immunol.* 1971. *13*: 57.

3 Kunkel, H.G., *Fed. Proc.* 1970. *29*: 55.

4 Cohn, M., Notani, G. and Rice, S.A., *Immunochemistry* 1969. *6*: 111.

5 Potter, M. and Lieberman, R., *J. Exp. Med.* 1970. *132*: 737.

6 Sirisinha, S. and Eisen, H.N., *Proc. Nat. Acad. Sci. US* 1971. *68*: 3130.

7 Eichmann, K., *Eur. J. Immunol.* 1972. *2*: 301.

8 Janeway, C.A., Jr. and Paul, W.E., *Eur. J. Immunol.* 1973. *3*: 340.

9 Granato, D., Braun, D.G. and Vassalli, P., *J. Immunol.* 1974. *113*: 417.

10 Rodkey, L.S., *J. Exp. Med.* 1974. *139*: 712.

11 McKearn, T.J., Stuart, F.P. and Fitch, F.W., *J. Immunol.* 1974. *113*: 1876.

12 Sakato, N. and Eisen, H., *J. Exp. Med.* 1975. *141*: 1411.

13 Cosenza, H. and Köhler, H., *Proc. Nat. Acad. Sci. US* 1972. *69*: 2701.

14 Jerne, N.K., *Ann. Inst. Pasteur* 1974. *125C*: 373.

15 Cosenza, H. and Köhler, H., *Science* 1972. *176*: 1027.

16 Chesebro, B. and Metzger, H., *Biochemistry* 1972. *11*: 766.

17 Liu, T.-Y. and Gotschlich, E., *J. Biol. Chem.* 1963. *238*: 1928.

18 Kennedy, J.C. and Axelrad, M.A., *Immunology* 1971. *20*: 253.

19 Jerne, N.K. and Nordin, A.A., *Science* 1963. *140*: 405.

20 Brown, P.B., Köhler, H. and Rowley, D.A., *J. Immunol.* 1975. *115*: 419.

21 Kluskens, L. and Köhler, H., *Proc. Nat. Acad. Sci. US* 1974. *71*: 5083.

22 Janeway, C.A., Jr., Sakato, N. and Eisen, H., *Proc. Nat. Acad. Sci. US* 1975. *72*: 2357.

23 Eichmann, K., *Eur. J. Immunol.* 1975. *5*: 511.

PRESENCE ON IDIOTYPE-SPECIFIC SUPPRESSOR T CELLS OF RECEPTORS THAT INTERACT WITH MOLECULES BEARING THE IDIOTYPE*

By FRANCES L. OWEN,‡ SHYR-TE JU, AND ALFRED NISONOFF

(From the Rosenstiel Basic Medical Sciences Research Center and the Department of Biology, Brandeis University, Waltham, Massachusetts 02154)

When immunized with keyhole limpet hemocyanin-*p*-azophenylarsonate (KLH-Ar),[1] all A/J mice produce anti-Ar antibodies of which a substantial proportion share a cross-reactive idiotype (CRI) (1). If, however, the mice are treated with rabbit anti-idiotypic (anti-id) antiserum before immunization, they produce anti-Ar antibodies lacking the CRI, although there is little or no effect on the total concentration of anti-Ar antibodies that are present (2, 3). Eichmann has shown that T cells from an idiotypically suppressed mouse can be used to transfer adoptively the suppressed state into a mildly irradiated syngeneic recipient (4). We have confirmed his results and have also obtained data indicating that B cells from an animal that is suppressed and then hyperimmunized are capable of transferring suppression, presumably through direct competition of a large number of B cells bearing nonidiotypic receptors with a relatively small number of B cells possessing receptors with the CRI in a nonimmunized animal (reference 5 and K. A. Ward, H. Cantor, and A. Nisonoff, unpublished data).

More recently, we have shown that T cells obtained from the spleen of a suppressed, hyperimmunized mouse form rosettes with A/J red blood cells (RBC) coated with Fab fragments of anti-Ar antibodies possessing the CRI (6, 7).[2] The responsible receptors on the T cells are synthesized endogenously and not passively adsorbed, as shown by experiments in which the receptors were removed by treatment with trypsin and reappeared upon standing in tissue culture (7).[2] In order to obtain an optimal number of idiotype-specific rosette-forming T cells, it is necessary to allow a rest period of 6–12 wk after immunization of the suppressed mice. At this time, the percentage of T cells that form rosettes generally varies between 6 and 10% of the total T-cell population. Specificity for the idiotype has been demonstrated by a variety of control experiments (6, 7).[2]

Since T cells from suppressed animals are capable of adoptively transferring suppression, it seemed possible that the rosette-forming population might include the suppressor T cells. The data to be presented here indicate that this is the case; the suppressor T-cell population can be depleted by removing those T cells which form idiotype-specific rosettes. Furthermore, the rosetted cells are

* Supported by grants AI-12895 and AI-12907 from the National Institutes of Health.

‡ Postdoctoral Fellow of the National Institutes of Health (grant no. AI-05192).

[1] *Abbreviations used in this paper:* anti-id, anti-idiotypic; Ar, *p*-azophenylarsonate; CFA, complete Freund's adjuvant; CRI, cross-reactive idiotype; KLH, keyhole limpet hemocyanin; PBS, phosphate-buffered saline, pH 7.0; RFC, rosette-forming cells; T(Ig⁻), T cells prepared by the method of Parish et al. (Materials and Methods).

[2] Owen, F. L., S-T. Ju, and A. Nisonoff. 1977. Binding to idiotypic determinants of large population of T cells in idiotypically suppressed mice. *Proc. Natl. Acad. Sci. U. S. A.* In press.

active as suppressors. The results do not establish whether all of the rosette-forming cells are suppressor cells.

Materials and Methods

Materials.   Adult A/J male mice were obtained from The Jackson Laboratory, Bar Harbor, Maine, or were first generation descendants of mice from that source. Complete Freund's adjuvant (CFA) was obtained from the Difco Laboratories, Detroit, Mich.; KLH from Calbiochem, La Jolla, Calif.; and *p*-aminophenylarsonic acid from Eastman Organic Chemicals Div., Eastman Kodak Co., Rochester, N. Y. The latter compound was recrystallized from a mixture of water and ethanol. Ficoll was obtained from Pharmacia Fine Chemicals, Inc., Piscataway, N. J. and Hypaque from Winthrop Laboratories, New York. Anti-Thy-1.2 antiserum was prepared by the method of Reif and Allen (8). The source of rabbit complement (C) and the method used for absorption are described elsewhere.[2] The activity and cytotoxicity are indicated by the data presented under Results. T cells were prepared from spleen by the method of Julius et al. (9), using nylon wool, or by a minor modification of the method of Parish et al. (10), in which spleen cells are coated with an IgG fraction of rabbit anti-mouse Fab and adsorbed to A/J RBC coated, using $CrCl_3$, with an IgG fraction of goat anti-rabbit Fab. Our modification consisted in the use of Fab fragments of rabbit anti-mouse Fab, rather than an IgG fraction, and of A/J RBC instead of sheep RBC. The B lymphocytes are adsorbed to the coated RBC while T cells, null cells, and glass-adherent cells remain free. The complexes of B cells and RBC are sedimented by centrifugation, which also removes free RBC. Purity of the T-cell preparations, estimated by treatment with anti-Thy-1.2 and C, is discussed under Results. The B cells were not used. Macrophages were not observed in rosettes.

Preparation of Anti-id Antibodies and Assays for CRI.   We have previously described the following procedures: preparation of KLH-Ar; specific purification of anti-Ar antibodies; preparation of rabbit anti-id antisera against anti-Ar antibodies of the A/J strain and proof of idiotypic specificity; radioimmunoassay for the CRI in unlabeled samples (1). The assay is based on a double antibody procedure, using 10 ng of ^{125}I-labeled specifically purified anti-Ar antibody as ligand, rabbit anti-id antiserum, and goat anti-rabbit Fc to precipitate the complexes; 0.25 mg of mouse IgG is also present in each test mixture. The presence of the idiotype in an unknown, unlabeled sample is determined through its capacity to inhibit the binding of the labeled ligand. Tests for idiotypic specificity of absorbed rabbit anti-id antiserum were carried out as described previously (1, 2). Total concentrations of anti-Ar antibodies were determined by a minor modification of the method of Klinman et al. (11), c.f., Ju et al. (12), in which the mouse antiserum is exposed to a polyvinyl surface coated with bovine serum albumin-Ar. This is followed by the addition of ^{125}I-labeled specifically purified rabbit anti-mouse IgG; the uptake of label provides a measure of the anti-Ar antibody content of the mouse serum. The method is standardized with sera containing known concentrations of anti-Ar antibodies, as determined by precipitin analysis. The useful range is 2–40 ng of anti-Ar antibody in 0.1 ml. All assays were carried out in duplicate.

Immunological Suppression of CRI.   This was accomplished by injecting two 0.5-ml portions of rabbit anti-id antiserum (total idiotype-binding capacity 60–80 μg) into male A/J mice, 6 wk of age. The injections were given 3 days apart. Starting 2 wk later, the mice were hyperimmunized with four injections of 0.5-mg portions of KLH-Ar in CFA over a period of 5 wk. The volume ratio of adjuvant to antigen was 1:1. These mice did not produce detectable quantities of the CRI but had serum titers of anti-Ar antibody between 1.0 and 6.4 mg/ml (mean, 3.9 mg/ml). The mice were allowed to rest for 8 wk before sacrifice and preparation of rosettes. The requirement for a rest period has been shown previously (6, 7).

Other Methods.   Proteins were labeled with ^{125}I by using chloramine-T (13). Fab fragments were prepared from nonspecific IgG or specifically purified anti-Ar antibodies by digestion with papain (14). They were purified by passage through DEAE-cellulose in 0.005 M phosphate buffer, pH 8.0; the Fab fragments are not retained on the column. Purity of the fragments was ascertained by Ouchterlony analysis. Fab fragments were coupled to A/J RBC with $CrCl_3$ by a modification of the method of Jandl and Simmons (15), as described elsewhere (6).[2] The Fab fragments were lightly iodinated with ^{125}I to permit quantitation of the amount coupled to the RBC; values ranged from 30 to 50 ng/10^6 RBC. Rosettes were prepared by mixing 10^7 lymphocytes with a 20-fold excess of RBC at room temperature in 0.2 ml of RPMI-1640 medium containing 5 μg/ml gentamycin, 3 $\times$

10^{-6} M glutamine, streptomycin, and penicillin. The mixture was centrifuged at 500 g for 5 min, then allowed to stand for 30 min at room temperature and 30 min in an ice bath. The cells were resuspended in the supernate and layered over a Ficoll-Hypaque solution (specific gravity, 1.077). The mixture was centrifuged for 15 min at 450 g. Free lymphocytes remained at the interface while rosettes and RBC were sedimented. Before this sedimentation, the percentage of rosette-forming cells (RFC) was determined on a 0.1-ml portion of the mixture by adding 0.1 ml of a solution containing 1% glutaraldehyde and 5% crystal violet in phosphate-buffered saline, pH 7.0 (PBS) (10). The dye selectively stains the lymphocytes. The rosettes were counted at a magnification of 630, using a Zeiss microscope fitted with a projection screen (Carl Zeiss, Inc., New York).

Results

The data presented in Tables I, II, and III were obtained by using splenic lymphocytes from 13 mice that had been suppressed with respect to the CRI, hyperimmunized with KLH-Ar, and allowed to rest for 9 wk before sacrifice. Lymphocytes were prepared from the 13 spleens and pooled.

The data in Table I show that 8.0% of the splenic lymphocytes, 5.0% of T cells prepared by passage through nylon wool, and 10.3% of T cells prepared by the method of Parish et al. (10) formed rosettes with A/J RBC coated with the Fab fragments of A/J anti-Ar antibodies possessing the CRI. No rosettes were detected when the lymphocytes were mixed with RBC coated with Fab fragments prepared from nonspecific A/J IgG.

That the receptors on the lymphocytes were reacting with idiotypic determinants is shown by the data on inhibition in Table I. 1 μl of serum containing 3.5 μg of anti-Ar antibody bearing the CRI inhibited rosette formation almost completely, whereas 5 μl of normal A/J serum, or 5 μl of A/J serum containing 18.5 μg of anti-Ar antibodies lacking the CRI, had a minimal effect on rosetting. The large percentages of lymphocytes forming idiotype-specific rosettes confirm previous observations with other suppressed mice (6, 7).

The data in Table II indicate that most of the rosette-forming lymphocytes are T cells. The results in the first three columns show the effectiveness of the anti-Thy-1.2 reagent, while the results in the last column of the table indicate that most of the rosette-forming lymphocytes are killed by anti-Thy-1.2 plus C. To obtain these lymphocytes, the rosettes were first separated from free lymphocytes by centrifuging through a Ficoll-Hypaque solution. The RBC were lysed when warmed to 37°C in the presence of the rabbit serum that was used as a source of C. This is attributable to the fragility of RBC treated with $CrCl_3$. The lymphocytes that had formed rosettes were not killed by the anti-Thy-1.2 reagent and only a small percentage were killed by exposure to C, whereas a mixture of the two reagents killed 92% of the cells. The number of cells killed was estimated by counting the number of lymphocytes remaining intact after each treatment, and which retained the ability to exclude trypan blue.

Table III presents data obtained after adoptive transfers of the T cells, prepared by the method of Parish et al. (10), from the pool of 13 spleens already described. Adoptive transfers were carried out with T cells, T cells depleted of idiotype-specific rosette-forming lymphocytes, and the rosettes themselves. Recipients were irradiated (200 R) adult A/J mice. Immunization consisted of four inoculations of 0.25 mg of KLH-Ar in CFA on days 3, 17, 24, and 31 after the adoptive transfer; a 1:1 volume-ratio of adjuvant to antigen solution was used. The immunized recipients were bled on day 35. In the control groups, which

TABLE I

Rosette-Forming Splenic T Lymphocytes in Suppressed, Hyperimmunized A/J Mice*

Red cell coat	Inhibitor	Source of lymphocytes		
		Spleen	T (nylon wool)	T(Ig⁻)‡
		% RFC		
Fab id§	None	8.0 (17/213)	5.0 (23/457)	10.3 (26/252)
Fab NIgG‖	None	0 (0/901)	0 (0/500)	0 (0/500)
Fab id	1.0 µl id serum¶	0.5 (3/602)	0.3 (1/593)	0.2 (1/503)
Fab id	5.0 µl normal A/J serum	6.3 (18/286)	3.8 (11/288)	9.6 (38/396)
Fab id	5.0 µl "suppressed" immune serum**	7.8 (85/1096)	3.9 (11/284)	10.0 (48/479)

* The mice were suppressed with respect to the CRI, then hyperimmunized with KLH-Ar as described in the text. They were allowed to rest for 9 wk before sacrifice. The spleens of 13 mice were pooled to carry out the experiments of Tables I, II, and III.

‡ Depleted of Ig-positive cells by the method of Parish et al. (see Materials and Methods). Data on sensitivity to anti-Thy-1.2 antibodies are given in Table II.

§ Fab fragments of anti-Ar antibodies possessing the CRI.

‖ Fab fragments of nonspecific A/J IgG.

¶ Containing 3.5 mg/ml of anti-Ar antibody possessing the CRI.

** Pooled serum from A/J mice, suppressed for the CRI then hyperimmunized. The serum contained 3.7 mg/ml of anti-Ar antibodies.

TABLE II

Sensitivity of Splenic Lymphocytes to Anti-Thy-1.2 Antiserum

Treatment of cells*	Spleen	T (nylon wool)	T(Ig⁻)‡	T rosettes§
	% Killed (number of viable cells are in parentheses)			
Anti-Thy-1.2	0 (211/200)‖	20 (120/150)	0 (255/250)	0 (254/250)
C	0 (213/200)	0 (150/150)	9 (229/250)	8 (230/250)
Anti-Thy-1.2 + C	51 (98/200)	74 (39/150)	82 (46/250)	92 (19/250)

* See Materials and Methods.

‡ See footnote ‡, Table I.

§ Rosettes were prepared from T cells depleted of Ig-positive cells according to Parish et al. (Materials and Methods). The procedures used caused lysis of the RBC, so that the cells counted were free lymphocytes derived from the rosettes (see text).

‖ The values represent the number of cells in a given volume of the hemocytometer chamber, with or without treatment, respectively. The control tube contained only RPMI-1640 medium and fetal calf serum. Final volumes were the same in each test. In some instances the number of cells counted in the treated samples were up to 6% higher than in the controls, owing to experimental error in counting.

received no cells, or received RBC coated with the idiotype but no lymphocytes (groups 8 and 9), the titers of anti-Ar antibodies ranged from 0.17 to 1.1 mg/ml with a mean of 0.45 mg/ml. These values are typical of titers obtained in irradiated recipients, which produce considerably lower concentrations of anti-Ar antibody than nonirradiated mice. Within a group there was no apparent correlation between the titer of anti-Ar antibody and the concentration of CRI per unit weight of antibody.

Transfer of 1×10^7 T cells (group 1) resulted in suppression of the CRI although normal titers of anti-Ar antibodies were formed by the recipients. A mixture with RBC coated with Fab fragments possessing the CRI, but under

TABLE III
Depletion of Suppressor Cells by Rosetting and Suppression by Rosettes*

Group no.	Cells transferred	No. of mice	Mean anti-Ar titer	Anti-Ar Ab required for 50% inhibition‡
			mg/ml	*ng*
1	10^7 T-enriched cells	5	1.3 (0.3–4.2)	3,000, >10,000, >25,000, >27,000, >105,000§
2	10^7 T-enriched cells + 2×10^8 RBC(id)‖	4	1.8 (0.7–3.8)	>17,500, >19,000, >52,000, >95,000
3	10^7 T-enriched cells depleted by rosetting¶	4	1.4 (0.2–5.1)	40, 69, 170, 200
4	5×10^5 T-enriched cells	9	0.5 (0.1–0.9)	56, 123, 125, 160, 260, 330, 700, 780, 840
5	5×10^5 T-enriched cells + 2×10^8 RBC(id)	4	1.0 (0.1–3.1)	160, 260, 370, 780
6	5×10^5 T-enriched cells as rosettes	9	1.8 (0.1–4.4)	4,200, >2,500, >22,500, >25,000, >25,000, >50,000, >57,500, >92,000, >118,000
7	5×10^5 T-enriched cells depleted by rosetting¶	10	0.4 (0.1–0.9)	25, 26, 31, 37, 38, 56, 63, 100, 130, 140
8	None	5	0.5 (0.1–1.1)	30, 47, 56, 56, 100
9	2×10^8 RBC(id)	5	0.4 (0.1–0.5)	17, 39, 42, 150, 163

* See footnote *, Table I, for a description of the source of donor cells. Rosettes were prepared with A/J RBC coated with Fab fragments of Anti-Ar antibodies possessing the CRI (see text). Recipient mice had received 200 R. Donors were suppressed with anti-id antiserum, hyperimmunized, and allowed to rest for 9 wk, as described in the text, before sacrifice.

‡ In the radioimmunoassay for CRI, using 10 ng of ^{125}I-labeled ligand.

§ Wherever the (>) symbol appears, 25 μl of serum was tested as inhibitor.

‖ RBC(id) refers to A/J RBC coated with Fab fragments of anti-Ar antibodies with CRI. In this group, the two types of cells were transferred as a mixture without centrifuging; i.e., not as rosettes.

¶ Using a 20:1 ratio of idiotype-coated RBC to lymphocytes.

conditions where rosettes did not form (i.e., without centrifuging to produce a pellet; group 2), had no effect on the suppressive activity of the T cells. However, when rosettes were allowed to form and were then removed by centrifugation (group 3), the remaining lymphocytes were not capable of suppressing the CRI. The adoptive transfer of 5×10^5 T cells did not induce the suppressed state (group 4), whereas transfer of the same number of rosette-forming T cells, as rosettes, was highly suppressive in each of the recipients (group 6). The adoptive transfer of 2×10^8 idiotype-coated RBC alone did not result in suppression.

Discussion

The experiments confirm our previous finding that substantial percentages of idiotype-specific rosette-forming lymphocytes can be produced under appropriate conditions (6, 7).[2] The system involves the anti-Ar antibodies of A/J mice, which exhibit strong intrastrain idiotypic cross-reactivity (1). The procedure consists, first, of suppressing the capacity to produce the CRI by administration of rabbit anti-id antiserum. Starting 2 wk later, the mice are hyperimmunized with KLH-Ar. High percentages of idiotype-specific rosette-forming T cells appear after a rest period of at least 6 wk; the percentage of RFC was much smaller when only 2 wk were allowed to elapse after the last inoculation of KLH-Ar (6, 7). Idiotype-specific RFC can also be induced in mice suppressed by adoptive transfer of lymphoid cells from suppressed syngeneic mice, rather than inoculation of anti-id antiserum (6).

Rosettes are formed with A/J RBC coated with Fab fragments of anti-Ar antibodies bearing the CRI. Evidence for idiotypic specificity included the failure to form rosettes with RBC coated with Fab fragments of nonspecific IgG; the inhibitory capacity of anti-Ar antiserum containing the CRI; and the absence of inhibition by anti-Ar antibodies lacking the CRI. Additional evidence, reported earlier (6, 7),[2] is the inhibitory capacity of anti-id antibody, present in the reaction mixture; the failure of anti-mouse IgG to cause inhibition; and the inhibition by free haptens, p-aminophenylarsonate or (p-azobenzenearsonic acid)-N-acetyl-L-tyrosine. It was also found that the receptors on lymphocytes responsible for rosette formation are synthesized by the cells and not passively adsorbed, and that most or all of the lymphocytes forming rosettes are T cells (6, 7).[2] The latter observation was confirmed in the present study (Table II). The data in Table I also indicate that T cells prepared by passage through nylon wool or by selective removal of B cells (10) are both capable of forming idiotype-specific rosettes.

The principal new observation is that the idiotype-specific rosette-forming lymphocytes include the suppressor cells. This was demonstrated in two ways (Table III). After removal of the rosette-forming lymphocytes by centrifugation, those lymphocytes remaining in the supernate showed a greatly reduced capacity to suppress the CRI upon adoptive transfer into mildly irradiated recipients. This cannot be attributed simply to the presence of idiotype-coated RBC in the mixture since the latter did not have suppressive activity (group 9, Table III). In addition, a mixture of 5×10^5 T cells plus RBC, prepared under conditions that do not permit rosette formation, was not suppressive, whereas the same number of rosetted cells caused suppression of the CRI after adoptive transfer (groups 5 and 6). Thus, the rosettes themselves were highly suppressive. The results therefore indicate that the rosette-forming T cells include the suppressor population. They do not, however, prove that all of the rosette-forming T cells are suppressor cells. This question is under investigation.

Another point of interest is the mechanism of formation of such large numbers of idiotype-specific T cells. We have previously shown[2] that injection of anti-id antiserum without subsequent hyperimmunization results in the formation of very small numbers of rosette-forming lymphocytes. The requirement for antigenic stimulation suggests that the agent which stimulates the multiplication of idiotype-specific suppressor T cells may be a complex of the idiotype with antigen. This possibility has been suggested by Eichmann (4). Although the mice were suppressed with respect to formation of the CRI, small amounts of idiotype might be produced which immediately form complexes with the excess of antigen present during hyperimmunization.

The fact that the suppressor cells interact with RBC coated with the idiotype suggests the possibility that T-cell-mediated suppression of idiotype occurs as a result of direct interaction between suppressor T cells and B lymphocytes bearing receptors with the CRI. This interpretation must be presented with some caution in view of the observation that allotype-specific suppression is mediated by T cells which interact with other T cells (16). The ability to generate large numbers of idiotype-specific T cells, and to isolate the suppressors, should facilitate the investigation of the mechanisms of induction of these cells and of suppression.

Summary

All A/J mice produce anti-p-azophenylarsonate (anti-Ar) antibodies, some of which share a cross-reactive idiotype. The idiotype can be suppressed by treatment with anti-idiotypic antiserum before immunization, although normal concentrations of anti-Ar antibodies are synthesized. We have previously reported that such suppressed mice, if hyperimmunized and then allowed to rest, contain up to 10% of splenic T cells which form rosettes with autologous RBC coated with Fab fragments of anti-Ar antibodies bearing the idiotype. Our present results indicate that the rosette-forming T cells include the idiotype-specific suppressor T-cell population. The suppressive activity is largely depleted by removal of the rosette-forming lymphocytes, and the rosettes themselves are highly suppressive. The data do not establish whether all of the idiotype-specific rosette-forming cells are suppressor cells. The system may provide a source of large numbers of suppressor cells for further study, and facilitate investigation of the mechanism of generation of idiotype-specific suppressor cells.

Received for publication 7 February 1977.

References

1. Kuettner, M. G., A. L. Wang, and A. Nisonoff. 1972. Quantitative investigations of idiotypic antibodies. VI. Idiotypic specificity as a potential genetic marker for the variable regions of mouse immunoglobulin polypeptide chains. *J. Exp. Med.* **135**:579.
2. Hart, D. A., L. L. Pawlak, and A. Nisonoff. 1972. Suppression of idiotypic specificities in adult mice by administration of antiidiotypic antibody. *J. Exp. Med.* **135**:1293.
3. Pawlak, L. L., D. A. Hart, and A. Nisonoff. 1973. Requirements for prolonged suppression of an idiotypic specificity in adult mice. *J. Exp. Med.* **137**:1442.
4. Eichmann, K. 1974. Idiotype suppression. II. Amplification of a suppressor T cell with antiidiotypic activity. *Eur. J. Immunol.* **5**:511.
5. Bangasser, S. A., A. A. Kapsalis, P. J. Fraker, and A. Nisonoff. 1975. Competition for antigen by cell populations having receptors with the same specificity, but of different idiotype. *J. Immunol.* **114**:610.
6. Ju, S-T., F. L. Owen, and A. Nisonoff. 1977. Structure and immunosuppression of a cross-reactive idiotype associated with anti-*p*-azophenylarsonate antibodies of strain A mice. *Cold Spring Harbor Symp. Quant. Biol.* **41**:in press.
7. Nisonoff, A., S-T. Ju, and F. L. Owen. 1977. Studies of structure and immunosuppression of a cross-reactive idiotype in strain A mice. *Transplant. Rev.* In press.
8. Reif, A. E., and J. M. Allen. 1966. Mouse thymus iso-antigens. *Nature (Lond.).* **209**:521.
9. Julius, M., E. Simpson, and L. A. Herzenberg. 1973. A rapid method for the isolation of functional thymus-derived murine lymphocytes. *Eur. J. Immunol.* **3**:645.
10. Parish, C. R., S. M. Kirov, N. Bowern, and R. V. Blanden. 1974. A one-step procedure for separating mouse T and B lymphocytes. *Eur. J. Immunol.* **4**:808.
11. Klinman, N. R., A. R. Pickard, N. H. Sigal, P. J. Gearhart, E. S. Metcalf, and S. K. Pierce. 1976. Assessing B cell diversification by antigen receptor and precursor cell analysis. *Ann. Immunol. (Paris)* **127C**:489.
12. Ju, S-T., A. Gray, and A. Nisonoff. 1977. Frequency of occurrence of idiotypes associated with anti-*p*-azophenylarsonate antibodies arising in mice immunologically suppressed with respect to a cross-reactive idiotype. *J. Exp. Med.* **145**:540.
13. Hunter, R. 1970. Standardization of the chloramine-T method of protein iodination. *Proc. Soc. Exp. Biol. Med.* **133**:989.
14. Porter, R. R. 1959. The hydrolysis of rabbit γ-globulin and antibodies with crystalline papain. *Biochem. J.* **73**:119.

15. Jandl, J. H., and R. L. Simmons. 1957. The agglutination and sensitization of red cells by metallic cations: interactions between multivalent metals and the red cell membrane. *Br. J. Haematol.* 3:19.

16. Herzenberg, L. A., K. Okumura, H. Cantor, V. L. Sato, F-W. Shen, E. A. Boyse, and L. A. Herzenberg. 1976. T-cell regulation of antibody responses: demonstration of allotype-specific helper T cells and their specific removal by suppressor T cells. *J. Exp. Med.* 144:330.

ANTIGEN- AND RECEPTOR-DRIVEN
REGULATORY MECHANISMS

VII. H-2-restricted Anti-Idiotypic Suppressor
Factor from Efferent Suppressor T Cells*

By MICHAEL H. DIETZ, MAN-SUN SY, BARUJ BENACERRAF, ALFRED
NISONOFF, MARK I. GREENE, AND RONALD N. GERMAIN

*From the Department of Pathology, Harvard Medical School, Boston, Massachusetts 02115; and the
Department of Biology, Brandeis University, Waltham, Massachusetts 02254*

Interactions between T lymphocyte subsets characterize virtually all well-studied models of antigen-specific immune suppression (1–11). In addition, soluble factors possessing antigen-binding sites and I region-coded determinants have been found to play major roles in these cell-cell interactions (reviewed in 12, 13). However, no clear consensus has been reached as to the precise phenotype of each participating T cell, the total number of cells involved in the overall pathway from initial antigen stimulation to actual suppression, or the importance of H-2 and/or variable region of immunoglobulin heavy chain (V_H)[1]-linked genetic restrictions at each step. At least some of the difficulty in resolving these issues arises because most studies to date appear to involve only a portion of a much larger sequence of cell interactions.

Over the past several years these laboratories have attempted to explore these issues by analyzing the sequence of T-T interactions in several T suppressor (Ts) models. Previous studies have defined in detail the production of antigen-specific T cell-derived suppressor factors (TsF) from antigen-induced first-order Ts (Ts_1) (we have termed this TsF "TsF_1") (14–17). In four different models, such TsF_1 have been shown to bear I-J-subregion-coded determinants and to function by inducing a second set of Ts, termed Ts_2, from resting T cell populations (2, 4, 17–21). This induction occurs across H-2 and background genetic differences (19, 22, 23). More-recent studies have shown that these TsF share serologically detectable idiotypic determinants with antibody of the same specificity (24, 25), and, in a model system involving suppression of delayed-type hypersensitivity (DTH) responses to the hapten azobenzenearsonate (ABA), that the Ts_2 induced by idiotypic ABA-TsF_1 are, in fact, anti-idiotypic (10). The present study continues this work by investigating the suppressive activity of such anti-idiotypic Ts_2. The data demonstrate that in contrast to Ts_1, Ts_2 are able to

* Supported by grants AI-14732, AI-16396, AI-12907, and AI-12895 from the National Institutes of Health, and by a grant from the Cancer Research Institute, Inc., New York.

[1] *Abbreviations used in this paper:* ABA, azobenzenearsonate; ABA-SC, ABA-coupled spleen cells; CRI, cross-reactive major idiotype of A/J anti-ABA antibodies; CRI-Ig, CRI-containing Ig; DNBS, dinitrobenzenesulfonate; DTH, delayed-type hypersensitivity; HBSS, Hanks' balanced salts solution; NP, 4-hydroxy-3-nitrophenyl acetyl; PBS, phosphate-buffered saline; TNP, trinitrophenyl; Ts, suppressor T cell(s); Ts_1, antigen-induced first-order Ts; Ts_2, TsF_1 induced second-order Ts; TsF, antigen-specific T cell-derived suppressor factors; TsF_1, soluble suppressor factor from Ts_1; TsF_2, soluble suppressor factor from Ts_2.

suppress already immune animals. Furthermore, the Ts_2 produce a second soluble factor, TsF_2, which lacks Ig-constant-region determinants, bears H-2-coded determinants, is anti-idiotypic, and fails to act across H-2 genetic differences. These results provide the basis for a hypothesis integrating a number of suppressor models into a single overall scheme.

Materials and Methods

Mice. A/J (H-2^a, IgH-1^e), A.By (H-2^b, Igh-1^e), B10.A (H-2^a, Igh-1^b) mice were obtained from The Jackson Laboratory, Bar Harbor, Maine. All animals in these experiments were 8–10 wk of age at the time of the experiment. All experimental groups consisted of at least four animals per group.

Preparation of Antigen and Antigen-coupled Cells. These procedures have been described in detail elsewhere (26). Briefly, a 40-mM solution of diazonium salt was prepared from *p*-arsanilic acid (Eastman Kodak Co., Rochester, N. Y.). The ABA solution was activated as previously described and conjugated to single-cell suspensions of erythrocyte-free spleen cells at a final concentration of 10 mM ABA. After washing in Hank's balanced salts solution (HBSS), the ABA-coupled spleen cells (ABA-SC) were used to induce DTH or Ts. Trinitrophenyl (TNP)-coupled spleen cells were prepared as described earlier (27).

Induction and Elicitation of DTH to ABA- to TNP-coupled Cells. To induce DTH to ABA or TNP, a total of 3×10^7 ABA-coupled syngeneic cells or TNP-coupled syngeneic spleen cells were injected subcutaneously into two separate sites on the dorsal flanks of mice. Challenge was performed four d later by injecting 30 μl of 10 mM diazonium salt of *p*-arsanilic acid or 30 μl of 10 mM 2,4,6-trinitrobenzene sulfonic acid, pH 7.4, (Eastman Kodak Co.) into the left footpad. 24 h after the footpad challenge, DTH reactivity was assessed by measuring the swelling of the footpad using a Fowler Micrometer (Schlesinger's for Tools, Brooklyn, N. Y.). The magnitude of the DTH was expressed as the increment of thickness of the challenged left footpad as compared with the untreated right footpad. Responses are given in units of 10^{-2} mm $\pm$ SEM.

Induction of Suppressor Cells and Preparation of Suppressor Factors. Normal A/J mice were injected intravenously with 5×10^7 ABA-SC. 7 d later, these mice served as donors of Ts_1 (10). Spleens from such animals were removed, and a single-cell suspension was prepared in HBSS. The cells were washed twice in HBSS and counted. To assay for the ability of such cells to inhibit ABA-specific DTH, 5×10^7 viable cells were injected intravenously into normal recipients, which were then primed subcutaneously with ABA-SC and challenged five d later. Production of TsF_1 was done precisely as described earlier (17). Briefly, 1×10^8–5×10^8 washed spleen cells in 1 ml of HBSS were subjected to alternate snap freezing at $-78°C$ and thawing at $37°C$. This was repeated four times and followed by centrifugation at 10,000 *g* for 90 min. The supernates were adjusted to 5×10^8 cell-equivalents/ml and frozen until use at $-40°C$. To test the ability of TsF_1 to inhibit ABA-specific DTH, the extract was injected intravenously into normal mice beginning at the time of immunization with 3×10^7 ABA-SC. 2×10^7 cell-equivalents of factor were administered each day for five successive d, at which time the animals were challenged and their footpad responses measured. To induce Ts_2, TsF_1 (2×10^7 cell-equivalents/d) was injected intravenously into normal A/J mice for 5 successive d (10). On day 7, spleen cells were assayed for Ts_2 activity by adoptive transfer to normal recipients, which were immunized and challenged as described above. TsF_2 was prepared from the spleens of TsF_1-treated animals in the same manner as TsF_1. TsF_2 was usually assayed by administering 2×10^7 cell-equivalents of extract/d per mouse on the day before and the day of footpad challenge of subcutaneously primed mice. As indicated in Results, however, Ts_1, TsF_1, Ts_2, and TsF_2 were also tested according to other schedules to assess afferent versus efferent suppressive activity.

Affinity Chromatography of TsF_2. Solid-phase immunoadsorbent columns were prepared and characterized as described earlier (17, 25). Soluble TsF_2 was fractionated on immunoadsorbents in the following manner. 5-ml plastic columns containing Sepharose 4B conjugated with the relevant protein were prepared and extensively washed with phosphate-buffered saline (PBS), pH 7.2, immediately before loading of the suppressor factors. The adsorption of the factor was performed at 4°C by allowing 2.4×10^8 cell-equivalents of suppressor factor in 1 ml to enter the gel matrix; the suppressor factor was then allowed to remain in the column for at least 60 min at 4°C. The column was then washed with at least five times its own void volume of PBS, pH 7.2. Such effluents were termed filtrates. Materials that remained adsorbed to the column were rapidly eluted with five bed-volumes of a glycine-HCl buffer, pH 2.8. The collected eluate was immediately neutralized to pH 7.0 with 1 N NaOH as the material emerged from the column. Both the filtrate and the eluate were concentrated to the original volume by negative-pressure dialysis at 4°C and were thereafter frozen at −40°C. Such materials were thawed and adjusted to the appropriate concentration immediately before use.

Purification of Suppressor Cells on Antigen-coated Plates. Specific suppressor cells were separated according to a modification of the method of Taniguchi and Miller (28), as reported earlier (29).

Antiserum Treatment. Anti-Thy 1.2 hybridoma antiserum was kindly provided by Dr. P. Lake, University College, London, England. 1×10^8 spleen cells from anifmls treated with TsF_1 were incubated with 1 ml of 1:20 dilution of anti-Thy 1.2 hybridoma antibody for 45 min at 0°C, washed once in HBSS, and then incubated again with 1 ml of a 1:10 dilution of Low-Tox rabbit complement (Cedarlane Laboratories Ltd., London, Ontario, Canada) for 30 min at 37°C. The cells were then washed twice in HBSS and passed over a Ficoll-Hypaque density gradient to remove the dead cells. The cells at the interface were recovered, extensively washed, counted (>90% viability as determined by trypan blue exclusion), and adjusted to the appropriate concentration for the preparation of TsF_2.

Statistical Analysis. Analysis of the significance of the differences in the data obtained was carried out using a Wang programmable computer. The means and SEM were expressed as well as the relevant P value obtained with the two-tailed Student's t test.

Results

Ts_1 and Ts_2 Differ in Their Ability to Suppress an Immunized Animal. Ts able to regulate the response of an immunized animal are frequently termed "efferent" suppressors and have been reported in several systems (30–32). This is in contrast to those Ts that act only when given before or at the time of initial antigen priming, the so-called "afferent" suppressors (31). Because Ts_1 arise before and stimulate Ts_2, the ability of these two cell types to suppress when transferred at different times during the ABA-DTH response was investigated. Ts_1 and Ts_2 cells were prepared as in Materials and Methods and transferred to sets of mice that were primed either on various days before or on the day of Ts transfer. All groups were challenged in the footpad 4 d after priming, and the DTH response assayed on day five. As shown in Fig. 1, both Ts_1 and Ts_2 suppress when given on the day of priming (afferent suppression). Ts_1 fail to decrease responses when transferred 2 or more d after priming. Ts_2, in contrast, could suppress the DTH responses of mice when transferred 2 or 3 d after priming, indicating that they could function in an efferent mode. These experiments also clearly indicate that in addition to differential receptor specificity, as reported earlier (10), Ts_1 and Ts_2 are functionally distinct suppressor T cell subsets.

The Ts_2 Capable of Suppressing When Transferred Late in an Evolving Immune Response are Anti-Idiotypic. In an earlier report (10) the Ts_2 induced by TsF_1 were shown to be anti-idiotypic by the demonstration that such cells adhered to plates coated with cross-reactive major idiotype of A/J anti-ABA antibodies (CRI)-containing mouse immunoglobulin (CRI-Ig). In that report, the anti-idiotypic cells were transferred to mice at the time of immunization. To insure that the Ts_2 that were able to act late in an ongoing immune response were the same anti-idiotypic Ts_2 described earlier, plate-separated T cells were transferred to immune mice 1 d before challenge (Fig. 2).

This experiment demonstrates that efferent suppressor cells were depleted from the whole T cell population by incubation on CRI-Ig-coated plates, but not on control

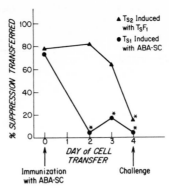

FIG. 1 Ts₁ and Ts₂ differ in their ability to s uppress an immunized animal. A/J mice were immunized with 3×10^7 ABA-SC subcutaneously on day 0. 5×10^7 Ts₁ (●), induced by intravenous injection of ABA-SC, or Ts₂ (▲), induced by TsF₁, were transferred intravenously into different groups of mice either on the day of immunization (day 0) or day 2, or day 3, or day 4. All mice were challenged 4 d after immunization with 30 μl of the diazonium salt of p-arsanilic acid as described in Materials and Methods. 24 h later, the degree of footpad swelling was determined. The percentage of suppression transferred was calculated according to the following formula:

$$\text{percent tolerance} = \left(\frac{\text{positive control} - \text{experimental value}}{\text{positive control} - \text{negative control}} \right) \times 100.$$

* Not significantly different from control.

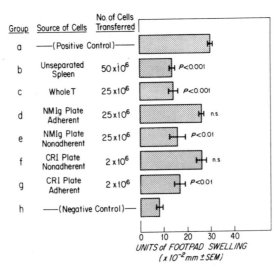

FIG. 2. The Ts₂ capable of suppressing when transferred late in an evolving immune response are anti-idiotypic. Ts₂ were obtained from A/J mice treated with 2×10^7 cell-equivalents of TsF₁ (intravenously) for 5 successive d. They were then incubated first on rabbit anti-mouse Ig plates to remove the immunoglobulin-bearing B cells. Part of the enriched T cell population was assayed for suppressor activity (group C) or subjected to further purification on plates coated either with normal A/J immunoglobulin (groups D and E) or anti-ABA antibodies containing CRI determinants (groups F and G). The various populations were transferred to different groups of A/J mice which had been immunized with 3×10^7 ABA-SC 3 d earlier to assay for efferent suppression. All the mice and the appropriate control animals were challenged with the diazonium salt 24 h after cell transfer and increase in footpad swelling was determined 24 h after challenge. The bars represent the mean footpad swelling of at least four mice/group.

Ig-coated plates. In addition, such cells were recovered only from CRI-Ig-coated plates. Therefore, the Ts$_2$ capable of inhibiting an ongoing immune response have anti-idiotypic receptors.

Ts$_2$ and a Soluble Suppressor Factor (TsF$_2$) Derived from Ts$_2$ Suppress Fully Immune Mice.   The preceding experiments, although demonstrating the ability of Ts$_2$ to act late after priming, did not prove that such Ts could inhibit the elicitation of DTH in a fully immune animal already able to give a response at the time of Ts transfer. The inability of Ts$_2$ to suppress when given on the day of challenge (Fig. 1) might indicate either that these cells cannot inhibit fully differentiated primed T cells or, alternatively, that additional time is necessary for optimal Ts$_2$ activity after transfer, either alone or in concert with additional cells of the recipient. To explore these issues, and to determine if Ts$_2$ acted via a soluble factor in analogy to Ts$_1$ and TsF$_1$, a series of experiments was carried out in which challenge was delayed until day 5, permitting Ts$_2$ or TsF$_2$ transfer beginning on day 4, a time at which DTH can be readily elicited. As shown in Fig. 3, either Ts$_2$, administered on day 4, or a soluble extract of Ts$_2$, termed TsF$_2$, given on days 4 and 5, markedly inhibited ABA DTH responses, indicating their ability to suppress the activity of already immune cells. Further, the source of TsF$_2$ is shown to be a T cell by the loss of TsF$_2$ activity from an anti-Thy 1.2 plus C-treated cell population.

Suppression by TsF$_2$ is Antigen Specific.   In an earlier report, it was shown that TsF$_1$ acted in an antigen-specific fashion (17). Because antigen-specific suppressor pathways may terminate with a nonspecific effector molecule (33–34), the antigen specificity of the TsF$_2$ was tested. The results in Fig. 4 show that TsF$_2$ was not able to suppress DTH to the hapten TNP. Hence, TsF$_2$ functions in an antigen-specific fashion.

Functional Characterization of TsF$_2$.   To further analyze the activity of TsF$_2$, a series of functional experiments comparing it to TsF$_1$ was carried out. In the experiment

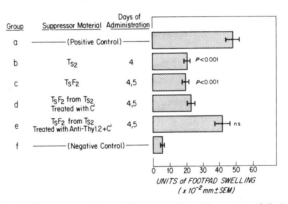

FIG. 3. Ts$_2$ and a soluble suppressor factor TsF$_2$ derived from Ts$_2$ suppress fully immune mice. Ts$_2$ were obtained from A/J mice as described in Materials and Methods. Some of the Ts$_2$ were assayed for suppressor activity by adoptive transfer (group B), whereas others were either not treated (group C) or treated with complement alone as control (group D) or with anti-Thy 1.2 antibody and complement (group E). After the antiserum treatment, viable cells were recovered by density gradient separation using Ficoll-Hypaque. These cells (groups C, D, and E) were then subjected to several freeze-thaws to produce TsF$_2$ as described in Materials and Methods. To assay for TsF$_2$ activity, A/J mice were immunized with 3×10^7 ABA-SC on day 0. On day 4 or 5, they were treated with 3×10^7 cell equivalents of TsF$_2$ (intravenously) for two successive d. All the animals and the appropriate controls were challenged on day 5 and the increased footpad swelling measured 24 h later. The bars represent the mean footpad swelling of at least four mice/group.

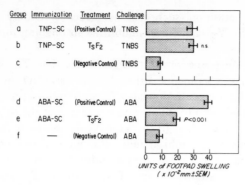

FIG. 4.   TsF₂ is antigen specific. A/J mice were immunized with either 3×10^7 ABA-SC or 3×10^7 TNP-SC subcutaneously on day 0. On days 3 and 4, they received 3×10^7 cell equivalents of TsF₂ obtained from animals injected with ABA-specific TsF₁ (groups B and E). All the animals and the appropriate controls were challenged on day 4 with the diazonium salt of p-arsanilic acid or TNBS in the footpad. 24 h later, the degree of footpad swelling was determined. The bars represent the mean footpad swelling of at least four mice/group.

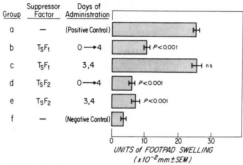

FIG. 5.   In contrast to TsF₁, which suppresses the afferent limb, TsF₂ suppresses the efferent limb of the immune response. 10^8 cell-equivalents of TsF₁ or TsF₂ were administered intravenously into different groups of A/J mice over 5-d periods (day 0 to day 4), beginning on the day of immunization with 3×10^7 ABA-SC (subcutaneously) (groups B and C), or only on the day before and the day of footpad challenge (days 3 and 4) (groups D and E). The mice were challenged with the diazonium salt on day 4. 24 h later, the degree of footpad swelling was determined.

depicted in Fig. 5, TsF₁ and TsF₂ were each administered to different sets of immunized A/J mice. They were either given over 5 d, beginning with the day of immunization and ending with the day of challenge, or they were administered only on the day before and the day of challenge. TsF₁ was able to suppress only when administered beginning with the day of immunization. TsF₂ was able to suppress when administered in either fashion. This pattern paralleled that seen with the cells, i.e., TsF₁ acts only in the afferent mode, whereas TsF₂ can act in the efferent mode.

A characteristic feature of TsF₁ studied in these laboratories is activity across H-2 differences. This contrasts markedly with data on antigen-specific TsF active in secondary humoral responses as reported by Tada et al. (35), and recent work showing that 4-hydroxy-3-nitrophenyl acetyl (NP)-specific efferent Ts failed to suppress across I-region differences (37). TsF₂ was thus tested for its ability to act across various genetic differences. Fig. 6 shows that TsF₁ from A/J mice was able to suppress DTH in the H-2-congenic strain A.BY (H-2ᵇ). On the other hand, TsF₂, produced from Ts₂ induced by the same lot of TsF₁, was unable to suppress the A.By strain when

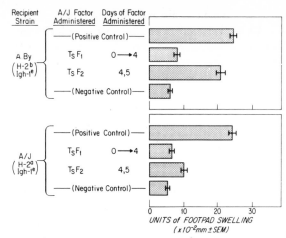

FIG. 6. H-2 restriction of ABA-TsF$_2$ activity. A total of 10^8 cell-equivalents of A/J-derived TsF$_1$ or 6×10^7 cell-equivalents of TsF$_2$ was administered on days 0–4 or 4 and 5 to A/J and A.BY mice, respectively. These mice were all immunized with ABA-syngeneic spleen cells subcutaneously on day 0 and challenged in the footpad with diazonium salt on day 5. Footpad swelling was determined 24 h later.

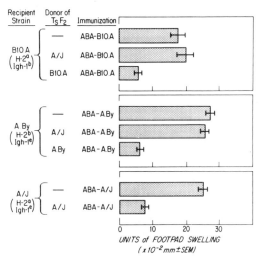

FIG. 7. Influence of H-2 and non-H-2 genes on the activity of ABA-TsF$_2$. ABA-TsF$_2$ was prepared as described in Materials and Methods from A/J, A.BY, and B10.A mice. 6×10^7 cell-equivalents of the indicated TsF$_2$ preparation was administered on days 3 and 4 to mice primed subcutaneously on day 0 with ABA-syngeneic spleen cells. On day 4, mice were challenged with diazonium salt in the footpad, and swelling determined 24 h later.

administered in the efferent mode. Thus, TsF$_2$, in contrast to TsF$_1$, cannot act across H-2 differences.

Additional experiments were then carried out to assess whether or not gene products of non-H-2 loci also played a role in restricting TsF$_2$ activity. Furthermore, it was important to insure that the inability of the A.By mice to be suppressed in the efferent mode by A/J TsF$_2$ was not because of some inherent defect in TsF$_2$ activity in A.By mice. Therefore, in the experiment depicted in Fig. 7, TsF$_2$ was produced not only in

A/J, but also in the H-2-congenic strain A.By and in B10.A, which, like A/J, is H-2^a but which differs in its background. The TsF$_2$ was produced in each of the three strains by administration of TsF$_1$ derived from the same strain. A/J-derived TsF$_2$ was only able to suppress the immune response in A/J mice. A.By and B10.A mice were nonetheless suppressed by syngeneic TsF$_2$. Thus, both H-2 and non-H-2 gene products play roles in genetically restricting the function of TsF$_2$.

Immunochemical Characterization of TsF$_2$. A/J TsF$_1$ has been shown to bear H-2-coded determinants, but no Ig constant region determinants (17, 25). In addition, TsF$_1$ binds to ABA-conjugated Sepharose 4B columns and is retained by an anti-CRI column. TsF$_2$ was similarly characterized immunochemically in a series of experiments, the results of which are shown in Table I. In these studies, the filtrates and acid eluates of TsF$_2$ passed over a variety of immunoadsorbents were tested in the efferent mode. TsF$_2$ from A/J (H-2^a) mice was retained by an anti-H-2^a column, but not an anti-H-2^q or by an anti-mouse Ig column. The activity retained on the anti-H-2^a absorbent could be recovered in the acid eluate. Furthermore, TsF$_2$ was retained by and eluted from a column to which CRI-Ig was coupled, but not to columns to which nonspecific immune A/J Ig was coupled nor to columns to which ABA was

TABLE I

Immunochemical Characterization of ABA-TsF$_2$

Experi-ment	Immunoadsorbent fraction*	Footpad size‡	Percent suppres-sion§
		$\times\ 10^{-2}\ mm\ \pm\ SEM$	
1	None (positive control)	25.2 ± 2.0	—
	RAMIg, filtrate	11.3 ± 1.1 ($P < 0.01$)	79
	RAMIg, eluate	29.8 ± 1.8 (NS)	0
	B10 anti-B10.A, filtrate	21.8 ± 2.1 (NS)	20
	B10 anti-B10.A, eluate	14.0 ± 1.5 ($P < 0.01$)	64
2	None (positive control)	23.2 ± 2	—
	Unfractionated ABA-TsF$_2$	5.0 ± 1.1 ($P < 0.001$)	100
	B10 anti-B10.A, filtrate	18.7 ± 0.9 (NS)	21
	B10 anti-B10.A, eluate	7.5 ± 0.2 ($P < 0.001$)	80
	(B10.D2 × B10.BR) anti-B10.6R, filtrate	5.5 ± 0.8 ($P < 0.001$)	86
	(B10.D2 × B10.BR) anti-B10.6R, eluate	23.0 ± 2.4 (NS)	0
3	None (positive control)	29.3 ± 3.0	—
	Unfractionated ABA-TsF$_2$	7.8 ± 1.2 ($P < 0.01$)	100
	A/J control Ig (CRI$^-$), filtrate	9.0 ± 0.9 ($P < 0.01$)	100
	A/J control Ig (CRI$^-$), eluate	30.5 ± 3.4 (NS)	0
	A/J anti-ABA Ig (CRI$^+$), filtrate	25.0 ± 1.4 (NS)	22
	A/J anti-ABA Ig (CRI$^+$), eluate	$16.5 \pm .9$ ($P < 0.01$)	65
	ABA-FγG, filtrate	13.8 ± 2.3 ($P < 0.07$)	79
	ABA-FγG, eluate	28.3 ± 1.7 (NS)	0

RAMIg, rabbit anti-mouse Ig; FγG, fowl gamma globulin.

* A/J ABA-TsF$_2$ was passed over indicated immunoadsorbent and filtrate, then acid eluate collected, and tested in the efferent mode. B10 anti-B10.A = anti-H-2^a, (B10.D2 × B10.BR); anti-B10.6R = anti-H-2K^q,I^q.

‡ P value is in comparison to positive control.

§ percent suppression = $\left(\dfrac{\text{positive control DTH} - \text{experimental DTH}}{\text{positive control DTH} - \text{negative control DTH}} \right) \times 100\%$.

coupled in the form of ABA-fowl gamma globulin. These experiments characterize TsF_2 as having H-2 determinants and anti-idiotypic specificity. It does not have anti-antigen binding activity.

Discussion

The studies reported above provide significant new information about the T cells and T cell-derived products involved in antigen-specific immune suppression of ABA DTH (10, 17, 25, 26, 29). The data indicate that antigen-specific Ts_1 and TsF_1 act in the afferent mode, before immunization, whereas Ts_2 and the soluble product of these anti-idiotypic cells, TsF_2, act in the efferent mode in fully immune animals. TsF_2 from the A/J strain has been characterized functionally and immunochemically. In contrast to TsF_1, TsF_2 fails to act across either H-2 or non-H-2 genetic differences, as shown by the observations that TsF_2 from A/J mice was not suppressive in A.By or B10.A mice. The active material lacks Ig constant region determinants and ABA-binding activity, but possesses H-2-coded determinants and has anti-idiotypic (anti-CRI) binding specificity. Similar idiotypic and anti-idiotypic TsF that suppress the idiotypic component of the humoral anti-ABA response have been reported previously (36).

Two major issues are raised by these results. The first involves the difference in mode of action between TsF_1 and TsF_2. Several earlier studies have demonstrated that some Ts act only in nonimmune animals to prevent priming, whereas other Ts can inhibit responses of immune animals (31, 37). The present report confirms these observations, in a model in which the direct relationship between the two cell types has been established. It is not clear if Ts_1 have any actual suppressive activity, or if they function only via Ts_2 induction. The time (24–48 h) needed for this inductive step would account for the inability of Ts_1 to function late in the immune response. Ts_2 clearly act in primed animals, but whether or not Ts_2 are actually effector Ts is not apparent. Their failure to suppress when given on the day of challenge may be simply a quantitative or homing problem, or may, in fact, reflect the necessity for Ts_2 to act on a third cell type which might be the actual effector suppressor cell. Whether or not the target of Ts_2 activity is another cell in the suppressor pathway or is the immune cell being suppressed, the specificity of action shown by the anti-idiotypic Ts_2 and the previous demonstration of the failure of such cells to suppress mice lacking the appropriate allotype linkage group of genes necessary for CRI expression indicate that the target cell bears idiotypic (CRI) determinants. Recent experiments have in fact identified a CRI^+ T cell distinct from the DTH effector cell that is required for Ts_2-mediated suppression (M.-S. Sy, unpublished observations).

The second major issue raised by this study involves the genetic restrictions evident in TsF_2, but not TsF_1 activity. Examining TsF obtained from Ts in the L-glutamic acid^{60}-L-alanine^{30}-L-tyrosine^{10}, L-glutamic acid^{50}-L-tyrosine^{50}, or ABA (M.-S. Sy, manuscript in preparation) models (19, 20) has not revealed the I-J restrictions of TsF function reported by Tada et al. (35). In each of these cases, TsF_1 activity across H-2 differences is quantitatively and qualitatively undistinguishable from that observed in H-2-identical animals. The finding that ABA-TsF_2 fails to act in H-2-different mice suggests strongly that one explanation for the apparent discrepancy in the earlier studies is that the TsF studied by Tada et al. (35) is a TsF_2 and not TsF_1, and that only such TsF_2 show H-2-restricted activity. This suggestion is consistent with the prolonged (4 wk) time period used to generate the keyhole limpet hemocyanin-specific Ts (38) and the ability of these cells to suppress primed T cell populations (39). It may be relevant that the suppressor cells identified by Owen et al. (40) after hyperimmunization and a prolonged rest period were anti-idiotypic in their specificity. We have recently proposed (41) an overall pathway of T cell-TsF activity based on this analysis that integrates the major features of several reported Ts models into a single scheme. In this model, antigen-specific, H-2-unrestricted TsF_1 induces Ts_2, which produces TsF_2. This material in turn acts in an H-2-restricted manner on antigen-primed idiotypic auxiliary Ts (Ts_3; T_{aux}) that mediate effector suppression.

Other groups have also demonstrated a role for H-2 gene products in Ts activity. Thus, Miller et al. (42) have shown that suppression requires identity between Ts donor and recipient at the H-2D locus in the dinitrophenyl-coupled cell model. Weinberger et al. (37) showed that NP-specific Ts_2 failed to act across I-region differences. Epstein and Cohn (43), exploring alloinduced Ts, and Jandinski et al. (44), studying TNP-modified, self-induced Ts active in humoral immune responses, observed H-2-restricted suppressor function. Rich and Rich (45) showed that TsF active in inhibiting mixed lymphocyte responses required I-C subregion identity between producer and acceptor (45). Moorehead (46, 47) found that dinitrobenzenesulfonate (DNBS)-induced TsF was also H-2K- or H-2-D-restricted in its interaction with its target cell. Recently, Yamuchi et al.[2] have found a sheep erythrocyte-specific TsF in the feedback suppression model which shows H-2 restriction. Thus, such H-2 genetic restrictions are common in suppressor systems, although the subregion involved appears to differ in the various examples, and the site in the pathway at which the genetic influence exists has not always been evaluated. In all of these studies, it is not yet clear if the restriction reflects the receptor specificity of the TsF itself, as is true of H-2-restricted cytoxic lymphocytes, the receptor specificity of the target cell, or some other interaction process controlled by H-2-linked genes.

A final point concerns the influence of background genes on TsF_2 activity. Several groups have demonstrated a role for non-H-2-, non-V_H-linked genes in Ts activity (48, 49). More recently, Weinberger et al. (37) and Eardley et al. (50) have reported on the critical role of V_H-linked genes in T-T interaction in Ts pathways. Based on the known anti-idiotypic specificity of TsF_2, and the finding that Ts_2 fail to act across allotype differences in congeneic strains of mice (50), it is most likely that the influence of non-H-2 genes on TsF_2 activity in the ABA model reflects a requirement for idiotype expression in the recipient to serve as the TsF_2 target. Thus, mapping studies currently in progress might be expected to reveal that the relevant non-H-2 genes controlling TsF_2 function are allotype-linked (V_H) genes. It should be pointed out that the requirement for idiotype expression to permit detection of Ts_2 activity can lead to detection of pseudorestrictions of TsF_1 activity. Sy et al. (51) have recently demonstrated that the apparent failure of ABA-TsF_1 (M.-S. Sy, manuscript in preparation) or CRI-coupled spleen cells to suppress CRI^- mice is not a result of a failure of Ts_2 induction by these materials because of V_H restrictions, but rather, of a failure of Ts activity because of the absence of an idiotypic target in these animals. Thus, the apparent V_H restriction of TsF_1 activity under such circumstances does not truly reflect the activity of the TsF itself but rather reflects genetic constraints on later

_____
[2] Yamuchi et al. Manuscript submitted for publication.

parts of the suppressor pathway. These findings must be kept in mind in evaluating any apparent restrictions of Ts interactions.

These experiments have extended our knowledge about the cells and factors involved in a single Ts model and pointed out the similarities and differences involved in T-T interactions in two successive stages of the ABA suppressor pathway. These observations have also been of help in understanding the relationship among the various suppressor systems in the literature. Future studies may reveal the basis for initiation of the suppressor pathway and the actual means by which Ts decrease immune responses.

Summary

Azobenzenearsonate (ABA)-specific T cell-derived suppressor factor (TsF$_1$) from A/J mice was used to induce second-order suppressor T cells (Ts$_2$). Comparison of suppressor T cells induced by antigen (Ts$_1$) with Ts$_2$ induced by TsF$_1$ revealed that Ts$_1$ were afferent suppressors active only when given at the time of antigen priming, and not thereafter, whereas Ts$_2$ could act when transferred at any time up to 1 d before antigen challenge for a delayed-type hypersentivity response. This was true even when the recipient could be shown to be fully immune before transfer of Ts$_2$, thus defining these cells as efferent suppressors. The anti-idiotypic specificity of the Ts$_2$ was demonstrated by the ability of Ts$_2$ to bind to idiotype (cross-reactive idiotype [CRI])-coated Petri dishes. A soluble extract from Ts$_2$ (TsF$_2$) was also capable of mediating efferent suppression that was functionally antigen- (ABA) specific. Comparison of TsF$_1$ with this new factor, TsF$_2$, revealed that both lack Ig-constant-region determinants, possess H-2-coded determinants, and show specific binding (to ABA and to CRI$^+$-Ig, respectively). TsF$_1$ acts in strains that differ with respect to H-2 and background genes, whereas TsF$_2$ shows H-2- and non-H-2-linked genetic restrictions. This existence of H-2 restriction of TsF$_2$ activity suggests that the apparent discrepancies in studies on H-2 restriction of TsF may be a result of the analysis of two separate classes of TsF, only one of which shows genetically restricted activity, thus unifying several models of suppressor cell activity.

We thank Ms. Ellen Morelock for her expert technical assistance and Ms. Terri Greenberg and Harriet Yake for their care in preparing this manuscript.

Received for publication 20 October 1980.

References

1. Feldmann, M., P. C. L. Beverley, J. Woody, and I. F. C. McKenzie. 1977. T-T interactions in the induction of suppressor and helper T cells: analysis of membrane phenotype of precursor and amplifier cells. *J. Exp. Med.* **145**:793.

2. Waltenbaugh, C., J. Thèze, J. A. Kapp, and B. Benacerraf. 1977. Immunosuppressive factor(s) specific for L-glutamic acid^{50}-L-tyrosine^{50} (GT). III. Generation of suppressor T cells by a suppressive extract derived from GT-primed lymphoid cells. *J. Exp. Med.* **146**:970.

3. Tada, T. 1977. Regulation of the antibody response by T cell products determines by different I subregions. *In* The Immune System: Genetics and Regulation, E. Sercarz, L. A. Herzenberg, and C. F. Fox, editors. Academic Press, Inc., New York. 345.

4. Germain, R. N., J. Thèze, J. A. Kapp, and B. Benacerraf. 1978. Antigen-specific T-cell-mediated suppression. I. Induction of L-glutamic acid-L-alanine-L-tyrosine specific suppressor T cells in vitro requires both antigen-specific T-cell-suppressor factor and antigen. *J. Exp. Med.* **147:**123.

5. Eardley, D. D., J. Hugenberger, L. McVay-Boudreau, F.-W. Shen, R. K. Gershon, and H. Cantor. 1978. Immunoregulatory circuits among T-cell sets. I. T-helper cells induce other T-cell sets to exert feedback inhibition. *J. Exp. Med.* **147:**1106.

6. Perry, L. L., B. Benacerraf, and M. I. Greene. 1978. Regulation of the immune response to tumor antigen. IV. Tumor antigen-specific suppressor factor(s) bear I-J determinants and induce suppressor T cells in vivo. *J. Immunol.* **121:**2144.

7. Araneo, B. A., R. L. Gowell, and E. E. Sercarz. 1979. Ir gene defect may reflect a regulatory imbalance. I. Helper T cell activity revealed in a strain whose lack of response is controlled by suppression. *J. Immunol.* **123:**961.

8. McDougal, J. S., F. W. Shen, and P. Elster. 1979. Generation of T helper cells in vitro. V. Antigen-specific Ly 1$^+$ T cells mediate the helper effect and induce feedback suppression. *J. Immunol.* **122:**435.

9. UytedeHaag, F., C. J. Heijnen, K. H. Pot, and R. E. Ballieux. 1979. T-T interactions in the induction of antigen-specific human suppressor T lymphocytes in vitro. *J. Immunol.* **123:**646.

10. Sy, M.-S., M. H. Dietz, R. N. Germain, B. Benacerraf, and M. I. Greene. 1980. Antigen- and receptor-driven regulatory mechanisms. IV. Idiotype-bearing I-J$^+$ suppressor T cell factors induce second-order suppressor cells which express anti-idiotypic receptors. *J. Exp. Med.* **151:**1183.

11. Weinberger, J. Z., R. N. Germain, B. Benacerraf, and M. E. Dorf. 1980. Hapten-specific T cell responses to 4-hydroxy-3-nitrophenyl acetyl. V. Role of idiotypes in the suppressor pathway. *J. Exp. Med.* **152:**161.

12. Germain, R. N., and B. Benacerraf. 1980. Helper and suppressor T cell factors. *Springer Sem. Immunopathol.* **3:**93.

13. Tada, T., and K. Okumura. 1980. The role of antigen specific T cell factors in the immune response. *Adv. Immunol.* **28:**1.

14. Kapp, J. A., C. W. Pierce, F. DelaCroix, and B. Benacerraf. 1976. Immunosuppressive factor(s) extracted from lymphoid cells of nonresponder mice primed with L-glutamic acid^{60}-L-alanine^{30}-L-tyrosine^{10} (GAT). I. Activity and antigenic specificity. *J. Immunol.* **116:**305.

15. Waltenbaugh, C., P. Debre, J. Thèze, and B. Benacerraf. 1977. Immunosuppressive factor(s) specific for L-glutamic acid^{50}-L-tyrosine^{50} (GT). I. Production, characterization, and lack of H-2 restriction for activity in recipient strain. *J. Immunol.* **118:**2073.

16. Greene, M. I., S. Fujimoto, and A. H. Sehon. 1977. Regulation of the immune response to tumor antigens. III. Characterization of thymic suppressor factor(s) produced by tumor-bearing hosts. *J. Immunol.* **119:**757.

17. Greene, M. I., B. A. Bach, and B. Benacerraf. 1979. Mechanisms of regulation of cell-mediated immunity. III. The characterization of azobenzenearsonate-specific suppressor T-cell-derived suppressor factors. *J. Exp. Med.* **149:**1069.

18. Thèze, J., C. Waltenbaugh, M. E. Dorf, and B. Benacerraf. 1977. Immunosuppressive factor(s) specific for L-glutamic acid^{50}-L-tyrosine^{50} (GT). II. Presence of I-J determinants on the GT. suppressive factor. *J. Exp. Med.* **146:**287.

19. Germain, R. N., J. Thèze, C. Waltenbaugh, M. E. Dorf, and B. Benacerraf. 1978. Antigen specific T-cell mediated suppression. II. *In vitro* induction by I-J coded L-glutamic acid^{50}-L-tyrosine^{50} (GT)-specific T cell suppressor factor (GT-T$_s$F) of suppressor T cells (T$_{s2}$) bearing distinct I-J determinants. *J. Immunol.* **121:**602.

20. Kapp, J. A. 1978. Immunosuppressive factor(s) extracted from lymphoid cells of nonre-

sponder mice primed with L-glutamic acid^{60}-L-alanine^{30}-L-tyrosine^{10} (GAT). IV. Lack of strain restriction among allogeneic, nonresponder donors and recipients. *J. Exp. Med.* **147:** 997.

21. Germain, R. N., and B. Benacerraf. 1978. Antigen-specific T cell-mediated suppression. III. Induction of antigen-specific suppressor T cells (T_{s2}) in L-glutamic acid^{60}-L-alanine^{30}-L-tyrosine^{10} (GAT) responder mice by nonresponder-derived GAT-suppressor factor (GAT-T_sF). *J. Immunol.* **121:**608.

22. Waltenbaugh, C., and B. Benacerraf. 1978. Specific suppressor extract stimulates the production of suppressor T cells. *In* Ir Genes and Ia Antigens. H. O. McDevitt, editor. Academic Press, Inc., New York. 549.

23. Pierres, M., B. Benacerraf, and R. N. Germain. 1979. Antigen-specific T cell-mediated suppression. VI. Properties of in vitro generated L-glutamic acid ^{60}-L-alanine^{30}-L-tyrosine^{10} (GAT)-specific T-suppressor factor(s) (GAT-T_sF) in responder mouse strains. *J. Immunol.* **123:**2756.

24. Germain, R. N., S.-T. Ju, T. J. Kipps, B. Benacerraf, and M. E. Dorf. 1979. Shared idiotypic determinants on antibodies and T cell derived suppressor factor specific for the random terpolymer L-glutamic acid ^{60}-L-alanine^{30}-L-tyrosine^{10}. *J. Exp. Med.* **149:**613.

25. Bach, B. A., M. I. Greene, B. Benacerraf, and A. Nisonoff. 1979. Mechanisms of regulation of cell-mediated immunity. IV. Azobenzenearsonate-specific suppressor factor(s) bear cross-reactive idiotypic determinants the expression of which is linked to the heavy-chain allotype linkage group of genes. *J. Exp. Med.* **149:**1084.

26. Bach, B. A., L. Sherman, B. Benacerraf, and M. I. Greene. 1978. Mechanisms of regulation of cell-mediated immunity. II. Induction and suppression of delayed-type hypersensitivity to azobenzenearsonate-coupled syngeneic cells. *J. Immunol.* **121:**1460.

27. Greene, M. I., M. Sugimoto, and B. Benacerraf. 1978. Mechanisms of regulation of cell-mediated immune responses. I. Effect of the route of immunization with TNP-coupled syngeneic cells on the induction and suppression of contact sensitivity to picryl chloride. *J. Immunol.* **120:**1604.

28. Taniguchi, M., and J. F. A. P. Miller. 1977. Enrichment of specific suppressor T cells and characterization of their surface markers. *J. Exp. Med.* **146:**1450.

29. Dietz, M. H., M.-S. Sy, M. I. Greene, A. Nisonoff, B. Benacerraf, and R. N. Germain. 1980. Antigen- and receptor-driven regulatory mechanisms. VI. Demonstration of cross-reactive idiotypic determinants on azobenzenearsonate specific suppressor T cells producing soluble suppressor factor(s). *J. Immunol.* **125:**2374.

30. Zembala, M., and G. L. Asherson. 1974. T cell suppression of contact sensitivity in the mouse. II. The role of soluble suppressor factor and its interaction with macrophages. *Eur. J. Immunol.* **4:**799.

31. Claman, H. N., S. D. Miller, M.-S. Sy, and J. W. Moorhead. 1980. Suppressive mechanisms involving sensitization and tolerance in contact allergy. *Immunol. Rev.* **50:**105.

32. Zembala, M., and G. L. Asherson. 1973. Depression of the T-cell phenomenon of contact sensitivity by T-cells from unresponsive mice. *Nature (Lond.).* **244:**227.

33. Asherson, G. L., and M. Zembala. 1974. T cell suppression of contact sensitivity in the mouse. III. The role of macrophages and the specific triggering of nonspecific suppression. *Eur. J. Immunol.* **4:**804.

34. Ptak, W., M. Zembala, M. Hanczakowski-Rewicka, and G. L. Asherson. 1978. Nonspecific macrophage suppressor factor: its role in the inhibition of contact sensitivity to picryl chloride by specific T suppressor factor. *Eur. J. Immunol.* **8:**645.

35. Tada, T., M. Taniguchi, and C. S. David. 1976. Properties of the antigen-specific suppressive T-cell factor in the regulation of antibody response of the mouse. IV. Special subregion assignment of the gene(s) that codes for the suppressive T-cell factor in the H-2 histocompatibility complex. *J. Exp. Med.* **144:**713.

36. Hirai, Y., and A. Nisonoff. 1980. Selective suppression of the major idiotypic component of an antihapten response by soluble T cell-derived factors with idiotypic or anti-idiotypic receptors *J. Exp. Med* **151**:1213.

37. Weinberger, J. Z., B. Benacerraf, and M. E. Dorf. 1980. Hapten-specific T-cell responses to 4-hydroxy-3-nitrophenyl acetyl. III. Interactions of effector suppressor T cells are restricted by I-A and Igh-V genes. *J. Exp. Med.* **151**:1413.

38. Takemori, T., and T. Tada. 1975. Properties of antigen-specific suppressive T-cell factor in the regulation of antibody response of the mouse. I. In vivo activity and immunochemical characterizations. *J. Exp. Med.* **142**:1241.

39. Taniguchi, M., K. Hayakawa, and T. Tada. 1976. Properties of antigen-specific suppressive T cell factor in the regulation of antibody response of the mouse. II. In vitro activity and evidence for the I region gene product. *J. Immunol.* **116**:542.

40. Owen, F. L., S.-T. Ju, and A. Nisonoff. 1977. Presence on idiotype-specific suppressor T-cells of receptors that interact with molecules bearing the idiotype. *J. Exp. Med.* **145**:1559.

41. Germain, R. N., and B. Benacerraf. Hypothesis: a single major pathway of T lymphocyte interactions in antigen specific immune suppression. *Scand. J. Immunol.* In press.

42. Miller, S. D., M.-S. Sy, and H. N. Claman. 1978. Genetic restrictions for the induction of suppressor T cells by hapten-modified lymphoid cells in tolerance to 1-fluoro-2,4-dinitro-benzene contact sensitivity. Role of the H-2D region of the major histocompatibility complex. *J. Exp. Med.* **147**:788.

43. Epstein, R., and M. Cohn. 1978. T-cell inhibition of humoral responsiveness. I. Experimental evidence for restriction by the K- and/or D-end of the H-2 gene complex. *Cell. Immunol.* **39**:110.

44. Jandinski, J. J., J. Li, P. J. Wettstein, J. A. Frelinger, and D. W. Scott. 1980. Role of self carriers in the immune response and tolerance. V. Reversal of trinitrophenyl-modified self-suppression of the B cell response by blocking of H-2 antigens. *J. Exp. Med.* **151**:133.

45. Rich, S. S., and R. R. Rich. 1976. Regulatory mechanisms in cell-mediated immune responses III. I-region control of suppressor cell interaction with responder cells in mixed lymphocyte reactions. *J. Exp. Med.* **143**:672.

46. Moorhead, J. W. 1977. Soluble factors in tolerance and contact sensitivity to DNFB in mice. II. Genetic requirements for suppression of contact sensitivity by soluble suppressor factor. *J. Immunol.* **119**:1773.

47. Moorhead, J. W. 1979. Soluble factors in tolerance and contact sensitivity to 2,4-dinitro-fluorobenzene in mice. III. Histocompatibility antigens associated with hapten dinitro-phenyl serve as target molecules on 2,4-dinitrofluorobenzene immune T-cells for soluble suppressor factor. *J. Exp. Med.* **150**:1432.

48. Taniguchi, M., T. Tada, and T. Tokuhisa. 1976. Properties of the antigen-specific suppressive T-cell factor in the regulation of antibody response of the mouse. III. Dual gene control of the T-cell-mediated suppression of the antibody response. *J. Exp. Med.* **144**:20.

49. Rich, S. S., F. M. Orson, and R. R. Rich. 1977. Regulatory mechanisms in cell-mediated immune responses. VI. Interaction of H-2 and Non-H-2 genes in elaboration of mixed leukocyte reaction suppresor factor. *J. Exp. Med.* **146**:1211.

50. Eardley, D. D., F. W. Shen, H. Cantor, and R. K. Gershon. 1979. Genetic control of immunoregulatory circuits: genes linked to the Ig locus govern communication between regulatory T-cell sets. *J. Exp. Med.* **150**:44.

51. Sy, M.-S., M. H. Dietz, A. Nisonoff, R. N. Germain, B. Benzcerraf, and M. I. Greene. 1980. Antigen- and receptor-driven regulatory mechanisms. V. The failure of idiotype-coupled spleen cells to induce unresponsiveness in animals lacking the appropriate V_H genes is caused by the lack of idiotype-matched targets. *J. Exp. Med.* **152**:1226.

Chapter **X**

The Molecular Biology of the Immune System

Chapter X

The Molecular Biology
of the Immune System

The mechanism whereby organisms can respond to the universe of antigens with specific antibody molecules has fascinated immunologists for decades. The capability of synthesizing virtually a limitless variety of protein molecules with a fixed and limited set of genetic elements poses serious questions for the one gene one polypeptide dogma of the classical molecular biologists. The development of recombinant DNA technology and the technology for producing monoclonal antibodies (derived from hybridomas) has allowed these problems to be addressed in a rigorous fashion. We can for the first time begin to ask questions as to just how many genes there are that encode antibody molecules, what their structures are and what controls their expression. Similarly, we can ask questions regarding the extent of somatically generated antibody diversity as well as the mechanisms that are responsible for it. This chapter concerns itself with the historical development of the molecular approaches to these problems and gives a brief summary of the virtual explosion of available information on the subject that has accumulated in the four years that have elapsed since the initial report from the laboratory of Dr. Susuma Tonegawa.

In a theoretical paper, Dryer and Bennett (1965) called attention to the fact that the requirements for the synthesis of antibodies posed problems for classical genetics. The inheritance and expression of the constant region (C) part of the molecule behaved as a single Mendelian character whereas the variable region (V) did not. The variable region which differs from molecule to molecule is, however, connected by normal peptide linkage to the constant part. Thus they propose that a novel mechanism must exist for synthesizing a single molecule in which the two different protein segments are joined together. They suggested that this occurred at the DNA level.

Support for the notion that protein or RNA splicing was not responsible for VC joining came from the work of Milstein (see the paper in this chapter by Kohler and Milstein) and Margulies et al. (1976). They showed that cells producing a single H chain (or L chain) could be fused to form a single hybrid cell and that the hybrid cell produced the parental chains only. Interchange between V and C parts did not take place. This result further supported the notion that rearrangement of DNA was responsible for joining segments encoding V and C regions together. After bringing these two DNA segments into proper alignment, a complete H and L chain could be produced. It was the pioneering work of Tonegawa and colleagues that showed that DNA rearrangements did indeed take place which could account for the expression of proper Ig molecules. Their initial report (Hozumi and Tonegawa), which is included in this chapter, provided the practical and theoretical framework for all of the molecular biology that follows. They showed that the V and C coding sequences existed in the DNA in non-continuous segments. These segments are disposed differently depending on whether or not the tissue from which the DNA is derived expresses immunoglobulin.

The nucleotide sequence of a light chain variable region coding segment (V_L) resident in the DNA of a cell not expressing Ig (germ-line or embryonic tissue) confirmed the fact that the C coding segment was not on a contiguous piece of DNA (this is presented in the paper of Tonegawa, et al.). The analysis of the complete nucleotide sequence of an embryonic V_L gene allowed other important conclusions to be drawn regarding the expression of immunoglobulin genes. Assuming (as they did correctly) that the embryonic V_L gene is identical in structure to the structure in germ line DNA, they concluded that the hypervariable segments seen in light chains are encoded on a contiguous piece of DNA and not as separate genetic elements which would be "free" to assort independently. Thus the variability seen amongst V_L protein sequences might have to be generated by alteration of the germ line sequence in somatic tissue (i.e. somatic mutation). This point will be addressed in more detail below. Further studies on V_L genes in cells expressing immunoglobulin (Brack et al., 1978) revealed that there was a separate non-contiguous DNA segment encoding part of the V_L protein sequence. This DNA segment is referred to as the J segment (for joining segment) and in the light chain, it encodes amino acid residues 99 through 112 (the end of the V sequence). In contrast to the variability seen within the V_L segment proper, the variability seen within, and at the boundaries of, the J segments occurs via a different mechanism. The full impact of the latter variability was realized only after the genes encoding V_L for the kappa class of light chains were studied.

The paper by Seidman, Max, and Leder, who investigated the genes encoding the kappa class of light chain, demonstrates that there are in fact 4 J segments encoding different amino acid sequences (an additional pseudo J segment was also found) that can associate independently with V_L segments. Thus, variability in sequence can be realized by combinatorial joining of different V genes with different J segments. In addition, as these authors further demonstrated, variation in sequence is generated at the junction between V and J segments. This comes about by a shift in the exact nucleotide at which the joining takes place, thus producing, in many cases, a different amino acid residue at position 99 (or 100). Structural studies indicate that this functional residue plays a key role in the antigen binding site.

Examination of the gene products of a given group of V_L genes revealed that indeed the same V gene could associate with different J segments and that junctional variability was also evident. As mentioned above, there is additional variation (from the germ line) in the sequence lying within the V gene proper (residues 1-99). Experiments measuring the number of V genes that could give rise to the group of proteins studied suggested strongly that there were more protein segments spanning residues 1-99 than genes in the germ line that could encode them. Thus, variation in protein sequence must arise via somatic mutation (Valbuena et al., 1978). With respect to the kappa light chains we can say with some assurance that different sequences can be assembled by combinational joining and junctional diversity, thereby multiplying the protein products elaborated by a single germ line V gene by about fifteen. The number of sequences that can be generated by somatic mutation from a given germ line gene is unknown as yet. A conservative estimate of 10 sequences so generated leads to a total of 150 different L chain sequences generated from a single germ line V_L gene. Estimates of the number of V_L genes give numbers approximating 300. Thus there would appear to be the potential in the mouse for generating a total of approximately 5×10^4 different kappa chains.

The mechanisms associated with the expression of the entire repertoire of V_L sequence expression are yet to be elucidated. It is clear that in order to achieve expression of a complete functional L chain, rearrangement of DNA sequence is essential. Examination of the sequence surrounding sites involved in rearrangement revealed two relevant facts. First, rearrangement is associated with deletion of the DNA between the V segment and the J segment. Second, there are nucleotide sequences existing downstream (to the 3' side) from the V segments that are homologous to sequences existing upstream from each J segment. It is reasonable to conclude that such sequences are involved in the deletion event, but that is yet to be demonstrated (Seidman et al., Sakano et al.). Since these rearrangements occur only in immunocompetent cells there must also exist a mechanism for controlling those events. Such a mechanism could account for the phenomenon of allelic exclusion. This phenomenon, in which only one of the parental chromosomes expresses its Ig product in any one lymphocyte, has been demonstrated in cells secreting immunoglobulin. The detailed mechanisms for achieving allelic exclusion are also yet to be elucidated.

The general features of the molecular biology of light chain expression are also evident in heavy chain expression. As described in the paper by Davis et al., DNA rearrangements also occur in association with H chain expression. Evidence is presented for the existence of separate V_H, J_H, and C_H segments[1]. As described in that paper by Sakano et al., 1980, an additional complexity became evident when the DNA sequences involved were studied in detail: two different types of rearrangements must occur in order to express a μ H chain or a γ_{2b} H chain. The first type is similar to that seen in L chain expression, i.e. the deletion resulting in the juxtaposition of V_H and J_H. The second type also involves a deletion juxtaposing the joined V_H-J_H segment and the $c\mu$ or $c\gamma_{2b}$. The J_H segments only exist adjacent to the $c\mu$ region. Thus, in order to express a class other than IgM a second deletion must take place allowing the V_H-J_H to be brought into juxtaposition with the appropriate constant region segment. Deletion of the genetic information encoding constant regions lying between μ and the one being expressed was independently suggested by Honjo et al. (1978) and supported by experiments of Rabbitts et al., 1980 and Corey et al., 1980. Those results together with the ones described above allow two important conclusions to be drawn. First, the order of constant regions genes on chromosome 12 of the mouse is most likely μ, γ, γ_3, γ_1, γ_{2b}, γ_{2a}, α and ε. Secondly, a cell must have the capability of expressing the μ heavy chain before it can express any other class of heavy chain.

Analysis of DNA sequences surrounding the various DNA segments involved in rearrangements revealed a feature regarding H chain expression not found in the case of L chain expression. The germ line DNA does not appear to encode necessarily the entire V_H sequence (minus J_H) in one contiguous piece of DNA. Instead there is an additional non-contiguous segment termed the D (for diversity) segment (Early et al., 1980). Examination of antibody protein sequences[1] demonstrated that V_H, D_H and J_H segments are independent and can join together in various combinations to produce an intact V_H protein sequence. As with L chain sequences this combinatorial joining allows for a large number of protein sequences to be generated from each germ line V_H segment. This is described in the paper by Schilling et al. (1980). Also reported in that paper is that base changes within V_H proper are also evident. Thus this mechanism too can give rise to variability of protein sequence. To date, we do not know how many D segments there are, what functional diversity is possible, or what the extent of mutational change is. Therefore, a guess as to how many V_H protein sequences can be produced by a mouse would likely be very inaccurate. Nonetheless, if we assume 100 V_H genes (residues 1-101), 5 D segments, 4 J segments plus an amino acid variability at each junction (5 possible residues), and six mutational variants for each germ line V_H segment, the total V_H chain sequences possible approximates to 10^5. Given the number of

----

[1] The homogeneous proteins studied in this report were the products of hybridomas selected for their ability to synthesize anti-dextran antibodies. Hybridoma technology, which enables the immortalization of spleen cells synthesizing antibodies, has become an invaluable tool for studying the biochemistry and molecular biology of immunoglobulin molecules. In addition, the technology has allowed the production of virtually limitless quantities of cells secreting antibodies specific for particular antigenic determinants. Thus, this important advance has provided a sort of technical watershed for immunologist and non-immunologist alike.

potential V_L sequences of 5×10^4, a total number of 10^9 - 10^{10} different antibody specificities could be achieved if all H and L chain combinations exist. Concomitant with the discovery that DNA rearrangements are responsible for control of Ig gene expression was the discovery that DNA encoding contiguous protein sequences need not itself be contiguous.[2] In the case of the L chain "gene" there are two introns, one between the hydrophobic leader and one between the V and C exons. In the case of the H chain "gene", introns also occur between V and C and also separate the hinge region and each of the domains of the constant region (see Sakano et al., 1979). Correct splicing of exons is obviously essential in order to produce a correct protein. Mutations in DNA that effect this process have been found and may account for various abnormal immunoglobulins produced in disease states (see Dunnick et al., 1980).

The ability to select an mRNA splicing pattern that results in alternate protein products being derived from a single RNA transcript provides a novel means for controlling gene expression. Several features of immunoglobulin expression appear to be controlled in this fashion. In a report by Rogers et al. (1980) evidence is presented that the two forms of IgM, the membrane bound form and the secreted form, are generated in this fashion. Two splice patterns, one for generating the carboxy terminus of the membrane-bound protein and one for generating the carboxy terminus of the secreted protein, exist. RNA produced by the former splice pattern encodes a protein with a hydrophobic tail used to anchor the IgM in the membrane. In contrast, mRNA resulting from the latter splice pattern excludes the information for this extra piece. We do not yet know what intra- or extracellular signals control the synthesis of these different forms of IgM.

It is known that cells can also express two H chain classes (μ and δ) containing the same V sequence simultaneously. The paper by Liu suggests that this too is the result of alternate splicing of a single transcript. Again, the controlling elements involved are not yet understood. As described in Chapter V, simultaneous expression of IgM and IgD appear to effect profoundly the immune reactivity of B lymphocytes.

The advances in molecular biology have had a dramatic impact on immunology. We can certainly expect that with the availability of homogeneous, antigen-specific populations of T cells, the genes encoding their receptors will be found and the chemical nature of those receptors elucidated. Similarly the isolation of genetic elements resident within the H-2 (and HLA) complex will help to elucidate the role of H-2 gene products and receptors in the immune response. What were heretofore intractable problems, e.g. thymic education, H-2 restriction, etc., may well reduce to the cloning and sequencing of pieces of DNA.

---

[2]DNA sequences transcribed into RNA can be deleted or spliced out such that their translation products need not appear in the cell. These sequences which do not encode a protein are referred to as intervening sequences or introns. The DNA sequences encoding proteins are referred to as encoding sequences or exons.

Papers Included in this Chapter

1. Hozumi, N. and Tonegawa, S. (1976) Evidence for somatic rearrangement of immunoglobulin genes coding for variable and constant regions. Proc. Natl. Acad. Sci. USA <u>73</u>, 3628.

2. Max, E.E., Seidman, J.G., and Leder, P. (1979) Sequences of five potential recombination sites encoded close to an immunoglobulin κ constant region gene. Proc. Natl. Acad. Sci. USA <u>76</u>, 3450.

3. Tonegawa, S., Maxam, A.M., Tizard, R., Bernard, O., and Gilbert, W. (1978) Sequence of a mouse germ-line gene for a variable region of an immunoglobulin light chain. Proc. Natl. Acad. Sci. USA <u>75</u>, 1485.

4. Davis, M.M., Calame, K., Early, P.W. Livant, D.W., Joho, R., Weissman, I.L., and Hood, L. (1980) An immunoglobulin heavy-chain gene is formed by at least two recombinational events. Nature <u>283</u>, 733.

5. Schilling, J., Clevinger, B., Davie, J.M., and Hood, L. (1980) Amino acid sequence of homogeneous antibodies to dextran and DNA arrangements in heavy chain V-region gene segments. Nature <u>283</u>, 35.

6. Kohler, G. and Milstein, C. (1975) Continuous cultures of fused cells secreting antibody of predefined specificity. Nature <u>256</u>, 495.

7. Liu, C.-P., Tucker, P.W., Mushinski, J.F., and Blattner, F.R. (1980) Mapping of heavy chain genes for mouse immunoglobulins M and D. Science <u>209</u>, 1348.

References Cited in this Chapter

Corey, S., Jackson, J., and Adams, J. (1980) Deletions in the constant region locus can account for switches in immunoglobulin heavy chain expression. Nature <u>285</u>, 450.

Dreyer, W.J. and Bennett, J.C. (1965) The molecular basis of antibody formation: A paradox. Proc. Natl. Acad. Sci. USA <u>54</u>, 864.

Dunnick, W., Rabbitts, T.H., and Milstein, C. (1980) An immunoglobulin deletion mutant with implications for heavy-chain switch and RNA splicing. Nature <u>286</u>, 669.

Early, P., Huang, H., Davis, M., Calame, K., and Hood., L. (1980) An immunoglobulin heavy chain variable region gene is generated from three segments of DNA: V_H, D, and J_H. Cell <u>19</u>, 981.

Honjo, T. and Katavka, T. (1978) Organization of immunoglobulin heavy chain genes and allelic deletion model. Proc. Natl. Acad. Sci. USA <u>75</u>, 2140.

Margulies, D.H., Kuehl, W.M., and Scharff, M.D. (1976) Somatic cell hybridization of mouse myeloma cells. Cell <u>8</u>, 405.

Rabbitts, T.H., Forster, A., Dunnick, W., and Bentley, D.L. (1980) The role of gene deletion in the immunoglobulin heavy chain switch. Nature <u>283</u>, 351.

Rogers, J., Early, P., Carter, C., Calame, K., Bond, M., Hood, L., and Well, R. (1980) Two mRNAs with different 3' ends encode membrane-bound and secreted forms of immunoglobulin μ chain. Cell <u>20</u>, 303.

Sakano, H., Huppi, K., Heinrich, G., and Tonegawa, S. (1979) Sequences at the somatic recombination sites of immunoglobulin light-chain genes. Nature 280, 288.

Sakano, H., Maki, R., Kurosawa, Y., Roeder, W., and Tonegawa, S. (1980) Two types of somatic recombination are necessary for the generation of complete immuno-globulin heavy-chain genes. Nature 286, 676.

Seidman, J.G., Max, E., and Leder, P. (1979) Sequences at the somatic recombination sites of immunoglobulin light-chain genes. Nature 280, 370.

Valbuena, O., Marcie, K.B., Weigert, M., and Perry, R.P. (1978) Multiplicity of germline genes specifying a group of related mouse κ chains with implications for the generation of immunoglobulin diversity. Nature 276, 780.

Evidence for somatic rearrangement of immunoglobulin genes coding for variable and constant regions

(κ-chain mRNA/restriction enzymes/RNA·DNA hybridization)

Nobumichi Hozumi and Susumu Tonegawa

Basel Institute for Immunology, 487, Grenzacherstrasse, CH-4058 Basel, Switzerland

Communicated by N. K. Jerne, July 2, 1976

ABSTRACT   A high-molecular-weight DNA from Balb/c mouse early embryo or from MOPC 321 plasmacytoma (a κ-chain producer) was digested to completion with *Bacillus amyloliquefaciens* strain H restriction enzyme (*Bam*H I). The resulting DNA fragments were fractionated according to size in preparative agarose gel electrophoresis. DNA fragments carrying gene sequences coding for the variable or constant region of κ chains were detected by hybridization with purified, ^{125}I-labeled, whole MOPC 321 κ mRNA and with its 3′-end half. The pattern of hybridization was completely different in the genomes of embryo cells and of the plasmacytoma. The pattern of embryo DNA showed two components, one of which (molecular weight = 6.0 million) hybridized with C-gene sequences and the other (molecular weight = 3.9 million) with V-gene sequences. The pattern of the tumor DNA showed a single component that hybridized with both V-gene and C-gene sequences and that is smaller (molecular weight = 2.4 million) than either of the components in embryo DNA. The results were interpreted to mean that the V_κ and C_κ genes, which are some distance away from each other in the embryo cells, are joined to form a contiguous polynucleotide stretch during differentiation of lymphocytes. Such joining occurs in both of the homologous chromosomes. Relevance of these findings with respect to models for V–C gene joining, activation of a specific V_κ gene, and allelic exclusion in immunoglobulin gene loci is discussed.

Both light and heavy chains of immunoglobulin molecules consist of two regions: the variable region (V region) and the constant region (C region) (1, 2). Uniqueness (i.e., one copy per haploid genome) of the genetic material coding for C region ("C gene") has been conjectured from normal Mendelian segregation of allotypic markers (3). Nucleic acid hybridization studies have confirmed this notion (4–10). Hybridization studies have also demonstrated that a group of closely related V regions are somatically generated from a few, conceivably even a single, germline gene(s) (V gene) (4–6). They did not, however, give us any reliable estimate of the *total* number of germ line V genes. However, given the enormous diversity of V regions, the existence of multiple germ line V genes seems likely. If this is so, a problem arises: how is the information in the V and C genes integrated to generate a contiguous polypeptide chain? Since V- and C-gene sequences exist in a single mRNA molecule as a contiguous stretch (11), such integration must take place at either the DNA or the RNA level. Integration at the RNA level could result from a "copy-choice" event during transcription or from joining of two RNA molecules after transcription.

We report here experimental evidence for possible joining of V and C sequences at the DNA level.

MATERIALS AND METHODS

Myeloma Tumors. Balb/c mouse plasmacytoma MOPC 321 was kindly provided by Dr. M. O. Potter. Tumors were maintained in mice as described (4).

Abbreviations: V and C genes, genes coding for variable and constant regions, respectively; M_r, molecular weight.

Bam H I Restriction Endonuclease. *Bacillus amyloliquefaciens* strain H, originally from Dr. F. Young, was obtained from Dr. T. Bickles at the Biozentrum, Basel. Cells were grown in L-broth. The *Bam*H I endonuclease was prepared according to the method of Wilson and Young (12), except that 10% glycerol was present in all buffers during the phosphocellulose step. Five milligrams of high-molecular-weight embryo or MOPC 321 tumor DNA in a buffer consisting of 6 mM Tris·HCl, pH 7.4, 6 mM MgCl₂, and 6 mM 2-mercaptoethanol was incubated with 10^4 units (1 unit is defined as the amount of the enzyme sufficient for digesting 1 μg of phage λ DNA in 60 min at 37° under the above conditions) of the purified enzyme at 37° for 4 hr. In order to investigate completeness of digestion we followed the following scheme. An aliquot of the reaction mixture containing mouse DNA was removed before incubation and mixed with a small amount (ratio of λ DNA to mouse DNA, 1:10) of phage λ DNA. This pilot mixture was incubated in parallel with the main reaction mixture under the conditions described above. After incubation the pilot mixture was electrophoresed in 0.9% agarose. DNA was visualized under ultraviolet light after staining with 1 μg/ml of ethidium bromide (13). Digestion of mouse DNA in the main reaction mixture was considered to be complete when the electrophoresis pattern of the λ DNA in the pilot mixture exhibited no indication of incomplete digestion.

Preparative Agarose Gel Electrophoresis. An apparatus conventionally used for separation of serum proteins was adapted for nucleic acids. Agarose (Sigma, electrophoresis grade) was melted in TA buffer (20 mM Tris-CH₃CO₂, pH 8.0, 18 mM NaCl, 2 mM Na₂EDTA, 20 mM NaCH₃CO₂) at a final concentration of 0.9% and cast in a Plexiglas tray (40 × 50 × 1 cm) with a central longitudinal partition. Agarose was allowed to solidify at room temperature. With the aid of a scalpel, 5-mm wide slots were made in the agarose at 10 cm from one end. The DNA digested with *Bam*H I was concentrated with 2-butanol by the procedures of Stafford and Bieber (14) and dialyzed against TA buffer. The DNA solution was supplemented with 10 × concentrated TA buffer and bromphenol blue. The mixture was warmed to 47°. Melted agarose was mixed with the warmed DNA solution such that the final mixture contained the original TA buffer, 0.01% dye, and 0.45% agarose. The mixture was transferred to the slot in the agarose sheet and was allowed to solidify at room temperature. In order to avoid overloading, the amount of the DNA was kept at less than 2.1 μg/mm² of the cross section of the gel. The gel was covered with plastic wrap and electrophoresis was carried out at 4° for 3 days. For the first 20 hr the electric current was kept at 0.75 mA/cm² of gel cross section; it was raised to 3 mA for the remainder of the run.

Extraction of DNA from Agarose Gel. After electrophoresis the gel was cut with a scalpel into 5-mm thick slices. Each slice was transferred to a test tube containing one volume of 5 M

NaClO$_4$ (15). The gel slices were melted by heating at 60° for 60 min with occasional shaking by hand. In order to remove dissolved agarose as well as NaClO$_4$, DNA was absorbed to a small column of hydroxyapatite (Bio-Gel, DNA grade) and then eluted with 0.4 M potassium phosphate, pH 6.9.

Hybridization. DNA eluted from hydroxyapatite was sonicated in a MSE sonicator to yield fragments 400–500 base pairs long. Purified, sonicated *Escherichia coli* DNA (Sigma type VIII) was added at a final concentration of 100 μg/ml. The mixture was dialyzed against water overnight, and DNA was precipitated with 2 volumes of ethanol. The precipitate was collected by centrifugation, dried under reduced pressure, and dissolved in 25 μl of a mixture consisting of 0.3 M NaCl, 0.03 M sodium citrate, 0.01 M Tris HCl, pH 7.5, 0.001 M Na$_2$EDTA, and 700–1500 cpm of ^{125}I-labeled mRNA or its 3′-end fragment. The hybridization mixture was covered with a thin layer of mineral oil and heated at 98–100° for 5 min. Annealing was carried out at 70° for 40–48 hr. The processing of the hybridization mixture has been described (4).

Preparation of High-Molecular-Weight DNA. DNA was purified from either 12- to 13-day-old Balb/c mouse embryos (Bomholtgard, Denmark) or MOPC 321 tumors according to procedures described by Gross-Bellard *et al.* (16).

Other Procedures. Methods for purification of κ-chain mRNA of MOPC 321, for isolation of 3′-end half fragment of the mRNA, and for RNA iodination have all been described (4, 6). The κ mRNA preparation used in the present work is the one described in Fig. 1 of ref. 6.

RESULTS

Detection of V- and C-gene sequences in DNA fragments

A straightforward procedure for quantitation of a specific sequence in a particular DNA preparation is to carry out hybridization either in liquid or preferentially on filter paper under conditions where labeled RNA is well in excess over DNA (RNA excess, RNA-driven hybridization) (17). For this procedure it is necessary that the RNA preparation is completely pure. The κ mRNA preparation used in the present experiments, although as pure as the present technology permits, does not meet this condition. From the "fingerprint" analysis of T1 RNase digest and from gel electrophoresis in 99% formamide, the purity of the κ mRNA preparation was estimated to be about 90% (4, 6). For this reason and also because of the difficulty of obtaining a large amount of purified κ mRNA, the RNA saturation procedure is not suitable for quantitation of immunoglobulin genes.

When the purpose is to detect a particular DNA sequence in relative rather than absolute terms, as in the present case, an alternative method is available which circumvents the above problems. In Fig. 1 a varying amount of MOPC 321 DNA was hybridized with ^{125}I-labeled κ mRNA of MOPC 321 or the half containing its 3′-end. The volume of hybridization mixtures, concentration of RNA, and time of incubation were constant. Under these conditions the level of ^{125}I-labeled RNA in the hybrid is a function of the concentration of the complementary DNA sequences. As shown in Fig. 1, the increment in the level of hybridization decreases as the concentration of complementary DNA sequences increases. Thus, when DNA is subjected to a fractionation procedure such as gel electrophoresis, and each fraction is assayed for a specific DNA sequence by this method, the shape of the resulting hybrid peak is flatter than expected from the actual distribution of complementary DNA sequences. (In principle, it is possible to correct each

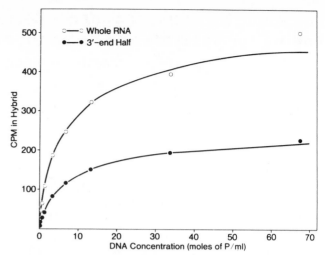

FIG. 1. Dependency of hybridization level on DNA concentration. ^{125}I-Labeled, whole κ mRNA of MOPC 321 (1300 cpm) or its 3′-end half fragment (600 cpm) was annealed with varying indicated amounts of denatured MOPC321 DNA in 0.3 M NaCl, 0.03 M sodium citrate (pH 7) at.70° for 42 hr. The volume of annealing mixture was 0.05 ml. Specific activity of the RNA was 6.5 × 10^7 cpm/μg. Each point represents an average of duplicate measurement. Intrinsic RNase-resistant counts, 22 cpm for whole RNA and 16 cpm for 3′-end half, are subtracted.

fraction for the nonlinearity of the assay by using Fig. 1 or a "standard curve". We have not done this, however, in Fig. 2.) This disadvantage is more than counterbalanced by the critical advantage; namely, that hybridization of minor RNA impurity does not overshadow that of the major RNA component. The method is particularly suitable where the impurity is distributed among many different RNA species, for in this case, DNA sequences complementary to the contaminating RNAs would be distributed among many different fractions, and hybridization of any particular RNA impurity would not stand out of the general background level. As previously shown, κ mRNA preparations have impurities of exactly this type (4, 6).

Since a RNA probe for V gene sequences is not available (the 3′-end half fragment is a RNA probe for C-gene sequences), V-gene sequences were determined indirectly from the difference in the two hybridization levels obtained by the whole molecule and the half-molecule containing the 3′-end (3′-half). For this purpose two RNA probes with identical specific activity were prepared; hybridization of whole molecules was carried out using twice as much input radioactivity as with 3′-end half-molecules. The final hybridization levels obtained with the whole molecules are just about twice those with the 3′-half in a wide range of DNA concentration (Fig. 1); this justified such an indirect measurement of V-gene sequences.

V- and C-gene sequences in embryo and myeloma DNA fragments generated with *Bam*H I endonuclease

High-molecular-weight embryo and MOPC 321 tumor DNAs were digested with *Bam*H I endonuclease, and the resulting DNA fragments were fractionated according to size by preparative agarose gel electrophoresis. DNA eluted from gel slices was assayed for V- and C-gene sequences by hybridization with ^{125}I-labeled whole or 3′-half κ mRNA of MOPC 321 under the condition described in the last section. The pattern of hybridization is shown in Fig. 2. With embryo DNA, two DNA components of molecular weight (M_r) 6.0 and 3.9 million hybridized with the whole RNA molecules, whereas only the 6.0 million M_r component hybridized with the 3′-half. The fact that only the 6.0 million M_r component exhibits any hybridization

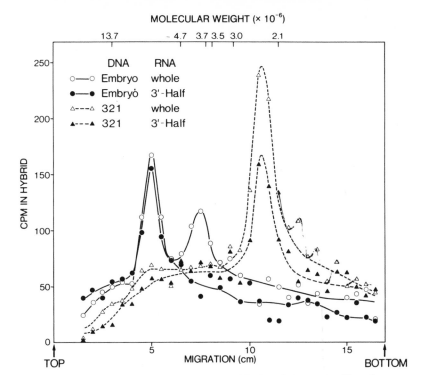

FIG. 2. Gel electrophoresis patterns of mouse DNA fragments, generated by *Bam*H I, carrying V- or C-gene sequences. ^{125}I-Labeled, whole κ mRNA of MOPC 321 (1220 cpm) or its 3′-end half fragment (600 cpm) was annealed with DNA extracted from gel slices. Intrinsic RNase-resistant counts are subtracted. Conditions of electrophoresis and hybridization are as described in *Materials and Methods*. Hybridization patterns with DNA of two different sources, embryo and MOPC 321 tumor, are superimposed. The molecular weight scale was obtained from phage λ DNA, digested by *Eco*RI (*E. coli*) endonuclease, which was electrophoresed in parallel with mouse DNAs.

above the background level with the 3′-end half indicates that not only the C-region sequences, but also the sequences corresponding to the untranslated region at the 3′-end of mRNA molecule, are in this component. Furthermore, since the 6.0 million M_r component hybridizes almost equally with the two RNA probes, it should not contain appreciable V gene sequences. Thus, the 3.9 million M_r component should contain V gene sequences as well as sequences corresponding to the untranslated region at the 5′-end of mRNA molecule.

Since V and C genes are in separate DNA fragments whose size is much larger than the size of either gene, they are probably some distance away from each other in the embryo genome. However, the possibility that the enzyme cleaved contiguously arranged V and C genes near the boundary is not entirely eliminated. Cleavage sites are possible in the nucleotides coding for the amino acids at positions 93–95 and 99–100 (18). However, the probability that either of these amino acid sequences provides the exact nucleotide sequence required is low.

The pattern of hybridization is completely different in the DNA from the homologous tumor (Fig. 2). Both RNA probes hybridized with a new DNA component of M_r 2.4 million. There are no indications that either of these RNA probes hybridizes with other DNA components above the general background level. These results indicate that both V and C genes, or the entire sequences represented in the mRNA molecule [except for the unlabeled poly(A)sequence], are contained in the 2.4 million M_r component in the tumor genome. The whole RNA hybridizes with this component nearly twice as well as does the 3′-end half, thereby supporting this notion. Hence, the $V_κ$ and $C_κ$ genes, which are most likely some distance away from each other in the embryo genome, are brought together in the plasma cells expressing this particular $V_κ$ gene, presumably to form a contiguous nucleotide stretch. The fact that

neither the 6.0 nor the 3.9 million M_r component exists in the plasma cell genome indicates that such rearrangement of immunoglobulin genes takes place in all of the homologous chromosomes.

DISCUSSION

We have shown that the pattern of *Bam*H I DNA fragments that carry immunoglobulin V- or C-gene sequences is completely different in the genomes of mouse embryo cells and of a murine plasmacytoma. The pattern of embryo DNA shows two components, one of which (M_r = 6.0 million) hybridizes with C-gene sequences and the other (M_r = 3.9 million) with V-gene sequences. The pattern of tumor DNA shows a single component that hybridizes with both V-gene and C-gene sequences and that is smaller (M_r = 2.4 million) than either of the components in embryo DNA. The straightforward interpretation of these results is that $V_κ$ and $C_κ$ genes, which are some distance away from each other in the embryo cells, are joined to form a contiguous polynucleotide stretch during differentiation of lymphocytes. Such joining occurs in both of the homologous chromosomes.

An alternative explanation of the results, namely, that accumulation of mutations or base modifications leading to either loss or gain of *Bam*H I sites generated the observed pattern difference, is not impossible. On this view, there would have to be a *Bam*H I site close to the V–C junction in embryo DNA. This *Bam*H I site would have to be lost by mutation or by base modification in the MOPC 321 tumor. By itself, such an alteration would cause the appearance of a single 9.9 million M_r component in the tumor. To achieve the M_r of the single component actually observed in the tumor (2.4 million), there would have to be new *Bam*H I sites created by mutation between the $V_κ$ gene and the nearest site on either side. Since there is no reason why there should be any selective pressures in-

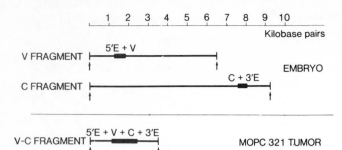

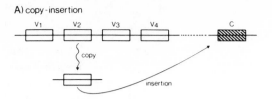

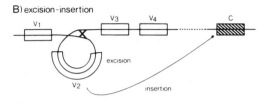

FIG. 3. Mouse DNA fragments carrying V_κ and C_κ genes. DNA fragments were generated by *Bam*H I restriction endonucleases. Arrows indicate *Bam*H I sites. 5′E and 3′E designate base sequences corresponding to untranslated regions of a κ-chain mRNA molecule at the 5′- and 3′-end, respectively. V and C designate base sequences corresponding to variable and constant regions, respectively. The relative position of these sequences within each fragment is deduced from the present results within the framework of either of the latter three models depicted in Fig. 4.

volving *Bam*H I sites, the occurrence of three base alterations would seem to be quite unlikely. Furthermore, such pattern changes are not unique to this particular combination of enzyme and DNA. With *Eco* RI enzyme and embryo and HOPC 2020 DNA (a λ-chain producer) we have obtained results that lead us to a similar conclusion (Hozumi and Tonegawa, unpublished).

There is also an alternative explanation for the absence of the embryonic DNA components in the tumor, namely, that the V–C gene joining took place in only one of the homologous chromosomes and that the other chromosome(s) has been lost during propagation of the tumor. In view of the known chromosome abnormalities of murine plasmacytomas, we cannot eliminate this trivial possibility.

Fig. 3 summarizes our interpretation of the experimental results. What is the mechanism by which the integration of V- and C-gene sequences is brought about? In the past, several models have been proposed, and some of these models are schematically illustrated in Fig. 4. In the "copy-insertion" model a specific V gene is duplicated and the copy is inserted at a site adjacent to a C gene (19). In this model, the embryonic sequence context of a V gene should be retained in the lymphocyte expressing that particular V gene. Since the embryonic DNA fragment carrying the MOPC 321 V gene does not seem to exist in the genome of this tumor, our results are clearly incompatible with this model. In the "excision-insertion" model, a specific V gene is excised into an episome-like structure, and this in turn is integrated adjacent to a C gene (20). In the "deletion" model, DNA in the interval between a particular V gene and the corresponding C gene loops out, is excised, and diluted out upon subsequent cell multiplication (21). In the "inversion" model, V genes and a C gene are arranged in a chromosome in opposite directions and a segment of chromosome between a particular V gene (inclusive) and a C gene (exclusive) is inverted. The latter three models are both consistent with the experimental results presented here.

It is easy to think of variants of these models. In one variant of the copy-insertion model, the C gene is copied and the copy is inserted next to the V gene. In another variant, both V and C genes are copied, and the copies inserted at a third location. These variants are just as incompatible with our results as is the copy-insertion model illustrated in Fig. 4. There are analogous variants of the "excision-insertion" model and they are just as compatible with the experimental results as the excision-insertion model illustrated in Fig. 4.

A "committed" bone marrow-derived (B) lymphocyte and

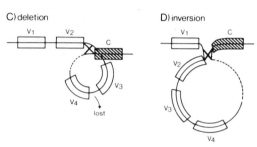

FIG. 4. Models for V–C gene joining at DNA level. See *text* for explanation.

a plasma cell produce antibody of only one specificity (22–24). In particular, it expresses only one light chain V gene. If there are multiple V_κ genes, there must exist a mechanism for the activation of one particular V_κ gene. In the light of the present findings, one intriguing possibility is that activation of a V gene is intimately coupled with its joining to a C gene. For instance, in the "excision-insertion" model, a promotor site may be created by the insertion of the excised V-DNA fragment. This would activate that particular V gene for transcription. In fact, on this view the current terminology of "V genes" and "C genes" is inappropriate. Rather, there are two segments of DNA, one specifying the V region and the other specifying the C region. The gene is *created* by joining.

For immunoglobulin loci, only one allele is expressed in any given lymphocyte (22). This is not the case with most autosomal genes. Our results suggest an interesting explanation for allelic exclusion—that is, that the two homologous chromosomes are, in any given plasma cell, homozygous. Homozygosis could result from the loss of one homologue followed by reduplication of the other or from somatic recombination between the centromere and the immunoglobulin locus (25, 26). The alternative view, that joining takes place in both chromosomes, presents a problem if there is more than one V segment. There is no intrinsic reason why the same V segment should be joined on both homologues.

The present results, when combined with those of our previous reports, demonstrate that both content and context of immunoglobulin genes are altered during differentiation of lymphocytes. Whether such a genetic event is common in other eukaryotic gene systems remains to be seen.

We thank Mr. G. R. Dastoornikoo, Ms. R. Schuller, and Ms. S. Stutz for their expert technical assistance. The "inversion model" was suggested by Dr. C. Steinberg.

1. Hilschmann, N. & Craig, L. C. (1965) *Proc. Natl. Acad. Sci. USA* **53,** 1403–1409.
2. Titani, K., Whitley, E., Avogardo, L. & Putnam, F. W. (1965) *Science* **149,** 1090–1092.
3. Milstein, C. & Munro, A. J. (1970) *Annu. Rev. Microbiol.* **24,** 335–358.
4. Tonegawa, S., Steinberg, C., Dube, S. & Bernardini, A. (1974) *Proc. Natl. Acad. Sci. USA* **71,** 4027–4031.
5. Tonegawa, S., Bernardini, A., Weimann, B. J. & Steinberg, C. (1974) *FEBS Lett.* **40,** 92–96.
6. Tonegawa, S. (1976) *Proc. Natl. Acad. Sci. USA* **73,** 203–207.
7. Faust, C. H., Diggelmann, H. & Mach, B. (1974) *Proc. Natl. Acad. Sci. USA* **71,** 2491–2495.
8. Rabbitts, T. H. (1974) *FEBS Lett.* **42,** 323–326.
9. Stavnezer, J., Huang, R. C. C., Stavnezer, E. & Bishop, J. M. (1974) *J. Mol. Biol.* **88,** 43–63.
10. Honjo, T., Packman, S., Swan, D., Nau, M. & Leder, P. (1974) *Proc. Natl. Acad. Sci. USA* **71,** 3659–3663.
11. Milstein, C., Brownlee, G. G., Cartwright, E. M., Javis, J. M. & Proudfoot, N. J. (1974) *Nature* **252,** 354–359.
12. Wilson, G. A. & Young, F. E. (1975) *J. Mol. Biol.* **97,** 123–125.
13. Sharp, P. A., Sugden, B. & Sambrook, J. (1973) *Biochemistry* **12,** 3055–3063.
14. Stafford, D. W. & Bieber, D. (1975) *Biochim. Biophys. Acta* **378,** 18–21.
15. Fukes, M. & Thomas, C. A., Jr. (1970) *J. Mol. Biol.* **52,** 395–397.
16. Gross-Bellard, M., Dudet, P. & Chambon, P. (1973) *Eur. J. Biochem.* **36,** 32–38.
17. Birnstiel, M. L., Sells, B. H. & Purdon, I. F. (1972) *J. Mol. Biol.* **63,** 21–39.
18. McKean, D., Potter, M. & Hood, L. (1973) *Biochemistry* **12,** 760–771.
19. Dreyer, W. J., Gray, W. R. & Hood, L. (1967) *Cold Spring Harbor Symp. Quant. Biol.* **32,** 353–367.
20. Gally, J. A. & Edelman, G. M. (1970) *Nature* **227,** 341–348.
21. Kabat, D. (1972) *Science* **175,** 134–140.
22. Pernis, B., Chiappino, G., Kelus, A. S. & Gell, P. G. H. (1965) *J. Exp. Med.* **122,** 853–876.
23. Mäkelä, O. (1967) *Cold Spring Harbor Symp. Quant. Biol.* **32,** 423–430.
24. Raff, M. C., Feldman, M. & De Petris, S. (1973) *J. Exp. Med.* **137,** 1024–1030.
25. Ohno, S. (1974) in *Chromosomes and Cancer,* ed. German, J. (John Wiley & Sons, New York) pp. 77–94.
26. Gally, J. A. (1967) *Cold Spring Harbor Symp. Quant. Biol.* **32,** 166–168.

Sequences of five potential recombination sites encoded close to an immunoglobulin κ constant region gene

(J segment/intervening sequence/palindromes/antibody diversity)

EDWARD E. MAX, J. G. SEIDMAN, AND PHILIP LEDER

Laboratory of Molecular Genetics, National Institute of Child Health and Human Development, National Institutes of Health, Bethesda, Maryland 20205

Contributed by Philip Leder, May 7, 1979

ABSTRACT   Immunoglobulin κ chain gene formation involves site-specific somatic recombination between one of several hundred germ-line variable region genes and a joining site (or "J segment") encoded close to the constant region gene. We have cloned and determined the nucleotide sequence of major portions of the recombination region of the mouse κ gene and discovered a series of five such J segments spread out along a segment of DNA 2.4 kilobases from the κ constant region gene. These J segments encode the 13 COOH-terminal amino acids of the variable region, probably including amino acids involved in the antigen combining site and in heavy/light chain contacts. The J segments also display striking sequence homology to one another in both their coding and immediately flanking sequences. Major elements of a short palindrome—CAC(^T_A)GTG—are preserved adjacent to the recombination sites of both variable and J region genes and constitute inverted repeats at both ends of the sequences to be joined. These palindromes can be written as a hypothetical stem structure that draws variable and J regions together, providing a possible molecular basis for the DNA joining event. Four of the J segments that we have discovered encode amino acid sequences already found in myeloma proteins. By altering the frame of recombination, we can account for additional light chain amino acid sequences, suggesting that the V/J joining event might generate antibody diversity somatically both by using different combinations of variable and J region genes and by using alternative joining frames.

Formation of a complete immunoglobulin gene involves a recombination event that seems to be critical to the development of immunoglobulin diversity as well as to the activation of immunoglobulin genes (1–4). Such recombination takes place between a germ-line variable (V) region gene and a distant segment of DNA, the J region, that encodes the 13 amino acids conventionally associated with the carboxy end of the V region. Evidently, the J segment contains two important signals: one for DNA recombination and one for RNA splicing. This has been demonstrated directly for mouse λ (5–7) and κ (8–10) light chain genes by using cloned segments of DNA derived from embryonic and antibody-producing myeloma cells. Even after recombination, however, light chain genes remain in three discrete coding segments separated from one another by two intervening sequences of DNA. Early studies of heavy chain genes suggest that this organization is a general pattern related to immunoglobulin domains (11, 12).

We have found that a J-region gene located 3.7 kilobases (kb) from the κ constant (C) region gene is joined in a myeloma cell to the germ-line gene corresponding to the V region of the MOPC-41 light chain (10). Except for the recombination event, the germ-line and expressed sequences are identical, suggesting that no further somatic alteration was involved in creating this gene (10). Moreover, we noted that the germ-line κ J-region had several interesting coding and flanking sequence homologies

to a λ J-region segment described by Bernard et al. (5). Here we extend our sequence information in the region surrounding this germ-line κ J segment. We have found five potential J-region genes, regularly spaced at intervals of 309–354 base pairs, on the 5′ side of the C region gene. Four of these correspond to amino acid sequences found in myeloma light chains. It is clear that, by altering the frame of recombination between germ-line V and J-region genes, other light chain sequences could be accounted for. Moreover, the amino acid sequence encoded by the light chain J and its heavy chain analogue might influence the character and conformation of the complementarity-determining region of an antibody (the antigen-combining site) (13). Various combinations of V and J regions could, therefore, be an important source of immunoglobulin diversity.

Finally, the five J-region sequences retain regions of homology that may be critical to both DNA recombination and RNA splicing. In particular, they retain major elements of the palindromic sequence CAC(^T_A)GTG that occurs in homologous positions adjacent to the recombination sites in germ-line κ and λ V-region genes. This sequence forms the basis for a hypothetical stem structure that may be drawn between all V- and J-region gene sequences thus far determined and that may represent an intermediate in the recombination between V- and J-region genes.

MATERIALS AND METHODS

Recombinant DNA Techniques. The cloned, 16-kb embryonic mouse κ C region fragment and derivative fragments used in these sequence studies have been described (9). All experiments were carried out in accordance with the National Institutes of Health Guidelines on Recombinant DNA Research at an EK2/P2 level of containment.

Sequencing Technique. The partial chemical degradation method of Maxam and Gilbert (14) was used for all sequencing. The restriction enzyme sites, sequencing strategy, overlaps, and extent of bidirectional sequencing are shown in Fig. 1. Our experience with thin (0.3 mm) gels suggests that sequences determined in a single direction are subject to <5% error. We have tried to use the most reliable portions of each run and have sequenced critical regions twice.

Computer Applications. The sequences have been set in format by using a program, PUBTRANS, written by Jacob V. Maizel, Jr. Homology and dyad symmetry searches were carried out by using the program of Korn and Queen (15).

RESULTS AND DISCUSSION

Sequence of Germ-Line DNA Fragment Containing Five Potential J-Region Genes and the κ C Region Gene. We have found (10) that the recombination event that formed the MOPC-41 light chain gene occurred 3.7-kb from the germ-line κ C region gene, adjacent to a J segment that carries codons for

Abbreviations: V, variable; kb, kilobase(s); C, constant.

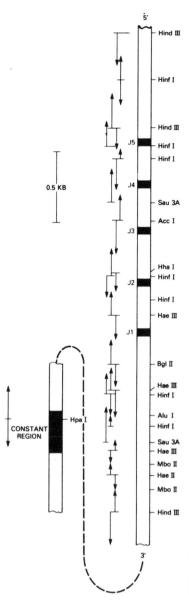

FIG. 1. Strategy for determining sequence of germ-line J- and C-region genes. DNA for sequencing was derived from a 6.5-kb *Eco*RI/*Bam*HI fragment cloned in pBR322. This plasmid was initially cleaved with *Hin*dIII, *Hin*dIII/*Xba* I/*Bam*HI, or *Hin*dIII/*Eco*RI and the resulting fragments were isolated by electrophoresis on polyacrylamide gels followed by electrophoretic elution. These fragments, as well as others generated by the indicated further cleavages, were end-labeled with ^{32}P by treatment with bacterial alkaline phosphatase followed by T4 polynucleotide kinase and [γ-^{32}P]ATP. The double-end-labeled fragments were cleaved at internal restriction sites to obtain single-end-labeled fragments which were subjected to the partial chemical degradation sequence analysis described by Maxam and Gilbert (14). Final electrophoresis on four separate 0.3-mm polyacrylamide/urea gels yielded autoradiograms that were readable up to about 250 base pairs from the labeled end.

gions are present in the 1.5-kb segment on the 3′ side of the last regularly occurring J segment. In addition, within 150 base pairs to the 5′ side of J5 there is a region of reiterated DNA that may represent a border of an otherwise unique sequence. This work has been done with BALB/c mice; analysis of κ light chain amino acid sequences (22) and mRNA precursors (23) in NZB myelomas suggests the existence of multiple J-region genes in NZB mice as well.

The Recombination Event: Clues Provided by Inverted Repeat Sequences Adjacent to the V- and J-Region Recombination Site. The structural homology shared among the five J-segments (Fig. 3) suggests a common ancestral gene. The homology in the coding sequence is striking. Excluding the J3 sequence, 10 of the 13 codons are unchanged except in a few third-base positions (see below). It is likely that the preserved coding sequence is of significance for the configuration of the immunoglobulin molecule, but it may play some further role in DNA or RNA joining as well. Two other regions of preserved homology are of interest. These are the G- and T-rich sequence on the 5′ side of each J segment (underlined in Fig. 3 and also partially preserved in the V-region genes) and the short palindrome—CAC(^T_A)GTG—present with some minor variations adjacent to each J-region recombination site (marked with an arrow in Fig. 3). The palindromic sequence is present adjacent to each germ-line V-region recombination site thus far sequenced (marked with an arrow in Fig. 3). These palindromes are complementary inverted repeat sequences and therefore can be visualized as forming an inversion loop or stem structure drawing the V and J sequences together. An example of such a stem structure, drawn between the J5 sequence and the VK41 sequence, is shown in Fig. 4. Similar structures can be drawn between each sequenced κ J and κ V gene as well as between the λ J and its V-region sequence (6, 7).

Although the significance of these palindromic sequences remains to be evaluated, some signal must ensure reasonably precise recombination between V- and J-region segments. If joining V and J does in fact involve the recognition of these short palindromes, it is significant that they do not become a part of the recombined sequence as does, for example, the integration core sequence of phage λ (24). From our studies on the sequences of the MOPC-41 recombinant and its precursors (10), we have deduced the specific nucleotides that were joined by the recombination event that formed the gene. These are shown by asterisks in Fig. 4. Both lie at the base of the hypothetical stem structure; the stem is eliminated by the recombination event. Different nucleotide pairs would be joined in the case of alternative frames of recombination (see below). The lower panel of Fig. 4 shows diagrammatically how both strands of the double helix might interact to form such a stem structure: in one region each strand interacts with its usual complementary partner, whereas in an adjacent region each strand interacts with a distal sequence of the same strand. Presumably, this structure, requiring the disruption of normal interactions of double-stranded DNA, would be thermodynamically unstable and would require special enzymes for formation. Given its completely hypothetical nature (e.g., the relative order of the V and J/C genes is not known), the major advantage of this structure would be in allowing recombination to be accomplished by several known enzymatic activities. Joining the DNA strands at the base of such a stem could be achieved by a nicking enzyme and DNA ligase, or a nicking-closing enzyme, or by bypassing the stem with a DNA polymerase. Any of these mechanisms would eliminate the stem while joining the appropriate V/J sequences. These short palindromic repeats are also reminiscent of the ends of translocatable drug-resistance elements (25), the insertion sequence ISI (26), and the integratable human virus Ad2 (27, 28).

amino acids 96–108 of the MOPC-41 light chain. We have now determined the sequence of extensive portions of the region surrounding this segment. The extent of the determined sequence (about 3 kb) and a major portion of the sequence itself is shown in Fig. 2. The sequence reveals a rather striking pattern: beginning with the J segment used to form the active MOPC-41 gene and repeating every 309–354 bases for the next 1.2 kb, a J-region segment appears, identifiable by its homology to the J expressed in MOPC-41. We number these regions J1 through J5, starting at the 3′ side (closest to the C region). Each J region encodes a 13 amino acid sequence; and four of the sequences—J5, J4, J2 and J1—correspond to J segments found in the myeloma light chains of MOPC-41, MOPC-21, MOPC-149, and MOPC-511, respectively (Table 1). The remaining sequence, J3, is discussed below.

No further J regions in the sequenced segment were detected by several computer searches for homology to the five known J regions. It is possible that other κ J regions are encoded in unsequenced portions of this fragment or in fragments not yet cloned. However, the latter seems unlikely because we have not been able to detect other J-containing *Eco*RI fragments by using this J-containing segment as a probe in *in situ* hybridization experiments (unpublished data). Furthermore, no other J re-

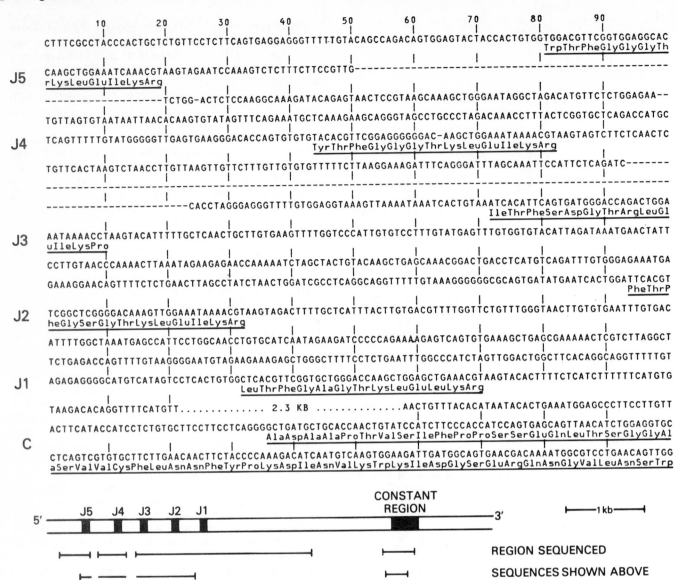

FIG. 2. Sequence of region containing κ J- and C-region genes. Coding regions of the genes are indicated by the amino acids below the nucleotide sequence. The dashes indicate undetermined nucleotides. The lengths of these unsequenced regions are estimated by restriction analysis. The diagram below indicates the extent of the sequence determined and shows which portion of the known sequence is given above. The J regions were identified by their homology with the J of the expressed MOPC-41 gene (J5). They are numbered from the J closest to the C region.

V/J Recombination as Mechanism for Generating Diversity. How can the set of five J-region sequences that we have found be reconciled with the existence of more than five different amino acid sequences in the J region of known κ chains? Table 1 lists the amino acid sequences corresponding to our germ line J regions as well as all the known BALB/c J regions derived from published amino acid sequence studies (29, 30). There are four J regions that correspond to germ-line genes and five that do not correspond to any germ-line J. It is apparent that each of these five non-germ-line sequences may be derived from one of the germ-line regions by changing the codon at position 96. We can explain the non-germ-line sequences by assuming that the recombination enzymes allow some flexibility in the frame of recombination. For instance, the recombination event that joined VK41 with J5 occurred between the second and third nucleotides of codon 95 (CCG):

$$\text{VK41} \qquad \text{J5} \qquad \text{recombined}$$
$$\underline{\text{CCT}}\ \underline{\text{CCC}} \times \underline{\text{TGG}}\ \underline{\text{TGG}} \rightarrow \underline{\text{CCG}}\ \underline{\text{TGG}}. \qquad [10]$$

If the same sequences recombined between the first and second nucleotides of codon 96, the resulting sequence (CCT CGG) would code for an arginine in position 96, corresponding to the non-germ-line J sequence shown as sequence 2 in Table 1. Table 2 indicates all the codons that could be generated at position 96 by shifting the recombination frame between VK41 and the four germ-line J-region genes. In addition to three of the known non-germ-line J regions, several other J regions that have not yet been observed could be generated. If this model is correct, these J sequences may be found as more light chains are sequenced. If we make an additional assumption—namely, that other VK genes have different codons at position 96—then all the known J-region amino acid sequences could be generated.

Role of the J Segment in the Antibody Molecule. From what is known of the three-dimensional structure of M603, a phosphocholine-binding IgA, it appears that amino acid substitutions in position 96 (which is a part of the complementarity-determining region) would alter the properties of the antigen-binding site (13). Furthermore, positions 96 and 98 of the J region are important for contacts between heavy and light chains, and this might also affect the configuration of the antigen-binding site (31). Analogous positions play a similar role

```
GluAspPheValAspTyrTyrCysLeuGlnTyrAlaSerSerPro        ━━━━━━▶
GAAGATTTTGTAGACTATTACTGTCTACAATATGCTAGTTCTCCTCCCACAGTGATACAAATCATAACTAAACCCCATGAAAGTAGAAA        VK41
                                         ------------------------------------------------------
```

```
GluAspPheGlySerTyrTyrCysGlnHisPheTrpSerThrPro        ━━━━━━▶
GAAGATTTTGGGAGTTATTACTGTCAACATTTTTGGAGTACTCCTCCCACAGTGATTCAAGCCATGACATAAACCATGCAGGGAAGCAGA        VK2
                                         ------------------------------------------------
```

```
GluAspPheGlySerTyrTyrCysGlnHisPheTrpSerAlaPro        ━━━━━━▶
GAGGATTTTGGGAGTTATTACTGTCAACATTTTTGGAGTGCTCCTCCCACAGTGATTCAAGCCATGACATAAACCATGCAGGGAAGCAGA        VK3
                                         ------------------------------------------------
```

```
                                         ◀━━━━━━  TrpThrPheGlyGlyGlyThrLysLeuGluIleLysArg
GGAGGGTTTTTGTACAGCCAGACAGTGGAGTACTACCACTGTGGTGGACGTTCGGTGGAGGCACCAAGCTGGAAATCAAACGTAAGTAG        J5
------------
```

```
                                                            *
                                         ◀━━━━━━  TyrThrPheGlyGlyGlyThrLysLeuGluIleLysArg
GCTCAGTTTTTGTATGGGGGTTGAGTTAAGGGACACCAGTGTG  TACACGTTCGGAGGGGGGACXAAGCTGGAAATAAAAACGTAAGTAG        J4
                                         TG
------------
```

```
                                         *          *  *              *                *
                                         ◀━━━━━━  IleThrPheSerAspGlyThrArgLeuGluIleLysPro
GGAGGG TTTTGT  GGAGGTAAAGTTAAAATAAATCACTGTAAATCACATTCAGTGATGGGACCAGACTGGAAATAAAAACCTAAGTAC        J3
------ ------
```

```
                                         *              *
                                         ◀━━━━━━  PheThrPheGlySerGlyThrLysLeuGluIleLysArg
GGCAGGTTTTTGTAAAGGGGGCGCAGTGATATGAATCACTG  GATTCACGTTCGGCTCGGGGACAAAGTTGGAAATAAAAACGTAAGTAG        J2
------------
```

```
                                         *              *              *
                                         ◀━━━━━━  LeuThrPheGlyAlaGlyThrLysLeuGluLeuLysArg
GGCAGGTTTTTGTAGAGAGGGGCATGTCATAGTCCTCACTGTGGCTCACGTTCGGTGCTGGGACCAAGCTGGAGCTGAAACGTAAGTAC        J1
------------
```

FIG. 3.   Nucleotide sequences at the recombination sites of several germ-line V- and J-region genes. K2 and K3 are closely related κ V regions (*Upper*) identified and cloned on the basis of cross hybridization to a cDNA probe specific for the expressed MOPC-149 gene (3). More extensive sequences are published elsewhere for K2 and K3 (3) and for VK41 (10). The J region (*Lower*) sequences of the present work have been aligned to facilitate comparison with each other and with the V regions. Homologous regions are underlined. The septanucleotide (boldface arrows) appearing (with minor variations) in the V regions as well as in the J regions may be important in the V/J recombination event and is discussed in the text. The asterisks identify amino acids that differ from the J5 sequence.

in the heavy chain (32). It seems, therefore, that the different J regions would affect the diversity not only of the primary structure of the antibody molecule but also of the antigen-combining site itself.

The Peculiar J3 Sequence Violates the GT/AG Excision Rule. The sequence encoded by J3 does not correspond to any published BALB/c myeloma sequence. It has alterations in three amino acid positions—99, 103, and 108—which are invariant in all known BALB/c J regions, as well as an aspartate in position 104, which also has not been observed among known BALB/c J regions. Moreover, all the other J-region genes end in the dinucleotide G-T, consistent with the splicing out of an intervening sequence beginning with a G-T and ending with an A-G. Almost all known intervening sequences have the form G-T. . .A-G (33–35). The J3 sequence, however, has a C-T instead of a G-T at the 5′ border of the intervening sequence. The significance of this alteration is not known, but if it interferes with RNA processing, the sequence would obviously not be expressed. If this were the case, it is further possible that the selective forces that act on an expressed sequence no longer maintain the functional sequence of J3. Hence, this J sequence may represent some intermediate stage in the evolutionary deterioration of a J gene.

Table 1.   Germ-line J-region gene translations and observed J-region amino acid sequences

Amino acid translation of germ-line BALB/c J-region genes	Amino acid sequences of J regions found in κ chains of BALB/c mice		
		96	108
J5 W T F G G G T K L E I K R*	1.	W T F G G G T K L E I K R*	
	2.	R - - - - - - - - - - - -	
J4 Y - - - - - - - - - - -*	3.	Y - - - - - - - - - - -*	
	4.	P - - - - - - - - - - - -	
J3 I - - S D - - R - - - -P*			
J2 F - - - S - - - - - - -*	5.	F - - - S - - - - - - -*	
	6.	W - - - S - - - - - - -	
	7.	I - - - S - - - - - - -	
J1 L - - - A - - - - L - -*	8.	L - - - A - - - - L - -*	
	9.	I - - - A - - - - L - -	

W = Trp, T = Thr, F = Phe, G = Gly, K = Lys, L = Leu, E = Glu, I = Ile, R = Arg, Y = Tyr, P = Pro, S = Ser, D = Asp, A = Ala. The myeloma proteins used as examples are identified and referenced as follows: 1, MOPC 41 (16); 2, MOPC 173 (17); 3, MOPC 21 (18); 4, MOPC 11 (19); 5, MOPC 149 (J. G. Seidman, unpublished results); 6, MOPC 321 (20); 7, X 24 (21); 8, MOPC 511 (E. Apella, personal communication); 9, TEPC 601 (21).
* Germ-line sequence.

Table 2.   Codon diversity created by alternative frames of recombination between germ-line, V-, and J-region genes

Germ-line genes	Junction sequence	Codons generated	
VK41	C C C	TGG - Trp*	(1)
		CGG - Arg*	(2)
J5	T G G	CCG - Pro*	(4)
		CCC - Pro*	(4)
VK41	C C C	TAC - Tyr*	(3)
		CAC - His	
J4	T A C	CCC - Pro*	(4)
VK41	C C C	TTC - Phe*	(5)
		CTC - Leu	
J2	T T C	CCC - Pro	
VK41	C C C	CTC - Leu*	(8)
		CCC - Pro	
J1	C T C		

* Accounts for a light chain of known sequence, corresponding to number in Table 1 as indicated by numbers in parentheses.

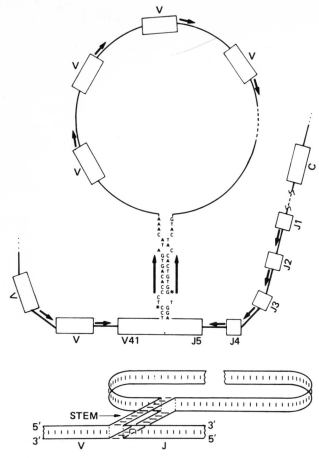

FIG. 4. Hypothetical stem structure formed between inverted repeats located next to germ-line V- and J-region genes. Above is a diagram of hypothetical intermediate in the chromosomal rearrangement κ V-, J-, and C-region genes. The distance between V- and J/C-region genes is not known and this is indicated by the broken line. It must be emphasized that the relative order of the V and J/C genes is also not known. However, preliminary evidence (unpublished results) suggests that the recombination event is accompanied by a deletion of sequences 5′ to the recombining J segment. Each V- and J-region gene is bordered by a palindrome that is also an inverted repeat sequence (boldface arrow) located on the 3′ side of the V genes and on the 5′ side of the J genes. Each of these sequences can be written as a complementary stem structure as shown in the figure in which the actual sequence of MOPC-41 recombinant is used as an example. The bases marked with asterisks are those actually joined to form the recombinant (10). The resulting amino acid codon sequence is indicated. Similar stem structures which draw V- and J-regions together can be written for each κ and λ V- and J-sequence now known (3, 5, 7). Below is a three-dimensional representation of the hypothetical stem intermediate. V- and J-coding sequences interact with their opposite strands in a normal DNA duplex, but the inverted repeat adjacent to the V gene is drawn as a stem interacting with its complement located on the same strand, presumably many thousands of bases away adjacent to a J sequence.

We are grateful to Dr. Jacob V. Maizel, Jr., for his assistance in establishing the computer programs used in these studies and to Dr. David Davies for his helpful comments regarding the potential significance of the J segment in the immunoglobulin molecule. We are also grateful to Terri Broderick for her expert assistance in the preparation of this manuscript.

1. Dreyer, W. J. & Bennett, J. C. (1965) *Proc. Natl. Acad. Sci. USA* **54**, 864–868.
2. Hozumi, N. & Tonegawa, S. (1976) *Proc. Natl. Acad. Sci. USA* **73**, 3628–3632.
3. Seidman, J. G., Leder, A., Edgell, M. H., Polsky, F., Tilghman, S. M., Tiemeier, D. C. & Leder, P. (1978) *Proc. Natl. Acad. Sci. USA* **75**, 3881–3885.
4. Rabbitts, T. H. & Forster, A. (1978) *Cell* **13**, 319–327.
5. Bernard, O., Hozumi, N. & Tonegawa, S. (1978) *Cell* **15**, 1133–1144.
6. Brack, C., Hirama, M., Lenhard-Schuller, R. & Tonegawa, S. (1978) *Cell* **15**, 1–14.
7. Tonegawa, S., Maxam, A. M., Tizard, R., Bernard, O. & Gilbert, W. (1978) *Proc. Natl. Acad. Sci. USA* **75**, 1485–1489.
8. Seidman, J. G., Leder, A., Nau, M., Norman, B. & Leder, P. (1978) *Science* **202**, 11–17.
9. Seidman, J. G. & Leder, P. (1978) *Nature (London)* **276**, 790–795.
10. Seidman, J. G., Max, E. E. & Leder, P. (1979) *Nature (London)*, in press.
11. Early, P. W., Davis, M. M., Kaback, D. B., Davidson, N. & Hood, L. (1979) *Proc. Natl. Acad. Sci. USA* **76**, 857–861.
12. Sakano, H., Rogers, J. H., Hüppi, I., Brack, C., Traunecker, A., Maki, R., Wall, R. & Tonegawa, S. (1979) *Nature (London)* **277**, 627–633.
13. Padlan, E. A., Davies, D. R., Rudikoff, S. & Potter, M. (1976) *Immunochemistry* **13**, 945–949.
14. Maxam, A. & Gilbert, W. (1977) *Proc. Natl. Acad. Sci. USA* **74**, 560–564.
15. Queen, C. L. & Korn, J. (1979) *Methods Enzymol.* **64**, in press.
16. Gray, W. F., Dreyer, W. J. & Hood, L. (1967) *Science* **155**, 465–467.
17. Schiff, C. & Fougereau, M. (1975) *Eur. J. Biochem.* **59**, 525–537.
18. Svasti, J. & Milstein, C. (1972) *Biochem. J.* **128**, 427–444.
19. Smith, G. P. (1978) *Biochem. J.* **171**, 337–347.
20. McKean, D., Potter, M. & Hood, L. (1973) *Biochemistry* **12**, 749–759.
21. Rao, D. N., Rudikoff, S. & Potter, M. (1978) *Biochemistry* **17**, 5555–5559.
22. Weigert, M., Gatmaitan, L., Loh, E., Schilling, J. & Hood, L. (1978) *Nature (London)* **276**, 785–790.
23. Perry, R. P., Kelley, D. E. & Schibler, U. (1979) *Proc. Natl. Acad. Sci. USA*, in press.
24. Landy, A. & Ross, W. (1977) *Science* **197**, 1147–1160.
25. Kleckner, N. (1977) *Cell* **11**, 11–23.
26. Ohtsubo, H. & Ohtsubo, E. (1978) *Proc. Natl. Acad. Sci. USA* **75**, 615–619.
27. Garon, C. F., Berry, K. W. & Rose, J. A. (1972) *Proc. Natl. Acad. Sci. USA* **69**, 2391–2395.
28. Wolfson, J. & Dressler, D. (1972) *Proc. Natl. Acad. Sci. USA* **69**, 3054–3057.
29. McKean, D. J. & Potter, M. (1979) *ICN–UCLA Symposium on T- and B-Cell Lymphocytes: Recognition and Function*, in press.
30. Kabat, E. A., Wu, T. T. & Bilofsky, H. (1979) *Sequences of Immunoglobulin Chains*, in press.
31. Padlan, E. A., Davies, D. R., Pecht, I., Givol, D. & Wright, C. (1976) *Cold Spring Harbor Symp. Quant. Biol.* **41**, 627–637.
32. Rao, D. N., Rudikoff, S., Krutzsch, H. & Potter, M. (1979) *Proc. Natl. Acad. Sci. USA*, **76**, 2890–2894.
33. Breathnach, R., Benoist, C., O'Hare, K., Gannon, F. & Chambon, P. (1978) *Proc. Natl. Acad. Sci. USA* **75**, 4853–4857.
34. Catterall, J. F., O'Malley, B., Robertson, M. A., Staden, R., Tanaka, Y. & Brownlee, G. G. (1978) *Nature (London)* **275**, 510–513.
35. Konkel, D., Tilghman, S. & Leder, P. (1978) *Cell* **15**, 1125–1132.

Sequence of a mouse germ-line gene for a variable region of an immunoglobulin light chain

(λ light chain/hypervariable region/DNA sequencing/interspersed noncoding sequences)

Susumu Tonegawa*, Allan M. Maxam[†], Richard Tizard[†], Ora Bernard*, and Walter Gilbert[†]

* Basel Institute for Immunology, 487 Grenzacherstrasse, CH-4005 Basel, Switzerland; and [†] Department of Biochemistry and Molecular Biology, Harvard University, Cambridge, Massachusetts 02138

Contributed by Walter Gilbert, January 9, 1978

ABSTRACT   We have determined the sequence of the DNA of a germ-line gene for the variable region of a mouse immunoglobulin light chain, the Vλ_{II} gene. The sequence confirms that the variable region gene lies on the DNA separated from the constant region. Hypervariable region codons appear in the germ-line sequence. A sequence for the hydrophobic leader, 19 amino acids that are cleaved from the amino terminus of the protein, appears near, but not continuous with, the light chain structural sequence: most of the leader sequence is separated from the rest of the gene by 93 bases of untranslated DNA.

Immunoglobulin molecules are synthesized by special differentiated cells which produce and export a protein specialized to bind an antigen. That binding site is a pocket between two subunits, a light chain and a heavy chain. The part of the light chain in contact with the antigen, the amino-terminal region, termed the variable region, differs extensively from one immunoglobulin to another; the carboxy-terminal half of the light chain, the constant region, is invariant and, in mice, falls into three classes, one κ and two λ. The variable region itself, though different in different immunoglobulins, shows certain consistencies in structure: three hypervariable regions embedded in a framework (1). The hypervariable regions span five to ten amino acids each around positions 30, 55, and 90 in the immunoglobulin chain and contain most of the variation. In the three-dimensional structure of an immunoglobulin, they are the segments in contact with antigen. Differences in sequence in framework regions provide a classification of light chains into different subgroups (2). Some of these differences are a change in only a single amino acid; others are more extensive. Estimates of the largest number of different variable region subgroups that can appear with single constant regions range from 10 to 100. (A similar structure holds for the heavy chains; the first 110 amino acids constitute a variable domain and the next 330, a series of constant domains.)

Hozumi and Tonegawa (3) showed, by restriction enzyme analysis and hybridization mapping techniques, that in the germ line, in embryonic tissue, the sequences coding for the variable region and those coding for the constant region of an immunoglobulin light chain lie separate from one another. These two genetic regions move during differentiation of lymphocyte precursors and come closer together to form a single transcription unit that will produce the final immunoglobulin molecule. Experiments with cloned immunoglobulin genes have since demonstrated this movement more directly and revealed a surprising structure for the active gene. Brack, Lenhard-Schuller, and Tonegawa (ref. 4; unpublished data) analyzed a gene isolated from a myeloma, by R-loop mapping

and by heteroduplex comparisons with embryonic genes. They showed that the variable and constant region genes move from distant positions in the embryonic DNA to less distant positions in the producing cell but do not become contiguous. The variable and constant region coding sequences in the myeloma are separated still by 1250 bases. Presumably the DNA rearranges to bring these two regions close enough to lie on a single messenger precursor. One hypothesizes that this precursor then loses the additional sequences as the messenger matures in the nucleus.

In this paper we report the sequence of a germ-line gene for a mouse λ light chain variable region. We determined the sequence of this DNA for two reasons: to establish that the DNA coded for a variable region truly in isolation from any constant portion, and to determine whether a germ-line gene contained hypervariable as well as framework sequences.

A leading model for the behavior of the hypervariable regions, suggested by Kabat and his coworkers (1, 5), is that hypervariable region DNA is extraneous to the germ-line gene and inserts, like a small version of phage lambda, into receptor sites in the gene. This model would explain why different immunoglobulins in different subgroups occasionally have identical hypervariable regions. If this model were true, we would expect a germ-line gene to be a framework surrounding attachment sites rather than hypervariable sequences. Alternative models for the production of different immunoglobulins within one subgroup envisage the germ-line variable gene as a complete, inherited sequence that could be expressed with a unique antigenic specificity, a germ-line idiotype. This specificity will then change by mutation in somatic cells. Either these changes are localized in hypervariable regions by some process that makes the DNA in these areas unusually labile or, alternatively, mutations might arise anywhere in the variable gene but appear only in the hypervariable areas because the selection pressure matching immunoglobulin to antigen will select primarily for mutations in the antigen combining site. Determination of the sequence of a germ-line immunoglobulin light chain gene answers some of these questions.

MATERIALS AND METHODS

The recombinant phage λgt-WES-Ig 13, containing a 4800-base-pair insert of mouse DNA carrying a Vλ gene (6, 7), was grown under P3-EK2 conditions in Basel. A total of 5 mg of purified phage DNA was used for the sequence determination in Cambridge. Milligram amounts of the phage DNA were cleaved with a mixture of *Eco*RI and *Hae* III restriction endonucleases and the three cleavage products of the mouse insert, 750, 1500, and 2500 base pairs long, were separated on a 6 × 200 × 400 mm slab gel polymerized from 5% (wt/vol) acrylamide/0.17% (wt/vol) bisacrylamide/50 mM Tris-borate (pH

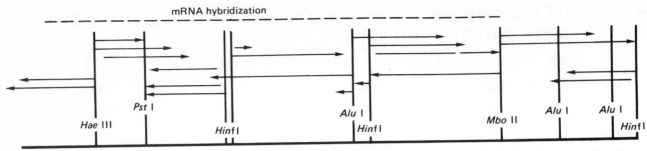

FIG. 1. Restriction endonuclease (7), mRNA hybridization (7), and DNA sequence map of the mouse embryonic Vλ_{II} region. The direction and the extent of the sequence determinations are shown by the arrows.

8.3)/1 mM EDTA, and extracted as described (8), except that the eluate was passed through a Millipore filter before ethanol precipitation.

The two larger *Hae* III fragments were then cleaved with *Hin*f, *Mbo* II, *Alu* I, or *Pst* I (9), end-labeled, and cleaved again, and their sequences were determined precisely as described (8) except for the following modifications. Tris·HCl buffer (50 mM, pH 9.5) was used instead of glycine/NaOH in the kinase reaction. One-tenth the carrier DNA (1μg) and tRNA (5μg) in all reactions for sequence determination produced sharper bands on autoradiograms. Addition of only sodium acetate, EDTA, and tRNA (no magnesium acetate) after the hydrazinolysis reactions and careful rinsing of the final ethanol precipitate to remove any residual sodium acetate eliminated occasional aberrations in the pyrimidine cleavage patterns. The "alternative guanine cleavage" (8) was used for guanine-only base-specificity.

For determination of long sequences, 200 nucleotides or so from the labeled end, products of sequencing reactions were loaded once or twice on the 20% gel described (8) and once on a 1.5 × 200 × 400 mm 10% gel polymerized from 9.5% (wt/vol) acrylamide/0.5% (wt/vol) bisacrylamide/7 M urea/100 mM Tris-borate (pH 8.3)/2 mM EDTA and electrophoresed at a high enough voltage (800–1100 V) and current to keep the surface of the gel at or above 50° during the run.

RESULTS AND DISCUSSION

Tonegawa *et al.* (6) isolated a derivative of phage lambda containing a 4800-base-pair insert of mouse DNA that hybridizes to the variable region half of a λ_I mRNA. A partial restriction map was constructed for the right third of this fragment, from a *Hae* III cut to the *Eco*RI end: the region in which hybridization to a messenger shows the putative variable region gene to lie (7). We chemically determined the sequence (8) of the relevant restriction fragments. In so doing, we could immediately establish the position of the variable region gene by comparison of the DNA sequence to the known protein sequences. Fig. 1 shows the restriction map of this area and indicates the experimental strategy that developed and checked

the sequence. Fig. 3 shows the sequence of some 750 base pairs around the variable region gene.

Identification of Variable Region. Although the original clone was identified by hybridization to a message for a λ_I immunoglobulin, the sequence on the DNA corresponds more closely to a λ_{II} variable region. λ_I is a minor group in the mouse; about 5% of the light chains have this variable region attached to a λ_I constant region. The sequences of 18 λ_I chains from mouse myelomas have been determined (2, 10–12). Twelve of these are identical and the six others differ by one, two, or three amino acids in the hypervariable regions. λ_{II} accounts for only 1–2% of mouse immunoglobulin (H. N. Eisen, personal communication); the sequence of only one example, MOPC 315, has been determined (13). The variable regions of λ_I and λ_{II} chains are very similar; their frameworks differ in seven places. [The two λ constant regions differ at about 29 places (13).] Fig. 2 schematically compares the protein sequence determined by the DNA and those of the λ_I and λ_{II} variable regions, while Fig. 3 shows the entire sequence. Specifically, at framework residues 16 (Glu or Gly), 19 (Thr or Ile), 62 (Ala or Val), 71 (Asn or Asp), 85 (Glu or Asp), and 87 (Ile or Met), the DNA sequence corresponds to a λ_{II} chain (underlined). Thus we believe that we have determined the sequence of the λ_{II} germ-line gene.

There is one exception to this framework agreement: at position 38, λ_I has Val and λ_{II} has Ile, while the DNA sequence at this point predicts Val. We regard the DNA sequence as unambiguous. Since the protein sequence suggests that the expressed gene in MOPC 315 has Ile at this point, we have an excellent candidate for a mutational change in a *framework* region. This interpretation would suggest that single amino acid changes in the framework regions of immunoglobulins are mutations and do not represent different germ-line genes, reducing sharply the number of different subgroups of κ chains that one might estimate to exist in the germ line.

The λ subgroups in the mouse may represent a degenerate example: there appear to be only single variable-region genes for λ_I and λ_{II} (restriction enzyme analysis by M. Hirama and S. Tonegawa, unpublished). Presumably the other possible λ frameworks have been lost in the mouse.

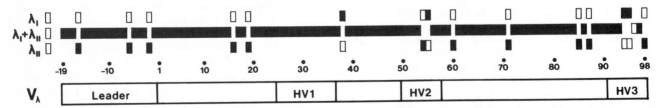

FIG. 2. Schematic comparison of the variable regions of mouse λ_I and λ_{II} immunoglobulin light chains with the DNA of the embryonic Vλ gene. The almost continuous bar in the middle represents an amino acid sequence common to most of the light chains, while boxes set apart from it mark positions at which they differ. Wherever the bar or a box is shaded, the sequence of that protein corresponds with the sequence of the DNA. Of 15 positions that distinguish the two variable regions, the DNA matches λ_{II} (MOPC 315) at 11 and λ_I (MOPC 104E) at 4. The actual sequences are given in Fig. 3.

```
CCCATACTAAGAGTTATATTATGTCTGTCTCACTGCCTGCTGCTGACCAATATTGAAAATAATAGACTTGGTTTGTGA
GGGTATGATTCTCAATATAATACAGACAGAGTGACGGACGACGACTGGTTATAACTTTTATTATCTGAACCAAACACT
```

λ_I   *MetAlaTrpIleSerLeuIleLeuSerLeuLeuAlaLeuSerSer*

λ_II   *MetAlaTrpThrSerLeuIleLeuSerLeuLeuAlaLeuCysSer*

MetAlaTrpThrSerLeuIleLeuSerLeuLeuAlaLeuCysSerGly   UGAPhe

```
ATTATGGCCTGGACTTCACTTATACTCTCTCTCCTGGCTCTCTGCTCAGGTCAGCAGCCTTTCTACACTGCAGTGGGTATGCAACAATACACATCTTGTCTCTGATTT
TAATACCGGACCTGAAGTGAATATGAGAGAGAGGACCGAGAGACGAGTCCAGTCGTCGGAAAGATGTGACGTCACCCATACGTTGTTATGTGTAGAACAGAGACTAAA
```

λ_I   *GlyAlaIleSerGlpAlaValValThrGlnGluSerAlaLeuThrThrSerProGlyGluThrValThrLeuThr*

λ_II   *GlyAlaSerSerGlpAlaValValThrGlnGluSerAlaLeuThrThrSerProGlyGlyThrValIleLeuThr*

AlaThrAspAspTrpIleSerTyrLeuPheAlaGlyAlaSerSerGlnAlaValValThrGlnGluSerAlaLeuThrThrSerProGlyGlyThrValIleLeuThr

```
GCTACTGATGACTGGATTTCTTACCTGTTTGCAGGAGCCAGTTCCCAGGCTGTTGTGACTCAGGAATCTGCACTCACCACATCACCTGGTGGAACAGTCATACTCACT
CGATGACTACTGACCTAAAGAATGGACAAACGTCCTCGGTCAAGGGTCCGACAACACTGAGTCCTTAGACGTGAGTGGTGTAGTGGACCACCTTGTCAGTATGAGTGA
```

      AsnThr   *Gly*   *Leu*   *Asn*   *Val*

λ_I   *CysArgSerSerThrGlyAlaValThrThrSerAsnTyrAlaAsnTrpValGlnGlnLysProAspHisLeuPheThrGlyLeuIleGlyGlyThrAsnAsnArgAla*

λ_II   *CysArgSerSerThrGlyAlaValThrThrSerAsnTyrAlaAsnTrpIleGlnGluLysProAspHisLeuPheThrGlyLeuIleGlyGlyThrSerAspArgAla*

CysArgSerSerThrGlyAlaValThrThrSerAsnTyrAlaAsnTrpValGlnGluLysProAspHisLeuPheThrGlyLeuIleGlyGlyThrSerAsnArgAla

```
TGTCGCTCAAGTACTGGGGCTGTTACAACTAGTAACTATGCCAACTGGGTTCAAGAAAAACCAGATCATTTATTCACTGGTCTAATAGGTGGTACCAGCAACCGAGCT
ACAGCGAGTTCATGACCCCGACAATGTTGATCATTGATACGGTTGACCCAAGTTCTTTTTGGTCTAGTAAATAAGTGACCAGATTATCCACCATGGTCGTTGGCTCGA
```
              HV1   **HV2**

λ_I   *ProGlyValProAlaArgPheSerGlySerLeuIleGlyAsnLysAlaAlaLeuThrIleThrGlyAlaGlnThrGluAspGluAlaIleTyrPheCysAlaLeuTrp*

λ_II   *ProGlyValProValArgPheSerGlySerLeuIleGlyAspLysAlaAlaLeuThrIleThrGlyAlaGlnThrGluAspAspAlaMetTyrPheCysAlaLeuTrp*

ProGlyValProValArgPheSerGlySerLeuIleGlyAspLysAlaAlaLeuThrIleThrGlyAlaGlnThrGluAspAspAlaMetTyrPheCysAlaLeuTrp

```
CCAGGTGTTCCTGTCAGATTCTCAGGCTCCCTGATTGGAGACAAGGCTGCCCTCACCATCACAGGGGCACAGACTGAGGATGATGCAATGTATTTCTGTGCTCTATGG
GGTCCACAAGGACAGTCTAAGAGTCCGAGGGACTAACCTCTGTTCCGACGGGAGTGGTAGTGTCCCCGTGTCTGACTCCTACTACGTTACATAAAGACACGAGATACC
```
                                                                                                 HV3

   Cys   *Arg*

λ_I   *TyrSerAsnHisTrp* | *ValPheGlyGlyGlyThrLysLeuThrValLeuGlyGlnPro* | *LysSerSerProSerValThrLeuPheProProSerSerGluGluLeuThr*

λ_II   *PheArgAsnHisPhe* | *ValPheGlyGlyGlyThrLysLeuThrValLeuGlyGlnPro* | *LysSerThrProThrLeuThrValPheProProSerSerGluGluLeuLys*

```
                 gtxttyggxggxggxacxaargtxacxgtxttrggxcarccxaartcxacxccxacxttracxgtxttyccxccxtcxtcxgargarttraar
                 ctx                                                                      agyagy         ctx
```

TyrSerThrHisPheHisAsnAspMetCysArgTrpGlySerArgThrArgThrLeuTrpTyrSerLeuThrThrIlePheLeuThrGlyLyrTyrMetSerLeuVal

```
TACAGCACCCATTTCCACAATGACATGTGTAGATGGGGAAGTAGAACAAGAACACTCTGGTACAGTCTCACTACCATCTTCTTAACAGGTGGCTACATGTCCCTAGTC
ATGTCGTGGGTAAAGGTGTTACTGTACACATCTACCCCTTCATCTTGTTCTTGTGAGACCATGTCAGAGTGATGGTAGAAGAATTGTCCACCGATGTACAGGGATCAG
```
   HV3   V|C   V|C

λ_I   *GluAsnLysAlaThrLeuValCysThrIleThrAspPheTyrProGlyVal....*

λ_II   *GluAsnLysAlaThrLeuValCysLeuIleSerAsnPheSerProGlySer....*

```
                 garaayaargcxacxttrgtxtgyttratytcxaayttytcxccxggxtcx
                 ctx             ctxataagy          agy           agy
```

CysSerLeuLeuLeuUAG

```
TGTTCTCTTTTACTATAGAGAAATTTATAAAAGCTGTTGTCTCGAGCAACAAAAAGTTTTATTCAACAAATTGTATAATAATTATGCCTTGATGACAAGCTTTGTTTA
ACAAGAGAAATGATATCTCTTTAAATATTTTCGACAACAGAGCTCGTTGTTTTTCAAAATAAGTTGTTTAACATATTATTAATACGGAACTACTGTTCGAAACAAAT
```

FIG. 3. DNA sequence of the mouse embryonic Vλ_II immunoglobulin light chain gene in direct comparison with all known mouse λ light chains. The nucleotide sequence of the gene coincides best with the amino acid sequence of the MOPC 315 λ_II light chain. Coding begins with the first amino acid of the hydrophobic leader, the Met at position 19, is interrupted at Ser 5 by a 93-base-pair intron (see text), resumes with Gly 4, and (with five single-base-change exceptions) continues from this point to Phe 98, where it ends abruptly. Positions which distinguish λ_I (MOPC 104E) and λ_II (MOPC 315) light chains are indicated (*), as are known substitutions (↑) in λ_I hypervariable regions (HV1, HV2, and HV3). Amino acid numbering begins with the cyclized glutamine (Glp) found at the amino terminus of mature light chains. [The Gln-Glu at 6–7 is predicted by the DNA, confirming a correction (10) for both λ_I and λ_II.] Underlined amino acids are encoded by the DNA below, while underlined DNA matches codons for at least one of the proteins above. V/C indicates variable-constant junctions based on the DNA (solid line) and on all light chain protein sequences (broken line).

Variable-Constant Junction. We hoped to find an isolated variable region on the DNA. The conventional assignment of the junction between variable and constant regions is at amino acid 112; however, the DNA sequence deviates from the protein

sequence sharply after amino acid 98. There is no agreement between λ light-chain amino acid codons and the DNA sequence over the next 240 bases past this point. We conclude that the embryonic DNA contains this variable region gene in iso-

HV1   T A C **T** A G T T G T A A C A G C C C C A G T A **C** T

HV2   A **A** T A G G **T G G** T A C C A G C A **A** C C G A G C **C** T

HV3   G C T C T A T G G T **A** - C **A** G C A C C C A T T T C

FIG. 4.   Double and triple DNA sequence homologies in the $V\lambda_{II}$ gene corresponding to immunoglobulin light chain hypervariable regions. The bottom strand of HV1 and top strands of HV2 and HV3 (Fig. 3) have been aligned 5′ to 3′, with the bases that could change to produce known λ_I and inferred λ_{II} amino acid substitutions shown in boldface.

lation from the constant region. No obvious element identifies the boundary as unusual; we cannot interpret a feature of the DNA sequence that shows how the structures are brought together. The identification of amino acid 112 as the variable-constant junction is based on analogy with human sequences. We do not know, for the mouse λ_{II} chains, that the junction should be there: we interpret our sequence as showing that the variable-constant junction of λ_{II} gene follows amino acid 98.

Hypervariable Regions. Sequences corresponding to hypervariable regions appear in this germ-line gene. The first hypervariable region in the DNA sequence fits both the λ_I and λ_{II} proteins (Fig. 3). The second hypervariable region corresponds to λ_{II} at position 54 (Ser rather than Asn) to λ_I at 55 (Asn rather than Asp) and to both elsewhere. However, at the third hypervariable region, the MOPC 315 protein deviates from our DNA sequence at three amino acids, two of which correspond to λ_I chains. Our finding that the hypervariable regions are in the germ-line DNA rules out models in which DNA is inserted into a pre-existing germ-line sequence. Differences among mouse λ light chains in the hypervariable regions are thus most likely due to the gene for the final protein having accrued one or several mutations before or during the antigen-driven expansion of the pool of precursor cells specialized to make that particular immunoglobulin.

There are no features in these hypervariable regions that are so dramatic as to define a mechanism by which these regions might be more labile than the rest of the DNA. We do not find extensive palindromes surrounding these hypervariable regions (14, 15). (Of the four palindromes suggested in ref. 14, we find only one in the second hypervariable region.) There is, however, a similarity in the DNA structure of the three hypervariable regions. Fig. 4 matches the bottom strand of the first hypervariable region with the top strand of the second and third. These similarities may be simply evolutionary relics, but it is not impossible that they could serve as a recognition site for an enzyme that would cleave or modify the DNA in order to make the sequence labile.

At three positions in the hypervariable regions and one in the framework, this $V\lambda_{II}$ gene deviates from λ_{II} (MOPC 315) expectations and agrees with the most common λ_I residues. This may reflect a past duplication that gave rise to both the λ_I and λ_{II} germ-line genes.

$V\lambda_{II}$ Gene Contains a Genetic Discontinuity—93 Nontranslated Bases in Leader Region. The light chain is synthesized as a longer precursor (16, 17), containing at its amino terminus a hydrophobic peptide that, presumably, is involved in moving this protein through the cell membrane (18, 19). As the protein matures, this precursor is cleaved at a glutamine (20), which cyclizes into a pyrrolidone carboxylic acid (10). Sequences of the 19-amino-acid long signal peptides for mouse λ light chains have been worked out by Schecter and his coworkers (20, 21) by translating purified immunoglobulin messenger in vitro. The DNA, however, codes for only four of the expected amino acids preceding the initial glutamine (Fig. 3) and then deviates entirely from the precursor protein. Ninety-three bases earlier in the DNA, we find a sequence for the first 15 amino acids of the light chain precursor, including

an ATG for the initial methionine. When this match was first made, only the λ_I precursor sequence was available, and that deviates from the DNA in three places. However, after the DNA sequence was established, Burstein and Schecter (ref. 20; personal communication) worked out the hydrophobic leader sequence for the λ_{II} light chain; that protein sequence agrees with the DNA at all points. Therefore the DNA sequence was a predictor of the protein sequence. The extra region of 93 bases cannot be translated into protein; it contains a stop signal in phase. This 93-base region is not an artifact of the cloning. It contains a unique Pst I cut that can be exhibited in uncloned embryonic DNA (M. Hirama and S. Tonegawa, unpublished).

We believe the functioning gene in the myeloma will consist of the precursor region followed first by the 93-base-long interspersed DNA, followed by the variable region gene, then by a 1250-base-long piece of noncoding DNA (4), and last, by the constant region gene. We call such an additional piece of DNA that arises within a gene an *intron* (for intragenic region or intracistron) and thus look upon the structure of this gene as leader(45)-intron(93)-variable (306)-intron(1250)-constant(348). The current level of analysis cannot exclude still more small introns within the constant region nor specify the exact location of the second intron.

Maturation. Does the sequence for the 93-base intron show us how this area can be skipped or excised from the message? There is only a short repetition at its ends, a CAGG sequence repeated in the correct phase. A four-base repeat, however, is not specific enough to be used by the RNA polymerase as a signal to stop and start up again on an RNA-priming model. The signals to eliminate the unwanted region may be contained in the base sequence and structure of the messenger. A hairpin structure could hold the point of splicing in its stem, but that would necessitate ligation from one chain across to the opposite side of the helix. Fig. 5 shows the one well-placed hairpin we find in the sequence; it is neither convincing nor impossible. An alternative structure would be to use a sequence elsewhere in the messenger as a template to bring into juxtaposition the bonds that are to be split and rejoined.

Genes in Pieces. We hypothesize that most higher cell genes will consist of informational DNA interspersed with silent sequences. The eukaryotic cistron is a transcription unit containing alternate regions to be excised from the messenger, the introns, and regions left to be expressed, exons. [There are many recent examples of this structure (4, 22–26); others are reviewed in ref. 27.] Thus the gene is a mosaic: sequences corresponding to expressed functions held in a matrix, the latter possibly 10 times larger than the coding components. This model immediately resolves two puzzles. Heterogeneous nuclear RNA would be the long transcription products from which the much smaller ultimate messengers are spliced. The extra DNA in higher cells would arise because the genes, the transcription units, are much larger than those for a single polypeptide product.

Assume that the splicing mechanism is general, independent of the gene structure, and recognizes simply a unique secondary structure in the RNA. Of what benefit then are the infilling sequences? They speed evolution. Single base changes not only

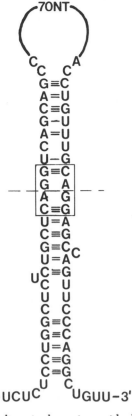

FIG. 5. Hypothetical hairpin structure in λ_{II} light chain precursor mRNA with the nontranslated sequences to be excised (intron) above the dashed line and coding sequences to be joined (exon) below. The base-paired sequences are from the ends of the 93-base noncoding segment in Fig. 3. Proper splicing of this precursor RNA would require a symmetric pair of cuts in the antiparallel sequences boxed in the stem, followed by ligation of the strands below to preserve one copy of the CAGG sequence. The strands to be joined, however, are not favorably oriented for ligation.

providing developmental control. A differentiation pathway can be defined by the appearance of a new splicing enzyme.

One striking interpretation, with this new picture, would be that the simultaneous expression of IgM and IgD with the same idiotype by a single cell (28) will turn out to be due to a V_H gene translocated near the C_H genes and transcribed together with C_μ and C_δ into a single V_H-μ-δ precursor, alternate splicings then providing the two products. The switch from IgM to IgG may be a new translocation of a V_H gene, or it may be a new processing of a V_H-μ-δ-γ precursor to produce a V_H-γ product.

We are grateful for the expert assistance of Rita Lenhard-Schuller. We thank I. Schecter for the communication of results before publication and P. Slonimsky, B. Blomberg, D. Wiley, and S. Harrison for discussions. W.G. is an American Cancer Society Research Professor. Part of this work was supported by National Institutes of Health Grant GM 09541-16.

can alter single amino acids, but now, if they occur at the boundaries of introns, can change the splicing to add or delete a string of amino acids. This generates a more rapid search through the space of protein molecules. Second, the splicing process need not be 100% efficient. For example, changes in silent positions, third base positions in codons, can alter both the pattern and the efficiency of splicing so that the product of a single transcription unit can be two polypeptide chains, one being the original gene product and the second, also synthesized at high frequency, the new product. Evolution can seek new solutions without destroying old. This resolves a classic problem: one thought that the organism had to create a second copy of an essential gene in order to mutate it to a new function. The intronic model eliminates any special duplication. The extra material, scattered widespread across the genome, can be called into action at any time. After a new gene function appears, there can be selective pressure for duplication. One consequence of the intron model is that the dogma of one gene, one polypeptide chain disappears. A gene, a contiguous region on DNA, now corresponds to one transcription unit, but that transcription unit can correspond to many polypeptide chains, of related or differing functions.

Since the gene is now spread out over a larger region of DNA, recombination between exons will be enhanced. Furthermore, if the exons correspond to identifiable functions put together by splicing to form a special combination in a finished protein, then recombination between these regions can sort their functions independently. In the λ light chain a hydrophobic precursor sequence is separated by an intron from its follower region; recombination could combine this leader sequence with some other protein. One might anticipate that middle repetitive sequences within introns will provide hot spots for recombination to reassort the exonic sequences.

Still another picture emerges if we hypothesize that specific new splicing patterns can be turned on by special gene products,

1. Wu, T. T. & Kabat, E. A. (1970) *J. Exp. Med.* **132**, 211–250.
2. Cohn, M., Blomberg, B., Geckeler, W., Raschke, W., Riblet, R. & Weigert, M. (1974) in *The Immune System; Genes, Receptors, Signals*, eds. Sercarz, E. E., Williams, A. R. & Fox, C. F. (Academic, New York), pp. 89–117.
3. Hozumi, N. & Tonegawa, S. (1976) *Proc. Natl. Acad. Sci. USA* **73**, 3628–3632.
4. Brack, C. & Tonegawa, S. (1977) *Proc. Natl. Acad. Sci. USA* **74**, 5652–5656.
5. Wu, T. T., Kabat, E. A. & Bilofsky, H. (1975) *Proc. Natl. Acad. Sci. USA* **72**, 5107–5110.
6. Tonegawa, S., Brack, C., Hozumi, N. & Schuller, R. (1977) *Proc. Natl. Acad. Sci. USA* **74**, 3518–3522.
7. Tonegawa, S., Brack, C., Hozumi, N. & Pirrotta, V. (1977) in *Cold Spring Harbor Symp. Quant. Biol.*, in press.
8. Maxam, A. M. & Gilbert, W. (1977) *Proc. Natl. Acad. Sci. USA* **74**, 560–564.
9. Roberts, R. J. (1977) in *Critical Reviews in Biochemistry*, ed. Fasman, G. D. (CRC Press, Cleveland, OH), pp. 123–164.
10. Weigert, M. G., Cesali, I. M., Yonkovich, S. J. & Cohn, M. (1970) *Nature* **228**, 1045–1047.
11. Apella, E. (1971) *Proc. Natl. Acad. Sci. USA* **68**, 590–594.
12. Cesari, I. M. & Weigert, M. (1973) *Proc. Natl. Acad. Sci. USA* **70**, 2112–2116.
13. Dugan, E. S., Bradshaw, R. A., Simms, E. S. & Eisen, H. N. (1973) *Biochemistry* **12**, 5400–5416.
14. Leder, P., Honjo, T., Seidman, J. & Swan, D. (1976) *Cold Spring Harbor Symp. Quant. Biol.* **41**, 855–862.
15. Wuilmart, C., Urbain, J. & Givol, D. (1977) *Proc. Natl. Acad. Sci. USA* **74**, 2526–2530.
16. Milstein, C., Brownlee, G., Harrison, T. M. & Mathews, M. B. (1972) *Nature New Biol.* **239**, 117–120.
17. Swan, D., Aviv, H. & Leder, P. (1972) *Proc. Natl. Acad. Sci. USA* **69**, 1967–1972.
18. Blobel, G. & Dobberstein, B. (1975) *J. Cell Biol.* **67**, 835–851.
19. Blobel, G. & Dobberstein, B. (1975) *J. Cell Biol.* **67**, 852–862.
20. Burstein, Y. & Schecter, I. (1977) *Biochem. J.* **165**, 347–354.
21. Burstein, Y. & Schecter, I. (1977) *Proc. Natl. Acad. Sci. USA* **74**, 716–760.
22. Berget, S. M., Moore, C. & Sharp, P. A. (1977) *Proc. Natl. Acad. Sci. USA* **74**, 3171–3175.
23. Chow, L. T., Gelinas, R. E., Broker, T. R. & Roberts, R. J. (1977) *Cell* **12**, 1–8.
24. Klessig, D. F. (1977) *Cell* **12**, 9–21.
25. Breathmark, R., Mandel, J. L. & Chambon, P. (1977) *Nature* **270**, 314–319.
26. Doel, M. T., houghton, M., Cook, E. A. & Carey, N. H. (1977) *Nucleic Acids Res.* **4**, 3701–3713.
27. Williamson, B. (1977) *Nature* **270**, 295–297..w
28. Rowe, D. S., Hug, K., Forni, L. & Permis, B. (1973) *J. Exp. Med.* **138**, 965–972.

An immunoglobulin heavy-chain gene is formed by at least two recombinational events

Mark M. Davis*, Kathryn Calame*, Philip W. Early*, Donna L. Livant*,
Rolf Joho† & Irving L. Weissman† & Leroy Hood*

* Division of Biology, California Institute of Technology, Pasadena, California 91125
† Laboratory of Experimental Oncology, Department of Pathology, Stanford Medical School, Stanford, California 94305

The events of B-cell differentiation can be reconstructed in part through an analysis of the organisation of heavy-chain gene segments in differentiated B cells. A mouse immunoglobulin α heavy-chain gene is composed of at least three noncontiguous germ-line DNA segments—a V_H gene segment, a J_H gene segment associated with the C_μ gene segment, and the C_α gene segment. These gene segments are joined together by two distinct types of DNA rearrangements—a V–J joining and a C_H switch.

THE antibody genes provide a unique opportunity for studying the molecular basis of one pathway of eukaryotic differentiation because the rearrangement of gene segments is correlated with the expression of antibody molecules. Antibody molecules are encoded by three unlinked gene families—λ, κ and heavy chain (H)[1]. The λ and κ families encode light (L) chains and the heavy-chain family encodes heavy chains. The light chains are encoded by three gene segments, V_L (variable), J_L (joining) and C_L (constant), which are separated in the genomes of cells undifferentiated with regard to antibody gene expression[2,3] During differentiation of the antibody-producing or B cell, the V_L and J_L gene segments are rearranged and joined together while the intervening DNA between the J_L and C_L gene segments remains unmodified[2-4]. This process of DNA rearrangement is termed V–J joining. During the expression of the rearranged gene in the differentiated B cell, the coding regions as well as the intervening DNA between the J_L and C_L gene segments are transcribed as part of a high molecular weight nuclear transcript. The intervening region is then removed by RNA splicing to produce a light-chain mRNA with contiguous V_L, J_L and C_L coding segments[5,6]. Recently, we demonstrated that the heavy chain contains three analogous gene segments, V_H, J_H and C_H, which undergo a similar type of V–J joining during B-cell differentiation[7-9]. Each antibody-producing cell synthesises only one $V_L J_L$ polypeptide sequence and one $V_H J_H$ sequence, which together form the antigen-binding (V) domain of the antibody molecule.

The heavy-chain genes seem to have a special role in the differentiation of the antibody-producing cell as reflected in a second phenomenon known as the switching of heavy-chain constant regions or C_H switching. Antibody molecules can be divided into five different immunoglobulin classes—IgM, IgD, IgG, IgA and IgE—which are determined by one of eight distinct heavy chain genes [that is, C_μ, C_δ, ($C_{\gamma 1}$, $C_{\gamma 2a}$, $C_{\gamma 2b}$, $C_{\gamma 3}$), C_α and C_ε]. Each class of immunoglobulin seems to be associated with unique functions such as complement fixation or the release of histamine from mast cells. The process of B-cell differentiation is complex and a variety of conflicting views exists on the transitional stages. It is generally agreed, however, that IgM is the earliest immunoglobulin class that is expressed in the differentiation of a B cell (see ref. 10). The immature B lymphocyte apparently has the capacity to differentiate along a variety of discrete pathways and produce progeny which may switch from IgM expression to the expression of any one of the other classes of immunoglobulins. The terminal stage of B-cell differentiation is the plasma cell, which is committed to synthesise and secrete large quantities of a single molecular species of antibody. Several studies suggest that the specificity or V domain does not change as different immunoglobulin classes

are expressed throughout this differentiation process[11-15]. During C_H switching, light-chain expression remains unaltered.

We are interested in studying the molecular mechanisms which permit a B cell (or its progeny) to express successive classes of antibody molecules. Here we demonstrate that an α heavy-chain gene derived from a terminally differentiated plasma cell is composed of three noncontiguous germ-line DNA segments. These gene segments are joined together by at least two distinct DNA rearrangements—a V–J joining and a C_H switch.

Organisation of the α heavy-chain gene is altered in differentiated as opposed to undifferentiated DNAs

We have constructed 'libraries' of recombinant phage[16] containing large (12–20 kilobases) inserts of mouse DNA in the vector Charon 4A (ref. 17). These libraries contain sufficient numbers of recombinants ($\sim 10^6$) for there to be a high probability that most single-copy sequences of a given genome are included. We have previously reported the construction of such a library from the DNA of the mouse IgA-producing myeloma tumour M603, and the subsequent isolation and characterisation of clones containing rearranged or differentiated genomic DNA. These clones, CH603α6 (α6) and CH603α125 (α125), have V_H and C_α gene segments on a single fragment of DNA[7,8]. The EcoRI restriction maps of these clones are shown in Fig. 1a. The α6 clone has three EcoRI fragments—one 7.3 kilobases long containing the V_H gene segment, a second of 5.1 kilobases containing the 5' portion of the C_α gene segment, and a third of 4.4 kilobases containing the 3' portion of the C_α gene segment. Approximately 6.8 kilobases of intervening DNA separate the V_H and C_α gene segments. The α125 clone also has three EcoRI fragments—one of 6.5 kilobases which is located on the 5' side of the 7.3- and 5.1-kilobase EcoRI fragments also described for α6.

The genomic clones were isolated using a cloned cDNA plasmid representing the entire heavy-chain mRNA of the IgA-producing tumour S107 (denoted S107 cDNA)[7]. The S107 V_H region is about 98% homologous to the M603 V_H region at the nucleotide level[9]. Analysis by the Southern blot procedure[18,19] of EcoRI-digested M603 DNA hybridised to the S107 cDNA probe demonstrates the presence of three EcoRI fragments corresponding to those described above in the α6 clone[7] Thus, the rearranged α6 clone is not an artefact of the cloning or isolation procedures.

When the S107 cDNA probe is used to examine a Southern blot of EcoRI-digested sperm or embryo DNA, a somewhat different pattern is observed[8]. In particular, the 5.1-kilobase

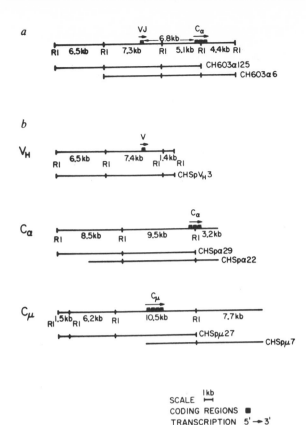

SCALE |—| 1kb
CODING REGIONS ■
TRANSCRIPTION 5' → 3'

Fig. 1 *a*, α Heavy-chain clones isolated from a genomic library of myeloma M603 DNA. CH603 α6 and CH603 α125 are two overlapping clones derived from a partial *Eco*RI library of M603 DNA, constructed using the phage Charon 4A (ref. 7). The position of the respective gene segments, as well as their direction of transcription, was determined by R-loop mapping and restriction analysis[7,8]. *b*, Germ-line V_H, C_μ and C_α clones isolated from a genomic library of sperm DNA from inbred BALB/c mice. Sperm DNA[33,34] was partially digested with restriction enzymes in two ways: (1) with a mixture of *Hae*III plus *Alu*I and (2) with *Eco*RI alone. After digestion in conditions designed to maximise the yield of 12–20-kilobase fragments, these fragments were selected on sucrose gradients. The *Hae*III/*Alu*I fragments were methylated with *Eco*RI methylase and blunt end ligated with synthetic *Eco*RI cleavage sites and then cleaved with *Eco*RI essentially as described previously[16] except that *Eco*RI linker ligations were done at 18 °C to lessen the endonuclease degradation. Both types of fragments were then ligated to the isolated arms of the bacteriophage Charon 4A (ref. 16) at 4 °C. The enzymatically recombined DNAs were packaged *in vitro* using the strains of Sternberg[35] and the protocol of Hohn[36]. The efficiency of packaging was 400,000 plaque forming units (PFU) per µg inserted DNA in the case of *Eco*RI partially digested DNA and 200,000 PFU per µg for *Hae*III/*Alu*I digestions. The background of non-recombinant Charon 4A was 25% and <5%, respectively. Approximately 500,000 *Eco*RI and 1,200,000 *Hae*III/*Alu*I clones were constructed and amplified. A library of ~500,000 clones provides a 90% chance of finding a given single copy sequence and a library of 1,200,000 clones a 99% chance[37]. The use of enzymes that recognise three different sequences reduces the possibility that a particular region of interest is lost from the library because it had too many or too few restriction sites to fall within the sucrose gradient size cut. Libraries were screened[7,16] with the cDNA clone (see text) S107 or M104E C_μ cloned cDNA probes identified by DNA sequence analysis[9,38]. The C_μ clone used extends from codon 300 to the 3'-untranslated region (~1,000 base pairs)[38]. Independent overlapping clones were obtained from the C_α and C_μ gene segments. Location of the coding regions and the direction of transcription were determined by DNA sequence analyses for the CHSp PC-3 clone, by heteroduplex analyses with the α6 clone for the CHSp α29 clone, and by R-loop mapping and restriction analyses of the CHSpµ27 clone[38]. The germ-line genomic clones will be denoted µ27 (CHSpµ27), V_H3 (CHSpV$_H$3), and α29 (CHSpα29).

*Eco*RI fragment containing most of the large intervening sequence and the 5' portion of the C_α gene segment is not present. This observation indicates that the α6 clone is the product of one or more DNA rearrangements which presumably occurred during B-cell differentiation[8].

The α gene is composed of at least three different germ-line segments of DNA

In view of the possibility that lymphocytes earlier in the M603 lineage might first have produced IgM molecules and later IgA molecules, we decided to investigate the possible contribution of germ-line C_μ sequences as well as V_H and C_α sequences to the myeloma α gene (Fig. 1*a*). We constructed several libraries of germ-line DNA (sperm) and proceeded to isolate clones containing V_H, C_μ and C_α gene segments using cloned cDNA probes (denoted S107 V_H, C_α and C_μ) for the corresponding coding regions and the screening procedure of Benton and Davis[20]. *Eco*RI restriction maps of several germ-line clones are shown in Fig. 1*b*. We chose sperm (germ-line) DNA for our undifferentiated genomic libraries to eliminate any possibility of DNA rearrangements which may occur in somatic tissues during embryogenesis. The C_α and V_H clones seem to be representative of germ-line sequence organisation because the *Eco*RI fragments containing the corresponding coding regions (in the clones) are identical in size to those found in the Southern blot analysis of undifferentiated DNA with C_α and S107 V_H cloned cDNA probes—9.5 and 7.4 kilobases, respectively (Fig. 2*a–d*). The V_H and C_α probes were derived from restriction fragments of the S107 cloned cDNA[7] extending from approximately the 5'-untranslated region to codon 108 (V_H) and from codons 108 to 274 (C_α). Figure 2*a* shows that a Southern blot of a germ-line

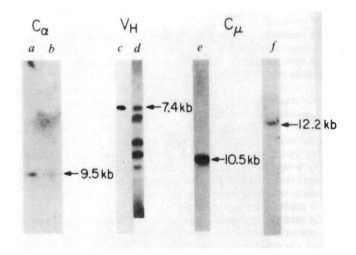

Fig. 2 Southern blot analyses of C_α, C_μ and S107 V_H coding regions in germ-line DNA. Approximately 3–20 µg of sperm DNA or 13-day embryo was digested with *Eco*RI and electrophoresed on a 0.7% neutral agarose gel for 10–12 h at 30–40 V. Gels were then blotted according to the procedures of Southern and Flavell[18,19] and hybridised with ^{32}P nick-translated cDNA[39]. Washing was for 1.5 h in 1 M NaCl, 1 M Tris *p*H 8, 0.1% SDS and 0.1% sodium pyrophosphate at 65 °C and for 1 h in 1×SSC, 0.1% SDS and 0.1% sodium pyrophosphate. Lanes *a* and *b* are α29 and embryo DNAs, respectively, hybridised to the C_α cDNA probe. Lane *c* is V_H3 DNA hybridised to the S107 cDNA probe. Lane *d* is sperm DNA hybridised to the S107 V_H cDNA probe. This probe cross-reacts with eight or nine closely related V_H gene segments[8]. Lanes *e* and *f* are µ27 and sperm DNAs, respectively, hybridised to the C_μ cDNA probe. Identical results in all cases were obtained with BALB/c 13-day embryo DNA or sperm DNA as has been found for a V_κ gene[34]. Sizes (given in kilobases) were determined by comparison with restriction fragments of PBR322 (ref. 40) or by the use of PBR322 multimers generated by *Bam*HI cleavage of PBR322 and limited ligation of the resulting monomer.

C_α clone (α29) digested with *Eco*RI and hybridised with the C_α cDNA probe yields a 9.5-kilobase fragment. A similar analysis of embryo DNA produces a band of identical size (Fig. 2b). Southern blots of a germ-line V_H clone (V_H3) and embryo DNA digested with *Eco*RI and hybridised to a S107 V_H cDNA probe gave, respectively, a 7.4-kilobase fragment (Fig. 2c) and several fragments including a 7.4-kilobase fragment (Fig. 2d). An important question is the relationship between the germ-line V_H3 and myeloma M603 V_H coding regions because there are at least eight V_H gene segments that hybridise to the S107 probe[8]. The germ-line V_H3 DNA sequence codes for the S107 V_H protein sequence[9] and, accordingly, differs from the M603 V_H region by a minimum of four base changes leading to three amino acid subsitutions[21]. Thus, the germ-line V_H3 and the myeloma M603 V_H gene segments may be encoded by two distinct germ-line V_H gene segments or the V_H3 gene segment may give rise to the M603 V_H gene segment by somatic mutation and selection. As we shall show subsequently, the V_H3 clone seems to be indistinguishable from the M603 clone in the V_H coding region and in more than 11-kilobases of 5′-flanking sequence by heteroduplex and restriction enzyme analyses. Therefore, our analyses of the localisation of V_H sequences in the myeloma α6 clone are valid because the germ-line V_H clone serves as a probe for the M603 V_H gene and its attendant 5′-flanking sequence.

The *Eco*RI fragment of the germ-line C_μ clone containing the C_μ coding region is smaller (10.5 kilobases; see Fig. 2e) than its counterpart seen on a Southern blot analysis of sperm DNA with a cloned μ cDNA probe (12.2 kilobases; see Fig. 2f). We believe this discrepancy arises from one (or more) deletion(s) in the DNA flanking the C_μ coding region during the propagation of the recombinant phage. In isolating μ clones from the M603 library, we obtained several 9.5–10-kilobase *Eco*RI fragments containing C_μ coding regions, whereas Southern blot analysis of M603 DNA with a cloned C_μ cDNA probe demonstrated a genomic fragment of 12.2 kilobases, as in the sperm DNA (data not shown). Attempts to isolate C_μ clones from a library of mouse liver DNA have led to similar results (N. Newell and F. Blattner, personal communication). We will present restriction enzyme data below which demonstrate that this apparent deletion in the μ clone does not affect our general conclusions.

The germ-line V_H, C_α and C_μ clones were compared with the myeloma α6 and α125 clones by heteroduplex analysis. Representative heteroduplexes from each of these comparisons show extensive homologies. The germ-line V_H clone shares approximately 11.6 kilobases of homology with the myeloma α125 clone (Fig. 3a). This homology extends from the 5′ end of the α125 clone up to and including the V_H coding region. The germ-line C_μ clone shows 5.0 kilobases of homology with the large intervening sequence of the myeloma α6 clone (Fig. 3b). Starting at its 3′ end, the germ-line C_α clone has about 6.4 kilobases of homology with the myeloma α6 clone (Fig. 3c). The heteroduplex measurements for these analyses are given in Table 1.

Fig. 3 Heteroduplex analyses of germ-line and somatic clones. The electron micrographs are shown on the left, tracings of these heteroduplexes in the middle and diagrammatic representations on the far right. a, Myeloma α125/germ-line V_H PC3. b, Myeloma α6/germ-line μ27. c, Myeloma α6/germ-line α22. Letters A–E indicate single-strand and double-strand regions for which measurements are given in Table 1. In corresponding line drawings, (r) and (l) refer to the right and left arms of the λ vector. Blocks indicate coding regions in these figures, CsCl purified phage particles were treated with 0.1 M NaOH for 10 min at 20 °C to lyse the phage and denature the DNA simultaneously. After neutralisation of the mixture, the DNA was allowed to reanneal in 50% (v/v) three times recrystallised formamide for 45 min at 20 °C before spreading for electron microscopy[41].

Table 1 Measurements of heteroduplex molecules

Heteroduplex	No. of Molecules	Distance in kilobases				
		A	B	C⁺	D	E
a Myeloma α125/ germ-line V_H3	35	4.5± 0.4	7.2± 0.7	11.6± 0.4		
b Myeloma α6/ germ-line μ27	30	4.8± 0.5	5.1± 0.5	5.0± 0.4	6.7± 0.4	7.5± 0.5
c Myeloma α6/ germ-line α22	26	9.3± 0.5	10.4± 0.9	6.4± 0.5	1.2± 0.2	

Measurements were standardised relative to two circular DNA molecules on the same grid [single-strand ΦX174 DNA (5,375 bases) and double-strand pBR322 DNA (4,365 base pairs)]. Letters refer to regions indicated in Fig. 3. The complete clone designations are given in Fig. 3. In heteroduplex *a*, C refers to the region of duplex between the germ-line V_H clone and the myeloma clone. A and B are the non-homologous single strands of these clones. Similarly, C for heteroduplexes *b* and *c* refers to the duplexes formed between myeloma and germ-line C_μ or C_α clones.

Comparative restriction analyses confirm and extend the heteroduplex data discussed above. A detailed restriction map for the M603 myeloma clones was obtained by double digestion with pairs of restriction enzymes and is shown in Fig. 4a. To compare the placement of these cleavage sites with those of the germ-line clones, the 5.1-kilobase restriction fragment of the α6 clone (Fig. 1a) that spans the region joining the germ-line C_α and C_μ sequences was subcloned. Using this fragment as a probe, detailed restriction comparisons of the myeloma α6 clone and the germ-line C_μ and C_α clones were made. Representative data are shown in Fig. 5 for these comparisons. Figure 5a, b and c represents HincII plus EcoRI digestions of the myeloma α6 5.1-kilobase RI fragment, the germ-line α29 clone and the germ-line C_μ clone, respectively. These digests were electrophoresed on an agarose gel, blotted onto a nitrocellulose filter and hybridised with the labelled 5.1-kilobase RI fragment. This enables homologous restriction fragments to be identified rapidly and precisely (arrows indicate identical fragments). Figure 5d, e and f shows a similar analysis using HindIII plus EcoRI digestions, respectively, of the myeloma 5.1-kilobase subclone, the germ-line α29 clone and the germ-line μ27 clone. These data, as well as additional restriction analyses of the germ-line V_H3 clone, demonstrate that 4 out of 4 restriction sites in the germ-line V_H3 clone, 10 out of 10 sites in germ-line α29 clone and 9 out of 10 sites in the germ-line μ27 clone corresponded exactly to those found in the myeloma α6 clone (Fig. 4). Not only do these restriction analyses independently confirm the heteroduplex results, but they also suggest that the component germ-line sequences of the α6 clone are very similar to their germ-line counterparts. Thus, the heteroduplex and restriction analyses demonstrate that the germ-line V_H gene segment and its 5'-flanking sequence, although not identical to its M603 counterpart, are very similar and may be used to analyse V_H gene segment organisation in the myeloma M603 clones.

DNA sequence analyses of the myeloma α6 clone and the germ-line μ27 clone have demonstrated that the heavy-chain gene family does have distinct V_H and J_H gene segments in the germ line[9]. Moreover, the germ-line J_H gene segment corresponding to that expressed in the myeloma α6 clone is associated with the germ-line C_μ gene[9]. This J_H gene segment contains the HhaI site that marks the end of the homology between the germ-line μ27 clone and the myeloma clones (Fig. 4). Accordingly, the distinct germ-line V_H and germ-line J_H gene segments are rearranged in the myeloma clones in a manner analogous to the V_L and J_L gene segments of myeloma light chain genes[2,3].

Although it seems unlikely, the M603 J_H gene segment and flanking sequences could have been fused to a germ-line C_μ clone as the result of some cloning artefact. To eliminate this possibility, we decided to demonstrate the germ-line association of the J_H flanking sequence and C_μ sequences by Southern blot analyses. On different slots of the same agarose gel, we electrophoresed mouse sperm DNA cleaved with either HincII or EcoRI, transferred these DNAs to a nitrocellulose filter and hybridised one lane of each digest with a C_μ probe and a second lane with the 5.1-kilobase EcoRI fragment from the intervening sequence of the myeloma α6 clone (Fig. 6). Figure 6a and b shows Southern blots of EcoRI-digested sperm DNA hybridised with the C_μ cDNA probe and the 5.1-kilobase EcoRI subclone from the α6 clone, respectively. In both lanes, a 12.2-kilobase band corresponding to the germ-line C_μ fragment can be identified. In a second digest (HincII) of mouse sperm DNA, a similar result is obtained with both the C_μ probe (Fig. 6c) and the 5.1-kilobase EcoRI subclone (Fig. 6d) hybridising to a 9.3-kilobase HincII fragment. The 9.5-kilobase EcoRI band in Fig. 6b and the 5.0-kilobase band in Fig. 6d correspond to C_α restriction fragments. In each case, one of the two bands from the 5.1-kilobase EcoRI probe co-migrated with the single C_μ DNA fragment. This analysis shows that at least part of the intervening sequence from the myeloma α6 clone is adjacent to the C_μ gene in the germ line. Thus, the apparent deletion in the germ-line μ27 clone is not a significant factor in our discussion.

A summary of these analyses is presented in Fig. 7. The myeloma α6 clone is composed of three noncontiguous germ-line gene segments: (1) a V_H gene segment and its 5'-flanking

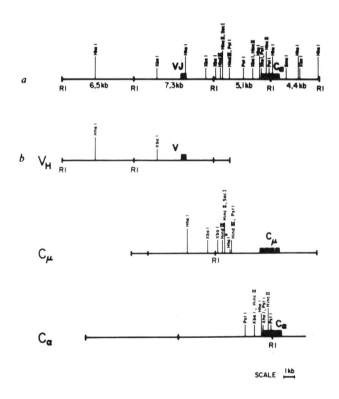

Fig. 4 *a*, Restriction map of myeloma clones α6 and α125. Specific cleavage sites were determined using double enzyme digestion and sizing by gel electrophoresis. Size standards were restriction digests of PBR322 and PBR322 multimers (see Fig. 2 legend). HindIII, HincII, PstI and SacI cleavage sites are only shown for the 5.1-kilobase EcoRI fragment. *b*, Restriction sites corresponding to those of the myeloma clones detected in the V_H, C_μ and C_α germ-line clones (Fig. 1). Restriction sites in regions corresponding to the 6.5- and 7.3-kilobase RI fragments of α6 and α125 were mapped by double digestion as above. All sites within the 5.1-kilobase RI fragment were compared by co-migration and blotting as described and illustrated in Fig. 5. The HhaI site marked by an asterisk in the C_μ clone was the one site found not to conform with those of the myeloma α6 clone.

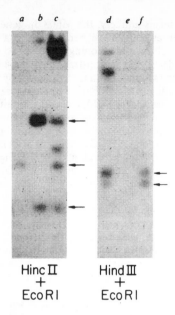

Fig. 5 Comparative restriction digests of the α6 myeloma, germ-line C_μ and C_α clones. Lanes *a–f* show parallel *Hinc*II + *Eco*RI and *Hind*III + *Eco*RI digests of the 5.1-kilobase *Eco*RI subclone of the α6 clone (Fig. 1*a*) and DNA from the germ-line α29 and μ27 clones (Fig. 1*b*). Samples were electrophoresed in 1% agarose, transferred to nitrocellulose and hybridised with nick-translated 5.1-kilobase α6 subclone DNA. The co-electrophoresis of restriction fragments allows a large number of different restriction sites to be compared rapidly. In this way, all 16 mapped sites in the 5.1-kilobase α6 subclone were compared with equivalent sites on the α29 and μ27 clones. Arrows show coincident bands.

sequence, (2) flanking sequences located 5′ to a germ-line C_μ gene which includes the J_H gene segment, and (3) the germ-line C_α gene segment with its flanking 3′ and 5′ sequences. Note that within the limits of the methods used here these three germ-line gene segments and their attendant flanking sequences apparently cover the entire myeloma α6 clone.

The α heavy-chain gene is formed by at least two recombinational events

Two distinct DNA rearrangements have occurred to form the M603 α gene—V–J joining and C_H switching (Fig. 8)[9]. The simplest interpretation of these observations is that the V_H gene segment is first joined to a J_H gene segment linked to the C_μ gene segment. This V–J joining, analogous to that which occurs in light chains, generates a rearranged μ gene that presumably leads to the expression of IgM molecules. V–J joining also commits an individual lymphocyte to the expression of a single V domain that remains invariant throughout subsequent steps of B-cell differentiation. Later, a C_H switch joins the V–J gene segment to the C_α gene segment to create a functional α gene and thus enables the differentiated lymphocyte to express IgA molecules. Thus, the myeloma α6 heavy-chain gene is assembled by two distinct and presumably independent DNA rearrangement events. Our heteroduplex, Southern blot and restriction mapping data show unequivocally that these two rearrangements occur at two distinct sites in the genome. These two sites of rearrangement are termed the V–J joining site and the C_H switch site, respectively.

These data are consistent with a differentiation pathway in which a B cell may switch from IgM to IgA synthesis while expressing the same V domain. We cannot establish that these two DNA rearrangement events occurred at different times, although this supposition is reasonable. We also cannot rule out the possibility of intermediate differentiation states in this B-cell lineage where IgG molecules were produced, although there is no direct evidence for such a stage.

The C_H switch may be explained by any one of several genetic models

Several mechanisms of C_H switching have been proposed[11,22–26], many of which are similar to those proposed for V–J joining[27]. These models can be categorised as involving either DNA rearrangements to replace one constant region with another (successive deletions, excision–insertions or inversions) or the differential processing of a large nuclear RNA transcript containing multiple heavy-chain constant-region genes[26].

The RNA processing model seems unlikely as a general mechanism for C_H switching at the level of antibody-secreting plasma cells. High molecular weight nuclear RNAs from three myeloma tumours hybridise only with a cDNA probe complementary to the class of immunoglobulin expressed in that tumour and not with probes from non-expressed immunoglobulin classes[28]. Moreover, cell fusion experiments hybridising two different myeloma cells demonstrate that the hybrid cells synthesise only parental $V_H C_H$ combinations[29], a result in conflict with the simple RNA processing model. Furthermore, neither $C_{\gamma 1}$ nor C_μ DNA probes hybridise on Southern blots with the M603 α clones (data not shown), contrary to one prediction of the RNA processing model.

On the other hand, the evidence presented here strongly supports DNA rearrangement as a fundamental element in the mechanism for heavy-chain switching. As summarised in Fig. 7, a very large segment of C_μ flanking sequence has been brought adjacent to C_α and V_H gene segments in the active α gene of myeloma tumour M603. The creation of the M603 α gene apparently requires two DNA rearrangements (Fig. 8). A C_H switching

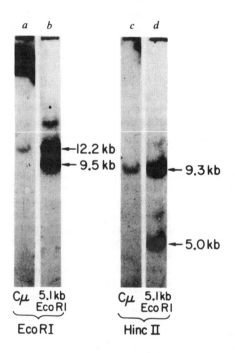

Fig. 6 Southern blots of the 5.1-kilobase *Eco*RI and C_μ fragments. Mouse sperm DNA was digested with *Eco*RI or *Hinc*II and 3 μg was loaded onto a 4 mm × 20 × 20 cm 0.7% agarose gel, electrophoresed at 40 V for 10 h and blotted as described. Probes used were either the 5.1-kilobase *Eco*RI fragment or a C_μ cDNA clone nick-translated to 4×10^8 c.p.m. per μg. Washing was as described in Fig. 3 legend except that lanes hybridised with the 5.1-kilobase *Eco*RI fragment were washed further in 10 mM NaCl, 10 mM Tris, 0.1% SDS and 0.1% NaPP$_i$ for 2 h at 68 °C to reduce the signal strength of weakly homologous repeats. Filters were exposed for 12 h with an intensifying screen at −80 °C. The faint band above the 12.2-kilobase band in *b* is a partial digestion product. All lanes shown were run on the same gel and blotted simultaneously, alignment being assisted by inclusion of pBR322 multimers as internal standards (see Fig. 2).

DNA rearrangement is not postulated in the RNA processing models[26]. The data presented here do not distinguish between the various types of DNA rearrangements proposed, but the hybridisation kinetics experiments of Honjo and Kataoka are consistent with a deletional mechanism for the C_H switch[25]. In addition, recent experiments suggest that V–J joining in mouse λ genes is accomplished by a deletional mechanism[4]. Thus, both types of DNA rearrangement, V–J joining and C_H switching, may arise through deletional mechanisms. If the deletional model is correct for either type of DNA rearrangement, the differentiation of B cells is irreversible because chromosomal information is lost with the excision of each deletional loop of chromosome.

Gene organisation studies may delineate distinct pathways of B-cell differentiation

It will be interesting to determine whether all joined α genes have the same switch site. Southern blot analyses of the DNA from a second IgA-producing myeloma tumour, H8, with the 5.1-kilobase *Eco*RI probe yield a restriction fragment pattern identical to that of M603 DNA (data not shown). In particular, the 5.1-kilobase *Eco*RI fragment (Fig. 1) which contains the switch site seems the same. These data suggest that the expressed α genes in both M603 and H8 myeloma tumours have the same C_H switch site. However, Southern blot analyses of several other closely related IgA-producing tumours do not show a 5.1-kilobase *Eco*RI fragment (M.M.D., unpublished observation). Thus, there may be multiple C_H switch sites for the C_α gene segment. As it seems that B cells producing IgM or IgG may switch to the production of IgA, perhaps distinct C_H switch points reflect distinct pathways of B-cell differentiation. It will also be interesting to determine the location and number of switch sites for other immunoglobulin classes. If each C_H gene segment has a unique site or set of sites for C_H switching, one may be able to trace the distinct pathways of B-cell differentiation by studying the sequence organisation of each functional heavy-chain gene.

DNA rearrangements of antibody gene segments lead to combinatorial amplification of immunoglobulin information

V–J joining and C_H switching are mediated by DNA rearrangements which display combinatorial properties that amplify the germ-line information encoding the antibody gene families. (1) The V and J gene segments of one antibody gene family may be joined in a combinatorial manner to generate diversity in the third hypervariable regions of both κ and heavy chains[4,8,9,30,31]. For example, mice may have at least 200 V_H and 5 J_H gene segments that may be joined combinatorially to generate 1,000 different $V_H J_H$ coding regions. (2) One V domain may be combinatorially switched among eight or more different C_H regions to carry out a variety of different effector functions

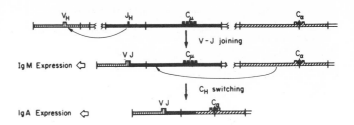

Fig. 8 Two types of DNA rearrangements leading to the creation of the myeloma α heavy-chain gene V–J joining and C_H switching. V–J joining indicates a DNA rearrangement that joins the V_H and J_H gene segments. Because the J_H segments seem to be associated with the germ-line C_μ gene[9], V–J joining permits a μ chain and IgM molecules to be expressed by the differentiating B cell. C_H switching denotes a DNA rearrangement that replaces the C_μ gene segment with a C_α gene segment. This second rearrangement presumably permits an α chain and IgA molecules to be expressed by the now fully differentiated lymphocyte.

that are directed at eliminating antigen or triggering defensive mechanisms such as complement fixation. Thus, each recognition (V) domain may be switched to many different effector (C) domains. The combinatorial properties of antibody gene segments and polypeptides therefore contribute to several fundamental aspects of the vertebrate immune response—V region diversity and the combinatorial switching of antigen-recognition (V) domains with a variety of different effector (C) domains during B-cell differentiation. Other complex eukaryotic gene families may use similar DNA combinatorial mechanisms for information amplification[32].

This work was supported by NSF grant PCM 76–81546, ACS grant IM56 and USPHS grant AI09072. M.M.D., P.W.E. and D.L.L. are supported by NIH training grant GM 07616. K.C. is supported by NIH Fellowship GM 05442. R.J. is a fellow of the Swiss National Foundation. I.L.W. is a faculty Research Awardee of the ACS. All experiments involving recombinant organisms were conducted in accordance with the revised NIH Guidelines on recombinant DNA, using P2, EK-2 or P3, EK-2 containment. We thank David Goldberg for his gift of PBR322 multimers, Tom Sargent for packaging extracts, Keichi Itakura for synthetic RI linkers, and David Anderson, Norman Davidson, Max Delbrück, Richard Flavell, Henry Huang, Tom Maniatis and Tom Sargent for helpful discussions.

Received 1 October 1979; accepted 22 January 1980.

1. Mage, R., Lieberman, R., Potter, M. & Terry, W. in *The Antigens* (ed. Sela, M.) 299–376 (Academic, New York, 1973).
2. Brack, C., Hirowa, A., Lenhard-Schueller, R. & Tonegawa, S. *Cell* **15**, 1–14 (1978).
3. Seidman, J. G., Max, E. E. & Leder, P. *Nature* **280**, 370–375 (1979).
4. Sakano, H., Huppi, K., Heinrich, G. & Tonegawa, S. *Nature* **280**, 288–294 (1979).
5. Gilmore-Herbert, M. & Wall, R. *Proc. natn. Acad. Sci. U.S.A.* **75**, 342–345 (1978).
6. Schibler, U., Marcu, K. B. & Perry, R. P. *Cell* **15**, 1495–1509 (1978).
7. Early, P. W., Davis, M. M., Kaback, D. B., Davidson, N. & Hood, L. *Proc. natn. Acad. Sci. U.S.A.* **76**, 857–861 (1979).
8. Davis, M., Early, P., Calame, K., Livant, D. & Hood, L. in *Eukaryotic Gene Regulation* (eds Axel, R. Maniatis, T. & Fox, C. F.) 393–406 (ICN UCLA Symp., Academic, New York, 1979).
9. Early, P. W., Huang, H. V., Davis, M. M., Calame, K. & Hood, L. *Cell* (submitted).
10. Raff, M. C. *Cold Spring Harb. Symp. quant. Biol.* **41**, 159–162 (1976).
11. Sledge, C., Fain, D. S., Black, B., Krieger, R. G. & Hood, L. *Proc. natn. Acad. Sci. U.S.A.* **73**, 923–927 (1976).
12. Fudenburg, H. H., Wang, A. C., Pink, J. R. L. & Levin, A. S. *Ann. N.Y. Acad. Sci.* **190**, 501–506 (1971).
13. Wang, A. C., Geisely, J. & Fudenberg, H. H. *Biochemistry* **12**, 528–534 (1973).
14. Levin, A. S., Fudenberg, H. H., Hopper, J. E., Wilson, S. & Nisonoff, A. *Proc. natn. Acad. Sci. U.S.A.* **68**, 169–171 (1971).
15. Wang, A. C., Wilson, S. K., Hopper, J. E., Fudenberg, H. H. & Nisonoff, A. *Proc. natn. Acad. Sci. U.S.A.* **66**, 337–343 (1970).
16. Maniatis, T. *et al.* *Cell* **15**, 687–701 (1978).
17. Blattner, F. R. *et al.* *Science* **196**, 161–169 (1977).
18. Southern, E. M. *J. molec. Biol.* **98**, 503–517 (1977).
19. Jeffreys, A. J. & Flavell, R. A. *Cell* **12**, 429–439 (1977).
20. Benton, W. D. & Davis, R. W. *Science* **196**, 180–182 (1977).
21. Hood, L. *et al. Cold Spring Harb. Symp. quant. Biol.* **41**, 817–836 (1976).
22. Smithies, O. *Science* **169**, 882–883 (1970).
23. Hood, L. *Fedn Proc.* **31**, 177–178 (1972).
24. Bevan, M. J., Parkhouse, R. M. E., Williamson, A. R. & Askonas, B. A. *Prog. Biophys. molec. Biol.* **25**, 131–162 (1972).

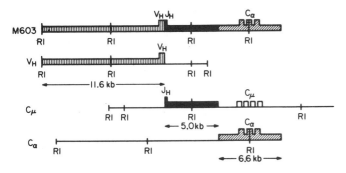

Fig. 7 Origins of the three germ-line components of the myeloma α heavy-chain gene. Various types of shading indicate homology by heteroduplex analyses and by restriction enzyme analyses.

25. Honjo, T. & Kataoka, T. *Proc. natn. Acad. Sci. U.S.A.* **75**, 2140–2144 (1978).
26. Rabbitts, T. H. *Nature* **275**, 291–296 (1978).
27. Tonegawa, S., Hozumi, V., Matthyssen, G. & Schuller, R. *Cold Spring Harb. Symp. quant. Biol.* **41**, 877–889 (1976).
28. Marcu, K. B., Schibler, U. & Perry, R. P. *Science* **204**, 1087–1088 (1979).
29. Schulman, M. J. & Kohler, G. *Nature* **274**, 917–919 (1978).
30. Max, E., Seidman, J. G. & Leder, P. *Proc. natn. Acad. Sci. U.S.A.* **76**, 3450–3454 (1979).
31. Schilling, J., Cleavinger, B., Davie, J. & Hood, L. *Nature* **283**, 35–40 (1980).
32. Hood, L., Huang, H. & Dreyer, W. J. *J. supramolec. Struct.* **7**, 531–559 (1977).
33. Witkin, S., Korngold, G. & Bendich, A. *Proc. natn. Acad. Sci. U.S.A.* **72**, 3295–3299 (1975).
34. Joho, R., Weissman, I. L., Early, P., Cole, J. & Hood, L. *Proc. natn. Acad. Sci. U.S.A.* (in the press).
35. Sternberg, N., Tiemeier, D. & Enquist, L. *Gene* **1**, 255–280 (1977).
36. Hohn, B. & Murray, K. *Proc. natn. Acad. Sci. U.S.A.* **74**, 3259–3263 (1977).
37. Clarke, L. & Carbon, J. *Cell* **9**, 91–99 (1976).
38. Calame, K., Rogers, J., Davis, M., Early, P., Livant, D., Wall, R. & Hood, L. *Nature* (submitted).
39. Maniatis, T., Jeffrey, A. & Kleid, D. G. *Proc. natn. Acad. Sci. U.S.A.* **72**, 1184–1188 (1975).
40. Sutcliffe, J. G. *Nucleic Acids Res.* **5**, 2721–2728 (1978).
41. Davis, R., Simon, M. & Davidson, N. *Meth. Enzym.* **21**D, 413–428 (1971).

Amino acid sequence of homogeneous antibodies to dextran and DNA rearrangments in heavy chain V-region gene segments

J. Schilling*, B. Clevinger†, J. M. Davie† & L. Hood*

* Division of Biology, California Institute of Technology, Pasadena, California 91125
† Department of Microbiology and Immunology, Washington University School of Medicine, St. Louis, Missouri 63110

The complete variable region sequences from ten antibodies and two myeloma proteins binding α-1,3 dextran have been determined. The diversity patterns of these homogeneous antibody molecules suggest that the variable regions of heavy chains are encoded by separate variable (V) and joining (J) gene segments. The most striking feature of these data is the extensive sequence variability of a region that we denote the D (diversity) segment which is located at the junction between the V and J segments in the centre of the third hypervariable region. The D segment diversity may arise from a novel somatic mutational mechanism or may be encoded by multiple D gene segments. For the first time, the amino acid sequence correlates of several V region idiotypes are determined.

STUDIES of immunoglobins at the protein and DNA levels have led to striking advances in our understanding of the organisation and evolution of antibody genes and mechanisms of antibody diversity[1-4]. There are three unlinked families or clusters of antibody genes—the λ and κ gene families encode light (L) chains and the heavy (H) gene family encodes heavy chains[5]. The κ or λ gene is composed of four distinct gene segments, leader (L), variable (V), joining (J) and constant (C), which are separated by intervening sequences in undifferentiated or germ-line DNA[6-12]. Heavy chain genes probably have a similar organisation of gene segments (P. Early, H. Huang, M. Davis and L.H., unpublished observations). During the differentiation of antibody-producing cells, the V_L and J_L gene segments are juxtaposed and together encode the variable or antigen-binding domain of the antibody molecule. (The term V or J segment will refer to the corresponding protein sequence; the term V or J gene segment will refer to the corresponding DNA coding regions.) Comparative amino acid sequence analyses of V regions showed that for both light and heavy chains, there are three regions of extreme variability termed hypervariable regions[13,14], which fold in three dimensions to constitute the walls of the antigen-binding site[15,16]. The diversity in the hypervariable regions of L chains has been most extensively studied and explained by a variety of genetic mechanisms. (1) There are

many germ-line V_κ (and J_κ) gene segments[5,10]. (2) A single V_κ gene segment may be joined to many different J_κ gene segments and conversely, a single J_κ gene segment may be joined to many different V_κ gene segments[17]. As the boundary of the V_L and J_L segments falls within the third hypervariable region, the combinatorial joining of these segments contributes to the diversity of antigen-binding sites. (3) Random single base mutations may occur throughout the V_L (and J_L) gene segments and those variants exhibiting altered and useful antigen-binding properties may be selected for clonal expansion. For example, 12 of 18 V_λ regions examined were identical to one another[18-20]. The six variant V_λ sequences differed from the 12 identical sequences by between one and three amino acid substitutions. All these substitutions occur in the hypervariable regions. These data suggest that a single V_λ gene is often expressed unchanged and less frequently is modified by somatic mutation to produce the variants. Recent DNA sequence analyses of germ-line and myeloma λ gene segments are consistent with this supposition[6,11,12]. The pattern of diversity wherein variant V region sequences exhibit few substitutions mostly concentrated in the hypervariable regions will be denoted λ-type diversity. (4) A site-specific recombinational mechanism has been postulated to explain amino acid sequence diversity at the junction of V_κ and J_κ gene segments[21-23]. This diversity is reflected as single codon substitutions at the N-terminal residue of the J_κ segment and as the insertion or deletion of single codons precisely at the V-J junction[23]. This pattern of diversity will be denoted the V-J junctional diversity.

We have analysed the diversity patterns in mouse antibodies directed against a simple antigenic-determinant, α-1,3 dextran, in order to determine the relative contributions of the diversification mechanisms discussed above. The hybridoma technique of Köhler and Milstein facilitates the analyses of antibody diversity by making large quantities of antigen-induced, homogeneous antibody molecules available[24]. This is accomplished by fusing normal antibody-secreting cells with mouse plasmacytoma cells[25]. The hybrid cells generate immortal cell lines that secrete large amounts of homogeneous antibody molecules of the two parental cell types. The light chains of the dextran antibodies are all of the λ type and appear to be very similar, if not identical, to one another. Hence, most of the diversity in the dextran system appears to reside in the

Table 1 RNA codons of positions 100 and 101 from V_H regions of myeloma and hybridoma proteins binding dextran*

V–J segments†	Proteins	Codons‡ 100	Codons‡ 101
		a b c	a b c
V_1–J_1	M104E	UAY	GAY
	J558	AGZ CGX	UAY
	Hdex2	AAY	UAY
	Hdex3	AGZ CGX	GAY
	Hdex7	GCX	GAY
	Hdex6	AGY UCX	CAY
V_2–J_1	Hdex8	UAY	GAY
V_3–J_1	Hdex9	AGZ CGX	UAY
V_4–J_1	Hdex10	GUX	AAY
V_1–J_2	Hdex4	AAZ	GAY
	Hdex5	AGY UCX	AAY
V_1–J_3	Hdex1	AAY	UAY

* These codon assignments are derived from the corresponding amino acid sequences

† The V and J segments are defined in Fig. 2.

‡ Y denotes pyrimidine; Z designates purine; and X denotes any one of the four bases.

heavy chains. We report here the complete amino acid sequences from the variable regions of 10 heavy chains derived from hybridoma antibodies to dextran. These data suggest that heavy chains have a D (diversity) segment located in the heart of the third hypervariable region. The D segment may be encoded by a separate germ-line gene segment or generated by a new type of somatic mutational mechanism.

Immunoglobulins binding dextran

Immunoglobulins binding dextran exhibit limited heterogeneity in several respects. First, preliminary amino acid sequence studies on two myeloma proteins binding α-1,3 dextran, M104E and J558, revealed that the V_L regions are identical[18,19] and that the V_H regions are identical through their first hypervariable regions[26]. The N-terminal sequences of V_H regions derived from normally-induced pooled serum antibodies to α-1,3 dextran also are identical to their myeloma counterparts[27]. Second, isoelectric focusing analyses of anti-dextran antibodies of the IgG class demonstrate limited heterogeneity with considerable sharing of bands between sera from different mice[28]. Moreover, the serum antibody light chains are identical to one another and to their myeloma counterparts by isoelectric focusing criteria[27]. Third, the light chains from the 10 hybridomas and 2 myeloma proteins reported here are all of the λ type and have identical isoelectric focusing patterns (J.S., unpublished observation). Thus, light chains from dextran antibodies appear to be very similar, if not identical, to one another. Therefore, most of the diversity in the antibodies to α-1,3 dextran must reside in the heavy chains. Fourth, the anti-α-1,3 dextran antibodies can be divided into several general categories by idiotypic analyses[29]. More than half of the anti-α-1,3 dextran antibodies from mice immunised with the branched dextran B1355 share a variable-region antigenic determinant or idiotype with both the M104E and J558 proteins. This idiotype is absent from non-dextran binding immunoglobulins and λ light chains and is denoted the 'cross-reactive' or IdX idiotype. Antibodies with the IdX idiotype are synthesised by multiple different antibody-producing clones. Additional idiotypes specific for either the M104E or the J558 immunoglobulins are present on a minor fraction of the

normally-induced antibodies. These idiotypes are denoted as individual or IdI idiotypes (for example, IdIM104E and IdIJ558). Another general category of dextran antibodies is that of those lacking the IdX determinant (IdX negative). The IdX-positive and IdX-negative antibodies to dextran are structurally very similar in that their light chains are identical by isoelectric focusing analysis and the heavy chains from both the IdX-positive and IdX-negative pools of serum antibodies are identical to the heavy chains of the M104E and J558 immunoglobulins for their N-terminal 27 and 53 residues, respectively[27]. These observations show that much of the V domain diversity in normal antibodies to α-1,3 dextran resides in the heavy chain beyond residue 53. Thus, we turned to the amino acid sequence analysis of homogeneous normal antibodies to α-1,3 dextran.

Diversity patterns in V_H regions from antibodies binding dextran

Figure 1 gives the complete amino acid sequences of V_H regions from 2 myeloma proteins and 10 hybridoma antibodies that bind dextran. The hypervariable regions and the variable region site of carbohydrate attachment are indicated. These V_H regions are compared to the published sequence of the myeloma M104E V_H region[30] with amino acid substitutions indicated by the one-letter code[31].

Two features are of interest in this set of V_H regions. First, each of these 12 V_H regions differs from all of the others by 1 to 13 amino acid substitutions (Fig. 1). We have not yet detected a pair of identical V_H regions in 12 attempts. Accordingly, the V_H sequence diversity in the antibody response to α-1,3 dextran is extensive. Second, patterns of variability divide the V_H regions from dextran antibodies into three distinct sections—V(1–99), V(100–101) and V(102–117). The highly variable section, V(100–101), separates the other two sections which are relatively conserved. Section V(100–101) exhibits nine different sequences in the 12 V_H regions (Fig. 2). Indeed, position 100 has eight alternatives and position 101 has four. In contrast, nine V_H regions have identical sequences for section V(1–99) and the remaining V_H regions differ by between two and seven residues from the others. Likewise, nine V_H regions have identical sequences for the section V(102–117), two have a second sequence (Hdex4 and Hdex5) and one has a third sequence (Hdex1). Thus, the relatively conserved sections V(1–99) and V(102–117) are separated by two highly variable residues at positions 100 and 101.

This same three-section pattern is seen when 20 completely sequenced V_H regions are aligned by homology with their anti-dextran counterparts (Fig. 3). The section homologous to V(1–99) can clearly be defined in each of these 20 V_H regions and each terminates precisely at the position homologous to 99 in the heavy chains binding dextran (Fig. 3). The homology relationships among the V(102–117) sections are also obvious. The V(102–117) sections from the various V_H regions are identical in size, apart from the levan-binding immunoglobulins, and exhibit extensive amino acid sequence homology to one another (56–100% sequence identity). Accordingly, sections homologous to V(1–99) and V(102–117) in the dextran V_H regions are clearly present in all of the other V_H regions examined to date. In contrast, there is no homology in the V(100–101) regions between these two sections because of extensive sequence and size variability (Fig. 3). The sections analogous to V(100–101) can be defined only as those residues remaining after the V(1–99) and V(102–117) sections have been assigned for each V region by homology.

Implications of V_H diversity patterns for gene organisation

The patterns of amino acid diversity in the V_H regions from antibodies binding α-1,3 dextran allow us to define tentatively V_H and J_H segments with properties similar to their light chain counterparts (Figs 1–3). The V(1–99) section appears to correspond to the V_H segment and V(102–117) section appears to

	Kabat–Wu Numbering	10	20	30	40	50	59	69	79	86	95	106		Idiotype Reactions	
	Sequential Numbering	10	20	30	40	50	60	70	80	90	100	110	Idx	IdI MI04	IdI J558
SOURCE	V_H REGION			hv1			hv2				hv3				
MYELOMA	MI04E						CHO						++++	++++	—
MYELOMA	J558						CHO			RY			++++	—	++++
HYBRIDOMA	Hdex 1						CHO			NYH V			++++	—	+
HYBRIDOMA	Hdex 2						CHO			NY			++++	—	+
HYBRIDOMA	Hdex 3						CHO			R			++++	—	—
HYBRIDOMA	Hdex 4						CHO			K-Y-Y-Q—L—			++++	—	—
HYBRIDOMA	Hdex 5						CHO			SNY-Y—Q—L—			++++	—	—
HYBRIDOMA	Hdex 6						CHO			SH			++++	—	—
HYBRIDOMA	Hdex 7						CHO			A			++++	++++	—
HYBRIDOMA	Hdex 8	T		V-HP		T-S	CHO		G				+	++++	—
HYBRIDOMA	Hdex 9				R	H-N-F				RY			++++	—	++++
HYBRIDOMA	Hdex 10					KK				VN			—	—	—
Substitution (Sequential)		16		34	39 40	52 54 55		63	73 77 78 80						

←————————————————— V segment —————————————————→ D ←–J segment–→

Idiotype Correlations   ⊢—⊣ Idx   ⊢—⊣ IdI

Fig. 1 The amino acid sequences and idiotypes from V regions of myeloma proteins and homogeneous hybridoma antibodies binding dextran. The myeloma protein M104E is used as a prototype sequence and variation from this sequence is indicated. The hypervariable regions are indicated by hv1, hv2 and hv3, respectively. CHO indicates the attachment of a carbohydrate moiety. The one-letter amino acid code of Dayhoff is used[31]. The V and J segments are indicated by arrows. The D segment includes positions 100 and 101. Sequential numbering denotes consecutive numbering of the V_H regions. The Kabat–Wu numbering is from ref. 36. Details of the complete heavy chain sequence of the M104E immunoglobulin are published elsewhere[30]. A similar strategy was used for the sequence analysis of the V_H regions from the hybridomas. The positions of amino acid substitutions in V segments are indicated. Eight hybrids were derived from individual BALB/c mice: Hdex6, 7, 8 and 9 were obtained at 5 or 6 days after a single injection of 100 μg dextran B1355 in complete Freund's adjuvant (CFA), and Hdex2, 3, 4 and 10 were obtained 1–3 days after hyperimmunisation with dextran in CFA followed 1–2 months later with 3 interperitoneal injections of 2×10^9 *Escherichia coli* at 2-day intervals. Hybrids Hdex1 and 5 were derived from a single (BAB-14 × BRVR)F$_1$ mouse at 7 days after the dextran–*E. coli* protocol. Hybrids were generated as described by Galfre *et al.*[41] using the HGPRT-deficient MPC-11 ($γ_{2b}κ$) line, 45.6TG1.7, (Hdex1, 3, 4, 5 and 10) or a non-secreting variety of the MOPC-21 ($γ_1κ$) line, NSI/1-Ag4-1, which synthesises but does not secrete κ chains (Hdex2, 6, 7, 8 and 9). Anti-dextran secreting hybrids were detected by direct binding of ^{125}I-dextran by culture supernates either in a tube-binding assay or after isoelectric focusing in polyacrylamide gels. Hybrids were cloned in soft agar[42] over feeder layers and grown in BALB/c mice as ascites tumours. The antibodies produced by the hybrids appeared representative of those found in the serum at the time of fusion. Idiotypes were determined as described by ref. 29. ++++ Indicates a reaction 10 times as strong as +; − indicates no reaction.

represent the J_H segment. The V(100–101) section will be called the D (diversity) segment and its origin will be discussed.

The extensive variability in these V_H regions at position 100 appears to mark the C-terminal end of the V_H segment, just as extensive variability at position 96 marks the C-terminal end of the V segment in kappa light chains[17]. The correspondence of V(1–99) to a V_H gene segment is confirmed by the DNA sequence analysis of a germ-line V_H gene segment which ends precisely at codon 99 (P. Early, H. Huang, M. Davis and L.H., unpublished observation).

The J_H segment includes at least positions 102–117. The C-terminal boundary of the J_H segment at position 117 has been defined by the DNA sequence analysis of an expressed V_H-J_H gene (H. Huang, P. Early, M. Davis and L.H., unpublished observations). This analysis demonstrates that the C_H region begins at a position homologous to residue 118 in the heavy chains from antibodies binding dextran, in agreement with the J_H and $C_{γ1}$ gene segment boundaries defined by DNA sequence analysis of a $C_{γ1}$ gene[32]. The N-terminal boundary of the J_H segment is unclear. Figure 2 shows three J_H sequences which differ from each other by two or three two-base substitutions (J_1, J_2 and J_3). Position 102 is the first position in the J_H segment at which linkage relationships among amino acid residues appear. A tryptophan at position 102 is invariably linked to valine, alanine and valine residues at positions 106, 109 and 113, respectively, in the J_1 segments, whereas the tyrosine at position 102 is linked to tyrosine, glutamine and leucine at these positions in the J_2 segments (Fig. 2). These linkage relationships imply that the J_1 and J_2 segments are encoded independently in the germ-line. The D segment comprises positions 100 and 101 and may or may not be a part of the J segment, depending on how V–J junctional diversity arises. These assignments of the V_H, J_H and D segment boundaries also are supported by the homology relationships exhibited among the other completely sequenced V_H regions (Fig. 3).

The D and J_H segments comprise almost the entire third hypervariable region (Fig. 2). This is in striking contrast to the $J_κ$ segment which just extends into the third hypervariable region by a single residue[17]. Moreover, the size variation seen in the D segment of heavy chains is far more extensive than that seen at the V–J boundary of kappa chains.

Rudikoff and coworkers have used more limited V_H region data to suggest the V_H segment extends from 1 to 98 and the J_H segment from 105 to 117. They conclude that positions 99 to 104 may be part of the V segment or, alternatively, part of the J segment[33]. These assignments are different from those made here.

Implications of V_H diversity patterns for mechanisms of antibody diversity

The V_H regions from dextran antibodies all differ from one another. Let us analyse possible contributions to this diversity by multiple germ-line gene segments, somatic mutation and the combinational joining of V_H and J_H segments.

Germ-line V_H and J_H gene segments. Four different amino acid sequences corresponding to the V_H gene segment are present in the 12 antibodies binding dextran and these are denoted V_1 to V_4 in Figs 1 and 2. The V_1 segment is expressed in nine different immunoglobulins binding dextran and, accordingly, appears to be encoded by a germ-line gene. This supposition follows from the improbability of generating the same V_H sequence many times by somatic mutation[4]. The V_2, V_3 and V_4 segments differ from V_1 by seven, four and two residue substitutions, respectively. The V_2, V_3 and V_4 segments may be encoded by distinct germ-line V_H gene segments or these sequences may arise from one or more somatic mutational mechanisms. However, the diversity patterns among the V_2, V_3 and V_4 segments do not resemble either of the two somatic mutational patterns described earlier for light chains, λ-type diversity and $V_κ$-$J_κ$ junctional diversity.

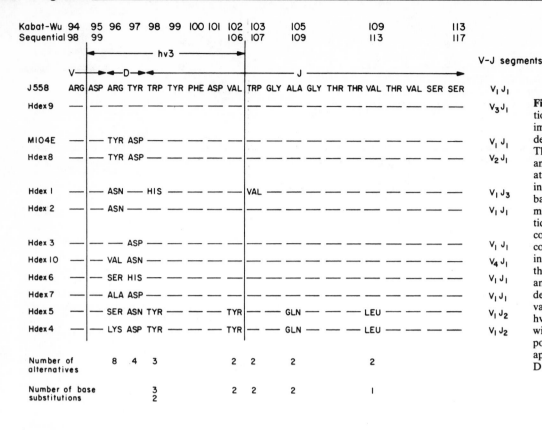

Fig. 2 The C-terminal portions of V_H regions from immunoglobulins binding dextran. See Fig. 1 legend. The number of alternative amino acids for each position at which variability exists are indicated. The number of base substitutions denotes the minimum number of nucleotide changes required to convert the most common codon at a particular position into the alternative codons at that position. The V_1 to V_4 and J_1 to J_3 segments are defined. The third hypervariable region is denoted by hv3. Three pairs of proteins with identical sequences at positions 100 and 101 are set apart from the rest. The V_H, D and J_H segments are defined in Fig. 1.

The three different J_H segments differ from one another by two to five residue alternatives (Fig. 2). Moreover, most of these J_H segment substitutions require two or even three base substitutions and repeats of two of these sequences have been seen. These observations strongly suggest that the three J_H sequences are directly encoded by three germ-line J_H gene segments. The J_H diversity noted in Fig. 3 appears to be significantly greater than the corresponding J_L diversity[17] and, accordingly, may be encoded by a larger number of germ-line J_H gene segments.

Combinational joining. The combinational joining of V_H and J_H gene segments to encode many distinct V_H regions provides yet another source of antibody diversity (Fig. 4). For example, identical J_H segments may be associated with differing V_H segments (such as M104E and Hdex8; J558 and Hdex9). Conversely, identical V_H segments may be associated with different J_H segments (for example, Hdex1, Hdex2 and Hdex4) (Fig. 1). Even more striking is the observation that the J_1 segment in V_H regions from antibodies binding dextran also occurs in the very different V_H regions from myeloma proteins binding phosphorylcholine (Fig. 3). Likewise, the same J_H segment may be found in heavy chains binding phosphorylcholine and galactan (for example, M167 and T601). Thus, the combinational joining of J_H and V_H gene segments appears to be an important source of diversity in these dextran-binding antibodies.

D segment diversity. This most striking diversity pattern in these V_H regions (Figs 2, 3). Diversity in the D segments may arise from one of three general mechanisms: (1) a novel somatic mutation mechanism possibly occurring as a result of the joining of V_H and J_H gene segments; (2) a large number of germ-line J_H or V_H gene segments which include the D segment; or (3) a third gene segment (D) encoding positions 100 and 101 which is joined with the V_H and J_H gene segments during the differentiation of an antibody-producing cell. Let us consider what constraints the diversity patterns of the V_H regions impose on each of these models.

Several observations argue against many germ-line V_H gene segments extending from codons 1 to 101. First, identical D segments may be associated with different V_H segments—M104E-Hdex8 and J558-Hdex9 (Fig. 1). Thus, the D segments

appear to associate independently with V_H segments. Second, the T15 V_H gene segment terminates at codon 99 (P. Early, H. Huang and L.H., unpublished observation). Accordingly, the D segment cannot be encoded as part of the V_H gene segment for the T15 V_H gene segment (Fig. 3). Identical D segments may be associated with different J_H segments—Hdex1-Hdex2 and X44-J539 (Fig. 3). A more striking example of this independent association of D and J_H segments is the observation that a J_1 segment is associated with very different D segments in immunoglobulins binding distinct haptens—T15 and M104E (Fig. 3). The independent segregation of D segments with both V_H and J_H segments suggests that the D coding regions are not merely germ-line extensions of V_H or J_H gene segments. In addition, homology comparisons among the V_H regions very clearly define the V_H and J_H segments (Fig. 3). In contrast, the D segments lack constancy both in size (0 to 7 residues) and amino acid sequence (20 different sequences in 25 different V_H regions compared) (Fig. 3). The V_H regions in Fig. 3 would require a very large number of germ-line gene segments if the V_H or J_H gene segments include the D segment.

The D segment diversity is quite distinct from the V–J junctional diversity of kappa chains in several regards. First, there are extensive size differences among the D segments whereas the κ junctional diversity shows at most the insertion or deletion of just a single codon[23]. Second, the two codons for the D segment associated with a particular $V_H(V_1)$ and a particular $J_H(J_1)$ segment exhibit three different nucleotide alternatives at three positions—100a, 100b and 101a (Table 1). This observation is important because it suggests that D segment diversity cannot be generated solely by recombination between the V_1 and J_1 gene segments as is suggested by the site-specific recombinational model of mouse kappa chains[21-23]. This model allows for only two alternatives at any nucleotide position. Accordingly, the extensive variability, the size differences and the presence of three bases at several nucleotide positions in the D segments constitute a distinctly new pattern of diversity that must be explained by a novel mechanism for somatic mutation or the presence of multiple D gene segments.

The striking base invariance at nucleotide positions 101b and

101c in the midst of the remarkable adjacent diversity raises several interesting possiblities (Table 1). Perhaps these invariant nucleotides form part of a recognition site for an enzyme that might mediate somatic mutation or join the D and J_H gene segments together. Alternatively, perhaps these invariant nucleotides are a part of the J_H gene segment.

If the D segment is encoded by a gene segment separate from the V_H and J_H gene segments, then heavy chain genes differ in a major organisational feature from light chain genes. Moreover, the presence of a putative D gene segment creates a second gene segment boundary (V_H–D and D–J_H) at which junctional diversity of the type noted in kappa chains can occur[21-23].

Antigen-binding diversity and V_H variability

The third hypervariable region forms an important part of the antigen-binding site[15,16]. Diversity in the dextran antibodies is concentrated in the third hypervariable region of heavy chains which includes the D segment and much of the J_H segment (Fig. 2). Studies on the binding specificites of the myeloma proteins M104E and J558 which differ only in the D segment have demonstrated that the antigen-binding properties of these two molecules are quite distinct[34,35]. Accordingly, the mechanism for producing D segment diversity does have an important role in modifying the antigen-binding properties of antibodies.

Variability in the third hypervariable region of the dextran antibodies arises from germ-line diversity (J_H segments), by combinational joining of V_H and J_H (and possibly D_H) segments and by the mechanism for producing D segment diversity. Of these, D segment diversity appears to be by far the most extensive and, presumably, important in generating a diversity of antigen-binding sites (Figs 2, 3).

One striking feature evident in Fig. 3 is that immunoglobulins binding particular haptens share similar V_H segments, exhibit D segments of similar size (for example, zero for levans, four for galactans, and so on), and to a lesser extent share similar J_H segments. Whether these correlations arise through some intrinsic properties of gene organisation, somatic mutational mechanisms or antigen selection is unknown.

Correlation of dextran idiotypes and protein structure

Several interesting conclusions can be drawn about the correlation of V_H structure and the various idiotypes of dextran-binding antibodies (Fig. 1). First, the individual idiotype appears to correlate with the D segment sequence. The V_H regions from the M104E and J558 proteins differ only by their D segments. As the light chains are identical, the individual idiotypes of M104E and J558 must be determined by positions 100 and 101. Moreover, the V_H region of Hdex9, which differs from that of J558 by four residues and is identical at positions 100 and 101, has the individual J558 idiotype. Likewise, the V_H region of Hdex8, which differs from that of M104E by seven residues and is identical in the D segment, also has the individual M104E idiotype. However, a single residue substitution can eliminate the individual idiotype as demonstrated by the observation that Hdex3 is lacking both the M104E and J558 individual idiotypes (Figs 1, 2). Partial and in one case complete IdI reactivity can be retained by D segments sharing one of two residues (Hdex1, Hdex2 and Hdex7). The overall conclusion is that the individual idiotypes of myeloma proteins M104E and J558 are determined by residues 100 and 101 and, accordingly, are either encoded by a third gene segment (D) or represent antigenic determinants generated by a novel somatic mutational mechanism. Second, the IdX idiotype correlates with several residues in the V_H segment. The V_H region from one IdX-negative protein, Hdex10, has been sequenced. Hdex10 differs from the two myeloma V_H regions by four residues—positions 54, 55, 100 and 101. The substitution at position 55 destroys the attachment site for the carbohydrate moiety, distinguishing it from the other

Protein	Specificity	V segment	D segment	J segment	References
		75 80 90 99	100 101	102 110 117	
M104E	1,3 Dextran	SSSTAYMQLNSLTSEDSAVYYCARD	YD	WYFDVWGAGTTVTVSS	32
Hdex8	"	---G--------------------	YD	----------------	This paper
J558	"	-------------------------	RY	----------------	"
Hdex9	"	--N---F------------------	RY	----------------	"
Hdex10	"	------------------------↓	VN	----------------	"
Hdex6	"	-------------------------	SH	----------------	"
Hdex3	"	-------------------------	R-	----------------	"
Hdex7	"	-------------------------	A-	----------------	"
Hdex2	"	-------------------------	NY	----------------	"
Hdex1	"	-------------------------	NY	H----V----------	"
Hdex5	"	-------------------------	SN	Y---Y--Q---L----	"
Hdex4	"	-------------------------	K-	Y---Y--Q---L----	"
T15,S63,S107,Y5236	Phosphorylcholine	-Q-IL-L-M-A-RA---T-I------	YYGSSY	----------------	46
H8	"	-Q-IL-L-M-A-RA---T-I------	---N--	----------------	"
W3207	"	-Q-IL-F-M-A-RA---T-I-----N	--KYDL	--V-------------	"
M603	"	-Q-IL-L-M-A-RA---T-I-----N	----T	----------------	47
M167	"	-Q-VL-L-M-A-RA---T-T---T--	ADYGDSYF	G---------------	48
M511	"	-Q-IL-L-M-A-RA---T-I------	GDYG-SY	----------------	"
A4	2,1 Levan	-K-SV-L-M-N-RA---TGI---TTG	[ ]	[ ]-AY--Q--L---	49
E109	"	-K-SVFL-M-N-RA---TGIH--TTG	[ ]	[ ]-AY--Q--L---	"
U61	"	-K-SV-L-M-N-RA---TGI---TTG	[ ]	[ ]-AY--Q--L-P-	"
A47N	"	-K-SV-L-M-N-RA---T-I---STG	[ ]	[ ]-AY--Q--L---	"
M315	DNP	-ENQFFLK-D-V-[ ]T-T----G-	NDH	L---Y--Q---L---	50
M21	Unknown	PKN-LFL-MT--R---T-M-----H	GNYPW	YAM-Y--Q--S-----	51
M173	Unknown	AKN-L-L-MSKVR---T-L-----S	PY	YAM-Y--Q--S-----	52
T601	Galactan	AKN-L-L-MSKVR---T-L-----G	GYY	G---------------	35
X44	"	AKN-L-L-MSKVR---T-L-----L	H---	G-AAY--Q--L----A	"
X24	"	AKN-L-L-MSKVR---T-L-----G	---	G---Y--Q--L-----	"
J539	"	AKNSL-L-MSKVR---T-L-----L	H--	G-NAY--Q--L----A	"

Fig. 3 A comparison of published heavy chain sequences including the C-terminal portions of V_H regions. The division of this region into V_H segments, D segments, and J_H segments is based on homology with the heavy chains binding dextran. The prototype sequence for the V_H and J_H segments is M104E. The prototypes for the D segments are the first sequence in each group. The J segments which do not extend to amino acid 117 are due to incomplete amino acid sequence data and not to shorter J segments.

11 V_H regions (Fig. 1). It is attractive to postulate that the IdX idiotype depends on amino acids at positions 54 and 55. This supposition is supported by the observation that Hdex8, which has greatly reduced IdX reactivity, has a substitution at position 54. Analysis of V_H regions from several additional IdX-negative antibody molecules are in progress to determine whether they also exhibit substitutions at positions 54 and 55 and to test the intriguing possibility that the IdX idiotype may be in part associated with a carbohydrate moiety.

Thus, idiotypes can apparently arise from V segments (IdX) or D segments (IdI). These data on the hybridoma antibodies binding α-1,3 dextran do allow us to assign structural correlates for the cross-reactive (V segment) and individual (D segment) idiotypes of several dextran antibodies. However, an idiotype localised to a particular region (for example, the IdIJ558 to positions 100 and 101) may require particular sequences in still other portions of the V region for its expression (for example, V and J segments like those found in dextran-binding antibodies). Accordingly, the interpretation of genetic mapping of idiotypes is complex. Indeed, one can never be certain what is being mapped—a V segment, a D segment, a J segment, the ability to produce certain somatic variants, or some combination of these.

In an historical context, note that Kabat and coworkers predicted the existence of minigenes for each hypervariable region based on an analysis of the V region amino acid sequences[13,36]. In addition, Capra and Kindt postulated the existence of multiple V coding elements based on an analysis of V_H sequence data and idiotypes[37,38]. If the D segments are encoded by distinct gene segments, they are 'minigenes' that encode the individual idiotypes of murine antibodies binding dextran.

The evolution of split-gene segments and DNA rearrangements

The V_H regions and V_L regions are homologous to one another at the protein level[31,39]. Both V_L and V_H are encoded by V and J gene segments of similar size. This homology implies that the split-gene strategy and the mechanism for rearranging split genes must have arisen before the divergence of light and heavy chains. Indeed, it is attractive to speculate that split genes and mechanisms for their rearrangement evolved before the emergence of the vertebrate immune system, possibly in families of genes coding membrane recognition functions. It will be interesting to determine whether there are other contemporary

multigene families using strategies of split genes and DNA rearrangment for the amplification of information[5]. This is discussed elsewhere[40].

The patterns of phenotypic diversity in the V_H regions from antibodies binding α-1,3 dextran have allowed us to make several important conclusions. We can define V_H and J_H segments which are homologous to their light chain counterparts. Diversity in a third region, the D segment, can be explained by a novel somatic mutational mechanism, the existence of a third germ-line (D) gene segment, or the less likely possibility of a large number of J_H gene segments. Antibody diversity appears to arise from several sources: multiple germ-line gene segments, combinational joining of V_H and J_H gene segments and the

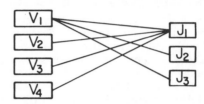

Fig. 4 A diagram indicating the combinational joining of V_H and J_H segments. V_1 to V_4 and J_1 to J_3 are defined in Fig. 2.

mechanism for producing D segment diversity which includes extensive nucleotide substitutions as well as the insertion or deletion of multiple residues. Correlations between the V_H sequences and the idiotypes of the antibodies to α-1,3 dextran are possible but there should be caution in the interpretation of genetic mapping studies which use idiotypes. Studies are in progress in our laboratories to expand our analyses of the phenotypic (protein) diversity in the dextran system and to analyse the gene segments encoding the dextran response.

This work was supported by USPHS Grants AI-10781, AI-15353, AI-11635, AI-05599 and CA-09118, NSF grant PCM76-09719, and National Institute of General Medical Sciences grant 5 T32 Gm 07616. We thank Dr Michael Hunkapiller and Ms Carol Oken for assistance in the sequence analysis of the J588 protein, Alex Shaffer for assistance with making hybridomas, Dr Daniel Hansburg for defining the dextran idiotypes and for helping in the early stages of this study, Philip Early for pointing out the possiblity of a D gene segment and Drs Douglas Fambrough and John Campbell for comments on the manuscript.

Received 27 July; accepted 31 October 1979.

1. Dreyer, W. J., Gray, W. R. & Hood, L. *Cold Spring Harb. Symp. quant. Biol.* **32**, 353–367 (1967).
2. Gally, J. A. & Edelman, G. M. *A. Rev. Genet.* **6**, 1–46 (1972).
3. Cohn, M. in *The Immune System, Genes, Receptors, Signals* (eds Sercarz, E. E., Williamson, A. R. & Fox, C. F.) 89–118 (Academic, New York, 1974).
4. Hood, L. *et al. Cold Spring Harb. Symp. quant. Biol.* **41**, 817–836 (1976).
5. Hood, L., Campbell, J. Y. & Elgin, S. C. R. *A. Rev. Genet.* **9**, 305–353 (1975).
6. Brack, C., Hirama, M., Lenhard-Schuller, R. & Tonegawa, S. *Cell* **15**, 1–14 (1978).
7. Seidman, J. G. *et al. Proc. natn. Acad. Sci. U.S.A.* **75**, 3881–3885 (1978).
8. Early, P. W., Davis, M. M., Kaback, D. B., Davidson, N. & Hood, L. *Proc. natn. Acad. Sci. U.S.A.* **76**, 857–861 (1979).
9. Smith, G. P., Hood, L. & Fitch, W. *A. Rev. Biochem.* **40**, 969–1012 (1971).
10. Seidman, J. G., Leder, A., Nau, M., Norman, B. & Leder, P. *Science* **202**, 11–17 (1978).
11. Tonegawa, S., Maxam, A. M., Tizard, R., Bernard, O. & Gilbert, W. *Proc. natn. Acad. Sci. U.S.A.* **75**, 1485–1489 (1978).
12. Bernard, O., Hozumi, N. & Tonegawa, S. *Cell* **15**, 1133–1144 (1978).
13. Wu, T. & Kabat, E. *J. exp. Med.* **132**, 211–250 (1970).
14. Capra, J. D. & Kehoe, J. M. *Proc. natn. Acad. Sci. U.S.A.* **71**, 845–849 (1974).
15. Amzel, L., Poljak, R., Saul, F., Varga, J. & Richard, F. *Proc. natn. Acad. Sci. U.S.A.* **71**, 1427–1430 (1974).
16. Padlan, E. A., Segal, D. M., Rudikoff, S., Potter, M., Spande, T. & Davies, D. R. *Nature, new Biol.* **245**, 165–167 (1973).
17. Weigert, M., Gatmaitan, L., Loh, E., Schilling, J. & Hood, L. *Nature* **276**, 785–790 (1978).
18. Weigert, M., Cesari, I. M., Yonkovich, S. J. & Cohn, M. *Nature* **228**, 1045–1047 (1970).
19. Appella, E. *Proc. natn. Acad. Sci. U.S.A.* **68**, 590–594 (1971).
20. Weigert, M. & Riblet, R. *Cold Spring Harb. Symp. quant. Biol.* **61**, 837–846 (1977).
21. Max, E. E., Seidman, J. G. & Leder, P. *Proc. natn. Acad. Sci. U.S.A.* (in the press).
22. Sakano, H., Hüppi, K., Heinrich, G. & Tonegawa, S. *Nature* **280**, 288–294 (1979).
23. Weigert, M. *et al. Nature* (in the press).
24. Köhler, G. & Milstein, C. *Nature* **256**, 495–497 (1975); *Eur. J. Immun.* **6**, 511–519 (1976).
25. Littlefield, J. N. *Science* **145**, 709–710 (1964).
26. Barstad, P. *et al. Eur. J. Immun.* **8**, 497–503 (1978).
27. Schilling, J., Hansburg, D., Davie, J. M. & Hood, L. *J. Immun.* **123**, 384–388 (1979).
28. Hansburg, D., Perlmutter, R. M., Briles, D. E. & Davie, J. M. *Eur. J. Immun.* **8**, 352–359 (1978).
29. Hansburg, D., Briles, D. E. & Davie, J. M. *J. Immun.* **117**, 569–575 (1976); **119**, 1406–1412 (1977).
30. Kehry, M., Sibley, C., Fuhrman, J., Schilling, J. & Hood, L. *Proc. natn. Acad. Sci. U.S.A.* **76**, 2932–2936 (1979).
31. Dayhoff, M. O. *Atlas of Protein Sequence and Structure 1972* (National Biomedical Research Foundation, Maryland, 1972).
32. Sokano, H. *et al. Nature* **277**, 627–633 (1979).
33. Rao, D. N., Rudikoff, S., Krutzach, H. & Potter, M. *Proc. natn. Acad. Sci. U.S.A.* **76**, 2890–2894 (1979).
34. Leon, M. A., Young, N. M. & McIntire, K. R. *Biochemistry* **9**, 1023–1030 (1970).
35. Lundblad, A. *et al. Immunochemistry* **9**, 535–544 (1972).
36. Kabat, E. A., Wu, T. T. & Bilofsky, H. *Variable Regions of Immunoglobulin Chains* (Bolt, Beranek & Newman, Cambridge, Mass., 1976); *Proc. natn. Acad. Sci. U.S.A.* **75**, 2429–2433 (1978).
37. Capra, J. D. & Kindt, T. *Immunogenetics* **1**, 417–427 (1975).
38. Kindt, J. G. & Capra, J. D. *Immunogenetics* **6**, 309–321 (1978).
39. Edelman, G. M. *et al. Proc. natn. Acad. Sci.* **63**, 78 (1969).
40. Hood, L., Huang, H. & Dreyer, W. J. *J. supramolec. Struct.* **7**, 531–559 (1977).
41. Galfre, G., Howe, S. C., Milstein, C., Butcher, G. W. & Howard, J. C. *Nature* **266**, 550–552 (1977).
42. Coffino, P., Baumel, R., Laskov, R. & Scharff, M. D. *J. cell. Physiol.* **79**, 429–440 (1972).
43. Hubert, J., Johnson, N., Barstad, P. & Hood, L. (in preparation).
44. Rudikoff, S. & Potter, M. *Biochemistry* **13**, 4033–4038 (1974).
45. Rudikoff, S. & Potter, M. *Proc. natn. Acad. Sci. U.S.A.* **73**, 2109–2112 (1976).
46. Vrana, M., Rudikoff, S. & Potter, M. *Proc. natn. Acad. Sci. U.S.A.* **75**, 1957–1961 (1978).
47. Francis, S. H., Leslie, R. G. Q., Hood, L. & Eisen, H. N. *Proc. natn. Acad. Sci. U.S.A.* **71**, 1123–1127 (1974).
48. Milstein, C., Adetugbo, K., Cowan, N. J. & Secker, D. S. *Prog. Immun.* **2**, 157–168 (1974).
49. Bourgeois, A., Fougereau, M. & Depreval, C. *Eur. J. Biochem.* **24**, 446–455 (1972).

Continuous cultures of fused cells secreting antibody of predefined specificity

THE manufacture of predefined specific antibodies by means of permanent tissue culture cell lines is of general interest. There are at present a considerable number of permanent cultures of myeloma cells[1,2] and screening procedures have been used to reveal antibody activity in some of them. This, however, is not a satisfactory source of monoclonal antibodies of predefined specificity. We describe here the derivation of a number of tissue culture cell lines which secrete anti-sheep red blood cell (SRBC) antibodies. The cell lines are made by fusion of a mouse myeloma and mouse spleen cells from an immunised donor. To understand the expression and interactions of the Ig chains from the parental lines, fusion experiments between two known mouse myeloma lines were carried out.

Each immunoglobulin chain results from the integrated expression of one of several V and C genes coding respectively for its variable and constant sections. Each cell expresses only one of the two possible alleles (allelic exclusion; reviewed in ref. 3). When two antibody-producing cells are fused, the products of both parental lines are expressed[4,5], and although the light and heavy chains of both parental lines are randomly joined, no evidence of scrambling of V and C sections is observed[4]. These results, obtained in an heterologous system involving cells of rat and mouse origin, have now been confirmed by fusing two myeloma cells of the same mouse strain,

The protein secreted (MOPC 21) is an IgG1 (κ) which has been fully sequenced[7,8]. Equal numbers of cells from each parental line were fused using inactivated Sendai virus[9] and samples containing 2×10^5 cells were grown in selective medium in separate dishes. Four out of ten dishes showed growth in selective medium and these were taken as independent hybrid lines, probably derived from single fusion events. The karyotype of the hybrid cells after 5 months in culture was just under the sum of the two parental lines (Table 1). Figure 1 shows the isoelectric focusing[10] (IEF) pattern of the secreted products of different lines. The hybrid cells (samples c–h in Fig. 1) give a much more complex pattern than either parent (a and b) or a mixture of the parental lines (m). The important feature of the new pattern is the presence of extra bands (Fig. 1, arrows). These new bands, however, do not seem to be the result of differences in primary structure; this is indicated by the IEF pattern of the products after reduction to separate the heavy and light chains (Fig. 1B). The IEF pattern of chains of the hybrid clones (Fig. 1B, g) is equivalent to the sum of the IEF pattern (a and b) of chains of the parental clones with no evidence of extra products. We conclude that, as previously shown with interspecies hybrids[4,5], new Ig molecules are produced as a result of mixed association between heavy and light chains from the two parents. This process is intracellular as a mixed cell population does not give rise to such hybrid molecules (compare m and g, Fig. 1A). The individual cells must therefore be able to express both isotypes. This result shows that in hybrid cells the expression of one isotype and idiotype does not exclude the expression of another: both heavy chain

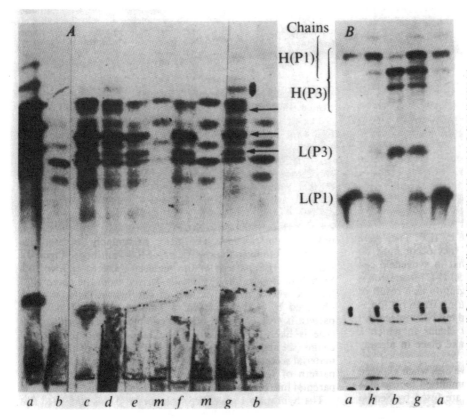

Fig. 1 Autoradiograph of labelled components secreted by the parental and hybrid cell lines analysed by IEF before (A) and after reduction (B). Cells were incubated in the presence of ^{14}C-lysine[14] and the supernatant applied on polyacrylamide slabs. A, pH range 6.0 (bottom) to 8.0 (top) in 4 M urea. B, pH range 5.0 (bottom) to 9.0 (top) in 6 M urea; the supernatant was incubated for 20 min at 37 °C in the presence of 8 M urea, 1.5 M mercaptoethanol and 0.1 M potassium phosphate pH 8.0 before being applied to the right slab. Supernatants from parental cell lines in: a, P1Bul; b, P3-X67Ag8; and m, mixture of equal number of P1Bul and P3-X67Ag8 cells. Supernatants from two independently derived hybrid lines are shown: c–f, four subclones from Hy-3; g and h, two subclones from Hy-B. Fusion was carried out[4,9] using 10^6 cells of each parental line and 4,000 haemagglutination units inactivated Sendai virus (Searle). Cells were divided into ten equal samples and grown separately in selective medium (HAT medium, ref. 6). Medium was changed every 3 d. Successful hybrid lines were obtained in four of the cultures, and all gave similar IEF patterns. Hy-B and Hy-3 were further cloned in soft agar[14]. L, Light; H, heavy.

Chains labels: H(P1), H(P3), L(P3), L(P1)

Lanes A: a b c d e m f m g b

Lanes B: a h b g a

and provide the background for the derivation and understanding of antibody-secreting hybrid lines in which one of the parental cells is an antibody-producing spleen cell.

Two myeloma cell lines of BALB/c origin were used. P1Bul is resistant to 5-bromo-2′-deoxyuridine[4], does not grow in selective medium (HAT, ref. 6) and secretes a myeloma protein, Adj PC5, which is an IgG2A (κ), (ref. 1). Synthesis is not balanced and free light chains are also secreted. The second cell line, P3-X63Ag8, prepared from P3 cells[2], is resistant to 20 μg ml^{-1} 8-azaguanine and does not grow in HAT medium.

isotypes ($\gamma 1$ and $\gamma 2a$) and both V_H and both V_L regions (idiotypes) are expressed. There are no allotypic markers for the C_κ region to provide direct proof for the expression of both parental C_κ regions. But this is indicated by the phenotypic link between the V and C regions.

Figure 1A shows that clones derived from different hybridisation experiments and from subclones of one line are indistinguishable. This has also been observed in other experiments (data not shown). Variants were, however, found in a survey of 100 subclones. The difference is often associated with changes

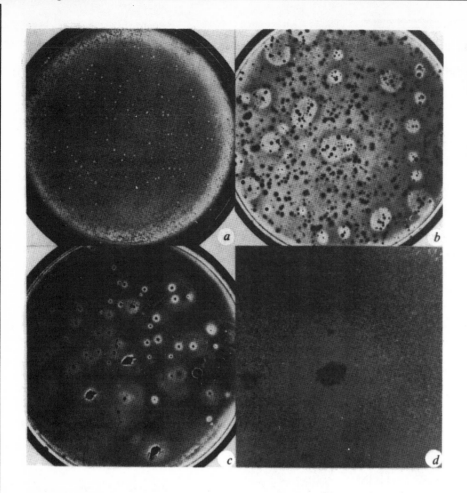

Fig. 2 Isolation of an anti-SRBC antibody-secreting cell clone. Activity was revealed by a halo of haemolysed SRBC. Direct plaques given by: *a*, 6,000 hybrid cells Sp-1; *b*, clones grown in soft agar from an inoculum of 2,000 Sp-1 cells; *c*, recloning of one of the positive clones Sp-1/7; *d*, higher magnification of a positive clone. Myeloma cells (10^7 P3-X67A g8) were fused to 10^8 spleen cells from an immunised BALB/c mouse. Mice were immunised by intraperitoneal injection of 0.2 ml packed SRBC diluted 1:10, boosted after 1 month and the spleens collected 4 d later. After fusion, cells (Sp-1) were grown for 8 d in HAT medium, changed at 1–3 d intervals. Cells were then grown in Dulbecco modified Eagle's medium, supplemented for 2 weeks with hypoxanthine and thymidine. Forty days after fusion the presence of anti-SRBC activity was revealed as shown in *a*. The ratio of plaque forming cells/total number of hybrid cells was 1/30. This hybrid cell population was cloned in soft agar (50% cloning efficiency). A modified plaque assay was used to reveal positive clones shown in *b–d* as follows. When cell clones had reached a suitable size, they were overlaid in sterile conditions with 2 ml 0.6% agarose in phosphate-buffered saline containing 25 μl packed SRBC and 0.2 ml fresh guinea pig serum (absorbed with SRBC) as source of complement. *b*, Taken after overnight incubation at 37 °C. The ratio of positive/total number of clones was 1/33. A suitable positive clone was picked out and grown in suspension. This clone was called Sp-1/7, and was recloned as shown in *c*; over 90% of the clones gave positive lysis. A second experiment in which 10^6 P3-X67Ag8 cells were fused with 10^8 spleen cells was the source of a clone giving rise to indirect plaques (clone Sp-2/3-3). Indirect plaques were produced by the addition of 1:20 sheep anti-MOPC 21 antibody to the agarose overlay.

in the ratios of the different chains and occasionally with the total disappearance of one or other of the chains. Such events are best visualised on IEF analysis of the separated chains (for example, Fig. 1*h*, in which the heavy chain of P3 is no longer observed). The important point that no new chains are detected by IEF complements a previous study[4] of a rat–mouse hybrid line in which scrambling of *V* and *C* regions from the light chains of rat and mouse was not observed. In this study, both light chains have identical C_κ regions and therefore scrambled V_L–C_L molecules would be undetected. On the other hand, the heavy chains are of different subclasses and we expect scrambled V_H–C_H to be detectable by IEF. They were not observed in the clones studied and if they occur must do so at a lower frequency. We conclude that in syngeneic cell hybrids (as well as in interspecies cell hybrids) *V*–*C* integration is not the result of cytoplasmic events. Integration as a result of DNA translocation or rearrangement during transcription is also suggested by the presence of integrated mRNA molecules[11] and by the existence of defective heavy chains in which a deletion of *V* and *C* sections seems to take place in already committed cells[12].

The cell line P3-X63Ag8 described above dies when exposed to HAT medium. Spleen cells from an immunised mouse also die in growth medium. When both cells are fused by Sendai virus and the resulting mixture is grown in HAT medium, surviving clones can be observed to grow and become established after a few weeks. We have used SRBC as immunogen, which enabled us, after culturing the fused lines, to determine the presence of specific antibody-producing cells by a plaque assay technique[13] (Fig. 2*a*). The hybrid cells were cloned in soft agar[14] and clones producing antibody were easily detected by an overlay of SRBC and complement (Fig. 2*b*). Individual clones were isolated and shown to retain their phenotype as almost all the clones of the derived purified line are capable of lysing SRBC (Fig. 2*c*). The clones were visible to the naked eye (for example, Fig. 2*d*). Both direct and indirect plaque

assays[13] have been used to detect specific clones and representative clones of both types have been characterised and studied.

The derived lines (Sp hybrids) are hybrid cell lines for the following reasons. They grow in selective medium. Their karyotype after 4 months in culture (Table 1) is a little smaller than the sum of the two parental lines but more than twice the chromosome number of normal BALB/c cells, indicating that the lines are not the result of fusion between spleen cells. In addition the lines contain a metacentric chromosome also present in the parental P3-X67Ag8. Finally, the secreted immunoglobulins contain MOPC 21 protein in addition to new, unknown components. The latter presumably represent the chains derived from the specific anti-SRBC antibody. Figure 3*A* shows the IEF pattern of the material secreted by two such Sp hybrid clones. The IEF bands derived from the parental P3 line are visible in the pattern of the hybrid cells, although obscured by the presence of a number of new bands. The pattern is very complex, but the complexity of hybrids of this type is likely to result from the random recombination of chains (see above, Fig. 1). Indeed, IEF patterns of the reduced material secreted by the spleen–P3 hybrid clones gave a simpler pattern of Ig chains. The heavy and light chains of the P3 parental line became prominent, and new bands were apparent.

The hybrid Sp-1 gave direct plaques and this suggested that it produces an IgM antibody. This is confirmed in Fig. 4 which shows the inhibition of SRBC lysis by a specific anti-IgM

Table 1 Number of chromosomes in parental and hybrid cell lines

Cell line	Number of chromosomes per cell	Mean
P3-X67Ag8	66,65,65,65,65	65
P1Bu1	Ref. 4	55
Mouse spleen cells	—	40
Hy-B (P1-P3)	112,110,104,104,102	106
Sp-1/7-2	93,90,89,89,87	90
Sp-2/3-3	97,98,96,96,94,88	95

antibody. IEF techniques usually do not reveal 19S IgM molecules. IgM is therefore unlikely to be present in the unreduced sample *a* (Fig. 3*B*) but μ chains should contribute to the pattern obtained after reduction (sample *a*, Fig. 3*A*).

The above results show that cell fusion techniques are a powerful tool to produce specific antibody directed against a predetermined antigen. It further shows that it is possible to isolate hybrid lines producing different antibodies directed against the same antigen and carrying different effector functions (direct and indirect plaque).

The uncloned population of P3–spleen hybrid cells seems quite heterogeneous. Using suitable detection procedures it should be possible to isolate tissue culture cell lines making different classes of antibody. To facilitate our studies we have used a myeloma parental line which itself produced an Ig. Variants in which one of the parental chains is no longer expressed seem fairly common in the case of P1–P3 hybrids (Fig. 1*h*). Therefore selection of lines in which only the specific antibody chains are expressed seems reasonably simple. Alternatively, non-producing variants of myeloma lines could be used for fusion.

We used SRBC as antigen. Three different fusion experiments were successful in producing a large number of antibody-producing cells. Three weeks after the initial fusion, 33/1,086

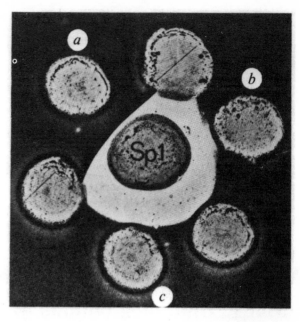

Fig. 4 Inhibition of haemolysis by antibody secreted by hybrid clone Sp-1/7-2. The reaction was in a 9-cm Petri dish with a layer of 5 ml 0.6% agarose in phosphate-buffered saline containing 1/80 (v/v) SRBC. Centre well contains 2.5 μl 20 times concentrated culture medium of clone Sp-1/7-2 and 2.5 μl mouse serum. *a*, Sheep specific anti-mouse macroglobulin (MOPC 104E, Dr Feinstein); *b*, sheep anti-MOPC 21 (P3) IgG1 absorbed with Adj PC-5; *c*, sheep anti-Adj PC-5 (IgG2a) absorbed with MOPC 21. After overnight incubation at room temperature the plate was developed with guinea pig serum diluted 1:10 in Dulbecco's medium without serum.

clones (3%) were positive by the direct plaque assay. The cloning efficiency in the experiment was 50%. In another experiment, however, the proportion of positive clones was considerably lower (about 0.2%). In a third experiment the hybrid population was studied by limiting dilution analysis. From 157 independent hybrids, as many as 15 had anti-SRBC activity. The proportion of positive over negative clones is remarkably high. It is possible that spleen cells which have been triggered during immunisation are particularly successful in giving rise to viable hybrids. It remains to be seen whether similar results can be obtained using other antigens.

The cells used in this study are all of BALB/c origin and the hybrid clones can be injected into BALB/c mice to produce solid tumours and serum having anti-SRBC activity. It is possible to hybridise antibody-producing cells from different origins[4,5]. Such cells can be grown *in vitro* in massive cultures to provide specific antibody. Such cultures could be valuable for medical and industrial use.

G. KÖHLER
C. MILSTEIN

*MRC Laboratory of Molecular Biology,
Hills Road, Cambridge CB2 2QH, UK*

Received May 14; accepted June 26, 1975.

[1] Potter, M., *Physiol. Rev.*, **52**, 631-719 (1972).
[2] Horibata, K., and Harris, A. W., *Expl Cell Res.*, **60**, 61-70 (1970).
[3] Milstein, C., and Munro, A. J., in *Defence and Recognition* (edit. by Porter, R. R.), 199-228 (MTP Int. Rev. Sci., Butterworth, London, 1973).
[4] Cotton, R. G. H., and Milstein, C., *Nature*, **244**, 42-43 (1973).
[5] Schwaber, J., and Cohen, E. P., *Proc. natn. Acad. Sci. U.S.A.*, **71**, 2203-2207 (1974).
[6] Littlefield, J. W., *Science*, **145**, 709 (1964).
[7] Svasti, J., and Milstein, C., *Biochem. J.*, **128**, 427-444 (1972).
[8] Milstein, C., Adetugbo, K., Cowan, N. J., and Secher, D. S., *Progress in Immunology*, II, 1 (edit. by Brent, L., and Holborow, J.), 157-168 (North-Holland, Amsterdam, 1974).
[9] Harris, H., and Watkins, J. F., *Nature*, **205**, 640-646 (1965).
[10] Awdeh, A. L., Williamson, A. R., and Askonas, B. A., *Nature*, **219**, 66-67 (1968).
[11] Milstein, C., Brownlee, G. G., Cartwright, E. M., Jarvis, J. M., and Proudfoot, N. J., *Nature*, **252**, 354-359 (1974).
[12] Frangione, B., and Milstein, C., *Nature*, **244**, 597-599 (1969).
[13] Jerne, N. K., and Nordin, A. A., *Science*, **140**, 405 (1963).
[14] Cotton, R. G. H., Secher, D. S., and Milstein, C., *Eur. J. Immun.*, **3**, 135-140 (1973).

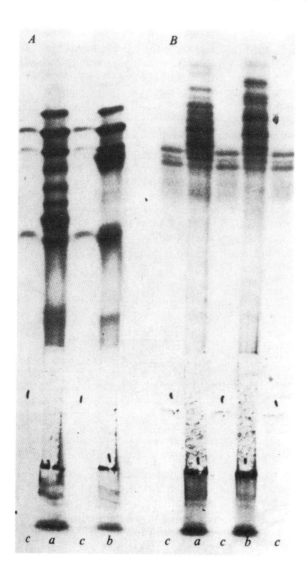

Fig. 3 Autoradiograph of labelled components secreted by anti-SRBC specific hybrid lines. Fractionation before (*B*) and after (*A*) reduction was by IEF. *p*H gradient was 5.0 (bottom) to 9.0 (top) in the presence of 6 M urea. Other conditions as in Fig. 1. Supernatants from: *a*, hybrid clone Sp-1/7-2; *b*, hybrid clone Sp-2/3-3; *c*, myeloma line P3-X67Ag8.

Mapping of Heavy Chain Genes for Mouse Immunoglobulins M and D

Chih-Ping Liu, Philip W. Tucker, J. Frederic Mushinski,

Frederick R. Blattner

In the normal mouse, B lymphocytes are constantly being produced from stem cells located in the bone marrow (*1*). Immature bone marrow cells bearing surface antibody of the immunoglobulin M (IgM) class migrate to the spleen and lymph nodes and divide to give more mature cells bearing both IgM and IgD (*2, 3*). On a given cell these share a common idiotype (*4*), which indicates that the two antibody molecules have the same or at least a very similar sequence in the variable (V) region. Double producer cells constitute a majority of splenic B lymphocytes (*2*), which after specific anti-

genic stimulation differentiate further into plasma cells capable of secreting antibody molecules. At some point in this differentiation, some B cells bearing both IgM and IgD can undergo a "class switch" leading to expression of one of the other antibody classes (IgG, IgA, or IgE). In mouse serum, IgM is plentiful but IgD is present at an extremely low level (*5*).

Both the switch to secretion and the class switch are achieved by nucleic acid rearrangements that lead to changes in the protein structure of the antibodies. There are several reports (*6*) indicating that the secreted IgM and the membrane IgM differ at their carboxyl termini. Two reports (*7*) described a molecular mechanism that could explain the structural changes accompanying secretion for IgM. They found two forms of messenger RNA (mRNA) that code for differing carboxyl terminal sequences on the μ heavy chain. One form coded for the

known sequence of the secreted IgM protein. The other coded for the same sequence, except that 20 amino acids at the carboxyl terminal were replaced by a new stretch of 41 amino acids having an extremely hydrophobic core of 26 residues; they proposed that this region could serve as a transmembrane anchor for the surface form of IgM. The gene segments coding for these alternative carboxyl terminals were found in the DNA. The coding sequence for the terminus of the secreted form was contiguous to the last constant (C) region domain ($C\mu4$) of the heavy chain, whereas the terminus for the membrane form was 1.85 kilobase pairs (kbp) on the 3' side of $C\mu4$. The authors proposed that expression of the two forms of the μ chain is controlled through alternate splicing of mRNA precursor of differing lengths, governed by the choice of RNA termination-polyadenylation sites.

The class switch is a more drastic change, involving complete replacement of one heavy chain C region (C_H) with another of different class. The mouse has at least eight C_H genes (μ, δ, γ_3, γ_1, γ_{2b}, γ_{2a}, α, and ϵ) that code, respectively, for the C_H regions of the eight classes (IgM, IgD, IgG3, IgG1, IgG2b, IgG2a, IgA, and IgE). Thus it seems difficult to invoke RNA splicing for such a switch since an RNA precursor of many hundred kilobases would be needed. Honjo and Kataoka (*8*) have proposed a DNA deletion model for the class switch in which the C_H genes in genomic DNA are arranged in tandem on the 3' side of the heavy chain variable region (V_H) genes. Expression of Ig class in this model is controlled by a somatic deletion of DNA so as to eliminate unexpressed intervening C_H genes. In this way, any V_H gene could ultimately be associated with any of the C_H genes in different cells without the need for splicing out huge intervening sequences. Honjo proposed that the $C\mu$ gene is first in the cluster and is followed by the four $C\gamma$ genes and the $C\alpha$ gene.

Results of recent studies (*9, 10*), in which the Southern hybridization meth-

Summary. A single DNA fragment containing both μ and δ immunoglobulin heavy chain genes has been cloned from normal BALB/c mouse liver DNA with a new λ phage vector Charon 28. The physical distance between the membrane terminal exon of μ and the first domain of δ is 2466 base pairs, with δ on the 3' side of μ. A single transcript could contain a variable region and both μ and δ constant regions. The dual expression of immunoglobulins M and D on spleen B cells may be due to alternate splicing of this transcript.

C.-P. Liu is a postdoctoral fellow in the Laboratory of Genetics, University of Wisconsin–Madison, Madison 53706. P. W. Tucker was an assistant professor, Department of Biochemistry, University of Mississippi Medical Center, Jackson 39216; he is now associate professor in the Department of Microbiology, University of Texas Southwestern Medical School, Dallas 75235. J. F. Mushinski is senior investigator in the Laboratory of Cell Biology, National Cancer Institute, Bethesda, Maryland 20205. F. R. Blattner is an associate professor of genetics in the Laboratory of Genetics, University of Wisconsin–Madison, Madison 53706.

od was used to probe the DNA of plasmacytomas, which are tumors of antibody-secreting cells, strongly support the deletion model and the gene order proposed by Honjo. Cloning of DNA from plasmacytomas has also revealed rearrangements that would be consistent with a deletion model for the class switch (*10*). Of course DNA cloning studies cannot easily discriminate between deletional mechanisms and other types of rearrangement.

The spacing and order of genes for C_H regions are just beginning to be determined by "chromosome walking experiments" in which a series of overlapping molecular clones that span the interval between C_H genes are isolated. Honjo (*11*) has evidence that $C\gamma_{2b}$ and $C\gamma_{2a}$ are about 20 kbp apart and Maki *et al.* (*12*) find a similar distance for $C\gamma_1$ to $C\gamma_{2b}$. We find that the J_H cluster is 8 kbp from $C\mu$ and that $C\mu$ is probably the first gene in the C_H cluster (*13*). Thus the indications from all available experiments strongly favor the gene order proposed by Honjo and support his hypothesis that genomic deletion mediates the class switch in antibody-secreting cells. However, it would be difficult to account for simultaneous double production of two classes sharing a single V gene if one of the C_H genes had to be deleted before the other could be activated. A different explanation must be sought for this aspect of antibody class expression.

A first step toward understanding the molecular mechanism of δ and μ chain expression would be to determine the structure of the $C\delta$ gene and its relationship to the $C\mu$ gene in germ-line DNA. In this article we present the results of mo-

lecular cloning of both $C\mu$ and $C\delta$ genes from BALB/c mouse DNA. We find that $C\delta$ is located only about 2 kbp on the 3′ side of the membrane terminal exon of the $C\mu$ gene. Thus $C\mu$ and $C\delta$ are much closer to one another in the chromosome than are the other C_H genes mapped to date. The distance, in fact, is shorter than the intron of approximately 8 kbp that separates the exonic cluster coding for the carboxyl end of variable portion of heavy chains (the J_H region) from the $C\mu$ gene (*13*). These results suggest a model with a single manageable transcript containing one V_H region and both $C\mu$ and $C\delta$ regions, which could be spliced in several ways to yield mRNA for μ or δ chains in membrane, secretory, and possibly other forms.

Materials and Methods

In the past, analysis of the $C\delta$ gene has been hindered by the lack of a suitable nucleic acid probe for hybridization. Recently, Finkelman *et al.* (*14*) identified two previously untyped plasmacytomas, TEPC 1017 and TEPC 1033, as producers of protein that reacts with a monoclonal antibody against IgD (*15*). They showed by several methods that this protein shares antigenic determinants with the IgD present on the membranes of splenic B lymphocytes. To obtain a $C\delta$ probe we prepared a complementary DNA (cDNA) clone of reverse-transcribed mRNA from TEPC 1017 (*16*) in plasmid vector pBR322. This probe, pδ54J, was initially used to screen shotgun collections (*13*) prepared from partially digested Eco RI fragments of

BALB/c liver DNA cloned into the λ bacteriophage vector Charon 4A (*17*). Due to the unfavorable distribution of Eco RI sites in the genomic DNA, we never succeeded in cloning a fragment containing both $C\mu$ and $C\delta$ in this vector.

A New Cloning Vector

For this project, a new cloning vector, Charon 28 (*18*), was developed (Fig. 1). The chief feature of this new vector is the ability to use Bam HI cloning sites in the vector to clone partial Mbo I digests of target mouse DNA. The six-base recognition site of Bam HI includes within it the four-base recognition site of Mbo I, and both enzymes cleave DNA to leave the same cohesive terminal (Fig. 1). Use of Mbo I to cut the target DNA permits generation of a wide spectrum of overlapping fragments, making it unlikely that any region of DNA would be impossible to clone.

Isolation of Genomic Cδ Clones

To prepare a Charon 28 shotgun collection, fragments obtained by partial Mbo I digestion of BALB/c DNA were fractionated by velocity sedimentation through NaCl gradients and cloned in Charon 28 (*19*). Screening by the megaplate method (*20*) with the pδ54J probe yielded several positive clones, and two, Ch28 257.3 and Ch28 257.4, were extensively characterized. We also characterized the $C\mu$ clone Ch4A142.7 obtained from a partial Eco RI fragment shotgun collection in Charon 4A (*13*).

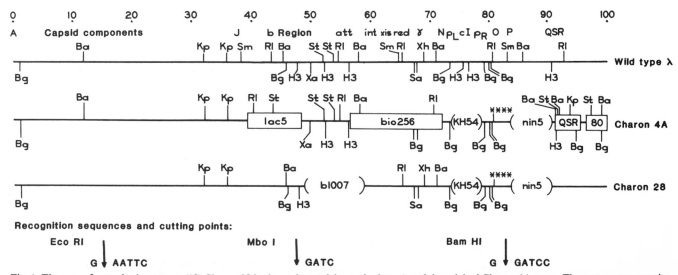

Fig. 1. The map of new cloning vector (*18*) Charon 28 is shown beneath bacteriophage λ and the original Charon 4A maps. The new vector permits cloning of Mbo I-cut target DNA into Bam HI sites in the vector. Charon 4A is suitable only for cloning Eco RI target fragments. The recognition specificities of the enzymes are shown at the bottom. Note that there is only one Eco RI site in Charon 28 DNA. A second Eco RI site expected between the Kpn I sites and the Bam site on the left arm of the vector was unexpectedly lost during construction of the vector. Abbreviations: *Ba*, Bam HI; *Bg*, Bgl II; *H3*, Hind III; *Kp*, Kpn I; *RI*, Eco RI; *Sa*, Sal I; *Sm*, Sma I; *St*, Sst I; *Xa*, Xba I; and *Xh*, Xho I.

Mapping the Cμ-Cδ Genomic Region

The structures of these clones are detailed in the maps in Fig. 2. The top line shows the entire 30-kbp segment of mouse DNA included in the three clones that contain the genes for Cμ and Cδ and the J_H cluster. The DNA inserts in the Charon phage clones are both in the u orientation (17), that is, the left vector arm is adjacent to the 3' end of the cloned immunoglobulin gene region. Since the Mbo I shotgun collection was not replicated in bulk before screening for single plaques, both of the Charon 28 clones (and the Charon 4A clone) result from independent cloning events. In the regions where they overlap, all three clones are identical when compared either by restriction mapping or by heteroduplex analysis in the electron microscope (data not shown); Ch28 257.3 includes a longer stretch of genomic DNA than Ch28 257.4 does. The left cloning points of these clones are very close, and we could not determine which clone overlaps which in that area, although they differ in that a Bam HI site is present in Ch28 257.3 and absent in Ch28 257.4 (see below). Analysis of all clones was helpful for defining the detailed restriction map shown in the bottom portion of Fig. 2. This map was obtained from measurements of 53 restriction fragments on agarose and acrylamide gels obtained by single and multiple digestion of Ch28 257.3 and Ch28 257.4. Some of these fragments were subcloned into pBR322 for further mapping, as indicated in Fig. 2. This was useful in some cases to resolve the order of closely spaced restriction sites. A computer was used to calculate the best-fit map by a least-squares method (21). The computed numerical locations of restriction sites are listed in (22).

The two Bam HI sites designated by an asterisk in Fig. 2 define the ends of the cloned segment in Ch28 257.3. The one at the left is definitely not in genomic DNA or in clones Ch28 257.4 or Ch4A142.7 and is probably the result of the fusion of Mbo I site into the Bam HI site of the vector. We do not know whether the Bam HI site at the right end of the clone is also a result of cloning or whether it is indeed in the genome.

One way that we established the order of Cμ and Cδ in genomic DNA was by analyzing the clones Ch4A142.7, and Ch28 257.4 by plaque hybridization with radioactive probes for Cμ (13, 23) and Cδ, with the vector as a negative control. We found positive hybridization signals for the Cμ probe with both genomic clones, but only Ch28 257.4 reacted with

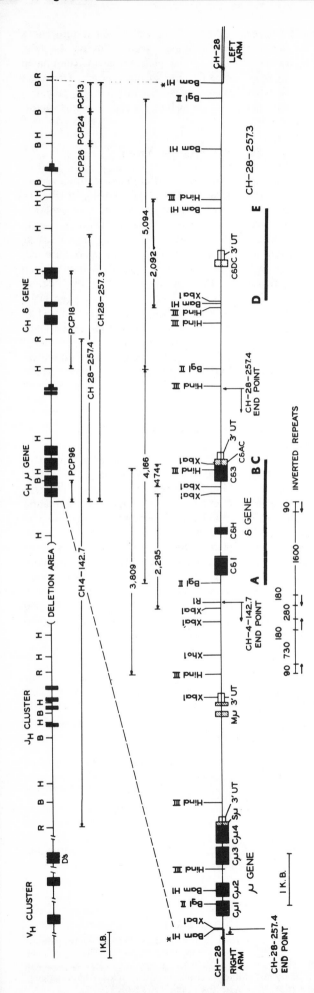

Fig. 2. Physical map of a 30-kbp portion of the Ig H chain locus from BALB/c mouse. The upper map shows the J_H cluster, the Cμ gene including the membrane-binding exon, and known portions of the Cδ gene drawn to scale. Restriction endonuclease sites Hind III, Bam HI, and Eco RI are designated H, B, and R in the upper portion. The V_H cluster and the cluster of D segments are believed to be located to the left and separated from J_H by an unknown distance. Arrows beneath the upper map show the recombinant DNA clones, derived from liver DNA of BALB/c mouse, that were used in this study. Subclones in pBR322 are designated with PCP. "Deletion area" designates a 2900-bp DNA region including repetitive DNA that is missing in clone Ch4A142.7 due to recombination during propagation of the clone in E. coli. An enlarged and more detailed map of the DNA of Ch28 257.3 is shown in the lower portion of the figure. DNA fragments showing positive hybridization to Cδ gene sequences are designated with arrows above the map showing fragment size in base pairs. The bars beneath the map show the two hybridizing regions that have homology with the Cδ probes used in Fig. 3; A, B, C, D, and E on the bars refer to the restriction sites. The specific positions of 8 exons within these fragments are taken from the accompanying article (26). The positions of inverted repeats are indicated beneath the map. End points of Ch4A142.7 and Ch28 257.4 on the Ch28 257.3 map are indicated with arrows. Bam HI sites designated with * may be due to cloning junctions in the vector and may be absent in genomic DNA.

Cδ. In light of the map (Fig. 2), this result shows that Cδ is on the 3′ side of Cμ in genomic DNA [see (24)].

Which Regions of Cδ Code for TEPC 1017 Messenger RNA?

The Southern hybridization technique (25) was used to identify more finely which fragments of Ch28 257.3 and Ch28 257.4 hybridized to Cμ and Cδ probes. In one series of experiments of this type, we used a probe made by direct ^{32}P-labeling of TEPC 1017 mRNA with polynucleotide kinase (Fig. 3). This RNA was purified by adsorption to oligo-(dT)-cellulose columns and sucrose gradient centrifugation; it corresponded in size with the major δ mRNA of the tumor. Other experiments done with the cloned cDNA in pδ54J gave identical results. However, the mRNA probe is most informative, since it is full length, unlike the cloned cDNA probe, and we would thus expect a signal for all exons in genomic DNA that are expressed strongly in the tumor. Both probes show a strong hybridization to a number of bands in the gels (Fig. 3). Interpretation of this pattern with respect to the restriction map can be summarized as follows. Two regions of hybridization were found about 3 kbp apart; they are underlined in Fig. 2. One of these regions is bounded on the left by the Bgl II site denoted A in Fig. 2 and on the right by Xba I site C. The other region is between Xba I site D and Bam HI site E. No hybridizing region was found in the interval of DNA between sites C and D in Fig. 2.

Detailed restriction mapping and DNA sequencing of the hybridized subfragments of Ch28 257.3 (26) showed that the two major C region domains of the TEPC 1017 δ mRNA plus the hingelike segment between them are coded in the leftmost of the hybridizing regions of Fig. 2, whereas the hybridizing region on the right includes an exon for the 26 amino acids at the carboxyl terminus of the TEPC 1017 δ chain. The position of the Cμ gene and its restriction map agreed with published data (27).

Distance from Cμ to Cδ

The placement of the Cδ1 and Cδ3 domains (Fig. 2) is based on sequence data derived from Ch28 257.4 and pδ54J DNA (26). The Hind III site B lies within the Cδ3 domain. DNA sequence analysis from this site in both cDNA and genomic DNA shows that the orientation of the Cδ gene is the same as that of the Cμ.

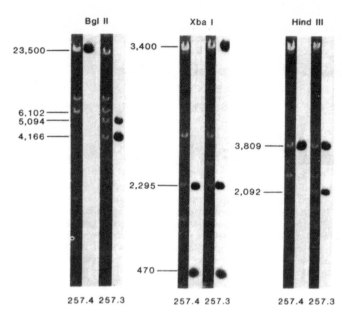

Fig. 3. DNA from Ch28 257.3 and Ch28 257.4 were digested with restriction enzymes Bgl II, Xba I, and Hind III, then subjected to Southern hybridization (25) and hybridized to ^{32}P-labeled δ mRNA from TEPC 1017. For each enzyme digestion, the photograph of agarose gel is on the left; the photograph of the x-ray autoradiograph, reduced to the same scale as the gel photograph, is on the right.

Placement of Cδ1 145 base pairs (bp) to the right of Bgl II site A is confirmed both by hybridization and sequence results. Thus we can estimate the distance from the carboxyl terminus of the membrane exon of Cμ to Cδ1 in the two genomic clones as 2466 bp (28). To extrapolate this result to the genome, additional confirmation is needed, since it is possible for a segment of DNA to be deleted or otherwise rearranged during propagation in *Escherichia coli*. [The region denoted "deletion area" between J_H and Cμ (Fig. 2) is an example of a highly reiterated segment that has shown this tendency (13).] We therefore did a series of Southern hybridization experiments with the pδ54J probe to determine the sizes of the Bgl II and Hind III DNA fragments in uncloned mouse liver DNA. We found both digests yielded fragments whose sizes agreed perfectly (± 10 percent) with the sizes found in the clones (data not shown). It would be quite un-

likely for three independent clones to have the same structure if serious cloning artifacts were present, but small deletions of regions of a few hundred base pairs could have gone undetected by our methods.

That no hybridization of polyadenylated TEPC 1033 mRNA, TEPC 1017 δ mRNA, or pδ54J cDNA has been observed to the left of the Bgl II site A is significant. Since only the 145 nucleotides separate that site from the first Cδ domain, we conclude that no part of the Cδ gene that is transcribed and transported to the cytoplasm of the two plasmacytomas can be coded by DNA located to the left of that Bgl II site. Thus, the exon is probably the first coding segment of Cδ, namely, Cδ1.

Inverted Repeats Near and Within Cδ

The structure of the DNA in Ch28 257.4 and Ch28 257.3 has been examined by electron microscopy. A characteristic double-foldback structure, frequently observed in single-stranded DNA, implies the presence of two pairs of DNA segments with inverted complementary sequences. An electron micrograph showing this structure and indicating the sizes of the DNA segments is shown in Fig. 4. When heteroduplexes between Ch28 257.4 and Ch4A142.7 were examined, the small loop could occasionally be observed at the end point of Ch4A142.7 DNA just to the left of Cδ1. Thus the central pair of inverted repeats is located near the immediate 5′ side of the Eco RI site defining the right end of the Ch4A142.7 clone (see Fig. 2). The orientation of the outer pair of inverted repeats was determined by the observa-

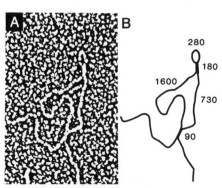

Fig. 4. Inverted repeats around the δ gene. (A) representative electron micrograph of the single strands of Ch28 257.4 DNA. (B) Interpretative drawing with the average measurements of each DNA segment. The orientation of the asymmetric double palindromic structure is indicated on Fig. 2.

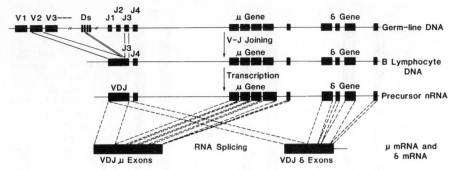

Fig. 5. Proposed model for dual expression of IgM and IgD in B lymphocytes. First, the joining of a V gene, a D segment, and a J region takes place in the DNA of B lymphocyte precursor cells to produce the V gene. The intervening DNA segments are deleted during the process. RNA precursor molecules are made and further processed by splicing out the intervening RNA sequence between J and μ and a sequence downstream of μ (to give rise to μ mRNA) or by splicing out intervening RNA sequences between J and δ including μ (to give rise to δ mRNA). Because of these two classes of mRNA carry identical V, D, and J, they are expected to have the same idiotype and antigenic specificity.

tion that the complete double-foldback structure could be seen in single-stranded DNA of plasmid DNA PCP13 (see Fig. 2), possible only if the entire structure was situated as shown. It is interesting that this assignment places one of the repeated DNA segments in the intron between the hinge and Cδ3, suggesting that the gene lies on the longer (1600-nucleotide) single-stranded DNA segment in Fig. 4. It would be interesting to speculate that these repeats may play a role in some aspect of Cδ expression.

Conclusions

We have shown that, in mouse liver DNA, genes for the μ and δ constant regions are in a cluster, with Cδ at the 3' side of Cμ. The distance between Cμ and Cδ is extremely short (2466 bp) and therefore far less than the distance from J_H to Cμ. The Cδ gene is closely linked to a pair of inverted repeats, one of which may be in an intron between Cδ domains.

The surface IgD and IgM molecules on a given lymphocyte show identical antigenic specificity and idiotype (3). The DNA deletion mechanism that successfully explains the class switch appears to be unsuited to explain this dual expression. Three other possible mechanisms for producing two classes of immunoglobulins with apparently identical V regions can be considered: (i) A long-lived μ mRNA molecule could persist after transition of the cell from μ to δ expression. (ii) There could be a duplication of the V region DNA leading to assembly of two complete genes. (iii) The dual expression could be determined by processing the mRNA precursor without DNA rearrangement. The first hypothesis appears to be eliminated by

the observation of cultured lymphoma cell lines (29) that produce both IgD and IgM of the same idiotype, even after long periods in culture. The hypothesis of V gene duplication and translocation goes back to the early history of immunology (30), but has not yet been supported experimentally. Our finding that the distance between μ and δ is only about 2.5 kbp, which is smaller than the J to Cμ segment known to be spliced, keeps the splicing argument as strong as any other. A schematic depiction of this proposal is shown in Fig. 5.

References and Notes

1. N. L. Warner, *Adv. Immunol.* **19**, 67 (1974).
2. U. Melchey, E. S. Vitetta, M. McWilliams, J. Philips-Quagliata, M. E. Lamm, J. W. Uhr, *J. Exp. Med.* **140**, 1427 (1974).
3. E. Abney and R. M. E. Parkhouse, *Nature (London)* **252**, 600 (1974).
4. R. L. Coffman and M. Cohn, *J. Immunol.* **118**, 1806 (1977).
5. E. Abney, I. R. Hunger, R. M. E. Parkhouse, *Nature (London)* **259**, 404 (1976); F. D. Finkelman, V. L. Woods, A. Berning, I. Scher, *J. Immunol.* **123**, 1253 (1979); A. Bargellesi, G. Corte, E. Cosulch, M. Ferrarini, *Eur. J. Immunol.* **9**, 490 (1979); J. Radl, J. Van Den Berg, C. M. Jol-Van Der Zijde, *J. Immunol.* **124**, 2513 (1980).
6. P. B. Williams, R. T. Kubo, H. M. Grey, *J. Immunol.* **121**, 2435 (1978); F. W. Alt, A. L. M. Bothwell, M. Knapp, E. Siden, E. Mather, M. Koshland, D. Baltimore, *Cell* **20**, 293 (1980); P. Singer, H. Singer, A. W. Williamson, *Nature (London)* **285**, 294 (1980).
7. J. Rogers, P. Early, C. Carter, K. Calame, M. Bond, L. Hood, R. Wall, *Cell* **20**, 303 (1980); P. Early, J. Rogers, M. Davis, K. Calame, M. Bond, R. Wall, L. Hood, *ibid.*, p. 313.
8. T. Honjo and T. Kataoka, *Proc. Natl. Acad. Sci. U.S.A.* **75**, 2140 (1978).
9. T. H. Rabbits, A. Forster, W. Dunnick, D. L. Bently, *Nature (London)* **283**, 351 (1980).
10. S. Cory and J. M. Adams, *Cell* **19**, 37 (1980); Y. Yaoita and T. Honjo, *Biomed. Res.*, in press; T. Kataoka, T. Kawakami, N. Takahashi, T. Honjo, *Proc. Natl. Acad. Sci. U.S.A.* **77**, 919 (1980); C. Coleclough, C. Cooper, R. P. Perry, *ibid.*, p. 1422.
11. T. Honjo, personal communication.
12. S. Tonegawa, personal communication.
13. N. Newell, J. E. Richards, P. W. Tucker, F. R. Blattner, *Science* **209**, 1128 (1980).
14. F. D. Finkelman, S. W. Kessler, J. F. Mushinski, M. Potter, *Fed. Proc. Fed. Am. Soc. Exp. Biol.* **39**, 481 (1980).
15. V. T. Oi, P. P. Jones, J. W. Goding, L. A. Herzenberg, L. A. Herzenberg, *Curr. Top. Microbiol. Immunol.* **81**, 115 (1978).
16. J. F. Mushinski, F. R. Blattner, J. D. Owens, F. D. Finkelman, S. W. Kessler, L. Fitzmaurice,

M. Potter, P. W. Tucker, *Proc. Natl. Acad. Sci. U.S.A.*, in press.
17. B. G. Williams and F. R. Blattner, *J. Virol.* **29**, 555 (1979); J. R. deWet, D. L. Daniels, J. L. Schroeder, B. G. Williams, K. Denniston-Thompson, D. D. Moore, F. R. Blattner, *ibid.* **33**, 401 (1980).
18. D. Rimm, D. Horness, J. Kucera, F. R. Blattner, in preparation. Charon 28 was constructed by D. Rimm, D. Horness, and J. Kucera, undergraduate students at the University of Wisconsin.
19. Mouse liver DNA was prepared by the procedures described in (20). Approximately 375 μg of liver DNA was digested with 90 units of Mbo I (New England Biolabs) at 37°C in 75 mM NaCl, 10 mM tris-HCl (pH 7.4), 10 mM MgCl₂, and 1 mM dithiothreitol. At 10, 20, and 30 minutes of digestion, one-third of the volume of the reaction sample was taken, and digestion was stopped by adding EDTA to a final concentration of 20 mM; the enzyme was inactivated by adding diethyl pyrocarbonate to a final concentration of 0.1 percent. The three samples were combined and layered onto a 12-ml 5 to 20 percent NaCl gradient and centrifuged in a Beckman SW41 rotor at 35,000 rev/min at 25°C for 4 hours. Twenty fractions were collected and 10 μl of each fraction was taken and subjected to electrophoresis in a 0.6 percent agarose gel. Fractions that contained fragments of 8 to 20 kb were pooled, and alcohol was added to a 70 percent concentration to precipitate DNA molecules. The DNA pellet was dissolved in 10 mM tris-HCl, pH 7.9, and 1 mM EDTA. These Mbo I DNA fragments were subsequently ligated with Charon 28 arms to generate recombinant DNA molecules. Charon 28 arms were prepared as follows: Charon 28 DNA at a concentration of 1 mg/ml was ligated at the cohesive ends with T4 DNA ligase, then digested with Bam HI enzyme. The digested DNA was loaded onto a 12-ml, 5 to 20 percent NaCl gradient and centrifuged at 35,000 rev/min at 25°C for 4 hours. Twenty fractions were collected; fractions containing the 32-kb arms were pooled, and DNA was precipitated with alcohol. The Charon 28 arm pellet was dissolved in 10 mM tris-HCl, pH 7.9, and 1 mM EDTA at a concentration of 1 mg/ml. The liver DNA fragments and the Charon 28 arms were mixed and ligated at a vector-to-target ratio of 3:1 (by weight). The overall DNA concentration was 650 μg/ml. Packaging of the ligated DNA in vitro was performed to generate viable recombinant phages by the method described in (20).
20. F. R. Blattner, A. E. Blechl, K. Denniston-Thompson, H. E. Faber, J. E. Richards, J. L. Slightom, P. W. Tucker, O. Smithies, *Science* **202**, 1279 (1978).
21. J. L. Schroeder and F. R. Blattner, *Gene* **4**, 167 (1978).
22. The restriction enzyme sites and their map coordinates (nucleotides) are: Bam HI*, 0; Xba I, 6; Bgl II, 516; Bam HI, 828; Hind III, 1,170; Hind III, 2,149; Xba I, 4,457; Hind III, 5,085; Xho I, 5,263; Xba I, 5,812; Xba I, 6,135; Eco RI, 6,173; Bgl II, 6,618; Xba I, 8,430; Xba I, 8,552; Hind III, 8,894; Xba I, 9,022; Hind III, 10,519; Bgl II, 10,784; Hind III, 11,675; Hind III, 11,966; Bam HI, 12,066; Xba I, 12,098; Bam HI, 13,891; Hind III, 14,058; Bam HI, 14,969; Bgl II, 15,878; and Bam HI*, 16,181.
23. U. Schibler, K. B. Marcu, R. Perry, *Cell* **15**, 1495 (1978); K. B. Marcu, U. Schibler, R. P. Perry, *Science* **204**, 1087 (1979).
24. Pollock *et al.* recently reported that the δ gene of the New Zealand Black (NZB) strain of mouse may be on the 5' end of the μ gene, since their serological data indicates that the NZB strain has the a allotype of δ whereas μ and the other constant regions of NZB are of the e allotype [R. R. Pollock, M. E. Dorf, M. F. Mescher, *Proc. Natl. Acad. Sci. U.S.A.* **77**, 4256 (1980)]. These data could be subjected to a variety of other interpretations. A cloning experiment on NZB DNA with the probes we have constructed might resolve this point at the molecular level.
25. E. M. Southern, *J. Mol. Biol.* **98**, 503 (1978). The DNA from Ch28 257.4 or from Ch28 257.3 was digested with various combinations of restriction enzymes, including Xba I, Bam HI, Hind III, Xho I, Eco RI, and Bgl II, and then subjected to electrophoresis through 0.8 percent agarose gels that were stained with ethidium bromide and photographed with ultraviolet illumination. Southern transfers were made of the replicate gels, and each nitrocellulose paper was hybridized with ³²P-labeled δ probe or ³²P-la-

beled μ probe, or in some cases, with δmRNA that was labeled with ^{32}P by using polynucleotide kinase as described by N. Maizels, *Cell* **9**, 431 (1976).

26. See the article by P. W. Tucker, C.-P. Liu, J. F. Mushinski, F. R. Blattner, *Science* **209**, 1353 (1980).

27. N. M. Gough *et al.*, *Proc. Natl. Acad. Sci. U.S.A.* **77**, 554 (1980).

28. To calculate this, we first calculated the distance from the Xba I site in the 3′ untranslated region of the μ membrane exon to the Bgl II site just to the left of Cδ1 by subtracting the coordinates (*22*) (6618 − 4457 = 2161). We then added to this value 145 bp for the Bgl II to Cδ1 distance and 160 bp for the distance from the Cμ exon terminator to the Xba I site (*7*). The result was 2466 nucleotide pairs.

29. S. Slavin and S. Strober, *Nature (London)* **272**, 624 (1978); M. R. Knapp, E. Severison-Grono-wicz, J. Schroeder, S. Strober, *J. Immunol.* **123**, 1000 (1979); M. Knapp, personal communication regarding the observation that IgM and IgD molecules on BCL-1 lymphoma cells show the same idiotype.

30. W. J. Dreyer and J. C. Bennett, *Proc. Natl. Acad. Sci. U.S.A.* **54**, 865 (1°65).

31. We thank M. Fiant and J. Richards for work on

heteroduplex and electromicroscopy, E. Lui for technical help on phage plaque purification, and W. Fiske and E. La Luzerne for manuscript preparation. All experiments were carried out in accordance with the NIH Guidelines on Recombinant DNA Research. Supported by grants GM 21812 (F.R.B.) and GM 06768 (C.P.L.) from the National Institute of General Medical Sciences, and by grant AI 16547 (P.W.T.) from the National Institute of Allergy and Infectious Diseases. This is paper number 2447 from the Laboratory of Genetics, University of Wisconsin.

16 July 1980

Notes